Treatments, Mechanisms, and Adverse Reactions of Anesthetics and Analgesics

Treatments, Mechanisms, and Adverse Reactions of Anesthetics and Analgesics

The Neuroscience of Pain, Anesthetics, and Analgesics

Edited by

Rajkumar Rajendram
Department of Medicine, King Abdulaziz Medical City, King Abdulaziz International Medical Research Center, Ministry of National Guard - Health Affairs, Riyadh, Saudi Arabia

College of Medicine, King Saud bin Abdulaziz University of Health Sciences, Riyadh, Saudi Arabia

Vinood B. Patel
Centre for Nutraceuticals, School of Life Sciences, University of Westminster, London, United Kingdom

Victor R. Preedy
King's College London, London, United Kingdom

Colin R. Martin
Institute for Health and Wellbeing, University of Suffolk, United Kingdom

Academic Press is an imprint of Elsevier
125 London Wall, London EC2Y 5AS, United Kingdom
525 B Street, Suite 1650, San Diego, CA 92101, United States
50 Hampshire Street, 5th Floor, Cambridge, MA 02139, United States
The Boulevard, Langford Lane, Kidlington, Oxford OX5 1GB, United Kingdom

Copyright © 2022 Elsevier Inc. All rights reserved.

No part of this publication may be reproduced or transmitted in any form or by any means, electronic or mechanical, including photocopying, recording, or any information storage and retrieval system, without permission in writing from the publisher. Details on how to seek permission, further information about the Publisher's permissions policies and our arrangements with organizations such as the Copyright Clearance Center and the Copyright Licensing Agency, can be found at our website: www.elsevier.com/permissions.

This book and the individual contributions contained in it are protected under copyright by the Publisher (other than as may be noted herein).

Notices

Knowledge and best practice in this field are constantly changing. As new research and experience broaden our understanding, changes in research methods, professional practices, or medical treatment may become necessary.

Practitioners and researchers must always rely on their own experience and knowledge in evaluating and using any information, methods, compounds, or experiments described herein. In using such information or methods they should be mindful of their own safety and the safety of others, including parties for whom they have a professional responsibility.

To the fullest extent of the law, neither the Publisher nor the authors, contributors, or editors, assume any liability for any injury and/or damage to persons or property as a matter of products liability, negligence or otherwise, or from any use or operation of any methods, products, instructions, or ideas contained in the material herein.

Library of Congress Cataloging-in-Publication Data
A catalog record for this book is available from the Library of Congress

British Library Cataloguing-in-Publication Data
A catalogue record for this book is available from the British Library

ISBN 978-0-12-820237-1

SET ISBN 978-0-12-821066-6

For information on all Academic Press publications
visit our website at https://www.elsevier.com/books-and-journals

Publisher: Nikki Levy
Acquisitions Editor: Natalie Farra
Editorial Project Manager: Timothy Bennett
Production Project Manager: Paul Prasad Chandramohan
Cover Designer: Miles Hitchen

Typeset by STRAIVE, India

Printed in the United States of America
Last digit is the print number: 10 9 8 7 6 5 4 3

Contents

Contributors   xxi
Preface   xxvii

Part I
Drugs and agents used in anesthesia and analgesia

1. Anesthesia for malignant hyperthermia susceptible patients

Calim Neder Neto,
Mariana Fontes Lima Neville, and
Helga Cristina Almeida da Silva

Introduction	3
Diseases associated with susceptibility to MH	4
Clinical manifestation—MH crisis	5
Diagnosis and differential diagnosis	6
Treatment of MH crisis	7
Anesthesia for patients susceptible to malignant hypertermia	9
Pre-operative care	9
Perioperative care	10
Applications to other areas	11
Other agents of interest	11
Mini-dictionary of terms	12
Key facts of malignant hyperthermia	12
Summary points	12
References	13

2. Use of dextran in regional anesthesia

Masahiko Tsuchiya

Introduction	15
History of dextran as an LA adjuvant	16
Adrenaline and dextran as LA adjuvant	16
Adverse effects of adjuvant adrenaline	16
Reduction of adrenaline toxicity by dextran addition	18
First appearance of dextran as LA adjuvant	18

Mechanism of dextran as LA adjuvant	18
Application of dextran as LA adjuvant for regional anesthesia	18
Suitability of LA-dextran mixture for compartment nerve block	18
LA-dextran mixture enhances analgesic effects of compartment nerve block	19
LA-dextran mixture used for compartment nerve block remains at injection site more than 24 h	19
Dextran inhibits unintended spread of injected LA	21
Summary of application of dextran as LA adjuvant	23
Future studies	23
Other agents	23
Mini-dictionary or terms	23
Key facts	24
Summary points	24
References	24

3. Intraperitoneal local anesthetic agents in the management of postoperative pain

Karlin Sevensma

Introduction	27
Anatomic considerations and agents	27
Laparoscopic cholecystectomy	27
Laparoscopic appendectomy	28
Gynecologic surgery	28
Instillation at hemidiaphragm for phrenic nerve pain prevention	28
Safety of intraperitoneal anesthetics	30
Applications to other areas	30
Other agents of interest	31
Mini-dictionary of terms	31
Key facts of intraperitoneal local anesthetic agents in the management of postoperative pain	31
Summary points	31
References	31

4. Automatic control of anesthesia via different vital signs

Jerry Chen, Maysam F. Abbod, and Jiann-Shing Shieh

Introduction	33
Anesthesia and its interpretation—Hypnosis, nociception, and neuromuscular relaxation	33
Clinical signs and physiological signals in the surgery	34
Modeling and control of anesthesia	35
Automatic control of hypnosis	35
Automatic control of analgesia	36
Automatic control of neuromuscular relaxation	36
The seven stages for automatic control of anesthesia	37
Applications to other areas	38
Other agents of interest	38
Key facts	39
Key facts of anesthetic agents	39
Key facts of Propofol	39
Mini-dictionary of terms	39
Summary points	40
References	40

5. Non-opioid based analgesia in otolaryngology

Yohanan Kim, Anthony Sanchez, and Khanh Nguyen

Introduction	43
Patient pain experience in otolaryngology	43
Oral cavity and oropharynx	44
Nasal surgery	44
Otologic surgery	44
Visceral/soft tissue	45
Oncologic surgery	45
Risk factors for greater pain experience	45
Non-opioid analgesia	45
Local anesthetics	45
Acetaminophen	46
NSAIDs	46
Gabapentinoids	47
Discussion	48
Applications to other areas	48
Other agents of interest	49
Key facts	49
Key facts of patient pain experience in otolaryngology	49
Key facts of pain medication use in otolaryngology	49
Mini-dictionary of terms	50
Summary points	50
References	50

Part II
Mechanisms of action of drugs

6. Buprenorphine: Mechanism and applications

Albert Lin and Anuj Aggarwal

Introduction	55
Buprenorphine pharmacology	55
Adverse effects	56
Clinical use of buprenorphine	57
Perioperative management	57
Applications to other areas	58
Other agents of interest	58
Methadone	58
Naltrexone	59
Key facts	59
Key facts of buprenorphine	59
Mini-dictionary of terms	59
Summary points	59
References	60

7. Caffeine as analgesic adjuvant

Thomas W. Weiser

Introduction	63
Caffeine: Mechanism of action and effects in preclinical models of pain	63
Caffeine's analgesic properties in humans	63
Pharmacokinetic properties in man in absence and presence of analgesics	64
Clinical efficacy data in acute pain trials	65
Caffeine and the ceiling effect of NSAIDs	66
Comparisons of caffeine-containing analgesics with active comparators	66
Caffeine and migraine	67
Caffeine and worsening of headache diseases?	67
Pain and functional impairment	68
Safety of caffeine and caffeine-containing analgesics	68
Effects attributable to caffeine	68
Effects attributable to the analgesic compound	69
Applications to other areas	69
Other agents of interest	70
Key facts of caffeine	70
Mini-dictionary of terms	70
Summary points	70
References	71

8. Chloroprocaine: Features and applications

Barbara Rupnik and Alain Borgeat

Introduction	73
History and controversies	73
Pharmacology overview	73
Structure and properties	74
Pharmacokinetics of 2-chloroprocaine	74
Mechanism of action	74
Clinical application	75
Adverse effects	77
Chloroprocaine 2%: Suggested application	77
Mini-dictionary of terms	77
Key facts of chloroprocaine: Features and applications	78
Summary points	78
References	78

9. Clonidine: Features and applications

Renato Santiago Gomez and Magda Lourenço Fernandes

Introduction	81
Pharmacological characteristics	81
Pharmacodynamics	81
Pharmacokinetics	82
Clinical applications in the perioperative period	82
Pre-anesthetic medication	82
Adjunct to regional anesthesia	83
Protection of the cardiovascular system	83
Controlled hypotension	84
Anti-shivering effect	85
Postoperative agitation	85
Applications to other areas	85
Other agents of interest	86
Conclusions	87
Key facts	87
Mini-dictionary of terms	87
Summary points	87
References	87

10. An excursion into secondary pharmacology of fentanyls with potential implications for drug design: σ_1 receptor

Piotr F.J. Lipiński, Edina Szűcs, Małgorzata Jarończyk, Piotr Kosson, Sándor Benyhe, Aleksandra Misicka, Jan Cz Dobrowolski, and Joanna Sadlej

The role of fentanyl	90
Secondary pharmacology of fentanyls	90

σ_1 receptor: A unique and intriguing protein	92
Affinity of fentanyls for σ_1R	94
Fentanyls and σ_1R: Insights from computations	95
Pharmacological relevance of fentanyls' interactions with σ_1R	96
Fentanyl structure as scaffold for dual μOR/σ_1R ligands	96
Successful realization of the dual μOR/σ_1R ligands concept	97
Applications to other areas	97
Other agents of interest	98
Key facts of σ_1 receptor	99
Mini-dictionary of terms	99
Summary points	99
References	99

11. Isoflurane: Mechanisms and applications

Lady Christine Ong Sio, Marina Varbanova, and Alexander Bautista

Introduction	101
Chemical structure and properties	102
Pharmacology and pharmacokinetics	102
Mechanism of action	102
Clinical effects	103
Central nervous system effects	104
Respiratory effects	104
Cardiac effects	104
Skeletal and smooth muscle effects	104
Hepatic effects	104
Renal effects	105
Obstetric effects	105
Adverse effects	105
Malignant hyperthermia	105
Teratogenic effects	105
Environmental effects	105
Metabolism	105
Drug-drug interactions	106
Opioids	106
Nitrous oxide	106
Neuromuscular blocking agents	106
Epinephrine	106
Calcium antagonists	106
Concomitant use of beta blockers	106
Application to other areas	106
Other agents of interest	106
Mini-dictionary of terms	107
Key facts of isoflurane	107
Summary points	107
References	107

viii Contents

12. The lidocaine patch: Features and applications: Post-thoracotomy pain and beyond

Alfonso Fiorelli, Pasquale Sansone, Caterina Pace, and Mario Santini

Introduction	109
Search strategy	110
Search results	110
Analysis of the results	116
Discussion	117
Conclusions	118
Key facts of lidocaine patch for controlling acute postoperative pain	118
Summary points	118
References	119

13. Memantine: Features and application in the management of chronic pain

Harsha Shanthanna

Introduction	121
Background: NMDA antagonists and pain	122
Memantine: Fundamental pharmacology	122
Memantine: Preclinical evidence of effect on pain	124
Memantine: Clinical evidence and literature review in chronic pain	124
Memantine: Considerations for clinical use	127
Applications to other areas	127
Other agents of interest	127
Conclusions	128
Key facts	128
Mini-dictionary of terms	128
Summary points	128
References	128

14. Midazolam: Mechanism and perioperative applications

Joe C. Hong

Introduction	131
Mechanism of action	132
Pharmacokinetics	132
Effects on the central nervous system	133
Effects on the respiratory system	134
Effects on the cardiovascular system	134
Clinical uses and dosage	134
Applications to other areas: Palliative care	135
Other agents of interest: Flumazenil	136
Key facts of midazolam	136
Mini-dictionary of terms	136

Summary points	137
References	137

15. Intravenous paracetamol: Features and applications

Ristiawan M. Laksono and Isngadi Ahmad Wagimin

Introduction	139
Chemical structure and pharmacokinetics	139
Paracetamol pharmacokinetics	140
Site of action of paracetamol	141
Inhibition of COX pathway	141
Modulation of endogenous cannabinoid system	141
Inhibition of nitric oxide	141
Pharmacokinetics of intravenous paracetamol in adult	142
The use of intravenous paracetamol in adult	142
Non-superiority of IV paracetamol	144
The role of paracetamol to control breakthrough pain	144
Paracetamol in obstetrics	144
Pharmacokinetics of intravenous paracetamol in children	145
The use of paracetamol in children	145
Side effect of intravenous paracetamol	146
Applications to other areas	146
Other agents of interest	147
Mini-dictionary of terms	147
Key facts of intravenous paracetamol	147
Summary points	148
References	148

16. Prilocaine: Mechanisms and application

Naresh Kumar Katari and Sreekantha B. Jonnalagadda

The journey of local anesthetic	151
Structure	152
Classification of anesthetic drugs	153
Prilocaine	153
Preparation method of prilocaine:	157
Pharmacology and mechanism action of local anesthetic agent—Prilocaine	157
Lipid solubility and protein binding:	158
Metabolism and excretion	158
Distribution	159
Elimination	160
Duration of anesthesia	160
Toxicity	160
Methemoglobinemia	160
Applications	161

Other applications of prilocaine 161
Other agents of interest 162
Mini-dictionary of terms 162
Key facts 162
Summary and conclusion 162
References 163

17. Sevoflurane: Features and uses in topical application for wound care

Manuel Gerónimo-Pardo

Introduction 165
Systemic effects of sevoflurane and haloethers as general anesthetics 165
Analgesic effects of sevoflurane and haloethers on the central nervous system 166
Peripheral analgesic effects of sevoflurane and haloethers 166
Off-label use of topical sevoflurane on painful chronic wounds 167
  Analgesic profile 167
  Analgesic effect on different types of wounds and pains 167
  Quality of life and opioid-sparing effect 172
Safety issues of topical sevoflurane 172
  Local adverse effects 172
  Systemic adverse effects for patients 172
  Systemic adverse effects for health workers 173
  Greenhouse effect 173
Dosages and methods of administration 173
Economic implications 173
Application to other areas 174
  Sevoflurane as an antimicrobial agent for infected wounds 174
  Sevoflurane as a prohealing agent for hard-to-heal wounds 175
  Topical sevoflurane for painful conditions different from wounds 175
Other agents of interest 175
Mini-dictionary of terms 176
Key facts of painful chronic wounds 177
Summary points 177
References 177

18. Tramadol as an analgesic

Ayman M. Mahmoud and Emad H.M. Hassanein

Introduction 181
Mechanism(s) of action 181
  Central analgesic mechanism 181
  Peripheral local anesthetic mechanism 181

  Effect of tramadol on other receptors 182
Pharmacokinetics 182
  Absorption 182
  Distribution 182
  Metabolism 182
  Excretion 182
Therapeutic uses 183
  Neuropathic pain 183
  Osteoarthritis pain 183
  Cancer pain 183
  Pain in the emergency department 184
  Acute myocardial infarction pain 184
  Postoperative pain 184
  Postoperative shivering 185
  Acute renal pain 185
  Tramadol as a local anesthetic 185
  Use/off-label use 185
Adverse effects and management of tramadol toxicity 185
Dependence, withdrawal, abuse, and tolerance 186
Contraindications 186
  Seizure disorders 186
  Liver disease 186
  Renal dysfunction 186
  Respiratory depression 187
  Suicidal tendency 187
Conclusion 187
Applications to other areas 187
Other agents of interest 187
Mini-dictionary of terms 187
Key facts 188
Summary points 188
Conflict of interest 188
References 188

Part III
Adverse effects, reactions, and outcomes

19. Long-term effects of anesthesia on the brain: an update on neurotoxicity

Rajkumar Rajendram, Vinood B. Patel, and Victor R. Preedy

Introduction 195
  Neurotransmitters signaling during cerebral development 196
  Neurotoxicity of anesthetic agents 197
  Historical overview 198

Effect of surgical stress on the developing brain	200
Cognitive and behavioral development in *Homo sapiens* after childhood surgery	200
The effect of anesthesia on neuronal networks	202
The long-term effects of excessive neurotransmitter modulation during CNS development	202
Studies in nonhuman primates	202
Application to other areas: The aging brain	203
Conclusion	204
Mini-dictionary of terms	205
Key facts of the United States Federal Drug Administration (FDA)	205
Key facts of anesthetic neurotoxicity	205
Key facts of postoperative cognitive dysfunction	205
Other agents of interest	206
Summary points	206
References	206

20. Breastfeeding and mother-baby dyad's competence following neuraxial labor analgesia

Roberto Giorgio Wetzl, Maria Lorella Gianni, Enrica Delfino, and Alessandra Consales

Introduction	211
Human rotational delivery and childbirth pain	211
From prayers to neuraxial analgesia: A brief history of childbirth pain relief	212
The analgesia-breastfeeding issue: Alleged mechanism of interference	213
Exclusive breastfeeding for the first 6 months of infant's life: Health impact	213
The analgesia-breastfeeding issue in the literature	213
Limitations of current knowledge	213
Conflicting study design, sampling strategies and enrolment criteria	213
Different accuracy in reporting labor neuraxial blockade techniques	213
The choice of confounders	215
Methodological flaws	215
Global literature overview	218
The analgesia-breastfeeding issue as a new research opportunity	218
Breastfeeding as a measure of dyadic neurological competence	218
Searching for objective, nonbiased indicators of breastfeeding initiation success	220

Measuring and grading breastfeeding initiation success	221
Conclusion	221
Applications to other areas	221
Other agents of interest	222
Mini-dictionary of terms	222
Key facts of labor analgesia state of art in guidelines and metaanalyses	222
Summary points	223
References	223

21. Mechanistic overview of how opioid analgesics promote constipation

Jesse J. Di Cello, Arisbel B. Gondin, Simona E. Carbone, and Daniel P. Poole

Introduction	227
The enteric nervous system: Master regulator of GI function	228
Opioid receptors and their endogenous ligands	229
Opioids inhibit GI function through actions on the ENS	230
How do opioids exert their inhibitory effects on GI motility to promote chronic constipation?	231
General effects of opioids on GI motility	231
Opioid-induced constipation is resistant to tolerance development	231
Application to other areas	232
Other agents of interest	232
Mini-dictionary of terms	232
Key facts of opioid-induced constipation	232
Summary points	233
References	233

22. Cognitive behavioral therapy for chronic pain and opioid use disorder

Marina G. Gazzola, Mark Beitel, Christopher J. Cutter, and Declan T. Barry

Introduction	235
Prevalence of chronic pain and OUD	235
Clinical complexity of patients with chronic pain and OUD	236
Provider consideration	236
Necessary pharmacological platform: Medication for opioid use disorder	237
Optimal medical management	237
Overview of CBT for chronic pain and OUD	238

CBT modules	238
Seven clinical problems or challenges	238
Clinical problem 1: Treating chronic pain as if it were acute pain	238
Clinical problem 2: Patient inactivity	238
Clinical problem 3: Decreased distress tolerance	238
Clinical problem 4: Catastrophizing	240
Clinical problem 5: Focusing on what's going wrong	240
Problem 6: Viewing substance use as "just happening" and "beyond their control"	241
Problem 7: Decreased assertiveness	241
Applications to other areas: Real-world delivery of CBT for chronic pain and OUD	241
Alternative therapies	242
Conclusions	242
Mini-dictionary of terms	242
Key facts of CBT for chronic pain and OUD	242
Summary points	242
References	243

23. Preoperative opioid and benzodiazepines: Impact on adverse outcomes

Martin Ingi Sigurðsson

Introduction	247
Definition of chronic opioid and BZD usage	247
Prevalence of chronic preoperative opioid and BZD usage	248
Characteristics of patient cohorts with preoperative chronic opioid and BZD usage	248
Perioperative outcomes of patients taking opioids and BZDs preoperatively	252
Inadequate perioperative analgesia	252
Increased risk of complications from analgesia	253
Other short-term outcomes	255
Long-term postoperative outcomes for patients	255
Application to other areas	256
Other agents of interest	256
Mini-dictionary of terms	258
Key facts	258
Summary points	258
References	259

24. Malignant hyperthermia syndrome and hydrogen sulfide signaling: Role of Kv7 channels

Mariarosaria Bucci, Valentina Vellecco, Antonio Mancini, and Giuseppe Cirino

Introduction	261
IVCT procedure	262
Hydrogen sulfide	262
Hydrogen sulfide and MH	263
Potassium channels as H_2S molecular targets in MHS subjects	264
The paradoxical depolarizing activity of Kv7 channel in MHS subjects	264
Protein posttranslational modifications: Persulfidation (S-sulphydration)	265
Persulfidation of Kv7 channels in MH syndrome	266
Applications to other areas	267
Hydrogen sulfide signaling and SKM: Beyond MH	267
Other agents of interest	268
Future perspectives	268
Mini-dictionary of terms	269
Key facts of "malignant hyperthermia syndrome and hydrogen sulfide signaling role of Kv7 channels"	269
Summary points	269
References	270

25. Problems with epidural catheter

Mustafa Kemal Arslantas

Introduction	273
Problems encountered while inserting an epidural catheter	273
Breakage of epidural catheters	274
Blocked epidural catheter	275
Epidural catheter removal difficulty	275
Spinal epidural hematoma	276
Spinal epidural abscess	277
Application to other areas	277
Other agents of interest	277
Mini-dictionary of terms	277
Key facts	277
Key facts of problems with epidural catheter	277
Summary points	278
References	278

xii Contents

26. Headache after neuraxial blocks: A focus on combined spinal-epidural anesthesia

Haluk Ozdemir and Reyhan Arslantas

Introduction	281
History of PDPH	281
The importance of PDPH	282
The mechanism of PDPH	282
The characteristics of PDPH	282
Combined spinal-epidural anesthesia	282
Combined spinal-epidural anesthesia for special patient groups	283
Effects of application techniques of CSEA on headache	283
Prophylaxis of PDPH after UDP in neuroaxial anesthesia	284
Management of PDPH	285
Supportive treatment	285
Medical management	285
Epidural blood patch	286
Alternative therapies	286
Conclusion	287
Application to other areas	287
Other agents of interest	287
Mini-dictionary of terms	287
Key facts	287
Key facts of headache after neuraxial blocks	287
Summary points	288
References	288

27. Liposomal bupivacaine, pain relief and adverse events

Hüseyin Oğuz Yılmaz and Alparslan Turan

Introduction	291
Liposomes and DepoFoam	292
Liposomal bupivacaine	292
Clinical pharmacology	294
Release characteristics and stability	294
Pharmacodynamics and mechanism of action	294
Pharmacokinetics	294
Safety and adverse events	297
Central nervous system toxicity	297
Cardiovascular system toxicity	298
Local adverse events	298
Maternal/fetal toxicity	299
Compatibility	299
Clinical uses and efficacy	299
Liposome bupivacaine versus placebo	300
Surgical site infiltration	300

Peripheral nerve blocks and perineural use	300
Neuraxial use	301
Administration	301
Application to other areas	301
Other agents of interest	301
Mini-dictionary of terms	303
Key facts of liposome bupivacaine	303
Summary points	304
References	304

28. Adverse events associated with analgesics: A focus on paracetamol use

Iwona Popiolek and Grzegorz Porebski

Introduction	309
Mechanism of toxicity	310
Clinical and laboratory manifestations of hepatotoxicity	311
Risk factors of liver injury	311
Management of acute poisoning	312
Chronic poisoning	312
Immediate hypersensitivity	313
Delayed hypersensitivity	313
Diagnosis and management of paracetamol-induced hypersensitivity reactions	314
Applications to other areas	314
Other agents of interest	315
Mini-dictionary of terms	315
Key facts of paracetamol-induced toxicity and hypersensitivity	316
Summary points	316
References	316

Part IV
Novel and nonpharmacological aspects and treatments

29. *Acronychia pedunculata* leaves and usage in pain

U.G. Chandrika and W.M.K.M. Ratnayake

Introduction	321
Traditional medicinal system in Sri Lanka	321
Acronychia pedunculata as a Sri Lankan medicinal plant	322
Chemistry and biological activities of *A. pedunculata*	322
Anti-inflammatory activity of *A. pedunculate* leaves	323

Identification of anti-inflammatory fractions
and active compounds of *A. pedunculate*
leaves   324
Analgesic activity of *A. pedunculate* leaves   325
Other areas of applications of *A. pedunculata*   325
Other agents of interest   325
Conclusion   326
Mini-dictionary of terms   326
Key facts   326
Summary   326
A list of other novel plant-based agents used
in analgesia or anesthesia   326
References   327

30. *Adansonia digitata* and its use in neuropathic pain: Prostaglandins and beyond

Aboyeji Lukuman Oyewole,
Abdulmusawwir O. Alli-Oluwafuyi,
Abdulrazaq Bidemi Nafiu, and Aminu Imam

Introduction   329
Brief description of *Adansonia digitata*   329
Distribution   330
Significance of *Adansonia digitata*   330
Relevance among local folks   330
Medicinal importance   331
Composition of *Adansonia digitata*
plant   332
Phytochemical constituents   332
Mineral components   332
Impact of *Adansonia digitata* on
pain   332
Effect on pain sensation   332
Effect on neuropathic pain   338
Potential mechanism of actions for *A.*
digitata extract, from prostaglandins and
beyond   340
Prostaglandin E2 pathway   340
Inflammatory and oxidative stress
pathways   341
Peripheral and central sensitization
pathways   341
Ions-mediated pathways   343
Applications to other areas   343
Metabolic diseases   343
Malaria   344
Neuroprotection   344
Antiviral   345
Other agents of interest   345
Mini-dictionary of terms   345
Key facts   346
Summary points   346
References   346

31. *Andrographis paniculata* standardized extract (ParActin) and pain

Rafael A. Burgos, Pablo Alarcón, and
Juan L. Hancke

Introduction   351
Andrographis paniculata: Classification
and composition   351
Preclinical antiinflammatory and analgesic
effects of *A. paniculata* and
andrographolide contained in ParActin   352
Mechanism of action of andrographolide, the
active principle of ParActin in inflammation
and pain   353
Effect of andrographolide on the NF-κB
pathway   353
Effects of Andrographolide on MAPK
and AP-1   355
Antioxidant effect of andrographolide and
role of the Nrf2/keap1 pathway   355
Pharmacokinetics and metabolism   357
Clinical pharmacology and side effects   357
Contraindications and potential interactions   357
Ongoing clinical trials in inflammation and
pain   358
Applications to other areas   358
Other agents of interest   358
Mini-dictionary of terms   358
Key facts   359
Summary points   359
References   359

32. Capsaicin: Features usage in diabetic neuropathic pain

Kongkiat Kulkantrakorn

Introduction   365
Capsaicin   365
Diabetic neuropathy   365
Clinical data of capsaicin in painful diabetic
neuropathy   366
Early clinical studies   366
Comparative studies   366
Modern-day studies   368
Low dose topical capsaicin   368
High dose topical capsaicin   368
Applications to other areas   369
Other agents of interest   369
Topical lidocaine   369
Vitamin and supplements   370
Mini-dictionary of terms   371
Key facts   371
Summary points   371
References   371

33. Cola nitida and pain relief

L.D. Adedayo, O. Bamidele, and
S.A. Onasanwo

Introduction on Cola nitida	375
Brief review on Cola nitida	375
Socio-cultural values and uses	375
Distribution of Cola nitida	376
Composition of Cola nitida	376
Uses of Cola nitida	376
Traditional uses of Cola nitida	377
Pharmacological potentials of Cola nitida	377
Pain and nociception	377
Causes of pain	377
Classification of pain	378
Nociceptive processing	378
Pathologic pain	378
Peripheral mechanisms of sensory transmission	378
Cola nitida as a therapeutic agent for pain relief	379
Application to other areas	381
Other agents of interest	381
Mini-dictionary	381
Key facts	382
Summary points	382
References	382

34. Analgesic effects of Ephedra herb and ephedrine alkaloids-free Ephedra herb extract (EFE)

Sumiko Hyuga, Shunsuke Nakamori,
Yoshiaki Amakura, Masashi Hyuga,
Nahoko Uchiyama, Yoshinori Kobayashi,
Takashi Hakamatsuka, Yukihiro Goda,
Hiroshi Odaguchi, and Toshihiko Hanawa

Introduction	385
Ephedra herb and ephedrine alkaloids	386
Novel active ingredients, herbacetin-glycosides, in Ephedra herb and the analgesic effect of herbacetin, an active metabolite of herbacetin-glycosides	386
Development of ephedrine alkaloids-free Ephedra Herb extract (EFE)	387
Adverse effects of EHE and safety of EFE	389
Analgesic effects of EHE and EFE on formalin-induced pain	389
Analgesic effects of EHE, EFE, ephedrine, and pseudoephedrine on formalin-induced pain	390
Analgesic effect of EFE on pain in arthritis model mouse	394

Reduction of capsaicin-induced pain via transient receptor potential vanilloid 1 (TRPV-1) by EHE	394
Application to other areas	395
Conclusions and perspectives	396
Other agents of interest	397
Mini-dictionary of terms	397
Key facts	397
Key facts of formalin test	397
Key facts of Ephedra herb macromolecule condensed-tannin (EMCT)	398
Summary points	398
References	398

35. Euphorbia bicolor (Euphorbiaceae) latex phytochemicals and applications to analgesia

Paramita Basu, Dayna L. Averitt, and
Camelia Maier

Introduction	402
Medicinal properties of the Euphorbia species	402
Medicinal properties of E. bicolor latex	403
Identification of E. bicolor latex phytochemicals	404
Evidence of pain-relieving properties of the E. bicolor latex phytochemicals	405
Development of E. bicolor phytochemicals as phytomedicines for pain management	413
Mini-dictionary of terms	413
Key facts about coumestans	414
Key facts about diterpenes	414
Key facts about flavonoids	414
Key facts about isoflavones	414
Summary points	414
References	415

36. Analgesic properties and mechanisms of action of Muntingia calabura extracts: A review

Zainul Aminuddin Zakaria, Tavamani Balan,
Mohd. Hijaz. Mohd. Sani, and
Nurfuzillah Abdul Rani

Introduction	419
Muntingia calabura L	420
Antinociceptive activity of Muntigia calabura leaves extracts	421
Mechanisms of action underlying the antinociceptive activity of Muntingia calabura leaves extracts	424
Conclusion	425

Applications to other areas	425
Other agents of interest	426
Mini-dictionary of terms	427
Key facts of traditional and complementary medicine	427
Summary points	428
References	428

37. Resolving neuroinflammation and pain with maresin 1, a specialized pro-resolving lipid mediator

Victor Fattori, Camila R. Ferraz, Fernanda Soares Rasquel-Oliveira, Tiago H. Zaninelli, Sergio M. Borghi, Rubia Casagrande, and Waldiceu A. Verri, Jr

Introduction	432
MaR1	434
Analgesic effects of MaR1	434
Clinical analgesic evidence of MaR1 and its precursors	436
Application to other areas	437
Other agents of interest	438
Mini-dictionary of terms	439
Key facts	439
Key facts of SPMs	439
Key facts of MaR1	439
Summary points	439
Funding	439
References	440

38. Therapeutic role of naringenin to alleviate inflammatory pain

Marília F. Manchope, Camila R. Ferraz, Sergio M. Borghi, Fernanda Soares Rasquel-Oliveira, Anelise Franciosi, Julia Bagatim-Souza, Amanda M. Dionisio, Rubia Casagrande, and Waldiceu A. Verri, Jr

Introduction	443
Naringenin actions on transient receptor potential (TRP) channels and inflammatory pain relief	444
Naringenin inhibits the production of endogenous cytokines that mediate inflammatory pain	446
Naringenin targets oxidative stress to reduce inflammatory pain	447
Analgesic pathways actively induced by naringenin	447
Naringenin modulates transcription factors and miRNA	448
Clinical applicability and safety	448

Applications to other areas	449
Other agents of interest	451
Mini-dictionary of terms	452
Key facts of inflammatory pain	452
Summary points	452
Funding	453
References	453

39. Analgesic properties of plants from the genus *Solanum* L. (Solanaceae)

F.J.R. Paumgartten, G.R. de Souza, A.J.R. da Silva, and A.C.A.X. De-Oliveira

Introduction	457
Solanum genus L. (Solanaceae)	459
Antinociceptive effects of *Solanum* spp. extracts	459
Antiinflammatory activity of *Solanum* spp. extracts	464
Type of analgesia produced by *Solanum* spp. extracts	465
Chemical constituents of active extracts	465
Other pharmaco-toxicological activities of interest	467
Application to other areas	467
Mini-dictionary of terms	467
Key facts of *Solanum* analgesia	468
Other agents of interest	468
Summary points	468
References	468

40. Dietary constituents act as local anesthetic agents: Neurophysiological mechanism of nociceptive pain

Mamoru Takeda and Yoshihito Shimazu

Introduction	473
Ascending pain pathway in trigeminal system	474
Sensory transduction and noxious transmission	475
Possible molecular targets for local anesthetic agents	476
Generator potential in the nociceptive terminals	476
Action potential in the nociceptive terminals	477
Peripheral mechanism for potential candidates of dietary constituents as local anesthetic agents	477
Modulation of generator potential by dietary constituents	477

xvi   Contents

Modulation of action potential by dietary constituents — 478
Functional significance for dietary constituents as local anesthetic agents — 479
Future direction — 480
Concluding remarks — 480
Mini-dictionary of terms — 480
Key facts — 482
Summary points — 482
References — 483

41. Pain response following prenatal stress and its modulation by antioxidants

Che Badariah Abd Aziz, Asma Hayati Ahmad, and Hidani Hasim

Introduction — 487
Oxidative stress — 488
Antioxidant — 490
Honey — 490
Protective effects of other antioxidants in the prenatally stressed offspring — 491
Resveratrol — 491
Spirulina — 492
Vitamins — 492
Applications to other areas — 493
Role on cardiovascular disease — 493
Role in diabetes mellitus — 493
Role as antimicrobial agent — 493
Other agents of interest — 493
Mini-dictionary of terms — 494
Key facts of prenatal stress — 495
Key facts of antioxidants — 495
Summary points — 495
References — 495

42. Physical activity and exercise in the prevention of musculoskeletal pain in children and adolescents

Pablo Molina-García, Patrocinio Ariza-Vega, and Fernando Estévez-López

Background — 499
Physical activity in the prevention of musculoskeletal pain — 499
Physical activity recommendations — 500
Physical fitness components in the prevention of musculoskeletal pain — 501
Cardiorespiratory fitness and aerobic/anaerobic training — 501
Muscular strength and resistance training — 501
Flexibility and mobility training — 502

Risk factors for MSKP and the preventive role of physical activity and exercise — 504
Biomechanical factors: An integrative approach — 504
Pediatric obesity — 505
Psychosocial factors — 506
Atypical brain development — 507
Applications to other areas — 507
Other agents of interest — 508
Mini-dictionary of terms — 508
Key facts — 508
Summary points — 508
References — 509

43. Linking aerobic exercise and childhood pain alleviation: A narrative

Tiffany Kichline, Adrian Ortega, and Christopher C. Cushing

Introduction — 513
Aerobic activity levels in youth with chronic pain — 513
Effect of aerobic exercise on pediatric chronic pain intensity — 514
Current literature on effect of aerobic exercise on chronic pediatric pain intensity — 514
Effect of aerobic activity across pediatric chronic pain conditions — 515
Individual differences and the effect of aerobic exercise — 517
Aerobic exercise and strength training in pediatric chronic pain — 517
Conclusion — 517
Application to other areas — 518
Other agents of interest — 518
Mini-dictionary of terms — 519
Key facts — 519
Summary points — 519
References — 519

44. Physical activity and exercise in the management of chronic widespread musculoskeletal pain: A focus on fibromyalgia

Thomas Davergne, Fernando Estévez-López, Ana Carbonell-Baeza, and Inmaculada C. Álvarez-Gallardo

Musculoskeletal pain — 523
Physical activity in musculoskeletal pain: A historical perspective — 523

Levels of physical activity in musculoskeletal pain 524
Interventions to enhance physical activity in musculoskeletal pain: Current evidence and novel approaches 524
Physical exercise in musculoskeletal pain 525
General advice to design physical exercise programs for chronic musculoskeletal pain 525
Physical activity in fibromyalgia 525
Aerobic exercise in fibromyalgia 526
Resistance training in fibromyalgia 536
Multicomponent exercise in fibromyalgia 536
Mind-body exercise in fibromyalgia 536
Exergames in fibromyalgia 540
General exercise recommendations 540
Effectiveness of physical exercise in fibromyalgia: A summary of the evidence 541
Applications to other areas 541
Other agents of interest 541
Mini-dictionary of terms 541
Key facts 541
Summary points 542
References 542

45. Spinal cord stimulation and limb pain

Timothy Sowder, Usman Latif, Edward Braun, and Dawood Sayed

History 545
Conventional tonic SCS 545
Dorsal root ganglion stimulation 546
High frequency 10 kHz SCS 546
Burst SCS 546
Ischemic pain 547
Patient selection 547
  Indications 547
  Failure of conservative treatment 547
  Psychological screening 547
  Medical comorbidities 548
  Screening trial 548
Outcomes 548
  Failed back surgical syndrome 548
  Complex regional pain syndrome 549
  Diabetic neuropathy 550
  Ischemic pain 550
Applications to other areas 550
Other agents of interest 550
Mini-dictionary of terms 551
Key facts of spinal cord stimulation and limb pain 551
Summary points 551
References 551

46. Effectiveness of neural mobilization on pain and disability in individuals with musculoskeletal disorders

Carlos Romero-Morales, César Calvo-Lobo, David Rodríguez-Sanz, Daniel López-López, Marta San Antolín, Victoria Mazoteras-Pardo, Eva María Martínez-Jiménez, Marta Losa-Iglesias, and Ricardo Becerro-de-Bengoa-Vallejo

Introduction 555
Headache and neck pain 557
Cervicobrachial pain 557
Median, cubital, and radial nerves entrapment syndromes 558
Low back pain 559
Lower limb and neurodynamic techniques 559
Clinical applications 560
Application to other areas 561
Other agents of interest 561
Mini-dictionary of terms 561
Key facts of musculoskeletal disorders 561
Summary points 561
References 562

47. Virtual reality and applications to treating neck pain

M. Razeghi, I. Rezaei, and S. Bervis

Introduction 565
Neck pain and VR-based rehabilitation 565
VR systems design in NP rehabilitation 566
Clinical efficacy of VR in neck disorders 569
  Pain 569
Kinematic impairments 570
Disability 570
Postural control and balance impairment 570
Factors affecting the efficacy of VR-based treatment in NP 570
VR application as an assessment tool in NP 571
Disadvantages of VR application in NP 571
Conclusions 572
Applications to other areas 572
Other agents of interest 572
Mini-dictionary of terms 572
Key facts of VR 572
Summary points 572
References 573

48. Virtual reality induced analgesia and dental pain

Elitsa Veneva, Ani Belcheva, and Ralitsa Raycheva

Introduction 575
Distraction as an approach to relieve injection discomfort 575

xviii Contents

Virtual reality devices in dental pain
management 576
Safety considerations 578
Effectiveness of VR in reduction of injection
discomfort 578
Applications to other areas 579
Other agents of interest 579
Mini-dictionary of terms 579
Key facts of the use of VR devices in dental
pain management 579
Summary points 579
References 580

49. Vibrotactile devices, DentalVibe, and local anesthesia

Elitsa Veneva, Ani Belcheva, and Ralitsa Raycheva

Introduction 583
Non-pharmacological approach to injection
discomfort 583
Melzack and Wall's gate control theory
of pain 583
Fear of dental injections 584
Reduction of discomfort during local
anesthesia 584
Vibrotactile devices for dental
use 584
Effectiveness in reduction of injection
discomfort 584
Evaluation of the efficacy of vibrotactile
devices 585
Applications to other areas 585
Other agents of interest 585
Mini-dictionary of terms 585
Key facts of vibrotactile devices for
dental use 585
Summary points 586
References 587

50. Cooled radiofrequency ablation as a treatment for knee osteoarthritis

Antonia F. Chen and Eric J. Moorhead

Introduction 589
Nonpharmacological and pharmacological
conservative treatments 589
Minimally invasive and surgical
approaches 590

Denervation treatment in knee
osteoarthritis 590
Cryoablation 590
Radiofrequency ablation 591
Clinical trials of radiofrequency ablation 592
Other studies of radiofrequency ablation 592
Summary of radiofrequency ablation 593
Cooled radiofrequency ablation 593
Clinical trials of cooled radiofrequency
ablation 593
Other studies of cooled radiofrequency
ablation 595
Summary of cooled radiofrequency
ablation 596
Applications to other areas 596
Other ablative procedures of interest 597
Mini-dictionary of terms 597
Key facts of cooled radiofrequency ablation 597
Summary points 597
References 598

51. Nonpharmacologic analgesic therapies: A focus on photobiomodulation, acustimulation, and cryoanalgesia (ice) therapy

Roya Yumul, Ofelia L. Elvir Lazo, and Paul F. White

Introduction 601
Photobiomodulation therapy 602
History 602
Mechanisms of PBMT 602
PBMT treatment parameters 602
How is PBMT administered? 605
PBMT side effects 607
PBMT clinical applications 608
Conclusion 609
Acustimulation 609
Acupuncture 609
Electroanalgesia 610
Cryoanalgesia (ice) therapy 610
Cryoanalgesia administration 611
Clinical applications 611
Conclusion 612
Other agents of interest 612
Mini-dictionary of terms 612
Key facts of nonpharmacologic analgesic
therapies 613
Summary points 613
References 613

52. New coping strategies and self-education for chronic pain management: E-health

Victoria Mazoteras-Pardo, Marta San Antolín, Daniel López-López, Ricardo Becerro-de-Bengoa-Vallejo, Marta Losa-Iglesias, Carlos Romero-Morales, David Rodríguez-Sanz, Eva María Martínez-Jiménez, and César Calvo-Lobo

Introduction	617
The epidemic of chronic pain	618
Effective chronic pain management	619
Pain detection	620
Tools to assess chronic pain	620
Current nonpharmacological interventions in chronic pain	621
E-health in chronic pain	623
Clinical applications	623
Application to other areas	624
Other agents of interest	624
Mini-dictionary of terms	624
Key facts of chronic pain	624
Summary points	625
References	625

53. Postoperative pain management: Truncal blocks in obstetric and gynecologic surgery

Pelin Corman Dincer

Introduction	629
Preparation for the truncal block	630
Patient selection	630
Truncal blocks of the abdominal wall	631
Quadratus lumborum block	633
Rectus sheath block	634
Iliohypogastric and ilioinguinal nerve block	635
Transversalis fascia plane block	636
Lumbar paravertebral block	637
Erector spina plane block	637
Conclusion	638
Application to other areas	638
Other agents of interest	638
Mini-dictionary of terms	638
Key facts	639
Key facts of truncal blocks in obstetric and gynecologic surgery	639
Summary points	639
References	639
Index	643

Contributors

Numbers in paraentheses indicate the pages on which the authors' contributions begin.

Maysam F. Abbod (33), Department of Electronic and Computer Engineering, Brunel University London, Uxbridge, United Kingdom

Che Badariah Abd Aziz (487), Department of Physiology, School of Medical Sciences, Universiti Sains Malaysia, Kubang Kerian, Kelantan, Malaysia

L.D. Adedayo (375), Neurophysiology Unit, Department of Physiology, College of Health Sciences, Bowen University, Iwo; Neuroscience and Oral Physiology Unit, Department of Physiology, Faculty of Basic Medical Sciences, University of Ibadan, Ibadan, Nigeria

Anuj Aggarwal (55), Stanford University, Department of Anesthesiology, Perioperative, and Pain Medicine, Redwood City, CA, United States

Asma Hayati Ahmad (487), Department of Physiology, School of Medical Sciences, Universiti Sains Malaysia, Kubang Kerian, Kelantan, Malaysia

Isngadi Ahmad Wagimin (139), Department of Anesthesiology and Intensive Care, Faculty of Medicine, University of Brawijaya, Kota Malang, East Java, Indonesia

Pablo Alarcón (351), Laboratory of Inflammation Pharmacology, Institute of Pharmacology and Morphophysiology, Universidad Austral de Chile, Valdivia, Chile

Abdulmusawwir O. Alli-Oluwafuyi (329), Biotranslational Research Group; Department of Pharmacology and Therapeutics, Faculty of Basic Clinical Sciences, University of Ilorin, Ilorin, Nigeria

Inmaculada C. Álvarez-Gallardo (523), Department of Physical Education, Faculty of Education Sciences, University of Cádiz, Cádiz, Spain

Yoshiaki Amakura (385), Department of Pharmacognosy, College of Pharmaceutical Sciences, Matsuyama University, Matsuyama, Ehime, Japan

Marta San Antolín (555,617), Department of Psychology, Universidad Europea de Madrid, Madrid, Spain

Patrocinio Ariza-Vega (499), Department of Physiotherapy, University of Granada, Granada, Spain

Mustafa Kemal Arslantas (273), Department of Anesthesiology and Reanimation, School of Medicine, Demiroglu Bilim University, Istanbul, Turkey

Reyhan Arslantas (281), Department of Anesthesiology and Reanimation, Taksim Training and Research Hospital, Istanbul, Turkey

Dayna L. Averitt (401), Department of Biology, Texas Woman's University, Denton, TX, United States

Julia Bagatim-Souza (443), Department of Pathology, Center of Biological Sciences, Londrina State University, Londrina, Paraná, Brazil

Tavamani Balan (419), Department of Pharmaceutical Technology, Faculty of Pharmacy and Health Sciences, University Kuala Lumpur Royal College of Medicine Perak, Ipoh, Perak, Malaysia

O. Bamidele (375), Neurophysiology Unit, Department of Physiology, College of Health Sciences, Bowen University, Iwo, Nigeria

Declan T. Barry (235), Department of Psychiatry, Yale School of Medicine; Pain Treatment Services, The APT Foundation; Child Study Center, Yale School of Medicine, New Haven, CT, United States

Paramita Basu (401), Department of Anesthesiology & Perioperative Medicine, Pittsburgh Center for Pain Research, and the Pittsburgh Project to End Opioid Misuse, University of Pittsburgh School of Medicine, Pittsburgh, PA, United States

Alexander Bautista (101), Department of Anesthesiology and Perioperative Medicine, University of Louisville, Louisville, KY, United States

Ricardo Becerro-de-Bengoa-Vallejo (555,617), Faculty of Nursing, Physiotherapy and Podiatry, Universidad Complutense de Madrid, Madrid, Spain

Mark Beitel (235), Department of Psychiatry, Yale School of Medicine; Pain Treatment Services, The APT Foundation; Child Study Center, Yale School of Medicine, New Haven, CT, United States

Ani Belcheva (575,583), Department of Pediatric Dentistry, Faculty of Dental Medicine, Medical University – Plovdiv, Plovdiv, Bulgaria

Sándor Benyhe (89), Institute of Biochemistry, Biological Research Centre, Hungarian Academy of Sciences, Szeged, Hungary

S. Bervis (565), Physical Therapy Department, School of Rehabilitation Sciences, Shiraz University of Medical Sciences, Shiraz, Iran

Alain Borgeat (73), Department of Anaesthesiology, Balgrist University Hospital, Zurich, Switzerland

Sergio M. Borghi (431, 443), Department of Pathology, Center of Biological Sciences, Londrina State University; Center for Research in Health Sciences, University of Northern Paraná, Londrina, Paraná, Brazil

Edward Braun (545), Department of Anesthesiology, University of Kansas Medical Center, Kansas City, KS, United States

Mariarosaria Bucci (261), Department of Pharmacy, School of Medicine and Surgery, University of Naples "Federico II", Naples, Italy

Rafael A. Burgos (351), Laboratory of Inflammation Pharmacology, Institute of Pharmacology and Morphophysiology, Universidad Austral de Chile, Valdivia, Chile

César Calvo-Lobo (555,617), Faculty of Nursing, Physiotherapy and Podiatry, Universidad Complutense de Madrid, Madrid, Spain

Simona E. Carbone (227), Drug Discovery Biology Theme, Monash Institute of Pharmaceutical Sciences, Monash University, Parkville, VIC, Australia

Ana Carbonell-Baeza (523), Department of Physical Education, Faculty of Education Sciences, University of Cádiz, Cádiz, Spain

Rubia Casagrande (431, 443), Department of Pharmaceutical Sciences, Center of Health Sciences, Londrina State University, Londrina, Paraná, Brazil

U.G. Chandrika (321), Department of Biochemistry, Faculty of Medical Sciences, University of Sri Jayewardenepura, Nugeggoda, Sri Lanka

Antonia F. Chen (589), Department of Orthopaedic Surgery, Brigham and Women's Hospital, Harvard Medical School, Boston, MA, United States

Jerry Chen (33), Department of Mechanical Engineering, Yuan Ze University, Taoyuan, Taiwan

Giuseppe Cirino (261), Department of Pharmacy, School of Medicine and Surgery, University of Naples "Federico II", Naples, Italy

Alessandra Consales (211), IRCCS Ca' Granda Ospedale Maggiore Policlinico Foundation, Neonatal Intensive Care Unit; Department of Clinical Sciences and Community Health, University of Milan, Milan, Italy

Christopher C. Cushing (513), Clinical Child Psychology Program, Schiefelbusch Institute for Life Span Studies, University of Kansas, Lawrence, KS, United States

Christopher J. Cutter (235), Department of Psychiatry, Yale School of Medicine; Pain Treatment Services, The APT Foundation; Child Study Center, Yale School of Medicine, New Haven, CT, United States

A.J.R. da Silva (457), Institute of Research on Natural Products, Federal University of Rio de Janeiro, Rio de Janeiro, RJ, Brazil

Helga Cristina Almeida da Silva (3), Department of Surgery, Discipline of Anaesthesiology, Pain and Intensive Care, Federal University of São Paulo, São Paulo, SP, Brazil

Thomas Davergne (523), Sorbonne Université, INSERM, Pierre Louis Institute of Epidemiology and Public Health (IPLESP), Paris, France

Enrica Delfino (211), Department of Anaesthesia, Intensive Care, and Out-Hospital Emergency, Aosta Valley Regional Hospital, Beauregard Hospital, Aosta, Italy

A.C.A.X. De-Oliveira (457), Department of Biological Sciences, National School of Public Health, Oswaldo Cruz Foundation, Rio de Janeiro, RJ, Brazil

Jesse J. Di Cello (227), Department of Physiology, Biomedicine Discovery Institute, Monash University, Clayton, VIC, Australia

Pelin Corman Dincer (629), Department of Oral & Maxillofacial Surgery, Faculty of Dentistry, Marmara University, Istanbul, Turkey

Amanda M. Dionisio (443), Department of Pathology, Center of Biological Sciences, Londrina State University, Londrina, Paraná, Brazil

Jan Cz Dobrowolski (89), National Medicines Institute, Warsaw, Poland

Ofelia L. Elvir Lazo (601), Department of Anesthesiology, Cedars-Sinai Medical Center, Los Angeles, CA, United States

Fernando Estévez-López (499, 523), Department of Pediatrics, Wilhelmina Children's Hospital, University Medical Center Utrecht, Utrecht University, Utrecht, the Netherlands

Victor Fattori (431), Department of Pathology, Center of Biological Sciences, Londrina State University, Londrina, Paraná, Brazil

Magda Lourenço Fernandes (81), Faculdade de Medicina da Universidade Federal de Minas Gerais, Belo Horizonte, Minas Gerais, Brazil

Camila R. Ferraz (431, 443), Department of Pathology, Center of Biological Sciences, Londrina State University, Londrina, Paraná, Brazil

Alfonso Fiorelli (109), Division of Thoracic Surgery, University of Campania "Luigi Vanvitelli", Caserta, Italy

Anelise Franciosi (443), Department of Pathology, Center of Biological Sciences, Londrina State University, Londrina, Paraná, Brazil

Marina G. Gazzola (235), Department of Psychiatry, Yale School of Medicine; Pain Treatment Services, The APT Foundation, New Haven, CT, United States

Manuel Gerónimo-Pardo (165), Department of Anesthesiology, Integrated Care Management of Albacete, Albacete, Spain

Maria Lorella Giannì (211), IRCCS Ca' Granda Ospedale Maggiore Policlinico Foundation, Neonatal Intensive Care Unit; Department of Clinical Sciences and Community Health, University of Milan, Milan, Italy

Yukihiro Goda (385), National Institute of Health Sciences, Kawasaki, Kanagawa, Japan

Renato Santiago Gomez (81), Faculdade de Medicina da Universidade Federal de Minas Gerais, Belo Horizonte, Minas Gerais, Brazil

Arisbel B. Gondin (227), Drug Discovery Biology Theme, Monash Institute of Pharmaceutical Sciences, Monash University, Parkville, VIC, Australia

Takashi Hakamatsuka (385), Division of Pharmacognosy, Phytochemisry, and Narcotics, National Institute of Health Sciences, Kawasaki, Kanagawa, Japan

Toshihiko Hanawa (385), Department of Kampo Medicine, Oriental Medicine Research Center of Kitasato University, Tokyo, Japan

Juan L. Hancke (351), Laboratory of Inflammation Pharmacology, Institute of Pharmacology and Morphophysiology, Universidad Austral de Chile, Valdivia, Chile

Hidani Hasim (487), Department of Chemical Pathology, School of Medical Sciences, Universiti Sains Malaysia, Kubang Kerian, Kelantan, Malaysia

Emad H.M. Hassanein (181), Department of Pharmacology & Toxicology, Faculty of Pharmacy, Al-Azhar University-Assiut Branch, Assiut, Egypt

Joe C. Hong (131), Department of Anesthesiology & Peri-Operative Medicine, David Geffen School of Medicine, University of California, Los Angeles, Ronald Reagan UCLA Medical Center, Los Angeles, CA, United States

Masashi Hyuga (385), Division of Biological Chemistry and Biologicals, National Institute of Health Sciences, Kawasaki, Kanagawa, Japan

Sumiko Hyuga (385), Department of Clinical Research, Oriental Medicine Research Center of Kitasato University, Tokyo, Japan

Aminu Imam (329), Neuroscience Unit, Department of Anatomy, Faculty of Basic Medical Sciences, University of Ilorin, Ilorin, Nigeria

Małgorzata Jarończyk (89), National Medicines Institute, Warsaw, Poland

Sreekantha B. Jonnalagadda (151), School of Chemistry & Physics, Westville Campus, University of KwaZulu-Natal, Durban, South Africa

Naresh Kumar Katari (151), Department of Chemistry, School of Science, GITAM deemed to be University, Hyderabad, Telangana, India

Tiffany Kichline (513), Clinical Child Psychology Program, University of Kansas, Lawrence, KS, United States

Yohanan Kim (43), Loma Linda University, Department of Otolaryngology—Head and Neck Surgery, Loma Linda, CA, United States

Yoshinori Kobayashi (385), Department of Pharmacognosy, School of Pharmacy, Kitasato University, Tokyo, Japan

Piotr Kosson (89), Toxicology Research Laboratory, Mossakowski Medical Research Institute, Polish Academy of Sciences, Warsaw, Poland

Kongkiat Kulkantrakorn (365), Division of Neurology, Department of Internal Medicine, Faculty of Medicine, Thammasat University, Klong Luang, Pathumthani, Thailand

Ristiawan M. Laksono (139), Department of Anesthesiology and Intensive Care, Faculty of Medicine, University of Brawijaya, Kota Malang, East Java, Indonesia

Usman Latif (545), Department of Anesthesiology, University of Kansas Medical Center, Kansas City, KS, United States

Albert Lin (55), Stanford University, Department of Anesthesiology, Perioperative, and Pain Medicine, Stanford, CA, United States

Piotr F.J. Lipiński (89), Department of Neuropeptides, Mossakowski Medical Research Institute, Polish Academy of Sciences, Warsaw, Poland

Daniel López-López (555,617), Research, Health and Podiatry Group, Department of Health Sciences, Faculty of Nursing and Podiatry, Universidade da Coruña, Ferrol, Spain

Marta Losa-Iglesias (555,617), Department of Nursing, Faculty of Health Sciences, Universidad Rey Juan Carlos, Alcorcón, Spain

Ayman M. Mahmoud (181), Physiology Division, Zoology Department, Faculty of Science, Beni-Suef University, Beni Suef, Egypt

Camelia Maier (401), Department of Biology, Texas Woman's University, Denton, TX, United States

Marília F. Manchope (443), Department of Pathology, Center of Biological Sciences, Londrina State University, Londrina, Paraná, Brazil

Antonio Mancini (261), Center of Biotechnologies, Cardarelli Hospital, Naples, Italy

Eva María Martínez-Jiménez (555,617), Faculty of Nursing, Physiotherapy and Podiatry, Universidad Complutense de Madrid, Madrid, Spain

Victoria Mazoteras-Pardo (555,617), Research Group "ENDOCU", Department of Nursing, Physiotherapy and Occupational Therapy, Faculty of Physiotherapy and Nursing of Toledo, University of Castilla-La Mancha, Ciudad Real, Spain

Aleksandra Misicka (89), Department of Neuropeptides, Mossakowski Medical Research Institute, Polish Academy of Sciences; Faculty of Chemistry, University of Warsaw, Warsaw, Poland

Pablo Molina-García (499), Biohealth Research Institute, Physical Medicine and Rehabilitation Service, Virgen de las Nieves University Hospital, Granada, Spain

Eric J. Moorhead (589), Global Clinical Affairs, Avanos Medical, Alpharetta, GA, United States

Abdulrazaq Bidemi Nafiu (329), Biotranslational Research Group, University of Ilorin, Ilorin; Departments of Human Physiology, Federal University Dutse, Dutse, Jigawa, Nigeria

Shunsuke Nakamori (385), Department of Pharmacognosy, School of Pharmacy, Kitasato University, Tokyo, Japan

Calim Neder Neto (3), Department of Surgery, Discipline of Anaesthesiology, Pain and Intensive Care, Federal University of São Paulo, São Paulo, SP, Brazil

Mariana Fontes Lima Neville (3), Department of Surgery, Discipline of Anaesthesiology, Pain and Intensive Care, Federal University of São Paulo, São Paulo, SP, Brazil

Khanh Nguyen (43), Loma Linda University, Department of Otolaryngology—Head and Neck Surgery, Loma Linda, CA, United States

Hiroshi Odaguchi (385), Oriental Medicine Research Center of Kitasato University, Tokyo, Japan

S.A. Onasanwo (375), Neuroscience and Oral Physiology Unit, Department of Physiology, Faculty of Basic Medical Sciences, University of Ibadan, Ibadan, Nigeria

Adrian Ortega (513), Clinical Child Psychology Program, University of Kansas, Lawrence, KS, United States

Aboyeji Lukuman Oyewole (329), Neurophysiology, Inflammation and Tropical Diseases Unit, Department of Physiology, Faculty of Basic Medical Sciences, University of Ilorin; Bioresearch Hub Laboratory; Biotranslational Research Group, University of Ilorin, Ilorin, Nigeria

Haluk Ozdemir (281), Department of Anesthesiology and Reanimation, Marmara University, Pendik Training and Research Hospital, Istanbul, Turkey

Caterina Pace (109), Division of Thoracic Surgery, University of Campania "Luigi Vanvitelli", Caserta, Italy

Vinood B. Patel (195), Department of Biomedical Science, School of Life Sciences, University of Westminster, London, United Kingdom

F.J.R. Paumgartten (457), Department of Biological Sciences, National School of Public Health, Oswaldo Cruz Foundation, Rio de Janeiro, RJ, Brazil

Daniel P. Poole (227), Drug Discovery Biology Theme, Monash Institute of Pharmaceutical Sciences, Monash University, Parkville, VIC, Australia

Iwona Popiolek (309), Toxicology Clinic, University Hospital in Krakow, Krakow, Poland

Grzegorz Porebski (309), Department of Clinical and Environmental Allergology, Jagiellonian University Medical College, Krakow, Poland

Victor R. Preedy (195), Faculty of Life Science and Medicine, King's College London, Franklin-Wilkins Building, London, United Kingdom

Rajkumar Rajendram (195), College of Medicine, King Saud bin Abdulaziz University for Health Sciences; Department of Medicine, King Abdulaziz Medical City, Ministry of National Guard—Health Affairs, Riyadh, Saudi Arabia; Department of Anaesthesia and Intensive Care, Stoke Mandeville Hospital, Buckinghamshire, United Kingdom

Nurfuzillah Abdul Rani (419), Department of Pharmaceutical Technology, Faculty of Pharmacy and Health Sciences, University Kuala Lumpur Royal College of Medicine Perak, Ipoh, Perak, Malaysia

Fernanda Soares Rasquel-Oliveira (431,443), Department of Pathology, Center of Biological Sciences, Londrina State University, Londrina, Paraná, Brazil

W.M.K.M. Ratnayake (321), Department of Pharmaceutical and Cosmetic Sciences, Faculty of Health Sciences, CINEC Campus, Malabe, Sri Lanka

Ralitsa Raycheva (575,583), Department of Social Medicine and Public Health, Faculty of Public Health, Medical University – Plovdiv, Plovdiv, Bulgaria

M. Razeghi (565), Physical Therapy Department, School of Rehabilitation Sciences, Shiraz University of Medical Sciences, Shiraz, Iran

I. Rezaei (565), Physical Therapy Department, School of Rehabilitation Sciences, Shiraz University of Medical Sciences, Shiraz, Iran

David Rodríguez-Sanz (555,617), Faculty of Nursing, Physiotherapy and Podiatry, Universidad Complutense de Madrid, Madrid, Spain

Carlos Romero-Morales (555,617), Physiotherapy Department, Faculty of Sport Sciences, Universidad Europea de Madrid, Madrid, Spain

Barbara Rupnik (73), Department of Anaesthesiology, Balgrist University Hospital, Zurich, Switzerland

Joanna Sadlej (89), Faculty of Mathematics and Natural Sciences, Cardinal Stefan Wyszyński University in Warsaw, Warsaw, Poland

Anthony Sanchez (43), Loma Linda University, Department of Otolaryngology—Head and Neck Surgery, Loma Linda, CA, United States

Mohd. Hijaz. Mohd. Sani (419), Department of Biomedical Sciences and Therapeutics, Faculty of Medicine and Health Sciences, University Malaysia Sabah, Kota Kinabalu, Sabah, Malaysia

Pasquale Sansone (109), Division of Thoracic Surgery, University of Campania "Luigi Vanvitelli", Caserta, Italy

Mario Santini (109), Division of Thoracic Surgery, University of Campania "Luigi Vanvitelli", Caserta, Italy

Dawood Sayed (545), Department of Anesthesiology, University of Kansas Medical Center, Kansas City, KS, United States

Karlin Sevensma (27), Department of Surgery, Metro Health University of Michigan Health, Wyoming, MI, United States

Harsha Shanthanna (121), Departments of Anesthesia and Health Research Methods, Evidence and Impact, McMaster University, Hamilton, ON, Canada

Jiann-Shing Shieh (33), Yuan Ze University, Taoyuan, Taiwan

Yoshihito Shimazu (473), Laboratory of Food and Physiological Sciences, Department of Life and Food Sciences, School of Life and Environmental Sciences, Azabu University, Sagamihara, Kanagawa, Japan

Martin Ingi Sigurðsson (247), Department of Anesthesiology and Critical Care, Landspitali—The National Hospital of Iceland, Reykjavik, Iceland

Lady Christine Ong Sio (101), Department of Anesthesiology and Perioperative Medicine, University of Louisville, Louisville, KY, United States

G.R. de Souza (457), Department of Biological Sciences, National School of Public Health, Oswaldo Cruz Foundation, Rio de Janeiro, RJ, Brazil

Timothy Sowder (545), Department of Anesthesiology, University of Kansas Medical Center, Kansas City, KS, United States

Edina Szűcs (89), Institute of Biochemistry, Biological Research Centre, Hungarian Academy of Sciences; Doctoral School of Theoretical Medicine, University of Szeged, Faculty of Medicine, Szeged, Hungary

Mamoru Takeda (473), Laboratory of Food and Physiological Sciences, Department of Life and Food Sciences, School of Life and Environmental Sciences, Azabu University, Sagamihara, Kanagawa, Japan

Masahiko Tsuchiya (15), Department of Anesthesiology, Osaka City University Graduate School of Medicine, Osaka; Department of Anesthesia, Kishiwada Tokushukai Hospital, Kishiwada, Japan

Alparslan Turan (291), Department of Outcomes Research, Anesthesiology Institute, Cleveland Clinic, Cleveland, OH, United States

Nahoko Uchiyama (385), Division of Pharmacognosy, Phytochemisry, and Narcotics, National Institute of Health Sciences, Kawasaki, Kanagawa, Japan

Marina Varbanova (101), Department of Anesthesiology and Perioperative Medicine, University of Louisville, Louisville, KY, United States

Valentina Vellecco (261), Department of Pharmacy, School of Medicine and Surgery, University of Naples "Federico II", Naples, Italy

Elitsa Veneva (575, 583), Department of Pediatric Dentistry, Faculty of Dental Medicine, Medical University – Plovdiv, Plovdiv, Bulgaria

Waldiceu A. Verri, Jr (431, 443), Department of Pathology, Center of Biological Sciences, Londrina State University, Londrina, Paraná, Brazil

Thomas W. Weiser (63), Medical Affairs, Sanofi-Aventis Deutschland GmbH, Main/Main, Germany

Roberto Giorgio Wetzl (211), European e-Learning School of Obstetric Anesthesia, Rome, Italy

Paul F. White (601), Department of Anesthesiology, Cedars-Sinai Medical Center, Los Angeles; The White Mountain Institute, The Sea Ranch, CA, United States

Hüseyin Oğuz Yılmaz (291), Department of Critical Care, Muğla Education and Research Hospital, Muğla, Turkey

Roya Yumul (601), Department of Anesthesiology, Cedars-Sinai Medical Center; Department of Anesthesiology, UCLA, Charles R. Drew University of Medicine and Science, Los Angeles, CA, United States

Zainul Aminuddin Zakaria (419), Department of Biomedical Sciences, Faculty of Medical and Health Sciences, University Putra Malaysia, Serdang, Selangor, Malaysia

Tiago H. Zaninelli (431), Department of Pathology, Center of Biological Sciences, Londrina State University, Londrina, Paraná, Brazil

Preface

The etiology of pain is complex and multifactorial. This complexity is magnified by the use of analgesics and local or general anesthetics. While analgesics can reduce pain, general anesthetics reduce consciousness and local anesthetics reduce localized pain. Analgesics may be used after surgery performed under anesthesia. However, anesthesia is not without risk. Depending on patient-related factors such as comorbid disease, anesthesia may be associated with significant morbidity or even mortality. Adverse events during or after anesthesia may necessitate the use of other pharmacological agents such as vasopressors, inotropes, sedatives, antiarrhythmics, or antiemetics.

The perception of pain itself results from a multifaceted interaction between illness beliefs, age, gender, time of onset, stress, socioeconomic status, and other factors. To a certain extent, one could argue that pain itself is helpful in treating disease. It can be considered the "sixth vital sign" as it indicates the need for assessment by a healthcare professional and the need for clinical investigations, diagnosis, and appropriate medication. The pain associated with myocardial infarction is a good example of this. One needs to consider though that some acute and chronic pain can lead to psychological distress and reduced quality of life. In the long term, chronic persistent pain can impact significantly on the family unit and many diseases present with pain. There are a plethora of pharmacological agents currently available for pain management. Furthermore, studies showing the beneficial effects of plant or natural extracts provide the foundation for further rigorous studies in clinical trials.

The neuroscience of pain in one condition may be relevant to understanding the pain observed in other conditions. The onset of pain, the cause of the pain, and the administration of analgesia or anesthesia is a continuing scientific spectrum. At each point, there is a firm scientific basis with established literature and also an ongoing drive to discover new facts and data. Hitherto, such material on the neuroscience of pain, anesthesia, and analgesia has been sporadic and/or written for the experts who specialize in narrow and focused areas. For example, the expert in the use of general anesthetics in the surgical setting may not necessarily be an expert in the molecular biology of neurons activated in the pain process. The cellular biologist may be aware of neither the science underpinning the provision of anesthesia nor the adverse outcomes that the anesthesiologist may encounter in the clinical setting. To address the aforementioned issues, the editors have compiled *The Neuroscience of Pain, Anesthetics, and Analgesics*.

The Neuroscience of Pain, Anesthetics, and Analgesics is divided into three books:

Book 1: *Features and Assessments of Pain, Anesthesia, and Analgesia*
Book 2: *Treatments, Mechanisms, and Adverse Reactions of Anesthetics and Analgesics*
Book 3: *The Neurobiology, Physiology, and Psychology of Pain*

This book, *Treatments, Mechanisms, and Adverse Reactions of Anesthetics and Analgesics*, is divided into four parts. *Part I, Drugs and Agents Used in Anesthesia and Analgesia*, covers topics on local anesthetic agents in the management of postoperative pain, use of dextran in regional anesthesia, and nonopioid-based analgesia. *Part II, Mechanisms of Action of Drugs*, covers agents such as Buprenorphine, Chloroprocaine, Caffeine, Clonidine, Fentanyl, Isoflurane, lidocaine, Memantine, Paracetamol, Sevoflurane, Tramadol, and Prilocaine. *Part III, Adverse Effects, Reactions, and Outcomes*, covers topics on the long-term effects of anesthesia on the brain, cognitive behavioral therapy for opioid use disorder, headache after neuroaxial blocks, and adverse events associated with Paracetamol. *Part IV, Novel and Nonpharmacological Aspects and Treatments*, covers topics on *Adansonia digitata* and use in neuropathic pain, *Cola nitida* and pain relief, Capsaicin for diabetic neuropathic pain, *Acronychia pedunculata* leaves for pain treatment, *Andrographis paniculata* extract for pain, and analgesic effects of Ephedra herb extract.

Each chapter has:

- *An Abstract (published online)*
- *Key Facts*
- *Mini-Dictionary of Terms*

xxvii

xxviii Preface

- *Applications to Other Areas*
- *Other Agents of Interest*
- *Summary Points*

The section *Key Facts* focuses on areas of knowledge written for the novice. The *Mini-Dictionary of Terms* explains terms that are frequently used in the chapter. The *Applications to Other Areas* describes other fields of neuroscience, pain science, analgesia, and anesthesia that the chapter may be relevant to. The *Summary Points* encapsulates each chapter in a succinct way.

The Neuroscience of Pain, Anesthetics, and Analgesics is designed for research and teaching purposes. It is suitable for neurologists and anesthesiologists as well as those interested in pain and its interlink with pain relief. It is valuable as a personal reference book and also for academic libraries that cover the neuroscience of pain, anesthetics, and analgesics. Contributions are from leading national and international experts including those from world-renowned institutions.

Rajkumar Rajendram, Victor R. Preedy, Vinood B. Patel, and Colin R. Martin
(Editors)

Part I

Drugs and agents used in anesthesia and analgesia

Chapter 1

Anesthesia for malignant hyperthermia susceptible patients

Calim Neder Neto, Mariana Fontes Lima Neville, and Helga Cristina Almeida da Silva

Department of Surgery, Discipline of Anaesthesiology, Pain and Intensive Care, Federal University of São Paulo, São Paulo, SP, Brazil

Abbreviations

MH	malignant hyperthermia
MHS	susceptible to malignant hyperthermia
Ca^{2+}	calcium ion
RyR1	ryanodine receptor type 1
DHPR	dihydropyridine receptor
CK	serum creatine kinase level
PACU	post-anaesthesia care unit
ETCO$_2$	end-tidal carbon dioxide
DIC	disseminated intravascular coagulation
VO$_2$	oxygen consumption
IVCT	in vitro contracture test
IV	intravenous
ICU	intensive care unit
O$_2$	oxygen
PaCO$_2$	partial pressure of carbon dioxide in arterial blood
BE	base excess
H$^+$	hydrogen ion

Introduction

Malignant hyperthermia (MH) is an inherited pharmacogenetic disorder with an autosomal dominant pattern, incomplete penetrance and variable expression, associated with several mutations (Rosenberg, Pollock, Schiemann, Bulger, & Stowell, 2015). The disease is expressed by a cellular hypermetabolic reaction when patients susceptible to Malignant Hyperthermia (SHM) are exposed to triggering agents such as halogenated volatile anaesthetics, depolarizing neuromuscular blocker (succinylcholine), or, rarely, heat/intense physical exercise (Fig. 1) (Hopkins, 2011; Rosenberg et al., 2015).

The first report of MH occurred in the 1960s (Ball, 2007; Denborough, Forster, Lovell, Maplestone, & Villiers, 1962). The original patient reported a positive family history, with 10 unexplained deaths during general anaesthesia in relatives, and previous instructions to avoid this anaesthesia (Ball, 2007; Denborough et al., 1962). When submitted to general anaesthesia with halothane to correct a fracture of the lower limb, he presented, minutes after starting the volatile drug, tachycardia, hypotension, sweating, and rapid consumption of soda lime—an indirect sign of increased carbon dioxide production (Ball, 2007; Denborough et al., 1962). Halothane was discontinued, blood transfusion and cooling of the patient with ice bags were performed, and the patient survived (Ball, 2007; Denborough et al., 1962).

The incidence of MH crisis is estimated between 1 case for every 10,000 to 250,000 anaesthesias (Rosenberg et al., 2015). Although the MH crisis can be triggered in the first anaesthesia, the episode usually occurs after the third exposure (Glahn et al., 2020). The MH crisis affects different ethnicities and occurs more frequently in male patients (2:1), compared to female patients (Rosenberg et al., 2015).

Malignant hyperthermia is a potentially fatal disease if not diagnosed and treated early and correctly (Rosenberg et al., 2015). The spread of knowledge about this disease, associated with the emergence of dantrolene sodium as a specific treatment, was responsible for reducing mortality from more than 80% in the 1980s to less than 5% today (Riazi,

Treatments, Mechanisms, and Adverse Reactions of Anesthetics and Analgesics. https://doi.org/10.1016/B978-0-12-820237-1.00001-6
Copyright © 2022 Elsevier Inc. All rights reserved.

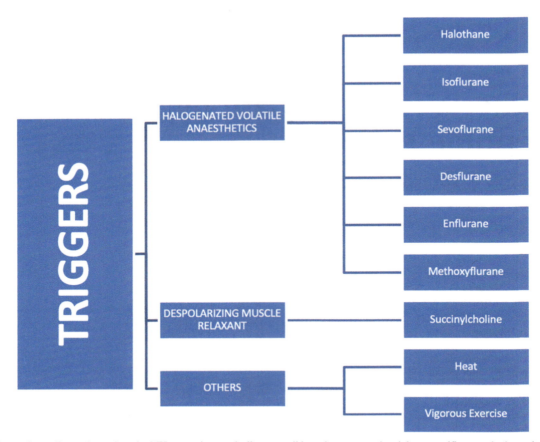

FIG. 1 Triggers for malignant hyperthermia. MH occurs in genetically susceptible patients exposed mainly to specific anaesthetics and more rarely to environmental stimuli.

Kraeva, & Hopkins, 2018; Rosenberg et al., 2015). The pathophysiology and molecular pathways involved in MH will be fully reviewed in another chapter of this book. Briefly, exposure to triggering agents leads to an abnormal increase in calcium levels in the skeletal muscle cell cytoplasm, thus exponentially increasing its contraction and metabolism (Hirshey Dirksen et al., 2011; Rosenberg et al., 2015). The regulation of this ion (Ca^{2+}) in the muscle cytoplasm is mainly linked to the type 1 ryanodine receptor (RyR1), activated by the dihydropyridine receptor (DHPR), and to the Ca^{2+}-ATPase system (Gonsalves et al., 2019; Hirshey Dirksen et al., 2011; Rosenberg et al., 2015). There are several mutations described in the gene encoding the RyR1 receptor on chromosome 19 that are related to SHM, as well as some mutations on chromosome 1, in the DHPR receptor gene (Gonsalves et al., 2019; Rosenberg et al., 2015).

Despite the rarity of the MH crisis, it is extremely important that anaesthesiologists be aware of its pathophysiology and the MHS associated diseases, in order to recognize the signs and symptoms of the acute crisis, its differential diagnosis and treatment, and how to conduct an anaesthesia in patients known to be susceptible to MH or at high risk, such as individuals with a positive family history for MH.

Diseases associated with susceptibility to MH

Susceptibility to MH is a concern in myopathies related to RyR1 receptor gene mutations, called "ryanodinopathies" (Bamaga et al., 2016; Litman, Griggs, Dowling, & Riazi, 2018; Riazi et al., 2018). It is estimated that up to 30% of patients with ryanodinopathies will also be susceptible to MH (Bamaga et al., 2016; Litman et al., 2018; Riazi et al., 2018). The main conditions that can be associated with *RyR1* gene variants are: Central Core Disease, King-Denborough Syndrome, Congenital Myopathy with Central or Internalized Nuclei, Statin Myopathies, Multiminicore Disease, Congenital Fiber-type Disproportion, Heat/Exercise-induced Exertional Rhabdomyolysis, Benign Samaritan Congenital Myopathy, Nemaline Rod Myopathy, Atypical Periodic Paralysis, and Idiopathic Hyper-CKemia (Bamaga et al., 2016; Litman et al., 2018; Riazi et al., 2018).

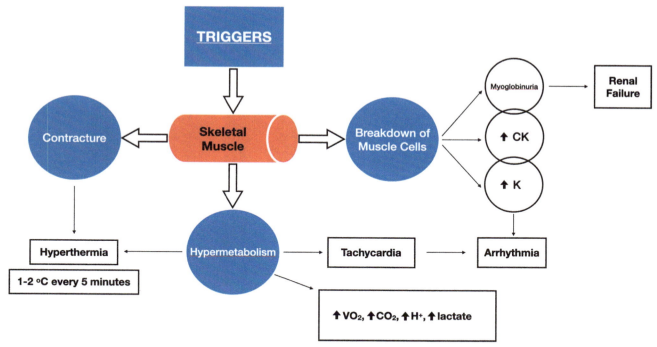

FIG. 2 Main clinical and laboratory findings of malignant hyperthermia crisis and associated physiopathologic processes. *CK*, serum creatine kinase level; *K*, serum potassium level; VO_2, oxygen consumption; CO_2, carbon dioxide; H^+, hydrogen.

Due to the possible correlation between these conditions and susceptibility to MH, patients should be considered as susceptible until the contrary is proven and, if they are submitted to anaesthesia, it must be performed without triggering agents (Litman et al., 2018; Riazi et al., 2018). Patients should ideally be referred to centres specialized in MH to proceed with the recommended diagnostic investigation (Litman et al., 2018; Riazi et al., 2018).

Clinical manifestation—MH crisis

There is a broad spectrum of clinical manifestations in the MH crisis. Symptoms and signs may start as soon as the patient is exposed to the triggering agent or even hours later (Fig. 2) (Ellis, Halsall, & Christian, 1990; Glahn et al., 2020; Riazi et al., 2018; Rosenberg et al., 2015).

Commonly, the earliest signs are tachycardia and hypercapnia, easily detected by monitoring the concentration of expired carbon dioxide ($ETCO_2$) (Glahn et al., 2020; Riazi et al., 2018; Rosenberg et al., 2015). The clinical findings result from the cellular hypermetabolism: hyperthermia, tachycardia, tachypnea, increased oxygen consumption (VO_2), metabolic acidosis, muscle contracture, and rhabdomyolysis, with resulting hyperkalemia and myoglobinuria (Glahn et al., 2020; Riazi et al., 2018; Rosenberg et al., 2015). Hyperthermia, despite being part of the name of the disease, may be absent or a late sign (Glahn et al., 2020; Riazi et al., 2018; Rosenberg et al., 2015). Patients can present dysfunction of multiple organs and systems, with cerebral oedema, cardiac arrhythmias due to hyperkalemia, acute renal failure due to myoglobinuria, acute pulmonary oedema, hepatitis, disseminated intravascular coagulation (DIC), and muscle compartment syndrome (Glahn et al., 2020; Riazi et al., 2018; Rosenberg et al., 2015).

Usually, the MH crisis is classified according to its clinical presentation, using the description proposed by Ellis et al. (1990) and Ranklev-Twetman (1990):

- Classical fulminant: rapid onset, exuberant, potentially fatal, muscular and metabolic clinical manifestations (typical crisis);
- Moderate: metabolic and muscular manifestations, without risk of death (abortive crisis);
- Mild: mild metabolic manifestations, without clinical manifestations (abortive crisis);
- Masseter rigidity or spasm with evidence of muscle injury;
- Masseter rigidity or spasm with metabolic alteration;
- Isolated masseter rigidity or spasm;

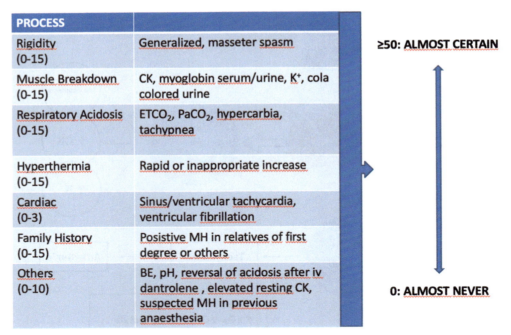

FIG. 3 Likelihood of the episode being a malignant hyperthermia crisis (Larach et al.). The clinical grading scale considers seven main processes that occur during a MH crisis. The total score of the scale composed by Larach et al. (1994) is used to predict the likelihood that an adverse event was a malignant hyperthermia episode.

- Unexplained perioperative death or cardiac arrest (atypical crisis);
- Other presentations: unexplained perioperative fever, rhabdomyolysis or acute kidney injury (atypical crisis).

In the 1990s, Larach et al. proposed a scale to quantify the likelihood of an event being a MH crisis (Larach et al., 1994). This classification may be used for epidemiological studies and studies of correlation with results of diagnostic tests, but it should not delay or even inhibit the institution of treatment at the time of the acute crisis (Fig. 3) (Larach et al., 1994).

Diagnosis and differential diagnosis

The diagnosis of MH crisis is based on clinical findings of hypermetabolism after the patient's exposure to any of the triggering agents (Glahn et al., 2010; Rosenberg et al., 2015). Patients may occasionally have muscle complaints in daily life, such as myalgia, cramps or intolerance to physical exertion (Santos et al., 2017).

Among the differential diagnoses at the moment of MH crisis, there are diseases that present with hypermetabolism or with thermoregulatory changes, specially the neuroleptic malignant syndrome, serotonin syndrome, status epilepticus, pheochromocytoma, drug intoxication (cocaine, ecstasy), infectious tetanus, delirium tremens, thyrotoxicosis, fatal catatonia, heatstroke, sepsis, stroke, and traumatic brain injury (Boushra, Miller, Koyfman, & Long, 2018; Simon, 1993, 1994).

The gold standard for the diagnosis of susceptibility to MH is the in vitro muscle contracture test after exposure to caffeine or halothane, according to European or North American protocols (Hopkins et al., 2015; Rosenberg et al., 2015; Rosenberg, Antognini, & Muldoon, 2002). The muscle sample is obtained by biopsy, preferably from the quadriceps thigh muscle, and the test must be performed within 5 h (Ellis et al., 1984).

To properly perform the test, the muscle sample must be maintained in a Krebs-Ringer solution at 37°C and oxygenated continuously with carbogen, submitted to supramaximal stimuli of 1 ms and frequency of 0.2 hertz (Ellis et al., 1984). The purpose of the test is to assess the contractile response of this muscle when exposed to halothane or caffeine (Ellis et al., 1984; Hopkins et al., 2015; Rosenberg et al., 2002, 2015). Thus, it is possible to classify patients as not susceptible (normal response: absence of contracture above the threshold) or susceptible (abnormal response, with contractures greater than the threshold when exposed to lower concentrations of agents) (Fig. 4) (Ellis et al., 1984; Hopkins et al., 2015; Rosenberg et al., 2002, 2015).

The use of genetic tests for the diagnosis of susceptibility to MH is restricted to the situations specified in the guidelines of the European MH Group (www.emhg.org) (Hopkins et al., 2015). There are several genes involved in this disease, with

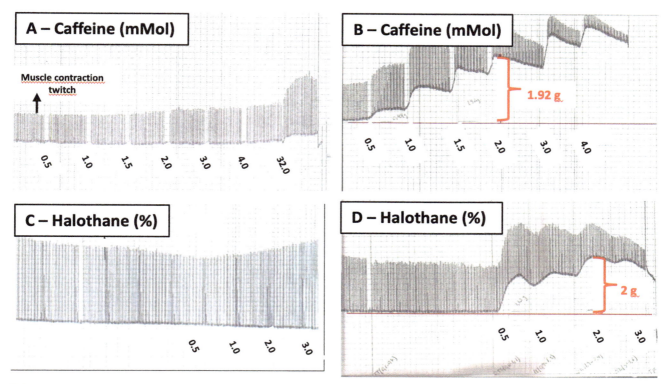

FIG. 4 In vitro contracture test (IVCT): examples of positive and negative tests, with caffeine or halothane. Example of IVCT, figure from author's portfolio. Graphics A and C are negative tests for caffeine and halothane respectively. Graphics B and D are positives tests. A test is considered positive with a contracture of at least 0.2 g before 2 mMol caffeine or 2% halothane.

several mutations described, and in many families the responsible genes/mutations have not yet been identified (Hopkins et al., 2015; Monnier, Procaccio, Stieglitz, & Lunardi, 1997; Rosenberg et al., 2002, 2015). Currently, the finding of a pathogenic mutation, previously proven to be associated with MH in the ryanodine or dihydropyridine genes, in a patient with a personal or family history of MH, is accepted as a positive genetic test and confirms the diagnosis of susceptibility to MH (Hopkins et al., 2015; Monnier et al., 1997; Rosenberg et al., 2002, 2015). However, a negative result in genetic tests does not exclude the diagnosis of MH susceptibility (Monnier et al., 1997; Rosenberg et al., 2015).

Treatment of MH crisis

Once the diagnosis of the MH crisis has been made, the patient's exposure to triggering agents should be stopped immediately, followed by the administration of the antidote dantrolene sodium (blockage of the excessive release of intracellular calcium) and the installation of clinical support measures, following a checklist to face the crisis and post-crisis moments (Hardy et al., 2018; Krivosic-Horber, 1990). There are several algorithms and guidelines for diagnosis and treatment of MH crisis, such as those from the European MH Group (www.emhg.org), the North American MH Group (www.mhaus.org), and several countries, as for example the British register of MH (www.ukmhr.ac.uk).

Immediate measures include patient hyperventilation with 100% oxygen inspired fraction and minute volume of 2–3 times the baseline with high flow, recognition of the situation as critical and call for help, change for total intravenous anaesthesia, informing the surgeon of the situation and finishing the surgery if possible, placement of the specific activated carbon filter between the patient and the anaesthesia circuit, and removal of the vaporizer from the anaesthesia station without wasting time changing the circuit and anaesthetic machine (Fig. 5) (Glahn et al., 2010; Riazi et al., 2018; Rosenberg et al., 2015).

Dantrolene is a lipophilic hydantoin derivative that acts by blocking RyR1 receptors (Flewellen, Nelson, Jones, Arens, & Wagner, 1983; Glahn et al., 2020; Krause, Gerbershagen, Fiege, Weißhorn, & Wappler, 2004). In initial doses of 2.4 mg/kg, it leads to a plasma concentration of 4.2 μg/mL, resulting in block of approximately 75% of skeletal muscle contraction (Flewellen et al., 1983). The elimination half-life is 12 h and the plasma concentration remains stable for up to 5 h after its administration (Krause et al., 2004). Dantrolene is metabolized by the liver and its metabolites are

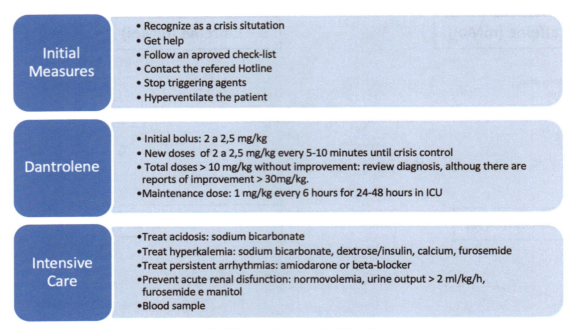

FIG. 5 Management of the malignant hyperthermia crisis. *ICU*, intensive care unit; *MH*, malignant hyperthermia.

excreted in the urine and bile (Dykes, 1975; Flewellen et al., 1983; Krause et al., 2004). There are two forms of presentation: the first and most available is in 20 mg/60 ml bottles (Dantrium) with final sodium dantrolene concentration of 0.33 mg/ml and pH 9.5, including mannitol in its composition; the second and more recent has presentation of 250 mg/5 ml bottles (Ryanodex) and is easier to dilute but not yet widely available (Glahn et al., 2020; Krause et al., 2004; Rosenberg et al., 2015). The dantrolene solution should be administered quickly as a bolus by a large vein, due to the possibility of phlebitis (Glahn et al., 2020; Krause et al., 2004). The initial dose of 2 to 2.5 mg/kg can be repeated every 5–10 min until the control of the acute crisis and improvement of clinical signs (control generally occurs below total doses of 10 mg/kg) (Glahn et al., 2010; Riazi et al., 2018). The MH crisis is considered to be controlled when the following criteria of clinical stability are present: $ETCO_2$ falling or normal, heart rate stable and without signs of arrhythmias, intravenous dantrolene already started, temperature decline and improvement of muscle rigidity (Larach et al., 2012). The main side effects of dantrolene are muscle weakness, phlebitis, respiratory distress, and gastrointestinal disorders (Flewellen et al., 1983).

Routine patient monitoring should be maintained (oximetry, electrocardiography, non-invasive blood pressure, and capnography), as well as central temperature and bladder catheterization to assess diuresis (Glahn et al., 2010; Riazi et al., 2018). Monitoring of invasive blood pressure, also used for serial collection of exams, and insertion of a central venous catheter should be considered (Glahn et al., 2010; Riazi et al., 2018). Serial laboratory tests are collected every 6 h: arterial blood gas, serum potassium, serum CK, serum/urine myoglobin, glucose, coagulation profile, and renal function (Glahn et al., 2010; Riazi et al., 2018; Rosenberg et al., 2015).

Active cooling must be installed in order to maintain the patient's temperature below 38.5°C and can be stopped when this value is reached (Rosenberg et al., 2015). Cooling measures include the administration of intravenous saline solution (2000–3000 mL) cooled to 4°C and the cooling of the body surface with wet and cold cloths or bags with ice, especially in neck, axilla, and groin regions (Glahn et al., 2010; Riazi et al., 2018; Rosenberg et al., 2015). The acute kidney injury is prevented with the maintenance of an urine output above 2 ml/kg/h, with the use of volemic expansion to achieve normovolemia and diuretics such as furosemide 0.5–1 mg/kg iv and mannitol 1 g/kg; it must be considered that each vial of dantrolene already has 3 g of mannitol (Glahn et al., 2010; Riazi et al., 2018; Rosenberg et al., 2015).

To avoid the appearance of life-threatening arrhythmias, hyperkalaemia should be treated quickly with sodium bicarbonate (if metabolic acidosis is present) or polarizing solution (10 units of regular insulin in 50 ml of glucose 50%) (Glahn et al., 2010; Riazi et al., 2018). In the presence of electrocardiographic alterations (peaked T wave, flattened P wave, or widened QRS complex), 10 to 30 mg/kg of calcium chloride should be administered to stabilize the myocyte membrane (Glahn et al., 2010). Cardiac arrhythmias generally stop when the hydro-electrolytic balance is re-established, but if they persist, standard antiarrhythmic agents such as amiodarone (3 mg/kg iv) or beta-blockers can be used (Glahn et al., 2010; Riazi et al., 2018). The use of calcium channel blockers as antiarrhythmic agents in MH crisis is contraindicated because, in association with dantrolene, they can lead the patient to hemodynamic collapse (Rosenberg et al., 2015).

After stabilization of the clinical condition, as evidenced by the decrease in $ETCO_2$, absence of arrhythmias, control of hyperthermia, and normal muscle tonus, patients must remain under intensive care for at least more 24 h with serial blood samples every 6 h to assess their evolution (Glahn et al., 2010). Maintenance dose of dantrolene (1 mg/kg 6/6 h) is necessary for 24–48 h, according to the evolution (Rosenberg et al., 2015).

After resolution of the condition and hospital discharge, the patient and family should be referred to a specialized centre to the complementary diagnostic tests and investigation of MH susceptibility (Glahn et al., 2010).

Anesthesia for patients susceptible to malignant hypertermia

The development of the acute event of MH results from two main factors, the susceptibility to MH and the exposure to the triggering agents (Hopkins, 2011; Rosenberg et al., 2015). As MH is a pharmacogenetic and potentially fatal disease if not diagnosed and treated early, until proven otherwise, all members of families where there is a history suggestive of malignant hyperthermia are considered as susceptible (Rosenberg et al., 2015). Possibly susceptible patients and those that have been already confirmed as susceptible, by the in vitro contracture test and/or molecular study, need an individualized anaesthetic preparation to avoid any contact with the MH triggering agents during elective and planned procedures. But the MH crisis can occur during anaesthesia in a patient with no previous history, a situation that requires a trained team to quickly start treatment and implement the safe anaesthesia protocol for MH (Hardy et al., 2018).

The patient with a previous suspected MH event or in high risk due to MH family history should ideally undergo consultation at a specialized centre to perform the in vitro contracture test and/or genetic test to confirm or discard susceptibility to MH before any elective anaesthetic procedure (Allen, Larach, & Kunselman, 1998; Ellis et al., 1984; Ørding et al., 1997; Urwyler, Deufel, McCarthy, & West, 2001). However, absence of the definitive MH diagnosis should not be a reason for refusing or cancelling the anaesthesia and surgery, especially when the patient does not have access to the ideal prior investigation or presents in an emergency situation or (Silva et al., 2019).

Anaesthesia free of triggering agents is considered safe for patients susceptible to MH (Scala et al., 2006; Silva et al., 2019; Wappler, 2010). However, even without the triggering agents, the patient susceptible to MH would still have a theoretical chance of 0.46% for developing MH, and the anaesthetic team would need to be skilled to diagnose and treat the MH crisis (Carr, Lerman, Cunliffe, McLeod, & Britt, 1995; Grinberg, Edelist, & Gordon, 1983).

The approach to safe anaesthesia of the patient definitely or possibly susceptible to MH has two stages, the pre-operative care and the perioperative care.

Pre-operative care

It is mandatory to obtain a detailed clinical history of MH patients and their families, with a focus on investigating previous events or complications during anaesthesia (Riazi et al., 2018; Rosenberg et al., 2015; Wappler, 2010). Personal complaints such as cramps, muscle pain, and exercise intolerance, or a history of neuromuscular diseases, are commonly found in ryanodinopathies and should motivate evaluation with a specialized neurologist (Litman et al., 2018; Wappler, 2010). Whenever necessary, it is possible to contact a specialized MH centre to clarify questions about the safe anaesthetic procedure. The websites of the European (www.emhg.org) and North-American MH groups (www.mhaus.org) provide the list of MH research centres around the world (Wappler, 2010).

The MH patient can be anxious and insecure about undergoing anaesthesia (Litman et al., 2018; Rosenberg et al., 2015; Wappler, 2010). The anaesthesiologist can reassure the patient, exposing the risks but stating that the entire perioperative period will be performed according to safe anaesthesia protocols for MH (Wappler, 2010). It should be emphasized that no drugs that trigger the MH crisis will be used (Litman et al., 2018; Rosenberg et al., 2015). Currently, specific complementary exams are not usually requested based only on the MH susceptibility (Wappler, 2010). The complementary investigation must be based on co-existing diseases, as it is recommended for patients without MH (Wappler, 2010).

In the past, it was believed that the use of oral dantrolene before surgery would provide protection against the development of the MH crisis, but this prophylactic use showed to be ineffective and is not anymore recommended (Krause et al., 2004). Additionally, the plasma level after oral dantrolene is not adequate to a MH crisis and the patient could present the side effects of this medication (Krause et al., 2004; Wappler, 2010).

Perioperative care

The patient should be as relaxed as possible to avoid stress as a possible trigger of the MH crisis (Rosenberg et al., 2015; Wappler, 2010). If there are no contraindications, preanaesthetic for anxiolysis could be prescribed, such as benzodiazepines and ketamine, which are safe for these patients (Wappler, 2010).

An important step in the care for MH susceptible patients is the correct "decontamination" of the anaesthesia machine so that it presents less than 5 ppm remnants of the halogenated volatile drug, a value considered safe because it does not trigger MH crises in pigs (Hopkins, 2011; Litman et al., 2018; Rosenberg et al., 2015; Wappler, 2010). Initially, vaporizers should be removed from the room and the removable parts of the anaesthesia system (soda lime, canister, silicone circuit, connectors) that may be contaminated with halogenated agents should be replaced (Fig. 6) (Rosenberg et al., 2015; Wappler, 2010). In addition, other triggering medications, such as succinylcholine, are removed from the operative room (Hopkins, 2011; Rosenberg et al., 2015; Wappler, 2010).

Each anaesthetic machine producer must determine the time required for each system to be cycled at a high flow of fresh gases (10 L/min) so that the system is clean of halogenated agents (Brünner, Pohl, & Grond, 2011; Crawford, Prinzhausen, & Petroz, 2007; Petroz & Lerman, 2002; Prinzhausen, Crawford, O'Rourke, & Petroz, 2006; Silva et al., 2019; Wappler, 2010; Whitty, Wong, Petroz, Pehora, & Crawford, 2009). Older devices required a time of 10 min to be clean while the most modern devices could require up to 120 min to do so (Brünner et al., 2011; Crawford et al., 2007; Petroz & Lerman, 2002; Prinzhausen et al., 2006; Silva et al., 2019; Wappler, 2010; Whitty et al., 2009). Currently, it is recommended, to facilitate the cleaning of the anaesthetic machine, the use of activated carbon filter in the inspiratory and expiratory branches of the anaesthesia circuit, to effectively accelerate the removal of the halogenated particles from the system and reducing the total time to less than 10 min (Gunter, Ball, & Than-Win, 2008; Müller-Wirtz et al., 2020; Rosenberg et al., 2015; Thoben et al., 2019).

The monitoring of the patient must be based on the surgical procedure and the clinical status of the patient, but the monitoring of the body core temperature and capnography are essential (Wappler, 2010). It is recommended to provide

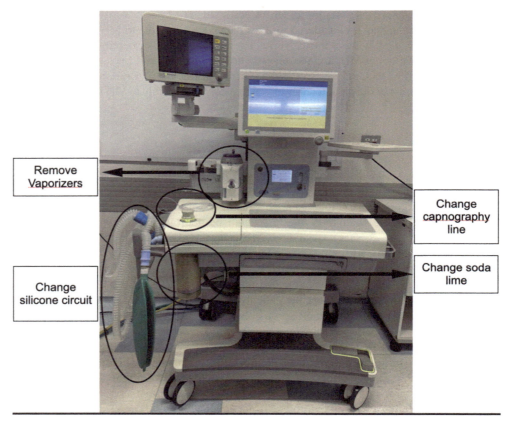

FIG. 6 Preparation of anaesthetic workstation for anaesthesia of MH patients. Course of action for preparing the anaesthetic machine for safe MH anaesthesia.

all the necessary material to manage a possible MH crisis, such as peripheral and central venous catheters, gastric and bladder probes, soda lime, tubes for collecting exams, syringes and needles of different sizes, ice, medications for critical events, and enough dantrolene for both treating the acute crisis and preventing recrudescence in the next 24–48 h (Silva et al., 2019; Wappler, 2010). The amount of dantrolene required for the first dose of 2.5 mg/kg in an adult of 70 kg would be 175 mg, and control of the crisis is usually achieved with up to 10 mg/kg, which implies having 700 mg of dantrolene (36 vials of 20 mg or three vials of 250 mg) for the acute crisis (Glahn et al., 2020).

General anaesthesia free of triggering agents or regional anaesthesia techniques associated with adequate sedation are safe in MH patients (Rosenberg et al., 2015; Wappler, 2010). Benzodiazepines, barbiturates, propofol, etomidate, ketamine, opioids, nitrous oxide, nondepolarizing neuromuscular blockers and local anaesthetics can be used (Rosenberg et al., 2015; Wappler, 2010). The end of surgery and the awakening of anaesthesia must be conducted in a smooth manner (Wappler, 2010). Patients susceptible to MH who have undergone anaesthesia with all appropriate care can safely perform the immediate postoperative period in the same post-anaesthetic recovery room (PACU) in which other patients are exhaling remnants of halogenated anaesthetics, since that the room is well ventilated (Barnes, Stowell, Bulger, Langton, & Pollock, 2015). The PACU care must follow the usual routine according to the patient's clinical status and surgical procedure, and nothing specific is necessary due to susceptibility to MH (Barnes et al., 2015). MH patients can safely undergo outpatient surgery without the need to stay overnight in the hospital and they can be discharged from the service 2 h after the end of anaesthesia free of triggering agents and without complications (Barnes et al., 2015; Brandom, 2009).

Applications to other areas

In this chapter we have reviewed main topics about anaesthesia for malignant hyperthermia (MH) patients. We have summarized key facts about MH, such as its definition as a hypermetabolic pharmacogenetic syndrome associated with halogenated agents and succinylcholine, the first report in 1960s, epidemiology, mortality, pathophysiology, and genetics. In the following section we described the most frequent diseases associated with susceptibility to MH, such as the central core disease myopathy, a disease allelic with MH. After, we detailed the cardinal clinical manifestations of a MH crisis including hypercarbia, tachycardia, muscle rigidity, acidosis, and hyperthermia. The differential diagnosis presented other causes of hyperthermia during anaesthesia that must be investigated. The section of diagnosis referred the guidelines for establishing the MH susceptibility confirmation by the gold standard in vitro contracture test, as well the situations where the molecular investigation can be employed. Treatment in the acute and post-acute phases was delineated, with emphasis in the specific antidote dantrolene. We explained the pre-operative and perioperative care during anaesthesia for patients susceptible to MH, underscoring the importance of cleaning the anaesthetic machine of all traces of halogenated agents.

Adequate knowledge about MH has important implications in the safety of the anaesthetic act, due to its high mortality and fast development. Delays in the diagnosis and correct treatment with dantrolene increase the morbidity and mortality of this condition. Therefore, adequate training of anaesthetic team is recommended.

Other agents of interest

In this chapter we have described anaesthetic drugs that trigger malignant hyperthermia (MH) in genetically susceptible patients. Although halothane was the first drug associated with MH, with faster beginning of hypermetabolism and more severe clinical manifestations, the other halogenated agents, as well the nondepolarizing neuromuscular blocker succinylcholine, can also trigger a MH crisis:

Desflurane
Enflurane
Halothane
Isoflurane
Methoxyflurane
Sevoflurane

A successful anaesthesia for malignant hyperthermia susceptible patients includes the avoidance of triggers cited above and the use of safe anaesthetics, such as the examples listed below:

Barbiturates: Thiopental
Benzodiazepines: Diazepam, lorazepam, and midazolam
Dexmedetomidine

Etomidate
Ketamine
Local anaesthetics: Bupivacaine, lidocaine, mepivacaine, prilocaine
Nitrous oxide, Xenon
Non depolarizing neuromuscular blockers: Rocuronium, vecuronium, pancuronium, atracurium, cisatracurium, mivacurium
Opioids: Alfentanil, fentanyl, remifentanil, sufentanyl
Propofol
Sugamadex

Mini-dictionary of terms

Disseminated Intravascular Coagulation (DIC): it is a systemic activation of blood coagulation with microvascular thrombosis leading to multiple organ dysfunctions. The consumption of clotting factors and platelets can result in severe hemorrhages.

Hypercapnia: or hypercarbia: It is a situation where there is abnormally elevated CO_2 levels in the blood/exhaled air.

Hyperkalaemia: increase in the level of plasmatic potassium above the normal value. This condition may lead to severe and fatal arrhythmias.

Malignant Hyperthermia triggering agents: agents/drugs that may precipitate a Malignant Hyperthermia crisis.

Post-Anaesthetic Recovery Room: a specific room where patients are monitored and stay while recovering after anaesthesia. It is in close proximity to the operating rooms and has dedicated staff.

Soda Lime: it is a chemical compound made of the mixture of calcium hydroxide and sodium hydroxide. It is used in granular form to remove the carbon dioxide from breathing gases of a closed breathing system.

Vaporizers: it is a medical device that is attached to an anaesthetic machine. It is responsible to deliver a chosen concentration of the volatile agent. It is used mainly in general anaesthesia.

Key facts of malignant hyperthermia

- Malignant Hyperthermia (MH) is a pharmacogenetic and potentially fatal disease.
- MH is expressed by a cellular hypermetabolism reaction after exposure to triggering agents.
- The incidence of MH crisis is estimated between 1 case for every 10,000 to 250,000 anaesthesias.
- There was a reduction of mortality from more than 80% in the 1980s to less than 5% nowadays with the appropriate diagnosis and treatment.
- The abnormal increase in calcium levels in the skeletal muscle cell cytoplasm exponentially increases muscle contraction and metabolism.
- Several mutations are described in the genes encoding the RyR1 and DHPR receptors.

Summary points

- Due to the possible correlation between some neuromuscular conditions and susceptibility to MH, these patients should be considered as susceptible until the contrary is proven.
- There is a broad spectrum of clinical manifestation of the MH crisis. Commonly, the earliest signs are tachycardia and hypercapnia.
- Outside of the crisis, the gold standard for the diagnosis of susceptibility to MH is the in vitro test of muscle contraction to exposure to caffeine and halothane.
- MH crisis: the patient's exposure to triggering agents should be stopped immediately, administer the dantrolene sodium antidote and clinical support measures.
- Patients susceptible to MH need an individualized anaesthetic preparation to avoid any contact with the MH triggering agents: correct "decontamination" of the anaesthesia machine, remove vaporizers and replace the removable parts of the anaesthesia system.

References

Allen, G. C., Larach, M. G., & Kunselman, A. R. (1998). The sensitivity and specificity of the caffeine halothane contracture test: A report from the north American malignant hyperthermia registry. *Anesthesiology, 88*, 579–588.

Ball, C. (2007). Unravelling the mystery of malignant hyperthermia. *Anaesthesia and Intensive Care, 3*(1), 26–31.

Bamaga, A. K., Riazi, S., Amburgey, K., Ong, S., Halliday, W., Diamandis, P., et al. (2016). Neuromuscular conditions associated with malignant hyperthermia in paediatric patients: A 25-year retrospective study. *Neuromuscular Disorders, 26*, 201–206.

Barnes, C., Stowell, K. M., Bulger, T., Langton, E., & Pollock, N. (2015). Safe duration of postoperative monitoring for malignant hyperthermia patients administered non-triggering anaesthesia: An update. *Anaesthesia and Intensive Care, 43*(1), 98–104.

Boushra, M. N., Miller, S. N., Koyfman, A., & Long, B. (2018). Consideration of occult infection and sepsis mimics in the sick patient without an apparent infectious source. *Journal of Emergency Medicine, 56*(1), 36–45.

Brandom, B. W. (2009). Ambulatory surgery and malignant hyperthermia. *Current Opinion in Anaesthesiology, 22*, 744–747.

Brünner, H. W., Pohl, S., & Grond, S. (2011). Washout of sevoflurane from the GE Avance and Amingo Carestation anesthetic machines. *Acta Anaesthesiologica Scandinavica, 55*, 1118–1123.

Carr, A. S., Lerman, J., Cunliffe, M., McLeod, M. E., & Britt, B. A. (1995). Incidence of malignant hyperthermia reactions in 2,214 patients undergoing muscle biopsy. *Canadian Journal of Anaesthesia, 42*, 281–286.

Crawford, M. W., Prinzhausen, H., & Petroz, G. C. (2007). Accelerating the washout of inhalational anesthetics from the Dräger Primus anesthetic workstation. *Anesthesiology, 106*, 289–294.

Denborough, M. A., Forster, J. F. A., Lovell, R. R. H., Maplestone, P. A., & Villiers, J. D. (1962). Anaesthetic deaths in a family. *British Journal of Anaesthesia, 34*, 395–396.

Dykes, M. H. (1975). Evaluation of a muscle relaxant: Dantrolene sodium (Dantrium). *Journal of the American Medical Association, 231*, 862–864.

Ellis, F. R., Halsall, P. J., & Christian, A. S. (1990). Clinical presentation of suspected malignant hyperthermia during Anaesthesia in 402 Probands. *Anaesthesia, 45*, 838–841.

Ellis, F. R., Halsall, P. J., Ørding, H., Fletcher, R., Rankley, E., Heffron, J. J. A., et al. (1984). A protocol for the investigation of malignant hyperpyrexia susceptibility. *British Journal of Anaesthesia, 56*, 1267–1269.

Flewellen, E. H., Nelson, T. E., Jones, W. P., Arens, J. F., & Wagner, D. L. (1983). Dantrolene dose–response in awake man: Implications for management of malignant hyperthermia. *Anesthesiology, 59*, 275–280.

Glahn, K. P. E., Bendixen, D., Girard, T., Hopkins, P. M., Johannsen, S., Rüffert, H., et al. (2020). Availability of dantrolene for the management of malignant hyperthermia crises: European Malignant Hyperthermia Group guidelines. *British Journal of Anaesthesia, 125*(2), 133–140.

Glahn, K. P. E., Ellis, F. R., Halsall, P. J., Müller, C. R., Snoeck, M. M., Urwyler, A., et al. (2010). Recognizing and managing a malignant hyperthermia crisis: Guidelines from the European Malignant Hyperthermia Group. *British Journal of Anaesthesia, 105*, 417–420.

Gonsalves, S. G., Dirksen, R. T., Sangkuhl, K., Pulk, R., Alvarellos, M., Vo, T., et al. (2019). Clinical Pharmacogenetics Implementation Consortium (CPIC) guideline for the use of potent volatile anesthetic agents and succinylcholine in the context of RYR1 or CACNA1S genotypes. *Clinical Pharmacology & Therapeutics, 105*(6), 1338–1344.

Grinberg, R., Edelist, G., & Gordon, A. (1983). Postoperative malignant hyperthermia episodes in patients who received "safe" anaesthetics. *Canadian Anaesthesists' Society Journal, 30*, 273–276.

Gunter, J. B., Ball, J., & Than-Win, S. (2008). Preparation of Dräger Fabius anesthesia machine for the malignant-hyperthermia susceptible patient. *Anesthesia & Analgesia, 107*, 1936–1945.

Hardy, J. B., Gouin, A., Damm, C., Compère, V., Veber, B., & Dureuil, B. (2018). The use of a checklist improves anaesthesiologists' technical and non-technical performance for simulated malignant hyperthermia management. *Anaesthesia Critical Care & Pain Medicine, 37*(1), 17–23.

Hirshey Dirksen, S. J., Larach, M. G., Rosenberg, H., Brandom, B. W., Parness, J., Lang, R. S., et al. (2011). Future directions in malignant hyperthermia research and patient care. *Anesthesia & Analgesia, 113*, 1108–1119.

Hopkins, P. M. (2011). Malignant hyperthermia: Pharmacology of triggering. *British Journal of Anaesthesia, 107*(1), 48–56.

Hopkins, P. M., Rüffert, H., Snoeck, M. M., Girard, T., Glahn, K. P. E., Ellis, F. R., et al. (2015). European Malignant Hyperthermia Group guidelines for investigation of malignant hyperthermia susceptibility. *British Journal of Anaesthesia, 115*(4), 531–539.

Krause, T., Gerbershagen, M. U., Fiege, M., Weißhorn, R., & Wappler, F. (2004). Dantrolene – A review of its pharmacology, therapeutic use and new developments. *Anaesthesia, 59*(4), 364–373.

Krivosic-Horber, R. (1990). Malignant hyperthermia. Treatment of the acute episode. *Acta Anaesthesiologica Belgica, 41*, 83–86.

Larach, M. G., Dirksen, S. J. H., Belani, K. G., Brandom, B. W., Metz, K. M., Policastro, M. A., et al. (2012). Creation of a guide for the transfer of care of the malignant hyperthermia patient from ambulatory surgery centers to receiving hospital facilities. *Anesthesia & Analgesia, 114*(1), 94–100.

Larach, M. G., Localio, A. R., Allen, G. C., Denborough, M. A., Ellis, F. R., Gronert, G. A., et al. (1994). A clinical grading scale to predict malignant hyperthermia susceptibility. *Anesthesiology, 86*, 771–779.

Litman, R. S., Griggs, S. M., Dowling, J. J., & Riazi, S. (2018). Malignant hyperthermia susceptibility and related diseases. *Anesthesiology, 128*, 159–167.

Monnier, N., Procaccio, V., Stieglitz, P., & Lunardi, J. (1997). Malignant hyperthermia susceptibility is associated with a mutation of the α1-subunit of the human dihydropyridine-sensitive L-type voltage-dependent calcium-channel receptor in skeletal muscle. *American Journal of Human Genetics, 60*, 1316–1325.

Müller-Wirtz, L. M., Godsch, C., Sessler, D. I., Volk, T., Kreuer, S., & Hüppe, T. (2020). Residual volatile anesthetics after workstation preparation and activated charcoal filtration. *Acta Anaesthesiologica Scandinavica, 64*, 759–765.

Ørding, H., Brancadoro, V., Cozzolino, S., Ellis, F. R., Glauber, V., Gonano, E. F., et al. (1997). In vitro contracture test for diagnosis of malignant hyperthermia following the protocol of the European MH Group: Results of testing patients surviving fulminant MH and un-related low-risk subjects. *Acta Anaesthesiologica Scandinavica, 41*, 955–966.

Petroz, G. C., & Lerman, J. (2002). Preparations of the Siemens KION anesthetic machine for patients susceptible to malignant hyperthermia. *Anesthesiology, 96*, 941–946.

Prinzhausen, H., Crawford, M. W., O'Rourke, J., & Petroz, G. C. (2006). Preparation of the Dräger Primus anesthetic machine for malignant hyperthermia-susceptible patients. *Canadian Journal of Anaesthesia, 53*, 885–890.

Ranklev-Twetman, E. (1990). Malignant hyperthermia: The clinical syndrome. *Acta Anaesthesiologica Belgica, 41*, 79–82.

Riazi, S., Kraeva, N., & Hopkins, P. M. (2018). Updated guide for the management of malignant hyperthermia. *Canadian Journal of Anesthesia, 65*(6), 709–721.

Rosenberg, H., Antognini, J. F., & Muldoon, S. (2002). Testing for malignant hyperthermia. *Anesthesiology, 96*, 232–237.

Rosenberg, H., Pollock, N., Schiemann, A., Bulger, T., & Stowell, K. (2015). Malignant hyperthermia: A review. *Orphanet Journal of Rare Diseases, 10*, 93.

Santos, J. M., Andrade, P. V., Galleni, L., Vainzof, M., Sobreira, C. F. R., Schmidt, B., et al. (2017). Idiopathic hyperCKemia and malignant hyperthermia susceptibility. *Canadian Journal of Anesthesia/Journal Canadien D'anesthésie, 64*(12), 1202–1210.

Scala, D., Di Martino, A., Cozzolino, S., Mancini, A., Bracco, A., Andria, B., et al. (2006). Follow-up of patients tested for malignant hyperthermia susceptibility. *European Journal of Anaesthesiology, 23*, 801–805.

Silva, H. C. A., Onari, E. S., Castro, I., Perez, M. V., Hortensi, A., & Amaral, J. L. G. (2019). Anesthesia for muscle biopsy to test susceptibility to malignant hyperthermia. *Brazilian Journal of Anesthesiology, 69*, 335–341.

Simon, H. B. (1993). Hyperthermia. *New England Journal of Medicine, 329*(7), 483–487.

Simon, H. B. (1994). Hyperthermia and heatstroke. *Hospital Practice, 29*(8), 65–80.

Thoben, C., Dennhardt, N., Krauß, T., Sümpelmann, R., Zimmermann, S., Rueffert, H., et al. (2019). Preparation of anaesthesia workstation for trigger-free anaesthesia. *European Journal of Anaesthesiology, 36*(11), 851–856.

Urwyler, A., Deufel, T., McCarthy, T., & West, S. (2001). Guidelines for molecular genetic detection of susceptibility to malignant hyperthermia. *British Journal of Anaesthesia, 86*, 283–287.

Wappler, F. (2010). Anesthesia for patients with a history of malignant hyperthermia. *Current Opinion in Anaesthesiology, 23*(3), 417–422.

Whitty, R. J., Wong, G. K., Petroz, G. C., Pehora, C., & Crawford, M. W. (2009). Preparation of the Dräger Fabius GS workstation for malignant hyperthermia-susceptible patients. *Canadian Journal of Anaesthesia, 56*, 497–501.

Chapter 2

Use of dextran in regional anesthesia

Masahiko Tsuchiya[a,b]

[a]*Department of Anesthesiology, Osaka City University Graduate School of Medicine, Osaka, Japan;* [b]*Department of Anesthesia, Kishiwada Tokushukai Hospital, Kishiwada, Japan*

Abbreviations

ASA	American Society of Anesthesiologists
AUC	area under concentration-time curve
BIS	bispectral index
LA	local anesthetic drug
MAC	minimum alveolar concentration
NRS	numerical rating scale

Introduction

More than half a century ago, local anesthetic drugs (LAs) were analyzed and extensively used, with various techniques and adjuvants to potentiate their effectiveness investigated and proposed (Swain, Nag, Sahu, & Samaddar, 2017). However, striking progress in general anesthesia thereafter, including safer agents and reliable monitoring techniques for vital signs, made it a versatile unique option for controlling surgical stress and pain. The shift to general anesthesia moved LAs to an auxiliary role for surgery, while a variety of previously used techniques to modify their effects have been largely forgotten. In recent years, opioid abuse has become a topic of public concern (Soffin, Lee, Kumar, & Wu, 2019), while the possibilities that regional anesthesia might reduce the risk of cancer recurrence in cancer surgery patients (Exadaktylos, Buggy, Moriarty, Mascha, & Sessler, 2006; Grandhi, Lee, & Abd-Elsayed, 2017) and may also be superior for postoperative recovery (Bugada, Ghisi, & Mariano, 2017) have been suggested. Presently, the advantages of regional anesthesia techniques and related adjuvant compounds for surgery are being reevaluated, including dextran, an LA adjuvant with a long history.

Recent advances in ultrasound technology have altered the mode of regional anesthesia and an interfascial compartment nerve block for the abdominal trunk can now be easily performed (Tsuchiya et al., 2012; Tsuchiya, Mizutani, Funai, & Nakamoto, 2016), with transversus abdominis plane and quadrates lumborum blocks typical applications. These new types of nerve blocks require accurate injection of a large amount of an LA into the targeted interfascial compartment. A well-performed interfascial compartment nerve block has potential to serve as a substitute for epidural anesthesia, the current gold standard method for surgical pain control. When antithrombotic or antiplatelet therapy is used, epidural anesthesia is frequently restricted due to an increased risk of spinal hematoma involved with spinal cord injury. Therefore, it is considered that trunk blocks, which are relatively safer in regard to hemorrhage-related nerve injury, will become increasingly important. However, there is risk of LA systemic toxicity when a large amount of the drug is given (Oda, 2019). Furthermore, when the block is performed by a single injection, insufficient analgesic duration for controlling postoperative pain is another weak point. These disadvantages must be solved for more widespread use in clinical practice of trunk blocks.

In this chapter, new applications for LAs, especially low-molecular weight dextran as an LA adjuvant, for resolving the disadvantages of trunk nerve blocks and improved potency and safety are discussed based on previous studies, as well as recent findings obtained by our research team.

Treatments, Mechanisms, and Adverse Reactions of Anesthetics and Analgesics. https://doi.org/10.1016/B978-0-12-820237-1.00061-2
Copyright © 2022 Elsevier Inc. All rights reserved.

History of dextran as an LA adjuvant

Adrenaline and dextran as LA adjuvant

Although various LAs have been developed and are widely used in anesthetic practice, ideal drugs with a longer duration of action, lower incidence of systemic toxicity, and better nerve fiber selectivity are anticipated. Adjuvants that can modify LA activity with potential analgesic effects or improved safety have been studied (Wiles & Nathanson, 2010), with several drugs and compounds proposed and examined. Adrenaline (epinephrine) is one of the oldest LA additives for such purposes and continues to be used. Dextran as well, which is composed of complex branched polysaccharides derived from sucrose with various lengths and weights (Fig. 1), has been tested as an LA adjuvant, though nearly forgotten in recent times.

Adverse effects of adjuvant adrenaline

A small amount of adrenaline can be added to an LA solution to prolong analgesia and reduce the potential danger of systemic LA toxicity (Murphy, Mather, Stanton-Hicks, Bonica, & Tucker, 1976). Added adrenaline induces vasoconstriction, and thus reduces systemic absorption of an administered LA and enhances the analgesia effects. Such an adjuvant technique is typically applied with lidocaine and seems to be optimal.

However, addition of adrenaline to a lidocaine solution does not always lead to good results. Lidocaine has potent vasodilation effects, which may enhance absorption of adrenaline into systemic circulation. Ueda et al. found that use of a lidocaine solution with a small amount of adrenaline significantly increased the plasma concentration of adrenaline in comparison with a pure adrenaline solution in cases of subcutaneous infiltration anesthesia (Fig. 2) (Ueda, Hirakawa, & Mori, 1985). An increase in adrenaline concentration in circulation under halothane anesthesia is known to cause severe cardiac dysrhythmia. Although recently introduced inhaled anesthesia agents such as desflurane and sevoflurane have substantial resistance to adrenaline-induced dysrhythmia, they are still affected by the risk of increased blood pressure and heart rate when a lidocaine-adrenaline mixture is used for regional anesthesia (Fig. 3). Furthermore, the presence of adrenaline decreases the threshold concentration of lidocaine to induce convulsions, thus increasing the systemic toxicity of lidocaine (Fig. 4) (Yokoyama, Goto, Ueda, Hirakawa, & Arakawa, 1993). Another study also demonstrated the toxic effect of adrenaline to lower lidocaine concentration leading to convulsions (Takahashi et al., 2006). Those latter results showed that concomitant administration of adrenaline with lidocaine elevates the concentration of extracellular lidocaine in the brain, a mechanism by which adrenaline may increase the central nervous system toxicity of lidocaine.

Together, these findings reveal that use of a lidocaine–adrenaline mixture for regional anesthesia has a substantial risk for co-administered adrenaline to increase the systemic toxicity of lidocaine, while lidocaine increases the concentration of

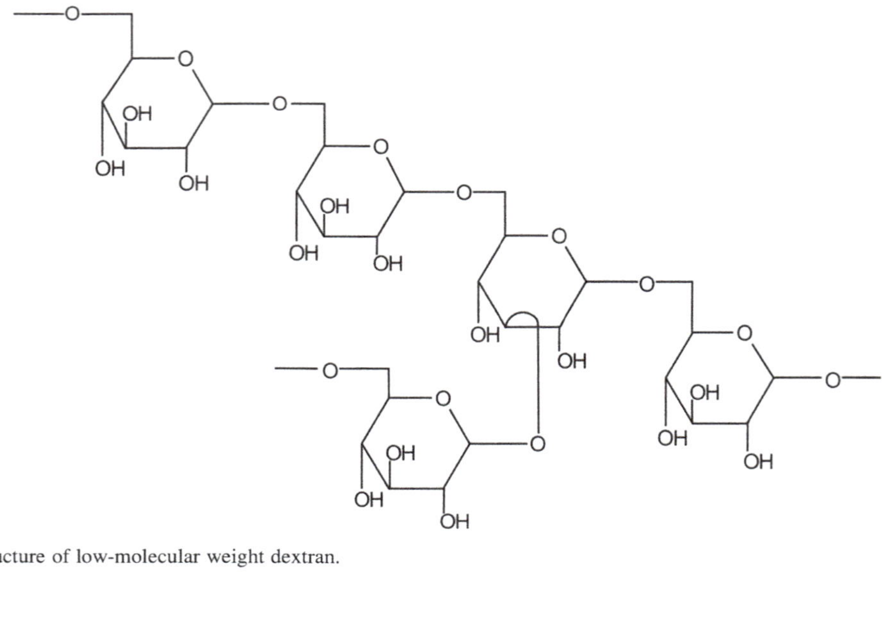

FIG. 1   Molecular structure of low-molecular weight dextran.

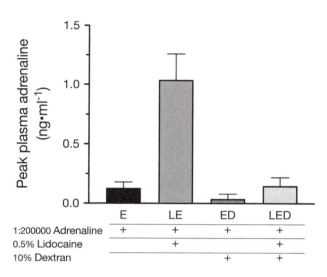

FIG. 2 Peak plasma adrenaline concentration following subcutaneous infiltrative injection of various adrenaline solutions in humans. E: 1:200,000 adrenaline in normal saline solution, LE: 1:200,000 adrenaline with 0.5% lidocaine in normal saline solution, ED: 1:200,000 adrenaline with 10% low-molecular weight dextran in saline solution, LED: 1:200,000 adrenaline with 0.5% lidocaine and 10% low-molecular weight dextran in saline solution. When adrenaline was subcutaneously injected with lidocaine (LE solution), the peak plasma adrenaline concentration showed an approximately 7-fold increase as compared with injection of pure adrenaline solution (E solution). However, the presence of low-molecular weight dextran in lidocaine adrenaline solution (LED solution) suppressed the toxic increase in plasma adrenaline concentration.

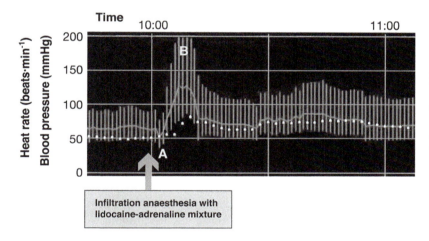

FIG. 3 Typical increases in blood pressure and heart rate with subcutaneous infiltration anesthesia using 20 ml of 0.5% lidocaine and 1:200,000 adrenaline in normal saline solution in patients receiving sevoflurane general anesthesia. (A) Soon after injection of a lidocaine-adrenaline mixture, blood pressure was slightly decreased, probably due to the β2-adrenergic effect of a small amount of absorbed adrenaline. (B) Thereafter, blood pressure and heart rate were greatly increased, mainly due to the α1- and β1-adrenergic effects of adrenaline.

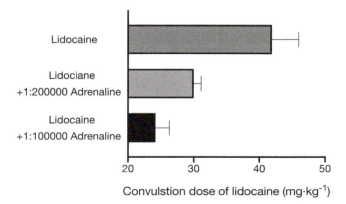

FIG. 4 Convulsion dose of plasma concentration of lidocaine. Male Wistar rats received continuous intravenous injections of three different lidocaine mixtures (1.5% lidocaine, 1.5% lidocaine with 1:200,000 adrenaline, 1.5% lidocaine with 1:100,000 adrenaline), then the total lidocaine dose from the beginning of infusion to onset of generalized convulsions was calculated. Addition of adrenaline to the lidocaine solutions decreased the threshold of lidocaine-induced convulsions in a dose-dependent manner, indicating that it increased lidocaine systemic toxicity.

18 **PART | I** Drugs and agents used in anesthesia and analgesia

adrenaline in plasma. Thus, use of a lidocaine solution with a small amount of adrenaline added as an adjuvant is not a completely safe technique, contrary to the understanding of many physicians. A decrease in adrenaline toxicity is an important aim for safe application of a lidocaine-adrenaline mixture for regional anesthesia.

Reduction of adrenaline toxicity by dextran addition

In a study by Ueda et al. that reported the toxicity of a lidocaine-adrenaline mixture, the authors also presented a novel solution to produce a safer combination (Ueda et al., 1985). Their investigation found that addition of low-molecular dextran (average molecular weight 40,000) into a lidocaine-adrenaline mixture suppressed absorption of adrenalin into circulation. The peak plasma adrenaline concentration was 0.15 ± 0.07 ng ml^{-1} with the lidocaine-adrenaline mixture with low-molecular weight dextran, while that was 1.04 ± 0.22 ng ml^{-1} with the standard lidocaine-adrenaline mixture (Fig. 2). Thus, they concluded that a lidocaine-adrenaline-dextran mixture was safe for application of lidocaine containing a small amount of adrenaline for regional anesthesia. Furthermore, those results demonstrated that dextran functions effectively as an adjuvant to improve safety when using an LA-adrenaline mixture. Thereafter, findings that dextran suppressed systemic absorption of lidocaine as well as adrenaline when used for subcutaneous infiltration anesthesia were confirmed by other researchers (Adams et al., 1988). Furthermore, their pharmacological results supported the possibility that dextran has good potential to enhance LA analgesic effects.

First appearance of dextran as LA adjuvant

The first description of dextran providing enhancement of analgesic effects in cases of regional anesthesia can be traced back to 1960 (Loder, 1960), more than two decades prior to the above mentioned studies. Thereafter, several studies and reports regarding the adjuvant effects of dextran for analgesia followed, some of which demonstrated prolongation of analgesia (Ito, Ichinohe, Shibukawa, Aida, & Kaneko, 2007; Kaplan, Miller Jr., & Gallagher Jr, 1975; Simpson, Hughes, & Long, 1982), whereas others failed to find potential effects for analgesia (Armstrong & Kingsnorth, 1986; Kingsnorth, Wijesinha, & Grixti, 1979). However, along with progress in general anesthesia techniques for surgery, LA adjuvants including dextran have gone rather unnoticed, thus the adjuvant effects of dextran remain inclusive.

Mechanism of dextran as LA adjuvant

Although several questions remain concerning the adjuvant effects of dextran when used with LAs, its mechanism of action has been studied in laboratory experiments. Dextran may form a water-soluble complex with LAs and it is known that an LA complex remains at the injection site longer than an unbound LA, probably because of the increase in viscosity associated with reduced diffusion of the complex (Aberg, Friberger, & Sydnes, 1978; Hassan, Renck, Lindberg, Akerman, & Hellquist, 1985). Additionally, addition of dextran was shown to alter the pH of an LA solution, which may further contribute to prolongation of action (Covino, 1986).

Application of dextran as LA adjuvant for regional anesthesia

Suitability of LA-dextran mixture for compartment nerve block

Because of possible interactions with LAs, the effectiveness of dextran as an LA adjuvant may greatly depend on the type of nerve block being employed, which could be related to the inconsistency of study results presented thus far. We consider that dextran may be more effective for an interfascial compartment block of the abdominal trunk, such as transversus abdominis plane block (Tsuchiya, Mizutani, & Ueda, 2019). Such trunk nerve block procedures are now widely available with the aid of ultrasound imaging, though require injection of a large amount of LA solution by unzipping of the interfascial compartment. Maintaining the amount of injected LA in the compartment for a longer period is essential to induce sufficient analgesia, thus it is quite reasonable to assume that the fluid retention properties of dextran will have a favorable impact.

A study of epidural anesthesia using a lidocaine-dextran mixture clearly demonstrated that dextran kept the solution at the injection site (Alkhawajah & Farag, 1992). Furthermore, use of a lidocaine-dextran mixture reduced vascular uptake of lidocaine from epidural space as compared to a lidocaine-saline standard mixture. Although the authors did not investigate analgesic effects, the effect of dextran to reduce systemic absorption of lidocaine from epidural space may prolong the duration and expand the area of epidural anesthesia.

LA-dextran mixture enhances analgesic effects of compartment nerve block

Based on the findings and considerations noted above, we conducted a study that confirmed low-molecular weight dextran as an effective adjuvant for LA that enhances its analgesic effects for a longer period in cases with an interfascial compartment block (Hamada et al., 2016). A brief outline of that study is presented following.

Patients with an American Society of Anesthesiologists (ASA) physical status of 1 or 2 and scheduled for a laparoscopic colectomy received a combination of two interfascial compartment blocks, a transversus abdominis plane block and rectus sheath block (Tsuchiya et al., 2012). They then received this two-block combination with either 0.2% levobupivacaine in a saline solution (control group, $n = 27$), or 0.2% levobupivacaine and 8% low-molecular weight dextran (average molecular weight 40,000) in a saline solution (dextran group, $n = 27$). Following anesthesia induction, a bilateral transversus abdominis plane block (0.2% levobupivacaine solution, 20 ml $\times$ 2) and rectus sheath block (0.2% levobupivacaine solution, 20 ml $\times$ 2) were performed under ultrasound guidance using the appropriate mixture for each group. During surgery, anesthesia was maintained with 1%–3% inhaled sevoflurane and 0.15–0.25 $\mu g\ kg^{-1}\ min^{-1}$ remifentanil to keep systolic blood pressure and heart rate within 70%–110% of their pre-anesthesia values, with a bispectral index (BIS) value from 40 to 60. Thirty minutes prior to completion of surgery, 200 μg of fentanyl was given by intravenous injection, and continuous intravenous infusions of fentanyl at 25 $\mu g\ h^{-1}$ and droperidol at 63 $\mu g\ h^{-1}$ were started for basic postoperative analgesia and antiemetic treatments and then continued until 24 h after surgery. As the main outcome measurements, changes in plasma concentration of levobupivacaine were periodically measured after completion of the trunk nerve block, while postoperative pain at rest was assessed using an 11-point numerical rating scale (NRS; 0, no pain, to 10, worst pain).

Patient age and body weight in the control group were 66 ± 10.3 years (mean $\pm$ standard deviation) and 60 ± 12.1 kg, respectively, and anesthesia time was 331 ± 66.6 min, while those in the dextran group were 67 ± 9.5 years, 59 ± 11.5 kg, and 325 ± 67.7 min, respectively, with no significant differences between the groups. In the control group, the plasma concentration of levobupivacaine rose quickly just after performing the nerve block and reached a maximum at 51 ± 30 min (T_{max}). In the dextran group, that rose in a more gradual manner with a significantly longer T_{max} value (73 ± 25 min, $P < 0.05$ vs. control group) (Fig. 5A, Table 1). The maximum concentration of levobupivacaine (C_{max}) in the control group was 1410 ± 322 ng ml^{-1}, while that in the dextran group was significantly lower at 1141 ± 287 ng ml^{-1} ($P < .05$). Also, the area under the plasma concentration-time curve (AUC) from 0 to 240 min was significantly lower in the dextran group ($172,484 \pm 50,502$ vs. $229,124 \pm 87,254$ ng min ml^{-1}, $P < .05$). Moreover, C_{max} values in 2 of the control group patients exceeded 1800 ng ml^{-1}, whereas that value was not exceeded in any of the dextran group patients. These results demonstrated that use of the dextran mixture resulted in lower C_{max} with a longer T_{max} value for the concentration of levobupivacaine in plasma, and consequently a lower AUC from 0 to 240 min as compared to the standard solution, indicating reduced absorption of levobupivacaine from the injected compartment into systemic circulation. Thus, we concluded that use of dextran reduces the risk of LA systemic toxicity.

As for analgesic effects, which could be of greater importance, NRS analysis demonstrated significantly better analgesia for a longer duration over the first postoperative 24 h in patients given the dextran mixture as compared to the control group given a standard solution (NRS 2 h after surgery, 1.5 ± 1.9 vs. 3.3 ± 2.7; 8 h, 1.5 ± 1.5 vs. 4.2 ± 2.7, 16 h, 2.1 ± 2.1 vs. 4.1 ± 2.5; 24 h, 2.1 ± 1.8 vs. 3.3 ± 2.0) (Fig. 5B).

Those results showed that low-molecular weight dextran as an LA adjuvant offers great clinical advantages, including enhancement of analgesia effect as well as reduction in LA toxicity risk, in patients undergoing a combined transversus abdominis plane block and rectus sheath block.

LA-dextran mixture used for compartment nerve block remains at injection site more than 24 h

We also investigated the adjuvant effects of dextran in patients undergoing a quadratus lumborum block, another type of interfascial compartment nerve block. Those findings showed that the LA-dextran mixture remained at the injection site for more than 24 h (Tsuchiya, Mizutani, & Ueda, 2018). Brief details of that study are presented following.

A quadratus lumborum block using a low-molecular weight dextran mixture with ropivacaine was applied in 18 patients undergoing open major abdominal surgery procedures, including hepatectomy ($n = 10$), total gastrectomy ($n = 3$), pancreaticoduodenectomy ($n = 2$), radical nephrectomy ($n = 1$), radical cystectomy ($n = 1$), and radical hysterectomy ($n = 1$).

Patient age and body weight were 67 ± 8.4 years and 72.6 ± 4.6 kg, respectively, while anesthesia time was 429 ± 123 min. The American Society of Anesthesiologists (ASA) physical status was 1 or 2. A bilateral quadratus lumborum block was performed following anesthesia induction. One hundred ml of 0.1% ropivacaine and 8% low-molecular weight dextran in saline solution with 2 mg of morphine was injected into interfascial space posterior of the quadratus lumborum muscle on each side, a so-called QLB2 nerve block. Anesthesia was maintained with desflurane and

FIG. 5 (A) Changes in levobupivacaine plasma concentration in patients receiving abdominal trunk nerve block (bilateral transversus abdominis plane block plus rectus sheath block) with 160 mg of levobupivacaine. The control group (□ n = 27) received 80 ml of 0.2% levobupivacaine in a saline solution, while the dextran group (■ n = 27) received 80 ml of 0.2% levobupivacaine and 8% low-molecular weight dextran in a saline solution. All patients underwent a laparoscopic colectomy. Data are expressed as the mean ± standard deviation. T_{max}, C_{max}, and $AUC_{0\text{-}240min}$ values for each group are shown in Table 1. Addition of dextran to the levobupivacaine solution was shown to suppress systemic absorption of levobupivacaine from the injection site, which may prolong analgesic effects and decrease the systemic toxicity of levobupivacaine. (B) NRS scores for postoperative pain in the same patients as A. Values are expressed as the mean ± standard deviation. NRS; 0, no pain, to 10, worst pain. Scores in the dextran group (■ n = 27) were significantly lower at each time point after surgery as compared to the control group (□ n = 27) ($P < .01$ at 2, 8, 16 h; $P = .035$ at 24 h).

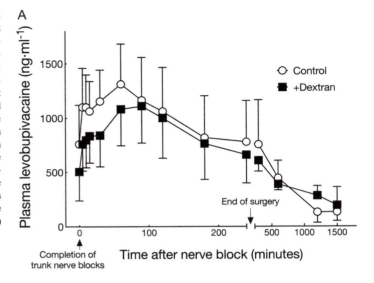

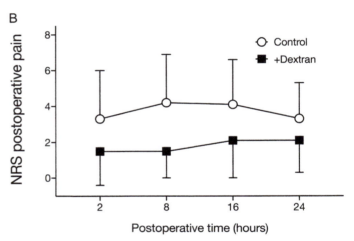

TABLE 1 Pharmacokinetic parameters for levobupivacaine plasma concentration in patients receiving 160 mg of levobupivacaine.

		Control group			95% confidence interval	Dextran group			95% confidence interval	P value
T_{max}	min	51	±	30	38–63	73	±	25	63–83	<.01
C_{max}	ng ml^{-1}	1410	±	322	1280–1540	1141	±	287	1017–1265	<.01
$AUC_{0\text{-}240min}$	ng ml^{-1} min	229,124	±	87,254	194,608–263,640	172,484	±	50,502	151,638–193,330	<.01

All patients received 160 mg of levobupivacaine as a trunk nerve block (bilateral transversus abdominis plane block plus rectus sheath block). The control group (n = 27) received 80 ml of 0.2% levobupivacaine in a saline solution, and the dextran group (n = 27) 80 ml of 0.2% levobupivacaine and 8% low-molecular weight dextran in a saline solution. Data are expressed as the mean ± standard deviation and were statistically analyzed using an unpaired t-test. Time courses of plasma levobupivacaine concentration in both groups are shown in Fig. 5A. T_{max}, time of maximum plasma concentration; C_{max}, maximum plasma concentration; $AUC_{0\text{-}240min}$, area under plasma concentration-time curve from 0 to 240 min. P values indicate comparisons between control and dextran groups. T_{max} was longer, and C_{max} and $AUC_{0\text{-}240min}$ were smaller in the dextran group, indicating that addition of dextran to the levobupivacaine solution suppressed systemic absorption of levobupivacaine from the injection site.

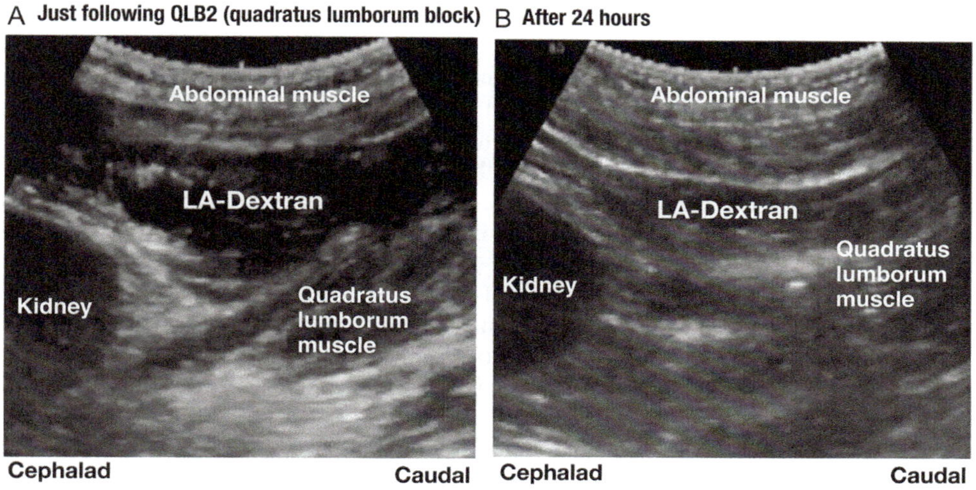

A **Just following QLB2 (quadratus lumborum block)** B **After 24 hours**

FIG. 6 (A) Sagittal ultrasound images obtained just after completion of a posterior quadrates lumborum block with 100 ml of 0.1% ropivacaine in 8% low-molecular weight dextran in a saline solution (LA-dextran). The injected local anesthetic mixture was found to be widespread throughout the targeted area over the interfascial space posterior to the quadrates lumborum muscle. (B) Sagittal ultrasound image obtained from same injection site at 24 h after initiation of quadrates lumborum block. Some of the local anesthetic mixture (LA-dextran) remained.

remifentanil, as well as intermittent administrations of rocuronium, to maintain systolic blood pressure and heart rate at 70%–110% of their preanesthesia values, as well as a bispectral index value of 40–60. Following surgery, flurbiprofen (50 mg) was given every 8 h, with acetaminophen (1000 mg) used as rescue treatment for pain.

In these patients, anesthesia was maintained with very low anesthetic concentrations; desflurane at a minimum alveolar concentration (MAC) of 0.6 ± 0.1 and remifentanil at 0.08 ± 0.04 $\mu g\,kg^{-1}\,min^{-1}$. No rescue drug was used in any case the first night after surgery and the NRS for pain was 2.2 ± 1.7. All patients successfully walked more than 20 m with less pain the next day. Prior to beginning rehabilitation approximately 24 h after the quadratus lumborum block, ultrasound examinations were performed and the results indicated that LA remained at the injected site (Fig. 6). No adverse effects including tissue necrosis over the area of injection were observed.

Those findings indicated that a bilateral quadratus lumborum block using 100 ml of 0.1% ropivacaine and an 8% low-molecular weight dextran mixture in each side reduced intraoperative requirements of desflurane and remifentanil, and also provided good postoperative analgesia without rescue analgesics, thus confirming the adjuvant effects of dextran for LA analgesia potency. Ultrasound imaging showing the LA-dextran mixture remaining at the injection site even after 24 h also supported the proposed mechanism of dextran as an adjuvant, in that the drug maintained the injection mixture at the injection site for an extended time, thus enabling long-lasting analgesic effects.

An LA-dextran mixture remaining for a long period at the injection site may be related to increased viscosity of the mixture, as discussed in the mechanism section above, which may be an additional beneficial effect for performing a nerve block. When a dextran mixture is injected into the wrong portion of parenchymal tissue or an area outside of the target interfascial compartment, extra high pressure could develop due to its high viscosity. A perceptible increase in injection pressure related to the site of injection can be of great help to avoid mis-injection of LA and perform an accurate compartment injection. This guidance effect via injection pressure may be related, at least in part, to the effectiveness of dextran when used as an adjuvant. In addition, it should be noted that feedback from injection pressure is especially beneficial for novice practitioners learning nerve block procedures.

Dextran inhibits unintended spread of injected LA

Based on the above results obtained with a quadratus lumborum block, we speculated that dextran could be used for an accurate nerve block in cases of single nerve block as well, because of its inhibition of unintended hazardous LA spreading in surrounding tissues. We performed a study of patients undergoing a mandibular nerve block performed at a site close to the oval foramen, from which the mandibular nerve appears, and confirmed a good adjuvant effect of dextran in those single nerve block cases (Tsuchiya, Mizutani, Yabe, Mori, & Ueda, 2019), with the details shown following.

Briefly, a mandibular nerve block was performed using a lateral extraoral approach with guidance from ultrasound imaging in 10 patients undergoing a parotidectomy under general anesthesia. Following anesthesia induction, the head was turned according to the surgical site with the mouth open, then a convex ultrasound transducer was placed just below and parallel to the zygomatic bone. Representative ultrasound images of the basal portion of the lateral pterygoid plate with this transducer position are shown in Fig. 7. Next, a 23-gauge nerve block needle was inserted towards the dorsal edge of the plate, close to the mandibular nerve. When the needle touched the plate edge, 3 ml of 0.3% ropivacaine and 7% low-molecular weight dextran in a saline solution were injected. Occasionally, the maxillary artery is shown in this section by ultrasound imaging and should not be traumatized. Anesthesia is maintained with desflurane and remifentanil to keep a bispectral index of 40–60.

Patient age and body weight were 60 ± 12 years and 69.5 ± 14.6 kg, respectively, while anesthesia time was 227 ± 92 min. Intraoperative haemodynamics were stably maintained by desflurane at 0.7 ± 0.1 MAC and remifentanil at 0.08 ± 0.03 $\mu g\ kg^{-1}\ min^{-1}$ without use of vasopressors. Eighteen hours after surgery, the NRS for pain was 1.2 ± 0.4 without use of a rescue drug, as compared to 2.7 ± 0.7 in our previous non-nerve block cases ($P < .01$). No side effects related to unintended spread of injected ropivacaine were noted.

Performance of a mandibular nerve block using a mixture of ropivacaine and low-molecular weight dextran provided good postoperative analgesia, as well as safety. These findings validated our speculation that such a dextran mixture enables an accurate nerve block and enhances analgesic LA potency without unintended spread of LA in single nerve block cases, the same as seen in patients who underwent an interfascial compartment nerve block.

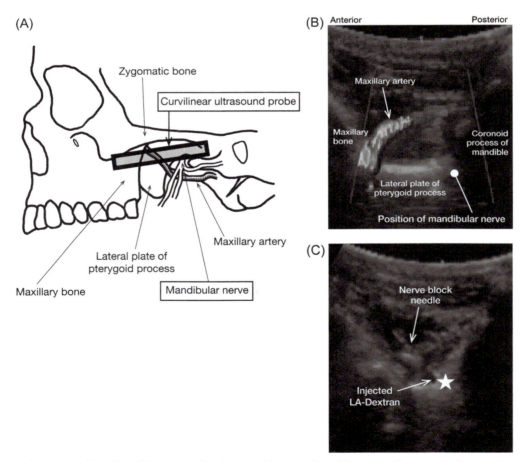

FIG. 7 (A) Anatomical relationships of mandibular nerve, lateral pterygoid plate, and maxillary artery. For a mandibular nerve block, the ultrasound probe is placed just below and parallel to the zygomatic bone. (B) Typical ultrasound image obtained at 4.5 MHz with curvilinear ultrasound probe placed at point depicted in (A). Color Doppler ultrasound imaging revealed the maxillary artery. The mandibular nerve could not be clearly seen with this view due to acoustic shadows from surrounding bone. (C) Ultrasound image obtained after injection of 3 ml of ropivacaine-dextran mixture (LA-dextran) to target point *(denoted by star)* close to mandibular nerve.

Summary of application of dextran as LA adjuvant

Together, these recent studies indicate that low-molecular weight dextran functions as an effective adjuvant for potentiation of LA analgesia in compartment nerve block cases, including transversus abdominis plane block, rectus sheath block, and quadratus lumborum block. Moreover, they show that an abdominal trunk block can be very effective and comparable to an epidural block, especially when performed with a large amount of LA-dextran mixture. In addition, use of an LA-dextran mixture may inhibit unintended spread of injected LA solution and increase accuracy of the nerve block. We also consider that the guidance effect of a dextran mixture for correct injection, resulting from the extra high pressure that develops when that mixture is injected into the wrong area instead of the target compartment, is a practical advantage. These characteristic functions are important considerations for use of dextran as an adjuvant. The mechanisms involved may be related to loose formation of an LA-dextran complex coupled with increased viscosity of the mixture and reduced diffusion from the injection site, resulting in maintaining the LA concentration in the same compartment as the target nerve.

Suppression of systemic absorption of LA by use of low-molecular weight dextran results in reduced risk of LA systemic toxicity (Oda, 2019; Tsuchiya, Mizutani, & Ueda, 2019). To decrease LA toxicity, use of a small dose, divided administration, or LAs with lower toxicity such as ropivacaine or levobupivacaine, as well as an aspiration test and ultrasound imaging to assess the safe location of the needle have been generally suggested As shown in results presented above, use of a dextran mixture is another effective method to lower the systemic toxicity of LA. This safety effect provides significant clinical advantages for dextran when used as an LA adjuvant. In addition, there is another effect of low-molecular weight dextran when added to a lidocaine-adrenaline solution, as it reduces the systemic toxicity of adrenaline by suppressing systemic adrenaline absorption. This is especially important for cases of minor surgery under regional anesthesia using a lidocaine-adrenaline solution.

Future studies

Based on studies conducted with a small number of patients demonstrating the adjuvant effects of dextran for use with LAs, large scale trials to confirm those results suggesting significant clinical advantages in terms of prolongation of analgesia are now required. In addition, most studies performed recently used low-molecular weight dextran. However, dextran is available in a variety of molecular weight types, and which is more effective or safer also requires future investigation.

Other agents

Liposomal bupivacaine has recently become commercially available in the United States for local infiltration anesthesia and brachial plexus block (Malik, Kaye, Kaye, Belani, & Urman, 2017; Skolnik & Gan, 2014). Although this agent has shown excellent potential, it is quite expensive and systemic or local toxicity occurring long after administration has yet to be determined. Adjuvant dextran was first used more than half a century ago, thus its long history of clinical use provides evidence of therapeutic safety. Furthermore, when used as an adjuvant for LAs, it is inexpensive and the approach is easily performed. Therefore, we consider that use of dextran as an LA adjuvant remains beneficial even when compared with newer LAs such as liposomal bupivacaine and both methods may be considered.

Mini-dictionary or terms

Adjuvant for local anesthetic drugs (LAs): Medical substance possessing an ability to modify the effects or functions of LAs.

Adrenaline: A typical LA adjuvant used to extend analgesic duration.

American Society of Anesthesiologists (ASA) physical status: Classifications of physiological status to predict operative risk prior to surgery, with patients classified as 1 and 2 considered to be safe for undergoing surgery.

Area under the concentration-time curve (AUC): Definite integral of a curve that shows variations in drug plasma concentration over time.

Bispectral index (BIS): Value used to monitor depth of general anesthesia, with 40–60 regarded as optimal.

Interfascial compartment nerve block: Series of nerve blocks consisting of a transversus abdominis plane block, rectus sheath block, quadratus lumborum block, and other similar nerve blocks frequently used for control of surgical pain.

Levobupivacaine: Amide LA consisting of a single enantiomer of bupivacaine hydrochloride.

Lidocaine: Typical amide LA used for all forms of regional anesthesia.

24 PART | I Drugs and agents used in anesthesia and analgesia

Local anesthetic systemic toxicity: Occasional lethal abnormality of the central nervous system and cardiac system due to a high blood concentration of LAs in vessels that develops by diffusion from the injection site or a mis-injection.

Low-molecular weight dextran: Complex of branched polysaccharides with average molecular weight of 40,000 that functions as an effective LA adjuvant used to potentiate analgesic effects with reduced LA systemic toxicity.

Minimum alveolar concentration (MAC): Concentration of volatile anesthetic in alveoli of the lung that has been shown to prevent movement in 50% of patients receiving invasive painful stimulation while under general anesthesia.

Numerical rating scale (NRS): Pain scale with 11 points, ranging from 0, no pain, to 10, worst pain.

Ropivacaine: Another single enantiomer amide LA with analgesic potency similar to that of bupivacaine.

Ultrasound guidance for nerve block: Imaging technique for performing nerve block safely that accurately visualizes the surrounding anatomy and placement of the needle at the target location in real time.

Key facts

- The AUC of plasma levobupivacaine from 0 to 240 min was $172,484 \pm 50,502$ ng min ml^{-1} in surgical patients undergoing a trunk nerve block with 160 mg of levobupivacaine in an 8% low-molecular weight dextran saline solution, while that was $229,124 \pm 87,254$ ng min ml^{-1} in patients receiving 160 mg of levobupivacaine in a saline solution ($P < .01$).
- NSR results demonstrated better postoperative analgesia in surgical patients undergoing a trunk nerve block with 0.2% levobupivacaine and 8% low-molecular weight dextran in a saline solution as compared to 0.2% levobupivacaine in a saline solution 8 h after surgery (1.5 ± 1.5 vs. 4.2 ± 2.7, $P < .01$).
- The mixture of LA and low-molecular weight dextran remained at the injection site for more than 24 h.
- When a mixture of LA and low-molecular weight dextran is injected into the wrong area instead of the target compartment, very high pressure can develop.
- In cases of subcutaneous infiltration anesthesia, the peak plasma adrenaline concentration was 0.15 ± 0.07 ng ml^{-1} when using lidocaine, adrenaline, and low-molecular weight dextran in a saline solution as compared to 1.04 ± 0.22 ng ml^{-1} when using the same concentration of lidocaine and adrenaline in a saline solution.

Summary points

- Use of an LA mixture with low-molecular weight dextran enhances analgesic duration and potency in surgical patients receiving a transversus abdominis plane block and other types of interfascial compartment nerve blocks, as well as a single nerve block.
- An interfascial compartment block of the abdominal trunk by use of a large amount of an LA mixture with low-molecular weight dextran is very effective and comparable to an epidural block.
- A mixture of LA with low-molecular weight dextran suppresses LA systemic absorption, thus reducing LA systemic toxicity.
- Use of low-molecular weight dextran in an LA mixture for nerve block analgesia increases block accuracy and performance.
- The mechanism of the effects of dextran used as an adjuvant is related to an increase in viscosity that causes reduced diffusion of the LA-dextran complex from the injection site.
- Addition of low-molecular weight dextran to a lidocaine-adrenaline solution reduces the systemic toxicity of adrenaline by inhibiting systemic adrenaline absorption.
- Low-molecular weight dextran functions effectively as an adjuvant to enhance LA analgesia and reduce the risk of LA systemic toxicity.

References

Aberg, G., Friberger, P., & Sydnes, G. (1978). Studies on the duration of local anaesthesia: A possible mechanism for the prolonging effect of dextran on the duration of infiltration anaesthesia. *Acta Pharmacologica et Toxicologica*, *42*(2), 88–92. https://doi.org/10.1111/j.1600-0773.1978.tb02174.x.

Adams, H. A., Biscoping, J., Kafurke, H., Muller, H., Hoffmann, B., Boerner, U., et al. (1988). Influence of dextran on the absorption of adrenaline-containing lignocaine solutions: A protective mechanism in local anaesthesia. *British Journal of Anaesthesia*, *60*(6), 645–650. https://doi.org/10.1093/bja/60.6.645.

Alkhawajah, A., & Farag, H. (1992). The effect of dextran on the pharmacokinetics of lignocaine during epidural anaesthesia. *The Journal of International Medical Research, 20*(2), 127–135. https://doi.org/10.1177/030006059202000205.

Armstrong, D. N., & Kingsnorth, A. N. (1986). Local anaesthesia in inguinal herniorrhaphy: Influence of dextran and saline solutions on duration of action of bupivacaine. *Annals of the Royal College of Surgeons of England, 68*(4), 207–208. https://www.ncbi.nlm.nih.gov/pubmed/2431648.

Bugada, D., Ghisi, D., & Mariano, E. R. (2017). Continuous regional anesthesia: A review of perioperative outcome benefits. *Minerva Anestesiologica, 83* (10), 1089–1100. https://doi.org/10.23736/S0375-9393.17.12077-8.

Covino, B. G. (1986, Jul). Pharmacology of local anaesthetic agents. *British Journal of Anaesthesia, 58*(7), 701–716. https://doi.org/10.1093/bja/58.7.701.

Exadaktylos, A. K., Buggy, D. J., Moriarty, D. C., Mascha, E., & Sessler, D. I. (2006). Can anesthetic technique for primary breast cancer surgery affect recurrence or metastasis? *Anesthesiology, 105*(4), 660–664. https://doi.org/10.1097/00000542-200610000-00008.

Grandhi, R. K., Lee, S., & Abd-Elsayed, A. (2017). The relationship between regional anesthesia and cancer: A metaanalysis. *The Ochsner Journal, 17*(4), 345–361. https://www.ncbi.nlm.nih.gov/pubmed/29230120.

Hamada, T., Tsuchiya, M., Mizutani, K., Takahashi, R., Muguruma, K., Maeda, K., et al. (2016). Levobupivacaine-dextran mixture for transversus abdominis plane block and rectus sheath block in patients undergoing laparoscopic colectomy: A randomised controlled trial. *Anaesthesia, 71*(4), 411–416. https://doi.org/10.1111/anae.13408.

Hassan, H. G., Renck, H., Lindberg, B., Akerman, B., & Hellquist, R. (1985). Effects of adjuvants to local anaesthetics on their duration. I. Studies of dextrans of widely varying molecular weight and adrenaline in rat infraorbital nerve block. *Acta Anaesthesiologica Scandinavica, 29*(4), 375–379. https://doi.org/10.1111/j.1399-6576.1985.tb02218.x.

Ito, E., Ichinohe, T., Shibukawa, Y., Aida, H., & Kaneko, Y. (2007). Anesthetic duration of lidocaine with 10% dextran is comparable to lidocaine with 1:160 000 epinephrine after intraosseous injection in the rabbit. *Oral Surgery, Oral Medicine, Oral Pathology, Oral Radiology, and Endodontics, 104* (3), e26–e31. https://doi.org/10.1016/j.tripleo.2007.03.008.

Kaplan, J. A., Miller, E. D., Jr., & Gallagher, E. G., Jr. (1975). Postoperative analgesia for thoracotomy patients. *Anesthesia and Analgesia, 54*(6), 773–777. https://doi.org/10.1213/00000539-197511000-00025.

Kingsnorth, A. N., Wijesinha, S. S., & Grixti, C. J. (1979). Evaluation of dextran with local anaesthesia for short-stay inguinal herniorraphy. *Annals of the Royal College of Surgeons of England, 61*(6), 456–458. https://www.ncbi.nlm.nih.gov/pubmed/496237.

Loder, R. E. (1960). A local-anaesthetic solution with longer action. *Lancet, 2*(7146), 346–347. https://doi.org/10.1016/s0140-6736(60)91485-9.

Malik, O., Kaye, A. D., Kaye, A., Belani, K., & Urman, R. D. (2017). Emerging roles of liposomal bupivacaine in anesthesia practice. *Journal of Anaesthesiology Clinical Pharmacology, 33*(2), 151–156. https://doi.org/10.4103/joacp.JOACP_375_15.

Murphy, T. M., Mather, L. E., Stanton-Hicks, M., Bonica, J. J., & Tucker, G. T. (1976). The effects of adding adrenaline to etidocaine and lignocaine in extradural anaesthesia I: Block characteristics and cardiovascular effects. *British Journal of Anaesthesia, 48*(9), 893–898. https://doi.org/10.1093/bja/48.9.893.

Oda, Y. (2019). Local anesthetic systemic toxicity: Proposed mechanisms for lipid resuscitation and methods of prevention. *Journal of Anesthesia, 33*(5), 569–571. https://doi.org/10.1007/s00540-019-02648-y.

Simpson, P. J., Hughes, D. R., & Long, D. H. (1982). Prolonged local analgesia for inguinal herniorrhaphy with bupivacaine and dextran. *Annals of the Royal College of Surgeons of England, 64*(4), 243–246. https://www.ncbi.nlm.nih.gov/pubmed/6178347.

Skolnik, A., & Gan, T. J. (2014). New formulations of bupivacaine for the treatment of postoperative pain: Liposomal bupivacaine and SABER-bupivacaine. *Expert Opinion on Pharmacotherapy, 15*(11), 1535–1542. https://doi.org/10.1517/14656566.2014.930436.

Soffin, E. M., Lee, B. H., Kumar, K. K., & Wu, C. L. (2019). The prescription opioid crisis: Role of the anaesthesiologist in reducing opioid use and misuse. *British Journal of Anaesthesia, 122*(6), e198–e208. https://doi.org/10.1016/j.bja.2018.11.019.

Swain, A., Nag, D. S., Sahu, S., & Samaddar, D. P. (2017). Adjuvants to local anesthetics: Current understanding and future trends. *The World Journal of Clinical Cases, 5*(8), 307–323. https://doi.org/10.12998/wjcc.v5.i8.307.

Takahashi, R., Oda, Y., Tanaka, K., Morishima, H. O., Inoue, K., & Asada, A. (2006). Epinephrine increases the extracellular lidocaine concentration in the brain: A possible mechanism for increased central nervous system toxicity. *Anesthesiology, 105*(5), 984–989. https://doi.org/10.1097/00000542-200611000-00020.

Tsuchiya, M., Mizutani, K., Funai, Y., & Nakamoto, T. (2016). In-line positioning of ultrasound images using wireless remote display system with tablet computer facilitates ultrasound-guided radial artery catheterization. *Journal of Clinical Monitoring and Computing, 30*(1), 101–106. https://doi.org/10.1007/s10877-015-9692-9.

Tsuchiya, M., Mizutani, K., & Ueda, W. (2018). Large volume of low concentration of local anesthetic dissolved with low-molecular weight dextran as adjuvant for ultrasound-guided posterior quadratus lumborum block greatly enhances and extends analgesic effects. *Minerva Anestesiologica, 84*(7), 876–878. https://doi.org/10.23736/S0375-9393.18.12653-8.

Tsuchiya, M., Mizutani, K., & Ueda, W. (2019). Adding dextran to local anesthetic enhances analgesia. *Journal of Anesthesia, 33*(1), 163. https://doi.org/10.1007/s00540-018-2573-x.

Tsuchiya, M., Mizutani, K., Yabe, M., Mori, T., & Ueda, W. (2019). Ultrasound-guided mandibular nerve block with local anesthetic and low-molecular weight dextran helps reduce anesthetic requirements for parotidectomy. *Minerva Anestesiologica, 85*(2), 202–203. https://doi.org/10.23736/S0375-9393.18.12966-X.

Tsuchiya, M., Takahashi, R., Furukawa, A., Suehiro, K., Mizutani, K., & Nishikawa, K. (2012). Transversus abdominis plane block in combination with general anesthesia provides better intraoperative hemodynamic control and quicker recovery than general anesthesia alone in high-risk abdominal surgery patients. *Minerva Anestesiologica, 78*(11), 1241–1247. https://www.ncbi.nlm.nih.gov/pubmed/23132262.

Ueda, W., Hirakawa, M., & Mori, K. (1985). Acceleration of epinephrine absorption by lidocaine. *Anesthesiology*, *63*(6), 717–720. https://doi.org/10.1097/00000542-198512000-00034.

Wiles, M. D., & Nathanson, M. H. (2010). Local anaesthetics and adjuvants – future developments. *Anaesthesia*, *65*(Suppl 1), 22–37. https://doi.org/10.1111/j.1365-2044.2009.06201.x.

Yokoyama, M., Goto, H., Ueda, W., Hirakawa, M., & Arakawa, K. (1993). Modification of intravenous lidocaine-induced convulsions by epinephrine in rats. *Canadian Journal of Anaesthesia*, *40*(3), 251–256. https://doi.org/10.1007/BF03037037.

Chapter 3

Intraperitoneal local anesthetic agents in the management of postoperative pain

Karlin Sevensma

Department of Surgery, Metro Health University of Michigan Health, Wyoming, MI, United States

Introduction

Finding ways to control pain postoperatively has long been a goal of the surgical team. The advent of laparoscopy has led to patients experiencing less pain than they would in traditional open surgeries, but there is still some pain associated with laparoscopic procedures (Gerges, Kanzai, & Jabbour-khoury, 2006). The use of local anesthetic intraperitoneally, the placement of that local anesthetic and the timing of placement with respect to dissection have been investigated and continue to be evaluated. If modifying the ways that local anesthetic is used can lead to better postoperative pain control and decrease the need for narcotic pain medication, this could be helpful in improving the patient experience of surgery. This chapter explores the different ways that local anesthetic is used within the peritoneal cavity during laparoscopic surgery. The usefulness of local anesthetic in the skin, muscle, or peritoneum via injection through the skin, though useful, is not the focus of this chapter.

Anatomic considerations and agents

Local anesthetic instilled within the peritoneal cavity can affect both the parietal peritoneum directly as well as the visceral peritoneum and viscera near the site of instillation. Therefore, the local anesthetic agent can act upon both the spinal nerves that innervate the parietal peritoneum and the visceral splanchnic nerves. In this regard, local anesthetic within the peritoneum can reduce both localized pain and referred visceral and somatic pain (Cha et al., 2012). Instillation of local anesthetic is limited by the volume of medication and the dose of local anesthetic that can be safely administered without risking toxicity. For this reason, local anesthetic is typically given in a certain location, adjacent to the organ that is to be removed or altered or near a structure that is targeted, such as the diaphragm and phrenic nerve. Another consideration is that local anesthetic will pool in certain areas of the body depending upon patient position. Common places for pooling in the supine patient are the subdiaphragmatic spaces and the pelvis. In addition to these areas, the folds of the peritoneal surface cause several other potential spaces for local anesthetic to pool: the right and left subhepatic spaces, the right and left paramesenteric spaces and the right and left paracolic gutters (Ioannidis et al., 2013). The type of local anesthetic is also a consideration, as different agents offer different duration of action and different side effect profiles. Lidocaine has been in use since 1948 but is not the favored local anesthetic for intraperitoneal use due to its short duration of action and its cardiotoxicity. Bupivacaine and ropivacaine have a longer duration of action and less cardiotoxicity, especially when used in their S-isomer forms (Ioannidis et al., 2013).

Laparoscopic cholecystectomy

Control of pain following laparoscopic cholecystectomy is multifaceted and typically requires a multimodality approach. The use of local anesthetic instilled into the abdomen at the time of surgery has been examined in several ways for laparoscopic cholecystectomy. Placement of the local anesthetic at the gallbladder fossa, over the liver and at one or both hemidiaphragms has been evaluated. Placement of the local anesthetic before pneumoperitoneum introduction, prior to dissection and after dissection has also been evaluated. A variety of local anesthetic agents and doses have been studied.

Initial studies of intraperitoneal local anesthetics for laparoscopic cholecystectomy were based upon gynecologic studies, and the doses of local anesthetic were typically quite small. These early studies did not always demonstrate a

---

Treatments, Mechanisms, and Adverse Reactions of Anesthetics and Analgesics. https://doi.org/10.1016/B978-0-12-820237-1.00002-8
Copyright © 2022 Elsevier Inc. All rights reserved.

28 PART | I Drugs and agents used in anesthesia and analgesia

significant effect on patient outcomes. Szem et al. examined using 10 mg of Bupivacaine applied over the liver and gall-bladder and at the hemidiaphragms prior to initiation of dissection. This technique demonstrated a decrease in pain scores in the first 6 h but no significant change in post-procedure analgesic use (Szem, Hydo, & Barie, 1996). Raetzell et al. used doses of 6.25 mg and 12.5 mg of bupivacaine applied over the hemidiaphragms and at the gallbladder fossa at the end of the procedure. Comparing that to saline instillation, they found no difference in postoperative pain scores or narcotic use (Raetzell, Maier, Schroder, & Wulf, 1995). Joris et al. performed a similar study, administering 10 mg of bupivacaine post-dissection at the right hemidiaphragm. No effect on pain was noted (Joris, Thiry, Paris, Weerts, & Lamy, 1995). A study performed in 2011 examined placement of 10 mg of levobupivacaine prior to dissection and found no effect on post-operative pain (Hilvering et al., 2011). Arguably, the dosing in these early studies was not enough to have an appreciable effect on post-operative pain.

Later studies with increased doses of intraperitoneal local anesthetic found both decreased pain scores as well as decreased narcotic consumption as compared to saline controls (Ahmed et al., 2008; Elhakim, Elkott, Ali, & Tahoun, 2000; Kucuk, Kadiogullari, Canoler, & Salvi, 2007; Labaille, Mazoit, Paqueron, Franco, & Benhamou, 2002; Yeh et al., 2014). In a study by Ahmed et al., bupivacaine 10 mg was compared to lidocaine 40 mg and plain saline irrigation vs no irrigation. Interestingly, even the group that had only saline irrigation of the diaphragmatic surface had decreased pain scores and required less postoperative pain medication (Ahmed et al., 2008). A study by Paulson et al. showed that placing 7.5 mg of bupivacaine around the liver before dissection and again after dissection increased same day discharge rates in laparoscopic cholecystectomy patients. The same study also compared patients who had just 7.5 mg of bupivacaine pre-dissection to patients with the same amount of bupivacaine administered postdissection and both groups had similar rates of same-day discharge (Paulson, Mellinger, & Baguley, 2003) (Table 1).

Laparoscopic appendectomy

The use of intraperitoneal local anesthetic has not been as widely studied for laparoscopic appendectomy as it has been for laparoscopic cholecystectomy. A standardized dose of local anesthetic as a whole number is also not as common in the literature, likely because most appendectomies are performed on children and teens and patient size varies significantly. Therefore, many studies examined a weight-based dose of local anesthetic.

In the studies that examine predissection placement of local anesthetic, lidocaine, levobupivacaine, and ropivacaine have been examined as weight-based regimens. The distribution of the local anesthetic was at the hemidiaphragms and in the peri-appendiceal area. These studies found a decrease in postoperative narcotic consumption that was significant (Colbert, O'Hanlon, Courtney, Quill, & Flynn, 1998; Kang & Kim, 2010; Kim et al., 2011; Thanapal et al., 2012). One of the studies also looked at serum cortisol levels as a marker for stress response to pain and found that cortisol levels were significantly lower in the patients receiving intraperitoneal local anesthetic (Thanapal et al., 2012). In a study examining post-dissection placement of 10 mg of bupivacaine at the appendiceal stump in adult patients, a decrease in pain scores and a decrease in postoperative narcotic consumption was noted (Sevensma, Schleichert, Schwickerath, Shoemaker, & Miller, 2018) (Table 2).

Gynecologic surgery

Local anesthetics instilled in the pelvis for gynecologic procedures predate the use of intraperitoneal local anesthetics in general surgery. The first study of this type was performed in 1991 and found that either pain scores or narcotic consumption or both were favorably affected by placing intraperitoneal local anesthetic (Ioannidis et al., 2013). The methods of placement varied to include the pelvis, the hemidiaphragms or both, all with significant decreases in pain. Haldane et al. describe a pouch of Douglas block for elective sterilization using a catheter technique to allow for re-dosing during the procedure (Gerges et al., 2006; Haldane, Stott, & McMenemin, 1998). The positive effects of intraperitoneal blockade are more pronounced in gynecologic surgery than in laparoscopic cholecystectomy, perhaps owing to the less traumatic nature and shorter duration of gynecologic cases (Ioannidis et al., 2013).

Instillation at hemidiaphragm for phrenic nerve pain prevention

Placement of local anesthetic at the hemidiaphragms has been shown in many studies to decrease shoulder pain following laparoscopy. The shoulder pain associated with laparoscopy is due to phrenic nerve injury and irritation caused by stretching the diaphragm with the pneumoperitoneum. Application of local anesthetic at the diaphragm has been compared to saline instillation and to nothing as a control. Interestingly, the instillation of plain saline at the hemidiaphragm appears to

TABLE 1 Laparoscopic cholecystectomy.

Author	Year	Agent(s) vs. saline	Dose	Timing	Location	Findings	Comments
Chundrigar	1993	bupivacaine	2.5 mg	Post-dissection	Gallbladder fossa	Decreased incidence of shoulder pain No change in narcotic consumption	
Raetzell	1995	Bupivacaine	6.25 mg 12.5 mg	Post-dissection	Subphrenic, gallbladder fossa	No difference in pain scores or morphine	Respiratory depression noted
Joris	1995	Bupivacaine	10 mg	Post-dissection	Right hemidiaphragm	No influence on pain	
Szem	1996	Bupivacaine	10 mg	Pre-dissection	Subphrenic, gallbladder, over liver	Decreased pain scores for first 6 h, no change morphine	
Elhakim	2000	Lidocaine	200 mg	Post-dissection	Right hemidiaphragm	Decreased pain scores & morphine consumption first 24 h	Decreased shoulder pain
Labaille	2002	Ropivacaine	100 mg 300 mg	Pre- and post-dissection	Subphrenic, gallbladder fossa, over liver	Decreased morphine consumption	No benefit to increasing dose to 300 mg
Paulson	2003	Bupivacaine	7.5 mg or 15 mg	Pre-and/or post-dissection	Peri-hepatic	Decreased length of stay	
Kucuk	2007	Bupivacaine vs Ropivacaine	100 mg 100 mg 150 mg	Post-dissection	Subphrenic, gallbladder fossa	Decreased morphine first 24 h	150 mg dose more effective than 100
Ahmed	2008	Bupivacaine vs lidocaine vs saline vs nothing	10 mg 40 mg	Post-dissection	Subphrenic, gallbladder fossa	Decreased pain scores & morphine consumption first 24 h	Even saline irrigation alone decreased pain scores
Pappas-Gogos	2008	Ropivacaine Saline irrigation	80 mg	Pre- vs post-dissection	Right hemidiaphragm	Decreased abdominal and shoulder pain scores	Pre-dissection group better than post-dissection
Gharaibeh	2000	Bupivacaine	2.5 mg	Post-dissection	Gallbladder fossa	Decreased incidence of shoulder pain	Suggests gallbladder fossa is a source of shoulder pain
Cha	2011	Ropivacaine	2 mg/kg In 100 mL saline	Pre-dissection	Subphrenic, gallbladder, over liver	Decreased narcotic consumption	Combined wound and intraperitoneal local anesthetic compared to no local anesthetic
Hilvering	2011	levobupivacaine	10 mg	Pre-dissection	Right hemidiaphragm and gallbladder	No influence on pain	Combined wound and intraperitoneal local anesthetic compared to no local anesthetic
Yeh	2014	Levoropivacaine	5 mg/kg in 200 mL saline	Pre-pneumo	Both hemidiaphragms prior to insufflation	Decreased pain score Decreased narcotic Decreased LOS	Combined wound and intraperitoneal local anesthetic compared to no local anesthetic

PART | I Drugs and agents used in anesthesia and analgesia

TABLE 2 Laparoscopic appendectomy.

Author	Year	Agent(s) vs. saline	Dose	Timing	Location	Findings	Comments
Kang	2010	Ropivacaine	2 mg/kg	Pre-dissection	Diaphragmatic and peri-appendiceal	Decreased fentanyl consumption	
Kim	2011	Lidocaine	3.5 mg/kg	Pre-dissection	Diaphragmatic and peri-appendiceal	Decreased pain scores and narcotic consumption	IV lidocaine had similar effect
Thanapal	2012	Ropivacaine Levo-bupivacaine	5 mL/kg 2 mL/kg	Pre-dissection	Peri-appendiceal	Decreased morphine consumption Lower cortisol levels	
Sevensma	2018	Bupivacaine	10 mg	Post-dissection	Appendiceal stump/cecum	Decreased pain score Decreased narcotic	Patient age > 18

have a positive effect on shoulder pain prevention postoperatively. The instillation of local anesthetic also appears to prevent some postoperative shoulder pain. Both saline and local anesthetic were better than nothing in preventing postoperative shoulder pain (Donatsky, Bjerrum, & Gogenur, 2013).

It would make sense that placing local anesthetic at the hemidiaphragms as early in the case as possible would give maximal effect. One study comparing predissection local anesthetic instillation to postdissection local anesthetic instillation found that pain scores were lower in the predissection group (Pappas-Gogos et al., 2008). Paulsen et al. found equivalency between predissection and postdissection groups (Paulson et al., 2003). The evidence for predissection infiltration rather than postdissection infiltration is not strong, and many studies have shown benefit of postdissection application of local anesthetic to the hemidiaphragm (Ahmed et al., 2008; Elhakim et al., 2000; Kucuk et al., 2007).

For patients undergoing laparoscopic cholecystectomy, placement of local anesthetic in the gallbladder fossa at the conclusion of surgery has been found to decrease shoulder pain, even when no local anesthetic was placed directly at the diaphragm. This effect was accomplished with relatively small doses of bupivacaine (2.5 mg). They found that placement of the local anesthetic in the gallbladder fossa at the end of the case decreased the incidence of shoulder pain in their patients. However, postoperative narcotic consumption was not affected (Chundrigar, Morris, Hedges, & Stamatakis, 1993; Gharaibeh & Al-Jaberi, 2000).

Safety of intraperitoneal anesthetics

Most published studies demonstrate no complications related to the use of intraperitoneal anesthetics. Using an appropriate dose of medication is important to avoid cardiotoxicity associated with overdose of local anesthetics, since they are absorbed and circulate systemically. One study did note significant respiratory depression in patients who received intraperitoneal Bupivacaine (Raetzell et al., 1995). This may have been due to narcotic use in the face of decreased need, and this finding was not replicated in other studies.

Applications to other areas

Although most work involving local anesthetic instillation is done in patients undergoing laparoscopy, there may be applications to open surgery as well. Application in thoracoscopic surgery and in other surgical subspecialties is also a possibility. As other minimally invasive techniques are developed, intraperitoneal local anesthetic use should be trialed in those procedures (Kahokehr, Sammour, Soop, & Hill, 2010).

Other agents of interest

Gas-humidifying devices have also been evaluated for their ability to aerosolize local anesthetic for intraperitoneal use. In a study by Greib et al., two basic types of humidifiers were evaluated for their ability to produce a sufficient concentration of local anesthetic in the intraperitoneal environment. Evaporation-based humidifiers were found to deliver an insufficient quantity of the medication to have an effect. Micro-vibration humidifiers, on the other hand, were able to aerosolize sufficient local anesthetic agent to be potentially useful. This study was designed to determine concentrations of anesthetic produced by these humidifying devices and did not evaluate potential effect on patient physiology (Greib et al., 2008). Some authors have tested aerosolized local anesthetic in vivo and have found it to be effective at reducing pain, though its use is not widespread (Colbert et al., 1998).

Mini-dictionary of terms

Intraperitoneal: within the peritoneal cavity; studies that involved injecting the peritoneum were excluded from this review.

Local anesthetic: medication that acts to block neurotransmission at the site of administration (Lidocaine, Bupivacaine and Ropivacaine are discussed in this chapter).

Key facts of intraperitoneal local anesthetic agents in the management of postoperative pain

- In studies involving laparoscopic appendectomy, the use of intraperitoneal local anesthetic agents has been shown to decrease postoperative pain scores and decrease the need for rescue pain medication compared with placebo.
- In studies involving laparoscopic cholecystectomy, the use of intraperitoneal local anesthetic agents has been shown to decrease pain scores of both abdominal and shoulder pain and decrease postoperative consumption of analgesics.

Summary points

- The instillation of local anesthetic into the peritoneal cavity at the time of various types of laparoscopic surgery has been found to improve postoperative pain as quantified by decreased time to discharge, decreased pain severity scores, and decreased amounts of rescue narcotic.
- No significant safety issues have been identified with the practice of intraperitoneal instillation of local anesthetic.
- Dosage, placement location, and timing of local anesthetic administration has been evaluated in a variety of settings, for different procedures and prior to and after dissection.
- Although more research is needed to standardize the practice of intraperitoneal local anesthetic use, it has been demonstrated as helpful as an adjunct in managing the pain associated with laparoscopic surgery and should be considered for all patients undergoing laparoscopic surgery.

References

Ahmed, B., Ahmed, A., Tan, D., Awad, Z., Al-Aali, A., Kilkenny, J., et al. (2008). Post-laparoscopic cholecystectomy pain: effects of Intraperitoneal local anethetics on pain control – A randomized prospective double-blind placebo-controlled trial. *The American Surgeon, 74*(3), 201–209.

Cha, S. M., Kang, H., Baek, C. W., Jung, Y. H., Koo, G. H., Kim, B. G., et al. (2012). Peritrocal and intraperitoneal ropivacaine for laparoscopic cholecystectomy: A prospective randomized, double-blind controlled trial. *Journal of Surgical Research, 175,* 251–258.

Chundrigar, T., Morris, R., Hedges, A., & Stamatakis, J. (1993). Intraperitoneal bupivacaine for effective pain relief after laparoscopic cholecystectomy. *Annals of the Royal College of Surgeons of England, 75,* 437–439.

Colbert, S., O'Hanlon, D. M., Courtney, D. F., Quill, D. S., & Flynn, N. (1998). Analgesia following appendicectomy – the value of peritoneal bupivacaine. *Canadian Journal of Anaesthesia, 45,* 729–734.

Donatsky, A., Bjerrum, F., & Gogenur, I. (2013). Intraperitoneal instillation of saline and local anesthesia for prevention of shoulder pain after laparoscopic cholecystectomy. *Surgical Endoscopy, 27,* 2283–2292.

Elhakim, M., Elkott, N., Ali, M., & Tahoun, H. (2000). Intraperitoneal lidocaine for postoperative pain after laparoscopy. *Acta Anaesthesiologica Scandinavica, 44,* 280–284.

Gerges, F., Kanzai, G., & Jabbour-khoury, S. (2006). Anesthesia for laparoscopy: A review. *Journal of Clinical Anesthesia, 18,* 67–78.

Gharaibeh, K., & Al-Jaberi, T. (2000). Bupivacaine istillation into the gallbladder bed after laparoscopic cholecystectomy: Does it decrease shoulder pain? *Journal of Laparoendoscopic and Advanced Surgical Techniques, 10*(3), 137–141.

Greib, N., Schlotterbeck, H., Dow, W., Joshi, G., Geny, B., & Diemunsch, P. (2008). An evaluation of gas humdifying devices as a means of intraperitoneal local anesthetic administration for laparoscopic surgery. *International Anesthesia Research Society, 107*, 2.

Haldane, G., Stott, S., & McMenemin, I. (1998). Pouch of Douglas block for laparoscopic sterilization. *Anaesthesia, 53*(6), 598–603.

Hilvering, B., Draaisma, W., van der Bilt, J., Valk, R., Kofman, K., & Consten, E. (2011). Randomized clinical trial of combined preincisional infiltration and intraperitoneal instillation of levobupivacaine for postoperative pain after laparoscopic cholecystectomy. *British Journal of Surgery, 98*, 784–789.

Ioannidis, O., Anastasilakis, C., Varnalidis, I., Paraskevas, G., Malakozis, S., Gatzos, S., et al. (2013). Intraperitoneal administration of local anesthetics in laparoscopic surgery: Pharmacological, anatomical, physiological and pathophysiological considerations. *Minerva Chirurgica, 68*, 599–612.

Joris, J., Thiry, E., Paris, P., Weerts, J., & Lamy, M. (1995). Pain after laparoscopic cholecystectomy: Characteristics and effect of intraperitoneal bupivacaine. *Anesthesia & Analgesia, 81*, 379–384.

Kahokehr, A., Sammour, T., Soop, M., & Hill, A. (2010). Intraperitoneal use of local anesthetic in laparoscopic cholecystectomy: Systematic review and metaanalysis of randomized controlled trials. *Journal of Hepato-Biliary-Pancreatic Sciences, 17*, 637–656.

Kang, H., & Kim, B. G. (2010). Intraperitoneal Ropivacaine for effective pain relief after laparoscopic appendectomy: A prospective, randomized, double-blind, placebo-controlled study. *The Journal of International Medical Research, 38*, 821–832.

Kim, T. H., Kang, H., Hong, J. H., Park, J. S., Baek, C. W., Kim, J. Y., et al. (2011). Intraperitoneal and intravenous lidocaine for effective pain relief after laparoscopic appendectomy: A prospective, randomized, double-blind. *Placebo-Controlled Study. Surgical Endoscopy, 25*, 3183–3190.

Kucuk, C., Kadiogullari, N., Canoler, O., & Salvi, S. (2007). A placebo-controlled comparison of bupivacaine and ropivacaine instillation for preventing postoperative pain avter laparoscopic cholecystectomy. *Surgery Today, 37*, 396–400.

Labaille, T., Mazoit, J., Paqueron, X., Franco, D., & Benhamou, D. (2002). The clinical efficacy and pharmacokinetics of intraperitoneal Ropivacaine for laparoscopic cholecystectomy. *Anesthesia and Analgesia, 94*, 100–105.

Pappas-Gogos, G., Tsimogiannis, K., Zikos, N., Nikas, K., Manataki, A., & Tsimoyiannis, E. (2008). Preincisional and intraperitoneal ropivacaine plus normal saline infusion for postoperative pain relief after laparoscopic cholecystectomy: A randomized double-blind controlled trial. *Surgical Endoscopy, 22*, 2036–2045.

Paulson, J., Mellinger, J., & Baguley, W. (2003). The use of intraperitoneal Bupivacaine to decrease length of stay in elective cholecystectomy patients. *The American Surgeon, 69*(4), 275–279.

Raetzell, M., Maier, C., Schroder, D., & Wulf, H. (1995). Intraperitoneal application of bupivacaine during laparoscopic cholecystectomy – Risk or benefit? *Anesthesia and Analgesia, 81*, 967–972.

Sevensma, K., Schleichert, T., Schwickerath, C., Shoemaker, A., & Miller, C. (2018). A randomized double blinded study to determine the effectiveness of utilizing intraperitoneal bupivicaine: Does it reduce post-operative opioid use following laparoscopic appendectomy? *American Journal of Surgery, 217*(3), 479–482.

Szem, J., Hydo, L., & Barie, P. (1996). A double-blinded evaluation of intraperitoneal bupivacaine vs saline for the reduction of postoperative pain and nausea after laparoscopic cholecystectomy. *Surgical Endoscopy, 10*, 44–48.

Thanapal, M. R., Tata, M. D., Tan, A. J., Subramaniam, T., Tong, J. M. G., Palayan, K., et al. (2012). Pre-emptive intraperitoneal local anaesthesia: An effective method in immediate post-operative pain management and metabolic stress response in laparoscopic Appendicectomy, a randomized, double-blinded. *Placebo-controlled Study. Journal of Surgery, 84*, 47–51.

Yeh, C. N., Tsai, C. Y., Cheng, C. T., Wang, S. Y., Liu, Y. Y., Chiang, K. C., et al. (2014). Pain relief from combined wound and intraperitoneal local anesthesia for patients who undergo laparoscopic cholecystectomy. *BMC Surgery, 14*, 28.

Chapter 4

Automatic control of anesthesia via different vital signs

Jerry Chen[a], Maysam F. Abbod[b], and Jiann-Shing Shieh[c]

[a]Department of Mechanical Engineering, Yuan Ze University, Taoyuan, Taiwan, [b]Department of Electronic and Computer Engineering, Brunel University London, Uxbridge, United Kingdom, [c]Yuan Ze University, Taoyuan, Taiwan

Abbreviations

AEP	Auditory Evoked Potential
ANS	Autonomic Nervous System
ApEn	Approximate Entropy
BIS	Bispectral Index
BP	Blood Pressure
CI	Complex Index
CNS	Central Nervous System
DFA	Detrended Fluctuation Algorithm
DoA	Depth of Anesthesia
ECG	Electrocardiography
EEG	Electroencephalography
EMD	Empirical Mode Decomposition
EMG	Electromyography
GABA	Gamma-aminobutyric Acid
HR	Heart Rate
IMF	Intrinsic Mode Functions
NMDA	N-methyl-D-aspartate
PD	Pharmacodynamics
PK	Pharmacokinetics
PVI	Pleth Variability Index
RE	Response Entropy
SAP	Systolic Arterial Pressure
SampEn	Sample Entropy
SE	State Entropy
SpO$_2$	Oxyhemoglobin Saturation
SSI	Surgical Stress Index

Introduction

Anesthesia and its interpretation—Hypnosis, nociception, and neuromuscular relaxation

The idea of anesthesia is to induce a temporary, reversible effect on subject that causes unconsciousness, amnesia, analgesia, akinesia and helps to stabilize cardiorespiratory and autonomic systems (De Hert, 2006; Evers & Crowder, 2009; Hedenstierna, 1995; Lenhardt, 2010; Neukirchen, Kienbaum, Warner, & Warner, 2008). The purpose of anesthesia is to ensure the surgery goes under the optimal conditions and keeps the patient from any adverse effect or mental trauma in both intra-operative and post-operative. To achieve this, anesthesia has to be functional in three major aspects: hypnotics, analgesics, and neuromuscular relaxation. While hypnotics caused the unconsciousness to prevent any intro-operative awareness or memorization; analgesics reduce or even eliminate the nociceptive reactions due to the painful stimulus, and neuromuscular relaxation induce paralysis for suppressing any muscle activity.

Treatments, Mechanisms, and Adverse Reactions of Anesthetics and Analgesics. https://doi.org/10.1016/B978-0-12-820237-1.00004-1
Copyright © 2022 Elsevier Inc. All rights reserved.

To measure hypnosis, the most widely used physiological signal is electroencephalography (EEG). One of the EEG based techniques such as Bispectral index (BIS) monitor define the depth of hypnosis by using the weighted sum of EEG sub parameters in both time domain and frequency domain, and high order spectral sub parameters) (Kaul & Bharti, 2002) to derive a one-dimensional scale from 0 to 100 where index of 100 represents the awake state and vice versa. Other common monitors for determine depth of hypnosis are like Entropy monitor which generate two numbers: State Entropy (SE) and Response Entropy (RE). The SE is calculated from the EEG activity in frequency range of 0.8–32 Hz. RE, however, is measured form the frequency range of 0.8–47 Hz.

For nociception or analgesia, relevant measurement is to use clinical signs caused by sympathetic response such as rapid increasing heart rate (HR) and blood pressure (BP); or other techniques like Surgical Stress Index (SSI), or Analgesic Nociception Index (ANI). SSI is derived from wave from amplitude of photoplethysmographic and heart beat intervals; it ranging from 0 (for very low stress level) to 100 (for very high stress level). In general anesthesia, SSI has positive correlation with nociceptive stimulus and negative correlation with antinociceptive medication (Huiku et al., 2007). ANI scale from 0 to 100 and has strong correlation with nociception, and it measure the influences of respiratory cycle on heart rate variability high frequency (0.15–15 Hz) for obtaining the parasympathetic tone activity (Jeanne, Logier, Jonckheere, & Tavernier, 2009).

Finally, neuromuscular relaxation/blockade could be monitored by clinical tests, qualitative evaluation and quantitative monitoring (Duțu et al., 2018). Clinical test includes the evaluation of respiratory parameters and muscle function (e.g., lifting, griping, etc.); qualitative evaluation assesses the visual or tactilely response of the stimulated muscle by using peripheral nerve stimulators (Naguib, Kopman, & Ensor, 2007) (e.g., train of four, double—burst, tetanic and post-tetanic count, etc.); quantitative monitoring records, quantifies the evoked muscle responses by numeric number while stimulating the peripheral nerve (Murphy, 2018).

Clinical signs and physiological signals in the surgery

Although the nature of awareness and pain is subjective, they could also be objectifying by clinical signs (i.e., sweating, lacrimation, papillary dilatation) and physiological signals (i.e., electromyography (EMG), electrocardiography (ECG), HR, BP, EEG, oxyhemoglobin saturation (SpO_2)). The signals presented during the surgery reflect the effects of the administered anesthesia agent on subject depending on the potency and the concentration of the drug and could be used as an indication for manipulating anesthesia.

Different symptoms and conditions through the three stages (i.e., induction, maintenance, and recovery) are as following:

During the induction period, the subjects are expected with paradoxical excitation, incoherent speech, purposeless movements or some degree of irregular respiratory pattern that could leads to apnea. Subjects gradually lose the corneal and eyelash reflexes, and they would be observed with decreasing muscle tone and reducing response to oral commands. EEG signal in this phase would show an increasing in the beta wave activity (13–25 Hz) (Evers & Crowder, 2009) compared with the prominent alpha activity (10 Hz) in the awake state.

In the maintenance period, the subject's level of anesthesia would mainly be accessed through the physiological signals (e.g., HR, BP) because the clinical sign becomes difficult to access. The physiological sign (e.g., EEG) starts to show the following indications: (1) Light state, signs of decreasing EEG beta activity, and increasing on both alpha (8–12 Hz) and delta activity (0–4 Hz) (Guignard, 2006; Vasella, Frascarolo, Spahn, & Magnusson, 2005); (2) Deep state, EEG is showing the characteristic of burst suppression (Chernik et al., 1990); and (3) Profound state, where the EEG becomes isoelectric. Other quantitative EEG-derived indexes range within the maintenances are as follow: BIS (scale from 0 to 100) is 40–60 (Ekman, Lindholm, Lennmarken, & Sandin, 2004; Gan et al., 1997; Gürses, Sungurtekin, Tomatir, & Dogan, 2004), RE (scale from 0 to 100) and SE (scale from 0 to 91) is 40–60.

Moreover, in the maintenance period, the unusual increasing of HR and BP are also indicating an inadequate state of anesthesia for the subject. Other clinical signs such as papillary dilatation, sweating, returning of muscle tone, lacrimation (Guignard, 2006), and other hormonal response such as catecholamine, corticosteroids (Chernik et al., 1990; Gelb, Leslie, Stanski, & Shafer, 2010) could also serve as the signs for inadequate state of anesthesia to the subject.

During the recovery period, subjects are expecting with clinical signs such as returning of spontaneous respiration, muscle tone, HR increases, BP increases, salivation, and tearing begins. As the skeletal-muscle tone returns, behaviors such as grimacing, swallowing, coughing, and defensive movements (e.g., hands reaching for the endotracheal or nasogastric tube) (Brown, Lydic, & Schiff, 2010). Physiological signal of EEG increases in alpha and beta activity until recover to the awake patterns.

Modeling and control of anesthesia

The design of model-based predictive controller for anesthesia not only has to consider unconsciousness, analgesia, and muscle relaxation mechanisms but also the Pharmacokinetics (PK) and Pharmacodynamics (PD) model as shown in Fig. 1. PK model cover the uptake and distribution of drug within the body and PD describe the dynamics to the effect-site concentration. PK model describe the interactions of administered agent and the three compartment: the rapidly equilibrating compartment (i.e., muscles, viscera), slowly equilibrating compartment (i.e., fat, bones), and central compartment (i.e., blood, brain, liver); PD model are transfer function for the effect-site concentration which also varies from patient to patient (Dumont, 2012; Sawaguchi et al., 2008).

The anesthesiologist is required to keep the patient in an adequate state of anesthesia in the majority of time in the surgery. That means the manipulation of anesthetic agents in the aspects of hypnosis, analgesia and neuromuscular blockade has to be done precisely to avoid any potential of under or overdosing of drug to the patient. Thus, an ideal automatic controller for anesthesia should at least be able to induce the patient rapidly with only a minimal overshoot and then keep the patient in an adequate state of anesthesia as the anesthesiologist did. However, this simple idea could be incredibly complicated for the implementation of that idea to the automatic controller plus the inherent variability for inter-patient and intra-operative. Most researches conducted for automatic control of anesthesia were focused on only one aspect at the same time, that is, it is either automatic control of hypnosis or nociception or neuromuscular relaxation.

Automatic control of hypnosis

The assessment of hypnosis is usually referred to depth of anesthesia (DOA). There are mainly two ways of methods for DOA assessment according to Kaul and Bharti (2002): subjective methods and objective methods. Subjective methods include autonomic response (i.e., hemodynamic changes, lacrimation, sweating, pupillary dilatation) and isolated forearm technique which is a subjective movement in response to a verbal command. On the other hand, objective methods are physiological signals such as spontaneous surface electromyogram (SEMG), lower esophageal contractility (LOC), heart rate variability (HRV), EEG, and evoked potentials.

Sometimes the physiological signal (e.g., EEG) could be easily affected by artifacts in the surgery it requires a signal preprocessing technique to filter out the noise from contaminated signal. Study (Liu, Chen, Fan, Abbod, & Shieh, 2016a) has noticed this problem and tried to find out the best way to get rid of this annoying phenomenon. After compared five different methods (i.e., median frequency, 95% spectral edge frequency, approximate entropy, sample entropy, and permutation entropy), they have concluded that permutation entropy has the better performances than other indices despite all the algorithm could discriminate awake from anesthesia state. Another approved technique for denoising the EEG signal are empirical mode decomposition was also studied (Liu, Chen, Fan, Abbod, & Shieh, 2016b). They first apply the EMD

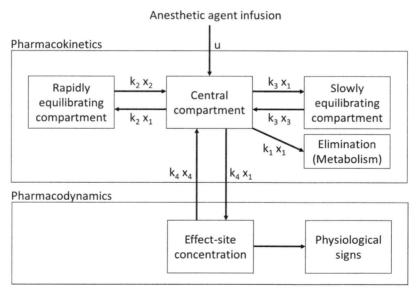

FIG. 1 An illustration of the transfer function of PK and PD. (u for drug infusing rate; k and x are the transfer function parameter that could be a function of patient's age and body weight (Sawaguchi, Furutani, Shirakami, Araki, & Fukuda, 2008).)

36 PART | I Drugs and agents used in anesthesia and analgesia

algorithm to decompose the signal into intrinsic mode functions than apply mean entropy for multiscale entropy to find the artifacts-free intrinsic mode function of the original signal. And the result shows a huge improve in the correlation coefficients between the prediction to DOA before and after noise filtering.

Due to the noise contamination, a mechanism for automatic control based not only on BIS but also other physiological signals (e.g., BP, HR) enhance monitor stabilization and drug operation of DoA. One study (Yu, Doctor, Fan, & Shieh, 2018) conducted such mechanism and choose BP as the second reference for automatic control of DOA. They compared four different control strategies: (a) use only BIS as the operating references; (b) use only BP as the operating references; (c) first use BIS as the operating references but when BIS lower than 50 the operating target switches to BP; (d) first use BIS as the operating references but when BIS is lost (namely, the reading becomes "−1") switch to use BP. Finally, with type-2 self-organizing fuzzy logic controllers (T2-SOFLCs) adapting both BIS and BP references (using BP as target set points when BIS drop lower than 50), the proposed controller improve the performance compare with BIS only control scheme. Since the controller could pick up another signal as new set point to proceed when one of the relying factor was affected by the noises or was lost through surgery, the multi-references controller scheme benefits the stableness on monitoring DoA. Furthermore, multi-references scheme reduces the intervention of specific anesthetic agents to the assessment in general, which broaden the drug-selection for monitoring and control of hypnosis in the practice.

Automatic control of analgesia

The evaluation of analgesia is based on the autonomic response to noxious stimulation. Several devices help to assess degree of analgesia by providing the analgesia/nociception balance. Among them all, the ANI are the newest and promising for monitoring and automatic controlling for analgesia. Similar to the contribution of BIS index in the assessment of hypnosis, ANI could be useful to sense degree of analgesia. However, measuring the analgesic state has not been as well-understood as the consciousness state since the ANI index still required more test on its suitability for higher number of patients and different types of surgeries. In José's research (Gonzalez-Cava et al., 2020), they evaluate the suitability of ANI as a guidance for adapting the decision of drug infusion. In the result, the using of minimum values of ANI with hemodynamic information has better result than non-specific traditional signs (i.e., heart rate and blood pressure) in the decision making correlation. Furthermore, the adaptation of ANI monitor in the decision making process could reduce the hemodynamic events by knowing the dose change in advanced.

The design of a closed-loop control scheme for automatic control of analgesia also encounter the problem of finding a proper feedback variable as a measuring to the analgesic state. One research (Casteleiro-Roca et al., 2018) proposed using drug infusion rate, and EMG as input variables to design a hybrid intelligent model for predicting the ANI index. They collect the response of patients regarding of the analgesic infusion rate and EMG measurement of the painful surgical stimuli to formulate an analgesic prediction model based on multi-layer perceptron artificial neural networks (MLP-ANN). Julien (De Jonckheere et al., 2013) designed a ANI based automatic analgesic drugs delivery devices by determining several decision rule based on the ANI analysis over 2000 patients during orthopedic surgical procedure under general anesthesia. Although the proposed drug delivery system has only been performed with small set of patients and larger set of patients are required to be validated, building up a proper delivery dose of analgesic drugs could reduce the risk for anesthetized patient to hemodynamic reactivity or post-operative hyperalgesia, and most importantly a well-design automatic controller for analgesia could reduce the labor of anesthesiologist in the surgery.

Automatic control of neuromuscular relaxation

To increase the quality of intubation and reduce airway injury, a useful neuromuscular monitoring for neuromuscular blocking agents are needed. An appropriate doses of neuromuscular blockade could improve the quality of endotracheal intubation and reduce the condition of dyspnea and surgical emphysema. Furthermore, the residual neuromuscular block in the post-operative is the signs or symptoms with some serious complications such as pharyngeal dysfunction, increased rick for aspiration pneumonia, acute respiratory events, etc. (Duțu et al., 2018). The monitoring techniques for neuromuscular monitoring have mechanomyography (MMG, the signal presented on the surface of muscle when it is contracted), EMG (the electrical activity produced by skeletal muscles), acceleromyography (AMG, a kind of piezoelectric myograph to measure the force produced by a muscle), kinemyography (KMG, the electrical signal generated by the bending of a piezoelectric sensor strip placed between the thumb and the index finger), phonomyography (PMG, an acoustic myography the which records the low-frequency sounds evoked by muscle contraction), compressomyography (CMG, the non-invasive neuromuscular transmission monitor measuring the pressure changes of a balloon resulted by hand contraction in response to electric stimulation of the ulnar nerve).

An automatic control of neuromuscular block during general anesthesia enhances the performance of neuromuscular blockade in the intra-operation. In Adamus & Belohlavek (2015), they presented a system of fuzzy control of neuromuscular block that minimize the workload of anesthetist and enhance the efficacy of muscle relaxant which enables the patient receive a minimum amount of drug for neuromuscular blockade. The drug absorption and metabolism is both inter- and intra-patient variations. It makes it difficult to the implementation of automatic control of neuromuscular relaxation. Research like (Merigo et al., 2018) proposed a PID control algorithm which adapt genetic algorithm for parameters tuning that aim to achieve the automatic regulation for neuromuscular blockade level. The optimal PID is tested in simulation on a database of patient models, and its performances not only satisfies the clinical specifications but also shows robustness result on the inter-patient variability test.

The seven stages for automatic control of anesthesia

The design of an automatic controller for anesthesia involves domain knowledge in both clinical aspect and engineering aspect. The complicated interactions and mechanisms within human body in terms of unconsciousness, amnesia, analgesia, and akinesia make it harder to determine the adequate general anesthesia. Fortunately, with mature development in electronics devices (e.g., computing resources, embedded system technology, etc.) and complicated signal processing algorithms, designing a multi-parameters monitor that is reliable for general anesthesia becomes more and more viable. The following seven stages as shown in Fig. 2 are proposed as the basic hierarchical system for measuring, interpreting, modeling, supervision, and controlling of the general anesthesia.

Stage 1 is signal decomposing. The vital signs collected during the surgery are never perfect, in fact, it is always contaminated by various noises presented in the surgery room. Sometimes, the noises even derive from the subjects themselves since there could be several physiological phenomena that are contributing the same one vital sign. The decomposition of signal separates the noises from the vital sign itself and disorganize the vital sign into the multiple fundamental part of the correlated physiological phenomena.

Stage 2 is the nonlinear analysis phase. The vital signs are most likely to be nonlinear and nonstationary throughout the surgery. Different from traditional methods of statistical analysis which assume the statistical properties of a signal are the

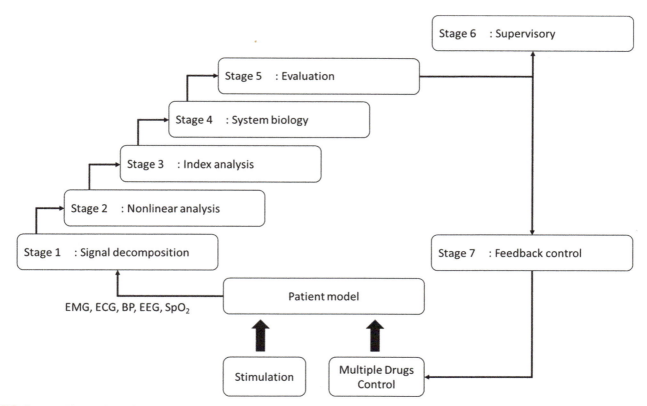

FIG. 2 A multistage hierarchical system for supervisory and control in general anesthesia.

38   PART | I Drugs and agents used in anesthesia and analgesia

same for the entire signal in the bigger scale, physiologic system is highly nonstationary and nonlinear. Thus, to obtain a better interpretation or evaluation for modeling anesthesia, non-stationary and nonlinear analysis such as approximate entropy (ApEn), sample entropy (SampEn), multi-scale entropy (MSE), and detrended fluctuation algorithm (DFA) are needed.

Stage 3 is an index analysis phase that transform the complex presentation from stage 2 to a human-friendly readable index or scalable indices/parameters for automatic control. Various techniques like BIS, AEP, CI, SSI, and PVI ranging from 0 to 100 for determine the adequate anesthesia. Those indices and their relations with anesthesia provides multiple indications or evidences for any decision-making on operating the anesthetic agents by either anesthesiologist or higher stage of automatic machine in the future.

Stage 4 is a system biology phase which model the multiple indications extracted in the previous stages and then propose the decision of whether to increase the dosage of anesthetic agents or to reduce it. In many cases, indices for anesthesia would be influenced by the factor along the surgery. Different types of administered agents could also possibly be the factor that contribute the affection. Hence, a platform or mechanism that integrates various physiologic parameters and capable of multi-input are required. Such mechanism could be Neuro-fuzzy or GA-fuzzy which are known of self-generating function for rules and adapting empirical regulations.

Stages 5 and 6 are evaluation and supervisory phases which allow anesthesiologist to review the adequate general anesthesia according to their clinical experience. The multistage hierarchical system could not only assist but reduce the anesthesiologist workload. Under the examination of anesthesiologist, the system is possible to enter the final feedback control for multi-drugs administration.

Stage 7 is the final feedback control phase that administer the control of multi-drugs. Due to the pharmacokinetic and pharmacodynamics effects of drugs in each subject, quantitative approaches are no longer useful for highly nonlinear human being system. Qualitative methods, on the other hand, have been increasingly adopted for such applications.

Applications to other areas

The majority of studies on automatic control of anesthesia have been working with intravenous anesthesia agents: Propofol. It is a milk-like appearance intravenous agents, which is also widely used in most cases of surgery and veterinary medicine for anesthesia. The agent often used with manual infusion pumps or computer-controlled infusion pumps. However, the exact mechanism of propofol action are still not cleared yet, only some evidences indicated that the infusing of propofol increases the primary inhibitory neurotransmitter gamma-aminobutyric acid (GABA) in the cerebral cortex (Kobayashi & Oi, 2017). One single bolus dose of propofol (2–2.5 mg/kg) would be affected enough to cause unconsciousness to majority of patients within 1 min (Langley & Heel, 1988). The hypnotic agent, propofol, possesses a pharmacokinetic characteristic of rapid onset and offset of drug effect and quick elimination by the metabolism (Schuttler & Schwilden, 2008). Despite the efficiency of propofol may present, the agent may provoke the fatal adverse effect, propofol infusion syndrome (PRIS), which is an acute refractory bradycardia that leads to asystole under one or more of the conditions as metabolic acidosis, hyperlipidemia, rhabdomyolysis, and enlarged or fatty liver (Cardone & Kam, 2007; Casserly et al., 2004). The provoking of PRIS are stated to be related with the direct inhibition of mitochondrial respiratory chain or impaired mitochondrial fatty acid metabolism intervened by propofol. And the successful management of PRIS depends on the awareness of the early signs like right bundle branch block (RBBB) with convex-curved ST elevation. Treatment options for PRIS are also little due to the refractory nature of the symptom, few successful treated cases involved using both transcutaneous pacing (TCP) and transvenous pacing (Abrahams, Reiter, Acker, & Sinson, 2002; Barclay, Williams, & Major, 1992; Cray, Robinson, & Cox, 1998; Culp, Augoustides, Ochroch, & Milas, 2004; Holzki, Aring, & Gillor, 2004; Parke et al., 1992; Withington, Decell, & Ayed, 2004; Wolf, Weir, Segar, Stone, & Shield, 2001), extracorporeal membrane oxygenation (ECMO) (Abrahams et al., 2002; Culp et al., 2004), hemodialysis or hemoperfusion with cardiorespiratory support (Machata, Gonano, Birsan, Zimpfer, & Spiss, 2005). Other common side effects caused by propofol would be irregular heart rate, low blood pressure, a burning sensation when injecting the agent, and apnea.

Other agents of interest

There are two classes of anesthetic agents: inhaled agents and intravenous agents. Common used inhaled anesthetics are a single gas (i.e., nitrous oxide) and volatile liquids (i.e., halothane, enflurane, isoflurane, desflurane, and sevoflurane). Non-opioid intravenous anesthetics include barbiturate (i.e., thiopental, methohexital), benzodiazepines (i.e., midazolam), propofol, ketamine, and etomidate. Each agent has a complicated molecular mechanism, and currently there are no ideal anesthetic drugs that could contribute all and only desired effects (e.g., hypnosis, amnesia, analgesia, immobility, etc.).

However, sensitivity of bispectral index (BIS) varies from types of administered anesthetic drug. Instances of nitrous oxide added general anesthetic regimen using propofol and remifentanil has proven to dramatically restrict the response of BIS on monitoring DoA during surgery (Coste, Guignard, Menigaux, & Chauvin, 2000). The low-dose ketamine alone does not correlate with the BIS readings (Friedberg, 1998; Suzuki, Edmonds, Tsueda, Malkani, & Roberts, 1998). Moreover, BIS reading comparison during sedation with sevoflurane, midazolam and propofol is also conducted in Ibrahim, Taraday, and Kharasch (2001); indication of propofol has quicker onset and offset for sedation during regional anesthesia while midazolam was more cardiostable is given by Khurana, Agarwai, Verma, and Gupta (2009); improved patient satisfaction with desflurane rather than propofol in the post-operative are also being studied in Luginbuhl, Wuthrich, Petersen-Felix, Zbinden, and Schnider (2003). Although BIS is easy to be affected by the conducted drugs, it helps to reduce the overall dose of anesthetic agents used during the surgery which may hastened the recovery time from anesthesia.

Key facts

Key facts of anesthetic agents

The introduction of neuromuscular blocking agents redefined the classes of anesthetic agents into narcosis, analgesia, muscle relaxation.

- Narcotics are useful as antitussive agents for airway procedures.
- The more powerful analgesics have the higher potential of it to caused addiction or undesirable side effects for patients.
- The neuromuscular blocking agent is most effective via intravenous or intramuscular injection contrast to orally.

Key facts of Propofol

- Evidence shows that propofol increases the primary inhibitory neurotransmitter gamma-aminobutyric acid (GBAA) in the cerebral cortex to induce hypnosis.
- One single bolus does of propofol (i.e., 2–2.5 mg/kg) is enough to unconsciousness to majority of patients within 1 min.
- The administration of propofol has a rare possibility to provoke the fatal propofol infusion syndrome which is an acute refractory bradycardia that leads to asystole.

Mini-dictionary of terms

Akinesia. The symptom of Loss the power of voluntary movement.
Amnesia. The symptom of partial or total loss of memory.
Analgesia. The inability to fell pain.
Barbiturate. A class of drug derived from barbituric acid and is used as a central nervous system depressant.
Benzodiazepine. A class of psychoactive drugs with structure of heterocyclic organic compounds often used as tranquilizers.
Desflurane. One of the anesthetic drugs that is a highly fluorinated methyl ethyl ether and is usually used for maintenance of general anesthesia.
Enflurane. A volatile organic liquid used as a general anesthetic.
Halothane. A volatile synthetic organic compound used as a general anesthetic.
Hypnosis. The state of induction which a person is in consciousness but apparently loses the power of voluntary action and is highly responsive to suggestion or direction.
Isoflurane. A general anesthetic that could be used to maintain anesthesia via inhalation.
Methohexital. A drug that acts as a central nervous system depressant with characteristics of short-acting and rapid onset of action.
Midazolam. A psychoactive medication for anesthesia, procedural sedation, trouble sleeping, and sever agitation.
Nitrous oxide. As known as laughing as or nitrous which is a colorless non-flammable gas with a slight metallic scent and taste and is often used in surgery and dentistry for anesthetic and pain reduction effects.
Propofol. A short-acting medication that results in a decreased level of consciousness and lack of memory for events. Often used in the starting and maintenance of general anesthesia.
Sevoflurane. A sweet-smelling, nonflammable inhalational anesthetic for induction and maintenance of general anesthesia.
Thiopental. An ultra-short-acting central nervous system depressant and has been used commonly in the induction phase of general anesthesia.

Summary points

- The study of anesthesia consists of hypnosis, nociception, and neuromuscular blockade.
- Indices for DOA are Bispectral index, State Entropy, Response Entropy.
- Indices for level of analgesia are Surgical Stress Index, Analgesic Nociception Index.
- Assessment of neuromuscular blockade is techniques such as mechanomyography, electromyography, acceleromyography, kinemyography, phonomyography, and compressomygraphy.
- Proper dosage of anesthetic agents for patients could avoid sever consequences on both intraoperative and postoperative.
- Within different period of the surgery (i.e., induction, maintenance, and recovery), different clinical signs and different characteristics of physiological signals are expected.
- The seven-stage for automatic control of anesthesia are: (1) Signal decomposing; (2) Nonlinear analysis; (3) Index analysis; (4) System biology; (5) Evaluation; (6) Supervisory; and (7) Feedback control.
- Propofol infusion syndrome (PRIS) could be fatal under one or more of the conditions as metabolic acidosis, hyperlipidemia, rhabdomyolysis, and enlarged or fatty liver.
- Propofol has quicker onset and offset for sedation compared to midazolam. But desflurane provide a better patient satisfaction in the post-operative than propofol did.

References

Abrahams, J. M., Reiter, G. T., Acker, M. A., & Sinson, G. P. (2002). Propofol. *Journal of Neurosurgery, 96*, 1160–1161.

Adamus, M., & Belohlavek, R. (2015). Fuzzy control of neuromuscular block during general anesthesia—System design, development and implementation. *International Journal of General Systems, 36*(6), 733–743.

Barclay, K., Williams, A. J., & Major, E. (1992). Propofol infusion syndrome in children. *British Medical Journal, 305*(6859), 953–954.

Brown, E. N., Lydic, R., & Schiff, N. D. (2010). General anesthesia, sleep, and coma. *The New England Journal of Medicine, 363*, 2638–2650.

Cardone, D., & Kam, P. C. A. (2007). Propofol infusion syndrome. *Anaesthesia, 62*, 690–701.

Casserly, B., O'Mahony, E., Timm, E. G., Haqqie, S., Eisele, G., & Urizar, R. (2004). Propofol infusion syndrome: An unusual cause of renal failure. *American Journal of Kidney Disease, 44*, 98–101.

Casteleiro-Roca, J. L., Jove, E., Gonzalez-Cava, J. M., Pérez, J. A. M., Calvo-Rolle, J. L., & Alvarez, F. B. (2018). Hybrid model for the ANI index prediction using Remifentanil drug and EMG signal. *Neural Computing and Applications, 32*, 1249–1258.

Chernik, D. A., Gillings, D., Laine, H., Hendler, J., Silver, J. M., Davidson, A. B., et al. (1990). Validity and reliability of the observer's assessment of alertness/sedation scale: Study with intravenous midazolam. *Journal of Clinical Psychopharmacology, 10*(4), 244–251.

Coste, C., Guignard, B., Menigaux, C., & Chauvin, M. (2000). Nitrous oxide prevents movement during orotracheal intubation without affecting BIS value. *Anesthesia & Analgesia, 91*(1), 130–135.

Cray, S. H., Robinson, B. H., & Cox, P. N. (1998). Lactic acidaemia and bradyarrhythmia in a child sedated with propofol. *Critical Care Medicine, 26*, 2087–2092.

Culp, K. E., Augoustides, J. G., Ochroch, A. E., & Milas, B. L. (2004). Clinical management of cardiogenic shock associated with prolonged propofol infusion. *Anesthesia and Analgesia, 99*, 221–226.

De Hert, S. G. (2006). Volatile anesthetics and cardiac function. *Seminars in Cardiothoracic and Vascular Anesthesia, 10*, 33–42.

De Jonckheere, J., Delecroix, M., Jeanne, M., Keribedj, A., Couturier, N., & Logier, R. (2013). Automated analgesic drugs delivery guided by vagal tone evaluation interest of the analgesia nociception index. In *Annual international conference of the IEEE EMBS, Osaka, Japan, 3–7.*

Dumont, G. A. (2012). Closed-loop control of anesthesia—A review. In *IFAC Symposium on Biological and Medical Systems* (pp. 373–378).

Duțu, M., Ivașcu, R., Tudorache, O., Morlova, D., Stanca, A., Negoiță, S., et al. (2018). Neuromuscular monitoring: An update. *Romanian Journal of Anaesthesia and Intensive Care, 25*(1), 55–60.

Ekman, A., Lindholm, M. L., Lennmarken, C., & Sandin, R. (2004). Reduction in the incidence of awareness using BIS monitoring. *Acta Anaesthesiologica Scandinavica, 48*, 20–26.

Evers, A. S., & Crowder, C. M. (2009). Mechanisms of anesthesia and consciousness. In P. G. Barash, B. F. Cullen, R. K. Stoelting, M. Cahalan, & M. C. Stock (Eds.), *Clinical anesthesia* (pp. 95–114). New York: Lippincott Williams & Wilkins.

Friedberg, B. L. (1998). The effect of a dissociative dose of ketamine on the bispectral index during propofol hypnosis. *Journal of Clinical Anesthesia, 11*(1), 4–7.

Gan, T. J., Glass, P. S., Windsor, A., Payne, F., Rosow, C., Sebel, P., et al. (1997). Bispectral index monitoring allows faster emergence and improved recovery from propofol, alfentanil, and nitrous oxide anesthesia. *Anesthesiology, 87*, 808–815.

Gelb, A. W., Leslie, K., Stanski, D. R., & Shafer, S. L. (2010). Monitoring the depth of anesthesia. In R. D. Miller, L. I. Eriksson, L. A. Fleisher, J. P. Wiener-Kronish, & W. L. Young (Eds.), *Miller's anesthesia* (pp. 1229–1265). New York: Churchill Livingstone.

Gonzalez-Cava, J. M., Arnay, R., León, A., Martín, M., Reboso, J. A., Calvo-Rolle, J. L., et al. (2020). Machine learning based method for the evaluation of the analgesia nociception index in the assessment of general anesthesia. *Computers in Biology and Medicine, 118*, 103645.

Guignard, B. (2006). Monitoring analgesia. *Best Practice & Research. Clinical Anaesthesiology, 20*(1), 161–180.

Gürses, E., Sungurtekin, H., Tomatir, E., & Dogan, H. (2004). Assessing propofol induction of anesthesia dose using bispectral index analysis. *Anesthesia & Analgesia, 98*(1), 128–131.

Hedenstierna, G. (1995). Contribution of multiple inert gas elimination technique to pulmonary medicine. 6. Ventilation-perfusion relationships during anaesthesia. *Thorax, 50*(1), 85–91.

Holzki, J., Aring, C., & Gillor, A. (2004). Death after re-exposure to propofol in a 3-year old child: A case report. *Paediatric Anaesthesia, 14*, 265–270.

Huiku, M., Uutela, K., Gils, N. V., Korhonen, I., Kymäläinen, M., Meriläinen, P., et al. (2007). Assessment of surgical stress during general aneaesthesia. *British Journal of Anaesthesia, 98*(4), 447–455.

Ibrahim, A. E., Taraday, J. K., & Kharasch, E. D. (2001). Bispectral index monitoring during sedation with Sevoflurane, Midazolam, and Propofol. *Anesthesiology, 95*, 1151–1159.

Jeanne, M., Logier, R., Jonckheere, J. D., & Tavernier, B. (2009). Validation of a graphic measurement of heart rate variability to assess analgesia/nociception balance during general anesthesia. In *Conference of IEEE EMBS. Minneapolis, Minnesota, USA, September 2–6*.

Kaul, H. L., & Bharti, N. (2002). Monitoring depth of anaesthesia. *Indian Journal of Anaesthesia, 46*(4), 323–332.

Khurana, P., Agarwai, A., Verma, R., & Gupta, P. (2009). Comparison of midazolam and propofol for BIS-guided sedation during regional anaesthesia. *Indian Journal of Anaesthesia, 53*(6), 662–666.

Kobayashi, M., & Oi, Y. (2017). Actions of propofol on neurons in the cerebral cortex. *Journal of Nippon Medical School, 84*(4), 165–169.

Langley, M. S., & Heel, R. C. (1988). Propofol. A review of its pharmacodynamic and pharmacokinetic properties and use as an intravenous anaesthetic. *Drugs, 35*, 334–372.

Lenhardt, R. (2010). The effect of anesthesia on body temperature control. *Frontiers in Bioscience (Scholar Edition), 2*, 1145–1154.

Liu, Q., Chen, Y. F., Fan, S. Z., Abbod, M. F., & Shieh, J. S. (2016a). A comparison of five different algorithms for EEG signal analysis in artifacts rejection for monitoring depth of anesthesia. *Biomedical Signal Processing and Control, 25*, 24–34.

Liu, Q., Chen, Y. F., Fan, S. Z., Abbod, M. F., & Shieh, J. S. (2016b). EEG artifacts reduction by multivariate empirical mode decomposition and multiscale entropy for monitoring depth of anesthesia during surgery. *Medical & Biological Engineering & Computing, 55*, 1435–1450.

Luginbuhl, M., Wuthrich, S., Petersen-Felix, S., Zbinden, A. M., & Schnider, T. W. (2003). Different benefit of bispectral index in desflurane and propofol anesthesia. *Acta Anaesthesiologica Scandinavica, 47*(2), 165–173.

Machata, A. M., Gonano, C., Birsan, T., Zimpfer, M., & Spiss, C. K. (2005). Rare but dangerous adverse effects of propofol and thiopental in intensive care. *Journal of Trauma, 58*, 643–645.

Merigo, L., Padula, F., Latronico, N., Mendonça, T., Paltenghi, M., Rocha, P., et al. (2018). Optimized PID tuning for the automatic control of neuromuscular blockade. *IFAC-PapersOnLine, 51*(4), 66–71.

Murphy, G. S. (2018). Neuromuscular monitoring in the perioperative period. *Anesthesia & Analgesia, 126*(2), 464–468.

Naguib, M., Kopman, A. F., & Ensor, J. E. (2007). Neuromuscular monitoring and postoperative residual curarisation: A meta-analysis. *British Journal of Anaesthesia, 98*(3), 302–316.

Neukirchen, M., Kienbaum, P., Warner, D. S., & Warner, M. A. (2008). Sympathetic nervous system: Evaluation and importance for clinical general anesthesia. *Anesthesiology, 109*, 1113–1131.

Parke, T. J., Stevens, J. E., Rice, A. S., Greenway, C. L., Bray, R. J., Smith, P. J., et al. (1992). Metabolic acidosis and fatal myocardial failure after propofol infusion in children: Five case reports. *British Medical Journal, 305*, 613–616.

Sawaguchi, Y., Furutani, E., Shirakami, G., Araki, M., & Fukuda, K. (2008). A model-predictive hypnosis control system under total intravenous anesthesia. *IEEE Transactions on Biomedical Engineering, 55*(3), 874–887.

Schuttler, J., & Schwilden, H. (2008). *Modern anesthetics* (p. 182). Berlin Heidelberg: Springer Verlag.

Suzuki, M., Edmonds, H. L., Tsueda, K., Malkani, A. L., & Roberts, C. S. (1998). Effect of ketamine on bispectral index and levels of sedation. *Journal of Clinical Monitoring and Computing, 14*(5), 373.

Vasella, F. C., Frascarolo, P., Spahn, D. R., & Magnusson, L. (2005). Antagonism of neuromuscular blockade but not muscle relaxation affects depth of anaesthesia. *British Journal of Anaesthesia, 94*(6), 742–747.

Withington, D. E., Decell, M. K., & Ayed, T. A. (2004). A case of propofol toxicity: Further evidence for a causal mechanism. *Pediatric Anaesthesia, 14*, 505–508.

Wolf, A., Weir, P., Segar, P., Stone, J., & Shield, J. (2001). Impaired fatty acid oxidation in propofol infusion syndrome. *Lancet, 357*, 606–607.

Yu, Y. N., Doctor, F., Fan, S. Z., & Shieh, J. S. (2018). An adaptive monitoring scheme for automatic control of anaesthesia in dynamic surgical environments based on bispectral index and blood pressure. *Journal of Medical Systems, 42*, 95.

Chapter 5

Non-opioid based analgesia in otolaryngology

Yohanan Kim, Anthony Sanchez, and Khanh Nguyen
Loma Linda University, Department of Otolaryngology—Head and Neck Surgery, Loma Linda, CA, United States

Abbreviations

APAP	Acetaminophen
COX	cyclooxygenase
IQR	interquartile range
IM	intramuscular
IV	intravenous
LA	local anesthetic(s)
mg	milligram
kg	kilogram
NSAID	non-steroidal anti-inflammatory drug
PHQ-9	9-item Patient Health Questionnaire
STOA	State Trait Operation Anxiety

Introduction

Pain control in the postsurgical setting is a multifaceted challenge that involves complex consideration of the experience of the surgeon, the perception of the patient, the physical toll of the surgical procedure, and the efficacy of the chosen analgesic regimen. In the ideal scenario, the physical demand of a given procedure is known and the patient's expectation is set appropriately. As a result of the surgeon's experience and knowledge, the appropriate analgesic regimen is prescribed, and the patient's post-surgical pain is well-controlled with medications and minimal to no adverse side effects.

In the recent opioid epidemic in the United States, it has become even more imperative to better understand this process in order to decrease overall opioid prescription and decrease the risk for diversion of opioid medication. The goal is to provide good patient care with effective pain control. As a result of increased attention to minimizing opioid medication prescription, there has been an increased scrutiny in physician prescribing patterns and interest in understanding patients' post-surgical pain experience and non-opioid alternatives for analgesia. This chapter will focus on recent clinical studies evaluating non-opioid based analgesia utilized in otolaryngology with treatment recommendations from recent findings.

Patient pain experience in otolaryngology

In order to effectively address patients' pain in the post-surgical setting, one important and necessary component is understanding the patient's pain experience during recovery and the actual analgesic requirement for a given procedure. This enables the surgeon to both provide good counsel for patients prior to surgery and allow the patients to have the appropriate expectation. Herein lies a challenge for the head and neck surgeon due to the diversity of surgical subsites. For the otolaryngologist, there is significant variation in the type of tissues and location of surgery including the cutaneous soft tissue; osseous work as in otologic procedures or facial trauma repair; mucosal surfaces for surgeries involving the oral cavity, larynx, and sinus. Additionally, surgery in each of these subsites also results in a myriad of functional limitations in the recovery period encompassing partial obstruction of normal respiratory pathway, alteration of normal swallowing mechanisms, alteration in normal speech and communication, and hearing difficulties among many others. It is not unexpected that studies have shown post-surgical pain is subsite dependent and thus varies by procedure (Sommer et al., 2009).

Treatments, Mechanisms, and Adverse Reactions of Anesthetics and Analgesics. https://doi.org/10.1016/B978-0-12-820237-1.00005-3
Copyright © 2022 Elsevier Inc. All rights reserved.

44 **PART | I** Drugs and agents used in anesthesia and analgesia

In selecting the optimal analgesic regimen, consideration of the primary surgical site is a reasonable first step in this complex algorithm (Table 1).

Oral cavity and oropharynx

Surgery involving the alimentary tract can severely affect patients' daily function and exacerbate the recovery process by limiting oral intake. This would include surgery on the tongue, any surface of the oral mucosa, and the oropharynx such as the palate and most commonly the tonsils. While data for each specific subsite is limited, most pain related studies have focused on one of the most common otolaryngologic surgery performed, namely tonsillectomy. It is well supported that post-surgical pain following tonsillectomy is often reported as moderate to severe and will often persist up to 2 weeks following surgery. (Kim et al., 2020) Additional data available include molar extractions that report maximum moderate to severe pain experience within the first 24 h of the procedure and a return to baseline by the seventh day (Isola, Matarese, Ramaglia, Cicciù, & Matarese, 2019). What remains unclear is whether patients' reported pain is truly the result of tissue injury and activation of nociceptive receptors vs discomfort from distorted anatomy due to the procedure and the resulting swelling. Regardless of the source, it is reasonable to counsel patients to anticipate a moderate level of pain lasting up to 7 days and sometimes longer following procedures involving the oral cavity and oropharynx.

Nasal surgery

Surgery involving the nasal sinus, nasal septum, nasal framework, and soft tissue is a unique set of procedures due to their effects on the upper airway. These include functional endoscopic sinus surgery, septoplasty, and rhinoplasty all of which often involves varying degrees of soft tissue as well as bony and cartilaginous structures that form the nose. As these procedures are often performed in the ambulatory setting, several studies have investigated patients' recovery course and the necessary analgesic requirement. The majority of patients report mild and at most moderate levels of pain during recovery with significant improvement by postoperative day 3 (Locketz et al., 2019). Various studies evaluating patients who underwent septoplasty and rhinoplasty report an overall moderate level of pain immediately after surgery with a quick trend toward mild pain within the first week of recovery (Sclafani, Kim, Kjaer, Kacker, & Tabaee, 2019).

Otologic surgery

Early studies evaluating pain by surgical sites found that patients who underwent otologic procedures reported the least amount of pain compared to other areas in the head and neck (Sommer et al., 2009). More recent studies have found that patients report a median pain score of 3 with an interquartile range (IQR) of 2–6 (Boyd et al., 2019). This corresponded with a median opioid use of four tablets of hydrocodone/acetaminophen (IQR 1–11.5). A separate study corroborated these findings and further reported that patients undergoing a transcanal procedure experience a shorter duration of pain compared to those who had a post auricular approach, and consumed less opioid medication (Qian, Alyono, Woods, Ali, & Blevins, 2019).

TABLE 1 Summary of surgical type and their associated pain severity and duration.

Surgical type/subsite	Severity	Duration
Oral cavity/oropharynx	Moderate to severe	>7 days
Nasal surgery	Mild to moderate	3–5 days
Ear surgery	Mild to moderate	3–5 days
Visceral/soft tissue	Mild to moderate	<3 days
Oncologic surgery	Moderate to severe	>7 days

Visceral/soft tissue

Surgery related to the salivary glands, lymph nodes, and thyroid/parathyroid tissue would fall under this category and there is a great level of interest due to the rising incidence of thyroid malignancy and increasing trend toward ambulatory endocrine surgery. An early prospective study evaluating patients' function following parotidectomy found that patients experience at most moderate pain with quick resolution to a mild level within the first week of surgery (Foghsgaard, Foghsgaard, & Homøe, 2007). Numerous studies evaluating patients' pain experience following endocrine surgery report similar findings of mild to moderate pain with minimal opioid consumption following surgery and quick resolution of pain (Nguyen et al., 2019; Papoian et al., 2020).

Oncologic surgery

Pain control following oncologic resection is a challenging endeavor. These procedures result in significant immediate alteration of normal anatomy, function, and covers multiple surgical sites when regional and free tissue transfers are required for reconstruction. While studies characterizing the exact degree of pain following these types of procedures are limited, there are numerous reports demonstrating that patients' often experience inadequate pain control (Hinther, Nakoneshny, Chandarana, Wayne Matthews, & Dort, 2018; Orgill, Krempl, & Medina, 2002). In such occasions, a multi-modality approach for pain control is more likely to achieve adequate analgesia and aide in patient recovery (Eggerstedt, Stenson, Ramirez, et al., 2019).

Risk factors for greater pain experience

Risk factors for greater postoperative pain have been described in numerous studies in many surgical subspecialties. In a systematic review by Yang et al. (2019), 33 studies representing 53,362 patients were evaluated to find predictors of poor postoperative pain control among adult patients undergoing a wide variety of surgeries. In this review, nine predictors of poor postoperative pain control were discovered. These included younger age, smoking, female sex, history of depressive symptoms, history of anxiety symptoms, sleep difficulties, higher body mass index, presence of preoperative pain, and use of preoperative analgesia. Factors which were not found to be a predictor included chronic pain, marital status, socioeconomic status, pain catastrophizing, education, and surgical history.

Anticipating which patients may have more post-operative pain than might be expected is of great utility to the otolaryngologist. While increased pain may serve as a sentinel warning of a complication of surgery or progression of the patient's disease, it may also simply be just that—post-operative pain. In an observational study done by Suffeda et al., patients who underwent otolaryngologic surgery were evaluated for depression and anxiety using the 9-item Patient Health Questionnaire (PHQ-9) for depression, and the 30-item State Trait Operation Anxiety (STOA) for anxiety. The pain catastrophizing score, which measures an exaggerated negative attitude toward pain was also given to all patients. Patients with higher depression and anxiety scores reported a greater postoperative pain experience compared to other patients who had similar procedures. Independently, it was also found that patients who required opioids in the recovery room after surgery generally had more post-operative pain as well. Interestingly, patients who had an exaggerated negative attitude toward pain as found on the pain catastrophizing survey were not found to have an increase in post-operative pain.

In another study by Sommer and colleagues which evaluated otolaryngologic surgical patients for post-operative pain, the most important predictor of pain was the surgical site. Surgery on the mouth, salivary glands, throat, or neck were found to be more painful post-operatively when compared to other areas such as the ear. In this study, pre-operative pain and the patient's pain catastrophizing scores were both found to associate with higher postoperative pain.

Non-opioid analgesia
Local anesthetics

Local anesthesia is a wide-spread and effective tool used in the specialty of otolaryngology. When injected these anesthetics can provide pain relief almost immediately and last for several hours. Local anesthetics (LA) work by reversibly binding to sodium channels present on peripheral nerves, inactivating them. This action decreases the rate of depolarization of the nerve, thus significantly decreasing the perception of pain. These medications are often formulated as a mixture containing epinephrine, which acts as a vasoconstrictor, decreasing skin-edge bleeding during surgery. This vasoactive effect is also thought to maintain a more localized concentration of the anesthetic in the area injected, providing longer-term anesthesia.

Many LA medications exist, with lidocaine and bupivacaine being perhaps the two most common ones used in otolaryngologic surgery and practice. The duration of action of these medications varies widely in the literature, with most sources claiming a longer duration of action with bupivacaine as opposed to lidocaine, lasting several hours. In otolaryngologic surgery, a local anesthetic is often injected into the skin and subcutaneous tissue several minutes before incision in an effort to decrease skin bleeding. Additional anesthetic might be injected around the incision at the end of the procedure, to provide some short-term post-operative pain relief.

Whether a patient is undergoing a procedure under general anesthesia or as an outpatient in the clinic, local anesthesia can sometimes be used to perform peripheral nerve blocks. These blocks are intended to anesthetize an entire area of the face or neck supplied by a single peripheral nerve. These blocks can be especially effective for anesthesia both during the procedure and afterwards. Examples of peripheral nerve blocks include inferior alveolar nerve blocks, infraorbital nerve blocks, and a field block for procedures on the auricle (Salam, 2004). These blocks can also be considered post-operatively if a patient develops significant pain, although the duration of effect will be limited.

Given the excellent efficacy of injectable local anesthetics, there is significant motivation to increase the duration of action of these medications. Liposomal bupivacaine was developed in an effort to accomplish this. Liposomes are structures consisting of a phospholipid bilayer, which in this case, are able to encapsulate the bupivacaine molecules. When injected into the tissue, the liposomes allow a slower and more controlled release of the anesthetic, leading to a longer duration of action (Chahar & Cummings 3rd, 2012). One review of the liposomal bupivacaine found that while time to effect was unchanged with the liposomal formation, duration of effect doubled (Chahar & Cummings 3rd, 2012). There is a relative lack of research regarding liposomal bupivacaine in head and neck surgery, and further studies are warranted.

Although serious adverse effects from injectable local anesthesia are rare, they can potentially be fatal. Some of these adverse effects include generalized tonic-clonic seizures, respiratory distress or failure, hypotension, and atrioventricular conduction delay leading to life threatening arrhythmias. It is important to note that these side effects are rate, especially when proper doses are administered. The most common side effects experienced are generally anxiety, pain at the injection site, and of course pain during the injection.

Acetaminophen

One of the most commonly used over-the-counter pain medications in the United States, acetaminophen (APAP) has been used for pain and fever relief for over a century. Though the exact mechanism of action of acetaminophen in treating pain is unknown, it is thought to act in the central nervous system as opposed to in the periphery. By inhibiting cyclooxygenase (COX) activities in the brain, acetaminophen can be useful in both treating pain and fevers; it has been proposed that the mechanism of action involves decreasing the amount of the COX enzyme, as opposed to inhibiting a COX receptor (Ghanem, Pérez, Manautou, & Mottino, 2016).

Acetaminophen is widely used in otolaryngology, either on its own or as one component of a more involved pain regimen. For example, in post-operative pediatric tonsillectomy patients, a regimen of alternating acetaminophen and ibuprofen has become the mainstay of pain control. With this alternating regimen, only a small proportion of patients and their parents report inadequate analgesia (Liu & Ulualp, 2015). APAP has also been found to decrease opioid use postoperatively in head & neck patients. Multimodal analgesia, which refers to a combination of non-opioid pain medications such as APAP, non-steroidal anti-inflammatory drugs, gabapentinoids, and local anesthetics is often applied in otolaryngology; this has been found to decrease opioid use even in head & neck free flap reconstruction both in the immediate postoperative period and at the time of discharge (Eggerstedt et al., 2019).

Acetaminophen also has multiple routes of administration, making it particularly useful in patients unable to tolerate oral intake. APAP may be given orally, rectally, intravenously, and in some cases, intramuscularly. It is available in pill, tablet, capsule, or liquid form, adding to its versatility. While often given intravenously in the operating room, it is largely accepted that oral acetaminophen is similarly effective. One particular study by Bhoja et al. (2020) specifically looked at the effectiveness of IV vs oral acetaminophen given within one hour prior to sinus surgery. The primary outcome was a pain score assessed 1 h post-operatively. They found that there was no significant difference in pain scores between patients who received acetaminophen via IV vs oral route.

NSAIDs

Nonsteroidal anti-inflammatory drugs (NSAID) are a class of medications that inhibit cyclooxygenase (COX) which suppresses prostaglandin production resulting in decreased inflammation and pain. The use of NSAIDs and non-opioid analgesia versus opioid analgesia has been well studied in the orthopedic and plastic surgery literature; and only recently these comparative studies have begun to surface in otolaryngology head and neck surgery. In part, this may be due to a reluctance among many otolaryngologists historically stems from a perceived fear concerning post-operative bleeding and hematoma formation in the richly vascular regions of the head and neck and its proximity to the airway which can result in catastrophic consequences.

Multiple studies have been done to demonstrate the safe use of NSAIDs. A recent systematic review looked at hematoma risks with NSAID use in various plastic surgery cases (Walker, Jones, Kratky, Chen, & Runyan, 2019). The meta-analysis included 15 studies with over 3000 patients and showed no statistical difference in the formation of hematomas with the use of ketorolac, ibuprofen or celecoxib. Another study performed on patients who underwent a thyroidectomy showed no statistical difference in hematoma formation between patients who did and did not receive ketorolac (Chin et al., 2011). However, one caveat is that postoperative ibuprofen has been shown to significantly increase the severity of post-tonsillectomy hemorrhage with at least a three-fold increased risk of requiring a blood transfusion in the pediatric population (Mudd et al., 2017). However, there was no increased risk of bleeding with ibuprofen use. This has led to a certain degree of controversy concerning the use of ibuprofen in post-tonsillectomy patients despite widespread acceptance of NSAID use after most ENT surgeries.

Older studies have looked at perioperative use of various NSAIDS, including ketorolac, dexketoprofen, lornoxicam, diclofenac, and rofexicob in patients who underwent septoplasty and rhinoseptoplasty. Generally pain scores were improved up to 24 h and they concluded NSAIDS were a safe means of decreasing immediate post-operative rescue analgesics, both opioid and non-opioid (Moeller, Pawlowski, Pappas, Fargo, & Welch, 2012; Ozer et al., 2012; Sener et al., 2008; Turan et al., 2002). However, these studies only followed patients for 24 h and did not evaluate on postoperative opiate use. In contrast, one recent randomized control trial compared outpatient opioid and non-opioid medication use in patients who underwent outpatient otolaryngologic surgeries including thyroidectomy, parathyroidectomy, functional endoscopic sinus surgery, septoplasty, endolaryngeal surgeries, and otologic surgeries (Nguyen et al., 2019). The study found that a course of oral ibuprofen as a primary postoperative analgesic in these patients was equally effective as compared to hydrocodone/acetaminophen and that the NSAID group reported significantly lower use of opioid medication.

While NSAID use on outpatient procedures have been studied more vigorously, there is a conspicuous lack of data on the effects of NSAID vs. opioid use on larger otolaryngologic surgeries, primarily cancer ablation and free flap reconstruction. As expected, patients who undergo these types of procedures are more likely to experience greater postoperative pain and often require both higher doses and longer durations of opioid use (Chua, Reddy, Lee, & Patt, 1999). Additionally, patients who have complete oncologic treatment often can develop chronic pain symptoms long after completion of treatment which predisposes them to extended opioid medication use and dependence. Therefore, multimodal approaches combining opioid and nonopioid analgesics are encouraged in the postoperative management of head and neck cancer patients. One retrospective study found significant reduction in morphine equivalents required per day in patients who were also treated with 200 mg celecoxib BID for at least 5 days postoperatively compared to patients treated with opioids only (Carpenter et al., 2018). There was no difference in the incidence of complications including hematoma formation, free flap failure, or wound dehiscence. While only celecoxib was used in this study, there is no reason to believe that other NSAIDs should not have similar benefits though additional research may need to be conducted to ensure no unwanted side effects compromise the health of the flap and the patient.

However, as widespread as the use of NSAIDs has become in the postoperative management of ENT procedures, they are not without potential side effects and should still be used judiciously. Providers must be cognizant of patients with a history of peptic ulcer disease, inflammatory bowel disease or kidney disease which may be exacerbated by the use of NSAIDs. Further, NSAIDs, with the exception of aspirin, have been shown to increase the risk of myocardial infarction and stroke (Kearney et al., 2006).

Gabapentinoids

Gabapentin has long been used in the management of neuropathic pain, particularly in the diabetic population, and although its exact mechanism remains unclear, there is some limited evidence that it may be useful in the management of postoperative pain. Multiple studies have noted significant improvement in pain and decreased opioid use with one-time preoperative gabapentin after outpatient ENT procedures in the immediate postoperative period up to 24 h (Lam, Choi, Wong, Irwin, & Cheung, 2015; Sanders & Dawes, 2016). Another randomized control trial observed a significant decrease in pain scores in patients who underwent head and neck oral mucosal surgeries with scheduled gabapentin use for 72 h (Townsend et al., 2018). However, there was no significant difference in opioid pain medication use or satisfaction with pain control.

Unfortunately, there is no current consensus regarding the dosage of gabapentin with the literature reporting doses from 75 to 1200 mg. Additional research regarding the optimal dose is required. Additionally, care needs to be taken when interpreting these studies as statistically significant results may not necessarily translate into clinically significant practice, especially given that the large bulk of the literature only followed the patients for 24 h or less.

Discussion

Due to the ongoing opioid epidemic, the matter of non-opioid pain management has risen in prominence. Unfortunately, despite the recent rise in interest and produced literature, there remains very little in the way of evidence-based guidelines. The management of pain in the acute postoperative period remains widely varied and inconsistent among practitioners. The current literature suggests that the various pain modalities previously discussed all have some efficacy in improving pain and decreasing opioid pain medication use. What is less apparent is which modality would be more effective, but this may also vary depending on the type of surgical procedure performed. As more prospective trials are conducted to provide additional insights where it is lacking, evidence-based guidelines will also be developed to help guide surgeons and clinicians. For now, we hope to inform practitioners on the available alternatives to opioid pain medications to help determine which modality may work best within the confines of their own practices (Table 2, Fig. 1).

Applications to other areas

In this chapter we have discussed the patient's pain experience with various otolaryngology procedures while also reviewing commonly used alternatives to opioid pain medication. While additional comparative studies are necessary to further determine the optimal pain management modalities, we believe that the proposed algorithm may be helpful in the management of other surgical subspecialties as well, outside of otolaryngology.

TABLE 2 Summary of non-opioid pain medication.

Drug class	Name	Route	Dose	Dose interval	Max daily dose	Half-life	Recommended use
Local Anesthetics	Lidocaine	Intradermal	To effect	30–60 min	3 mg/kg (7 mg/kg with epinephrine)	1.5 h	Intraoperatively, immediately prior to procedure, or in the acute postoperative period
	Bupivacaine	Intradermal	To effect	120–140 min	2 mg/kg	3 h	
	Prilocaine	Intradermal	To effect	30–90 min	6 mg/kg	2 h	
Acetaminophen		Oral, rectal, IV	650–1000 mg	4–6 h	4000 mg	2 h	Acute postoperative period
NSAIDs	Ibuprofen	Oral	200–800 mg	4–6 h	3200 mg	2 h	Acute postoperative period
	Celecoxib	Oral	100–400 mg	12 h	800 mg	11 h	
	Naproxen	Oral	250–500 mg	6–8 h	1500 mg	15 h	
	Ketorolac	IV or IM	30 mg	N/A	N/A	6 h	Single dose
Gabapentinoids	Gabapentin	Oral	300–1200 mg	12 h	3600 mg	6 h	Single dose preoperatively or scheduled in the acute postoperative period

h, hours; mg, milligram; kg, kilogram; IV, intravenous; IM, intramuscular.

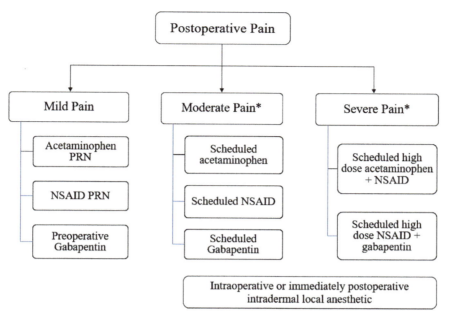

FIG. 1 Proposed algorithm for expected postoperative pain. *Supplemented by judicious use of opiate pain medication as needed.

Other agents of interest

Corticosteroids have an anti-inflammatory effect and decrease both edema and pain. Single dose intraoperative dexamethasone has been shown to decrease postoperative pain and an earlier return to a regular diet in the pediatric post-tonsillectomy patients (Steward, Grisel, & Meinzen-Derr, 2011). However, due to their systemic side effects, corticosteroids are not routinely used in most otolaryngologic procedures. Further research is required to determine if intraoperative steroids decrease postoperative pain in other otolaryngologic procedures.

Key facts
Key facts of patient pain experience in otolaryngology

- Patients' pain experience following surgery is site-dependent
- Patients can expect pain experience following outpatient surgery to be mild to moderate with resolution within the first week after surgery
- Patients should be counseled to expect surgery involving the mouth and oropharynx to result in more moderate to severe pain and can take more than 7 days to improve

Key facts of pain medication use in otolaryngology

- Local anesthetics are especially useful prior to and during procedures as they are very effective at eliminating pain, but are short acting.
- Acetaminophen is a well-tolerated medication which is effective for mild and moderate pain, making it especially useful post-operatively in otolaryngologic procedures.
- Historically, otolaryngologists have hesitated in the use of NSAIDs due to a perceived bleeding risk, however they have gained wider acceptance and now are staple in non-opioid pain management.
- Gabapentin is less widely utilized among otolaryngologists, likely due to the lack of familiarity with this medication, but may have a role in adjunct pain management.

50 PART | I Drugs and agents used in anesthesia and analgesia

Mini-dictionary of terms

Liposomal bupivacaine. A long acting, injectable local anesthetic most often used after cutaneous procedures.

Acetaminophen. A common analgesic with antipyretic properties.

NSAID (nonsteroidal anti-inflammatory drug). Large class of non-opioid analgesics that act on the COX-1 and COX-2 pathways.

Gabapentinoids. A class of medications that are derivatives of the inhibitory neurotransmitter, gamma-Aminobutyric acid, and are often used for neuropathic pain.

Summary points

- Patient's pain experience following otolaryngology surgery is dependent upon surgical site.
- Same-day otolaryngology surgeries result in mild to moderate pain.
- Pain following surgery should resolve within 7 days following surgery.
- Surgery involving the mouth and oropharynx can be more severe and last more than 7 days.
- Patients' should be counseled that discomfort following surgery may be related to impairment of normal function and may not improve with analgesic medication.
- Alternative and non-medical approaches to managing discomfort should be provided to patients as part of the post-surgical care instructions.
- NSAIDs are quite effective as the mainstay pain regimen in most outpatient otolaryngologic procedures.
- Gabapentin is a neuropathic agent and may have its uses in postoperative pain management, but the exact role in the management of acute postoperative pain remains unclear and additional investigation is necessary.

References

Bhoja, R., Ryan, M. W., Klein, K., Minhajuddin, A., Melikman, E., Hamza, M., et al. (2020). Intravenous vs oral acetaminophen in sinus surgery: A randomized clinical trial. *Laryngoscope Investigative Otolaryngology, 5*(3), 348–353. https://doi.org/10.1002/lio2.375.

Boyd, C., Shew, M., Penn, J., Muelleman, T., Lin, J., Staecker, H., et al. (2019). Postoperative opioid use and pain management following otologic and neurotologic surgery. *The Annals of Otology, Rhinology, and Laryngology, 129*(2), 175–180. https://doi.org/10.1177/0003489419883296.

Carpenter, P. S., Shepherd, H. M., McCrary, H., Torrecillas, V., Kull, A., Hunt, J. P., et al. (2018). Association of celecoxib use with decreased opioid requirements after head and neck cancer surgery with free tissue reconstruction. *JAMA Otolaryngology. Head & Neck Surgery, 144*(11), 988–994. https://doi.org/10.1001/jamaoto.2018.0284.

Chahar, P., & Cummings, K. C., 3rd. (2012). Liposomal bupivacaine: A review of a new bupivacaine formulation. *Journal of Pain Research, 5*, 257–264. https://doi.org/10.2147/JPR.S27894.

Chin, C. J., Franklin, J. H., Turner, B., Sowerby, L., Fung, K., & Yoo, J. H. (2011). Ketorolac in thyroid surgery: Quantifying the risk of hematoma. *Journal of Otolaryngology - Head & Neck Surgery, 40*(3), 196–199.

Chua, K. S., Reddy, S. K., Lee, M. C., & Patt, R. B. (1999). Pain and loss of function in head and neck cancer survivors. *Journal of Pain and Symptom Management, 18*(3), 193–202. https://doi.org/10.1016/s0885-3924(99)00070-6.

Eggerstedt, M., Stenson, K. M., Ramirez, E. A., et al. (2019). Association of perioperative opioid-sparing multimodal analgesia with narcotic use and pain control after head and neck free flap reconstruction. *JAMA Facial Plastic Surgery, 21*(5), 446–451. https://doi.org/10.1001/jamafacial.2019.0612.

Foghsgaard, S., Foghsgaard, J., & Homøe, P. (2007). Early post-operative morbidity after superficial parotidectomy: A prospective study concerning pain and resumption of normal activity. *Clinical Otolaryngology, 32*(1), 54–57. https://doi.org/10.1111/j.1365-2273.2007.01315.x.

Ghanem, C. I., Pérez, M. J., Manautou, J. E., & Mottino, A. D. (2016). Acetaminophen from liver to brain: New insights into drug pharmacological action and toxicity. *Pharmacological Research, 109*, 119–131. https://doi.org/10.1016/j.phrs.2016.02.020.

Hinther, A., Nakoneshny, S. C., Chandarana, S. P., Wayne Matthews, T., & Dort, J. C. (2018). Efficacy of postoperative pain management in head and neck cancer patients. *Journal of Otolaryngology - Head & Neck Surgery, 47*(1), 29. https://doi.org/10.1186/s40463-018-0274-y.

Isola, G., Matarese, M., Ramaglia, L., Cicciù, M., & Matarese, G. (2019). Evaluation of the efficacy of celecoxib and ibuprofen on postoperative pain, swelling, and mouth opening after surgical removal of impacted third molars: A randomized, controlled clinical trial. *International Journal of Oral and Maxillofacial Surgery, 48*(10), 1348–1354. https://doi.org/10.1016/j.ijom.2019.02.006.

Kearney, P. M., Baigent, C., Godwin, J., Halls, H., Emberson, J. R., & Patrono, C. (2006). Do selective cyclo-oxygenase-2 inhibitors and traditional non-steroidal anti-inflammatory drugs increase the risk of atherothrombosis? Meta-analysis of randomised trials. *British Medical Journal (Clinical Research Ed.), 332*(7553), 1302–1308. https://doi.org/10.1136/bmj.332.7553.1302.

Kim, M., Kacker, A., Kutler, D. I., Tabaee, A., Stewart, M. G., Kjaer, K., et al. (2020). Pain and opioid analgesic use after otorhinolaryngologic surgery. *Otolaryngology – Head and Neck Surgery*, 194599820933223 (Advance online publication) https://doi.org/10.1177/0194599820933223.

Lam, D. M., Choi, S. W., Wong, S. S., Irwin, M. G., & Cheung, C. W. (2015). Efficacy of pregabalin in acute postoperative pain under different surgical categories: A meta-analysis. *Medicine, 94*(46). https://doi.org/10.1097/MD.0000000000001944, e1944.

Liu, C., & Ulualp, S. O. (2015). Outcomes of an alternating ibuprofen and acetaminophen regimen for pain relief after tonsillectomy in children. *The Annals of Otology, Rhinology, and Laryngology, 124*(10), 777–781. https://doi.org/10.1177/0003489415583685.

Locketz, G. D., Brant, J. D., Adappa, N. D., Palmer, J. N., Goldberg, A. N., Loftus, P. A., et al. (2019). Postoperative opioid use in sinonasal surgery. *Otolaryngology – Head and Neck Surgery, 160*(3), 402–408. https://doi.org/10.1177/0194599818803343.

Moeller, C., Pawlowski, J., Pappas, A. L., Fargo, K., & Welch, K. (2012). The safety and efficacy of intravenous ketorolac in patients undergoing primary endoscopic sinus surgery: A randomized, double-blinded clinical trial. *International Forum of Allergy & Rhinology, 2*(4), 342–347. https://doi.org/10.1002/alr.21028.

Mudd, P. A., Thottathil, P., Giordano, T., Wetmore, R. F., Elden, L., Jawad, A. F., et al. (2017). Association between ibuprofen use and severity of surgically managed posttonsillectomy hemorrhage. *JAMA Otolaryngology. Head & Neck Surgery, 143*(7), 712–717. https://doi.org/10.1001/jamaoto.2016.3839.

Nguyen, K. K., Liu, Y. F., Chang, C., Park, J. J., Kim, C. H., Hondorp, B., et al. (2019). A randomized single-blinded trial of ibuprofen- versus opioid-based primary analgesic therapy in outpatient otolaryngology surgery. *Otolaryngology – Head and Neck Surgery, 160*(5), 839–846. https://doi.org/10.1177/0194599819832528.

Orgill, R., Krempl, G. A., & Medina, J. E. (2002). Acute pain management following laryngectomy. *Archives of Otolaryngology – Head & Neck Surgery, 128*(7), 829–832. https://doi.org/10.1001/archotol.128.7.829.

Ozer, A. B., Erhan, O. L., Keles, E., Demirel, I., Bestas, A., & Gunduz, G. (2012). Comparison of the effects of preoperative and intraoperative intravenous application of dexketoprofen on postoperative analgesia in septorhinoplasty patients: Randomised double blind clinical trial. *European Review for Medical and Pharmacological Sciences, 16*(13), 1828–1833.

Papoian, V., Handy, K. G., Villano, A. M., Tolentino, R. A., Hassanein, M. T., Nosanov, L. S., et al. (2020). Randomized control trial of opioid- versus nonopioid-based analgesia after thyroidectomy. *Surgery, 167*(6), 957–961. https://doi.org/10.1016/j.surg.2020.01.011.

Qian, Z. J., Alyono, J. C., Woods, O. D., Ali, N., & Blevins, N. H. (2019). A prospective evaluation of postoperative opioid use in otologic surgery. *Otology & Neurotology, 40*(9), 1194–1198. https://doi.org/10.1097/MAO.0000000000002364.

Salam, G. A. (2004). Regional anesthesia for office procedures: Part I. Head and neck surgeries. *American Family Physician, 69*(3), 585–590.

Sanders, J. G., & Dawes, P. J. (2016). Gabapentin for perioperative analgesia in otorhinolaryngology-head and neck surgery: Systematic review. *Otolaryngology – Head and Neck Surgery, 155*(6), 893–903. https://doi.org/10.1177/0194599816659042.

Sclafani, A. P., Kim, M., Kjaer, K., Kacker, A., & Tabaee, A. (2019). Postoperative pain and analgesic requirements after septoplasty and rhinoplasty. *Laryngoscope, 129*(9), 2020–2025. https://doi.org/10.1002/lary.27913.

Sener, M., Yilmazer, C., Yilmaz, I., Bozdogan, N., Ozer, C., Donmez, A., et al. (2008). Efficacy of lornoxicam for acute postoperative pain relief after septoplasty: A comparison with diclofenac, ketoprofen, and dipyrone. *Journal of Clinical Anesthesia, 20*(2), 103–108. https://doi.org/10.1016/j.jclinane.2007.09.009.

Sommer, M., Geurts, J. W., Stessel, B., Kessels, A. G., Peters, M. L., Patijn, J., et al. (2009). Prevalence and predictors of postoperative pain after ear, nose, and throat surgery. *Archives of Otolaryngology – Head & Neck Surgery, 135*(2), 124–130. https://doi.org/10.1001/archoto.2009.3.

Steward, D., Grisel, J., & Meinzen-Derr, J. (2011). Steroids for improving recovery following tonsillectomy in children. *The Cochrane Database of Systematic Reviews, 8*(2).

Townsend, M., Liou, T., Kallogjeri, D., Schoer, M., Scott-Wittenborn, N., Lindburg, M., et al. (2018). Effect of perioperative gabapentin use on postsurgical pain in patients undergoing head and neck mucosal surgery: A randomized clinical trial. *JAMA Otolaryngology. Head & Neck Surgery, 144*(11), 959–966. https://doi.org/10.1001/jamaoto.2018.0282.

Turan, A., Emet, S., Karamanlioğlu, B., Memiş, D., Turan, N., & Pamukcu, Z. (2002). Analgesic effects of rofecoxib in ear-nose-throat surgery. *Anesthesia and Analgesia, 95*(5). https://doi.org/10.1097/00000539-200211000-00039.

Walker, N. J., Jones, V. M., Kratky, L., Chen, H., & Runyan, C. M. (2019). Hematoma risks of nonsteroidal anti-inflammatory drugs used in plastic surgery procedures: A systematic review and meta-analysis. *Annals of Plastic Surgery, 82*(6S Suppl 5), S437–S445. https://doi.org/10.1097/SAP.0000000000001898.

Yang, M., Hartley, R. L., Leung, A. A., Ronksley, P. E., Jetté, N., Casha, S., et al. (2019). Preoperative predictors of poor acute postoperative pain control: A systematic review and meta-analysis. *BMJ Open, 9*(4). https://doi.org/10.1136/bmjopen-2018-025091, e025091.

Part II

Mechanisms of action of drugs

Chapter 6

Buprenorphine: Mechanism and applications

Albert Lin[a] and Anuj Aggarwal[b]

[a]Stanford University, Department of Anesthesiology, Perioperative, and Pain Medicine, Stanford, CA, United States, [b]Stanford University, Department of Anesthesiology, Perioperative, and Pain Medicine, Redwood City, CA, United States

Abbreviations

CNS	central nervous system
IV	intravenous
NMDA	N-methyl-D-aspartate
OMT	opioid maintenance therapy
ORL1	opioid receptor-like 1
PCA	patient controlled analgesia
PO	per os (by mouth)
PRN	pro re nata (as needed)
SNRI	serotonin/norepinephrine reuptake inhibitor

Introduction

Buprenorphine is a potent, long-acting partial opioid agonist that has been used to treat chronic pain as well as opioid dependence. Due to the combination of its' unique pharmacodynamic and pharmacokinetic properties, it can decrease the ability of non-potent opioids to work as effectively (morphine, codeine, oxycodone, hydrocodone, and methadone) which is often the anesthesiologist's primary concern and experience with buprenorphine. Potent opioids (sufentanil, fentanyl, and hydromorphone) will still work at higher doses. The μ-receptor blocking effect remains present for 24–36 h after the last dose. Reinduction after surgery requires the patient stop all opioids and experience withdrawal before restarting buprenorphine; a challenging prospect given the combined stressors of acute on chronic pain in the setting of a possible opioid use disorder. This can be a very challenging time for some patients and has been associated with relapse to opioid misuse, heroin use, opioid overdose, and death. In addition, though buprenorphine's partial agonism adds safety in cases of overdose, dangers of CNS depression and respiratory depression remain, especially in combination with other CNS depressants. Here, we will review the pharmacology and clinical applications of buprenorphine for the anesthesiologist, addressing concerns and unique challenges in the perioperative period as well as its' use outside of the perioperative period (Table 1).

Buprenorphine pharmacology

Buprenorphine is a partial μ-opioid receptor agonist with high affinity, low intrinsic activity, and slow dissociation (half-life of association/dissociation of 2–5 h). These properties account for its prolonged therapeutic effect and for its ability to displace other μ-agonists when used for OMT (McCance-Katz, Sullivan, & Nallani, 2010). In addition, buprenorphine increases μ-receptor expression on membrane surfaces (Zaki, Keith, Brine, Carroll, & Evans, 2000). Buprenorphine is a κ-receptor antagonist which may have clinical effects in treating depression (Karp, Butters, & Begley, 2014; Mello & Mendelson, 1985). It is also an agonist for ORL1; activation in the dorsal horn has been shown to be analgesic while cerebral activation appears to blunt antinociception. This suggests that μ-receptor mediated antinociception can be reduced via ORL1 activation (Khhroyan, Polgar, Jiang, Zaveri, & Toll, 2009; Mello & Mendelson, 1985; Spagnolo, Calo, & Polgar, 2008). The supraspinal component of buprenorphine-induced antinociception does not appear to be mediated via μ-opioid

Treatments, Mechanisms, and Adverse Reactions of Anesthetics and Analgesics. https://doi.org/10.1016/B978-0-12-820237-1.00006-5
Copyright © 2022 Elsevier Inc. All rights reserved.

56 PART | II Mechanisms of action of drugs

TABLE 1 Buprenorphine preparations currently available in the United States.

Trade name (product)	Dose form	Doses
Sublocade® (Buprenorphine)	Extended release, subcutaneous	100 mg/0.5 cc, 300 mg/1.5 cc
Butrans® (Buprenorphine)	Transdermal patch	5/7.5/10/15/20 µg/h[a]
Suboxone® (Buprenoprhine + naloxone)	Sublingual tab and film	2/0.5 mg; 4/1 mg, 8/2 mg[b]
Zubsolv® (Bupre + naloxone)	Sublingual	1.4/0.36 mg, 2.9/0.71 mg, 5.7/1.4 mg, 8.6 mg/2.1 mg, 11.4/2.9 mg[c]
Bunavail® (Bupre + naloxone)	Mucoadhesive buccal film	2.1/0.3 mg, 4.2/0.7 mg, 6.3/1 mg
Buprenex® (Buprenorphine)	IV/IM	0.3 mg[d]
Probuphine® (Buprenorphine HCl)	Transdermal implant	80 mg (for 6 months)[e]
Belbuca® (Buprenorphine)	Buccal film	75/150/300/450/600/750/900 mcg[f]

[a]Per manufacturer, 5 µg/h patch = 10–20 mg PO morphine/24 h.
[b]Reduce patient to 30 MME before initiating, package insert recommends 75 µg for <30 mg MME, 150 µg for 30–89 mg MME, 300 µg for 90–160 mg MME, and consider consideration of other analgesic agent greater than 160 mg MME. 600/750/900 µg are for titration from lower doses of buprenorphine only.
[c]Per manufacturer, 1.4 mg buprenorphine Zubsolv® is equivalent to Suboxone® 2 mg sublingual tablets.
[d]Per manufacturer, 0.3 mg of IV buprenorphine equivalent to 10 mg of IV morphine.
[e]Per manufacturer, peak concentration in 12 h after implantation with steady state achieved in approximately 1 month after implantation, patient should prior to implantation by on a stable dose ≤8 mg of buprenorphine. Per the clinical trials study, dosing at steady state was equivalent to 8 mg/day of sublingual buprenorphine as determined by plasma concentration.
[f]Comparative bioavailability of Belbuca (buccal) is 45%–65%, Butrans (patch) 15%, and Suboxone/Bunavail/Zubsolv 29% ± 10%.
The table summarizes available buprenorphine formulations as of 2020 in the United States with available dosings, highlighting the versatility of buprenorphine but further compounding challenges to comprehensive understanding of buprenorphine pharmacology.

receptors, but rather through activation of other unique receptors, evidenced by studies showing that peripheral administration of naloxone antagonizes buprenorphine's dose response curve, whereas supraspinal intracerebroventricular administration does not (Ding & Raffa, 2009).

Buprenorphine is a highly lipid soluble molecule and is 96% bound to α- and β-globulins in plasma. Metabolism is mediated primarily through CYP3A4 N-dealkylation, and the major metabolite is norbuprenorphine (Khanna & Sivaram, 2015; Brewster, Humphrey, & McLeavy, 1981). Both buprenorphine and norbuprenorphine undergo rapid hepatic glucuronidation via UGT2B7 and UGT1A1 to buprenorphine-3-glucuoronide and norbuprnorphine-3-glucuronide, both of which can exceed parent drug levels. All metabolites except norbuprenorphine-3-glucuronide have been shown to have analgesic properties (Brown, Holtzman, Kim, & Kharasch, 2011). Buprenorphine is eliminated primarily in stool, with 10%–30% elimination in urine in conjugated form. Assuming that bioavailability of intravenous buprenorphine is 100%, the relative bioavailability of intramuscular and sublingual administration are 70% and 29%, respectively. Sublingual buprenorphine has an onset of action of 30–60 min and peaks at 1–4 h. The duration of action is 6–12 h with doses less than 4 mg and 24–72 h with doses greater than 16 mg; the prolonged terminal half-life may be secondary to enterohepatic recirculation (Kuhlman, Lalani, & Malluilo, 1996; Nath, Upton, & Everhart, 1999; Schuh & Johanson, 1999). Intravenous buprenorphine has an onset time of approximately 5–15 min and a duration of 6–9 h (Heel, 1979). Secondary to rapid glucoridination, buprenorphine and norbuprenorphine are rapidly conjugated; this, in addition to the fact that buprenorphine and norbuprenorphine do not inhibit CYP enzymes at therapeutic doses, lead to a safe drug interaction profile (Zhang, Ramamoorthy, Tyndale, & Sellers, 2003). The pharmacokinetic profile of buprenorphine and the incidence of adverse events do not seem to change in geriatric patients, patients with renal failure on dialysis, or patients with liver failure (Boger, 2006; Seripa, Pilotto, Panza, Matera, & Pilotto, 2010).

Sublingual administration of buprenorphine is done in conjunction with naloxone to counter potential intravenous misuse in a ratio of 4:1 as naloxone has poor bioavailability (3%) in the sublingual formulation. Withdrawal symptoms after stopping long-term buprenorphine treatment usually manifest within 3–5 days.

Adverse effects

Respiratory depression with buprenorphine monotherapy is rare because of a ceiling dose effect with a documented incidence of 1%–11%. Elderly patients, obese patients, and patients with sleep apnea or pre-existing neuromuscular disease are the most susceptible. However, when used in combination with other central nervous depressants such as benzodiazepines, buprenorphine can cause fatal respiratory depression (Dahan, Yassen, & Romberg, 2006; McCance-Katz et al., 2010). The incidence of constipation with buprenorphine use is much lower at 1%–5% compared to full μ-agonists. Furthermore, buprenorphine does not cause spasms of the sphincter of Oddi (Cuer, Dapoigny, & Ajmi, 1989; Shipton, 2005). Cognitive function in comparative studies showed favorability towards buprenorphine vs methadone and similar effect profiles when compared to placebo (Shmygalev et al., 2011; Soyka et al., 2005). In addition, unlike other potent opioids, buprenorphine does not appear to exert immunosuppressive effects—though most studies on this topic have been conducted in animals—nor does it depress the production of sex hormones that can lead to hypogonadism (Aloisi et al., 2011; Sacerdote, 2008). Due to its pharmacokinetic parameters, higher doses with prolonged administration of opioid antagonists is often needed for reversal.

Clinical use of buprenorphine

Buprenorphine is extensively used for OMT, with a Cochrane review of 13 studies concluding that buprenorphine is an effective treatment for the treatment of opioid dependence. Furthermore, buprenorphine is an effective detoxifying agent for opioid dependence that is equal to, if not better than, methadone (Gowing, Ali, & White, 2006). Buprenorphine's lower abuse potential and better safety profile make it more appealing for OMT when compared to methadone, which has increased drug–drug interactions and often needs to be administered through special clinics. With regards to pain management, buprenorphine has been found to be efficacious in treating moderate to severe pain, with studies showing that buprenorphine provides equivalent analgesic effects but decreased adverse effects and increased patient compliance when compared to morphine or fentanyl (Bohme & Likar, 2003). It has also been shown to be effective in treating neuropathic pain by potentially blocking secondary hyperalgesia from central sensitization. Furthermore, preclinical studies showed buprenorphine to be effective in treating bone pain, heat pain, pain related to nerve growth factor injections, and cold pressor pain—a significantly wider range of pain phenotypes when compared to fentanyl, which was only effective for cold pressor pain (Wolff et al., 2012). Studies have also demonstrated that administering other opioids to patients receiving buprenorphine therapy is safe and effective.

There has been great concern regarding buprenorphine use in the perioperative setting out of concern for possible analgesic ceiling effects; however, no analgesic ceiling effects have been seen when additional full μ-receptor agonists are given at therapeutic doses (Butler, 2013). Conversion calculations of buprenorphine to morphine suggest a ratio of 1:110–115 to achieve the same degree of analgesia, though clinical judgement is needed as buprenorphine does not neatly convert to other opioids using morphine equivalence calculations (Sittl, Likar, & Poulsten-Nautrup, 2005). Different formulations have different bioavailability, which make conversion even more challenging. Finally, there is concern regarding buprenorphine's high affinity at the μ-receptor causing displacement of adjuvant opioid agents; however, buprenorphine occupies fewer receptors (leading to a receptor reserve for additive μ-agonists) and increases μ-receptor expression with long-term use and has similar affinity to sufentanil and hydromorphone (Brown et al., 2011; Volpe et al., 2011).

Perioperative management

Discontinuation of buprenorphine can be challenging, and the perioperative period compounds the difficulty of discontinuing buprenorphine with conflicting guidelines and recommendations. Tapering of buprenorphine is recommended for patients with stable psychosocial parameters and with strong motivations to discontinue; however, studies disagree on the optimal taper duration (Mauger, Fraser, & Gill, 2014). Additionally, multiple studies have concluded that amongst the adverse effects of buprenorphine, discontinuation for any reason and being out of treatment was the major predictor of death (Soyka, 2015). Furthermore, studies have demonstrated that patients often desire to remain on buprenorphine out of fear of withdrawing or relapsing to illicit opioid use. One meta-analysis found that rates of relapse to illicit opioid use exceeded 50% in every included study at 1 month following buprenorphine cessation (Bentzley, Barth, Back, & Book, 2015). Discontinuation of buprenorphine in the perioperative setting poses dangers and challenges to patients that may not be immediately apparent to the surgical or anesthesia team taking care of the patient on the day of surgery.

For patients on high doses of buprenorphine (daily dose greater than 10 mg) who are to undergo surgeries with a high degree of anticipated post-operative pain, consideration should be made to delay surgery such that the daily dose of buprenorphine can be tapered down to 8 mg per day at least 72 h prior to surgery. This decision should be made by the surgeon and anesthesiologist in conjunction with the patient's buprenorphine prescriber. In general, most guidelines suggest decreasing the dosage of buprenorphine by 2 mg per day every 2–3 days until the 8 mg per day dosing goal is reached. Post-operatively, patients should receive their usual outpatient dosing of buprenorphine in addition to supplemental pain medications as needed (Book, Myrick, Malcom, & Strain, 2007; Sen et al., 2016). Evidence for management comes largely from cases studies suggesting that patients are most likely to benefit from divided dosing of home buprenorphine with additional buprenorphine given as needed (2 mg q4–6h prn) for breakthrough pain. The discussion and plan should be made in conjunction with the patient's buprenorphine prescriber. Multimodal techniques have not been studied in patients on opioid maintenance therapy, but extensive guidelines exist for multimodal pain management in the immediate preoperative and intraoperative settings (Wenzel, Schwenk, Baratta, & Viscusi, 2016). For patients taking less than 2–4 mg of buprenorphine a day, they can be treated similarly to other patients on high dose opioid regimens (methadone, fentanyl patch, etc.) (Li, Schmiesing, & Aggarwal, 2020).

In the postoperative period, higher doses of opioid agonists may be necessary. Opioids with higher μ-activity such as sufentanil, fentanyl, or hydromorphone should be considered while codeine and hydrocodone should be avoided. In addition, sublingual or IV buprenorphine can be added to this regimen to provide additional analgesia (Bullingham, McQuay, Dwyer, Allen, & Moore, 1981). Another strategy is to divide the patient's total daily dose of buprenorphine into three or four doses spaced throughout the day (Childers & Arnold, 2012; Johnson, Fudala, & Payne, 2005). Furthermore, supplemental buprenorphine (2–4 mg every 6–8 h) can be added as a PRN medication, based on case reports showing that this dosing is effective for managing post-surgical pain.

Applications to other areas

This chapter focused on the clinical applications of buprenorphine in the outpatient and perioperative settings. Maintaining continuity of pain care between the two care settings is of particular importance. No less important, however, is creating continuity of care between outpatient clinics and emergency rooms where many patients with opioid use disorder are often first diagnosed and treated. Several studies have demonstrated the role that buprenorphine can play in the emergency setting for treating opioid withdrawal as well as opioid addiction. For example, buprenorphine has been shown to be safe and effective for treating opioid withdrawal symptoms while decreasing ED visits compared to supportive care alone. ED-initiated buprenorphine was also associated with increased patient engagement in outpatient addiction treatment programs and reduced illicit opioid use in patients who received outpatient follow-up after being managed in the ED (Cisewski, Santos, Koyfman, & Long, 2019; D'Onofrio et al., 2017). In patients who present to an ED with symptoms suggestive of opioid withdrawal, buprenorphine 4 mg can be administered, followed by a 1–2 h observation period. If the patient remains symptomatic, an additional dose should be given. If the patient's symptoms resolve, the patient should be discharged with a three-day supply of buprenorphine equal to the total dose required to achieved symptom resolution in the ED. This dose is typically between 8 and 16 mg (Cisewski et al., 2019). The patient should have outpatient follow-up within 2–3 days. Of note, if opioid withdrawal symptoms are inadvertently precipitated through buprenorphine use, treatment should consist of non-opioid agents such as clonidine or benzodiazepines. Buprenorphine's high potency prevents the therapeutic benefits from additional opioid administration.

Other agents of interest

In this chapter we have described in detail the pharmacology of buprenorphine and how its partial agonism and long effective half-life lead to unique clinical opportunities in chronic pain and OMT, along with unique challenges in the perioperative period. Other long-acting opioid formulations—such as fentanyl patches and methadone—are often used for similar indications and are useful comparisons. Unlike methadone, buprenorphine does not have a myriad of receptor targets allowing for fewer drug–drug interactions; however, for this very reason, patients may not report similar analgesic efficacy with buprenorphine compared to methadone, and may need other agents that can mimic methadone's other analgesic effects mediated through NMDA antagonism and SNRI properties. Fentanyl patches offer the benefit of full opioid agonism but have variable pharmacokinetic properties and long half-lives, which when combined with their full agonism can lead to a higher incidence of respiratory depression and sedation compared to buprenorphine.

Methadone

Methadone is a long-acting opioid agonist that primary binds to the μ-opioid receptor. It is a well-established long-term treatment for opioid addiction and has been shown to increase treatment adherence and reduce opioid use when compared with placebo or non-pharmacologic treatments.[60] Methadone has also been associated with decreased mortality from opioid use disorder and with reduced criminal behavior. Common side effects include constipation, fatigue, and peripheral edema. Notably, methadone has been linked to QTc prolongation and torsades de pointes, especially in patients with a history of arrhythmias and/or structural heart disease. Special consideration should be given to patients who have a QTc greater than 450 ms and/or concurrently take other QTc prolonging medications. Methadone also has a greater risk for drug overdose when compared to buprenorphine given that it is a full agonist. Prescribers should bear in mind the relatively long (and variable) half-life of 8–60 h and slowly titrate doses, especially when initiating therapy. For the treatment of opioid use disorder, methadone is usually started at 20–30 mg daily and titrated in increments of 5–10 mg every 2–3 days. Further titrations beyond 80–120 mg daily should occur no more frequently than on a weekly basis.

Naltrexone

Naltrexone is a competitive opioid antagonist that shows the highest affinity for μ-receptors. It is used to prevent relapse in opioid use disorder by blocking the euphoric effects of illicit opioid use. Naltrexone can be taken orally or subcutaneously, with the latter route usually dosed at 380 mg per injection every 4 weeks. Both formulations have been found to be effective for treating opioid dependence when compared to placebo. Of note, naltrexone can cause immediate opioid withdrawal symptoms if patients are not tapered off other opioids prior to naltrexone initiation. Therefore, the initiation of naltrexone should be conducted under medical supervision.

Key facts
Key facts of buprenorphine

- Buprenorphine is a high-affinity partial agonist for the μ-opioid receptor, a kappa-receptor antagonist, and an ORL1 agonist.
- Sublingual buprenorphine has an onset of action of 30–60 min and peaks at 1–4 h. Duration of action is 6–12 h at <4 mg dosing and 24–72 h at >16 mg dosing.
- Sublingual buprenorphine is often combined with naloxone in a 4:1 ratio (e.g., 8 mg buprenorphine to 2 mg naloxone) to discourage abuse. Naloxone has only 3% bioavailability when administered sublingually.
- Respiratory depression with buprenorphine monotherapy is a rare occurrence, with a documented incidence of 1%–11%. The risk is elevated in elderly patients, obese patients, and patients with sleep apnea or neuromuscular disease.
- Despite the concerns surrounding buprenorphine's possible analgesic ceiling effects, the concomitant use of other opioid agents has been shown to be both safe and effective.

Mini-dictionary of terms

Norbuprenorphine. An active metabolite of buprenorphine formed via hepatic N-dealkylation by CYP3A4.
Opioid maintenance therapy. The use of long-acting opioid agents such as buprenorphine and methadone to maintain abstinence from opioid abuse. The goal of OMT is to relieve the cravings associated with opioid addiction, to block the euphoric effects of dependent opioids, and to relieve withdrawal symptoms.
Partial agonist. A drug that binds to and activates a given receptor, but only with partial efficacy relative to a full agonist.
Buprenorphine induction. A medically supervised process in which a patient transitions from an abused opioid to a dose of buprenorphine. The abused opioid is often held until the patient experiences withdrawal symptoms prior to the initiation of buprenorphine.
Opioid use disorder. A DSM-V disorder defined as a problematic pattern of opioid use leading to clinically significant problems or distress. Also known as "opioid addiction."

Summary points

- Buprenorphine is a potent, long-acting partial opioid agonist that is most commonly used for the treatment of opioid use disorder and chronic pain.
- The major metabolite of buprenorphine is norbuprenorphine, which is formed via CYP3A4-mediated N-dealkylation. Both buprenorphine and norbuprenorphine undergo hepatic glucuronidation to buprenorphine-3-glucuronide and norbuprenorphine-3-glucuronide, respectively. All metabolites except norbuprenorphine-3-glucuronide have analgesic properties.

60 PART | II Mechanisms of action of drugs

- The relative bioavailabilities of buprenorphine when administered intramuscularly or sublingually are 70% and 29%, respectively.
- The rate of respiratory depression and constipation are much lower with buprenorphine when compared to full agonists.
- Buprenorphine has been demonstrated to be an effective treatment for opioid dependence that is equal, if not better, than methadone. Buprenorphine has also been shown to be efficacious in moderate to severe pain.
- No analgesic ceiling effects have been demonstrated following the concurrent use of buprenorphine and other mu-receptor agonists.
- Buprenorphine should be continued in the perioperative setting. For patients taking doses exceeding 10 mg per day, consideration should be given to tapering down the dose to approximately 8 mg per day at least 72 h prior to surgery.
- In the postoperative period, higher doses of opioid agonists—such as sufentanil, fentanyl, or hydromorphone—should be considered, while avoiding codeine and hydrocodone. Sublingual or IV buprenorphine can be added to the patient's pain control regimen.

References

Aloisi, A. M., Ceccarelli, I., Carlucci, M., Suman, A., Sindaco, G., Mameli, S., … Pari, G. (2011). Hormone replacement therapy in morphine-induced hypogonadic male chronic pain patients. *Reproductive Biology and Endocrinology, 9*, 26.

Bentzley, B. S., Barth, K. S., Back, S. E., & Book, S. W. (2015). Discontinuation of buprenorphine maintenance therapy: Perspectives and outcomes. *Journal of Substance Abuse Treatment, 52*, 48–57.

Boger, R. H. (2006). Renal impairment: A challenge for opioid treatment? The role of buprenorphine. *Palliative Medicine, 20*(Suppl 1), S17–S23.

Bohme, K., & Likar, R. (2003). Efficacy and tolerability of a new opioid analgesic formulation, buprenorphine transdermal therapeutic system (TDS), in the treatment of patients with chronic pain. A randomized, double-blind, placebo-controlled study. *Pain Clinic, 15*(2), 193–202.

Book, S. W., Myrick, H., Malcom, R., & Strain, E. C. (2007). Buprenorphine for postoperative pain following general surgery in a buprenorphine-maintained patient. *The American Journal of Psychiatry, 164*(6), 979.

Brewster, D., Humphrey, M. J., & McLeavy, M. A. (1981). Biliary excretion, metabolism and enterohepatic circulation of buprenorphine. *Xenobiotica, 11*, 189–196.

Brown, S. M., Holtzman, M., Kim, T., & Kharasch, E. (2011). Buprenorphine metabolites, buprenorphine-3-glucoronide and norbuprenorphine-3-glucuronide, are biologically active. *Anesthesiology, 115*(6), 1251–1260.

Bullingham, R. E., McQuay, H. J., Dwyer, D., Allen, M. C., & Moore, R. A. (1981). Sublingual buprenorphine used postoperatively: Clinical observations and preliminary pharmacokinetic analysis. *British Journal of Clinical Pharmacology, 12*(2), 117–122.

Butler, S. (2013). Buprenorphine—Clinically useful but often misunderstood. *Scandinavian Journal of Pain, 4*, 148–152.

Childers, J. W., & Arnold, R. M. (2012). Treatment of pain in patients taking buprenorphine for opioid addiction #221. *Journal of Palliative Medicine, 15*(5), 613–614.

Cisewski, D. H., Santos, C., Koyfman, A., & Long, B. (2019). Approach to buprenorphine use for opioid withdrawal treatment in the emergency setting. *The American Journal of Emergency Medicine, 37*(1), 143–150.

Cuer, J. C., Dapoigny, M., & Ajmi, S. (1989). Effects of buprenorphine on motor activity of the sphincter of Oddi in man. *European Journal of Clinical Pharmacology, 36*(2), 203–204.

Dahan, A., Yassen, A., & Romberg, R. (2006). Buprenorphine induces ceiling in respiratory depression but not in analgesia. *British Journal of Anaesthesia, 96*(5), 627–632.

Ding, Z., & Raffa, R. B. (2009). Identification of an additional supraspinal component to the analgesic mechanism of action of buprenorphine. *British Journal of Pharmacology, 157*, 831–843.

D'Onofrio, G., Chawarski, M. C., O'Connor, P. G., Pantalon, M. V., Busch, S. H., Owens, P. H., et al. (2017). Emergency department-initiated buprenorphine for opioid dependence with continuation in primary care: Outcomes during and after intervention. *Journal of General Internal Medicine, 32*(6), 660–666.

Gowing, L., Ali, R., & White, J. (2006). Buprenorphine for the management of opioid withdrawal. *Cochrane Database of Systematic Reviews, 2*, CD002025.

Heel, R. (1979). Buprenorphine: A review of its pharmacological properties and therapeutic efficiency. *Drugs, 17*, 81–110.

Johnson, R. E., Fudala, P. J., & Payne, R. (2005). Buprenorphine: Considerations for pain management. *Journal of Pain and Symptom Management, 29*(3), 297–326.

Karp, J. F., Butters, M. A., & Begley, A. E. (2014). Safety, tolerability, and clinical effect of low-dose buprenorphine for treatment-resistant depression in midlife and older adults. *The Journal of Clinical Psychiatry, 75*(8), 785–793.

Khanna, I., & Sivaram, P. (2015). Buprenorphine—An attractive opioid with underutilized potential in treatment of chronic pain. *Journal of Pain Research, 8*, 859–870.

Khhroyan, T. V., Polgar, W. E., Jiang, F., Zaveri, N. T., & Toll, L. (2009). Nociception/orphanin FQ receptor activation attenuates antinociception induced by mixed nociception/orphanin FQ/mu-opioid receptor agonists. *The Journal of Pharmacology and Experimental Therapeutics, 331*(3), 949–953.

Kuhlman, J. J., Lalani, S., & Malluilo, J. (1996). Human pharmacokinetics of intravenous, sublingual, and buccal buprenorphine. *Journal of Analytical Toxicology, 20*, 369–378.

Mauger, S., Fraser, R., & Gill, K. (2014). Utilizing buprenorphine-naloxone to treat illicit and prescription-opioid dependence. *Neuropsychiatric Disease and Treatment, 10,* 587–598.

McCance-Katz, E. F., Sullivan, L. E., & Nallani, S. (2010). Drug interaction of clinical importance among the opioids, methadone and buprenorphine, and other frequently prescribed medications: A review. *The American Journal on Addictions, 19*(1), 4–16.

Mello, N. K., & Mendelson, J. K. (1985). Behavioral pharmacology of buprenorphine. *Drug and Alcohol Dependence, 14*(3–4), 283–303.

Nath, R. P., Upton, R. A., & Everhart, E. T. (1999). Buprenorphine pharmacokinetics: Relative bioavailability of sublingual tablet and liquid formulations. *Journal of Clinical Pharmacology, 39*(6), 619–623.

Sacerdote, P. (2008). Opioid-induced immunosuppression. *Current Opinion in Supportive and Palliative Care, 2*(1), 14–18.

Schuh, K. J., & Johanson, C. E. (1999). Pharmacokinetic comparison of the buprenorphine sublingual liquid and tablet. *Drug and Alcohol Dependence, 56* (1), 55–60.

Sen, S., Arulkumar, S., Cornett, E., Gayle, J., Flower, R., Fox, C., et al. (2016). New pain management options for the surgical patient on methadone and buprenorphine. *Current Pain and Headache Reports, 20*(3), 16.

Seripa, D., Pilotto, A., Panza, F., Matera, M. G., & Pilotto, A. (2010). Pharmacogenetics of cytochrome P450 (CYP) in the elderly. *Ageing Research Reviews, 9*(4), 457–474.

Shipton, E. (2005). Safety and tolerability of buprenorphine. In K. Budd, & R. Raffa (Eds.), *Buprenorphine—The unique opioid analgesic* (pp. 92–101). Stuttgart: Thieme Verlag Kg.

Shmygalev, S., Damm, M., Weckbecker, K., Berghaus, G., Petzke, F., & Sabatowski, R. (2011). The impact of long-term maintenance treatment with buprenorphine on complex psychomotor and cognitive function. *Drug and Alcohol Dependence, 117*(2–3), 190–197.

Sittl, R., Likar, R., & Poulsten-Nautrup, B. (2005). Equipotent doses of transdermal fentanyl and transdermal buprenorphine in patients with cancer and non-cancer pain. Results of a retrospective cohort study. *Clinical Therapeutics, 27,* 225–237.

Soyka, M. (2015). New developments in the management of opioid dependence: Focus on sublingual buprenorphine-naloxone. *Substance Abuse and Rehabilitation, 6*(6), 1–14.

Soyka, M., Hock, B., Kagerer, S., Lehnert, R., Limmer, C., & Kuefner, H. (2005). Less impairment on one portion of a driving-relevant psychomotor battery in buprenorphine-maintained than in methadone maintained patients: Results of a randomized clinical trial. *Journal of Clinical Psychopharmacology, 25*(5), 490–493.

Spagnolo, B., Calo, G., & Polgar, W. E. (2008). Activities of mixed NOP and mu-opioid receptor ligands. *British Journal of Pharmacology, 153*(3), 609–619.

Volpe, D. A., McMahon Tobin, G. A., Mellon, D. R., Katki, A. G., Parker, R. J., Colatsky, T., et al. (2011). Uniform assessment and ranking of opioid mu receptor binding constants for selected opioid drugs. *Regulatory Toxicology and Pharmacology, 59*(3), 385–390.

Wenzel, J. T., Schwenk, E. S., Baratta, J. L., & Viscusi, E. R. (2016). Managing opioid-tolerant patients in the perioperative surgical home. *Anesthesiology Clinics, 34*(2), 287–301.

Wolff, R. F., Aune, D., Truyers, C., Hernandez, A. V., Misso, K., Riemsma, R., et al. (2012). Systemic review of efficacy and safety of buprenorphine versus fentanyl or morphine in patients with chronic moderate to severe pain. *Current Medical Research and Opinion, 28*(5), 833–845.

Zaki, P. A., Keith, D. E., Brine, G. A., Carroll, F. I., & Evans, C. J. (2000). Ligand-induced changes in surface mu-opioid receptor number: Relationship to G protein activation? *The Journal of Pharmacology and Experimental Therapeutics, 282*(3), 1127–1134.

Zhang, W., Ramamoorthy, Y., Tyndale, R. F., & Sellers, E. M. (2003). Interaction of buprenorphine and its metabolite norbuprenorphine with cytochromes p450 in vitro. *Drug Metabolism and Disposition, 31*(6), 768–772.

Chapter 7

Caffeine as analgesic adjuvant

Thomas W. Weiser

Medical Affairs, Sanofi-Aventis Deutschland GmbH, Frankfurt/Main, Germany

Abbreviations

APC   fixed dose combination of aspirin, paracetamol, and caffeine
EFSA   European Food Safety Authority
EMA   European Medicines Agency
NSAID   Nonsteroidal anti-inflammatory drug
OTC   over-the-counter
WHO   World Health Organization

Introduction

Caffeine-containing food and beverages are consumed since centuries in many regions of the world, with coffee and tea being the most prominent examples. Caffeine possesses mild stimulant effects, which probably contribute to its widespread consumption. Besides that, caffeine is used as a combination partner with non-narcotic analgesics to treat acute pain like headache or toothache since around the beginning of the 20th century. From today's perspective it is hard to tell which intentions and observations led to the development of these combinations.

Nowadays, we are in the comfortable situation to understand caffeine's pharmacological mechanism(s) of action in more detail, and to have a rich basis of state-of-the-art clinical trials which have investigated the effects of caffeine in the context of analgesia.

Caffeine alone is used for treating some few pain conditions, which will be briefly discussed in this chapter. The main part of this chapter will discuss the effects of caffeine as a co-analgesic (i.e., in combination with analgesics of a variety of chemical classes acting on different pharmacological targets). Such combinations are available OTC in many countries of the world, and for some indications—like acute migraine attacks—their efficacy is in the range of specialized prescription-only treatments.

Caffeine: Mechanism of action and effects in preclinical models of pain

Caffeine (1,3,7-trimethylxanthine) occurs in various plant species, and has been isolated by the German chemist Friedlieb Ferdinand Runge in 1819. Its main pharmacological action is the inhibition of adenosine receptors, with low selectivity for the four subtypes described so far (A_1-, A_{2A}-, A_{2B}-, A_3-receptors; Jacobson & Müller, 2016). Data from preclinical pain models have been described in detail by Sawynok (2011): Caffeine induced antinociception at doses of 35–100 mg/kg in rodents. In combination with aspirin, paracetamol, or morphine, the antinociceptive actions of these agents were augmented. The interaction with carbamazepine and tricyclic antidepressants, which are used to treat neuropathic pain in humans, appears to be more complex (with inhibition of antinociceptive effects with low doses of caffeine). Since the focus of this chapter is the use of caffeine for the treatment of pain in humans, preclinical data will not be discussed further.

Caffeine's analgesic properties in humans

Caffeine itself appears to possess analgesic properties on its own. In clinical trials investigating efficacy of caffeine-containing analgesics which included caffeine-arms, trends of Diamond, Balm, and Freitag (2000) and Weiser, Richter, Hegewisch, Muse, and Lange (2018), and even superiority over placebo has been demonstrated (Diener, Pfaffenrath, Pageler, Peil, & Aicher, 2005). A strong cup of coffee is recommended by some physicians as homemade remedy for treating headache (although the actual caffeine content of coffee can vary considerably, EFSA NDA Panel (EFSA Panel on Dietetic Products, Nutrition and Allergies), 2015).

In some indications, caffeine was shown to be effective. Headache as a sequel of spinal lumbar puncture is relatively common (Basurto, Osorio, & Bonfill, 2015). Oral, as well as intravenous caffeine are effective treatments for this headache.

Relatively uncommon, but burdensome for patients affected, is hypnic headache, which is defined in the "International Classification of Headache Disorders 3rd edition" as "Frequently-recurring headache attacks developing only during sleep, causing wakening and lasting for up to 4hours, without characteristic associated symptoms and not attributed to other pathology" (IHS CLASSIFICATION ICHD-3, n.d.). Caffeine has been shown to be effective both as acute, as well as prophylactic treatment (Silva-Néto, Santos, & Peres, 2019).

Besides that, caffeine is widely used as an adjuvant in fixed-dose combinations with analgesic agents. This chapter will therefore focus on the properties of caffeine as a co-analgesic.

Pharmacokinetic properties in man in absence and presence of analgesics

Caffeine has co-analgesic properties and is typically combined with one or two WHO stage 1 analgesic compounds (like aspirin, ibuprofen, and paracetamol). Its pharmacokinetic properties, as well as metabolism, have been investigated extensively by others and will not be discussed in detail here. Oral caffeine is absorbed completely and rapidly, reaching maximum plasma levels about 0.5 h after intake, with half-life times in the range of few hours (Arnaud, 2011). These values fit well to the pharmacokinetic properties of many WHO stage 1 analgesics.

In many textbooks on pharmacology, as well as in many review articles, one finds the hypothesis that caffeine might accelerate and augment pain relief by pharmacokinetic properties, i.e., by accelerating absorption of analgesic compounds. This appears not to be the case under fasted conditions (after at least 8 h abstinence from food, with only tap water allowed to drink), which are usually applied in pharmacokinetic studies. One should bear in mind that these studies usually are run for regulatory purposes (i.e., obtaining the market authorization for a generic medicinal product based on the bioequivalence with an already registered product containing the same active ingredient(s)). Health authorities (like the EMA) recommend to run such trials under strict fasted conditions to reduce complexity of clinical bioequivalence trials (CHMP, 2010). Since analgesics are rarely taken under such conditions (i.e., after >8 h of fasting), the relevance of these data for the every-day situation is negligible.

The combination of 250 mg aspirin and 200 mg paracetamol is absorbed equally fast in the absence or presence of 50 mg caffeine (Weiser & Weigmann, 2019), and comparable findings were reported for 400 mg ibuprofen in the presence or absence of 100 mg caffeine. When taken after a standard breakfast, however, 400 mg ibuprofen was absorbed faster in the presence of 100 mg caffeine in comparison to 400 mg ibuprofen (as lysinate) without caffeine (Weiser, Schepers, Mück, & Lange, 2019; Fig. 1). Moreover, higher peak plasma levels were reached (Fig. 2). This is of particular interest, since fast releasing oral analgesic formulations (like ibuprofen as lysinate, sodium salt, or soft gel capsule) are discussed to act stronger and faster compared to standard oral formulations (Moore, Derry, Straube, Ireson-Paine, & Wiffen, 2014). Direct head-to-head comparisons, however, did not find significant faster onsets of action for the treatment of headache

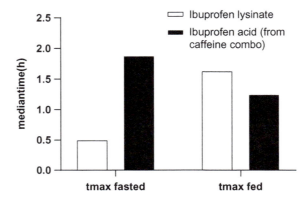

FIG. 1 Effect of caffeine on speed of uptake of ibuprofen. Median times to reach maximum ibuprofen plasma concentration (tmax) after intake of 400 mg as lysine salt, or as combination with 100 mg caffeine under fasted and fed conditions. *Data from Weiser, T., Schepers, C., Mück, T., & Lange, R. (2019). Pharmacokinetic properties of ibuprofen (IBU) from the fixed-dose combination IBU/caffeine (400/100 mg; FDC) in comparison with 400 mg IBU as acid or lysinate under fasted and fed conditions—Data from 2 single-center, single-dose, randomized crossover studies in healthy volunteers.* Clinical Pharmacology in Drug Development, 8, 742–753.

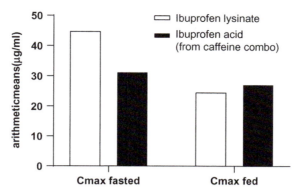

FIG. 2 Effect of caffeine on the plasma concentration of ibuprofen. Mean maximum ibuprofen plasma concentrations (Cmax) after intake of 400 mg as lysine salt, or as combination with 100 mg caffeine under fasted and fed conditions. *Data from Weiser, T., Schepers, C., Mück, T., & Lange, R. (2019). Pharmacokinetic properties of ibuprofen (IBU) from the fixed-dose combination IBU/caffeine (400/100 mg; FDC) in comparison with 400 mg IBU as acid or lysinate under fasted and fed conditions—Data from 2 single-center, single-dose, randomized crossover studies in healthy volunteers.* Clinical Pharmacology in Drug Development, 8, 742–753.

(Heintze & Fuchs, 2015), or pain after extraction of third molars (Kyselovič, Koscova, Lampert, & Weiser, 2020) for these fast releasing formulations.

Thus, it can be assumed that on non-empty stomach caffeine indeed can accelerate the absorption of analgesic compounds (at least in the case of ibuprofen), which contributes to the faster and stronger pain relief of caffeine-containing analgesics in comparison to their counterparts without caffeine.

Clinical efficacy data in acute pain trials

The duration of "acute pain" is not well defined and depends very much on the location and etiology of the pain event. Acute pain after tooth extraction might last for hours to 1 or 2 days, a single migraine episode might last from 4 to 72 h (IHS CLASSIFICATION ICHD-3, n.d.), and acute low back pain can last up to 6 weeks.

In many countries, caffeine-containing analgesics are available OTC, and in this context the label allows duration of use for few days only. Thus, from the patients' perspective this restriction can be expected to define the meaning of "acute pain" in terms of duration (i.e., pain with a maximum duration of few days), and, moreover, it reflects the duration of the most often treated pain forms in self-medication (i.e., tension-type headache and migraine attacks; Radtke & Neuhauser, 2009).

In the case of migraine, and certainly other acute pain forms as well, patients desire fast and complete pain relief (Lipton, Hamelsky, & Dayno, 2002). A high response rate is desirable to increase the patients' chance for successful treatment.

An adjuvant (like caffeine) supports efficacy of the analgesic ingredient without having pronounced effects on its own. Thus, the analgesics with or without caffeine can be expected to act qualitatively similar (but quantitatively different).

Under this assumption, caffeine will positively affect all clinical efficacy readouts: It will improve mean pain reduction, it will induce earlier onset of action (Fig. 3), and it will provide a higher number of patients reporting a given pain relief and

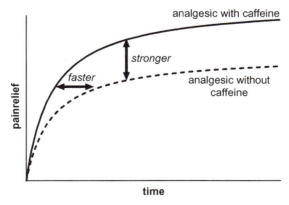

FIG. 3 Co-analgesics and analgesia over time. Simulation of the effect of a co-analgesic (like caffeine) on the time-effect curve of an acute pain analgesic. Efficacy of the combination is higher at every timepoint, which results in a faster and more pronounced pain relief compared to the analgesic alone.

66 PART | II Mechanisms of action of drugs

not needing rescue medication. This is exemplified by the data from a recent clinical trial investigating ibuprofen 400 mg plus caffeine 100 mg, corresponding doses of ibuprofen and caffeine, and placebo in the dental extraction model (Weiser et al., 2018): Compared to ibuprofen alone, mean pain reduction over 0–8 h was 30%–40% higher, median time to onset of meaningful pain relief was 36% shorter, responder-rate (percentage of patients achieving more than 50% pain reduction over 0–6 h) was 40% higher, and intake of rescue medication or a second analgesic dose was 50% lower with the combination.

The beneficial effect of caffeine as a co-analgesic has been analyzed by Derry, Derry, and Moore (2012). Data from 19 studies (with altogether 7238 patients) were identified, and 14 trials in headache, dysmenorrhea, and postsurgical pain were analyzed. In headache, responder rates for analgesic plus caffeine were 80%, with analgesic alone 73%. In dysmenorrhea the figures were 50% and 39%, and in postsurgical pain 60% and 52%, respectively. Altogether, 5%–10% more patients achieved relevant pain relief with the analgesic plus caffeine, compared to the analgesic alone. The adjuvant effect of caffeine has also been confirmed in a large clinical trial in tension type headache and migraine: Here, the combination aspirin/paracetamol/caffeine (500/400/100 mg) acted faster, reduced pain more effectively, and had a higher responder rate compared with aspirin/paracetamol alone (Diener et al., 2005).

The trials described above investigated analgesics' effects in acute pain which usually ceases within hours or a few days. By definition, acute (low) back pain can last up to 6 weeks. A recent study investigated the efficacy of ibuprofen 400 mg plus caffeine 100 mg with ibuprofen and placebo (3 doses daily for 6 days) in acute upper and lower back pain (Predel, Ebel-Bitoun, Lange, & Weiser, 2019). Somewhat surprisingly, in terms of pain reduction all treatment arms were not significantly different from each other. Based on these study data it is hard to judge whether or not caffeine might offer additional benefit in this pain condition.

Acute benefits of caffeine when given in addition to opioids for the treatment of cancer pain have been reported (Suh et al., 2013). Thus, caffeine's adjuvant effects can be considered not to be limited to WHO step 1 analgesics.

Caffeine and the ceiling effect of NSAIDs

NSAIDs have shallow dose-response curves, and above a certain dose their efficacy for acute pain relief cannot be increased by increasing the dose. This "ceiling effect" is well described for ibuprofen. This compound is available OTC up to doses of 400 mg in many countries, whereas doses of 600 or 800 mg are available with a prescription. Efficacy for acute pain relief levels at 400 mg: In clinical acute pain trials investigating ibuprofen of doses from 400 to 600 or 800 mg in more than 1000 patients, higher doses than 400 mg did not provide more acute pain relief (Kellstein et al., 2000; Laska et al., 1986; Seymour, Ward-Booth, & Kelly, 1996), but ibuprofen with caffeine does so (Weiser et al., 2018; Fig. 3). Higher doses of ibuprofen (600–800 mg) provide stronger anti-inflammatory efficacy, but this appears not to be relevant for many acute pain forms.

Comparisons of caffeine-containing analgesics with active comparators

In clinical efficacy trials caffeine-containing analgesics were not only compared with the respective analgesics without caffeine (as well as placebo), but in several cases also with active comparators.

In a trial on tension-type headache and migraine, Diener et al. (2005) compared aspirin/paracetamol/caffeine (500/400/100 mg) with the same dose of aspirin/paracetamol, aspirin alone (1000 mg), paracetamol alone (1000 mg), caffeine (100 mg) and placebo. The caffeine-containing combination was superior to all other treatment arms (including onset of action, pain reduction, and responder-rate).

In two trials on acute migraine attacks, aspirin/paracetamol/caffeine (500/500/130 mg) were compared with 50 mg sumatriptan and placebo (Goldstein et al., 2005), and with 400 mg ibuprofen and placebo (Goldstein, Silberstein, Saper, Ryan Jr., & Lipton, 2006). In both studies, the caffeine-containing combination was superior to the active comparators.

In a review and meta-analysis, Petersen, Kops, and Heintze (2017) compared response rates of WHO stage 1 analgesics in migraine treatment with 100 mg sumatriptan as reference compound. Sumatriptan showed a response rate (percentage of pain-free patients at 2 h after intake) of 32%, followed by aspirin/paracetamol/caffeine (500/400–500/100–130 mg) with 31%. Response rates were 28% for ibuprofen (400 mg), 23% for aspirin (900–1000 mg), 18% for paracetamol (1000 mg), and 11% for placebo, respectively (Fig. 4).

Thus, a rich body of evidence demonstrates the superior efficacy of caffeine-containing analgesics.

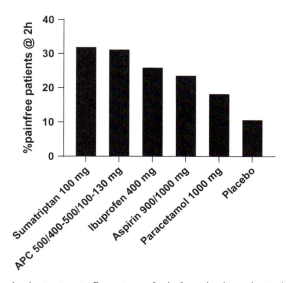

FIG. 4 Comparison of efficacy of various migraine treatments. Percentages of pain-free migraine patients with different acute treatments investigated in randomized clinical trials. The responder rate of APC came very close to that of the maximum used dose of sumatriptan. *Data from Petersen, K. U., Kops, M., & Heintze, K. (2017). WHO step 1 analgesics in the treatment of (migraine) headache.* Pharmakon, 1, 69–74.

Caffeine and migraine

In migraine, another mechanism of action adds to caffeine's co-analgesic effect. During a migraine attack, meningeal blood vessels are dilatated, which contributes to abnormal high activation of the sensory nerves innervating them and projecting to the trigeminal nucleus caudalis, which is involved in nociceptive processing (Just, Arndt, Weiser, & Doods, 2006). Triptans are agonists at peripheral serotonin 1B/D receptor, and their activation induces constriction of meningeal blood vessels, and thereby pain relief in migraine. Other compounds, like ergotamine, act similar (although ergotamine acts on other receptor systems as well; Dahlöf & Massen Van Den Brink, 2012). Caffeine itself also induces vasoconstriction (Blaha, Benes, Douville, & Newell, 2007), and caffeine-induced constriction of meningeal blood vessels will add another facet to its use as supplement to analgesics for treating migraine.

Caffeine and worsening of headache diseases?

Habitual caffeine intake can lead to the development of tolerance, i.e., the person consuming caffeine on a regular basis adapts to caffeine's pharmacological effects. Upon cessation of consumption, withdrawal symptoms might occur, with headache being one of them. This headache is defined as "headache developing within 24 hours after regular consumption of caffeine in excess of 200 mg/day for more than 2 weeks, which has been interrupted. It resolves spontaneously within 7 days in the absence of further consumption" (IHS CLASSIFICATION ICHD-3, n.d.). People affected by this headache either cope it with ingestion of caffeine-containing foods or beverages, or (in the longer run) will reduce their caffeine consumption. No reports in literature databases were identified which suggest that analgesics are used to treat this condition.

Dietary caffeine is discussed as a risk factor for the chronification of migraine. In a meta-analysis by Buse, Greisman, Baigi, and Lipton (2019), only one case-control study was identified which described a higher risk for chronification with high dietary intake of caffeine in the time period preceding the occurrence of chronic headache, and therefore the authors considered evidence for the risk to be relatively weak.

It has also been discussed whether the use of caffeine-containing analgesics might promote medication-overuse headache (i.e., worsening of a headache condition due to intake of acute pain treatments on more than 10–15 days per month for a period of more than 3 months; IHS CLASSIFICATION ICHD-3, n.d.). This has been ruled out by Bigal et al. (2008). Caffeine-containing analgesics did not show higher risks for the development of chronic migraine headache. Higher risks were only observed for analgesics containing barbiturates or opioids (i.e., with substances which can induce dependence by themselves; Fig. 5).

Thus, the hypothesis that caffeine-containing acute pain analgesics bear higher risks for headache chronification are not supported by scientific data.

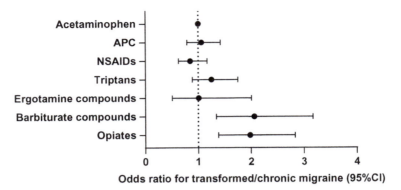

FIG. 5 Role of different analgesics for chronification of migraine. Association of medication used to treat migraine in 2005 and development of transformed (synonym with chronic) migraine in 2006. Paracetamol use was set as reference to 1. Use of medication not containing barbiturates or opioids did not show higher risks for chronification. *Data from Bigal, M. E., Serrano, D., Buse, D., Scher, A., Stewart, W. F., & Lipton, R. B. (2008). Acute migraine medications and evolution from episodic to chronic migraine: A longitudinal population-based study. Headache, 48, 1157–1158.*

Pain and functional impairment

Impairment of cognitive function and daily activities has been reported for, e.g., tension-type headache (Smith, 2016), or migraine (Farmer et al., 2001). Effective analgesia restored mental performance and the ability to perform daily activities.

Especially in migraine impairment of daily activities is prominent, and in consequence is an endpoint in many clinical migraine trials. Lipton et al. (1998) investigated the combination of aspirine/paracetamol/caffeine in three migraine trials, and reported little or no functional disability 2 h after onset of treatment in 59% of the verum, and only 34% in the placebo group. Qualitatively similar results were reported by Diener et al. (2005) for a slightly different aspirine/paracetamol/caffeine combination in a study investigating patients suffering from migraine or tension-type headache (Fig. 6).

Thus, successful analgesia will not only reduce acute pain, but in addition can restore function (including mental capabilities) to pre-pain levels.

Safety of caffeine and caffeine-containing analgesics
Effects attributable to caffeine

Many foods and beverages contain caffeine, and possible health risks and benefits have been discussed widely. Two thorough analyses on that topic were published within the last years: The EFSA came to the conclusion that in adults, lifelong intake of 400 mg caffeine per day is safe (200 mg per day in pregnant women; EFSA NDA Panel (EFSA

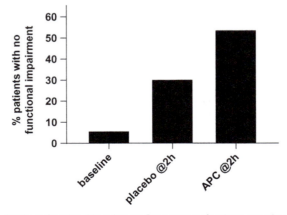

FIG. 6 Effect of a caffeine-containing analgesic on functional impairment. Percentages of patients reporting functional impairment before, and 2 h after begin of treatment with placebo or APC (500/400/100 mg) in a trial investigating tension-type and migraine headache patients. *Data from Diener, H. C., Pfaffenrath, V., Pageler, L., Peil, H., & Aicher B. (2005) The fixed combination of acetylsalicylic acid, paracetamol and caffeine is more effective than single substances and dual combination for the treatment of headache: A multicentre, randomized, double-blind, single-dose, placebo-controlled parallel group study. Cephalalgia, 25, 776–887.*

Panel on Dietetic Products, Nutrition and Allergies), 2015). A Canadian research team published an update of the assessment done by "Health Canada" and came to similar conclusions, namely that 400 mg per day are safe for adults, and 300 mg per day for pregnant women (Wikoff et al., 2017). Thus, the occasional intake of an acute pain analgesic containing 100–130 mg caffeine can be expected to be negligible in terms of safety. Postdural headache is usually treated with caffeine doses of 300–500 mg in a clinical setting and should therefore pose no special risk on the patients (Basurto et al., 2015).

Despite this relatively high safety, some patients are concerned because of special (seeming) risks or tolerability issues. Although cardiovascular safety has been shown in many investigations, and was subject to the above mentioned safety assessments, patients often fear that caffeine might pose a risk to health because of its effect on blood pressure. Indeed, caffeine doses of a few hundred milligrams can increase blood pressure, but this effect is relatively small and transient. Daniels, Molé, Shaffrath, and Stebbins (1998) observed an increase of systolic/diastolic blood pressure of about 10/5 mmHg after ingestion of caffeine doses of 6 mg/kg; and Papakonstantinou et al. (2016) measured increases of about 5/3 mmHg after ingestion of 2.3 mg/kg caffeine. A caffeine dose of 100 mg corresponds to about 1.3 mg/kg (Weiser et al., 2019) and can be expected to induce even smaller changes. On the other hand a short period of physical exercise (e.g., climbing the stairs to up to the third floor) raises systolic and diastolic blood pressure in sport students by up to 14/9 mmHg (Raschka, Müller-Nalbach, Rühl, & Koch, 2002). Thus, the blood pressure effects of light physical exercise exceed those of a caffeine-dose found in a caffeine-containing analgesic.

Sleep disturbance is another concern associated with caffeine-containing analgesics, and some patients fear that they will spend the night awake after taking a caffeine-containing analgesic in the evening. Clinical data, however, show a different picture: In a clinical efficacy trial on back pain, the fixed-dose combination of 400 mg ibuprofen and 100 mg caffeine was administered 3 times daily for 6 days (Predel et al., 2019). Of the 256 patients treated with this combination, only 2 reported sleep disorder as adverse event, i.e., less than 1% (with the same frequency in the placebo arm).

Effects attributable to the analgesic compound

Many WHO stage 1 analgesics can be used in self-medication (e.g., ibuprofen at daily doses of up to 1200 mg to treat acute pain for few days), as well as under a doctor's surveillance, where the same substance can be administered at daily doses of up to 2400 mg for the treatment of chronic pain for an extended period of time. Safety information from this usage will also be captured by the drug labels for the self-medication products, and therefore the perception of actual risks of such products might be biased.

To gain more information on the actual risks under conditions more relevant for self-medication, Moore et al. (2015) analyzed data from acute pain studies with close to 20,000 patients. This analysis showed that adverse event rates for the analgesics in general were not different from placebo, underpinning the safety and tolerability of these analgesics when used for treating acute pain events.

On the other hand, from the scientific point of view it appears reasonable to assume that the combination of two active ingredients has a higher risk to induce adverse events than a single compound alone. In the trials published by Weiser et al. (2018) and Predel et al. (2019), AE frequencies were higher for the combination ibuprofen/caffeine versus ibuprofen alone, however, the assessment of tolerability was not statistically different between the treatments (with 91%–93% of patients assessing tolerability as "good," "very good," or "excellent," irrespective of whether their medication contained caffeine or not). In both studies, only 1.4%, and 0.8%, respectively, of patients treated with the combination discontinued the trial due to adverse events. Thus, tolerability of caffeine-containing acute-pain analgesics appears not to be an issue for the patients.

Applications to other areas

The concept of combining pharmacological agents with different modes of action is not new, and was thoroughly described by the Swiss professor Emil Bürgi as early as 1938 (Bürgi, 1938). Interestingly, in his book he cited a cuneiform Babylonian inscription describing the combination approach dating back to 4000 years BC.

Based on today's scientific knowledge, combining compounds with different pharmacological effects make a lot of sense. As mentioned before, WHO stage 1 analgesics like ibuprofen or paracetamol have shallow dose-response curves, and their analgesic effects cannot be increased above a certain level by raising the dose. On the other hand, combinations of ibuprofen and paracetamol are stronger analgesics than each of the compounds alone (since both have different pharmacological modes of action; Mehlisch, Aspley, Daniels, Southerden, & Christensen, 2010). Combining analgesic compounds with caffeine follows the same rule. Thus, one interesting approach for development of new analgesics might be to search the databases for analgesic compounds with different modes of action, and check their potential for combination based on pharmacodynamic, pharmacokinetic, safety, and tolerability aspects. Such an approach could provide faster and stronger acting acute pain analgesics with even higher responder rates than the preparations known today—the clinical development, however, would be very demanding.

Other agents of interest

Caffeine's co-analgesic (and probably also its own analgesic) effects are mainly due to the blockade of adenosine receptors. Thus, other inhibitors of adenosine receptors might be promising co-analgesics as well.

Besides the naturally occurring adenosine receptor antagonists, synthetic compounds have been synthetized in the past, with often higher affinity and specificity for the different receptor subtypes (Jacobson & Müller, 2016). Such compounds might be even more promising combination partners for WHO stage 1 analgesics than caffeine.

Thus, caffeine is not only a useful compound, may it be as single agent or as co-analgesic, to provide better relief for patients suffering from pain. This compound might also be the starting point for the development of more effective analgesics in the future.

Key facts of caffeine

- has analgesic properties on its own, which are employed for the treatment of, e.g., post-dural puncture headache.
- (co-)analgesic effects can mostly be attributed to its inhibition of adenosine receptors.
- increases efficacy of acute pain analgesics, i.e., caffeine-containing combinations act faster, stronger, and in a higher number of patients than their counterparts without caffeine.
- in combination with aspirin and paracetamol was the first OTC drug to be registered for the treatment of migraine attacks in the United States.
- as co-analgesic helps to overcome the ceiling effect, e.g., of 400 mg ibuprofen for the treatment of acute pain.

Mini-dictionary of terms

Co-analgesic (or analgesic adjuvant). Compound without pronounced analgesic properties, which augments efficacy when combined with an analgesic.
Adenosine receptor. Class of receptors expressed in many different tissues, including nociceptive neurons, being activated by adenosine.
Tension-type headache. Most abundant primary headache. Pathophysiology not well understood, can in most cases successfully treated with WHO stage 1 analgesics.
Migraine. Primary headache disorder with complex neurological and vascular pathophysiology.

Summary points

- Caffeine is an inhibitor of adenosine receptors, and possesses analgesic efficacy in animal models, as well as in humans.
- Under fasted conditions, caffeine does not accelerate speed of absorption of ibuprofen, but does so under non-fasted conditions.
- In head-to-head comparison efficacy trials, caffeine-containing analgesic acted faster, stronger, and in higher numbers of patients then analgesics without caffeine.
- Caffeine-containing analgesics' efficacy comes close to prescription-only therapy (e.g., sumatriptan) in acute migraine.
- In headache and migraine trials, caffeine-containing analgesics improved functional outcomes, e.g., return to daily activities.
- Epidemiological studies showed no increased risk for caffeine-containing analgesics for the chronification of headache disorders, demonstrating that the risk for development of medication-overuse headache is not higher than for analgesics without caffeine.
- Caffeine-containing analgesics for the treatment of acute pain show high safety and tolerability (like their counterparts without caffeine).

References

Arnaud, M. J. (2011). Pharmacokinetics and metabolism of natural methylxanthines in animal and man. In B. B. Fredholm (Ed.), *Methylxanthines* (pp. 35–93). Heidelberg, Dordrecht, London, New York: Springer.

Basurto, O. X., Osorio, D., & Bonfill, C. X. (2015). Drug therapy for treating post-dural puncture headache. *Cochrane Database of Systematic Reviews, 7*, CD007887.

Bigal, M. E., Serrano, D., Buse, D., Scher, A., Stewart, W. F., & Lipton, R. B. (2008). Acute migraine medications and evolution from episodic to chronic migraine: A longitudinal population-based study. *Headache, 48*, 1157–1158.

Blaha, M., Benes, V., Douville, C. M., & Newell, D. W. (2007). The effect of caffeine on dilated cerebral circulation and on diagnostic CO_2 reactivity testing. *Journal of Clinical Neuroscience, 14*, 464–467.

Bürgi, E. (1938). *Die Arzneikombinationen*. Berlin: Verlag von Julius Springer.

Buse, D. C., Greisman, J. D., Baigi, K., & Lipton, R. B. (2019). Migraine progression: A systematic review. *Headache, 59*, 306–338.

CHMP. (2010). *Guideline on the investigation of bioequivalence.* https://www.ema.europa.eu/en/investigation-bioequivalence#current-effective-version-section. (Accessed 1 July 2020).

Dahlöf, C., & Massen Van Den Brink, A. (2012). Dihydroergotamine, ergotamine, methysergide and sumatriptan – Basic science in relation to migraine treatment. *Headache, 52*, 707–714.

Daniels, J. W., Molé, P. A., Shaffrath, J. D., & Stebbins, C. L. (1998). Effects of caffeine on blood pressure, heart rate, and forearm blood flow during dynamic leg exercise. *Journal of Applied Physiology, 85*, 154–159.

Derry, C. J., Derry, S., & Moore, R. A. (2012). Caffeine as an analgesic adjuvant for acute pain in adults. *Cochrane Database of Systematic Reviews*, CD009281.

Diamond, S., Balm, T. K., & Freitag, F. G. (2000). Ibuprofen plus caffeine in the treatment of tension-type headache. *Clinical Pharmacology and Therapeutics, 68*, 312–319.

Diener, H. C., Pfaffenrath, V., Pageler, L., Peil, H., & Aicher, B. (2005). The fixed combination of acetylsalicylic acid, paracetamol and caffeine is more effective than single substances and dual combination for the treatment of headache: A multicentre, randomized, double-blind, single-dose, placebo-controlled parallel group study. *Cephalalgia, 25*, 776–887.

EFSA NDA Panel (EFSA Panel on Dietetic Products, Nutrition and Allergies). (2015). Scientific opinion on the safety of caffeine. *EFSA Journal, 13* (4102), 1–120.

Farmer, K., Cady, R., Bleiberg, J., Reeves, D., Putnam, G., O'Quinn, S., et al. (2001). Sumatriptan nasal spray and cognitive function during migraine: Results of an open-label study. *Headache, 41*, 377–384.

Goldstein, J., Silberstein, S. D., Saper, J. R., Elkind, A. H., Smith, T. R., Gallagher, R. M., et al. (2005). Acetaminophen, aspirin, and caffeine versus sumatriptan succinate in the early treatment of migraine: Results from the ASSET trial. *Headache, 45*, 973–982.

Goldstein, J., Silberstein, S. D., Saper, J. R., Ryan, R. E., Jr., & Lipton, R. B. (2006). Acetaminophen, aspirin, and caffeine in combination versus ibuprofen for acute migraine: Results from a multicenter, double-blind, randomized, parallel-group, single-dose, placebo-controlled study. *Headache, 46*, 444–453.

Heintze, K., & Fuchs, W. (2015). Effects of food on pharmacokinetics of immediate release oral formulations. *British Journal of Clinical Pharmacology, 80*, 1239.

IHS CLASSIFICATION ICHD-3. https://ichd-3.org/. (Accessed 1 July 2020).

Jacobson, K. A., & Müller, C. E. (2016). Medicinal chemistry of adenosine, P2Y and P2X receptors. *Neuropharmacology, 104*, 31–49.

Just, S., Arndt, K., Weiser, T., & Doods, H. (2006). Pathophysioly of migraine: A role for neuropeptides. *Drug Discovery Today, 3*, 327–333.

Kellstein, D. E., Lipton, R. B., Geetha, R., Koronkiewicz, K., Evans, F. T., Stewart, W. F., et al. (2000). Evaluation of a novel solubilized formulation of ibuprofen in the treatment of migraine headache: A randomized, double-blind, placebo-controlled, dose-ranging study. *Cephalalgia, 20*, 233–243.

Kyselovič, J., Koscova, E., Lampert, A., & Weiser, T. (2020). A randomized, double-blind, placebo-controlled trial of ibuprofen lysinate in comparison to ibuprofen acid for acute postoperative dental pain. *Pain and therapy, 9*, 249–259.

Laska, E. M., Sunshine, A., Marrero, I., Olson, N., Siegel, C., & McCormick, N. (1986). The correlation between blood levels of ibuprofen and clinical analgesic response. *Clinical Pharmacology and Therapeutics, 40*, 1–7.

Lipton, R. B., Hamelsky, S. W., & Dayno, J. M. (2002). What do patients with migraine want from acute migraine treatment? *Headache, 42*(Suppl 1), S3–S9.

Lipton, R. B., Stewart, W. F., Ryan, R. E., Jr., Saper, J., Silberstein, S., & Sheftell, F. (1998). Efficacy and safety of acetaminophen, aspirin, and caffeine in alleviating migraine headache pain: Three double-blind, randomized, placebo-controlled trials. *Archives of Neurology, 55*, 210–217.

Mehlisch, D. R., Aspley, S., Daniels, S. E., Southerden, K. A., & Christensen, K. S. (2010). A single-tablet fixed-dose combination of racemic ibuprofen/paracetamol in the management of moderate to severe postoperative dental pain in adult and adolescent patients: A multicenter, two-stage, randomized, double-blind, parallel-group, placebo-controlled, factorial study. *Clinical Therapeutics, 32*, 1033–1049.

Moore, R. A., Derry, S., Straube, S., Ireson-Paine, J., & Wiffen, P. J. (2014). Faster, higher, stronger? Evidence for formulation on efficacy for ibuprofen in acute pain. *Pain, 155*, 14–21.

Moore, R. A., Wiffen, P. J., Derry, S., Maguire, T., Roy, Y. M., & Tyrrell, L. (2015). Non-prescription (OTC) oral analgesics for acute pain—An overview of cochrane reviews. *Cochrane Database of Systematic Reviews, 11*, CD010794.

Papakonstantinou, E., Kechribari, I., Sotirakoglou, K., Tarantilis, P., Gourdomichali, T., Michas, G., et al. (2016). Acute effects of coffee consumption on self-reported gastrointestinal symptoms, blood pressure and stress indices in healthy individuals. *Nutrition Journal, 15*, 26.

Petersen, K. U., Kops, M., & Heintze, K. (2017). WHO step 1 analgesics in the treatment of (migraine) headache. *Pharmakon, 1*, 69–74.

Predel, H. G., Ebel-Bitoun, C., Lange, R., & Weiser, T. (2019). A randomized, placebo- and active-controlled, multi-country, multi-center parallel group trial to evaluate the efficacy and safety of a fixed-dose combination of 400 mg ibuprofen and 100 mg caffeine compared with ibuprofen 400 mg and placebo in patients with acute lower back or neck pain. *Journal of Pain Research, 12*, 2771–2783.

Radtke, A., & Neuhauser, H. (2009). Prevalence and burden of headache and migraine in Germany. *Headache, 49*, 79–89.

Raschka, C., Müller-Nalbach, M., Rühl, T., & Koch, H. J. (2002). Circadian rhythm of heart rate, blood pressure and lactate during stair climbing with different external weight loads. *Schweizerische Zeitschrift für Sportmedizin und Sporttraumatologie, 50*, 151–154.

Sawynok, J. (2011). Methylxanthines and pain. In B. B. Fredholm (Ed.), *Methylxanthines* (pp. 312–331). Heidelberg, Dordrecht, London, New York: Springer.

Seymour, R. A., Ward-Booth, P., & Kelly, P. J. (1996). Evaluation of different doses of soluble ibuprofen and ibuprofen tablets in postoperative dental pain. *The British Journal of Oral & Maxillofacial Surgery, 34*, 110–114.

Silva-Néto, R. P., Santos, P. E. M. S., & Peres, M. F. P. (2019). Hypnic headache: A review of 348 cases published from 1988 to 2018. *Journal of the Neurological Sciences, 15*, 103–109.

Smith, P. A. (2016). Acute tension-type headaches are associated with impaired cognitive function and more negative mood. *Frontiers in Neurology, 7*, 42.

Suh, S. Y., Choi, Y. S., Oh, S. C., Kim, Y. S., Cho, K., Bae, W. K., et al. (2013). Caffeine as an adjuvant therapy to opioids in cancer pain: A randomized, double-blind, placebo-controlled trial. *Journal of Pain and Symptom Management, 46*, 474–482.

Weiser, T., Richter, E., Hegewisch, A., Muse, D. D., & Lange, R. (2018). Efficacy and safety of a fixed-dose combination of ibuprofen and caffeine in the management of moderate to severe dental pain after third molar extraction. *European Journal of Pain, 22*, 28–38.

Weiser, T., Schepers, C., Mück, T., & Lange, R. (2019). Pharmacokinetic properties of ibuprofen (IBU) from the fixed-dose combination IBU/caffeine (400/100 mg; FDC) in comparison with 400 mg IBU as acid or lysinate under fasted and fed conditions—Data from 2 single-center, single-dose, randomized crossover studies in healthy volunteers. *Clinical Pharmacology in Drug Development, 8*, 742–753.

Weiser, T., & Weigmann, H. (2019). Effect of caffeine on the bioavailability and pharmacokinetics of an acetylsalicylic acid-paracetamol combination: Results of a phase I study. *Advances in Therapy, 36*, 597–607.

Wikoff, D., Welsh, B. T., Henderson, R., Brorby, G. P., Britt, J., Myers, E., et al. (2017). Systematic review of the potential adverse effects of caffeine consumption in healthy adults, pregnant women, adolescents, and children. *Food and Chemical Toxicology, 109*, 585–648.

Chapter 8

Chloroprocaine: Features and applications

Barbara Rupnik and Alain Borgeat

Department of Anaesthesiology, Balgrist University Hospital, Zurich, Switzerland

Abbreviations

2-CHP	2-chloroprocaine
CSF	cerebrospinal fluid
EDTA	ethylenediaminetetraacetic acid
LA	local anesthetic
NSAID	nonsteroidal anti-inflammatory drug
TNS	transient neurological symptoms
VGSC	voltage-gated sodium channel

Introduction

History and controversies

2-Chloroprocaine was first introduced into clinical practice in 1950s and was widely employed due to its favorable pharmacological profile (Ravindran, Bond, Tasch, Gupta, & Luerssen, 1980). However, in the beginning of the 1980s case reports emerged, describing various neurological sequelae after neuraxial anesthesia where 2% or 3% 2-chloroprocaine was used. In all reported cases either a lumbar epidural or a caudal anesthesia was attempted. Inadvertent intrathecal injection of a large dose of 2-chloroprocaine was suspected in the majority of the cases but not in all. Symptoms described were ranging from lower extremity motor weakness, impaired temperature sensation and tactile discrimination with retained vibration, pressure sensation and proprioception in the lumbar and sacral dermatomes, and the inability to void in the immediate postanesthetic phase to adhesive arachnoiditis, clinically presenting as an incomplete cauda equina syndrome, weeks after the performed neuraxial anesthetic with 3% 2-chloroprocaine (Ravindran et al., 1980; Reisner, Hochman, & Plumer, 1980). Persisting fecal and urinary incontinence, perineal numbness, no sensation during sexual intercourse, positive Babinski reflex and dysesthesia of the lower extremities were reported a year or more after the neuraxial anesthesia performed with 2-chloroprocaine (Moore, Spierdijk, van Kleef, Coleman, & Love, 1982). As possible reasons for these complications a direct neurotoxicity of large doses of 2-chloroprocaine intrathecally and the sodium bisulfite, antioxidant added to the formulations of 2-chloroprocaine, were discussed (Moore et al., 1982; Ravindran et al., 1980; Reisner et al., 1980)-Following these reports 2-chloroprocaine was withdrawn from the market in the early 1980s (Moore et al., 1982).

A series of volunteer studies was performed in 2004, where a low dose of preservative- and antioxidant-free 2-chloroprocaine was injected intrathecally. In this series the incidence of TNS was very low and no permanent neurological injuries were reported. Thereafter, 2-chloroprocaine was gradually reintroduced into clinical practice of neuraxial as well as peripheral regional anesthesia.

Pharmacology overview

2-chloroprocaine belongs to an ester class of LA with a $p\mathrm{K_a}$ of 8.97 (Rebel & Schell, 2015). It shares a basic structure, common to all local anesthetic molecules, consisting of a lipophilic aromatic ring, intermediate chain (in the case of 2-CHP an ester bond), and terminal amine (Becker & Reed, 2006). 2-Chloroprocaine has a fast onset of action, with sensory and motor block onset time of 4.3 ± 2.4 and 7.1 ± 3.7 min, respectively, in peripheral nerve blocks (Saporito, Anselmi, Borgeat, & Aguirre, 2016). Onset time for spinal anesthesia is 8–10 min, depending on the dose administered. Duration of action for peripheral nerve block is 105 ± 26 and 91 ± 25 min for sensory and motor blocks, respectively, when 30 mL of

---

Treatments, Mechanisms, and Adverse Reactions of Anesthetics and Analgesics. https://doi.org/10.1016/B978-0-12-820237-1.00008-9
Copyright © 2022 Elsevier Inc. All rights reserved.

3% 2-CHP are applied (Saporito et al., 2016). Duration of spinal anesthesia is dose-dependent, with 98 ± 20 and 132 ± 23 min to complete sensory regression for 30 and 60 mg of applied 2-CHP, respectively (Smith, Kopacz, & McDonald, 2004).

Structure and properties

2-Chloroprocaine hydrochloride (benzoic acid, 4-amino-2-chloro-2-(diethylamino) ethyl ester, monohydrochloride; $C_{13}H_{20}Cl_2N_2O_2$) is a salt of chloroprocaine with a molecular weight of 307.21 g/mol (Fig. 1). It is a derivative of procaine (Gonter & Kopacz, 2005). Its melting point is 177 °C. It is soluble in water (Foldes & Mc, 1952). At 25 °C the pH of preservative- and antioxidant-free 2% 2-chloroprocaine (2-CHP) is 3.34 and pH of the 3% 2-CHP is 3.31. pH of commercially available solutions is adjusted between 2.7 and 4.0 with the addition of sodium hydroxide or hydrochloric acid.

2-CHP is available either preservative-free or with the addition of EDTA and methylparaben. Sodium bicarbonate (NaHCO$_3$) can alkalize the solution of 2-CHP to physiological range (pH 7.4). A hypobaric solution of 2-CHP can be made by a 1:1 dilution with sterile water. Plain 2% and 3% 2-CHP are hyperbaric relative to CSF at 37 °C. A solution with added NaHCO$_3$ is denser relative to CSF than the plain solution of 2-CHP, whereas the addition of epinephrine does not change the density of the plain 2-CHP. When small volumes of NaHCO$_3$ or epinephrine were added, no precipitation of the solution was observed (Na & Kopacz, 2004).

Pharmacokinetics of 2-chloroprocaine

2-Chloroprocaine is metabolized in plasma through ester linkage hydrolysis by the enzymes pseudocholinesterase and liver esterase to two inactive metabolites 2-chloro-para-aminobenzoic acid (CABA) and diethylaminoethanol (DEAE). CABA undergoes acetylation to N-acetyl-CABA (Krohg & Jellum, 1981). The mean in vitro half-life ($t_{1/2}$) of 2-CHP is 21 and 25 s, for men and women, respectively. In vivo $t_{1/2}$, measured after epidural application, is 3.1 min (Kuhnert et al., 1986). The highest tissue concentrations are found in liver, lungs, heart, and brain. Metabolites are excreted in urine (Krohg & Jellum, 1981; Kuhnert, Kuhnert, & Reese, 1982). As the spinal fluid contains no pseudocholinesterase, metabolism and elimination of 2-CHP depends on the diffusion to the epidural space and vascular absorption.

Mechanism of action

Local anesthetics produce a sensory and motor blockade by binding to voltage-gated sodium channels (VGSC), which causes a reversible and concentration dependent reduction in peak inward sodium current, attenuating the action potential and blocking its conduction along the axon (Catterall, 2012; Shah, Votta-Velis, & Borgeat, 2018) The action of LA is determined by their pK_a value, degree of lipophilicity, intrinsic vasoactivity, and degree of protein binding (Lirk, Picardi, & Hollmann, 2014). LA with pK_a values close to the physiological pH have a shorter onset of action, with the exception of 2-chloroprocaine, which has the highest pK_a of all local anesthetics and the shortest onset of action (Table 1). High concentration of 2-CHP formulations and consequently a high concentration gradient across the lipid bilayer has been proposed as mechanism for the short onset time seen in 2-CHP nerve blocks (Rebel & Schell, 2015).

Lipophilicity is related to potency of local anesthetics as this determines its ability to diffuse across lipid membranes. 2-Chloroprocaine has a low lipid solubility and thereby a low potency. Intrinsic vasoconstrictive properties, observed in amide-linked LA, appear to prolong the duration of LA action. Ester-linked LA, however, have been shown to have intrinsic vasodilatative properties in vitro (Willatts & Reynolds, 1985). The degree of protein binding of 2-CHP, which also determines the duration of action, is not known (Rebel & Schell, 2015).

In vitro studies showed a bacteriostatic effect on cultures of *E. coli*, *S. aureus*, and Enterobacter Aerogenes (Foldes & Mc, 1952).

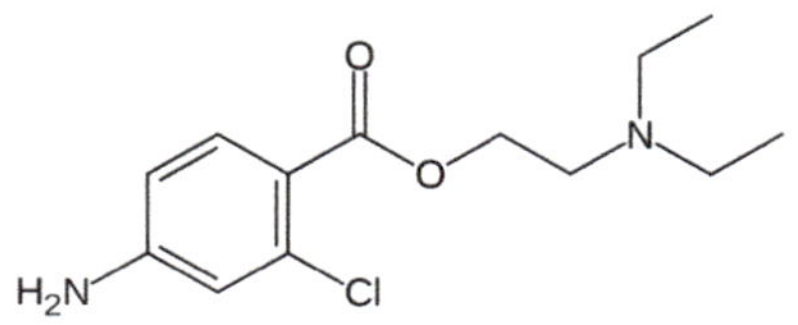

FIG. 1 Chemical structure of chloroprocaine 2%.

TABLE 1 Comparative pharmacology of local anesthetics.

Classification and compounds	pK_a	% nonionized at pH 7.4
Esters		
Procaine	8.9	3
Chloroprocaine	8.7	5
Tetracaine	3.5	7
Amides		
Lidocaine	7.9	24
Mepivacaine	7.6	39
Prilocaine	7.9	24
Bupivacaine	8.1	17
Levobupivacaine		
Ropivacaine	8.1	17

Clinical application

In various European countries and recently in the USA a preservative-free formulation of 2-CHP 1% has been approved for intrathecal use for surgery expected to last less than 40 min (Goldblum & Atchabahian, 2013). 2-Chloroprocain is also used in peripheral nerve blocks for surgeries of short duration.

With its short onset of action (5–10 min) and fast recovery (70–150 min) when used for spinal anesthesia, 2-chloroprocaine has an advantageous pharmacodynamic profile in day case surgery (Goldblum & Atchabahian, 2013; Pollock, 2012). Duration of action of spinal anesthesia depends on the dose of 2-chloroprocaine applied. For a surgery lasting 40–60 min, 30 mg of plain 2-chloroprocaine has been suggested, 40–45 mg for surgery up 70 min and 60 mg for a 90-min procedure (Goldblum & Atchabahian, 2013). Kouri et al. reported that 40 mg of 2-chloroprocaine showed a reliable spinal anesthesia lasting 60 min and discharge time of 120 min (Kouri & Kopacz, 2004). The addition of fentanyl appears to prolong the surgical block and does not seem to decrease time to discharge (Vath & Kopacz, 2004).

Compared to low dose bupivacaine used for day case surgery, 2-chloroprocaine has been shown to have a more predictable recovery profile. In addition bupivacaine showed an excessive primary block failure in the doses used for ambulatory surgery, making it an unreliable choice for this type of surgery (Valanne, Korhonen, Jokela, Ravaska, & Korttila, 2001; Nair, Abrishami, Lermitte, & Chung, 2009). A meta-analysis done by our group showed a significant reduction in motor and sensory block regression with 2-chloroprocaine compared to the low-dose bupivacaine, which translates to a shorter time to ambulation with a possibly faster discharge leading to a favorable cost profile, an important aspect in day case surgery (Table 2).

2-Chloroprocaine seems to have an advantageous safety profile in terms of most common complications after spinal anesthesia when compared with other local anesthetics employed for surgery of short duration. Urinary retention and spontaneous voiding play an important role in day case surgery. Bupivacaine as well as lidocaine have been known to cause bladder overdistension with postoperative urinary retention. In a study comparing intrathecal 2-chloroprocaine and lidocaine, problems with micturition requiring a bladder catheterization were reported only in the lidocaine group. There was no evidence for postoperative urinary retention where 2-chloroprocaine was used (Breebaart, Teune, Sermeus, & Vercauteren, 2014).

3% chloroprocaine has been used for peripheral nerve blocks in day-case surgery. When compared to 1.5% mepivacaine in popliteal sciatic block, 3% chloroprocaine showed a significantly shorter onset time and block duration. Sensory onset time when 90 mg of 3% chloroprocaine were used was 4–6 min and motor block onset time was 7–10 min. Duration of motor block was 90–115 min (Saporito et al., 2016). Its favorable pharmacodynamic profile might lead to a reduced anesthesia-controlled time, shorter recovery room length of stay, translating into reduced perioperative costs (Gonano et al., 2009; Saporito et al., 2016).

In a study on volunteers Marica et al. have studied pain on intradermal injection, comparing chloroprocaine combined with sodium bicarbonate, lidocaine alone, lidocaine combined with sodium bicarbonate and normal saline combined with

TABLE 2 Details of studies comparing low dose bupivacaine to chloroprocaine 2%.

Articles	Modified Jadad score	Drugs used	Setting	Primary outcome	Secondary outcomes urinary and TNS details
Lacasse et al. (2011)	12	40 mg plain 2% CP vs. 7.5 mg hyperbaric 0.75% bupivacaine	Randomized double blind study	Time until discharge	Peak block height (+time to reach it), duration of the sensory (regression to light touch) and motor blocks (Bromage's score), the length of stay in the PACU, the time until ambulation, and until micturition
			General/ urological and gynecologic surgery <60 min		FU: check for side effects after 24 h and 7 days Urinary: time to spontaneous voiding = discharge criteria. No urinary retention
					TNS: one case per groups developed TNS at 24 h
Maes, Laubach, and Poelart (2016)	11	40 mg 2% 2-CP vs 40 mg 2% 2-CP with 1 μg sufentanil vs. 7.5 mg hyperbaric 0.5% bupivacaine with 1 μg sufentanil	Elective low-risk cesarean section in healthy parturients	Complete regression of motor blockade (modified Bromage scale = 1)	Sensory block, maternal blood pressure, heart rate, SaO_2, presence of nausea and vomiting, pain, Apgar score, baby admission to NICU
			Randomized, single blind, controlled study		No FU details, no urinary or TNS details reported
					CEA started after complete motor blockade resolution
Yoos and Kopacz (2005)	11	40 mg 2% 2-CP vs 7.5 mg hyperbaric 0.5% bupivacaine	Randomized, double blind, crossover, volunteer study	Complete sensory resolution	Regression of motor block (Bromage's score), simulated discharge criteria: block resolution (down to S2), time to voiding, time to ambulation, residual bladder volume
			Pain assessment: tourniquet and electrical stimulation		FU: for 72 h and 6 m: Check for side effects
					Urinary<: bladder US to all; spontaneous voiding = discharge criteria. After voiding again bladder US
					No urinary retentions
					TNS: no TNS at 72 h/6 m
Teunkens et al. (2016)	13	40 mg plain 1% 2-CP, 40 mg 1% plain lidocaine, or 7.5 mg 0.5% plain bupivacaine	Prospective, double blind, randomized controlled trial	Time to complete recovery of sensory block (return to S5)	Time to recovery from motor block (Bromage score), failure rates, incidence of hypotension/bradycardia, postoperative pain, first mobilization, voiding and discharge times, and the incidence of TNS
			Outpatient knee arthroscopy		FU: 24 h after surgery (daily phone in the case of TNS until problem resolved)
					Urinary: voiding time registered, not discharge criteria. If retention was suspected: US and if ≤500 mL: single catheterization
					TNS: no TNS reported

2-CP, 2-chloroprocaine; *FU*, follow up; *h*, hours; *m*, months; *NICO*, neonatal intensive care unit; *TNS*, transient neurologic symptoms.

sodium bicarbonate. Chloroprocaine alone or combined with sodium bicarbonate has shown least pain on injection (Marica, O'Day, Janosky, & Nystrom, 2002). There was no difference in pain scores between these two solutions, confirming findings of previous studies that have postulated than pH alone cannot explain pain felt on intradermal injection (Marica et al., 2002; Fitton, Ragbir, & Milling, 1996; Prien, 1994; Skidmore, Patterson, & Tomsick, 1996). Chloroprocaine can be considered as an alternative to lidocaine for local infiltration anesthesia.

Adverse effects

Compared to other short acting local anesthetics used for spinal anesthesia, 2-chloroprocaine has a lower incidence of transient neurological symptoms (TNS) and urinary retention (Boublik, Gupta, Bhar, & Atchabahian, 2016; Goldblum & Atchabahian, 2013; Manassero & Fanelli, 2017). In a study analyzing equal doses of 2-chloroprocaine and lidocaine intrathecally in patients undergoing knee arthroscopy, Casati et al. found no TNS in the 2-chloroprocaine group compared to five patients in the lidocaine group (Casati, Fanelli, Danelli, et al., 2007). Another study comparing lidocaine and 2-chloroprocaine for spinal anesthesia in short urological procedures reported four patients with TNS in the lidocaine group, whereas none were reported in the 2-chloroprocaine group. However, this study describes the only transient incomplete cauda equina syndrome reported in a patient receiving intrathecal 2-chloroprocaine since its reintroduction in 2004. The exact etiology of the cauda equina symptoms in this patient remains unclear as the patient refused further examination. In a retrospective study reviewing 601 spinal anesthetics performed with chloroprocaine no TNS were reported (Hejtmanek & Pollock, 2011). An author of this study reported four TNS in more than 4000 spinal anesthetics performed with 2-chloroprocaine between 2004 and 2012 (Goldblum & Atchabahian, 2013).

In a retrospective review of 122 patients nausea necessitating treatment with ondansetron has been reported in 11 patients receiving intrathecal 2-chloroprocaine, with or without the addition of intrathecal fentanyl. Two patients receiving 2-chloroprocaine with fentanyl intrathecally have complained of pruritus (Yoos & Kopacz, 2005).

Rebound pain has been described after spinal anesthesia as well as after single shot peripheral nerve blocks and continuous infusions of local anesthetics. It is defined as severe burning or dull aching pain occurring 8–24 h after the block wears off, lasting approximately 2 h and typically not responding adequately to opiate analgesia (Henningsen, Sort, Moller, & Herling, 2018; Lavand'homme, 2018; Williams, Bottegal, Kentor, Irrgang, & Williams, 2007; Williams, Ibinson, Mangione, et al., 2015). Pathophysiology of rebound pain is poorly understood and is beyond the scope of this chapter. We have noticed this phenomenon in our study, where a single shot popliteal sciatic block with 2-chloroprocaine 3% was performed for minor ambulant foot surgery (Saporito et al., 2016). A favorable postoperative pain outcome was achieved with a multimodal pain therapy beginning intraoperatively with the application of paracetamol and NSAID, followed by tramadol at the end of surgery.

Since preservative free formulation has been introduced in 2004 only one case of TNS has been described. We have found no reports of long-term and lasting adverse effects of perineurally or intrathecally applied chloroprocaine.

Chloroprocaine 2%: Suggested application

- **Ambulatory surgery:** Any type of surgery lasting less than 1 h fulfilling the criteria for ambulatory surgery.
- **Infiltration:** Due to the lack of pain during application. This drug can be used for procedure of short duration (e.g., esthetic surgery).
- **C-section:** May be an alternative due to its very short onset of action.

Mini-dictionary of terms

TNS. Pain in the buttocks, and lower extremities after spinal anesthesia, starting a few hours after the anesthetic and lasting from a few hours up to 10 days.

Day-case surgery. Synonymous with day-case or outpatient surgery, is surgery that does not require an overnight hospital stay.

Baricity of local anesthetics. The ratio of the density of local anesthetic and CSF at 37 °C.

Half-life ($t_{1/2}$) of a drug. The estimated time it takes for the maximum concentration of a drug to be reduced to 50% of the maximal concentration.

pK_a. Negative base-10 logarithm of the acid dissociation constant (K_a) of a solution. It indicates the strength of an acid, i.e., the lower the pK_a value, the stronger the acid.

Key facts of chloroprocaine: Features and applications

- 2-Chloroprocaine is an ester class local anesthetic with a short onset and short duration of action.
- After a series of case reports about neurological complications after neuraxial anesthesia performed with chloroprocaine, it has largely disappeared from clinical use.
- Neurological complications had been ascribed to a direct neurotoxicity of large doses of 2-chloroprocaine intrathecally and the sodium bisulfite, antioxidant added to the formulations of 2-chloroprocaine.
- In 2004 a series of volunteer studies were performed with preservative- and antioxidant-free 2-chloroprocaine injected intrathecally where a very low incidence of TNS and no permanent neurological complications were reported.
- In the recent years the popularity of chloroprocaine has increased especially for spinal anesthesia in ambulatory surgery.
- Since its reintroduction into clinical practice only one case of TNS has been described.
- Compared to lidocaine for spinal anesthesia, chloroprocaine has a lower incidence of TNS and urinary retention.
- Compared to low-dose bupivacaine for spinal anesthesia, chloroprocaine has shown a more reliable block characteristics and shorter recovery time.

Summary points

- Chloroprocaine is an ester class local anesthetic with a short onset and duration of action.
- Since the introduction of preservative- and antioxidant-free formulation of chloroprocaine one case of TNS and no permanent neurological sequelae have been described.
- Chloroprocaine has been used in spinal anesthesia for surgery up to 90 min duration.
- Chloroprocaine has been used in peripheral nerve blocks for short ambulatory surgery.
- Chloroprocaine has a low incidence of TNS and urinary retention.
- Due to its short recovery time chloroprocaine is advantageous in spinal anesthesia for ambulatory surgery.

References

Becker, D. E., & Reed, K. L. (2006). Essentials of local anesthetic pharmacology. *Anesthesia Progress, 53*, 98–108 (quiz 109–110).

Boublik, J., Gupta, R., Bhar, S., & Atchabahian, A. (2016). Prilocaine spinal anesthesia for ambulatory surgery: A review of the available studies. *Anaesthesia Critical Care & Pain Medicine, 35*, 417–421.

Breebaart, M. B., Teune, A., Sermeus, L. A., & Vercauteren, M. P. (2014). Intrathecal chloroprocaine vs. lidocaine in day-case surgery: Recovery, discharge and effect of pre-hydration on micturition. *Acta Anaesthesiologica Scandinavica, 58*, 206–213.

Casati, A., Fanelli, G., Danelli, G., et al. (2007). Spinal anesthesia with lidocaine or preservative-free 2-chlorprocaine for outpatient knee arthroscopy: A prospective, randomized, double-blind comparison. *Anesthesia and Analgesia, 104*, 959–964.

Catterall, W. A. (2012). Voltage-gated sodium channels at 60: Structure, function and pathophysiology. *The Journal of Physiology, 590*, 2577–2589.

Fitton, A. R., Ragbir, M., & Milling, M. A. (1996). The use of pH adjusted lignocaine in controlling operative pain in the day surgery unit: A prospective, randomised trial. *British Journal of Plastic Surgery, 49*, 404–408.

Foldes, F. F., & Mc, N. P. (1952). 2-Chloroprocaine: A new local anesthetic agent. *Anesthesiology, 13*, 287–296.

Goldblum, E., & Atchabahian, A. (2013). The use of 2-chloroprocaine for spinal anaesthesia. *Acta Anaesthesiologica Scandinavica, 57*, 545–552.

Gonano, C., Kettner, S. C., Ernstbrunner, M., Schebesta, K., Chiari, A., & Marhofer, P. (2009). Comparison of economical aspects of interscalene brachial plexus blockade and general anaesthesia for arthroscopic shoulder surgery. *British Journal of Anaesthesia, 103*, 428–433.

Gonter, A. F., & Kopacz, D. J. (2005). Spinal 2-chloroprocaine: A comparison with procaine in volunteers. *Anesthesia and Analgesia, 100*, 573–579.

Hejtmanek, M. R., & Pollock, J. E. (2011). Chloroprocaine for spinal anesthesia: A retrospective analysis. *Acta Anaesthesiologica Scandinavica, 55*, 267–272.

Henningsen, M. J., Sort, R., Moller, A. M., & Herling, S. F. (2018). Peripheral nerve block in ankle fracture surgery: A qualitative study of patients' experiences. *Anaesthesia, 73*, 49–58.

Kouri, M. E., & Kopacz, D. J. (2004). Spinal 2-chloroprocaine: A comparison with lidocaine in volunteers. *Anesthesia and Analgesia, 98*, 75–80. table of contents.

Krohg, K., & Jellum, E. (1981). Urinary metabolites of chloroprocaine studied by combined gas chromatography–mass spectrometry. *Anesthesiology, 54*, 329–332.

Kuhnert, B. R., Kuhnert, P. M., Philipson, E. H., Syracuse, C. D., Kaine, C. J., & Yun, C. H. (1986). The half-life of 2-chloroprocaine. *Anesthesia and Analgesia, 65*, 273–278.

Kuhnert, B. R., Kuhnert, P. M., & Reese, A. L. (1982). Urinary metabolites of chloroprocaine. *Anesthesiology, 56*, 483–484.

Lacasse, M. A., Roy, J.-D., Forget, J., Vandenbroucke, F., Seal, R. F., Beaulieu, D., … Massicotte, L. (2011). Comparison of bupivacaine and 2-chloroprocaine for spinal anesthesia for outpatient surgery: A double-blind randomized trial. *Canadian Journal of Anesthesia, 58*, 384–391.

Lavand'homme, P. (2018). Rebound pain after regional anesthesia in the ambulatory patient. *Current Opinion in Anaesthesiology, 31*, 679–684.

Lirk, P., Picardi, S., & Hollmann, M. W. (2014). Local anaesthetics: 10 essentials. *European Journal of Anaesthesiology, 31*, 575–585.

Maes, S., Laubach, M., & Poelart, J. (2016). Randomised controlled trial of spinal anaesthesia with bupivacaine or 2-chloroprocaine during caesarean section. *Acta Anaesthesiologica Scandinavica, 60*, 642–649.

Manassero, A., & Fanelli, A. (2017). Prilocaine hydrochloride 2% hyperbaric solution for intrathecal injection: A clinical review. *Local and Regional Anesthesia, 10*, 15–24.

Marica, L. S., O'Day, T., Janosky, J. E., & Nystrom, E. U. (2002). Chloroprocaine is less painful than lidocaine for skin infiltration anesthesia. *Anesthesia and Analgesia, 94*, 351–354. table of contents.

Moore, D. C., Spierdijk, J., van Kleef, J. D., Coleman, R. L., & Love, G. F. (1982). Chloroprocaine neurotoxicity: Four additional cases. *Anesthesia and Analgesia, 61*, 155–159.

Na, K. B., & Kopacz, D. J. (2004). Spinal chloroprocaine solutions: Density at 37 degrees C and pH titration. *Anesthesia and Analgesia, 98*, 70–74. table of contents.

Nair, G. S., Abrishami, A., Lermitte, J., & Chung, F. (2009). Systematic review of spinal anaesthesia using bupivacaine for ambulatory knee arthroscopy. *British Journal of Anaesthesia, 102*, 307–315.

Pollock, J. E. (2012). Intrathecal chloroprocaine—Not yet "safe" by US FDA parameters. *International Anesthesiology Clinics, 50*, 93–100.

Prien, T. (1994). Intradermal anaesthesia: Comparison of several compounds. *Acta Anaesthesiologica Scandinavica, 38*, 805–807.

Ravindran, R. S., Bond, V. K., Tasch, M. D., Gupta, C. D., & Luerssen, T. G. (1980). Prolonged neural blockade following regional analgesia with 2-chloroprocaine. *Anesthesia and Analgesia, 59*, 447–451.

Rebel, A., & Schell, R. (2015). Faust's anesthesiology review, 4th ed. *Anesthesia and Analgesia, 120*, 953.

Reisner, L. S., Hochman, B. N., & Plumer, M. H. (1980). Persistent neurologic deficit and adhesive arachnoiditis following intrathecal 2-chloroprocaine injection. *Anesthesia and Analgesia, 59*, 452–454.

Saporito, A., Anselmi, L., Borgeat, A., & Aguirre, J. A. (2016). Can the choice of the local anesthetic have an impact on ambulatory surgery perioperative costs? Chloroprocaine for popliteal block in outpatient foot surgery. *Journal of Clinical Anesthesia, 32*, 119–126.

Shah, J., Votta-Velis, E. G., & Borgeat, A. (2018). New local anesthetics. *Best Practice & Research. Clinical Anaesthesiology, 32*, 179–185.

Skidmore, R. A., Patterson, J. D., & Tomsick, R. S. (1996). Local anesthetics. *Dermatologic Surgery, 22*, 511–522 (quiz 523–514).

Smith, K. N., Kopacz, D. J., & McDonald, S. B. (2004). Spinal 2-chloroprocaine: a dose-ranging study and the effect of added epinephrine. *Anesthesia and Analgesia, 98*, 81–88. table of contents.

Teunkens, A., Vermeulen, K., Van Gerven, E., Fieuws, S., Van der Velde, M., & Rex, S. (2016). Comparison of 2-chloroprocaine, bupivacaine, and lidocaine for spinal anesthesia in patients undergoing knee arthroscopy in an outpatient setting: A double-blind randomized controlled trial. *Regional Anesthesia and Pain Medicine, 41*, 576–583.

Valanne, J. V., Korhonen, A. M., Jokela, R. M., Ravaska, P., & Korttila, K. K. (2001). Selective spinal anesthesia: a comparison of hyperbaric bupivacaine 4 mg versus 6 mg for outpatient knee arthroscopy. *Anesthesia and Analgesia, 93*, 1377–1379. table of contents.

Vath, J. S., & Kopacz, D. J. (2004). Spinal 2-chloroprocaine: the effect of added fentanyl. *Anesthesia and Analgesia, 98*, 89–94. table of contents.

Willatts, D. G., & Reynolds, F. (1985). Comparison of the vasoactivity of amide and ester local anaesthetics. An intradermal study. *British Journal of Anaesthesia, 57*, 1006–1011.

Williams, B. A., Bottegal, M. T., Kentor, M. L., Irrgang, J. J., & Williams, J. P. (2007). Rebound pain scores as a function of femoral nerve block duration after anterior cruciate ligament reconstruction: Retrospective analysis of a prospective, randomized clinical trial. *Regional Anesthesia and Pain Medicine, 32*, 186–192.

Williams, B. A., Ibinson, J. W., Mangione, M. P., et al. (2015). Research priorities regarding multimodal peripheral nerve blocks for postoperative analgesia and anesthesia based on hospital quality data extracted from over 1,300 cases (2011-2014). *Pain Medicine, 16*, 7–12.

Yoos, J. R., & Kopacz, D. J. (2005). Spinal 2-chloroprocaine for surgery: An initial 10-month experience. *Anesthesia and Analgesia, 100*, 553–558.

Chapter 9

Clonidine: Features and applications

Renato Santiago Gomez and Magda Lourenço Fernandes

Faculdade de Medicina da Universidade Federal de Minas Gerais, Belo Horizonte, Minas Gerais, Brazil

Abbreviations

AMI   acute myocardial infarction
ADHD attention deficit hyperactivity disorder
CNS   central nervous system

Introduction

Clonidine is an old drug, synthesized in 1962, in Germany (Sanchez Munoz, De Kock, & Forget, 2017). Although it was emerged from research on nasal decongestants, it was initially marketed as an antihypertensive medication, due to its important hypotensive effect (Dollery et al., 1976). In the following decades, several studies tried to clarify the properties and the exact mechanism of action of this agent. From the description of its sedative effects, the use of clonidine has expanded to other indications, especially in anesthesia and in the perioperative period.

In addition to the sedative effect, other properties associated with clonidine favored the expansion of its use: analgesia, antiemesis, reduced bleeding, reduced anesthetic induction time, hemodynamic and hormonal stability, reduced oxygen consumption, renal protection, anesthetic-sparing effect, anxiolysis, sedation, anti-tremors, reduced recovery time, and myocardial protection (Sanchez Munoz et al., 2017).

However, the unique pharmacodynamic profile, centered on a receptor associated with multiple actions, demands care in clinical handling to achieve the desired therapy. Furthermore, the therapeutic effect is maximum within a narrow range of plasma concentration and individual variations in pharmacokinetics may interfere with this effect. This requires that clonidine therapy be individualized and carefully monitored to avoid and correct possible failures and possible adverse effects (Frisk-Holmberg, 1983).

The purpose of this chapter was to describe the pharmacological characteristics and clinical applications of clonidine in the perioperative period, aiming to guide its best use, based on scientific evidence.

Pharmacological characteristics

Clonidine acts as a partial agonist of presynaptic receptors and thus can interfere in the peripheral regulation of noradrenaline release. It then decreases the neuronal sympathetic tone efferent to the heart, kidneys, and peripheral vasculature. On the other hand, it increases vagal tone in the cardiovascular system, resulting in a reduction in total peripheral resistance, renal vascular resistance, and blood pressure (Lowenthal, Matzek, & Macgregor, 1988).

Pharmacodynamics

Clonidine is a imidazoline centrally acting alpha2-adrenergic agonist, as is noradrenaline. Alpha-adrenoreceptor stimulation occurs both in the central nervous system (CNS) and in the periphery, with greater affinity for presynaptic receptors (Frisk-Holmberg, 1983). Pre-synaptic stimulation of alpha-2 receptors is coupled via protein G and other effectors, including inhibition of adenylate cyclase and its effects on the potassium and calcium channels. Thus, the drug's action restricts to the release of norepinephrine (Sanchez Munoz et al., 2017).

Treatments, Mechanisms, and Adverse Reactions of Anesthetics and Analgesics. https://doi.org/10.1016/B978-0-12-820237-1.00009-0
Copyright © 2022 Elsevier Inc. All rights reserved.

There are three subtypes of alpha-adrenergic receptors: alpha-2a, alpha-2b, and alpha-2c. Stimulation of these different subtypes has particular effects, namely (Nguyen et al., 2017):

- Sedation, analgesia and sympatholysis: alpha-2a receptor.
- Vasoconstriction and anti-tremor mechanism: alpha-2b receptor.
- Withdrawal to stimuli, contraction of the muscles of the extremities, stimulus to blink, variation in blood pressure and breathing patterns: alpha-2c receptor.
- Central action of clonidine: all centrally located alpha-2 subtypes, resulting in different manifestations.

In the CNS, clonidine produces its effects by stimulating alpha2-adrenergic receptors in the brain stem. It acts especially on the locus coeruleus, a region of the pons-medulla that receives innervations from the anterior portion of the brain, the area responsible for wakefulness. The sedative effects of clonidine result from the stimulation of alpha-2 receptors in this area, activating inhibitory neurons, which results in decreased output of sympathetic stimuli from the CNS (Nguyen et al., 2017).

The stimulation of adrenoreceptors in the spinal vasomotor center is considered as the main mechanism of the hypotensive effect (Frisk-Holmberg, 1983). The decrease in the output of stimuli from the spinal cord to the peripheral nerves results in peripheral vasodilation, in addition to the decrease in blood pressure, heart rate, and cardiac output (Nguyen et al., 2017).

Clonidine also modifies the potassium channels in the neurons of the CNS, causing hyperpolarization of their cell membranes. It is possible that this is the mechanism of action by which its administration is associated with reducing the need for anesthetics (Nguyen et al., 2017). The analgesic effects of clonidine are related to the stimulation of pre- and post-synaptic alpha-2 adrenoreceptors in peripheral or medullary locations. It inhibits the release of spinal substance P, decreasing the transmission of nociceptive neurons and preventing harmful stimuli from being transmitted to the brain (Nguyen et al., 2017). In addition to this analgesic effect, clonidine also improves the quality of analgesia in regional anesthesia. Acting on alpha-2 receptors in the spinal cord and brain stem and interfering in the conduction of painful fibers A and C, it increases the conductance to potassium and the duration of action of the local anesthetic. It also promotes local vasoconstriction and thus prolongs the duration of the anesthetic's effect (Prabhakar et al., 2019).

Noradrenaline is also known to play an important role in the reconsolidation of memory. In view of the activation of presynaptic alpha-2 receptors and the consequent decrease in the release of norepinephrine, experimental studies suggest that clonidine may interfere in the reconsolidation of memory to cocaine, with potential use to reduce the relapse of drug addicts (Denny & Unterwald, 2019).

Pharmacokinetics

Clonidine is a highly lipophilic drug with a high volume of distribution. It is rapidly absorbed after oral administration, with a maximum plasma concentration time between 1.5 and 2 h and a half-life of approximately 8–12 h (Sanchez Munoz et al., 2017). The therapeutic dose of clonidine generally ranges from 0.2 to 0.6 mg day^{-1}. The maximum plasma concentrations resulting from these dosages reach approximately 2 µg L^{-1} (Lowenthal et al., 1988). After intravenous administration, clonidine exhibits a multiphase pharmacokinetic profile, with rapid but extensive plasma distribution. Being highly fat-soluble, it readily penetrates extravascular sites, as well as the CNS. This results in a large apparent volume of distribution, of approximately 2 L kg^{-1} (Lowenthal et al., 1988).

The degree to which clonidine is bound to serum albumin *in vitro* varies within the range of 20%–40%. There appears to be no change in plasma protein binding in patients with end-stage renal disease. Clonidine easily crosses the placenta and its concentrations are the same in maternal serum and cord blood (Lowenthal et al., 1988). Its elimination occurs mainly by renal mechanisms, and the main metabolite formed, *p*-hydroxyclonidine, represents less than 10% of the concentration of the drug eliminated unchanged (Lowenthal et al., 1988).

Clinical applications in the perioperative period

The clinical use of clonidine in the perioperative period has grown and diversified from the record of its multiple benefits. However, there are controversies about the real advantages of this drug in some of these indications. In addition, the advent of a more specific alpha2-adrenergic agonist, dexmedetomidine, also limited the use of clonidine in certain settings.

Pre-anesthetic medication

Pre-medication with oral clonidine have several objectives, such as sedation, attenuation of hemodynamic responses, and decreased incidence of nausea and vomiting. As pre-anesthetic medication, in adults with ASA I or II physical status, clonidine at a dose of 0.15 mg, increased hemodynamic stability and decreased the requirement for anesthetics to induce general anesthesia. In comparison with midazolam, at the dose of 7.5 mg, it still adjusted the stress response, preventing an increase in plasma levels of the adrenocorticotropic hormone. Clonidine did not delay postoperative recovery suggesting its use may be superior to midazolam in these patients (Paris et al., 2009). Similarly, another study evaluated that a dose of 0.2 mg produced a significant reduction in anxiety, compared to benzodiazepines, however, with a dose of 0.3 mg, persistent hypotension was seen in the postoperative period, being its use advised against at this dosage (Carabine, Wright, & Moore, 1991).

In adult patients who received a dose of 100 µg of clonidine orally before surgery, attenuation of the hemodynamic responses produced by pneumoperitoneum was observed during laparoscopic cholecystectomy. In comparison with the placebo group, these patients maintained greater hemodynamic stability after intubation, during pneumoperitoneum and also extubation. They also presented a lower incidence of postoperative nausea and vomiting, resulting in better patient satisfaction and good cost-benefit ratio (Masud et al., 2017).

Adjunct to regional anesthesia

Adjuvants are drugs that work synergistically with local anesthetics to improve the quality and effectiveness of regional anesthesia. They can act by several ways, interfering with the onset of action, the duration and/or quality of analgesia and the potential adverse effects related to anesthesia (Prabhakar et al., 2019). Clonidine is a non-opioid drug widely used as an adjunct to regional anesthesia, although the results are sometimes controversial, both in terms of efficacy and dosage.

To compare the efficacy of clonidine at a dose of 30–300 µg, as an adjunct to various types of anesthetic blocks, a meta-analysis included 20 randomized studies with 1054 patients. It was found that clonidine prolonged postoperative analgesia, sensory and motor block, however clonidine increased the risk of arterial hypotension, orthostatic hypotension or fainting, and sedation (Pöpping et al., 2009).

In obstetrics, the use of clonidine in the neuroaxis, associated with local anesthetic, for the purpose of analgesia in the postoperative period of cesarean section was evaluated in a recent meta-analysis, which included 18 randomized clinical trials. The results showed that, compared to placebo, clonidine improved postoperative analgesia, decreasing the consumption of analgesics in the first 24 h and/or the time required for the first dose of analgesics. On the other hand, its use resulted in an increase in the incidence of intraoperative hypotension and sedation, without, however, negatively altering the pH of the umbilical artery or Apgar index of the fetus. The effect of the drug on intraoperative bradycardia, nausea and vomiting, pruritus, and postoperative sedation was inconclusive in this analysis. The intrathecal clonidine dose ranged from 30 to 150 µg, while the epidural clonidine dose ranged from 75 to 400 µg. Based on these findings, the authors argue that clonidine may be a useful analgesic supplement in women undergoing cesarean section under neuraxial anesthesia (Allen et al., 2018). Despite the heterogeneity of studies in some of its aspects, including the dosage, the meta-analysis supports the use of clonidine as an option for analgesia in obstetrics.

In pediatrics, clonidine is also one of the well-known adjuvants of anesthetic blocks, especially epidural block. The effectiveness of this indication was evaluated in a recent meta-analysis of randomized clinical trials, which aimed to assess the efficacy and adverse effects of clonidine, in comparison with other adjuvants, including morphine. As a result, it was found that longer duration of postoperative analgesia and motor block in the group that received clonidine, compared to other adjuvants. The clonidine group also had a lower incidence of nausea and vomiting in the postoperative period, while other events such as bradycardia, hypotension, and urinary retention were not different. Despite the heterogeneity of the studies, including the dose of clonidine (1–3 µg kg^{-1}), the authors argued that this drug by epidural route, in comparison with other adjuvants, provides longer duration of analgesia and a lower incidence of nausea and postoperative vomiting (Yang, Yu, & Zhang, 2018). However, some controversy persists about the best dosage by caudal epidural route, according to the type of surgery. In potentially more painful surgeries, such as hypospadias correction, there was no difference in analgesic consumption and pain scale, with the use of 1, 2, or 3 µg kg^{-1} clonidine, associated with bupivacaine, compared to use of bupivacaine without this adjuvant. However, with the use of 3 µg kg^{-1} the children were more sedated, which represented greater comfort (Bonisson et al., 2019).

Protection of the cardiovascular system

One of the most relevant aspects of the discussion on the applications of clonidine in the perioperative period, refers to its potential protective effects on the cardiovascular system, reducing the risk of myocardial infarction, stroke, and even death. Considering the moderating action of sympathetic activity, research has evaluated the beneficial effects of clonidine and other alpha-2 agonist agents in patients undergoing cardiac or non-cardiac surgery.

With regard to non-cardiac surgery, a large, randomized, blinded study compared a low dose of clonidine $(0.2\ mg\ day^{-1})$ with placebo, administered immediately before surgery and maintained for 72 h. 10,010 patients were evaluated, with or without risk for atherosclerotic disease. As a result, it was found that clonidine did not reduce the occurrence of death or acute myocardial infarction (AMI) within 30 days. On the other hand, the use of clonidine increased the risk of clinically relevant hypotension and non-fatal cardiac arrest (Devereaux et al., 2014).

Other studies have made a similar analysis, including patients undergoing cardiac surgery. A Cochrane meta-analysis published in 2018 compiled some of this research to assess the effectiveness and safety of alpha-2 agonists in reducing mortality and cardiac complications. Twenty-one clinical trials with clonidine, 24 with dexmedetomidine and two with mivazerol, used in adults undergoing cardiac or non-cardiac surgery, were included. Based on the results, the authors reaffirmed that, with regard to non-cardiac surgery, there is evidence of moderate to high quality, that these agents do not prevent death, AMI or stroke in the perioperative period. The important adverse effects associated with the use of alpha-2 agonists, such as hypotension and bradycardia, were also highlighted. A similar result was also seen in the analysis of studies that evaluated cardiac surgery. No changes were found in the risk of mortality or AMI, and an increased risk of bradycardia has also been reported. As for the risk of stroke, the results were inconclusive. Based on the results of the meta-analysis, the authors advocated that there are no convincing reasons for the use of alpha-2 adrenergic agonists in order to reduce the risks of death in the perioperative period of non-cardiac or cardiac surgery and that the use of such drugs is associated with important adverse effects, especially hypotension and bradycardia (Duncan et al., 2018).

Thus, it is clear that the adverse effects of clonidine and other alpha-2 agonist agents are limiting the use of these agents in the perioperative period. For this reason, a recent meta-analysis specifically evaluated the occurrence of such effects in patients without risk of cardiovascular event, who underwent non-cardiac surgery under general anesthesia. Fifty-six studies (4868 patients) who received clonidine or dexmedetomidine were included, compared to placebo. The incidence of serious adverse events was assessed up to 72 h after surgery. With regard to clonidine, the analysis revealed that the incidence of intraoperative hypotension was significantly higher compared to placebo. As for postoperative hypotension, no significant difference was observed with clonidine. Likewise, undesirable bradycardia, requiring correction with medication, has been reported, but with no significant difference between clonidine and placebo. Hypertension also did not differ and tachycardia did not have enough data for analysis. On the contrary, dexmedetomidine was associated with a higher risk of bradycardia and hypotension in the intraoperative period, which persisted even after treatment interruption, while related to a lower incidence of hypertension and tachycardia. It was concluded that there is a risk of hypotension and bradycardia associated with the use of alpha-2 agonists in the perioperative period. The authors advocated that, in relation to clonidine, the quality of the evidence is low or there is a lack of evidence for clinical recommendations. Regarding dexmedetomidine, the evidence favours its clinical use to avoid spikes in hypertension and tachycardia during prolonged surgery, under general anesthesia, in patients without risk of cardiovascular disease. However, administration by continuous infusion during the intraoperative period and postoperative monitoring are recommended to mitigate the risk of hypotension and bradycardia (Demiri et al., 2019).

Based on these findings, we can conclude that the routine use of clonidine in the perioperative period with the aim of promoting myocardial protection is not indicated, and it is reserved for scenarios that require strict control of blood pressure and heart rate. In addition, its use is safer in patients without cardiovascular comorbidities and dosages should certainly be differentiated according to the situation and therapeutic objective.

Controlled hypotension

In view of the good sympathetic control exercised by alpha-2 agonists, one of the most consolidated indications of clonidine in the intraoperative period would be in surgeries that require controlled hypotension, in order to reduce bleeding and improve surgical conditions. This is the case, for example, with some spine, ear, nose, throat, and plastic surgeries. In such scenario, clonidine seems to favour hypotensive anesthesia even when used orally.

In endoscopic nasal polyp surgery, it was found that the amount of bleeding and mean arterial pressure were significantly lower in patients who received oral clonidine 0.2 mg preoperatively (Tugrul et al., 2016).

In spine surgery, bleeding reduction was evaluated in a randomized, blinded study that included 120 adult patients, aiming to compare clonidine, dexmedetomidine, and placebo. Clonidine was administered orally (0.2 mg 90 min before surgery) and dexmedetomidine by continuous infusion (bolus of $0.5\ \mu g\ kg^{-1}$ and maintenance of $0.25\ \mu g\ kg^{-1}\ h^{-1}$). The results showed a significant reduction in intraoperative blood loss in patients who received clonidine or dexmedetomidine, compared to the control group, with a more significant reduction in the clonidine group. The authors concluded that dexmedetomidine or clonidine are equally effective in maintaining hemodynamic stability and intraoperative bleeding in patients undergoing spine surgery. However, they pointed out that the reduction in bleeding was more significant with clonidine (Janatmakan et al., 2019).

In theory, dexmedetomidine would be an option preferable to clonidine to promote intraoperative hypotension, in view of its more specific effect and the ease of dose adjustment. However, a randomized and controlled study that aimed to compare these two drugs in 94 patients undergoing endoscopic sinusectomy, showed no difference in the bleeding score between these two drugs. The authors pointed out that the more prolonged effect of clonidine, associated with its low cost, may make it a preferable option as an adjunct to hypotensive anesthesia (Escamilla et al., 2019).

Anti-shivering effect

One of the most worrying complications of the surgical-anesthetic procedure is hypothermia and consequent postoperative tremors. Perioperative hypothermia is associated with increased oxygen consumption and increased risk of cardiovascular events, infections and bleeding. Despite measures to prevent heat loss, such as thermal mattresses and blowers and heating of venous fluids, hypothermia is frequent, affecting more than half of the patients in the anesthetic recovery room (Lewis et al., 2015). Tremors represent a way for the body to produce heat to compensate for hypothermia, however they cause great discomfort for the patient and can increase surgical pain.

Alpha-2 agonist drugs are potentially effective in controlling tremors. This benefit was studied in a meta-analysis that investigated the effects of alpha-2 agonists clonidine or dexmedetomidine in preventing postoperative tremors in patients undergoing surgery under general anesthesia. Twenty studies were included with 1401 adult subjects who received an alpha-2 agonist compared to another alpha-2 agonist or placebo. Thirteen of these studies compared clonidine with a control, while seven compared dexmedetomidine with a control. Doses, methods, and time of administration varied between studies, as well as the form of administration (oral, bolus venous, or continuous venous). The high risk of bias in the studies was highlighted. As a result, it was found that the use of alpha-2 agonists significantly reduced the risk of tremors, compared to a placebo or control. When considering clonidine and dexmedetomidine separately, such evidence was found for both drugs. However, due to heterogeneity (I2 = 80%), the authors considered the evidence of low quality, especially in clonidine studies. As for adverse effects, the analysis suggested that sedation and bradycardia were significantly more common with dexmedetomidine compared to placebo, while sedation was not significant with clonidine. There were no reports of major cardiovascular complications, such as death from AMI or stroke. The authors concluded that there is evidence that clonidine and dexmedetomidine can reduce postoperative tremors, with more intense residual sedation in patients who received dexmedetomidine. However, they pointed out that the quality of this evidence was very low (Lewis et al., 2015).

Also in patients undergoing neuraxial anesthesia, the use of alpha-2 agonists can be useful for controlling the tremor. Randomized study that included 90 patients undergoing spinal anesthesia, compared tramadol (1 mg kg^{-1}), clonidine (1 µg kg^{-1}) and dexmedetomidine (0.5 µg kg^{-1}) on this effect. It was concluded that dexmedetomidine was better than tramadol or clonidine to promote sedation and tremor control, however with a higher occurrence of hypotension and bradycardia (Venkatraman et al., 2018).

Postoperative agitation

Agitation during awakening from general anesthesia is a relatively common occurrence in children, especially when sevoflurane is used. In view of its sedative effects, clonidine is one of the drugs considered to control such event. In a recent meta-analysis, the efficacy of some of these drugs was compared in children who underwent eye surgery. The results showed that clonidine or dexmedetomidine, as well as other drugs (ketamine, propofol, fentanyl, midazolam, sufentanil, and remifentanil) were superior to placebo. However, the effects of dexmedetomidine were considered to be superior to other drugs (Tan et al., 2019).

Applications to other areas

The antihypertensive effect of clonidine was discovered by chance, more than 40 years ago, when it was used for this purpose. But nowadays it is rarely prescribed as an antihypertensive, except in exceptional cases, such as those refractory to other classes of antihypertensive drugs most used. In hypertension the initial dose is 0.1 mg twice a day, with the usual dose reaching up to 0.8 mg/day. However, clonidine started to be investigated in several other clinical indications for which its use has expanded, mainly in the field of psychiatry. In these cases, the dosage varies greatly depend on the type of disorder and the intensity of the symptoms (Musini, Pasha, Gill, & Wright, 2017).

Clonidine has been used successfully to quickly suppress the signs and symptoms of opiate withdrawal. It helps opiate detoxification, reducing sympathetic hyperactivity, controlling tachycardia, hypertension, sweating, restlessness, and relieving insomnia. It is also used to make alcohol abstinence possible, with superiority compared to placebo demonstrated in several studies. It also proved to be more effective in smoking cessation than placebo (Naguy, 2016).

Recently, alpha-2 agonists have also been used to treat some other disorders such as attention deficit hyperactivity disorder (ADHD), post-traumatic stress disorder and Tourette's syndrome (Chiu & Campbell, 2018). Considered as a non-stimulant, these drugs are options in individuals who cannot tolerate the dopaminergic side effects of stimulants, in patients who require symptomatic coverage for 24 h or for others who are at risk of abuse or diversion of stimulants. Both clonidine and guanfacine are approved for use in the United States, in combination with some stimulant, for children who experience break-through symptoms with a stimulant alone (Mattingly & Anderson, 2016).

In Tourette's syndrome, the use of alpha-2-adrenoreceptor agonists such as clonidine is better established when associated with ADHD. In this condition, clonidine has a low risk of adverse reactions and promising efficacy. Sedation and hypotension can often limit higher dosages, which are sometimes necessary for adequate control of tics. Compared to clonidine, guanfacine has long lasting effects so that less dosages per day are needed. In addition, sedation and hypotension are less relevant (Roessner et al., 2013).

Menopausal syndrome, characterized by vasomotor changes, including hot flashes and night sweats, affects up to 70% of women in menopause and impairs their quality of life. Hyperactivity of the sympathetic nervous system, mediated through alpha-2 adrenergic receptors, is an important factor responsible for such changes. Clonidine has been used to control these symptoms, although with weak evidence, based on descriptive studies and expert opinion (Naguy, 2016).

A recent systematic review suggested that clonidine may be an effective and low-cost pharmacological option for individuals with behavioral disorders related to Autism Spectrum Disorder. It seems especially useful in cases where little or no benefit has been obtained with other pharmacological interventions combined with multidisciplinary approaches. Although this recommendation is based on case reports and two cross-sectional studies, the authors highlight greater benefit in young people with whom they struggle with hyperactivity, hyperexcitation, impulsivity, difficulty sleeping, and aggression (Banas & Sawchuk, 2020).

Other agents of interest

In addition to clonidine, another agonist of alpha-2 adrenergic receptors that deserves to be highlighted due to the spread of its use in the last two decades is dexmedetomidine. This drug was approved for use in the United States in 1999, for use as an analgesic and sedative in intensive care. However, such properties have expanded their use to various stages of the perioperative period (Afonso & Reis, 2012).

Dexmedetomidine is a relatively selective alpha-2 adrenergic receptor agonist, with a wide range of pharmacological properties. It is 8–10 times more selective for the alpha-2 receptor than clonidine. It has low affinity for beta adrenergic, muscurinic, dopaminergic, and serotonergic receptors. It binds to the alpha-2 receptors of locus ceruleus and spinal cord, promoting sedation and analgesia, respectively. The greater affinity for the alpha-2 receptor selectively leads to bradycardia, vasodilation, and hypotension (Naaz & Ozair, 2014).

Dexmedetomidine exhibits linear kinetics up to 24 h, in a continuous venous dose of 0.2–0.7 µg/kg/h. Its volume of distribution is about 118 L and the protein binding 94%. Its oral bioavailability is low due to extensive first-pass metabolism. However, sublingual and intranasal administration increases the bioavailability (84%), which gives it a potential role in sedation in children and premedication (Naaz & Ozair, 2014).

Unlike other sedatives or anesthetics, dexmedetomidine induces minimal respiratory depression, even when higher doses are used. Thus, it can be used safely in patients with spontaneous breathing, which favors its use in situations such as craniotomy with the patient awake and intubation awake. Its applications as premedication, as an adjunct to general or regional anesthesia and as a postoperative sedative and analgesic resemble benzodiazepines, but it seems more beneficial from the point of view of side effects (Lee, 2019).

Several randomized controlled trials have shown that patients treated with dexmedetomidine have significantly less delirium in the intensive care unit compared to patients treated with lorazepam, midazolam or propofol. However, in these studies, dexmedetomidine was compared with modulators of GABA receptors, which are well known for increasing the incidence of delirium. Similarly to clonidine, it has been used to prevent agitation on awakening in children undergoing general anesthesia (Lee, 2019).

Conclusions

Clonidine acts on alpha-2 receptors with wide distribution, both central and peripheral, exhibiting different clinical effects. Its highly lipophilic pharmacological profile with a large volume of distribution demands specific care, with the dosage, to avoid undesirable effects. Therefore, treatment should be instituted individually and monitored, with a view to its hemodynamic effects related to decreased sympathetic activity.

Whether by oral administration, venous or associated with nerve blocks, clonidine remains used in several clinical situations in the perioperative period. Among such indications, the control of nausea, vomiting, and tremors stands out. In addition, its use favors better hemodynamic control during anesthesia, either to prevent hypertension or to promote controlled hypotension. Despite this benefit, the continued use of clondine or even dexmedetomidine, aiming to protect the cardiovascular system in the perioperative period, does not seem to be supported by current evidence.

As an adjunct to local anesthetics, aiming to obtain postoperative analgesia, clonidine stands out as a good option, especially in children, although controversies over the best dose remain. Also in these patients, clonidine is an option to reduce agitation upon awakening from anesthesia.

Finally, despite the advent of dexmedetomidine, an alpha-2 agonist that has a more specific pharmacological profile, there is no evidence to abandon the use of clonidine. In well-indicated clinical situations, it can be an excellent cost-benefit agent.

Key facts

Clonidine is well-known drug alpha2-adrenergic agonist. Whether by oral administration, venous or associated with nerve blocks, it is used in several clinical situations in the perioperative period. Among such indications, the control of nausea, vomiting and tremors stands out. In addition, its use favours better hemodynamic control during anesthesia, either to prevent hypertension or to promote controlled hypotension.

Mini-dictionary of terms

Antiemesis. A drug used to treat nausea and vomiting.
Anxiolysis. A drug that decreased the anxiety.
Agonist. A drug that has an affinity.
Sympatholysis. An effect of reducing the actions of the sympathetic nervous system.
Adjuvant. A drug that enhances the effect of another.

Summary points

1. Clonidine is a imidazoline centrally acting apha2-adrenergic agonist, as is noradrenaline. Alpha-adrenoreceptor stimulation occurs both in the CNS and in the periphery, with greater affinity for presynaptic receptors.
2. Clinical clonidine effects: sedative, analgesic, antiemesis, reduced bleeding, reduced anesthetic induction time, hemodynamic and hormonal stability, reduced oxygen consumption, renal protection, anesthetic-sparing effect, anxiolysis, sedation, anti-tremors, reduced recovery time and myocardial protection.
3. Perioperative use of clonidine
   - Pre-anesthetic medication
   - Adjunct to regional anesthesia
   - Protection of the cardiovascular system
   - Controlled hypotension
   - Anti-shivering effect
   - Control of postoperative agitation

References

Afonso, J., & Reis, F. (2012). Dexmedetomidine: Current role in anesthesia and intensive care. *Revista Brasileira de Anestesiologia, 62*(1), 118–133.

Allen, T. K., et al. (February 2018). The impact of neuraxial clonidine on postoperative analgesia and perioperative adverse effects in women having elective caesarean section—A systematic review and meta-analysis. *British Journal of Anaesthesia, 120*(2), 228–240. 1471-6771 https://www.ncbi.nlm.nih.gov/pubmed/29406172.

Banas, K., & Sawchuk, B. (2020). Clonidine as a treatment of behavioural disturbances in autism spectrum disorder: A systematic literature review. *Journal of Canadian Academy of Child and Adolescent Psychiatry, 29*(2), 110–120.

Bonisson, A. C. M., et al. (2019). Combination of clonidine-bupivacaine in caudal epidural anesthesia for hypospadias surgery in children: Prospective, randomized, blind study. *Revista Brasileira de Anestesiologia*, 1806-907X. *69*(1), 27–34. January–February 2019 https://www.ncbi.nlm.nih.gov/pubmed/30482552.

Carabine, U. A., Wright, P. M., & Moore, J. (July 1991). Preanaesthetic medication with clonidine: A dose-response study. *British Journal of Anaesthesia*, 0007-0912. *67*(1), 79–83. https://www.ncbi.nlm.nih.gov/pubmed/1859765.

Chiu, S., & Campbell, K. (2018). *Clonidine for the treatment of psychiatric conditions and symptoms: A review of clinical effectiveness, safety, and guidelines [Internet].* Ottawa (ON): Canadian Agency for Drugs and Technologies in Health. PMID: 30303668.

Demiri, M., et al. (December 2019). Perioperative adverse events attributed to α2-adrenoceptor agonists in patients not at risk of cardiovascular events: Systematic review and meta-analysis. *British Journal of Anaesthesia*, 1471-6771. *123*(6), 795–807. https://www.ncbi.nlm.nih.gov/pubmed/31623842.

Denny, R. R., & Unterwald, E. M. (September 2019). Clonidine, an α2 adrenergic receptor agonist, disrupts reconsolidation of a cocaine-paired environmental memory. *Behavioural Pharmacology*, 1473-5849. *30*(6), 529–533. https://www.ncbi.nlm.nih.gov/pubmed/31386639.

Devereaux, P. J., et al. (April 2014). Clonidine in patients undergoing noncardiac surgery. *The New England Journal of Medicine*, 1533-4406. *370*(16), 1504–1513. https://www.ncbi.nlm.nih.gov/pubmed/24679061.

Dollery, C. T., et al. (January 1976). Clinical pharmacology and pharmacokinetics of clonidine. *Clinical Pharmacology and Therapeutics*, 0009-9236. *19*(1), 11–17. https://www.ncbi.nlm.nih.gov/pubmed/1245090.

Duncan, D., et al. (March 2018). Alpha-2 adrenergic agonists for the prevention of cardiac complications among adults undergoing surgery. *Cochrane Database of Systematic Reviews*, 1469-493X. *3*, CD004126. Disponível em https://www.ncbi.nlm.nih.gov/pubmed/29509957.

Escamilla, Y., et al. (November 2019). Randomized clinical trial to compare the efficacy to improve the quality of surgical field of hypotensive anesthesia with clonidine or dexmedetomidine during functional endoscopic sinus surgery. *European Archives of Oto-Rhino-Laryngology*, 1434-4726. *276*(11), 3095–3104. https://www.ncbi.nlm.nih.gov/pubmed/31363901.

Frisk-Holmberg, M. (February 1983). Clinical pharmacology of clonidine. *Chest*, 0012-3692. *83*(2 Suppl), 395–397. https://www.ncbi.nlm.nih.gov/pubmed/6822135.

Janatmakan, F., et al. (February 2019). Comparing the effect of clonidine and dexmedetomidine on intraoperative bleeding in spine surgery. *Anesthesia and Pain Medicine*, 2228-7523. *9*(1), e83967. https://www.ncbi.nlm.nih.gov/pubmed/30881906.

Lee, S. (2019). Dexmedetomidine: Present and future directions. *Korean Journal of Anesthesiology, 72*(4), 323–330.

Lewis, S. R., et al. (August 2015). Alpha-2 adrenergic agonists for the prevention of shivering following general anaesthesia. *Cochrane Database of Systematic Reviews*, 1469-493X. *8*, CD011107. https://www.ncbi.nlm.nih.gov/pubmed/26256531.

Lowenthal, D. T., Matzek, K. M., & Macgregor, T. R. (May 1988). Clinical pharmacokinetics of clonidine. *Clinical Pharmacokinetics*, 0312-5963. *14*(5), 287–310. https://www.ncbi.nlm.nih.gov/pubmed/3293868.

Masud, M., et al. (October 2017). Role of oral clonidine premedication on intra-operative haemodynamics and PONV in laparoscopic cholecystectomy. *Mymensingh Medical Journal*, 2408-8757. *26*(4), 913–920. https://www.ncbi.nlm.nih.gov/pubmed/29208884.

Mattingly, G. W., & Anderson, R. H. (2016). Optimizing outcomes in ADHD treatment: From clinical targets to novel delivery systems. *CNS Spectrums, 21*(S1), 45–59.

Musini, V. M., Pasha, P., Gill, R., & Wright, J. M. (2017). Blood pressure lowering efficacy of clonidine for primary hypertension. *Cochrane Database of Systematic Reviews, 9*, CD008284.

Naaz, S., & Ozair, E. (2014). Dexmedetomidine in current anaesthesia practice—A review. *Journal of Clinical and Diagnostic Research, 8*(10). GE01-4.

Naguy, A. (2016). Clonidine use in psychiatry: Panacea or panache. *Pharmacology, 98*(1–2), 87–92.

Nguyen, V., et al. (June 2017). Alpha-2 agonists. *Anesthesiology Clinics*, 1932-2275. *35*(2), 233–245. https://www.ncbi.nlm.nih.gov/pubmed/28526145.

Paris, A., et al. (July 2009). Effects of clonidine and midazolam premedication on bispectral index and recovery after elective surgery. *European Journal of Anaesthesiology*, 1365-2346. *26*(7), 603–610. https://www.ncbi.nlm.nih.gov/pubmed/19367170.

Pöpping, D. M., et al. (August 2009). Clonidine as an adjunct to local anesthetics for peripheral nerve and plexus blocks: A meta-analysis of randomized trials. *Anesthesiology*, 1528-1175. *111*(2), 406–415. https://www.ncbi.nlm.nih.gov/pubmed/19602964.

Prabhakar, A., et al. (December 2019). Adjuvants in clinical regional anesthesia practice: A comprehensive review. *Best Practice & Research. Clinical Anaesthesiology*, 1878-1608. *33*(4), 415–423. https://www.ncbi.nlm.nih.gov/pubmed/31791560.

Roessner, V., Schoenefeld, K., Buse, J., Bender, S., Ehrlich, S., & Münchau, A. (2013). Pharmacological treatment of tic disorders and Tourette syndrome. *Neuropharmacology, 68*, 143–149.

Sanchez Munoz, M. C., De Kock, M., & Forget, P. (May 2017). What is the place of clonidine in anesthesia? Systematic review and meta-analyses of randomized controlled trials. *Journal of Clinical Anesthesia*, 1873-4529. *38*, 140–153. https://www.ncbi.nlm.nih.gov/pubmed/28372656.

Tan, D., et al. (August 2019). Effect of ancillary drugs on sevoflurane related emergence agitation in children undergoing ophthalmic surgery: A Bayesian network meta-analysis. *BMC Anesthesiology*, 1471-2253. *19*(1), 138. https://www.ncbi.nlm.nih.gov/pubmed/31370793.

Tugrul, S., et al. (2016). Effect of the premedication with oral clonidine on surgical comfort in patients undergoing fess due to advanced nasal polyposis: A randomized double blind clinical trial. *American Journal of Otolaryngology*, 1532-818X. *37*(6), 538–543. November–December 2016 https://www.ncbi.nlm.nih.gov/pubmed/27720506.

Venkatraman, R., et al. (2018). A prospective, randomized, double-blinded control study on comparison of tramadol, clonidine and dexmedetomidine for post spinal anesthesia shivering. *Revista Brasileira de Anestesiologia*, 1806-907X. *68*(1), 42–48. January–February 2018 https://www.ncbi.nlm.nih.gov/pubmed/28546012.

Yang, Y., Yu, L. Y., & Zhang, W. S. (2018). Clonidine versus other adjuncts added to local anesthetics for pediatric neuraxial blocks: A systematic review and meta-analysis. *Journal of Pain Research*, 1178-7090. *11*, 1027–1036. https://www.ncbi.nlm.nih.gov/pubmed/29910631.

Chapter 10

An excursion into secondary pharmacology of fentanyls with potential implications for drug design: σ_1 receptor

Piotr F.J. Lipiński[a], Edina Szűcs[b,c], Małgorzata Jarończyk[d], Piotr Kosson[e], Sándor Benyhe[b], Aleksandra Misicka[a,f], Jan Cz Dobrowolski[d], and Joanna Sadlej[g]

[a]*Department of Neuropeptides, Mossakowski Medical Research Institute, Polish Academy of Sciences, Warsaw, Poland;* [b]*Institute of Biochemistry, Biological Research Centre, Hungarian Academy of Sciences, Szeged, Hungary;* [c]*Doctoral School of Theoretical Medicine, University of Szeged, Faculty of Medicine, Szeged, Hungary;* [d]*National Medicines Institute, Warsaw, Poland;* [e]*Toxicology Research Laboratory, Mossakowski Medical Research Institute, Polish Academy of Sciences, Warsaw, Poland;* [f]*Faculty of Chemistry, University of Warsaw, Warsaw, Poland;* [g]*Faculty of Mathematics and Natural Sciences, Cardinal Stefan Wyszyński University in Warsaw, Warsaw, Poland*

Abbreviations

(+)-MR200	[(+)-methyl (1R,2S)-2-{[4-(4-chlorophenyl)-4-hydroxypiperidin-1-yl]methyl}-1-phenylcyclopropanecarboxylate]
4-IBP	N-(1-benzyl-4-piperidyl)-4-iodo-benzamide
5-HT$_{1A}$ receptor	serotonin 1A receptor
5-HT$_{2A}$ receptor	serotonin 2A receptor
Anavex 2–73	1-(2,2-Diphenyltetrahydro-3-furanyl)-N,N-dimethylmethanamine
BD-1047	N'-[2-(3,4-dichlorophenyl)ethyl]-N,N,N'-trimethylethane-1,2-diamine
BD-1063	1-[2-(3,4-dichlorophenyl)ethyl]-4-methylpiperazine
BiP	binding immunoglobulin protein
E-52862	4-(2-((5-methyl-1-(naphthalen-2-yl)-1H-pyrazol-3-yl)oxy)ethyl)morpholine
ER	endoplasmic reticulum
H1 (receptor)	histamine 1 receptor
hERG	human *Ether-à-go-go*-Related Gene; the abbreviation is usually used to denote a potassium ion channel whose subunit is encoded by this gene
I$_2$	I2-imidazoline binding sites
IC$_{50}$	half maximal inhibitory concentration
I$_{hERG}$	electrical current associated with the hERG channel
K$_i$	inhibition constant
OCT1	organic cation transporter 1
PDB	Protein Data Bank
PRE-084	2-morpholin-4-ylethyl 1-phenylcyclohexane-1-carboxylate
RC-106	(E)-4-benzyl-1-(3-(naphthalen-2-yl)but-2-en-1-yl)piperidine
RMSD	root-mean-square deviation
δOR	δ-opioid receptor
κOR	κ-opioid receptor
μOR	μ-opioid receptor
σ_1R	σ_1 receptor
σ_2R	σ_2 receptor

Treatments, Mechanisms, and Adverse Reactions of Anesthetics and Analgesics. https://doi.org/10.1016/B978-0-12-820237-1.00010-7
Copyright © 2022 Elsevier Inc. All rights reserved.

The role of fentanyl

Among the most potent and useful analgesic agents, one must certainly name fentanyl (*N*-phenyl-*N*-[1-(2-phenylethyl) piperidin-4-yl]propanamide, see Fig. 1 for structures of fentanyl and all mentioned derivatives thereof). The compound is a high-affinity and selective μ-opioid receptor (μOR) agonist. Even though the μOR affinities of fentanyl and morphine are at a similar nanomolar level (Volpe et al., 2011), fentanyl is a much more potent analgesic, with a faster onset of action, mainly due to its higher lipophilicity. Depending on the type of administration, fentanyl has been found to be 50–100 times more potent than morphine and approximately 0.1 mg of parenteral fentanyl gives analgesia equal to that elicited by 10 mg of parenteral morphine (Vardanyan & Hruby, 2014). The indications for fentanyl use include perioperative pain management, but also induction and maintenance of anesthesia (alone or in combination with other agents). Furthermore, fentanyl is used for the treatment of cancer and non-cancer (severe) chronic pain.

Fentanyl is a parent compound for a larger family of synthetic opioids: 4-anilidopiperidines or "fentanyls," as we shall refer to them hereafter. This group comprises numerous derivatives (Vardanyan & Hruby, 2014) that were synthesized in the course of intensive studies on the structure-activity relationships of opioid analgesics. Interesting examples of fentanyl analogues include ultrapotent μOR agonists, like carfentanil (up to even 10,000-times more potent than morphine in some analgesic tests) or ohmefentanyl (7000-times more potent than morphine), and also ultra-short acting analgesics, like alfentanil or remifentanil. While the majority of fentanyl derivatives has been of significance only from the research point of view, a few received regulatory agencies' approval and have been used in human (alfentanil, remifentanil, and sufentanil) or veterinary (carfentanil) medicine.

Regrettably, fentanyls have also enjoyed popularity as substances of abuse. Like all μOR agonists, they produce euphoria and addiction. Their high potencies, as well as relatively easy and cheap syntheses in clandestine laboratories, make them very popular on the black market. Being very potent and having a narrow difference between a euphoric and a lethal dose, fentanyls are easy to overdose, which results in death due to respiratory depression. Fentanyl and its derivatives are the single major cause of drug overdose deaths in the United States. According to the statistics from 2018, opioids were responsible for almost 47,000 of drug overdose deaths and among these more than 50% were caused by fentanyls (Center for Disease Control and Prevention, 2019). The problem of fentanyl deaths has also been noted in European countries (Mounteney, Giraudon, Denissov, & Griffiths, 2015).

Secondary pharmacology of fentanyls

Drugs often bind at more than one molecular target. It is not rare that apart from having a certain high-affinity main target, a medicinal substance has a plethora of minor targets to which it binds with moderate or low (but noticeable) affinities. These secondary targets (or off-targets) contribute to drugs' pharmacological and toxicological profiles by being responsible for both some favorable therapeutic effects (ones additional to the on-target action) as well as adverse drug reactions. Hence, in modern drug discovery, analysis of off-targets has become a standard practice. The main motivation for this is associated with safety considerations but there are also attempts to discover drugs with intended promiscuity, that is to say, "selectively non-selective" compounds, since the multitarget profile of action may be beneficial in many cases (Morphy & Rankovic, 2005).

Secondary pharmacology of fentanyl and its derivatives does not seem to be thoroughly explored, at least in terms of available affinity data for molecular targets other than μOR. On the one hand, this may be partially due to the fact that fentanyls' practical use had been firmly established before the awareness of the importance of secondary pharmacology for drug safety became widespread. On the other hand, fentanyl and the approved analogues are administered in very low doses that give rise to very low blood concentrations (of the order of a few nmol/L) in which medicinally relevant action via some low affinity off-targets is unlikely. Furthermore, pharmaceutical fentanyls are considered to have a relatively high safety margin (in medicinal settings), and their adverse effects are typical "on-target" side-effects associated with the μ-opioid receptors. For all the above reasons digging for some low affinity secondary binding sites of fentanyls might not have been considered important.

Still, in our opinion, investigations of fentanyls' off-targets are justified, in particular given the deplorable popularity of this group of compounds as drugs of abuse. In drug addicts, with developed opioid tolerance, the tissue levels of abused fentanyls may be much higher than those observed in the patients. Presumably, they might even be high enough to saturate a certain off-target site population. In such a case, at least in principle, an off-target activity could become relevant from the point of view of pharmacodynamics or toxicity.

An example of a binding site for which this is indeed likely to happen was recently provided by Tschirhart, Li, Guo, and Zhang (2019). The authors investigated in vitro the interactions of fentanyl and norfentanyl with the hERG potassium ion

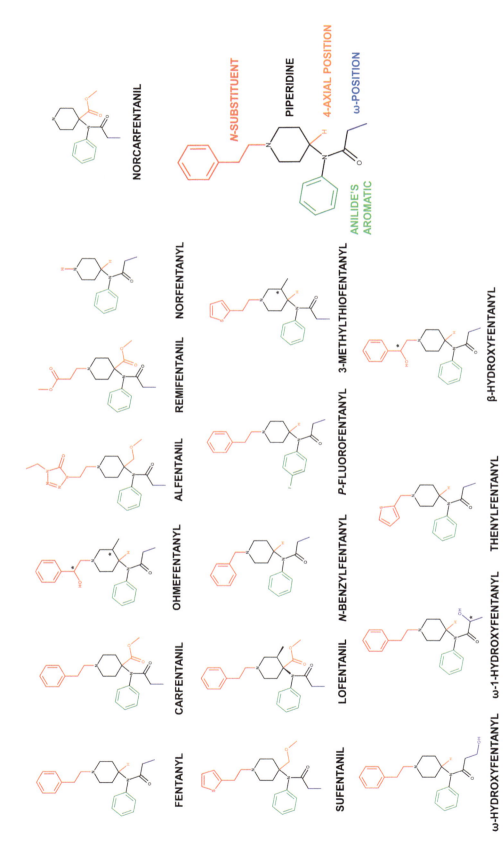

FIG. 1 Structures of fentanyl and its derivatives considered in the chapter.

channel, which plays an important role in the electrical activity of the heart. While norfentanyl exhibited no significant inhibitory action on the channel, fentanyl was shown (using a whole-cell voltage-clamp method) to decrease the I_{hERG} current with $IC_{50} = 0.3$ µM if a voltage protocol mimicking a human ventricular action potential was applied. By a kind of competitive experiment and by using cells expressing hERG mutants, the authors were also able to find hints in favor of this inhibitory action being associated with binding to the typical hERG inhibitors binding site. While the obtained hERG inhibition IC_{50} value is higher than the concentrations seen in patients treated with opioids, it overlaps with the upper range of drug blood concentrations found post-mortem in cases of fentanyl-related deaths (up to 1.14 µM, Martin, Woodall, & McLellan, 2006). As hERG inhibition is associated with (potentially fatal) cardiotoxicity, further investigation of the in vivo significance of these findings is warranted.

Apart from these recent hERG data, there are a few other reports mentioning the affinity of fentanyls for some secondary targets. Fentanyl was found to have low micromolar affinities for serotonin receptors, including 5-HT$_{1A}$ ($K_i = 2.1$ µM) and 5-HT$_{2A}$ ($K_i = 1.3$ µM) (Martin, Introna, & Aronstam, 1991; Rickli, Liakoni, Hoener, & Liechti, 2018). Similar micromolar affinity for the 5-HT$_{1A}$ receptor was also determined for sufentanil (7) (IC_{50} values between 2.2 and 5.5 µM) and alfentanil was shown to have no appreciable binding (Martin et al., 1991). According to Tao et al. binding to 5-HT$_{1A}$ receptors might contribute to the lethality of fentanyl overdose (Tao, Karnik, Ma, & Auerbach, 2003). Other authors found that fentanyl exhibits micromolar affinity for I_2-imidazoline binding sites ($K_i = 5.5$ or 8.6 µM) (Dardonville et al., 2006) and the histamine H1 receptor ($IC_{50} \sim 20$ µM) (Vasudevan, Moore, Schymura, & Churchill, 2012). A recent paper examined the inhibitory activity of several opioids on the organic cation transporter 1 (OCT1) function and it turned out that fentanyl and sufentanil are weak OCT1 inhibitors with IC_{50} values of 46 and 19 µM, respectively (Meyer et al., 2019).

Last but not least, it is to be noted that fentanyls, despite being µOR-selective opioids, have moderate to high affinities for other opioid receptors. Fentanyl itself binds to the µ receptors with low nanomolar affinity ($K_i \sim 1$ to a few nM), and to the δ- and κ- receptors with affinities in the middle nanomolar range (e.g., Maguire et al. give K_i values of 180 and 290 nM, for the δ- and κ-opioid receptors, respectively (Maguire et al., 1992)). However, there are examples, lofentanil being one of them, in the cases of which µ-selectivity is much lower. This compound was reported to have subnanomolar K_i values of 0.023, 0., and 0.60 for µ-, δ-, and κ-opioid receptors, respectively (Maguire et al., 1992).

In the course of our research on fentanyls, we turned our attention to the possibility that these compounds might bind to the σ$_1$ receptor—one of the most interesting and rather mysterious molecular targets that modern pharmacology and medicinal chemistry have in their areas of interest.

σ$_1$ receptor: A unique and intriguing protein

In the early studies, the σ binding sites had been mistakenly taken for a type of opioid receptors. As the research on them advanced, it became clear that there are two σ subtypes, σ$_1$ and σ$_2$ receptors, and that they do not have much in common with the opioid receptors. In fact, the σ$_1$ receptors are utterly unique in mammalian receptorome.

This $\sim$25-kDa protein (223 amino acids) shares no homology with other mammalian proteins (Hayashi, Tsai, Mori, Fujimoto, & Su, 2011). The closest homologue is the yeast sterol isomerase ERG2P, but σ$_1$R does not have any enzymatic function. The protein structure is unique in having an unusual fold, a single-pass transmembrane topology, and two-helices that seem to be partially embedded in the membrane.

Furthermore, the σ$_1$ receptors seem to be more than typical receptors (receiving and transducing signals). According to the dominant opinion (Schmidt & Kruse, 2019), σ$_1$R are rather ligand-operated chaperones that modulate cellular signaling pathways and functions. In cells, σ$_1$R localizes mainly in the mitochondria-associated membrane of the endoplasmic reticulum (ER) where it is anchored by a single transmembrane segment. The receptor is bound to the binding immunoglobulin protein (BiP) which has an important role in protein folding, oligomerization, and sorting. Activation of σ$_1$R upon binding of agonists or upon decreasing ER calcium concentrations is assumed to cause the receptor to unbind from BiP. σ$_1$R can then translocate to different organelles and engage in interactions with a variety of proteins in the ER, mitochondrial, nuclear, or even plasma membranes. Almost 50 proteins of divergent structures were reported to be bound by σ$_1$R and for many of them it was speculated that such interaction could be physiologically relevant. σ$_1$R was proposed to directly or indirectly regulate ion channels (Morales-Lázaro, González-Ramírez, & Rosenbaum, 2019), kinases and receptors, including some G-protein coupled receptors, for example, the µ-opioid receptor (Kim et al., 2010) or dopamine receptors (Almansa & Vela, 2014).

The involvement of σ$_1$R in protein–protein interactions is regulated by small molecular ligands (Fig. 2) that bind the receptor. Compounds which promote these interactions are termed "σ$_1$R agonists," and those that decrease them are "σ$_1$R antagonists." Several endogenous substances have been proposed to be endogenous ligands of σ$_1$R (progesterone, N,N-dimethyltryptamine, D-erythro-sphingosine, choline) but conclusive settlements as to this problem are lacking (Schmidt &

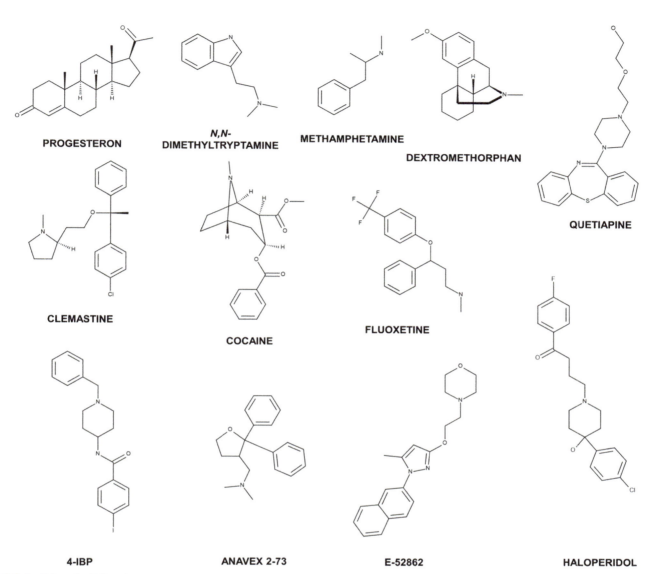

FIG. 2 Selected σ₁R ligands.

Kruse, 2019). Nonetheless, the diversity of endogenous and exogenous compounds that are known to bind to σ₁R with moderate to high affinities is truly notable and the promiscuity is another interesting feature of σ₁R. The ligands include butyrophenones, phenothiazines, thioxanthenes, steroids, benzomorphans, N-substituted piperidines, alkylamines, and others (Almansa & Vela, 2014; Narayanan, Bhat, Mesangeau, Poupaert, & McCurdy, 2011). Importantly, some approved or investigational medicines (e.g., dextromethorphan, clemastine, quetiapine, fluoxetine, haloperidol, etc.), whose intended (main) targets are distinct from σ₁R, have an appreciable affinity for the receptor. Furthermore, a few substances of abuse (e.g., cocaine or methamphetamine) are known to interact with σ₁R.

The above mentioned substances are non-selective σ₁R binders. The variety of selective ligands of this receptor have also been discovered in the med-chem programs (Weber & Wünsch, 2017) motivated by the possible beneficial effects of σ₁R modulation in the treatment of schizophrenia, depression, drug addiction, neurodegenerative diseases, neuropathic pain, cancers, and in cardioprotection. Some of the selective σ₁R ligands even made their way to clinical trials, but so far none have been approved. To the best of our knowledge, two of them are currently being considered in the trials: Anavex 2–73 (mixed muscarinc/σ₁R agonist for Alzheimer's disease treatment) and E-52862 (selective σ₁R antagonist for the treatment of neuropathic pain).

Affinity of fentanyls for $\sigma_1 R$

Taking into account the broad involvement of $\sigma_1 R$ in cellular physiology, the molecular promiscuity of the receptor and the fact that many fentanyls resemble known $\sigma_1 R$ ligands, we were then curious if fentanyls have an appreciable affinity for $\sigma_1 R$. All the more that for the parent compound that is for fentanyl, two older papers reported it to be a moderate (IC_{50} = 354 nM, Largent, Wikström, Gundlach, & Snyder, 1987) or a weak (K_i > 1 µM, Chen, Smith, Cahill, Cohen, & Fishman, 1993) $\sigma_1 R$ ligand. As to the other fentanyls, no information on $\sigma_1 R$ affinity had been available before our undertaking of the problem. To address this issue, we tested fentanyl and 11 commercially available derivatives for $\sigma_1 R$ affinity by a competition binding assay with a selective radioligand (Lipiński, Szűcs, et al., 2019). The structures of the tested derivatives are given in Fig. 1. The results of the assay are presented in Table 1 along with affinity data for µOR (Lipiński, Kosson, et al., 2019).

In agreement with the previous reports mentioned, the parent fentanyl (1) turned out to exhibit a rather low $\sigma_1 R$ affinity with K_i of 3718 nM. However, some modest structural modifications can result in significant enhancement of $\sigma_1 R$ binding. Shortening the N-phenethyl chain by just one $-CH_2-$ unit (N-benzylfentanyl) gives an affinity improvement 15-times over and a submicromolar K_i of 241 nM. A conservative H/F exchange at the *para*-position of anilide's aromatic (*p*-fluorofentanyl) brings about a 10-times affinity increase (K_i = 370.30 nM) compared to the parent. Similar binding strength is seen in the case of 3-methylthiofentanyl (K_i = 387.78 nM), in which introduction of a methyl group in the piperidine position C3 is accompanied by replacement of phenyl to 2-thienyl in the N-chain. Introduction of a hydroxyl group at β-, ω-, or ω-1-hydroxy positions of the fentanyl structure is highly unfavorable for binding as derivatives with these modifications exhibit no more than 50% radioligand displacement at 10 µM. Similarly unfavorable for $\sigma_1 R$ binding seems 4-axial substitution. Norcarfentanil, remifentanil and alfentanil inhibited no more than 35% of the radioligand binding at 10 µM. Still, this modification cannot be entirely adverse to binding, as sufentanil was shown to have an affinity (K_i = 1552.88 nM) better than that of the parent.

On comparing these data to µOR affinities (having in mind the only approximate nature of such comparisons), it turns out that fentanyl and sufentanil bind to $\sigma_1 R$ by more than three orders of magnitude weaker than to µOR. Slightly less selective seem *p*-fluorofentanyl and 3-methylthiofentanyl, for which this difference would be roughly two orders of magnitude. A derivative with no selectivity is N-benzylfentanyl, in the case of which the affinity measures are on a similar middle nanomolar level in both receptors. Alfentanil, remifentanil, and β-hydroxyfentanyl are totally µOR-selective.

TABLE 1 Affinities of fentanyl and its derivatives for $\sigma_1 R$ and µOR.

	$\sigma_1 R$	µOR
	$K_i \pm SEM$ (nM)	$IC_{50} \pm SD$ (nM)
fentanyl	3718.06 ± 0.96	1.23 ± 0.14
N-benzylfentanyl	240.81 ± 0.78	489.7 ± 28.6
p-fluorofentanyl	370.30 ± 0.83	0.48 ± 0.03
3-methylthiofentanyl	387.78 ± 0.84	1.10 ± 0.10
β-hydroxyfentanyl	>10,000	2.81 ± 0.13
ω-hydroxyfentanyl	>10,000	97.7 ± 5.8
ω-1-hydroxyfentanyl	>10,000	489.0 ± 40.6
norcarfentanil	>10,000	295.1 ± 1.3
remifentanil	>10,000	0.60 ± 0.08
alfentanil	>10,000	38.9 ± 2.8
sufentanil	1552.88 ± 0.86	0.40 ± 0.03
thienylfentanyl	886.24 ± 0.81	245.5 ± 12.9

SEM, standard error of the mean; *SD*, standard deviation.
The values taken from Lipiński, P. F. J., Szűcs, E., Jarończyk, M., Kosson, P., Benyhe, S., Misicka, A., … Sadlej, J. (2019). Affinity of fentanyl and its derivatives for the σ 1-receptor. *MedChemComm, 10*, 1187–1191; Lipiński, P., Kosson, P., Matalińska, J., Roszkowski, P., Czarnocki, Z., Jarończyk, M., … Sadlej, J. (2019). Fentanyl family at the mu-opioid receptor: Uniform assessment of binding and computational analysis. *Molecules, 24*, 740.

Looking at this from a different angle, it is to be noted that some trends found in structure-affinity relationships for fentanyls at σ₁R are opposite to those at μOR. First, N-benzylfentanyl has an affinity for μOR almost 400-times inferior to its parent fentanyl, while in the case of σ₁R it is the shorter derivative that binds significantly better. Second, 4-axially substituted derivatives which are usually potent μOR binders (like alfentanil and remifentanil) have very poor affinity for σ₁R. On the other hand, another congener of this group, sufentanil exhibits slightly improved σ₁R binding compared to the parent, so the 4-axial substitution does not seem to be entirely adverse. On the other hand, similarly to what is seen with μOR, p-F and 3-Me substitutions are also beneficial in the case of σ₁R. The improvement of affinity is even greater than in μOR.

Fentanyls and σ₁R: Insights from computations

Taking advantage of the recently solved σ₁R crystal structures, we attempted modeling the interactions of fentanyls with the receptor binding site to see what might be a probable structural basis for the obtained trends in affinity. Fentanyl and two submicromolar σ₁R binders, that is N-benzylfentanyl and p-fluorofentanyl, were subject to molecular docking and molecular dynamics simulations.

According to computations, all these three derivatives assume a binding mode (Figs. 3 and 4) that is fairly similar to the one found for a high-affinity ligand, 4-IBP (N-(1-benzyl-4-piperidyl)-4-iodo-benzamide) in the crystal structure (PDB accession code: 5HK2 (Schmidt et al., 2016)). The features characteristic for this binding mode are:

- ionic interaction of protonated piperidine's nitrogen with E172,
- positioning of the anilide's ring towards α4 and α5 helices,
- location of the N-substituent towards the bottom of β-barrel (close to D126).

Aside from the single ionic interaction, the remaining contacts seen in the predicted binding modes are of apolar character (Fig. 3).

Overall, the binding pose of the shorter derivative, N-benzylfentanyl is well-aligned on the crystallographic binding pose of the high-affinity ligand, 4-IBP (Fig. 4). The pharmacophoric features are closely positioned in both cases, with the displacement measure RMSD (root mean square deviation of atomic positions) for N-benzyl rings being 1.89 Å, for the other ring pair 1.37 Å and for the protonated nitrogens 1.20 Å. If N-phenethyl is present in the structure, the additional single bond and perhaps lack of stacking with Trp164 contribute to the lower affinity of the parent fentanyl. In the case of p-fluorofentanyl, the factor that seem responsible for the submicromolar affinity are the interactions of the halogen with Tyr206. This interaction model also seems suitable for explaining the very low affinity of hydroxylated derivatives

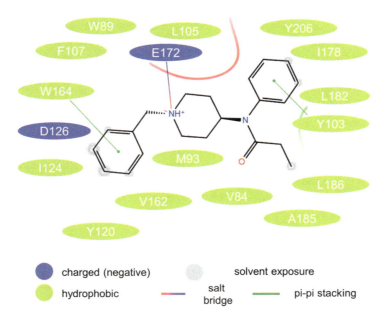

FIG. 3 Interactions of N-benzylfentanyl with the σ₁R as found by molecular modeling.

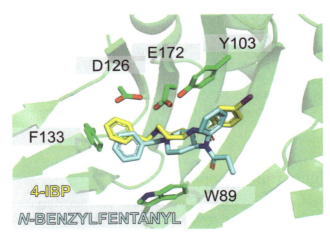

FIG. 4 Superposition of binding poses of *N*-benzylfentanyl and 4-IBP.

(no H-bonding partners reachable by OH-groups in β-, ω- or ω-1-positions) or of those with 4-axial substituents (no place for accommodation of the substituents without changing the binding mode or creating strain).

Pharmacological relevance of fentanyls' interactions with $\sigma_1 R$

There comes the question of whether the experimentally determined $\sigma_1 R$ affinities of fentanyls can have any practical pharmacological significance. In the case of the parent fentanyl, it seems not. Drug levels around the $\sigma_1 R$ K_i of fentanyl are not reachable with the doses in normal medicinal use. Perhaps in only the most severe instances of overdose, one could observe some significant $\sigma_1 R$ occupation. Still, it is worth noting that cocaine, whose effects have been many times studied in the context of $\sigma_1 R$, exhibits a similar affinity for this receptor, with K_i values of a few micromoles (Su, 2019). In the cases of *p*-fluorofentanyl and 3-methylthiofentanyl, which are not in medicinal use but appear as street drugs, the difference between $\sigma_1 R$ and μOR affinities (about two orders of magnitude) is smaller, but still a large one, and on assuming similar abuse patterns as in the case of fentanyl, concentrations that could give some relevant $\sigma_1 R$ binding are not likely to be reached.

Potentially the most interesting situation is that with *N*-benzylfentanyl. This derivative has similar affinities for both $\sigma_1 R$ and μOR. On the one hand, being a rather weak opioid, it has no medicinal use and it does not frequently (if ever) appear as the main constituent of designer drugs. On the other hand, the compound has been found every now and then as an impurity in the samples of clandestinely made fentanyl (Coleman, 2007). Furthermore, its presence was established in the urine samples of some opioid users (Daniulaityte, Carlson, Juhascik, Strayer, & Sizemore, 2019) and the substance was reported to be involved in some cases of fentanyl-analogue-related deaths (Daniulaityte et al., 2019). Hence, given the role of $\sigma_1 R$ in drug addiction and stimulant action (Katz, Hong, Hiranita, & Su, 2016), it would be interesting to see whether co-administration of fentanyl and *N*-benzylfentanyl somehow influences the euphoria, tolerance, and addiction caused by fentanyl.

Yet another issue deserves mention. Very modest structural modifications of the fentanyl scaffold, like p-fluorination or deletion of one methylene unit, are able to confer appreciable $\sigma_1 R$ affinity. The non-pharmaceutical fentanyl analogues that are produced in clandestine laboratories often include such structural motifs, as well as motifs found in high affinity $\sigma_1 R$ ligands. It cannot be excluded that some of them could exhibit physiologically relevant $\sigma_1 R$ binding. In consequence, for full elucidation of pharmacological profile, it would be useful to assay for $\sigma_1 R$ affinity the "street" fentanyls, including the novel ones that appear now and then.

Fentanyl structure as scaffold for dual μOR/$\sigma_1 R$ ligands

Looking at the $\sigma_1 R$ affinities of fentanyls from the perspective of medicinal chemistry and novel analgesic drug discovery, we came to entertain the idea that fentanyl or its derivatives might be starting points for creating high-affinity mixed μOR/$\sigma_1 R$ ligands. Based on what is known about the role of the $\sigma_1 R$ in modulating pain, such compounds could have promising analgesic properties, perhaps also beneficial in the management of neuropathic pain.

σ_1R antagonists have been proposed as therapeutic agents for the treatment of neuropathic pain and as adjuvants to opioid therapy (Almansa & Vela, 2014; Zamanillo, Romero, Merlos, & Vela, 2013). There is much preclinical evidence in favor of these propositions that is further corroborated by the positive results of Phase 2 clinical trials in which E-52862 was evaluated for the management of oxaliplatin-induced neuropathy (Bruna et al., 2018). In a short overview, it is to be said that σ_1R antagonists, while having no antinociceptive action in classical models of acute nociception, are able to inhibit pain in sensitizing and neuropathic pain models. Furthermore, upon coadministration with opioids, σ_1R antagonists are able to enhance their analgesic effects without increasing their propensity to evoke the typical side-effects or tolerance and rewarding.

The mere idea of developing dual $\mu OR/\sigma_1R$ ligands was put forward earlier. Prezzavento et al. (2017) discovered that phenazocine enantiomers that had been known to be good binders of μ-opioid and σ_1-receptor (Carroll et al., 1992) exhibit antagonistic σ_1R activity. They proved that the analgesic action of phenazocines is associated with σ_1R antagonism. In their in vivo work, a σ_1R agonist PRE-084 was able to significantly reduce the antinociceptive actions of both enantiomers at doses that did not interfere with the opioid analgesia exerted by fentanyl and morphine. Thus they suggested that the mixed $\mu OR/\sigma_1R$ profile might also explain the significant in vivo analgesia observed not only for $(-)$-phenazocine (a subnanomolar μOR agonist), but also for $(+)$-phenazocine (a middle nanomolar μOR agonist). In the latter case, σ_1R antagonism would be responsible for potentiation of the antinociception evoked by (moderate) activation of μOR. That such potentiation is possible had been shown many times before, but Prezzavento et al. (2017) were, to the best of our knowledge, the first to demonstrate such an effect upon administration of a single (but dual-acting) agent. Their work was concluded by hypothesizing that the development of μOR agonist/σ_1R antagonist dual ligands should yield compounds with increased analgesic action and fewer side effects (Prezzavento et al., 2017).

Successful realization of the dual $\mu OR/\sigma_1R$ ligands concept

The validity of our speculations that fentanyl structure could be used for creating dual μOR agonist/σ_1R antagonist ligands, and that such compounds would have favorable analgesic properties, has been recently confirmed by two works performed independently of our efforts.

With the intent of finding μOR agonist/σ_1R antagonist compounds, a group from ESTEVE Pharmaceuticals S.A. dealt with 4-aryl-1-oxa-4,9-diazaspiro[5.5]undecane derivatives (Fig. 5) with N-arylethyl substituent (García et al., 2020). This class of compounds can be perceived as a relatively close derivative of fentanyl structure, with an extended and constrained fragment corresponding to propionanilide moiety in fentanyls. In an effort to explore the structure-activity relationships, the authors obtained a set of 52 analogues with diversified μOR and σ_1R binding properties both with respect to their binding strength and their mutual ratios. The compound shown in Fig. 5 as **1** emerged as the most interesting one since it turned out to have quite a high and balanced affinity at the intended targets ($K_i \mu OR = 175$ nM, $K_i \sigma_1R = 58$ nM), μOR agonist/σ_1R antagonist functional profile, as well as selectivity against the anti-targets and good physicochemical properties. It was further tested in vivo in the mouse paw pressure pain test where it had potent analgesic activity, comparable to that found for oxycodone, even though being about 10-times weaker a μOR ligand than oxycodone. The compound was also found to elicit local, peripheral analgesia after the ipl administration, which was abolished by a σ_1R agonist PRE-084. Moreover, the dual ligand caused less constipation than oxycodone at equianalgesic doses. The authors remark that this compound was chosen for a lead optimization program that resulted in a clinical candidate to be disclosed soon.

A further group of fentanyl-related compounds was investigated in another paper (Xiong et al., 2020). The authors hybridized N-arylpropionamide (of fentanyl) and 4-benzyl piperidine (of RC-106, a pan-sigma ligand) and performed SAR analysis of this hybrid framework (Fig. 5) by obtaining and assaying for receptor affinities 22 novel derivatives. The most promising of them (Fig. 5, compound **2**), exhibited very strong binding at both μOR ($K_i = 1.9$ nM) and σ_1R ($K_i = 2.1$ nM). Moreover, the substance had dose-dependent analgesic effects in Phase II of the formalin test as well as it was found to be as effective (against neuropathic pain) as E-52862 in the chronic constriction injury model.

Applications to other areas

Apart from the σ_1 receptor, there is another σ-binding site: the σ_2 receptor (σ_2R). Recently, σ_2R was identified as transmembrane protein 97 (TMEM97). Several propositions as to potential therapeutic use of σ_2R ligands have been put forward. They include, for example, therapy of cancers (as single agents, as delivery agents or as sensitizers of cancer therapeutics) and neurological, inflammatory, and autoimmune diseases (Zeng & Mach, 2017). Many σ_1R ligands have appreciable affinity for σ_2R. In light of this, it may be interesting to investigate affinities of fentanyls for σ_2R. All the more that

98 PART | II Mechanisms of action of drugs

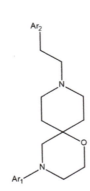

4-ARYL-1-OXA-4,9-DIAZASPIRO[5.5]UNDECANES WITH N-ARYLETHYL SUBSTITUENT

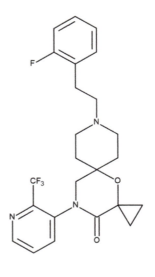

1

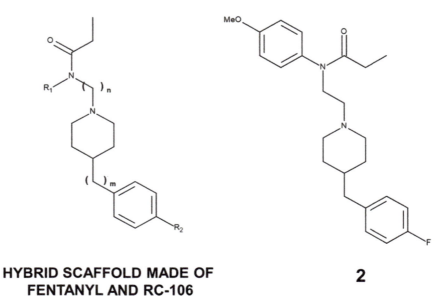

HYBRID SCAFFOLD MADE OF FENTANYL AND RC-106

2

FIG. 5 Mixed μOR agonist/σ_1R antagonist ligands.

there exists a prediction from a QSAR model derived from a large dataset of σ_2R ligands that fentanyl would bind σ_2R with K_i in middle nanomolar range (Rescifina et al., 2017).

Other agents of interest

In this chapter we have described fentanyls in the context of their affinity for σ_1R. Other opioid ligands have also been tested for binding to this receptor (Largent et al., 1987). These include, for example:

Morphine.
Naloxone.
Levorphanol.
Levallorphan.

Dextrometorphan.
Meperidine.
Cyclazocine.
Pentazocine.

Morphine and naloxone were found to be essentially non-binders ($IC_{50} > 100,000$ nM). Very low binding was observed also for levorphanol ($IC_{50} > 10,000$ nM). In the case of levallorphan and meperidine, the reported IC_{50} values were of a few micromoles (1890 and 2290 nM, respectively). Dextrometorphan has a submicromolar affinity ($IC_{50} = 810$ nM). ($\pm$)-Cyclazocine and ($\pm$)-pentazocine exhibit rather strong $\sigma_1 R$ binding, their reported IC_{50} values being 102 and 25 nM, respectively.

Key facts of σ_1 receptor

- The protein is utterly unique in mammalian receptorome, with respect to sequence, structure, and function.
- Currently, it is thought to be a ligand-operated molecular chaperone.
- It binds a number of structurally diverse compounds, including medicinal substances and drugs of abuse.
- Selective $\sigma_1 R$ antagonists have been proposed as potential therapeutics in several areas.
- $\sigma_1 R$ antagonist, E-52862, is considered in clinical trials (indications: neuropathic pain and potentiation of analgesic action of opioids).

Mini-dictionary of terms

Chaperones. Proteins that help in proper folding/unfolding of other proteins.
Docking. A computational technique for predicting how (in terms of structure) a ligand binds to its target.
IC_{50}. Half maximal inhibitory concentration; a kind of affinity measure. The lower it is, the stronger a compound binds.
K_i. Inhibition constant; a kind of affinity measure. The lower it is, the stronger a compound binds.

Summary points

- 12 commercially available fentanyls were investigated for $\sigma_1 R$ affinity.
- The parent fentanyl exhibits a micromolar affinity ($K_i = 3178$ nM).
- Modest structural changes as deleting one methylene unit from N-substituent (N-benzylfentanyl, $K_i = 241$ nM) or introduction of fluorine in the anilide ring (p-fluorofentanyl, $K_i = 370$ nM) yield submicromolar affinities.
- Dual $\sigma_1 R/\mu OR$ ligands based on fentanyl scaffold could be analgesic agents with interesting and beneficial properties.
- Recent papers report fentanyl-related high affinity $\sigma_1 R$ antagonists/μOR agonists dual ligands with potent analgesic activity.

References

Almansa, C., & Vela, J. M. (2014). Selective sigma-1 receptor antagonists for the treatment of pain. *Future Medicinal Chemistry, 6*, 1179–1199.

Bruna, J., Videla, S., Argyriou, A. A., Velasco, R., Villoria, J., Santos, C., et al. (2018). Efficacy of a novel sigma-1 receptor antagonist for oxaliplatin-induced neuropathy: A randomized, double-blind, placebo-controlled phase IIa clinical trial. *Neurotherapeutics, 15*, 178–189.

Carroll, F. I., Abraham, P., Parham, K., Bai, X., Zhang, X., Brine, G. A., et al. (1992). Enantiomeric N-substituted N-normetazocines: A comparative study of affinities at .sigma., PCP, and .mu. opioid receptors. *Journal of Medicinal Chemistry, 35*, 2812–2818.

Center for Disease Control and Prevention. (2019). *Synthetic opioid overdose data.*

Chen, J. C., Smith, E. R., Cahill, M., Cohen, R., & Fishman, J. B. (1993). The opioid receptor binding of dezocine, morphine, fentanyl, butorphanol and nalbuphine. *Life Sciences, 52*, 389–396.

Coleman, J. J. (2007). *Fentanyl analogs in street drugs.* Prescription Drug Research Center LLC.

Daniulaityte, R., Carlson, R. R., Juhascik, M. P., Strayer, K. E., & Sizemore, I. E. (2019). Street fentanyl use: Experiences, preferences, and concordance between self-reports and urine toxicology. *International Journal of Drug Policy, 71*, 3–9.

Dardonville, C., Fernandez-Fernandez, C., Gibbons, S.-L., Ryan, G. J., Jagerovic, N., Gabilondo, A. M., et al. (2006). Synthesis and pharmacological studies of new hybrid derivatives of fentanyl active at the µ-opioid receptor and I2–imidazoline binding sites. *Bioorganic & Medicinal Chemistry, 14*, 6570–6580.

García, M., Virgili, M., Alonso, M., Alegret, C., Fernández, B., Port, A., et al. (2020). 4-Aryl-1-oxa-4,9-diazaspiro[5.5]undecane derivatives as dual µ-opioid receptor agonists and σ 1 receptor antagonists for the treatment of pain. *Journal of Medicinal Chemistry, 63*, 2434–2454.

Hayashi, T., Tsai, S.-Y., Mori, T., Fujimoto, M., & Su, T.-P. (2011). Targeting ligand-operated chaperone sigma-1 receptors in the treatment of neuro-psychiatric disorders. *Expert Opinion on Therapeutic Targets, 15*, 557–577.

Katz, J. L., Hong, W. C., Hiranita, T., & Su, T.-P. (2016). A role for sigma receptors in stimulant self-administration and addiction. *Behavioural Pharmacology, 27*, 100–115.

Kim, F. J., Kovalyshyn, I., Burgman, M., Neilan, C., Chien, C.-C., & Pasternak, G. W. (2010). σ 1 receptor modulation of G-protein-coupled receptor signaling: Potentiation of opioid transduction independent from receptor binding. *Molecular Pharmacology, 77*, 695–703.

Largent, B. L., Wikström, H., Gundlach, A. L., & Snyder, S. H. (1987). Structural determinants of sigma receptor affinity. *Molecular Pharmacology, 32*, 772–784.

Lipiński, P., Kosson, P., Matalińska, J., Roszkowski, P., Czarnocki, Z., Jarończyk, M., et al. (2019). Fentanyl family at the mu-opioid receptor: Uniform assessment of binding and computational analysis. *Molecules, 24*, 740.

Lipiński, P. F. J., Szűcs, E., Jarończyk, M., Kosson, P., Benyhe, S., Misicka, A., et al. (2019). Affinity of fentanyl and its derivatives for the σ 1-receptor. *MedChemComm, 10*, 1187–1191.

Maguire, P., Tsai, N., Kamal, J., Cometta-Morini, C., Upton, C., & Loew, G. (1992). Pharmacological profiles of fentanyl analogs at μ, δ and κ opiate receptors. *European Journal of Pharmacology, 213*, 219–225.

Martin, D. C., Introna, R. P., & Aronstam, R. S. (1991). Fentanyl and sufentanil inhibit agonist binding to 5-HT1A receptors in membranes from the rat brain. *Neuropharmacology, 30*, 323–327.

Martin, T. L., Woodall, K. L., & McLellan, B. A. (2006). Fentanyl-related deaths in Ontario, Canada: Toxicological findings and circumstances of death in 112 cases (2002-2004). *Journal of Analytical Toxicology, 30*, 603–610.

Meyer, M. J., Neumann, V. E., Friesacher, H. R., Zdrazil, B., Brockmöller, J., & Tzvetkov, M. V. (2019). Opioids as substrates and inhibitors of the genetically highly variable organic cation transporter OCT1. *Journal of Medicinal Chemistry, 62*, 9890–9905.

Morales-Lázaro, S. L., González-Ramírez, R., & Rosenbaum, T. (2019). Molecular interplay between the Sigma-1 receptor, steroids, and ion channels. *Frontiers in Pharmacology, 10*. https://doi.org/10.3389/fphar.2019.00419.

Morphy, R., & Rankovic, Z. (2005). Designed multiple ligands. An emerging drug discovery paradigm. *Journal of Medicinal Chemistry, 48*, 6523–6543.

Mounteney, J., Giraudon, I., Denissov, G., & Griffiths, P. (2015). Fentanyls: Are we missing the signs? Highly potent and on the rise in Europe. *The International Journal on Drug Policy, 26*, 626–631.

Narayanan, S., Bhat, R., Mesangeau, C., Poupaert, J. H., & McCurdy, C. R. (2011). Early development of sigma-receptor ligands. *Future Medicinal Chemistry, 3*, 79–94.

Prezzavento, O., Arena, E., Sánchez-Fernández, C., Turnaturi, R., Parenti, C., Marrazzo, A., et al. (2017). (+)-and (−)-Phenazocine enantiomers: Evaluation of their dual opioid agonist/σ1antagonist properties and antinociceptive effects. *European Journal of Medicinal Chemistry, 125*, 603–610.

Rescifina, A., Floresta, G., Marrazzo, A., Parenti, C., Prezzavento, O., Nastasi, G., et al. (2017). Development of a Sigma-2 receptor affinity filter through a Monte Carlo based QSAR analysis. *European Journal of Pharmaceutical Sciences, 106*, 94–101.

Rickli, A., Liakoni, E., Hoener, M. C., & Liechti, M. E. (2018). Opioid-induced inhibition of the human 5-HT and noradrenaline transporters in vitro: Link to clinical reports of serotonin syndrome. *British Journal of Pharmacology, 175*, 532–543.

Schmidt, H. R., & Kruse, A. C. (2019). The molecular function of σ receptors: Past, present, and future. *Trends in Pharmacological Sciences, 40*, 636–654.

Schmidt, H. R., Zheng, S., Gurpinar, E., Koehl, A., Manglik, A., & Kruse, A. C. (2016). Crystal structure of the human σ1 receptor. *Nature, 532*, 527–530.

Su, T.-P. (2019). Non-canonical targets mediating the action of drugs of abuse: Cocaine at the Sigma-1 receptor as an example. *Frontiers in Neuroscience, 13*. https://doi.org/10.3389/fnins.2019.00761.

Tao, R., Karnik, M., Ma, Z., & Auerbach, S. B. (2003). Effect of fentanyl on 5-HT efflux involves both opioid and 5-HT 1A receptors. *British Journal of Pharmacology, 139*, 1498–1504.

Tschirhart, J. N., Li, W., Guo, J., & Zhang, S. (2019). Blockade of the human ether A-go-go–related gene (hERG) potassium channel by fentanyl. *Molecular Pharmacology, 95*, 386–397.

Vardanyan, R. S., & Hruby, V. J. (2014). Fentanyl-related compounds and derivatives: Current status and future prospects for pharmaceutical applications. *Future Medicinal Chemistry, 6*, 385–412.

Vasudevan, S. R., Moore, J. B., Schymura, Y., & Churchill, G. C. (2012). Shape-based reprofiling of FDA-approved drugs for the H1 histamine receptor. *Journal of Medicinal Chemistry, 55*, 7054–7060.

Volpe, D. A., McMahon Tobin, G. A., Mellon, R. D., Katki, A. G., Parker, R. J., Colatsky, T., et al. (2011). Uniform assessment and ranking of opioid μ receptor binding constants for selected opioid drugs. *Regulatory Toxicology and Pharmacology, 59*, 385–390.

Weber, F., & Wünsch, B. (2017). Medicinal chemistry of σ1 receptor ligands: Pharmacophore models, synthesis, structure affinity relationships, and pharmacological applications. In F. J. Kim, & G. W. Pasternak (Eds.), *Sigma proteins: Evolution of the concept of sigma receptors* (pp. 51–79). Springer International Publishing.

Xiong, J., Jin, J., Gao, L., Hao, C., Liu, X., Liu, B.-F., et al. (2020). Piperidine propionamide as a scaffold for potent sigma-1 receptor antagonists and mu opioid receptor agonists for treating neuropathic pain. *European Journal of Medicinal Chemistry, 191*, 112144.

Zamanillo, D., Romero, L., Merlos, M., & Vela, J. M. (2013). Sigma 1 receptor: A new therapeutic target for pain. *European Journal of Pharmacology, 716*, 78–93.

Zeng, C., & Mach, R. H. (2017). The evolution of the sigma-2 (σ2) receptor from obscure binding site to bona fide therapeutic target. In S. Smith, & S. Tsung-Ping (Eds.), *Sigma receptors: Their role in disease and as therapeutic agents* (pp. 49–61). Springer Nature.

Chapter 11

Isoflurane: Mechanisms and applications

Lady Christine Ong Sio, Marina Varbanova, and Alexander Bautista
Department of Anesthesiology and Perioperative Medicine, University of Louisville, Louisville, KY, United States

Abbreviations

MAC	minimum alveolar concentration
GABA-A	gamma aminobutyric acid
AEA	*N*-arachidonoylethanolamine or endocannabinoid anandamide
atm	atmosphere
CBF	cerebral blood flow
CO_2	carbon dioxide
CSF	cerebrospinal fluid
EEG	electroencephalogram
COPD	chronic obstructive pulmonary disease
CYP	cytochrome P450 enzyme
CYP2E1	cytochrome P450 family 2 subfamily E member 1
ED	effective dose

Introduction

The introduction of inhalational agents in clinical practice has enabled safe practice in the conduct of general anesthesia. The discovery of ether and chloroform in the nineteenth century marked the beginning of general anesthesia. General anesthesia is defined as a drug-induced reversible depression of the central nervous system, leading to a loss of response to external stimuli. An ideal agent should provide a reversible state of hypnosis, amnesia, analgesia, and immobility with autonomic and sensory blockade. Inhalation agents are able to provide these components in varying clinically relevant concentrations.

The introduction of inhaled anesthetics in clinical practice began with the successful use of nitrous oxide in 1844 in the field of dental anesthesia, followed by the discovery of the properties of ether in 1846 and chloroform in 1847. However, the introduction of inhaled anesthetics came 80 years later (Eger, 1993).

Halothane was the first inhaled anesthetic to have been successfully marketed due its potency, nonflammability, and acceptable anesthetic profile (Halsey, 1981). However, less-than-ideal cardiovascular effect, hepatic toxicity, and substantial degree of metabolism prompted research to synthesize new volatile anesthetics, which are equal to or even better than halothane (Terrell, 2008).

A methyl ethyl ether derivative, enflurane, was introduced in 1973. Unlike halothane, hepatotoxicity in enflurane was less likely. However, its side effects included metabolism to inorganic fluoride and central nervous system stimulation at high concentrations.

Isoflurane is one of the series of fluorinated methyl ethyl ethers that was originally synthesized by Ross Terrell in 1965 (Prys-Roberts, 1981). However, it was released for general use in the USA and Canada 16 years later.

Desflurane, a totally fluorinated methyl ethyl ether, was introduced in 1992, and in 1994, sevoflurane, a totally fluorinated methyl isopropyl ether was introduced. To date, isoflurane, desflurane, and sevoflurane are the most common inhaled anesthetics used in the United States, along with nitrous oxide. Balancing the pharmacokinetic benefits of these inhaled anesthetics while minimizing risks remain a challenge to the anesthesiologist.

This book chapter aims to review the chemical structure and properties of isoflurane, discuss its pharmacodynamics and pharmacokinetics, and its application in clinical practice today.

Treatments, Mechanisms, and Adverse Reactions of Anesthetics and Analgesics. https://doi.org/10.1016/B978-0-12-820237-1.00011-9
Copyright © 2022 Elsevier Inc. All rights reserved.

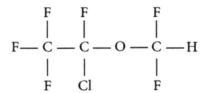

FIG. 1 Structure of isoflurane.

Chemical structure and properties

Isoflurane is a halogenated volatile agent, 2-chloro-2-(difluoromethoxy)-1,1,1-trifluoro-ethane (Fig. 1). It is a structural isomer of enflurane, with three fluorine atoms on the terminal ethyl carbon, giving isoflurane its molecular stability. Hence, isoflurane is resistant to chemical or biological reactions. Isoflurane does not react with metal and is stable in soda lime and ultraviolet light (DiFazio, Brown, Ball, Heckel, & Kennedy, 1972).

Pharmacology and pharmacokinetics

A partition coefficient pertains to the ratio of the concentration of a compound in one medium to that in another at equilibrium. Thus, the blood-gas partition coefficient of an inhaled anesthetic describes how the latter will partition itself between two phases after equilibrium has been reached. Therefore, the higher the blood-gas coefficient, the higher the uptake of the gas in the blood and therefore, induction will be slower. Table 1 lists the different blood-gas coefficients of inhaled anesthetics. Isoflurane has a moderately high blood-gas coefficient of 1.4, meaning if the gas is in equilibrium, concentration in the blood will be 1.4 times higher than that in the alveoli, with consequent slow uptake and induction of general anesthesia.

Isoflurane has high fat solubility that is associated with prolonged emergence, particularly after prolonged exposure due to its accumulation in tissues. Hence, this property of isoflurane limits its use in short procedures. If the anticipated procedure is of long duration, the use of isoflurane has more value because it is inexpensive, and it is the most potent of the volatile fluorinated volatile anesthetics.

It is supplied as a colorless bottled liquid that readily evaporates at standard temperature and pressure. Samples stored for 5 years in indirect sunlight in clear and colorless glass bottles showed an unchanged composition by gas chromatography. It is delivered via a vaporizer that is mounted on an anesthesia machine.

Mechanism of action

The mechanism of action of all inhalational anesthetics is poorly understood. The dominant molecular mechanisms by which volatile anesthetics depress neural functions both in the brain and in the spinal levels are not clearly defined. It has been demonstrated in rodent studies that isoflurane binds to GABA-A and glycine receptors, which contributes to depressant actions on the spinal ventral horn, thus producing immobility (Grasshoff & Antkowiak, 2006).

TABLE 1 Properties of inhaled anesthetics.

Inhaled anesthetic	Blood-gas coefficient	MAC (%)
Desflurane	0.45	6
Enflurane	1.8	1.7
Halothane	2.3	0.75
Isoflurane	1.4	1.4
Nitrous oxide	0.47	104
Sevoflurane	0.65	2

Isoflurane has been found to affect the endocannabinoid system by reducing plasma endocannabinoid anandamide (AEA), which may affect systemic inflammatory responses during surgery and promotes stress reduction after loss of consciousness (Weis et al., 2010).

Minimum alveolar concentration (MAC) is the concentration at 1 atmosphere (atm) of inhaled anesthetic, which prevents movement in response to surgical stimulation in 50% of patients (Campagna, Miller, & Forman, 2003). The MAC of isoflurane is 1.28% as shown in Table 1.

Clinical effects

Table 2 summarizes the effects of isoflurane on different organs.

TABLE 2 Clinical effects of isoflurane.

CNS effects	
Cerebral Metabolic Rate	Decrease
Cerebral Blood Flow	Increase
Intracranial Pressure	Increase
CSF formation	No Effect
EEG frequency	Decrease
Seizures	Decrease
Evoked potentials	Increase latency Decrease amplitude
Respiratory effects	
Respiratory Rate	Increase
Tidal volume	Decrease
Functional residual capacity	Decrease
Dead space	Increase
Cardiac effects	
Arterial blood pressure	Decrease
Systemic vascular resistance	Decrease
Heart rate	Increase
Cardiac output	No effect
Skeletal and smooth muscle effects	
Relaxation of skeletal/smooth muscle	Increase
Uterine smooth muscle	Relaxation
Hepatic effects	
Hepatic circulation	Vasodilatation
Hepatic artery blood flow	No effect
Renal effects	
Renal blood flow	Decrease
Glomerular filtration rate	Decrease
Urine output	Decrease

Central nervous system effects

Isoflurane produces a dose-dependent cerebral vasodilation with a reduction of cerebral metabolic rate. It increases cerebral blood flow (CBF) and intracranial pressure by blunting cerebral autoregulation. However, at <1 MAC values, there is only a modest decrease in CBF and CO_2 responsiveness. At low or high concentrations, isoflurane causes no change in the rate of cerebrospinal fluid (CSF) formation and in resistance to resorption of CSF (Stoelting & Hillier, 2006).

At doses approaching 1 MAC, the EEG frequency decreases. With isoflurane, burst suppression appears on EEG at about 1.5 MAC and electrical silence happens at 2 MAC. Isoflurane does not evoke seizure activity on EEG (Stoelting & Hillier, 2006).

Respiratory effects

Isoflurane is usually administered following an intravenous induction agent due to its pungent odor. The sole use of isoflurane alone can precipitate coughing due to activation of airway reflexes even in apneic periods. This effect is usually prominently observed in smokers and in patients with reactive airway diseases, such as asthma and chronic obstructive pulmonary disease (COPD).

The respiratory effect of isoflurane is dose-dependent. Spontaneous ventilation is usually preserved with an associated increase in rate and decrease in tidal volume. The respiratory depressant effect may progressively shift the respiratory carbon dioxide response curve to the right, leading to a blunted ventilatory response to hypercapnia. As with other inhalational agents, isoflurane blunts the hypoxic pulmonary drive. Isoflurane can also reverse hypoxic pulmonary vasoconstriction, which may result in hypoxia, due to a concomitant increase in ventilation and perfusion mismatch following perfusion of a poorly ventilated lung (Stoelting & Hillier, 2006).

Cardiac effects

Isoflurane produces a dose-dependent reduction in arterial blood pressure due to peripheral vasodilation while preserving cardiac output. During high isoflurane concentrations, progressive tachycardia can be seen with its use. Isoflurane does not sensitize the myocardium to produce arrhythmias. Coronary arterial vasodilation can lead to a coronary artery steal phenomenon; however, this does not appear to be clinically significant (Stoelting & Hillier, 2006).

Administering isoflurane to susceptible patients can prolong the QT interval. This effect may be exacerbated by the patient's disease conditions or concomitant perioperative medications (e.g., congenital long QT syndrome or taking drugs that can prolong the QT interval). Isolated cases of cardiac arrhythmia associated with QT prolongation have been reported. There are very rare reports of torsade de pointes.

Skeletal and smooth muscle effects

Similar to all inhalational anesthetics, isoflurane induces dose-dependent relaxation of both skeletal and smooth muscles by inhibiting nicotinic acetylcholine receptors. The degree of skeletal muscle relaxation contributed by isoflurane alone may be insufficient to prevent movement during surgery. Hence, the use of a neuromuscular blocking agent is needed to potentiate its effect.

Isoflurane on smooth muscles in the gastrointestinal and genitourinary tracts can cause postoperative nausea, emesis, ileus, and urinary retention.

Hepatic effects

Isoflurane causes vasodilation of hepatic circulation while maintaining total hepatic blood flow and hepatic artery blood flow. Unlike halothane, isoflurane effects on the liver is mild and self-limited, as with other volatile anesthetics. Trifluoroacetic acid is a potential metabolite of isoflurane, and a condition similar to halothane hepatitis, albeit extremely low, has been reported in humans (Pawson & Sandra, 2008).

Renal effects

Dose-related decreases in renal blood flow, glomerular filtration rate, and urine output have been observed in volatile anesthetics, including isoflurane, and are due to effects on systemic blood pressure and cardiac output (Stoelting & Hillier, 2006).

Obstetric effects

Smooth muscle relaxation contributed by isoflurane may be beneficial if uterine relaxation is needed to facilitate removal of a retained placenta or products of conception. However, uterine relaxation produced by volatile anesthetics may contribute to blood loss secondary to uterine atony. Isoflurane, as any other inhaled anesthetic, readily crosses the placenta and enters the fetus, but it is rapidly exhaled by the newborn infant (Stoelting & Hillier, 2006).

Adverse effects

The most frequent adverse events of isoflurane (incidence >10%) are white blood cell count increase, agitation, breath holding, cough, nausea, and chills/shivering.

Malignant hyperthermia

Isoflurane shares the same risk for malignant hyperthermia with the rest of the volatile agents. It is very important to avoid its use in susceptible and at-risk patients (Stoelting & Hillier, 2006).

Teratogenic effects

Occupational exposure to inhalational anesthetics has often been associated with health hazards and reproductive toxicity, but the available evidence is weak and comes mostly from epidemiological studies that have been criticized. Studies based on registered data generally showed no association between occupational exposure to inhalational anesthetics and reproductive effects. Animal studies also showed a lack of carcinogenicity, organ toxicity, and reproductive effects with trace concentrations, as observed in operating rooms. Nonetheless, it is good practice to limit levels of exposure (Burm, 2003).

Environmental effects

Inhalational agents have been shown to undergo little metabolism during clinical applications and evaporate almost completely to the environment. This contributes to the greenhouse effect and their potential to destroy the stratospheric ozone layer. Isoflurane has an average lifetime of 3.2 years, despite its global warming potential of 510 (Schoenenberger, 2015).

Volatile anesthetics such as desflurane, enflurane, and isoflurane all have CHF_2 moiety in their structures, such that its degradation by strong bases contributes to carbon monoxide formation. However, sevoflurane and halothane form negligible amounts of carbon monoxide (Baxter, Garton, & Kharasch, 1998). Factors influencing carbon monoxide production from inhaled anesthetics include: (1) dryness of carbon monoxide absorbent; (2) high temperature of carbon dioxide absorbent during low gas flows; (3) prolonged high fresh gas flows, leading to desiccation of carbon dioxide absorbent; and (4) type of carbon dioxide absorbent (Baxter & Kharasch, 1997) (Baxter et al., 1998).

Metabolism

The liver is the major site of metabolism for most inhaled anesthetics. The presence of hepatic CYP2E1 is important in the oxidative metabolism of isoflurane, resulting in the release of fluoride and chloride ions to form reactive intermediates that react with water to form carboxylic acids. Enzyme induction with drugs such as phenobarbital or phenytoin increases the separation of fluoride ions from isoflurane; however, the resulting plasma concentrations of fluoride remain less than with other inhaled anesthetics (Mazze, Woodruff, & Heerdt, 1982).

An association with fulminant hepatic necrosis is very low in isoflurane, unlike in halothane (Miller, 2015). Disease states that decrease hepatic blood flow and drug delivery, i.e., cirrhosis, can alter the metabolism of isoflurane.

The kidneys also contain the CYP enzyme, including CYP2E1 that catalyze both phase 1 and phase 2 reactions. It is an additional site where inhaled anesthetic metabolism occurs. Resultant serum fluoride concentrations have been implicated with fluoride-induced nephrotoxicity; however, studies have shown that this is not clinically significant with isoflurane.

Drug-drug interactions

Opioids

The minimum alveolar concentration (MAC) for isoflurane is reduced by concomitant administration of intravenous anesthetics, such as opioids and benzodiazepines. Opioids such as fentanyl and its analogs, when combined with isoflurane, may lead to a synergistic decrease in blood pressure and respiratory rate.

Nitrous oxide

N_2O decreases the MAC of isoflurane.

Neuromuscular blocking agents

Isoflurane decreases the required doses of neuromuscular blocking agents. It potentiates all commonly used muscle relaxants, the effect being most profound with the nondepolarizing type. Therefore, less than the usual amounts of such agents should be used. In general, anesthetic concentrations of isoflurane at equilibrium reduce the ED_{95} of succinylcholine, atracurium, pancuronium, rocuronium, and vecuronium by approximately 25%–40% or more.

Epinephrine

Isoflurane is similar to sevoflurane in the sensitization of the myocardium to the arrhythmogenic effect of exogenously administered epinephrine. The dose of epinephrine producing ventricular extrasystoles in 50% of humans anesthetized with 1.25 MAC isoflurane is 6.7 micrograms per kilogram when epinephrine is injected submucosally. This would equal 47 mL of a 1:100,000 epinephrine-containing solution in a 70-kg man.

Calcium antagonists

Isoflurane may lead to marked hypotension in patients treated with calcium antagonists.

Concomitant use of beta blockers

Concomitant use of beta blockers may exaggerate the cardiovascular effects of inhalational anesthetics, including hypotension and negative inotropic effects.

Application to other areas

- Safer inhalational anesthetics for the pediatric population have been the subject of many researches worldwide. A study by Tao, et al. in 2016 has shown that isoflurane, but not desflurane, is more likely to induce developmental neurotoxicity in the context of multiple exposures to anesthesia (Tao et al., 2016).
- A retrospective case series performed by Shankar et al. in 2005 in children aged 1–16 years with severe asthma requiring invasive mechanical ventilation has shown that isoflurane improves arterial pH and reduced partial pressure of arterial carbon dioxide, leading to improvement and rapid weaning from mechanical ventilation (Shankar, Churchwell, & Deshpande, 2006).

Other agents of interest

The current role of volatile agents in the setting of critical care medicine has been a subject of interest. Expanding the use of volatile anesthetics as a means of sedation in the intensive care unit has attracted researchers on their likely benefits. A trial conducted by Kong and colleagues in 1989 compared isoflurane to midazolam in 60 ICU patients. Several trials have also been conducted and have shown faster patient recovery. Its potential for end-organ protective properties has also been of research interest. However, like sedatives, volatile agents can induce deep levels of sedation, not to mention respiratory depressant effects and reduction of patient mobility. Further research is needed to study their long-term effects on mortality, length of mechanical ventilation, as well as early and long-term cognitive effects. To date, a low number of studies and subjects limit their applications in the ICU (Jerath et al., 2015).

Mini-dictionary of terms

Cytochrome P450 (CYP P450): A group of enzymes that are particularly important in humans in that it participates in several oxidation-reduction reactions to metabolize thousands of exogenous and endogenous chemicals, hormone synthesis and breakdown, and drug metabolism. Its peak absorption wavelength is measured at 450 nm by spectrophotometry, hence the term "P450". Human enzymes are found in the mitochondria and endoplasmic reticulum of cells, and so far, 57 human genes have been identified to code for various P450 enzymes. Many drugs either induce the biosynthesis of an enzyme (enzyme induction) or directly inhibit CYP activity (enzyme inhibition); thus, it is essential to know how one drug affects another.

ED_{95}: The dose required to achieve the desired effect in 95% of the population.

GABA receptors: Neurotransmitter gamma-aminobutyric acid (GABA) is the chief inhibitory compound in the human central nervous system. When GABA binds to GABA receptors, it changes its shape to allow chloride ions to enter the neuron and as an effect, the neuron reduces its excitability or is said to be "inhibited". Understanding GABA receptors is important and is the target of many anesthetic effects.

Minimum alveolar concentration (MAC): The concentration at 1 atm of inhaled anesthetic, which prevents movement in response to surgical (pain) stimulation in 50% of patients. It can be used to compare the potency of inhaled anesthetics. Certain physiological and pathological states can affect MAC, thereby increasing or decreasing the anesthetic required to prevent movement. Human studies have determined that MAC peaks at 6 months of age, after which it progressively decreases at the rate of 4% every decade over the age of 40 years.

Key facts of isoflurane

- Isoflurane has myocardial protective properties, limiting infarct size and improving functional recovery from myocardial ischemia. The mechanism for this protection mimics ischemic preconditioning and involves the opening of adenosine triphosphate-dependent potassium channels.
- The benefits of isoflurane pretreatment are not limited to the cardiovascular system. There is evidence indicating that the inhibitory effects of isoflurane on hepatic cancer aggressiveness may be mediated by regulation of the PI3K/AKT-induced NF-κB signaling pathway. These findings indicate that isoflurane may be regarded as a preferable anesthetic and as an additional antitumor agent for clinical treatment of patients with hepatic carcinoma.

Summary points

- Isoflurane, like all inhalational agents, decreases the cerebral metabolic rate of oxygen and maintains a dose-dependent increase in cerebral flood flow.
- Isoflurane's pungency limits its use as a sole inhalational induction agent; however, its low cost and easy titratability makes it favorable for use in the operating room.
- In general, the mechanism of action of isoflurane and volatile anesthetics is poorly understood. It has been demonstrated that the binding of isoflurane to GABA-A and glycine receptors as well its effect on the endocannabinoid system may promote isoflurane's depressant actions, inflammatory responses to surgery, and stress reduction.
- Isoflurane affects the central nervous system, respiratory, cardiovascular, hepatic and renal systems, as well as skeletal and smooth muscles.
- Adverse effects include predisposition to malignant hyperthermia in susceptible patients and toxic environmental byproducts.

References

Baxter, P. J., Garton, K., & Kharasch, E. D. (1998). Mechanistic aspects of carbon monoxide formation from volatile anesthetics. *Anesthesiology, 89*(4), 929–941. https://doi.org/10.1097/00000542-199810000-00018.

Baxter, P. J., & Kharasch, E. D. (1997). Rehydration of desiccated Baralyme prevents carbon monoxide formation from desflurane in an anesthesia machine. *Anesthesiology, 86*(5), 1061–1065. https://doi.org/10.1097/00000542-199705000-00009.

108 PART | II Mechanisms of action of drugs

Burm, A. G. (2003). Occupational hazards of inhalational anaesthetics. *Best Practice & Research. Clinical Anaesthesiology, 17*(1), 147–161. https://doi.org/10.1053/bean.2003.0271.

Campagna, J. A., Miller, K. W., & Forman, S. A. (2003). Mechanisms of actions of inhaled anesthetics. *The New England Journal of Medicine, 348*(21), 2110–2124. https://doi.org/10.1056/NEJMra021261.

DiFazio, C. A., Brown, R. E., Ball, C. G., Heckel, C. G., & Kennedy, S. S. (1972). Additive effects of anesthetics and theories of anesthesia. *Anesthesiology, 36*(1), 57–63. https://doi.org/10.1097/00000542-197201000-00010.

Eger, E. (1993). *Desflurane(Suprane): A compendium and reference. Vol. 1* (p. 119). Nutley, NJ: Anaquest.

Grasshoff, C., & Antkowiak, B. (2006). Effects of isoflurane and enflurane on GABAA and glycine receptors contribute equally to depressant actions on spinal ventral horn neurones in rats. *British Journal of Anaesthesia, 97*(5), 687–694. https://doi.org/10.1093/bja/ael239.

Halsey, M. J. (1981). Investigations on isoflurane, sevoflurane and other experimental anaesthetics. *The British Journal of Anaesthesia, 53*(Suppl 1), 43S–47S. Retrieved from https://www.ncbi.nlm.nih.gov/pubmed/7016154.

Jerath, A., Ferguson, N. D., Steel, A., Wijeysundera, D., Macdonald, J., & Wasowicz, M. (2015). The use of volatile anesthetic agents for long-term critical care sedation (VALTS): Study protocol for a pilot randomized controlled trial. *Trials, 16*, 560. https://doi.org/10.1186/s13063-015-1083-5.

Mazze, R. I., Woodruff, R. E., & Heerdt, M. E. (1982). Isoniazid-induced enflurane defluorination in humans. *Anesthesiology, 57*(1), 5–8. https://doi.org/10.1097/00000542-198207000-00002.

Miller, R. D. (2015). *Miller's anesthesia* (8th ed.). Philadelphia, PA: Churchill Livingston/Elsevier.

Pawson, P. F., & Sandra. (2008). Anesthetic agents. In J. Maddison, S. Page, & D. Church (Eds.), *Small animal clinical pharmacology* (pp. 83–112). W.B. Saunders.

Prys-Roberts, C. (1981). Isoflurane. *British Journal of Anaesthesia, 53*(12), 1243–1245. https://doi.org/10.1093/bja/53.12.1243.

Schoenenberger, M. K. V. T. S. R. M. R. D. H. M. H. F. (2015). Modern inhalation anesthetics: Potent greenhouse gases in the global atmosphere. *Geophysical Research Letters, 42*(5), 1606–1611.

Shankar, V., Churchwell, K. B., & Deshpande, J. K. (2006). Isoflurane therapy for severe refractory status asthmaticus in children. *Intensive Care Medicine, 32*(6), 927–933.

Stoelting, R. K., & Hillier, S. (2006). *Pharmacology and physiology in anesthetic practice*. Philadelphia: Lippincott Williams and Wilkins.

Tao, G., Xue, Q., Luo, Y., Li, G., Xia, Y., & Yu, B. (2016). Isoflurane is more deleterious to developing brain than desflurane: The role of the Akt/GSK3beta signaling pathway. *BioMed Research International, 2016*, 7919640. https://doi.org/10.1155/2016/7919640.

Terrell, R. C. (2008). The invention and development of enflurane, isoflurane, sevoflurane, and desflurane. *Anesthesiology, 108*(3), 531–533. https://doi.org/10.1097/ALN.0b013e31816499cc.

Weis, F., Beiras-Fernandez, A., Hauer, D., Hornuss, C., Sodian, R., Kreth, S., ... Schelling, G. (2010). Effect of anaesthesia and cardiopulmonary bypass on blood endocannabinoid concentrations during cardiac surgery. *British Journal of Anaesthesia, 105*(2), 139–144. https://doi.org/10.1093/bja/aeq117.

Chapter 12

The lidocaine patch: Features and applications: Post-thoracotomy pain and beyond

Alfonso Fiorelli, Pasquale Sansone, Caterina Pace, and Mario Santini

Division of Thoracic Surgery, University of Campania "Luigi Vanvitelli", Caserta, Italy

Abbreviations

FEV1	forced expiratory volume 1 s
FVC	forced vital capacity
GPE	global perceived effect
ICU	intensive care unit
IV	intravenous
LEP	laser-evoked potential
LOHS	length of hospital stay
MeSH	Medical Subject Headings
NRS	numeric rating scale
PCA	patient-controlled analgesia
POHD	post-operative days
POHS	post-operative hours
RCT	randomized control trial
VAS	Visual Analogue Scale
VRS	verbal rating pain scale
WOMAC	Western Ontario and McMaster Universities Osteoarthritis Index

Introduction

Acute postoperative pain is present in more than 80% of patients undergoing surgery. Approximately 75% of patients define it as moderate, severe, or extreme, and only 50% report an adequate postoperative pain relief (Chou et al., 2016; Sansone et al., 2015). An inadequate pain control is strongly associated with poor outcomes and increased of morbidity and mortality, especially in patients undergoing thoracotomy. Post-thoracotomy pain reduces coughing and the mobilization of secretions, favoring respiratory complications, such as atelectasis and pneumonia (Fiorelli et al., 2016). Unfortunately, the ideal strategy for controlling acute postoperative pain is still debate. Systemic opioid administration continues to be the main route, but it may cause side effects especially in patients with multiple comorbidities (Wenk & Schug, 2011). Regional techniques (Hogan & Abram, 1997; Manion & Brennan, 2011), such as epidural analgesia, selective nerve or muscle blocks may fail, and are sometimes contraindicated or not feasible for anatomic reasons. Thus, current evidences (Helander et al., 2017; Tan, Law, & Gan, 2015; Wick, Grant, & Wu, 2017) suggest that the administration of two or more analgesics with different mechanisms of action is more effective than monotherapy as it reduces the level of pain while minimizing the dose and adverse opioid effects.

Lidocaine has been widely used as part of a multimodal technique. There are many routes of administration, and the choice depends upon several factors (i.e., site of surgery, duration of action, potential adverse effects, etc.). Local infiltration is associated with short-lasting effect, while systemic side effects, and delay wound healing may occur with continuous local infusion (Brower & Johnson, 2003). Lidocaine patches have been used with success in the treatment of postherpetic neuralgia and a variety of neuropathic pain syndromes (Demant et al., 2015; Derry, Wiffen, Moore, & Quinlan,

Treatments, Mechanisms, and Adverse Reactions of Anesthetics and Analgesics. https://doi.org/10.1016/B978-0-12-820237-1.00012-0
Copyright © 2022 Elsevier Inc. All rights reserved.

2014; Goddard & Reaney, 2018; Martini et al., 2018; Sansone et al., 2017), but its effectiveness for controlling acute postoperative pain remains unclear.

Thus, we therefore conduct this review aiming to evaluate whether the Lidocaine 5% patch as adjunct to standard analgesic treatments may confer benefits in reduction postoperative pain, opioid consumption, and length of hospital stay.

Search strategy

A literature review was carried out using MEDLINE, PubMed, Scopus, Google Scholar, and Cochrane database until the end of March 2020. The free text and the MeSH terms were: lidocaine, Lidoderm, lidocaine patch, postoperative, postthoracotomy, analgesia, and acute pain. We recruited additional papers from bibliographies of the selected papers. Papers not written in English language, abstracts, case reports, letters, reviews, and metaanalysis, papers published from the same groups, and those evaluating the role of Lidocaine patch for controlling not acute postoperative pain (i.e., chronic or neuropathic pain) were excluded. The papers were judged relevant, and included in this review by agreement between two independent literature reviewers (AF and MS).

Search results

A total of 365 papers were identified from the initial search. Of these, 355 were excluded, as (i) they were duplicates or abstract or case series; (ii) did not evaluate central topics of interest; (iv) presented wrong outcomes or protocol, and (v) were not written in English language. Thus, our review included a total of 10 papers based on PRISMA guideline (Liberati et al., 2009) (Fig. 1). The characteristics of the identified studies were summarized in Table 1.

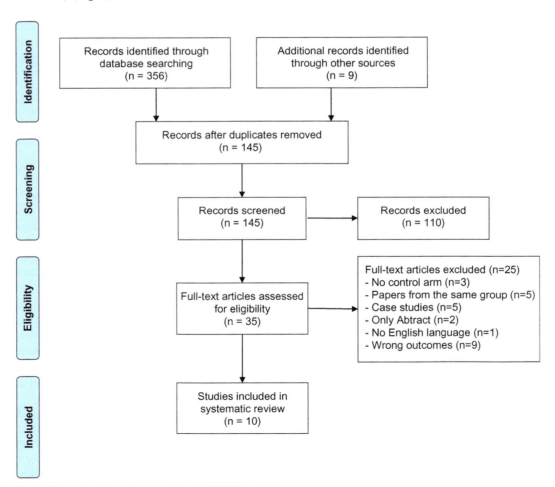

FIG. 1 Flowchart of the systematic review according to PRISMA guideline.

TABLE 1 Selected studies

Authors	Study design	Surgery	Patients	Intervention	Outcomes	Results	Limits	Conclusions
Liu et al. (2019)	Retrospective	Minimally invasive or traditional cardiothoracic surgical procedures	Lidocaine group: 72 patients vs. Control group: 79 patients	Lidocaine patch was placed near surgical incision All patients postoperatively received standard care (opioid and non-opioid analgesics)	Pain scores at 12 h ICU stays LOHS Cumulative opioid and non-opioid consumption	No difference ($P=0.58$) No difference ($P=0.86$) No difference ($P=0.47$) No difference ($P>0.05$)	Retrospective design Low baseline pain scores (only 20% had sternotomy)	The Lidocaine patch did not provide any advantages over pain control, analgesic consumption, ICU and LHOS
Fiorelli et al. (2019)	RCT	Lobectomy via thoracotomy	Lidocaine group: 45 patients vs. Control group: 45 patients	Lidocaine or placebo patch were placed at the level of thoracotomy for three days before surgery All patients postoperatively received IV morphine PCA	Pain evaluation at rest and after coughing at 6;12;24;48;72 POHs (VAS scores) Cumulative morphine consumption Recovery of respiratory function (FEV1 and FVC) LOHS Alteration of peripheral painful pathways measured with N2 and P2 LEPs Adverse effects	Lidocaine group had a significant reduction in VAS scores at rest ($P=0.013$) and after coughing ($P=0.015$) Significant reduction in Lidocaine group ($P=0.001$) Better recovery of FEV1 ($P=0.025$) and FVC ($P=0.037$) in Lidocaine group No difference Significant alteration of LEPs in control group but not in Lidocaine group No difference	Small sample size No sophisticated strategy to quantify the pain	Pre-emptive use of Lidocaine patches significantly reduce post-thoracotomy pain, and morphine consumption

Continued

TABLE 1 Selected studie—cont'd

Authors	Study design	Surgery	Patients	Intervention	Outcomes	Results	Limits	Conclusions
Lau et al. (2018)	RCT	Non-oncological Gynecological open surgery	Lidocaine group: 14 patients vs. Control group: 14 patients	Lidocaine or placebo patch were applied along wound incision All patients postoperatively received IV morphine PCA	Pain scores at rest and on movement at 24 and at 48 POHS	Lidocaine group had lower postoperative pain scores at rest ($P=0.024$) but not on movement ($P=0.4$) at 24 POHS. No difference were found for postoperative pain scores at rest ($P=0.3$) and on movement at 48 ($P=0.7$) POHS	Spirometry performed at bedside Quality of recovery and patient satisfaction not evaluated	The Lidocaine patch 5% provided significant pain relief at rest
					Cumulative morphine consumption	Lower in lidocaine group ($P=0.065$)		
					Spirometry (FEV1 and FVC)	No difference		
					LHOS	No difference		
					Adverse effect	None		
Sadigursky et al. (2017)	RCT	Total knee arthroplasty	Lidocaine group: 22 patients vs. Control group: 23 patients	Lidocaine patch was placed on surgical incision All patients received routine postoperative analgesia with oral analgesic and IV opioid	Pain at 24 POH; 48 POH; 72 POH; 14 POD; 28 POD (VAS scale)	Lidocaine group had lower postoperative pain scores at all time-points	No blind study After discharge patients themselves reported the amount of analgesic drugs used	Lidocaine patches significantly reduced pain, opioid pain rescue load and systemic side effects
					Cumulative analgesic and opioid consumption	Lidocaine group presented smaller consumption only at 72POH ($P=0.001$) and 28 POD ($P=0.013$)		
					Postoperative functional data (WOMAC Questionnaire)	No significant difference ($P=0.42$)		
					Side effects due to opioid (nausea)	Significant lower in lidocaine group		

						Adverse effects	None		
Vrooman et al. (2015)	RCT	Robotic cardiac surgery	Lidocaine group: 39 patients vs. Control group: 39 patients	Lidocaine or placebo patch were externally placed around surgical incisions All patients received IV fentanyl PCA		Pain evaluation at 3;7;30;90;180 POD (VAS scores and phone interview) Cumulative analgesic and opioid consumption Patient satisfaction Adverse effects to Lidocaine patch	No significant difference ($P=0.27$) No significant difference ($P=0.87$) No significant difference ($P=0.17$) None	Different pain measurements	Lidocaine patches were ineffective in reducing acute and chronic pain
Khanna et al. (2012)	Retrospective	Total knee arthroplasty	Lidocaine group: 31 patients Control group: 22 patients	Lidocaine patch was placed on surgical incision All patients received standard analgesic treatment		Postoperative pain (VAS score) during 11 PODs LOHS Patient satisfaction Side effects	No difference No difference 94% of patients were satisfied of lidocaine patch None	Retrospective nature Small sample size Patients were monitored only during their hospital stay but no during rehabilitation stay More female subjects were in the lidocaine patch 5% group	The Lidocaine patch 5% did not provide significant additional pain relief .
Kwon et al. (2012)	RCT	Laparoscopic vaginal hysterectomy and myomectomy	Lidocaine group: 20 patients vs. Control group: 20 patients	Lidocaine or placebo patch were applied on each port site for the first 36 postoperative hours All patients postoperatively received IV fentanyl		Postoperative pain at rest and during ambulation (VAS and VRS scores) Postoperative analgesic requirements Patient satisfaction Side effects	Pain scores at rest were lower in Lidocaine group No difference was found for pain during ambulation No difference No difference No difference	Small sample size	The Lidocaine patch 5% provided significant pain relief

Continued

TABLE 1 Selected studie—cont'd

Authors	Study design	Surgery	Patients	Intervention	Outcomes	Results	Limits	Conclusions
Kim (2011)	RCT	Percutaneous endoscopic lumbar discectomy	Lidocaine group: 50 patients vs. Control group: 50 patients	Lidocaine or placebo patch was applied 1 h before operation Preoperative skin infiltration of lidocaine, and intraoperative IV fentanyl were administered in all patients	Postoperative pain (NRS scores) Patient satisfaction Side effects	NRS scores were lower in Lidocaine group ($P<0.05$) Higher in lidocaine group ($P<0.05$) No difference	Placebo patch was not similar to lidocaine patch Pain intensity was different according to the degree of spinal disc herniation Perception of pain was lowered postoperatively compared to radicular pain originating from herniated nucleus pulposus	The Lidocaine patch 5% provided significant pain relief and patient satisfaction
Habib et al. (2009)	RCT	Radical retropubic prostatectomy	Lidocaine group: 36 patients Placebo Group: 34 patients	Lidocaine patch applied in the first 24 POH In all cases postoperative analgesia was provided using IV morphine PCA	Postoperative pain at rest and after coughing (VAS score) were evaluated at 6; 12 and 24 POHS Quality of postoperative pain Interference of pain with mood, walking, and deep breathing. Postoperative Morphine consumption LOHS Side effects	Lidocaine group had lower VAS scores on coughing over all time periods ($P=0.0001$), and at rest for up to 6 h ($P=0.0003$). Better in lidocaine group ($P=0.037$) Reduced in Lidocaine group ($P<0.05$) No difference No difference None	Small sample size	Lidocaine patch 5% was associated with reduced pain scores and better quality of pain control

Saber et al. (2009)	RCT	Laparoscopic ventral hernia repair	Lidocaine group: 15 patients vs. Control group: 15 patients	Lidocaine or placebo patch were externally placed approximately to the location of the underlying intra-abdominal mesh	Postoperative pain at discharge; at 10–14 days; at 1–2 months (VRS rates)	Lidocaine group had lower postoperative pain scores at discharge ($P=$ 0.0067) but not at 10–14 days ($P=0.12$); at 1–2 months ($P=0.83$)	Small sample size No blind study	The Lidocaine patch 5% provided significant pain relief at discharge
					Cumulative analgesic and morphine consumption	No significant difference		
				All patients received standard postoperative analgesia	LOHS	No significant difference ($P=0.14$)		
					Side effects	none		

Analysis of the results

Liu, Wai, and Nunez (2019) retrospectively evaluated 91 patients undergoing minimally invasive or traditional cardiothoracic surgical procedures, and admitted to ICU. Patients were splitted into two groups based whether they received lidocaine 5% patch (Lidocaine Group $n = 47$) or standard care alone (control group; $n = 44$). Lidocaine patch was placed near sternotomy and/or thoracotomy sites. Pain was evaluated 12 h after placement of the first lidocaine patch using VAS; cumulative opioid and non-opioid consumption, length of ICU and hospital stay. The comparison of the two groups showed no significant differences regarding VAS scores ($P = 0.58$); cumulative opioid and non-opioid consumption ($P > 0.05$), and ICU stay ($P = 0.86$); and LOHS ($P = 0.47$). No adverse effects related to Lidocaine patch occurred. In theory, the effects of Lidocaine patch could be less evident as the most of patients (80%) received a minimally invasive approach associated with low intensity of pain.

Fiorelli et al. (2019) analyzed a total of 90 patients undergoing lobectomy for lung cancer via thoracotomy. Patients were randomly assigned to receive lidocaine 5% patch ($n = 45$) or a placebo ($m = 45$). The patch was placed at level of the thoracotomy for 3 days before surgery. Postoperative analgesia was induced in all cases with IV morphine PCA. Pain at rest and after coughing were evaluated with VAS at 6;12;24;48;72 POHs. Cumulative morphine consumption, recovery of respiratory function measured by FEV1 and FVC, LOHS, and peripheral painful pathways measured with N2 and P2 with LEP were also evaluated. Lidocaine group was associated with a significant reduction in pain intensity at rest ($P = 0.013$) and after coughing ($P = 0.015$), in morphine consumption ($P = 0.001$); a better recovery of FEV1 ($P = 0.025$) and of FVC ($P = 0.037$), and no significant changes in N2 and P2 LEPs until 6 months after operation. No adverse effects related to Lidocaine patch occurred. Small sample size and the use of not sophisticated strategy as VAS to quantify the pain could affect the results.

Lau, Li, Lee, and Chan (2018) evaluated 28 patients undergoing elective non-oncological gynecological surgery via open incision. Patients were randomly assigned to lidocaine 5% patch ($n = 14$) or placebo patch ($n = 14$). Patches were applied along surgical incision for 24 h. Postoperative pain at rest and on movement were measured with VAS at 24 POHs, and at 48 POHS. Cumulative morphine consumption, recovery of respiratory function measured by FEV1 and FVC, and LOHS were also evaluated. The lidocaine group had lower postoperative pain scores at rest ($P = 0.024$), but not on movement at 24 h ($P = 0.4$). No significant differences were found for pain scores at rest ($P = 0.39$) and on movement ($P = 0.75$). Compared to placebo, Lidocaine may slightly lower cumulative morphine consumption over time, despite it was not significant ($P = 0.065$). Spirometry tests showed no significant difference among two groups ($P = 0.09$). No adverse effects related to Lidocaine patch occurred. Spirometry was performed at bedside. Yet, quality of recovery and patient satisfaction were not evaluated.

Sadigursky et al. (2017) analyzed 48 patients undergoing total knee arthroplasty. Patients were randomly assigned to lidocaine 5% patch in adjunction to routine post-operative analgesia ($n = 24$) or to routine post-operative analgesia alone ($n = 24$). Pain levels were measured with VAS from the immediate postoperative to the end of a 28-day interval. Cumulative opioid consumption and functional data measured with the WOMAC index were also evaluated. Lidocaine group had lower postoperative pain scores at all-time points and smaller analgesics and morphine consumption at 72 POH ($P = 0.001$), and at 28 POD ($P = 0.01$). No significant difference was found regarding the WOMAC index. The study was not blind as control group did not receive placebo patch. There was no control on the analgesic consumption after discharge as that was reported by the patients themselves.

Vrooman et al. (2015) analyzed a total of 78 patients undergoing robotic cardiac surgery. Patients were randomly assigned to lidocaine 5% patch or identical-appearing placebo patches. Patch was applied around each incision 12 h/day until pain resolved, or for 6 months. Supplemental opioid was provided by IV PCA or orally. Pain was initially evaluated with a VAS, and subsequently by telephone. Patients satisfaction was simultaneously recorded with GPE scale. No differences were found between two study groups regarding post-operative pain scores ($P = 0.27$), analgesic and morphine consumption ($P = 0.87$); and patient satisfaction ($P = 0.17$). Lidocaine patch was not associated to specific side effects. The different scale for measuring pain and the low post-operative pain due to minimally invasive approach could affect the results.

Khanna, Peters, and Singh (2012) retrospectively evaluated the data of 53 consecutive patients undergoing knee arthroplasty and admitted to rehabilitation unit. Of these, 31 received the lidocaine patch 5%, and 22 patients served as the control group. Post-operative pain was daily measured with the VAS starting on rehabilitation day 1 until day 11. Patient's functional capacity and satisfaction were evaluated using the WOMAC index. The authors found a statistically significant difference in VAS scores on day 3 ($P < 0.05$) between the two groups, with the control group demonstrating better pain relief. However, both groups reported similar pain improvements by the end of their hospital stay. Of the 31 patients receiving lidocaine patch, most of them (94%) reported satisfaction, and attributed pain relief to its use (WOMAC score). Lidocaine patch was not associated with side effects. The retrospective nature of the study, the small sample size, patients monitored only during their hospital stay but not during rehabilitation stay, and the higher rates of female in Lidocaine group are all limits that could affect the results.

Kwon et al. (2012) evaluated 40 patients undergoing laparoscopic gynecologic surgery. Patients were randomly assigned to Lidocaine 5% patch ($n = 20$) or to placebo patch ($n = 20$). The patch was placed at the four port sites, and changed every 12 h for 36 h after surgery. Postoperative pain was evaluated using the VAS and the Prince Henry and 5-point VRS at 1, 6, 12, 24, and 36 h after surgery. Additional dose of analgesics postoperatively administered and patient's satisfaction were also measured. The VAS score for wound pain was lower in the lidocaine patch group at 1 and 6 h after surgery than the control group ($P = 0.005$ and $P < 0.0005$, respectively). The VAS scores for postoperative pain were lower in the lidocaine patch group at rest 1 h after surgery ($P = 0.045$). The 5-point VRS score for postoperative pain was lower in the lidocaine patch group at 6 and 12 h after surgery ($P = 0.015$ and $P = 0.035$, respectively) than in the control group. Post-operative pain during ambulation, and the analgesics consumption were similar between two groups. Lidocaine patch was not associated with side effects.

Kim (2011) evaluated 100 patients undergoing a single level percutaneous endoscopic lumbar discectomy at L4–L5. Patients were randomly assigned to 5% Lidocaine patch ($n = 50$) or placebo patch ($n = 50$). The patches were placed at the site of surgical site 1 h before the surgery and removed after the patient entered the operating room. Pain was measured at each stage during the surgery by patient's NRS. The authors found that the NRS scores at the stages of needle insertion ($P < 0.05$), skin incision ($P < 0.05$), serial dilation and insertion of working channel ($P < 0.05$), and subcutaneous suture ($P < 0.05$) were significantly lower in the Lidocaine group than the control group. Postoperative operator's and patients' satisfaction scores were also significantly higher in Lidocaine group than in the control group. There were no adverse effects in both groups. Placebo patch was not similar to lidocaine patch; the pain intensity before operation were different based on the degree of spinal disc herniation and the perception of pain was lowered postoperatively compared to radicular pain originating from herniated nucleus pulposus. All these limits could affect the results.

Habib et al. (2009) evaluated 70 patients undergoing radical retropubic prostatectomy. Patients were randomly assigned to lidocaine 5% patch ($n = 36$) or to placebo patch ($n = 34$). The patch was applied on each side of the surgical incision for 24 h. Postoperative pain at rest and on movement were measured with VAS scores at 6; 12; and 24 POHs. Postoperative morphine consumption and LOHS were also evaluated. The authors found that the Lidocaine patch group reported significantly less pain on coughing over all time periods ($P = 0.0001$), and at rest for up to 6 h ($P = 0.0003$). Lidocaine compared to placebo group rated that the quality of postoperative pain control was significantly better ($P = 0.037$) as the pain interfered with their walking, deep breathing, and mood significantly less ($P < 0.05$). Postoperative morphine consumption and LHOS were similar between the groups. No adverse effects related to lidocaine patch were found.

Saber, Elgamal, Rao, Itawi, and Martinez (2009) analyzed 35 patients undergoing laparoscopic hernia ventral repair. Patients were randomly assigned to lidocaine 5% patch ($n = 15$) in adjunction to routine post-operative analgesia or to routine post-operative analgesia alone ($n = 15$). Patch was applied on the anterior abdominal wall corresponding to the placement site of the underlying mesh for 12 h daily for three successive days. Postoperative morphine consumption and LOHS were also evaluated. Postoperative pain was measured with the VRS at discharge; at 10–14 days; at 1–2 months. Analgesics consumption and LOHS were also evaluated. The authors found that lidocaine group had lower postoperative pain scores at discharge ($P = 0.0067$), but not at 10–14 days ($P = 0.12$); at 1–2 months ($P = 0.83$). No significant differences were found regarding cumulative consumption of analgesic and of morphine, and of LOHS. The study was not blind as control group did not receive placebo patch.

Discussion

The results of these reviews are contrast. Among 10 selected papers (Sadigursky et al., 2017; Fiorelli et al., 2019; Habib et al., 2009; Khanna et al., 2012; Kim, 2011; Kwon et al., 2012; Lau et al., 2018; Liu et al., 2019; Saber et al., 2009; Vrooman et al., 2015), 7 (Sadigursky et al., 2017; Fiorelli et al., 2019; Habib et al., 2009; Khanna et al., 2012; Kim, 2011; Lau et al., 2018; Saber et al., 2009) showed that Lidocaine patch as adjunct to standard analgesic treatment significantly reduced postoperative pain scores. However, these results were associated with a significant reduction of cumulative analgesic consumption in only two studies (Sadigursky et al., 2017; Fiorelli et al., 2019), while no studies (Sadigursky et al., 2017; Fiorelli et al., 2019; Habib et al., 2009; Khanna et al., 2012; Kim, 2011; Lau et al., 2018; Saber et al., 2009) found a significant reduction of LOHS in Lidocaine group. By contrast, three studies (Kwon et al., 2012; Liu et al., 2019; Vrooman et al., 2015) found that the Lidocaine patch did not provide any significant advantages over pain control, analgesic consumption, and LOHS. The formulation delivers sufficient amounts of lidocaine to block sodium channels reducing both generation and conduction of peripheral pain impulses in damaged or dysfunctional nociceptors underlying patch site application, but insufficient amounts to block sodium channels on large myelinated A-ß sensory fibers (Gammaitoni, Alvarez, & Galer, 2003). Since patients undergoing total knee arthroplasty (Kwon et al., 2012) or thoracotomy (Liu et al., 2019; Vrooman et al., 2015) experienced deep pain at the level of the bone or the intercostal space, the lidocaine 5% patch most likely could not penetrate to provide adequate pain relief at this level. The use of lidocaine patch was safe, as no specific side effects was noted in all analyzed studies. Each patch contains approximately 700 mg of lidocaine, but only 5% of this dose is absorbed systemically, with a mean maximum plasma concentration about 10% of the antiarrhythmic dose (Fiorelli et al., 2015; Gammaitoni et al., 2003).

The results of the present analysis are difficult to compare for the following limitations. (i) All selected studies analyzed different surgical populations including patients undergoing cardio-thoracic; orthopedic; gynecological; urological, and general surgery procedures. An additional bias was that surgery was performed using a minimally invasive approach in some studies or a traditional approach in others. (ii) The evaluation of post-operative pain was not standardized since it was measured in different timing of postoperative period and with different methods (i.e., VAS, VRS). (iii) Postoperative analgesia was different among the studies including IV morphine or fentanyl PCA or oral administration of non-opioid analgesic. Yet, some studies did not evaluate the amount of additional analgesics postoperatively administered. (iv) All RCT studies included a small sample with less than 50 patients randomized to each arm; and two of these were no blinded as the control group did not receive a placebo patch.

Conclusions

This systematic review suggests contrast results on the useful of lidocaine 5% patch for controlling acute postoperative pain. The Lidocaine seems to not provide any advantages over deep acute post-operative pain control. Yet, it was not associated with reduction of analgesic consumptions neither with lower LOHS. However, lidocaine 5% patch remains a safe procedure as no specific side effects was found. Additional large, well designed studies are needed to confirm these findings.

Key facts of lidocaine patch for controlling acute postoperative pain

- Acute postoperative pain is present in more than 80% of patients undergoing surgery, and only 50% report an adequate postoperative pain relief.
- An inadequate pain control is strongly associated with poor outcomes and increased of morbidity and mortality, especially in patients undergoing thoracotomy.
- The administration of two or more analgesics is more effective than monotherapy as it reduces the level of pain while minimizing the dose and adverse opioid effects.
- Lidocaine patches have been used with success in the treatment of post-herpetic neuralgia and a variety of neuropathic pain syndromes, but its effectiveness for controlling acute post-operative pain remains unclear.
- We reviewed 10 papers in order to evaluate whether the Lidocaine 5% patch as adjunct to standard analgesic treatments may confer benefits in reduction postoperative pain, opioid consumption, and length of hospital stay.
- The results were difficult to compare due to the different surgical populations, and the lack of a standard protocol for measuring post-operative pain and for postoperative analgesia.
- Lidocaine patch seemed not provide any significant benefits.
- Seven papers showed that Lidocaine patch as adjunct to standard analgesic treatment significantly reduced postoperative pain scores while three papers found no significant benefits.
- Lidocaine patch was associated with a significant reduction of morphine or other analgesic consumption in only two papers while no papers reported a lower length of hospital stay in patients treated with Lidocaine patch.
- Lidocaine patch was safe, as no specific side effects was noted in all analyzed studies.

Summary points

- This chapter focus on the efficacy of Lidocaine 5% patch as adjunction to standard analgesic treatments for controlling acute post-operative pain is still debate.
- A systematic review of literature was performed and 10 papers were selected to investigate this issue.
- Lidocaine patch seemed not provide any significant benefits for reducing deep, acute postoperative pain, opioid consumption, and length of hospital stay.
- It was a safe procedure as no specific side effects was found in all studies.
- Different surgical populations, and the lack of a standard protocol are the main limits of this analysis.
- Future larger and well-designed studies are needed to corroborate these results.

References

Brower, M. C., & Johnson, M. E. (2003 May-June). Adverse effects of local anesthetic infiltration on wound healing. *Regional Anesthesia and Pain Medicine, 28*(3), 233–240.

Chou, R., Gordon, D. B., de Leon-Casasola, O. A., Rosenberg, J. M., Bickler, S., Brennan, T., et al. (2016). Management of postoperative pain: A clinical practice guideline from the American Pain Society, the American Society of Regional Anesthesia and Pain Medicine, and the American Society of Anesthesiologists' Committee on Regional Anesthesia, Executive Committee, and Administrative Council. *The Journal of Pain, 17*(2), 131–157.

Demant, D. T., Lund, K., Finnerup, N. B., Vollert, J., Maier, C., Segerdahl, M. S., et al. (2015). Pain relief with lidocaine 5% patch in localized peripheral neuropathic pain in relation to pain phenotype: A randomised, double-blind, and placebo-controlled, phenotype panel study. *Pain, 156*(11), 2234–2244.

Derry, S., Wiffen, P. J., Moore, R. A., & Quinlan, J. (2014 July 24). Topical lidocaine for neuropathic pain in adults. *Cochrane Database of Systematic Reviews, 7*, CD010958.

Fiorelli, A., Izzo, A. C., Frongillo, E. M., Del Prete, A., Liguori, G., Di Costanzo, E., et al. (2016). Efficacy of wound analgesia for controlling post-thoracotomy pain: A randomized double-blind study†. *European Journal of Cardio-Thoracic Surgery, 49*(1), 339–347.

Fiorelli, A., Mazzella, A., Passavanti, B., Sansone, P., Chiodini, P., Iannotti, M., et al. (2015). Is pre-emptive administration of ketamine a significant adjunction to intravenous morphine analgesia for controlling postoperative pain? A randomized, double-blind, placebo-controlled clinical trial. *Interactive Cardiovascular and Thoracic Surgery, 21*(3), 284–290.

Fiorelli, A., Pace, C., Cascone, R., Carlucci, A., De Ruberto, E., Izzo, A. C., et al. (2019). Preventive skin analgesia with lidocaine patch for management of post-thoracotomy pain: Results of a randomized, double blind, placebo controlled study. *Thoracic Cancer, 10*(4), 631–641.

Gammaitoni, A. R., Alvarez, N. A., & Galer, B. S. (2003). Safety and tolerability of the lidocaine patch 5%, a targeted peripheral analgesic: A review of the literature. *Journal of Clinical Pharmacology, 43*(2), 111–117.

Goddard, J. M., & Reaney, R. L. (2018). Lidocaine 5%-medicated plaster (Versatis) for localised neuropathic pain: Results of a multicentre evaluation of use in children and adolescents. *British Journal of Pain, 12*(3), 189–193.

Habib, A. S., Polascik, T. J., Weizer, A. Z., White, W. D., Moul, J. W., ElGasim, M. A., et al. (2009). Lidocaine patch for postoperative analgesia after radical retropubic prostatectomy. *Anesthesia and Analgesia, 108*(6), 1950–1953.

Helander, E. M., Menard, B. L., Harmon, C. M., Homra, B. K., Allain, A. V., Bordelon, G. J., et al. (2017). Multimodal analgesia, current concepts, and acute pain considerations. *Current Pain and Headache Reports, 21*(1), 3.

Hogan, Q. H., & Abram, S. E. (1997 January). Neural blockade for diagnosis and prognosis. A review. *Anesthesiology, 86*(1), 216–241.

Khanna, M., Peters, C., & Singh, J. R. (2012). Treating pain with the lidocaine patch 5% after total knee arthroplasty. *PM & R: The Journal of Injury, Function, and Rehabilitation, 4*(9), 642–646.

Kim, K. H. (2011 June). Use of lidocaine patch for percutaneous endoscopic lumbar discectomy. *The Korean Journal of Pain, 24*(2), 74–80.

Kwon, Y. S., Kim, J. B., Jung, H. J., Koo, Y. J., Lee, I. H., Im, K. T., et al. (2012). Treatment for postoperative wound pain in gynecologic laparoscopic surgery: Topical lidocaine patches. *Journal of Laparoendoscopic & Advanced Surgical Techniques. Part A, 22*(7), 668–673.

Lau, L. L., Li, C. Y., Lee, A., & Chan, S. K. (2018). The use of 5% lidocaine medicated plaster for acute postoperative pain after gynecological surgery: A pilot randomized controlled feasibility trial. *Medicine (Baltimore), 97*(39), e12582.

Liberati, A., Altman, D. G., Tetzlaff, J., Mulrow, C., Gøtzsche, P. C., Ioannidis, J. P., et al. (2009). The PRISMA statement for reporting systematic reviews and meta-analyses of studies that evaluate health care interventions: Explanation and elaboration. *Journal of Clinical Epidemiology, 62*(10), e1–34.

Liu, M., Wai, M., & Nunez, J. (2019). Topical lidocaine patch for postthoracotomy and poststernotomy pain in cardiothoracic intensive care unit adult patients. *Critical Care Nurse, 39*(5), 51–57.

Manion, S. C., & Brennan, T. J. (2011 July). Thoracic epidural analgesia and acute pain management. *Anesthesiology, 115*(1), 181–188.

Martini, A., Del Balzo, G., Schweiger, V., Zanzotti, M., Picelli, A., Parolini, M., et al. (2018). Efficacy of lidocaine 5% medicated plaster (VERSATIS®) in patients with localized neuropathic pain poorly responsive to pharmacological therapy. *Minerva Medica, 109*(5), 344–351.

Saber, A. A., Elgamal, M. H., Rao, A. J., Itawi, E. A., & Martinez, R. L. (2009). Early experience with lidocaine patch for postoperative pain control after laparoscopic ventral hernia repair. *International Journal of Surgery, 7*(1), 36–38.

Sadigursky, D., de Castro Oliveira, M., Macedo, G., Roque Paim Costa, F., Lemos Azi, M., & Figueiredo Alencar, D. (2017). Effectiveness of lidocaine patches for pain treatment after total knee arthroplasty. *Medical Express, 4*(6), M170603.

Sansone, P., Pace, M. C., Passavanti, M. B., Pota, V., Colella, U., & Aurilio, C. (2015 July-August). Epidemiology and incidence of acute and chronic Post-Surgical pain. *Annali Italiani di Chirurgia, 86*(4), 285–292.

Sansone, P., Passavanti, M. B., Fiorelli, A., Aurilio, C., Colella, U., De Nardis, L., et al. (2017 May). Efficacy of the topical 5% lidocaine medicated plaster in the treatment of chronic post-thoracotomy neuropathic pain. *Pain Management, 7*(3), 189–196. https://doi.org/10.2217/pmt-2016-0060 (Epub 2017 Mar 3).

Tan, M., Law, L. S., & Gan, T. J. (2015). Optimizing pain management to facilitate enhanced recovery after surgery pathways. *Canadian Journal of Anaesthesia, 62*(2), 203–218.

Vrooman, B., Kapural, L., Sarwar, S., Mascha, E. J., Mihaljevic, T., Gillinov, M., et al. (2015). Lidocaine 5% patch for treatment of acute pain after robotic cardiac surgery and prevention of persistent incisional pain: A randomized, placebo-controlled, double-blind trial. *Pain Medicine, 16*(8), 1610–1621.

Wenk, M., & Schug, S. A. (2011). Perioperative pain management after thoracotomy. *Current Opinion in Anaesthesiology, 24*, 8–12.

Wick, E. C., Grant, M. C., & Wu, C. L. (2017). Postoperative multimodal analgesia pain management with nonopioid analgesics and techniques: A review. *JAMA Surgery, 152*(7), 691–697.

Chapter 13

Memantine: Features and application in the management of chronic pain

Harsha Shanthanna

Departments of Anesthesia and Health Research Methods, Evidence and Impact, McMaster University, Hamilton, ON, Canada

Abbreviations

AMPA	α-amino-3-hydroxy-5-methyl-4-isoxazolepropionic acid
CNS	central nervous system
FDA	Food and Drug Administration
FIQ	fibromyalgia impact questionnaire
fMRI	functional magnetic resonance imaging
IC50	half-maximal inhibitory concentration
MG^{2+}	Magnesium
NMDA	*N*-methyl-D-aspartate
PLP	phantom limb pain
RCT	randomized controlled trial

Introduction

Chronic pain is very common and affects a large proportion of the population, causing personal distress, limiting functions, and affecting quality of life, as well as increasing health care costs (Mills, Nicolson, & Smith, 2019). Conditions causing chronic pain are noted as the most common cause of disability and burden worldwide (Disease, Injury, & Prevalence, 2017). In the United States, an estimated 100 million adults were affected by chronic non-cancer pain in 2012, with associated treatment costs ranging from $560 to $635 billion (Gaskin & Richard, 2012). In the same year, a National Health Interview Survey reported 126.1 million adults as having some pain in the previous 3 months, with 25.3 million adults (11.2%) suffering from daily pain and 23.4 million (10.3%) reporting a lot of pain (Nahin, 2015). Population studies from other countries have reported similar figures, with moderate or severe chronic pain as one in five in Europe (Breivik, Collett, Ventafridda, Cohen, & Gallacher, 2006), and 10.4%–14.3% in the United Kingdom—estimated based on the results of four studies (Fayaz, Croft, Langford, Donaldson, & Jones, 2016). Chronic pain can occur after an injury or surgery, although a major proportion of the elderly population suffer from chronic pain resulting from musculoskeletal issues such as arthritis and degenerative conditions (British Pain Society (2012); Fayaz et al., 2016; Hooten, Cohen, & Rathmell, 2015).

The management of chronic pain is often challenging and complex. As many to most conditions do not have primary therapy focused on a particular pathology, treatment is symptomatic and directed toward symptom relief and empowering individuals to manage their pain rather than treat it. Therapeutic options are associated with inconsistent results and often only modest improvement in symptoms (Chaparro, Smith, Moore, Wiffen, & Gilron, 2013; Derry et al., 2017; Reid, Eccleston, & Pillemer, 2015; Turk, Wilson, & Cahana, 2011). Although opioids are strong analgesics, their efficacy for chronic pain conditions is questionable (Els et al., 2017). Moreover, they have significant potential for long-term adverse effects for the patient as well societal harms due to diversion. There is a need to advance mechanism-based pain understanding leading to diagnosis (Vardeh, Mannion, & Woolf, 2016) and management of chronic pain conditions (Woolf & Max, 2001), whenever appropriate. In that direction, it becomes important to consider medications that influence the activity of *N*-methyl-D-aspartate (NMDA) receptors. Among the known NMDA receptor antagonists, ketamine is widely recognized for its analgesia potential and its use during surgery. Memantine is an oral NMDA antagonist with a potential for analgesic effect and a reasonably safe profile. It is the first NMDA antagonist to have been approved for the therapy of moderate to severe Alzheimer's disease, and has received approvals from the Food and Drug Administration

Treatments, Mechanisms, and Adverse Reactions of Anesthetics and Analgesics. https://doi.org/10.1016/B978-0-12-820237-1.00013-2
Copyright © 2022 Elsevier Inc. All rights reserved.

(FDA) and the European Medicines Agency in the 2000s. Typically it is started as a 5 mg/day dose and increased over weeks to a 20 mg/day dose. It has been shown to be effective both as monotherapy (Kishi et al., 2017) and also in addition to other medications effective in Alzheimer's (Tariot et al., 2004). In this chapter, we review the background, relevant pharmacology, and preclinical and clinical evidence of its effect on pain. We also provide considerations for clinical use in appropriate populations.

Background: NMDA antagonists and pain

Generation of pain signals by any injury, both within the nociceptive and neuropathic pathways, leads to secondary changes of peripheral and central sensitization. In acute conditions, there is development of *physiological sensitization* processes, such as hyperalgesia (decreased threshold of nociceptor neurons), and allodynia (nociceptive activation by non-noxious stimuli) (Woolf, 2011). These have the potential to turn into *pathological and persistent pain* due to neuroplasticity. Central to these changes is the release of glutamate, which is the major excitatory transmitter within the central nervous system (CNS). Glutamate release is associated with damage to neural structures contributing to neuropathic pain. Among the important classes of receptors that facilitate the actions of glutamate are ionotropic receptors that gate ion channels. These include (α-amino-3-hydroxy-5-methyl-4-isoxazolepropionic acid) AMPA, kainate, and NMDA receptors. Of the three, NMDA receptors play an important role in the neuroplasticity and neurodegeneration processes, including the generation and sustenance of pain signals (Woolf, 2011; Woolf & Thompson, 1991). They also have unique properties: they control a cation channel that is highly permeable to monovalent ions and calcium; simultaneous binding of glutamate and glycine is required for efficient activation; and at resting membrane potential these channels are blocked by extracellular magnesium and open only on simultaneous depolarization and agonist binding. Hence, NMDA antagonists have the potential to modify pain signals, thereby decreasing the chances of developing persisting pain and also the potential to decrease the intensity and duration of chronic pain (Aiyer, Mehta, Gungor, & Gulati, 2018; Petrenko, Yamakura, Baba, & Shimoji, 2003). NMDA receptors are present both within the peripheral tissues as well as within the CNS. There are eight different medications with a potential for NMDA blockade that are presently used in clinical practice. They include ketamine, amantadine, memantine, methadone, dextromethorphan, carbamazepine, phenytoin, and magnesium. Of these, ketamine has been well recognized for its analgesic properties. However, its use is limited because of it being parenteral in preparation and use, apart from associated psychological and systemic side effects.

Memantine: Fundamental pharmacology

Memantine (1-amino-3,5-dimethyladamantane) was first synthetized in 1968 by researchers at Eli Lilly as an agent to lower blood sugar levels (Fig. 1). Although it was registered for use in Germany for a variety of neurological conditions, it was not until the 1980s when its NMDA-antagonistic property was discovered (Bormann, 1989; Parsons, Danysz, & Quack, 1999).

Memantine is a low to moderate affinity, uncompetitive open channel (it can enter the channel and block current flow only after channel opening) NMDA receptor antagonist. Unlike many other compounds that carry NMDA blocking properties even under physiological transmission, memantine exerts its action by inhibiting calcium (Ca^{2+}) influx that occurs under pathological conditions or in excessive NMDA activation. The half maximal inhibitory concentration (IC50) for

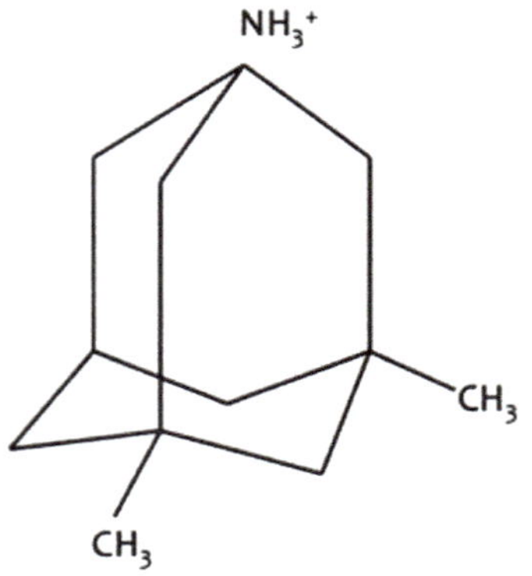

1 amino-3,5-dimethyl-adamantane

FIG. 1 Chemical structure of memantine.

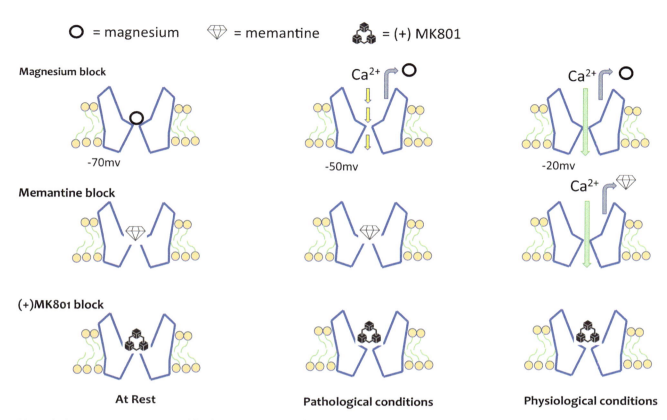

N-methyl D Aspartate receptor blockade kinetics: with synaptic depolarisation both magnesium and memantine demonstrate rapid unblocking kinetics and are able to leave the NMDA receptor channel, although memantine demonstrates a much stronger voltage-dependency (physiological blockade); whereas (+) MK801, another non-competitive NMDA antagonist does not demonstrate such kinetics and is poorly tolerated. Adapted from: Parsons CG, DanyszW, Quack G. Memantineis a clinically well tolerated N-methyl-D-aspartate (NMDA) receptor antagonist--a review of preclinical data. Neuropharmacology. 1999 Jun;38(6):735-67.

FIG. 2 NMDA receptor blocking kinetics of magnesium, memantine and (+) MK801.

NMDA blockade has been observed to be 1 μM (Johnson & Kotermanski, 2006). At higher concentrations memantine can have effects on many receptors including nicotinic acetylcholine receptors, serotonin receptors, sigma-1 receptors, and voltage-activated sodium channels, and hence can influence the uptake of serotonin and dopamine (Johnson & Kotermanski, 2006). It also has rapid blocking/unblocking kinetics, rapidly leaving the NMDA receptor channel during normal physiological activation. These properties have made it clinically useful and safe (Fig. 2). Experimental research indicates that it binds to the same channel site as Magnesium (Mg^{2+}); magnesium can actually decrease its NMDA antagonistic effect and mutations that influence Mg^{2+} binding also influence memantine action (Johnson & Kotermanski, 2006; Majláth, Török, Toldi, & Vécsei, 2016).

In humans, an orally administered dose is 100% bioavailable, undergoes minimal metabolism, and exhibits a terminal elimination half-life of 60–80 h (75% or greater of the dose is eliminated intact in the urine). It is 45% plasma protein bound. It has been observed to not inhibit cytochrome P-450 isoenzymes in vitro, and is minimally affected by food, sex, or age (Johnson & Kotermanski, 2006). The metabolites of memantine, N-3,5-dimethyl-gludantan, isomeric 4- and 6-hydroxy-memantine, and 1-nitroso-3,5-dimethyl-adamantane, have not been observed to display any NMDA antagonistic properties. The administration of daily dose in the range of 5–30 mg/day can result in a serum concentration of 0.025–0.529 μM and cerebrospinal fluid concentration of 0.05–0.31 μM (Kornhuber & Quack, 1995). Memantine does not have any major drug interactions. As it is renally excreted, its doses need to be appropriately reduced in patients with renal impairment (Periclou, Ventura, Rao, & Abramowitz, 2006). It must also be used with caution in patients with severe liver disease (Buvanendran & Kroin, 2008).

Memantine: Preclinical evidence of effect on pain

There is mixed evidence to suggest that memantine has analgesic properties in preclinical studies. As its analgesic effect is due to its NMDA-blocking properties, results from studies depend on the model used and the relative contribution of NMDA pathways and their post-translational effects (Pickering & Morel, 2018). Present evidence is also unclear about the doses needed to achieve sufficient NMDA blockade to provide meaningful analgesia response, either directly or in surrogate animal models. Most electrophysiological models have shown weak effects on acute somatic pain, when compared to other stronger NMDA antagonists such as MK-801, which retained its effect even with high intensity stimulation. Although memantine (20 mg/kg), ketamine (2 mg/kg), and MK-801 (0.2 mg/kg) reduced wind-up of single motor unit reflexes of 50%, 51%, and 46% of controls, respectively, it was less effective with acute nociceptive pinch reflexes (80% of controls), in comparison to ketamine and MK-801, which were equieffective (Jones, McClean, Parsons, & Headley, 2001; Parsons et al., 1999). But memantine has shown effects on responses that are usually associated with secondary sensitization either in acute pain or chronic neuropathic pain conditions. Memantine has shown effects against hyperalgesia in both nociceptic (Neugebauer, Kornhuber, Lücke, & Schaible, 1993) and neuropathic models (Carlton & Hargett, 1995; Eisenberg, LaCross, & Strassman, 1995). In macaques, very large doses of memantine were observed to reduce mechanical allodynia induced by nerve ligation (Parsons et al., 1999). Relatively recently, Morel et al. demonstrated that memantine initiated before surgery had better effect than when initiated post-surgery. In adult male Sprague-Dawley rats, a pre-emptive protocol was started 4 days before surgery and continued for 2 days post-surgery and compared to a postoperative protocol that included 7-day treatment beginning on the day of surgery. Tests for tactile allodynia, mechanical hyperalgesia, and spatial memory were conducted before and after surgery. Control animals (without memantine treatment) with spinal nerve ligation displayed nociception, impaired memory, and increased expression of the phosphorylated residues. While postoperative memantine had no beneficial effect, pre-emptive memantine prevented the development of post-surgical nociception and impairment of spatial memory, and did not increase the expression of phosphorylated residues at spinal, insular, and hippocampal levels (Morel et al., 2013).

Memantine: Clinical evidence and literature review in chronic pain

The use of memantine for chronic pain is relatively recent. Since the year 2000, a small number of reviews have attempted to summarize the evidence of memantine and its use, either for chronic pain conditions or to prevent chronic pain (Aiyer et al., 2018; Collins, Sigtermans, Dahan, Zuurmond, & Perez, 2010; Hewitt, 2000; Kurian, Raza, & Shanthanna, 2019; Pickering & Morel, 2018). There have been three recent clinical reviews looking into the use of memantine in chronic pain. Loy et al. published a systematic review on the use of memantine specific to the treatment of phantom limb pain (PLP) (Loy, Britt, & Brown, 2016). Their criteria allowed inclusion of all types of reports and included eight articles in their assessment. While studies looking at chronic PLP did not observe any improvements, use of memantine in acute PLP indicated some benefit, as observed by one prospective study, two case series and a case report. Pickering and Morel reviewed literature specific to neuropathic pain conditions (Pickering & Morel, 2018), and included all designs including case reports. The types of chronic pain population included in these reports were quite similar to the review by Kurian et al. (2019). Pickering and Morel concluded that at present results are inconclusive. Although they do not recommend its routine use, they note that consideration for its use can be made as an adjunct, especially in the initial phases of treatment in patients with opioid tolerance. More recently, Kurian et al. published a systematic review and meta-analysis of comparative trials in 2019 (Kurian et al., 2019). They looked within Medline, Embase, and Cochrane Central Registry of Controlled Trials (CENTRAL) for any comparative studies published until February 2019 and retrieved 15 trials of which 13 trials used memantine to treat chronic pain and the remaining two trials looked at preventing the development of chronic pain after breast surgery and amputations. The study with the largest sample size was published only as an abstract and included 170 patients of chronic pelvic pain (Dimitrakov, Chitalov, & Dechev, 2009). Other chronic pain populations considered were fibromyalgia, complex regional pain syndrome, chronic headache, PLP, postherpetic neuralgia, and neuropathic pain conditions. Eleven studies had compared memantine to placebo and five trials had a crossover design. Memantine was used in titrated but varying doses in studies. The maintenance dose for chronic pain treatment ranged from 20 to 55 mg/day, and the dosage for preventing chronic pain was 30 mg/day. In their pooled effect on pain relief, memantine did not affect pain scores in chronic pain. However, memantine was observed to decrease pain intensity in patients who were treated to prevent chronic pain; memantine group ($n = 30$) compared to the control group ($n = 29$): mean difference $= -1.02$ units (95% confidence interval (CI) -1.38 to -0.66)–low-quality evidence (Fig. 3). Among adverse effects, memantine was observed to cause more dizziness than placebo; relative risk $= 4.90$ (95% CI 1.26–18.99)–moderate quality of evidence (Fig. 4). Effects of memantine beyond the subjective outcome of pain relief have been observed in complex regional pain syndrome

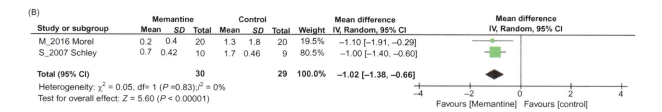

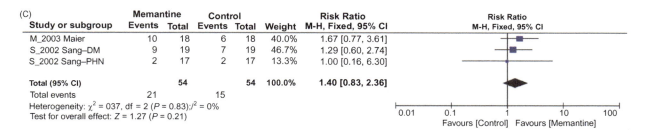

Analyses of pain treatment and prevention with memantine in chronic pain patients. (a) Comparison of mean differences with memantine and control for chronic pain treatment; (b) comparison of mean differences with memantine and control for prevention of persistent pain after surgery; and (c) comparison of pain success with memantine and control for chronic pain treatment. CI: Confidence Interval; DN: Diabetic Neuropathy; IV: Inverse Variance; M-H, Mantel-Haenszel test; PHN: Postherpetic Neuralgia; SD: Standard Deviation

With permission from: European Journal of Pain, Volume: 23, Issue: 7, Pages: 1234-1250, First published: 08 March 2019, DOI: (10.1002/ejp.1393)

FIG. 3 Effectiveness of memantine for chronic pain treatment or prevention.

and fibromyalgia. In a randomized controlled trial (RCT) study by Gustin et al., 20 patients with complex regional pain syndrome were randomized to receive a combination therapy of either morphine plus memantine or morphine plus placebo. Pain scores at rest and with movement were recorded apart from objective signals using functional magnetic resonance imaging (fMRI). The combination therapy was effective in reducing pain at rest and movement, apart from improving disability. Activation of signals within the contralateral primary somatosensory (cS1) and anterior cingulate cortex areas were significantly reduced during movement of the affected hand indicating its association to analgesic effect (Gustin et al., 2010). Fayed et al. assessed the effects of memantine versus placebo in 25 patients with fibromyalgia. At 6 months' follow-up, memantine-treated patients showed decreased clinical global impression scores and fibromyalgia impact questionnaire (FIQ) scores, and the placebo group actually reported less pain induced by sphygmomanometer. All patients were observed for cerebral metabolites as well as fMRI signals. Patients in the memantine group showed increased glutamate and its associated ratios with creatinine and glutamine in the posterior cingulate cortical areas indicating localized metabolic effects.

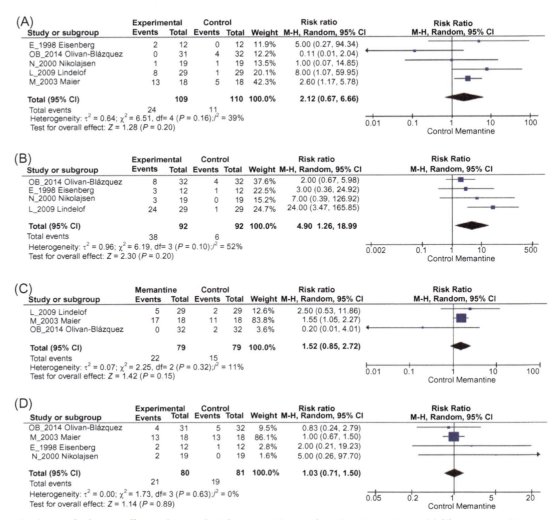

Analyses of adverse effects observed with memantine in chronic pain patients. (a) Nausea–vomiting observed with memantine compared to control group; (b) dizziness observed with memantine compared to control group; (c) drowsiness observed with memantine compared to control group; (d) and headache observed with memantine compared to control group. CI: Confidence Interval; IV: Inverse Variance; M-H, Mantel-Haenszel test; SD: Standard Deviation

With permission from: European Journal of Pain, Volume: 23, Issue: 7, Pages: 1234-1250, First published: 08 March 2019, DOI: (10.1002/ejp.1393)

FIG. 4 Potential adverse effects of memantine observed in studies using it for chronic pain.

The same group also demonstrated significant increases in creatine and choline in the right posterior insula and a correlation between choline and the FIQ scores in the posterior insula (Fayed et al., 2014). In another study on PLP patients by Schwenkreis et al., both pain scores and neural correlates of intracortical facilitation and inhibition were affected in both placebo and treatment groups and there was no correlation between pain score changes and intracortical neural correlates (Schwenkreis et al., 2003). More recently, Shanthanna et al. conducted a factorial design feasibility trial on patients undergoing thoracoscopic lung resection surgeries, to prevent persisting pain. Memantine was administered for 1 month after an initial infusion of ketamine up to 24 h. As this was a pilot study, no clinical inferences can be made, but based on the limited observation no particular benefit of NMDA blockade was observed compared to placebo. Although another protocol on the use of memantine for neuropathic pain in ketamine-responsive patients has been published (Pickering et al., 2014), its final report is not yet reported (clinicaltrials.gov, n.d.). Based on preclinical observations (Eisenberg et al.,

1995; Morel et al., 2013) and limited but available clinical evidence (Kurian et al., 2019), the analgesic effects of memantine are more evident when used as a prophylaxis or in early stages, rather than in established neuropathies (Eisenberg, Kleiser, Dortort, Haim, & Yarnitsky, 1998). Unfortunately, the use of memantine has not been reported in experimental models of acute pain in humans, as indicated by a recent review by Thompson et al. (2019). It is also not clear if only certain chronic pain populations are responsive to memantine. Although, Kurian et al. (2019) did not find any trials apart from Sang, Booher, Gilron, Parada, and Max (2002). Kavirajan notes an RCT of 527 diabetic neuropathy patients in which memantine was compared to placebo, with no difference in pain management. However, the reference indicated for this review is not clear and does not lead to any study reporting (Kavirajan, 2009). Apart from classical neuropathic pain conditions, such as PLP or postherpetic neuralgia, several studies have looked at its use in fibromyalgia (Olivan-Blazquez et al., 2014).

Memantine: Considerations for clinical use

Based on the available evidence from preclinical and clinical studies, memantine should not be considered as a first-line option for analgesia in neuropathic or non-neuropathic conditions. However, consideration for its use can be made in pain conditions refractory to analgesics normally used in clinical practice. It is better to use it in early stages or during the potential phase of transition from acute to chronic pain, such as during the development of persistent post-surgical pain. However, it is not clear if memantine is appropriate as a treatment for postsurgical pain. Suzuki (2009) recommends that memantine be used as a long-term therapy after initial ketamine administration, as was considered by Shanthanna et al. (2020). Such patient populations can include amputations and breast resections in which there is a high risk of neuropathic pain. If selecting patients based on their response to ketamine, a known parenteral analgesic, is not possible, memantine should be started at a low dose of 5 or 10 mg/day and titrated up over the next 3–4 weeks. The maintenance dose usually ranges between 20 and 40 mg/day. Although a majority of studies for pain patients indicate a dose of 20 mg/day, some studies have used up to 40 mg/day (Gustin et al., 2010; Lindelof & Bendtsen, 2009). Being an adjunct, it is important that a continuous dosing is employed, and it is not used as rescue medication. It may also be challenging to discern the actual clinical benefit in pain patients whose pain may vary on a day-to-day basis. Hence, it is important to note if there are gains in pain relief and decide to continue on a longer-term basis only if appropriate. A trial of therapy in suitable patients should be at least 4–8 weeks so as to achieve and establish consistent therapeutic levels. It is important to advise patients on potential adverse effects. These primarily include dizziness, headache, fatigue, and lethargy (Kavirajan, 2009). In the systematic review by Kurian et al., dizziness was observed to be frequent with memantine with an absolute risk increase of 35% (Kurian et al., 2019). In patients who develop side effects, one must consider decreasing the dose or stopping the medication as dizziness can lead to falls and secondary injuries.

Applications to other areas

As memantine has established evidence in elderly patients suffering from neurocognitive disease states such as Alzheimer's and vascular dementia, it should be considered to treat chronic pain in such patients, even if they are not on primary therapy. Spengos et al. report a patient with memory and cognitive decline who was put on treatment with memantine as the patient could not tolerate donezepil. The patient also suffered from chronic daily headaches. To their surprise, stabilization of memantine treatment at a dose of 10 mg/day worked effectively to control the patient's chronic headache (Spengos, Theleritis, & Paparrigopoulos, 2008). Memantine may also have synergistic effects or additional benefits in patients with substance abuse disorders, especially opioid use disorder and chronic pain, as both share the NMDA pathway.

Other agents of interest

The NMDA antagonist dextromethorphan has been studied in several surgeries for postoperative pain relief as a component of multimodal analgesia in some surgical populations such as dental extractions, colon surgery, and hysterectomy. Nelson et al. report potential for decrease in pain in diabetic neuropathy with dextromethorphan compared to placebo (Nelson, Park, Robinovitz, Tsigos, & Max, 1997). However, its efficacy is still not well established for routine clinical use (Suzuki, 2009; Weinbroum, Rudick, Paret, & Ben-Abraham, 2000). Although there continues to be a lot of enthusiasm and effort in bringing newer agents belonging to the same group (Gilron & Dickenson, 2014), none have been cleared for use in chronic pain practice. For example, neramexane is a newer non-competitive NMDA antagonist with potency and properties similar to memantine. Although, phase Ib clinical trials for chronic pain showed positive results, phase

128 PART | II Mechanisms of action of drugs

II results indicated no superiority to existing treatments (Malyshkin, Medvedev, Danysz, & Bespalov, 2005; Rammes, 2009).

Conclusions

As a drug belonging to NMDA antagonist group, memantine has definitely shown promise for analgesia and a good safety profile. The efficacy of memantine for chronic pain conditions is still not proven. Future studies should look into the effect of memantine on specific populations, especially those in acute to chronic pain transition. It may also be useful to assess its benefit on pain in elderly patients who are already on memantine therapy for other reasons.

Key facts

- Memantine is an oral NMDA antagonist
- Preclinical studies indicate that memantine may be useful in chronic pain conditions
- Present evidence for clinical use of memantine is limited

Mini-dictionary of terms

Chronic pain: Pain that persists for at least 12 weeks.
Neuropathic pain: Pain resulting from somatosensory nerve-related disease or injury.
Non-competitive blockade: Blockade by a non-competitive antagonist prevents the action of an agonist by binding to a non-agonist site to prevent activation.
Alzheimer's disease: Alzheimer's disease, the most common cause of dementia, is an irreversible, progressive disorder that causes brain cells to degenerate and die. It results in a continuous decline in thinking, behavioral and social skills that disrupts a person's ability to function independently.
Phantom limb pain: Painful sensation that is felt in an area of the body where an amputation has occurred. Although the limb is gone, the patient still experiences painful sensation.
fMRI: Functional magnetic resonance imaging is a non-invasive imagine technique that measures brain activity.

Summary points

- NMDA receptors are involved in the pain neurotransmission and neuroplasticity.
- Memantine is an oral NMDA receptor blocker approved for use in Alzheimer's disease.
- Preclinical studies indicate that memantine has analgesic potential.
- Existing evidence on the clinical use of memantine for chronic pain conditions is limited and mixed.
- Existing studies suggest that it may have a greater potential if used during the translation from acute to chronic pain.
- Although memantine has a very good safety profile, risk of dizziness was observed to be higher than placebo.
- The use of memantine for chronic pain is off-label.

References

Aiyer, R., Mehta, N., Gungor, S., & Gulati, A. (2018). A systematic review of NMDA receptor antagonists for treatment of neuropathic pain in clinical practice. *The Clinical Journal of Pain, 34*(5), 450–467. https://doi.org/10.1097/ajp.0000000000000547.

Bormann, J. (1989). Memantine is a potent blocker of N-methyl-D-aspartate (NMDA) receptor channels. *European Journal of Pharmacology, 166*(3), 591–592. https://doi.org/10.1016/0014-2999(89)90385-3.

Breivik, H., Collett, B., Ventafridda, V., Cohen, R., & Gallacher, D. (2006). Survey of chronic pain in Europe: Prevalence, impact on daily life, and treatment. *European Journal of Pain, 10*(4), 287–333. https://doi.org/10.1016/j.ejpain.2005.06.009.

British Pain Society. (2012). *National pain audit final report 2010–2012*. Retrieved from https://www.britishpainsociety.org/static/uploads/resources/files/members_articles_npa_2012_1.pdf.

Buvanendran, A., & Kroin, J. S. (2008). Early use of memantine for neuropathic pain. *Anesthesia & Analgesia, 107*(4), 1093–1094.

Carlton, S. M., & Hargett, G. L. (1995). Treatment with the NMDA antagonist memantine attenuates nociceptive responses to mechanical stimulation in neuropathic rats. *Neuroscience Letters, 198*(2), 115–118. https://doi.org/10.1016/0304-3940(95)11980-b.

Chaparro, L. E., Smith, S. A., Moore, R. A., Wiffen, P. J., & Gilron, I. (2013). Pharmacotherapy for the prevention of chronic pain after surgery in adults. *The Cochrane Database of Systematic Reviews, 2013*(7). https://doi.org/10.1002/14651858.CD008307.pub2, CD008307.

clinicaltrials.gov n.d. https://clinicaltrials.gov/ct2/show/NCT01602185?term=memantine%2C+pickering&cond=neuropathic+pain&draw=2&rank=3. Retrieved from https://clinicaltrials.gov/ct2/show/NCT01602185?term=memantine%2C+pickering&cond=neuropathic+pain&draw=2&rank=3.

Collins, S., Sigtermans, M. J., Dahan, A., Zuurmond, W. W., & Perez, R. S. (2010). NMDA receptor antagonists for the treatment of neuropathic pain. *Pain Medicine*, *11*(11), 1726–1742.

Derry, S., Wiffen, P. J., Kalso, E. A., Bell, R. F., Aldington, D., Phillips, T., et al. (2017). Topical analgesics for acute and chronic pain in adults—An overview of Cochrane reviews. *Cochrane Database of Systematic Reviews*, *5*(5). https://doi.org/10.1002/14651858.CD008609.pub2, CD008609.

Dimitrakov, J. D., Chitalov, Z., & Dechev, I. (2009). Memantine in the alleviation of symptoms of chronic pelvic pain syndrome: A randomized, double-blind, placebo-controlled trial [abstract 339]. *The Journal of Urology*, *11*(2), 100–101.

Disease, G. B. D., Injury, I., & Prevalence, C. (2017). Global, regional, and national incidence, prevalence, and years lived with disability for 328 diseases and injuries for 195 countries, 1990-2016: A systematic analysis for the global burden of disease study 2016. *Lancet*, *390*(10100), 1211–1259. https://doi.org/10.1016/S0140-6736(17)32154-2.

Eisenberg, E., Kleiser, A., Dortort, A., Haim, T., & Yarnitsky, D. (1998). The NMDA (N-methyl-D-aspartate) receptor antagonist memantine in the treatment of postherpetic neuralgia: A double-blind, placebo-controlled study. *European Journal of Pain*, *2*(4), 321–327.

Eisenberg, E., LaCross, S., & Strassman, A. M. (1995). The clinically tested N-methyl-D-aspartate receptor antagonist memantine blocks and reverses thermal hyperalgesia in a rat model of painful mononeuropathy. *Neuroscience Letters*, *187*(1), 17–20. https://doi.org/10.1016/0304-3940(95)11326-r.

Els, C., Jackson, T. D., Hagtvedt, R., Kunyk, D., Sonnenberg, B., Lappi, V. G., et al. (2017). High-dose opioids for chronic non-cancer pain: An overview of cochrane reviews. *The Cochrane Database of Systematic Reviews*, *10*(10). https://doi.org/10.1002/14651858.CD012299.pub2, CD012299.

Fayaz, A., Croft, P., Langford, R. M., Donaldson, L. J., & Jones, G. T. (2016). Prevalence of chronic pain in the UK: A systematic review and meta-analysis of population studies. *BMJ Open*, *6*(6). https://doi.org/10.1136/bmjopen-2015-010364, e010364.

Fayed, N., Olivan-Blazquez, B., Herrera-Mercadal, P., Puebla-Guedea, M., Perez-Yus, M. C., Andres, E., et al. (2014). Changes in metabolites after treatment with memantine in fibromyalgia. A double-blind randomized controlled trial with magnetic resonance spectroscopy with a 6-month follow-up. *CNS Neuroscience & Therapeutics*, *20*(11), 999–1007.

Gaskin, D. J., & Richard, P. (2012). The economic costs of pain in the United States. *The Journal of Pain*, *13*(8), 715–724. https://doi.org/10.1016/j.jpain.2012.03.009.

Gilron, I., & Dickenson, A. H. (2014). Emerging drugs for neuropathic pain. *Expert Opinion on Emerging Drugs*, *19*(3), 329–341. https://doi.org/10.1517/14728214.2014.915025.

Gustin, S. M., Schwarz, A., Birbaumer, N., Sines, N., Schmidt, A. C., Veit, R., et al. (2010). NMDA-receptor antagonist and morphine decrease CRPS-pain and cerebral pain representation. *Pain*, *151*(1), 69–76.

Hewitt, D. J. (2000). The use of NMDA-receptor antagonists in the treatment of chronic pain. *Clinical Journal of Pain*, *16*(2 Suppl), S73–S79.

Hooten, W. M., Cohen, S. P., & Rathmell, J. P. (2015). Introduction to the symposium on pain medicine. *Mayo Clinic Proceedings*, *90*(1), 4–5. https://doi.org/10.1016/j.mayocp.2014.11.007.

Johnson, J. W., & Kotermanski, S. E. (2006). Mechanism of action of memantine. *Current Opinion in Pharmacology*, *6*(1), 61–67. https://doi.org/10.1016/j.coph.2005.09.007.

Jones, M. W., McClean, M., Parsons, C. G., & Headley, P. M. (2001). The in vivo relevance of the varied channel-blocking properties of uncompetitive NMDA antagonists: Tests on spinal neurones. *Neuropharmacology*, *41*(1), 50–61. https://doi.org/10.1016/s0028-3908(01)00041-7.

Kavirajan, H. (2009). Memantine: A comprehensive review of safety and efficacy. *Expert Opinion on Drug Safety*, *8*(1), 89–109. https://doi.org/10.1517/14740330802528420.

Kishi, T., Matsunaga, S., Oya, K., Nomura, I., Ikuta, T., & Iwata, N. (2017). Memantine for Alzheimer's disease: An updated systematic review and meta-analysis. *Journal of Alzheimer's Disease*, *60*(2), 401–425. https://doi.org/10.3233/jad-170424.

Kornhuber, J., & Quack, G. (1995). Cerebrospinal fluid and serum concentrations of the N-methyl-D-aspartate (NMDA) receptor antagonist memantine in man. *Neuroscience Letters*, *195*(2), 137–139. https://doi.org/10.1016/0304-3940(95)11785-u.

Kurian, R., Raza, K., & Shanthanna, H. (2019). A systematic review and meta-analysis of memantine for the prevention or treatment of chronic pain. *European Journal of Pain*, *23*(7), 1234–1250. https://doi.org/10.1002/ejp.1393.

Lindelof, K., & Bendtsen, L. (2009). Memantine for prophylaxis of chronic tension-type headache—A double-blind, randomized, crossover clinical trial. *Cephalalgia*, *29*(3), 314–321.

Loy, B. M., Britt, R. B., & Brown, J. N. (2016). Memantine for the treatment of phantom limb pain: A systematic review. *Journal of Pain & Palliative Care Pharmacotherapy*, *30*(4), 276–283. https://doi.org/10.1080/15360288.2016.1241334.

Majláth, Z., Török, N., Toldi, J., & Vécsei, L. (2016). Memantine and kynurenic acid: Current neuropharmacological aspects. *Current Neuropharmacology*, *14*(2), 200–209. https://doi.org/10.2174/1570159x14666151113123221.

Malyshkin, A. A., Medvedev, I. O., Danysz, W., & Bespalov, A. Y. (2005). Anti-allodynic interactions between NMDA receptor channel blockers and morphine or clonidine in neuropathic rats. *European Journal of Pharmacology*, *519*(1–2), 80–85. https://doi.org/10.1016/j.ejphar.2005.07.003.

Mills, S. E. E., Nicolson, K. P., & Smith, B. H. (2019). Chronic pain: A review of its epidemiology and associated factors in population-based studies. *British Journal of Anaesthesia*, *123*(2), e273–e283. https://doi.org/10.1016/j.bja.2019.03.023.

Morel, V., Etienne, M., Wattiez, A. S., Dupuis, A., Privat, A. M., Chalus, M., et al. (2013). Memantine, a promising drug for the prevention of neuropathic pain in rat. *European Journal of Pharmacology*, *721*(1–3), 382–390.

Nahin, R. L. (2015). Estimates of pain prevalence and severity in adults: United States, 2012. *The Journal of Pain*, *16*(8), 769–780. https://doi.org/10.1016/j.jpain.2015.05.002.

Nelson, K. A., Park, K. M., Robinovitz, E., Tsigos, C., & Max, M. B. (1997). High-dose oral dextromethorphan versus placebo in painful diabetic neuropathy and postherpetic neuralgia. *Neurology*, *48*(5), 1212–1218. https://doi.org/10.1212/wnl.48.5.1212.

Neugebauer, V., Kornhuber, J., Lücke, T., & Schaible, H. G. (1993). The clinically available NMDA receptor antagonist memantine is antinociceptive on rat spinal neurones. *Neuroreport, 4*(11), 1259–1262. https://doi.org/10.1097/00001756-199309000-00012.

Olivan-Blazquez, B., Herrera-Mercadal, P., Puebla-Guedea, M., Perez-Yus, M. C., Andres, E., Fayed, N., et al. (2014). Efficacy of memantine in the treatment of fibromyalgia: A double-blind, randomised, controlled trial with 6-month follow-up. *Pain, 155*(12), 2517–2525.

Parsons, C. G., Danysz, W., & Quack, G. (1999). Memantine is a clinically well tolerated N-methyl-D-aspartate (NMDA) receptor antagonist—A review of preclinical data. *Neuropharmacology, 38*(6), 735–767.

Periclou, A., Ventura, D., Rao, N., & Abramowitz, W. (2006). Pharmacokinetic study of memantine in healthy and renally impaired subjects. *Clinical Pharmacology and Therapeutics, 79*(1), 134–143. https://doi.org/10.1016/j.clpt.2005.10.005.

Petrenko, A. B., Yamakura, T., Baba, H., & Shimoji, K. (2003). The role of N-methyl-D-aspartate (NMDA) receptors in pain: A review. *Anesthesia and Analgesia, 97*(4), 1108–1116. https://doi.org/10.1213/01.ane.0000081061.12235.55.

Pickering, G., & Morel, V. (2018). Memantine for the treatment of general neuropathic pain: A narrative review. *Fundamental & Clinical Pharmacology, 32*(1), 4–13. https://doi.org/10.1111/fcp.12316.

Pickering, G., Pereira, B., Morel, V., Tiberghien, F., Martin, E., Marcaillou, F., et al. (2014). Rationale and design of a multicenter randomized clinical trial with memantine and dextromethorphan in ketamine-responder patients. *Contemporary Clinical Trials, 38*(2), 314–320.

Rammes, G. (2009). Neramexane: A moderate-affinity NMDA receptor channel blocker: New prospects and indications. *Expert Review of Clinical Pharmacology, 2*(3), 231–238.

Reid, M. C., Eccleston, C., & Pillemer, K. (2015). Management of chronic pain in older adults. *BMJ, 350*, h532. https://doi.org/10.1136/bmj.h532.

Sang, C. N., Booher, S., Gilron, I., Parada, S., & Max, M. B. (2002). Dextromethorphan and memantine in painful diabetic neuropathy and postherpetic neuralgia—Efficacy and dose-response trials. *Anesthesiology, 96*(5), 1053–1061. https://doi.org/10.1097/00000542-200205000-00005.

Schwenkreis, P., Maier, C., Pleger, B., Mansourian, N., Dertwinkel, R., Malin, J. P., et al. (2003). NMDA-mediated mechanisms in cortical excitability changes after limb amputation. *Acta Neurologica Scandinavica, 108*(3), 179–184.

Shanthanna, H., Turan, A., Vincent, J., Saab, R., Shargall, Y., O'Hare, T., et al. (2020). N-methyl-D-aspartate antagonists and steroids for the prevention of persisting post-surgical pain after thoracoscopic surgeries: A randomized controlled, factorial design, international, multicenter pilot trial. *Journal of Pain Research, 13*, 377–387. https://doi.org/10.2147/JPR.S237058.

Spengos, K., Theleritis, C., & Paparrigopoulos, T. (2008). Memantine and NMDA antagonism for chronic migraine: A potentially novel therapeutic approach? *Headache, 48*(2), 284–286. https://doi.org/10.1111/j.1526-4610.2007.01016.x.

Suzuki, M. (2009). Role of N-methyl-D-aspartate receptor antagonists in postoperative pain management. *Current Opinion in Anaesthesiology, 22*(5), 618–622.

Tariot, P. N., Farlow, M. R., Grossberg, G. T., Graham, S. M., McDonald, S., & Gergel, I. (2004). Memantine treatment in patients with moderate to severe Alzheimer disease already receiving donepezil: A randomized controlled trial. *JAMA, 291*(3), 317–324. https://doi.org/10.1001/jama.291.3.317.

Thompson, T., Whiter, F., Gallop, K., Veronese, N., Solmi, M., Newton, P., et al. (2019). NMDA receptor antagonists and pain relief: A meta-analysis of experimental trials. *Neurology, 92*(14), e1652–e1662. https://doi.org/10.1212/wnl.0000000000007238.

Turk, D. C., Wilson, H. D., & Cahana, A. (2011). Treatment of chronic non-cancer pain. *Lancet, 377*(9784), 2226–2235. https://doi.org/10.1016/s0140-6736(11)60402-9.

Vardeh, D., Mannion, R. J., & Woolf, C. J. (2016). Toward a mechanism-based approach to pain diagnosis. *The Journal of Pain, 17*(9 Suppl), T50–T69. https://doi.org/10.1016/j.jpain.2016.03.001.

Weinbroum, A. A., Rudick, V., Paret, G., & Ben-Abraham, R. (2000). The role of dextromethorphan in pain control. *Canadian Journal of Anaesthesia, 47*(6), 585–596. https://doi.org/10.1007/bf03018952.

Woolf, C. J. (2011). Central sensitization: Implications for the diagnosis and treatment of pain. *Pain, 152*(3 Suppl), S2–15. https://doi.org/10.1016/j.pain.2010.09.030.

Woolf, C. J., & Max, M. B. (2001). Mechanism-based pain diagnosis: Issues for analgesic drug development. *Anesthesiology, 95*(1), 241–249. https://doi.org/10.1097/00000542-200107000-00034.

Woolf, C. J., & Thompson, S. W. (1991). The induction and maintenance of central sensitization is dependent on N-methyl-D-aspartic acid receptor activation; implications for the treatment of post-injury pain hypersensitivity states. *Pain, 44*(3), 293–299.

Chapter 14

Midazolam: Mechanism and perioperative applications

Joe C. Hong

Department of Anesthesiology & Peri-Operative Medicine, David Geffen School of Medicine, University of California, Los Angeles, Ronald Reagan UCLA Medical Center, Los Angeles, CA, United States

Abbreviations

ASA-PS	American Society of Anesthesiologist physical status
CBF	cerebral blood flow
CMRO$_2$	cerebral metabolic rate of oxygen
CNS	central nervous system
CYP3A4	cytochrome P450 3A4 isoenzyme
CYP3A5	cytochrome P450 3A5 isoenzyme
EEG	electroencephelogram
GABA	gamma-aminobutyric acid
GABA$_A$	GABA type A receptor
GCSE	generalized convulsive status epilepticus
ICP	intracranial pressure
PONV	postoperative nausea and vomiting

Introduction

In 1959, chlordiazepoxide (Librium) became the first benzodiazepine to be patented. Since then, several benzodiazepines have been synthesized with the goal of producing an increase in potency and a more rapid onset of anxiolysis, sedation, anterograde amnesia, and hypnosis. In 1963, diazepam (Valium) was synthesized followed by lorazepam (Ativan) in 1971. Although effective, these older benzodiazepines were poorly water-soluble. Thus, parenteral preparations of these compounds required propylene glycol which often caused venous irritation on injection. Midazolam, synthesized in 1976, quickly became the benzodiazepine of choice for use in procedural sedation (Wick, 2013). The main advantages of midazolam were its water solubility, more rapid onset of action, and a relatively shorter duration of action. Because midazolam is water-soluble at low pH as a hydrochloride salt, it does not require propylene glycol for solubility and thus does not produce venous irritation on injection. Once in the bloodstream, the increase in pH results in the closure of the 7-membered 1,4-diazepine ring (see Fig. 1). Ring closure markedly increases the lipid solubility, speeds up central nervous system (CNS) uptake, and restores the pharmacologic property of midazolam at the GABA$_A$ receptor.

The unique feature of having both water and lipid solubility makes midazolam a versatile drug for various medical situations. In addition to the intravenous (IV) route of administration, oral, intramuscular, intranasal, and subcutaneous routes are possible with midazolam. This opens multiple facets for its use. Young children presenting for medical procedures, an individual experiencing status epilepticus, and an elderly adult in a palliative care facility are just some examples of patients that do not routinely have IV access but can still benefit from alternative routes of midazolam administration.

Midazolam's predictable pharmacokinetics makes it the safest benzodiazepine for procedural sedation. Its anxiolytic, sedative, and amnestic properties markedly increase patient satisfaction in the perioperative period. The relatively fast onset and offset compared to other benzodiazepine allows midazolam to be titrated to effect. The availability of a safe and effective reversal agent (flumazenil) further improves the safety profile of midazolam.

Treatments, Mechanisms, and Adverse Reactions of Anesthetics and Analgesics. https://doi.org/10.1016/B978-0-12-820237-1.00014-4
Copyright © 2022 Elsevier Inc. All rights reserved.

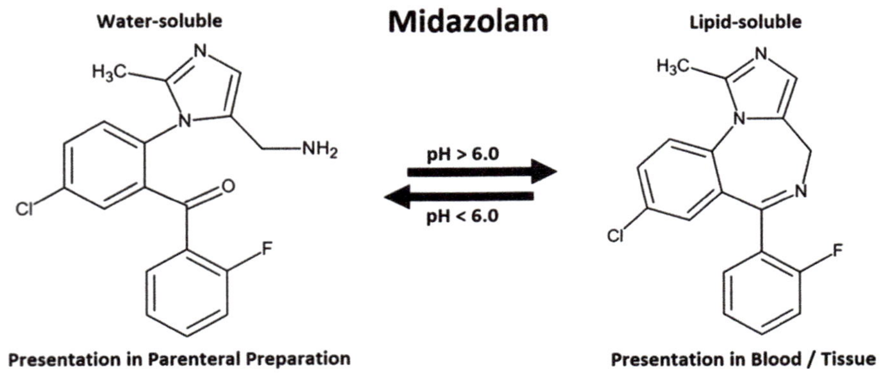

FIG. 1 Midazolam structure with relationship to presentation and function. Midazolam is water-soluble in parental preparation as a hydrochloride salt due to its open diazepine ring. Upon administration, the rise in tissue pH causes closure of the 7-membered 1,4-diazepine ring, restoring its lipid solubility and pharmacologic property of at the $GABA_A$ receptor.

Mechanism of action

Midazolam is the most commonly administered benzodiazepine in the perioperative setting due to its effectiveness at rapidly achieving procedural sedation, hypnosis, anxiolysis, and anterograde amnesia (Winsky-Sommerer, 2009). Benzodiazepines potentiate the neural inhibitory effect of gamma-aminobutyric acid (GABA) on the $GABA_A$ receptor. The mechanism of action for midazolam is mediated via its interaction with the $GABA_A$ receptor. Benzodiazepines including midazolam exert their effect by binding to the $GABA_A$ receptor at a site that is distinct from the binding site of GABA on the $GABA_A$ receptor. Specifically, midazolam binds at the junction of the α and γ subunits on the extracellular side of the transmembrane $GABA_A$ receptor (Sigel & Steinmann, 2012). Midazolam binding does not directly activate the chloride channel. Rather, the binding of midazolam to $GABA_A$ receptor potentiates the effect of GABA-induced chloride conductance, resulting in hyperpolarization of the postsynaptic membrane. In other words, benzodiazepines are not direct agonists but are rather positive allosteric modulators at the $GABA_A$ receptor. The chloride-mediated inhibitory postsynaptic potential decreases the probability of neurotransmission and is responsible for the clinical effects of midazolam.

Pharmacokinetics

The pharmacokinetic properties of midazolam have been well documented (Dundee, Halliay, Harper, et al., 1984). As with other benzodiazepines, the quick onset of midazolam is primarily due to its high lipid solubility which allows rapid penetration of the CNS. As such, benzodiazepines as a class of medications, have a high volume of distribution. Cessation of clinical effects is primarily due to the rapid redistribution of midazolam to inactive tissue compartments followed by a relatively slower hepatic microsomal oxidation and glucuronide conjugation. Of the common clinically utilized benzodiazepines, midazolam has the highest lipid solubility, thus resulting in the fastest onset of action. Midazolam also has the highest plasma clearance and thus the shortest elimination half-life compared to lorazepam and diazepam. The relatively faster onset and higher clearance of midazolam classify it as a short-acting benzodiazepine, allowing for easier titration in the perioperative period (see Table 1).

TABLE 1 Pharmacokinetic properties of benzodiazepines.

	Elimination half-life (h)	Plasma clearance (mL/kg/min)	Volume of distribution (L/kg)	Protein binding (%)
Midazolam	1.7–3.5	5.8–9.0	1.1–1.7	94–98
Lorazepam	11–22	0.8–1.5	0.8–1.3	88–92
Diazepam	20–50	0.2–0.5	0.7–1.7	98–99
Flumazenil	0.7–1.3	13–17	0.9–1.9	40–50

Elimination half-life, plasma clearance, volume of distribution, and protein binding of commonly used benzodiazepines. Pharmacokinetic parameters from Dundee et al. (1984).

Midazolam is available as oral, intramuscular, and intravenous (IV) preparations. Off-label routes of administration include subcutaneous, intranasal, and rectal. Pharmacokinetic studies on orally administered midazolam (Thummel, O'Shea, Paine, et al., 1996) have demonstrated a significant first-pass effect with less than 50% bioavailability after oral ingestion. Cmax after oral ingestion occurs in 30–80 min. When administered intravenously, peak effects are observed within 5 min. When administered intramuscularly, peak effects are observed within 15 min.

Distribution to peripheral tissue compartments is primarily responsible for the cessation of clinical effects of midazolam. Distribution occurs immediately after IV administration, with a distribution half-life of 6–15 min (Allonen, Ziegler, & Klotz, 1981). The rapid clearance of midazolam is aided by the imidazole ring which undergoes oxidation at a much faster rate than diazepam and lorazepam which do not have this ring structure (Dundee et al., 1984). Compared to lorazepam and diazepam, hepatic clearance of midazolam is up to 5–10 times faster respectively. Because the plasma clearance of lorazepam and diazepam is much slower, midazolam should ideally be the only benzodiazepine used for perioperative sedation or for continuous infusion for ICU sedation (Gommers & Bakker, 2008). Metabolism occurs in the liver via CYP3A4 and CYP3A5. These cytochrome p450 isoenzymes hydroxylates midazolam mainly into 1-hydroxymidazolam as well as minor metabolites of 4-hydroxymidazolam and 1,4-hydroxymidazolam. These three metabolites are then conjugated by glucuronic acid to form water-soluble compounds for renal elimination. Patients with mutations of CYP3A4 will have a delay in metabolism resulting in a prolonged duration of action (MacKenzie & Hall, 2017).

Midazolam pharmacokinetics is affected by age, obesity, liver, and renal disease. Patients with renal dysfunction may experience prolonged sedation due to the accumulation of active metabolites. 1-hydroxymidazolam, which is the major metabolite of midazolam hydroxylation, retains much of the sedative activities of the parent midazolam (Bauer, Ritz, Haberthur, et al., 1995). Patients with marked liver disease or cirrhosis can have a reduced ability to metabolize midazolam. The elimination half-life of midazolam in patients with liver cirrhosis has been shown to be increased to upwards of 3.9 h. This compares with the elimination half-life of 1.6 h in patients without cirrhosis (MacGilchrist, Birnie, Cook, et al., 1986). Patients with both renal and liver disease often have concomitant hypoalbuminemia. Because midazolam is highly protein-bound, the relative unbound fraction of midazolam in these patients also possess a risk for more profound and prolong effects. Careful titration is therefore paramount in these patients (Franken, Masman, de Winter, et al., 2017). On the other hand, chronic alcohol users have an increased metabolism of midazolam to the extent of remaining healthy liver parenchyma. Ethanol causes an induction of the same cytochrome p450 isoenzymes which metabolize benzodiazepines. Obesity increases the volume of distribution of midazolam resulting in slower elimination. Lastly, the elderly possess higher body composition of adipose tissue compared to lean tissue, thus increasing the volume of distribution and slows elimination. The decrease in baseline liver and renal organ function in the elderly also slows both hepatic metabolism and renal elimination of metabolites (Franken, de Winter, van Esch, et al., 2016).

Effects on the central nervous system

GABA is an inhibitory neurotransmitter in the CNS. The effect of GABA binding on the $GABA_A$ receptor causes an increase in chloride ion conductance, resulting in hyperpolarization of the postsynaptic neuron. Benzodiazepines including midazolam also bind to $GABA_A$ receptors, but at a site that is distinct from GABA. When benzodiazepines are bound to $GABA_A$ receptors, a conformational change occurs which potentiates the chloride conductance normally mediated by GABA binding, resulting in enhanced neuronal inhibition (Mohler, Fritschy, & Rudolph, 2002). In addition to the clinical effects of anxiolysis, sedation, anterograde amnesia, and hypnosis; the enhanced neuronal inhibition by benzodiazepines also produced anticonvulsant effects.

Benzodiazepines including midazolam causes a dose dependent decrease in cerebral metabolic rate of oxygen ($CMRO_2$) and a concomitant decrease in cerebral blood flow (CBF). The ratio of $CMRO_2$/CBF is not altered (Forster, Juge, & Morel, 1982). The potential beneficial effect of decreased CBF and intracranial pressure (ICP) must be weighed against the respiratory depressant effect of benzodiazepines which tends to do the opposite. Respiratory depression can lead to an increase in partial pressure of CO_2 resulting in an increase in CBF and ICP. Unlike barbiturates or propofol, there is a ceiling effect with benzodiazepines as escalating doses will cause incremental EEG slowing but will not produce an isoelectric EEG. This observation puts limitations on midazolam's neuroprotective effect. Patients with elevated intracranial pressure (ICP) exhibit little to no change with midazolam administration. Despite not having a well demonstrated neuroprotective effect, benzodiazepines including midazolam are potent therapeutics for alcohol withdrawal seizures, local anesthetic-induced seizures, and status epilepticus (Alshehri, Abulaban, Bokhari, et al., 2017).

Effects on the respiratory system

Benzodiazepines causes a dose dependent depression in central respiratory drive. There are three separate mechanisms responsible for this phenomenon. First, benzodiazepines can cause relaxation of the upper airway tone resulting in airway obstruction (Norton, Ward, Karan, et al., 2006). Midazolam should therefore be minimized or even avoided in patients with known obstructive sleep apnea. Secondly, benzodiazepines can cause a suppression of ventilatory drive in response to carbon dioxide tension. This response is synergistic with opioids. Patients receiving a co-administration of benzodiazepines and opioids should be carefully monitored to avoid apnea (Tverskoy, Fleyshman, Ezry, et al., 1989). Lastly, benzodiazepines can also suppress the ventilatory drive in response to hypoxemia (Alexander & Gross, 1988). Careful respiratory monitoring with pulse oximetry is warranted in patients receiving benzodiazepines in the perioperative period. Capnography is highly recommended when multiple classes of respiratory depressants are likely to be co-administered with benzodiazepines.

Effects on the cardiovascular system

Benzodiazepines including midazolam produces mild hemodynamic effects. There is a small decrease in arterial blood pressure primarily mediated by a minor decrease in systemic vascular resistance. Cardiac output shows little change (Ruff & Reves, 1990). Care should be taken in hypovolemic patients as midazolam and diazepam have been shown to depress the baroreceptor reflexes (Coote, 2006). Ablation of the baroreceptor reflex can impede compensatory mechanisms leading to a more significant drop in blood pressure. As such, maintaining a euvolemic state in the perioperative period is of importance.

Clinical uses and dosage

Midazolam is the most commonly administered premedication in the perioperative setting. For procedural sedation, the clinical effects of midazolam are highly desirable and include hypnosis, anxiolysis, and anterograde amnesia. When properly titrated, patients may seem arousable but lack the recall of intraoperative events (Bauer, Dom, Ramirez, et al., 2004). Favorable pharmacokinetics of rapid onset and short duration of action allows midazolam to be titrated. In addition to sedative-hypnotic properties, midazolam has shown to have beneficial effects on postoperative nausea and vomiting (PONV).

Midazolam is available in oral and parenteral preparations. Common indications, doses, and routes of administration are listed in Table 2. For procedural premedication, IV midazolam is commonly administered to adult patients presenting with procedural anxiety. For American Society of Anesthesiologists Physical Status (ASA-PS) I and II patients, an initial 1–2 mg IV can be given with incremental doses of 0.5–1 mg titrated to the desired level of sedation. However, downward dose adjustments should be made for the elderly, ASA-PS III and higher patients, and patients with hepatic and or renal impairments (Klotz, 2009). The level of patient anxiety, the length, and the type of procedure should also be considered in determining the initial bolus dose and subsequent titration doses. For pediatric patients, neonates rarely require a sedative premedication as separation anxiety does not occur until after 6 months of age. When IV access is available, a bolus dose of

TABLE 2 Midazolam for procedural premedication.

	Route	Initial dose	Dose adjustment
Pediatrics	PO	0.25–0.5 mg/kg	Decrease dose for renal and/or hepatic impairment
Pediatrics	IM	–0.15 mg/kg	Decrease dose for renal and/or hepatic impairment
Pediatrics (<5 years)	IV	0.05–0.1 mg/kg	Decrease dose for renal and/or hepatic impairment
Pediatrics (5–12 years)	IV	0.025–0.05 mg/kg	Decrease dose for renal and/or hepatic impairment
Adult	IV	1–2 mg	Decrease dose for ASA-PS III and above, the elderly, and for patients with renal and/or hepatic impairments

Common Midazolam doses and routes of administration for procedural premedication.

0.05–0.1 mg/kg IV for patients from 6 months to 5 years of age is suitable for anxiolysis and induction of amnesia. For 5–12 years of age, the initial IV dose should be lowered to 0.025–0.05 mg/kg. Since most pediatric patients presenting for a medical procedure or surgery may not be willing to have IV access established, oral midazolam is a viable alternative. For oral route of administration, 0.25–0.5 mg/kg oral midazolam can be given with the resulting sedation and anxiolysis occurring in 15–20 min (Manso, Guittet, & Vandenhende, 2019). Although oral and IV routes of administration are most prevalent, alternative routes of administration can also be employed. A single intramuscular dose of 0.1–0.15 mg/kg midazolam can be used for pediatric patients. Other routes that have been described include intranasal and rectal routes of administration.

In addition to procedural sedative effects, midazolam has been shown to mitigate the incidence of PONV. In a meta-analysis (Grant, Kim, Page, et al., 2016) looking at a broad variety of surgical procedures requiring general anesthesia and high-risk patient characteristics for PONV, the administration of IV midazolam was associated with a significant decrease in PONV and the need of a rescue antiemetic. Furthermore, the authors found no association to delayed postoperative recovery.

Benzodiazepines are the first line treatment for generalized convulsive status epilepticus (GCSE). The initial pharmacologic therapy for GCSE hinges upon the presence or absence of IV access. When IV access is available, lorazepam 0.1 mg/kg IV is the drug of choice (Treiman, Meyers, Walton, et al., 1998). Although IV midazolam is also efficacious in stopping GCSE, the longer duration of action of lorazepam makes it a more effective agent to prevent breakthrough seizures once the benzodiazepine begins to wear off. When IV access is not available, midazolam 10 mg IM or 10 mg intranasally is the drug of choice (Arya, Kothari, Zang, et al., 2015). Although not well studied, intranasal or buccal midazolam are absorbed more rapidly than IM midazolam which suggests that these alternative routes of administration may be superior (Hirsch, 2012). As such, buccal and intranasal midazolam is often prescribed for outpatients for interruption of GCSE that can be administered at home during an emergency without IV access and without medical personnel.

Midazolam is the only benzodiazepine suitable for induction of general anesthesia. The slower onset and the markedly longer duration of action of lorazepam and diazepam makes them ill-suited for induction of general anesthesia. For midazolam, the onset of general anesthesia occurs 30–90 s after an induction dose of 0.2–0.4 mg/kg IV. The half-time to observed EEG effects after an IV bolus is between 2 and 3 min (Breimer, Hennis, Burm, et al., 1990). Although midazolam can be used for induction of general anesthesia, its onset time is still slower compared with propofol, etomidate, and ketamine. The major disadvantage of midazolam as an induction agent for general anesthesia is the substantially longer return to consciousness. After induction with midazolam, return to consciousness may take from 10 to 30 min. This is markedly longer than the aforementioned IV induction agents which take between 3 and 12 min to return to consciousness (Fields, Rosbolt, & Cohn, 2009).

Applications to other areas: Palliative care

Palliative medicine is a complex interdisciplinary practice involving social, ethical, and medical experts. The intention of this section is to provide a brief overview of the role of midazolam in palliative care. It is not a guide for the management of these patients.

Terminally ill patients may experience multiple unpleasant symptoms near the end of life. Refractory symptoms of terminal dyspnea, pain, anxiety, insomnia, and agitated delirium are all too common. When these symptoms cannot be adequately relieved, palliative sedation has often been used an as an option of last resort (Lo & Rubenfeld, 2005). Medical management of these symptoms is multifaceted and usually requires a combination of drugs for palliation. Benzodiazepines are considered first-line drugs, with midazolam as the most commonly used benzodiazepine for palliative therapy (Lux, Protus, & Kimbrel, 2017). Midazolam suites this purpose well with its sedative, anxiolytic, and anticonvulsant properties. Its short duration of action also allows for ease of titration for patients undergoing continuous infusions. Its ability to be given via multiple routes, including IM or subcutaneously without significant pain on injection, lends to its over flexibility in various care settings.

Palliative sedation may be indicated when a patient is experiencing unbearable and intractable symptoms from their disease process. Initiation of palliative sedation can be started with an infusion of midazolam at 0.5–1 mg/h. This can be titrated upwards with a range of 1–20 mg/h as the upper range of effective dose. Boluses of midazolam with a range of 1–5 mg IV can also be used concurrently and titrated to the endpoint of calm and comfort (Bodnar, 2017). If additional comfort measures are necessary, consider adding an antipsychotic such as haloperidol once the midazolam dose reaches >30 mg/24 h (Howard, Twycross, Shuster, et al., 2014). For management of terminal dyspnea, opioids are typically co-administered with midazolam to help manage the anxiety associated with the dyspnea. When midazolam is used in

136 PART | II Mechanisms of action of drugs

conjunction with morphine, it has been shown to provide superior dyspnea relief compared to either agents when used alone (Navigante, Cerchietti, Castro, et al., 2006).

Other agents of interest: Flumazenil

Flumazenil is a competitive antagonist at the benzodiazepine binding site on the $GABA_A$ receptor. It is indicated for the reversal of the sedative-hypnotic effects of benzodiazepines. Flumazenil binds at the junction of the α and γ subunits on the extracellular side of the $GABA_A$ receptor near the benzodiazepine binding site. Binding of flumazenil causes neither activation nor inhibition. Instead, it is a neutral allosteric modulator that prevents benzodiazepines from binding (Kucken, Teissere, Seffinga-Clark, et al., 2003).

Reversal of benzodiazepine intoxication is typically performed by administering 0.2 mg IV flumazenil every 1–2 min until adequate reversal of sedation. Relative to other benzodiazepines, flumazenil is rapidly metabolized by the liver and quickly eliminated, even more so than midazolam. Therefore, there exists a potential that flumazenil can be rapidly cleared with the resumption of benzodiazepine intoxication. This is especially true if the intoxication is due to long-acting benzodiazepines such as lorazepam or diazepam. As such, careful monitoring is necessary to prevent the return of benzodiazepine intoxication (Sivilotti, 2016). An infusion of flumazenil at 0.5–1 µg/kg/minute can be used to mitigate this concern (Kleinberger, Grimm, Laggner, et al., 1985). Side effects of benzodiazepine reversal with flumazenil are limited. Unlike diazepam, there is no venous irritation on injection. Similar to other benzodiazepines, metabolism is via CYP3A4 and CYP3A5 without active metabolites. End products of metabolism are conjugated to glucuronic acid and renally eliminated. In patients with a physical dependence on benzodiazepines, rapid reversal with flumazenil may potentially cause withdrawal symptoms including seizures (Ngo, Anthony, Samuel, et al., 2007).

Key facts of midazolam

- Midazolam is water soluble in pharmaceutical preparations. Upon administration, an increase in pH causes closure of the 7-membered 1,4-diazepine ring, restoring the pharmacologic property of midazolam at the $GABA_A$ receptor.
- Midazolam is a positive allosteric modulator at the $GABA_A$ receptor. Midazolam binds to the $GABA_A$ receptor and potentiates the effect of GABA-induced chloride conduction resulting in hyperpolarization of the postsynaptic neuron.
- Midazolam's high lipid solubility accounts for its fast onset and also for its short duration of action. High lipid solubility results in fast CNS penetrance. High lipid solubility also means midazolam redistributes to inert peripheral compartments resulting in the cessation of the clinical effects.
- Midazolam causes a concomitant decrease in both $CMRO_2$ and CBF. The $CMRO_2$ to CBF ratio does not change. Although EEG slowing is shown, midazolam has a ceiling effect as escalating doses do not produce an isoelectric EEG.
- Midazolam can be used for the treatment of GCSE and is the treatment of choice to be administered at home without IV access and without trained medical personnel. Midazolam 10 mg IM or intranasally can be utilized in this setting.
- Flumazenil is an allosteric modulator at the $GABA_A$ receptor. It binds $GABA_A$ receptor to but does not increase the chloride conduction. Thus, it functions as a competitive antagonist to benzodiazepine binding at the $GABA_A$ receptor.

Mini-dictionary of terms

American Society of Anesthesiologist physical status. A classification system which categories a patient's health status endorsed by the American Society of Anesthesiologists. There are six categories of classification with increasing levels of comorbid conditions. Class I = healthy. Class II = mild systemic disease. Class III = severe systemic disease. Class IV = severe systemic disease with constant threat to life. Class V = moribund patient who is not expected to survive without the operation. Class VI = brain-dead individual for organ donation.

Capnography. The monitoring of the concentration or partial pressure of carbon dioxide in respiratory gas. Capnography monitors in the medical setting are used to detect adverse respiratory events such as hypoventilation.

C_{max}. C_{max} is an abbreviation for the maximum or peak serum concentration after a drug is administered. C_{max} is determined by the total dose administered and the apparent volume of distribution of the medication within the subject.

Cytochrome p450 isoenzymes. Cytochrome p450 is a family of isoenzymes responsible for the biotransformation of many drugs mainly via oxidation. Although the majority of these isoenzymes are located in the liver, extrahepatic metabolism also occurs in the lungs, kidneys, and gastrointestinal tract. On the cellular level, the isoenzymes are membrane proteins located within the smooth endoplasmic reticulum.

Propylene glycol. An organic compound with the chemical formula $CH_3CH(OH)CH_2OH$. It is used in pharmaceutical preparations as a solvent to solubilize lipid-soluble drugs of various presentations including oral, injectable, and topical formulations. When used to dissolve intravenous preparations (as with benzodiazepines), the intravenous injection can cause venous irritation. Other rare side effects are dose-dependent and may include CNS depression, bradycardia, QRS and T wave abnormalities on ECG, arrhythmia, seizures, lactic acidosis and hemolysis.

Summary points

- Midazolam is the most commonly utilized benzodiazepine in the perioperative period. It produced anxiolysis, sedation, anterograde amnesia, and hypnosis.
- Midazolam has fast onset and short duration of action compared to other benzodiazepines.
- Midazolam can be administered via multiple routes of administration. In pediatric patients presenting for surgery, sedation and anxiolysis can be achieved 15–20 min after an oral administration of 0.25–0.5 mg/kg midazolam.
- Midazolam is metabolized via CYP450 isoenzymes with active metabolized renally eliminated. The elderly and patients with hepatic and/or renal dysfunction may experience longer durations of action. Chronic alcohol consumption causes CYP450 induction, resulting in cross-tolerance.
- Midazolam can cause respiratory depressant effects especially given in conjunction with other sedatives or analgesics in the perioperative setting. When used in conjunction with opioids and especially with propofol, capnography monitoring is necessary to detect adverse effects of respiratory depression.
- Benzodiazepine overdose can be reversed with flumazenil. Because of flumazenil's relatively shorter duration of action compared to other benzodiazepines, repeated administration or an infusion may be necessary to prevent the recurrence of overdose symptoms.

References

Alexander, C. M., & Gross, J. B. (1988). Sedative doses of midazolam depress hypoxic ventilatory responses in humans. *Anesthesia and Analgesia, 67,* 377–382.

Allonen, H., Ziegler, G., & Klotz, U. (1981). Midazolam kinetics. *Clinical Pharmacology and Therapeutics, 30,* 653–661.

Alshehri, A., Abulaban, A., Bokhari, R., et al. (2017). Intravenous versus non-intravenous benzodiazepines for the cessation of seizures: A systematic review and meta-analysis of randomized controlled trials. *Academic Emergency Medicine, 24,* 875–883.

Arya, R., Kothari, H., Zang, Z., et al. (2015). Efficacy of nonvenous medications for acute convulsive seizures: A network meta-analysis. *Neurology, 85,* 1859–1868.

Bauer, K. P., Dom, P. M., Ramirez, A. M., et al. (2004). Preoperative intravenous midazolam: Benefits beyond anxiolysis. *Journal of Clinical Anesthesia, 16,* 177–183.

Bauer, T. M., Ritz, R., Haberthur, C., et al. (1995). Prolonged sedation due to accumulation of conjugated metabolites of midazolam. *Lancet, 346,* 145–147.

Bodnar, J. (2017). A review of agents for palliative sedation/continuous deep sedation: Pharmacology and practical applications. *Journal of Pain & Palliative Care Pharmacotherapy, 31,* 16–37.

Breimer, L. T., Hennis, P. J., Burm, A. G., et al. (1990). Quantification of the EEG effect of midazolam by aperiodic analysis in volunteers: Pharmacokinetic/pharmacodynamic modelling. *Clinical Pharmacokinetics, 18,* 245–253.

Coote, J. H. (2006). Landmarks in understanding the central nervous control of the cardiovascular system. *Experimental Physiology, 92,* 3–18.

Dundee, J. W., Halliay, N. J., Harper, K. W., et al. (1984). Midazolam: A review of its pharmacological properties and therapeutic use. *Drugs, 28,* 519–543.

Fields, A. M., Rosbolt, M. B., & Cohn, S. M. (2009). Induction agents for intubation of the trauma patient. *The Journal of Trauma, 67,* 867–869.

Forster, A., Juge, O., & Morel, D. (1982). Effects of midazolam on cerebral blood flow in human volunteers. *Anesthesiology, 56,* 453–455.

Franken, L. G., de Winter, B. M. C., van Esch, H. J., et al. (2016). Pharmacokinetic consideration and recommendations in palliative care, with focus on morphine, midazolam, and haloperidol. *Expert Opinion on Drug Metabolism & Toxicology, 12,* 669–680.

Franken, L. G., Masman, A. D., de Winter, B. M. C., et al. (2017). Hypoalbuminemia and decreased midazolam clearance in terminally ill adult patients, an inflammatory effect? *British Journal of Clinical Pharmacology, 83,* 1701–1712.

Gommers, D., & Bakker, J. (2008). Medications for analgesia and sedation in the intensive care unit: An overview. *Critical Care, 12*(Suppl 3), S4.

Grant, M. C., Kim, J., Page, A. J., et al. (2016). The effect of intravenous midazolam on postoperative nausea and vomiting: A meta-analysis. *Anesthesia and Analgesia, 122,* 656–663.

Hirsch, L. J. (2012). Intramuscular versus intravenous benzodiazepines for prehospital treatment of status epilepticus. *The New England Journal of Medicine, 366,* 659–660.

Howard, P., Twycross, R., Shuster, J., et al. (2014). Benzodiazepines. *Journal of Pain and Symptom Management, 47,* 955–964.

Kleinberger, G., Grimm, G., Laggner, A., et al. (1985). Weaning patients from mechanical ventilation by benzodiazepine antagonist Ro 15-1788. *Lancet, 2,* 268–269.

Klotz, U. (2009). Pharmacokinetics and drug metabolism in the elderly. *Drug Metabolism Reviews, 41,* 67–76.

138 PART | II Mechanisms of action of drugs

Kucken, A. M., Teissere, J. A., Seffinga-Clark, J., et al. (2003). Structural requirements for imidazobenzodiazepine binding to GABA(A) receptors. *Molecular Pharmacology, 63*, 289–296.

Lo, B., & Rubenfeld, G. (2005). Palliative sedation in dying patients: "We turn to it when everything else hasn't worked". *Journal of the American Medical Association, 294*, 1810–1816.

Lux, M. R., Protus, B. M., & Kimbrel, J. (2017). A survey of hospice and palliative care physicians regarding palliative sedation practices. *The American Journal of Hospice & Palliative Care, 34*, 217–222.

MacGilchrist, A. J., Birnie, G. G., Cook, A., et al. (1986). Pharmacokinetics and pharmacodynamics of intravenous midazolam in patients with severe alcoholic cirrhosis. *Gut, 27*, 190–195.

MacKenzie, M., & Hall, R. (2017). Pharmacogenomics and pharmacogenetics for the intensive care unit: A narrative review. *Canadian Journal of Anaesthesia, 64*, 45–64.

Manso, M. A., Guittet, C., & Vandenhende, F. (2019). Efficacy of oral midazolam for minimal and moderate sedation in pediatric patients: A systematic review. *Paediatric Anaesthesia, 29*, 1094–1106.

Mohler, H., Fritschy, J. M., & Rudolph, U. (2002). A new benzodiazepine pharmacology. *The Journal of Pharmacology and Experimental Therapeutics, 300*, 2–8.

Navigante, A. H., Cerchietti, L. C., Castro, M. A., et al. (2006). Midazolam as adjunct therapy to morphine in the alleviation of severe dyspnea perception in patients with advanced cancer. *Journal of Pain and Symptom Management, 31*, 38–47.

Ngo, A. S., Anthony, C. R., Samuel, M., et al. (2007). Should a benzodiazepine antagonist be used in unconscious patients presenting to the emergency department? *Resuscitation, 74*, 27–37.

Norton, J. R., Ward, D. S., Karan, S., et al. (2006). Differences between midazolam and propofol sedation on upper airway collapsibility using dynamic negative airway pressure. *Anesthesiology, 104*, 1155–1164.

Ruff, R., & Reves, J. G. (1990). Hemodynamic effects of lorazepam-fentanyl anesthetic induction for coronary artery bypass surgery. *Journal of Cardiothoracic Anesthesia, 4*, 314–317.

Sigel, E., & Steinmann, M. E. (2012). Structure, function, and modulation of GABA(A) receptors. *The Journal of Biological Chemistry, 287*, 40224–40231.

Sivilotti, M. L. (2016). Flumazenil, naloxone and the 'coma cocktail'. *British Journal of Clinical Pharmacology, 81*, 428–436.

Thummel, K. E., O'Shea, D., Paine, M. F., et al. (1996). Oral first-pass elimination of midazolam involves both gastrointestinal and hepatic CYP3A-mediated metabolism. *Clinical Pharmacology and Therapeutics, 59*, 491–502.

Treiman, D. M., Meyers, P. D., Walton, N. Y., et al. (1998). A comparison of four treatments for generalized convulsive status epilepticus. Veterans affairs status epilepticus cooperative study group. *The New England Journal of Medicine, 339*, 792–798.

Tverskoy, M., Fleyshman, G., Ezry, J., et al. (1989). Midazolam-morphine sedative interaction in patients. *Anesthesia and Analgesia, 68*, 282–285.

Wick, J. Y. (2013). The history of benzodiazepines. *The Consultant Pharmacist, 28*, 538–548.

Winsky-Sommerer, R. (2009). Role of GABA$_A$ receptors in the physiology and pharmacology of sleep. *The European Journal of Neuroscience, 29*, 1779–1794.

Chapter 15

Intravenous paracetamol: Features and applications

Ristiawan M. Laksono and Isngadi Ahmad Wagimin

Department of Anesthesiology and Intensive Care, Faculty of Medicine, University of Brawijaya, Kota Malang, East Java, Indonesia

Abbreviations

AM404	N-arachidinoyl-phenolamine
$\mathbf{C_{max}}$	maximum concentration
COVID-19	corona virus disease 2019
COX	cyclo-oxygenase
$\mathbf{CSF_{max}}$	maximum cerebrospinal fluid concentration
FAAH	fatty acidamide hydrolase
IV	intravenous
NAPQI	N-acetyl-p-benzoquinone imine
NMDA	N-methyl-D-aspartate
NO	nitric oxide
NSAID	non-steroidal anti-inflammatory drug
PGE2	prostaglandin E2
VAS	visual analog scale
VRS	visual rating score

Introduction

Paracetamol, N-acetyl-p-aminophenol, which is also known as acetaminophen, is a compound that has antipyretic and analgesic effects. This drug is freely available on the market and used to treat pain and fever. Paracetamol can be given per oral, per rectal, or intravenously. Pharmacologically, paracetamol has analgesic and antipyretic effects but has very small peripheral anti-inflammatory properties (Pacifici & Allegaert, 2014; Parker et al., 2018). Intravenous paracetamol administration can be used as a preemptive analgesic for major oncology operations (Laksono, Isngadi, & Murti, 2019). Generally, IV paracetamol used for immediate analgesic and antipyretic effects. Intravenous paracetamol also used as multimodal analgesia with opioid-sparing effect.

Chemical structure and pharmacokinetics

Chemically, paracetamol is phenol, and like other phenols, it is easily oxidized. Paracetamol is a low molecular weight compound (Fig. 1) (Graham, Davies, Day, Mohamudally, & Scott, 2013). This drug is a very weak acid ($p\mathrm{K_a}$ 9.7), therefore, it is in a non-ionic form in physiological pH condition (Craig, 1990). The partition coefficient between octanol and water is 3.2 and in this range, passive diffusion through cell membranes can occur. After intravenous administration, paracetamol binding with plasma protein can be neglected, it was concluded that paracetamol is distributed throughout the body without being bound to tissue (Gazzard, Ford-Hutchinson, Smith, & Williams, 1973; Prescott, Speirs, Critchley, Temple, & Winney, 1989).

The peak plasma concentration (C_{max}) of paracetamol after administration of a therapeutic dose (1 g) intravenously or orally with fast absorption is around 20 mg/L (130 μM) to 30 mg/L (200 μM). Plasma paracetamol levels of 30 mg/L can be considered as therapeutic range (Graham et al., 2013).

Treatments, Mechanisms, and Adverse Reactions of Anesthetics and Analgesics. https://doi.org/10.1016/B978-0-12-820237-1.00015-6
Copyright © 2022 Elsevier Inc. All rights reserved.

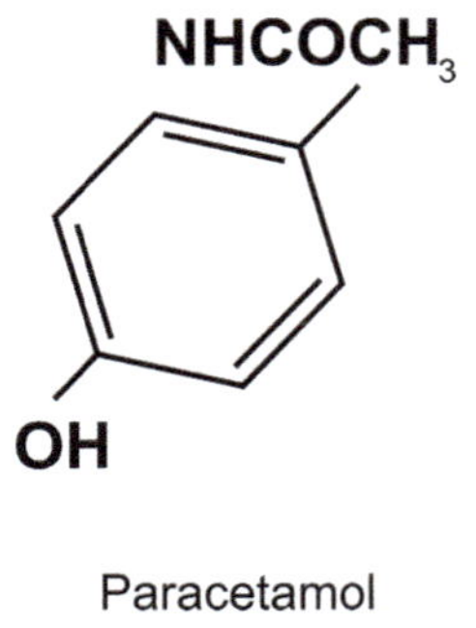

NHCOCH₃

Paracetamol

FIG. 1 Chemical structure of paracetamol.

Paracetamol pharmacokinetics

After oral administration, the drug is quickly absorbed by the gastrointestinal (GI) tract. Its systemic bioavailability is dose-dependent and ranges from 70% to 90%. The rate of absorption after oral administration is greatly affected by the rate of gastric emptying. When the stomach is full, the absorption of oral acetaminophen will be inhibited. The distribution volume of paracetamol is around 0.9 L/kg. Paracetamol is bound by red blood cells around 10%–20%. The half-life of the drug in plasma is 1.9–2.5 h. Total clearance from the body is 4.5–5.5 mL/kg/min. After giving a therapeutic dose of paracetamol to healthy people, most of the paracetamol undergo conjugation in the liver by phase II enzymes to form glucuronide derivatives (paracetamol-glu) and sulfate derivatives (Paracetamol-sul) (Fig. 2). Early pharmacokinetic studies showed that paracetamol given as a therapeutic dose would be excreted at 55% as glucuronide, 30% as sulfate, and 4% excreted as a product of oxidative metabolism. The metabolic half-life ($t_{1/2}$) of paracetamol is 1.5–2.5 h but can be extended after an overdose of paracetamol (Caparrotta, Antoine, & Dear, 2018; Forrest, Clements, & Prescott, 1982; Prescott, 1980). A small dose of paracetamol will be oxidized by phase I reactions that are catalyzed by several oxidases (cytochrome P450, CYP450), nicotinamide adenine dinucleotide phosphate oxidase, and oxygen (Prescott, 1980). This phase I oxidative reaction metabolizes 2%–10% paracetamol into a reactive intermediate, N-acetyl-p-benzoquinone imine (NAPQI), which causes hepatotoxic effects of paracetamol. This metabolite becomes inactive after being reduced with glutathione and excreted in urine as cysteine (about 4%) and as a conjugate of mercapturic acid (about 4%) (Dasgupta & Karowski, 2020). An overdose of acetaminophen can cause acute liver necrosis due to depletion of glutathione and damage due to reactive N-acetyl-p-benzoquinone-imine (NAPQI) metabolism. This damage can be prevented by early administration of sulfhydryl compounds such as methionine and N-acetylcysteine (Forrest et al., 1982).

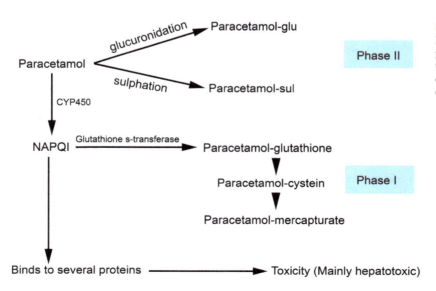

FIG. 2 Paracetamol metabolism through liver. Phase II including conjugation of paracetamol to sulphation and glucuronidation. Phase I forming NAPQI that binds to glutathione S-transferase (GSH) and changed into paracetamol mercapturate. NAPQI that does not bind to GSH will causing toxicity.

Site of action of paracetamol

Historically, paracetamol was originally categorized as a non-steroidal anti-inflammatory drugs (NSAIDs). Several investigations comparing its mechanism of action with classical NSAIDs, such as acetylsalicylic acid, which inhibits the cyclooxygenase pathway, turned out that paracetamol proved ineffective as an anti-inflammatory drug (Högestätt et al., 2005).

Paracetamol act in the central and peripheral nervous system. This drug has a mechanism of action similar to nonsteroidal anti-inflammatory drugs (NSAIDs), i.e, resembling selective COX-2 inhibitors. Paracetamol is a weaker analgesic than NSAIDs or selective COX-2 inhibitors but more preferred because of its better tolerance. Aside from the similarities with NSAIDs, the mechanism of action of paracetamol is uncertain, but currently accepted mechanism of action of paracetamol is that paracetamol works by inhibiting the COX-1 and COX-2 isoenzyme, thus resulting in inhibition of the formation of phenoxyl radicals from tyrosine residues and inhibits the formation of prostaglandins (PG). Paracetamol is thought to have selectivity for COX-2. The selectivity against COX-2 is shown by poor antiplatelet activity and good gastrointestinal tolerance of paracetamol. Paracetamol also inhibits other peroxidase enzymes, including myeloperoxidase, this is different from non-selective NSAIDs and selective COX-2 inhibitors. Paracetamol has a central and peripheral effect, the analgesic effect of paracetamol is reduced by several endogenous neurotransmitter inhibitors including the serotonergic system, opioids, and cannabinoids (Graham et al., 2013).

Inhibition of COX pathway

Paracetamol can penetrate the blood–brain barrier and is distributed homogeneously throughout the central nervous system. Paracetamol has been shown to produce selective inhibition of the COX pathway in the brain, while inhibition by paracetamol is not present in peripheral tissues as in the stomach (Högestätt et al., 2005). The site of action for paracetamol in the central nervous system was initially discovered by Flower and Vane (1972) who found that paracetamol produced a stronger inhibition of PG synthesis and other related factors in the microsomal fraction of rabbits' brains than in the dog's spleen. The central site of action of paracetamol is significantly proven when the administration of systemic paracetamol inhibits the increase in central PGE2 production, which is caused by the administration of pyrogen or pain stimulation (Feldberg, Gupta, Milton, & Wendlandt, 1973). Furthermore, this drug act through the central (cyclooxygenase, serotonergic neuronal descent pathway, L-arginine/nitric oxide pathway, and cannabinoid system) antinociception process and "redox" mechanism. The antipyretic effect of acetaminophen is largely due to the effective inhibition of cyclooxygenase-2 in the hypothalamus, thus preventing the stimulation of heat-regulating centers. Decreased production of prostaglandins in neurons and endothelial blood vessels have a role as analgesics. Acetaminophen has no anti-inflammatory or anti-thrombotic activity (Dasgupta & Krasowski, 2020).

Modulation of endogenous cannabinoid system

Another mechanism of action of paracetamol as an analgesic is by involving modulation of the endogenous cannabinoid system. This mechanism might explain the "relaxing" or "calming" effects that have been reported in some patients using paracetamol. This effect is characterized by the uniqueness of well-being, relaxation, and calmness. It was suspected that after deacetylation of paracetamol to p-aminophenol, it is conjugated with arachidonic acid by fatty acidamide hydrolase (FAAH) enzymes to form N-arachidinoyl-phenolamine (AM404) in the brain, spinal cord, and dorsal ganglia fibers (Högestätt et al., 2005). The structures of AM404, anandamide, and N-arachidonoylethanolamine (AEA), the latter being endogenous cannabinoid neurotransmitters. Because they have similar structures, AM404 is a weak agonist of type 1 and 2 cannabinoid receptors (CB1 and CB2), and anandamide membrane transporter inhibitors (AMT), which cause an increase in endogenous cannabinoid levels. AM404 is also a potential activator of the vanilloid subtype 1 receptor (TRPV1). Paracetamol directly activates the TRPV1 receptor and indirectly activates the CB1 cannabinoid receptor by increasing endogenous levels of anandamide. Both of these mechanisms are in thermoregulation and pain centers in the brain. Thus the mechanism of action of paracetamol in the central nervous system in addition to COX inhibition mechanism, also through the vanilloid and cannabinoid signaling pathways (Ghanem, Pérez, Manautou, & Mottino, 2016).

Inhibition of nitric oxide

Inhibition of the formation of nitric oxides (NO) may also be a mechanism of action of paracetamol as an analgesic. The L-arginine/NO pathway is activated by P substances and NMDA receptors for NO synthesis, which is an important neurotransmitter in the nociceptive process of the spinal cord (Jóźwiak-Bebenista & Nowak, 2014). The mechanism of action

142 **PART | II** Mechanisms of action of drugs

of paracetamol through the central nerve may also occur through the stimulatory effect on the descending pathway of serotoninergic. Research conducted on healthy volunteers who are stimulated by pain with electrical stimulation of the median nerve showed that the analgesic effect of paracetamol is completely blocked in groups of subjects treated with paracetamol combined with tropisetron or granisetron (5-HT3 receptor antagonists) (Jóźwiak-Bebenista & Nowak, 2014).

The anti-nociceptive effect of paracetamol is reduced by serotonin inhibitors (5-hydroxytryptamine, 5-HT3), endogenous opioids, endogenous cannabinoids, and possibly acetylcholine (Graham et al., 2013). Due to the reduced analgesic activity of paracetamol caused by 5-HT3 receptor antagonists, tropisetron, and granisetron, the maintenance or activation of several serotonergic systems is very important to produce a good analgesic effect of paracetamol (Graham et al., 2013).

The mechanism of action of paracetamol is very complex, working at all levels of pain stimulus conduction from tissue receptors through the spinal cord to the thalamus and brain cortex, where the pain sensation is generated. Possibly, paracetamol works peripherally and centrally by inhibition of COX activity in the peripheral and central mechanism of action through (cyclooxygenase, serotonergic neuronal descent pathway, L-arginine/nitrite oxide pathway, and cannabinoid system) antinociception and "redox" mechanism (Jóźwiak-Bebenista & Nowak, 2014).

The combination of paracetamol therapy with NSAIDs will produce a better analgesic effect than paracetamol alone because this combination will produce a synergistic effect and additive effect (Graham et al., 2013). While the combination therapy between paracetamol and opioid can increase the efficacy and can cause an "opioid-sparing effect" so that the administration of intravenous paracetamol can reduce the need for opiates for acute pain therapy (Graham et al., 2013).

Pharmacokinetics of intravenous paracetamol in adult

Intravenous paracetamol has widely used across the world. It was approved by US FDA in November 2010 for the treatment of acute pain, fever in 2 years old children and older, and adults. This drug has been produced in several country with various brand names (Table 1). Fews were used in many experiments and widely distributed in several country.

Pain that appears postoperatively is often relieved using opioids. Opioids are known to have side effects of slowing down gastric emptying so that it interferes with the absorption of multimodal analgesic drugs used by the oral route. Paracetamol is a multimodal analgesic that can be combined with opioids, but the use of the oral route has lower efficacy than the intravenous route. Plasma concentrations in patients taking oral paracetamol showed a decrease when added with morphine, but there was an unexpected increase when the use of morphine was stopped. This shows the accumulation of paracetamol which was absorbed by the small intestine due to the cessation of the use of morphine. The use of IV paracetamol, when added to morphine treatment as multimodal analgesia, showed a small change in plasma concentrations. The pharmacokinetics of IV paracetamol have lower variability than oral paracetamol when used together with morphine (Raffa et al., 2018).

Administration of drugs by intravenous route gives higher concentration in the body. Several studies show this results. Administration of 1000 mg of Paracetamol preoperatively had blood plasma concentration (C_{max}) of 19.3 mg/L when administered intravenously and 13 mg/L when administered orally at the 30 min (van der Westhuizen, Kuo, Reed, & Holder, 2011). A total of 96% and 67% of patients achieved maximal plasma concentrations in IV and oral group respectively. Also, it was reported that paracetamol concentrations were positively correlated with body weight, so a dose of 1.5 g was needed for patients with 50–75 kg weight and 2 g for patients weighing more than 75 kg. The maximal concentration (C_{max}) of paracetamol in plasma is higher than the oral and rectal routes (Singla et al., 2012). The study was conducted on healthy adult men administered with 1000 mg of paracetamol via IV and oral routes, and 1300 mg via rectal route. Paracetamol IV had maximum cerebrospinal fluid concentration (CSF_{max}) higher than oral and rectal paracetamol which is 5.94, 3.72 and 4.13 μg/mL, respectively.

Studies comparing the effectiveness of IV to oral paracetamol administration for orthopedic surgery pain were carried out by Brett et al. (2012). The results of these studies indicate IV paracetamol had a higher plasma concentration in the early period of surgery than oral paracetamol. Rescue analgesia and length of stay in IV group were also lower than oral group.

The use of intravenous paracetamol in adult

Antipyretic effect of 1000 mg paracetamol on endotoxins induced volunteers with a median age of 33 years was observed by Peacock et al. (2011). A total of 45 and 36 subjects received IV and oral treatment of paracetamol included in modified intention-to-treat analysis populations. The intravenous route shows a significant temperature decrease in 30–90 min compared to the oral route. Improvement of the body temperature of 0.3 °C in IV groups can be considered in using this route to obtain a faster effect and in the case of patients unable to receive drugs via oral route. There are two subjects in the IV route

TABLE 1 List of intravenous paracetamol brands and manufacturers.

Brands	Manufacturer	Reference
Ofirmev	Cadence Pharmaceutical	Achuff et al. (2019), Meyering, Stringer, and Hysell (2017), Kayhan, Sanli, Ozgul, Kirteke, and Yologlu (2018), Peacock, Breitmeyer, Pan, Smith, and Royal (2011), Cook et al. (2016), and Wininger et al. (2010)
Perfalgan	Bristol-Myers Squibb	Bektas et al. (2009), Brett, Barnett, and Pearson (2012), Flint et al. (2017), Kumpulainen et al. (2007), Raffa et al. (2018), Eken et al. (2014), and Singla et al. (2012)
IV Paracetamol	Actavis	Chiam, Weinberg, and Bellomo (2015)
IV Paracetamol	Pfizer	Chiam et al. (2015)
Acetaminophen	Fresenius Kabi Austria GmbH	Nahum, Weissbach, Kaplan, and Kadmon (2020)
Cetapain	Darya-Varia	*MIMS Indonesia: Petunjuk Konsultasi* (2019)
Farmadol	Fahrenheit	*MIMS Indonesia: Petunjuk Konsultasi* (2019)
Nofebril	Yarindo Farmatama	*MIMS Indonesia: Petunjuk Konsultasi* (2019)
Paracetamol Phapros	Phapros	*MIMS Indonesia: Petunjuk Konsultasi* (2019)
Pehamol	Phapros	*MIMS Indonesia: Petunjuk Konsultasi* (2019)
Pireta	Mahakam Beta Farma	*MIMS Indonesia: Petunjuk Konsultasi* (2019)
Pyrexin	Meprofarm	*MIMS Indonesia: Petunjuk Konsultasi* (2019)
Sanmol infusion	Sanbe	*MIMS Indonesia: Petunjuk Konsultasi* (2019)
Tamoliv	Kalbe	*MIMS Indonesia: Petunjuk Konsultasi* (2019)
Paracetamol Uni-Pharma	Uni-Pharma pharmaceutical	*MIMS Indonesia: Petunjuk Konsultasi* (2019)

Several brands of IV paracetamol are distributed in several countries.

that had mild adverse events. Both routes had similar safety, no hepatic adverse event (AE) nor AE due to endotoxin administration reported.

Pathan et al. (2016) conducted a study comparing the effectiveness of morphine, diclofenac (NSAID), and paracetamol in reducing renal colic pain. This study reported that diclofenac is more effective reducing >50% numerical pain rating scale (NRS) score and safer than morphine, and there was no difference between paracetamol and morphine. Diclofenac and paracetamol are statistically significant in reducing >50% NRS score in patients with ureteric calculi. Rescue analgesia in the diclofenac and paracetamol groups was lower than in the morphine group. Yang, Qiu, and Wei (2016) commented on this result that the use of NSAIDs could cause an increased risk in coronary events, so it requires careful consideration before applying it to patients. The use of morphine in relieving renal colic pain has a high side effect and often requires further analgesia. Whereas, the use of paracetamol in reducing renal colic pain can be an option because of its lower side effects and similar efficacy to diclofenac.

Intravenous paracetamol can also be administered as preemptive analgesia during electroconvulsive therapy (ECT). This therapy needs analgesia so that the side effect can be minimized and the patients can continue the therapies. A study by Karaaslan, Akbas, Ozkan, and Zayman (2019) investigated IV paracetamol vs IV ibuprofen effectivity as preemptive analgesia for ECT patients. IV paracetamol can reduce the duration of seizures in patients compared to ibuprofen and placebo. There were patients in the placebo, ibuprofen, and paracetamol groups who experienced nausea and vomiting, but pruritus occurred in the placebo and ibuprofen group only and none in the paracetamol group. However, in the study by Karaaslan et al. (2019) there were no significant differences in the number of patients experiencing headaches and myalgia between groups. Another study involving 126 patients undergoing ECT reported a significant difference in patients experiencing headaches between the placebo group and paracetamol (Isuru et al., 2017). The group of patients who received 1000 mg of paracetamol treatment 2 h before anesthesia experiencing headaches was 36%, while the placebo group was 71%. The use of IV paracetamol as analgesia in ECT patients can be a choice to reduce the side effect of the therapy.

Meyering et al. (2017) reported the use of intravenous paracetamol as multimodal analgesia with prochlorperazine and diphenhydramine to treat headaches in the emergency department. The results showed that the addition of IV paracetamol was better than prochlorperazine and diphenhydramine alone in reducing the visual analog scale (VAS) score and rescue analgesia. Reduction of VAS at 49.2 min post-IV paracetamol administration reach clinically significant result.

Non-superiority of IV paracetamol

Several studies explain the use of IV paracetamol in partial/total knee arthroplasties does not provide significant results in reducing pain. This was obtained from measurements of VAS, opioid rescue, and pain scores. Studies by Hickman et al. (2018) prove intravenous use of paracetamol is not superior to oral paracetamol in 486 patients with total hip arthroplasty and total knee arthroplasty. Patients were given 1000 mg of oral paracetamol before surgery did not have a significant difference compared with intraoperative intravenous paracetamol in morphine milligram equivalent (MME), length of stay in the PACU and hospital, as well as in reducing post-operative nausea and vomiting. Similar results appear from the use of IV paracetamol for the management of postoperative pain in shoulder arthroplasties (Patterson et al., 2018).

The study of IV paracetamol by Greenberg et al. (2018) in 131 craniotomy patients showed that 24 h postoperative opioid-free patients were twice as likely as placebo treatments. A study by Eken, Serinken, Elicabuk, Uyanik, and Erdal (2014) reported the use of intravenous paracetamol similar to morphine in alleviating acute mechanical low back pain based on VAS scores and verbal rating scores (VRS) at 30 min, better than dexketoprofen group. The adverse events in the paracetamol group were lower than those in the morphine group and the reduction in pain score at 30 min was similar to morphine. Another study reported a decrease in pain intensity for patients undergoing lumbar disk surgery with dexketoprofen as a postoperative pain adjunct. Tunali et al. (2013) shows that dexketoprofen is more effective than paracetamol, but it was explained that it was caused by the type of surgery with more severe pain in the paracetamol group.

Barnaby et al. (2019) conducted a study comparing the effectiveness of intravenous hydromorphone and paracetamol use in patients with acute pain in the Emergency Department (ED). Hydromorphone showed the ability to reduce the NRS score in the first 5 min better than paracetamol. The number of patients needed rescue analgesia before 1 h was similar between hydromorphone and paracetamol groups, but this number was lower in the hydromorphone group at 1 h after drug administration. The number of patients experiencing adverse events in the form of nausea and vomiting in the hydromorphone group was higher than that of paracetamol, although it was not statistically significant. Another study reported that the use of paracetamol as an adjunct to hydromorphone decreased the NRS score on first-hour treatment but was not clinically significant (Chang et al., 2019). The study was conducted in patients over 65 years with acute severe pain in the ED. Elderly patients often experience under treatment of pain and changes in their ability to absorb drugs due to aging, so that paracetamol was used as an alternative drug candidate for elderly patients.

The role of paracetamol to control breakthrough pain.

Breakthrough pain is defined as discomfort experienced by a patient that arises between the time the drug is given for pain therapy with normal doses, usually characterized as acute and severe (Medical Dictionary for the Health Professions and Nursing, 2018). There are several patients that aren't become pain-free after given postoperative analgesic therapy. Awan and Durrani (2015), conducted a study of patients who had received postoperative pain therapy. The study found that 40% of patients experienced mild pain and 39% of patients had moderate pain within 24 h after surgery, while 56% of patients experienced mild pain and 33% of patients had moderate pain within 48 h after surgery. Mild to moderate pain that appears after surgery in patients who have received pain therapy is thought to be a breakthrough pain that appears between the time of administration of analgesic doses of therapy. Paracetamol given between the time of administration of ketorolac tromethamine analgesic can reduce the number of patients who experience moderate pain to severe pain after surgery. The use of paracetamol between doses of ketorolac tromethamine therapy effectively reduces pain intensity but fails to produce complete analgesia (Bangash & Durrani, 2018).

Paracetamol in obstetrics

Paracetamol has been used as first-line therapy for pain and fever in pregnant women. Paracetamol clearance increased in pregnant women, so that pregnant women will more quickly experience a decrease in the analgesic effect of paracetamol.

Intravenous paracetamol given preoperatively for cesarean section has a morphine-sparing effect. This effect also occurs in oral administration. If the administration of paracetamol is combined with NSAID, it will produce a synergistic effect so that this therapy combination becomes more potent and more effective (Allegaert & van den Anker, 2017). However, paracetamol administration in pregnant women should be taken carefully, because it shows the occurrence of prenatal ductus arteriosus closure so that the safety of using paracetamol during pregnancy is questioned. So that the use of paracetamol in pregnant women after 6 months of gestation must be limited (Becquet, Bonnet, Ville, Allegaert, & Lapillonne, 2018).

Paracetamol can be found in low levels in breast milk, with milk:plasma ratio of 0.76. If paracetamol levels $<0.1\%$ of the dose given are found in breast milk, then breastfeeding for the baby does not need to be stopped. Since it is known that paracetamol has a morphine-sparing effect, paracetamol administration has clear benefits as part of a multimodal analgesia therapy protocol (Allegaert & van den Anker, 2017). The administration of intravenous paracetamol to patients undergoing elective cesarean section with regional anesthesia techniques has been proven to reduce the need for opioids and increase patient comfort. So it is recommended the use of paracetamol for patients who performed elective cesarean section with regional anesthesia techniques (Luttrel, Gayden, & Pellegrini, 2016).

Pharmacokinetics of intravenous paracetamol in children

The use of IV paracetamol in preterm infants treated in intensive care units is widely used across the country. Infants had lower paracetamol clearance than adults. The main metabolic pathway of paracetamol in infants are by sulfation, whereas glucuronidation is low, and after that, paracetamol will be metabolized into NAPQI which is hepatotoxic. A study by Flint et al. (2017) conducted in 24–32 gestational weeks gestational ages neonates underwent a central venous catheter placement procedure in the first 7 days, showed the concentration of paracetamol-glucuronide metabolites increased in proportion to gestational age. The concentration of paracetamol-sulfate predominates metabolites in extremely preterm, preterm, and full-term neonates (Cook et al., 2016; Flint et al., 2017). Glucuronidation increase with the increased age of the neonates. Administration of IV paracetamol did not show saturation of sulfation with increasing the dose to 20 mg/kg body weight (Flint et al., 2017). Weight and postnatal ages have a significant effect on the formation clearance of glucuronidation and oxidative pathway metabolites (Cook et al., 2016). These results indicate the presence of glucuronidation at 24–32 gestational weeks (Flint et al., 2017). The increase of glucuronidation in infants can alleviate the formation of NAPQI.

Hammer et al. (2020) conducted a study of neonates and infants treated with low-dose and high-dose IV paracetamol to treat acute postoperative pain. A total of 215 subjects were randomized to the high dose, low dose, and 2 control groups consisting of neonates (<28 days), consisting of extreme preterm (≥28 to <32 weeks gestational age), preterm (≥32 to <37 weeks gestational age), full-term (≥37 to ≤40 weeks gestational age); younger infants (28 weeks to <6 months); intermediate-age infants (>6 to <12 months); and older infants (>12 to <24 months) each. Neonates and younger infants have a higher median half-life than intermediates and older infants, whereas intermediates and older infants have similar median half-life to adults. Clearance in infants is higher than adults and the C_{max} in neonates is lower than in infants. The efficacy of IV paracetamol did not significantly differ between groups. The number of patients in the study group who needed rescue medication was lower than placebo. Pharmacokinetics of neonates and infants receiving intravenous paracetamol 12.5 and 15 mg/kg body weight, respectively, are similar to 1000 mg/kg doses in adults every 6 h.

The use of paracetamol in children

Intravenous paracetamol was licensed for use in children in Europe since 2005. A total of 105 respondents from the Association of Pediatric Anesthetists of Great Britain and Ireland Linkmen and the British Pediatric Pain Traveling Club, followed a survey conducted by Wilson-Smith and Morton (2009). The study reported 45% of respondents used IV paracetamol for infants less than 1 year, 24% of them were preterm neonates <32 weeks postconceptional age (PCA). In the >1–12 months gestational age group, more than 50% use a total daily dose of 60 mg/kg/day whereas in the group of >36–44 weeks PCA uses a dose of 30 mg/kg/day. More than 70% of the respondents using the dose above UK licensed dose (30 mg/kg) and 50% of them are in groups of >36–44 weeks PCA. When compared to the pharmacokinetic study of paracetamol using propacetamol by (Allegaert et al., 2004), the majority of respondent in the previous study have used the appropriate total daily dose, i.e., 60 mg/kg/day for >36 weeks PCA, 50 mg/kg/day for 32–36 weeks PCA, and 40 mg/kg/day for <32 weeks PCA.

Intravenous propacetamol has an effectiveness similar to dexibuprofen for treating upper respiratory tract infection (URTI) patients aged 6–14 years accompanied by fever (Choi et al., 2018). Propacetamol is a prodrug of paracetamol, which through the process of hydrolyzation changed into paracetamol. Based on the study of Choi et al. (2018), the

146 PART | II Mechanisms of action of drugs

propacetamol IV group experienced fewer adverse events than the control group. The body temperature of patients receiving IV propacetamol treatment reached <38 °C at 0.5 h while the control group takes 1 h. Based on this study, the use of IV propacetamol may be an option if antipyretic administration is needed immediately or cannot be done by oral route.

Intravenous paracetamol can be used for treatment in children that requires effective and rapid analgesics and antipyretics. Paracetamol works through several binding mechanisms with receptors in the central nervous system in reducing pain (Sharma & Mehta, 2014). Therefore, the concentration of paracetamol in cerebrospinal fluid is a good criterion for the effectiveness of paracetamol. Kumpulainen et al. (2007) conducted a study that measured peak concentrations of paracetamol in CSF in 32 children underwent lower part of the body surgery with spinal anesthesia. At 5 min after administration, paracetamol was already detected in CSF though the concentrations below 1 mg/L. The peak concentration of paracetamol in CSF was detected at 57 min with a concentration of 18 mg/L. Gender showed a correlation with CSF concentration, girls showed higher concentrations of paracetamol in CSF than boys. The highest plasma concentration of 33 mg/L was observed at 19 min. These results appear to be higher than in adults. Based on Singla et al. (2012), maximum paracetamol concentration in CSF in adult was reached after 2 h with a concentration of 5.94 μg/mL, while the maximum concentration of paracetamol in plasma at 15 min was 21.6 μg/mL.

The use of IV paracetamol for placement of peripherally inserted central catheter in preterm infants <32 weeks gestational ages showed no superiority to sucrose (Roofthooft et al., 2017). This is based on the Premature Infant Pain Profile (PIPP) and the COMFORTneo score, which is not statistically significant. However, in this study, infants treated with sucrose were not randomized and assessments were not blinded to the groups.

Side effect of intravenous paracetamol

Chiam et al. (2015) explained a review of the side effects of IV paracetamol. Some AEs that appear in some clinical trials are hemodynamic change, whereas some AE including liver, renal functions, asthma, thrombocytopenia, and anaphylaxis are more likely occur in the chronic or overdose use. Nausea and vomiting are occurring in the use of paracetamol in general rather than intravenous administration (Hickman et al., 2018). Transient hypotension was seen in several clinical trial reports. Hypotensive events can increase length of stay and also morbidity. Some studies indicate this effect through outcomes in the form of decreased systolic blood pressure (SBP), diastolic blood pressure (DBP), and mean arterial pressure (MAP). They suggest that these changes are due to the content of mannitol in intravenous paracetamol preparations. Some brands of IV paracetamol contain close to 4 g of mannitol every 1 g/100 mL infusion. To obtain a total daily dose of 4 g/day paracetamol, requires 400 mL infusion which is equivalent to mannitol of 15.4–15.64 g. The amount exceeds mannitol prescribed for diuresis and oliguria (Chiam et al., 2015). However, based on further studies, the hypotension effect on IV paracetamol proclaimed comes from paracetamol itself and not from mannitol (Chiam, Weinberg, Bailey, McNicol, & Bellomo, 2016). This was observed in decreased SBP, DBP, and MAP in the first 20 min of infusion in the IV paracetamol group but not in the saline and mannitol groups.

In another study, Chiam, Bellomo, Churilov, and Weinberg (2018) observed the effect of infusion of 1000 mg IV paracetamol in cardiac surgery patients. The results showed a significant decrease in SBP, DBP, and MAP compared to patients who were administered with saline at 30 min after infusions. The effect of transient hypotension on IV paracetamol was also observed in pediatric cardiac patients in the PICU. The patients median age were 8.8 months with a total of 608 patients included. A total of 5% of patients experience a decrease in MAP below 15%, and 20% have a decrease in MAP below 10%, which seems to be significantly observed starting at 30 min. This decrease correlates with age, baseline MAP, and skin temperature (Achuff et al., 2019). Hypotensive effect of IV paracetamol was also seen in critically ill children with septic shock on inotropic support that observed from the decreased MAP and SBP (Nahum et al., 2020).

The use of intravenous paracetamol also needs to consider its effectiveness compared to the oral route. IV paracetamol in Australia cost $ 3.30/1000 mg, whereas paracetamol tablet cost$ 1.24/500 mg, and for suppository cost $ 0.026/500 mg. IV paracetamol reaches its maximum concentration in plasma faster, eventually, it gives the same result of pain reduction by oral route (Chiam et al., 2015). IV paracetamol would be more appropriate for critically ill, postsurgical patients, or patients with difficulties for orally administered drugs.

Applications to other areas

Paracetamol is widely known for its ability as an analgesic and antipyretic. But, several studies show that Paracetamol has another benefit. A study by Garrido, Arancibia, Campos, and Valenzuela (1991) showed that unchanged paracetamol molecules might act as antioxidants by trapping free radicals and inhibiting peroxidation in hepatocytes. The same result was previously described by DuBois, Hill, and Burk (1983) that paracetamol inhibits peroxidation in the vitamin E deficient hepatocyte. Those studies were done in rats isolated hepatocytes. Sogut, Solduk, Gokdemir, and Kaya (2015) conducted a clinical study showing that Intravenous paracetamol can prevent DNA damage due to increased reactive oxygen species (ROS) in lymphocytes subject to trauma.

Zhao et al. (2017) reported that paracetamol could reduce cognitive dysfunction induced by using lipopolysaccharides (LPS). It was found that paracetamol prevented short-term memory deficits, attenuated microglial activation in hippocampal DG regions, prevented superoxide dismutase (SOD) decrease, attenuated malondialdehyde (MDA) increase in the hippocampi that was induced by LPS administration intracerebroventricularly (i.c.v.). It also found that paracetamol administration i.c.v. can down regulate GSK3β by phosphorylating GSK3β, alleviate BDNF expression in the hippocampus, and reduce neuron apoptosis, so that paracetamol might ameliorate learning and memory abilities.

Chronic ingestion of paracetamol of 20 mg/kg body weight given to streptozotocin-induced diabetes mice can normalize blood sugar levels (Blough & Wu, 2011). It also can prevent hyperglycemia in animal with a high-fat diet. This is thought to be due to the ability of paracetamol in improving insulin production/secretion. Paracetamol with a dose of 30 mg/kg also shows the ability to maintain elevated blood sugar levels due to the aging process. It also reported that paracetamol can improve the function and structure of skeletal muscles, possess cardioprotective, and neuroprotective which are thought to be due to their ability as antioxidants (Blough & Wu, 2011).

Other agents of interest

Another drug that is often available over the counter (OTC) and is prescribed as the first-line analgesic and antipyretic is ibuprofen. Ibuprofen can also be used together with paracetamol to produce a synergistic effect to accelerate the analgesic effect. Ibuprofen is classified as an NSAID and works by binding to COX-1 and COX-2, thus inhibiting the formation of inflammatory prostanoids. Ibuprofen can reduce fever in children with slightly better abilities compared with paracetamol but not statistically significant (Narayan, Cooper, Morphet, & Innes, 2017). The use of Ibuprofen in reducing fever has an equivalent ability to paracetamol for clinical use. Based on studies by Jayawardena and Kellstein (2017), ibuprofen (7.5 mg/kg) has an antipyretic effect for febrile children, which is more favorable than paracetamol (10–15 mg/kg) because it has a faster onset, longer and greater temperature reduction. However, both drugs have a similar safety profile.

Intravenous ibuprofen can be used as an alternative to multimodal analgesia to IV paracetamol for postoperative bariatric surgery (Kayhan et al., 2018). It didn't reduce opioid consumption but can reduce the severity of pain. Southworth, Peters, Rock, and Pavliv (2009) reported the use of ibuprofen as postoperative analgesia in abdominal and orthopedic surgery can reduce pain with rest and movement and reduce the incidence of pyrexia. Morphine consumption is also reduced in the use of ibuprofen at doses of 400 mg and 800 mg q6h.

Mini-dictionary of terms

Patent Ductus arteriosus. Openings in blood vessels that lead from the heart that do not close due to developmental defect and often occur in preterm infants.

Visual analog scale. Psychometric instrument developed for pain measurement.

C_{max}. Maximum drug concentration in blood plasma after administration until second dose administration.

CSF_{max}. Maximum drug concentration in the cerebrospinal fluid after drug administration.

Propacetamol. Prodrug of paracetamol by esterification and carboxylic acid diethylglycine which has analgesic and antipyretic properties.

Key facts of intravenous paracetamol

- Paracetamol was first made as a drug in 1877 and was used clinically for the first time in 1893.
- Paracetamol infusion was first introduced and used for patients in the hospital in 1985 and is indicated when the patient cannot be administered enterally.
- The recommended maximum daily dose used in adult patients is 3–4 g and when used in high doses it can cause toxicity including liver failure.
- Intravenous paracetamol can provide higher speed for achieving plasma drug concentrations (15 min after intravenous administration) and significantly causes paracetamol levels in plasma to be higher than the oral and rectal routes.
- Paracetamol is the activator of several acute respiratory distress syndrome (ARDS) related protein in COVID-19 patients, so the use of that drugs in large doses to patients already at risk of respiratory failure should be taken cautiously.

148 PART | II Mechanisms of action of drugs

- Paracetamol is recommended as the first choice for treating pain in COVID-19 patients compared to NSAIDs because of its minimal side effects and thought not to mask the symptoms of the disease.

Summary points

- This chapter explains the clinical application of intravenous paracetamol as an analgesic and antipyretic with minimal side effect
- Intravenous paracetamol often used when an immediate analgesic effect needed
- Intravenous paracetamol as multimodal analgesia has opioid-sparing effect
- Administration of IV paracetamol give higher C_{max} and CSF_{max} than those oral and rectal routes, since its mechanism of action is in the central nerve system, IV paracetamol become favorable than the other routes
- It was found that this drug can cause transient hypotension
- It is suggested that unchanged paracetamol molecules give antioxidant effect

References

Achuff, B. J., Moffett, B. S., Acosta, S., Lasa, J. J., Checchia, P. A., & Rusin, C. G. (2019). Hypotensive response to IV acetaminophen in pediatric cardiac patients. *Pediatric Critical Care Medicine*, *20*(6), 527–533.

Allegaert, K., & van den Anker, J. N. (2017). Perinatal and neonatal use of paracetamol for pain relief. *Seminars in Fetal & Neonatal Medicine*, *22*(5), 308–313.

Allegaert, K., Anderson, B. J., Naulaers, G., de Hoon, J., Verbesselt, R., Debeer, A., … Tibboel, D. (2004). Intravenous paracetamol (propacetamol) pharmacokinetics in term and preterm neonates. *European Journal of Clinical Pharmacology*, *60*(3). https://doi.org/10.1007/s00228-004-0756-x. 191–7.

Awan, H., & Durrani, Z. (2015). Postoperative pain management in the surgical wards of a tertiary care hospital in Peshawar. *JPMA. The Journal of the Pakistan Medical Association*, *65*(4), 358–361.

Bangash, A. A., & Durrani, Z. (2018). Effectiveness of acetaminophen in control of breakthrough pain: Randomized controlled trial. *JPMA. The Journal of the Pakistan Medical Association*, *68*(7), 994–1001.

Barnaby, D. P., Chertoff, A. E., Restivo, A. J., Campbell, C. M., Pearlman, S., White, D., et al. (2019). Randomized controlled trial of intravenous acetaminophen versus intravenous hydromorphone for the treatment of acute pain in the emergency department. *Annals of Emergency Medicine*, *73*(2), 133–140.

Becquet, O., Bonnet, D., Ville, Y., Allegaert, K., & Lapillonne, A. (2018). Paracetamol/acetaminophen during pregnancy induces prenatal ductus arteriosus closure. *Pediatrics*, *142*(1), e20174021.

Bektas, F., Eken, C., Karadeniz, O., Goksu, E., Cubuk, M., & Cete, Y. (2009). Intravenous paracetamol or morphine for the treatment of renal colic: A randomized, placebo-controlled trial. *Annals of Emergency Medicine*, *54*(4). https://doi.org/10.1016/j.annemergmed.2009.06.501. 568–74.

Blough, E. R., & Wu, M. (2011). Acetaminophen: Beyond pain and fever-relieving. *Frontiers in Pharmacology*, *2*, 72.

Brett, C. N., Barnett, S. G., & Pearson, J. (2012). Postoperative plasma paracetamol levels following oral or intravenous paracetamol administration: A double-blind randomised controlled trial. *Anaesthesia and Intensive Care*, *40*(1), 166–171.

Caparrotta, T. M., Antoine, D. J., & Dear, J. W. (2018). Are some people at increased risk of paracetamol-induced liver injury? A critical review of the literature. *European Journal of Clinical Pharmacology*, *74*(2), 147–160.

Chang, A. K., Bijur, P. E., Ata, A., Campbell, C., Pearlman, S., White, D., et al. (2019). Randomized clinical trial of intravenous acetaminophen as an analgesic adjunct for older adults with acute severe pain. *Academic Emergency Medicine: Official Journal of the Society for Academic Emergency Medicine*, *26*(4), 402–409.

Chiam, E., Bellomo, R., Churilov, L., & Weinberg, L. (2018). The hemodynamic effects of intravenous paracetamol (acetaminophen) vs normal saline in cardiac surgery patients: A single center placebo controlled randomized study. *PLoS One*, *13*(4), e0195931.

Chiam, E., Weinberg, L., Bailey, M., McNicol, L., & Bellomo, R. (2016). The haemodynamic effects of intravenous paracetamol (acetaminophen) in healthy volunteers: A double-blind, randomized, triple crossover trial. *British Journal of Clinical Pharmacology*, *81*(4), 605–612.

Chiam, E., Weinberg, L., & Bellomo, R. (2015). Paracetamol: A review with specific focus on the haemodynamic effects of intravenous administration. *Heart, Lung and Vessels*, *7*(2), 121–132.

Choi, S. J., Moon, S., Choi, U. Y., Chun, Y. H., Lee, J. H., Rhim, J. W., et al. (2018). The antipyretic efficacy and safety of propacetamol compared with dexibuprofen in febrile children: A multicenter, randomized, double-blind, comparative, phase 3 clinical trial. *BMC Pediatrics*, *18*(1), 201.

Cook, S. F., Roberts, J. K., Samiee-Zafarghandy, S., Stockmann, C., King, A. D., Deutsch, N., et al. (2016). Population pharmacokinetics of intravenous paracetamol (acetaminophen) in preterm and term neonates: Model development and external evaluation. *Clinical Pharmacokinetics*, *55*(1), 107–119.

Craig, P. N. (1990). Drug compendium. In C. Hansch, P. G. Sammes, & J. B. Taylor (Eds.), *Comprehensive medicinal chemistry* (p. 245). Oxford: Pergamon.

Dasgupta, A., & Krasowski, M. D. (2020). *Analgesics. Therapeutic drug monitoring data*. Academic Press. https://doi.org/10.1016/B978-0-12-815849-4.00014-1.

DuBois, R. N., Hill, K. E., & Burk, R. F. (1983). Antioxidant effect of acetaminophen in rat liver. *Biochemical Pharmacology*, *32*(17), 2621–2622.

Eken, C., Serinken, M., Elicabuk, H., Uyanik, E., & Erdal, M. (2014). Intravenous paracetamol versus dexketoprofen versus morphine in acute mechanical low back pain in the emergency department: A randomised double-blind controlled trial. *Emergency Medicine Journal : EMJ, 31*(3), 177–181.

Feldberg, W., Gupta, K. P., Milton, A. S., & Wendlandt, S. (1973). Effect of pyrogen and antipyretics on prostaglandin acitvity in cisternal c.s.f. of unanaesthetized cats. *The Journal of Physiology, 234*(2), 279–303.

Flint, R. B., Roofthooft, D. W., van Rongen, A., van Lingen, R. A., van den Anker, J. N., van Dijk, M., et al. (2017). Exposure to acetaminophen and all its metabolites upon 10, 15, and 20mg/kg intravenous acetaminophen in very-preterm infants. *Pediatric Research, 82*(4), 678–684.

Flower, R. J., & Vane, J. R. (1972). Inhibition of prostaglandin synthetase in brain explains the anti-pyretic activity of paracetamol (4-acetamidophenol). *Nature, 240*(5381), 410–411.

Forrest, J. A., Clements, J. A., & Prescott, L. F. (1982). Clinical pharmacokinetics of paracetamol. *Clinical Pharmacokinetics, 7*(2), 93–107.

Garrido, A., Arancibia, C., Campos, R., & Valenzuela, A. (1991). Acetaminophen does not induce oxidative stress in isolated rat hepatocytes: Its probable antioxidant effect is potentiated by the flavonoid silybin. *Pharmacology & Toxicology, 69*(1), 9–12.

Gazzard, B. G., Ford-Hutchinson, A. W., Smith, M. J., & Williams, R. (1973). The binding of paracetamol to plasma proteins of man and pig. *The Journal of Pharmacy and Pharmacology, 25*(12), 964–967.

Ghanem, C. I., Pérez, M. J., Manautou, J. E., & Mottino, A. D. (2016). Acetaminophen from liver to brain: New insights into drug pharmacological action and toxicity. *Pharmacological Research, 109*, 119–131.

Graham, G. G., Davies, M. J., Day, R. O., Mohamudally, A., & Scott, K. F. (2013). The modern pharmacology of paracetamol: Therapeutic actions, mechanism of action, metabolism, toxicity and recent pharmacological findings. *Inflammopharmacology, 21*(3), 201–232.

Greenberg, S., Murphy, G. S., Avram, M. J., Shear, T., Benson, J., Parikh, K. N., et al. (2018). Postoperative intravenous acetaminophen for craniotomy patients: A randomized controlled trial. *World Neurosurgery, 109*, e554–e562.

Hammer, G. B., Maxwell, L. G., Taicher, B. M., Visoiu, M., Cooper, D. S., Szmuk, P., et al. (2020). Randomized population pharmacokinetic analysis and safety of intravenous acetaminophen for acute postoperative pain in neonates and infants. *Journal of Clinical Pharmacology, 60*(1), 16–27.

Hickman, S. R., Mathieson, K. M., Bradford, L. M., Garman, C. D., Gregg, R. W., & Lukens, D. W. (2018). Randomized trial of oral versus intravenous acetaminophen for postoperative pain control. *American Journal of Health-System Pharmacy, 75*(6), 367–375.

Högestätt, E. D., Jönsson, B. A., Ermund, A., Andersson, D. A., Björk, H., Alexander, J. P., et al. (2005). Conversion of acetaminophen to the bioactive N-acylphenolamine AM404 via fatty acid amide hydrolase-dependent arachidonic acid conjugation in the nervous system. *The Journal of Biological Chemistry, 280*(36), 31405–31412.

Isuru, A., Rodrigo, A., Wijesinghe, C., Ediriweera, D., Premadasa, S., Wijesekara, C., & Kuruppuarachchi, L. (2017). A randomized, double-blind, placebo-controlled trial on the role of preemptive analgesia with acetaminophen [paracetamol] in reducing headache following electroconvulsive therapy [ECT]. *BMC Psychiatry, 17*(1), 275. https://doi.org/10.1186/s12888-017-1444-6.

Jayawardena, S., & Kellstein, D. (2017). Antipyretic efficacy and safety of ibuprofen versus acetaminophen suspension in febrile children: Results of 2 randomized, double-blind, single-dose studies. *Clinical Pediatrics, 56*(12), 1120–1127.

Jóźwiak-Bebenista, M., & Nowak, J. Z. (2014). Paracetamol: Mechanism of action, applications and safety concern. *Acta Poloniae Pharmaceutica, 71*(1), 11–23.

Karaaslan, E., Akbas, S., Ozkan, A. S., & Zayman, E. P. (2019). Effects of preemptive intravenous paracetamol and ibuprofen on headache and myalgia in patients after electroconvulsive therapy: A placebo-controlled, double-blind, randomized clinical trial. *Medicine, 98*(51), e18473.

Kayhan, E. G., Sanli, M., Ozgul, U., Kirteke, R., & Yologlu, S. (2018). Comparison of intravenous ibuprofen and acetaminophen for postoperative multimodal pain management in bariatric surgery: A randomized controlled trial. *Journal of Clinical Anesthesia, 50*, 5–11.

Kumpulainen, E., Kokki, H., Halonen, T., Heikkinen, M., Savolainen, J., & Laisalmi, M. (2007). Paracetamol (acetaminophen) penetrates readily into the cerebrospinal fluid of children after intravenous administration. *Pediatrics, 119*(4), 766–771.

Laksono, R. M., Isngadi, I., & Murti, A. H. (2019). Intravenous paracetamol as a preemptive analgesia to reduce postoperative pain after major oncologic surgery. *Anaesthesia, Pain & Intensive Care, 23*(1), 28–32.

Luttrel, H. M., Gayden, J., & Pellegrini, J. (2016). Effectiveness of intravenous acetaminophen administration in the postoperative pain management of the cesarean section patient. *Anasthesia eJournal, 4*(1), 43–45.

Medical Dictionary for the Health Professions and Nursing. (2018, February 15). Retrieved from http://medicaldictionary.thefreedictionary.com/breakthrough+pain.

Meyering, S. H., Stringer, R. W., & Hysell, M. K. (2017). Randomized trial of adding parenteral acetaminophen to prochlorperazine and diphenhydramine to treat headache in the emergency department. *The Western Journal of Emergency Medicine, 18*(3), 373–381.

MIMS Indonesia: Petunjuk Konsultasi. (2019). Jakarta: MIMS Pharmacy Guide.

Nahum, E., Weissbach, A., Kaplan, E., & Kadmon, G. (2020). Hemodynamic effects of intravenous paracetamol in critically ill children with septic shock on inotropic support. *Journal of Intensive Care, 8*, 14.

Narayan, K., Cooper, S., Morphet, J., & Innes, K. (2017). Effectiveness of paracetamol versus ibuprofen administration in febrile children: A systematic literature review. *Journal of Paediatrics and Child Health, 53*(8), 800–807.

Pacifici, G. M., & Allegaert, K. (2014). Clinical pharmacology of paracetamol in neonates: A review. *Current Therapeutic Research, Clinical and Experimental, 77*, 24–30.

Parker, S. L., Saxena, M., Gowardman, J., Lipman, J., Myburgh, J., & Roberts, J. A. (2018). Population pharmacokinetics of intravenous paracetamol in critically ill patients with traumatic brain injury. *Journal of Critical Care, 47*, 15–20.

Pathan, S. A., Mitra, B., Straney, L. D., Afzal, M. S., Anjum, S., Shukla, D., et al. (2016). Delivering safe and effective analgesia for management of renal colic in the emergency department: A double-blind, multigroup, randomised controlled trial. *Lancet, 387*(10032), 1999–2007.

Patterson, D. C., Cagle, P. J., Jr., Poeran, J., Zubizarreta, N., Mazumdar, M., Galatz, L. M., et al. (2018). Effectiveness of intravenous acetaminophen for postoperative pain management in shoulder arthroplasties: A population-based study. *Journal of Orthopaedic Translation, 18*, 119–127.

Peacock, W. F., Breitmeyer, J. B., Pan, C., Smith, W. B., & Royal, M. A. (2011). A randomized study of the efficacy and safety of intravenous acetaminophen compared to oral acetaminophen for the treatment of fever. *Academic Emergency Medicine: Official Journal of the Society for Academic Emergency Medicine, 18*(4), 360–366.

Prescott, L. F. (1980). Kinetics and metabolism of paracetamol and phenacetin. *British Journal of Clinical Pharmacology, 10*(Suppl 2), 291S–298S.

Prescott, L. F., Speirs, G. C., Critchley, J. A., Temple, R. M., & Winney, R. J. (1989). Paracetamol disposition and metabolite kinetics in patients with chronic renal failure. *European Journal of Clinical Pharmacology, 36*(3), 291–297.

Raffa, R. B., Pawasauskas, J., Pergolizzi, J. V., Jr., Lu, L., Chen, Y., Wu, S., et al. (2018). Pharmacokinetics of Oral and intravenous paracetamol (acetaminophen) when co-administered with intravenous morphine in healthy adult subjects. *Clinical Drug Investigation, 38*(3), 259–268.

Roofthooft, D. W. E., Simons, S. H. P., van Lingen, R. A., Tibboel, D., van den Anker, J. N., Reiss, I. K. H., et al. (2017). Randomized controlled trial comparing different single doses of intravenous paracetamol for placement of peripherally inserted central catheters in preterm infants. *Neonatology, 112*, 150–158.

Sharma, C. V., & Mehta, V. (2014). Paracetamol: Mechanisms and updates. *Continuing Education in Anaesthesia, Critical Care and Pain, 14*(4), 153–158.

Singla, N. K., Parulan, C., Samson, R., Hutchinson, J., Bushnell, R., Beja, E. G., et al. (2012). Plasma and cerebrospinal fluid pharmacokinetic parameters after single-dose administration of intravenous, oral, or rectal acetaminophen. *Pain Practice, 12*(7), 523–532.

Sogut, O., Solduk, L., Gokdemir, M. T., & Kaya, H. (2015). Impact of single-dose intravenous paracetamol on lymphocyte DNA damage and oxidative stress in trauma patients. *Biomedical Research, 26*(1), 23–30.

Southworth, S., Peters, J., Rock, A., & Pavliv, L. (2009). A multicenter, randomized, double-blind, placebo-controlled trial of intravenous ibuprofen 400 and 800 mg every 6 hours in the management of postoperative pain. *Clinical Therapeutics, 31*(9), 1922–1935.

Tunali, Y., Akçil, E. F., Dilmen, O. K., Tutuncu, A. C., Koksal, G. M., Akbas, S., ... Yentur, E. (2013). Efficacy of intravenous paracetamol and dexketoprofen on postoperative pain and morphine consumption after a lumbar disk surgery. *Journal of Neurosurgical Anesthesiology, 25*(2). https://doi.org/10.1097/ANA.0b013e31827464af. 143–7.

van der Westhuizen, J., Kuo, P. Y., Reed, P. W., & Holder, K. (2011). Randomised controlled trial comparing oral and intravenous paracetamol (acetaminophen) plasma levels when given as preoperative analgesia. *Anaesthesia and Intensive Care, 39*(2), 242–246.

Wilson-Smith, E. M., & Morton, N. S. (2009). Survey of i.v. paracetamol (acetaminophen) use in neonates and infants under 1 year of age by UK anesthetists. *Paediatric Anaesthesia, 19*(4), 329–337.

Wininger, S. J., Miller, H., Minkowitz, H. S., Royal, M. A., Ang, R. Y., Breitmeyer, J. B., & Singla, N. K. (2010). A randomized, double-blind, placebo-controlled, multicenter, repeat-dose study of two intravenous acetaminophen dosing regimens for the treatment of pain after abdominal laparoscopic surgery. *Clinical Therapeutics, 32*(14). https://doi.org/10.1016/j.clinthera.2010.12.011. 2348–69.

Yang, L., Qiu, S., & Wei, Q. (2016). Re: Delivering safe and effective analgesia for Management of Renal Colic in the emergency department: A double-blind, multigroup, randomised controlled trial. *European Urology, 70*(4), 702.

Zhao, W. X., Zhang, J. H., Cao, J. B., Wang, W., Wang, D. X., Zhang, X. Y., et al. (2017). Acetaminophen attenuates lipopolysaccharide-induced cognitive impairment through antioxidant activity. *Journal of Neuroinflammation, 14*(1), 17.

Chapter 16

Prilocaine: Mechanisms and application

Naresh Kumar Katari[a] and Sreekantha B. Jonnalagadda[b]

[a]*Department of Chemistry, School of Science, GITAM deemed to be University, Hyderabad, Telangana, India;* [b]*School of Chemistry & Physics, Westville Campus, University of KwaZulu-Natal, Durban, South Africa*

Abbreviations

ANS	autonomic nervous system
CNS	central nervous system
CV	cardiovascular
IV	intravenous
LAs	local anesthetics
MIC	minimum inhibitory concentration
NLCs	nanostructured lipid carriers
PE	premature ejaculation
PNS	peripheral nervous system
SLNs	solid lipid nanoparticles
TNS	transient neurologic symptoms

Prilocaine is a general local anesthetic drug (LAs) used as pain relief treatment. Claes Tegner and Nils Löfgren first invented the drug. The molecular structure of the prilocaine contains 2-amide functional group, and the IUPAC name of the compound is *N*-(2-methylphenyl)-2-(propylamino)propanamide. The chemical structure of the prilocaine is shown in Fig. 1, having similarities to lignocaine (lidocaine, xylocaine) and xylocaine. It has significant advantages compared to other local anesthetic agents, i.e., long duration of anesthesia, no need of adrenaline, >50% less toxic, but its absorption is slower. The main side effect of prilocaine is methemoglobinemia. It is a type of condition in the form of hemoglobin iron oxidized to ferric (Fe^{3+}) state leading to rapid oxygen desaturation. This effect rarely occurs when a higher dose of the anesthesia drug is administered to the patient. Methemoglobinemia may impact on life forbidding situations, including coma, seizures, shock, respiratory disorders, etc. The long-term damage includes hypoxic encephalopathy and myocardial infarction, and even death. Methylene blue or ascorbic acid is the promising drug to treat the methemoglobinemia as a nominal dose of 1 mg/kg.

Prilocaine is a toluene derivative having amino amide linkage and acts as a LA. It is shown therapeutic action against nerves system and stabilize the neuronal membrane by preferential binding and inhibit the depolarization of sodium channel gate voltage. The results indicate that decreasing the membrane permeability the amino amide derivative of prilocaine, in this *N*-propyl-DL-alanine and 2-methylaniline has to be combined to appear amide bond. Table 1 summarizes the physicochemical and biological properties of prilocaine drug.

The local anesthesia defines as a loss of sensation in a body affected part, which is produced through the misery of excitation in nerve endings and provides the loss of sensation without making the loss of consciousness. The primary effect of the LAs takes place at the time of depolarization phases of the action potential. These effects include decreasing the rate of depolarization.

The journey of local anesthetic

In the 18th century, the mission of the LA drugs started with natural products. Cocaine is the first isolated LA drug from coca leaves. But it took more than 25 years to employ in clinical practices (Ruetsch, Bonibc, & Borgeatac, 2001). In 1884, the first clinical trial of cocaine was performed by administrating on the eye (Davison, 1965). Uses of the first LA drug quickly spread worldwide, but at the toxic effect of this drug caused many deaths. From this, synthetic organic chemistry played a crucial role in developing and synthesis of various anesthetic medications for pain relief (Hadeed & Siegel, 1989). In 1943, Lofgren developed lignocaine (lidocaine) which exhibited new local anesthetic property. In next, the procaine-like

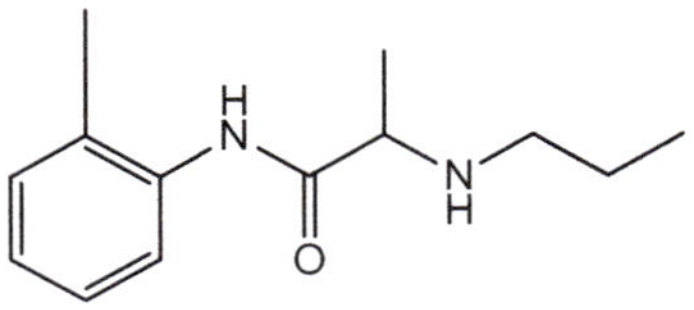

152 PART | II Mechanisms of action of drugs

FIG. 1 Chemical structure of prilocaine.

drugs are the ester derivatives of para-aminobenzoic acid. Lignocaine is an amide derivative of diethylamino acetic acid. After observing the clinical activity of lignocaine, various amide derivatives such as prilocaine, bupivacaine and etidocaine were designated for medical treatment as local anesthetics. Each drug has its specific pharmacological action. Between 1891 and 1972, new amino ester local anesthetics like tropocaine, eucaine, holocaine, orthoform, benzocaine, and tetracaine were synthesized. In addition to these drugs, amino amide local anesthetics like nirvaquine, procaine, chloroprocaine, cinchocaine, lidocaine, mepivacaine, prilocaine, efocaine, bupivacaine, etidocaine, and articaine have also been developed (Ruetsch et al., 2001). All these drugs have shown pharmacological activity against pain in a different mode of action. Still, most of these drugs exert toxicity on the central nervous system (CNS) and cardiovascular (CV) toxicity.

Structure

Local anesthetic agents are having a common standard molecular form like lipophilic aromatic ring linked to a hydrophilic a tertiary amine side chain having either an ester or amide bond (Fig. 2) (Perrin, Bull, & Black, 2019). Based on linking bond in general categorized into two different groups, i.e., amides and esters. Tertiary amines are most injectable LAs and few secondary amines (like prilocaine, hexylcaine). From the structure of LAs, the lipophilic part is the most significant portion in the construction of the molecule. All local anesthetics are amphipathic, and form both lipophilic and hydrophilic state. The amino derivative of ethyl alcohol or acetic acid is a hydrophilic part, LAs without hydrophilic component are not acceptable or suitable for injection. The important thing is the nature of the linkage to state the various properties of LAs, like biotransformation and others. Amide linkages are fewer sustainable to hydrolysis than ester linkages, and this impacts on the time interval and hydrophilic group affect the onset action.

TABLE 1 Physicochemical and biological properties of the prilocaine.

Melting point	37.5–38
Boiling point	159
Flash Point	134.3
Density	134.3 g/mL
Molecular weight	220.16
pK_a	7.9
Percent base (RN) at pH 7.4	25
Approximate onset of action (min)	2–4
Partition Coefficient	0.9
Protein binding	55%
Metabolism	Hepatic and renal
Half-life	10–150 min
Elimination	Renal exclusion
Onset	Fast
Relative potency	2
Duration	Moderate

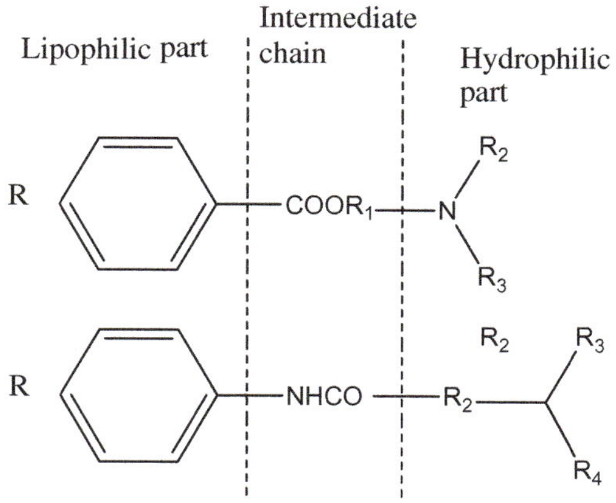

FIG. 2 General structure of the ester and amide local anesthetics (Perrin et al., 2019).

LAs are weak bases because the due existence of the amino group and occur in solution partially as the free base and cation, termed the conjugate base. Due to weak base property, they are poorly soluble in water and reacts with the acid to form salts and unstable in air. These local anesthetics exist in ionized and unionized equilibrium forms. The Henderson-Hasselbach equation describes the ratios of these in the weak base (Peck, Hill, & Williams, 2008).

$$pK_a = pH - \log[\text{base}]/[\text{conjugate acid}]$$

pK_a is the pH where 50% of drugs are in ionized, and 50% is in unionized form. LAs are less emulsifiable weak bases at $pK_a > 7.4$. Generally, they exhibit highly in the ionized form at pH (pH -7.4). The base form in the drugs have a high response to the onset of action, and deionized form of the compound disperses most willingly across the nerve sheath.

Low pK_a values of LAs like mepivacaine, etidocaine, prilocaine, and lidocaine reflect their quick onset of action. Nearly 35% of these agents are in base form at tissue pH -7.4. In distinction, other drugs such as procaine, bupivacaine, tetracaine, where pK_a is 8.1–8.6, exist mainly in the cationic form (85%–95%) at pH 7.4 and with that, the onset of action is slow when compared to others (Covino & Donald, 1981).

Classification of anesthetic drugs

In the past few decades, LAs are widely used to deliver analgesia by reducing the potential of the PNS and CNS to generate pain signals. Anesthetic drugs are an essential part of all typical surgical process and play a key role in the patient's body control from pains. In general, the LAs drugs are classified into three types, i.e., esters, amides, quinoline compounds.

1. Ester compounds: butacaine, cocaine, ethyl aminobenzoate (benzocaine), hexylcaine, piperocaine, tetracaine.
2. Esters of para-aminobenzoic acids: chloroprocaine, procaine, propoxycaine.
3. Amides: articaine, bupivacaine, dibucaine, etidocaine, lidocaine, mepivacaine, prilocaine, ropivacaine.
4. Quinolines: centbucridine.

These drugs are commonly utilized in anesthesiology, ophthalmology, and otolaryngology and also the medication for oncological and chronic pain. The primary function of these anesthetics is to minimize the pain during and after medical treatment or processes like surgery. Mechanism of LAs is associated with a disturbance of electrochemical progressions in the nerve or nerve fibers. They also act on the voltage-gated sodium channels, and LAs interactions through calcium, potassium, and hyperpolarization-gated ion channels, ligand-gated channels, and G protein-coupled receptors. Amino esters can break down through hydrolysis by pseudocholinesterase contained in the plasma and aminoamides to undergo hepatic biotransformation via aromatic hydroxylation, N-dealkylation, and amide hydrolysis (Heavner, 2007).

Prilocaine

Prilocaine was first introduced as a LA agent in 1960 for the infiltration anesthesia, peripheral nerve block and peridural. Prilocaine is a secondary amide analogue of lidocaine and having longer duration rapid onset action. Because of rapid tissue uptake action, biotransformation, and quick fall of prilocaine plasma level make prilocaine as a good anesthetic, and it is generally used as intravenous regional anesthesia. These factors increase the safety uptake of the drug and make prilocaine usage as a popular regional anesthesia procedure (Petersen, Luck, Kristensen, & Mikkelsen, 1977). But the toxicity levels shown similar symptoms and signs related to a LAs overdose. Prilocaine drug having same anesthetic effect compares to lidocaine. Both the molecular forms (charged and uncharged) of the anesthetic drug are essential in the neural blockade; Prilocaine has a better onset of action than other LAs of pK_a less than 7.9 (Prilocaine pK_a is 7.9). The advantageous property is its lower systemic toxicity, due to slow absorption rate from the injection site, more excellent tissue distribution, and more rapid metabolism than lidocaine (Wildsmith, 1985). The main reason for the toxicity levels in LAs is an intravenous (IV) injection and extravascular administration of an unnecessary higher dose. Toxicity in CNS follows the quick IV administration is correlated with the inherent anesthetic power of the LAs (Concepcion & Covino, 1984). The toxicity potentials of LAs will be affected by various aspects, i.e., absorption rate, tissue distribution and metabolism of the drugs. Both prilocaine and lidocaine have a per cent less toxicity effect because of the slow absorption rate. LAs similar property in inherent anesthetic strength and rapid IV toxicity. But compared to lidocaine, prilocaine has approximately more than 60 undergo enzymatic degradation, mainly in the liver. Prilocaine undergoes the maximum space in hepatic uptake and complete metabolism before the renal excretion (Stewart, Rogers, Mahaffey, Witherspoon, & Woods, 1963). When we compare LAs strength and duration interval, prilocaine redistributes faster than mepivacaine or lidocaine. Due to the lower levels of prilocaine in blood, it shows a tendency to form the least vasodilation than lidocaine. In dental treatment procedures, the patient requires local anesthesia to minimize and prevent the pain. Because of less vasodilation and toxicity compared to lidocaine, Prilocaine amide group LA drugs are widely used in dentistry from the 1960s (Mclure & Rubin, 2005). The solubility of prilocaine in the aqueous nature leads to quick tissue elimination through the bloodstream. It is generally co-administered with a vasoconstrictor to extend the anesthetic effect interval in patients. An undissociated form of prilocaine is high lipophilic and penetrable via the biological membrane (Bragagni, Maestrelli, Mennini, Ghelardini, & Mura, 2010). Prilocaine, in combination with felypressin shown 77.8%–80% of anesthetic effect after maxillary infiltration and with onset action among 2–3 min and duration ranges are from 24.8 to 61.6 min (Petersen et al., 1977).

The liposomal formulation approach shows the therapeutic effects of drugs. In general liposomes are colloidal phospholipid vesicles showing safe, fruitful drug carrier systems (Grant & Bansinath, 2001). Liposomal formulation drugs extensively applied in topical drug delivery system. This new approach shows the evolution of a new drug delivery system, with improved curative efficacy. M. C Volpato et al. performed the liposomal prilocaine anesthetic efficacy in maxillary infiltration anesthesia. Results indicate that in maxillary infiltration liposomal prilocaine exhibits better anesthetic efficiency compared to plane prilocaine in combination with felypressin (Petersen, Luck, Kristensen, & Mikkelsen, 1977). Another study showed that liposomes carrier prepared by combination prilocaine drug with hydroxypropyl-β-cyclodextrin. The anesthetic effect of developed liposomal formulation determined by in vivo on Guinea pigs based on the test of dorsal-muscle contraction (Lim, MacLeod, Ries, & Schwarz, 2007). The prepared all liposomal formulations shown higher values of encapsulation efficacy, from 82% and 85% ranges (Table 2) in the form of base drug and loading drug techniques. From these outcomes, the binary filling method shown new platform to the drug with cyclodextrins in the liposome. This also suggested an exceptional strategy to improve the therapeutic action of prilocaine, and this efficacy represents an excellent choice to replace with liposomes in place of vasoconstrictors for the maintenance of the drug-anesthetic effect (Bragagni et al., 2010).

TABLE 2 Properties of liposomes loaded with prilocaine as hydrochloride salt or prilocaine-liposomal complex (Bragagni et al., 2010).

Liposomes	EE %	Deformab. $n = 5$	Particle size (nm) $n = 5$	P.I. $n = 5$	Z-potential (mV) $n = 5$
Empty	–	1.09	431 ± 44	0.2 ± 0.024	$+34.6 \pm 4.9$
PRL.HCl 1% aqueous phase	85	1.04	331 ± 34	0.18 ± 0.015	$+59.6 \pm 8.2$
PRL-HPβCD 1% aqueous phase	83	1.04	479 ± 51	0.19 ± 0.016	$+36.0 \pm 5.9$
PRL 1% in lipophilic phase	84	1.07	433 ± 38	0.27 ± 0.026	$+32.5 \pm 4.2$
PRL.HCl 2% aqueous phase	82	1.43	506 ± 55	0.16 ± 0.025	$+55.4 \pm 7.4$
PRL 2% in lipophilic phase	82	1.09	367 ± 32	0.33 ± 0.029	$+25.1 \pm 5.5$
PRL + PRL-HPβCD double load 1 + 1%	84	1.11	324 ± 35	0.28 ± 0.023	$+28.7 \pm 4.7$

PRL, prilocaine.

Topical anesthesia showed various applications in minimizing pain, anxiety by needle injections and anesthesia during the surface skin surgery process. Local systemic anesthetics available in the form of gels, creams, ointments etc. deliver prilocaine, lidocaine, and benzocaine etc. noninvasively (Merck & Co. Inc, 1996). Topical local anesthetics via skin are convenient for straightforward application, less adverse action, long-range sustainable drug release. The topological anesthetic product is a eutectic mixture of LAs contains 2.5% prilocaine and 2.5% lidocaine (EMLA cream, Astra Pharmaceuticals). Eutectic mixture supplies adequate absorption of active substances, those improve the analgesic activity (Lener, Bucalo, & Kist, 1997).

You, Yuan, and Chen (2017) designed a lidocaine and prilocaine coloaded nanodrug distribution system for the topical anesthetic analgesic therapy and compared its effect with solid lipid nanoparticles (SLNs) and nanostructured lipid carriers (NLCs). The results showed that the combination of both lidocaine and prilocaine exhibit higher efficiency than single loaded drug system. SLNs with the co-loaded drug have a small particle size compared with NLCs and co-loaded drug and shown better ex vivo skin infiltration ability. Lidocaine/prilocaine NLCs shown strong in vivo anesthesia analgesic effect than SLN system. This research indicates that both SLNs and NLCs promise dual drug systems in topological anesthetic therapy with various advantages.

Yamashita et al. (2019) examined the effect of two different LAs preparation through dental treatment and studied the way different LAs show activity on the autonomic nervous system (ANS) during impacted mandibular third molar extraction. It was observed that the low-frequency/high-frequency rate was increasing during extraction compared with lidocaine/adrenaline. These findings suggest different circulatory dynamics in tooth removal with two different local anesthetics preparations, i.e. decreasing the parasympathetic nervous activity with lidocaine/adrenaline and developing sympathetic nervous activity with prilocaine/felypressin.

Kesici, Demirci, and Kesici (2019), reported that prilocaine and bupivacaine also show a significant effect in antimicrobial activity against *Escherichia coli*, *Staphylococcus aureus*, and *Pseudomonas aeruginosa*. The broth micro-dilution technique was used for determination of minimum inhibitory concentration (MIC) on Mueller Hinton broth (Oxoid, UK) using 96-well microplates. The study showed that prilocaine has more antimicrobial effect than bupivacaine (Table 3). This study exposed that the antimicrobial effect abilities would also consider as understanding in the choice of LAs agents to treat the problem of an infection that potency grows during local anesthetic infiltration and leads to morbidity (Kesici et al., 2019).

Tulaci, Arslan, Tulaci, and Yazici (2020) showed that usage of LAs prilocaine in combination with tramadol as nasal packing removal for effecting in pain and anxiety levels in patients. Tramadol is a centrally acting opioid having a sedative effect, nasal absorption capacity and less risk of addiction. This study reveals that prilocaine, in combination with tramadol, has more analgesic and anxiolytic effect during nasal packing removal.

Premature ejaculation (PE) is one of the common sexual dysfunction in men (Laumann, Paik, & Rosen, 1999). Prilocaine/lidocaine bases both have low melting point solids, and the combination of these two drugs contain a melting point lower than room temperature. Aerosol development from lidocaine/prilocaine mixture affords a dosage distribution of 7.5 mg lidocaine base with 2.5 mg of prilocaine base per metered dose actuation. The application of a developed topical aerosol spray on glams penis 15 min earlier interaction extends ejaculation time, and it showed the improvement of sexual satisfaction in men and their partner (Henry & Morales, 2003).

TABLE 3 The results of various concentrations of anesthetic agents against *S. aureus, P. aeruginosa,* and *E. coli* strains and distribution of MIC values (Kesici et al., 2019).

Prilocaine (mg/mL)	0.078	0.156	0.313	0.625	1.25	2.50	5.00	10.00	20.00
S. aureus ATCC 29213	+	+	+	+	+	+	+	+	−[a]
P. aeruginosa ATCC 27853	+	+	+	+	+	+	+	+	+
E. coli ATCC 25922	+	+	+	+	+	+	+	−	−
Bupivacaine (mg/mL)	0.078	0.156	0.313	0.625	1.25	2.50	5.00	10.00	20.00
S. aureus ATCC 29213	+	+	+	+	+	+	+	+	+
P. aeruginosa ATCC 27853	+	+	+	+	+	+	+	+	+
E. coil ATCC 25922	+	+	+	+	+	+	+	+	−[a]

[a]MIC values.

156 **PART | II** Mechanisms of action of drugs

TABLE 4 Aspects of clinical studies of intrathecal prilocaine (Black et al., 2011; Camponovo et al., 2010; de Weert et al., 2000; Gebhardt, Herold, Weiss, Samakas, & Schmittner, 2012; Martinez-Bourio et al., 1998; Ostgaard et al., 2000; Ozden et al., 2014; Reisli et al., 2003).

References	Number of subjects	Prilocaine dose	Solution	Motor block duration	Void time
Gebhardt et al. (2012)	120	10/20/30	2% H	NA	173/193/211
Ozden et al. (2014)	50 (25)	15	0.5% H	134	153
Black et al. (2011)	48 (22)	20 (+20 μg fentanyl)	2% P	After 2 h, 19/22 had bromage score of 0	205
Reisli et al. (2003)	30	40	2% P	76.8	NA
Camponovo et al. (2010)	90 (30)	40/60	2% H/2% P	92/118	195/218
Martinez-Bourio et al. (1998)	200 (100)	68.6	5% H	NA	NA
Ostgaard et al. (2000)	50	80	2% P	197	NA
de Weert et al. (2000)	69 (34)	80	2% P	Time until onset of regression	NA

Study on prilocaine has mainly described the rate of transient neurologic symptoms (TNS) compared with other LAs and the onset action and time of sensorial and motor block in the patient getting prilocaine intrathecal (Boublik, Gupta, Bhar, & Atchabahian, 2016). The use of LAs in case of spinal anesthesia would deliver quick recovery duration. Few reported rates of TNS are summarized in Table 4.

Gebhardt et al. (2012) described the ideal dose form for short dosage spinal anesthesia with hyperbaric prilocaine for the patients in perianal surgery. Different dosages (10, 20, 30 mg) were conducted in 116 patients, and they found encouraging results. The authors state that all three treatments have shown sufficient anesthesia for the procedures, dose 10 mg was proposed for perianal surgery because of the motor block nonappearance and quicker retrieval time.

Kaban et al. (Ozden et al., 2014) did the comparison studies of hyperbaric prilocaine and bupivacaine in the form of block determination in day-case spinal anesthesia by using 30 mg of 0.5% hyperbaric prilocaine with 20 μg fentanyl and 7.5 mg of 0.5% bupivacaine with 20 μg fentanyl. They conclude that prilocaine 30 mg + 20 μg fentanyl shown quicker sensory block determination and shorter discharge time compared with 5 mg of 0.5% bupivacaine +20 μg fentanyl.

Black, Newcombe, Plummer, McLeod, and Martin (2011) sought the comparison studies of prilocaine and bupivacaine with fentanyl in the use of spinal anesthesia for ambulatory arthroscopic surgery of the knee. At 2-h time set prilocaine (20 mg) with fentanyl (μg) found extreme sensory block level and regression of sensory block to L4, and also motor block at 1 h and 2 h, and levels of hemodynamic stability. Prilocaine blocking levels completely determined in 86% of the patients. The advantages included prior ambulation, decreasing hazard of urinary retention, and shorter time to discharge the patient. They conclude that mixture of prilocaine and fentanyl is an excellent alternative to the small prescription bupivacaine and fentanyl, for spinal anesthesia in ambulatory arthroscopic knee surgery and they stress that it signifies an exceptional alternate to lidocaine in the ambulatory care. And studies have shown prilocaine in combination with hyperbaric solution showing the earlier action to motor block onset and less interval of medical block and signifying its advances for the ambulatory setting. Spinal anesthesia by 60 mg or 40 mg of 2% hyperbaric prilocaine is similar to 60 mg of 2% plain prilocaine in the form of onset of sensory block at T10 (Camponovo, Fanelli, Ghisi, Cristina, & Fanelli, 2010).

Spinal or epidural anesthesia conducted incessantly with definitive in aged patients suffering transurethral resection of the prostate. Continuous spinal anesthesia had an additional quick onset of action and produced the extra active sensory and motor blockade and had to be a better quicker retrieval time. Prilocaine looks up to be harmless and secure LA in medication for any continuous spinal anesthesia or continuous epidural anesthesia. The effect of prilocaine in continuous spinal anesthesia and constant epidural anesthesia depends on hemodynamic constancy and quality of anesthesia, and patient recovery rate undergoes transurethral resection of the prostate gland (Reisli, Celik, Tuncer, Yosunkaya, & Otelcioglu, 2003).

In 2005, Zaric and Pace (2009) carried out a systematic review on prilocaine effect on rate of TNS. The rate of TNS with different intrathecal agents includes 1437 patients in 14 clinical trials, and they report that spinal anesthesia was

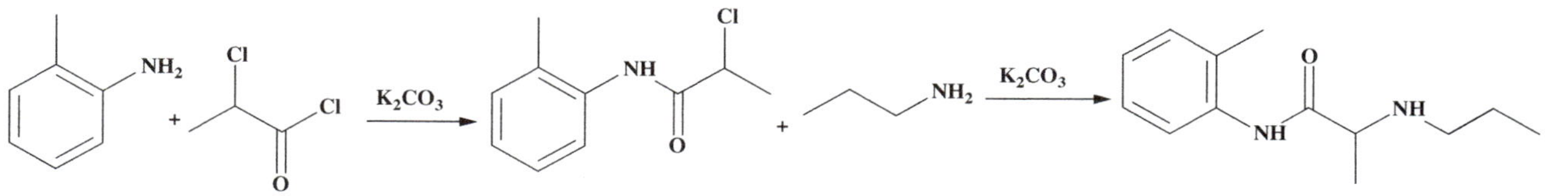

FIG. 3 Synthetic route of prilocaine local anesthetic (*Astra Apotekernes Kem Fab., Br. Pat. 839*.943, 1958).

suggestively greater in lidocaine than bupivacaine, prilocaine or procaine. But compared to data of four agents to prilocaine, they found that risk assessments are developing more in TNS with lidocaine versus prilocaine (95% CI: 2.66 to 22.52) (de Weert et al., 2000; Martinez-Bourio et al., 1998; Ostgaard, Hallaraker, Ulveseth, & Flaatten, 2000). These studies are suggesting that TNS has shown very sporadic with prilocaine as compared with lidocaine. Prilocaine showed no other side effects and safe medication, and low TNS rate.

Preparation method of prilocaine:

A preparation method (*Astra Apotekernes Kem Fab., Br. Pat. 839*.943, 1958) of prilocaine base is through the reaction of 2-chloropropionic acid and o-toluidine in the presence of 4-(4,6-dimethoxy-1,3,5-triazine-2-)-4-ethylmorpholine chloride to give *N*-(2-methylphenyl)-2-chloropropionamide, add *N*-propylamine in acetone solvent to intermediate. The crude product of prilocaine base is reacted under potassium carbonate. The prilocaine base is refined with a mixed solvent of n-heptane and ethanol to obtain medicinal Grade prilocaine base, and the synthetic route is shown in Fig. 3.

Pharmacology and mechanism action of local anesthetic agent—Prilocaine

LAs showed their pharmacological action on the nerve membrane and stops the generation of the action potential in nerve axons by blocking the inflow of sodium between voltage-gated sodium channels in the nerve axon membrane. Anesthetic agents bind to a specific site in the sodium channels and prevent conformational fluctuations and lead to damage conduction. As extra receptors are occupied, gradual decreasing the rate and degree of depolarization till conduction flops. The application of LAs on the nerve membrane, there is a specific route in the disappearance of sensations, it based on the various sensitivities of the nerve involvement. First pain fibers are blocked than pressure, temperature, and motor kinetics.

Many theories have been developed from the past few decades to describe the mechanism of action of LAs, i.e., acetylcholine, calcium displacement, and surface charge theories. The acetylcholine theory suggests that acetylcholine involved in nerve condition, in accumulation to its role as a neurotransmitter at nerve synapse (Dettbarn, 1967). The calcium displacement theory reveals that LA nerve block was done by movement of calcium from the membrane site, and it controls the permeability to sodium (Goldman & Blaustein, 1996). The surface charge theory states that the LAs show the effect through the nerve membrane binding and change in electrical potential at the surface membrane (Wei, 1969).

Diffusion and binding are the two main factors involve in the activity of LAs, (a) drug diffusion by nerve sheath and (b) binding to receptor location in the ion channel. In vivo, expression of LAs depends on the constructive relationship among the molecular weight of LAs and lipophilicity, duration of action, potency, protein binding, and harmfulness and inverse relationship between rapidity of onset action. In vitro appearance of LA activity based on additional aspects, such as injection spot, quantity, essential vasoactivity, and design. The appearance of sensory against motor block differs and is dependent on several aspects, including the drug agent and kind of block performed.

Previous reports and history demonstrated that two views influence the mechanism of action of general LAs, i.e., the Meyer–Overton observation and the Unitary Hypothesis. Meyer–Overton hypothesis separately represent that the anesthetic agent's potency linked to lipid solubility and coming to Unitary Hypothesis stands that all LAs are chemically varied agents and they develop their anesthetic effect by the same (unknown) mechanism. In 1984, Franks and Lieb indicate that the maximum number of general anesthetics inhibited the lipid-free enzyme firefly luciferase, specified that they turn directly on protein (Harrison, 2006). LA agents are shown reversible blocking action potential at sodium channel level by inter interrupting axonal conduction. All LA actions are the non-specified effort on the nerve to perform sodium channel. Compound with more lipophilic/hydrophobic nature contains the high potent, long-term, and high toxicity (Catterall & Mackie, 2006). The hydrophobic part in the sodium channel pore that contains a binding affinity for the lipophilic/hydrophobic group of the LAs. Binding positions occurs with sodium gate or pore in an open or stimulated position, and

158 PART | II Mechanisms of action of drugs

anesthetics are needed sources to their binding site. Once anesthetic agents stabilize the sodium channel at inactive state, the nerve cannot repolarize. The strength and bond between the hydrophobic groups affect the therapeutic action shown its action (Hondeghem & Miller, 1989). The anesthetic agent's metabolism is similar to the duration of effect and toxicity. Ester linked LAs are commonly short-acting site. Plasma cholinesterase rapidly deactivates them LAs (amide linked) are metabolized in the liver part via the cytochrome P450 enzyme system. Greater than 50% of amide contained LAs are bound to Alpha1 acid glycoprotein (AAG), a fluid originates in plasma. Instability levels of anesthetic agents are showing an effect on metabolism (Rosenberg, Veering, & Urmey, 2004). The other important feature in LAs is pH plays a very significant character in LAs activity. When anesthetic agents are inoculated on regular tissue, there is a quick equilibrium of the pH, permitting the unprotonated compound to diffuse across the cell membrane. The cationic form of LAs binds with sodium channel or sodium port; the ionization effect must take place after passage via a membrane to be shown anesthetic effect. The cationic or anionic charged form of these LAs will not diffuse by the membrane; it states that pH will affect the ability of LAs to get down the cell membrane (Hondeghem & Miller, 1989; Harrison, 2006; Rosenberg et al., 2004). Another important thing is nerve fibers contain different liability in the effect of LAs. These differences are mainly based on fiber diameter and myelination. Fibers like minor diameter unmyelinated fibers for example, type C pain fibers are more sensitive in blocking the effects of anesthetic and others like heavily myelinated, thin fibers like type A motor fibers stand fewer blocking effect (Hondeghem & Miller, 1989). The minor fibers are accountable for the pain and autonomic activity. These fibers are effected more quickly by anesthetic agents and block the membrane site. This results in pain reduction without the blockage of neuronal activity. Nerve cells are bounded with lipoprotein medium that splits or separates the gap from the extracellular space. Within this membrane exist intramembrane and transmembrane proteins that act as pores and enzymes for passive and active transport of vital chemicals and elements. The presence of an active sodium transport system results in a transmembrane potential by the maintenance of an ionic gradient. (Altman, Smith-Coggins, & Ampel, 1985).

Lipid solubility and protein binding:

The aromatic ring in the prilocaine directs its nature of solubility. As nerve membranes are mostly lipids, the drug with high lipid empathy indicates the greatest potency. Receptor sites and sodium channels are substantially protein, and tremendously protein-binding molecules bind securely to the active site. The other important thing is protein binding generates a bank, to remove the unbound drug from active sites by vascular uptake. Longer duration of action shows the more percentage of protein binding (Yagiela, 2010). Lipid solubility is the primary step to determine the intrinsic anesthetic potency. Biochemical composition in nerve membrane gives a logical theory or explanation of the relationship between lipid solubility and anesthetic potency. In vitro studies of isolated nerves define the relationship among the partition coefficient of LAs and small concentration essential for conduction blockade (Truant & Takman, 1959). For example, among the amino acid agents, mepivacaine and prilocaine are lesser lipid-soluble and weaker amide agents compared to etidocaine. Etidocaine is the most lipophilic and potent anesthetic. In vivo studies in man specify the relationship between lipid solubility and anesthetic effectiveness is not precise as in an isolated nerve.

Prilocaine impairs the conduction in neural membrane tissue on the site of action, in the central nervous system, the nerve excitation takes place when the anesthetic concentration rises rapidly. This is nonsensical because prilocaine blocks the neuronal conduction. LAs also showed an effect on potassium channels, which may itself produce neuronal excitation. Sometimes the patient may experience agitation, dizziness, involuntary muscle activity, disorientation, etc. when anesthetics are administered in higher concentration than maximum dose or injected directly into the vasculature.

Prilocaine stabilize the nerve membrane by ionic fluxes are essential for the initiation and conduction impulses and shows the LAs effect. Prilocaine acts on a sodium channel on the neuronal membrane and minimizes the spread of seizure activity and propagation.

Metabolism and excretion

In plasma ester, LAs are hydrolyzed via pseudocholinesterase to p-aminobenzoic acid and few other derivatives. Formed by-products or derivatives further undergo the biodegradation in the liver, and a minute amount of drug will be eliminated unchanged. And amide anesthetics are metabolized in the liver mainly by CYP3A4 and CYP1A2 isoforms. More than 50% amide linked local anesthetics are bound to Alpha 1 acid glycoprotein, it was found in plasma. The metabolism rate depends on liver blood flow and function of the liver, and slow liver blood flow can hold back amide metabolism and possible for increasing the toxicity levels (Yagiela, 2010). In vivo metabolic studies of prilocaine appears that the hydrolysis takes place at amide linkage. The enzymatic activity of the liver responsible for the in vivo hydrolysis was located exclusively in the microsome fraction.

Prilocaine is metabolized equally in both liver and kidney and eliminated by the kidney in the form of urine. And it is not metabolized through plasma esterases. Prilocaine hydrolysis through amides forms (or yields) O-toluidine and N-propylalanine. Two of these compounds further undergo ring hydroxylation. The moiety of O-toluidine developed methemoglobin, in both in vitro and in vivo characterization (Fig. 5). Methemoglobin is cured by methemoglobin reductase, and this procedure may enhance by intravenous methylene blue (1–2 mg/kg). The metabolism of prilocaine occurs in both kidney and liver, and the hepatic and renal dysfunction may possess prilocaine function. Compared with other LAs, prilocaine plasma binding capacity may depend on the concentration of the drug. At 0.5–1.0 mg/mL of prilocaine it is 55% protein bound.

Prilocaine crosses the blood-brain barrier and placental barrier apparently through passive distribution. Applications include acidosis and use of CNS stimulants, and sedatives disturb the CNS levels of prilocaine essential to develop the systemic effects in the body. In the rhesus monkey, arterial blood levels of 20 mg/mL have initiated to be a threshold for convulsive activity.

Akerman et al. reports made that the hydrolysis rate gradually reduced by many substituents around the amide linkage. The presented substitution showing the great influence on the enzymatic hydrolysis seems to be situated adjacent carbon in the side chain. Increasing the substituent size on nitrogen increased the hydrolysis rate of secondary amines, i.e., xylidide and toluidine derivatives (Akerman, Ross, & Telc, 2009). The in vitro nerve blocking activity was associated with directly to the intravenous toxicity and impartially straight to the dermal toxicity.

Distribution

When the LAs are injected to the target site of organ, distribution within the organ plays a crucial role in gaining clinical pharmacological effect. Prilocaine administered intravenously, and steady-state volume of distribution is 0.7 to 4.4 L/Kg (mean 2.6, ±1.3 SD, $n = 13$). The higher delivery rate of prilocaine exhibited lesser plasma absorption of the drug, compared to the same volumes of lidocaine and prilocaine administered (Fig. 4).

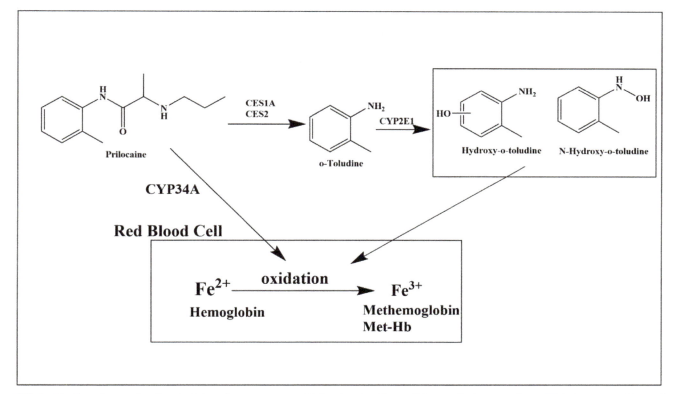

FIG. 4 The flow diagram describes the LAs pathway to receptors binding location. Extracellular anesthetics occurs in equilibria between exciting and nonexciting forms. The charged cations enter poorly into lipid membrane, intracellular access is realized by the uncharged form way. Reequilibration of intracellular results to the development of the high active charged species, it may bind to receptor at the internal atrium of the sodium channel (hydrophobic pathway). *(Copied from google source https://basicmedicalkey.com/local-anesthetics/.)*

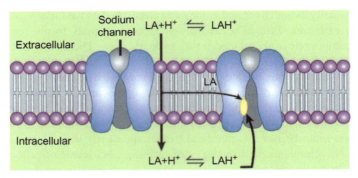

FIG. 5 Metabolic pathway of prilocaine.

Elimination

The elimination and half-life time of the prilocaine takes nearly from 10 min to 2.3 h (mean 70, ±48 SD, $n = 13$). The total clearance is 18–64 mL/min/kg (mean 38, ±15 SD, $n = 13$). After metabolism, drug will be eliminated through urine. In the elimination process, Hepatic extraction, and biotransformation both will play a key role in amino amide linkage local anesthetics.

Duration of anesthesia

Anesthesia duration is mainly linked with the degree of protein binding of several LAs. Blockade conduction maintains the LAs interactions with protein-membrane or receptor located in the sodium channel of the nerve membrane. Anesthetic agents who have a greater affinity to bind to the receptor site for an extended period, it forms a long duration of conduction blockade. When we use prilocaine for infiltration injection in dental treatment, the onset duration of anesthesia is less than 2 min with a regular period of soft tissue anesthesia is nearly 2 h. From electrical stimulation investigation, prilocaine hydrochloride (4% Citanest) simple dental-injection supply approximately 10 min of onset duration of anesthesia in maxillary infiltration injections. Use 4% Citanest plain dental injection for inferior alveolar nerve block onset action shows average less than 3–4 min with an approximate onset action on soft tissue 2.5 h.

Toxicity

The toxicity factor of LAs is seen in body organs, which depends on the sodium channels for appropriate function, includes central nervous system, methemoglobinemia and heart. The CNS is the most diplomatic to LA effect than the cardiac system and generally distinct signs and symptoms for first toxicity.

The toxicity levels of the cardiovascular system are mainly in three stages. The first stage contains hypertension and tachycardia. The second stage is related to myocardial depression and hypotension. The final stage involves peripheral vasodilatation, severe hypotension, and a variety of arrhythmias such as sinus bradycardia, conduction blocks, ventricular tachyarrhythmias, and asystole.

The CNS is the maximum affected organism in the body in local anesthesia. Tinnitus, blurred vision, dizziness, tongue paresthesia, and circumoral numbness are the main CNS symptoms. Excitative signals include nervousness, agitation, restlessness, and muscle twitching are the result of the blockade of inhibitory pathways. The first symptom development of CNS depression with slurred speech, drowsiness, unconsciousness, and then respiratory arrest.

Methemoglobinemia

Prilocaine breaks down (metabolized) in both kidney and plasma, where one of the metabolites lead to methemoglobinemia. Methemoglobinemia states that oxidized amount of hemoglobin in which iron in heme is oxidized to the ferric (Fe^{3+}) form. Indications of methemoglobinemia might show effects instantly, or sometimes it might be late some hours after the introduction and categorized through a cyanotic skin discoloration or irregular coloring of the blood. Methylene blue is the first line to treat the disorder methemoglobinemia. Methylene blue stimulates the methemoglobin enzymatic deduction through NADPH-methemoglobin reductase and also reduced to leuco methylene blue that, in turn, reacts with methemoglobin (Frey & Kehrer, 1999).

Applications

Different anesthetic agents are used to conduct anesthesia are considered as potential active agents for familial malignant hyperthermia. The critical factor is safety and effect of prilocaine stands on the exact dosage, adequate precautions, and the use of emergency conditions. These days minor and major surgical processes are performed by regional anesthesia in out-patient clinics to minimize or diagnostic and therapeutic reasons. Prilocaine and lidocaine both have the best ranking in using LA agents in all major and minor surgeries (US Food and Drug Administration). Eutectic mixture of prilocaine and lidocaine (EMLA) has been showing a significant effect to reduce pain from needle sticks. EMLA cream having 2.5% prilocaine and 2.5% of lidocaine, an emulsion in which the oil phase is a eutectic mixture of prilocaine and lidocaine in 1:1 weight ratio. The EMLA mixture contains a melting point less than room temperature, because of this both LAs are in the form of liquid phase than crystal. The usage of EMLA cream on the skin, it affords dermal anesthesia through manumit of EMLA ointment into skin epidermal and dermal layers and the growth of prilocaine and lidocaine in the vicinity of cutaneous pain receptors and nerve endings. Both lidocaine and prilocaine are the amide local anesthetics and control nerve membranes by inhibiting the ionic fluxes for initiation and conduction impulses. The dermal application of EMLA may cause transient, local blanching followed by a transient, local redness or erythema. Aerosol development from lidocaine/prilocaine mixture delivers dose distribution of 7.5 mg lidocaine base and 2.5 mg of prilocaine base per metered dose actuation. The usages of a developed topical aerosol spray to glams penis 15 min earlier interaction prolongs ejaculation period, and it showed improvement of sexual gratification in males and their partner (Henry & Morales, 2003). EMLA used safely in infants older than 3 months of age, and it was not producing any other local or systemic adverse effects applied to a skin area. The developed cream has shown that effective alternative to conservative infiltration anesthesia for dermabrasion and artificial surgical process.

Prilocaine HCI Dental Injection (4% Citanest Plain), is a sterile, non-pyrogenic isotonic solution acts as an anesthetic agent and is administered by injection. It produces anesthesia in dentistry by infiltration techniques or a nerve block. 4% Citanest Plain Dental propose to usage in maxillary infiltration anesthesia process, the affected part (painful) in the body shown action in 15 min later the injection. Prilocaine hydrochloride (4%) is suitable for the short time anesthesia process in anterior maxillary teeth. Therefore, prilocaine has a better clinical performance to provide rapid dental anesthesia and earlier teeth extraction than lidocaine.

Prilocaine in combination with neostigmine in IV regional anesthesia has reduced sensory and motor block onset action and increases the excellence of anesthesia by extending the period of the first analgesic condition. Neostigmine toxic effect found that typically related to systemic absorption but did not require in therapy. Adding neostigmine to LAs in IVRA shown increasing the quality rate of anesthesia (Turan, Karamanlýoglu, Memis, Kaya, & Pamukçu, 2002). In a few cases, mostly in high-risk patient treatment, adrenaline may cause to hypertensive crisis, arrhythmia, and uncontrolled diabetes. Prilocaine is the best choice as least vasodilator and the best option for patients whom vasoconstrictor is contraindicated. Prilocaine, in combination with felypressin, could be the right choice for patients who have a contraindication to the use of lidocaine with adrenaline.

Other applications of prilocaine

The EMLA cream (lidocaine 2.5% + prilocaine 2.5% eutectic mixture) is specified as a topical anesthetic used for regular undamaged skin and genital mucous membrane for normal small surgeries and also in infiltration anesthesia.

The use of prilocaine and lidocaine in pregnancy has been studied in rats and the results shown reproduction studies with prilocaine reveals that no indication of impaired fertility or damage in fetus compared to lidocaine. But animal reproduction investigations are not every time equal to studies on humans. Therefore the use of EMLA cream would be utilized during pregnancy when it exactly required. Reproduction studies have investigated in rats by administration of an aqueous mixture covering lidocaine HCl and prilocaine HCl (1:1 ratio). The quantity at 40 mg/kg respectively observed that no teratogenic, embryotoxic or fetotoxic effect.

The uses EMLA cream in kids below 7 years has revealed that less advantageous compared to older children and adults. This investigation demonstrated that the importance of sensitive and mental condition care of younger children undertaking medical or surgical therapy. Using EMLA cream in young children, particularly newborns below 3 months of age, necessary to follow precautions and care also needed to avoid side effects.

Other agents of interest

The authors have no conflict of interest.

Mini-dictionary of terms

ANS   autonomic nervous system
CNS   central nervous system
CV   cardiovascular
IV   intravenous
LAs   local anesthetics
MIC   minimum inhibitory concentration
NLCs   nanostructured lipid carriers
PE   premature ejaculation
PNS   peripheral nervous system
SLNs   solid lipid nanoparticles
TNS   transient neurologic symptoms

Key facts

- Prilocaine is an amide derivative, and it shows it therapeutic action against nerve system. It stabilizes the neuronal membrane by preferential binding and inhibits the depolarization of sodium channel gate voltage, and it decreases the membrane permeability.
- Prilocaine stabilize the nerve membrane by ionic fluxes are essential for the initiation and conduction impulses and shows the LA effect.
- Prilocaine has higher rate TNS compared with other LAs onset action, time of sensorial and motor block in the patient getting prilocaine intrathecal. The use of LAs in case of spinal anesthesia would deliver quick recovery duration by using prilocaine.
- Prilocaine impairs the conduction in neural membrane tissue on the site of action, in the central nervous system, the nerve excitation takes place when the anesthetic concentration rises rapidly.
- Due to the rapid tissue uptake action, biotransformation, and quick fall of prilocaine plasma level make prilocaine as a good anesthetic, and it is generally used as an intravenous regional anesthetic drug.
- The main advantages of the prilocaine are a long duration of anesthesia, no need for adrenaline, and less toxic compared to other LAs.
- The metabolism of prilocaine takes place in both liver and kidney.
- The use of prilocaine in dental treatment infiltration injection, the onset duration of anesthesia is less than 2 min with a regular period of soft tissue anesthesia is nearly 2 h.
- Prilocaine shows a better clinical performance to provide rapid dental anesthesia and earlier teeth extraction than lidocaine.
- High doses of prilocaine lead a methemoglobinemia; it is an oxidized amount of hemoglobin in which iron in heme is oxidized to the ferric (Fe^{3+}).
- Prilocaine, in combination with lidocaine, has been widely using as a topological anesthetic cream.

Summary and conclusion

Prilocaine is a secondary amide analogue of lidocaine and having a longer duration of rapid onset action. Prilocaine acts on a sodium channel on the neuronal membrane and minimizes the spread of seizure activity and propagation. Prilocaine shows its therapeutic action against nerves system and stabilize the neuronal membrane by preferential binding and inhibit the depolarization of sodium channel gate voltage. Because of fast tissue uptake action, biotransformation prilocaine plasma levels made that prilocaine as a good anesthetic agent in intravenous regional anesthesia medication. With the lowest levels of prilocaine in blood, it shows a tendency to form the least vasodilation than lidocaine. Prilocaine solubility nature in the aqueous atmosphere may lead to fast tissue elimination via the bloodstream. The main property is its lower systemic toxicity, due to slow absorption rate from the injection site, excellent tissue distribution and more rapid metabolism than lidocaine. The liposomes which prepared from prilocaine in combination with other compounds like felypressin and cyclodextrin showed a greater anesthetic efficiency compared with prilocaine. The eutectic mixture of local anesthetics contains 2.5% prilocaine and 2.5% lidocaine (EMLA cream) is a topical anesthetic ointment, it supplies adequate absorption of active substances, those improve analgesic activity. The high dosage of prilocaine may lead to methemoglobinemia condition in the form of hemoglobin iron oxidized to ferric (Fe^{3+}) state, leading to rapid oxygen desaturation. The use of methylene blue or ascorbic acid is the promising drug to treat the methemoglobinemia as a nominal dose of 1 mg/kg.

References

Akerman, S. B. A., Ross, S. B., & Telc, A. (2009). The enzymatic hydrolysis of a series of aminoacylanilides in relation to their nerve blocking effect and toxicity. *Acta Pharmacologica et Toxicologica, 32*(1–2), 88–96.

Altman, R. S., Smith-Coggins, R., & Ampel, L. L. (1985). Local anesthetics. *Annals of Emergency Medicine, 14*, 1209–I217.

Astra Apotekernes Kem Fab., Br. Pat. 839.943. (1958). .

Black, A. S., Newcombe, G. N., Plummer, J. L., McLeod, D. H., & Martin, D. K. (2011). Martin spinal anesthesia for ambulatory arthroscopic surgery of the knee: A comparison of low-dose prilocaine and fentanyl with bupivacaine and fentanyl. *British Journal of Anesthesia, 106*(2), 183–188.

Boublik, J., Gupta, R., Bhar, S., & Atchabahian, A. (2016). Prilocaine spinal anesthesia for ambulatory surgery: A review of the available studies. *Anesthesia Critical Care & Pain Medicine, 35*(6), 417–421.

Bragagni, M., Maestrelli, F., Mennini, N., Ghelardini, C., & Mura, P. (2010). Liposomal formulations of prilocaine: Effect of complexation with hydroxypropyl-ß-cyclodextrin on drug anesthetic efficacy. *Journal of Liposome Research, 20*(4), 315–322.

Camponovo, C., Fanelli, A., Ghisi, D., Cristina, D., & Fanelli, G. (2010). A prospective, doubleblinded, randomized, clinical trial comparing the efficacy of 40 mg and 60 mg hyperbaric 2% prilocaine versus 60 mg plain 2% prilocaine for intrathecal anesthesia in ambulatory surgery. *Anesthesia and Analgesia, 111*, 568–572.

Catterall, W. A., & Mackie, K. (2006). *Local anesthetics. Goodman and Gillman's the pharmacologic basis of therapeutics* (11th ed.). Columbus, OH: McGraw Hill Chapter 14.

Concepcion, M., & Covino, B. G. (1984). Rational use of local anesthetics. *Drugs, 27*, 256–270.

Covino, B. G., & Donald, B. (1981). Giddon, pharmacology of local anesthetic agents. *Journal of Dental Research, 60*(8), 1454–1459.

Davison, M. H. A. (1965). In *The evolution of anesthesia* (pp. 117–119). Altrincham, UK: John Sherratt & Son.

de Weert, K., Traksel, M., Gielen, M., Slappendel, R., Weber, E., & Dirksen, R. (2000). The incidence of transient neurological symptoms after spinal anesthesia with lidocaine compared to prilocaine. *Anesthesia, 55*, 1020–1024.

Dettbarn, W. D. (1967). The acetylcholine system in peripheral nerve. *Annals of the New York Academy of Sciences, 144*, 483–50.3.

Frey, B., & Kehrer, B. (1999). Toxic methaemoglobin concentrations in premature infants after application of a prilocaine containing cream and peridural prilocaine. *European Journal of Pediatrics, 158*, 785–788.

Gebhardt, V., Herold, A., Weiss, C., Samakas, A., & Schmittner, M. D. (2012). Dosage finding for low-dose spinal anesthesia using hyperbaric prilocaine in patients undergoing perianal outpatient surgery. *Acta Anaesthesiologica Scandinavica, 57*(2), 249–256.

Goldman, D. E., & Blaustein, M. P. (1996). Ions, drugs and the axon membrane. *Annals of the New York Academy of Sciences, 137*, 967.

Grant, G. J., & Bansinath, M. (2001). Liposomal delivery systems for local anesthetics. *Regional Anesthesia and Pain Medicine, 26*, 61–63.

Hadeed, A. J., & Siegel, S. R. (1989). Maternal cocaine use during pregnancy:effect on the newborn infant. *Pediatrics, 84*, 205.

Harrison, N. L. (2006). General anesthetics: Mechanisms of action. In H. C. Hemmings, & P. M. Hopkins (Eds.), *Foundations of Anesthesia* (pp. 287–294). (2nd ed.). Philadelphia: Mosby Elsevier.

Heavner, J. E. (2007). Local anesthetics. *Current Opinion in Anaesthesiology, 20*, 336–342.

Henry, R., & Morales, A. (2003). Topical lidocaine–prilocaine spray for the treatment of premature ejaculation: A proof of concept study. *International Journal of Impotence Research, 15*, 277–281.

Hondeghem, L. M., & Miller, R. D. (1989). Local anesthetics. In *Basic and clinical pharmacology* (pp. 315–322). East Norwalk, CT: Appleton & Lange.

Kesici, S., Demirci, M., & Kesici, U. (2019). Bacterial inhibition efficiency of prilocaine and bupivacaine. *International Wound Journal, 16*(5), 1185–1189.

Laumann, E. O., Paik, A., & Rosen, C. (1999). Sexual dysfunction in the United States: Prevalence and predictors. *JAMA, 281*, m537–m545.

Lener, E. V., Bucalo, B. D., & Kist, D. A. (1997). Topical anesthetic agents in dermatologic surgery: A review. *Dermatologic Surgery, 23*, 673–683.

Lim, T., MacLeod, B. A., Ries, C. R., & Schwarz, S. (2007). The quaternary lidocaine derivative, QX-314, produces long-lasting local anesthesia in animal models in vivo. *Anesthesiology, 2*, 305–311.

Martinez-Bourio, R., Arzuaga, M., Quintana, J., Aguilera, L., Aguirre, J., Saez-Eguilaz, J. L., et al. (1998). Incidence of transient neurologic symptoms after hyperbaric sub-arachnoid anesthesia with 5% lidocaine and 5% prilocaine. *Anesthesiology, 88*, 624–628.

Mclure, H. A., & Rubin, A. P. (2005). Review of local anesthetic agents. *Minerva Anestesiologica, 71*, 59–74.

Merck & Co. Inc (1996). *The merck index.* New Jersey: Merck & Co. Inc.

Ostgaard, G., Hallaraker, O., Ulveseth, O. K., & Flaatten, H. (2000). A randomized study of lidocaine and prilocaine for spinal anesthesia. *Acta Anaesthesiologica Scandinavica, 44*, 436–440.

Ozden, G. K., Dilek, Y., Taylan, A., Sayin, M. M., Duray, S., & Haluk, G. (2014). Spinal anesthesia with hyperbaric prilocaine in day-case perianal surgery: Randomised controlled trial. *The Scientific World Journal, 608372*, 6.

Peck, T. E., Hill, S. A., & Williams, M. (2008). *Pharmacology for anesthesia and intensive care* (3rd ed.). Cambridge University Press Medicine.

Perrin, S. L., Bull, C., & Black, B. S. (2019). Local anesthetic drugs. *Anesthesia and Intensive Care Medicine*, j. mpaic.12.003.

Petersen, J. K., Luck, H., Kristensen, F., & Mikkelsen, L. (1977). A comparison of four commonly used local analgesics. *International Journal of Oral Surgery, 6*, 51–59.

Reisli, R., Celik, J., Tuncer, S., Yosunkaya, A., & Otelcioglu, S. (2003). Anesthetic and haemodynamic effects of continuous spinal versus continuous epidural anesthesia with prilocaine. *European Journal of Anaesthesiology, 20*, 26–30.

Rosenberg, P. H., Veering, B. T., & Urmey, W. F. (2004). Maximum recommended doses of local anesthetics: A multifactorial concept. *Regional Anesthesia and Pain Medicine, 29*, 564–575.

Ruetsch, Y. A., Bonibc, T., & Borgeatac, A. (2001). From cocaine to ropivacaine: The history of local anesthetic drugs. *Current Topics in Medicinal Chemistry, 1*, 175–182.

Stewart, D. M., Rogers, W. P., Mahaffey, J. E., Witherspoon, S., & Woods, E. F. (1963). Effect of local anesthetics on the cardiovascular system of the dog. *Anesthesiology, 24*, 620–624.

Truant, A. P., & Takman, B. (1959). Differential physical-chemical and neuropharmacologic properties of local anesthetic agents. *Anesthesia and Analgesia, 38*, 478–484.

Tulaci, K. G., Arslan, E., Tulaci, R. G., & Yazici, H. (2020). Effect of prilocaine and its combination with tramadol on anxiety and pain during nasal packing removal. *European Archives of Oto-Rhino-Laryngology, 277*(5), 1385–1390.

Turan, A., Karamanlýoglu, B., Memis, D., Kaya, G., & Pamukçu, Z. (2002). Intravenous regional anesthesia using prilocaine and neostigmine. *Anesthesia and Analgesia, 95*(5).

Wei, L. Y. (1969). Role of surface dipoles on axon membrane. *Science, 163*, 280–282.

Wildsmith, J. A. (1985). Prilocaine—An underutilized local anesthetic regional anesthesia. *The Journal of Neural Blockade in Obstetrics, Surgery & Pain Control, 10*, 155–159.

Yagiela, J. A. (2010). Local Anesthetics. In J. A. Yagiela, F. J. Dowd, & B. S. Johnsonet al.*Pharmacology and therapeutics for dentistry* (p. 256). (6th ed.). St Louis (MO): Mosby Elsevier.

Yamashita, K., Kibe, T., Shidou, R., Kohjitani, A., Nakamura, N., & Sugimura, M. (2019). Difference in the effects of lidocaine with adrenaline and prilocaine with felypressin on the autonomic nervous system during extraction of the impacted mandibular third molar: A randomized control trial. *Journal of Oral and Maxillofacial Surgery, 78*(2), 215.e1–215.e8.

You, P., Yuan, R., & Chen, C. (2017). Design and evaluation of lidocaine- and prilocaine-coloaded nanoparticulate drug delivery systems for topical anesthetic analgesic therapy: A comparison between solid lipid nanoparticles and nanostructured lipid carriers. *Drug Design, Development and Therapy, 11*, 2743–2752.

Zaric, D., & Pace, N. L. (2009). Transient neurologic symptoms (TNS) following spinal anesthesia with lidocaine versus other local anesthetics. *Cochrane Database of Systematic Reviews, 2*.

Chapter 17

Sevoflurane: Features and uses in topical application for wound care

Manuel Gerónimo-Pardo

Department of Anesthesiology, Integrated Care Management of Albacete, Albacete, Spain

Abbreviations

EMLA	eutectic mixture of local anesthetics
GWP$_{20}$	twenty-year global warming potential
MRSA	methicillin-resistant *Staphylococcus aureus*
NRS	numerical rating scale
NSAID	nonsteroidal antiinflammatory drug
TRP	transient receptor potential
TRPM3	transient receptor potential melastatin 3

Introduction

Chronic wounds can negatively affect across all areas of patients' quality of life. Superinfection can associate exudate and bad odor, for instance. However, patients report pain as the worst symptom by far (Briggs, Nelson, & Martyn-St James, 2012).

Nonsteroidal antiinflammatory drugs (NSAIDs) and opioids are the most prescribed systemic painkillers for intense pain. However, patients suffering from painful hard-to-heal wounds need to take painkillers for a long time. Considering that patients suffering from these wounds are elderly and fragile, the risk of developing severe side effects is very high (Luchting & Azad, 2019), which globally contributes to worsening the health problem.

Local analgesics could be a great option to avoid unwanted effects derived from systemic analgesics, but a paucity of widely accepted alternatives exists. A eutectic mixture of local anesthetics (EMLA), namely, lidocaine and prilocaine, has proven to significantly reduce the pain from sharp debridement of leg ulcers. However, it requires to be applied for at least 30 min before debridement, which is inconvenient in overloaded workplaces; besides, it lacks approval for persistent daily pain. Ibuprofen-releasing foam dressings were intended to relieve persistent pain, but evidence on their usefulness is still inconclusive (Briggs et al., 2012).

Therefore, the current therapeutic armamentarium is lacking a topical analgesic drug that is effective against both somatic and neuropathic persistent pain associated with wounds of different etiologies and, at the same time, able to control incidental pain caused by dressing and cleansing. Ideally, such a drug should not favor local infections or interfere with ulcer healing.

This chapter is focused on how current evidence strongly suggests that sevoflurane could become such an ideal drug.

Systemic effects of sevoflurane and haloethers as general anesthetics

Sevoflurane is a highly fluorinated (seven fluorines) derivative from methyl isopropyl ether. It was synthesized in the 1960s and later introduced in routine clinical anesthetic practice in 1992, wherein it was presented as a highly volatile nonflammable liquid. Chlorine-free sevoflurane and desflurane are markedly less pungent than other chlorine-containing haloethers, namely, methoxyflurane, enflurane, and isoflurane (Fig. 1; Wang, Ming, & Zhang, 2020).

So far, sevoflurane has been used in millions of anesthetic procedures in many countries around the world, showing a positive benefit-risk ratio over a broad spectrum of patients (Brioni, Varughese, Ahmed, & Bein, 2017).

Treatments, Mechanisms, and Adverse Reactions of Anesthetics and Analgesics. https://doi.org/10.1016/B978-0-12-820237-1.00017-X
Copyright © 2022 Elsevier Inc. All rights reserved.

166 PART | II Mechanisms of action of drugs

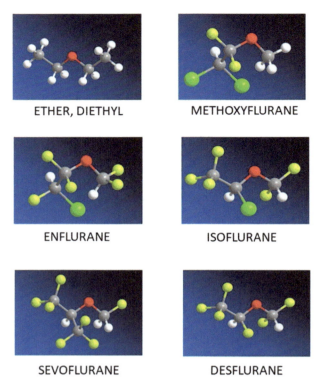

FIG. 1 Molecular structures of diethyl ether and haloethers. *White balls*: hydrogen. *Gray balls*: carbon. *Red balls*: oxygen. *Yellow balls*: fluorine. *Green balls*: chlorine. *(Courtesy of María Luisa González Moral, MD.)*

A great number of different molecular targets for sevoflurane has been identified, but the precise molecular mechanisms underlying their anesthetic effects have not yet been fully understood (Perouansky, Pearce, Hemmings, & Franks, 2020). Besides hypnosis, sevoflurane also produces vasodilatory and bronchodilatory effects mediated through smooth muscle relaxation (Thorlacius & Bodelsson, 2004).

Analgesic effects of sevoflurane and haloethers on the central nervous system

In addition to their well-known hypnotic effects, experiments conducted on animal models strongly suggest that sevoflurane and haloethers also exert an analgesic/anesthetic effect on the central nervous system. In rats with a transected spinal cord, sevoflurane administered by inhalation at anesthetic doses was found to produce a profound analgesic effect at the level of the spinal cord, measured as a reduction in nociceptive C response (Wang, Deng, You, Liu, & Zhao, 2005). Experiments in which sevoflurane was injected intrathecally showed a complete anesthetic effect. For instance, a subarachnoid injection of neat sevoflurane produced reversible, dose-dependent motor and sensitive blocks in dogs (García-Fernández et al., 2005). In a rat model, a subarachnoid injection of a lipidic-emulsified formulation of sevoflurane induced spinal anesthesia as indicated by the successful motor and sensitive blockade developed. Isoflurane and enflurane, other haloethers, elicited the same response, indicating that this is a class effect (Guo et al., 2014).

Peripheral analgesic effects of sevoflurane and haloethers

Drug repurposing of sevoflurane as a topical analgesic had a serendipitous origin; the author intended to study the in vitro antimicrobial properties of liquid sevoflurane when a drop accidentally touched the author's lip, producing an unexpected immediate and complete anesthetic effect on the lip (and a mild headache). This personal experience was quite surprising considering that classic anesthetic texts stated that haloethers were devoid of any clinically relevant peripheral analgesic effects (Koblin, 2005). Nevertheless, a further thorough search in the literature revealed some interesting papers showing a real analgesic peripheral effect of haloethers when they were administered topically, intradermally, and subcutaneously.

Forearms of volunteers were covered with cotton soaked in isoflurane or sevoflurane, and then covered with plastic to prevent drugs from volatilization; compared to contralateral forearms that served as controls, a mild analgesic effect was found after provoking mechanical and electrical noxious stimuli (Fassoulaki, Sarantopoulos, Karabinis, & Derveniotis, 1998; Fassoulaki, Skouteri, Siafaka, & Sarantopoulos, 2005).

An intradermal injection of microdroplets encapsulating different drugs, including diethyl ether and the haloethers methoxyflurane, enflurane, and isoflurane, produced reversible local anesthesia in a rat model of pain caused by a tail clamp and electrical stimulus (Haynes & Kirkpatrick, 1985).

A subcutaneous injection of neat enflurane and isoflurane produced a concentration-dependent cutaneous analgesic effect, which was of intermediate intensity compared to that produced by lidocaine and prilocaine in a rat model of pain caused by a pin-prick test (Chu et al., 2008). The authors concluded that inhaled anesthetics had a direct analgesic effect on skin.

In the author's opinion, haloethers do not attain sufficient concentration at nerve endings to noticeably block nervous transmission when administered by inhalation. On the contrary, local administration exposes the nerve endings to high enough concentrations to completely block nervous transmission.

Concerning the molecular mechanism of peripheral action, it remains unknown. This is logical, as haloethers were believed to lack any important peripheral effects (Koblin, 2005); therefore, studies focused on assessing such effects have not yet been conducted. It would be a hard-to-reach task, as cutaneous pain is mediated by a great variety of receptors (Benarroch, 2015). Studies conducted to identify the molecular mechanisms underlying the hypnotic effect found that haloethers act on some of these receptors as voltage-gated Na^+, K^+, and Ca^{2+} channels (Koblin, 2005) and more recently on sensory transient receptor (TRP) channels (Caterina & Pang, 2016). TRP channels are involved in different skin biological sensory processes, such as acute pain and acute itch, and in some pathophysiological processes, such as inflammatory pain, neuropathic pain, and different forms of allodynia and hyperalgesia (Caterina & Pang, 2016), which make them an interesting target to explain the clinical effects of topical sevoflurane on painful wounds. As stated, it has been recently found that sevoflurane inhibits the thermosensitive nociceptor ion channel transient receptor potential melastatin 3 (TRPM3) (Kelemen et al., 2020). Clearly, further research in this field is needed to ascertain, if possible, the precise peripheral mechanism of analgesic effects.

Off-label use of topical sevoflurane on painful chronic wounds

Since the first report in 2011 supporting drug repurposing of sevoflurane as a topical analgesic (Gerónimo-Pardo, Martínez-Monsalve, & Martínez-Serrano, 2011), 20 clinical cases (Table 1) and 6 small series of cases (Table 2) have been reported, accounting for approximately 2000 applications in 270 patients. Such a body of evidence makes it possible to summarize the key features of the usual analgesic pattern.

Analgesic profile

The analgesic effect was very reliable since it appeared in all but 1 of the 270 treated patients; the exception was a young male patient suffering from a pressure ulcer in the context of a neuromuscular disease of unknown etiology (Table 2, series #2).

Topical sevoflurane typically produced a rapid, intense, and long-lasting analgesic effect (Fig. 2).

Latency was very short, usually less than 10 min (Tables 1 and 2), which is very convenient; although not assessed, topical sevoflurane seems to be a better alternative to perform debridements than EMLA, especially in overloaded workplaces.

The analgesic effect was usually very intense, which allowed to perform the majority of debridements planned using sevoflurane as the only analgesic agent (Tables 1 and 2). The reported sharp decrease in pain intensity from more than 8 points to less than 1 in a few minutes or even a few seconds deserves to be highlighted (Table 1, cases #07, #09, #10, #18, #19).

The analgesic effect usually lasted for 8–12 h, which makes sevoflurane suitable for treating persistent pain, but the duration of analgesia was quite variable as pain relief ranged from a few hours (Table 2, series #5) to more than 24 hours (Table 1, cases #07, #18, #19).

Analgesic effect on different types of wounds and pains

So far, the analgesic effect has been mainly reported not only for leg ulcers but also for diabetic foot, neoplastic wounds, pressure ulcers, and postoperative wounds (Tables 1 and 2). Remarkably, some of the patients successfully treated suffered

TABLE 1 Topical sevoflurane for painful wounds: Case reports.

Order and references	Patient and wound	Effectiveness	Safety	Comment
#01/Gerónimo-Pardo et al. (2011)	76-year-old woman with a painful left leg ulcer (and a painless right one) of somatic characteristics	Systemic analgesics, epidural analgesia, and EMLA had failed; opioids toxicity. Immediate (2 min), intense (from 8 to 4 NRS points), and long-lasting (12 h) analgesia for 16 days with sevoflurane (10 mL/d)	Pruritus. No systemic effects	First case reporting topical sevoflurane as a rescue analgesic. Ulcer healed after 16 days
#02/Martínez-Monsalve and Gerónimo-Pardo (2011)	73-year-old man with an extremely painful ischemic leg ulcer	Previous opioids toxicity. Intense and long-lasting analgesia for 3 days with sevoflurane (10 mL/d), allowing cleansing and even sharp debridement (complete anesthesia in 10 min)	Mild discomfort during sharp debridement	Patient died from heart attack (antecedent of severe ischemic cardiopathy) deemed unrelated to sevoflurane
#03/Rueda-Martínez, Gerónimo-Pardo, Martínez-Monsalve, and Martínez Serrano (2014)	43-year-old immunocompromised man with a postoperative wound infected by resistant *P. aeruginosa* and sensitive *S. aureus*	Wound closed after 4 weekly irrigations with sevoflurane (5 mL). Immediate (<2 min) and long-lasting (8–10 h) analgesic effect, intense enough to allow for debridement	Neither local nor systemic adverse effects	Sevoflurane successfully tried as the only antimicrobial agent against resistant bacteria. Mood and physical conditions improved
#04/Ferrara, Domingo-Chiva, Selva-Sevilla, Campos-García, and Gerónimo-Pardo (2016)	61-year-old man with recurrent frontal epidural abscess; culture positive for *E. coli*	Five previous surgical drainages Epidural abscess disappeared and wound healed after 8 weeks of daily sevoflurane (5 mL). Analgesic effect also evidenced, but details not provided	Neither local nor systemic adverse effects	Sevoflurane successfully tried as the only antimicrobial agent. New surgery (craniotomy) avoided
#05/Imbernón et al. (2016)	73-year-old woman with several painful leg ulcers; culture positive for MRSA	Rapid (10 min), intense, and long-lasting (8 h) analgesic effect with sevoflurane (10 mL)	Erythema and itching on wounds edges	Most of the ulcers healed in 21 days (associated systemic antibiotics)
#06/Fernández-Ginés, Cortiñas-Sáenz, Fernández-Sánchez, and Morales-Molina (2017)	62-year-old woman with a painful leg ulcer. Last will due to terminal oncologic disease prevented invasive measures	Adverse effects due to high dosage of ineffective systemic analgesics Rapid (3 min), intense (allowed cleansings), and long-lasting analgesic effects with sevoflurane (6 mL/12 h) at home for 35 days; marked reduction in analgesic consumption and associated adverse effects	Irritative itching on wound edges	Trend to ulcer healing. Patient regained quality of life for her last days
#07/Amores-Valenciano, Navarro-Carrillo, Romero-Cebrián, and Gerónimo-Pardo (2018)	84-year-old man with leg ulcers due to congestive heart failure. Pain was somatic and neuropathic	Quality-of-life-interfering adverse effects due to high dosage of ineffective systemic analgesics. Immediate (<1 min), intense (from 10 to 2 NRS points), and long-lasting (at least 24 h, until next dose) analgesic effects with sevoflurane, treated for 12 months	Neither local nor systemic adverse effects	Ulcers progressively healed. Patient regained quality of life
#08/Fernández-Ginés, Cortiñas-Sáenz, et al. (2019)	39-year-old male with painful inguinal ulcers secondary to invasive penile carcinoma. Pain was somatic and neuropathic	Adverse effects from an ineffective high dose of opioids (800 mg/d of equivalent morphine). Rapid (4 min), intense (from 9 to 3 NRS points), and long-lasting (>24 h, until next daily doses) analgesic effects with sevoflurane (40 mL then increased to 120 mL/d due to tumoral progression), treated for 100 days; marked reduction in analgesic consumption and associated adverse effects	Reddened itchy zone surrounding the application	Highest dose reported. Sevoflurane stopped after being referred to the palliative care unit for terminal sedation

#09/Gerónimo-Pardo and Jiménez-Roldán (2019)	81-year-old woman with a vasculopathic leg ulcer. Pain was somatic and neuropathic	Effective systemic analgesics, but they were discontinued due to adverse effects Rapid (5 min), complete (from 10 to 0 NRS points), and long-lasting (8–10 h) analgesic effect with sevoflurane (5 mL), for 3 months	Neither local nor systemic adverse effects	Sevoflurane discontinued after ulcer healing
#10/Padilla-del Rey, Gerónimo-Pardo, García-Fernández, and Cartagena-Sevilla (2019)	51-year-old man with diabetic foot. Pain was mainly neuropathic	Foot amputation was planned due to failure of all analgesic measures, including invasive techniques provided in a pain clinic Rapid (10 min), intense (from 9 to 1 NRS points), and long-lasting (14 h) analgesic effect with sevoflurane (10 mL) for 40 days (ulcer healed)	Neither local nor systemic adverse effects	Amputation avoided in extremis because sevoflurane was tried the evening before Remarkable change in prognosis and in the subsequent quality of life
#11/Quintana-Castanedo, Recarte-Marín, Pérez-Jerónimo, Conde-Montero, and de la Cueva-Dobao (2019)	28-year-old diabetic woman with extremely painful ulcerative necrobiosis lipoidica	Sevoflurane produced analgesia applied once at usual dose (1 mL/cm^2), but analgesic details not provided. Further analgesic effect after punch grafting	Not reported	Possible synergistic role of sevoflurane to prepare wound bed to better receive punch grafting
#12/Lledó-Carballo et al. (2019)	81-year-old woman with a chronic leg ulcer; culture positive for *E. cloacae*	Intense (from 7 to 3 NRS points) analgesic effect for three weekly applications at usual dose (1 mL/cm^2)	Neither local nor systemic adverse effects	Devitalized wound bed improved for punch grafting after sevoflurane. Possible synergistic role of sevoflurane to prepare wound bed to better receive punch grafting
#13/Gencay (2019)	76-year-old woman with a pressure ulcer; *P. aeruginosa* in culture	Culture negative after 30 days with sevoflurane at usual dose (1 mL/cm^2), trend to epithelization	Not reported	Further sevoflurane planned
#14/Navarro-Buendía, Satorres-Pérez, Jiménez-Ruíz, and Alberdi-Bellón (2020)	86-year-old woman with venous leg ulcers; *P. aeruginosa* in culture. Pain was somatic and neuropathic	Systemic analgesics had failed Details of dose and analgesic effect not provided, except for reducing pain from 9 to 6 NRS points. Treated for 2 months	Neither local nor systemic adverse effects	Sevoflurane tried both as an analgesic and as an antimicrobial agent. Great improvement in quality of life
#15/Losa-Palacios, Achaerandio-de Nova, Restrepo-Pérez, and Gerónimo-Pardo (2019)	41-year-old homeless man with an infected postoperative elbow wound containing osteosynthesis material	Wound covered with plastic; sevoflurane (4–5 mL) was injected and then aspirated after 4–5 min. Wound healed in 35 days without antibiotics, osteosynthesis material preserved; also, ambulatory negative-pressure wound therapy for 10 days	Neither local nor systemic adverse effects	Treatment applied when possible in an uncooperative patient
#16/Losa-Palacios, Achaerandio-de Nova, and Gerónimo-Pardo (2020)	83-year-old man with a postoperative wound containing osteosynthesis material infected by MRSA	Wound covered with plastic; sevoflurane (20 mL) was injected and then aspirated after 5 min. Analgesic effect intense enough to allow for sharp debridement. Sevoflurane applied every 2–4 days plus guided antibiotics for 2 months, until wound closure	Not reported	Successful initial evolution with sevoflurane, while antibiotics were useless against infecting agents. Further surgeries avoided

Continued

TABLE 1 Topical sevoflurane for painful wounds: Case reports—cont'd

Order and references	Patient and wound	Effectiveness	Safety	Comment
#17/Castillo-Carrión, Liria-Sánchez, and Gerónimo-Pardo (2020)	88-year-old, tetraparetic woman with a painless pressure ulcer; culture positive for *P. aeruginosa*	Flap surgery was planned Ulcer improved with sevoflurane at home (5 mL daily) and healed after 10 weeks	Neither local nor systemic adverse effects	In this case, sevoflurane was tried as a prohealing agent, and a surgery was avoided in a frail patient
#18/Fernández-Ginés, Cortiñas-Sáenz, Selva-Sevilla, and Gerónimo-Pardo (2020)	87-year-old oncologic woman with a painful leg ulcer. Pain was somatic and neuropathic	Systemic analgesics toxicity. Leg amputation refused Rapid (3 min), intense (from 9 to 2 NRS points), and long-lasting (>24 h) analgesic effect with daily sevoflurane (5 mL) at home for 35 days	Local erythema and itch, well tolerated	Ulcer healed. Noticeable improvement in quality of life until death due to gynecologic tumoral progression
#19/Pinar-Sánchez and Gerónimo-Pardo (2020)	88-year-old diabetic woman with bilateral painful leg ulcers. Pain was somatic and neuropathic. Culture positive for MRSA and *P. aeruginosa*	Systemic analgesics had failed and produced adverse effects Immediate (in seconds), intense (from 8 to 0 NRS points), and long-lasting (>24 h) analgesic effect with daily sevoflurane (20 mL) first admitted (21 days), then at home (15 days). Culture negative after 4 days. Ulcer healed	Mild and transient burning sensation at wound edges	Able to rest at night from the first day, noticeable improvement in quality of life
#20/Achaerandio-de Nova, Losa-Palacios, San Martín-Martínez, and Gerónimo-Pardo (2020)	74-year-old diabetic and obese woman with a postoperative skin wound infected by MRSA after knee arthroplasty	Wound was covered by plastic and sevoflurane (5 mL) was injected and then aspirated after 5–10 min. Sevoflurane daily plus guided antibiotics for 6 weeks, until wound closure	Neither local nor systemic adverse effects	Successful initial evolution with sevoflurane, while antibiotics were useless against the infecting agent. Further surgeries avoided

Reports appear in chronological order.
Created by the author.

TABLE 2 Topical sevoflurane for painful wounds: Series of cases.

Order and reference	Patients and wounds	Effectiveness	Safety
#1/Martínez-Monsalve and Gerónimo-Pardo (2013)	Sevoflurane applied to 9 outpatients suffering from chronic venous ulcers, treated in a weekly basis	Significant ($P<.001$) before/after pain reduction sevoflurane (from 7.4 ± 0.5 to 2.1 ± 0.6 NRS points for the first 9 cleansings; from 7.2 ± 1.3 to 1.1 ± 0.6 NRS points for further 67 pooled debridements) Four ulcers healed after 4, 18, 54, and 154 weeks High mean patients' satisfaction (8.8 ± 0.7 arbitrary points out of 10)	One-third suffered from itching or erythema
#2/Fernández-Ginés, Cortiñas-Sáenz, Mateo-Carrasco, et al. (2017)	Sevoflurane applied to 10 outpatients with venous leg ulcers or pressure ulcers; comparison with 5 patients under standard analgesic care	One patient failed sevoflurane. For the rest, clinically significant reduction in pain intensity from 9 points to 2 or less from the 1st day to the 90th day, paralleled with a reduction in morphine consumption from 100 mg/d at the start to 20 mg/d from days 30 to 90. Nearly no changes in control group	Mild reddening and pruritus in 4 patients One death at day 64 in the sevoflurane group was deemed unrelated to treatment (history of previous ischemic cardiomyopathy)
#3a/Imbernón-Moya et al. (2017a) #3b/Imbernón-Moya et al. (2017b) #3c/Imbernón-Moya et al. (2018)	Sevoflurane applied to 30 outpatients suffering from venous ulcers, treated thrice weekly	Short latency (mean 3.9 ± 1.5 min) Intense analgesic effect at cleaning from nearly 9 VAS points at baseline to less than 4 points at the 1st cleaning, and less than 1 point at 12th Long-lasting effects (mean 12 ± 2.9 h, range 8–18 h) All patients significantly reduced analgesic consumption. Significant improvement in quality of life (Charing Cross Venous Ulcer Questionnaire: 83 ± 14 points at baseline, 50 ± 14 points at the 12th visit) and functional capacity (Barthel: 82 ± 13 points at baseline, increased to 91 ± 12 points at the 12th visit) Significant reduction in ulcer size to the half (8.4 ± 9.7 cm^2 at baseline, 4.2 ± 5.4 cm^2 at the 12th visit)	Mild and transient pruritus ($n=5$), erythema ($n=2$), heat ($n=3$), irritative dermatitis ($n=1$) No systemic effects
#4/López-Riascos, Rivas-Ramírez, and Carrillo-Torres (2019)	Sevoflurane applied to 30 patients with venous or pressure ulcers to perform sharp debridement	Sevoflurane doses: 2.1 mL/cm^2. Short latency (<5 min) in 93%. Mild pain in 67% during debridement and 97% post procedure	No adverse effects
#5/ Martínez-Monsalve, Selva-Sevilla, and Gerónimo-Pardo (2019)	Sevoflurane applied to 152 inpatients to perform sharp debridement of ulcers of vascular (72%; venous, arterial, and mixed), systemic and infectious etiologies	Short latency (median 2 min). Intense, long-lasting analgesic effect (Fig. 2) Sharp debridement fully achieved in a high percentage of patients (93%), most of them (85%) after a single irrigation High median patients' global satisfaction (8 arbitrary points out of 10)	Transient slight of mild pruritus at the wound edges in 34% Rebound pain in 7%
#6/Fernández-Ginés, Cortiñas-Sáenz, Agudo-Ponce, et al. (2020)	Sevoflurane applied to 39 critically ill patients with painful pressure ulcers, comparison with 73 patients under iv opioids	Persistent pain (24 matched pairs): pain was significantly more reduced with sevoflurane from the first to the fourth week Debridement (21-sevoflurane; 19-opioids): shorter latency for sevoflurane, similar effectiveness, postprocedural analgesia significantly longer for sevoflurane (7.3 ± 2.3 vs 1.6 ± 1.4 h)	Slight burning sensation, erythema, and pruritus with sevoflurane in 3 patients. No systemic effects for sevoflurane (50% for opioids)

Reports appear in chronological order.
Created by the author.

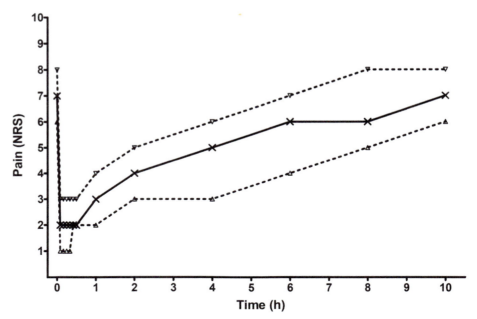

FIG. 2 Typical pain response to topical sevoflurane on painful wounds. Median values of pain over time corresponding to 152 patients suffering from painful vascular ulcers who were treated with topical sevoflurane to perform debridements. Latency was very short (2 min); the analgesic effect was intense (from 7 to 2 NRS points of pain) and most debridements could be performed; pain intensity progressively returned to baseline over the 10 h assessed. *NRS*, numerical rating scale. *(From Martínez-Monsalve, A., Selva-Sevilla, C., & Gerónimo-Pardo, M. (2019). Analgesic effectiveness of topical sevoflurane to perform sharp debridement of painful wounds. Journal of Vascular Surgery, 69, 1532–1537, with permission.)*

from neuropathic pain (Table 1, cases #07, #08, #09, #10, #18). Sevoflurane would be expected to act on all kind of wounds, regardless of their etiology, provided the free nerve endings could be reached.

Quality of life and opioid-sparing effect

Quality of life was noticeably improved for 30 patients suffering from painful venous leg ulcers, which received a course of topical sevoflurane thrice weekly for a month (Imbernón-Moya et al., 2017a). Remarkably, quality of death was also improved for palliative oncologic patients suffering from painful neoplastic or leg ulcers (Table 1, cases #06, #18); the latter discarded pain-driven suicide ideation.

Quality of life improved due to the analgesic effect and also due to the reduction or disappearance of adverse effects caused by systemic analgesics after tapering their dosages (Tables 1 and 2). The opioid-sparing effect (Table 1, cases #02, #06, #07, #08, #18) is remarkably considering the current worldwide opioid epidemic context (Furlan, Harvey, & Chadha, 2020); also, contrary to opioids, neither tolerance nor dependence phenomena have been reported in patients treated for several months with topical sevoflurane (Table 1, cases #08, #09; Table 2, series #2).

Safety issues of topical sevoflurane

Local adverse effects

The most frequently reported unwanted effect is mild and transient itching at wound edges, which is usually well tolerated (Tables 1 and 2).

Erythema is also frequently reported, but in the author's opinion, it is more a desirable effect that an unwanted one, since it reflects the vasodilatory action of sevoflurane on small vessels and capillaries, which may be beneficial to cicatrization (see below).

Systemic adverse effects for patients

In the author's experience, health workers are mainly concerned about the possibility of the patient falling asleep, but patients are protected by the physiopathology of chronic wounds and sevoflurane volatility. As blood supply to chronic wounds is usually impaired, drugs are slowly absorbed into the systemic circulation, yielding low venous concentrations. Once venous blood reaches alveolar circulation, volatile sevoflurane crosses from the alveolar capillaries to the alveoli to be exhaled with normal respiration, which was defined as "pulmonary first pass" (Gerónimo-Pardo & Cortiñas-Sáenz, 2018), thus making the risk for hypnosis extremely low; in fact, it has not been reported so far.

Systemic adverse effects for health workers

Occupational exposure to topical sevoflurane is a field for research, since only a small research has been conducted so far. Three patients were treated with 10–20 mL of sevoflurane in a big closed room. Not surprisingly, levels measured in the passive badges attached to the nurse's white coat were higher than those in badges placed around the room, but all levels were below the limits established in the few countries that have been regulating sevoflurane exposure so far (Fernández-Ginés, Selva-Sevilla, Cortiñas-Sáenz, & Gerónimo-Pardo, 2019). Nevertheless, as occupational exposure to sevoflurane in the operating room has been advocated as a cause of unspecific unwanted symptoms, such as fatigue, dizziness, discomfort, or headache, health workers are advised to use some safety measures to minimize occupational exposition to topical sevoflurane, such as treating patients in well-ventilated rooms (Imbernón-Moya, Ortiz-de Frutos, Sanjuan-Alvarez, & Portero-Sanchez, 2018).

Greenhouse effect

All inhalational anesthetics have a potential for greenhouse effect. Ether and CO_2 share a 20-year global warming potential (GWP_{20}) of 1, but values for sevoflurane, isoflurane, and desflurane are higher (349, 1401, and 3714, respectively) (Ryan & Nielsen, 2010), thus making sevoflurane the less harmful haloether. Obviously, the lesser the dose applied to a wound the better it is for the planet.

Dosages and methods of administration

The first dose reported was 1 mL/cm^2 (Table 1, cases #01, #02; Table 2, series #1), but the author reliably knows it was an arbitrary decision; therefore, physicians should be free to adapt doses to every individual wound. As a general rule, the cleaner the wound bed the lesser the dose required.

Sometimes new doses are required to progress in debridement as the analgesic effect was restrained to superficial layers due to volatilization (Table 1, case #2). Covering the wound with plastic to inject sevoflurane into the created greenhouse would keep it retained until further aspiration (Fig. 3), allowing for a more intense and deeper analgesic effect (and lessening room pollution). As for the case illustrated in Fig. 3, the "greenhouse technique" has proven useful in enhancing the antimicrobial effect in other postoperative wounds with infected osteosynthesis material (Table 1, cases #15, #16, #20).

Concerning easy-to-handle, nonliquid new formulations, the only case reported so far involved a marathon runner suffering from plantar fasciitis who was successfully treated with a 0.5% gel formulation (Fernández-Ginés, Cortiñas-Sáenz, Navajas-Gómez de Aranda, & Sierra-García, 2018). Had the gel proved useful to wounds, some advantages would arise, such as easier application and lower occupational exposure; besides, a 0.5% formulation will be more respectful of the environment than will a 100% formulation.

Economic implications

A hospital-restricted 250-mL bottle of sevoflurane costs approximately €75 to the hospital pharmacy (expressed in 2020 current prices), thus making the usual daily dose of 5–10 mL more expensive (€1.5–3) than several systemic analgesics taken together.

However, sevoflurane treatment proved to be cheaper than usual care when other direct costs were considered. In a preliminary study including out-hospital patients referred to a pain clinic for painful nonrevascularizable vascular ulcers, 30 patients receiving the best analgesic treatment were compared to 20 patients also treated with sevoflurane. After a year, the median costs for the sevoflurane group were more than 10-fold lower than those for the comparator, mainly due to savings in costs derived from less frequentation to primary care and emergency departments as well as less admittances to hospitals.

Finally, also considering the significantly better analgesic response, topical sevoflurane proved to be cost-effective (C. Selva-Sevilla, F. D. Fernández-Ginés, M. Cortiñas-Sáenz, & M. Gerónimo Pardo, personal communication, February 22, 2019). Decision policy makers should be interested in promoting further research in this field to confirm these promising results.

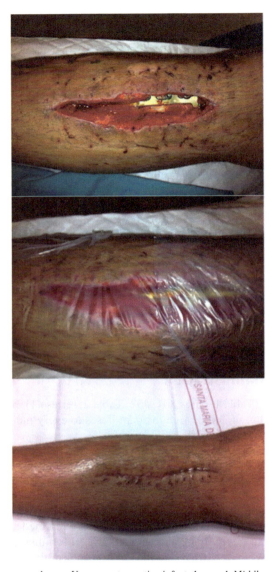

FIG. 3 Wound covered with plastic to make a greenhouse. *Upper*: postoperative infected wound. *Middle*: sevoflurane volatilization can be prevented by injecting the drug into a plastic greenhouse, which favors a longer and closer contact with the wound bed and the osteosynthesis material to enhance both analgesic and antimicrobial effects. *Down*: the wound healed after 15 days without removing the osteosynthesis material. *(Courtesy of Francisco Saura-Sánchez, MD.)*

Application to other areas

Sevoflurane as an antimicrobial agent for infected wounds

In vitro, ether and haloethers exhibited antimicrobial properties as nonconventional antibiotics (Kristiansen & Amaral, 1997), yet the mechanism of antimicrobial action has not been elucidated. On the basis of their solvent properties, a possible solvent effect on cell membranes has been theorized (Martínez-Serrano, Gerónimo-Pardo, Martínez-Monsalve, & Crespo-Sánchez, 2017), which would make them suitable for treating microorganisms resistant to conventional antibiotics.

In vivo, topical ether was successfully used 100 years ago as an antimicrobial agent for infected wounds (DeTarnowsky, 1914; Distaso & Bowen, 1917; Souligoux, 1913; Waterhouse, 1915) and even for acute peritonitis (Morestin, 1913; Sklass, 1913). Now, topical sevoflurane has also proven to be effective for treatment on wounds infected by both gram-positive and gram-negative organisms (Table 1, cases #03, #04, #12 to #16, #19, #20), including MRSA (Table 1, cases #16, #19, #20) and multidrug-resistant *P. aeruginosa* (Table 1, case #03). However, controversy exists as some authors argue against sevoflurane exhibiting a real antimicrobial effect on infected wounds (Fernández-Ginés, Cortiñas-Sáenz, Navajas-Gómez de Aranda, Yoldi Bocanegra, & Sierra-García, 2017). On the basis of successful personal experiences (Table 1, cases #03, #04, #15, #16, #19, #20), the author finds the benefit-risk favorable to try sevoflurane for infected wounds, especially when dealing with resistant organisms.

Sevoflurane as a prohealing agent for hard-to-heal wounds

Some treated ulcers healed after application of topical sevoflurane (Tables 1 & 2), but attributing a direct prohealing effect to sevoflurane is still hazardous. Sevoflurane-driven pain relief allowed patients suffering from venous leg ulcers to regain their normal life activities and improved their mood (Table 2, series #3a), which could have played an important indirect role in wound healing. Conversely, topical sevoflurane was specifically applied as a prohealing agent on an 88-year-old tetraparetic patient suffering from a painless, hard-to-heal pressure ulcer, scheduled for plastic surgery. The ulcer healed after 10 weeks, and neither pain relief nor regaining daily activities played any role in this case, but a confounding antimicrobial effect could not be discarded since the ulcer was superinfected by *P. aeruginosa* (Table 1, case #17).

Apart from the analgesic and antimicrobial effects as confounding factors, sevoflurane could promote wound healing mediated by its well-known vasodilatory effect (Thorlacius & Bodelsson, 2004), which could explain the occasionally reported hyperemic appearance (Table 1, cases #05, #08, #18). Maybe the supply of nutrients to the wound would be increased through dilated vessels (Castillo-Carrión et al., 2020), thus favoring healing. In line with this, a 2% ether solution was found to cause a rapid growth of granulation tissue on acute war wounds; the most striking histological feature was a marked production of fibroblasts and new blood vessels (Distaso & Bowen, 1917). Similarly, the granulation tissue of wounds treated with topical sevoflurane proliferates rapidly and exhibits a characteristic wine-red color (Fig. 4).

The vasodilatory effect would make sevoflurane especially useful for ulcers of vasculopathic etiology (Gerónimo-Pardo & Jiménez-Roldán, 2019), and another interesting advantage comes from the possibility of increasing the rate of success of skin grafting, as sevoflurane would prepare the wound bed in a nonaggressive manner to better receive the graft (Quintana-Castanedo et al., 2019).

Topical sevoflurane for painful conditions different from wounds

Mucosa are plenty of easy-to-access free nerve endings, as the author's lip proved, making conditions affecting mucosa suitable to be treated. For instance, sevoflurane could be useful for conjunctival diseases; in line with this, ether was reported to be a painless topical antiviral agent to treat herpetic keratitis (Kronenberg, 1950).

The author accumulates experience with some novel treatments. For instance, male patients occasionally suffer from severe agitation in the postanesthetic recovery room due to glans irritation caused by bladder catheters; glans irritation immediately ceases after an injection of 1 mL of sevoflurane into the urethra through a small catheter, allowing patients to rest, which is a key feature regarding the immediate postoperative period (Gerónimo-Pardo, 2021). In addition, a patient suffering from an extremely painful chronic anal fissure experienced an immediate, complete, and sustained analgesic effect after applying 3 mL of sevoflurane into the rectum; being flatulent for 1–2 h was the only unwanted effect. The fissure healed after 1 month of applications every two or three days (Cifuentes-Tébar, Rueda-Martínez, Selva-Sevilla, & Gerónimo-Pardo, 2021). Similarly, rectal sevoflurane has been shown to be effective in treating pain and itching caused by hemorrhoids.

New formulations such as gels or patches would allow the expansion of indications of haloethers to painful medical conditions coursing with intact skin, as exemplified (Fernández-Ginés et al., 2018); the potential for development is enormous.

Other agents of interest

Sevoflurane is a halogenated ether derivative. After the communication of M. Souligoux (1913) about his pioneering experiences in treating infected wounds with topical ether (the author does not know the reason why Mr. Souligoux decided to try ether), this new method of treatment spread from France to other countries, such as the United Kingdom (Waterhouse, 1915) and the United States (DeTarnowsky, 1914). In spite of lacking systemic antibiotics and all modern facilities that are available nowadays, a pool of patients suffering from infections was reported to have been healed thanks to ether irrigations. Infected open wounds were very frequent in the context of the First World War (Distaso & Bowen, 1917; Waterhouse,

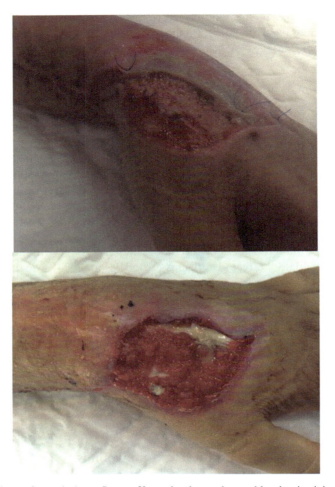

FIG. 4 Characteristic granulation tissue after topical sevoflurane. *Upper*: hand wound caused by abrasive injury showing a postoperative coverture defect. *Lower*: abundant wine-red granulation tissue spreading over the wound bed and covering tendons after 10 days of irrigations with 10 mL of sevoflurane. *(Courtesy of Sergio Losa-Palacios, MD.)*

1915), and septic arthritis were also successfully treated (Waterhouse, 1915), but the main interest was focused on the treatment of intraabdominal infections, where many "miraculous" healings were achieved (Waterhouse, 1915). Although less interesting for those authors, an analgesic effect for topical ether was also reported (Waterhouse, 1915). The author does not know why reports on this topical ether stopped a few years later. Maybe the information was lost after the First World War, or maybe the availability of modern systemic antibiotics turned this technique obsolete. In any way, nowadays, ether could be a valid topical alternative for the treatment of complicated wounds in developing countries, unable to get more expensive anesthetics, since it could be easily manufactured in the same country (Chang, Goldstein, Agarwal, & Swan, 2015).

Mini-dictionary of terms

Haloethers: A family of inhalational anesthetics obtained after halogenation of ether with either fluorine or chlorine.
Chronic leg ulcers: Ulcers that remained open for a long time, usually more than 4 weeks, regardless of the etiology, although leg ulcers are the most frequent ones.
Hard-to-heal ulcers: Ulcers that showed resistance to man-healing alternatives.
Somatic pain: Pain of an inflammatory nature that originated in damaged tissues and was conducted by the nervous system.
Neuropathic pain: Pain that originated in damaged nervous structures, typically in nerves or free nerve endings; thus, the nervous system serves as an origin of the pain as well as a conductor.

Refractory pain: Pain resistant to conventional systemic analgesics or even invasive techniques. It is frequently neuropathic in nature.

Key facts of painful chronic wounds

- Chronic wounds are a major health problem as patients can suffer from persistent pain and recurrent episodes of wound infection.
- Patients suffering from chronic wounds report pain as the worst symptom, which greatly interferes with their quality of life.
- Patients suffering from chronic wounds are usually elderly and fragile, which exposes them as being at high risk for developing adverse effects caused by systemic analgesics, mainly NSAIDs and opioids.
- There is a paucity of effective local analgesics both for persistent pain and pain associated with cleansing and dressing changes.
- In addition to increased pain, wound infection negatively affects patients' quality of life due to exudate and bad odor.
- Nonhealing wounds can be colonized or infected by resistant organisms after repeated antibiotic cycles.

Summary points

- This chapter is focused on sevoflurane, which is a fluorinated ether derivative widely used to provide general anesthesia by inhalation.
- The precise hypnotic mechanism of the action of sevoflurane and all other haloethers on the central nervous system remains elusive, as well as the analgesic mechanism on the spinal cord.
- Sevoflurane exhibits a good benefit-risk profile; in addition to hypnosis, other systemic effects include vasodilation and bronchodilation mediated by smooth muscle relaxation.
- Sevoflurane is formulated as a volatile liquid, which allows for its topical use in wounds.
- Sevoflurane irrigation on painful wounds produces a rapid, intense, and long-lasting analgesic/anesthetic effect; as for a hypnotic effect, the precise analgesic mechanism of action has not yet been elucidated.
- Current evidence suggests that topical sevoflurane may also have beneficial effects as an antimicrobial agent on infected wounds, which could be mediated by its solvent properties, and as a prohealing agent on chronic wounds, which could be mediated by its vasodilatory properties.
- As for inhalation, the benefit-risk profile seems to be beneficial for topical use as well, as reported local adverse effects on wounds are usually mild and transient, and no systemic effects have been reported.
- So far, topical sevoflurane has proven effective for both somatic and neuropathic pain and for wounds of diverse etiology.
- A preliminary economic evaluation strongly suggests that sevoflurane could be a cost-effective agent as a topical analgesic for out-patient treatment of painful nonrevascularizable leg ulcers.
- New formulations of sevoflurane could allow the expansion of its use as a topical analgesic to painful conditions other than wounds.

References

Achaerandio-de Nova, A., Losa-Palacios, S., San Martín-Martínez, A., & Gerónimo-Pardo, M. (2020). Topical sevoflurane for superficial MRSA infection after knee arthroplasty. Case report. *Revista Chilena de Ortopedia y Traumatología, 61*, 083–089 (Spanish).

Amores-Valenciano, P., Navarro-Carrillo, A., Romero-Cebrián, A., & Gerónimo-Pardo, M. (2018). Topical sevoflurane for rescue analgesia in refractory pain due to chronic venous ulcers. *Emergencias, 30*, 133–138 (Spanish).

Benarroch, E. E. (2015). Ion channels in nociceptors: Recent developments. *Neurology, 84*, 1153–1164.

Briggs, M., Nelson, E. A., & Martyn-St James, M. (2012). Topical agents or dressings for pain in venous leg ulcers. *Cochrane Database of Systematic Reviews, 11*. CD001177.

Brioni, J. D., Varughese, S., Ahmed, R., & Bein, B. (2017). A clinical review of inhalation anesthesia with sevoflurane: From early research to emerging topics. *Journal of Anesthesia, 31*, 764–778.

Castillo-Carrión, P., Liria-Sánchez, P. J., & Gerónimo-Pardo, M. (2020). Off-label topical sevoflurane as an alternative treatment for a pressure ulcer in a frail patient. *Gerokomos, 31*, 268–270 (Spanish).

Caterina, M. J., & Pang, Z. (2016). TRP channels in skin biology and pathophysiology. *Pharmaceuticals (Basel), 9*, 77.

Chang, C., Goldstein, E., Agarwal, N., & Swan, K. G. (2015). Ether in the developing world: Rethinking an abandoned agent. *BMC Anesthesiology, 15*, 149.

178 PART | II Mechanisms of action of drugs

Chu, C. C., Wu, S. Z., Su, W. L., Shieh, J. P., Kao, C. H., Ho, S. T., et al. (2008). Subcutaneous injection of inhaled anesthetics produces cutaneous analgesia. *Canadian Journal of Anaesthesia, 55*, 290–294.

Cifuentes-Tébar, J., Rueda-Martínez, J. L., Selva-Sevilla, C., & Gerónimo-Pardo, M. (2021). Analgesic effectiveness and improvement in quality of life after using topical sevoflurane for an extremely painful anal fissure. *Journal of Coloproctology, 41*, 206–209.

DeTarnowsky, G. (1914). Sulphuric ether lavage in infections: The Souligoux-Morestin method. *JAMA, 62*, 281–282.

Distaso, A., & Bowen, T. R. (1917). Auto-disinfection of wounds by the use of ether solution. *British Medical Journal, 1*, 259–261.

Fassoulaki, A., Sarantopoulos, C., Karabinis, G., & Derveniotis, C. (1998). Skin application of isoflurane attenuates the responses to a mechanical and an electrical stimulation. *Canadian Journal of Anaesthesia, 45*, 1151–1155.

Fassoulaki, A., Skouteri, I., Siafaka, I., & Sarantopoulos, C. (2005). Local application of volatile anesthetics attenuates the response to a mechanical stimulus in humans. *Canadian Journal of Anaesthesia, 52*, 951–957.

Fernández-Ginés, F. D., Cortiñas-Sáenz, M., Agudo-Ponce, D., Navajas-Gómez de Aranda, A., Morales-Molina, J. A., Fernández-Sánchez, C., et al. (2020). Pain reduction of topical sevoflurane vs intravenous opioids in pressure ulcers. *International Wound Journal, 17*, 83–90.

Fernández-Ginés, F. D., Cortiñas-Sáenz, M., Fernández-Sánchez, C., & Morales-Molina, J. A. (2017). Topical sevoflurane: A new palliative therapeutic option for skin ulcers. *Medicina Paliativa, 24*, 104–108 (Spanish).

Fernández-Ginés, F. D., Cortiñas-Sáenz, M., Mateo-Carrasco, H., Navajas-Gómez de Aranda, A., Navarro-Muñoz, E., Rodríguez-Carmona, R., et al. (2017). Efficacy and safety of topical sevoflurane in the treatment of chronic skin ulcers. *American Journal of Health-System Pharmacy*, e176–e182.

Fernández-Ginés, D., Cortiñas-Sáenz, M., Navajas-Gómez de Aranda, A., Navas-Martinez, M. C., Morales-Molina, J. A., Sierra-García, F., et al. (2019). Palliative analgesia with topical sevoflurane in cancer-related skin ulcers: A case report. *European Journal of Hospital Pharmacy, 26*, 229–232.

Fernández-Ginés, F. D., Cortiñas-Sáenz, M., Navajas-Gómez de Aranda, A., & Sierra-García, F. (2018). Topical sevoflurane: Analgesic management in a marathon runner with plantar fasciitis. *Sport Sciences for Health, 14*, 459–462.

Fernández-Ginés, F. D., Cortiñas-Sáenz, M., Navajas-Gómez de Aranda, A., Yoldi Bocanegra, R., & Sierra-García, F. (2017). Reply: To chronic venous ulcer treatment with topical sevoflurane by Imbernón et al. *International Wound Journal, 14*, 591.

Fernández-Ginés, D., Cortiñas-Sáenz, M., Selva-Sevilla, C., & Gerónimo-Pardo, M. (2020). Sevoflurane topical analgesia for intractable pain with suicidal ideation. Case report. *BMJ Supportive & Palliative Care*. https://doi.org/10.1136/bmjspcare-2019-002023.

Fernández-Ginés, D., Selva-Sevilla, C., Cortiñas-Sáenz, M., & Gerónimo-Pardo, M. (2019). Occupational exposure to sevoflurane following topical application to painful wounds. *La Medicina del Lavoro, 110*, 363–371.

Ferrara, P., Domingo-Chiva, E., Selva-Sevilla, C., Campos-García, J., & Gerónimo-Pardo, M. (2016). Irrigation with liquid sevoflurane and healing of a postoperative, recurrent epidural infection: A potential cost-saving alternative. *World Neurosurgery, 90*, 702.e1–702.e5.

Furlan, A. D., Harvey, A. M., & Chadha, R. (2020). Warning from Canada: Latin America, South Africa and India may face an opioid epidemic in the coming years. *Journal of Global Health, 10*(1), 010324.

García-Fernández, J., Parodi, E., García, P., Matute, E., Gómez-de Segura, I. A., Cediel, R., et al. (2005). Clinical actions of subarachnoid sevoflurane administration in vivo: A study in dogs. *British Journal of Anaesthesia, 95*, 530–534.

Gencay, I. (2019). Topical sevoflurane: An alternative treatment for pressure ulcers. *Journal of the College of Physicians and Surgeons–Pakistan, 29* (Suppl. 2), S92–S94.

Gerónimo-Pardo, M., & Cortiñas-Sáenz, M. (2018). Analgesic efficacy of topical sevoflurane on wounds. *Revista de la Sociedad Española del Dolor, 25*, 106–111 (Spanish, English).

Gerónimo-Pardo, M., & Jiménez-Roldán, C. (2019). Sevoflurane as a topical analgesic/anesthetic on painful chronic wounds: A new example of drug repositioning. *Actualidad en Farmacología y Terapéutica, 17*, 131–134 (Spanish).

Gerónimo-Pardo, M., Martínez-Monsalve, A., & Martínez-Serrano, M. (2011). Analgesic effect of topical sevoflurane on venous leg ulcer with intractable pain. *Phlébologie, 40*, 95–97.

Gerónimo-Pardo, M. (2021). Topical sevoflurane for immediate and long-lasting relief of extremely severe postoperative agitation caused by urethral irritation (Spanish/English). *Actas Urológicas Españolas, 45*, 175–176.

Guo, J., Zhou, C., Liang, P., Huang, H., Li, F., Chen, X., et al. (2014). Comparison of subarachnoid anesthetic effect of emulsified volatile anesthetics in rats. *International Journal of Clinical and Experimental Pathology, 7*, 8748–8755.

Haynes, D. H., & Kirkpatrick, A. F. (1985). Ultra-long-duration local anesthesia produced by injection of lecithin-coated methoxyflurane microdroplets. *Anesthesiology, 63*, 490–499.

Imbernón, A., Blázquez, C., Puebla, A., Churruca, M., Lobato, A., Martínez, M., et al. (2016). Chronic venous ulcer treatment with topical sevoflurane. *International Wound Journal, 13*, 1060–1062.

Imbernón-Moya, A., Ortiz-de Frutos, J., Sanjuan-Alvarez, M., & Portero-Sanchez, I. (2018). Application of topical sevoflurane before cleaning painful skin ulcers. *Actas Dermo-Sifiliográficas, 109*, 447–448 (Spanish).

Imbernón-Moya, A., Ortiz-de Frutos, F. J., Sanjuan-Alvarez, M., Portero-Sanchez, I., Merinero-Palomares, R., & Alcazar, V. (2017a). Pain, quality of life, and functional capacity with topical sevoflurane application for chronic venous ulcers: A retrospective clinical study. *EJVES Short Reports, 36*, 9–12.

Imbernón-Moya, A., Ortiz-De Frutos, F. J., Sanjuan-Alvarez, M., Portero-Sanchez, I., Merinero-Palomares, R., & Alcazar, V. (2017b). Healing of chronic venous ulcer with topical sevoflurane. *International Wound Journal, 14*, 1323–1326.

Imbernón-Moya, A., Ortiz-De Frutos, F. J., Sanjuan-Alvarez, M., Portero-Sanchez, I., Merinero-Palomares, R., & Alcazar, V. (2018). Pain and analgesic drugs in chronic venous ulcers with topical sevoflurane use. *Journal of Vascular Surgery, 68*, 830–835.

Kelemen, B., Lisztes, E., Vladár, A., Hanyicska, M., Almássy, J., Oláh, A., et al. (2020). Volatile anaesthetics inhibit the thermosensitive nociceptor ion channel transient receptor potential melastatin 3 (TRPM3). *Biochemical Pharmacology, 174*, 113826.

Koblin, D. D. (2005). Mechanism of action. In M. D. Miller (Ed.), *Miller's anesthesia* (6th ed., pp. 105–130). Philadelphia, PA: Churchill Livingstone/Elsevier.

Kristiansen, J. E., & Amaral, L. (1997). The potential management of resistant infections with non-antibiotics. *The Journal of Antimicrobial Chemotherapy, 40*, 319–327.

Kronenberg, B. (1950). A further report on the use of ether in the treatment of herpetic keratitis. *New York State Journal of Medicine, 50*, 2825–2826.

Lledó-Carballo, A., Horcajada-Reales, C., Crespo-Moreno, A., Martínez-López, M. E., Esteban-Garrido, E., Toral-Morillas, M., et al. (2019). Approach and treatment of a venous vascular ulcer with topical sevoflurane and autologous micrografts. *Heridas y Cicatrización, 3*, 24–28 (Spanish).

López-Riascos, S. D., Rivas-Ramírez, C., & Carrillo-Torres, O. (2019). Locally applied sevoflurane: An analgesic and therapeutic alternative in difficulty chronic skin ulcers. *Revista Mexicana de Anestesiología, 42*, 268–274 (Spanish).

Losa-Palacios, S., Achaerandio-de Nova, A., & Gerónimo-Pardo, M. (2020). Conservative multimodal management of osteosynthesis material in surgical wounds with polymicrobial superinfection, including methicillin-resistant *Staphylococcus aureus*. Clinical case. *Revista Española de Cirugía Ortopédica y Traumatología, 64*, 125–129 (Spanish).

Losa-Palacios, S., Achaerandio-de Nova, A., Restrepo-Pérez, M., & Gerónimo-Pardo, M. (2019). Uncooperative patient with an infected elbow osteosynthesis: Alternative management with topical sevoflurane. *Wounds, 31*, E68–E72.

Luchting, B., & Azad, S. C. (2019). Pain therapy for the elderly: Is opioid-free an option? *Current Opinion in Anaesthesiology, 32*, 86–91.

Martínez-Monsalve, A., & Gerónimo-Pardo, M. (2011). Analgesic effect of topical sevoflurane on ischaemic wound in a patient with ischaemic cardiopathy and respiratory failure secondary to morphine. *Heridas y Cicatrización, 6*, 46–49 (Spanish).

Martínez-Monsalve, A., & Gerónimo-Pardo, M. (2013). The analgesic effect of topical sevoflurane on painful varicose ulcers in ambulatory patients. *Heridas y Cicatrización, 1*, 16–19 (Spanish).

Martínez-Monsalve, A., Selva-Sevilla, C., & Gerónimo-Pardo, M. (2019). Analgesic effectiveness of topical sevoflurane to perform sharp debridement of painful wounds. *Journal of Vascular Surgery, 69*, 1532–1537.

Martínez-Serrano, M., Gerónimo-Pardo, M., Martínez-Monsalve, A., & Crespo-Sánchez, M. D. (2017). Antibacterial effect of sevoflurane and isoflurane. *Revista Española de Quimioterapia, 30*, 84–89.

Morestin, H. (1913). Eventration. Ulceration and rupture of the bag. From the omentum. Beginning peritonitis. Intervention. Washing the peritoneum with ether. Cure of eventration. Healing (French). *Bulletin et mémoires de la Société des chirurgiens de Paris*, (séance du 12 février), 284–287.

Navarro-Buendía, G. A., Satorres-Pérez, M., Jiménez-Ruíz, L., & Alberdi-Bellón, M. (2020). Efects of topical sevoflurane in the treatment of vascular ulcers. Report of a clinical case. *Hospital a Domicilio, 4*, 45–50.

Padilla-del Rey, M. L., Gerónimo-Pardo, M., García-Fernández, M. R., & Cartagena-Sevilla, J. (2019). Painful diabetic foot ulcer amputation avoided thanks to topical sevoflurane. *Revista de la Sociedad Española del Dolor, 26*, 253 (Spanish/English).

Perouansky, M., Pearce, R. A., Hemmings, H. C., & Franks, N. P. (2020). Inhaled anesthetics: Mechanisms of action. In M. Gropper, L. Eriksson, L. Fleisher, J. Wiener-Kronish, N. Cohen, & K. Leslie (Eds.), *Miller's anesthesia* (9th ed., pp. 487–508). Philadelphia, PA: Churchill Livingstone/Elsevier.

Pinar-Sánchez, J., & Gerónimo-Pardo, M. (2020). Topical sevoflurane as rescue therapy in a frail patient suffering from leg ulcers causing refractory pain and superinfected by methicillin-resistant *Staphylococcus aureus* and sensitive *Pseudomonas aeruginosa*. Case report. *Medicina Paliativa, 27*, 141–145 (Spanish).

Quintana-Castanedo, L., Recarte-Marín, L., Pérez-Jerónimo, L., Conde-Montero, E., & de la Cueva-Dobao, P. (2019). Ulcerative necrobiosis lipoidica diabeticorum successfully treated with topical sevoflurane and punch grafting. *International Wound Journal, 16*, 1234–1236.

Rueda-Martínez, J. L., Gerónimo-Pardo, M., Martínez-Monsalve, A., & Martínez Serrano, M. (2014). Topical sevoflurane and healing of a post–operative wound superinfected by multi–drug–resistant *Pseudomonas aeruginosa* and sensitive *Staphylococcus aureus* in an immunocompromised patient. *Surgical Infections, 15*, 843–846.

Ryan, S. M., & Nielsen, C. J. (2010). Global warming potential of inhaled anesthetics: Application to clinical use. *Anesthesia and Analgesia, 111*, 92–98.

Sklass, A. (1913). *The use of ether in cases of generalized peritonitis* (Thesis). Lyon: Université de Lyon (French).

Souligoux, C. (1913). On the use of ether in infections (French). *Bulletin et mémoires de la Société des chirurgiens de Paris*, (séance du 19 février), 293–296.

Thorlacius, K., & Bodelsson, M. (2004). Sevoflurane promotes endothelium-dependent smooth muscle relaxation in isolated human omental arteries and veins. *Anesthesia and Analgesia, 99*, 423–428.

Wang, Y., Deng, X., You, X., Liu, S., & Zhao, Z. (2005). Involvement of GABA and opioid peptide receptors in sevoflurane-induced antinociception in rat spinal cord. *Acta Pharmacologica Sinica, 26*, 1045–1048.

Wang, Y., Ming, X. X., & Zhang, C. P. (2020). Fluorine-containing inhalation anesthetics: Chemistry, properties, and pharmacologies. *Current Medicinal Chemistry*. https://doi.org/10.2174/0929867326666191003155703.

Waterhouse, H. F. (1915). The employment of ether in surgical therapeusis, with special reference to its use in septic peritonitis, pyogenic arthritis, and gunshot wounds. *British Medical Journal, 1*, 233–237.

Chapter 18

Tramadol as an analgesic

Ayman M. Mahmoud[a] and Emad H.M. Hassanein[b]

[a]*Physiology Division, Zoology Department, Faculty of Science, Beni-Suef University, Beni Suef, Egypt,* [b]*Department of Pharmacology & Toxicology, Faculty of Pharmacy, Al-Azhar University-Assiut Branch, Assiut, Egypt*

Introduction

Tramadol is an atypical opioid and one of the commonly prescribed analgesics. Tramadol differs from typical opioid drugs such as morphine as it modulates the reuptake of monoamines (noradrenaline and serotonin (5HT)) at presynaptic terminals (Raffa, Friderichs, Reimann, et al., 1992). In 1962, studies have reported the potent analgesic effects of tramadol. In 1977, tramadol was approved in Germany and the Food and Drug Administration (FDA) has approved it in 1995. Importantly, extensive clinical studies reported the high efficacy of tramadol for the treatment of moderate to severe pain (Bono & Cuffari, 1997). Chemically, tramadol is 2-(dimethyl amino)-methyl)-1-(3′-methoxyphenyl) cyclohexanol and is a 4-phenyl-piperidine analogue of the opioid codeine. Commercially available tramadol is present in a mixture of both dextro (+) and levo (−) enantiomers (Fig. 1). Tramadol hydrochloride is actively soluble in either water or ethanol (Grond & Sablotzki, 2004). The natural origin of tramadol is still a great area of interest. It has been identified in *Nauclea latifolia* and it accounts <0.00002% (w/w) in its roots and bark (Kusari, Tatsimo, Zühlke, et al., 2014). However, most of the marketed tramadol is obtained from chemical synthesis as it is characterized by a simple reaction and has significant economic outcomes (Alvarado, Guzmán, Díaz, et al., 2005).

Tramadol is commercially available with 164 brands worldwide and is manufactured and marketed by 108 pharmaceutical companies. It is available in different dosage forms, including oral preparation (capsule, tablet and syrup) and topical preparation (cream, ointment and gel). Additionally, parenteral preparation of tramadol is also available and some formulae contains other drugs such as acetaminophen (Subedi, Bajaj, Kumar, et al., 2019).

Mechanism(s) of action

Central analgesic mechanism

In spite of the decades passed since its discovery, the exact mechanism of action of tramadol remains a great area of research interest. Some of its established effects are the ability to inhibit noradrenaline and 5HT reuptake and activate the μ-opioid receptors (MOR). It is present in a racemic mixture of the positive and negative enantiomers; both have significant synergism.

Tramadol activates the MOR weakly as compared with that of the morphine while its metabolite *O*-desmethyl tramadol (M1) showed more potent analgesic action (Ide, Minami, Ishihara, et al., 2006; Lai, Ma, Porreca, et al., 1996). Notably, (+)-tramadol has a greater affinity for MOR and inhibits 5HT uptake more efficient than the (−)-tramadol. On the other hand, (−)-tramadol inhibits noradrenaline uptake more potently than (+)-tramadol (Raffa et al., 1992; Raffa, Friderichs, Reimann, et al., 1993). Therefore, the potent analgesic effect of tramadol is mediated by its action on the central nervous system (CNS).

Peripheral local anesthetic mechanism

Tramadol has a local anesthetic effect mediated by its blocking action on K^+ channels (Mert, Gunes, Guven, et al., 2003).

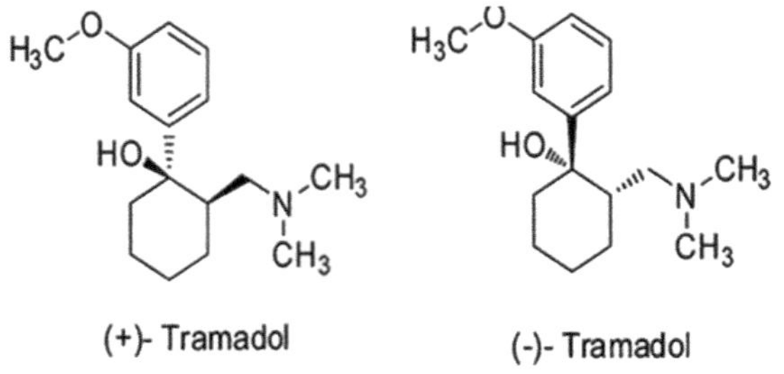

182 PART | II Mechanisms of action of drugs

FIG. 1 Isomers of tramadol.

Effect of tramadol on other receptors

Tramadol has effects on receptors rather than MOR such as cholinergic muscarinic; M1 and M3 (Shiraishi, Minami, Uezono, et al., 2001), cholinergic nicotinic $\alpha7$ (Shiraishi, Minami, Uezono, et al., 2002) and serotonergic; 5-HT2C receptors (Ogata, Minami, Uezono, et al., 2004). However, the impact of these receptors on the analgesic effect of tramadol is still unclear.

Pharmacokinetics

Absorption

Rapid absorption of tramadol was observed after oral administration and it reached a peak serum concentration after 2 h (Gillen, Haurand, Kobelt, et al., 2000). The oral bioavailability of tramadol is 70% following a single dose owing its first-pass effect while its 90%–100% bioavailability possibly due to saturation in phase I metabolism (Dayer, Desmeules, & Collart, 1997).

Distribution

The distribution of tramadol occurs rapidly around the whole body with a plasma protein binding of about 20%. The half-life of tramadol is 5.1 h after a single oral dose of 100 mg while that of the potent metabolite M1 is 9 h (Dayer et al., 1997).

Metabolism

Tramadol undergoes phase I and phase II metabolism. In phase I, tramadol is metabolized to M1, having both MOR and noradrenaline activities, by the cytochrome P450 isoenzyme CYP2D6. Since M1 has a potent analgesic effect, the pharmacological activity of tramadol depends greatly on CYP2D6. However, CYP2D6 genes are very polymorphic and hence the plasma levels of tramadol can either be reduced or increased leading to the difference in responses between patients receiving tramadol. Therefore, tramadol pharmacokinetics is highly dependent on the individual genotype (Gan, Ismail, Wan Adnan, et al., 2007). CYP2B6 and CYP3A4 are also involved and catalyze the production of other metabolites such as the inactive metabolite N-desmethyltramadol (M2). M1 and M2 metabolites undergo phase II metabolism producing N-didesmethyltramadol (M3), N,N,O-tridesmethyltramadol (M4), and N,O-didesmethyltramadol (M5) secondary metabolites which are further converted to inactive compounds through sulfation or glucuronidation (Subrahmanyam, Renwick, Walters, et al., 2001). Of note, M1 has a greater affinity for MOR than that of tramadol and hence contributes greatly in the central opioid-derived analgesic mechanism (Lai et al., 1996).

Excretion

Tramadol is excreted mainly through the kidney as demonstrated using radioactive isotopes where approx. 90% of the radio-labeled tramadol was excreted in the urine and only residual activity recovered in the feces. About 10%–30% of tramadol is excreted as unchanged, whereas 60% is excreted as a metabolite (Lintz, Barth, Osterloh, et al., 1986).

Therapeutic uses

Tramadol is an effective analgesic prescribed to relieve moderate to severe pain. It is recommended as a suitable drug for step 2 of the world health organization (WHO) Analgesic Ladder (Tassinari, Drudi, Rosati, et al., 2011). The wide range and high effectiveness of tramadol against different forms of pain were documented in the 2016 US drug spending report as a one of the top 20 prescribed medications (IMS Institute for Healthcare Informatics, 2016).

Several studies have reported the potent analgesic action of tramadol against different acute and chronic pain such as the cancer pain (Bono & Cuffari, 1997), and chronic noncancer pain, including osteoarthritis (Roth, 1998), low back pain (Schnitzer, Gray, Paster, et al., 2000), diabetic neuropathy (Harati, Gooch, Swenson, et al., 1998), and polyneuropathy (Sindrup, Andersen, Madsen, et al., 1999). Tramadol could also be the drug of choice in the pain involves monoaminergic and opioid system such as chronic pain, depression, or obsessive-compulsive disorder (Reeves & Cox, 2008; Rojas-Corrales, Gibert-Rahola, & Mico, 2007; Shapira, Keck Jr., Goldsmith, et al., 1997).

Neuropathic pain

Neuropathic pain can occur in injured neurons as a result of nerve injury and along nociceptive and descending modulatory pathways in the CNS. This pain is accompanied with various conditions such as diabetes, surgery and cancer (Cohen & Mao, 2014). Neuropathic pain may occur in the absence of a noxious stimulus or be induced by the presence of sensory stimuli. Numbness and allodynia clinical phenomena are usually associated with neuropathic pain and the patients suffer burning sensations, abnormal sensitivity to normally painless stimuli, or oversensitivity to painful sensory stimuli. Notably, the response to drugs differs from one patient to another and this is attributed to the exact pattern which vary between people and disease (Helfert, Reimer, Höper, et al., 2015). The patients with chronic neuropathic pain may suffer moderate or severe pain for many years with a significant loss of quality of life and increased healthcare costs (Andrew, Derry, Taylor, et al., 2014). Tramadol has effectively relieved the neuropathic pain as demonstrated in randomized and quasi-randomized controlled trials compared the effectiveness of tramadol versus placebo and other pain medication, or no treatment in people of both sexes and all ages with neuropathic pain of all degrees of severity (Duhmke, Cornblath, & Hollingshead, 2004). A recent study evaluated the combination of gabapentin and tramadol in three different algesiometric assays in streptozotocin-induced diabetes in mice and concluded that this combination significantly produced a dose-dependent antinociception with a significant synergism that is attributed to the multimodal analgesic action of tramadol and gabapentin (Miranda, Sierralta, Aranda, et al., 2018).

Osteoarthritis pain

Osteoarthritis (OA) pain is one of the common causes of chronic pain thought to be increased with age. The most affected joints by OA are hips, knees and hands (Eitner, Hofmann, & Schaible, 2017). Administration of tramadol effectively relieved OA pain immediate as well as extended-release (ER) formulation (Fishman, Kistler, Ellerbusch, et al., 2007). A combination of tramadol with paracetamol produced an effective synergism with different mechanisms that do not overlap in the preclinical studies. The synergism between tramadol and paracetamol significantly exhibited a more rapid pain action than that of tramadol alone as well as more persistent pain relief than that observed for paracetamol only (Schnitzer, 2003). Additionally, a chronic low back/osteoarthritic pain study showed that the drug combination can also be used similarly to codeine/acetaminophen combinations in treating benign chronic pain (Schnitzer, 2003). The safety profile of the tramadol/acetaminophen combination is at least as favorable as that of codeine/acetaminophen and is well-tolerated with long-term use (Schnitzer, 2003).

Cancer pain

Cancer is debilitating disease and has been estimated to cause over eight million deaths per annum (approximately 13% of deaths worldwide) (International Agency for Research on Cancer and WHO, 2014). Cancer pain may be one of the most feared symptoms associated with the disease. In the early and intermediate stages of cancer, 40% of the patients suffer moderate to severe pain and increased to 90% in patients with advanced stages (Paley, Johnson, Tashani, et al., 2015). There are several drugs effective in relieving cancer pain, but it is inadequately controlled in up to 50% of cancer patients and remains unsatisfactory. Therefore, the development of ideal (safe, effective and inexpensive) analgesic becomes a great area of research interest and challenge. Managing cancer pain in older adults could be complex and challenging (Guerard & Cleary, 2017). Several studies demonstrated the effectiveness of tramadol and other nonopioid against cancer pain where

184   PART | II Mechanisms of action of drugs

tramadol exhibited a relatively good relief from the pain associated with cancer (Grond & Sablotzki, 2004; Grond, Zech, & Lynch, 1992; Li, Zhang, Xiao, et al., 2018).

Pain in the emergency department

The pin is one of the reasons requires emergency department (ED) admission. In spite of the large available analgesics, there is no ideal analgesic suitable for patients in ED (Motov & Khan, 2008). Several scholars focused to improve the quality of pain management as the consumption of the opioids dramatically raised by 400% and their clinical application in ED elevated by more than 60% (Pletcher, Kertesz, Kohn, et al., 2008). Overconsumption of opioids was strongly correlated with hypoxia and hypotension and the high rate of deaths from opioids abuse (Alexander, Kruszewski, & Webster, 2012). Therefore, alternative therapeutic strategies to overcome these problems become urgent to decrease the need for rescue opioid prescriptions in EDs (Butti, Bierti, Lanfrit, et al., 2017). Interestingly, a combination of tramadol with paracetamol significantly decreased the intravenous morphine requirement and resulted in strong pain relief and hence the patient become satisfied (Bouida, Beltaief, Msolli, et al., 2019). Therefore, tramadol is a promising alternative for opioids in patients admitted to the ED.

Acute myocardial infarction pain

Myocardial infarction (MI) is the events of heart attack caused by the formation of plaques in the interior walls of the arteries leading to decrease blood flow to the heart and injuring heart muscles owing to a deficiency in oxygen supply. The most common characteristic symptoms of MI include chest pain (left arm to the neck), shortness of breath, sweating, nausea, vomiting and others (Lu, Liu, Sun, et al., 2015). Evidences proved the effectiveness of tramadol in the management of MI associated pain (Abdi & Basgut, 2016; Sunshine, 1994).

Postoperative pain

Surgery is usually associated with acute pain of moderate to severe intensity named postoperative pain. In a general postoperative, pain intensity is generally decreased to mild pain over the first 24–48 h after surgery with analgesics, but a proportion of patients may suffer moderate or severe pain beyond this period (Sommer, de Rijke, van Kleef, et al., 2008). Strong opioid analgesics may be prescribed immediately to relief the postoperative acute pain and their use is characterized by high efficiency and quick-acting. However, they are associated with adverse effects and therefore risk/benefit ratio should be weighed, especially with long time administration (Capdevila, Biboulet, & Barthelet, 1998). As a result, the clinical application of opioids is limited for relieving a subacute pain (1–3 months after surgery) (Hegmann, Weiss, Bowden, et al., 2014) and these encourage scholars to find adequate and effective analgesics for the control of persistent postoperative pain. As an alternative to opioids, tramadol may be prescribed because it possessed a dual mechanism of analgesic action mediated through binding to MOR and inhibition of norepinephrine and 5HT reuptake (Raffa et al., 1992). Several studies documented the advantages of tramadol over other analgesics in postoperative pain. In this context, tramadol has been used as the first choice for effective management for the relief of persistent postoperative pain of musculoskeletal disorder, as well as various types of chronic pain (Chung, Lee, Seo, et al., 2007). In another study, the analgesic effect of tramadol was compared to acetaminophen and tramadol effectively treated postoperative pain after lumbar disc surgery (Yilmaz, Sarihasan, Kelsaka, et al., 2015). On the other hand, acetaminophen alone didn't provide sufficient analgesia (Yilmaz et al., 2015). In postoperative pain after microdiscectomy surgery, Dogar and Khan concluded that a combination of low-dose tramadol (1 mg/kg) and acetaminophen has equivalent analgesic efficacy with a decrease in nausea and vomiting incidence when compared with the higher dose of tramadol (1.5 mg/kg) and acetaminophen combination (Dogar & Khan, 2017). In patients with maxillofacial trauma and who had undergone maxillofacial surgery, intravenous tramadol has been used for the treatment of postoperative pain (Degala & Nehal, 2018).

Postoperative pain in children is a common complication after surgery. Inadequate or lacking control of the postoperative pain associated with the physiological, metabolic and psychological impact on the child. Notably, the use of local analgesics and anesthetics may be associated with intraoperative comfort. Therefore, adjuvant drugs are urgently added to the regimen in order to expand and enhance the block quality and increase the analgesia duration (Goudarzi, Kamali, Yazdi, et al., 2019). Drug combinations may be the best choice for postoperative pain in children. Opioids are associated with severe undesirable adverse effects, whereas tramadol can be used effectively for postoperative pain. Administration of tramadol prior surgery can help reducing pain afterward. In comparison with opioids, tramadol administration is associated with a lower risk for respiratory or hemodynamic depression and therefore it may be the best analgesic choice for children in the perioperative period (Schnabel, Reichl, Meyer-Frießem, et al., 2015). The addition of dexmedetomidine, tramadol and neostigmine to lidocaine 1.5% prolonged the duration of postoperative analgesia in the lower abdominal pain surgery among children (Gerbershagen, Aduckathil, van Wijck, et al., 2013). In addition, a combination of tramadol with paracetamol effectively relieved postoperative pain after major abdominal surgery in children (Ali, Sofi, & Dar, 2017). The combination of tramadol with local anesthetics resulted in an effectively increased duration of postoperative analgesia in the children (Shrestha & Bista, 2005). However, tramadol should not be prescribed for the age below 16 years and extended release formulation of tramadol should be avoided for children below 18 years and in the tonsils or adenoids surgery (Afshari & Ghooshkhanehee, 2009).

Postoperative shivering

At least 10% and as many as 66% of patients subjected to surgical operations experienced shivering during the course of recovery. Shivering is an irregular, spontaneous, continuous constriction of muscle activities for more than 15 min. Physiologically, shivering increased the rate of metabolism and maintain normal body temperature. However, shivering may result in severe adverse events and complications, including increased oxygen consumption, troubling cardiopulmonary function, elevation of the intracranial and intraocular pressures and increased risk of myocardial ischemia and ultimate to severe pain and patient discomfort (Kranke, Eberhart, Roewer, et al., 2002). Although the use of opioids showed beneficial effects, the severe adverse effects represent the main limitation of their application (Tie, Su, He, et al., 2014). Tramadol has shown a high efficacy in controlling post spinal shivering in comparison with clonidine (Alijanpour, Banihashem, Maleh, et al., 2016). The efficacy of tramadol is increased through the combination with other drugs such as paracetamol and dexketoprofen (Fornasari, Allegri, Gerboni, et al., 2017), and it is assumed to be a good applicant to decrease intra- and postanesthetic shivering (Mahesh & Kaparti, 2014).

Acute renal pain

Renal colic negatively affected the quality of life and caused anxiety and/or depression (Diniz, Schor, & Blay, 2006). Therefore, rapid and effective analgesia is crucial in renal colic management at the ED. Tramadol was introduced as an effective analgesic with less adverse effect in comparison with opioids in ureteral calculi. A double-blind study showed the effectiveness of tramadol against renal colic pain (Stankov, Schmieder, Zerle, et al., 1994). The intravenous administration of tramadol/diclofenac sodium showed superior pain relief in comparison with buscopan/diclofenac (Al-Jasmawy, Muhi, & Kadhim, 2015).

Tramadol as a local anesthetic

Clinically, tramadol delivers a local anesthetic action for minor skin surgery (Altunkaya, Ozer, Kargi, et al., 2003) and was used as an adjuvant to local anesthesia in brachial plexus block (Shin, Ju, Jang, et al., 2017) and to support the anesthesia during long operations (Ege, Calisir, Al-Haideri, et al., 2018). In upper extremity surgery, tramadol prolonged the duration of both sensory and motor blocks and analgesia, and shortened the time to onset of sensory and motor blocks (Shin et al., 2017).

Use/off-label use

Premature ejaculation (PE) is a male sexual problem simply defined by a short ejaculatory latency (McMahon, Althof, Waldinger, et al., 2008). Tramadol is prescribed off-label for the treatment of PE with a mechanism not fully understood yet (Wong & Malde, 2013). The beneficial effects of tramadol in PE were reported in 11 clinical trials, evaluated by 6 systematic reviews, 3 of which pooled data in a meta-analysis (Abdel-Hamid, Andersson, Waldinger, et al., 2016). Patients who received tramadol "on a knife's edge" are exquisitely sensitive to develop other sexual dysfunctions (Abdel-Hamid et al., 2016) which were attributed to the long-term action of tramadol on the monoaminergic system (Barber, 2011).

Adverse effects and management of tramadol toxicity

In general, tramadol accompanied toxicity is not regarded as lethal toxicity and is relatively safe when compared with the classical opioids (WHO, 2020). Tramadol adverse effects can be categorized into most frequent, less common and rare adverse effects. The most frequent adverse effects being nausea, dizziness, drowsiness, fatigue, sweating, vomiting,

186 **PART | II** Mechanisms of action of drugs

and dry mouth (Grond & Sablotzki, 2004). Less common adverse effects include diarrhea and cardiovascular complications (tachycardia and postural hypotension) whereas the rare adverse effects are represented as respiratory depression, seizures, tremors, bradycardia, psychosis and anxiety (WHO, 2020). Coadministration of tramadol with other drugs or alcohol significantly participates in tramadol intoxication (WHO, 2020). Notably, the symptoms of tramadol intoxication start at doses above 500 mg and after 4 h of administration (Spiller, Gorman, Villalobos, et al., 1997). Overdoses of tramadol are associated with lethargy, insomnia, nausea, nervousness, seizures, coma, hypertension, an increase in heart rate, and serotonin syndrome (Grond & Sablotzki, 2004). The effects of tramadol overdose could be managed by (i) respiratory support (ii) opioid antagonists such as naloxone or nalmefene, and (iii) anticonvulsants like benzodiazepines in case of seizures. Serotonin syndrome is attributed to serotine reuptake inhibition related mechanism that may be lethal if it left untreated. It can be effectively treated by discontinuing tramadol and the administration of benzodiazepines or cyproheptadine (Hassamal, Miotto, Dale, et al., 2018). Of note, tramadol is listed as Category C in the pregnancy risk drug by FDA and is therefore should not be prescribed for pregnant women (Källén & Reis, 2015).

Dependence, withdrawal, abuse, and tolerance

Animal studies showed that tramadol might produce physical dependence with mild withdrawal symptoms while other studies didn't report these findings (Cha, Song, Lee, et al., 2014; Epstein, Preston, & Jasinski, 2006). As compared to morphine, human studies showed the low potential of dependence with tramadol because of its lower affinity to the MOR (Epstein et al., 2006). Experimental studies reported the low abuse potential of tramadol compared with morphine (Epstein et al., 2006). In addition, clinical studies demonstrated reinforcing effects of tramadol in mild form and less than that of morphine (Epstein et al., 2006). The WHO expert committee reported the relative safety of tramadol (dependence and abuse potential) when compared with morphine (WHO, 2020). Potential abuse to tramadol increased in persons with a history of drug abuse (WHO, 2020). Since tramadol possesses a dual mechanism of action and inhibitory action on 5HT and noradrenaline, neurotransmitters that have a key role in mood, its potential for addiction should be considered (Barber, 2011).

Regarding tolerance, the chronic administration of tramadol has been assessed and compared with classic opioids. In neuropathic pain rat models, chronic administration of tramadol didn't cause tolerance (Kayser, Besson, & Guilbaud, 1991; Tsai, Sung, Chang, et al., 2000). In a mouse model of acetic acid-induced pain, tolerance has been developed with morphine, but not with tramadol (Miranda & Pinardi, 1998). In line with these findings, clinical trials showed the low tolerance and dependence associated with tramadol (Friderichs, Felgenhauer, Jongschaap, et al., 1978; Preston, Jasinski, & Testa, 1991). In Rhesus monkeys, tramadol exhibited a lower intense reinforcing effect than that was observed with codeine or pentatozine (Yanagita, 1978). Furthermore, tramadol induced conditioned place preference (CPP) associated with increased dopamine in the nucleus accumbens of rats (Sprague, Leifheit, Selken, et al., 2002; Tzschentke, Bruckmann, & Friderichs, 2002). Based on the findings of different studies, the effects of tramadol should be monitored by an expert.

Contraindications

Seizure disorders

The overdose of tramadol increases the risk of seizures, and is therefore should not be administered with a history of seizures and other neurologic disorders (Murray, Carpenter, Dunkley, et al., 2019).

Liver disease

Since liver diseases have a significant impact on the metabolism of drugs, tramadol should be administered with caution in patients with liver fibrosis and cirrhosis because the levels of tramadol and its active metabolite M1 change as the elimination altered (Subedi et al., 2019).

Renal dysfunction

Kidney diseases affect the elimination of tramadol and M1. The excretion rate of tramadol and M1 decreases in the case of renal impairment (Afshari & Ghooshkhanehee, 2009).

Respiratory depression

Respiratory depression may occur in patients with acute alcohol intoxication upon the use of tramadol, effects that could be managed by naloxone (Michaud, Augsburger, Romain, et al., 1999).

Suicidal tendency

The use of tramadol in a patient suffered from emotional disturbances as well as with the patent with a history of misuse of tranquilizers, or alcoholics should be carefully considered because it may lead to suicidal or addicts and may ultimate to death (Barbera, Fisichella, Bosco, et al., 2013).

Conclusion

Preclinical studies and clinical trials have shown the analgesic efficacy of tramadol against both acute and chronic pain. Tramadol is an effective analgesic against neuropathic, inflammatory, postoperative, OA, renal colic, postoperative shivering, cancer, and acute MI pain. The chronic administration of tramadol is associated with weak dependence, tolerance, signs of abuse and adverse effects when compared with the classical opioids. The analgesic effect of tramadol is reasonably safe and has shown antidepressant-like activity in several preclinical studies. The analgesic efficacy of tramadol increases when combined with paracetamol and other analgesics. However, the use of tramadol should be avoided during pregnancy and should be carefully prescribed in patients with a history of seizures, liver disease and renal impairment.

Applications to other areas

Tramadol works as a local anesthetic where it delivers a local anesthetic action for minor skin surgery. It has been also used as an adjuvant to local anesthesia in brachial plexus block and to support the anesthesia during long operations. Tramadol is prescribed off-label for the treatment of premature ejaculation with a mechanism not fully understood yet. However, chronic administration of tramadol may increase the risk of sexual dysfunctions due to its long-term action on the monoaminergic system. Several preclinical studies have shown the antidepressant-like activity of tramadol. Therefore, it might be prescribed for chronic pain in patients suffering depression, pending further preclinical and clinical investigations to determine it's the mechanism(s) underlying its antidepressant-like.

Other agents of interest

The analgesic efficacy of tramadol increases when combined other analgesics such as paracetamol. A combination of gabapentin and tramadol produces a dose-dependent antinociception effect with a significant synergism that is attributed to the multimodal analgesic action of these agents. The addition of dexmedetomidine, tramadol and neostigmine to lidocaine 1.5% prolonged the duration of postoperative analgesia in the lower abdominal pain surgery among children. In upper extremity surgery, tramadol prolonged the duration of both sensory and motor blocks, analgesia, and shortened the time to onset of sensory and motor blocks.

Mini-dictionary of terms

Atypical opioids. Are a group of pain-relieving drugs that have a different potency to the μ-opioid receptors, different adverse effects and abuse potential as compared to the classical opioids.

Neuropathic pain. Is the pain caused by a disease or damage to the somatosensory nervous system neuropathic pain.

Osteoarthritis (OA) pain. Is one of the common causes of chronic pain thought to be increased with age. The most affected joints by OA are hips, knees and hands.

Postoperative pain. Is an acute pain of moderate to severe intensity usually associated with surgery.

Postoperative shivering. Is a common complication of anesthesia that occurs in patients subjected to surgical operations. Shivering is an irregular, spontaneous, continuous constriction of muscle activities for more than 15 min.

Key facts

- Tramadol is a centrally acting synthetic opioid analgesic.
- Tramadol was created by a German drug company in 1962, tested for 15 years in Germany and brought to the foreign market in 1977.
- Until 1995, tramadol wasn't available in the US but now is quite popular.
- Tramadol is well-tolerated when used for pain but can cause some common side effects.
- Tramadol may carry a lower risk than other drugs of abuse when used for medical purposes.
- Tramadol is recommended as a suitable drug for step 2 of the world health organization (WHO) Analgesic Ladder.

Summary points

- This chapter focuses on the analgesic effect and mechanism of tramadol.
- Tramadol is an atypical opioid and one of the most common effective analgesics that is prescribed to relieve different forms of acute and chronic pain.
- Tramadol binds weakly to the μ-opioid receptor (MOR) and differs from typical opioid drugs as it modulates the reuptake of the monoamines noradrenaline and serotonin.
- As compared to morphine, human studies showed the low potential of dependence with tramadol because of its lower affinity of to the MOR.
- Chronic administration of tramadol is associated with weak dependence, tolerance and signs of abuse.
- Tramadol has less adverse effects when compared with the classical opioids.
- The analgesic efficacy of tramadol increases when combined with paracetamol and other analgesics.
- Tramadol should be avoided during pregnancy and should be carefully prescribed in patients with a history of seizures, liver disease and renal impairment.

Conflict of interest

None.

References

Abdel-Hamid, I. A., Andersson, K. E., Waldinger, M. D., et al. (2016). Tramadol abuse and sexual function. *Sexual Medicine Reviews*, *4*(3), 235–246.

Abdi, A., & Basgut, B. (2016). An evidence-based review of pain management in acute myocardial infarction. *Journal of Cardiology & Clinical Research*, *4*(4), 1067.

Afshari, R., & Ghooshkhanehee, H. (2009). Tramadol overdose induced seizure, dramatic rise of CPK and acute renal failure. *JPMA. The Journal of the Pakistan Medical Association*, *59*(3), 178.

Alexander, G. C., Kruszewski, S. P., & Webster, D. W. (2012). Rethinking opioid prescribing to protect patient safety and public health. *JAMA*, *308*(18), 1865–1866.

Ali, S., Sofi, K., & Dar, A. Q. (2017). Comparison of intravenous infusion of tramadol alone with combination of tramadol and paracetamol for postoperative pain after major abdominal surgery in children. *Anesthesia, Essays and Researches*, *11*(2), 472–476.

Alijanpour, E., Banihashem, N., Maleh, P. A., et al. (2016). Prophylactic effect of oral clonidine and tramadol in postoperative shivering in lower abdominal surgery. *Open Journal of Anesthesiology*, *6*(9), 137–147.

Al-Jasmawy, H. O., Muhi, A. I., & Kadhim, A. A. (2015). Efficacy of combined tramadol with diclofenac in comparison with monotherapy treatment using buscopan diclofenac or tramadol in renal pain control. *Medical Journal of Babylon*, *12*(4), 1070–1076.

Altunkaya, H., Ozer, Y., Kargi, E., et al. (2003). Comparison of local anaesthetic effects of tramadol with prilocaine for minor surgical procedures. *British Journal of Anaesthesia*, *90*(3), 320–322.

Alvarado, C., Guzmán, Á., Díaz, E., et al. (2005). Synthesis of tramadol and analogous. *Journal of the Mexican Chemical Society*, *49*(4), 324–327.

Andrew, R., Derry, S., Taylor, R. S., et al. (2014). The costs and consequences of adequately managed chronic non-cancer pain and chronic neuropathic pain. *Pain Practice : The Official Journal of World Institute of Pain*, *14*(1), 79–94.

Barber, J. (2011). Examining the use of tramadol hydrochloride as an antidepressant. *Experimental and Clinical Psychopharmacology*, *19*(2), 123–130.

Barbera, N., Fisichella, M., Bosco, A., et al. (2013). A suicidal poisoning due to tramadol. A metabolic approach to death investigation. *Journal of Forensic and Legal Medicine*, *20*(5), 555–558.

Bono, A. V., & Cuffari, S. (1997). Effectiveness and tolerance of tramadol in cancer pain. A comparative study with respect to buprenorphine. *Drugs*, *53* (Suppl 2), 40–49.

Bouida, W., Beltaief, K., Msolli, M. A., et al. (2019). Effect on morphine requirement of early administration of oral acetaminophen vs. acetaminophen/ tramadol combination in acute pain. *Pain Practice : The Official Journal of World Institute of Pain*, *19*(3), 275–282.

Butti, L., Bierti, O., Lanfrit, R., et al. (2017). Evaluation of the effectiveness and efficiency of the triage emergency department nursing protocol for the management of pain. *Journal of Pain Research, 10*, 2479–2488.

Capdevila, X., Biboulet, P., & Barthelet, Y. (1998). Postoperative analgesia. Specificity in the elderly. *Annales Françaises d'Anesthèsie et de Rèanimation, 17*(6), 642–648.

Cha, H. J., Song, M. J., Lee, K. W., et al. (2014). Dependence potential of tramadol: Behavioral pharmacology in rodents. *Biomolecules & Therapeutics, 22*(6), 558–562.

Chung, J.-Y., Lee, J.-J., Seo, H.-Y., et al. (2007). Effect of tramadol/acetaminophen combination drug in acute pain after spinal surgery. *Journal of Korean Society of Spine Surgery, 14*(3), 137–143.

Cohen, S. P., & Mao, J. (2014). Neuropathic pain: Mechanisms and their clinical implications. *BMJ (Clinical Research Ed.), 348*, f7656.

Dayer, P., Desmeules, J., & Collart, L. (1997). Pharmacology of tramadol. *Drugs, 53*(Suppl 2), 18–24.

Degala, S., & Nehal, A. (2018). Comparison of intravenous tramadol versus ketorolac in the management of postoperative pain after oral and maxillofacial surgery. *Oral and Maxillofacial Surgery, 22*(3), 275–280.

Diniz, D. H., Schor, N., & Blay, S. L. (2006). Stressful life events and painful recurrent colic of renal lithiasis. *The Journal of Urology, 176*(6 Pt 1), 2483–2487 (discussion 2487).

Dogar, S. A., & Khan, F. A. (2017). Tramadol-paracetamol combination for postoperative pain relief in elective single-level microdisectomy surgery. *Journal of Neurosurgical Anesthesiology, 29*(2), 157–160.

Duhmke, R. M., Cornblath, D. D., & Hollingshead, J. R. (2004). Tramadol for neuropathic pain. *The Cochrane Database of Systematic Reviews, 6*, CD003726.

Ege, B., Calisir, M., Al-Haideri, Y., et al. (2018). Comparison of local anesthetic efficiency of tramadol hydrochloride and lidocaine hydrochloride. *Journal of Oral and Maxillofacial Surgery : Official Journal of the American Association of Oral and Maxillofacial Surgeons, 76*(4), 744–751.

Eitner, A., Hofmann, G. O., & Schaible, H. G. (2017). Mechanisms of osteoarthritic pain. Studies in humans and experimental models. *Frontiers in Molecular Neuroscience, 10*, 349.

Epstein, D. H., Preston, K. L., & Jasinski, D. R. (2006). Abuse liability, behavioral pharmacology, and physical-dependence potential of opioids in humans and laboratory animals: Lessons from tramadol. *Biological Psychology, 73*(1), 90–99.

Fishman, R. L., Kistler, C. J., Ellerbusch, M. T., et al. (2007). Efficacy and safety of 12 weeks of osteoarthritic pain therapy with once-daily tramadol (tramadol Contramid OAD). *Journal of Opioid Management, 3*(5), 273–280.

Fornasari, D., Allegri, M., Gerboni, S., et al. (2017). A "novel" association to treat pain: Tramadol/dexketoprofen. The first drug of a "new pharmacological class". *Acta Bio-Medica : Atenei Parmensis, 88*(1), 17–24.

Friderichs, E., Felgenhauer, F., Jongschaap, P., et al. (1978). Pharmacological studies on analgesia, dependence on and tolerance of tramadol, a potent analgetic drug (author's transl). *Arzneimittel-Forschung, 28*(1a), 122–134.

Gan, S. H., Ismail, R., Wan Adnan, W. A., et al. (2007). Impact of CYP2D6 genetic polymorphism on tramadol pharmacokinetics and pharmacodynamics. *Molecular Diagnosis & Therapy, 11*(3), 171–181.

Gerbershagen, H. J., Aduckathil, S., van Wijck, A. J., et al. (2013). Pain intensity on the first day after surgery: A prospective cohort study comparing 179 surgical procedures. *Anesthesiology, 118*(4), 934–944.

Gillen, C., Haurand, M., Kobelt, D. J., et al. (2000). Affinity, potency and efficacy of tramadol and its metabolites at the cloned human mu-opioid receptor. *Naunyn-Schmiedeberg's Archives of Pharmacology, 362*(2), 116–121.

Goudarzi, T. H., Kamali, A., Yazdi, B., et al. (2019). Addition of dexmedetomidine, tramadol and neostigmine to lidocaine 1.5% increasing the duration of postoperative analgesia in the lower abdominal pain surgery among children: A double-blinded randomized clinical study. *Medical Gas Research, 9*(3), 110–114.

Grond, S., & Sablotzki, A. (2004). Clinical pharmacology of tramadol. *Clinical Pharmacokinetics, 43*(13), 879–923.

Grond, S., Zech, D., Lynch, J., et al. (1992). Tramadol—A weak opioid for relief of cancer pain. *Pain Clinic Utrecht, 5*(4), 241.

Guerard, E. J., & Cleary, J. F. (2017). Managing cancer pain in older adults. *Cancer Journal (Sudbury, Mass.), 23*(4), 242–245.

Harati, Y., Gooch, C., Swenson, M., et al. (1998). Double-blind randomized trial of tramadol for the treatment of the pain of diabetic neuropathy. *Neurology, 50*(6), 1842–1846.

Hassamal, S., Miotto, K., Dale, W., et al. (2018). Tramadol: Understanding the risk of serotonin syndrome and seizures. *The American Journal of Medicine, 131*(11), 1382.e1381–1382.e1386.

Hegmann, K. T., Weiss, M. S., Bowden, K., et al. (2014). ACOEM practice guidelines: Opioids for treatment of acute, subacute, chronic, and postoperative pain. *Journal of Occupational and Environmental Medicine, 56*(12), e143–e159.

Helfert, S. M., Reimer, M., Höper, J., et al. (2015). Individualized pharmacological treatment of neuropathic pain. *Clinical Pharmacology and Therapeutics, 97*(2), 135–142.

Ide, S., Minami, M., Ishihara, K., et al. (2006). Mu opioid receptor-dependent and independent components in effects of tramadol. *Neuropharmacology, 51*(3), 651–658.

IMS Institute for Healthcare Informatics. (2016). *Medicines use and spending in the US: A review of 2015 and outlook to 2020.* Parsippany, NJ: IMS Institute for Healthcare Informatics. https://static1.squarespace.com/static/54d50ceee4b05797b34869cf/t/5711197b45bf21650748e8ad/1460738430435/IMS+Health+2015.pdf.

International Agency for Research on Cancer and WHO. (2014). *GLOBOCAN 2012: Estimated cancer incidence, mortality and prevalence worldwide in 2012.* Lyon: IRAC. http://globocan.iarc.fr/Default.aspx.

Källén, B., & Reis, M. (2015). Use of tramadol in early pregnancy and congenital malformation risk. *Reproductive Toxicology (Elmsford, N.Y.), 58*, 246–251.

Kayser, V., Besson, J. M., & Guilbaud, G. (1991). Effects of the analgesic agent tramadol in normal and arthritic rats: Comparison with the effects of different opioids, including tolerance and cross-tolerance to morphine. *European Journal of Pharmacology, 195*(1), 37–45.

Kranke, P., Eberhart, L. H., Roewer, N., et al. (2002). Pharmacological treatment of postoperative shivering: A quantitative systematic review of randomized controlled trials. *Anesthesia and Analgesia, 94*(2), 453–460. table of contents.

Kusari, S., Tatsimo, S. J., Zühlke, S., et al. (2014). Tramadol—A true natural product? *Angewandte Chemie (International ed. in English), 53*(45), 12073–12076.

Lai, J., Ma, S. W., Porreca, F., et al. (1996). Tramadol, M1 metabolite and enantiomer affinities for cloned human opioid receptors expressed in transfected HN9.10 neuroblastoma cells. *European Journal of Pharmacology, 316*(2–3), 369–372.

Li, P., Zhang, Q., Xiao, Z., et al. (2018). Activation of the P2X(7) receptor in midbrain periaqueductal gray participates in the analgesic effect of tramadol in bone cancer pain rats. *Molecular Pain, 14*, 1744806918803039.

Lintz, W., Barth, H., Osterloh, G., et al. (1986). Bioavailability of enteral tramadol formulations. 1st communication: Capsules. *Arzneimittel-Forschung, 36*(8), 1278–1283.

Lu, L., Liu, M., Sun, R., et al. (2015). Myocardial infarction: Symptoms and treatments. *Cell Biochemistry and Biophysics, 72*(3), 865–867.

Mahesh, T., & Kaparti, L. (2014). A randomised trial comparing efficacy, onset and duration of action of pethidine and tramadol in abolition of shivering in the intra operative period. *Journal of Clinical and Diagnostic Research : JCDR, 8*(11), Gc07–09.

McMahon, C. G., Althof, S., Waldinger, M. D., et al. (2008). An evidence-based definition of lifelong premature ejaculation: Report of the International Society for Sexual Medicine Ad Hoc Committee for the definition of premature ejaculation. *BJU International, 102*(3), 338–350.

Mert, T., Gunes, Y., Guven, M., et al. (2003). Differential effects of lidocaine and tramadol on modified nerve impulse by 4-aminopyridine in rats. *Pharmacology, 69*(2), 68–73.

Michaud, K., Augsburger, M., Romain, N., et al. (1999). Fatal overdose of tramadol and alprazolam. *Forensic Science International, 105*(3), 185–189.

Miranda, H. F., & Pinardi, G. (1998). Antinociception, tolerance, and physical dependence comparison between morphine and tramadol. *Pharmacology, Biochemistry, and Behavior, 61*(4), 357–360.

Miranda, H. F., Sierralta, F., Aranda, N., et al. (2018). Synergism between gabapentin-tramadol in experimental diabetic neuropathic pain. *Fundamental & Clinical Pharmacology, 32*(6), 581–588.

Motov, S. M., & Khan, A. N. (2008). Problems and barriers of pain management in the emergency department: Are we ever going to get better? *Journal of Pain Research, 2*, 5–11.

Murray, B. P., Carpenter, J. E., Dunkley, C. A., et al. (2019). Seizures in tramadol overdoses reported in the ToxIC registry: Predisposing factors and the role of naloxone. *Clinical toxicology (Philadelphia, Pa.), 57*(8), 692–696.

Ogata, J., Minami, K., Uezono, Y., et al. (2004). The inhibitory effects of tramadol on 5-hydroxytryptamine type 2C receptors expressed in Xenopus oocytes. *Anesthesia and Analgesia, 98*(5), 1401–1406. table of contents.

Paley, C. A., Johnson, M. I., Tashani, O. A., et al. (2015). Acupuncture for cancer pain in adults. *The Cochrane Database of Systematic Reviews, 2015*(10), Cd007753.

Pletcher, M. J., Kertesz, S. G., Kohn, M. A., et al. (2008). Trends in opioid prescribing by race/ethnicity for patients seeking care in US emergency departments. *JAMA, 299*(1), 70–78.

Preston, K. L., Jasinski, D. R., & Testa, M. (1991). Abuse potential and pharmacological comparison of tramadol and morphine. *Drug and Alcohol Dependence, 27*(1), 7–17.

Raffa, R. B., Friderichs, E., Reimann, W., et al. (1992). Opioid and nonopioid components independently contribute to the mechanism of action of tramadol, an 'atypical' opioid analgesic. *The Journal of Pharmacology and Experimental Therapeutics, 260*(1), 275–285.

Raffa, R. B., Friderichs, E., Reimann, W., et al. (1993). Complementary and synergistic antinociceptive interaction between the enantiomers of tramadol. *The Journal of Pharmacology and Experimental Therapeutics, 267*(1), 331–340.

Reeves, R. R., & Cox, S. K. (2008). Similar effects of tramadol and venlafaxine in major depressive disorder. *Southern Medical Journal, 101*(2), 193–195.

Rojas-Corrales, M. O., Gibert-Rahola, J., & Mico, J. A. (2007). Role of atypical opiates in OCD. Experimental approach through the study of 5-HT(2A/C) receptor-mediated behavior. *Psychopharmacology, 190*(2), 221–231.

Roth, S. H. (1998). Efficacy and safety of tramadol HCl in breakthrough musculoskeletal pain attributed to osteoarthritis. *The Journal of Rheumatology, 25*(7), 1358–1363.

Schnabel, A., Reichl, S. U., Meyer-Frießem, C., et al. (2015). Tramadol for postoperative pain treatment in children. *The Cochrane Database of Systematic Reviews, 2015*(3), Cd009574.

Schnitzer, T. (2003). The new analgesic combination tramadol/acetaminophen. *European Journal of Anaesthesiology. Supplement, 28*, 13–17.

Schnitzer, T. J., Gray, W. L., Paster, R. Z., et al. (2000). Efficacy of tramadol in treatment of chronic low back pain. *The Journal of Rheumatology, 27*(3), 772–778.

Shapira, N. A., Keck, P. E., Jr., Goldsmith, T. D., et al. (1997). Open-label pilot study of tramadol hydrochloride in treatment-refractory obsessive-compulsive disorder. *Depression and Anxiety, 6*(4), 170–173.

Shin, H. W., Ju, B. J., Jang, Y. K., et al. (2017). Effect of tramadol as an adjuvant to local anesthetics for brachial plexus block: A systematic review and meta-analysis. *PLoS One, 12*(9), e0184649.

Shiraishi, M., Minami, K., Uezono, Y., et al. (2001). Inhibition by tramadol of muscarinic receptor-induced responses in cultured adrenal medullary cells and in Xenopus laevis oocytes expressing cloned M1 receptors. *The Journal of Pharmacology and Experimental Therapeutics, 299*(1), 255–260.

Shiraishi, M., Minami, K., Uezono, Y., et al. (2002). Inhibitory effects of tramadol on nicotinic acetylcholine receptors in adrenal chromaffin cells and in Xenopus oocytes expressing alpha 7 receptors. *British Journal of Pharmacology, 136*(2), 207–216.

Shrestha, B. R., & Bista, B. (2005). Tramadol along with local anaesthetics in the penile block for the children undergoing circumcision. *Kathmandu University Medical Journal (KUMJ)*, *3*(1), 26–29.

Sindrup, S. H., Andersen, G., Madsen, C., et al. (1999). Tramadol relieves pain and allodynia in polyneuropathy: A randomised, double-blind, controlled trial. *Pain*, *83*(1), 85–90.

Sommer, M., de Rijke, J. M., van Kleef, M., et al. (2008). The prevalence of postoperative pain in a sample of 1490 surgical inpatients. *European Journal of Anaesthesiology*, *25*(4), 267–274.

Spiller, H. A., Gorman, S. E., Villalobos, D., et al. (1997). Prospective multicenter evaluation of tramadol exposure. *Journal of Toxicology. Clinical Toxicology*, *35*(4), 361–364.

Sprague, J. E., Leifheit, M., Selken, J., et al. (2002). In vivo microdialysis and conditioned place preference studies in rats are consistent with abuse potential of tramadol. *Synapse (New York, N.Y.)*, *43*(2), 118–121.

Stankov, G., Schmieder, G., Zerle, G., et al. (1994). Double-blind study with dipyrone versus tramadol and butylscopolamine in acute renal colic pain. *World Journal of Urology*, *12*(3), 155–161.

Subedi, M., Bajaj, S., Kumar, M. S., et al. (2019). An overview of tramadol and its usage in pain management and future perspective. *Biomedicine & Pharmacotherapy = Biomedecine & Pharmacotherapie*, *111*, 443–451.

Subrahmanyam, V., Renwick, A. B., Walters, D. G., et al. (2001). Identification of cytochrome P-450 isoforms responsible for cis-tramadol metabolism in human liver microsomes. *Drug Metabolism and Disposition: The Biological Fate of Chemicals*, *29*(8), 1146–1155.

Sunshine, A. (1994). New clinical experience with tramadol. *Drugs*, *47*(Suppl 1), 8–18.

Tassinari, D., Drudi, F., Rosati, M., et al. (2011). The second step of the analgesic ladder and oral tramadol in the treatment of mild to moderate cancer pain: A systematic review. *Palliative Medicine*, *25*(5), 410–423.

Tie, H. T., Su, G. Z., He, K., et al. (2014). Efficacy and safety of ondansetron in preventing postanesthesia shivering: A meta-analysis of randomized controlled trials. *BMC Anesthesiology*, *14*, 12.

Tsai, Y. C., Sung, Y. H., Chang, P. J., et al. (2000). Tramadol relieves thermal hyperalgesia in rats with chronic constriction injury of the sciatic nerve. *Fundamental & Clinical Pharmacology*, *14*(4), 335–340.

Tzschentke, T. M., Bruckmann, W., & Friderichs, E. (2002). Lack of sensitization during place conditioning in rats is consistent with the low abuse potential of tramadol. *Neuroscience Letters*, *329*(1), 25–28.

WHO. (2020). *WHO expert committee on drug dependence: Forty-second report.*

Wong, B. L., & Malde, S. (2013). The use of tramadol "on-demand" for premature ejaculation: A systematic review. *Urology*, *81*(1), 98–103.

Yanagita, T. (1978). Drug dependence potential of 1-(m-methoxyphenyl)-2-dimethylaminomethyl-cyclohexan-1-ol hydrochloride (tramadol) tested in monkeys. *Arzneimittel-Forschung*, *28*(1a), 158–163.

Yilmaz, M. Z., Sarihasan, B. B., Kelsaka, E., et al. (2015). Comparison of the analgesic effects of paracetamol and tramadol in lumbar disc surgery. *Turkish Journal of Medical Sciences*, *45*(2), 438–442.

Part III

Adverse effects, reactions, and outcomes

Chapter 19

Long-term effects of anesthesia on the brain: an update on neurotoxicity

Rajkumar Rajendram[a,b,c], Vinood B. Patel[d], and Victor R. Preedy[e]

[a]*College of Medicine, King Saud bin Abdulaziz University for Health Sciences, Riyadh, Saudi Arabia,* [b]*Department of Medicine, King Abdulaziz Medical City, Ministry of National Guard—Health Affairs, Riyadh, Saudi Arabia,* [c]*Department of Anaesthesia and Intensive Care, Stoke Mandeville Hospital, Buckinghamshire, United Kingdom,* [d]*Department of Biomedical Science, School of Life Sciences, University of Westminster, London, United Kingdom,* [e]*Faculty of Life Science and Medicine, King's College London, Franklin-Wilkins Building, London, United Kingdom*

Abbreviations

CNS	central nervous system
FDA	Federal Drug Administration, United States of America
GA	general anesthesia
GABA	gamma-aminobutyric acid
ISPOCD1	International multicenter study on long-term POCD 1
NMDA	*N*-methyl-D-aspartate
POCD	postoperative cognitive dysfunction
POD	postoperative delirium

Introduction

It has been more than a century and a half since "Ether Day" (i.e., the first public demonstration of modern intraoperative anesthesia in 1846 by William Morton (Schlich, 2017)). The field of anesthesia has advanced significantly in this time. In the early 20th century, the medical specialty of anesthesia began as a branch of surgery and over the past 50 years, or so, this field has grown exponentially.

Today, in the third decade of the 21st century, anesthetists seem to be able to provide anesthesia to all patient groups regardless of age, comorbidity, and frailty almost effortlessly. Every year millions of painful, complex procedures are performed worldwide. None would be possible without general anesthesia (GA).

This has four components; amnesia, hypnosis, analgesia, and paralysis (Fig. 1). To achieve sufficient depth of anesthesia (i.e., to completely ablate all awareness and sensation of pain), the brain may be subjected to any of a broad range of classes of anesthetic agents, either alone or as 'cocktails'.

Anesthesia is necessary to ensure patients' comfort during unpleasant interventions. Patients undergoing GA typically receive one or more anesthetic agents intravenously (e.g., benzodiazepines, barbiturates, propofol, ketamine, etomidate, etc.) and/or via inhalation (i.e., volatile or gaseous anesthetics e.g., nitrous oxide, halothane, isoflurane, sevoflurane, desflurane, etc.): Fig. 2. However, in the 1970s, animal studies raised concerns that anesthetic and sedative drugs can damage the developing brain (Wilkinson, Mark, Wilson, & Patel, 1973). Since then, the long-term outcomes after GA have been scrutinized in a range of in vitro and in vivo studies.

Recent data from human studies also suggest an association between exposure to anesthesia in early childhood and subsequent neurodevelopmental deficit, but causation has not been established (Zaccariello et al., 2019). While concerns have also been raised regarding anesthetic exposure of the developing brain in utero, there is currently little to no evidence in humans to support these concerns.

Despite the lack of robust data and the uncertainty over this issue, the United States Federal Drug Administration (FDA) has issued an alert about the neurotoxic effects of GA (*FDA Drug Safety Communication: FDA Approves Label Changes*

Treatments, Mechanisms, and Adverse Reactions of Anesthetics and Analgesics. https://doi.org/10.1016/B978-0-12-820237-1.00019-3
Copyright © 2022 Elsevier Inc. All rights reserved.

196 PART | III Adverse effects, reactions, and outcomes

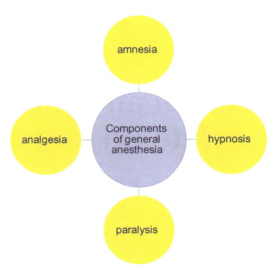

FIG. 1 Schematic representation of the components of general anesthesia. In the schematic representation anesthesia refers to a temporary, medically-induced state of lack of sensation or awareness. Analgesia refers to the prevention or relief of pain. Paralysis refers to muscle relaxation. Amnesia refers to loss of memory. Hypnosis refers to lack of awareness and unconsciousness (compiled by the authors).

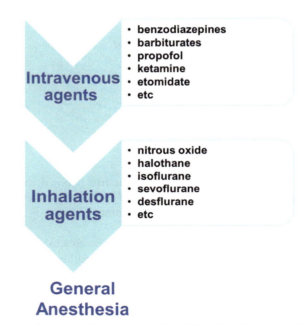

FIG. 2 Schematic representation of the typical agents used in general anesthesia. The schematic representation illustrates that the state of general anesthesia may be induced and maintained by intravenous agents and inhalational agents (compiled by the authors).

for Use of General Anesthetic and Sedation Drugs in Young Children | FDA, n.d.-a.; *FDA Drug Safety Communication: FDA Review Results in New Warnings about Using General Anesthetics and Sedation Drugs in Young Children and Pregnant Women* | FDA, n.d.-b). The possibility that these agents may be neurotoxic is of the utmost importance to patients, parents, and the clinicians who care for them. This chapter reviews the literature suggesting the toxicity of anesthetics to the central nervous system (CNS), the proposed pathogenesis, and the clinical data and the limitations of studies in humans.

Neurotransmitters signaling during cerebral development

All facets of neuronal development are defined by an army of neurotransmitters. The activities of these neurotransmitters are maintained in a fine balance by a plethora of neuromodulators. Indeed, most aspects of neuronal migration, differentiation, maturation and synaptogenesis (i.e., the main processes of CNS development) are controlled by two major neurotransmitters (i.e., gamma-aminobutyric acid (GABA) and glutamate) (Jevtovic-Todorovic, 2016; Jevtovic-Todorovic et al., 2003).

The formation of trillions of contacts (i.e., synapses) between neurons by dendritic branching is known as synaptogenesis. The vast majority of synaptogenesis and the formation of fundamental neuronal networks occurs during CNS development. However, synapses and neuronal networks are very pliable. Their perpetual remodeling occurs throughout life (i.e., new synapses are ceaselessly formed while others are pruned away). It is truly remarkable that purposeful neuronal circuits and ordered neuronal maps result from this seemingly chaotic process.

This developmental synaptogenesis is directed by activity-dependent remodeling. So, neuronal impulses and communication between neurons are needed for timely, precise, targeted synaptogenesis (Dobbing & Sands, 1979; Jevtovic-Todorovic, 2016). All aspects of synaptogenesis are tightly controlled by glia. These cells actively participate in intercellular signaling to provide an appropriate milieu for neuronal interactions (Allen & Barres, 2005).

Synaptic signaling and electrical activity are extremely important during synaptogenesis. Indeed, in the early phases of synaptogenesis, the major inhibitory neurotransmitter (GABA) is, somewhat paradoxically, excitatory (Ben-Ari, 2002). So, unphysiological modulation of neurotransmitter signaling may be toxic to the developing CNS (Jevtovic-Todorovic, 2016).

Thus, neuronal communication and activity are crucial for synaptogenesis and the stable receptor structures required for cognitive and behavioral development (Jevtovic-Todorovic, 2016).

The effect of general anesthetics on the brain must be considered in this context. General anesthetics are very potent transient inhibitors of neuronal communication (Hudetz, 2012). Thus, the final common endpoint of their administration is amnesia, analgesia, hypnosis, and paralysis (i.e., anesthesia; Fig. 1). Anesthesia is achieved by unphysiologically 'toning down' or even abolishing neuronal communication.

Despite significant advances in understanding in recent years, the precise mechanisms of action of general anesthetic agents remain elusive (Hemmings et al., 2019). However, anesthetics generally enhance inhibitory synaptic transmission and/or inhibit excitatory synaptic transmission (Hemmings et al., 2019). General anesthetics are powerful modulators of GABA and glutamate (Hemmings et al., 2019). Thus, a significant functional imbalance of these neurotransmitters occurs during GA. The agents which enhance $GABA_A$ mediated inhibitory neurotransmission include both intravenous anesthetics (e.g., barbiturates, benzodiazepines, propofol, and etomidate), and inhalational anesthetics (e.g., sevoflurane, desflurane) (Franks, 2008; Hirota, Roth, Fujimura, Masuda, & Ito, 1998; Jevtovic-Todorovic, 2016; Nishikawa & Harrison, 2003). However, a handful of intravenous anesthetics [e.g., phencyclidine (PCP) and ketamine] and inhalational anesthetics, nitrous oxide and xenon inhibit excitatory neurotransmission by blocking N-methyl-D-aspartate (NMDA) receptors (Franks, Dickinson, de Sousa, Hall, & Lieb, 1998; Hemmings et al., 2019; Jevtovic-Todorovic, 2016; Jevtović-Todorović et al., 1998; Lodge & Anis, 1982).

However, relatively recent research has recognized that clinically relevant doses and combinations of general anesthetics may be neurotoxic in adult and immature brains. This emerging data suggests that these agents are potentially deleterious to humans at the extremes of age in vivo (Berger et al., 2015; Zaccariello et al., 2019). Indeed, maintaining a fine balance in neurotransmitter release and their receptor activation during critical stages of synaptogenesis is clearly important.

Neurotoxicity of anesthetic agents

At birth the mammalian CNS is only partially developed. It undergoes a period of frenetic synaptogenesis in the postnatal period (Dobbing & Sands, 1979). This is the 'brain growth spurt' (Dobbing & Sands, 1979). The timing of the brain growth spurt varies between species. In rats, it occurs from 1 to 2 days before birth to 1–2 weeks after birth. In non-human primates, it begins around the second trimester in utero and ends 5 weeks postpartum. In *Homo sapiens* it begins in utero from the sixth gestational month and ends in the early stages of infancy (Dobbing, 1974; Dobbing & Sands, 1979).

Trillions of synaptic connections form during the brain growth spurt (Jevtovic-Todorovic, 2016). Each neuron vastly expands its dendritic surfaces to accept contact with incoming axons (Jevtovic-Todorovic, 2016). Neurons must establish their final destination and form meaningful connections in the early stages of CNS development (Jevtovic-Todorovic, 2016). These carefully targeted connections are the morphological and physiological foundations of functional neuronal circuits (Jevtovic-Todorovic, 2016). Neurons without useful connections are redundant. Such neurons are destined to die by apoptosis (i.e., programmed cell death) (Jevtovic-Todorovic, 2016; Jevtovic-Todorovic et al., 2003). The apoptosis which occurs naturally during the normal development of the mammalian CNS is a tightly controlled physiological process.

198 **PART | III** Adverse effects, reactions, and outcomes

Unphysiological changes in the synaptic milieu may initiate apoptosis inappropriately. A disturbance of the balance between glutamatergic and GABAergic neurotransmission by excessive suppression of neuronal activity at a crucial stage of CNS development could be a generic signal for neuronal apoptosis (Jevtovic-Todorovic, 2016).

The aim of GA is to induce unconsciousness and loss of pain sensation. So, it is important to question whether anesthetic agents could initiate widespread neuronal apoptosis. This is a particularly emotive issue because millions of children worldwide receive GA every year. The vast majority of these children only receive a single exposure for a relatively minor surgical intervention (Jevtovic-Todorovic, 2016). However, sadly some children require multiple operations and/or are deeply sedated in intensive care units for prolonged periods (Jevtovic-Todorovic, 2016). This latter, the more fragile cohort is therefore exposed to several anesthetic agents at a critical phase of CNS development (Jevtovic-Todorovic, 2016).

Over 70 years ago, it was noted that when children required GA for an operation the postoperative behavioral changes in those under 2 years of age were worse than those in older children (Levy, 1945). At that time these postoperative neurocognitive effects were attributed to the emotional and physical trauma of having surgery (Levy, 1945). As mentioned earlier, the possibility that anesthetic agents may be toxic to the developing brain was not systematically investigated until relatively recently. However, emerging data strongly suggest that general anesthetic agents are toxic to developing animal brains and imply that these agents could be detrimental to the developing human brain in vivo.

Historical overview

The side effects of GA on the brain were first described in the 1970s. Diethyl ether and halothane were found to induce brain injury if diffused directly into the brain (Wilkinson et al., 1973). Then, exposure of rats during early development (but not adult rats) to environmental concentrations of halothane in operating rooms was found to affect synapses and reduce the ability to learn (Quimby, Aschkenase, Bowman, Katz, & Chang, 1974; Quimby, Katz, & Bowman, 1975). It was subsequently established that rats were more sensitive to halothane during the second trimester in utero (i.e., during organogenesis). The offspring of pregnant rats exposed to halothane during this period demonstrated persistent learning deficits as adults (Smith, Bowman, & Katz, 1978).

The topic GA-induced neurotoxicity was pushed into the spotlight when it was reported that alcohol or ketamine can trigger extensive neuroapoptosis, and long-term spatial learning and memory disorders in postnatal day 7 rats (Ikonomidou et al., 1999, 2000). The developing CNS is most vulnerable to general anesthetics (i.e., NMDA antagonists or GABA agonists) during the 'brain growth spurt' (Dobbing, 1974). The CNS is much less vulnerable in the later stages of synaptogenesis (Rizzi, Carter, Ori, & Jevtovic-Todorovic, 2008; Yon, Daniel-Johnson, Carter, & Jevtovic-Todorovic, 2005).

Anesthetic agents cause significant widespread apoptosis of developing neurons i n several mammalian species (Jevtovic-Todorovic, 2016; Jevtovic-Todorovic et al., 2003). For example, the landmark study by Jevtovic-Todorovic demonstrated that exposure of postnatal day 7 rats to midazolam, isoflurane and nitrous oxide for 6 h caused neuronal apoptosis in several regions of the brain (Jevtovic-Todorovic et al., 2003).

Silver staining can characterize patterns of cells death. This technique was used to examine the extent of neurodegeneration in different regions of the brain of young rats (7 days old) subjected to anesthesia (Jevtovic-Todorovic et al., 2003). The effects of 6 h exposure to a cocktail of midazolam, nitrous oxide, and isoflurane on indices of neurodegeneration was staggering: up to 68 fold increases in the laterodorsal region of the thalamic nuclei [Table 1; (Jevtovic-Todorovic et al., 2003)]: There are of course more sophisticated measures of apoptosis and neurodegenerative processes but the study admirably demonstrates the degree to which anesthetics can impact on different brain regions.

This exposure to a "cocktail" that is commonly used by anesthetists was also associated with prolonged impairment in learning and memory. An explosion of research showing that other anesthetic agents induced apoptosis of neurons during CNS development followed in the wake of these studies.

Subsequent animal studies demonstrated that GA caused extensive acute injury and long-term cognitive impairments (Abdelghani, Thakore, Kaphle, Lasky, & Kheir, 2019; Brambrink et al., 2010; Creeley et al., 2014; Pearn et al., 2018; Zimin et al., 2018). Even worse, epigenetic modulation resulted in adverse effects on the memory of the offspring of rodents exposed to GA as neonates (Dalla Massara et al., 2016; Ju et al., 2019; Palanisamy et al., 2011). Taken together, these preclinical data support the notion that exposure of the developing brain to GA may have long-term adverse neurodevelopmental impact.

In 2016, the FDA issued a warning that prolonged or repeated exposure to GA in children younger than 3 years or in pregnant women during their third trimester may have neurodevelopmental effects (*FDA Drug Safety Communication: FDA Review Results in New Warnings about Using General Anesthetics and Sedation Drugs in Young Children and Pregnant Women | FDA*, 2016). The FDA also advised a change in labelling of anesthetic and sedative agents (*FDA Drug Safety Communication: FDA Review Results in New Warnings about Using General Anesthetics and Sedation Drugs in*

TABLE 1 Fold increases in neurodegeneration in response to anesthesia compared to controls.

60–70-fold increase in neurodegeneration

Thalamic nuclei: Laterodorsal

50–60-fold increase in neurodegeneration

Subiculum

Thalamic nuclei: Anteroventral

Retrosplenial cortex

Thalamic nuclei: Anteromedial

40–50-fold increase in neurodegeneration

Thalamic nuclei: Parafascicularis

Globus pallidus

30–40-fold increase in neurodegeneration

Neocortex (layers II and IV): Occipital

Thalamic nuclei: Reuniens

Thalamic nuclei: Paraventricular

Neocortex (layers II and IV): Parietal

Hypothalamus: Ventromedial

Nucleus accumbens

Mammillary complex

Cingulate cortex 3

20–30-fold increase in neurodegeneration

Diagonal band of Broca

Thalamic nuclei: Anterodorsal

Medial septal nucleus

Rostral caudate nucleus

Amygdaloid nuclei: Medial

Amygdaloid nuclei: Cortical

Neocortex (layers II and IV): Temporal

Hypothalamus: Anterior

Amygdaloid nuclei: Basolateral

Hypothalamus: Dorsomedial

Hippocampus, rostral CA1

10–20-fold increase in neurodegeneration

Neocortex (layers II and IV): Frontal

Rats aged 7-days-old were administered a combination of midazolam, nitrous oxide, and isoflurane. This cocktail of anesthetic agents is typically used clinically in pediatric anesthesia. Rats were anesthetized for 6 h. The degree of neurodegeneration in sections of the brain was assessed by counting the extent of silver staining (argyrophilic profiles). In the original article the fold increases are presented on a regional basis. To emphasize the magnitude of the changes we have ranked data in terms of fold changes. Fold increases relate to the number of times the density of degenerating neurons have increased compared to control brains. In all cases $P < 0.001$. The data only pertains only regions where the neurodegeneration rate was greater than 15-fold (Jevtovic-Todorovic et al., 2003). (Data adapted from Jevtovic-Todorovic, V., Hartman, R. E., Izumi, Y., Benshoff, N. D., Dikranian, K., Zorumski, C. F., et al. (2003). Early exposure to common anesthetic agents causes widespread neurodegeneration in the developing rat brain and persistent learning deficits. *Journal of Neuroscience, 23*(3), 876–882. doi:10.1523/jneurosci.23-03-00876.2003.)

Young Children and Pregnant Women| FDA, 2016) and then approved these label changes in 2017 (*FDA Drug Safety Communication: FDA Approves Label Changes for Use of General Anesthetic and Sedation Drugs in Young Children*|FDA, 2017). These regulatory actions heightened the debate over this controversial issue. That exposure of human neonates to GA can cause long-term neurodevelopmental impairment is not yet definitively proven. As yet, there is no definite clinical evidence establishing a causal link between early life GA exposure to long-term neurobehavioral abnormalities. The warning issued by the FDA has been questioned by authors highlighting the limitations of the currently available data (Warner, Shi, & Flick, 2018). Moreover, subsequent clinical trials have not suggested any causal association between adverse long-term neurodevelopmental outcomes and exposure to GA in early life (McCann et al., 2019; Warner et al., 2018).

The translation of preclinical data to humans is always difficult. However, specific issues weaken the power of preclinical data in this field. For example, in clinical trials, the indication for anesthesia (i.e., surgery etc.) and/or early social stress (e.g., maternal separation and social isolation (Reinwald et al., 2018)), may have confounding effects.

Effect of surgical stress on the developing brain

In clinical practice anesthesia is only administered to allow surgery or another invasive procedure to be performed and reduce the stress of the procedure. Surgery is associated tissue injury, pain, and neuronal stimulation. Thus, the purported neurotoxicity of GA during CNS development requires confirmation in the context of surgical stimulation.

In comparison to anesthesia alone, anesthesia and noxious stimulation exacerbated neuronal apoptosis and worsened long-term cognitive impairments (Shu et al., 2012). These observations shatter the supposition that surgical stimulation may 'protect' subjects from the neurotoxic effects of anesthesia. However, a prospective study showed that the effect of midazolam was weaker than that of surgery on the hippocampal development of preterm infants who had surgery (Duerden et al., 2016). Noncardiac surgery in preterm infants (<42 weeks) was an independent risk factor for cognitive impairments at the age of 3–6 years (Gano et al., 2015). These observations suggest that the impact of surgery on the cognitive performance of infants can outweigh the effects of anesthesia.

Preclinical models show that ketamine reduces cell death in the brains of rats exposed to inflammatory pain on postnatal days 1–4 (Rovnaghi, Garg, Hall, Bhutta, & Anand, 2008). However, exposure of these rats to ketamine without noxious stimuli caused acute neural injury and long-term cognitive impairment (Dong, Rovnaghi, & Anand, 2016; Chaoxuan Dong, Rovnaghi, & Anand, 2014). In conjunction, these observations suggest that assessing the effect of GA on the developing brain without surgery probably overestimates the neurotoxic effects of GA. Regardless, it is important to consider the evidence from studies of the effect of GA on *Homo sapiens*.

Cognitive and behavioral development in *Homo sapiens* after childhood surgery

Data suggest that exposure to GA may have long-term cognitive effects on children were first reported over 30 years ago (Backman & Kopf, 1986). They described an increased incidence of postoperative cognitive impairment after children were anesthetized with ketamine and halothane for a minor procedure (removal of congenital nevi). They found that some children developed regressive behavioral changes that lasted up to 18 months postprocedure. Children under 3 years old were the most sensitive.

Clinical investigations are still embryonic, but the emerging data suggest that exposure of very young children to anesthesia may have detrimental effects on cognition and behavior. A retrospective cohort study of 5357 children found that the risk of learning disability in adolescents was increased if they had received more than one general anesthetic as a child under 4 years of age (Wilder et al., 2009). Alarmingly, the cognitive scores of the subset exposed to anesthesia under the age of 4 years were two standard deviations below predicted. Thus, early exposure to GA may prevent achievement of full academic potential.

An increased risk of behavioral disabilities in later life has also been associated with procedures performed in premature infants (e.g., premature infants with patent ductus arteriosus requiring surgical closure (Chorne, Leonard, Piecuch, & Clyman, 2007)). However, the severity of their disease (i.e., necessitating surgical intervention) may have contributed to this observation.

In Iowa 12%–14% of otherwise healthy children exposed to GA at a very young age performed below the 5th percentile on academic achievement tests (Block et al., 2012). While there are several possible confounding factors, this result is concerning. After adjustment for comorbidities, children who had more than one exposure to anesthesia under the age of 2 years had a significantly higher risk of developing hyperactivity problems (Block et al., 2012).

One study examined the impact of general anesthesia on Attention-Deficit/Hyperactivity Disorder (ADHD) (Sprung et al., 2012). The study included the analysis of over 5000 children who received general anesthesia or no general anesthesia before aged 2 years. Cases of ADHD identified before19 years of age. The study showed that those repeatedly exposed to general anesthesia at an early age were significant at risk of ADHD (Fig. 3). Those with no exposure to general anesthesia had an accumulative incidence of ADHD of 7%. Those exposed to single episodes of general anesthesia had an accumulative incidence of ADHD of approximately 11%.while those who received 2 or more episodes the incidence was approximately 18% (Sprung et al., 2012).

The Mayo Anesthesia Safety in Kids (MASK) Study found that anesthesia exposure before the age of 3 years was not associated with deficits in the primary outcome of general intelligence (Warner, Zaccariello, et al., 2018). However, recently published data, extracted from the MASK study and analysed by factor and cluster analyses, showed that multiple exposures to GA before the age of 3 years were associated with specific deficits in neuropsychological tests (Zaccariello et al., 2019).

When interpreting these data it is important to consider the complexity of children's development. Age of exposure seems to be a very important determinant of the vulnerability to the potential adverse effects of anesthesia on cognition. However, the subtlety of the neurocognitive assessment influences the ability to diagnose deficits at any given age (Ing et al., 2012; Ing, DiMaggio, Malacova, et al., 2014; Ing, DiMaggio, Whitehouseet al., 2014). For example, the outcomes of studies of 10-year-old children exposed to anesthesia when under 3 years old depends on the assessment tools which were used.

Some studies reported that the academic achievement scores of children exposed to anesthesia were not significantly different to those of unexposed children (Ing, DiMaggio, Malacova, et al., 2014). However, the exposed children had subtle but significant impairment in language abilities and increased risk of disability in cognition (Ing, DiMaggio, Malacova, et al., 2014). These deficits were detected even in children who had only had a single exposure to anesthesia (Ing, DiMaggio, Malacova, et al., 2014). These results suggest that the choice of assessment tools and outcome measures are crucially important in the ability to capture anesthesia-induced developmental neurotoxicity.

Thus, the few clinical studies available to date associate exposure of neonates and infants to surgery and GA with significant and prolonged neurocognitive and behavioral effects. However, all of the data are retrospective. So, none of the studies conducted to date could control for any confounding factors.

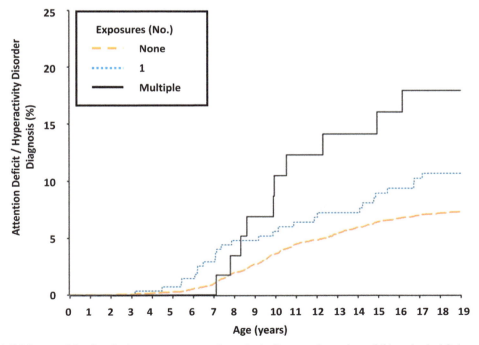

FIG. 3 Attention-deficit/hyperactivity disorder in response to general anesthesia. Data pertains to those children that had 0, 1, or multiple procedures under general anesthesia. Those with no exposure to general anesthesia had an accumulative incidence of ADHD of 7%. Those exposed to single episodes of general anesthesia had an accumulative incidence of ADHD of approximately 11%. While in those who received 2 or more episodes the incidence was approximately 18% For further details see Sprung et al. (2012).

202 PART | III Adverse effects, reactions, and outcomes

It is impossible to conduct a rigorous randomized, double-blinded prospective clinical study to address this question. So, investigators in this field can only obtain observational data in *Homo sapiens*. Therefore, it is of the utmost importance to continue to conduct well-designed observational clinical studies to predict the risk of cognitive and behavioral complications after GA and identify strategies to reduce the risk.

The effect of anesthesia on neuronal networks

The morphological changes described above can be detected on histological assessment. They represent substantial changes in neuronal structure and networks. However, some data suggest that the function of the remaining, apparently 'unaffected' neurons may also be suboptimal (Jevtovic-Todorovic, 2016; Jevtovic-Todorovic et al., 2003). Thus, subtle changes may persist in the 'survivors' long after the grossly damaged neurons have been removed.

For example, exposure of postnatal day 7 rats to GA causes a long-term impairment of synaptic transmission that can be detected in the hippocampus from postnatal day 27 to day 33 (Jevtovic-Todorovic et al., 2003). Despite robust short-term potentiation, long-term potentiation is significantly impaired. This suggests that the exposure to GA results in persistent disturbance of neuronal circuits in the young hippocampus.

A deficit in long-term potentiation was confirmed on examination of synaptic transmission using patch clamp recordings of evoked inhibitory post synaptic current and evoked excitatory post synaptic current (Sanchez et al., 2011). The mechanisms for the long-lasting changes in synaptic communication after anesthesia remain unknown. Some data suggest that anesthetics impair axon targeting. This may be mediated by inhibition of axonal growth cone collapse, resulting in lack of proper response to guidance cues, thus causing axon targeting errors (Mintz, Barrett, Smith, Benson, & Harrison, 2013). As the hippocampus is critical to learning and the development of memory, this observation has significant repercussions.

The long-term effects of excessive neurotransmitter modulation during CNS development

The studies described above demonstrate that exposure to anesthesia depletes neurones (Nikizad, Yon, Carter, & Jevtovic-Todorovic, 2007; Rizzi et al., 2008) and results in prolonged impairment of synapses (Briner et al., 2010, 2011; Head et al., 2009; Lunardi, Ori, Erisir, & Jevtovic-Todorovic, 2010). These effects seem to translate into lasting behavioral effects. For example, the cognitive development of control animals was faster than the development of animals exposed to general anesthetics at the peak of synaptogenesis (Fredriksson, Archer, Alm, Gordh, & Eriksson, 2004; Fredriksson, Pontén, Gordh, & Eriksson, 2007). The gap widened into adulthood. The impact of anesthesia on the incidence of ADHD has already been described (Fig. 3).

Intravenous general anesthetics like propofol or thiopental do not alter mouse behavior when administered alone. However, when administered in combination with ketamine on postnatal day 10, they alter behavior later in young adulthood (Fredriksson et al., 2007). Ketamine seems to be particularly neurotoxic. Administration of ketamine alone at early stages of brain development in rats causes later deficits in habituation, memory and learning (Fredriksson et al., 2004). Experimental animal data suggest that cognitive deficits are more profound with "cocktails" of anesthetics (Fredriksson et al., 2004, 2007; Jevtovic-Todorovic et al., 2003). This is alarming because anesthetic agents are frequently combined in clinical practice.

It is plausible that neuronal apoptosis induced by anesthesia is at least partially responsible for some the neurocognitive changes observed in these studies. Data suggest that during vulnerable periods, multiple, short exposures to anesthesia may significantly impair structural and cognitive development (Han et al., 2015; Zou et al., 2009). Yet, causality is difficult to establish.

Observations from rodent studies translate poorly to *Homo sapiens*. Data from rodent models cannot be directly extrapolated to clinical practice. So, to minimize the effect of differences between species, the effect of GA on nonhuman primates has been studied.

Studies in nonhuman primates

Exposure to ketamine anesthesia for 3 h induced neuronal apoptosis in postnatal day 5 rhesus monkeys (Slikker et al., 2007). Exposure to 1.5% isoflurane for 5 h induced neuronal apoptosis in postnatal day 6 rhesus monkeys (Brambrink et al., 2010). Postnatal day 7 rhesus monkeys developed prolonged cognitive impairments after exposure to ketamine anesthesia for 24 h. Although ketamine is not commonly used to sedate patients in intensive care units, exposure of critically ill patients to sedative doses of other anesthetic agents (e.g., propofol, midazolam, and/or an opiate) for more than 24 h is common. The primates exposed to ketamine exhibited long-term disturbances in all important aspects of cognitive development (Paule et al., 2011). These included motivation, learning and psychomotor speed (Paule et al., 2011). These effects occurred despite an absence of physiological or metabolic derangement.

In rhesus monkeys, as in rats, neonatal exposure to anesthetic agents alone or in combination induces neuronal and glial apoptosis and long-term cognitive impairment (Brambrink et al., 2012, 2010; Creeley et al., 2014; Paule et al., 2011; Slikker et al., 2007; Zimin et al., 2018; Zou et al., 2011). For example caspase 3 a marker of apoptosis increases in different regions of the developing mammalian brain in response to inhalational anesthetics (Fig. 4) Furthermore, GA affected myelin formation in the developing brains of nonhuman primates (Zhang et al., 2019). Sevoflurane exposure in early life disrupted folate metabolism and inhibited oligodendrocyte development in nonhuman primates (Zhang et al., 2019). Nonhuman primates are very similar to *Homo sapiens* in terms of timing and duration of the brain growth spurt and the complexity of the CNS. Thus, these data suggest that translation to *Homo sapiens* is likely. However, the experimental setting is fraught with challenges.

Thus, in recent years, considerable effort has been put into elucidating the mechanisms of anesthesia-induced apoptosis of developing neurons. It has been found to involve several cascades of cellular events that ultimately lead to neuronal deletion and impairment of accurate synaptogenesis.

In this chapter we have reviewed the data suggesting that anesthetic agents are toxic to the developing brain. However, exposure to surgery and GA is also associated with postoperative delirium (POD) and postoperative cognitive dysfunction (POCD) in frail, elderly patients (Berger et al., 2015; Pappa, Theodosiadis, Tsounis, & Sarafis, 2017).

Application to other areas: The aging brain

In 1998, an international multicenter study on the long-term POCD (ISPOCD1) found that 1 week after major noncardiac surgery POCD affected over 25% of elderly patients. After 3 months 10% of patients were still affected by POCD (Moller et al., 1998).

However, the clinical data suggesting that anesthesia is relevant to the POCD in elderly *Homo sapiens* are even less conclusive than the observations in children. The pathophysiology of POD and POCD is unclear and many questions remain unanswered (Berger et al., 2015). However, some preclinical data support the suggestion that exposure to GA may cause neurocognitive side effects in the elderly.

In a rodent model of POCD, exposure to isoflurane and nitrous oxide caused cognitive deficits that lasted for weeks (Culley, Baxter, Yukhananov, & Crosby, 2003). However, the POCD seemed to be transient in all cases. A later study found that the effect of isoflurane was greater than that of nitrous oxide (Culley, Baxter, Crosby, Yukhananov, & Crosby, 2004). Furthermore, two characteristic features of Alzheimer's disease (tauopathy and excessive plaque formation) are influenced by general anesthetics (Tang & Eckenhoff, 2013). Thus, it is possible that there is an interaction between exposure to GA and the pathogenesis of Alzheimer's disease may be relevant to POCD. It is also plausible that the pathogenesis of POCD may involve neuronal apoptosis similar to that which occurs when young children are exposed to general anesthesia during the brain growth spurt.

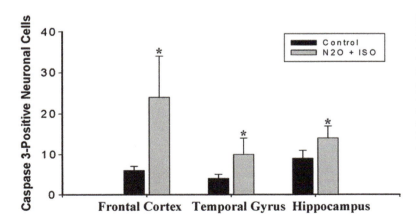

FIG. 4 Caspase 3 positivity in different regions of the brain in response to nitrous oxide gas and isoflurane. Key: N20, Nitrous oxide. ISO, isoflourane. Data as mean + SD ($n = 5$). * indicates $P < 0.05$. Exposure of infant monkeys was with 70% N_2O and 1% isoflurane. For further details see Zou et al. (2011). *(From Zou, X., Liu, F., Zhang, X., Patterson, T. A., Callicott, R., Liu, S., et al. (2011). Inhalation anesthetic-induced neuronal damage in the developing rhesus monkey.* Neurotoxicology and Teratology, 33(5), 592–597. https://doi.org/10.1016/j.ntt.2011.06.003 *with permission.)*

However, the clinical data are far from conclusive. For example, the parameters relevant to exposure to GA investigated in the ISPOCD1 study, only the duration of anesthesia was found to be a risk factor for early POCD. However, duration of anesthesia was not a risk factor for late POCD. Furthermore, in clinical practice the duration of anesthesia is fundamentally dependent on the duration of surgery. Moreover, the ISPOCD1 study and the subsequent ISPOCD studies failed to demonstrate any association of POCD with the type of anesthesia or the administration of any specific anesthetic agents (Funder & Steinmetz, 2012).

Conclusion

The association between exposure of children to GA and the risk of adverse long-term neurodevelopmental is complex and summarized in Fig. 5. The influence of genetic factors is unknown (Fig. 5). Interrelationships and outcomes have been hotly debated since the turn of the century. The FDA recently entered the fray, pouring fuel on the fire. Yet, despite monumental international efforts, causality has not been definitively determined and the pathogenesis remains unclear. It is impossible to conduct randomized clinical trials to resolve this issue. So, well-designed observation studies are required to define the risk factors for these adverse postoperative effects on cognition and to develop strategies for their prevention.

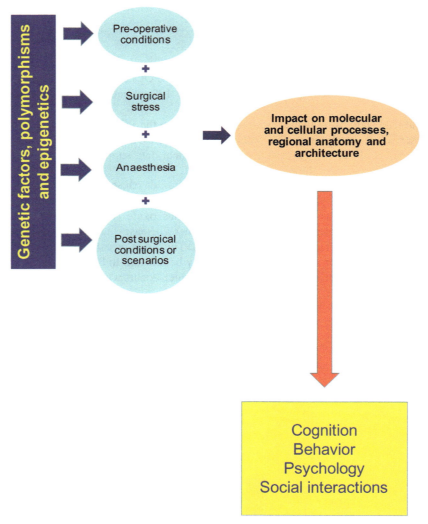

FIG. 5 Summary of the effects of anesthesia on the brain. In this schematic figure consideration must be given to the fact that the impact of anesthesia on the brain will also include the effects of preoperative conditions, surgical stress and postsurgical conditions or scenarios. There will also be interplay with genetic profiles, polymorphisms and epigenetics (compiled by the authors).

Mini-dictionary of terms

Apoptosis: programmed cell death. There are a variety of biochemical events that will lead to apoptosis, including DNA fragmentation. Apoptosis is distinct from necrosis. For example, apoptosis may affect single cells or clusters whereas necrosis will impact of contiguous (continuous) cells.

General anesthesia: General anesthesia is a medically induced coma. In this state protective reflexes are lost. It is induced by the administration of one or a 'cocktail' of general anesthetic agents.

General anesthetic agents: medications that induce a state of anesthesia when administered at an appropriate dose. These are four classes of medications which are used as anesthetic agents. These include intravenous and inhalational anesthetics, analgesics, and neuromuscular blocking drugs.

Neurotoxicity: toxin-induced damage to the central and/or peripheral nervous system.

Neurotransmitter: Chemicals made by a nerve cell to specifically transmit a message. These chemicals are released from a nerve cell across a synapse with a target cell. The recipient can be another cell in a muscle, nerve or gland cell.

Oligodendrocyte: A large central nervous system glial cell. Some oligodendrocytes produce the myelin sheath insulating neuronal axons.

Post-operative cognitive dysfunction (POCD): The response of cognitive processes to surgical stress or interventional procedures that require anesthesia. POCD is one component of a spectrum of disorders and conditions that arise after surgical stress. Virtually all body systems are impacted by surgical stress in one way or another. In terms of POCD it has been suggested that 10–30% of patients are affected at or after discharge.

Synaptogenesis: The formation of contacts (i.e., synapses) between neurons.

Key facts of the United States Federal Drug Administration (FDA)

The FDA was founded over a hundred years ago in 1906 though under another name.

The FDA mission is *"The Food and Drug Administration is responsible for protecting the public health by ensuring the safety, efficacy, and security of human and veterinary drugs, biological products, and medical devices; and by ensuring the safety of our nation's food supply, cosmetics, and products that emit radiation."*

The FDA is involved in regulating, controlling or overseeing aspects of medical devices, drugs (veterinary and human), food, tobacco and other health related domains.

FDA regulates both prescription and nonprescription drugs: the latter are also called "over the counter drugs".

The time to time the FDA will recall drugs including anesthetics on the grounds of safety.

The **Anesthetic and Analgesic Drug Products Advisory Committee (AADPAC)** functions under the umbrella of the FDA.

Key facts of anesthetic neurotoxicity

Anesthesia is necessary to ensure patients' comfort during unpleasant interventions.

In the 1970s, animal studies raised concerns that anesthetic agents may be neurotoxic.

The long-term outcomes after GA have been scrutinized in a range of in vitro and in vivo studies.

The pathophysiology of anesthesia-induced neurotoxicity is complex and poorly understood.

The United States Federal Drug Administration (FDA) has issued an alert about the neurotoxic effects of GA.

Recent data from human studies also suggest an association between exposure to anesthesia in early childhood and subsequent neurodevelopmental deficit, but causation has not been established.

Key facts of postoperative cognitive dysfunction

One week after major noncardiac surgery over 25% of elderly patients are affected by POCD. After 3 months 10% of patients are still affected by POCD.

The pathophysiology of POCD is unclear.

Preclinical data suggest that exposure to GA may cause neurocognitive side effects in the elderly.

The clinical studies investigating the relationship between anesthesia and POCD in elderly *Homo sapiens* are inconclusive.

Other agents of interest

No new anesthetics have been developed for over 40 years. Historically, the discovery of novel anesthetic agents has generally involved serendipity as much as strategy (McKinstry-Wu et al., 2015). Drug development thereafter has generally involved modification of the existing agents, rather than the invention of new chemotypes (McKinstry-Wu et al., 2015). The safety profile of agents such as remimazolam (a midazolam analogue) and methoxycarbonyl etomidate are likely to be similar to their parent molecules. To develop new types of anesthetic agents that are safer and more effective a new paradigm of anesthetic drug discovery is required. In this context the ability of to screen the ability of thousands of molecules to bind to a single protein target has recently become possible with high-throughput analyses (McKinstry-Wu et al., 2015). Apoferritin, a soluble protein, binds anesthetic agents in proportion to their potency (McKinstry-Wu et al., 2015). Indeed, high throughput techniques have identified 2600 compounds from a library of 350,000 that prevented 1-aminoanthracene from binding to apoferritin (McKinstry-Wu et al., 2015). Further studies suggested that two of these novel agents could effectively induce anesthesia in mice (McKinstry-Wu et al., 2015). However, many further stages of drug development will be required before these agents can be introduced into routine clinical practice (McKinstry-Wu et al., 2015). The preclinical and clinical trials of these, and any other potential candidate anesthetic agents, must certainly evaluate their potential for neurotoxicity.

Summary points

Anesthesia-induced neurotoxicity may be particularly damaging to the developing brain.

Generalized neuronal apoptosis is an important mechanism of anesthesia-induced developmental neurotoxicity;

Long-lasting impairments in neuronal communication and faulty formation of neuronal circuits may result from exposure to anesthesia at a young age;

In animals and possibly in humans, exposure to anesthesia at extremes of brain age may cause prolonged reduction of cognitive and behavioral performance.

The mechanisms of neurotoxicity of anesthetics to the developing brain may be relevant to postoperative cognitive dysfunction in the elderly.

References

Abdelghani, R., Thakore, S., Kaphle, U., Lasky, J. A., & Kheir, F. (2019). Radial endobronchial ultrasound-guided transbronchial cryobiopsy. *Journal of Bronchology & Interventional Pulmonology*. https://doi.org/10.1097/LBR.0000000000000566.

Allen, N. J., & Barres, B. A. (2005). Signaling between glia and neurons: Focus on synaptic plasticity. *Current Opinion in Neurobiology*, *15*(5), 542–548. https://doi.org/10.1016/j.conb.2005.08.006.

Backman, M. E., & Kopf, A. W. (1986). Iatrogenic effects of general anesthesia in children: Considerations in treating large congenital nevocytic nevi. *The Journal of Dermatologic Surgery and Oncology*, *12*(4), 363–367. https://doi.org/10.1111/j.1524-4725.1986.tb01921.x.

Ben-Ari, Y. (2002). Excitatory actions of gaba during development: The nature of the nurture. *Nature Reviews. Neuroscience*, *3*(9), 728–739. https://doi.org/10.1038/nrn920.

Berger, M., Nadler, J. W., Browndyke, J., Terrando, N., Ponnusamy, V., Cohen, H. J., et al. (2015). Postoperative cognitive dysfunction: Minding the gaps in our knowledge of a common postoperative complication in the elderly. *Anesthesiology Clinics*, *33*(3), 517–550. https://doi.org/10.1016/j.anclin.2015.05.008.

Block, R. I., Thomas, J. J., Bayman, E. O., Choi, J. Y., Kimble, K. K., & Todd, M. M. (2012). Are anesthesia and surgery during infancy associated with altered academic performance during childhood? *Anesthesiology*, *117*(3), 494–503. https://doi.org/10.1097/ALN.0b013e3182644684.

Brambrink, A. M., Back, S. A., Riddle, A., Gong, X., Moravec, M. D., Dissen, G. A., et al. (2012). Isoflurane-induced apoptosis of oligodendrocytes in the neonatal primate brain. *Annals of Neurology*, *72*(4), 525–535. https://doi.org/10.1002/ana.23652.

Brambrink, A. M., Evers, A. S., Avidan, M. S., Farber, N. B., Smith, D. J., Zhang, X., et al. (2010). Isoflurane-induced neuroapoptosis in the neonatal rhesus macaque brain. *Anesthesiology*, *112*(4), 834–841. https://doi.org/10.1097/ALN.0b013e3181d049cd.

Briner, A., De Roo, M., Dayer, A., Muller, D., Habre, W., & Vutskits, L. (2010). Volatile anesthetics rapidly increase dendritic spine density in the rat medial prefrontal cortex during synaptogenesis. *Anesthesiology*, *112*(3), 546–556. https://doi.org/10.1097/ALN.0b013e3181cd7942.

Briner, A., Nikonenko, I., De Roo, M., Dayer, A., Muller, D., & Vutskits, L. (2011). Developmental stage-dependent persistent impact of propofol anesthesia on dendritic spines in the rat medial prefrontal cortex. *Anesthesiology*, *115*(2), 282–293. https://doi.org/10.1097/ALN.0b013e318221fbbd.

Chorne, N., Leonard, C., Piecuch, R., & Clyman, R. I. (2007). Patent ductus arteriosus and its treatment as risk factors for neonatal and neurodevelopmental morbidity. *Pediatrics*, *119*(6), 1165–1174. https://doi.org/10.1542/peds.2006-3124.

Creeley, C. E., Dikranian, K. T., Dissen, G. A., Back, S. A., Olney, J. W., & Brambrink, A. M. (2014). Isoflurane-induced apoptosis of neurons and oligodendrocytes in the fetal rhesus macaque brain. *Anesthesiology*, *120*(3), 626–638. https://doi.org/10.1097/ALN.0000000000000037.

Culley, D. J., Baxter, M. G., Crosby, C. A., Yukhananov, R., & Crosby, G. (2004). Impaired acquisition of spatial memory 2 weeks after isoflurane and isoflurane-nitrous oxide anesthesia in aged rats. *Anesthesia and Analgesia, 99*(5), 1393–1397. table of contents https://doi.org/10.1213/01.ANE.0000135408.14319.CC.

Culley, D. J., Baxter, M., Yukhananov, R., & Crosby, G. (2003). The memory effects of general anesthesia persist for weeks in young and aged rats. *Anesthesia and Analgesia, 96*(4), 1004–1009. table of contents https://doi.org/10.1213/01.ane.0000052712.67573.12.

Dalla Massara, L., Osuru, H. P., Oklopcic, A., Milanovic, D., Joksimovic, S. M., Caputo, V., et al. (2016). General anesthesia causes epigenetic histone modulation of c-Fos and brain-derived neurotrophic factor, target genes important for neuronal development in the immature rat hippocampus. *Anesthesiology, 124*(6), 1311–1327. https://doi.org/10.1097/ALN.0000000000001111.

Dobbing, J. (1974). The later growth of the brain and its vulnerability. *Obstetrical and Gynecological Survey, 29*(7), 468–470. https://doi.org/10.1097/00006254-197407000-00012.

Dobbing, J., & Sands, J. (1979). Comparative aspects of the brain growth spurt. *Early Human Development, 3*(1), 79–83. https://doi.org/10.1016/0378-3782(79)90022-7.

Dong, C., Rovnaghi, C. R., & Anand, K. J. S. (2014). Ketamine affects the neurogenesis of rat fetal neural stem progenitor cells via the PI3K/Akt-p27 signaling pathway. *Birth Defects Research. Part B, Developmental and Reproductive Toxicology, 101*(5), 355–363. https://doi.org/10.1002/bdrb.21119.

Dong, C., Rovnaghi, C. R., & Anand, K. J. S. (2016). Ketamine exposure during embryogenesis inhibits cellular proliferation in rat fetal cortical neurogenic regions. *Acta Anaesthesiologica Scandinavica, 60*(5), 579–587. https://doi.org/10.1111/aas.12689.

Duerden, E. G., Guo, T., Dodbiba, L., Chakravarty, M. M., Chau, V., Poskitt, K. J., et al. (2016). Midazolam dose correlates with abnormal hippocampal growth and neurodevelopmental outcome in preterm infants. *Annals of Neurology, 79*(4), 548–559. https://doi.org/10.1002/ana.24601.

FDA Drug Safety Communication: FDA approves label changes for use of general anesthetic and sedation drugs in young children | FDA (2017). Retrieved November 5, 2020, from https://www.fda.gov/drugs/drug-safety-and-availability/fda-drug-safety-communication-fda-approves-label-changes-use-general-anesthetic-and-sedation-drugs.

FDA Drug Safety Communication: FDA review results in new warnings about using general anesthetics and sedation drugs in young children and pregnant women | FDA (2016). Retrieved November 5, 2020, from https://www.fda.gov/drugs/drug-safety-and-availability/fda-drug-safety-communication-fda-review-results-new-warnings-about-using-general-anesthetics-and.

Franks, N. P. (2008). General anaesthesia: From molecular targets to neuronal pathways of sleep and arousal. *Nature Reviews. Neuroscience, 9*(5), 370–386. https://doi.org/10.1038/nrn2372.

Franks, N. P., Dickinson, R., de Sousa, S. L., Hall, A. C., & Lieb, W. R. (1998). How does xenon produce anaesthesia? *Nature, 396*(6709), 324. https://doi.org/10.1038/24525.

Fredriksson, A., Archer, T., Alm, H., Gordh, T., & Eriksson, P. (2004). Neurofunctional deficits and potentiated apoptosis by neonatal NMDA antagonist administration. *Behavioural Brain Research, 153*(2), 367–376. https://doi.org/10.1016/j.bbr.2003.12.026.

Fredriksson, A., Pontén, E., Gordh, T., & Eriksson, P. (2007). Neonatal exposure to a combination of N-methyl-D-aspartate and gamma-aminobutyric acid type A receptor anesthetic agents potentiates apoptotic neurodegeneration and persistent behavioral deficits. *Anesthesiology, 107*(3), 427–436. https://doi.org/10.1097/01.anes.0000278892.62305.9c.

Funder, K. S., & Steinmetz, J. (2012). Post-operative cognitive dysfunction - lessons from the ISPOCD studies. *Trends in Anaesthesia and Critical Care, 2*(3), 94–97. https://doi.org/10.1016/j.tacc.2012.02.009.

Gano, D., Andersen, S. K., Glass, H. C., Rogers, E. E., Glidden, D. V., Barkovich, A. J., et al. (2015). Impaired cognitive performance in premature newborns with two or more surgeries prior to term-equivalent age. *Pediatric Research, 78*(3), 323–329. https://doi.org/10.1038/pr.2015.106.

Han, T., Hu, Z., Tang, Y., Shrestha, A., Ouyang, W., & Liao, Q. (2015). Inhibiting Rho kinase 2 reduces memory dysfunction in adult rats exposed to sevoflurane at postnatal days 7-9. *Biomedical Reports, 3*(3), 361–364. https://doi.org/10.3892/br.2015.429.

Head, B. P., Patel, H. H., Niesman, I. R., Drummond, J. C., Roth, D. M., & Patel, P. M. (2009). Inhibition of p75 neurotrophin receptor attenuates isoflurane-mediated neuronal apoptosis in the neonatal central nervous system. *Anesthesiology, 110*(4), 813–825. https://doi.org/10.1097/ALN.0b013e31819b602b.

Hemmings, H. C. J., Riegelhaupt, P. M., Kelz, M. B., Solt, K., Eckenhoff, R. G., Orser, B. A., et al. (2019). Towards a comprehensive understanding of anesthetic mechanisms of action: A decade of discovery. *Trends in Pharmacological Sciences, 40*(7), 464–481. https://doi.org/10.1016/j.tips.2019.05.001.

Hirota, K., Roth, S. H., Fujimura, J., Masuda, A., & Ito, Y. (1998). GABAergic mechanisms in the action of general anesthetics. *Toxicology Letters, 100–101*, 203–207. https://doi.org/10.1016/s0378-4274(98)00186-6.

Hudetz, A. G. (2012). General anesthesia and human brain connectivity. *Brain Connectivity, 2*(6), 291–302. https://doi.org/10.1089/brain.2012.0107.

Ikonomidou, C., Bittigau, P., Koch, C., Genz, K., Stefovska, V., Hörster, F., et al. (2000). Ethanol-induced apoptotic neurodegeneration and fetal alcohol syndrome. *Science (New York, N.Y.), 287*(5455), 1056–1060. https://doi.org/10.1126/science.287.5455.1056.

Ikonomidou, C., Bosch, F., Miksa, M., Bittigau, P., Vöckler, J., Dikranian, K., et al. (1999). Blockade of NMDA receptors and apoptotic neurodegeneration in the developing brain. *Science (New York, N.Y.), 283*(5398), 70–74. https://doi.org/10.1126/science.283.5398.70.

Ing, C., DiMaggio, C. J., Malacova, E., Whitehouse, A. J., Hegarty, M. K., Feng, T., et al. (2014). Comparative analysis of outcome measures used in examining neurodevelopmental effects of early childhood anesthesia exposure. *Anesthesiology, 120*(6), 1319–1332. https://doi.org/10.1097/ALN.0000000000000248.

Ing, C., DiMaggio, C., Whitehouse, A., Hegarty, M. K., Brady, J., von Ungern-Sternberg, B. S., et al. (2012). Long-term differences in language and cognitive function after childhood exposure to anesthesia. *Pediatrics, 130*(3), e476–e485. https://doi.org/10.1542/peds.2011-3822.

Ing, C., DiMaggio, C. J., Whitehouse, A. J. O., Hegarty, M. K., Sun, M., von Ungern-Sternberg, B. S., et al. (2014). Neurodevelopmental outcomes after initial childhood anesthetic exposure between ages 3 and 10 years. *Journal of Neurosurgical Anesthesiology*, 26(4), 377–386. https://doi.org/10.1097/ANA.0000000000000121.

Jevtovic-Todorovic, V. (2016). General anesthetics and neurotoxicity: How much do we know? *Anesthesiology Clinics*, 34(3), 439–451. W.B. Saunders https://doi.org/10.1016/j.anclin.2016.04.001.

Jevtovic-Todorovic, V., Hartman, R. E., Izumi, Y., Benshoff, N. D., Dikranian, K., Zorumski, C. F., et al. (2003). Early exposure to common anesthetic agents causes widespread neurodegeneration in the developing rat brain and persistent learning deficits. *Journal of Neuroscience*, 23(3), 876–882. https://doi.org/10.1523/jneurosci.23-03-00876.2003.

Jevtović-Todorović, V., Todorović, S. M., Mennerick, S., Powell, S., Dikranian, K., Benshoff, N., et al. (1998). Nitrous oxide (laughing gas) is an NMDA antagonist, neuroprotectant and neurotoxin. *Nature Medicine*, 4(4), 460–463. https://doi.org/10.1038/nm0498-460.

Ju, L. S., Yang, J. J., Xu, N., Li, J., Morey, T. E., Gravenstein, N., et al. (2019). Intergenerational effects of sevoflurane in young adult rats. *Anesthesiology*, 131(5), 1092–1109. https://doi.org/10.1097/ALN.0000000000002920.

Levy, D. M. (1945). Psychic trauma of operations in children: And a note on combat neurosis. *American Journal of Diseases of Children*, 69(1), 7–25. https://doi.org/10.1001/archpedi.1945.02020130014003.

Lodge, D., & Anis, N. A. (1982). Effects of phencyclidine on excitatory amino acid activation of spinal interneurones in the cat. *European Journal of Pharmacology*, 77(2–3), 203–204. https://doi.org/10.1016/0014-2999(82)90022-x.

Lunardi, N., Ori, C., Erisir, A., & Jevtovic-Todorovic, V. (2010). General anesthesia causes long-lasting disturbances in the ultrastructural properties of developing synapses in young rats. *Neurotoxicity Research*, 17(2), 179–188. https://doi.org/10.1007/s12640-009-9088-z.

McCann, M. E., de Graaff, J. C., Dorris, L., Disma, N., Withington, D., Bell, G., et al. (2019). Neurodevelopmental outcome at 5 years of age after general anaesthesia or awake-regional anaesthesia in infancy (GAS): An international, multicentre, randomised, controlled equivalence trial. *Lancet (London, England)*, 393(10172), 664–677. https://doi.org/10.1016/S0140-6736(18)32485-1.

McKinstry-Wu, A. R., Bu, W., Rai, G., Lea, W. A., Weiser, B. P., Liang, D. F., … Eckenhoff, R. G. (2015). Discovery of a novel general anesthetic chemotype using high-throughput screening. *Anesthesiology*, 122(2), 325–333. https://doi.org/10.1097/ALN.0000000000000505.

Mintz, C. D., Barrett, K. M. S., Smith, S. C., Benson, D. L., & Harrison, N. L. (2013). Anesthetics interfere with axon guidance in developing mouse neocortical neurons in vitro via a γ-aminobutyric acid type A receptor mechanism. *Anesthesiology*, 118(4), 825–833. https://doi.org/10.1097/ALN.0b013e318287b850.

Moller, J. T., Cluitmans, P., Rasmussen, L. S., Houx, P., Rasmussen, H., Canet, J., et al. (1998). Long-term postoperative cognitive dysfunction in the elderly ISPOCD1 study. ISPOCD investigators. International study of post-operative cognitive dysfunction. *Lancet (London, England)*, 351(9106), 857–861. https://doi.org/10.1016/s0140-6736(97)07382-0.

Nikizad, H., Yon, J.-H., Carter, L. B., & Jevtovic-Todorovic, V. (2007). Early exposure to general anesthesia causes significant neuronal deletion in the developing rat brain. *Annals of the New York Academy of Sciences*, 1122, 69–82. https://doi.org/10.1196/annals.1403.005.

Nishikawa, K., & Harrison, N. L. (2003). The actions of sevoflurane and desflurane on the gamma-aminobutyric acid receptor type A: Effects of TM2 mutations in the alpha and beta subunits. *Anesthesiology*, 99(3), 678–684. https://doi.org/10.1097/00000542-200309000-00024.

Palanisamy, A., Baxter, M. G., Keel, P. K., Xie, Z., Crosby, G., & Culley, D. J. (2011). Rats exposed to isoflurane in utero during early gestation are behaviorally abnormal as adults. *Anesthesiology*, 114(3), 521–528. https://doi.org/10.1097/ALN.0b013e318209aa71.

Pappa, M., Theodosiadis, N., Tsounis, A., & Sarafis, P. (2017). Pathogenesis and treatment of post-operative cognitive dysfunction. *Electronic Physician*, 9 (2), 3768–3775. https://doi.org/10.19082/3768.

Paule, M. G., Li, M., Allen, R. R., Liu, F., Zou, X., Hotchkiss, C., et al. (2011). Ketamine anesthesia during the first week of life can cause long-lasting cognitive deficits in rhesus monkeys. *Neurotoxicology and Teratology*, 33(2), 220–230. https://doi.org/10.1016/j.ntt.2011.01.001.

Pearn, M. L., Schilling, J. M., Jian, M., Egawa, J., Wu, C., Mandyam, C. D., et al. (2018). Inhibition of RhoA reduces propofol-mediated growth cone collapse, axonal transport impairment, loss of synaptic connectivity, and behavioural deficits. *British Journal of Anaesthesia*, 120(4), 745–760. https://doi.org/10.1016/j.bja.2017.12.033.

Quimby, K. L., Aschkenase, L. J., Bowman, R. E., Katz, J., & Chang, L. W. (1974). Enduring learning deficits and cerebral synaptic malformation from exposure to 10 parts of halothane per million. *Science*, 185(4151), 625–627. https://doi.org/10.1126/science.185.4151.625.

Quimby, K. L., Katz, J., & Bowman, R. E. (1975). Behavioral consequences in rats from chronic exposure to 10 ppm halothane during early development. *Anesthesia and Analgesia*, 54(5), 628–633. https://doi.org/10.1213/00000539-197509000-00017.

Reinwald, J. R., Becker, R., Mallien, A. S., Falfan-Melgoza, C., Sack, M., Clemm von Hohenberg, C., et al. (2018). Neural mechanisms of early-life social stress as a developmental risk factor for severe psychiatric disorders. *Biological Psychiatry*, 84(2), 116–128. https://doi.org/10.1016/j.biopsych.2017.12.010.

Rizzi, S., Carter, L. B., Ori, C., & Jevtovic-Todorovic, V. (2008). Clinical anesthesia causes permanent damage to the fetal Guinea pig brain. *Brain Pathology (Zurich, Switzerland)*, 18(2), 198–210. https://doi.org/10.1111/j.1750-3639.2007.00116.x.

Rovnaghi, C. R., Garg, S., Hall, R. W., Bhutta, A. T., & Anand, K. J. S. (2008). Ketamine analgesia for inflammatory pain in neonatal rats: A factorial randomized trial examining long-term effects. *Behavioral and Brain Functions*, 4. https://doi.org/10.1186/1744-9081-4-35.

Sanchez, V., Feinstein, S. D., Lunardi, N., Joksovic, P. M., Boscolo, A., Todorovic, S. M., et al. (2011). General anesthesia causes long-term impairment of mitochondrial morphogenesis and synaptic transmission in developing rat brain. *Anesthesiology*, 115(5), 992–1002. https://doi.org/10.1097/ALN.0b013e3182303a63.

Schlich, T. (2017). The history of anaesthesia and the patient-reduced to a body? *Lancet (London, England)*, 390(10099), 1020–1021. https://doi.org/10.1016/S0140-6736(17)32362-0.

Shu, Y., Zhou, Z., Wan, Y., Sanders, R. D., Li, M., Pac-Soo, C. K., et al. (2012). Nociceptive stimuli enhance anesthetic-induced neuroapoptosis in the rat developing brain. *Neurobiology of Disease, 45*(2), 743–750. https://doi.org/10.1016/j.nbd.2011.10.021.

Slikker, W. J., Zou, X., Hotchkiss, C. E., Divine, R. L., Sadovova, N., Twaddle, N. C., et al. (2007). Ketamine-induced neuronal cell death in the perinatal rhesus monkey. *Toxicological Sciences : An Official Journal of the Society of Toxicology, 98*(1), 145–158. https://doi.org/10.1093/toxsci/kfm084.

Smith, R. F., Bowman, R. E., & Katz, J. (1978). Behavioral effects of exposure to halothane during early development in the rat. Sensitive period during pregnancy. *Anesthesiology, 49*(5), 319–323. https://doi.org/10.1097/00000542-197811000-00004.

Sprung, J., Flick, R. P., Katusic, S. K., Colligan, R. C., Barbaresi, W. J., Bojanić, K., et al. (2012). Attention-deficit/hyperactivity disorder after early exposure to procedures requiring general anesthesia. *Mayo Clinic Proceedings, 87*(2), 120–129. https://doi.org/10.1016/j.mayocp.2011.11.008.

Tang, J. X., & Eckenhoff, M. F. (2013). Anesthetic effects in Alzheimer transgenic mouse models. *Progress in Neuro-Psychopharmacology & Biological Psychiatry, 47*, 167–171. https://doi.org/10.1016/j.pnpbp.2012.06.007.

Warner, D. O., Shi, Y., & Flick, R. P. (2018). Anesthesia and neurodevelopment in children: Perhaps the end of the beginning. *Anesthesiology, 128*(4), 700–703. Lippincott Williams and Wilkins https://doi.org/10.1097/ALN.0000000000002121.

Warner, D. O., Zaccariello, M. J., Katusic, S. K., Schroeder, D. R., Hanson, A. C., Schulte, P. J., et al. (2018). Neuropsychological and behavioral outcomes after exposure of young children to procedures requiring general anesthesia: The mayo anesthesia safety in kids (MASK) study. *Anesthesiology, 129* (1), 89–105. https://doi.org/10.1097/ALN.0000000000002232.

Wilder, R. T., Flick, R. P., Sprung, J., Katusic, S. K., Barbaresi, W. J., Mickelson, C., et al. (2009). Early exposure to anesthesia and learning disabilities in a population-based birth cohort. *Anesthesiology, 110*(4), 796–804. https://doi.org/10.1097/01.anes.0000344728.34332.5d.

Wilkinson, H. A., Mark, V. H., Wilson, R., & Patel, P. (1973). The toxicity of general anesthetics diffused directly into the brain. *Anesthesiology, 38*(5), 478–481. https://doi.org/10.1097/00000542-197305000-00011.

Yon, J.-H., Daniel-Johnson, J., Carter, L. B., & Jevtovic-Todorovic, V. (2005). Anesthesia induces neuronal cell death in the developing rat brain via the intrinsic and extrinsic apoptotic pathways. *Neuroscience, 135*(3), 815–827. https://doi.org/10.1016/j.neuroscience.2005.03.064.

Zaccariello, M. J., Frank, R. D., Lee, M., Kirsch, A. C., Schroeder, D. R., Hanson, A. C., et al. (2019). Patterns of neuropsychological changes after general anaesthesia in young children: Secondary analysis of the mayo anesthesia safety in kids study. *British Journal of Anaesthesia, 122*(5), 671–681. https:// doi.org/10.1016/j.bja.2019.01.022.

Zhang, L., Xue, Z., Liu, Q., Liu, Y., Xi, S., Cheng, Y., et al. (2019). Disrupted folate metabolism with anesthesia leads to myelination deficits mediated by epigenetic regulation of ERMN. *eBioMedicine, 43*, 473–486. https://doi.org/10.1016/j.ebiom.2019.04.048.

Zimin, P. I., Woods, C. B., Kayser, E. B., Ramirez, J. M., Morgan, P. G., & Sedensky, M. M. (2018). Isoflurane disrupts excitatory neurotransmitter dynamics via inhibition of mitochondrial complex I. *British Journal of Anaesthesia, 120*(5), 1019–1032. https://doi.org/10.1016/j.bja.2018.01.036.

Zou, X., Liu, F., Zhang, X., Patterson, T. A., Callicott, R., Liu, S., et al. (2011). Inhalation anesthetic-induced neuronal damage in the developing rhesus monkey. *Neurotoxicology and Teratology, 33*(5), 592–597. https://doi.org/10.1016/j.ntt.2011.06.003.

Zou, X., Patterson, T. A., Sadovova, N., Twaddle, N. C., Doerge, D. R., Zhang, X., et al. (2009). Potential neurotoxicity of ketamine in the developing rat brain. *Toxicological Sciences : An Official Journal of the Society of Toxicology, 108*(1), 149–158. https://doi.org/10.1093/toxsci/kfn270.

Chapter 20

Breastfeeding and mother-baby dyad's competence following neuraxial labor analgesia

Roberto Giorgio Wetzl[a], Maria Lorella Gianni[b,c], Enrica Delfino[d], and Alessandra Consales[b,c]

[a]European e-Learning School of Obstetric Anesthesia, Rome, Italy, [b]IRCCS Ca' Granda Ospedale Maggiore Policlinico Foundation, Neonatal Intensive Care Unit, Milan, Italy, [c]Department of Clinical Sciences and Community Health, University of Milan, Milan, Italy, [d]Department of Anaesthesia, Intensive Care, and Out-Hospital Emergency, Aosta Valley Regional Hospital, Beauregard Hospital, Aosta, Italy

Abbreviations

BIS	breastfeeding initiation success
BNBAS	Brazelton neonatal behavioral assessment scale
ENNS	early neonatal neurobehavioral scale
IBFAT	infant breastfeeding assessment tool
LATCH	latch, audible, type, comfort, help
NACS	neurologic and adaptive capacity score
PIBBS	preterm infant breastfeeding behavioral scale

Introduction

Over the last two decades, neuraxial labor analgesia has been increasingly proposed to laboring and delivering women as an effective and safe pain relief strategy (Anim-Somuah, Smyth, Cyna, & Cuthbert, 2018). Nowadays, authoritative international guidelines recognize to laboring and delivering women the right to receive it, even at their own request (WHO, 2018).

At the same time, after a long-lasting decline started after World War II and culminated in the 1970s, breastfeeding has eventually earned the status of "normative standard for infant feeding and nutrition" (Kramer & Kakuma, 2012). UNICEF (2013) suggests that an exclusive breastfeeding support strategy could have a huge impact on global health.

The revival of breastfeeding has raised concerns in the scientific community about the compatibility of the increasingly diffused pharmacological pain relief protocols with breastfeeding support strategies such as the UNICEF/WHO Baby-Friendly Hospital Initiative (2009).

Human rotational delivery and childbirth pain

As a consequence of two evolutionary adaptations due to bipedalism and brain enlargement, human childbirth is a uniquely complex (and sometimes complicated) process (Fig. 1). To overcome the hindrance of a mid-plane pelvic stricture (bispinous diameter) the fetus has to first flex their head and then rotate it and the shoulder girdle, while descending through a twisted, tight birth canal (rotational birth), finally emerging in occiput-anterior position (Gruss & Schmitt, 2015). In evolutionary terms, this made also advantageous for a human mother to be assisted during childbirth, particularly in the final stage of delivery (Trevathan, Smith, & McKenna, 1999).

The severe pain experienced by most women while laboring and delivering has to be evaluated within this context. First-stage pain is due to distension of the cervix and lower uterine segment, it has a colicky, visceral character, and is transmitted to T10-L1 spinal level. The severe, somatic, second-stage pain is due to stretching of cervical, vaginal, and perineal soft structures and is transmitted to S2-S4 spinal level (Bonica, 1979).

Childbirth pain persistence throughout evolution undoubtedly demonstrates its physiologic role: it alerts the woman about childbirth imminence, delays expulsive efforts protecting the mother and fetus from birth traumatism, activates

Treatments, Mechanisms, and Adverse Reactions of Anesthetics and Analgesics. https://doi.org/10.1016/B978-0-12-820237-1.00020-X
Copyright © 2022 Elsevier Inc. All rights reserved.

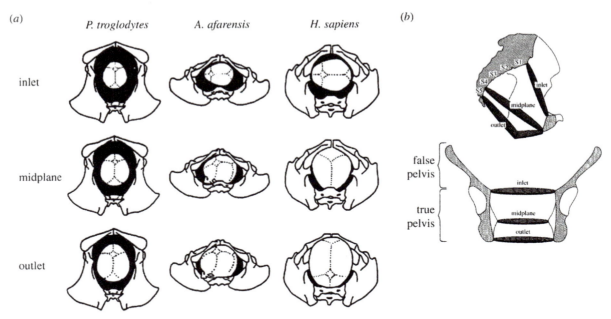

FIG. 1 Two stages of evolution of birth mechanism. (A) From left to right: occiput-posterior birth in chimpanzee (*P. troglodytes*), laterally-oriented head position throughout birth in Australopytechine (*A. afarensis*) and occiput-anterior birth in modern humans (*Homo sapiens*). (B) Configuration of human birth canal. *(From Gruss, L., & Schmitt, D. (2015). The evolution of the human pelvis: Changing adaptations to bipedalism, obstetrics and thermoregulation. Philosophical Transactions of the Royal Society of London. Series B, Biological Sciences, 370(1663), 20140063. doi:10.1098/rstb.2014.0063.)*

dyad's bonding through a hormonal cascade (Regalia, Bestetti, & Fumagalli, 2014). However, when it exceeds the woman's coping capacity, labor pain can prevent mother and baby from recovering in between contractions, block cervical dilation (*cervical dystocia*), or hindrance the progression of the fetal presenting part. In these cases, it has to be recognized as childbirth pathology and treated accordingly.

Although some Scholars denied the existence of labor pain (Dick-Read, 1933), others reported evidence of great interindividual variability in pain perception and coping ability among women of different socio-economic and cultural backgrounds. Melzack (1984) found that labor pain's intensity was referred to as severe to intolerable by 61.3% of nulliparas and by 46.3% of multiparas.

From prayers to neuraxial analgesia: A brief history of childbirth pain relief

Recipes for childbirth pain relief can be traced back up to the beginning of history: Sumerian tablets collected in the Vorderasiatisches Museum in Berlin contain prayers asking for liberation from childbirth pain. However, labor analgesia had been criticized, mainly on ethical grounds, for centuries, until, in 1847, Simpson first introduced diethyl ether as labor analgesic and, a few years later, the mighty Queen Victoria herself asked for analgesia during the birth of her last two children.

In the modern era, systemic techniques of labor analgesia were adopted, first inhalational and then parenteral. Thereafter, regional techniques were introduced: peripheral nervous blocks in the 1920s and neuraxial blockades (subarachnoid, epidural, and combined spinal-epidural blockades) after World War II, mainly in the Western world. Nowadays, worldwide, the most administered neuraxial drugs are local anesthetics (mainly bupivacaine and ropivacaine) combined with opioids (mainly fentanyl and sufentanil) in low-dosage/high-volume mixtures (mobile neuraxial analgesia). If a facility lacks resources to perform neuraxial blockades, it is always possible to administer parenteral opioids (to date, the safest alternative to neuraxial blockade is the short-acting opioid remifentanil). Nowadays, despite doubts raised by the lack of standardized research, labor analgesia is a widespread practice. In 2018 the World Health Organization (WHO, 2018), acknowledging the safety of modern analgesic techniques, recommended the use of pharmacological labor analgesia (epidural or parenteral opioid administration) "for healthy pregnant women requesting pain relief during labor, depending on a woman's preferences." This way, current international guidelines recognize only to the delivering woman the right to evaluate her own coping ability and legitimately request labor pain relief.

The analgesia-breastfeeding issue: Alleged mechanism of interference

Lately, the introduction of less aggressive analgesic techniques has curbed concerns about the potential impact of neuraxial labor analgesia on labor progression. A Cochrane metaanalysis stated that studies performed after 2005 demonstrate that low-dose/high-volume combined neuraxial labor analgesia does not affect instrumental or cesarean delivery rate, although it may be associated with a prolonged second stage of labor and increased oxytocin administration (Anim-Somuah et al., 2018). However, a new issue regarding the newborn arose in the 1990s. Many articles by pediatricians and midwives associate neuraxial labor analgesia with a potential negative impact on breastfeeding initiation. The hypothesized mechanism is that analgesic drugs may make the mother and/or the baby dysphoric, and/or interfere with the delicate hypothalamic-pituitary hormonal interplay supporting breastfeeding. Particularly, a negative role was hypothesized for opioids, which anesthetists had over time added to analgesic mixtures to avoid the motor block previously determined by high-dosage local anesthetics. In pharmacological terms, although centrally acting opioids (fentanyl and sufentanil) can reduce the hypothalamic-pituitary incretion of oxytocin (efferent branch of the milk ejection reflex elicited by stimulation of the mother's nipple), they also increase the anterior-pituitary incretion of the most important lactogenic hormone, prolactin, and thus lactocytes' milk production. Anyway, many articles in the literature (mainly by anesthetists) observed that, nowadays, dosage and route of administration of neuraxial drugs may abate the potential negative impact on milk ejection reflex. This is particularly true in the early peripartum period (until 60–72 h), when progesterone is still blocking milk production.

Exclusive breastfeeding for the first 6 months of infant's life: Health impact

According to WHO, exclusive breastfeeding for the first 6 months of an infant's life (and continued breastfeeding for 2 years or more, together with complementary feeding starting at 6 months) represents the best way to guarantee a healthy psycho-physic development to the nursling. Indeed, an ever-growing body of evidence demonstrates that depriving infants of the benefits of breastfeeding expose them to an increased risk of morbidity including gastrointestinal infections, diarrhea, necrotizing enterocolitis, respiratory infections and, later on in life, asthma. Formula-feeding increases the risk of otitis media, caries, malocclusion, obesity, and leukemia. Notably, risk of sudden infant death syndrome appears more consistent in formula-fed infants.

As for mothers, breastfeeding immediately after delivery contributes to reducing the risk of postpartum hemorrhage. In the long term, mothers who don't breastfeed or breastfeed only for a short period of time are more likely to develop breast or ovarian cancer, type II diabetes, hypertension, and cardiovascular disorders.

The analgesia-breastfeeding issue in the literature

Limitations of current knowledge

Although labor analgesia has dramatically evolved over the years, the absence of a standardized research methodology has limited the comparison of findings between studies (Szabo, 2013). Since many of them lack the necessary thoroughness, over time this situation has contributed to create many prejudices against analgesia, that still impact clinical practice (French, Cong, & Chung, 2016).

Conflicting study design, sampling strategies and enrolment criteria

In such delicate fields as childbirth and breastfeeding, it is almost impossible to conduct randomized controlled trials, due to crucial ethical issues. Although some Authors keep proposing experimental study designs (Reynolds, 2011), neuraxial analgesia can no longer be considered an experimental intervention and, therefore, to measure its alleged side effects observational studies appear more appropriate. However, only very few studies are "real world" cohort studies and, generally, the sampling criteria and enrolment criteria adopted are conflicting. Moreover, in case-control or descriptive studies, it is very difficult to avoid selection bias, given the numerous confounders to consider (Table 1).

Different accuracy in reporting labor neuraxial blockade techniques

High-dose analgesia with both local anesthetics and opioids can increase the risk of operative delivery (Anim-Somuah et al., 2018) and negatively impact both mother's ability to meet the newborn's needs and newborn's ability to express such needs and effectively suck on the breast. The main concern with local anesthetics is to lower the concentration of

TABLE 1 Conflicting sampling and enrolment strategies.

Study design and selection bias issues		Comments
Wilson et al. (2010)	Control group selected sequentially matching laboring women to study group cases	Selection bias possible as many existing confounders were not considered
Lee et al. (2017)	No control group without neuraxial analgesia	The inference could only be made about breastfeeding side-effects that are due to different drugs' mixture and dosages
Marzan Chang and Heaman (2005)	Study groups not formed randomly, but self-selecting to have neuraxial labor analgesia	Preexisting differences between study groups could account for the differences found in the dependent variables
Ransjö-Arvidson et al. (2001)	Any analgesia (pudendal block with mepivacaine, and/or pethidine, and/or and neuraxial analgesia) versus no analgesia	It seems to address more the right to receive labor analgesia rather than technical side effects of analgesia on breastfeeding
Dozier et al. (2013)	Epidural analgesia categorically defined as "any anesthetic agent(s) in the epidural or peridural space; this included bolus doses, continuous labor infusions, and patient-controlled pumps, and excluded local, spinal, or general anesthesia"	It is not specified if in the "no epidural group" parenteral pethidine or nitrous oxide administration is also considered in addition to the "no analgesia at all" group
Torvaldsen, Roberts, Simpson, Thompson, and Ellwood (2006)	Incorrect definition of study predictor (epidural). All women delivering vaginally with epidural analgesia also had pethidine administered	It is well known that parenteral pethidine can severely compromise both neurobehavioral and breastfeeding indexes
Marzan Chang and Heaman (2005)	Exclusive neuraxial labor analgesia versus a control group with no analgesia at all	Study predictors correctly defined
Different generalizability due to population enrolment criteria		**Comments**
Jordan, Emery, Bradshaw, Watkins, and Friswell (2005)	Parous women excluded	In parous women, past delivery and breastfeeding experiences can influence coping strategies for both labor pain and breastfeeding difficulties
Wieczorek, Guest, Balki, Shah, and Carvalho (2010)	Nulliparous women excluded	Nulliparous women more frequently choose epidural and show associated delivery and breastfeeding problems
Marzan Chang and Heaman (2005)	Parity considered only as a confounding factor	It correctly reflects the "real world" clinical setting
Wiklund, Norman, Uvnäs-Moberg, Ransjö-Arvidson, and Andolf (2009)	Induced labor not taken into account as obstetric confounder	Women needing labor induction more often require analgesia due to prolonged and painful labors
Jordan et al. (2009)	Induced labor included and taken into account as predictor	Drugs used to induce labor may interact with hormonal cascade sustaining breastfeeding (acting as dopamine agonists, prostaglandins suppress prolactin incretion and exogenous oxytocin may interfere with endogenous patterns of oxytocin incretion and sensitivity of oxytocin receptors)
Jonas et al. (2009)	Intra labor oxytocin administration and cumulative dosage described	Exogenous oxytocin administration influences the efferent arm of let-down reflex
Baumgarder, Muehl, Fischer, and Pribbenow (2003)	The peripartum oxytocin administration not considered	Oxytocin administration may reflect the course of delivery and influence breastfeeding initiation

TABLE 1 Conflicting sampling and enrolment strategies—cont'd

Different generalizability due to population enrolment criteria		Comments
Jonas, Nissen, Ransjö-Arvidson, Matthiesen, and Uvnäs-Moberg (2008)	Operative (instrumental or cesarean) deliveries excluded	More women requiring neuraxial analgesia have troubled deliveries and need operative interventions
Halpern et al. (1999)	Operative deliveries (instrumental or cesarean) included and taken into account as a study confounder	Otherwise, neuraxial analgesia could be considered a causal factor of operative deliveries
Dozier et al. (2013)	Women only selected if they initiate to breastfeed during hospital stay ("ever breastfeeding once")	The choice expressed during hospital stay could reflect a communicative style and compliance with team's policy, rather than the mother's real will
Torvaldsen, Roberts, Simpson, Thompson, & Ellwood (2006), Henderson, Dickinson, Evans, McDonald, and Paech (2003), and Wiklund et al. (2009)	Not differentiating between women intending or not to breastfeed	The proportion of breastfeeding success could be influenced by the different proportion of women not intending to breastfeed in neuraxial labor analgesia groups (sometimes, women choosing neuraxial analgesia have already decided not to breastfeed)
Henderson et al. (2003)	Educational level considered and taken into account as a study confounder	Generally, low educational level associated with lower breastfeeding commitment
Lee et al. (2017)	Single language population selected (generally, English speaking women, mainly Caucasian)	Often non-Caucasian women are economically and culturally more committed to breastfeed
Bell, White-Traut, and Medoff-Cooper (2010)	Focus only on newborns (separated from the mother in the first hour of life to evaluate the endpoint, suction skills from a bottle containing formula-milk)	The separation breaks the dyadic interaction essential to establish the breastfeeding skills. Only after the first suction session, mothers willing to breastfeed were assisted to breastfeed

administered drugs (to avoid maternal motor block or neonatal "floppy" muscular tone), while with opioids low cumulative dosage and short half-life seem the most important prerequisites (the main burden being on the baby). This led to the now prevailing low-dosage/high-volume combined analgesia techniques (Wang, Sun, & Huang, 2017). However, studies not performed by anesthetists are often inaccurate in reporting the technique used (Table 2).

The choice of confounders

So many confounders must be taken into account that comparing findings among studies about the relationship between childbirth analgesia and breastfeeding initiation is very difficult. The most renowned ones are: parity, mode of delivery, mother's age, gestational age, ethnicity, educational level, pregnancy and breastfeeding support practices, socio-economic status, work perspectives, social support, healthcare professionals' skills, facility breastfeeding policy, skin-to-skin contact, rooming-in, personality trait, motivation to breastfeed (Table 3).

Methodological flaws

Many studies do not even consider having to check their study's statistical power to be able to state what they claim (Jonas et al., 2009). Other studies are not powered enough to detect a real difference, if there ever was one (Baumgarder et al., 2003; Beilin et al., 2005; Bell et al., 2010). Mostly, observational studies only perform bivariate analyses; nonetheless Authors use associations to support cause-effect relationships (Wiklund et al., 2009).

TABLE 2 Different accuracy in reporting labor neuraxial blockade techniques.

Methods of reporting neuraxial blockade techniques		Breastfeeding results
Sepkoski, Lester, Ostheimer, and Brazelton (1992)	Single local anesthetic analgesia (anesthetics cumulative dosage and venous and arterial umbilical blood level of drugs reported)	Poor suction score in infants delivered with high-dosage bupivacaine
Radzyminski (2003)	Low dosage/high volume mixture of local anesthetics and opioids (reported venous cord blood level of drugs)	Reporting umbilical blood levels of bupivacaine 10 times lower than Sepkoski's, she does not find any significant difference in breastfeeding behaviors at birth or at 24 h of age (suction score, neurodevelopmental score and cord blood level monitored)
Jonas et al. (2009)	Very low dosage of local anesthetics and opioids (cumulative dosage of administered drugs reported)	No significant difference in duration of skin-to-skin before breastfeeding session, duration of breastfeeding session, number of breastfeeding sessions
Henderson et al. (2003)	Analgesia drugs reported only per neuraxial technique (combined spinal-epidural) and drug protocol	Nulliparous women who choose neuraxial analgesia are more likely to breastfeed for shorter duration. Those findings may reflect a copying style, rather than an effect of epidural analgesia: the level of pain tolerance in labor might predict women's ability to cope with breastfeeding difficulties after birth and vice versa
Wiklund et al. (2009)	Analgesia drugs reported only per type (even if the study was conducted as a single-center research)	Although in the Discussion section it is affirmed that "it is not possible to estimate how much epidural administration affects the success of breastfeeding," since "it was not possible to conduct a full-scale multivariate analysis," in Abstract it is stated that: (1) significantly fewer babies of mothers with epidural suckled within the first 4 h of life [odds ratio (OR) 3.79]; (2) the babies were also more often given artificial milk during hospital stay (OR 2.19); and (3) fewer were fully breastfed at discharge (OR 1.79)
Baumgarder et al. (2003)	Neither analgesia technique nor anesthetics and/or opioids concentration or dosage reported (even if the study was conducted as a single-center research)	Neuraxial analgesia had a negative impact on breastfeeding in the first 24 h even though it did not inhibit the percentage of breastfeeding attempts in the first hour

TABLE 3 The choice of confounders to be considered.

	Confounder	Breastfeeding results and comments
Agboado, Michel, Jackson, and Verma (2010)	Ethnicity, parity and in-hospital infant feeding practices are independent predictors of breastfeeding cessation at 6 months. Mode of delivery, timing of breastfeeding initiation, marital and socio-economic status do not predict breastfeeding cessation	White ethnicity was associated with shorter breastfeeding duration than nonwhite; nulliparous women were associated with shorter breastfeeding duration than parous; breastfeeding initiation after 1 h, formula in hospital, and teats in hospital were associated with shorter breastfeeding duration than breastfeeding initiation <1 h, no formula in hospital, and no teats in hospital
Henderson et al. (2003) and Wilson et al. (2010)	Maternal age considered and taken into account as study confounder	Older nulliparous women breastfeed longer although they are also more likely to ask for labor analgesia

TABLE 3 The choice of confounders to be considered—cont'd

	Confounder	Breastfeeding results and comments
Jonas et al. (2008)	Specific personality trait in mothers who ask for epidural analgesia	Compared to women receiving no medical interventions or oxytocin, women receiving epidural analgesia did not score lower on anxiety and aggression-hostility and higher on socialization if tested with a personality inventory 2 days after birth. However, after 2 days after birth and until 6 months postpartum the pattern reverses. Exogenously administered oxytocin influenced some of the scales in a dose-dependent manner in a positive direction
Tamminen, Verronen, Saarikoski, Göransson, and Tuomiranta (1983)	Mode of delivery	Infants born by cesarean sections with general anesthesia or instrumental deliveries have more difficulties in starting to breastfeed. Once started, breastfeeding duration is not affected
Patel, Liebling, and Murphy (2003)	Method of operative delivery in the second stage of labor (instrumental versus cesarean section)	The emergent cesarean section in the second stage of labor does not influence initiation or duration of exclusive breastfeeding
Lee et al. (2017)	Maternal commitment to breastfeed (evaluated through a breastfeeding motivational measurement scale)	In women committed to breastfeed, labor epidural solutions containing fentanyl concentrations as high as 2 µg/mL do not influence breastfeeding rates at 6 weeks postpartum
Halpern et al. (1999)	In-hospital and out-of-hospital breastfeeding support practices (immediate skin-to-skin contact, rooming-in, no teats proposed, no formula discharge pack offered, in-hospital breastfeeding support by International Board Certified Lactation Consultants, perinatal Breastfeeding Clinic available and free of charge 6/7, and phone counseling 24/24)	In a hospital that strongly promotes breastfeeding, epidural labor analgesia with local anesthetics and opioids does not impede breastfeeding success
Zuppa et al. (2014)	Full rooming-in (versus partial rooming-in: only from 10 am to 8 pm)	Multivariate analysis indicates that epidural analgesia is an independent factor which interferes with breastfeeding, but full rooming-in seems to be a protective factor against the risk determined by epidural analgesia itself
Patel et al. (2003)	Hospital stay length after an emergent cesarean section in the second stage of labor	A longer hospital stay may mean more support and determine a higher breastfeeding rate at discharge
Wieczorek et al. (2010)	Anticipated off-work time	The longer median off-work time (12 months) could explain why after a week from birth 100% of enrolled women were still breastfeeding. In Canadian population, the breastfeeding proportion was so high that verifying study endpoints was almost impossible
Jordan et al. (2009)	Index of income or social status (e.g., Townsend index of multiple deprivation)	A higher deprivation index hindered breastfeeding. However, the global findings of the study have to be evaluated in light of high intention of exclusively bottle-feeding of enrolled population: 18% at ward admission and 45% at discharge

Global literature overview

A recent review of the Academy of Breastfeeding Medicine (Martin, Vickers, Landau, Reece-Stremtan, and Academy of Breastfeeding Medicine, 2018) considered meaningful on the topic only the studies by Wilson et al. (2010) in nulliparous and Lee et al. (2017) in parous women. Considering the Halpern et al. (1999) study results together with the aforementioned studies, it could be argued that, from a strictly pharmacological point of view, current evidence reassuringly debunks the risk of neuraxially administered drugs being detrimental for breastfeeding initiation. However, it cannot be excluded that the psychological profile of women requesting analgesia is associated with motivational differences.

The analgesia-breastfeeding issue as a new research opportunity

In the 1980s, the fear of HIV and cytomegalovirus breast milk transmissibility had caused all milk banks around the world to close. Back then, formula feeding seemed the wonderful, unrestrained future of infant nutrition. This trend was inverted when Lucas (Lucas & Cole, 1990) demonstrated the protective effect of breast milk against necrotizing enterocolitis, and when, also thanks to associations of mothers (e.g., La Leche League), the scientific medical world rediscovered the importance of breastfeeding. Doubts on the potential impact of neuraxial labor analgesia on breastfeeding thus started.

Initially, research on the analgesia-breastfeeding issue prompted studies using neurobehavioral tools that only indirectly can evaluate breastfeeding endpoints (Table 4). In particular, although some researchers demonstrated a correlation between some neurobehavioral synthetic scales and breastfeeding competences (Marzan Chang & Heaman, 2005; Radzyminski, 2005), these tools were deeply flawed (Brockhurst, Littleford, & Halpern, 2000; Camann & Brazelton, 2000). Subsequently, many researchers (mainly anesthetists) focused their attention directly on the dyad's breastfeeding behavior, considering the analgesia-breastfeeding issue as a new, truly enriching study opportunity (Albani et al., 1999; Beilin et al., 2005; Chen, Li, Wang, & Wang, 2008; Gizzo et al., 2012; Halpern et al., 1999; Lee et al., 2017; Wilson et al., 2010). The challenge was now to find objectively measurable outcomes to quantify the clinical burden of labor analgesia on the dyad.

Breastfeeding as a measure of dyadic neurological competence

Breastfeeding is the most natural and physiological thing to do for both mother and newborn. However, it is actually a very complex process, which requires great skills and harmonic interplay of the parties involved. Both from a hormonal and neurological point of view, the ejection reflex, i.e., the key dyadic reflex sustaining breastfeeding, is a true orgasm. From the dyadic perspective, effective breastfeeding is a sensitive and specific indicator of well-being and, in a way, assessing it could even be considered a proper on-field neurological examination (Radzyminski, 2005). The majority of the innate, adaptive reflexes that allow breastfeeding are integrated at around 6 months of age, except the swallowing reflex, which persists throughout adulthood. This may offer a valid argument in favor of the WHO recommendation to exclusively breastfeed infants for at least the first 6 months of life, since apparently that is what evolution favored to guarantee species survival (Table 5).

TABLE 4 Tools to measure the effects of neuraxial labor analgesia on the newborn.

	Pros	Cons
Apgar score	Standardized score designed to assess newborn's overall health at birth	Low sensitivity (only modified by severe alterations of newborn's physiology)
Umbilical cord blood gas analysis	Objective evaluation of newborn's acid-base status	Low specificity (cannot differentiate between maternal, fetal and utero-placental factors)
Neurobehavioral analytic tests (BNBAS, ENNS): they are qualitative tests and only describe neonatal behavior	Noninvasive, useful for follow-up	High sensitivity (may detect alterations of ultimately no clinical relevance) (Camann & Brazelton, 2000)
Neurobehavioral synthetic tests (NACS): they are quantitative tests and evaluate neonatal behavior with a score	Noninvasive, useful for follow-up	Low sensitivity (cannot detect subtle neurobehavioral effects) (Brockhurst, Littleford, & Halpern, 2000)

BNBAS, Brazelton neonatal behavioural assessment scale (Als, Tronick, Lester, & Brazelton, 1977); *ENNS*, early neonatal neurobehavioral scale (Scanlon, Brown, Weiss, & Alper, 1974); *NACS*, neurologic and adaptive capacity score (Amiel-Tison et al., 1982).

TABLE 5 The complex neurologic background of mother-baby's breastfeeding abilities.

Step of dyadic interaction	Behavior description	Neurologic background and evidence
Consciousness (ability to react to and interact with outer environment)	In order to start breastfeeding successfully, first of all the newborn has to be *awake, quiet, and responsive*	To breastfeed, the neonate should be in a quiet, alert state, eyes open (the 4th neurodevelopmental state according to Brazelton). Moreover, the transition from a neurobehavioral state to another has to be smooth (Als et al., 1977)
Breast crawl (instinctive behavior that guides the baby to the mother's breast)	Immediately after birth, mother and infant mutually interact expressing *dyadic* behaviors. Breastfeeding can successfully start only if dyadic behaviors are left undisturbed. An interruption of even just 20 min can compromise the infant's ability to carry out the breast crawl instinct *Skin-to-skin* contact is the practice of leaving undisturbed the naked newborn on the mother's naked breasts right after birth. Generally, newborns display innate adaptive reflexes which lead to the latching on the mother's nipple and sucking highly nutritive and immune-active colostrum in about 50 min. They also stabilize their body temperature and respiratory homeostasis, and reduce oxygen and glucose consumption, and basal metabolic rate	Widström et al. (2011) have codified 9 phases of spontaneous behavior during early skin-to-skin mother-baby interaction: crying, relaxation, awakening, activity, crawling, resting, familiarization, suckling, and sleeping. This sequence demonstrates that newborns are able to express early self-regulation capacities. A Cochrane metaanalysis (Moore, Bergman, Anderson, & Medley, 2016) has confirmed the importance of early skin-to-skin contact to develop a strong mother-infant interaction and increase the probability of a longer duration of breastfeeding. Moreover, immediate skin-to-skin contact not only optimizes mother-infant short-term interaction, but also 1 year later, in terms of maternal sensitivity, child self-regulation, dyadic mutuality and reciprocity (Bystrova et al., 2009)
Searching (ability to localize the nipple)	The baby turns their head to localize the nipple and open their mouth (*gape response*) toward a tactile stimulus on their cheek or lips	The *rooting reflex* appears at 32 weeks of gestational age (wGA) and is integrated between 4 and 6 months of baby's life, but persists longer in breastfed infants
Latch-on (ability to grasp and retain breast tissue in the mouth). It requires the capacity to inhibit protective reflexes	The newborn is fed "at breast" and not "at nipple" because we now know that most of the mammary gland is situated beneath the nipple. Putting in their mouth a significant portion of mother breast, the tip of the nipple is positioned optimally at 6–8 mm from the hard-soft palate junction. In addition to using their tongue, the baby has to use their lips to create an effective seal around the breast	The *tongue thrust reflex* (28 wGA) allows grasping and securing the breast inside oral cavity. It is integrated at 6 months of life. The *pharyngeal (gag) reflex* (18 wGA) is elicited by something reaching the mid-posterior tongue. It also protects the nipple from the pressure of the baby's gums during suction. During latch-on and sucking, the baby has to be able to prevent eliciting the *pharyngeal (gag) reflex*
Sucking (or ability to compress the areolar tissue and express the milk from the ducts). It requires muscle strength sufficient to squeeze the mammary gland	Once correctly latched on the breast, the newborn begins a series of coordinated tongue and mandible movements. At first, the tongue compresses the nipple against the palate, causing it to elongate, due to its elastic nature (Burton, Deng, McDonald, & Fewtrell, 2013). A subsequent downward motion of the posterior portion of the tongue, aided by the seal offered by cheeks, lips and chin, increases the negative pressure inside the oral cavity. The nipple evenly expands and the nipple duct diameter increases, facilitating milk to flow into the oral cavity. Mandibular movements and tongue peristaltic waves push the milk	The *sucking reflex* (15–18 wGA) is fully integrated between 6 and 12 months. Proper functioning of *cranial nerves* and sufficient *muscular tone* are both necessary for the strong, coordinate facial and tongue movements that guarantee the suckle ability, while an effective lip seal will create the negative pressure necessary to retain the breast within the mouth. An adequate muscle tone is also necessary to support the weight of the maternal breast on the lower mandible

Continued

TABLE 5 The complex neurologic background of mother-baby's breastfeeding abilities—cont'd

Step of dyadic interaction	Behavior description	Neurologic background and evidence
	bolus posteriorly, triggering the swallow reflex, while the soft palate elevates making contact with the posterior pharyngeal wall, thus protecting the airways	
Mother's let-down reflex (mutual interplay ability)	The sucking movements of the newborn have to be rhythmic and strong enough to elicit the milk ejection reflex in the mother. Let-down requires an active tone of the newborns' neck flexor and extensor muscles. These muscles allow to maneuver the head maintaining a spatial relationship with the body and latch-on to the breast	The *let-down reflex* is a nipple-hypothalamic-pituitary maternal reflex favoring milk ejection. Once latched on the breast, newborn's active muscular tone allows maintaining attachment by preventing the head from falling forward with gravity and breaking suction (Radzyminski, 2005)
Swallowing (ability to coordinate different vital functions)	Sucking, swallowing and breathing are closely interconnected both from an anatomical and functional point of view and are controlled by overlapping cranial nerves. Unlike adults, infants' base of the tongue, soft palate and epiglottis are in close proximity to one another, facilitating both safety and coordination. A 1:1:1 ratio has been described as the optimal relationship between sucking, swallowing and breathing (Isaac & Choi, 2019)	The *swallowing reflex* (9–14 wGA) is elicited by liquid reaching sensory receptors on the tongue, soft palate and back of the mouth. To prevent chocking and aspiration the newborn can count on two main protective reflexes. The *pharyngeal (gag) reflex* and the *cough reflex*. The latter protects the airways but it is observed at birth in only 25%–50% of term newborns. However, 90% of infants present it by 1–2 months of age

A successful initiation of breastfeeding is a very significant indicator of a newborn's well-being and dyadic functional integrity. If present, it could testify the substantial harmlessness of the whole peripartum period, including pain management strategy. Conversely, since breastfeeding is a multifactorial process involving two subjects, it can be very difficult to point out exactly what went wrong in case of failure.

Labor analgesia could theoretically interfere with breastfeeding at many levels: it might depress neonatal alertness, compromise innate reflexes or reduce muscular tone, impairing breast crawl, nipple localization, latch-on or sucking. It has been speculated that labor analgesia could affect maternal bonding with the newborn, and even interfere with oxytocin, prolactin and cortisol incretion, impairing breastfeeding initiation or duration.

Searching for objective, nonbiased indicators of breastfeeding initiation success

While designing a study, choosing an already existent endpoint makes comparison with other studies' findings easier (Wetzl, Giannì, Delfino, & Consales, 2020). However, on our topic, studies led by different healthcare professionals use different endpoints.

Anesthetists mostly evaluate indexes of breastfeeding initiation success (BIS) too general to be clinically useful: "breastfeeding initiation rate" (Wilson et al., 2010), "breastfeeding difficulty" as assessed by women on postpartum day 1 (Beilin et al., 2005), "any breastfeeding" proportion at hospital discharge (Albani et al., 1999; Halpern et al., 1999), or "full breastfeeding," a compound index obtained considering WHO's "exclusively" and "predominantly" breastfeeding together (Volmanen, Valanne, & Alahuhta, 2004).

Other professionals (obstetricians, midwives, pediatricians, neonatologists and International Board Certified Lactation Consultants—IBCLCs) use neonatal or maternal breastfeeding skill indexes, which in most cases show too great an interobserver variability to be scientifically valid. The most renowned breastfeeding efficacy measurement tools are two synthetic scales, the IBFAT [infant breastfeeding assessment tool (Matthews, 1988)] and the LATCH [latching-audible-type-comfort-help (Jensen, Wallace, & Kelsay, 1994)]. The PIBBS [preterm infant breastfeeding behavior scale (Nyqvist, Rubertsson, Ewald, & Sjödén, 1996)] is an analytic test limited to selected groups of dyads (Lober, Dodgson, & Kelly, 2020). Notwithstanding their limitations, these scales are largely used as research endpoints (IBFAT, Riordan, Gross, Angeron, Krumwiede, & Melin, 2000; LATCH, Marzan Chang & Heaman, 2005; PIBBS, Radzyminski, 2005).

In the Baby-Friendly Hospital Initiative, the UNICEF/WHO's endpoint is the "proportion of women with exclusive breastfeeding throughout the hospital stay." This index verifies adherence of a facility to the guidelines but, from a scientific point of view, the results thus obtained may be subjected to a selection bias, because of healthcare providers' strong commitment and mothers' dedication to the goal of exclusive breastfeeding at a Baby-Friendly hospital. Moreover, such an index does not consider the time of hospital discharge as a possible confounder nor measure the efficacy of breastfeeding initiation.

Measuring and grading breastfeeding initiation success

Dewey (Dewey, Nommsen-Rivers, Heinig, & Cohen, 2003) proposed three main indicators of breastfeeding initiation failure in physiological childbirths:

1. Poor suckling capacity (IBFAT score <10, max. 12).
2. Delayed onset of lactation (>72 h).
3. Excessive weight loss (>10%) at 72 h.

Searching for an index to measure dyad's breastfeeding competence, the suckling capacity can be excluded, since it reflects only neonatal behavior in the process. Also, onset of lactation can be considered a restricted index, highlighting only maternal competence in breastfeeding initiation. Conversely, weight loss at *nadir*, typically reached between 60 and 72 h after birth (Flaherman et al., 2015), can be considered the most objective measure of dyadic breastfeeding performance.

Combining the WHO/UNICEF exclusive breastfeeding index with Dewey's weight loss at *nadir*, a study (Wetzl et al., 2019) defined BIS as "exclusive breastfeeding during the entire hospital stay, associated with a postnatal weight loss <7% at 60–72 h from birth." This index evaluates as objectively as possible the dyadic interaction and allows grading BIS.

Wetzl study on neuraxial labor analgesia (low-dosage/high-volume mixture of sufentanil and ropivacaine) enrolled an unselected cohort of women in an Italian Baby-Friendly hospital and performed a full-scale multivariate analyses on 50 peripartum confounders. Following the WHO suggestion (WHO, 2018), the study predictors were the "a priori" choice for analgesia and the request for analgesia "*as a last resort*." Only the "a priori" group was identified as at risk of breastfeeding initiation failure, whereas women who asked for analgesia "*as a last resort*" showed the same success odds as women who "*successfully coped*" with pain. It seemed that more than labor analgesia itself, what really mattered was what lay behind that choice: mothers' psychological characteristics, cultural background, education, work perspectives, and family support. If a woman, who is motivated and feels supported, wants to breastfeed, she breastfeeds, regardless of whether or not she received analgesia.

Considering the value of <7% as the index of physiologic weight loss at *nadir*, neonatal weight loss also has a predictive value in terms of exclusive breastfeeding at 6 months (Delfino, unpublished data).

Conclusion

From the 1990s, the perspective on the relationship between neuraxial labor analgesia and breastfeeding initiation success has substantially changed. While a growing body of evidence has demonstrated that the impact of neuraxial blockade on labor and delivery course has become minimal, a new potential issue has arisen: the alleged interference of analgesia on breastfeeding, core behavior in mammal species survival. Initially, the alleged interference was considered as probably due to administration of opioid-containing mixtures. When new endpoints were introduced to measure the practical impact of neuraxial blockade on breastfeeding behavior, the pharmacological-issue lost its credibility, whereas the maternal motivational-issue currently offers the most significant aspect to elaborate on in future on-field research.

Applications to other areas

- BIS can be considered a valuable index of efficacy and safety also in case of postoperative analgesia after cesarean delivery. Worldwide, cesarean delivery rates are steadily increasing and postoperative analgesia impacts both mother and baby in the crucial time of their first interaction. Therefore, analgesia has to be considered within in the frame of a strategy strongly supporting mother-baby bonding (rooming-in). Key aspects to take into account are: favoring the neuraxial route to administer drugs to reduce the neonatal burden of analgesics; using the lowest cumulative dosage to avoid interference with breastfeeding; adopting a multimodal approach combining opioids with local anesthetic in low dosage/high volume mixtures to reduce maternal side-effects (nausea, urinary retention, pruritus, sedation,

constipation); and targeting the pain relief level to steps of a surgical fast-track process (time of spontaneous urination, early mobilization, oral alimentation, bowel function).
- Choosing a true multidisciplinary approach targeted at "real world" health interdisciplinary endpoints (breastfeeding success, positive experience of childbirth) instead to mono-disciplinary, indirect indexes of well-being (mother satisfaction for analgesia, length of first suction session, suction strength measured by a nutritive sucking apparatus) is a strategy that can be extended to the whole field of pain relief clinical practice and research.

Other agents of interest

- *Pethidine*: Belsey et al. (1981) demonstrated that peripartum administration of pethidine to mothers affected newborn's neurobehavioral state. Neurobehavioral alterations at BNBAS peaked at 7 days postpartum and could still be seen at 6 weeks of life. Pharmacokinetics supports these findings given that in newborns pethidine half-life is 2–7 times longer than in adults (3–6 h), while nor-pethidine's reaches 20–36 h (vs 15 h). This significantly increases the risk of neurotoxicity of nor-pethidine, which, especially in the premature infant, can cause convulsive seizures.
- *Mepivacaine*: Scanlon et al. (1974) observed that neonates born to mothers who had received mepivacaine or lidocaine during labor were "awake, but floppy." The reduction in muscle tone at ENNS was fleeting in case of lidocaine use, whereas it lasted longer if mepivacaine had been administered. Pharmacokinetics confirms that in newborns mepivacaine has a half-life of 9 h, while lidocaine of just 3 h (Brown et al., 1975).
- *Propofol*: Using the ENNS, Celleno et al. (1989) demonstrated that propofol used in anesthesia for cesarean delivery caused more noticeable neurobehavioral alterations in neonates than thiopentone. Propofol seems to have a slower elimination in newborns than in their mothers, due to an immature glucuronidation process. The reported generalized irritability in 25% newborns exposed to propofol lasted for the first 4 h of life, interfering in the critical time of neonatal early adaptation to extra-uterine life (golden hours).

Mini-dictionary of terms

Exclusive breastfeeding: Infant receives only breast-milk (including expressed breast-milk or milk from a wet nurse) and nothing else, except for oral rehydrating solutions, medicines, and vitamins and minerals, if necessary.
Formula feeding: Infant is fed only on a breast-milk substitute.
Baby-Friendly Hospital Initiative: UNICEF/WHO initiative intending to defend, promote and support breastfeeding initiation in hospital settings. Among 10 proposed steps to successful breastfeeding there are respect for recommendations of the international code of marketing of breast-milk substitutes (WHO, 1981), immediate mother-baby skin-to-skin contact after childbirth, mother-baby rooming-in throughout the hospital stay and support to exclusive breastfeeding until hospital discharge.

Key facts of labor analgesia state of art in guidelines and metaanalyses

- After having strongly recommended to avoid the neuraxial labor analgesia as "one of the most striking examples of the medicalization of normal birth" (WHO, 1996), nowadays, WHO guidelines "for a positive childbirth experience" recommend neuraxial labor analgesia "for healthy pregnant women requesting pain relief during labor, depending on a woman's preferences" (WHO, 2018).
- WHO also reports that "women expressing an a priori desire for a neuraxial analgesia to help with a pain-free labor, to alleviate the fear of pain and/or to remain in control during labor" could behave differently from "other women (who) request a neuraxial analgesia as a last resort, when the level of pain and/or sense of control over the labor was overwhelming and unmanageable" (WHO, 2018).
- WHO guidelines even recommend as an alternative option "parenteral opioids, such as fentanyl, diamorphine and pethidine, for healthy pregnant women requesting pain relief during labor, depending on a woman's preferences" (WHO, 2018).
- According to a Cochrane metaanalysis, although neuraxial labor analgesia was reported to be associated with an increased instrumental delivery rate in studies performed before 2005, this is no longer true when studies following that date are considered (Anim-Somuah et al., 2018).
- The same Cochrane metaanalysis also admits that between women with neuraxial analgesia and women with parenteral opioids analgesia "there was no difference between cesarean section rates" (Anim-Somuah et al., 2018).

Summary points

- The complex, prolonged, and sometimes difficult mechanism of human childbirth prompts midwife assistance to delivery.
- Despite its physiologic role in human childbirth, labor pain may exceed the parturient coping capacity.
- Women have been asking for labor pain relief since the dawn of time, but have been only recently recognized the right to receive it, even at their own request.
- Neuraxial analgesia is considered a safe pain-relief strategy for the parturient, but concerns have been raised about the analgesic burden on the baby.
- Nowadays, breastfeeding has earned the status of normative standard for infant feeding and nutrition.
- Among proposed methods to evaluate neonatal well-being after neuraxial analgesia, the ones directly evaluating breastfeeding outcomes are the most reliable.
- Due to lack of standardization and different confounders considered, findings of studies on possible labor analgesia adverse effects on breastfeeding are conflicting.
- Mother and newborn's competences necessary to successfully breastfeed are so complex that objective measurement of breastfeeding efficacy could be considered a real world on-field neurological examination.
- A promising index of breastfeeding initiation success to consider is the combination of exclusive-breastfeeding proportion and weight loss at *nadir*.
- Maternal motivation to breastfeed seems to be the key, being able to nullify drug adverse effects or unfavorable birth circumstances.

References

Agboado, G., Michel, E., Jackson, E., & Verma, A. (2010). Factors associated with breastfeeding cessation in nursing mothers in a peer support programme in Eastern Lancashire. *BMC Pediatrics, 10,* 3. https://doi.org/10.1186/1471-2431-10-3.

Albani, A., Addamo, P., Renghi, A., Voltolin, G., Peano, L., & Ivani, G. (1999). The effect on breastfeeding rate of regional anesthesia technique for cesarean and vaginal childbirth. *Minerva Anestesiologica, 65*(9), 625–630.

Als, H., Tronick, E., Lester, B., & Brazelton, T. (1977). The Brazelton neonatal behavioral assessment scale (BNBAS). *Journal of Abnormal Child Psychology, 5*(3), 215–231. https://doi.org/10.1007/bf00913693.

Amiel-Tison, C., Barrier, G., Shnider, S., Levinson, G., Hughes, S., & Stefani, S. (1982). A new neurologic and adaptive capacity scoring system for evaluating obstetric medications in full-term newborns. *Anesthesiology, 56*(5), 340–350. https://doi.org/10.1097/00000542-198205000-00003.

Anim-Somuah, M., Smyth, R., Cyna, A., & Cuthbert, A. (2018). Epidural versus non-epidural or no analgesia for pain management in labour. *Cochrane Database of Systematic Reviews, 5*(5). https://doi.org/10.1002/14651858.CD000331.pub4.

Baumgarder, D., Muehl, P., Fischer, M., & Pribbenow, B. (2003). Effect of labor epidural anesthesia on breast-feeding of healthy full-term newborns delivered vaginally. *The Journal of the American Board of Family Practice, 16*(1), 7–13. https://doi.org/10.3122/jabfm.16.1.7.

Beilin, Y., Bodian, C., Weiser, J., Hossain, S., Arnold, I., Feierman, D. E., … Holzman, I. (2005). Effect of labor epidural analgesia with and without fentanyl on infant breast-feeding: A prospective, randomized, double-blind study. *Anesthesiology, 103*(6), 1211–1217. https://doi.org/10.1097/00000542-200512000-00016.

Bell, A., White-Traut, R., & Medoff-Cooper, B. (2010). Neonatal neurobehavioral organization after exposure to maternal epidural analgesia in labor. *Journal of Obstetric, Gynecologic, and Neonatal Nursing, 39*(2), 178–190. https://doi.org/10.1111/j.1552-6909.2010.01100.x.

Belsey, E. M., Rosenblatt, D. B., Lieberman, B. A., Redshaw, M., Caldwell, J., Notarianni, L., … Beard, R. W. (1981). The influence of maternal analgesia on neonatal behaviour: I. Pethidine. *BJOG, 88,* 398–406.

Bonica, J. (1979). Peripheral mechanism and pathways of parturition pain. *British Journal of Anaesthesia, 51,* 3–9S.

Brockhurst, N., Littleford, J., & Halpern, S. (2000). The neurologic and adaptive capacity score: A systematic review of its use in obstetric anesthesia research. *Anesthesiology, 92*(1), 237–246. https://doi.org/10.1097/00000542-200001000-00036.

Brown, W. U., Bell, G. C., Lurie, A. O., Weiss, B., Scanlon, J. W., & Alper, M. H. (1975). Newborn blood levels of lidocaine and mepivacaine in the first postnatal day following maternal epidural anesthesia. *Anesthesiology, 42,* 698–707.

Burton, P., Deng, J., McDonald, D., & Fewtrell, M. (2013). Real-time 3D ultrasound imaging of infant tongue movements during breast-feeding. *Early Human Development, 89*(9), 635–641. https://doi.org/10.1016/j.earlhumdev.2013.04.009.

Bystrova, K., Ivanova, V., Edhborg, M., Matthiesen, A., Ransjö-Arvidson, A., Mukhamedrakhimov, R., … Widström, A. (2009). Early contact versus separation: Effects on mother-infant interaction one year later. *Birth, 36*(2), 97–109. https://doi.org/10.1111/j.1523-536X.2009.00307.x.

Camann, W., & Brazelton, T. (2000). Use and abuse of neonatal neurobehavioral testing. *Anesthesiology, 92*(1), 3–5. https://doi.org/10.1097/00000542-200001000-00006.

Celleno, D., Capogna, G., Tomassetti, M., Costantino, P., Di Feo, G., & Nisini, R. (1989). Neurobehavioral effects of propofol following elective caesarean section. *British Journal of Anaesthesia, 62,* 649–654.

Chen, Y. M., Li, Z., Wang, A. J., & Wang, J. M. (2008). Effect of labor analgesia with ropivacaine on the lactation of paturients. *Zhonghua Fu Chan Ke Za Zhi, 43,* 502–505 (in Chinese).

Dewey, K., Nommsen-Rivers, L., Heinig, M., & Cohen, R. (2003). Risk factors for suboptimal infant breastfeeding behavior, delayed onset of lactation, and excess neonatal weight loss. *Pediatrics, 112*(3 Pt 1), 607–619. https://doi.org/10.1542/peds.112.3.607.

Dick-Read, G. (1933). *Natural childbirth*. London: Heineman.

Dozier, A., Howard, C., Brownell, E., Wissler, R., Glantz, J., Ternullo, S., … Lawrence, R. (2013). Labor epidural anesthesia, obstetric factors and breastfeeding cessation. *Maternal and Child Health Journal, 17*(4), 689–698. https://doi.org/10.1007/s10995-012-1045-4.

Flaherman, V., Schaefer, E., Kuzniewicz, M., Li, S., Walsh, E., & Paul, I. (2015). Early weight loss nomograms for exclusively breastfed newborns. *Pediatrics, 135*(1), e16–e23. https://doi.org/10.1542/peds.2014-1532.

French, C. A., Cong, X., & Chung, K. S. (2016). Labor epidural analgesia and breastfeeding: A systematic review. *Journal of Human Lactation, 33*(3), 507–520. https://doi.org/10.1177/0890334415623779.

Gizzo, S., Di Gangi, S., Saccardi, C., Patrelli, T., Paccagnella, G., Sansone, L., … Nardelli, G. (2012). Epidural analgesia during labor: Impact on delivery outcome, neonatal well-being, and early breastfeeding. *Breastfeeding Medicine, 7*, 262–268. https://doi.org/10.1089/bfm.2011.0099.

Gruss, L., & Schmitt, D. (2015). The evolution of the human pelvis: Changing adaptations to bipedalism, obstetrics and thermoregulation. *Philosophical Transactions of the Royal Society of London. Series B, Biological Sciences, 370*(1663), 20140063. https://doi.org/10.1098/rstb.2014.0063.

Halpern, S., Levine, T., Wilson, D., MacDonell, J., Katsiris, S., & Leighton, B. (1999). Effect of labor analgesia on breastfeeding success. *Birth, 26*(2), 83–88. https://doi.org/10.1046/j.1523-536x.1999.00083.x.

Henderson, J., Dickinson, J., Evans, S., McDonald, S., & Paech, M. (2003). Impact of intrapartum epidural analgesia on breast-feeding duration. *Australian and New Zealand Journal of Obstetrics and Gynaecology, 43*(5), 372–377. https://doi.org/10.1046/j.0004-8666.2003.t01-1-00117.x.

Isaac, N., & Choi, E. (2019). Infant anatomy and physiology for feeding. In S. Hetzel Campbell, J. Lauwers, R. Mannel, & B. Spencer (Eds.), *Interdisciplinary lactation care*. Burlington, MA: Jones & Bartlett Learning (chapter 3).

Jensen, D., Wallace, S., & Kelsay, P. (1994). LATCH: A breastfeeding charting system and documentation tool. *Journal of Obstetric, Gynecologic, and Neonatal Nursing, 23*(1), 27–32. https://doi.org/10.1111/j.1552-6909.1994.tb01847.x.

Jonas, K., Johansson, L. M., Nissen, E., Ejdebäck, M., Ransjö-Arvidson, A. B., & Uvnäs-Moberg, K. (2009). Effects of intrapartum oxytocin administration and epidural analgesia on the concentration of plasma oxytocin and prolactin, in response to suckling during the second day postpartum. *Breastfeeding Medicine, 4*(2), 71–82. https://doi.org/10.1089/bfm.2008.0002.

Jonas, K., Nissen, E., Ransjö-Arvidson, A., Matthiesen, A., & Uvnäs-Moberg, K. (2008). Influence of oxytocin or epidural analgesia on personality profile in breastfeeding women: A comparative study. *Archives of Women's Mental Health, 11*(5–6), 335–345. https://doi.org/10.1007/s00737-008-0027-4.

Jordan, S., Emery, S., Bradshaw, C., Watkins, A., & Friswell, W. (2005). The impact of intrapartum analgesia on infant feeding. *BJOG, 112*(7), 927–934. https://doi.org/10.1111/j.1471-0528.2005.00548.x.

Jordan, S., Emery, S., Watkins, A., Evans, J., Storey, M., & Morgan, G. (2009). Associations of drugs routinely given in labour with breastfeeding at 48 hours: Analysis of the Cardiff Births Survey. *BJOG, 116*, 1622–1629 (discussion 1630-2) https://doi.org/10.1111/j.1471-0528.2009.02256.x.

Kramer, M. S., & Kakuma, R. (2012). Optimal duration of exclusive breastfeeding (review). *Cochrane Database of Systematic Reviews*, (8), CD003517. https://doi.org/10.1002/14651858.CD003517.pub2.

Lee, A., McCarthy, R., Toledo, P., Jones, M., White, N., & Wong, C. (2017). Epidural labor analgesia-fentanyl dose and breastfeeding success: A randomized clinical trial. *Anesthesiology, 127*(4), 614–624. https://doi.org/10.1097/ALN.0000000000001793.

Lober, A., Dodgson, I. E., & Kelly, L. (2020). Using the Preterm Infant Breastfeeding Behavior Scale (PIBBS) with late preterm infants. *Clinical Lactation*. https://doi.org/10.1891/CLINLACT-D-20-00001.

Lucas, A., & Cole, T. (1990). Breast milk and neonatal necrotising enterocolitis. *Lancet, 336*(8730), 1519–1523. https://doi.org/10.1016/0140-6736(90)93304-8.

Martin, E., Vickers, B., Landau, R., Reece-Stremtan, S., & Academy of Breastfeeding Medicine. (2018). ABM clinical protocol #28: Peripartum analgesia and anesthesia for the breastfeeding mother. *Breastfeeding Medicine, 13*(3), 164–171.

Marzan Chang, Z., & Heaman, M. (2005). Epidural analgesia during labor and delivery: Effects on the initiation and continuation of effective breastfeeding. *Journal of Human Lactation, 21*(3), 305–314 (quiz 315-9, 326) https://doi.org/10.1177/0890334405277604.

Matthews, M. (1988). Developing an instrument to assess infant breastfeeding behaviour in the early neonatal period. *Midwifery, 4*(4), 154–165. https://doi.org/10.1016/s0266-6138(88)80071-8.

Melzack, R. (1984). The myth of painless childbirth (the John Bonica lecture). *Pain, 19*, 321–337.

Moore, E., Bergman, N., Anderson, G., & Medley, N. (2016). Early skin-to-skin contact for mothers and their healthy newborn infants. *Cochrane Database of Systematic Reviews, 11*(11), CD003519. https://doi.org/10.1002/14651858.CD003519.pub4.

Nyqvist, K., Rubertsson, C., Ewald, U., & Sjödén, P. (1996, September). Development of the Preterm Infant Breastfeeding Behavior Scale (PIBBS): A study of nurse-mother agreement. *Journal of Human Lactation, 12*(3), 207–219. https://doi.org/10.1177/089033449601200318.

Patel, R., Liebling, R., & Murphy, D. (2003). Effect of operative delivery in the second stage of labor on breastfeeding success. *Birth, 30*(4), 255–260. https://doi.org/10.1046/j.1523-536x.2003.00255.x.

Radzyminski, S. (2003). The effect of ultra low dose epidural analgesia on newborn breastfeeding behaviors. *Journal of Obstetric, Gynecologic, and Neonatal Nursing, 32*(3), 322–331. https://doi.org/10.1177/0884217503253440.

Radzyminski, S. (2005). Neurobehavioral functioning and breastfeeding behavior in the newborn. *Journal of Obstetric, Gynecologic, and Neonatal Nursing, 34*(3), 335–341. https://doi.org/10.1177/0884217505276283.

Ransjö-Arvidson, A., Matthiesen, A., Lilja, G., Nissen, E., Widström, A., & Uvnäs-Moberg, K. (2001). Maternal analgesia during labor disturbs newborn behavior: Effects on breastfeeding, temperature, and crying. *Birth, 28*(1), 5–12. https://doi.org/10.1046/j.1523-536x.2001.00005.x.

Regalia, A., Bestetti, G., & Fumagalli, S. (2014). Il dolore. In R. Spandrio, A. Regalia, & G. Bestetti (Eds.), *Fisiologia della nascita. Dai prodromi al postpartum*. Roma: Carocci Editore (chapter 8).

Reynolds, F. (2011). Labour analgesia and the baby: Good news is no news. *International Journal of Obstetric Anesthesia, 20*(1), 38–50. https://doi.org/10.1016/j.ijoa.2010.08.004.

Riordan, J., Gross, A., Angeron, J., Krumwiede, B., & Melin, J. (2000). The effect of labor pain relief medication on neonatal suckling and breastfeeding duration. *Journal of Human Lactation, 16*(1), 7–12.

Scanlon, J., Brown, W., Weiss, J., & Alper, M. (1974). Neurobehavioral responses of newborn infants after maternal epidural anesthesia. *Anesthesiology, 40*(2), 121–128. https://doi.org/10.1097/00000542-197402000-00005.

Sepkoski, C., Lester, B., Ostheimer, J., & Brazelton, T. (1992). The effects of maternal epidural anesthesia on neonatal behavior during the first month. *Developmental Medicine and Child Neurology, 34*(12), 1072–1080. https://doi.org/10.1111/j.1469-8749.1992.tb11419.x.

Szabo, A. (2013). Intrapartum neuraxial analgesia and breastfeeding outcomes: Limitations of current knowledge. *Anesthesia & Analgesia, 166*(2), 399–405. https://doi.org/10.1213/ANE.0b013e318273f63c.

Tamminen, T., Verronen, P., Saarikoski, S., Göransson, A., & Tuomiranta, H. (1983). The influence of perinatal factors on breast feeding. *Acta Paediatrica Scandinavica, 72*(1), 9–12. https://doi.org/10.1111/j.1651-2227.1983.tb09655.x.

Torvaldsen, S., Roberts, C., Simpson, J., Thompson, J., & Ellwood, D. (2006). Intrapartum epidural analgesia and breastfeeding: A prospectic cohort study. *International Breastfeeding Journal, 1*, 24. https://doi.org/10.1186/1746-4358-1-24.

Trevathan, W., Smith, E., & McKenna, J. (1999). Evolutionary obstetrics. In *Evolutionary medicine* (pp. 183–207). New York: Oxford University Press.

UNICEF. (2013). *Improving child nutrition. The achievable imperative for global progress.* New York: United Nations Children's Fund (UNICEF).

UNICEF/WHO. (2009). *Baby Friendly Hospital initiative. Revised, updated and expanded for integrated care. Section III. Breastfeeding promotion and support in a baby-friendly hospital. A 20 hour course for maternal staff.* Geneva, Switzerland: World Health Organization Document Production Services.

Volmanen, P., Valanne, J., & Alahuhta, S. (2004). Breast-feeding problems after epidural analgesia for labour: A retrospective cohort study of pain, obstetrical procedures and breast-feeding practices. *International Journal of Obstetric Anesthesia, 13*(1), 25–29. https://doi.org/10.1016/S0959-289X(03)00104-3.

Wang, Y., Sun, S., & Huang, S. (2017). Effects of epidural labor analgesia with low concentrations of local anesthetics on obstetric outcomes: A systematic review and meta-analysis of randomized controlled trials. *Anesthesia and Analgesia, 124*(5), 1571–1580. https://doi.org/10.1213/ANE.0000000000001709.

Wetzl, R. G., Delfino, E., Peano, L., Gogna, D., Vidi, Y., Vielmi, F., ... Arioni, C. (2019). A priori choice of neuraxial labour analgesia and breastfeeding initiation success: A community-based cohort study in an Italian baby-friendly hospital. *BMJ Open, 9*(3), e025179. https://doi.org/10.1136/bmjopen-2018-025179.

Wetzl, R. G., Giannì, M., Delfino, E., & Consales, A. (2020). Planning the teamwork, endpoints, and design of an observational study on neuraxial labor analgesia and breastfeeding success: Advantages, difficulties, and drawbacks of a truly multi- and interdisciplinary approach. *SAGE Research Cases.* https://doi.org/10.4135/9781529744101.

WHO. (1981). *International code of marketing of breast-milk substitutes.* Geneva, Switzerland: WHO. https://apps.who.int/iris/handle/10665/40382.

WHO. (1996). *Safe motherhood. Care in normal birth. A practical guide.* Geneva, Switzerland: WHO. WHO/FRH/HSM/96.24.

WHO (2018). WHO recommendations: Intrapartum care for a positive childbirth experience. Geneva, Switzerland: World Health Organization. License: CC BY-NC-SA 3.0 IGO.

Widström, A., Lilja, G., Aaltomaa-Michalias, P., Dahllöf, A., Lintula, M., & Nissen, E. (2011). Newborn behaviour to locate the breast when skin-to-skin: A possible method for enabling early self-regulation. *Acta Paediatrica, 100*(1), 79–85. https://doi.org/10.1111/j.1651-2227.2010.01983.x.

Wieczorek, P., Guest, S., Balki, M., Shah, V., & Carvalho, J. (2010). Breastfeeding success rate after vaginal delivery can be high despite the use of epidural fentanyl: An observational cohort study. *International Journal of Obstetric Anesthesia, 19*(3), 273–277. https://doi.org/10.1016/j.ijoa.2010.02.001.

Wiklund, I., Norman, M., Uvnäs-Moberg, K., Ransjö-Arvidson, A., & Andolf, E. (2009). Epidural analgesia: Breast-feeding success and related factors. *Midwifery, 25*(2), e31–e38. https://doi.org/10.1016/j.midw.2007.07.005.

Wilson, M., MacArthur, C., Cooper, G., Bick, D., Moore, P., Shennan, A., & on Behalf of COMET Study Group UK. (2010). Epidural analgesia and breastfeeding: A randomised controlled trial of epidural techniques with and without fentanyl and a non-epidural comparison group. *Anesthesia, 65*(2), 145–153. https://doi.org/10.1111/j.1365-2044.2009.06136.x.

Zuppa, A. A., Alighieri, G., Riccardi, R., Cavani, M., Iafisco, A., Cota, F., & Romagnoli, C. (2014). Epidural analgesia, neonatal care and breastfeeding. *Italian Journal of Pediatrics, 40*, 82. https://doi.org/10.1186/s13052-014-0082-6.

Chapter 21

Mechanistic overview of how opioid analgesics promote constipation

Jesse J. Di Cello[a], Arisbel B. Gondin[b], Simona E. Carbone[b], and Daniel P. Poole[b]

[a]*Department of Physiology, Biomedicine Discovery Institute, Monash University, Clayton, VIC, Australia,* [b]*Drug Discovery Biology Theme, Monash Institute of Pharmaceutical Sciences, Monash University, Parkville, VIC, Australia*

Abbreviations

βarr2	beta arrestin 2
CNS	central nervous system
DOR	delta opioid receptor
ENS	enteric nervous system
ERK	extracellular signal-regulated kinases
GI	gastrointestinal
GIRK	G protein-coupled inwardly rectifying potassium
GPCR	G protein-coupled receptor
GRK	G protein-coupled receptor kinase
IPAN	intrinsic primary afferent neuron
KOR	kappa opioid receptor
MOR	mu opioid receptor
OBD	opioid-induced bowel dysfunction
OIC	opioid-induced constipation
PAMORA	peripherally acting mu opioid receptor antagonist

Introduction

Opioids derived from the poppy plant (*Papaver somniferum*) have been historically used for their pain-relieving and hedonic properties. Today, opioids are still the mainstay treatment for moderate to severe pain which is a testament to both their analgesic efficacy and the lack of good alternatives. In the US alone, more than 250 million opioid prescriptions are written each year which has led to the highly publicized "opioid epidemic" (Crockett et al., 2019). Although their analgesic efficacy is unparalleled, the administration of opioids often leads to adverse side-effects including withdrawal, addiction, analgesic tolerance and respiratory depression, with the latter causing overdose-associated death. Opioids produce analgesia via the activation of the mu opioid receptor (MOR) in the central nervous system (CNS). In addition, opioids activate MOR at several sites in the periphery, including the gastrointestinal (GI) tract. The activation of MOR in the GI tract is associated with the development of several adverse side-effects, collectively known as opioid-induced bowel dysfunction (OBD). OBD includes, but is not limited to, esophageal dysfunction, gastroparesis, postoperative ileus, and opioid-induced constipation (OIC). OIC is the most prevalent form of OBD, affecting up to 80% of patients (Abramowitz et al., 2013). OIC is defined by the Rome IV criteria as new or worsening constipation symptoms when initiating, increasing or changing opioid therapy (Lacy et al., 2016). In a multinational survey conducted in 2009 by Bell and colleagues, patients suffering from OIC reported a significantly lower quality of life and increased sickness-related absences from work, highlighting both the social and economic burden (Bell, Annunziata, & Leslie, 2009). Furthermore, OIC is a major cause of patient noncompliance with opioid treatment, with approximately a third of patients either lowering the dosage or completely discontinuing treatment leading to inadequate or absence of pain relief (Nee et al., 2018). Although the first line of treatment for OIC is laxatives, approximately 50% of patients are unresponsive to these which makes alternative pharmacotherapies a more suitable option for relieving constipation (Christensen, Olsson, From, & Breivik, 2016). Patients suffering laxative refractory OIC are usually administered peripherally acting mu opioid receptor antagonists (PAMORAs). PAMORAs

Treatments, Mechanisms, and Adverse Reactions of Anesthetics and Analgesics. https://doi.org/10.1016/B978-0-12-820237-1.00021-1
Copyright © 2022 Elsevier Inc. All rights reserved.

block the binding of opioids to peripheral tissues without crossing the blood brain barrier, thereby blocking OBD while retaining CNS-associated analgesia. Despite their effectiveness, use of PAMORAs can be limited by cost and several side-effects including nausea, emesis and diarrhea (Thapa, Kappus, Hurt, & Diamond, 2019). Therefore, there is a need for new pharmacotherapies which can relieve GI side-effects of opioid analgesics. In this chapter, we provide an overview of the fundamental molecular and physiological mechanisms underlying OIC.

The enteric nervous system: Master regulator of GI function

The GI tract is intrinsically innervated by the enteric nervous system (ENS), which extends from the distal esophagus through to the rectum. The ENS is a division of the autonomic nervous system and contains a similar number of neurons to that present in the spinal cord. The ENS is organized into a dense network of highly interconnected ganglia containing enteric neurons and glial cells. There are two major ganglionated plexuses, which are defined both anatomically and functionally (Fig. 1). The **myenteric plexus** lies between the circular and longitudinal smooth muscle layers of the *muscularis externa* and is principally responsible for the control and coordination of mixing and propulsion of digestive contents. The **submucosal plexus** is located within the submucosal layer and is primarily involved in the regulation of secretomotor function to control water and electrolyte balance. The ENS is capable of detecting and responding to changes to the local environment of the GI tract, such as distension of the gut wall. The organization of the ENS into reflex circuits allows for autonomous control of GI function, with influence from the CNS. These circuits include ascending excitatory pathways that

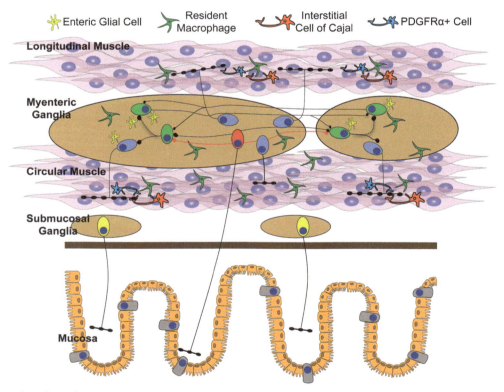

FIG. 1 The myenteric pathways involved in coordinating GI motility. Intrinsic primary afferent neurons (IPANs, *red neuron*) project to the intestinal mucosa to detect mechanical distortion or the presence of chemical stimuli in the lumen. IPANs initiate the motility reflex through the coordinated activation of ascending excitatory and descending inhibitory neural pathways. Both ascending and descending interneurons *(green)* receive input from IPANs, and provide synaptic output to excitatory and inhibitory motoneurons, respectively *(light blue)*. Enterochromaffin cells (EC cells) located in the epithelial lining of the mucosa *(gray)* release serotonin following physical or chemical changes in the lumen. Serotonin activates 5-HT receptors expressed by mucosally-projecting IPANs *(red neuron)*. Excitatory motoneurons release acetylcholine as their major neurotransmitter to produce an oral contraction, whereas inhibitory motoneurons release nitric oxide and VIP to promote aboral relaxation. Submucosal neurons *(yellow)* innervate cells of the epithelial layer and vasculature and their main function is to regulate intestinal secretion and blood flow. Opioid receptors are expressed by neurons of the ENS and are primarily localized to myenteric motor neurons. Other cell types which mediate myenteric responses or contribute to neuromuscular transmission include enteric glia, resident muscularis macrophages, interstitial cells of Cajal (ICC) and PDGFRα+ cells, as indicated. There is limited evidence for expression of opioid receptors by these cell types. *(Modified from Spencer, N. J., Dinning, P. G., Brookes, S. J., & Costa, M. (2016). Insights into the mechanisms underlying colonic motor patterns.* The Journal of Physiology, 594*(15), 4099–4116.)*

promote contractions proximal to GI contents and descending inhibitory pathways that relax the smooth muscle and facilitate movement of contents along the gut. Other reflex patterns are involved in mixing digesta to facilitate mechanical and enzymatic breakdown, and in the coordination of secretomotor activity. There are several distinct neuronal subtypes defined by their functional role, electrophysiological properties, morphology, projections and the transmitters and neurochemical markers that they contain. These include intrinsic primary afferent neurons ("IPANs" or "sensory" neurons) which detect mechanical or chemical stimulation of the mucosa and are capable of initiating and propagating a reflex response. Ascending and descending interneurons connect sensory neurons to the excitatory and inhibitory motor neurons, respectively. These motor neurons innervate neuroeffector cells that are coupled to smooth muscle cells, allowing indirect control of muscle activity. The control of motility and secretion is an essential process that must be tightly regulated. Factors that perturbate the activity of any of these neuron types, such as exposure to opioid drugs, can lead to dysmotility or to defective secretory function, resulting in diarrhea or constipation. Each neuronal subset expresses a unique complement of cell surface receptors, enabling them to respond to distinct transmitters. This diversity can be targeted pharmacologically to allow for selective modulation of the ENS.

Opioid receptors and their endogenous ligands

The opioid system is most commonly known for controlling pain, reward and addictive behaviors, but it also modulates several essential physiological processes including respiratory and GI function. Opioids exert their biological actions through the activation of three main opioid receptor subtypes: MOR, delta (DOR) and kappa (KOR). The precursor molecules pro-opiomelanocortin, pro-enkephalin and pro-dynorphin are proteolytically cleaved, generating the three classes of biologically active opioid peptides: endorphins, enkephalins and dynorphins, respectively. The endogenous peptides derived from these precursors all have the N-terminal tetrapeptide sequence Tyr-Gly-Gly-Phe and activate the different opioid receptor subtypes with varying affinities: endorphins have similar affinity for MOR and DOR, enkephalins have a slightly higher affinity for DOR and dynorphins have higher affinity for KOR (Janecka, Fichna, & Janecki, 2004; Thompson, Canals, & Poole, 2014).

Opioid receptors are members of the large family of G protein-coupled receptors (GPCRs) that transduce intracellular signals through heterotrimeric G proteins, composed of alpha (α), beta (β) and gamma (γ) subunits. The canonical signaling pathway for opioid receptors involves the recruitment of inhibitory G proteins that regulate second messenger production and ion channel conductivity. This ultimately leads to suppression of neuronal excitability (Al-Hasani & Bruchas, 2011). Translocation of the $G\alpha_{i/o}$ subunit inhibits adenylate cyclase activity and subsequent cyclic adenosine monophosphate (cAMP) production. The $G\beta\gamma$ subunit promotes the opening of G protein-coupled inwardly-rectifying potassium (GIRK) channels which results in neuronal hyperpolarization in postsynaptic nerve terminals (Fig. 2). The $G\beta\gamma$ subunit also inhibits voltage-gated calcium channels which suppress neurotransmitter release (Fig. 2) (Standifer & Pasternak, 1997). G protein signaling also activates downstream signaling molecules including, but not limited to, extracellular signal-regulated kinases (ERKs) (Belcheva, Wong, & Coscia, 2000; Duraffourd, Kumala, Anselmi, Brecha, & Sternini, 2014).

Upon prolonged or repeated agonist exposure, GPCRs undergo phosphorylation by G protein-coupled receptor kinases (GRKs), which increases the affinity of the receptor for the scaffold protein beta arrestin (Gurevich & Gurevich, 2019). Beta arrestins block further interaction of the receptor with the G protein, resulting in termination of its signaling in a process known as receptor desensitization. In addition, beta arrestins promote receptor internalization by interacting with the cellular trafficking machinery (Tian, Kang, & Benovic, 2014) and can also act as molecular scaffolds to support noncanonical, G protein-independent signaling (Defea, 2008; Luttrell & Gesty-Palmer, 2010; Peterson & Luttrell, 2017).

Opioid receptors and their cognate peptide ligands (e.g., endorphins) are expressed widely throughout the CNS where they modulate pain transmission, euphoria, sedation and respiratory control. The endogenous opioid system is also present in the GI tract (Galligan & Sternini, 2017; Sternini, Patierno, Selmer, & Kirchgessner, 2004). Enteric neurons that produce opioid neurotransmitters innervate the smooth muscle and mucosa and modulate GI motility and secretion (Poole et al., 2011; Thompson et al., 2014). Additionally, opioid peptides are also expressed in other cell types in the GI tract including enteroendocrine and immune cells (Collins & Verma-Gandhu, 2006; Kokrashvili et al., 2009). Immune-derived opioids produced during intestinal inflammation promote targeted peripheral analgesia and represent a potential mechanism that may be harnessed therapeutically (Carbone & Poole, 2020). Thus, in addition to pain perception, opioids and their respective receptors play a pivotal role in the homeostatic regulation of GI function.

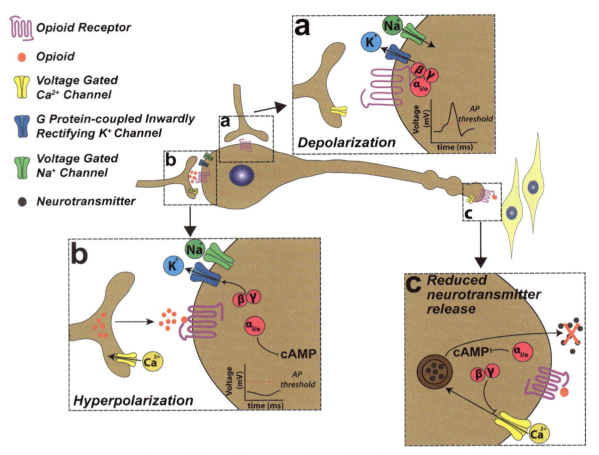

FIG. 2 Established molecular mechanisms underlying opioid receptor modulation of enteric neurotransmission. (A) At the postsynaptic level, the opening of Na⁺ channels *(green)* results in neuronal depolarization. After the action potential is triggered, K⁺ channels open *(blue)* which leads to hyperpolarization and the restoration of the membrane potential to its resting state. (B) The activation of opioid receptors results in the dissociation of the heterotrimeric G protein into Gα$_{i/o}$ and βγ subunits. The Gα$_{i/o}$ subunit inhibits adenylate cyclase and subsequent cAMP production. The βγ subunit directly activates K⁺ channels leading to hyperpolarization and reduced neuronal excitability. (C) At the presynaptic level, the βγ subunit directly inhibits Ca^{2+} channels *(yellow)* and the associated influx of Ca^{2+} ions into the nerve terminal, ultimately suppressing neurotransmitter release.

Opioids inhibit GI function through actions on the ENS

MOR is the primary target of clinically important opioid analgesics, including morphine. These drugs suppress nociceptive signaling through interaction with MOR expressed by primary afferent neurons and their peripheral terminals, by neurons of the dorsal horn of the spinal cord, and in brain regions involved in the processing of nociceptive input and the sensation of pain. Opioid receptors are established therapeutic targets for digestive disease and are the drivers of adverse effects of opioids on gut function. Distribution and functional studies indicate that MOR is extensively expressed by neurons located within ascending and descending pathways in the ENS. This has almost exclusively been determined in laboratory species, with very limited information available about the distribution of MOR in the human ENS despite their established clinical importance (DiCello et al., 2020; Lay et al., 2016; Lupp, Richter, Doll, Nagel, & Schulz, 2011). The constipating effects of opioid analgesics are mediated through on-target actions at MOR expressed by enteric neurons. Evidence to support this conclusion comes from the absence of morphine-induced delays in GI transit in MOR-deficient mice (Roy, Liu, & Loh, 1998) and from studies demonstrating that the inhibitory effects of opioids on neurogenic contractions are retained in isolated segments of the GI tract where central influence is effectively removed (DiCello et al., 2019). Furthermore, opioid-induced constipation can be diminished by peripherally restricted MOR antagonists (PAMORAs: e.g., nalmedine, methylnaltrexone, Naloxegol), and the peripherally-restricted MOR agonist loperamide is a commonly used antidiarrheal. Notably, the general effect of opioids on GI motility is consistent across species, suggesting that the distribution within functional neuronal subclasses is highly conserved.

How do opioids exert their inhibitory effects on GI motility to promote chronic constipation?

General effects of opioids on GI motility

The activation of opioid receptors on enteric neurons alters cellular signaling via the modulation of ion channel conductance (Galligan & Akbarali, 2014). Opioids influence ion channel conductance through a G protein-dependent mechanism (Johnson, 1990; Karras & North, 1979). Fig. 2 illustrates established opioid receptor signaling and ion channel modulation in enteric neurons. Altered enteric neuron excitability and suppressed neurotransmitter release adversely impacts bowel movements including dysmotility and inhibited transit.

GI motility is essential for survival and is tightly controlled to ensure effective absorption or secretion of water and electrolytes, and the absorption of nutrients. Smooth muscle layers provide the mechanical apparatus for a series of coordinated motility patterns, collectively known as peristaltic contractions. Peristalsis is achieved by the coordinated interplay of myenteric pathways (Spencer et al., 2016). Ultimately, the suppression of enteric neuronal activity accounts for the observed physiological and pharmacological effects of opioids on GI motility. By limiting the neural activity of inhibitory neurons, opioid receptor agonists enhance circular muscle tone (termed disinhibition). This leads to spasmodic contractions of the muscle and disrupts the coordinated peristaltic movements needed to effectively expel contents (Wood & Galligan, 2004). Moreover, the activation of opioid receptors on excitatory pathways inhibits the neurogenic contractions needed to propel contents. In addition to disrupting peristalsis, DiCello et al. (2019) established that the pharmacological activation of either MOR or DOR in the mouse colon disrupts cyclical propulsive motor events that are highly conserved across mammalian species, including humans. These actions are likely to contribute to constipation. Opioids also inhibit submucosal neurons which suppresses the movement of water and electrolytes across the lumen, thereby leading to dry stools and more severe constipation.

Opioid-induced constipation is resistant to tolerance development

The effectiveness of pain relief by opioids is significantly diminished over time, with increasing doses required to achieve the equivalent effect. This is known as analgesic tolerance and is a significant limitation to the long-term use of opioids. In direct contrast, tolerance to opioids does not develop in enteric pathways controlling motility. This leads to sustained, intractable constipation that persists for the duration of the treatment period (Crockett et al., 2019). This constipation is resistant to traditional osmotic, stimulant or bulk-forming laxatives and can be so severe that it impacts patient compliance with treatment regimens (Christensen et al., 2016). This clinical observation also occurs in laboratory animals and the proposed mechanism for its occurrence may be due to the lack of tolerance development in enteric neurons of the colon (Ross, Gabra, Dewey, & Akbarali, 2008).

Conceptually, the distinct cell type-specific regulation of MOR in nociceptive and enteric pathways has been attributed to unique functional roles for the scaffolding protein, beta arrestin 2 (βarr2). βarr2 interacts with opioid receptors to terminate canonical G protein-mediated signaling while still promoting the activation of other downstream signaling pathways. Early studies showed that βarr2 knockout mice presented enhanced and prolonged analgesia after chronic morphine treatment with reduced adverse effects including respiratory depression, constipation and tolerance (Bohn et al., 1999; Bohn, Lefkowitz, & Caron, 2002; Raehal, Walker, & Bohn, 2005). Consequently, recent efforts have focused on the development of opioid receptor agonists that are biased toward the G protein pathway and away from βarr2-dependent signaling in the hope to enhance analgesic efficacy while also diminishing adverse effects. This has led to the development of compounds such as oliceridine (TRV130) (DeWire et al., 2013) or PZM21 (Manglik et al., 2016) that have shown effective analgesia with improved side-effect profiles in rodents. However, subsequent re-examination of these agonists by other groups indicate that respiratory depression and constipation may persist in rodents (Altarifi et al., 2017; Hill et al., 2018; Kudla et al., 2019). Although initially promising, oliceridine (TRV130) was not approved by the FDA due to the relative risk to benefit. Schmid et al. (2017) have suggested that a higher degree of bias away from βarr2 recruitment might improve the therapeutic window by separating analgesia from respiratory depression even further. However, a recent study reassessing the signaling profile of these novel compounds has reported that they present a similar efficacy to the partial agonist buprenorphine and proposed that low intrinsic efficacy might underlie the improved side-effect profiles of these novel agonists (Gillis et al., 2020). However, the underlying theory supporting the development of G protein-biased opioid receptor ligands has been challenged. Mice that express MOR with key C-terminal phosphorylation sites required for effective beta arrestin recruitment removed exhibit respiratory depression, constipation, and opioid withdrawal following opioid administration (Kliewer et al., 2019). Furthermore, a recent study demonstrated that opioid-induced respiratory depression and constipation in βarr2 knockout mice were equivalent to that in wildtype mice (Kliewer et al., 2020), directly contradicting earlier research.

Application to other areas

In this chapter we have provided a basic overview of the mechanisms underlying opioid-induced constipation. We have described the autonomic pathways that regulate the neural control of GI physiology, the molecular biology of opioid receptors, the prolonged inhibitory actions of opioids on enteric pathways, and highlighted new pharmacological approaches to develop safer opiate analgesics. In the following section, we briefly describe non-GI side-effects associated with opioid use. In marked contrast to the prolonged inhibitory effects of opioids in the ENS, tolerance usually develops to their analgesic actions. This is problematic since higher doses are required to maintain analgesic efficacy. However, as described in the section above, this leads to greater or prolonged GI complications, including constipation. Another common side-effect which is now considered a major health burden in first world countries is the development of physical dependence and addiction to opioids. The highly addictive nature of opioids may initially arise from escalating dosages in the clinical setting to help maintain their analgesic profile. Finally, opioid use can potentially lead to respiratory depression and overdose. Therefore, careful consideration of the patient's clinical history and an appropriate opioid regimen is critical for reducing the likelihood of these adverse side-effects.

Other agents of interest

In this chapter we have highlighted the mechanisms underlying OIC. PAMORAs that are used to treat severe cases of OIC include:

Naloxegol
Methylnaltrexone
Naldemedine

Potential opportunities to treat or avoid OIC and other bowel disorders have been described in detail (Canals, Poole, Veldhuis, Schmidt, & Bunnett, 2019; Carbone & Poole, 2020) and include:

Enkephalinase inhibitors
Allosteric modulators of opioid receptors
pH-responsive opioid agonists
Preclinical agents that have shown fewer GI side-effects

Mini-dictionary of terms

Morphine: The main alkaloid of opium that acts via MOR to produce potent analgesia. Although morphine is routinely administered for moderate to severe pain, its use is limited by adverse side-effects including OIC.
Constipation: A common digestive problem whereby patients pass hard, dry stools. Patients with severe constipation have fewer than three bowel movements per week and often suffer from GI pain.
Enteric nervous system (ENS): The intrinsic nervous system of the GI tract and often referred to as the "second brain" because of its autonomy and dense neuron population (similar number of neurons in the spinal cord). It controls local GI functions including motility, secretion and blood flow. Enteric neurons may also relay and receive information from the CNS via parasympathetic and sympathetic connections.
Tolerance: The reduction in the response to a drug following its repeated use. Most patients develop tolerance to the analgesic properties of opioids, whereas some side-effects including constipation persist throughout the treatment regimen. Thus, tolerance to opioids is a system-dependent phenomenon.
Beta arrestins: A small family of scaffolding proteins which regulate GPCR signaling.

Key facts of opioid-induced constipation

- OBD is a collection of adverse GI side-effects, with the most common being OIC.
- OIC affects up to 80% of patients administered clinical opioids.
- Half of patients suffering from OIC are unresponsive to laxatives.

- Opioids exert their effects through specific GPCRs. Clinical opiates mainly target the MOR subtype.
- All opioid receptor subtypes are expressed by enteric neurons, and their activation leads to the inhibition of GI motility and secretion.
- The CNS and ENS have different tolerance profiles to the actions of opioids, which is reflected by the clinical effects of opioid drugs.

Summary points

- Colonic movement and secretion are controlled and coordinated by the ENS located exclusively within the gut wall.
- Opioid receptors belong to the seven transmembrane GPCR superfamily and comprise of three classical subtypes; MOR, DOR and KOR.
- Opioid receptors which are targets of opioid analgesics are also expressed by enteric neurons. Regulation of these receptors differs markedly between the ENS and pain pathways.
- Opioids suppress both enteric neuronal excitability and neurotransmitter release, leading to disruption of the coordinated movement of bowel contents and secretomotor function, ultimately leading to constipation.
- Tolerance does not develop to the inhibitory actions of opioids on ENS function in the colon, resulting in sustained constipation that persists until opioid administration is ceased.

References

Abramowitz, L., Beziaud, N., Labreze, L., Giardina, V., Causse, C., Chuberre, B., et al. (2013). Prevalence and impact of constipation and bowel dysfunction induced by strong opioids: A cross-sectional survey of 520 patients with cancer pain: DYONISOS study. *Journal of Medical Economics, 16* (12), 1423–1433.

Al-Hasani, R., & Bruchas, M. R. (2011). Molecular mechanisms of opioid receptor-dependent signaling and behavior. *Anesthesiology, 115*(6), 1363–1381.

Altarifi, A. A., David, B., Muchhala, K. H., Blough, B. E., Akbarali, H., & Negus, S. S. (2017). Effects of acute and repeated treatment with the biased mu opioid receptor agonist TRV130 (oliceridine) on measures of antinociception, gastrointestinal function, and abuse liability in rodents. *Journal of Psychopharmacology, 31*(6), 730–739.

Belcheva, M. M., Wong, Y. H., & Coscia, C. J. (2000). Evidence for transduction of mu but not kappa opioid modulation of extracellular signal-regulated kinase activity by G(z) and G(12) proteins. *Cellular Signalling, 12*(7), 481–489.

Bell, T., Annunziata, K., & Leslie, J. B. (2009). Opioid-induced constipation negatively impacts pain management, productivity, and health-related quality of life: Findings from the National Health and Wellness Survey. *Journal of Opioid Management, 5*(3), 137–144.

Bohn, L. M., Lefkowitz, R. J., & Caron, M. G. (2002). Differential mechanisms of morphine antinociceptive tolerance revealed in (beta)arrestin-2 knockout mice. *The Journal of Neuroscience, 22*(23), 10494–10500.

Bohn, L. M., Lefkowitz, R. J., Gainetdinov, R. R., Peppel, K., Caron, M. G., & Lin, F. T. (1999). Enhanced morphine analgesia in mice lacking beta-arrestin 2. *Science, 286*(5449), 2495–2498.

Canals, M., Poole, D. P., Veldhuis, N. A., Schmidt, B. L., & Bunnett, N. W. (2019). G-protein-coupled receptors are dynamic regulators of digestion and targets for digestive diseases. *Gastroenterology, 156*(6), 1600–1616.

Carbone, S. E., & Poole, D. P. (2020). Inflammation without pain: Immune-derived opioids hold the key. *Neurogastroenterology and Motility, 32*(2), e13787.

Christensen, H. N., Olsson, U., From, J., & Breivik, H. (2016). Opioid-induced constipation, use of laxatives, and health-related quality of life. *Scandinavian Journal of Pain, 11*, 104–110.

Collins, S., & Verma-Gandhu, M. (2006). The putative role of endogenous and exogenous opiates in inflammatory bowel disease. *Gut, 55*(6), 756–757.

Crockett, S. D., Greer, K. B., Heidelbaugh, J. J., Falck-Ytter, Y., Hanson, B. J., Sultan, S., et al. (2019). American Gastroenterological Association Institute Guideline on the medical management of opioid-induced constipation. *Gastroenterology, 156*(1), 218–226.

Defea, K. (2008). Beta-arrestins and heterotrimeric G-proteins: Collaborators and competitors in signal transduction. *British Journal of Pharmacology, 153*(Suppl. 1), S298–S309.

DeWire, S. M., Yamashita, D. S., Rominger, D. H., Liu, G., Cowan, C. L., Graczyk, T. M., et al. (2013). A G protein-biased ligand at the mu-opioid receptor is potently analgesic with reduced gastrointestinal and respiratory dysfunction compared with morphine. *The Journal of Pharmacology and Experimental Therapeutics, 344*(3), 708–717.

DiCello, J. J., Carbone, S. E., Saito, A., Rajasekhar, P., Ceredig, R. A., Pham, V., et al. (2020). Mu and delta opioid receptors are coexpressed and functionally interact in the enteric nervous system of the mouse colon. *Cellular and Molecular Gastroenterology and Hepatology, 9*(3), 465–483.

DiCello, J. J., Saito, A., Rajasekhar, P., Sebastian, B. W., McQuade, R. M., Gondin, A. B., et al. (2019). Agonist-dependent development of delta opioid receptor tolerance in the colon. *Cellular and Molecular Life Sciences, 76*(15), 3033–3050.

Duraffourd, C., Kumala, E., Anselmi, L., Brecha, N. C., & Sternini, C. (2014). Opioid-induced mitogen-activated protein kinase signaling in rat enteric neurons following chronic morphine treatment. *PLoS One, 9*(10), e110230.

Galligan, J. J., & Akbarali, H. I. (2014). Molecular physiology of enteric opioid receptors. *American Journal of Gastroenterology Supplements, 2*(1), 17–21. https://doi.org/10.1038/ajgsup.2014.5.

Galligan, J. J., & Sternini, C. (2017). Insights into the role of opioid receptors in the GI tract: Experimental evidence and therapeutic relevance. *Handbook of Experimental Pharmacology, 239*, 363–378.

Gillis, A., Gondin, A. B., Kliewer, A., Sanchez, J., Lim, H. D., Alamein, C., et al. (2020). Low intrinsic efficacy for G protein activation can explain the improved side effect profiles of new opioid agonists. *Science Signaling, 13*(625), eaaz3140.

Gurevich, V. V., & Gurevich, E. V. (2019). GPCR signaling regulation: The role of GRKs and arrestins. *Frontiers in Pharmacology, 10*, 125. https://doi.org/10.3389/fphar.2019.00125.

Hill, R., Disney, A., Conibear, A., Sutcliffe, K., Dewey, W., Husbands, S., et al. (2018). The novel mu-opioid receptor agonist PZM21 depresses respiration and induces tolerance to antinociception. *British Journal of Pharmacology, 175*(13), 2653–2661.

Janecka, A., Fichna, J., & Janecki, T. (2004). Opioid receptors and their ligands. *Current Topics in Medicinal Chemistry, 4*(1), 1–17.

Johnson, S. M. (1990). Opioid inhibition of cholinergic transmission in the guinea-pig ileum is independent of intracellular cyclic AMP. *European Journal of Pharmacology, 180*(2–3), 331–338.

Karras, P. J., & North, R. A. (1979). Inhibition of neuronal firing by opiates: Evidence against the involvement of cyclic nucleotides. *British Journal of Pharmacology, 65*(4), 647–652.

Kliewer, A., Gillis, A., Hill, R., Schmidel, F., Bailey, C., Kelly, E., et al. (2020). Morphine-induced respiratory depression is independent of beta-arrestin2 signalling. *British Journal of Pharmacology, 177*(13), 2923–2931.

Kliewer, A., Schmiedel, F., Sianati, S., Bailey, A., Bateman, J. T., Levitt, E. S., et al. (2019). Phosphorylation-deficient G-protein-biased mu-opioid receptors improve analgesia and diminish tolerance but worsen opioid side effects. *Nature Communications, 10*(1), 367.

Kokrashvili, Z., Rodriguez, D., Yevshayeva, V., Zhou, H., Margolskee, R. F., & Mosinger, B. (2009). Release of endogenous opioids from duodenal enteroendocrine cells requires Trpm5. *Gastroenterology, 137*(2), 598–606 (606 e591-592).

Kudla, L., Bugno, R., Skupio, U., Wiktorowska, L., Solecki, W., Wojtas, A., et al. (2019). Functional characterization of a novel opioid, PZM21, and its effects on the behavioural responses to morphine. *British Journal of Pharmacology, 176*(23), 4434–4445.

Lacy, B. E., Mearin, F., Chang, L., Chey, W. D., Lembo, A. J., Simren, M., et al. (2016). Bowel disorders. *Gastroenterology, 150*(6), 1393–1407.e5. https://doi.org/10.1053/j.gastro.2016.02.031.

Lay, J., Carbone, S. E., DiCello, J. J., Bunnett, N. W., Canals, M., & Poole, D. P. (2016). Distribution and trafficking of the mu-opioid receptor in enteric neurons of the Guinea pig. *American Journal of Physiology. Gastrointestinal and Liver Physiology, 311*(2), G252–G266.

Lupp, A., Richter, N., Doll, C., Nagel, F., & Schulz, S. (2011). UMB-3, a novel rabbit monoclonal antibody, for assessing mu-opioid receptor expression in mouse, rat and human formalin-fixed and paraffin-embedded tissues. *Regulatory Peptides, 167*(1), 9–13.

Luttrell, L. M., & Gesty-Palmer, D. (2010). Beyond desensitization: Physiological relevance of arrestin-dependent signaling. *Pharmacological Reviews, 62*(2), 305–330.

Manglik, A., Lin, H., Aryal, D. K., McCorvy, J. D., Dengler, D., Corder, G., et al. (2016). Structure-based discovery of opioid analgesics with reduced side effects. *Nature, 537*(7619), 185–190.

Nee, J., Zakari, M., Sugarman, M. A., Whelan, J., Hirsch, W., Sultan, S., et al. (2018). Efficacy of treatments for opioid-induced constipation: Systematic review and meta-analysis. *Clinical Gastroenterology and Hepatology, 16*(10), 1569–1584 (e1562).

Peterson, Y. K., & Luttrell, L. M. (2017). The diverse roles of arrestin scaffolds in G protein-coupled receptor signaling. *Pharmacological Reviews, 69*(3), 256–297.

Poole, D. P., Pelayo, J. C., Scherrer, G., Evans, C. J., Kieffer, B. L., & Bunnett, N. W. (2011). Localization and regulation of fluorescently labeled delta opioid receptor, expressed in enteric neurons of mice. *Gastroenterology, 141*(3), 982–991. e918.

Raehal, K. M., Walker, J. K., & Bohn, L. M. (2005). Morphine side effects in beta-arrestin 2 knockout mice. *The Journal of Pharmacology and Experimental Therapeutics, 314*(3), 1195–1201.

Ross, G. R., Gabra, B. H., Dewey, W. L., & Akbarali, H. I. (2008). Morphine tolerance in the mouse ileum and colon. *The Journal of Pharmacology and Experimental Therapeutics, 327*(2), 561–572.

Roy, S., Liu, H. C., & Loh, H. H. (1998). Mu-opioid receptor-knockout mice: The role of mu-opioid receptor in gastrointestinal transit. *Brain Research. Molecular Brain Research, 56*(1–2), 281–283.

Schmid, C. L., Kennedy, N. M., Ross, N. C., Lovell, K. M., Yue, Z., Morgenweck, J., et al. (2017). Bias factor and therapeutic window correlate to predict safer opioid analgesics. *Cell, 171*(5), 1165–1175. e1113.

Spencer, N. J., Dinning, P. G., Brookes, S. J., & Costa, M. (2016). Insights into the mechanisms underlying colonic motor patterns. *The Journal of Physiology, 594*(15), 4099–4116.

Standifer, K. M., & Pasternak, G. W. (1997). G proteins and opioid receptor-mediated signalling. *Cellular Signalling, 9*(3–4), 237–248.

Sternini, C., Patierno, S., Selmer, I. S., & Kirchgessner, A. (2004). The opioid system in the gastrointestinal tract. *Neurogastroenterology and Motility, 16* (Suppl. 2), 3–16.

Thapa, N., Kappus, M., Hurt, R., & Diamond, S. (2019). Implications of the opioid epidemic for the clinical gastroenterology practice. *Current Gastroenterology Reports, 21*(9), 44.

Thompson, G. L., Canals, M., & Poole, D. P. (2014). Biological redundancy of endogenous GPCR ligands in the gut and the potential for endogenous functional selectivity. *Frontiers in Pharmacology, 5*, 262.

Tian, X., Kang, D. S., & Benovic, J. L. (2014). Beta-arrestins and G protein-coupled receptor trafficking. *Handbook of Experimental Pharmacology, 219*, 173–186.

Wood, J. D., & Galligan, J. J. (2004). Function of opioids in the enteric nervous system. *Neurogastroenterology and Motility, 16*(Suppl. 2), 17–28.

Chapter 22

Cognitive behavioral therapy for chronic pain and opioid use disorder

Marina G. Gazzola[a,b], Mark Beitel[a,b,c], Christopher J. Cutter[a,b,c], and Declan T. Barry[a,b,c]

[a]*Department of Psychiatry, Yale School of Medicine, New Haven, CT, United States,* [b]*Pain Treatment Services, The APT Foundation, New Haven, CT, United States,* [c]*Child Study Center, Yale School of Medicine, New Haven, CT, United States*

Abbreviations

CBT	cognitive behavioral therapy
LTOT	long-term opioid therapy
MOUD	medication for opioid use disorder
OUD	opioid use disorder
SUD	substance use disorder

Introduction

In this chapter, we describe cognitive-behavioral therapy (CBT) for chronic pain and opioid use disorder (OUD). We first provide a context for understanding the prevalence and importance of addressing these interrelated chronic medical conditions. We then summarize findings regarding the clinical complexity of patients with OUD and chronic pain and the challenges faced by them and their providers. We provide a brief primer on OUD diagnosis and medication for OUD (MOUD), the necessary pharmacological platform for implementing CBT. We review the modules of CBT for chronic pain and OUD and describe the key clinical deficits that they are designed to address. After summarizing the evidence supporting this approach, we offer some information about implementation of this approach in real-world settings.

Prevalence of chronic pain and OUD

Chronic pain and opioid use disorder (OUD) are prevalent and interrelated chronic medical conditions. Chronic pain, defined as noncancer pain lasting most days over at least 3 months often limits patient mobility, autonomy, and quality of life. The Centers for Disease Control estimates that 20% of adult Americans experience chronic pain and 8% have "high-impact" chronic pain that limits their life or work activities on most days (Dahlhamer, Lucas, Zelaya, et al., 2016). More than 2 million individuals in the United States have OUD (or opioid addiction),[a] a chronic relapsing condition that is associated with high rates of mortality; in 2019, opioids were the most common cause of the greater than 70,000 overdose deaths (Ahmad, Rossen, & Sutton, 2020).

In the United States, the prescribing of opioids for pain, especially noncancer chronic pain, increased precipitously in the 1990s and 2000s (Rummans, Burton, & Dawson, 2018). It reached a peak of more than 255,000,000 prescriptions in 2012, approximating to 81 prescriptions per 100 persons nationally (Jones, Haug, Stitzer, & Svikis, 2000). During this time period, the prevalence of OUD and opioid-related harms also dramatically increased (Kolodny et al., 2015). In 2016, the Centers for Disease Control and Prevention (CDC) issued guidelines that discouraged the use of opioids as a first-line strategy for managing chronic pain (Dowell, Haegerich, & Chou, 2016). However, millions of patients with chronic pain still receive long-term opioid therapy (LTOT; over 160,000,000 prescriptions were written for opioids in 2018, representing 51.4 per 100 people) (Jones et al., 2000). Although the OUD-related risk associated with LTOT was once downplayed, a systematic review found the rates of "opioid addiction" among patients on LTOT ranged from 8% to 12% (Vowles et al., 2015). Additionally, rates of opioid "misuse" ranged from 22% to 29% (Vowles et al., 2015). The presence of opioid misuse

---

a. The terms substance use disorders and addition are used interchangeably in this chapter.

Treatments, Mechanisms, and Adverse Reactions of Anesthetics and Analgesics. https://doi.org/10.1016/B978-0-12-820237-1.00022-3
Copyright © 2022 Elsevier Inc. All rights reserved.

increases the risk of eventually having an OUD, and many patients with OUD report transitioning from use of prescription opioids to heroin (Jones, 2013). There appears to be a dose-response relationship between both prescription opioid duration and amount (morphine equivalent daily dose) and risk of developing OUD (Barry et al., 2018; Bohnert, Logan, Ganoczy, & Dowell, 2016; Edlund et al., 2014).

Conversely, among patients with OUD, chronic pain is prevalent, with estimates ranging from 37% to more than 60% (Barry, Beitel, Garnet, et al., 2009; Barry, Beitel, Joshi, & Schottenfeld, 2009; Dhingra et al., 2013; Dunn, Finan, Tompkins, Fingerhood, & Strain, 2015; Jamison, Kauffman, & Katz, 2000; Rosenblum et al., 2003). The most common primary locations for pain among this patient population are in the back, legs, shoulder or total body, and most report an accident or nerve damage as the cause of their pain (Barry et al., 2013; Jamison et al., 2000). Many patients with chronic pain and OUD report using alcohol and nonmedical or illicit use of opioids and benzodiazepines to relieve pain. Notably, many report lifetime utilization of medical treatments for pain relief (e.g., prescribed analgesics, physical therapy, nerve-block) as well as a variety of complementary and integrated health approaches (e.g., acupuncture, herbal medication, prayer, counseling, self-help groups, yoga, or hypnosis), including body-based methods (heat therapy, ice therapy, massage, using a TENS unit, or seeing a chiropractor) (Barry et al., 2012, 2013; Dunn et al., 2015; Jamison et al., 2000).

Patients may experience opioid-induced hyperalgesia because of prolonged opioid exposure (Roeckel, Le Coz, Gavériaux-Ruff, & Simonin, 2016). While some studies have noted that hyperalgesia can occur due to opioid withdrawal (Angst & Clark, 2006), this is unlikely to be a significant contributor to the extensive pain history many patients seeking treatment for OUD experience. Some patients with untreated OUD and chronic pain may be concerned that if they seek treatment for OUD, they will be asked to reduce their intake of short-acting opioid analgesics, illicit opioids, or other substances, and their pain may worsen. However, most patients entering MOUD with chronic pain report interest in receiving onsite treatment of both chronic pain and OUD (Barry, Beitel, et al., 2010; Barry, Irwin, et al., 2010; Beitel et al., 2016).

Clinical complexity of patients with chronic pain and OUD

Compared to those with each chronic medical condition alone, patients with both chronic pain and OUD often have higher rates of psychiatric comorbidity, including anxiety, mood, substance use, and personality disorders (Barry et al., 2016; Haller & Acosta, 2010; Higgins, Smith, & Matthews, 2020; Hser et al., 2017). Despite these high rates of co-occurring psychopathology, patients with chronic pain and OUD seeking MOUD report low rates of mental health care, which may also complicate their treatment (Barry et al., 2016). They also have higher rates of sleep disturbance (Baldassarri et al., 2020; Peles, Schreiber, & Adelson, 2006, 2009). Patients with both chronic pain and SUDs also have an elevated risk of opioid overdose (Dunn et al., 2010). Pain flares may prompt patients to increase opioid use to manage pain (Griffin et al., 2016) and are associated with increased opioid cravings among patients seeking treatment for OUD (Tsui et al., 2016). As noted previously, patients with chronic pain and OUD report a variety of licit (e.g., alcohol, tobacco) and illicit (e.g., benzodiazepines, cannabis[b]) substances to manage their pain (Barry, Beitel, Joshi, et al., 2009).

Provider consideration

While substance use disorders and chronic pain have been traditionally treated by separate providers, clinicians in both areas recognize similar clinical challenges to treating patients with comorbid pain and OUD. Clinicians perceive that they need more training in both pain and OUD treatment (Barry, Beitel, et al., 2010; Barry, Irwin, et al., 2010; Beitel et al., 2017; Jamison, Scanlan, Matthews, Jurcik, & Ross, 2016), which may be a result of the historically limited exposure to addiction medicine during medical training (Wood, Samet, & Volkow, 2013). Providers also perceive high clinical complexity among patients with OUD and chronic pain (Barry, Beitel, et al., 2010; Barry, Irwin, et al., 2010; Beitel et al., 2017), which is reflected in the aforementioned high rates of psychiatric and medical comorbidities that may often go untreated (Barry et al., 2016). Both addiction counselors and primary care physicians report difficulties coordinating drug counseling and pain treatment given the lack of available integrated treatments and referrals for these complex patients (Barry et al., 2008; Barry, Beitel et al., 2010; Barry, Irwin, et al., 2010; Beitel et al., 2017). Despite these challenges, counselors report positive views of and interest in receiving training on psychosocial interventions for chronic pain and OUD, including CBT (Barry et al., 2008; Oberleitner et al., 2016). In one study, when asked about their attitudes toward nonpharmacologic approaches to pain, addiction counselors viewed CBT to have the highest perceived treatment efficacy and were most willing to refer patients to CBT compared to other methods such as progressive muscle relaxation training, stress management training, or physical therapy (Oberleitner et al., 2016).

---

b. We recognize that legalization of cannabis for medical and recreational purposes has occurred to varying degrees depending on the reader's location.

Necessary pharmacological platform: Medication for opioid use disorder

Medication for opioid use disorder (MOUD) is an evidence-based pharmacotherapy for OUD that involves methadone (full agonist), buprenorphine (partial agonist), and naltrexone (antagonist). MOUD confers a decrease in mortality, morbidity, and infectious disease risk (e.g., Hepatitis C and HIV) (Johnson et al., 2020; National Academies of Sciences, Engineering, and Medicine, 2019; Sordo et al., 2017). In the United States, methadone and buprenorphine (usually in combination with naloxone) are the most frequently used MOUD. While methadone is generally provided in opioid treatment programs (which involve considerable local, state, and federal oversight), buprenorphine/naloxone can be prescribed for OUD in office-based settings by licensed providers who have a DEA-waiver following completion of specialized training. Naltrexone, a long-acting, injectable, opioid antagonist, is also approved for the treatment of OUD, but may not be suitable for patients with chronic pain receiving LTOT, as it can precipitate withdrawal and requires a minimum 7–10-day abstinence period prior to initiation. As such, this chapter focuses on opioid agonist therapy as the pharmacologic platform for optimal medical management of comorbid OUD and chronic pain.

Established protocols exist for MOUD induction, maintenance, and taper. Methadone is typically initiated at a 10–30 mg dose that is tapered by 5–10 mg to reach an optimal dose between 60 and 120 mg per day on an individualized basis (Substance Abuse and Mental Health Services Administration, 2020). Buprenorphine, a partial agonist which should be initiated after a patient is experiencing some opioid withdrawal symptoms to prevent precipitating withdrawal, is similarly tapered, with a typical initial dose of 2–4 mg that is increased by 2 mg to a maintenance dose ranging from 4 to 24 mg (Substance Abuse and Mental Health Services Administration, 2020). MOUD involving methadone or buprenorphine is usually taken once daily; while some patients may report temporary pain relief from MOUD, the key purposes of MOUD are to attenuate opioid withdrawal and craving and provide an opioid blockade (i.e., illicit opioids do not produce euphoria). While split-dosing may be used to improve analgesia, this may be easier to do with buprenorphine than methadone. Additional research is needed to establish optimal dosing strategies (including formulation and dose) to improve both pain and substance use disorder outcomes among patients on MOUD.

Optimal medical management

At our research-based clinic, we offer patients integrated treatment targeting both chronic pain and OUD, which we term "optimal medical management." Optimal medical management comprises four interrelated components: (1) MOUD; (2) nonopioid pharmacotherapy to address pain and comorbid psychiatric conditions; (3) psychotherapy; and (4) other non-pharmacological pain management interventions (NPMIs), including exercise and relaxation training. Optimal medical management is guided by the interdisciplinary approach to pain management, the gold-standard approach for addressing chronic pain, which is steeped in the biopsychosocial model (Gatchel, McGeary, McGeary, & Lippe, 2014). The psychotherapy that we offer patients in routine clinical practice for chronic pain and OUD includes CBT and other evidence-based approaches for addressing pain or SUDs [e.g., acceptance and commitment therapy (Hughes, Clark, Colclough, Dale, & McMillan, 2017), mindfulness-based strategies (Veehof, Trompetter, Bohlmeijer, & Schreurs, 2016)[c]]. CBT is a particularly promising nonpharmacological platform for integrated treatment of OUD and chronic pain because of its efficacy in treating SUDs (Carroll & Onken, 2005; Dutra et al., 2008; Magill et al., 2019; Moore et al., 2016), chronic pain (Burns et al., 2015; Ehde, Dillworth, & Turner, 2014; Williams, Fisher, Hearn, & Eccleston, 2020), and associated psychiatric disorders (Barry et al., 2016; Cuijpers, Cristea, Karyotaki, Reijnders, & Huibers, 2016; Driessen & Hollon, 2010; Kaczkurkin & Foa, 2015) both alone and when combined with medications. Three pilot studies (Currie, Hodgins, Crabtree, Jacobi, & Armstrong, 2003; Ilgen et al., 2011; Morasco et al., 2016) and two randomized clinical trials (Ilgen et al., 2016, 2020) have found support for the efficacy of in-person delivered CBT for SUDs and chronic pain. Our team has also found support for the efficacy of CBT for OUD and chronic when offered in conjunction with MOUD involving methadone (Barry et al., 2019) and buprenorphine (Barry & Schottenfeld, 2020). One additional pilot study found that an online CBT-informed self-help pain management intervention combined with onsite behavioral therapy at a methadone maintenance treatment program was associated with reduced medication misuse and pain interference, though the attrition rate was 45% (Wilson et al., 2018).

---

c. Drs. Barry, Beitel, and Cutter are the Founding Director, Research Director, and Clinic Director, respectively, of the APT Foundation Pain Treatment Services, a clinical-research program that is dedicated to the assessment and treatment of co-occurring OUD and chronic pain.

Overview of CBT for chronic pain and OUD

CBT for chronic pain and OUD comprises 10–12 sessions, lasting approximately 45 min, that are designed to provide patients with coping skills to better manage these interrelated chronic medical conditions. The seven modules are designed to improve coping skills to address seven common clinical problems or challenges experienced by patients with chronic pain and OUD (see Table 1). A core task of CBT is to interrupt the "Pain-OUD Dysfunction Cycle," whereby patients manage pain or associated distress by using opioids, which although may bring immediate relief, results in several negative consequences, which fuel poor pain management. Therapists attempt to make patients aware of the thoughts, feelings, and behaviors, which are associated with poor pain management, and to manage them without recourse to illicit opioids (Fig. 1).

CBT modules

In the 12-session version, the first session contains assessments for substance use and chronic pain, along with a description of the integrated focus of treatment. Sessions 2–11 each center around a separate CBT module, shown in Table 1. The structure of each appointment is the same: (1) check-in regarding past-week MOUD adherence, substance use, pain, coping, and any "homework" and setting the agenda for the current session, (2) teaching or reinforcing coping skills, and (3) establish assignments to practice the coping skills learned between sessions. The final session serves to review the coping skills the patient learned throughout the counseling period and to formulate an overall plan for how to cope in challenging situations.

Seven clinical problems or challenges
Clinical problem 1: Treating chronic pain as if it were acute pain

Many patients conceptualize their chronic pain the way they view acute pain. While the goal of treating acute pain is to relieve a patient's suffering and then solve the source of the pain, many patients with chronic pain have had chronic pain for years, despite interventions aimed at solving the root of their pain. Some patients and clinicians may expect that eliminating pain is the first step to improving quality of life. This may contribute to frustration by both the patient and provider and may lead to promotion of inappropriate treatments and behaviors. An assumption of CBT is that improving nonpain aspects of a patient's experience, such as increased mobility or engaging in nondrug-related pleasurable activities, pain will become less central to their lives. The Psychoeducation module of CBT reviews the differences between acute and chronic pain, appropriate treatment approaches and targets, and emphasizes the importance of self-management.

Clinical problem 2: Patient inactivity

Many patients with chronic pain are inactive and may fear engaging in physical activity. For example, in one study, patients on MOUD with methadone who reported chronic pain were significantly less likely to be physically active (Beitel et al., 2016). A common barrier to exercise in people with chronic pain is fear that being active could increase their pain (Boutevillain, Dupeyron, Rouch, Richard, & Coudeyre, 2017; Martinez-Calderon, Flores-Cortes, Morales-Asencio, & Luque-Suarez, 2019). However, deconditioning due to prolonged avoidance of physical activity can actually worsen the pain. Continued avoidance and passivity decrease patient activation and engagement in activities that provide pleasure and distract from pain. Behavioral Activation is a CBT module that promotes engaging in both physical activities and pleasurable activities that do not involve substance use. Components of this module include identifying and trying to relieve barriers to exercise; setting exercise goals for patients; engaging in graded, time-limited activity such as walking, swimming, or stretching; monitoring exercise with tools like pedometers and accelerometers; and discussing nondrug related pastimes that bring joy to patients. Patients are prescribed weekly exercise and nonsubstance-related activities to perform on their own. While patients in methadone maintenance treatment who have chronic pain express low interest in participating in exercise groups, we have found that novel, engaging, low-to-moderate impact onsite exercise groups (e.g., exergames involving the Wii Fit) are feasible; following a single session, patients report acute improvements in pain intensity, stress, depression, anxiety, and happiness (Dimeola et al., 2021).

Clinical problem 3: Decreased distress tolerance

Many patients with comorbid OUD and chronic pain have difficulties tolerating distress. This may contribute to nonmedical opioid use and return to illicit substance use. Two CBT modules target increasing distress tolerance: Relaxation

TABLE 1 Modules for CBT for chronic pain and OUD.

Clinical problems	Module	Activities	Examples
Acute model of pain	Psychoeducation	• Discussing the biopsychosocial approach to pain management • Distinguishing between acute and chronic pain and their management • Discussing relationship between opioid use disorder and chronic pain • Explaining the role of conditioning and exercise in promoting functioning • Establishing a framework for the biological, psychological, and social factors shaping self-management of opioid use disorder and chronic pain	• Recognition of the importance of self-management
Inactivity	Exercise and behavioral activation	• Examining and mitigating barriers to exercise • Graded, time-limited, and controlled engagement in physical activity • Discussing nondrug-related pleasurable activities • Setting exercise goals • Monitoring exercise using tools like a pedometer or accelerometer	• Walking • Swimming • Stretching
Decreased distress tolerance	Relaxation training	• Imagery • Progressive muscular relaxation • Diaphragmatic breathing	• Deep breathing when waiting in a line at the store
	Coping with cravings or pain flare-ups	• Self-soothing • Distraction • Positive self-talk • Thinking through consequences • Urge surfing • Mindfulness	• "I can get through this" • Talking to a friend • "I've done this before"
Catastrophizing	Cognitive restructuring	• Identification of errors in thinking surrounding OUD and pain • Changing thinking patterns surrounding opioid use and pain	• Recognizing specific thoughts (e.g., "My pain will never get better") • Replace maladaptive thoughts with adaptive thoughts
Focusing only on what's wrong	Resilience training	• Recognizing positive events • Coping with negative events	• Appreciating nice weather, supportive partner • Acknowledge patient achievements, like starting treatment • Keeping a list of positive events (e.g., gratitude, accomplishments)
Viewing substance use as "beyond my control"	Functional analysis of behavior	• Identifying substance use and pain flare-up relapse precursors and consequences • Reviewing future plans	• Tracking precursors to substance use • Recognizing thoughts can affect behaviors
Lack of assertiveness	Assertiveness training	• Discussing communication styles • Practicing assertive communication toward both opioid use and pain	• Refusal skills for substances • Communication with clinicians about patient needs

Modified from Barry, D. T., Beitel, M., Cutter, C. J., Fiellin, D. A., Kerns, R. D., Moore, B. A., ... Ginn, J. (2019). An evaluation of the feasibility, acceptability, and preliminary efficacy of cognitive-behavioral therapy for opioid use disorder and chronic pain. *Drug and Alcohol Dependence, 194*, 460–467.

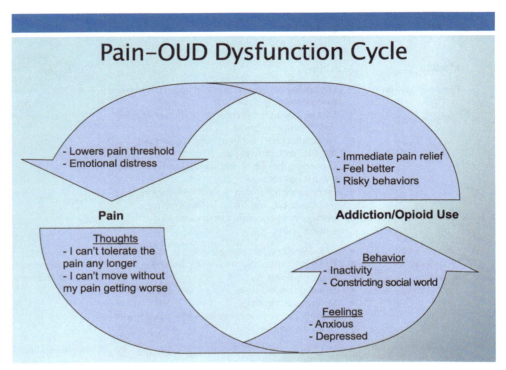

FIG. 1 The Pain-OUD Dysfunction Cycle describes how patients may use opioids for immediate pain relief, but face negative consequences as a result.

Training and Distress Coping Skills. Relaxation Training focuses on breathing techniques, visualization, and progressive muscular relaxation, while Distress Coping Skills address distress related to pain episodes (e.g., pain flare-up) and OUD (e.g., cravings) through behaviors like distraction, positive self-talk, reviewing negative consequences, and urge surfing. Combined, these two modules provide patients with a menu of options for how to handle challenging situations on their own.

Clinical problem 4: Catastrophizing

Many patients with comorbid pain and OUD engage in pain catastrophizing, which describes feeling stuck, magnification of pain, and feelings of helplessness in the face of pain going forward. For such patients, pain signals a dire outcome, which contributes to inactivity by causing patients to avoid activities they fear might worsen pain. Among patients with chronic pain, pain catastrophizing is an important predictor of pain intensity, interference, and overall functioning and promotes anxiety, depression, and lower quality of life (Craner, Sperry, Koball, Morrison, & Gilliam, 2017; Geisser, Robinson, Keefe, & Weiner, 1994; Quartana, Campbell, & Edwards, 2009; Severeijns, Vlaeyen, van den Hout, & Weber, 2001). Pain acceptance, on the other hand, is associated with decreased interference and improved functioning (Craner et al., 2017; Kratz, Ehde, Bombardier, Kalpakjian, & Hanks, 2017; McCracken, 1998). Similarly, in patients with OUD and chronic pain, improved pain outcomes are positively associated with pain acceptance and negatively associated with pain catastrophizing (Garnet et al., 2011; Mun et al., 2019). CBT clinicians help patients recognize and address pain catastrophizing and other "thinking errors" that negatively influence behaviors, mood, and feelings including stress (i.e., cognitive restructuring). Both the Coping Skills module (see Problem 3 above) and the Resilience Training module (see Problem 5 below) can also help patients address these cognitive errors. These modules, in part, are designed to assist patients in fostering reality-bound thoughts, develop positive coping statements, and engage in self-soothing with an ultimate goal of decreasing stress, which places the person at risk of poor pain management and relapse to nonmedical opioid use.

Clinical problem 5: Focusing on what's going wrong

For patients with OUD and chronic pain, who may also be struggling with other medical or psychiatric problems, it is easy to focus on aspects of their life that are not going well and ignore positive events. This contributes to feelings of helplessness

Behavioral Chains and Triggers	Thoughts and Feelings	Behavior	Positive Consequences	Negative Consequences
What sets me up to use?	What was I thinking? What was I feeling?	What did I do then?	What positive thing(s) happened?	What negative thing(s) happened?
What I plan to do differently next time:				

FIG. 2 Example of functional analysis worksheet.

and pessimism, and can worsen depressive symptoms. For example, a study of patients receiving methadone maintenance therapy found that those with either current or lifetime chronic pain had higher levels of pessimism than those without a history of chronic pain (Beitel et al., 2012). The module Resilience Training involves helping patients recognize positive events, including parts of their lives (e.g., nice weather, supportive spouse) or things they have done (e.g., started treatment, attended clinic) and savor the accompanying emotions (e.g., gratitude, accomplishment). Resilience Training also involves problem solving to cope with negative events (e.g., inclement weather, arguments with spouse); patients are guided to tease apart what role they may have played, what they can learn, and what short-term and long-term solutions may be available to them. Patients are trained to become aware of small positive and negative events that they may have ignored in the past.

Problem 6: Viewing substance use as "just happening" and "beyond their control"

Patients may feel powerless in the face of their OUD and chronic pain. For example, many patients may not recognize that certain patterns of behaviors, thoughts, or emotions may increase their risk of relapse (either to nonmedical opioid use or a pain flare-up). Functional Analysis helps patients to learn more and analyze their own behavioral patterns so that they are better able to identify precursors to and consequences of relapse (Haynes & O'Brien, 1990). An example of functional analysis work is in Fig. 2.

Problem 7: Decreased assertiveness

This module focuses on styles of communication and the likelihood of whether they produce the desired effect. Passive, aggressive, and passive-aggressive styles of communication are associated with personality disorders, which are common among patients on MOUD with co-occurring chronic pain and OUD (Barry, Beitel, Garnet, et al., 2009; Barry, Beitel, Joshi, et al., 2009; Barry et al., 2016) and may account for some of the clinical problems reported by providers who treat these patients (Barry et al., 2008; Barry, Beitel, et al., 2010; Barry, Irwin, et al., 2010; Beitel et al., 2017). Patients are taught to recognize different communication styles and to role-play being assertive. For example, regarding pain, patients might identify what they want to convey to their physician at their next visit, and role-play with the CBT therapist the anticipated interaction, until the patient is communicating in an assertive manner. In case of illicit opioid use, the therapist might role-play assertive drug-refusal strategies.

Applications to other areas: Real-world delivery of CBT for chronic pain and OUD

In busy methadone maintenance clinics or office-based practices, providers may not have the resources to deliver the full-course of CBT outlined above. We have found support for the feasibility, acceptability, and initial efficacy of offering specific CBT modules delivered in open-group format (e.g., psychoeducation, relaxation training) (Barry et al., 2014; Dimeola et al., 2021). We have also found that it is feasible to train addiction counselors to deliver the psychoeducation component of CBT with goal setting involving exercise (Butner et al., 2018). For providers who wish to try a brief course of CBT with a patient with chronic pain and OUD, we suggest that they might (1) review the psychoeducation module; (2) assess and recommend paced physical exercise; and (3) assess and address cognitive-errors or other barriers that interfere with exercise engagement. If you have more time, you then might offer a review of the additional skills and together with your patient select one or more additional modules.

Alternative therapies

In addition to CBT, nonpharmacologic and nonopioid treatments derived from a biopsychosocial approach for chronic pain are recommended by numerous institutional guidelines and systemic reviews (Dowell et al., 2016; Kamper et al., 2015; McDonagh et al., 2020; National Academies of Sciences, Engineering and Medicine, Health and Medicine Division, Board on Global Health; Board on Health Sciences Policy, Global Forum on Innovation in Health Professional Education, & Forum on Neuroscience and Nervous System Disorders, 2019). Evidence-based nonpharmacologic treatments shown to consistently better function for chronic pain include CBT, acupuncture, multidisciplinary rehabilitation, exercise, massage, and mind-body practices (e.g., meditation, relaxation, or mindfulness-based stress reduction) (National Academies of Sciences, Engineering, and Medicine et al., 2019; Skelly et al., 2020), although the exact effect and duration vary based on the therapy and location of pain. Patients also view multidisciplinary pain management positively and endorse decreases in pain-related factors in studies of such programs (Nees et al., 2020).

Conclusions

Chronic pain and OUD are interrelated and prevalent chronic medical conditions, which are associated with high levels of psychiatric distress among patients and frustration because of perceived lack of expertise and training by providers. In conjunction with MOUD, CBT is a promising approach to examine and address multiple clinical problems that accompany these medical conditions. For those providers, who do not have the time or resources to offer a full-course of individual CBT, a brief version conducted either individually or in groups is recommended. More research is needed to examine further the effectiveness of CBT for chronic pain and OUD in individual and group formats as part of interdisciplinary approach involving MOUD and perhaps other NPMIs.

Mini-dictionary of terms

Opioid use disorder: A chronic medical condition involving opioid use patterns that interfere with functioning.
Chronic pain: Noncancer pain lasting at least 3 months.
Cognitive behavioral therapy: A psychotherapy approach focused on thought patterns and how they impact actions and emotions.
Catastrophizing: Feeling stuck, magnification of pain, and feelings of helplessness in the face of pain.
Opioid agonist: A medication that acts on the same receptors as heroin, oxycodone, and other pain-relieving medications with similar chemical composition.

Key facts of CBT for chronic pain and OUD

- Chronic pain and opioid use disorder (OUD) are common and often co-occur.
- A growing number of patients diagnosed with OUD report that index opioid use involved opioid analgesics.
- Medication for opioid use disorder is a necessary pharmacological platform for treating chronic pain and OUD.
- Cognitive behavioral therapy (CBT) is a well-established treatment for psychiatric disorders and has been used for pain and opioid use disorder separately and in patients who have both conditions.
- CBT for chronic pain and OUD teaches patients coping skills to help better manage their pain and cravings for opioids without using illicit substances.

Summary points

- This chapter discusses treating patients with comorbid chronic pain and opioid use disorder (OUD) with a promising psychotherapy, cognitive behavioral therapy (CBT).
- Compared to those with each chronic medical condition alone, patients with both chronic pain and OUD may be more clinically complicated, with higher rates of psychiatric comorbidity.
- Comorbid pain and OUD are best treated by an interdisciplinary approach combining medication for opioid use disorder with nonopioid pharmacotherapy, psychotherapy, and other nonpharmacological pain management interventions.
- CBT is designed to improve coping skills to address common clinical problems experienced by patients with chronic pain and OUD.
- CBT interrupts the "Pain-OUD Dysfunction Cycle," in which patients use opioids for immediate pain relief, but face ensuing negative consequences.

References

Ahmad, F. B., Rossen, L., & Sutton, P. (2020). *Provisional drug overdose death counts.* Retrieved from https://www.cdc.gov/nchs/nvss/vsrr/drug-overdose-data.htm.

Angst, M. S., & Clark, J. D. (2006). Opioid-induced hyperalgesia: A qualitative systematic review. *Anesthesiology, 104*(3), 570–587. https://doi.org/10.1097/00000542-200603000-00025.

Baldassarri, S. R., Beitel, M., Zinchuk, A., Redeker, N. S., Oberleitner, D. E., Oberleitner, L. M. S., ... Barry, D. T. (2020). Correlates of sleep quality and excessive daytime sleepiness in people with opioid use disorder receiving methadone treatment. *Sleep & Breathing.* https://doi.org/10.1007/s11325-020-02123-z.

Barry, D. T., Beitel, M., Cutter, C. J., Fiellin, D. A., Kerns, R. D., Moore, B. A., ... Ginn, J. (2019). An evaluation of the feasibility, acceptability, and preliminary efficacy of cognitive-behavioral therapy for opioid use disorder and chronic pain. *Drug and Alcohol Dependence, 194*, 460–467.

Barry, D. T., Beitel, M., Cutter, C. J., Joshi, D., Falcioni, J., & Schottenfeld, R. S. (2010). Conventional and nonconventional pain treatment utilization among opioid dependent individuals with pain seeking methadone maintenance treatment: A needs assessment study. *Journal of Addiction Medicine, 4*(2), 81–87. https://doi.org/10.1097/ADM.0b013e3181ac913a.

Barry, D. T., Beitel, M., Garnet, B., Joshi, D., Rosenblum, A., & Schottenfeld, R. S. (2009). Relations among psychopathology, substance use, and physical pain experiences in methadone-maintained patients. *The Journal of Clinical Psychiatry, 70*(9), 1213–1218. https://doi.org/10.4088/JCP.08m04367.

Barry, D. T., Beitel, M., Joshi, D., & Schottenfeld, R. S. (2009). Pain and substance-related pain-reduction behaviors among opioid dependent individuals seeking methadone maintenance treatment. *The American Journal on Addictions, 18*(2), 117–121. https://doi.org/10.1080/10550490902772470.

Barry, D. T., Bernard, M. J., Beitel, M., Moore, B. A., Kerns, R. D., & Schottenfeld, R. S. (2008). Counselors' experiences treating methadone-maintained patients with chronic pain: A needs assessment study. *Journal of Addiction Medicine, 2*(2), 108–111. https://doi.org/10.1097/ADM.0b013e31815ec240.

Barry, D. T., Cutter, C. J., Beitel, M., Kerns, R. D., Liong, C., & Schottenfeld, R. S. (2016). Psychiatric disorders among patients seeking treatment for co-occurring chronic pain and opioid use disorder. *The Journal of Clinical Psychiatry, 77*(10), 1413–1419. https://doi.org/10.4088/JCP.15m09963.

Barry, D. T., Irwin, K. S., Jones, E. S., Becker, W. C., Tetrault, J. M., Sullivan, L. E., ... Fiellin, D. A. (2010). Opioids, chronic pain, and addiction in primary care. *The Journal of Pain, 11*(12), 1442–1450.

Barry, D. T., Marshall, B. D. L., Becker, W. C., Gordon, A. J., Crystal, S., Kerns, R. D., ... Edelman, E. J. (2018). Duration of opioid prescriptions predicts incident nonmedical use of prescription opioids among U.S. veterans receiving medical care. *Drug and Alcohol Dependence, 191*, 348–354. https://doi.org/10.1016/j.drugalcdep.2018.07.008.

Barry, D. T., Savant, J. D., Beitel, M., Cutter, C. J., Moore, B. A., Schottenfeld, R. S., & Fiellin, D. A. (2012). Use of conventional, complementary, and alternative treatments for pain among individuals seeking primary care treatment with buprenorphine-naloxone. *Journal of Addiction Medicine, 6*(4), 274–279. https://doi.org/10.1097/ADM.0b013e31826d1df3.

Barry, D. T., Savant, J. D., Beitel, M., Cutter, C. J., Moore, B. A., Schottenfeld, R. S., & Fiellin, D. A. (2013). Pain and associated substance use among opioid dependent individuals seeking office-based treatment with buprenorphine-naloxone: A needs assessment study. *The American Journal on Addictions, 22*(3), 212–217. https://doi.org/10.1111/j.1521-0391.2012.00327.x.

Barry, D. T., Savant, J. D., Beitel, M., Cutter, C. J., Schottenfeld, R. S., Kerns, R. D., ... Keneally, N. (2014). The feasibility and acceptability of groups for pain management in methadone maintenance treatment. *Journal of Addiction Medicine, 8*(5), 338.

Barry, D., & Schottenfeld, R. S. (2020). *Cogntive-behavioral therapy with buprenophrine/naloxone: For chronic pain and opioid use disorder: A randomized controlled trial.* (manuscript in preparation).

Beitel, M., Oberleitner, L., Kahn, M., Kerns, R. D., Liong, C., Madden, L. M., ... Barry, D. T. (2017). Drug counselor responses to patients' pain reports: A qualitative investigation of barriers and facilitators to treating patients with chronic pain in methadone maintenance treatment. *Pain Medicine, 18*(11), 2152–2161. https://doi.org/10.1093/pm/pnw327.

Beitel, M., Savant, J. D., Cutter, C. J., Peters, S., Belisle, N., & Barry, D. T. (2012). Psychopathology and pain correlates of dispositional optimism in methadone-maintained patients. *The American Journal on Addictions, 21*(Suppl. 1), S56–S62. https://doi.org/10.1111/j.1521-0391.2012.00293.x.

Beitel, M., Stults-Kolehmainen, M., Cutter, C. J., Schottenfeld, R. S., Eggert, K., Madden, L. M., ... Barry, D. T. (2016). Physical activity, psychiatric distress, and interest in exercise group participation among individuals seeking methadone maintenance treatment with and without chronic pain. *The American Journal on Addictions, 25*(2), 125–131.

Bohnert, A. S. B., Logan, J. E., Ganoczy, D., & Dowell, D. (2016). A detailed exploration into the association of prescribed opioid dosage and overdose deaths among patients with chronic pain. *Medical Care, 54*(5), 435–441. https://doi.org/10.1097/MLR.0000000000000505.

Boutevillain, L., Dupeyron, A., Rouch, C., Richard, E., & Coudeyre, E. (2017). Facilitators and barriers to physical activity in people with chronic low back pain: A qualitative study. *PLoS One, 12*(7), e0179826. https://doi.org/10.1371/journal.pone.0179826.

Burns, J., Nielson, W. R., Jensen, M. P., Heapy, A., Czlapinski, R., & Kerns, R. D. (2015). Specific and general therapeutic mechanisms in cognitive-behavioral treatment for chronic pain. *Journal of Consulting and Clinical Psychology, 83*(1), 1–11. https://doi.org/10.1037/a0037208.

Butner, J. L., Bone, C., Ponce Martinez, C. C., Kwon, G., Beitel, M., Madden, L., ... Barry, D. T. (2018). Training drug counselors to deliver a brief psychosocial intervention for chronic pain among patients in opioid agonist treatment: A pilot investigation. *Substance Abuse, 39*, 199–205.

Carroll, K. M., & Onken, L. S. (2005). Behavioral therapies for drug abuse. *The American Journal of Psychiatry, 162*(8), 1452–1460. https://doi.org/10.1176/appi.ajp.162.8.1452.

Craner, J. R., Sperry, J. A., Koball, A. M., Morrison, E. J., & Gilliam, W. P. (2017). Unique contributions of acceptance and catastrophizing on chronic pain adaptation. *International Journal of Behavioral Medicine, 24*(4), 542–551. https://doi.org/10.1007/s12529-017-9646-3.

Cuijpers, P., Cristea, I. A., Karyotaki, E., Reijnders, M., & Huibers, M. J. (2016). How effective are cognitive behavior therapies for major depression and anxiety disorders? A meta-analytic update of the evidence. *World Psychiatry, 15*(3), 245–258.

Currie, S. R., Hodgins, D. C., Crabtree, A., Jacobi, J., & Armstrong, S. (2003). Outcome from integrated pain management treatment for recovering substance abusers. *The Journal of Pain, 4*(2), 91–100.

Dahlhamer, J., Lucas, J., Zelaya, C., et al. (2016). *Prevalence of chronic pain and high-impact chronic pain among adults—United States, 2016.*

Dhingra, L., Masson, C., Perlman, D. C., Seewald, R. M., Katz, J., McKnight, C., … Portenoy, R. K. (2013). Epidemiology of pain among outpatients in methadone maintenance treatment programs. *Drug and Alcohol Dependence, 128*(1–2), 161–165.

DiMeola, K. A., Haynes, J., Barone, M., Beitel, M., Madden, L. M., Cutter, C. J., … Barry, D.T. (2021). A pilot investigation of nonpharmacological pain management intervention groups in methadone maintenance treatment. *Journal of Addiction Medicine.* https://doi.org/10.1097/ADM.0000000000000877.

Dowell, D., Haegerich, T. M., & Chou, R. (2016). CDC guideline for prescribing opioids for chronic pain—United States, 2016. *JAMA, 315*(15), 1624–1645. https://doi.org/10.1001/jama.2016.1464.

Driessen, E., & Hollon, S. D. (2010). Cognitive behavioral therapy for mood disorders: Efficacy, moderators and mediators. *The Psychiatric Clinics of North America, 33*(3), 537–555. https://doi.org/10.1016/j.psc.2010.04.005.

Dunn, K. E., Finan, P. H., Tompkins, D. A., Fingerhood, M., & Strain, E. C. (2015). Characterizing pain and associated coping strategies in methadone and buprenorphine-maintained patients. *Drug and Alcohol Dependence, 157*, 143–149.

Dunn, K. M., Saunders, K. W., Rutter, C. M., Banta-Green, C. J., Merrill, J. O., Sullivan, M. D., … Von Korff, M. (2010). Opioid prescriptions for chronic pain and overdose: A cohort study. *Annals of Internal Medicine, 152*(2), 85–92. https://doi.org/10.7326/0003-4819-152-2-201001190-00006.

Dutra, L., Stathopoulou, G., Basden, S. L., Leyro, T. M., Powers, M. B., & Otto, M. W. (2008). A meta-analytic review of psychosocial interventions for substance use disorders. *The American Journal of Psychiatry, 165*(2), 179–187. https://doi.org/10.1176/appi.ajp.2007.06111851.

Edlund, M. J., Martin, B. C., Russo, J. E., DeVries, A., Braden, J. B., & Sullivan, M. D. (2014). The role of opioid prescription in incident opioid abuse and dependence among individuals with chronic noncancer pain: The role of opioid prescription. *The Clinical Journal of Pain, 30*(7), 557–564. https://doi.org/10.1097/ajp.0000000000000021.

Ehde, D. M., Dillworth, T. M., & Turner, J. A. (2014). Cognitive-behavioral therapy for individuals with chronic pain: Efficacy, innovations, and directions for research. *American Psychologist, 69*(2), 153–166.

Garnet, B., Beitel, M., Cutter, C. J., Savant, J., Peters, S., Schottenfeld, R. S., & Barry, D. T. (2011). Pain catastrophizing and pain coping among methadone-maintained patients. *Pain Medicine, 12*(1), 79–86. https://doi.org/10.1111/j.1526-4637.2010.01002.x.

Gatchel, R. J., McGeary, D. D., McGeary, C. A., & Lippe, B. (2014). Interdisciplinary chronic pain management: Past, present, and future. *American Psychologist, 69*(2), 119.

Geisser, M. E., Robinson, M. E., Keefe, F. J., & Weiner, M. L. (1994). Catastrophizing, depression and the sensory, affective and evaluative aspects of chronic pain. *Pain, 59*(1), 79–83. https://doi.org/10.1016/0304-3959(94)90050-7.

Griffin, M. L., McDermott, K. A., McHugh, R. K., Fitzmaurice, G. M., Jamison, R. N., & Weiss, R. D. (2016). Longitudinal association between pain severity and subsequent opioid use in prescription opioid dependent patients with chronic pain. *Drug and Alcohol Dependence, 163*, 216–221. https://doi.org/10.1016/j.drugalcdep.2016.04.023.

Haller, D. L., & Acosta, M. C. (2010). Characteristics of pain patients with opioid-use disorder. *Psychosomatics, 51*(3), 257–266. https://doi.org/10.1016/S0033-3182(10)70693-9.

Haynes, S. N., & O'Brien, W. H. (1990). Functional analysis in behavior therapy. *Clinical Psychology Review, 10*(6), 649–668.

Higgins, C., Smith, B. H., & Matthews, K. (2020). Comparison of psychiatric comorbidity in treatment-seeking, opioid-dependent patients with versus without chronic pain. *Addiction (Abingdon, England), 115*(2), 249–258. https://doi.org/10.1111/add.14768.

Hser, Y.-I., Mooney, L. J., Saxon, A. J., Miotto, K., Bell, D. S., & Huang, D. (2017). Chronic pain among patients with opioid use disorder: Results from electronic health records data. *Journal of Substance Abuse Treatment, 77*, 26–30. https://doi.org/10.1016/j.jsat.2017.03.006.

Hughes, L. S., Clark, J., Colclough, J. A., Dale, E., & McMillan, D. (2017). Acceptance and commitment therapy (ACT) for chronic pain: A systematic review and meta-analyses. *The Clinical Journal of Pain, 33*(6), 552–568. https://doi.org/10.1097/ajp.0000000000000425.

Ilgen, M. A., Bohnert, A. S., Chermack, S., Conran, C., Jannausch, M., Trafton, J., & Blow, F. C. (2016). A randomized trial of a pain management intervention for adults receiving substance use disorder treatment. *Addiction, 111*, 1385–1393.

Ilgen, M. A., Coughlin, L. N., Bohnert, A. S., Chermack, S., Price, A., Kim, H. M., … Blow, F. C. (2020). Efficacy of a psychosocial pain management intervention for men and women with substance use disorders and chronic pain: A randomized clinical trial. *JAMA Psychiatry.* https://doi.org/10.1001/jamapsychiatry.2020.2369.

Ilgen, M. A., Haas, E., Czyz, E., Webster, L., Sorrell, J. T., & Chermack, S. (2011). Treating chronic pain in veterans presenting to an addictions treatment program. *Cognitive and Behavioral Practice, 18*(1), 149–160.

Jamison, R. N., Kauffman, J., & Katz, N. P. (2000). Characteristics of methadone maintenance patients with chronic pain. *Journal of Pain and Symptom Management, 19*(1), 53–62.

Jamison, R. N., Scanlan, E., Matthews, M. L., Jurcik, D. C., & Ross, E. L. (2016). Attitudes of primary care practitioners in managing chronic pain patients prescribed opioids for pain: A prospective longitudinal controlled trial. *Pain Medicine, 17*(1), 99–113. https://doi.org/10.1111/pme.12871.

Johnson, W. D., Rivadeneira, N., Adegbite, A. H., Neumann, M. S., Mullins, M. M., Rooks-Peck, C., … Sipe, T. A. (2020). Human immunodeficiency virus prevention for people who use drugs: Overview of reviews and the ICOS of PICOS. *The Journal of Infectious Diseases, 222*(Suppl. 5), S278–S300. https://doi.org/10.1093/infdis/jiaa008.

Jones, C. M. (2013). Heroin use and heroin use risk behaviors among nonmedical users of prescription opioid pain relievers—United States, 2002-2004 and 2008-2010. *Drug and Alcohol Dependence, 132*(1–2), 95–100. https://doi.org/10.1016/j.drugalcdep.2013.01.007.

Jones, H. E., Haug, N. A., Stitzer, M. L., & Svikis, D. S. (2000). Improving treatment outcomes for pregnant drug-dependent women using low-magnitude voucher incentives. *Addictive Behaviors, 25*(2), 263–267.

Kaczkurkin, A. N., & Foa, E. B. (2015). Cognitive-behavioral therapy for anxiety disorders: An update on the empirical evidence. *Dialogues in Clinical Neuroscience*, *17*(3), 337–346.

Kamper, S. J., Apeldoorn, A. T., Chiarotto, A., Smeets, R. J. E. M., Ostelo, R. W. J. G., Guzman, J., & van Tulder, M. W. (2015). Multidisciplinary biopsychosocial rehabilitation for chronic low back pain: Cochrane systematic review and meta-analysis. *BMJ*, *350*, h444. https://doi.org/10.1136/bmj.h444.

Kolodny, A., Courtwright, D. T., Hwang, C. S., Kreiner, P., Eadie, J. L., Clark, T. W., & Alexander, G. C. (2015). The prescription opioid and heroin crisis: A public health approach to an epidemic of addiction. *Annual Review of Public Health*, *36*, 559–574. https://doi.org/10.1146/annurev-publhealth-031914-122957.

Kratz, A. L., Ehde, D. M., Bombardier, C. H., Kalpakjian, C. Z., & Hanks, R. A. (2017). Pain acceptance decouples the momentary associations between pain, pain interference, and physical activity in the daily lives of people with chronic pain and spinal cord injury. *The Journal of Pain*, *18*(3), 319–331. https://doi.org/10.1016/j.jpain.2016.11.006.

Magill, M., Ray, L., Kiluk, B., Hoadley, A., Bernstein, M., Tonigan, J. S., & Carroll, K. (2019). A meta-analysis of cognitive-behavioral therapy for alcohol or other drug use disorders: Treatment efficacy by contrast condition. *Journal of Consulting and Clinical Psychology*, *87*(12), 1093.

Martinez-Calderon, J., Flores-Cortes, M., Morales-Asencio, J. M., & Luque-Suarez, A. (2019). Pain-related fear, pain intensity and function in individuals with chronic musculoskeletal pain: A systematic review and meta-analysis. *The Journal of Pain*, *20*(12), 1394–1415. https://doi.org/10.1016/j.jpain.2019.04.009.

McCracken, L. M. (1998). Learning to live with the pain: Acceptance of pain predicts adjustment in persons with chronic pain. *Pain*, *74*(1), 21–27. https://doi.org/10.1016/S0304-3959(97)00146-2.

McDonagh, M. S., Selph, S. S., Buckley, D. I., Holmes, R. S., Mauer, K., Ramirez, S., ... Chou, R. (2020). AHRQ comparative effectiveness reviews. In *Nonopioid pharmacologic treatments for chronic pain*. Rockville, MD: Agency for Healthcare Research and Quality (US).

Moore, B. A., Fiellin, D. A., Cutter, C. J., Buono, F. D., Barry, D. T., Fiellin, L. E., ... Schottenfeld, R. S. (2016). Cognitive behavioral therapy improves treatment outcomes for prescription opioid users in primary care buprenorphine treatment. *Journal of Substance Abuse Treatment*, *71*, 54–57. https://doi.org/10.1016/j.jsat.2016.08.016.

Morasco, B. J., Greaves, D. W., Lovejoy, T. I., Turk, D. C., Dobscha, S. K., & Hauser, P. (2016). Development and preliminary evaluation of an integrated cognitive-behavior treatment for chronic pain and substance use disorder in patients with the hepatitis C virus. *Pain Medicine*, *17*, 2280–2290.

Mun, C. J., Beitel, M., Oberleitner, L., Oberleitner, D. E., Madden, L. M., Bollampally, P., & Barry, D. T. (2019). Pain catastrophizing and pain acceptance are associated with pain severity and interference among methadone-maintained patients. *Journal of Clinical Psychology*, *75*(12), 2233–2247. https://doi.org/10.1002/jclp.22842.

National Academies of Sciences, Engineering, and Medicine. (2019). *Medications for opioid use disorder save lives*. Washington, DC: The National Academies Press.

National Academies of Sciences, Engineering, and Medicine, Health and Medicine Division; Board on Global Health, Board on Health Sciences Policy, Global Forum on Innovation in Health Professional Education, & Forum on Neuroscience and Nervous System Disorders. (2019). The National Academies Collection: Reports funded by National Institutes of Health. In C. Stroud, S. M. Posey Norris, & L. Bain (Eds.), *The role of nonpharmacological approaches to pain management: Proceedings of a workshop*. Washington, DC: National Academies Press (US).

Nees, T. A., Riewe, E., Waschke, D., Schiltenwolf, M., Neubauer, E., & Wang, H. (2020). Multidisciplinary pain management of chronic back pain: Helpful treatments from the patients' perspective. *Journal of Clinical Medicine*, *9*(1). https://doi.org/10.3390/jcm9010145.

Oberleitner, L. M., Beitel, M., Schottenfeld, R. S., Kerns, R. D., Doucette, C., Napoleone, R., ... Barry, D. T. (2016). Drug counselors' attitudes toward nonpharmacologic treatments for chronic pain. *Journal of Addiction Medicine*, *10*(1), 34–39. https://doi.org/10.1097/ADM.0000000000000177.

Peles, E., Schreiber, S., & Adelson, M. (2006). Variables associated with perceived sleep disorders in methadone maintenance treatment (MMT) patients. *Drug and Alcohol Dependence*, *82*(2), 103–110. https://doi.org/10.1016/j.drugalcdep.2005.08.011.

Peles, E., Schreiber, S., & Adelson, M. (2009). Documented poor sleep among methadone-maintained patients is associated with chronic pain and benzodiazepine abuse, but not with methadone dose. *European Neuropsychopharmacology*, *19*(8), 581–588. https://doi.org/10.1016/j.euroneuro.2009.04.001.

Quartana, P. J., Campbell, C. M., & Edwards, R. R. (2009). Pain catastrophizing: A critical review. *Expert Review of Neurotherapeutics*, *9*(5), 745–758. https://doi.org/10.1586/ern.09.34.

Roeckel, L.-A., Le Coz, G.-M., Gavériaux-Ruff, C., & Simonin, F. (2016). Opioid-induced hyperalgesia: Cellular and molecular mechanisms. *Neuroscience*, *338*, 160–182. https://doi.org/10.1016/j.neuroscience.2016.06.029.

Rosenblum, A., Joseph, H., Fong, C., Kipnis, S., Cleland, C., & Portenoy, R. K. (2003). Prevalence and characteristics of chronic pain among chemically dependent patients in methadone maintenance and residential treatment facilities. *JAMA*, *289*(18), 2370–2378. https://doi.org/10.1001/jama.289.18.2370.

Rummans, T. A., Burton, M. C., & Dawson, N. L. (2018). How good intentions contributed to bad outcomes: The opioid crisis. *Mayo Clinic Proceedings*, *93*(3), 344–350. https://doi.org/10.1016/j.mayocp.2017.12.020.

Severeijns, R., Vlaeyen, J. W. S., van den Hout, M. A., & Weber, W. E. J. (2001). Pain catastrophizing predicts pain intensity, disability, and psychological distress independent of the level of physical impairment. *The Clinical Journal of Pain*, *17*(2), 165–172. Retrieved from https://journals.lww.com/clinicalpain/Fulltext/2001/06000/Pain_Catastrophizing_Predicts_Pain_Intensity.9.aspx.

Skelly, A. C., Chou, R., Dettori, J. R., Turner, J. A., Friedly, J. L., Rundell, S. D., ... Ferguson, A. J. R. (2020). AHRQ comparative effectiveness reviews. In *Noninvasive nonpharmacological treatment for chronic pain: A systematic review update*. Rockville, MD: Agency for Healthcare Research and Quality (US).

Sordo, L., Barrio, G., Bravo, M. J., Indave, B. I., Degenhardt, L., Wiessing, L., ... Pastor-Barriuso, R. (2017). Mortality risk during and after opioid substitution treatment: Systematic review and meta-analysis of cohort studies. *BMJ*, *357*, j1550. https://doi.org/10.1136/bmj.j1550.

Substance Abuse and Mental Health Services Administration. (2020). *Medications for opioid use disorder. Treatment improvement protocol (TIP) series 63*. Rockville, MD: Substance Abuse and Mental Health Services Administration. Publication No. PEP20-02-01-006.

Tsui, J. I., Lira, M. C., Cheng, D. M., Winter, M. R., Alford, D. P., Liebschutz, J. M., ... Samet, J. H. (2016). Chronic pain, craving, and illicit opioid use among patients receiving opioid agonist therapy. *Drug and Alcohol Dependence, 166*, 26–31. https://doi.org/10.1016/j.drugalcdep.2016.06.024.

Veehof, M. M., Trompetter, H. R., Bohlmeijer, E. T., & Schreurs, K. M. G. (2016). Acceptance- and mindfulness-based interventions for the treatment of chronic pain: A meta-analytic review. *Cognitive Behaviour Therapy, 45*(1), 5–31. https://doi.org/10.1080/16506073.2015.1098724.

Vowles, K. E., McEntee, M. L., Julnes, P. S., Frohe, T., Ney, J. P., & van der Goes, D. N. (2015). Rates of opioid misuse, abuse, and addiction in chronic pain: A systematic review and data synthesis. *Pain, 156*(4), 569–576. https://doi.org/10.1097/01.j.pain.0000460357.01998.f1.

Williams, A. C. C., Fisher, E., Hearn, L., & Eccleston, C. (2020). Psychological therapies for the management of chronic pain (excluding headache) in adults. *Cochrane Database of Systematic Reviews, 8*(8), CD007407. https://doi.org/10.1002/14651858.CD007407.pub4.

Wilson, M., Finlay, M., Orr, M., Barbosa-Leiker, C., Sherazi, N., Roberts, M. L. A., ... Roll, J. M. (2018). Engagement in online pain self-management improves pain in adults on medication-assisted behavioral treatment for opioid use disorders. *Addictive Behaviors, 86*, 130–137.

Wood, E., Samet, J. H., & Volkow, N. D. (2013). Physician education in addiction medicine. *JAMA, 310*(16), 1673–1674. https://doi.org/10.1001/jama.2013.280377.

Chapter 23

Preoperative opioid and benzodiazepines: Impact on adverse outcomes

Martin Ingi Sigurðsson

Department of Anesthesiology and Critical Care, Landspitali—The National Hospital of Iceland, Reykjavik, Iceland

Abbreviations

ASA	American Society of Anesthesiology
BZD	benzodiazepine
CDC	Centre for Disease Control
NSAID	nonsteroidal antiinflammatory drug

Introduction

In the past decade, the number of individuals using opioid and benzodiazepines (BZDs) chronically has increased substantially. This increase seems to be across all age categories and independent of underlying individual morbidity and frailty. This increase has been associated with an increase in adverse outcomes and mortality for the chronic use of these medications.

Following this, it is rational that there has also been a rapid increase in the number of patients taking opioids and BZDs that present for various types of surgery. This poses a unique set of challenges for the perioperative management of this group of patients. This includes the awareness of the increased risk of several perioperative complications, ensuring adequate monitoring and an individualized approach to mitigate the risk of adverse outcomes. This chapter will review the available literature on the epidemiology and outcomes of patients who take opioids and BZDs chronically, highlight current management strategies and identify areas of future research and ongoing challenges in the perioperative management of these patients.

Since the beginning of the 21st century, chronic usage of opioids for noncancer pain has risen dramatically (Frenk, Porter, & Paulozzi, 1999–2012; Jain, Phillips, Weaver, & Khan, 2018). Additionally, there has been an alarming trend in the incidence of harmful use and abuse of opioids as well as opioid-related deaths (Dufour, Joshi, Pasquale, et al., 2014; Seth, Scholl, Rudd, & Bacon, 2018). This has resulted in the initiation of multiple programs to reduce the chronic usage of opioids via clinician education and legislation changes. While the results for the initial years are promising (Von Korff, Dublin, Walker, et al., 2016), multiple challenges still exist.

Simultaneously, there has been a rise in the prescription of BZDs as well as BZD-associated complications (Bachhuber, Hennessy, Cunningham, & Starrels, 2016; Olfson, King, & Schoenbaum, 2015). Overall, there seems to have been less concern about the potential harmful effects of chronic BZD usage, but this is likely to change in the upcoming years (Lembke, Papac, & Humphreys, 2018).

Given this rise in the chronic usage of opioids and BZDs in the global population, a large increase has been seen in the number of patients presenting for surgery who chronically use opioids and BZDs. This patient cohort presents unique challenges to the perioperative physician, both in the pre, peri and postoperative period. This chapter will touch on some of these challenges. Please note that this chapter will not discuss short-term usage of these medications for patients undergoing surgery, including usage in the perioperative and immediate postoperative period.

Definition of chronic opioid and BZD usage

There is some variability in definitions of chronic opioid and BZD usage, that affects the epidemiology and scope of the problem substantially. The Centre for Disease Control and prevention (CDC) defines chronic opioid usage as the usage of

---

Treatments, Mechanisms, and Adverse Reactions of Anesthetics and Analgesics. https://doi.org/10.1016/B978-0-12-820237-1.00023-5
Copyright © 2022 Elsevier Inc. All rights reserved.

opioids most days for 3 months or more (Dowell, Haegerich, & Chou, 2016). This allows for a simple definition that is easy to use both clinically as well as in research, but depends only on duration of usage and is insensitive to amount of medication given.

Another definition of chronic tailored for the care of surgical patients was published by Schoenfeld, Belmont, Blucher, et al. (2018). This definition separates patients into groups based on their preoperative opioid consumption into nonopioid users, acute usage (30 days or less), exposed group (received an opioid prescription in the last 12 months but noncontinuous use), intermediate sustained group (continuous usage for less than 6 months) and chronic sustained group (continuous usage for more than 6 months). This definition is potentially more beneficial for surgical cohorts, since it allows for separating patients that initiate opioid consumption immediately preceding surgery, is relatively easy to use and provides some quantification of usage. This definition however does not include a quantification of the opioid usage (Schoenfeld et al., 2018).

Two other definitions, designed for noncancer pain, include an absolute dose of opioids allowing more granularity in the definition of chronic opioid usage. The definition published by Svendsen et al. includes three definitions with varying sensitivity for chronic opioid use, a wide definition (usage of more than 4500 mg morphine equivalents for at least 9 months); an intermediary definition (usage of more than 9000 mg morphine equivalents for 12 months or more) and a strict definition (usage of more than 18,000 mg morphine equivalents for more than 12 months with at least 10 filled prescriptions) (Svendsen, Skurtveit, Romundstad, Borchgrevink, & Fredheim, 2012). Edlund et al. (2014) also published a classification for chronic opioid usage in noncancer pain. This definition separates duration of usage into six groups based on a combination of duration [acute (90 days or less) vs chronic (more than 90 days)] and amount low-dose (average daily dose of 1–36 mg morphine equivalents), medium dose (average daily dose of 36–120 mg morphine equivalents) and high-dose (average daily dose of more than 120 mg morphine equivalents) (Edlund et al., 2014). While these definitions allows a higher sensitivity in the definition of chonic opioid usage and take into account the absolute amount of opioid used, they are very complex to incorporate into research protocols and do not translate easily into clinical practice.

There is even less consensus on the definition of chronic BZD usage, but several studies have defined chronic BZD usage as usage of a BZD for more than 90 days (Airagnes, Lemogne, Renuy, et al., 2019; Carrier, Roberge, Courteau, & Vanasse, 2016).

Prevalence of chronic preoperative opioid and BZD usage

The prevalence of chronic opioid usage differs substantially based on the type of surgery and definitions used. A large cross-sectional study analyzing patterns of preoperative opioid prescription in a large surgical cohort from Michigan defined preoperative opioid usage as any self-reported usage by patients on preoperative assessment (Hilliard, Waljee, Moser, et al., 2018). Using this definition, 23.1% of all individuals reported any preoperative usage of opioids, most commonly hydrocodone. There was a large variability in the prevalence of chronic usage depending on type of surgery, ranging from intrathoracic (15.8%) to the spine or spinal cord (57.1%) surgery (Table 1; Hilliard et al., 2018). Oleisky et al. compared the incidence of preoperative chronic opioid consumption depending on different definitions in a population of over 2000 patients presenting for spine surgery (Oleisky, Pennings, Hills, et al., 2019). This study found that using the CDC criteria, the prevalence of chronic opioid consumption was 29%. Using the more detailed Schoenfeld criteria, 4.8% of patients were acutely using opioids in the 30 days preceding surgery; 53.8% were exposed to opioids in the 12 months preceding surgery; 7.5% had intermediate sustained opioid use in the 6 months preceding surgery and 12.3% had chronic sustained usage for more than 6 months (Oleisky et al., 2019).

Sigurdsson et al. studied the incidence and outcomes of patients presenting for surgery with a preexisting opioid usage defined as filling a prescription for opioids within 6 months preceding surgery, excluding those that only filled prescription in the last 30 days prior to surgery. In this nationwide cohort, 23.9% of the patients met this definition (Sigurdsson, Helgadottir, Long, et al., 2019).

Fewer data exist regarding chronic preoperative BZD usage in surgical cohorts. Hozack et al. found that 12% out of 290 patients presenting for minor upper extremity orthopedic surgery had used BZDs in the 6 months preceding surgery (Hozack, Rivlin, Lutsky, et al., 2019). Similarly, in a the nationwide surgical cohort from Iceland, 13.6% of a cohort of 41,170 surgical cases had filled a prescription for a BZD in the 6 months preceding surgery, and of these, 6.2% of the patients had filled a prescription for both BZDs and opioids preceding surgery (Sigurdsson et al., 2019).

Characteristics of patient cohorts with preoperative chronic opioid and BZD usage

Several authors have described the association between patient and procedural variables and preoperative chronic opioid and BZD usage. Sigurdsson et al. described the cohort taking opioids and BZDs preoperatively, defining preexisting usage

TABLE 1 Type and location of procedures for patients with self-identified preoperative opioid use compared with those with no opioid use.

Characteristic	Study group			OR (95% CI)	P value
	Overall (n = 34,186)	Opioid use (n = 7894)	No opioid use (n = 26,292)		
Body area					
Intrathoracic	1552 (4.6)	245 (15.8)	1307 (84.2)	1 [Reference]	NA
Head	3715 (11.1)	745 (20.1)	2970 (80.0)	1.35 (1.14–1.57)	<.001
Neck	4151 (12.4)	804 (19.4)	3347 (80.6)	1.29 (1.10–1.50)	.001
Thorax	2170 (6.5)	364 (16.8)	1806 (83.2)	1.08 (0.90–1.28)	.40
Spine or spinal cord	1472 (4.4)	840 (57.1)	632 (42.9)	7.09 (5.98–8.41)	<.001
Abdomen					
Upper	3299 (9.8)	766 (23.2)	2533 (76.8)	1.62 (1.38–1.89)	<.001
Lower	4969 (14.8)	962 (19.4)	3999 (80.6)	1.29 (1.10–1.50)	.001
Perineum	3496 (10.4)	727 (20.8)	2769 (79.2)	1.41 (1.19–1.64)	<.001
Pelvis (except hip)	125 (0.4)	53 (42.4)	72 (57.6)	3.95 (2.69–5.74)	<.001
Upper leg (except knee)	1581 (4.7)	566 (35.8)	1015 (64.1)	2.97 (2.51–3.52)	<.001
Knee or popliteal	1933 (5.8)	401 (20.7)	1532 (79.2)	1.40 (1.17–1.66)	<.001
Lower leg	772 (2.3)	309 (40.0)	463 (60.0)	3.56 (2.92–4.34)	<.001
Shoulder or axilla	1856 (5.5)	323 (17.3)	1533 (82.8)	1.12 (0.94–1.35)	.21
Upper arm and elbow	246 (0.7)	89 (36.2)	157 (63.8)	3.02 (2.25–4.06)	<.001
Forearm, wrist, hand	1359 (4.0)	347 (25.5)	1012 (74.5)	1.83 (1.52–2.20)	<.001
Other	893 (2.7)	285 (32.2)	608 (68.4)	2.50 (2.06–3.04)	<.001

From Hilliard, P. E., Waljee, J., Moser, S., et al. (2018). Prevalence of preoperative opioid use and characteristics associated with opioid use among patients presenting for surgery. *JAMA Surgery, 153*(10), 929–937.

as filling a prescription in the last 6 months preceding surgery (excluding the last 30 days preop) (Sigurdsson et al., 2019). They compared patients taking opioids, BZDs or both with the cohort that had not filled any prescriptions preoperatively (Table 2). Contrasted with those who did not fill any prescriptions for opioids and BZDs, patients who had filled a prescription for opioids only were older and had a higher burden of comorbidity, including a higher incidence of congestive heart failure and chronic obstructive lung disease. They also had a higher frailty risk score and a higher ASA physical status classification. Similarly, patients who had filled a prescription for BZD only were also older and had a higher rate of chronic conditions, including ischemic heart disease and reduced renal function. They also had a higher ASA physical status

250   **PART | III** Adverse effects, reactions, and outcomes

TABLE 2 Patient characteristics and medication usage for patients who filled a preoperative prescription (within 6 months excluding those who only filled in the last 30 days preoperatively) for opioids only, benzodiazepines only or both medications.

Characteristic	No. (%)				P value[a]
	Neither opioids nor benzodiazepines (n = 28,956)	Opioids only (n = 7460)	Benzodiazepines only (n = 3121)	Both opioids and benzodiazepines (n = 2633)	
Female sex	16,098 (55.6)	3935 (52.7)	2163 (69.3)	1810 (68.7)	<.001
Age, mean (SD), years	55 (19)	57 (17)	63 (16)	61 (15)	<.001
Comorbidities					
Ischemic heart disease	3171 (11.0)	1038 (13.9)	534 (17.1)	506 (19.2)	<.001
Congestive heart failure	895 (3.1)	348 (4.7)	170 (5.4)	192 (7.3)	<.001
Hypertension	2561 (8.8)	870 (11.7)	428 (13.7)	428 (16.3)	<.001
Diabetes	895 (3.1)	343 (4.6)	117 (3.7)	137 (5.2)	<.001
COPD	622 (2.1)	290 (3.9)	189 (6.1)	255 (9.7)	<.001
eGFR < 60 mL/min per 1.73 m^2	4736 (16.4)	1278 (17.1)	671 (21.5)	491 (18.6)	
Malignant neoplasm	3100 (10.7)	1001 (13.4)	569 (18.2)	514 (19.5)	<.001
Benign neoplasm	3571 (12.3)	1159 (15.5)	551 (17.7)	593 (22.5)	<.001
Organic psychiatric disorder	508 (1.8)	135 (1.8)	63 (2.0)	74 (2.8)	.002
Substance abuse disorder	831 (2.9)	396 (5.3)	213 (6.8)	391 (14.8)	<.001
Alcohol abuse	614 (2.1)	299 (4.0)	154 (4.9)	273 (10.4)	<.001
Opioid abuse	57 (0.2)	56 (0.8)	15 (0.5)	67 (2.5)	<.001
Hypnotic and sedative abuse	117 (0.4)	81 (1.1)	57 (1.8)	122 (4.6)	<.001
Schizophrenia	144 (0.5)	26 (0.3)	39 (1.2)	23 (0.9)	<.001
Mood disorder	888 (3.1)	368 (4.9)	298 (9.5)	393 (14.9)	<.001
Anxiety disorder	591 (2.0)	232 (3.1)	207 (6.6)	279 (10.6)	<.001
Personality disorder	203 (0.7)	84 (1.1)	62 (2.0)	81 (3.1)	<.001
Frailty score class[b]					
Low	25,753 (88.9)	6284 (84.2)	2565 (82.2)	1894 (71.9)	<.001
Intermediate	2879 (9.9)	1044 (14.0)	498 (16.0)	662 (25.1)	<.001
High	324 (1.1)	132 (1.8)	58 (1.9)	77 (2.9)	<.001
Elixhauser comorbidity index >0[c]	4019 (13.9)	1183 (15.9)	679 (21.8)	542 (20.6)	<.001
Prescribed medications					
Anticoagulant	3357 (11.6)	1189 (15.9)	570 (18.3)	548 (20.8)	<.001

TABLE 2 Patient characteristics and medication usage for patients who filled a preoperative prescription (within 6 months excluding those who only filled in the last 30 days preoperatively) for opioids only, benzodiazepines only or both medications—cont'd

Characteristic	Neither opioids nor benzodiazepines ($n = 28,956$)	Opioids only ($n = 7460$)	Benzodiazepines only ($n = 3121$)	Both opioids and benzodiazepines ($n = 2633$)	P value
Antiplatelet	2214 (7.6)	730 (9.8)	399 (12.8)	338 (12.8)	<.001
Cardiovascular	12,513 (43.2)	4057 (54.4)	2046 (65.6)	1773 (67.3)	<.001
Hormone	5246 (18.1)	2377 (31.9)	1008 (32.3)	1235 (46.9)	<.001
Musculoskeletal	9157 (31.6)	4652 (62.4)	1226 (39.3)	1631 (61.9)	<.001
Antidepressant	4730 (16.3)	2119 (28.4)	1605 (51.4)	1562 (59.3)	<.001
Respiratory	6861 (23.7)	2514 (33.7)	1260 (40.4)	1331 (50.6)	<.001
ASA physical status classification[d]					
I	8743 (31.4)	1601 (22.3)	355 (11.8)	244 (9.6)	<.001
II	13,611 (48.9)	3912 (54.6)	1724 (57.4)	1403 (55.3)	
III	4725 (17.0)	1453 (20.3)	812 (27.0)	773 (30.5)	
IV	626 (2.3)	185 (2.6)	106 (3.5)	104 (4.1)	
V	108 (0.4)	14 (0.2)	9 (0.3)	11 (0.4)	
Primary mode of anesthesia					
General	21,878 (78.7)	5598 (78.3)	2258 (75.6)	1943 (76.7)	.005
Other	67 (0.2)	18 (0.3)	8 (0.3)	5 (0.2)	
Regional and neuroaxial	5871 (21.1)	1532 (21.4)	720 (24.1)	584 (23.1)	

Abbreviations: *ASA*, American Society of Anesthesiology; *COPD*, chronic obstructive pulmonary disease; *eGFR*, estimated glomerular filtration rate.
[a]*P values reflect the statistical comparison between all 4 groups.*
[b]*The 3 frailty score categories consist of patients with a total frailty score of less than 5 (low), 5 to 15 (intermediate), and higher than 15 (high).*
[c]*The Elixhauser comorbidity index is a severity index to quantify various patient comorbidities from multiple chronic diseases into a single number that can be used to assess and correct for patient comorbidity burden.*
[d]*The ASA physical status classification is a classification system commonly used by anesthesiologists to assess the fitness of patients presenting for surgery. It has five categories ranging from healthy person (I) to a patient predicted to lose life or limb in the next 24 h without the operation (V), and an additional category (VI) for a person presenting for organ donation.*
From Sigurdsson, M. I., Helgadottir, S., Long, T. E., et al. (2019). Association between preoperative opioid and benzodiazepine prescription patterns and mortality after noncardiac surgery. *JAMA Surgery, 154*(8), e191652.

classification and a higher risk of frailty. Alarmingly, patients taking medications from both classes had the highest risk of underlying comorbid conditions, including ischemic heart disease, congestive heart failure and reduced kidney function (Sigurdsson et al., 2019). Hilliard et al. studied the patient and procedural characteristics of patients self-reporting opioid usage prior to surgery in a large surgical cohort. This found that patients reporting opioid use were more likely to have obstructive sleep apnea, a higher ASA score and a higher comorbidity burden measured by Charlson comorbidity index (Hilliard et al., 2018).

Taken together, the epidemiology suggests that a substantial portion of population presenting for surgery are chronically taking opioids or BZDs, and that this very population is already at an elevated risk of complications after surgery due to underlying comorbidities.

Perioperative outcomes of patients taking opioids and BZDs preoperatively

Patients with preoperative chronic opioid and BZDs usage are at increased risk of adverse effects during the perioperative period, both in the immediate postoperative phase as well as long-term. Fig. 1 provides an overview of the potential risks to this population during the perioperative period and a framework of potential strategies to mitigate these risks.

Inadequate perioperative analgesia

Patients chronically taking opioids that subsequently develop acute pain, such as during surgery, are at increased risk of inadequate analgesia in the perioperative period. A study using acute pain assessment on a numeric rating scale following surgery in over 50,000 patients found that the odds of experiencing severe pain (a pain score of seven or more on a 0–10 scale) were 1.25 (95% CI: 1.09–1.54) for patients taking preoperative opioids, after adjustment for other variables associated with experiencing severe pain (female gender, country, younger age and a longer duration of surgery) (Schnabel, Yahiaoui-Doktor, Meissner, Zahn, & Pogatzki-Zahn, 2020). There might be several underlying reasons. Some of the patient characteristics of patients that chronically use opioids are shared with the population that experiences a higher postoperative pain, such as preoperative diagnosis of depression, anxiety and younger age (Carroll, Angst, & Clark, 2004). However, patients chronically using opioids will additionally likely develop opioid tolerance, where a gradual increase in the dose of opioid is needed for the same analgesic effect. This is due to a cellular and receptor adaption at multiple levels, including the opioid receptor, downstream targets, neurons and synapses (Mehta & Langford, 2009). Therefore, any acute pain in a patient with a preexisting opioid tolerance is likely to require a higher dosage of opioids for adequate

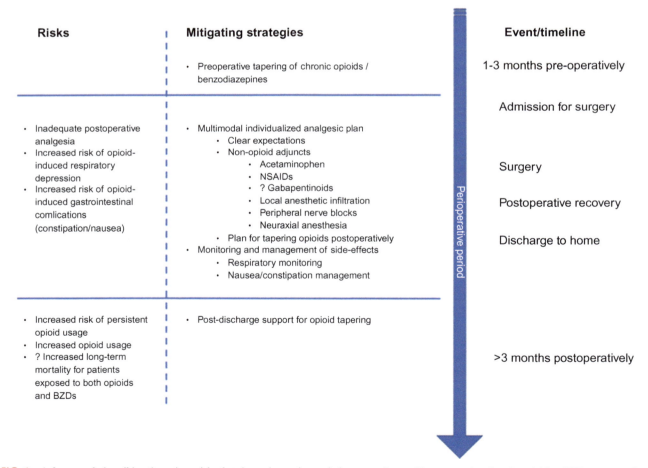

FIG. 1 A framework describing the unique risks that the perioperative period poses patients with preoperative chronic opioid or BZD use, as well as potential mitigating strategies.

relief. An additional challenge is the risk of opioid induced hyperalgesia, defined as an increased sensitivity to painful stimuli, resulting from opioid use (Lee, Silverman, Hansen, Patel, & Manchikanti, 2011). Opioid induced hyperalgesia can present in the postoperative period in the form of pain that is unaffected or even made worse by administration of opioids. This differential diagnosis needs to be considered in the patient whose pain does not reduce or even increases after opioid administration, as it requires a complete shift in management strategy.

Several strategies have been formed to improve the perioperative management of acute pain in the patient with underlying chronic opioid usage. An honest discussion to manage patient expectations regarding perioperative pain and pain management is the foundation of any such effort, and some clinicians identify patients at risk and refer them to a pain specialist prior to large surgeries to generate a thorough individualized perioperative plan.

Continuing home dosage of opioids, ideally with the same agent and route of administration, is essential to avoid opioid withdrawal that complicates acute pain management. A combination of medications including of nonopioid adjuncts such as acetaminophen and nonsteroidal antiinflammatory agents is warranted (Mehta & Langford, 2006), and gabapentinoids can be considered. Additionally, usage of NMDA antagonists such as ketamine or methadone can be considered for patients with preexisting chronic opioid consumption. Furthermore, the utilization of wound infiltration with local anesthetics, peripheral nerve block as well as neuraxial analgesia should be sought whenever feasible (Chou, Gordon, de Leon-Casasola, et al., 2016). In fact the assessment of the benefits and risk of invasive procedures such as neuraxial analgesia should take into account that such procedures might be especially beneficial in the population chronically taking opioids. Taken together, a careful thought should be taken to design and implement a perioperative pain plan that takes into account existing usage, the extent of surgery and duration of recovery.

Several health systems have additionally initiated a preoperative opioid tapering (https://anes-conf.med.umich.edu/opioidtaper/, accessed on 3rd September 2020), but results from studies on such initiatives are lacking (Aronson, Westover, Guinn, et al., 2018). Clinicians interested in assisting patients to perform a preoperative opioid tapering can use decision support via smartphone apps and computer algorithms (Fig. 2) but are also encouraged to seek expert help with pharmacists or physicians with experience in opioid tapering to ensure success.

It is challenging to determine if preoperative BZD usage affects postoperative pain control is challenging, given the crossover between the common indications for BZD usage (depression, anxiety and sleep difficulties) and risk factors for poor postoperative pain control (younger age, female gender, smoking, depression, anxiety, sleep difficulties) (Yang, Hartley, Leung, et al., 2019). For now, any causal effect of BZDs on worse acute pain following surgery remains undetermined.

Increased risk of complications from analgesia

Differential opioid tolerance is defined as the differential rate and speed of tolerance development for various effects and side-effects of opioids (Hayhurst & Durieux, 2016). It is important to keep this in mind when treating patients with opioid tolerance, as tolerance to the analgesic effects of opioids builds relatively fast compared to the build-up of tolerance for some of the side effects. Among the side-effects with delayed build-up of tolerance compared to the build-up of tolerance to analgesic effect is tolerance for respiratory depression. This has been experimentally demonstrated (Paronis & Woods, 1997), but is additionally supported by the identification of preexisting usage of opioids as a risk factor for opioid-induced respiratory depression (Gupta et al., 2018). Differential opioid tolerance explains how patients with chronic opioid usage are more likely to suffer from respiratory complications from opioid usage in the acute setting, such as after surgery. This warrants vigilance in prescribing and extended monitoring for opioid-induced respiratory depression in this patient cohort.

While the isolated usage of BZDs is generally considered to be relatively safe, concurrent usage of opioids and BZDs potentiates the respiratory depression from opioids (Boon, van Dorp, Broens, & Overdyk, 2020; Jann, Kennedy, & Lopez, 2014). Therefore, patients chronically taking BZDs that are exposed to opioids in the perioperative period are at increased risk of respiratory complications. Patients who are chronically taking both opioids and BZDs are likely at an even higher risk of respiratory complications, given the combination of differential opioid tolerance and the likelihood of increased postoperative opioid usage in the background of chronic BZD usage. The increased risk of respiratory depression warrants special considerations, including vigilant monitoring for respiratory complications, careful titration of opioids and the usage of nonopioid adjuncts whenever possible.

Another aspect of differential opioid tolerance that warrants discussion is opioid-induced constipation. Tolerance to this common side-effect generates very slowly, resulting in a very high ratio of opioid-induced constipation for the cohort of patients using opioids chronically (Muller-Lissner, Bassotti, Coffin, et al., 2017). This can potentially slow the recovery of patients after surgery, in particular after gastrointestinal surgery.

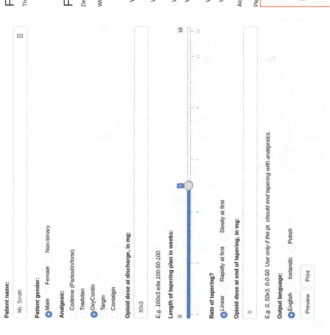

FIG. 2 An example of custom-made opioid tapering plan from www.reduceplan.com. This initiative supports patients and treating clinicians in tapering opioids and benzodiazepines, for example prior to surgery. Successful tapering of medications could potentially reduce complications associated with chronic opioid use in the immediate perioperative and postoperative period (www.reduceplan.com, accessed on September 3, 2020).

Other short-term outcomes

Sigurdsson et al. reported several other short-term outcomes of surgical patients chronically taking opioids, BZDs or both medications (Sigurdsson et al., 2019). When 30-day mortality was compared between each group and a propensity-score matched control group taking neither medication, adjusting for patient- and procedural differences, no difference was found in the group taking preoperative opioids only (1.3% vs 1.0%) or BZD only (1.9% vs 1.2%). However, the group that filled prescription for both medications preoperatively had a higher 30-day mortality than the matched control group (3.2% vs 1.8%). While unadjusted length of stay was longer for the group taking BZDs only or both opioids and BZDs, the length of stay was similar for all group and propensity-score matched controls in the adjusted assessment. While these results establish a correlation only and not causation, they highlight a group of patients (taking both opioids and BZDs preoperatively) that warrants a particular attention in the postoperative period extending past discharge from hospital.

Long-term postoperative outcomes for patients

Several authors have studied postoperative opioid consumption in patients taking opioids chronically prior to surgery. Oleisky et al. compared the rates of prescription filling for opioids in roughly 2300 patients following spine surgery based on different definitions of preoperative chronic opioid consumption (Fig. 3; Oleisky et al., 2019). This revealed that

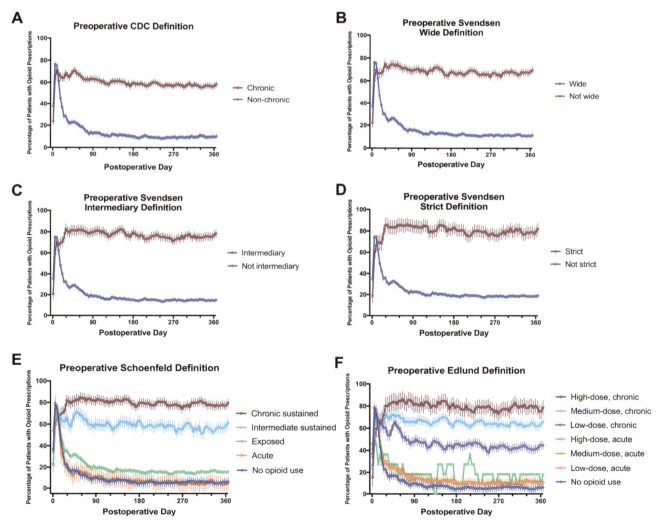

FIG. 3 The percentage of patients filling an opioid prescription for the first postoperative year following spine surgery in groups separated by preexisting chronic opioid usage by various definitions. *(From Oleisky, E. R., Pennings, J. S., Hills, J., et al. (2019). Comparing different chronic preoperative opioid use definitions on outcomes after spine surgery.* The Spine Journal, 19(6), 984–994.)

depending on definition of chronic opioid use, 50%–80% of all patients with chronic opioid use continued to fill prescriptions in the year following surgery, compared with less than 20% of those with acute exposure or no preoperative exposure to opioids. There did not appear to be an increase in the chronically prescribed opioids with time after an initial surge of increased opioid prescription filling associated with the immediate postoperative period (Fig. 3). Similarly in the cohort described by Sigurdsson et al., 43%, 23%, or 66% of patients who had filled a preoperative prescription for opioids, BZDs or both, respectively, filled a prescription for an opioid more than 3 months after surgery, compared with 12% of those who were exposed to neither medication (Sigurdsson et al., 2019). In this study, there was additionally an increased long-term mortality for the group of patients who filled prescriptions for both opioids and BZDs preoperatively, compared with a propensity-score matched control group (1.41; 95% CI, 1.22–1.64) (Fig. 4). Additionally, when patients were separated into quartiles based on the cumulative amount of defined daily dose consumed preoperatively, there was evidence of a dose-response relationship, with the majority of the mortality signal originating from patients in third and fourth quartile.

Application to other areas

Modalities that support reduced chronic opioid and BZD usage overall will also benefit the subset of this population that presents for surgery. For opioids, there are very limited interventions shown to be effective to reduce chronic usage on an individual basis (Eccleston, Fisher, Thomas, et al., 2017). For both medication classes however, strategies to alter the prescription practices of clinicians have been shown to be effective to reduce chronic usage (Black, Khor, & Demirkol, 2020; Tannenbaum, Martin, Tamblyn, Benedetti, & Ahmed, 2014).

Similarly, modalities to monitor patients on chronic opioids and BZDs in the perioperative period could prove helpful to minimize the risk of additional harm from these medications. Most importantly, there is a need to monitor these patients closely for respiratory depression for the first few postoperative days, either in hospital or at home. There is a considerable development in monitoring devices for inpatient use, including wearable devices that can be used outside the postanesthesia recovery room (Ayad, Khanna, Iqbal, & Singla, 2019), but few devices have been developed for use in patients discharged to home.

The most important unanswered question is whether a preoperative intervention to assist patient in reducing or tapering off chronic opioids and BZDs will improve their perioperative care and reduce the chances of postoperative complications (Hah, Bateman, Ratliff, Curtin, & Sun, 2017). Such interventions might benefit from the occurrence of surgery as a motivator for patients and clinicians to review the indications for chronic opioid and BZD treatment and overall treatment strategy. Furthermore, tapering medications preoperatively might lead to a permanent reduction or discontinuation of these medications if successful. Such preoperative tapering programs are currently operational for patients on chronic opioids. Currently it is impossible to comment on causality for the associations found between chronic usage of opioids and BZDs and adverse outcomes. Therefore, intervention trials with a control arm, such as trials of preoperative medication tapering, are desperately needed. Such trials will hopefully shed further light on successful strategies to tackle the problem of overuse of opioids and BZDs overall, as well as provide answers for the best approach to this complex and expanding group of patients.

Other agents of interest

Several other medication classes have the potential to reduce complications in patients with chronic opioid and BZD usage. Gabapentinoids were incorporated into acute postoperative pain regimens around 2010, with the hope that they would reduce the amount of opioids used postoperatively. If successful, these medications could potentially improve acute pain control for patients chronically on opioids or BZDs while the reduction in opioids used might reduce the likelihood of adverse outcomes. However, while studies in limited subsets of patients have been promising, a more general approach to usage of gabapentinoids was not successful in showing a reduction in the absolute amount of opioids used postoperatively (Hah, Mackey, Schmidt, et al., 2018). A recent metanalysis summarizing an almost decade of studies on gabapentinoids, found the effects of gabapentinoids for management of acute pain to be minimal (Verret, Lauzier, Zarychanski, et al., 2020). In this metaanalysis, results from 117 trials incorporating 9060 patients found that on average, patients randomized to gabapentinoids used on average of 7.9 mg less morphine equivalent doses in the first 24 h postoperatively compared with controls (Verret et al., 2020). Additionally, the metaanalysis identified an increased risk of adverse events in patients receiving gabapentinoids, but no change in opioid-related adverse events (Verret et al., 2020). Taken together, there is little justification for routine inclusion of gabapentinoids in acute pain regiments. It should be noted, however, that results for patients using opioids or BZDs chronically might differ from findings that include also patients not exposed to these medications, so there might be a role in their inclusion for individual patients.

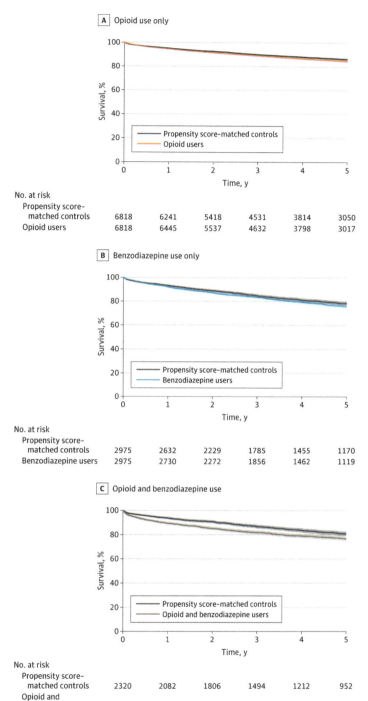

FIG. 4 Long-term survival in a mixed nationwide surgical cohort of patients based on preoperative exposure to opioids or benzodiazepines (BZDs). The figures show the postoperative survival of patients who filled a prescription for (A) opioids only (B) BZDs only or (C) both medications, compared with a propensity-score matched control group matched for patient and procedure characteristics. *(From Sigurdsson, M. I., Helgadottir, S., Long, T. E., et al. (2019). Association between preoperative opioid and benzodiazepine prescription patterns and mortality after noncardiac surgery. JAMA Surgery, 154(8), e191652.)*

258 PART | III Adverse effects, reactions, and outcomes

The other potential agent is liposomal bupivacaine, that provides a 72 h duration of action after single delivery (Balocco, Van Zundert, Gan, Gan, & Hadzic, 2018). This medication, if successfully used for infiltration or peripheral nerve blockage, could potentially allow for a prolonged pain relief not mediated by opioids. To date, database registries have identified cohorts with reduced postoperative opioid consumption following peripheral nerve blocks with liposomal bupivacaine (Asche, Dagenais, Kang, Ren, & Maurer, 2019), while others have found no difference in inpatient opioid consumption (Pichler, Poeran, Zubizarreta, et al., 2018). These studies however also include both patients with and without preoperative chronic opioid usage.

Mini-dictionary of terms

Chronic opioid/BZD use: Definitions vary, but generally chronic use refers to usage of the medications every day (or almost every day) for 90 days or more. Other definitions provide a higher sensitivity (requiring a longer duration of use) or account for amount of medication used.

Tolerance: The requirement of a higher dose of medication to achieve the same biological effect. Tolerance to opioids generally refers to the phenomena of requiring higher dose of medication to achieve the same analgesic effect, but tolerance to other effects of the medication also builds.

Differential opioid tolerance: Difference in the rate of tolerance build up for various effects and side-effects of opioids. Most importantly, tolerance to gastrointestinal and respiratory side effects builds more slowly than tolerance to the analgesic effects.

Opioid-induced hyperalgesia: Increased sensitivity to painful stimulus resulting from opioid use.

Persistent opioid use: Continuous usage of opioids intended past the expected duration of use in the acute setting, such as postoperatively. Definitions vary, but most commonly this refers to continued usage of opioids after 90 days postoperatively.

Key facts

- There is an alarming global increase in the number of patients who take opioids or BZDs chronically.
- Out of all patients presenting for surgery, roughly 20%–30% of patients take opioids chronically, and 10%–15% of patients take BZDs chronically.
- Patients who take opioids and BZDs are likely to have a higher burden of comorbidities such as heart failure and chronic obstructive lung disease. These diseases themselves render the patients more susceptible to complications from surgery.
- Acutely, patients with preexisting chronic usage of opioids are more likely to suffer from inadequate postoperative analgesia. Additionally, they are likely to require higher doses of opioids postoperatively and due to differential opioid tolerance, they are more likely to suffer from opioid-induced respiratory depression. One study found that postoperative 30-day mortality of patients with preoperative use of both BZDs and opioids was higher than matched controls.
- For the patient with chronic opioid usage, individualized plan for postoperative management of pain with a clear and honest discussion to manage expectations should be performed preoperatively depending on medication history and planned surgery. Agents used to support perioperative analgesia could include nonopioid analgesics such as acetaminophen, nonsteroidal antiinflammatory drugs, ketamine or methadone. Furthermore, inclusion of local anesthetics, either via infiltration, peripheral nerve blockade or in the form of neuraxial analgesia should be considered whenever feasible.
- Long-term, patients with preexisting usage of opioids or BZDs are more likely to have persistent usage of opioids more than 3 months postoperatively.
- An interesting ongoing and future aspect of caring for this patient population is the development of clinics and resources to support preoperative tapering of opioids and BZDs, in the hope that these adverse outcomes can be affected.

Summary points

- Preoperative chronic usage of opioids or BZDs is common in the surgical population.
- The population with chronic opioid or BZD usage has a comorbidity burden that renders them susceptible to complications.
- Patients with preoperative opioid usage are at risk for both inadequate analgesia as well as respiratory depression from additional opioids given in the early postoperative phase.

- Patients with preoperative opioid or BZD usage are more likely to use opioids persistently past 3 months postoperatively.
- Patients who take both opioids and BZDs preoperatively have a higher 30-day and long-term mortality than matched controls.
- The identified associations cannot establish causality, the results from interventions aiming at preoperative tapering hoping to reduce postoperative complications associated with chronic opioid and BZD us will hopefully provide guidance for the management of this population.

References

Airagnes, G., Lemogne, C., Renuy, A., et al. (2019). Prevalence of prescribed benzodiazepine long-term use in the French general population according to sociodemographic and clinical factors: Findings from the CONSTANCES cohort. *BMC Public Health, 19*, 566.

Aronson, S., Westover, J., Guinn, N., et al. (2018). A perioperative medicine model for population health: An integrated approach for an evolving clinical science. *Anesthesia and Analgesia, 126*, 682–690.

Asche, C. V., Dagenais, S., Kang, A., Ren, J., & Maurer, B. T. (2019). Impact of liposomal bupivacaine on opioid use, hospital length of stay, discharge status, and hospitalization costs in patients undergoing total hip arthroplasty. *Journal of Medical Economics, 22*, 1253–1260.

Ayad, S., Khanna, A. K., Iqbal, S. U., & Singla, N. (2019). Characterisation and monitoring of postoperative respiratory depression: Current approaches and future considerations. *British Journal of Anaesthesia, 123*, 378–391.

Bachhuber, M. A., Hennessy, S., Cunningham, C. O., & Starrels, J. L. (2016). Increasing benzodiazepine prescriptions and overdose mortality in the United States, 1996-2013. *American Journal of Public Health, 106*, 686–688.

Balocco, A. L., Van Zundert, P. G. E., Gan, S. S., Gan, T. J., & Hadzic, A. (2018). Extended release bupivacaine formulations for postoperative analgesia: An update. *Current Opinion in Anaesthesiology, 31*, 636–642.

Black, E., Khor, K. E., & Demirkol, A. (2020). Responsible prescribing of opioids for chronic non-cancer pain: A scoping review. *Pharmacy, 8*(3), 150.

Boon, M., van Dorp, E., Broens, S., & Overdyk, F. (2020). Combining opioids and benzodiazepines: Effects on mortality and severe adverse respiratory events. *Annals of Palliative Medicine, 9*, 542–557.

Carrier, J. D., Roberge, P., Courteau, J., & Vanasse, A. (2016). Predicting chronic benzodiazepine use in adults with depressive disorder: Retrospective cohort study using administrative data in Quebec. *Canadian Family Physician, 62*, e473–e483.

Carroll, I. R., Angst, M. S., & Clark, J. D. (2004). Management of perioperative pain in patients chronically consuming opioids. *Regional Anesthesia and Pain Medicine, 29*, 576–591.

Chou, R., Gordon, D. B., de Leon-Casasola, O. A., et al. (2016). Management of postoperative pain: A clinical practice guideline from the American Pain Society, the American Society of Regional Anesthesia and Pain Medicine, and the American Society of Anesthesiologists' Committee on Regional Anesthesia, Executive Committee, and Administrative Council. *The Journal of Pain, 17*, 131–157.

Dowell, D., Haegerich, T. M., & Chou, R. (2016). CDC guideline for prescribing opioids for chronic pain—United States, 2016. *JAMA, 315*, 1624–1645.

Dufour, R., Joshi, A. V., Pasquale, M. K., et al. (2014). The prevalence of diagnosed opioid abuse in commercial and medicare managed care populations. *Pain Practice, 14*, E106–E115.

Eccleston, C., Fisher, E., Thomas, K. H., et al. (2017). Interventions for the reduction of prescribed opioid use in chronic non-cancer pain. *Cochrane Database of Systematic Reviews, 11*, CD010323.

Edlund, M. J., Martin, B. C., Russo, J. E., DeVries, A., Braden, J. B., & Sullivan, M. D. (2014). The role of opioid prescription in incident opioid abuse and dependence among individuals with chronic noncancer pain: The role of opioid prescription. *The Clinical Journal of Pain, 30*, 557–564.

Frenk, S. M., Porter, K. S., & Paulozzi, L. J. (1999–2012). Prescription opioid analgesic use among adults: United States. *NCHS Data Brief, 2015*, 1–8.

Gupta, K., Prasad, A., Nagappa, M., Wong, J., Abrahamyan, L., & Chung, F. F. (2018). Risk factors for opioid-induced respiratory depression and failure to rescue: A review. *Current Opinion in Anaesthesiology, 31*, 110–119.

Hah, J. M., Bateman, B. T., Ratliff, J., Curtin, C., & Sun, E. (2017). Chronic opioid use after surgery: Implications for perioperative management in the face of the opioid epidemic. *Anesthesia and Analgesia, 125*, 1733–1740.

Hah, J., Mackey, S. C., Schmidt, P., et al. (2018). Effect of perioperative gabapentin on postoperative pain resolution and opioid cessation in a mixed surgical cohort: A randomized clinical trial. *JAMA Surgery, 153*, 303–311.

Hayhurst, C. J., & Durieux, M. E. (2016). Differential opioid tolerance and opioid-induced hyperalgesia: A clinical reality. *Anesthesiology, 124*, 483–488.

Hilliard, P. E., Waljee, J., Moser, S., et al. (2018). Prevalence of preoperative opioid use and characteristics associated with opioid use among patients presenting for surgery. *JAMA Surgery, 153*, 929–937.

Hozack, B. A., Rivlin, M., Lutsky, K. F., et al. (2019). Preoperative exposure to benzodiazepines or sedative/hypnotics increases the risk of greater filled opioid prescriptions after surgery. *Clinical Orthopaedics and Related Research, 477*, 1482–1488.

Jain, N., Phillips, F. M., Weaver, T., & Khan, S. N. (2018). Pre-operative chronic opioid therapy: A risk factor for complications, readmission, continued opioid use and increased costs after one- and two-level posterior lumbar fusion. *Spine, 43*(19), 1331–1338.

Jann, M., Kennedy, W. K., & Lopez, G. (2014). Benzodiazepines: A major component in unintentional prescription drug overdoses with opioid analgesics. *Journal of Pharmacy Practice, 27*, 5–16.

Lee, M., Silverman, S. M., Hansen, H., Patel, V. B., & Manchikanti, L. (2011). A comprehensive review of opioid-induced hyperalgesia. *Pain Physician, 14*, 145–161.

Lembke, A., Papac, J., & Humphreys, K. (2018). Our other prescription drug problem. *The New England Journal of Medicine, 378*, 693–695.

Mehta, V., & Langford, R. M. (2006). Acute pain management for opioid dependent patients. *Anaesthesia, 61*, 269–276.

Mehta, V., & Langford, R. (2009). Acute pain management in opioid dependent patients. *Reviews in Pain, 3*, 10–14.

Muller-Lissner, S., Bassotti, G., Coffin, B., et al. (2017). Opioid-induced constipation and bowel dysfunction: A clinical guideline. *Pain Medicine, 18*, 1837–1863.

Oleisky, E. R., Pennings, J. S., Hills, J., et al. (2019). Comparing different chronic preoperative opioid use definitions on outcomes after spine surgery. *The Spine Journal, 19*, 984–994.

Olfson, M., King, M., & Schoenbaum, M. (2015). Benzodiazepine use in the United States. *JAMA Psychiatry, 72*, 136–142.

Paronis, C. A., & Woods, J. H. (1997). Ventilation in morphine-maintained rhesus monkeys. II: Tolerance to the antinociceptive but not the ventilatory effects of morphine. *The Journal of Pharmacology and Experimental Therapeutics, 282*, 355–362.

Pichler, L., Poeran, J., Zubizarreta, N., et al. (2018). Liposomal bupivacaine does not reduce inpatient opioid prescription or related complications after knee arthroplasty: A database analysis. *Anesthesiology, 129*, 689–699.

Schnabel, A., Yahiaoui-Doktor, M., Meissner, W., Zahn, P. K., & Pogatzki-Zahn, E. M. (2020). Predicting poor postoperative acute pain outcome in adults: An international, multicentre database analysis of risk factors in 50,005 patients. *Pain Reports, 5*, e831.

Schoenfeld, A. J., Belmont, P. J., Jr., Blucher, J. A., et al. (2018). Sustained preoperative opioid use is a predictor of continued use following spine surgery. *The Journal of Bone and Joint Surgery. American Volume, 100*, 914–921.

Seth, P., Scholl, L., Rudd, R. A., & Bacon, S. (2018). Overdose deaths involving opioids, cocaine, and psychostimulants—United States, 2015-2016. *MMWR. Morbidity and Mortality Weekly Report, 67*, 349–358.

Sigurdsson, M. I., Helgadottir, S., Long, T. E., et al. (2019). Association between preoperative opioid and benzodiazepine prescription patterns and mortality after noncardiac surgery. *JAMA Surgery, 154*, e191652.

Svendsen, K., Skurtveit, S., Romundstad, P., Borchgrevink, P. C., & Fredheim, O. M. (2012). Differential patterns of opioid use: Defining persistent opioid use in a prescription database. *European Journal of Pain, 16*, 359–369.

Tannenbaum, C., Martin, P., Tamblyn, R., Benedetti, A., & Ahmed, S. (2014). Reduction of inappropriate benzodiazepine prescriptions among older adults through direct patient education: The EMPOWER cluster randomized trial. *JAMA Internal Medicine, 174*, 890–898.

Verret, M., Lauzier, F., Zarychanski, R., et al. (2020). Perioperative use of gabapentinoids for the management of postoperative acute pain: A systematic review and meta-analysis. *Anesthesiology, 133*, 265–279.

Von Korff, M., Dublin, S., Walker, R. L., et al. (2016). The impact of opioid risk reduction initiatives on high-dose opioid prescribing for patients on chronic opioid therapy. *The Journal of Pain, 17*, 101–110.

Yang, M. M. H., Hartley, R. L., Leung, A. A., et al. (2019). Preoperative predictors of poor acute postoperative pain control: A systematic review and meta-analysis. *BMJ Open, 9*, e025091.

Chapter 24

Malignant hyperthermia syndrome and hydrogen sulfide signaling: Role of Kv7 channels

Mariarosaria Bucci[a], Valentina Vellecco[a], Antonio Mancini[b], and Giuseppe Cirino[a]

[a]*Department of Pharmacy, School of Medicine and Surgery, University of Naples "Federico II", Naples, Italy,* [b]*Center of Biotechnologies, Cardarelli Hospital, Naples, Italy*

Abbreviations

3-MST	3 mercaptosulfutransferase
BKCa	large-conductance Ca^{2+}-activated K^+ channel
CACNA1S	calcium channel, voltage-dependent, L type, alpha-1s subunit
cAMP	cyclic adenosine monophosphate
CBS	cystathionine beta synthase
cGMP	cyclic guanosine monophosphate
CSE	cystathionine gamma lyase
DiBac4	bis-(1,3-dibutylbarbituric acid) trimethine oxono
EMHG	European Malignant Hyperthermia Group
IbTX	iberiotoxin
IVCT	in vivo contracture test
K$_{ATP}$	ATP-activated K channels
Kv7	voltage-dependent K channels
MH	malignant hyperthermia
MHAUS	MH Association of the United States
miRNA	microRNA
PDE	phosphodiesterase
PHSKM	primary human skeletal muscle
PTMs	posttranslational modification
RyR1	ryanodine receptor-1
SKM	skeletal muscle

Introduction

Despite the genetic basis of malignant hyperthermia (MH) syndrome are well established, and the scientific community hopes for DNA screening as the primary diagnostic approach. The diagnosis still needs an invasive procedure based on a classical bioassay, called In Vitro Contracture Test (IVCT) (Hopkins et al., 2015). IVCT is performed on a skeletal muscle biopsy harvested from the *vastus* group of the quadriceps that assesses in vitro, the skeletal muscle contractility. The assay consists in challenging the tissue with two different triggers: halothane and caffeine and the diagnosis for MH susceptibility depend upon the tissue response observed. In recent years, studies on the molecular basis of the MH syndrome have proposed RyR1 receptor function as a crucial element in the appearance of MH. More than 300 mutations of RyR1 decoding gene have been discovered, also associated with mutations in the CACNA1S receptor (Gomez, Holford, & Yamaguchi, 2016; Rosenberg & Rueffert, 2011). Unfortunately, only a minority of MH susceptible subjects display RyR1 gene mutations and, a causative role for MH susceptibility, has been identified for about 1/10 of the whole RyR1 mutations discovered (Ibarra Moreno et al., 2019). Thus, the IVCT procedure even though invasive and based on the phenotypic response, is still

Treatments, Mechanisms, and Adverse Reactions of Anesthetics and Analgesics. https://doi.org/10.1016/B978-0-12-820237-1.00024-7
Copyright © 2022 Elsevier Inc. All rights reserved.

262 PART | III Adverse effects, reactions, and outcomes

"The gold standard for MH diagnosis" for both European group of MH (EMHG) and the MH Association of the United States (MHAUS).

IVCT procedure

Briefly, under regional anesthesia bundles of 15–25 mm length and 2–3 mm thickness is dissected from *vastus* group of the quadriceps and quickly preserved in Krebs-Ringer solution oxygenated with carbogen (95% oxygen and 5% carbon dioxide mixture). The tissues are transported in a laboratory (in less than 30 min) and mounted in an isolated organ bath chamber filled with oxygenated Krebs-Ringer solution at 37°C. Bundles are tied on both ends and mounted vertically in the experimental tissue baths (15 mL volume) at 37°C so that the lower end is fixed and the upper end is connected to an isometric transducer with a resting tension of 2 mN (milliNewton; 0.2 g tension). Following 5 min, the bundles are stimulated with a 1–2 ms supramaximal stimulus at a frequency of 0.2 Hz. During electrical stimulation the muscle is stretched slowly to $150 \pm 10\%$ of initial length; this new length is considered the *optimal length*. The muscle is then allowed to stabilize at the *optimal length* for at least 20 min before starting the IVCT. Then the signal is acquired, amplified and continuously recorded by a data acquisition software system. Accordingly, to EMHG guidelines (Hopkins et al., 2015), the diagnosis depends on the increased resting tension of the tested muscle specimen: the threshold value for susceptibility is established as an increase of ≥ 0.2 g above the lowest resting tensions. A negative diagnosis (MHN subject) can be done when both caffeine, at 3 mM or more, and halothane, at 0.44 mmol/L or more, do not increase the resting tension more than 0.2 g. A positive diagnosis (MHS subject) is made when caffeine, at a concentration of 2 mM or less, and halothane, at 0.44 mmol/L or less, increase the resting tension more than 0.2 g (Fig. 1). Recently, EMHG has further detailed the diagnosis according to the susceptibility of the bundle to both triggers, only to caffeine, and only to halothane (Hopkins et al., 2015).

Hydrogen sulfide

The finding that ICTV is still today the gold standard for MH diagnosis strongly suggests that, over RyR1 mutations, other mechanisms are involved in the MH syndrome susceptibility. From here, the necessity to open to "a wider vision" on alternative/additive mechanisms that could concur to the excessive contractility of skeletal muscle typical of MH susceptible subjects. In this view, in the last 5 years, a growing interest has been focused on the potential involvement of the endogenous gasotransmitter hydrogen sulfide in the skeletal muscle physiopathology. Carbon monoxide, nitric oxide and hydrogen sulfide constitute a family of endogenous mediators that share the characteristic to be a gas. They are produced in our body by several enzymes with a half-life that ranges from seconds to minutes. Being gaseous molecules, they freely permeate through cell membranes without specific transporters reaching molecular targets relatively far from the site of their biosynthesis. Among them, H_2S is the latest discovered, firstly as an endogenous neuromodulator in the central nervous system (Abe & Kimura, 1996; Eto & Kimura, 2002; Miles & Kraus, 2004). The production of H_2S in mammalian tissues has been known for a long time but it was largely ignored and just considered a metabolic end-product. Three main enzymes are responsible for H_2S production (Fig. 2): cystathionine beta synthase (CBS), cystathionine gamma lyase (CSE or CGL or CTH) and 3-mercaptosulfutransferase (3-MST). CBS and CSE are pyridoxal-dependent enzymes and they require the amino acid L-cysteine as substrate (Kimura, 2014), while 3-MST is a pyridoxal-independent enzyme and it needs 3-mercaptopyruvate as substrate (Modis, Asimakopoulou, Coletta, Papapetropoulos, & Szabo, 2013; Fig. 2). All enzymes are constitutively expressed and appear to be tissue-specific (preferentially) (Cirino, Vellecco, & Bucci, 2017; Szabo et al., 2013). The presence of H_2S signaling in skeletal muscle (SKM) has been recently assessed. SKM expresses the three constitutive H_2S-generating enzymes in rat (Du, Li, Yang, Li, & Jin, 2013) and humans (Vellecco et al., 2016) that actively produce basal, detectable levels of H_2S. The role of hydrogen sulfide as an endogenous mediator has been extensively studied, it has neuroprotective actions, exerts vasodilatory properties, and also it has a protective role in gut and liver function (Fiorucci & Distrutti, 2016; Wallace & Wang, 2015). Taking advantage from its gaseous nature that allows it to travel through the cells, H_2S has many different molecular targets, in particular, it activates several subtypes of potassium channels, including ATP-activated K channels (K_{ATP}) (Fitzgerald et al., 2014; Medeiros et al., 2012), large-conductance Ca^{2+}-activated K^+ channel (BK_{Ca}) and subtypes 7 of voltage-dependent K channels (Kv7) (Hedegaard et al., 2014; Martelli et al., 2013). H_2S also acts as a pan inhibitor of phosphodiesterase (PDEs), a family of 11 isoenzymes that metabolize cyclic nucleotides cAMP and cGMP (Bucci et al., 2010). By blocking these enzymes H2S increases cAMP and cGMP levels prolonging their action (Bucci et al., 2010, 2012).

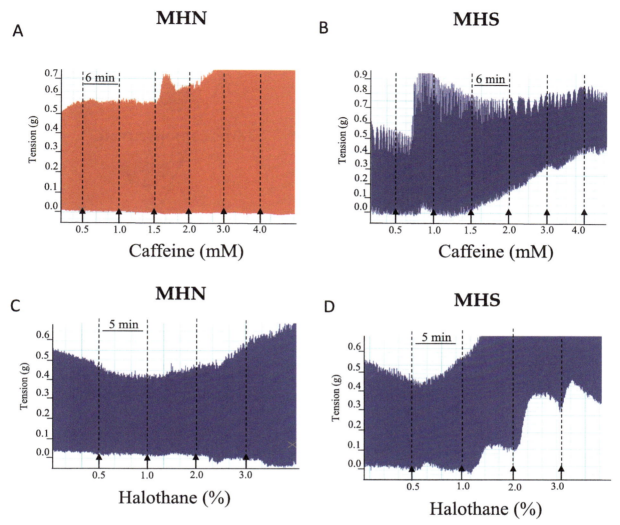

FIG. 1 IVCT original traces for MHS diagnosis. (A) Following 20 min of muscle bundle stabilization, caffeine has been added to the organ bath with a flow of 5 mL/min at progressive concentration of 0.5, 1.0, 1.5, 2.0, 3.0, 4.0 (6 min of contact for each concentration). The absence of any contracture from the resting tension provides an MHN diagnosis for caffeine. (B) Following 20 min of muscle bundle stabilization, caffeine has been added to the organ bath with a flow of 5 mL/min at progressive concentration of 0.5, 1.0, 1.5, 2.0, 3.0, 4.0 (6 min of contact for each concentration). The progressive increase in resting tension above 0.2 g (2 mN) provides an MHS diagnosis for caffeine. (C) Following 20 min of stabilization, the muscle bundle has been exposed to progressive halothane concentration of 0.5%, 1%, 2%, and 3% (5 min of contact for each concentration). The absence of any contracture from the resting tension provides an MHN diagnosis for halothane. (D) Following 20 min of stabilization, the muscle bundle has been exposed to progressive halothane concentration of 0.5%, 1%, 2%, and 3% (5 min of contact for each concentration). The progressive increase in resting tension ≥ 0.2 g (2 mN) provides a MHS diagnosis for halothane.

Hydrogen sulfide and MH

Which is the link between H_2S signaling and MH? Why and how this gasotransmitter should contribute and/or interfere with skeletal muscle hypercontractility, on which MH susceptibility diagnosis is based? As described above, in IVCT procedure the triggers used to perform MH diagnosis are caffeine and halothane. Both drugs share with H_2S some molecular targets: caffeine is known as a nonselective inhibitor of PDEs, and halothane mechanism of action involves potassium channels activation. From this evidence, we rose the hypothesis for the H_2S involvement in human MH.

Evaluation of H_2S-generating enzymes in muscle bundles reveals a significantly higher level of H_2S in MHS compared to MHN. Molecular analysis shows that the main source of H_2S in MHS patients is the CBS enzyme, that results strongly overexpressed, in terms of both protein and mRNA, compared to MHN (Vellecco et al., 2016). This evidence suggests that CBS-derived H_2S is the main source of the "pathological" high levels of H_2S measured in MHS subjects. However, circulating plasma levels of H_2S are not significantly different between MHN and MHS, implying that increased level of H_2S

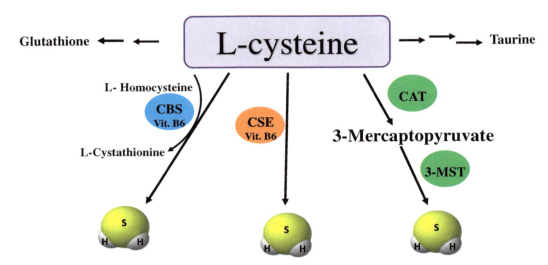

FIG. 2 Simplified scheme for H_2S biosynthesis. L-cysteine acts as a substrate leading to the synthesis of three major final products: H_2S, taurine, and glutathione. *CAT*, cysteine aminotransferase; *CBS*, cystathionine β-synthase; *CSE*, cystathionine γ-lyase; *3-MST*, 3-mercaptopyruvate sulfur transferase.

in MHS skeletal muscle is a local phenomenon, not detectable by measuring H_2S in plasma. This finding leads to two considerations: the first is that plasma levels of H_2S cannot be used as a diagnostic marker for MH susceptibly; this assumption does not surprise since, considering that H_2S is a potent endogenous vasodilator, the elevated concentration of SKM-derived H_2S should cause a reduction in blood pressure and there is no clinical evidence of a hypotensive phenotype among MHS subjects. The second, and more relevant consideration, is that the high amount of H_2S generates within the SKM may interact at different molecular level, by inducing posttranslational modification (see page 10) further contributing to the development of the MH syndrome.

Potassium channels as H_2S molecular targets in MHS subjects

Potassium channels are the first recognized molecular targets of H_2S, and up to now, the H_2S-K^+ channels the molecular interaction is the most extensively characterized. In SKM, the subtype voltage-gated Kv7.1-Kv7.5 (or KCNQ1-5) potassium channels, have been demonstrated to induce myogenesis and to protect SKM from statin-induced toxicity (Iannotti et al., 2010; Roura-Ferrer, Solé, Martínez-Mármol, Villalonga, & Felipe, 2008), even though, among Kv7 subtypes, Kv7.4 shows the highest degree of expression in both murine and human skeletal muscle cells (Iannotti, Barrese, Formisano, Miceli, & Taglialatela, 2013). However, western blot and Rt-PCR analysis of SKM did not show any significant difference between MHN and MHS biopsies for all Kv7 subtypes including Kv7.4 channels expression (Vellecco et al., 2020). Conversely, functional studies have shown that Kv7 channels are involved in MH. Indeed, the exposure of MHS biopsies to retigabine, a selective activator of Kv7 channels, increases the muscle contracture over 0.2 g, mimicking the classical IVCT triggers i.e., caffeine and halothane, while no effect is observed in MHN bundle (Fig. 3; Vellecco et al., 2020). To get a proof of a causative role for H_2S, the IVCT bioassay has been performed by incubating negative samples (MHN tissue) with NaHS followed by exposure with retigabine. In this condition, retigabine caused a significant increase in muscle contracture that mimics the MHS response confirming that the H_2S high concentration within the SKM of MH patients leads to activation of Kv7 channels (Vellecco et al., 2020). Kv7 is not the only potassium channels involved in MHS hypercontractility, K_{ATP} channels are also involved. Indeed, incubation of MHS bundles with the selective K_{ATP} channels blocker, glibenclamide, significantly reduces the halothane-induced increase in tension (Fig. 4; Vellecco et al., 2016). The evidence furnished by the classical pharmacological bioassay allows to formulate a hypothesis but does not give information on the molecular mechanism(s) of the phenomenon observed. Therefore, further studies on potassium channels and MHS susceptibility are essential.

The paradoxical depolarizing activity of Kv7 channel in MHS subjects

Primary human SKM (PHSKM) cells, starting from MHN (PHSKMN) and MHS (PHSMKS) biopsies, have been developed and characterized to further study potassium channels role, specifically to study the hyperpolarizing (relaxing)

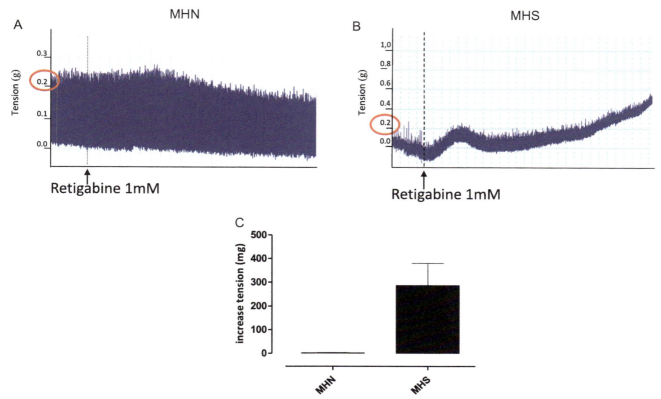

FIG. 3 Retigabine triggers a contractile response in biopsies of MH susceptible (MHS), but not in MH negative (MHN), subjects. (A) Representative trace of skeletal muscle bundle obtained from MHN subjects incubated with retigabine (1 mM) for 15 min. (B) Representative trace of skeletal muscle bundle obtained from MHS subjects incubated with retigabine for 15 min. (C) Effect of retigabine incubation, expressed as increased tension, in MHN and MHS biopsies. The value enclosed in *red circle* indicates the established threshold for susceptibility diagnosis. *(From Vellecco, V., Martelli, A., Bibli, I. S., Vallifuoco, M., Manzo, O. L., Panza, E., et al. (2020). Anomalous Kv7 channel activity in human malignant hyperthermia syndrome unmasks a key role for H2S and persulfidation in skeletal muscle. British Journal of Pharmacology, 177(4), 810–823, with permission.)*

or depolarizing (contracting) effects. One simple and direct method involving the use of a dye has been used. PHSKM cells have been incubated for 1 h with DiBac4, a membrane potential-sensitive dye and the relative-fluorescence changes, linked to hyperpolarizing (relaxing) or depolarizing (contracting) effects of openers and blockers of different subtypes of K^+ channels have been measured (Vellecco et al., 2020). Fig. 5 shows the responses of PHSKMN cells following incubation with selective activators or blockers of different K^+ channel subtypes i.e., K_{ATP}, large-conductance Ca^{2+}-activated and Kv7 channels. All the activators induce hyperpolarization while the blockers induce depolarization. When the same experiment has been performed on PHSKMS cells the pattern of the response is deeply modified. The retigabine effect is switched to depolarizing activity implying an anomalous behavior of Kv7 channels in PHSKMS cells. Conversely, cromakalim and NS1619, selective openers of K_{ATP} and BK_{Ca} channels respectively, induced a more intense hyperpolarizing activity. This effect most likely, compensates the paradoxical depolarizing activity of Kv7. Finally, the proof of concept of the causative role of augmented levels of H_2S for paradoxical depolarizing activity comes from PHSKMN cells. As shown in Fig. 5C, incubation with NaHS and then exposure to retigabine triggers a paradoxical depolarizing activity with an inversion analog to that observed in PHSKMS cells. This result has been also confirmed in vitro by using human tissues as described above. Indeed, by using IVCT procedure, incubation of MHN tissue with NaHS followed by exposure with retigabine leads to a significant increase of skeletal muscle contracture switching a negative response into a positive one (Vellecco et al., 2020).

Protein posttranslational modifications: Persulfidation (S-sulphydration)

Proteins posttranslational modifications (PTMs) are a heterogeneous group of transient molecular events in which some specific "sensitive" amino acids, within the protein structure, are available for a covalent and/or enzymatic reaction. PTMs refer to: (i) modification of an existing functional group; (ii) attachment with an endogenous molecule. In both cases, the effect is a modulation of proteins structure and function with consequent effect on cell signaling (Li et al., 2016;

266 PART | III Adverse effects, reactions, and outcomes

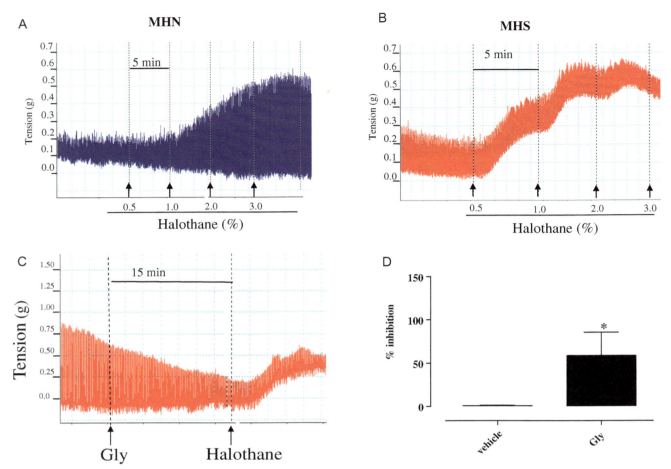

FIG. 4 Glibenclamide reduces halothane-induced contraction in MH susceptible (MHS) subjects. (A) Representative trace of skeletal muscle bundle from MHN subjects incubated with halothane during IVCT. Following progressive halothane concentration of 0.5%, 1%, 2%, and 3% (5 min of contact each concentration) no increase in tension is observed, providing a MHN diagnosis for halothane (B) Representative trace of skeletal muscle bundle from MHS subjects incubated with halothane during IVCT. Following progressive halothane concentration of 0.5%, 1%, 2%, and 3% (5 min of contact each concentration), the increase of resting tension above 0.2 g provides MHS diagnosis. (C) Representative trace of skeletal muscle bundle from MHS subjects incubated for 15 min with KATP potassium channel blocker glibenclamide (10 mM) followed by halothane challenge (2%) (D) glibenclamide significantly inhibits halothane-induced contraction in MHS biopsies. *(This figure was originally published in Vellecco, V., Mancini, A., Ianaro, A., Calderone, V., Attanasio, C., Cantalupo, A., et al. (2016). Cystathionine β-synthase-derived hydrogen sulfide is involved in human malignant hyperthermia. Clinical Science (London, England), 130, 35–44. https://portlandpress.com/clinsci/article-abstract/130/1/35/71299/Cystathionine-synthase-derived-hydrogen-sulfide-is?redirectedFrom=fulltext.)*

Qiyao et al., 2017). The PTMs include phosphorylation, glutathionylation, palmitoylation and other physiological modifications of proteins. Among gasotransmitters, S-nitrosylation is the most studied PTM: the attachment of a NO group to SH moiety of L-cysteine sensitive residues has been demonstrated to be involved in a wide spectrum of cell signaling and disorders of protein. S-nitrosylation contribute to the pathophysiology of several diseases i.e., pulmonary hypertension and cancer (Anand & Stamler, 2012; Nakamura & Lipton, 2013). In this context, recent literature has shown that PTMs are induced also by H_2S, named S-sulfhydration (or persulfidation) (Paul & Snyder, 2012; Zhang, Du, Tang, Huang, & Jin, 2017) affecting a wide range of physiological and/or pathological processes, such as cell proliferation/survival, mitochondrial bioenergetics, vascular reactivity and atherogenesis, oxidative stress and inflammation (Bibli et al., 2018; Módis et al., 2016).

Persulfidation of Kv7 channels in MH syndrome

The data discussed so far concerning the involvement of the hydrogen sulfide pathway in MH syndrome can be summarized as follows: (i) in SKM of patients susceptible to MH there is an increased CBS-derived H_2S production; (ii) increased level

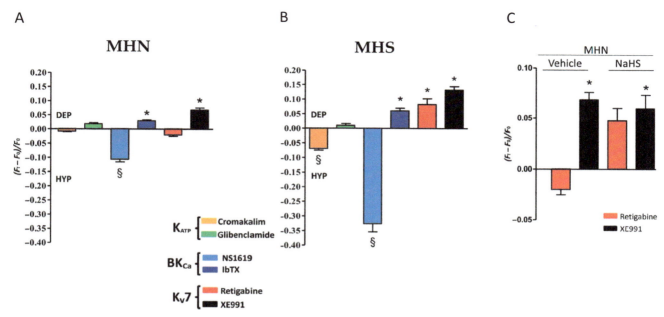

FIG. 5 Activation of Kv7 channels induces a paradoxical depolarizing response in primary human skeletal muscle cells derived from MHS biopsies (PHSKMS). (A) PHSKMN have been exposed to activators and blockers of K_{ATP} (cromakalim and glibenclamide), BKCa (NS1619 and IbTX), and Kv7 (retigabine and XE991) and the membrane potential evaluated. As expected, potassium channel activators evoke a hyperpolarizing (HYP) effect, while the potassium channel blockers induce a depolarizing (DEP) effect. (B) PHSKMS have been exposed to activators and blockers of K_{ATP}, BKCa, and Kv7 and the membrane potential evaluated. The pattern of the response results significantly modified revealing a paradoxical depolarizing response to retigabine. (C) PHSKMN have been preincubated with NaHS and then exposed to retigabine and XE991 for membrane potential evaluation. In this condition, retigabine exerts a paradoxical depolarizing activity with a similar trend observed in PHSKMS. No difference has been detected between the two groups, following XE991. *(From Vellecco, V., Martelli, A., Bibli, I. S., Vallifuoco, M., Manzo, O. L., Panza, E., et al. (2020). Anomalous Kv7 channel activity in human malignant hyperthermia syndrome unmasks a key role for H2S and persulfidation in skeletal muscle. British Journal of Pharmacology, 177(4), 810–823, with permission.)*

of H_2S in MHS skeletal muscle is a local phenomenon, not detectable by measuring H_2S in plasma; (iii) Potassium channels, specifically Kv7 and K_{ATP} subtypes, are involved in the anomalous behavior observed in SKM of patients susceptible to MH. But how H_2S affect K^+ channels activity is not clear. Indeed, there is no difference in the expression of Kv7 channels between MHN and MHS biopsies (Vellecco et al., 2020). One feasible hypothesis is that H_2S alters the activity of Kv7 channels, most likely by inducing PTMs. Based on prediction software Dianna 1.1 (the unified software for L-cysteine state, di-sulfide bond partner prediction and ternary cysteine classification) Kv7.4 channels possess only one highly nucleophilic cysteine (L-Cys87) and specifically modified biotin switch assay has clearly revealed that exclusively in MHS biopsies Kv7.4 is persulfidated (Vellecco et al., 2020). This latter result completes the possible mechanism indicating that CBS-derived H_2S, through Kv7 channels persulfidation, leads to an anomalous depolarizing effect. This molecular event translates into a pathological hypercontractility of SKM typical of MH syndrome (Fig. 6). It is worthy to consider that, being a gas, H_2S could interact with many other targets and induce persulfidation; in this regard almost 100 sensitive L-cysteine residues has been identified in RYR1 molecular structure that could be target of persulfidation modifying the channel activity (Chaube et al., 2014; Sun, Wang, Miyagi, Hess, & Stamler, 2013).

Applications to other areas
Hydrogen sulfide signaling and SKM: Beyond MH

The increased levels of CBS-derived H_2S found in skeletal muscle of MH susceptible patients unveils a broader issue: how H_2S, by acting on its multiple targets, could contribute to SKM physiology? Not much is available in literature, however, some clues can be reached from disorders in which H_2S-genering enzymes are malfunctioning.

For more than 50 years it is known that the lack and/or impairment of CBS activity is cause of hyperhomocysteinemia (HHcy) (Finkelstein, Mudd, Irreverre, & Laster, 1964; Kanwar, Manaligod, & Wong, 1976; Morris et al., 2017). HHcy is a risk factor for several human diseases and patients with HHcy manifest some characteristic features (early thrombotic events, cognitive decline) included in SKM dysfunctions. High levels of homocysteine (Hcy) in cerebrospinal fluids have

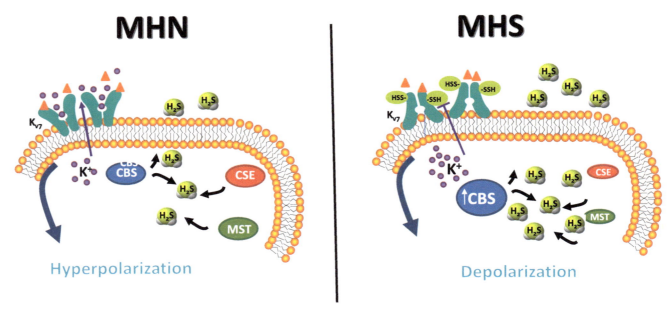

FIG. 6 Schematic cartoon summarizing the main findings discussed. The scheme depicts the effects of retigabine (▲) in human SKM of MHN and MHS MHN: CBS, CSE and 3-MST produce physiological amount of H_2S. retigabine, a selective activator of Kv7 channels, induces potassium efflux from the cells with a consequent hyperpolarizing effect without affecting the resting condition. MHS: CBS protein expression is increased and H_2S levels are augmented. H_2S induces persulfidation of Kv7 channels modifying channel activity. As shown, the persulfidated Kv7 channels display a paradoxical depolarizing effect leading to muscle contraction.

also been found in patients affected by multiple sclerosis and amyotrophic lateral sclerosis, neurological pathologies that induce muscle degeneration (Zoccolella et al., 2012). However, considering the data available, it is not clear if the HHcy-derived deleterious effect on SKM is due to a direct toxic effect of high levels Hcy. Indeed, HCY competes with L-cysteine transport, with consequent reduction not only in L-cysteine content but also in L-cysteine-derived antioxidant metabolites such as taurine and glutathione (see Fig. 2), or to a reduced levels of H_2S with consequent minor activation of H_2S molecular targets. Another clue for a role of H_2S in SKM performance comes from the Chronic Fatigue Syndrome, a serious debilitating disease with unknown etiology characterized by postexertional malaise and muscle and joint pain, coupled to dysregulation of body temperature and impairment of cognitive and autonomic function (Collatz, Johnston, Staines, & Marshall-Gradisnik, 2016). The hypothesis raised, that could explain the variety of the symptoms apparently unrelated, is a systemic dysregulation of H_2S signaling/metabolism (Dix Lemle, 2009) as well as of nitric oxide and urea metabolism (Monro & Basant, 2018). PDEs inhibition is the other most studied molecular mechanism for H_2S biological activity (Bucci et al., 2010, 2012). Many are the PDE isoforms expressed in SKM, that by hydrolysing cAMP and cGMP modulate the intracellular signaling (Nio et al., 2017; Tetsi et al., 2017). Interestingly, it has been demonstrated that in *mdx mice*, a murine model of Duchenne Muscular Dystrophy (DMD), treatment with association of PDE5 and PDE4 inhibitors, sildenafil and piclamilast, displayed antifibrotic effect, reduction of respiratory muscle weakness and cardiac dysfunction (Nio et al., 2017; Percival et al., 2012). The beneficial effect of sildenafil has been found also in a clinical study performed in boys (8–13 years of age) with DMD (Nelson et al., 2014). It is intriguing that, already in 1964, Nichol showed that in hind leg muscle of dystrophic mice there was a strong reduction in the SH-group compared to healthy mice (Nichol, 1964) suggesting thiol deficiency as a tissue feature of muscular dystrophy. All the information presented strongly suggest that H_2S could participate to SKM function, as it happens in many other districts of the body, by acting on K channels and PDE.

Other agents of interest
Future perspectives

The discovery that MH susceptibility, over RyR1 genetic mutations, involves a posttransduction mechanism that impacts on potassium channels activity, brings a novel and fresh impulse to the research on this rare disease. Indeed, this raises the possibility not only for the development of an alternative, less invasive method of MH diagnosis, but also for further gain inside into the molecular mechanisms of excitation-contraction coupling. Below, are summarized some considerations and new perspective for a possible alternative diagnostic procedure.

(i) In IVCT the triggers (caffeine and halothane) are used at extremely high concentration (up to 20 mM), even for an in vitro test. The exact molecular mechanisms through which both triggers induce contraction in MHS biopsies is still not clear. The use of the Kv7 selective activator retigabine, as an alternative trigger in the IVCT, could make the diagnosis easier and more reliable. Indeed, in the MHS biopsies retigabine elicits a contraction over 0.2 g from basal tension already at 1 mM. The lower concentration of retigabine as well as the presence of a specific target, compared to classical IVCT triggers, may ensure more specificity of action reducing ambiguous response.

(ii) The finding that SKM of MHS subjects displays elevated levels of H_2S compared to MHN implies a possibility to mitigate the sampling invasiveness requested for IVCT. Indeed, a needle biopsy should be enough to evaluate H_2S content. This measurement, together with a genetic testing may help to define a less invasive diagnostic procedure.

(iii) MHS patients display increased levels of H_2S within SKM, not detectable by measuring H_2S in plasma. Thus, "potential MHS subject" cannot be found by evaluating H_2S blood content. The discovery of new plasma and/or urine marker(s) related to H_2S signaling/metabolism could overcome this hurdle. It is known that catabolism of endogenous H_2S occurs mainly through exhalation and excretion in the urine as thiosulfate. This latter compound is considered the major end-product of H_2S catabolism and a nonspecific marker of entire-body H_2S production (Wang, 2012). Therefore, evaluation of the urinary level of thiosulfate, as a stable metabolite of H_2S, could become a marker of MH susceptibility, being an indirect index of the total (SKM included) H_2S content. The evaluation of thiosulfate as possible marker is simple and inexpensive since requires only the recruitment of known susceptible patients.

MiRNAs are small noncoding RNAs involved in the fine-tuning gene expression by accelerating degradation of mRNA and/or by inhibiting translation (Bartel, 2009). The identification of the microRNAs (miRNAs) that regulate mRNA expression of the H_2S-generating enzymes (CBS, CSE and 3-MST) is another diagnostic possibility. Again this could be achieved by using both urine and plasma specimens of known susceptible patients to identify a pattern of H_2S-related miRNA characteristic of MH susceptibility.

Mini-dictionary of terms

H_2S-generating enzymes: Group of three enzymes (CBS, CSE, 3MST) belonging to transsulfuration pathway that actively produce H_2S.

Posttranslational mechanism: Heterogeneous group of transient molecular events in which some specific "sensitive" amino acids, within the protein structure, are available for a covalent and/or enzymatic reaction.

Persulfidation (S-sulfhydration): Posttranslational event performed by H_2S on "sensitive" L-cysteine within protein structure that modifies the protein biological functions.

Retigabine: Selective opener of Kv7 potassium channel, known in therapy as antianticonvulsant used as an adjunctive treatment for partial epilepsy.

Glibenclamide: Selective blocker of K_{ATP} potassium channels by inhibiting regulatory subunit sulfonylurea receptor 1 (SUR1), known in therapy for type 2 diabetes.

Key facts of "malignant hyperthermia syndrome and hydrogen sulfide signaling role of Kv7 channels"

- In Vitro Contracture Test (IVCT) is the gold standard procedure for MH susceptibility diagnosis.
- IVCT is performed on a skeletal muscle biopsy to assess in vitro the skeletal muscle contractility following the challenge with two different triggers: halothane and caffeine.
- The presence of H_2S signaling in SKM has been recently assessed. Human SKM expresses the three constitutive H_2S-generating enzymes, namely CBS, CSE and 3-MST.
- MHS biopsies display increased levels of H_2S compared to MHN due to an overexpression of CBS. The increase in CBS-derived H_2S is confined within the SKM and it is not detectable in plasma.
- The selective Kv7 activator retigabine induces hyperpolarization (resting condition) in SKM biopsy of MHN patients while in MHS biopsies displays a paradoxical depolarizing activity (contraction).
- In SKM biopsy of MHS patients Kv7 potassium channels are persulfidated.

Summary points

- In MHS subjects there are augmented levels of H_2S within the skeletal muscle.

270 PART | III Adverse effects, reactions, and outcomes

- CBS is overexpressed in SKM of MHS subjects.
- The increased level of H_2S in MHS skeletal muscle is a local phenomenon since plasma levels are not affected.
- Potassium channels, specifically Kv7 and K_{ATP} subtypes, are involved in anomalous behavior observed in SKM of patients susceptible to MH.
- Persulfidation is the molecular mechanism through which H_2S modifies Kv7 channels activity switching the hyperpolarizing response (resting condition) to depolarizing (contraction).

References

Abe, K., & Kimura, H. (1996). The possible role of hydrogen sulfide as an endogenous neuromodulator. *The Journal of Neuroscience, 16*, 1066–1071.

Anand, P., & Stamler, J. S. (2012). Enzymatic mechanisms regulating protein S-nitrosylation: Implications in health and disease. *Journal of Molecular Medicine (Berlin), 90*, 233–244.

Bartel, D. P. (2009). MicroRNAs: Target recognition and regulatory functions. *Cell, 136*(2), 215–233.

Bibli, S. I., Hu, J., Sigala, F., Wittig, I., Heidler, J., Zukunft, S., et al. (2018). Cystathionine γ lyase sulfhydrates the RNA binding protein HuR to preserve endothelial cell function and delay atherogenesis. *Circulation, 139*(1), 101–114.

Bucci, M., Papapetropoulos, A., Vellecco, V., Zhou, Z., Pyriochou, A., Roussos, C., et al. (2010). Hydrogen sulfide is an endogenous inhibitor of phosphodiesterase activity. *Arteriosclerosis, Thrombosis, and Vascular Biology, 30*, 1998–2004.

Bucci, M., Papapetropoulos, A., Vellecco, V., Zhou, Z., Zaid, A., Giannogonas, P., et al. (2012). cGMP-dependent protein kinase contributes to hydrogen sulfide-stimulated vasorelaxation. *PLoS One, 7*(12), e53319.

Chaube, R., Hess, D. T., Wang, Y. J., Plummer, B., Sun, Q. A., Laurita, K., et al. (2014). Regulation of the skeletal muscle ryanodine receptor/Ca^{2+} release channel RyR1 by S-palmitoylation. *Journal of Biological Chemistry, 289*, 8612–8619.

Cirino, G., Vellecco, V., & Bucci, M. (2017). Nitric oxide-hydrogen sulfide: The gas paradigm of the vascular system. *British Journal of Pharmacology, 174*, 4021–4031.

Collatz, A., Johnston, S. C., Staines, D. R., & Marshall-Gradisnik, S. M. (2016). A systematic review of drug therapies for chronic fatigue syndrome/myalgic encephalomyelitis. *Clinical Therapeutics, 38*, 1263–1271.

Dix Lemle, M. (2009). Hypothesis: Chronic fatigue syndrome is caused by dysregulation of hydrogen sulfide metabolism. *Medical Hypothesis, 72*, 108–109.

Du, J., Li, W., Yang, C., Li, Q., & Jin, H. (2013). Hydrogen sulfide is endogenously generated in rat skeletal muscle and exerts a protective effect against oxidative stress. *Chinese Medicine Journal, 126*, 930–936.

Eto, K., & Kimura, H. (2002). A novel enhancing mechanism for hydrogen sulfide-producing activity of cystathionine beta-synthase. *Journal of Biological Chemistry, 277*(8), 42680–42685. https://doi.org/10.1074/jbc.M205835200.

Finkelstein, J. D., Mudd, S. H., Irreverre, F., & Laster, I. (1964). Homocystinuria due to cystathionine synthetase deficiency: The mode of inheritance. *Science, 46*(3645), 785–787.

Fiorucci, S., & Distrutti, E. (2016). Targeting the transsulfuration-H2S pathway by FXR and GPBAR1 ligands in the treatment of portal hypertension. *Pharmacological Research, 111*, 749–756. https://doi.org/10.1016/j.phrs.2016.07.040.

Fitzgerald, R., DeSantiago, B., Lee, D. Y., Yang, G., Kim, J. Y., & Foster, D. B. (2014). H2S relaxes isolated human airway smooth muscle cells via the sarcolemmal K(ATP) channel. *Biochemical and Biophysical Research Communications, 446*, 393–398.

Gomez, A. C., Holford, T. W., & Yamaguchi, N. (2016). Malignant hyperthermia-associated mutations in the S2-S3 cytoplasmic loop of type 1 ryanodine receptor calcium channel impair calcium-dependent inactivation. *American Journal of Physiology. Cell Physiology, 311*(5), C749–C757.

Hedegaard, E. R., Nielsen, B. D., Kun, A., Hughes, A. D., Krøigaard, C., Mogensen, S., et al. (2014). KV 7 channels are involved in hypoxia-induced vasodilatation of porcine coronary arteries. *British Journal of Pharmacology, 171*, 69–82.

Hopkins, P. M., Rüffert, H., Snoeck, M. M., Girard, T., Glahn, K. P., Ellis, F. R., et al. (2015). European Malignant Hyperthermia Group guidelines for investigation of malignant hyperthermia susceptibility. *British Journal of Anaesthesia, 115*, 531–539.

Iannotti, F. A., Barrese, V., Formisano, L., Miceli, F., & Taglialatela, M. (2013). Specification of skeletal muscle differentiation by repressor element-1 silencing transcription factor (REST)-regulated Kv7.4 potassium channels. *Molecular Biology of the Cell, 24*, 274–284.

Iannotti, F. A., Panza, E., Barrese, V., Viggiano, D., Soldovieri, M. V., & Taglialatela, M. (2010). Expression, localization, and pharmacological role of Kv7 potassium channels in skeletal muscle proliferation, differentiation, and survival after myotoxic insults. *Journal of Pharmacology and Experimental Therapeutics, 332*, 811–820.

Ibarra Moreno, C. A., Hu, S., Kraeva, N., Schuster, F., Johannsen, S., Rueffert, H., et al. (2019). An assessment of penetrance and clinical expression of malignant hyperthermia in individuals carrying diagnostic ryanodine receptor 1 gene mutations. *Anesthesiology, 131*(5), 983–991. https://doi.org/10.1097/ALN.0000000000002813.

Kanwar, Y. S., Manaligod, J. R., & Wong, P. W. (1976). Morphologic studies in a patient with homocystinuria due to 5, 10-methylenetetrahydrofolate reductase deficiency. *Pediatric Research, 10*(6), 598–609.

Kimura, H. (2014). Production and physiological effect of hydrogen sulfide. *Antioxidants & Redox Signaling, 20*, 783–793.

Li, Q., Uygun, B. E., Geerts, S., Ozer, S., Scalf, M., Gilpin, S. E., et al. (2016). Proteomic analysis of naturally-sourced biological scaffolds. *Biomaterials, 75*, 37–46.

Martelli, A., Testai, L., Breschi, M. C., Lawson, K., McKay, N. G., Miceli, F., et al. (2013). Vasorelaxation by hydrogen sulphide involves activation of Kv7 potassium channels. *Pharmacological Research, 70*, 27–34.

Medeiros, J. V., Bezerra, V. H., Lucetti, L. T., Lima-Júnior, R. C., Barbosa, A. L., & Tavares, B. M. (2012). Role of KATP channels and TRPV1 receptors in hydrogen sulfide-enhanced gastric emptying of liquid in awake mice. *European Journal of Pharmacology, 693*, 57–63.

Miles, E. W., & Kraus, J. P. (2004). Cystathionine β-synthase: Structure, function, regulation, and location of homocystinuria-causing mutations. *Journal of Biological Chemistry, 279*, 29871–29874.

Modis, K., Asimakopoulou, A., Coletta, C., Papapetropoulos, A., & Szabo, C. (2013). Oxidative stress suppresses the cellular bioenergetic effect of the 3-mercaptopyruvate sulfurtransferase/hydrogen sulfide pathway. *Biochemical and Biophysical Research Communications, 433*(4), 401–407. https://doi.org/10.1016/j.bbrc.2013.02.131.

Módis, K., Ju, Y., Ahmad, A., Untereiner, A. A., Altaany, Z., Wu, L., et al. (2016). S-sulfhydration of ATP synthase by hydrogen sulfide stimulates mitochondrial bioenergetics. *Pharmacological Research, 113*, 116–124.

Monro, J. A., & Basant, K. P. (2018). A molecular neurobiological approach to understanding the aetiology of chronic fatigue syndrome (myalgic encephalomyelitis or systemic exertion intolerance disease) with treatment implications. *Molecular Neurobiology, 55*, 7377–7388.

Morris, A. A., Kožich, V., Santra, S., Andria, G., Ben-Omran, T. I., Chakrapani, A. B., et al. (2017). Guidelines for the diagnosis and management of cystathionine beta-synthase deficiency. *Journal of Inherited Metabolic Disease, 40*(1), 49–74.

Nakamura, T., & Lipton, S. A. (2013). Emerging role of protein-protein transnitrosylation in cell signalling pathways. *ARS Journal, 18*, 239–249.

Nelson, M. D., Rader, F., Tavyev, J., Tang, X., Nelson, S. F., Miceli, M. C., et al. (2014). PDE5 inhibition alleviates functional muscle ischemia in boys with Duchenne muscular dystrophy. *Neurology, 82*, 2085–2091.

Nichol, C. J. (1964). Sulfhydryl and disulphide concentration in distrophic mouse muscle. *Canadian Journal of Biochemistry, 42*, 1643–1645.

Nio, Y., Tanaka, M., Hirozane, Y., Muraki, Y., Okawara, M., Hazama, M., et al. (2017). Phosphodiesterase 4 inhibitor and phosphodiesterase 5 inhibitor combination therapy has antifibrotic and anti-inflammatory effects in MDX mice with Duchenne muscular dystrophy. *FASEB Journal, 31*, 1–14.

Paul, B. D., & Snyder, S. H. (2012). H$_2$S signaling through protein sulfhydration and beyond. *Nature Reviews. Molecular Cell Biology, 13*, 499–507.

Percival, J. M., Whitehead, N. P., Adams, M. E., Adamo, C. M., Beavo, J. A., & Froehner, S. C. (2012). Sildenafil reduces respiratory muscle weakness and fibrosis in the mdx mouse model of Duchenne muscular dystrophy. *The Journal of Pathology, 228*, 77–87.

Qiyao, L., Shortreed, M. R., Wenger, C. D., Frey, B. L., Schaffer, L. V., Scalf, M., et al. (2017). Global post-translational modification discovery. *Journal of Proteome Research, 16*, 1383–1390.

Rosenberg, H., & Rueffert, H. (2011). Clinical utility gene card for malignant hyperthermia. *European Journal of Human Genetics, 19*(6).

Roura-Ferrer, M., Solé, L., Martínez-Mármol, R., Villalonga, N., & Felipe, A. (2008). Skeletal muscle Kv7 (KCNQ) channels in myoblast differentiation and proliferation. *Biochemical and Biophysical Research Communications, 369*, 1094–1097.

Sun, Q. A., Wang, B., Miyagi, M., Hess, D. T., & Stamler, J. S. (2013). Oxygen coupled redox regulation of the skeletal muscle ryanodine receptor/Ca^{2+} release channel (RyR1): Sites and nature of oxidative modification. *Journal of Biological Chemistry, 288*, 22961–22971.

Szabo, C., Coletta, C., Chao, C., Módis, K., Szczesny, B., Papapetropoulos, A., et al. (2013). Tumor derived hydrogen sulfide, produced by cystathionine-β-synthase, stimulates bioenergetics, cell proliferation, and angiogenesis in colon cancer. *Proceedings of the National Academy of Sciences of the United States of America, 110*, 12474–12479.

Tetsi, L., Charles, A. L., Paradis, S., Lejay, A., Talha, S., Geny, B., et al. (2017). Effects of cyclic nucleotide phosphodiesterases (PDEs) on mitochondrial skeletal muscle functions. *Cellular and Molecular Life Sciences, 74*, 1883–1893.

Vellecco, V., Mancini, A., Ianaro, A., Calderone, V., Attanasio, C., Cantalupo, A., et al. (2016). Cystathionine β-synthase-derived hydrogen sulfide is involved in human malignant hyperthermia. *Clinical Science (London, England), 130*, 35–44.

Vellecco, V., Martelli, A., Bibli, I. S., Vallifuoco, M., Manzo, O. L., Panza, E., et al. (2020). Anomalous Kv7 channel activity in human malignant hyperthermia syndrome unmasks a key role for H$_2$S and persulfidation in skeletal muscle. *British Journal of Pharmacology, 177*(4), 810–823.

Wallace, J. L., & Wang, R. (2015). Hydrogen sulfide-based therapeutics: Exploiting a unique but ubiquitous gasotransmitter. *Nature Reviews. Drug Discovery, 14*, 329–345.

Wang, R. (2012). Physiological implications of hydrogen sulfide: A whiff exploration that blossomed. *Physiological Reviews, 92*(2), 791–896.

Zhang, D., Du, J., Tang, C., Huang, Y., & Jin, H. (2017). H2S-induced sulfhydration: Biological function and detection methodology. *Frontiers in Pharmacology, 8*, 608.

Zoccolella, S., Tortorella, C., Iaffaldano, P., Direnzo, V., D'Onghia, M., Paolicelli, D., et al. (2012). Elevated plasma homocysteine levels in patients with multiple sclerosis are associated with male gender. *Journal of Neurology, 259*, 2105–2110.

Chapter 25

Problems with epidural catheter

Mustafa Kemal Arslantas

Department of Anesthesiology and Reanimation, School of Medicine, Demiroglu Bilim University, Istanbul, Turkey

Abbreviations

ASA American Society of Anesthesiologists
CSH cerebrospinal fluid
SEH spinal epidural hematoma

Introduction

The epidural block has widespread clinical applications in obstetrics, surgery, acute postoperative pain management, and chronic pain relief. Epidural infusions of local anesthetics and opioids via catheter are used to provide analgesia for days if necessary, to relieve postoperative pain after labor analgesia and major surgeries (e.g., thoracic, abdominal, and lower extremity). However, due to some problems encountered while using the epidural catheter, it may not function or develop various complications. This section reviews the problems encountered while inserting, using, and removing epidural catheters and their solutions.

Problems encountered while inserting an epidural catheter

The epidural space contains spinal nerve roots, excessive venous plexus, arteries, loose areolar connective tissue, and fat (Jiang, Shi, & Xu, 2015). However, recent studies suggest that there are meningo-vertebral ligaments between the lumbosacral spinal canal wall and the surrounding dura, which divide the epidural space into compartments of different sizes and shapes. The relationship between these structures and the epidural catheter may contribute to epidural catheter failure, irregular distribution of anesthesia, intravascular migration, and epidural hemorrhage.

Intravascular migration: Despite the appropriate technique and atraumatic placement, epidural catheters may migrate into epidural blood vessels. Intravascular migration of the epidural catheter has been reported more frequently in obstetric anesthesia than nonobstetric surgery; it may be due to physiological changes associated with pregnancy and labor.

During the initial placement of an epidural catheter in obstetric neuraxial analgesia, blood vessel puncture has a reported incidence of 6%–8% (Morrison & Buchan, 1990; Pan, Bogard, & Owen, 2004). Catheter migration can occur during and after insertion and may not always be related to the previous vessel puncture (Crosby, 1990; Dickson & Doyle, 1999; Jeon, Lee, Yoon, Kim, & Lee, 2013). Researchers have conducted several studies to reveal the causes of intravascular catheter migration. A systematic review (Mhyre, Greenfield, Tsen, & Polley, 2009) evaluates the evidence of the strategies proposed to minimize intravascular catheter migration during epidural catheter placement in pregnant women and recommends:

- Patient positioning in the lateral position.
- Distension of the epidural space with fluid before threading the catheter.
- Using a single rather than multiorifice catheter.
- Using the wired embedded polyurethane rather than nylon catheter.
- Limiting the depth of catheter insertion to 6 cm or less.

Determining the epidural catheter's intravascular placement is extremely important to prevent serious complications such as systemic local anesthetic intoxication.

There are several ways to determine the epidural catheter's migration during initial placement, including careful aspiration of the epidural catheter resulting in blood return (spontaneously or by aspiration) and tachycardia after using an epinephrine-containing epidural test dose. Moore and Batra (1981) proposed 45 mg of lidocaine with 15 µg of epinephrine

---

Treatments, Mechanisms, and Adverse Reactions of Anesthetics and Analgesics. https://doi.org/10.1016/B978-0-12-820237-1.00025-9
Copyright © 2022 Elsevier Inc. All rights reserved.

as the epidural test dose to avoid the consequences of intravascular injection of a critical amount of local anesthetic or opioid.

There is reasonable evidence for detecting intravascular misplacement of the epidural catheter as follows (Guay, 2006):

- In nonpregnant adult patients: Observation of an increase in SBP $\geq$15 mmHg or either an increase in SBP $\geq$15 mmHg or an increase in HR $\geq$10 bpm after the injection of 10 or 15 µg of epinephrine.
- In pregnant women: Signs of sedation, drowsiness, or dizziness within 5 min after the injection of 100 µg of fentanyl.
- In children: Increase in SBP $\geq$15 mmHg after the injection of 0.5 µg/kg of epinephrine.

Some case reports or case series have been published, showing that the epidural test dose not only fails to identify catheter misplacement but can also cause a serious side effect (Guay, 2006; McLean, Rottman, & Kotelko, 1992; Seidman & Marx, 1988). Therefore, more studies are needed to determine the best strategies for detecting intravascular catheter misplacement. Small bolus doses of drugs should be chosen. The problems such as catheter occlusion, failure of analgesia, or regression of the block should make the clinician consider the possibility of catheter misplacement (Bush & Kramer, 1993).

Subarachnoid migration: Most inadvertent dural punctures are typically detected during placement from cerebrospinal fluid (CSF) return from the epidural needle or aspiration from the catheter. Besides, a positive epidural test dose-response, or an exaggerated response to the administration of local anesthetic via the epidural catheter also suggests intrathecal migration of the epidural catheter (Betti, Carvalho, & Riley, 2017). The epidural catheter's apparent intrathecal migration can even be detected approximately 3 h to 4 days after insertion (Betti et al., 2017; Jaeger & Madsen, 1997). The dural holes made by Tuohy or spinal needles in the combined spinal-epidural anesthesia ease catheter penetration through the dura mater, and the risk of subarachnoid migration may increase. The dura mater may partially cut by the Tuohy needle during epidural placement, and that the epidural catheter later may migrate intrathecally through a partial dural tear. The risk of epidural catheter migration is higher, especially in multiple dural punctures with spinal needle or the formation of the hole with Tuohy needle than single dural punctures (Holmstrom, Rawal, Axelsson, & Nydahl, 1995).

Subarachnoid migration of the epidural catheter, if unnoticed, can lead to life-threatening complications. This complication is usually detected by clinical signs such as respiratory depression, motor blockade, and an unexpected drop in blood pressure. Consideration should be taken when injecting drugs through the epidural catheter under general anesthesia. Because significant fluctuations in blood pressure depend on anesthesia's depth, bleeding and fluid management will hide most of these symptoms, especially in major surgeries. In the case of continuous injection through the epidural catheter, even if previous bolus infusions are successful, aspiration tests should be performed each time before bolus injection, and vital signs should be carefully monitored after bolus injection (Ida, Sumida, & Kawaguchi, 2019).

Catheter migration to other locations: Apart from intravascular and subarachnoid migration when placing an epidural catheter, migration to surrounding tissues (e.g., intrathoracic, transforaminal, etc.) may occur very rarely. Even with experienced anesthetists can lead to a catheter's misplacement, particularly under suboptimal conditions, such as deviating from the patient's standard appropriate positioning. The Tuohy needle can enter directly into the pleural space after leaving the "relatively tight paravertebral tissue" and the resulting "loss of resistance" feeling. A puncture of the thoracic wall and thoracic catheter migration may result despite checking the loss of resistance, aspiration test, and test dose injection (Giest, Strauss, & Schaarschmidt, 2011; Shime, Shigemi, Hosokawa, & Miyazaki, 1991). Some methods have been described for evaluating the catheter position in the epidural space. These techniques can evaluate the catheter position's visibility or migration at the thoracic level by postoperative chest radiography using radio-opaque epidural catheters or dye (Shime et al., 1991).

Failure to obtain the expected sensory and hemodynamic response despite drug administered through the catheter may indicate catheter misplacement.

Breakage of epidural catheters

Epidural catheter breakage is a rare condition reported by case reports. Due to the user or manufacturer errors, severe damage, and catheter breakage during epidural catheter insertion or removal may result. If the epidural catheter encounters resistance while it is being advanced through the needle, then the needle should not be further advance, and further angulation and rotation should be avoided. Breaking the catheter into two fragments when placing an epidural catheter usually occurs when the catheter is pulled through the Tuohy needle due to the catheter being cut by the sharp tip of the needle. To avoid this complication, the needle and the catheter should remove as one set. Minimal force should be used while advancing and removing the catheter through the Tuohy needle (Hobaika, 2008).

The manufacturing process of epidural catheters and inherent to the materials used could predispose to catheter breakage. However, comparative studies on the tensile strength of catheter materials used have resulted in conflicting findings. The researchers find that nylon or polyurethane epidural catheters were more resistant to tensile strength than

Teflon or polyethylene epidural catheters (Nishio, Sekiguchi, Aoyama, Asano, & Ono, 2001). The other study reported similar high tensile strength with polyurethane epidural catheters compared with clear nylon and radiopaque nylon epidural catheters (Ates, Yucesoy, Unlu, Saygin, & Akkas, 2000).

Management of broken epidural catheter includes radiological evaluation (e.g., radiography, magnetic resonance imaging, computed tomography with multislice thin sections and three-dimensional or multiplanar reconstruction) neurosurgical consultation. A surgical extraction is considered if there is a possibility that the retained epidural catheter might cause neurological problems (Mayorga-Buiza, Gabella, Marquez-Rivas, & Rivero, 2012).

Blocked epidural catheter

Failure to inject drugs into the catheter is a common cause of epidural catheter problems. The epidural catheter's intraluminal obstruction can be caused by a blood clot (Hung & Hsieh, 2016), a manufacturing defect (Kulkarni, Pai, Shah, & Joshi, 2012), or an improper catheter connection and injector assembly (Gupta, Singh, & Kachru, 2001).

When a vascular puncture occurs with an epidural catheter, if the blood flowing to the catheter is not taken into account, the epidural catheter can be rapidly blocked by a blood clot. General practice is to withdraw the catheter until blood cannot be aspirated. However, it alone cannot prevent the clogging of this catheter. The blood in the catheter must be removed before it coagulates. The catheter should be flushed immediately with 0.9% saline and then slowly retracted to avoid possible catheter occlusion (Kumar, 1990). Before the epidural catheter is placed, a saline-filled syringe should be prepared, and if blood is aspirated after the catheter is inserted, the catheter should be flushed immediately. Injecting 1/3 of the total initial dose of local anesthesia or advancing a 3 to 5 cm catheter into the epidural space before the epidural catheter is inserted can reduce vascular puncture incidence (Hung & Hsieh, 2016).

Some problems with the connector assembly components and the filter can also lead to drug injection failure. Manufacturing errors (Singh & Solanki, 2003) or user errors (Gupta et al., 2001) can cause them. Whenever a failure to inject drugs into the catheter is encountered, the catheter's position in the connector assembly must be checked.

Epidural catheter removal difficulty

The epidural catheter is usually easily removed by gently pulling it out and is rarely difficult. The removal difficulties can be due to kinking, coiling, knotting, and looping of the catheter.

The catheter's kinking is mostly due to the physical properties of the material from which the catheter is made (Bhakta, Olteanu, & Zaheer, 2017; Toledano & Tsen, 2014). Some materials that catheters are made can be affected by body temperature, causing the catheter to kink more easily (Tandon, Pandey, & Pandey, 2013). Kink can be in the epidural area or under the skin, or outside the skin (Sexton, 2012). The skin fixation device used with some rigid catheters may cause kink formation 1–1.5 cm proximal to the catheter's entry into the skin (Bhakta et al., 2017). Also, kink may occur due to the practitioner's technique and due to excessive force. In the cases reported in the literature, it was revealed that the catheter was kinked in the examinations made as a result of the difficulty when injecting drugs through the catheter (Aslanidis, Fileli, & Pyrgos, 2010). Computed tomography and magnetic resonance impedance examination can help visualize the course of the kinking of the epidural catheter (Aslanidis et al., 2010).

Another problem encountered while removing the epidural catheter is the knotting of the catheter. The incidence of knotting is unknown exactly, but some of the authors reported one case in the 21,000 to 30,000 epidural insertions. Some authors state that the possibility of knotting increases with epidural catheters inserted more than 5 cm into the epidural space (Park, Cho, & Lee, 2017). Most cases of the knotted epidural catheter in the literature involved obstetric patients with lumbar epidural anesthesia (Macfarlane & Paech, 2002). Forcing the catheter to remove from entrapment can lead to serious complications like breakage, hematoma formation, and neurological deficit (Folk, Joye, Duc Jr., & Bailey, 2000). The authors made several recommendations to facilitate the removal of the epidural catheter. Gently steady traction of the epidural catheter in the insertion position is recommended because the retraction forces during removal are minimal when insertion and removal positions are similar. If this fails, continuous traction at the different positions can be attempted, and in the case of a knot, it can be reduced in size enough to allow pulling. Trying it after a while may affect catheter movement and then facilitate catheter removal (Folk et al., 2000). Epidural catheters that cannot be removed using these strategies may need to be removed by surgical methods under local or general anesthesia. If the catheter cannot be removed despite all efforts, radiological imaging should be performed. While removing the catheter, it may break, leaving a segment in the epidural space. If the broken epidural catheter causes complications, neurosurgical consultation should be made to decide whether it can be surgically removed (Blanchard et al., 1997).

Spinal epidural hematoma

The incidence of spinal epidural hematoma (SEH) is not fully known. However, according to a study of 3.7 million patients published in recent years, the nonobstetric population incidence was 18.5 per 100,000 epidural catheterizations instead of 0.6 per 100,000 in the obstetric population (Rosero & Joshi, 2016). SEH risk depends on the epidural catheterization technique used, difficult or traumatic (bloody) placement, patient comorbidities, coagulation defects, usage of anticoagulant drugs, and the surgical procedure performed. The most common symptoms of CSH are progressive motor and sensory block, back pain, and bowel/bladder dysfunction (Lee, Posner, Domino, Caplan, & Cheney, 2004). If SEH is detected due to anticoagulant medication, an emergency reversing agent(s) appropriate for the anticoagulant drug should be used. If an antiplatelet drug is used, an urgent platelet transfusion should be done. For possible decompressive surgery, urgent neurosurgical consultation should be obtained.

Assure the appropriate interval between the catheter placement or removal and the anticoagulant medications. There are some guidelines published by anesthesia associations on the timing of neuraxial anesthesia during antithrombotic therapy (Narouze et al., 2018; Table 1).

TABLE 1 The timing of epidural catheter removal during antithrombotic therapy.

	The interval from dose to catheter removal	
Anticoagulant	Before	After
Warfarin	Verify normal INR	No delay
IV Heparin (unfractionated)	4 to 6 h and verify normal aPTT	2 h
Low molecular weight heparin (LMWH)		
Therapeutic (subcutaneous)	$\geq$24 h	$\geq$4 h
Prophylactic (subcutaneous)	$\geq$12 h	$\geq$4 h
Factor X-a inhibitors		
Fondaparinux	$\geq$4 days	$\geq$8 h
New anticoagulants *Recommend an interval of 5 half-lives between the last dose of the drug		
Dabigatran, rivaroxaban, apixaban, edoxaban	$\geq$3 days	$\geq$6 h
Antiplatelet medication platelet P2Y12 receptor blockers		
Clopidogrel, ticlopidine	Neuraxial catheters can be maintained for 1 or 2 days if no loading dose will be administered	Without loading dose: immediate With loading dose: 6 h
Prasugrel, ticagrelor	Neuraxial catheters should not be maintained after administration	Without loading dose: immediate With loading dose: 6 h
Platelet GP IIb/IIIa inhibitors	Avoid while catheter is in place	Contraindicated 4 weeks post-op
Antiplatelet agents		
Aspirin	No restrictions	No restrictions
Clopidogrel	Avoid while catheter is in place	0 h; 6 h if loading dose
Prasugrel	Avoid while catheter is in place	6 h
Ticagrelor	Avoid while catheter is in place	6 h
Ticlopidine	Avoid while catheter is in place	6 h

The recommendations in this table are based on the guidelines of the ASRA fourth edition (Narouze, S., Benzon, H. T., Provenzano, D., Buvanendran, A., De Andres, J., Deer, T., Rauck, R., & Huntoon, M. A. (2018). Interventional spine and pain procedures in patients on antiplatelet and anticoagulant medications (second edition): Guidelines from the American Society of Regional Anesthesia and Pain Medicine, the European Society of Regional Anaesthesia and Pain Therapy, the American Academy of Pain Medicine, the International Neuromodulation Society, the North American Neuromodulation Society, and the World Institute of Pain. *Regional Anesthesia and Pain Medicine, 43*(3), 225–262. https://doi.org/10.1097/AAP.0000000000000700).

Spinal epidural abscess

Spinal epidural abscess is a suppurative infection of the central nervous system that can have serious consequences such as rare but permanent complications and even death. Studies have reported abscess rates of 5.4 to 7.2 per 100,000 epidural catheterizations in nonobstetric patients; this rate is much lower in obstetric patients (Rosero & Joshi, 2016). Rapid diagnosis and correct treatment are important as they can prevent complications and provide treatment in many cases. Clinical diagnosis can be made with back pain, neurological deficits (paralysis, motor weakness, radiculopathy, bowel or bladder dysfunction, etc.), and signs of infection (fever, leukocytosis, increased erythrocyte sedimentation rate, etc.). Modern imaging techniques such as computed tomography and especially magnetic resonance imaging greatly help the diagnosis. *Staphylococcus aureus* is the causative agent in most cases of spinal epidural abscess caused by pyogenic bacteria (Park et al., 2014). An empirical vancomycin regimen and a third or fourth-generation cephalosporin (Grade 2C) are recommended in these patients. Once the infecting organism(s) have been identified in the cultures made, treatment regimens should be simplified and directed to this pathogen. In addition to antibiotics, a surgical procedure is used for aspiration or drainage of the abscess in treating a spinal epidural abscess (Tuchman, Pham, & Hsieh, 2014).

Application to other areas

Peripheral nerve block techniques include continuous blockage for pain management with a catheter such as an epidural catheter. Insertion of a catheter adjacent to a nerve or plexus allows the nerve block to be extended by intermittent injection or continuous infusion. Problems similar to those with epidural catheters may be encountered during the use of these techniques. Although it is safer than neuraxial blocks, the problems and solution suggestions described in this section are valid for these applications.

Other agents of interest

Other neuraxial anesthesia techniques other than epidural catheter

- *Epidural anesthesia*: A catheter is not required and can be used for intraoperative anesthesia, and postoperative analgesia with local anesthetics administered a single bolus dose from the needle.
- *Spinal anesthesia*: It provides the rapid onset of more intense anesthesia using single doses of local anesthetics and opioids compared to epidural anesthesia.

Mini-dictionary of terms

Catheter: A long, very thin tube used to inject fluid to the epidural space.
Neuraxial anesthesia: The administration of medication into the subarachnoid or epidural space around the nerves of the central nervous system to produce anesthesia and analgesia, such as spinal anesthesia, caudal anesthesia, and epidural anesthesia.
Tuohy: A hollow hypodermic needle, very slightly curved at the end, suitable for inserting epidural catheters.

Key facts
Key facts of problems with epidural catheter

- Epidural infusions of local anesthetics and opioids via catheter are used to provide analgesia. However, due to some problems encountered while using the epidural catheter, it may not function or develop various complications.
- Intravascular migration of the epidural catheter has been reported more frequently in obstetric anesthesia than nonobstetric surgery; it may be due to physiological changes associated with pregnancy and labor.
- The epidural test dose not only fails to identify catheter misplacement but can also cause a serious side effect.
- Subarachnoid migration of the epidural catheter, if unnoticed, can lead to life-threatening complications.
- Breaking the catheter into two fragments when placing an epidural catheter usually occurs when the catheter is pulled through the Tuohy needle due to the catheter being cut by the sharp tip of the needle. To avoid this complication, the needle and the catheter should remove as one set.
- Forcing the catheter to remove from entrapment can lead to serious complications like breakage, hematoma formation, and neurological deficit.

278 PART | III Adverse effects, reactions, and outcomes

- Assure the appropriate interval between the catheter placement or removal and the anticoagulant medications.
- Rapid diagnosis and correct spinal epidural abscess treatment are important as they can prevent complications and provide treatment in many cases.

Summary points

- This chapter reviews the problems encountered while inserting, using, and removing epidural catheters and their solutions.
- The evidence of the strategies proposed to minimize intravascular catheter migration during epidural catheter placement in pregnant women and recommends:
  - Patient positioning in the lateral position.
  - Distension of the epidural space with fluid before threading the catheter.
  - Using a single rather than multiorifice catheter.
  - Using the wired embedded polyurethane rather than nylon catheter.
  - Limiting the depth of catheter insertion to 6 cm or less.
- Current evidence suggests that the small bolus doses of drugs should be chosen to avoid the consequences of intravascular injection of a critical amount of local anesthetic or opioid.
- Subarachnoid migration of the epidural catheter, if unnoticed, can lead to life-threatening complications. This complication is usually detected by clinical signs such as respiratory depression, motor blockade, and an unexpected drop in blood pressure.
- If the epidural catheter encounters resistance while it is being advanced through the needle, then the needle should not be further advance, and further angulation and rotation should be avoided.
- Management of broken epidural catheter includes radiological evaluation (e.g., radiography, magnetic resonance imaging, computed tomography with multislice thin sections and three-dimensional or multiplanar reconstruction) and neurosurgical consultation.
- The epidural catheter's intraluminal obstruction can be caused by a blood clot, a manufacturing defect, an improper catheter connection, and an injector assembly.
- The removal difficulties can be due to kinking, coiling, knotting, and looping of the catheter.
- The most common spinal epidural hematoma symptoms are progressive motor and sensory block, back pain, and bowel/bladder dysfunction.
- Clinical diagnosis of spinal epidural abscess can be made with back pain, neurological deficits (paralysis, motor weakness, radiculopathy, bowel or bladder dysfunction, etc.), and signs of infection (fever, leukocytosis, increased erythrocyte sedimentation rate, etc.).

References

Aslanidis, T., Fileli, A., & Pyrgos, P. (2010). Management and visualization of a kinked epidural catheter. *Hippokratia, 14*(4), 294–296. https://www.ncbi.nlm.nih.gov/pubmed/21311644. https://www.ncbi.nlm.nih.gov/pmc/articles/PMC3031330/pdf/hippokratia-14-294.pdf.

Ates, Y., Yucesoy, C. A., Unlu, M. A., Saygin, B., & Akkas, N. (2000). The mechanical properties of intact and traumatized epidural catheters. *Anesthesia and Analgesia, 90*(2), 393–399. https://doi.org/10.1213/00000539-200002000-00029.

Betti, F., Carvalho, B., & Riley, E. T. (2017). Intrathecal migration of an epidural catheter while using a programmed intermittent epidural bolus technique for labor analgesia maintenance: A case report. *A & A Case Reports, 9*(12), 357–359. https://doi.org/10.1213/XAA.0000000000000616.

Bhakta, P., Olteanu, D. S., & Zaheer, H. (2017). 90 degrees kinking of Vygon epidural catheter. *Journal of Anesthesia, 31*(5), 796. https://doi.org/10.1007/s00540-017-2321-7.

Blanchard, N., Clabeau, J. J., Ossart, M., Dekens, J., Legars, D., & Tchaoussoff, J. (1997). Radicular pain due to a retained fragment of epidural catheter. *Anesthesiology, 87*(6), 1567–1569. https://doi.org/10.1097/00000542-199712000-00036.

Bush, D. J., & Kramer, D. P. (1993). Intravascular migration of an epidural catheter during postoperative patient-controlled epidural analgesia. *Anesthesia and Analgesia, 76*(5), 1150–1151. https://doi.org/10.1213/00000539-199305000-00042.

Crosby, E. T. (1990). Epidural catheter migration during labour: An hypothesis for inadequate analgesia. *Canadian Journal of Anaesthesia, 37*(7), 789–793. https://doi.org/10.1007/BF03006538.

Dickson, M. A., & Doyle, E. (1999). The intravascular migration of an epidural catheter. *Paediatric Anaesthesia, 9*(3), 273–275. https://doi.org/10.1046/j.1460-9592.1999.00333.x.

Folk, J. W., Joye, T. P., Duc, T. A., Jr., & Bailey, M. K. (2000). Epidural catheters: The long and winding road. *Southern Medical Journal, 93*(7), 732–733. https://www.ncbi.nlm.nih.gov/pubmed/10923969.

Giest, J., Strauss, J., & Schaarschmidt, K. (2011). Case report: Failed epidural puncture for insertion of a catheter. *Anästhesiologie, Intensivmedizin, Notfallmedizin, Schmerztherapie, 46*(1), 8–11. https://doi.org/10.1055/s-0030-1270553 (Kasuistik: Perioperative Schmerztherapie—Fehllage eines thorakalen Periduralkatheters nach Punktion in Seitenlage).

Guay, J. (2006). The epidural test dose: A review. *Anesthesia and Analgesia, 102*(3), 921–929. https://doi.org/10.1213/01.ane.0000196687.88590.6b.

Gupta, S., Singh, B., & Kachru, N. (2001). "Blocked" epidural catheter: Another cause. *Anesthesia and Analgesia, 92*(6), 1617–1618. https://doi.org/10.1097/00000539-200106000-00058.

Hobaika, A. B. (2008). Breakage of epidural catheters: Etiology, prevention, and management. *Revista Brasileira de Anestesiologia, 58*(3), 227–233. https://doi.org/10.1590/s0034-70942008000300005.

Holmstrom, B., Rawal, N., Axelsson, K., & Nydahl, P. A. (1995). Risk of catheter migration during combined spinal epidural block: Percutaneous epiduroscopy study. *Anesthesia and Analgesia, 80*(4), 747–753. https://doi.org/10.1097/00000539-199504000-00017.

Hung, K. C., & Hsieh, S. W. (2016). Epidural catheter blockage caused by a blood clot: Is it time to change our practice? *Journal of Clinical Anesthesia, 35*, 205–206. https://doi.org/10.1016/j.jclinane.2016.08.003.

Ida, M., Sumida, M., & Kawaguchi, M. (2019). Intraoperative subarachnoid migration of the epidural catheter used for continuous infusion leading to delayed detection. *Journal of Clinical Anesthesia, 55*, 115. https://doi.org/10.1016/j.jclinane.2019.01.002.

Jaeger, J. M., & Madsen, M. L. (1997). Delayed subarachnoid migration of an epidural arrow FlexTip plus catheter. *Anesthesiology, 87*(3), 718–719. https://doi.org/10.1097/00000542-199709000-00050.

Jeon, J., Lee, I. H., Yoon, H. J., Kim, M. G., & Lee, P. M. (2013). Intravascular migration of a previously functioning epidural catheter. *Korean Journal of Anesthesiology, 64*(6), 556–557. https://doi.org/10.4097/kjae.2013.64.6.556.

Jiang, H., Shi, B., & Xu, S. (2015). An anatomical study of lumbar epidural catheterization. *BMC Anesthesiology, 15*, 94. https://doi.org/10.1186/s12871-015-0069-x.

Kulkarni, P. K., Pai, V. A., Shah, R. P., & Joshi, S. R. (2012). Intraluminal obstruction of epidural catheter due to manufacturing defect. *Journal of Anaesthesiology Clinical Pharmacology, 28*(2), 280. https://doi.org/10.4103/0970-9185.94935.

Kumar, C. M. (1990). Prevention of obstruction of epidural catheter by blood clot. *Canadian Journal of Anaesthesia, 37*(6), 709–710. https://doi.org/10.1007/BF03006508.

Lee, L. A., Posner, K. L., Domino, K. B., Caplan, R. A., & Cheney, F. W. (2004). Injuries associated with regional anesthesia in the 1980s and 1990s: A closed claims analysis. *Anesthesiology, 101*(1), 143–152. https://doi.org/10.1097/00000542-200407000-00023.

Macfarlane, J., & Paech, M. J. (2002). Another knotted epidural catheter. *Anaesthesia and Intensive Care, 30*(2), 240–243. https://doi.org/10.1177/0310057X0203000223.

Mayorga-Buiza, M. J., Gabella, F., Marquez-Rivas, J., & Rivero, M. (2012). Broken epidural catheter. *Anaesthesia, 67*(12), 1407. https://doi.org/10.1111/anae.12078.

McLean, B. Y., Rottman, R. L., & Kotelko, D. M. (1992). Failure of multiple test doses and techniques to detect intravascular migration of an epidural catheter. *Anesthesia and Analgesia, 74*(3), 454–456. https://doi.org/10.1213/00000539-199203000-00021.

Mhyre, J. M., Greenfield, M. L., Tsen, L. C., & Polley, L. S. (2009). A systematic review of randomized controlled trials that evaluate strategies to avoid epidural vein cannulation during obstetric epidural catheter placement. *Anesthesia and Analgesia, 108*(4), 1232–1242. https://doi.org/10.1213/ane.0b013e318198f85e.

Moore, D. C., & Batra, M. S. (1981). The components of an effective test dose prior to epidural block. *Anesthesiology, 55*(6), 693–696. https://doi.org/10.1097/00000542-198155060-00018.

Morrison, L. M., & Buchan, A. S. (1990). Comparison of complications associated with single-holed and multi-holed extradural catheters. *British Journal of Anaesthesia, 64*(2), 183–185. https://doi.org/10.1093/bja/64.2.183.

Narouze, S., Benzon, H. T., Provenzano, D., Buvanendran, A., De Andres, J., Deer, T., et al. (2018). Interventional spine and pain procedures in patients on antiplatelet and anticoagulant medications (second edition): Guidelines from the American Society of Regional Anesthesia and Pain Medicine, the European Society of Regional Anaesthesia and Pain Therapy, the American Academy of Pain Medicine, the International Neuromodulation Society, the North American Neuromodulation Society, and the World Institute of Pain. *Regional Anesthesia and Pain Medicine, 43*(3), 225–262. https://doi.org/10.1097/AAP.0000000000000700.

Nishio, I., Sekiguchi, M., Aoyama, Y., Asano, S., & Ono, A. (2001). Decreased tensile strength of an epidural catheter during its removal by grasping with a hemostat. *Anesthesia and Analgesia, 93*(1), 210–212. https://doi.org/10.1097/00000539-200107000-00041.

Pan, P. H., Bogard, T. D., & Owen, M. D. (2004). Incidence and characteristics of failures in obstetric neuraxial analgesia and anesthesia: A retrospective analysis of 19,259 deliveries. *International Journal of Obstetric Anesthesia, 13*(4), 227–233. https://doi.org/10.1016/j.ijoa.2004.04.008.

Park, K. H., Cho, O. H., Jung, M., Suk, K. S., Lee, J. H., Park, J. S., et al. (2014). Clinical characteristics and outcomes of hematogenous vertebral osteomyelitis caused by gram-negative bacteria. *The Journal of Infection, 69*(1), 42–50. https://doi.org/10.1016/j.jinf.2014.02.009.

Park, J. T., Cho, D. W., & Lee, Y. B. (2017). Knotting of a cervical epidural catheter in the patient with post-herpetic neuralgia: A rare complication. *Journal of Lifestyle Medicine, 7*(1), 41–44. https://doi.org/10.15280/jlm.2017.7.1.41.

Rosero, E. B., & Joshi, G. P. (2016). Nationwide incidence of serious complications of epidural analgesia in the United States. *Acta Anaesthesiologica Scandinavica, 60*(6), 810–820. https://doi.org/10.1111/aas.12702.

Seidman, S. F., & Marx, G. F. (1988). Epinephrine test dose is not warranted for confirmation of intravascular migration of epidural catheter in a parturient. *Canadian Journal of Anaesthesia, 35*(1), 104–106. https://doi.org/10.1007/BF03010560.

Sexton, A. J. (2012). Double kinking of a thoracic catheter within the epidural space. *Anaesthesia and Intensive Care, 40*(4), 732–733. https://www.ncbi.nlm.nih.gov/pubmed/22813519.

Shime, N., Shigemi, K., Hosokawa, T., & Miyazaki, M. (1991). Intrathoracic migration of an epidural catheter. *Journal of Anesthesia, 5*(1), 100–102. https://doi.org/10.1007/s0054010050100.

Singh, B., & Solanki, S. S. (2003). Blocked epidural catheter: Time to look beyond the catheter. *Anaesthesia, 58*(4), 400–401. https://doi.org/10.1046/j.1365-2044.2003.03095_24.x.

Tandon, M., Pandey, C. K., & Pandey, V. K. (2013). Epidural catheter kinking over the scapular margins. *Indian Journal of Anaesthesia, 57*(3), 318–319. https://doi.org/10.4103/0019-5049.115596.

Toledano, R. D., & Tsen, L. C. (2014). Epidural catheter design: History, innovations, and clinical implications. *Anesthesiology, 121*(1), 9–17. https://doi.org/10.1097/ALN.0000000000000239.

Tuchman, A., Pham, M., & Hsieh, P. C. (2014). The indications and timing for operative management of spinal epidural abscess: Literature review and treatment algorithm. *Neurosurgical Focus, 37*(2). https://doi.org/10.3171/2014.6.FOCUS14261, E8.

Chapter 26

Headache after neuraxial blocks: A focus on combined spinal-epidural anesthesia

Haluk Ozdemir[a] and Reyhan Arslantas[b]

[a]*Department of Anesthesiology and Reanimation, Marmara University, Pendik Training and Research Hospital, Istanbul, Turkey,* [b]*Department of Anesthesiology and Reanimation, Taksim Training and Research Hospital, Istanbul, Turkey*

Abbreviations

ACTH	adrenocorticotropic hormone
CSE	combined spinal-epidural
CSEA	combined spinal-epidural anesthesia
CSF	cerebrospinal fluid
EBP	epidural blood patch
EDA	epidural anesthesia
GONB	greater occipital nerve block
NTN	needle-through-needle
PDPH	postdural puncture headache
RCTs	randomized controlled trials
SPGB	sphenopalatine ganglion block
UDP	unintentional dural puncture

Introduction

Postdural puncture headache (PDPH) is a well-recognized complication of spinal anesthesia since its first introduction into clinical practice by Bier in 1898. Combined spinal-epidural anesthesia (CSEA) is an anesthetic modality that combines spinal and epidural techniques. By using CSEA, some of the disadvantages of spinal and epidural anesthesia (EDA) can be reduced or eliminated, and the advantages of both techniques can be maintained. CSEA is increasingly becoming popular in obstetric anesthesia and high-risk surgeries. Since the technique contains an intentional dural puncture as part of its spinal anesthesia component, concerns relating to PDPH arises.

History of PDPH

Spinal anesthesia was introduced into clinical practice by German surgeon Karl August Bier in August 1898. After the first successful cocainization of the subarachnoid space and operations on five patients, Bier himself and his assistant Hildebrandt has practiced the same procedure voluntarily to each other as a "self-experiment to investigate further some of the problems he encountered with the first administration in patients." Among these, six of them, including Bier and Hildebrandt, had developed headaches and were both forced to stay in bed for several days (Bier, 1899). In his 1989 pioneering report, Bier concluded that spinal anesthesia would only be successful when there is at least some reflux of cerebrospinal fluid (CSF) and PDPH. Interestingly, with these still valid statements today, he also correctly predicted that the headache was caused by the loss of CSF (Bier, 1899).

The medical literature of the early 1900s has numerous applications of spinal anesthesia using large spinal needles. Although headache was reported to be a complication in 50% of subjects, it was said to resolve within 24 h (Lee, 1979). The use of increasingly small-gauge needles and the increasing popularity of noncutting, pencil-point needles has dramatically reduced the incidence of PDPH following neuraxial anesthesia (Hoskin, 1998).

Treatments, Mechanisms, and Adverse Reactions of Anesthetics and Analgesics. https://doi.org/10.1016/B978-0-12-820237-1.00026-0
Copyright © 2022 Elsevier Inc. All rights reserved.

The importance of PDPH

A surprising fact is that, though seemingly rare and can be reasonably managed, PDPH is a leading cause of lawsuits (Davies, Posner, Lee, Cheney, & Domino, 2009). This may reflect a change in patients' expectations; the disappointment patients experience when this complication occurs is a disabling condition that has potential financial, social, and psychological effects (Gaiser, 2013). This also gives the background as to why there has been a significant investigation into the diagnosis and management of PDPH and why the prevention and treatment of PDPH are always current issues (Gaiser, 2017).

The studies showed that PDPH is not only an acute occurrence; it may become chronic. After an unintentional dural puncture (UDP) with a 17-gauge Tuohy needle, the chronic headache was reported in 28% of patients, of whom nearly 20% experienced serious disability from headaches (Webb et al., 2012). In another study with obstetric patients, MacArthur et al. reported a 23% incidence of headache lasting longer than 6 weeks in parturients who had sustained a dural puncture with a large-gauge needle. This difference was found to be significantly greater when compared with matched controls (MacArthur, Lewis, & Knox, 1993).

The mechanism of PDPH

The exact mechanism of PDPH remains unknown. The postulated cause is the decreased CSF pressure due to the loss of CSF in the epidural space through the dural puncture site. The decrease in the CSF pressure creates a loss of the cushioning effect provided by the intracranial fluid. The resulting traction placed on intracranial pain-sensitive structures elicits pain (Morewood, 1993). Another possible cause is the distension of the cerebral blood vessels. By a sudden drop in the CSF pressure, the intracranial vessels' vasodilation occurs for maintaining a constant intracranial volume under Monroe Kelly's hypothesis, resulting in pathophysiology similar to a vascular headache (Grant, Condon, Hart, & Teasdale, 1991). The beneficial effects of vasoconstrictor drugs such as caffeine and theophylline in PDPH treatment support this mechanism (Basurto Ona, Osorio, & Bonfill Cosp, 2015).

The characteristics of PDPH

According to the International Headache Society, PDPH is any headache that develops within 5 days of dural puncture and is not better accounted for by any other cause (Headache Classification Committee of the International Headache Society (IHS) The international classification of headache disorders, 3rd edition, 2018).

Postdural puncture's typical headache is often diffuse or dull and worsens within 15 min after sitting or standing up. Stiffness of the neck, nausea, tinnitus, hyperacusis, and photophobia generally accompanies. It is important to consider other causes of headache when assessing the situation since headache is a widespread symptom. There may be other causes of headache like tension headache, migraine, sinusitis, meningitis, cortical sinus thrombosis, subarachnoid hemorrhage, brain tumor, subdural hematoma, or cerebral infarction in the patients who have had a spinal, epidural, or a combined spinal-epidural (CSE) block.

Combined spinal-epidural anesthesia

In the history of anesthesia, epidural and spinal blocks are highly accepted and extensively used techniques in various surgical interventions and pain relief worldwide. But both techniques have well-known drawbacks, which are especially important in obstetric and high-risk patient groups. CSEA is a comparatively new technique that claimed to combine the advantages of both epidural and spinal techniques, including faster onset due to the spinal component, more reliable anesthesia with the ability to improve an adequate block, extending the block when necessary, providing postoperative analgesia, minimal motor and sensory blockade, improved mobilization and improved patient comfort (Rawal, Van Zundert, Holmstrom, & Crowhurst, 1997). Experience with low-dose CSEA showed that the incidence of maternal hypotension and the need for vasopressors was reduced in women undergoing operative delivery (Ghazi & Raja, 2007). Additionally, compared to EDA, the dose of anesthetics required is reduced, and toxicity from local anesthetics is eliminated (van de Velde, Teunkens, Hanssens, van Assche, & Vandermeersch, 2001).

Combined spinal-epidural anesthesia for special patient groups

CSEA, with all these features, is becoming widely popular in critical patients undergoing surgery and women during pregnancy, who are considered particularly at increased risk for PDPH. UDP reported to be occurring in 0.4%–6% of parturients who were given EDA are associated with acute severe positional headache in approximately 70%–80% of these parturients. (Amorim, Gomes de Barros, & Valenca, 2012; Choi, Ahn, & Kim, 2006; Kuczkowski, 2004; Paech, Banks, & Gurrin, 2001).

Studies have demonstrated that young age and female gender increase the risk of PDPH. Parturients pose the greatest risk population because they fulfill these risk criteria. Its high incidence may be attributed to increased estrogen levels, which influence the cerebral vessels' tone, thereby increasing the vascular distension in response to CSF hypotension (Kuczkowski, 2004; Liu, Kalarickal, & Rosinia, 2012). In vaginal delivery, increased CSF pressure during second-stage labor can increase the risk for CSF leak and increase dural opening size, which may increase the PDPH (Wu et al., 2006). Postpartum women may be at particular risk because of reduced CSF volumes and reduced CSF production during diuresis (Cook, 2000).

The effect of obesity on PDPH is a controversial issue. Obesity may pose technical difficulties risks for determining landmarks for CSEA application and may increase UDP risk. In a study, 99 patients who had an accidental dural puncture were compared to find the relation between their body mass indexes and the development of PDPH. Those patients with a body mass index (BMI) greater than $30 \, kg/m^2$ had a 25% incidence of PDPH. The patients with a BMI less than $30 \, kg/m^2$ had a 45% PDPH. It is accepted that increased epidural pressure during obesity decreases the leak from dural puncture cites and has been considered as a protective factor (Faure, Moreno, & Thisted, 1994).

Effects of application techniques of CSEA on headache

The CSE technique involves the intentional subarachnoid blockade and epidural catheter placement during the same procedure. Two different techniques are used to apply CSEA. The separate needle technique uses two different lumbar interspaces (e.g., epidural catheter placement at L1–2 followed by a subarachnoid block at L3–4). The needle-through-needle technique (NTN), also called as single space technique, consists of the insertion of an epidural needle into the lumbar epidural space, followed by the passage of the tip of a spinal needle through this epidural like an introducer. After spinal injection and the spinal needle's withdrawal, a catheter is threaded through the epidural needle for use after the spinal analgesia wears off. The NTN technique is the most widely reported CSE technique in the literature and is likely to be the most frequently used worldwide (Blanshard & Cook, 2004).

It is commonly accepted that the incidence of PDPH is related to the size of the dural lesion. Although not all patients who experienced a dural puncture develop a headache, after a puncture with an epidural needle, over 50% of patients will develop PDPH, whereas the risk of PDPH after dural perforation with a spinal needle varies between 1.5% and 11% depending on the type of needle that has been used (Choi et al., 2003). Angle et al. showed that epidural needle gauge is the most important factor associated with CSF leak. In this cadaveric study conducted in an in vitro model, with an 18-gauge Tuohy needle, the puncture angle between 30 and 90 degrees produced leak reductions, and the CSF leakage tended to be less with a cutting needle of the same gauge. But these results were not reached a statistical significance (Angle et al., 2003). A systematic review showed that parallel or longitudinal to the long axis of the spinal cord rather than a perpendicular insertion of a cutting or beveled needle would result in a statistically significant lower incidence of PDPH. The rationale or mechanisms for this finding is not clear. Still, it appears that the reduction of PDPH risk with the parallel insertion of the needle is similar to the reduced incidence seen with the use of pencil-point needles (Richman et al., 2006).

A high incidence of PDPH remains a major concern in the parturient, which may be as high as 10%–15% even when 25-to-26-gauge spinal needles are used (Cesarini, Torrielli, Lahaye, Mene, & Cabiro, 1990; Naulty et al., 1990). However, several researchers commented on the very low incidence or deficiency of PDPH following CSE block. Dennison reported only two PDPH in 400 patients (0.5%) who received CSE for cesarean delivery (Dennison, 1987). Kumar reported similar results in 300 patients (Kumar, 1987). Interestingly, in a CSE study where the single space NTN technique was compared with the separate spaces technique, only one PDPH was noted, and it occurred in the separate spaces group (Lyons, Macdonald, & Mikl, 1992). Brownridge (1991) had concerns for separate needle technique, that damage to the spinal needle tip might occur during its passage through the extradural needle. When this, combined with a directional change of the needle just before entering the dura, could traumatize the dura and increase the likelihood of spinal headache. He also had concerns about any delay which might occur because of difficulty in threading the extradural catheter. In his opinion, the patient needs undivided attention and reassurance since the onset of spinal anesthesia often leads to patient anxiety, and symptoms associated with hypotension need to be recognized and treated without delay. His preference was to have the epidural in place before administering the spinal anesthetic at a separate place.

Rawal et al. (1997) suggest that CSE's many factors reduce the likelihood of PDPH, presumably by minimizing CSF leak. The use of fine spinal needles (26–32 G); meticulous dural penetration because the Tuohy needle acts as an introducer; deflection of the spinal needle by the Tuohy needle tip leading to the reduced overlap of the dura mater and arachnoid mater; and the effect of the epidural opioid. A retrospective review of the use of the NTN technique for the CSEA in more than 6000 patients at Queen Charlotte's Hospital in London indicated a PDPH incidence of 0.13% in patients where the 27-gauge Whitacre spinal needle was passed no more than twice. These reports suggest that the risk of PDPH in parturients, particularly if spinal or epidural opioids are used as part of the technique, is very low or nonexistent (Cox et al., 1995). The Tuohy needle in the epidural space serves as an introducer and allows a meticulous puncture of the dura. Multiple attempts at identifying the subarachnoid space can thus be avoided. The technique allows the use of the finest diameter spinal needles. The risk of CSF leakage through the dura is decreased because of the increased pressure in the epidural space that results from the presence of an epidural catheter and the administration of epidural local anesthetics and opioids. This can be expected to splint the dura against the arachnoid membranes. With a single-segment CSE technique, the spinal needle is deflected somewhat as it exits through the Tuohy needle, approaching the dura at an angle. The holes in the dura and subarachnoid are less likely to overlap, reducing CSF leakage risk (Stacey, Watt, Kadim, & Morgan, 1993).

Norris et al. (1994) compared epidural block versus CSE block for labor analgesia. They concluded that intentional dural puncture with a small-gauge pencil-point needle during the CSE block induction does not increase the risk of postpartum headache when compared to epidural analgesia. Furthermore, patients who had epidural analgesia ($n = 388$) were significantly more likely to suffer an unintended dural puncture with the epidural needle than those who received CSE block ($n = 536$). Their two possible explanation for this unexpected finding was that the women they choose for CSE are in early labor. The women they most often chose for EA are well into the more painful active phase of labor (>5 cm cervical dilation). It's possible that these women more likely to move during induction of neuraxial anesthesia and, as a result of this, are more prone to dural puncture. The second explanation was while inducing CSEA; if one is uncertain of the epidural needle's location, one can insert the spinal needle and look for CSF. Obtaining CSF through the spinal needle is a sign for the anesthetist not to advance the epidural needle (Norris et al., 1994). Notwithstanding, obtaining CSF from the spinal needle inserted through the epidural needle may not be possible. Maneuvering the spinal or the epidural needle while attempting to puncture the dura may increase the risk of accidental dural puncture with the epidural needle. The intentional or UDP may increase the incidence of headaches (Norris et al., 1994). According to Balestrieri (2003), the risk of PDPH is highest in that subset of patients in whom placement of the epidural is technically difficult, either because of obesity, scoliosis, or for some other reason. In these patient groups, the CSE technique allows identifying the epidural space by the spinal needle placement with a return of CSF. While the lack of CSF return does not imply a failure to locate the epidural space, the presence of CSF return through the spinal needle is a reliable indicator of successful placement.

In a 2019 study by Doo et al., it was shown that the epidural needle's paramedian deviation might affect dural puncture during the NTN CSE technique and the wrong passage of the spinal needle through the Tuohy curve instead of the back hole may contribute to the failure of dural puncture (Doo et al., 2019).

Prophylaxis of PDPH after UDP in neuroaxial anesthesia

Neuraxial anesthesia is considered safe, but UDP is the most common complication. In the last three decades, research has focused on the treatment and prophylactic modalities of PDPH. A systematic review and meta-analyses of 2013 for determining whether the incidence of the accidental (unintentional) dural puncture in laboring women can be reduced included 54 articles, 13 nonrandomized controlled trials, and 41 randomized controlled trials (RCTs), reporting on a total of 98,869 patients. Their meta-analyses of non-RCTs suggested a lower risk for an UDP when using liquid versus air to identify the epidural space. In their analysis of RCTs, this comparison did not reveal a significant difference. CSE analgesia, patient position, type of the catheter, the Tuohy needle's gauge, use of ultrasound, and training grade had no effects on the incidence of UDP or PDPH, neither in non-RCTs nor in RCTs (Heesen, Klohr, Rossaint, Van De Velde, & Straube, 2013).

Today, there's no clear consensus exists on how to prevent patients from PDPH best. The most common action following UDP is still to place an epidural catheter (Baysinger, Pope, Lockhart, & Mercaldo, 2011). Another method of managing a PDPH is the intrathecal insertion of epidural catheters (Ayad, Demian, Narouze, & Tetzlaff, 2003). But the safety and efficacy of intrathecal catheters are still controversial (Lambert, 2004; Rosenblatt, Bernstein, & Beilin, 2004). Two meta-analyses showed reduced PDPH and EBP requirements but failed to show statistical significance after intrathecal epidural catheter placements (Apfel et al., 2010; Heesen et al., 2013).

Russell compared intrathecal catheter and epidural catheters after UDP with a 16 G or 18 G epidural needle in a randomized study of 115 women. The intrathecal catheters were placed in the same space, and the epidural catheters in another space and left there at least 24 h after delivery. His study did not show intrathecal catheters' superiority over epidural catheters in PDPH (intrathecal 72% vs epidural 62%). The result of Russel's study established that intrathecal catheter placement following UDP does not prevent headache; however, he showed a significantly greater requirement for two or more additional attempts to establish neuraxial analgesia associated with repeating the epidural and a 9% risk of second dural puncture. More than one-third of women suffered complications in the repeat epidural catheter group, a rate three times greater than the intrathecal catheter placement group (Russell, 2012).

Although RCTs did not reveal a significant difference, another much-debated but inconclusive issue in regional anesthesia among the anesthetists is the medium used in loss of resistance technique: air or saline. In a study by Aida, Taga, Yamakura, Endoh, and Shimoji (1998), the epidural block was performed with air-filled ($n = 1812$) or saline-filled ($n = 1918$) syringes in patients with acute or chronic pain. The study revealed no difference in the incidence of UDP (2.2% in both groups). But the incidence of headache was significantly higher in the air-filled syringe group (34% vs 10%). To prevent headaches, the smallest amount of air should be used to identify the epidural space. In another study carried out by Segal and Arendt (2010) included 929 parturients. The study results gave no difference among the mediums chosen for loss of resistance technique in determining epidural space.

Epidural blood patch (EBP) is an attractive option in the treatment of PDPH, especially in high-risk groups after UDP, even before PDPH symptoms occur. In these patients, recited epidural catheter in CSE may allow the use of an early EBP. But some observational studies support delaying therapeutic EPB for 24–48 h or more (Safa-Tisseront et al., 2001). This approach seems rationale symptoms may have begun to resolve a fibroblastic response at the site of the puncture present within 24 h. Anyway, and the effect of the local anesthetic dissipated. It's known that CSF is procoagulant, but local anesthetics interfere with coagulation by inhibiting platelet aggregation and granule release. Since the symptoms of PDPH occurs in 30%–40% of patients after dural puncture and the lack of data from randomized trials and facing a parturient with an incapacitating headache, Paech suggested that it is reasonable to consider conducting an early EBP, especially the PDPH is persistent after aggressive conservative therapy (Paech, 2005).

Management of PDPH

The basis of PDPH management is interfering with the CSF leakage. Headache management can be accomplished by replacing the lost CSF, closing the puncture, and controlling cerebral vasodilation. Once a diagnosis of PDPH has been reached, severity, onset, and duration assessment of PDPH should be done, and the available treatment modalities and their success rates should be discussed with the patients realistically. Reassuring in this stage seems useful since over 85% of the cases resolve within 6 weeks without any treatment. It is wise to adhere to institutional treatment protocols for PDPH if there exists one.

Supportive treatment

Although it is still popular, clinicians should not routinely recommend rest after a lumbar puncture to prevent PDPH. There was no evidence suggesting that routine bed rest after a dural puncture is beneficial for avoiding PDPH onset in a systemic review comparing bed rest and immediate mobilization after a postdural puncture (Arevalo-Rodriguez, Ciapponi, Roque i Figuls, Munoz, & Bonfill Cosp, 2016).

The recommendation of bed rest results in patient discomfort after particular operations such as Cesarean section and may lead to complications such as venous stasis in people with risk factors. The role of fluid supplementation as a prophylactic measure for PDPH remains unclear, and there is no point in administering fluids to the appropriately hydrated patients (Arevalo-Rodriguez et al., 2016). Abdominal binders may alleviate the PDPH by raising the intraabdominal pressure, which is transmitted to the epidural space. Today they are no longer used in clinical practice due to patient discomfort. Prone position has also been advocated as it raises the intra-abdominal pressure, but the difficulty in patients' compliance decreases its usage.

Medical management

Many drug therapies have proven to be effective in the treatment of PDPH. Acetaminophen and nonsteroid antiinflammatory drugs have been used extensively for initial symptomatic relief of PDPH. These drugs, along with opioids, combination drugs such as butalbital-acetaminophen-caffeine, caffeine-acetaminophen, and antiemetics, may help control the symptoms and reduce the need for more aggressive therapy; unfortunately, they may not provide complete relief.

Caffeine is a safe option and has shown effectiveness in treating PDPH. It can be considered for the treatment of PDPH in patients in which simple analgesics are ineffective and, EBP option is rejected by the patient or considered contraindicated. Caffeine is a CNS stimulant that acts by blockade of adenosine receptors leading to increased cerebral arterial vasoconstriction, which decreased cerebral blood flow. Caffeine also increases CSF production by activating sodium-potassium pumps (Baratloo et al., 2016). It was shown that intravenous caffeine sodium benzoate dosed at 500 mg in 1000 mL of normal saline during and after spinal anesthesia minimizes rates of PDPH in patients compared with patients who received solely 1000 mL of normal saline. A similar pain-reducing effect of caffeine has been seen in patients with PDPH in doses of 300 mg taken per oral when assessed with a pain scale before caffeine administration, 4 h after caffeine administration, and 24 h after caffeine administration.

In comprehensive meta-analyses (Basurto Ona et al., 2015), it was shown that there was an increase in the proportion of participants who reported improvement in theophylline pain scores compared to traditional treatment. Gabapentin, hydrocortisone, and theophylline have been shown to decrease pain severity scores. In this study, no conclusive results were reached for other drugs such as sumatriptan, adrenocorticotropic hormone (ACTH), pregabalin, and cosyntropin. As the authors have stated, these conclusions should be interpreted with care due to lack of information for an accurate appraisal of the risk of bias, the small sample sizes, and their limited generalizability since nearly half of the participants were postpartum women 30s.

Epidural blood patch

It has long been accepted that the gold standard in the treatment of PDPH is an EBP and has been seen to provide pain relief in around 61%–98% of cases (Kwak, 2017).

The exact mechanism by which the EBP relieves PDPH is still not precisely explained. The postulated mechanism for its effectiveness is that it initially increases pressure in the lumbar neuraxial canal and compressing the thecal sac. The increased pressure translocates to CSF from the spinal canal to the cranium (Beards, Jackson, Griffiths, & Horsman, 1993), which leads to immediate pain relief in most patients. The mass effect created by increased CSF volume reverses the mechanical traction over the pain-sensitive areas. Also, it leads to the reversal of temporary central venous dilation that contributes to pain (Gaiser, 2017; Kroin et al., 2002). Another theory is that it may plug the puncture's dural leak and rapid healing when the injected blood clots.

The technique was first introduced by Gormley (1960), who observed that "bloody taps were associated with reduced a reduced headache rate and postulated that blood might act as a sealing material for dural puncture." The technique consists of injecting a small amount of autologous blood into the epidural space to stop the leak of CSF fluid. The immediate pain relief after injection is typical. The time of EBP and the amount of blood to be given has long been controversial issues. Since most PDPH is self-resolving, EBP can be reserved for patients who have resistant headaches to conservative treatments for 10 days. The expedited symptom resolution usually necessitates aggressive management and sooner EBPs in today's clinical practice. There are several contraindications for the procedure, like infection at the injection site, coagulopathies, and anticoagulation. The most common complication of EBP is back pain, minor and self-limiting at the injection site. The associated rare complications are meningitis, arachnoiditis, epidural or subdural hematomas, abscesses, facial nerve palsy, and cauda equina syndrome (Booth, Pan, Thomas, Harris, & D'Angelo, 2017). In a randomized, blinded clinical study, it was found that injecting 15, 20, and 30 mL amount of autologous blood revealed 61%, 73%, and 67% of partial headache relief, respectively. It was concluded from this study that 20 mL is optimal for therapeutic EBP. The EBP injection within 48 h of dural puncture was less effective than EBP done later (Paech, Doherty, Christmas, Wong, & Epidural Blood Patch Trial Group, 2011).

Alternative therapies

Greater occipital nerve block (GONB) and bilateral trans nasal sphenopalatine ganglion block (SPGB) have been offered as alternative treatment modalities for PDPH. In GONB greater occipital nerve was blocked using local anesthetic and dexamethasone, and pain relief was achieved by interruption of pain signals to the trigeminal and greater occipital nerves (Niraj, Kelkar, & Girotra, 2014). It was shown in a prospective audit that complete or partial resolution of PDPH had been seen in 67% of patients. SPGB involves a cotton tip applicator saturated with a local anesthetic, applied to the posterior pharynx via the naris where the sphenopalatine ganglion is located (Cohen et al., 2018). A retrospective review showed that the women with PDPH treated with SPGB had a rapid onset of pain relief and a lower complication rate than those who received EBP.

Conclusion

CSEA is a neuraxial anesthesia technique that provides fast, deep, and reliable anesthesia with its spinal component and has the ability to expand anesthesia and analgesia through an epidural catheter. Besides, CSEA, like other neuraxial techniques, can cause headaches after dural puncture. The incidence of PDPH is related to the size of the dural lesion. Headache management can be accomplished by replacing the lost CSF, closing the puncture, and controlling cerebral vasodilation. EBP provides pain relief for PDPH by initially increasing pressure in the lumbar neuraxial canal and compressing the thecal sac. The mass effect created by increased CSF volume reverses the mechanical traction over the pain-sensitive areas. Further studies are needed to determine the true incidence of spinal headache after a combined technique; if the incidence of headache is confirmed to be very low, the combined technique may be better than either spinal or EDA alone.

Application to other areas

PDPH, which can develop after CSEA, can also be seen after other therapeutic and diagnostic applications such as spinal anesthesia and lumbar puncture. The basis of PDPH management is interfering with the CSF leakage.

Other agents of interest

Other neuraxial anesthesia techniques other than combine spinal-epidural anesthesia

- *Epidural anesthesia:* A catheter is not required and can be used for intraoperative anesthesia and postoperative analgesia with local anesthetics administered a single bolus dose from the needle.
- *Spinal anesthesia:* It provides the rapid onset of more intense anesthesia using single doses of local anesthetics and opioids compared to EDA.

Mini-dictionary of terms

Catheter: A long, very thin tube used to inject fluid to the epidural space.

Neuraxial anesthesia: The administration of medication into the subarachnoid or epidural space around the nerves of the central nervous system to produce anesthesia and analgesia, such as spinal anesthesia, caudal anesthesia, and EDA.

Tuohy: A hollow hypodermic needle, very slightly curved at the end, suitable for inserting epidural catheters.

Tinnitus: The hearing ringing, buzzing, or other sounds without an external cause.

Hyperacusis: Heightened sensitivity to sound, with aversive or pained reactions to normal environmental sounds.

Photophobia: Abnormal visual intolerance to light.

Epidural blood patch: Application of an epidural blood patch using 20 mL of freshly drawn blood.

Cosyntropin: A synthetic derivative of adrenocorticotropic hormone (ACTH).

Key facts

Key facts of headache after neuraxial blocks

- Continuous spinal epidural block for anesthesia and pain management provides rapid onset of motor and sensory block. These effects can be prolonged by administering an intermittent or continuous infusion of local anesthetic with an epidural catheter.
- PDPH is a well-recognized complication of spinal anesthesia.
- The incidence of PDPH is related to the size of the dural lesion.
- Postdural puncture's typical headache is often diffuse or dull and worsens within 15 min after sitting or standing up. Stiffness of the neck, nausea, tinnitus, hyperacusis, and photophobia generally accompanies.
- Young age and female gender increase the risk of PDPH.
- There is no difference among the mediums chosen for the loss of resistance technique in determining epidural space.
- Headache management can be accomplished by replacing the lost CSF, closing the puncture, and controlling cerebral vasodilation.

Summary points

- This chapter reviews the PDPH is one of the most common complications of neuraxial anesthesia.
- The postulated cause is the decreased CSF pressure due to the loss of CSF in the epidural space through the dural puncture site. The decrease in the CSF pressure creates a loss of the cushioning effect provided by the intracranial fluid.
- Acetaminophen and nonsteroid antiinflammatory drugs, along with opioids, combination drugs such as butalbital-acetaminophen-caffeine, caffeine-acetaminophen, and antiemetics, may help control the symptoms and reduce the need for more aggressive therapy; unfortunately, they may not provide complete relief.
- Caffeine is a safe option and has shown effectiveness in treating PDPH.
- EBP is an attractive option in the treatment of PDPH, especially in high-risk groups after UDP, even before PDPH symptoms occur.
- CSE techniques have been reported to be associated with a low incidence of PDPHs and as a result of a low incidence of EBPs.

References

Aida, S., Taga, K., Yamakura, T., Endoh, H., & Shimoji, K. (1998). Headache after attempted epidural block: The role of intrathecal air. *Anesthesiology, 88* (1), 76–81. https://doi.org/10.1097/00000542-199801000-00014.

Amorim, J. A., Gomes de Barros, M. V., & Valenca, M. M. (2012). Post-dural (post-lumbar) puncture headache: Risk factors and clinical features. *Cephalalgia, 32*(12), 916–923. https://doi.org/10.1177/0333102412453951.

Angle, P. J., Kronberg, J. E., Thompson, D. E., Ackerley, C., Szalai, J. P., Duffin, J., et al. (2003). Dural tissue trauma and cerebrospinal fluid leak after epidural needle puncture: Effect of needle design, angle, and bevel orientation. *Anesthesiology, 99*(6), 1376–1382. https://doi.org/10.1097/00000542-200312000-00021.

Apfel, C. C., Saxena, A., Cakmakkaya, O. S., Gaiser, R., George, E., & Radke, O. (2010). Prevention of postdural puncture headache after accidental dural puncture: A quantitative systematic review. *British Journal of Anaesthesia, 105*(3), 255–263. https://doi.org/10.1093/bja/aeq191.

Arevalo-Rodriguez, I., Ciapponi, A., Roque i Figuls, M., Munoz, L., & Bonfill Cosp, X. (2016). Posture and fluids for preventing post-dural puncture headache. *Cochrane Database of Systematic Reviews, 3*. https://doi.org/10.1002/14651858.CD009199.pub3, CD009199.

Ayad, S., Demian, Y., Narouze, S. N., & Tetzlaff, J. E. (2003). Subarachnoid catheter placement after wet tap for analgesia in labor: Influence on the risk of headache in obstetric patients. *Regional Anesthesia and Pain Medicine, 28*(6), 512–515. https://doi.org/10.1016/s1098-7339(03)00393-6.

Balestrieri, P. J. (2003). The incidence of postdural puncture headache and combined spinal-epidural: Some thoughts. *International Journal of Obstetric Anesthesia, 12*(4), 305–306. https://doi.org/10.1016/S0959-289X(03)00067-0.

Baratloo, A., Rouhipour, A., Forouzanfar, M. M., Safari, S., Amiri, M., & Negida, A. (2016). The role of caffeine in pain management: A brief literature review. *Anesthesia and Pain Medicine, 6*(3). https://doi.org/10.5812/aapm.33193, e33193.

Basurto Ona, X., Osorio, D., & Bonfill Cosp, X. (2015). Drug therapy for treating post-dural puncture headache. *Cochrane Database of Systematic Reviews, 7*. https://doi.org/10.1002/14651858.CD007887.pub3, CD007887.

Baysinger, C. L., Pope, J. E., Lockhart, E. M., & Mercaldo, N. D. (2011). The management of accidental dural puncture and postdural puncture headache: A North American survey. *Journal of Clinical Anesthesia, 23*(5), 349–360. https://doi.org/10.1016/j.jclinane.2011.04.003.

Beards, S. C., Jackson, A., Griffiths, A. G., & Horsman, E. L. (1993). Magnetic resonance imaging of extradural blood patches: Appearances from 30 min to 18 h. *British Journal of Anaesthesia, 71*(2), 182–188. https://doi.org/10.1093/bja/71.2.182.

Bier, A. (1899). Versuche über Cocainisirung des Rückenmarkes. *Deutsche Zeitschrift für Chirurgie, 51*(3), 361–369. https://doi.org/10.1007/BF02792160.

Blanshard, H. J., & Cook, T. M. (2004). Use of combined spinal-epidural by obstetric anaesthetists. *Anaesthesia, 59*(9), 922–923. https://doi.org/10.1111/j.1365-2044.2004.03918.x.

Booth, J. L., Pan, P. H., Thomas, J. A., Harris, L. C., & D'Angelo, R. (2017). A retrospective review of an epidural blood patch database: The incidence of epidural blood patch associated with obstetric neuraxial anesthetic techniques and the effect of blood volume on efficacy. *International Journal of Obstetric Anesthesia, 29*, 10–17. https://doi.org/10.1016/j.ijoa.2016.05.007.

Brownridge, P. (1991). Spinal anaesthesia in obstetrics. *British Journal of Anaesthesia, 67*(5), 663. https://doi.org/10.1093/bja/67.5.663.

Cesarini, M., Torrielli, R., Lahaye, F., Mene, J. M., & Cabiro, C. (1990). Sprotte needle for intrathecal anaesthesia for caesarean section: Incidence of postdural puncture headache. *Anaesthesia, 45*(8), 656–658. https://doi.org/10.1111/j.1365-2044.1990.tb14392.x.

Choi, D. H., Ahn, H. J., & Kim, J. A. (2006). Combined low-dose spinal-epidural anesthesia versus single-shot spinal anesthesia for elective cesarean delivery. *International Journal of Obstetric Anesthesia, 15*(1), 13–17. https://doi.org/10.1016/j.ijoa.2005.05.009.

Choi, P. T., Galinski, S. E., Takeuchi, L., Lucas, S., Tamayo, C., & Jadad, A. R. (2003). PDPH is a common complication of neuraxial blockade in parturients: A meta-analysis of obstetrical studies. *Canadian Journal of Anaesthesia, 50*(5), 460–469. https://doi.org/10.1007/BF03021057.

Cohen, S., Levin, D., Mellender, S., Zhao, R., Patel, P., Grubb, W., et al. (2018). Topical sphenopalatine ganglion block compared with epidural blood patch for postdural puncture headache management in postpartum patients: A retrospective review. *Regional Anesthesia and Pain Medicine, 43*(8), 880–884. https://doi.org/10.1097/AAP.0000000000000840.

Cook, T. M. (2000). Combined spinal-epidural techniques. *Anaesthesia, 55*(1), 42–64. https://doi.org/10.1046/j.1365-2044.2000.01157.x.

Cox, M., Lawton, G., Gowrie-Mohan, S., Priest, T., Arnold, A., & Morgan, B. (1995). Ambulatory extradural analgesia. *British Journal of Anaesthesia, 75*(1), 114–115. https://doi.org/10.1093/bja/75.1.114.

Davies, J. M., Posner, K. L., Lee, L. A., Cheney, F. W., & Domino, K. B. (2009). Liability associated with obstetric anesthesia: A closed claims analysis. *Anesthesiology, 110*(1), 131–139. https://doi.org/10.1097/ALN.0b013e318190e16a.

Dennison, B. (1987). Combined subarachnoid and epidural block for caesarean section. *Canadian Journal of Anaesthesia, 34*(1), 105–106. https://doi.org/10.1007/BF03007703.

Doo, A. R., Shin, Y. S., Choi, J. W., Yoo, S., Kang, S., & Son, J. S. (2019). Failed dural puncture during needle-through-needle combined spinal-epidural anesthesia: A case series. *Journal of Pain Research, 12*, 1615–1619. https://doi.org/10.2147/JPR.S178640.

Faure, E., Moreno, R., & Thisted, R. (1994). Incidence of postdural puncture headache in morbidly obese parturients. *Regional Anesthesia, 19*(5), 361–363. https://www.ncbi.nlm.nih.gov/pubmed/7848940.

Gaiser, R. R. (2013). Postdural puncture headache: A headache for the patient and a headache for the anesthesiologist. *Current Opinion in Anaesthesiology, 26*(3), 296–303. https://doi.org/10.1097/ACO.0b013e328360b015.

Gaiser, R. R. (2017). Postdural puncture headache: An evidence-based approach. *Anesthesiology Clinics, 35*(1), 157–167. https://doi.org/10.1016/j.anclin.2016.09.013.

Ghazi, A., & Raja, Y. (2007). Combined low-dose spinal-epidural anaesthesia versus single-shot spinal anaesthesia for elective caesarean delivery. *International Journal of Obstetric Anesthesia, 16*(1), 90–91. https://doi.org/10.1016/j.ijoa.2006.07.003.

Gormley, J. B. (1960). Treatment of postspinal headache. *Anesthesiology, 21*, 565–566.

Grant, R., Condon, B., Hart, I., & Teasdale, G. M. (1991). Changes in intracranial CSF volume after lumbar puncture and their relationship to post-LP headache. *Journal of Neurology, Neurosurgery, and Psychiatry, 54*(5), 440–442. https://doi.org/10.1136/jnnp.54.5.440.

(2018). Headache Classification Committee of the International Headache Society (IHS) The international classification of headache disorders, 3rd edition. *Cephalalgia, 38*(1), 1–211. https://doi.org/10.1177/0333102417738202.

Heesen, M., Klohr, S., Rossaint, R., Van De Velde, M., & Straube, S. (2013). Can the incidence of accidental dural puncture in laboring women be reduced? A systematic review and meta-analysis. *Minerva Anestesiologica, 79*(10), 1187–1197. https://www.ncbi.nlm.nih.gov/pubmed/23857441.

Hoskin, M. F. (1998). Spinal anaesthesia—The current trend towards narrow gauge atraumatic (pencil point) needles. Case reports and review. *Anaesthesia and Intensive Care, 26*(1), 96–106. https://doi.org/10.1177/0310057X9802600115.

Kroin, J. S., Nagalla, S. K., Buvanendran, A., McCarthy, R. J., Tuman, K. J., & Ivankovich, A. D. (2002). The mechanisms of intracranial pressure modulation by epidural blood and other injectates in a postdural puncture rat model. *Anesthesia and Analgesia, 95*(2), 423–429. table of contents https://doi.org/10.1097/00000539-200208000-00035.

Kuczkowski, K. M. (2004). Post-dural puncture headache in the obstetric patient: An old problem. New solutions. *Minerva Anestesiologica, 70*(12), 823–830. https://www.ncbi.nlm.nih.gov/pubmed/15702063.

Kumar, C. M. (1987). Combined subarachnoid and epidural block for caesarean section. *Canadian Journal of Anaesthesia, 34*(3 (Pt 1)), 329–330. https://doi.org/10.1007/BF03015181.

Kwak, K. H. (2017). Postdural puncture headache. *Korean Journal of Anesthesiology, 70*(2), 136–143. https://doi.org/10.4097/kjae.2017.70.2.136.

Lambert, D. (2004). Cautious optimism on reducing spinal headache with spinal catheters. *Regional Anesthesia and Pain Medicine, 29*(3), 299. author reply 299–301 https://doi.org/10.1016/j.rapm.2003.12.022.

Lee, J. A. (1979). Arthur Edward James Barker 1850–1916. British poineer of regional analgesia. *Anaesthesia, 34*(9), 885–891. https://doi.org/10.1111/j.1365-2044.1979.tb08541.x.

Liu, H., Kalarickal, P. L., & Rosinia, F. (2012). A case of paradoxical presentation of postural postdural puncture headache. *Journal of Clinical Anesthesia, 24*(3), 255–256. https://doi.org/10.1016/j.jclinane.2011.04.020.

Lyons, G., Macdonald, R., & Mikl, B. (1992). Combined epidural/spinal anaesthesia for caesarean section. Through the needle or in separate spaces? *Anaesthesia, 47*(3), 199–201. https://doi.org/10.1111/j.1365-2044.1992.tb02117.x.

MacArthur, C., Lewis, M., & Knox, E. G. (1993). Accidental dural puncture in obstetric patients and long term symptoms. *BMJ, 306*(6882), 883–885. https://doi.org/10.1136/bmj.306.6882.883.

Morewood, G. H. (1993). A rational approach to the cause, prevention and treatment of postdural puncture headache. *CMAJ, 149*(8), 1087–1093. https://www.ncbi.nlm.nih.gov/pubmed/8221447.

Naulty, J. S., Hertwig, L., Hunt, C. O., Datta, S., Ostheimer, G. W., & Weiss, J. B. (1990). Influence of local anesthetic solution on postdural puncture headache. *Anesthesiology, 72*(3), 450–454. https://doi.org/10.1097/00000542-199003000-00010.

Niraj, G., Kelkar, A., & Girotra, V. (2014). Greater occipital nerve block for postdural puncture headache (PDPH): A prospective audit of a modified guideline for the management of PDPH and review of the literature. *Journal of Clinical Anesthesia, 26*(7), 539–544. https://doi.org/10.1016/j.jclinane.2014.03.006.

Norris, M. C., Grieco, W. M., Borkowski, M., Leighton, B. L., Arkoosh, V. A., Huffnagle, H. J., et al. (1994). Complications of labor analgesia: Epidural versus combined spinal epidural techniques. *Anesthesia and Analgesia, 79*(3), 529–537. https://doi.org/10.1213/00000539-199409000-00022.

Paech, M. (2005). Epidural blood patch—Myths and legends. *Canadian Journal of Anesthesia, 52*(1), R47. https://doi.org/10.1007/BF03023087.

Paech, M., Banks, S., & Gurrin, L. (2001). An audit of accidental dural puncture during epidural insertion of a Tuohy needle in obstetric patients. *International Journal of Obstetric Anesthesia, 10*(3), 162–167. https://doi.org/10.1054/ijoa.2000.0825.

Paech, M. J., Doherty, D. A., Christmas, T., Wong, C. A., & Epidural Blood Patch Trial Group. (2011). The volume of blood for epidural blood patch in obstetrics: A randomized, blinded clinical trial. *Anesthesia and Analgesia, 113*(1), 126–133. https://doi.org/10.1213/ANE.0b013e318218204d.

Rawal, N., Van Zundert, A., Holmstrom, B., & Crowhurst, J. A. (1997). Combined spinal-epidural technique. *Regional Anesthesia, 22*(5), 406–423. https://doi.org/10.1016/s1098-7339(97)80026-0.

Richman, J. M., Joe, E. M., Cohen, S. R., Rowlingson, A. J., Michaels, R. K., Jeffries, M. A., et al. (2006). Bevel direction and postdural puncture headache: A meta-analysis. *The Neurologist, 12*(4), 224–228. https://doi.org/10.1097/01.nrl.0000219638.81115.c4.

Rosenblatt, M. A., Bernstein, H. H., & Beilin, Y. (2004). Are subarachnoid catheters really safe? *Regional Anesthesia and Pain Medicine, 29*(3), 298. author reply 299–301 https://doi.org/10.1016/j.rapm.2003.12.009.

Russell, I. F. (2012). A prospective controlled study of continuous spinal analgesia versus repeat epidural analgesia after accidental dural puncture in labour. *International Journal of Obstetric Anesthesia, 21*(1), 7–16. https://doi.org/10.1016/j.ijoa.2011.10.005.

Safa-Tisseront, V., Thormann, F., Malassine, P., Henry, M., Riou, B., Coriat, P., et al. (2001). Effectiveness of epidural blood patch in the management of post-dural puncture headache. *Anesthesiology, 95*(2), 334–339. https://doi.org/10.1097/00000542-200108000-00012.

Segal, S., & Arendt, K. W. (2010). A retrospective effectiveness study of loss of resistance to air or saline for identification of the epidural space. *Anesthesia and Analgesia, 110*(2), 558–563. https://doi.org/10.1213/ANE.0b013e3181c84e4e.

Stacey, R. G., Watt, S., Kadim, M. Y., & Morgan, B. M. (1993). Single space combined spinal-extradural technique for analgesia in labour. *British Journal of Anaesthesia, 71*(4), 499–502. https://doi.org/10.1093/bja/71.4.499.

van de Velde, M., Teunkens, A., Hanssens, M., van Assche, F. A., & Vandermeersch, E. (2001). Post dural puncture headache following combined spinal epidural or epidural anaesthesia in obstetric patients. *Anaesthesia and Intensive Care, 29*(6), 595–599. https://doi.org/10.1177/0310057X0102900605.

Webb, C. A., Weyker, P. D., Zhang, L., Stanley, S., Coyle, D. T., Tang, T., et al. (2012). Unintentional dural puncture with a Tuohy needle increases risk of chronic headache. *Anesthesia and Analgesia, 115*(1), 124–132. https://doi.org/10.1213/ANE.0b013e3182501c06.

Wu, C. L., Rowlingson, A. J., Cohen, S. R., Michaels, R. K., Courpas, G. E., Joe, E. M., et al. (2006). Gender and post-dural puncture headache. *Anesthesiology, 105*(3), 613–618. https://doi.org/10.1097/00000542-200609000-00027.

Chapter 27

Liposomal bupivacaine, pain relief and adverse events

Hüseyin Oğuz Yılmaz[a] and Alparslan Turan[b]

[a]Department of Critical Care, Muğla Education and Research Hospital, Muğla, Turkey; [b]Department of Outcomes Research, Anesthesiology Institute, Cleveland Clinic, Cleveland, OH, United States

Abbreviations

Bup.HCl	bupivacaine hydrochloride
C_{max}	maximum observed plasma concentration
ECG	electrocardiogram
FDA	United States Food and Drug Administration
LB	liposomal bupivacaine
MLV	multilamellar vesicle
MVL	multivesicular liposome
$\mathbf{p}K_a$	negative base-10 logarithm of the acid dissociation constant (K_a)
SD	standard deviation
T_{max}	time to attain C_{max}
ULV	unilamellar vesicle

Introduction

Optimal pain control during and after surgery increases quality of life and satisfaction of patients, improves clinical outcomes, fastens patient discharge and healing, decreases postoperative chronic pain states, and ultimately decrease the economic burden (Gan, 2017). Opioids have been the main drug of choice for controlling postoperative pain with worrisome side effects such as respiratory depression, nausea and vomiting, potential of addiction etc. Multimodal analgesia was introduced to decrease the need for opioids by inhibiting the nociceptive stimulus in multiple points of the nociception pathway starting from the site of injury to the central nervous system and it is widely utilized for management of postoperative pain. This approach includes combinations of local anesthetics, acetaminophen, nonsteroidal antiinflammatory drugs, steroids, gabapentinoids, and opioids (Beverly, Kaye, Ljungqvist, & Urman, 2017). In order to decrease the need for opioids, perioperative administration of local anesthetics as wound infiltration or nerve blocks are commonly used as a part of multimodal analgesia regimen (Chou et al., 2016).

Bupivacaine hydrochloride is the most common local anesthetic used in the perioperative period. However, when used as a single injection, duration of action is relatively short (2–8 h) and does not cover the whole length of postoperative phase. Additives such as epinephrine, clonidine, dexmedetomidine, dexamethasone may prolong the duration of action but still does not meet the need (Biyani, Chhabra, Baidya, & Anand, 2014). Postoperative repeated infiltration of the local anesthetic is impractical. Continuous infusion of bupivacaine via catheters and management of therapy requires significant expertise, time, and effort (Balocco, Van Zundert, Gan, Gan, & Hadzic, 2018). Continuous infusion can also cause various complications such as catheter dislocation, intravascular migration, infection, and decrease the mobility of the patients (Heimburger, McClave, Gramlich, & Merritt, 2010). Additionally, high drug plasma concentrations related to high doses, rapid systemic absorption or intravascular injection of bupivacaine can cause systemic toxicity.

In search of relatively longer acting, easy to use, uncomplicated and safe local anesthetic for regional anesthesia/analgesia, liposomal local anesthetic formulations were developed. Sustained-release multivesicular liposomal (MVL)

Treatments, Mechanisms, and Adverse Reactions of Anesthetics and Analgesics. https://doi.org/10.1016/B978-0-12-820237-1.00027-2
Copyright © 2022 Elsevier Inc. All rights reserved.

bupivacaine formulation was developed and approved by the US Food and Drug Administration (FDA) in 2011 for postoperative analgesia (Golf, Daniels, & Onel, 2011; Kim, Kim, & Murdanhe, 2012). This novel drug structure continuously releases the active molecule (bupivacaine HCL) over 72 h after a single injection and promise a longer duration of analgesia and less opioid use after surgery. Encapsulation of bupivacaine into a liposomal carrier is expected to provide a safe and reliable method to extend duration of activity and improve postoperative pain management.

However, despite the promising results of primary studies and theoretical advantages, extended duration of activity with the use of liposomal bupivacaine (LB) in the perioperative period conclude sparse results. While safety profile is still promising, superior efficacy compared to plain bupivacaine is unclear and requires more trials. This chapter summarizes the structure, pharmacology, and safety of the formulation, along with its efficacy in various clinical settings and methods.

Liposomes and DepoFoam

Liposomes are artificial nanostructures consisting of an aqueous core surrounded by a lipid bilayer. They were discovered in 1960s (Bangham, 1989; Bangham, Standish, & Watkins, 1965) and became ideal carriers for drugs to improve their absorption, metabolism, half-life, and toxicity characteristics. Liposomes are biocompatible and biodegradable. They bind to the active drug molecule into or on to their membrane and are expected to deliver the drug without rapid degradation to minimize side effects. Only a fraction of the active component is bioavailable at any time, thus they can deliver drugs to specific targets in high concentrations without systemic drug toxicity (Allen & Cullis, 2013; Bozzuto & Molinari, 2015; Çağdaş, Sezer, & Bucak, 2014). In vivo behavior of the liposomes is affected by the liposome size, structure, and composition. Liposomes smaller than 120 nm are rapidly absorbed by the tissue and distributed systematically, whereas larger ones stays in injection site for a longer period of time (Grant & Bansinath, 2001).

Liposome structure may be unilamellar, multilamellar, or multivesicular. Distinct from conventional liposomes, multivesicular liposomes (MVLs), contain multiple nonconcentric aqueous chambers (vesicles), and they are much larger in size. The basic morphological differences between unilamellar, multilamellar, and MVLs are illustrated in Fig. 1 (Ye, Asherman, Stevenson, Brownson, & Katre, 2000).

MVLs were first reported by Kim et al. (Kim, Turker, Chi, Sela, & Martin, 1983). MVLs are made of phospholipids, cholesterol, and triglycerides, and the composition of these liposomes must contain at least one amphipathic lipid and one neutral lipid. Under the light microscope, MVLs are spheroids with granular internal structure. Somewhat analogous to embryonic blastulae, the walls between internal aqueous compartments are single bilayers rather than two bilayer in close contact (Kim et al., 1983). In detail, there is an outermost bilayer membrane and internal structure divided into numerous compartments with bilayer septa. Like bubbles in gas-liquid foam, these compartments (vesicles) are polyhedral in shape. Vesicles are coordinated with each other in tetrahedral junctions and triolein molecules stabilizes these junctions (Mantripragada, 2002). MVL technology was named as DepoFoam by Pacira Pharmaceuticals in San Diego, CA, United States.

The soap bubble like, tightly packed, nonconcentric nature of MVLs, with polyhedral vesicles and tetrahedral junctions, increases stability of the structure and extends drug release. Active molecule is released from the MVLs by three mechanisms: membrane breakdown, membrane reorganization, and diffusion. During the release of the drug, diameter of the internal vesicles increases over time by coalescence. While a breach in membranes of the outermost vesicles results in drug delivery to the tissue, the breaches in the membranes of internal vesicles results drug redistribution within the MVL, by internal fusion and division, rather than drug release to medium (Mantripragada, 2002). The particles can be much larger and allow larger drug volumes to be trapped within the particle in a small volume of injection further extending the drug release to several weeks.

Liposomal bupivacaine

Encapsulating LAs in liposomes was first proposed in 1979 (Grant & Bansinath, 2001). Currently only available sustained release bupivacaine formulation on the market is EXPAREL, a DepoFoam technology used to encapsulate bupivacaine by Pacira Pharmaceuticals Inc. The active component of the sustained-release drug formulation "bupivacaine" is an amide-type, highly lipid soluble, long-acting local anesthetic. The concentration of bupivacaine in the solution is 13.3 mg/mL. Distinct from plain solution (bupivacaine hydrochloride), bupivacaine is encapsulated in DepoFoam as phosphate salt and concentration is expressed as a free base. Therefore, MVL bupivacaine and bupivacaine HCL are not bioequivalent even in same mg doses and not interchangeable. This should be taken into consideration when comparing different formulations. For example, 266 mg of bupivacaine base in DepoFoam is chemically equivalent to 300 mg of bupivacaine HCl (Bergese, Ramamoorthy, et al., 2012).

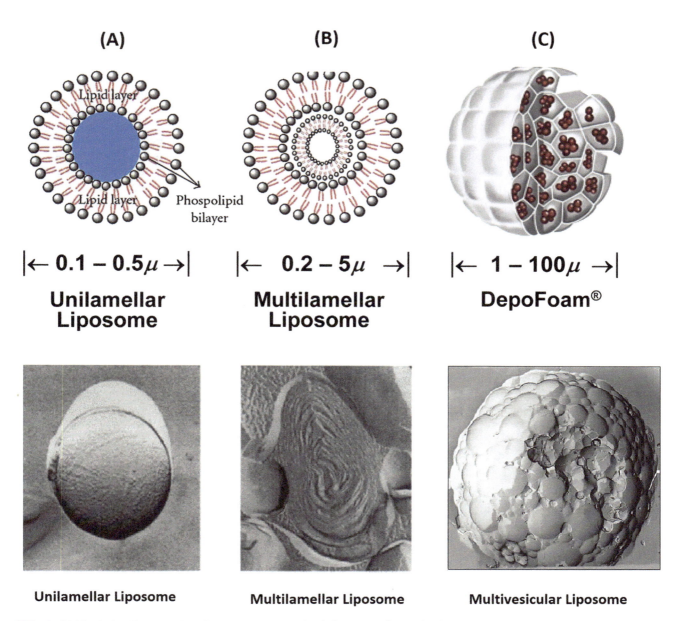

FIG. 1 Multivesicular liposome (Depofoam) versus conventional liposomes. Schematic diagram and electron microscope photographs of (A) unilamellar liposomes (unilamellar vesicles or ULV) consist of a single concentric lipid bilayer enclosing an aqueous volume; (B) multilamellar liposomes (multilamellar vesicles or "MLV") have numerous concentric membranes; (C) multivesicular liposome (MVL) (DepoFoam) contains multiple nonconcentric internal aqueous compartments and typically have a larger diameter of 1–100 mm. These electron micrographs show that DepoFoam particles are structurally distinct from MLV and ULV. *(From Ye, Q., Asherman, J., Stevenson, M., Brownson, E., & Katre, N. V. (2000). DepoFoam™ technology: A vehicle for controlled delivery of protein and peptide drugs.* Journal of Controlled Release, *64(1–3), 155–166. doi:https://doi.org/10.1016/S0168-3659(99)00146-7, with permission.)*

The size of the liposome particles in the formulation ranges from 24 to 31 μm. They are suspended in 0.9% sodium chloride solution. Vials do not contain any preservative solution. Inactive ingredients, which constitutes the structure of the liposomes are cholesterol, 4.7 mg/mL; 1, 2-dipalmitoyl-*sn*-glycero-3 phospho-rac-(1-glycerol), 0.9 mg/mL; tricaprylin, 2.0 mg/mL; and 1, 2-dierucoylphosphatidylcholine, 8.2 mg/mL. The pH of the solution is in the range of 5.8–7.4 (Pacira Pharmaceuricals Inc, 2018).

Clinical pharmacology

Release characteristics and stability

Release characteristics of the multivesicular LB formulation play a pivotal role on its suggested effect. In vitro and in vivo release characteristics of microparticles can be assesses by microdialysis methods (Manna et al., 2019; McAlvin et al., 2014). In a micro dialysis model, only 10% of the bupivacaine in MVLs was released at 48 h. At the same time 99% of the plain bupivacaine HCL (generic) formulation was released. Release of bupivacaine from liposomal formulation was completed in 800 h (Fig. 2A) (McAlvin et al., 2014). Higher temperature, higher agitation speed, and higher concentrations of albumin/plasma proteins increase the release rate. In opposite, extreme alterations from drug pK_a (8.1) decreases the release rate of bupivacaine from MVL (Manna et al., 2019). There are three phases in drug release (Fig. 2B). (1) Initial burst, results from the readily available dissolved bupivacaine in the solution or rapidly dissolved from liposome surface. Approximately, 3% of the bupivacaine in the solution is in free (nonencapsulated) form; (2) lag phase represents slow diffusion of bupivacaine through layers of the lipid membranes. Gradual coalescence of lipid vesicles are important and determines the speed of this phase; (3) secondary phase, is related with the erosion of MVL (Manna et al., 2019). This release behavior is observed and confirmed in clinical studies as a bimodal distribution pattern with a first peak in the 1st hour and a second peak between 12 and 36 h (Hu, Onel, Singla, Kramer, & Hadzic, 2013) (Fig. 2C).

Pharmacodynamics and mechanism of action

After release from the MVL, bupivacaine reversibly binds to the intracellular portion of voltage-dependent sodium channels, prevents the influx of sodium, increases the excitation threshold of the neurons, inhibits the depolarization of nerve membrane, and thereby inhibits the conduction of the stimuli. The degree of nerve blockage depends on the concentration and volume of bupivacaine administered.

The onset of action of LB is similar to plain bupivacaine HCL. In healthy volunteers, subcutaneous infiltration of LB bupivacaine causes substantial analgesia in 5 min (Apseloff, Onel, & Patou, 2013). The onset of action for effective analgesia in the clinical setting is longer. For example, after brachial plexus block with 133 mg LB, patients lose sensitivity to cold, pinprick, and light touch within 30 min (Patel et al., 2020). It can be mixed with plain bupivacaine HCl to enhance faster onset of action (Gadsden & Long, 2016). Possibly due to lower concentrations of bupivacaine in the infiltration site, the duration of motor blockage with LB is shorter compared to plain bupivacaine (Viscusi, Candiotti, Onel, Morren, & Ludbrook, 2012).

Pharmacokinetics

Pharmacokinetic characteristics of MVL bupivacaine in various clinical scenarios is well defined (Hu et al., 2013). Pharmacokinetic properties from various experiments are summarized in Table 1 (Bramlett et al., 2012; Golf et al., 2011; Gorfine et al., 2011; Viscusi et al., 2012).

Systemic absorption of bupivacaine following its release from MVL bupivacaine mimics plain bupivacaine HCL. The absorption rate is highest in intercostal nerve blocks followed by caudal/epidural injections > brachial plexus block > femoral and sciatic nerve blocks. The addition of vasoconstrictors such as epinephrine decreases local blood flow and slows the systemic absorption and blood concentration of local anesthetics. This effect is more pronounced for short-acting agents; therefore it is less likely to change absorption and duration of action for sustained release formulations such as LB (Brummett & Williams, 2011).

Plasma levels of bupivacaine follow a bimodal distribution after injection of LB. Systemic absorption of the readily available free dose (3%) demonstrates a first-order short-term release pattern and constitutes the first peak within 1 h. Thereafter, zero-order prolonged-release maintains the drug concentration, provides prolonged analgesia and the second peak occurs between 12 and 36 h depending on the delivered dose and site of administration. Plasma levels of bupivacaine can be significantly high even after 96 h (Fig. 2C) (Hu et al., 2013; Viscusi et al., 2012).

Bupivacaine is an amide-type LA and is metabolized primarily in the liver. A decrease in hepatic clearance, with severe hepatic disease or a decrease in hepatic blood flow, can increase plasma drug concentration and cause toxicity. Bupivacaine and its inactive metabolites are mainly excreted by kidneys. Only 6% of bupivacaine is excreted unchanged in the urine. Urinary excretion is affected by urinary blood flow urinary pH and acidifying the urine increases the excretion (Babst & Gilling, 1978). The carrier, MVL, undergoes a lipid degradation and clearance in the injection site. A significant portion of liposome component may be detectable at the injection site exceeding 21 days (Mantripragada, 2002).

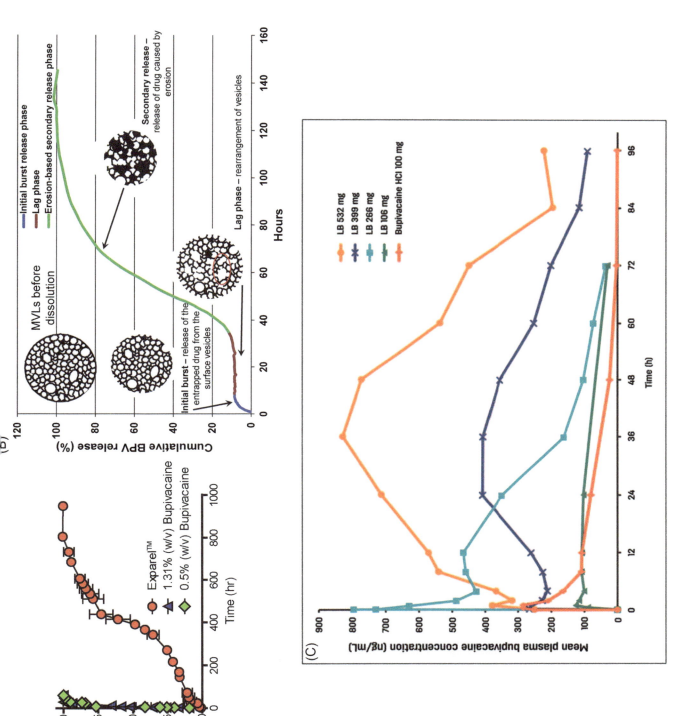

FIG. 2 Release characteristics of liposomal bupivacaine. (A) Cumulative release of bupivacaine from Exparel. Release of free bupivacaine HCL 1.31% and 0.5% is also shown. (B) Schematic detailing the possible release mechanism of bupivacaine from the MVLs. (C) Plasma bupivacaine concentration versus time for liposomal bupivacaine 106, 266, 399, and 532 mg, and bupivacaine HCl 100 mg. (With permission from (A) McAlvin, J. B., Padera, R. F., Shankarappa, S. A., Reznor, G., Kwon, A. H., Chiang, H. H., … Kohane, D. S. (2014). Multivesicular liposomal bupivacaine at the sciatic nerve. Biomaterials, 35 (15), 4557–4564. doi:https://doi.org/10.1016/j.biomaterials.2014.02.015. (B) Manna, S., Wu, Y., Wang, Y., Koo, B., Chen, L., Petrochenko, P., … Zheng, J. (2019). Probing the mechanism of bupivacaine drug release from multivesicular liposomes. Journal of Controlled Release, 294(December 2018), 279–287. doi:https://doi.org/10.1016/j.jconrel.2018.12.029. (C) Hu, D., Onel, E., Singla, N., Kramer, W. G., & Hadzic, A. (2013). Pharmacokinetic profile of liposome bupivacaine injection following a single administration at the surgical site. Clinical Drug Investigation, 33(2), 109–115. doi:https://doi.org/10.1007/s40261-012-0043-z.)

TABLE 1 Summary of pharmacokinetic parameters following single doses of liposome bupivacaine via surgical site infiltration, interscalene brachial plexus block and epidural administration.

	Surgical site infiltration							Interscalene brachial plexus block	Epidural administration			
	Bunionectomy[a]		Hemorrhoidectomy[b]	Total knee arthroplasty[c]				Shoulder arthroplasty[d]	Healthy volunteers[e]			
Parameters $n = 26$	LB 106 mg (8 mL) $n = 25$	LB 266 mg (20 mL) $n = 25$	LB 133 mg (60 mL) $n = 24$	LB 266 mg (60 mL) $n = 26$	LB 399 mg (60 mL) $n = 21$	LB 532 mg (60 mL) $n = 34$	Bup. HCl 150 mg (60 mL) $n = 15$	LB 133 mg (10 mL) $n = 8$	LB 89 mg (20 mL) $n = 8$	LB 155 mg (20 mL) $n = 7$	LB 266 mg (20 mL) $n = 6$	Bup. HCl 50 mg (20 mL)
C_{max}, mean (SD), ng/mL	166	867	262 (277)	340 (107)	500 (173)	935 (371)	205 (111)	209 (12.4)	120 (47)	134 (54)	250 (64)	300 (78)
T_{max}, median (min-max), h	2.0 (0.5–24)	0.5 (0.25–36)	12.0 (0.5–37.0)	24.0 (0.5–49.0)	24.0 (0.22–48.0)	36 (2.25–71.0)	24.0 (0.6–48.0)	48 (3–74)	7 (0.5–16)	24 (2–24)	24 (8–48)	0.7 (0.5–0.9)
$AUC_{0-tlast}$, mean (SD), h I ng/mL	5864 (2038)	16,867 (7868)						11,484 (8615)	4064 (1325)	6387 (1708)	13,198 (3996)	1960 (414)
$AUC_{0-\infty}$, mean (SD), h I ng/mL	7105 (2283)	18,289 (7569)	7826 (3317)	17,370 (8540)	27,630 (12,939)	60,174 (25117)	7460 (4118)	11,590 (8603)	4151 (1312)	6565 (1679)	13,954 (4336)	1961 (414)
$T_{1/2}$, mean (SD), h	34.1 (17.0)	24 (39)	13.4 (4.0)	17.1 (6.8)	18.8 (5.1)	16.9 (4.8)	10.7 (3.9)	11 (5)	16 (6)	14 (5)	19 (8)	6 (1)

AUC_∞, area under the plasma concentration time curve from time zero to infinity; *Bup,* bupivacaine; C_{max}, maximum plasma drug concentration; *LB,* liposome bupivacaine; *SD,* standard deviation; T_{max}, time to maximum drug concentration.
[a]*Golf et al. (2011).*
[b]*Gorfine, Onel, Patou, and Krivokapic (2011).*
[c]*Bramlett, Onel, Viscusi, and Jones (2012).*
[d]*Patel et al. (2020).*
[e]*Viscusi et al. (2012).*

Safety and adverse events

All local anesthetics including bupivacaine can cause systemic and local adverse events (AEs). Major safety concerns for bupivacaine formulations are local anesthetic systemic toxicity, particularly central nervous system and cardiovascular system toxicities, which can be fatal. The response, side effect/toxicity as our concern, depends on administration rate/speed of the drug, and concentration of the drug at the affected tissue. Systemic AEs and toxicity are dose-related and often caused by high plasma concentrations of bupivacaine. Accidental intravascular injection, administration of high doses, rapid absorption from the injection site, diminished tolerance can cause high plasma concentrations and systemic toxicity (Bourne, Wright, & Royse, 2010).

The MVL bupivacaine formulation has been designed to deliver higher bupivacaine doses with a single injection in a slow-release pattern, avoiding high plasma concentrations and toxicity while providing enough sensory block at any given time. In addition, only free form of bupivacaine can diffuse to brain and heart muscles, bupivacaine-loaded liposomes cannot penetrate to blood-brain barrier or the heart, thus even in accidental IV injection, tissue concentration will not be high enough to trigger serious AEs.

Patient-related factors which alter plasma protein binding, cardiac output, local blood flow, metabolism, and elimination of LA (i.e., systemic disease which decreases protein production, pregnancy, advanced age, acid-base status, surgical stress, competition with other drugs for protein binding sites, and flow dynamics) may diminish individual tolerance and increase toxicity (Bourne et al., 2010; Richard, Rickert, et al., 2011).

Single-dose LB infiltration is well tolerated up to 750 mg. The most common (>10%) postoperative AEs observed with MVL bupivacaine are nausea, pyrexia, pruritus, constipation, vomiting, and dizziness. Reported side effects are mostly mild to moderate intensity (Dasta, Ramamoorthy, Patou, & Sinatra, 2012; Hu et al., 2013; Naseem et al., 2012; Viscusi, Sinatra, Onel, & Ramamoorthy, 2014). Incidences of AEs were lower with lower doses of both plain and MVL formulations (i.e. <266 mg liposome bupivacaine caused fewer AEs). The side effect profile of LB is similar to bupivacaine hydrochloride even in higher doses up to threefold. Most of the common AEs are possibly due to surgery itself or opioid rescue (Ilfeld et al., 2015). Most often, life-threatening acute systemic toxicity is not seen until the blood level of bupivacaine is sufficiently elevated, therefore clinically relevant systemic toxicity is rare in therapeutic doses of LB (Aggarwal, 2018).

In preclinical in vivo dose-finding studies, the maximum tolerated intravenous dose of LB was sixfold higher than plain bupivacaine hydrochloride (4.5 mg/kg vs 0.75 mg/kg), while the maximum tolerated intraarterial dose was 15-fold higher (1.5 mg/kg vs 0.1 mg/kg) (Joshi, Patou, & Kharitonov, 2015). Additionally, bupivacaine dose within liposomes can be much larger than a given dose of plain bupivacaine hydrochloride and still produce equal or lower maximum plasma concentration. For example, threefold higher intravenous or epidural doses of liposome bupivacaine compared to bupivacaine hydrochloride, produced a similar maximum plasma concentration (C_{max}) in animal models (Joshi et al., 2015). These animal data cannot be directly extrapolated to humans; nevertheless, they suggest a favorable safety profile for LB even after accidental intravenous injections. In humans, systemic toxicity, in particular, central nervous system and cardiovascular system toxicities are expected to occur at plasma concentrations of 2000 and 4000 ng/mL, respectively (Bardsley, Gristwood, Baker, Watson, & Nimmo, 1998; Jorfeldt et al., 1968). Single or repeated high doses of LB revealed plasma concentrations well below these thresholds in various clinical settings (Table 1).

Central nervous system toxicity

Free bupivacaine is highly lipid-soluble and potent and can cross the blood-brain barrier. Central nervous system is particularly sensitive to overdoses and selectively blocking the inhibition of the excitatory pathways leads to toxicity.

Liposome bupivacaine increases the toxic dosages of bupivacaine required to induce seizure in vivo. For example, when compared to plain bupivacaine, rabbits required more than twice the dose of LB to produce seizures (Boogaerts et al., 1993). Similarly, when compared to the same doses of LB, plain bupivacaine solution is associated with more frequent neurological symptoms in dogs (Richard, Rickert, et al., 2011). Consistent with this safety profile in animal models, the overall incidence of central nervous system related AEs in LB-treated patients were 21% after peripheral nerve blocks (Ilfeld, 2017), and 14.5% after surgical site infiltrations (Viscusi et al., 2014) which are similar to comparable doses of plain bupivacaine HCL and placebo. Dizziness, headache, somnolence, lethargy, and hypesthesia are the most common AEs with mild to moderate intensity. No treatment-related severe neurological AEs were reported in clinical studies exploring the safety of LB (Ilfeld et al., 2015; Viscusi et al., 2014). Between 2012 and 2016, 66 central nervous system related AEs were reported to the FDA AEs reporting system, including 12 seizures (Aggarwal, 2018).

Cardiovascular system toxicity

With increasing concentrations of plasma bupivacaine, central nervous system toxicity will be followed by cardiovascular toxicity. In animal models, serious cardiovascular toxicity requires two- to fourfold higher plasma bupivacaine concentration compared to the concentration which produces seizures (Groban, 2003). These extremely high concentrations of plasma bupivacaine can often be seen following accidental intravascular injection of bupivacaine HCl. Bupivacaine blocks sodium, potassium, and L-type calcium channels in the sarcolemma of cardiomyocytes, impair oxidative phosphorylation and aerobic metabolism in mitochondria. Initial symptoms of cardiac toxicity are consistent with sympathetic stimulation (i.e., hypertension, tachycardia, ventricular arrhythmias) and progress to cardiovascular depression. Higher blood concentrations of bupivacaine inhibits baroreflex sensitivity, depress cardiac conductivity, reduce contractility, depress automaticity, reduce the duration of the refractory period, and can cause systemic vasodilatation. This effect may lead to atrioventricular block, ventricular arrhythmias, cardiovascular collapse, and cardiac arrest (Boogaerts et al., 1993; Bourne et al., 2010; Malik, Kaye, Kaye, Belani, & Urman, 2017; Neal et al., 2010). Even after inadvertent intravascular injection of MVL bupivacaine, liposomes do not cross the blood-brain barrier or enter cardiac cells, and release of bupivacaine from MVLs is quite slow as described above, therefore it is less likely to cause such incidents with higher doses of LB administered.

Cardiovascular toxicity of bupivacaine formulations is harder to study in humans. The toxic plasma bupivacaine concentration is not well defined but 4000 µg/mL is an accepted threshold (Bergese, Onel, Morren, & Morganroth, 2012). In clinical studies, single doses of LB (150, 300, 450, 600, 750 mg) by subcutaneous infiltration in healthy volunteers (Naseem et al., 2012), and surgical site infiltration after total knee arthroplasty (Bergese, Onel, et al., 2012) did not produce clinically significant changes in QT/QTc interval, heart rate, or PR interval. Repeated dosing of liposome bupivacaine in healthy volunteers also did not produce any meaningful change in ECG parameters, and no cardiac toxicity was observed (Rice, Heil, & Biernat, 2017). The most frequently reported cardiac AEs in clinical studies are bradycardia and sinus tachycardia with mild to moderate severity and the liposome bupivacaine demonstrated a similar cardiac safety profile compared to bupivacaine HCl (Ilfeld et al., 2015; Viscusi et al., 2014). In an evaluation of pharmacovigilance data between 2012 and 2016, 64 cardiovascular-related AEs were reported, including 4 life-threatening bradycardia and 13 cardiovascular arrests. However, the relation of reported adverse cardiovascular events to treatment was not available (Aggarwal, 2018).

Local adverse events

High concentrations and prolonged exposure to LAs (i.e., continuous infusion and controlled release formulations) (Padera, Bellas, Tse, Hao, & Kohane, 2008) can enhance myotoxicity (Hussain et al., 2018), neurotoxicity (Hogan, 2008) and cause clinically significant morbidity. When given in sufficiently large doses, liposome bupivacaine has the potential to cause local tissue injury both by providing prolonged exposure to bupivacaine and by the large multivesicular liposome structure itself.

Inflammation and wound healing

Minimal to mild granulomatous inflammation is the main reported histologic finding as a local reaction after LB infiltration in animal models. Granulomatous inflammation is more severe and lasts longer in pathological specimens compared to plain bupivacaine HCl; however, this reaction was considered as a normal reaction to foreign bodies and did not cause any functional deficit (Richard et al., 2012; Richard, Ott, et al., 2011; Richard, Rickert, et al., 2011). In a rat model, MVL bupivacaine caused more and longer-lasting local tissue inflammation compared to either 0.5% or 1.31% bupivacaine HCl (McAlvin et al., 2014). On histologic examination, a mixed infiltrate consisting of macrophages with foamy cytoplasm, a smaller population of lymphocytes, and lesser neutrophils were seen. Foamy macrophages were likely to be reflecting the uptake of the liposome structure (McAlvin et al., 2014; Richard, Ott, et al., 2011), and were not seen in the bupivacaine HCl infiltrated rats. Prolonged inflammation in preclinical studies arise concerns for wound healing in humans. Despite evidence from animal studies, there have been no reports of granulomatous inflammation in humans. In a retrospective review of 10 clinical trials of LB compared to bupivacaine HCL, wound healing scores, wound scarring, and long-term healing were similar between LB and bupivacaine HCL groups (Baxter, Bramlett, Onel, & Daniels, 2013). The most commonly reported local AEs were pain, swelling, and erythema at the site of injection which were also seen with bupivacaine HCl. Liposome bupivacaine does not appear to adversely affect wound healing after surgery.

Muscle injury

Bupivacaine can cause calcium-induced apoptosis of muscle cells. High concentrations (25 mg/mL) (Richard, Ott, et al., 2011), and high doses (30 mg/kg) (Richard, Rickert, et al., 2011) of LB caused mild muscle injury and did not cause a functional deficit in animal models. In rat models, LB-induced myotoxicity was similar to that of 0.5% bupivacaine HCl, but it was less than 1.31% bupivacaine HCl. To date, no AEs related to muscle injury have been reported in human studies.

Nerve injury

Bupivacaine can cause axonal damage with high concentrations, or prolonged exposure to the drug (Hogan, 2008; Yang, Abrahams, Hurn, Grafe, & Kirsch, 2011). The proposed cellular mechanism of neurotoxicity is elevated intracellular calcium level induced apoptosis, mitochondrial damage, loss of axonal microtubules, and generation of free oxygen radicals. Bupivacaine can also decrease local blood flow and neural perfusion and enhance nerve ischemia.

Administration of LB did not cause any nerve injury in the brachial plexus of rabbits and dogs (Richard et al., 2012), and the sciatic nerve of rats (McAlvin et al., 2014). Pathological findings in peripheral nerves were similar at short term (3–4 days) and long term (14–15 days) and did not differ between groups of 0.5% bupivacaine HCl, 1.33% bupivacaine HCL, or LB (1.33%) formulation. To further evaluate the safety, LB was injected intraneural-extra fascicularly to sciatic nerves of pigs (Damjanovska et al., 2015). Low doses of (4 mL) 0.5% bupivacaine HCL, LB 1.33% and 0.9% NaCl were infiltrated and compared with extraneural injections. Intraneural injection of LB resulted in a longer sensory blockade but did not cause persistent motor or sensory deficit. Intraneural injection of both LB or bupivacaine HCL did not cause histological nerve injury (Damjanovska et al., 2015). In clinical studies, there is no report of nerve injury after LB administration to date.

The safety of neuraxial use of LB was also questioned with epidural (Joshi et al., 2015) and intrathecal administration (Joshi et al., 2015; Malinovsky et al., 1997) to dogs and rabbits. Epidural administration of 40 mg LB did not cause any histological damage to the spinal cord or adjacent tissues. Animals that received LB or empty MVL (as placebo) had minimal chronic inflammation, presented as a few large foamy macrophages in histologic specimens of the adipose tissue around nerves (Joshi et al., 2015). In clinical studies, epidural administration of LB did not cause clinically evident neurotoxicity (Boogaerts et al., 1994; Viscusi et al., 2012).

Maternal/fetal toxicity

There aren't any studies of liposome bupivacaine in pregnant women. Animal reproduction studies have been done for plain bupivacaine HCl. Subcutaneous administration of bupivacaine HCL to rats and rabbits during organogenesis, at doses equivalent to 1.5–1.6 times the maximum recommended human dose of liposomal bupivacaine (266 mg), was associated with embryo-fetal deaths (Pacira Pharmaceuricals Inc, 2018). Based on bupivacaine data liposome bupivacaine is labeled as pregnancy category C.

Compatibility

To maintain the integrity of liposomes, local anesthetics other than bupivacaine should not be admixed within the syringe or co-administered in sequence to the same site of infiltration with LB. Lidocaine has a greater affinity to MVLs than bupivacaine, thus it may lead to rapid release of unencapsulated free bupivacaine causing inadvertently high bupivacaine concentrations (Joshi et al., 2015). If lidocaine is planned to be used concomitantly, it is recommended to wait at least 20 min after lidocaine infiltration, prior to infiltration of MVL bupivacaine to the same site. If bupivacaine HCl will be used in addition to LB, the additional dose of plain bupivacaine dose should not exceed 50% of the liposomal dose (Pacira Pharmaceuricals Inc, 2018; Vandepitte et al., 2017).

Contact with surfactants and antiseptics may disrupt the MVL structure, so antiseptics (povidone-iodine) applied to the infiltration site should be allowed to dry before administration of LB (Pacira Pharmaceuricals Inc, 2018).

Clinical uses and efficacy

Currently, available LB formulation has been approved only for single-dose surgical site infiltrations, transversus abdominis plane blocks (TAP), and interscalene brachial plexus nerve blocks for postoperative local/regional analgesia (FDA, 2015) (Pacira Pharmaceuricals Inc, 2018). The maximum approved dose is 266 mg. Promising theoretical advantages of the LB formulation have also been studied for off the label indications.

Liposome bupivacaine versus placebo

Surgical infiltration of MVL bupivacaine has provided superior analgesia compared to placebo in initial phase II and phase III trials including surgical site infiltration for bunionectomy (Golf et al., 2011) and hemorrhoidectomy (Gorfine et al., 2011). MVL bupivacaine also provided superior analgesia compared to placebo with single-injection femoral nerve block for total knee arthroplasty (Hadzic et al., 2016). These studies demonstrated the analgesic efficacy of liposome bupivacaine.

Surgical site infiltration

Multiple randomized controlled trials have been conducted to compare MVL bupivacaine to plain bupivacaine or conventional agents for surgical site infiltration and several reviews have been published. For bunionectomy, there are no clinical trials comparing MVL bupivacaine with an active comparator. Following hemorrhoidectomy, in two active-controlled trials, while one had negative findings (NCT00744848, not published) (Gabriel & Ilfeld, 2019), the other reported improved postoperative analgesia, and less opioid use with 266 mg LB compared to plain bupivacaine HCL (Haas, Onel, Miller, Ragupathi, & White, 2012).

For breast surgery, liposome bupivacaine infiltrated into implant pockets did not provide clinically important pain reduction compared to bupivacaine HCL (Nadeau, Saraswat, Vasko, Elliott, & Vasko, 2016; Smoot, Bergese, Onel, Williams, & Hedden, 2012). Less opioid consumption was reported with LB use (Smoot et al., 2012).

Regarding abdominal surgery, infiltration of LB to the surgical wound after colectomy was associated with less opioid use compared to multimodal analgesia regimens (i.e., intravenous patient-controlled analgesia) without any LA infiltration (Raman, Lin, & Krishnan, 2018). However, when compared to local infiltration of bupivacaine HCl as an active comparator, infiltration of LB did not provide any additional benefit (Knudson, Dunlavy, Franko, Raman, & Kraemer, 2016). Additional randomized controlled trials did not demonstrate better postoperative analgesia with infiltration of LB compared to bupivacaine HCl for inguinal hernia repair, laparoscopic urologic surgery, and robotic sacrocolpopexy (Gabriel & Ilfeld, 2019).

Periarticular infiltration of LB for knee arthroplasty mostly provided no additional benefit over infiltration of conventional local analgesics (Ilfeld, Gabriel, & Eisenach, 2018). Additionally, periarticular injection of LB resulted in poorer pain control compared to bupivacaine HCl in the immediate postoperative period (4–8 h) after total knee arthroplasty and shoulder arthroplasty (Abildgaard, Chung, Tokish, & Hattrup, 2019). Only 2 of the 13 randomized controlled trials reported benefit with periarticular infiltration of LB (Ilfeld et al., 2018). In the PILLAR study (Mont, Beaver, Dysart, Barrington, & Del Gaizo, 2018), infiltration of LB to knee joint with a special technique was reported to provide better analgesia and less opioid consumption compared to bupivacaine HCl. The authors of the PILLAR study concluded that, non-superior findings of the previous studies were mostly due to inappropriate infiltration technique, and suggested a specific technique (Mont et al., 2018). Regarding shoulder arthroplasty, LB in addition to interscalene nerve block with plain bupivacaine provided similar postoperative analgesia compared to interscalene alone with slightly fewer AEs (Sabesan et al., 2017).

Peripheral nerve blocks and perineural use

Administration of LB for peripheral nerve blocs and field/fascial blocks provided more promising results compared to surgical site infiltration of LB.

Femoral nerve block with LB provided prolonged block in healthy volunteers (Ilfeld, Malhotra, Furnish, Donohue, & Madison, 2013). Femoral nerve block with LB (266 mg) was effective after total knee arthroplasty with modestly lower pain scores and less opioid use, with a similar AE profile compared to placebo (Hadzic et al., 2016). The "modest" decrease in pain scores was attributed to the lack of sciatic and obturator nerve blocks. The effectiveness of femoral nerve block with LB for postoperative analgesia has yet to be compared with plain bupivacaine HCl to drive a conclusion.

Interscalene brachial plexus block with LB has been studied for postoperative analgesia after shoulder arthroplasty. In a phase III multicenter RCT (Patel et al., 2020), a single injection interscalene brachial plexus block with LB (133 or 266 mg) compared to placebo, provided superior analgesia throughout 48 h after surgery, decreased opioid consumption, and was approved by FDA for this indication. In another trial, combined use of liposome bupivacaine (133 mg, 10 mL) with bupivacaine HCl (5 mL, 0.25%) for interscalene brachial plexus block modestly lowered patients worst pain scores within 1 week after major shoulder surgery compared to bupivacaine HCl only. However, the functionality of the extremity, sleep duration, sensory and motor characteristics of the block, postoperative opioid consumption, and incidence of AEs did not differ between groups (Vandepitte et al., 2017).

Thoracic surgery is extremely painful. High thoracic epidural catheterization is difficult to perform and not safe enough. Safer and easier analgesic methods are required. In patients who underwent lung resection, intercostal nerve block with LB provided good pain control for at least 72 h (Rice et al., 2015). In a large retrospective study, intercostal nerve block with LB was compared with high thoracic epidural analgesia for video-assisted thoracotomy and the length of stay was 1 day shorter in the LB group (Mehran et al., 2017). Considering the available evidence, intercostal nerve block was considered as a safe and effective alternative to high thoracic epidural analgesia, and may be used as a rescue analgesia technique if thoracic epidural analgesia fails (Campos, Seering, & Peacher, 2019).

The transversus abdominis plane block is an effective method for postoperative analgesia after abdominal surgeries. The effectiveness of LB for TAP block was demonstrated with a placebo-controlled study in a small cohort of inguinal hernia repair. Transversus abdominis plane block with LB was associated with better pain control and lower opioid requirements up to 72 h compared to TAP block with plain bupivacaine HCl after robot-assisted hysterectomies and donor nephrectomies (Hutchins, Kesha, Blanco, Dunn, & Hochhalter, 2016). TAP block with LB also provided effective analgesia after major open abdominal surgery, and it was non-inferior to continuous lumbar epidural analgesia with bupivacaine HCl (Ayad et al., 2016). Transversus abdominis plane block with LB may be considered as an option in abdominal surgeries combined with other multimodal analgesia components (Malik et al., 2017).

Neuraxial use

Epidural administration of liposome bupivacaine was first studied in 1997 with unilamellar liposomes and provided longer sensory block compared to bupivacaine HCl (Boogaerts et al., 1993). More recently, epidural administration of 266 mg MVL bupivacaine in healthy volunteers resulted in a longer duration of sensory block and lesser motor block compared to 50 mg plain bupivacaine HCl (Viscusi et al., 2012). No further neuraxial studies have been conducted yet. Considering the chance of avoiding potential complications related to epidural catheters, prolonged sensory block with a single epidural injection of LB is promising for postoperative analgesia after major surgery. But further studies are required to drive any conclusion for epidural or intrathecal use.

Administration

MVLs stay precisely where they are placed and do not readily diffuse within tissue as easily as free bupivacaine HCL. Therefore, it requires more injections to ensure adequate coverage. The manufacturer recommends a special "moving needle technique" with slow injection. LB should be injected in multiple injections consisting of 1–2 mL aliquots to an entire surgical area with a 25 gauge or larger bore needle. The maximum allowed dose of LB is 266 mg (20 mL) for a single injection. It can be expanded with normal (0.9%) saline or ringer lactate solution up to 300 mL depending on the size of the surgical area (Pacira Pharmaceuricals Inc, n.d.). It can be mixed with plain bupivacaine HCL (up to 50% of the milligram dose of LB) prior to administration to enhance a faster onset of action and cover the analgesic need in the immediate perioperative period (Gadsden & Long, 2016; Vandepitte et al., 2017).

Application to other areas

Besides postoperative analgesia, prolonged sensory block with LB may also be useful in chronic pain conditions. For example, LB has the potential to provide significant relief for 2–3 days before definitive treatment with radiofrequency ablation for facet joint arthropathy. It may also be considered to be used in chronic headaches, migraines, scalp blocks, and trigger point injectios (Malik et al., 2017).

Other agents of interest

To provide prolonged postoperative analgesia a number of biomaterials are being studied (Balocco et al., 2018; Coppens et al., 2019; He, Qin, Huang, & Ma, 2020; McAlvin et al., 2014). Some examples are:

- Sucrose acetate isobutyrate extended-release bupivacaine (SABER-bupivacaine) is a biocompatible and biodegradable compound and depot formulation of bupivacaine which is designed to provide prolonged analgesia for 72 h. It has a high concentration of bupivacaine base (13.2%, 660 mg bupivacaine base in 5 mL).

- HTX-011 contains two active ingredients. Bupivacaine and low-dose meloxicam are incorporated into a biodegradable polymer (Biochromomer). The polymer undergoes hydrolysis in the tissues over 3 days and releases active components. Currently not FDA approved.
- Neosaxitoxin is a neurotoxin derivative, found in shellfish that reversibly blocks voltage-gated sodium channels. It blocks neural conduction, impeding the propagation of nerve impulses. It has less affinity to peripheral sodium channels in peripheral nerves compared to cardiac sodium channels and therefore has less potential for cardiotoxicity and does not cross the blood-brain barrier. Currently not FDA approved.

However, none of these are on the market yet. Currently, the only available sustained release formulation for prolonged analgesia is EXPAREL. Active component of sustained-release formulations is usually bupivacaine. Therefore, adverse effects are mostly identical to bupivacaine HCL, but it requires higher doses to induce local anesthetic systemic toxicity. Table 2 summarizes the properties of some investigational local anesthetics which are being developed for prolonged perioperative analgesia.

TABLE 2 Summary points and side effects of other agents of interest and liposomal bupivacaine.

	EXPAREL (on the market)	POSIMIR[a] (investigational)	HTX-011[b] ZYNRELEF (investigational)	Neosaxitoxin[c] (investigational)
Delivery system/carrier	DEPOFOAM, multivesicular liposome	SABER, sucrose acetate isobutyrate + benzyl alcohol	Biochronomer, biodegradable polymer	No commercial preparations available yet. Plain solutions and liposomal formulations are being investigated
Active drug, concentration, dose	Bupivacaine base, 1.33%, 266 mg/20 mL	Bupivacaine base, 13.2%, 660 mg/5 mL	Bupivacaine/ meloxicam, 400 mg/12 mg, 12 mL	Can be used alone or in combination 30 µg Neosaxitoxin (3 µg/mL) Bupivacaine 0.2% Epinephrine 5 µg/mL
Claimed duration of analgesia	72 h	72 h	72 h	48 h More than 7 days when embedded in liposomes
Technique	Surgical site infiltration, local regional block	Needle free instillation directly to the surgical site	Needle free instillation directly to the surgical site	Surgical site infiltration
C_{max}	<900 ng/mL	<900 ng/mL	<600 mg	170 /mL <40 pg/mL when coadministered with epinephrine
Common side effects	Similar to bupivacaine HCL, mild to moderate intensity	Similar to bupivacaine HCL, mild to moderate intensity	AEs profile like that of saline placebo and bupivacaine HCl	No difference in serious adverse events compared with bupivacaine
	Nausea Pyrexia Pruritus Constipation Vomiting Dizziness Headache	Nausea Pyrexia Pruritus Constipation Vomiting Dizziness Headache Metallic taste	Nausea Constipation Vomiting Dizziness	With 40 mg NeoSaxitoxin Perioral numbness (48%) Tingling (52%) None reported in patients under general anesthesia

TABLE 2 Summary points and side effects of other agents of interest and liposomal bupivacaine—cont'd

	EXPAREL (on the market)	POSIMIR (investigational)	HTX-011 ZYNRELEF (investigational)	Neosaxitoxin (investigational)
Local anesthetic systemic toxicity (LAST)	Predictable pharmacokinetics Safe C_{max} values No systemic toxicity	Stable release rate Reassuring C_{max} values No evidence of LAST	Safe C_{max} values No evidence of LAST	No cardiotoxicity Less hypotension compared with Tetradoxin
Wound healing	Unaffected	Unaffected	Unaffected	Unaffected
Nerve injury	Similar to bupivacaine HCL	Similar to bupivacaine HCL	Similar to bupivacaine HCL	Not neurotoxic
Muscle injury	Similar to Bupivacaine HCL	Similar to bupivacaine HCL	Similar to bupivacaine HCL	NA
Chondrolysis	Less chondrotoxic than bupivacaine HCl, Intraarticular injection is not recommended by the manufacturer	Not reported any	Not reported any	NA
Other	Prolonged granulomatous inflammation (foreign body reaction), healed without deficit	Bruise-like discoloration Prolonged granulomatous inflammation (foreign body reaction)		LD 50: 5 mcg/kg (rats)
Special point of interest	Low C_{max}, even after inadvertent intravascular injection Should not be mixed with lidocaine	Extremely low risk of inadvertent Intravascular injection	Extremely low risk of inadvertent Intravascular injection	Therapeutic index improves when mixed with bupivacaine and/or epinephrine

Novel agents for prolonged postoperative local/reginal analgesia are not limited to these in the table. Various active agents and carriers are currently being investigated (King, Beutler, Kaye, & Urman, 2017; Santamaria, Woodruff, Yang, & Kohane, 2017).
[a]*DURECT CORPORATION (2020).*
[b]*Lachiewicz et al. (2020).*
[c]*King et al. (2017).*

Mini-dictionary of terms

Multivesicular liposome: A liposome structure that contains multiple nonconcentric aqueous chambers. It is the carrier of bupivacaine and provides prolonged analgesia.
Bimodal distribution: Two peaks of the plasma concentration of a specified drug.

Key facts of liposome bupivacaine

- LB is an extended-release formulation of bupivacaine based on MVL technology which is designed to release bupivacaine slowly for up to 72 h following a single injection.
- The maximum allowed dose of LB is 266 mg (20 mL) for a single injection. It can be expanded with normal (0.9%) saline or Ringer lactate solution up to 300 mL depending on the size of the surgical area.
- Liposome bupivacaine structure cannot cross BBB and does release the encapsulated drug slowly. Therefore, accidental intravascular injection is less likely to cause central nervous system or cardiovascular toxicity.

304 PART | III Adverse effects, reactions, and outcomes

- Local anesthetics other than bupivacaine HCL should not be admixed within the syringe or co-administered in sequence to the same site of infiltration with LB.
- Skin disinfectants should be allowed to dry before administration of LB.

Summary points

- Bupivacaine is the most used local anesthetic for perioperative analgesia. However, when applied as a single injection, the duration of action is not enough, repeated infiltration is impractical, and catheter techniques require time and expertise.
- The safety and side effect profile of LB is similar to bupivacaine hydrochloride even in higher doses up to threefold.
- Single and repeated infiltration of LB in clinically effective doses produces maximum plasma concentration of bupivacaine below expected thresholds of systemic toxicity.
- When compared to placebo, administration of LB for surgical site infiltration, TAP block, femoral nerve block, and interscalene nerve block effectively decreases postoperative pain and opioid consumption.
- Current evidence does not clearly demonstrate the superiority of LB to plain bupivacaine hydrochloride.
- LB is not considered as cost-effective in the majority of trials. Cost-effectiveness should be assessed in a case-by-case basis. For example, it may be considered as a safe and effective alternative to thoracic epidural analgesia for thoracic surgery. However, it is not approved yet for this particular indication.

References

Abildgaard, J. T., Chung, A. S., Tokish, J. M., & Hattrup, S. J. (2019). Clinical efficacy of liposomal bupivacaine. *JBJS Reviews*, *7*(7). https://doi.org/10.2106/JBJS.RVW.18.00192, e8.

Aggarwal, N. (2018). Local anesthetics systemic toxicity association with exparel (bupivacaine liposome)—A pharmacovigilance evaluation. *Expert Opinion on Drug Safety*, *17*(6), 581–587. https://doi.org/10.1080/14740338.2017.1335304.

Allen, T. M., & Cullis, P. R. (2013). Liposomal drug delivery systems: From concept to clinical applications. *Advanced Drug Delivery Reviews*, *65*(1), 36–48. https://doi.org/10.1016/j.addr.2012.09.037.

Apseloff, G., Onel, E., & Patou, G. (2013). Time to onset of analgesia following local infiltration of liposome bupivacaine in healthy volunteers: A randomized, single-blind, sequential cohort, crossover study. *International Journal of Clinical Pharmacology and Therapeutics*, *51*(5), 367–373. https://doi.org/10.5414/CP201775.

Ayad, S., Babazade, R., Elsharkawy, H., Nadar, V., Lokhande, C., Makarova, N., … Turan, A. (2016). Comparison of transversus abdominis plane infiltration with liposomal bupivacaine versus continuous epidural analgesia versus intravenous opioid analgesia. *PLoS One*, *11*(4). https://doi.org/10.1371/journal.pone.0153675, e0153675.

Babst, C. R., & Gilling, B. N. (1978). Bupivacaine: A review. *Anesthesia Progress*, *25*(3), 87.

Balocco, A. L., Van Zundert, P. G. E., Gan, S. S., Gan, T. J., & Hadzic, A. (2018). Extended release bupivacaine formulations for postoperative analgesia: An update. *Current Opinion in Anaesthesiology*, *31*(5), 636–642. https://doi.org/10.1097/ACO.0000000000000648.

Bangham, A. (1989). The 1st description of liposomes – A citation classic commentary on diffusion of univalent ions across the lamellae of swollen phospholipids. *Current Contents/Life Sciences*, (13), 14. Retrieved from http://garfield.library.upenn.edu/classics1989/A1989T670000001.pdf.

Bangham, A. D., Standish, M. M., & Watkins, J. C. (1965). Diffusion of univalent ions across the lamellae of swollen phospholipids. *Journal of Molecular Biology*, *13*(1), 238–252. https://doi.org/10.1016/S0022-2836(65)80093-6.

Bardsley, H., Gristwood, R., Baker, H., Watson, N., & Nimmo, W. (1998). A comparison of the cardiovascular effects of levobupivacaine and rac-bupivacaine following intravenous administration to healthy volunteers. *British Journal of Clinical Pharmacology*, *46*(3), 245–249. https://doi.org/10.1046/j.1365-2125.1998.00775.x.

Baxter, R., Bramlett, K., Onel, E., & Daniels, S. (2013). Impact of local administration of liposome bupivacaine for postsurgical analgesia on wound healing: A review of data from ten prospective, controlled clinical studies. *Clinical Therapeutics*, *35*(3), 312–320. e5 https://doi.org/10.1016/j.clinthera.2013.02.005.

Bergese, S. D., Onel, E., Morren, M., & Morganroth, J. (2012). Bupivacaine extended-release liposome injection exhibits a favorable cardiac safety profile. *Regional Anesthesia and Pain Medicine*, *37*(2), 145–151. https://doi.org/10.1097/AAP.0b013e31823d0a80.

Bergese, S. D., Ramamoorthy, S., Patou, G., Bramlett, K., Gorfine, S. R., & Candiotti, K. A. (2012). Efficacy profile of liposome bupivacaine, a novel formulation of bupivacaine for postsurgical analgesia. *Journal of Pain Research*, *5*, 107–116. https://doi.org/10.2147/JPR.S30861.

Beverly, A., Kaye, A. D., Ljungqvist, O., & Urman, R. D. (2017). Essential elements of multimodal analgesia in enhanced recovery after surgery (ERAS) guidelines. *Anesthesiology Clinics*, *35*(2), e115–e143. https://doi.org/10.1016/j.anclin.2017.01.018.

Biyani, G., Chhabra, A., Baidya, D. K., & Anand, R. K. (2014). Adjuvants to local anaesthetics in regional anaesthesia – Should they be used? Part I: Pros. *Trends in Anaesthesia and Critical Care*. https://doi.org/10.1016/j.tacc.2013.12.002.

Boogaerts, J., Declercq, A., Lafont, N., Benameur, H., Akodad, E. M., Dupont, J.-C., & Legros, F. J. (1993). Toxicity of bupivacaine encapsulated into liposomes and injected intravenously. *Anesthesia & Analgesia*, *76*(3), 553–555. https://doi.org/10.1213/00000539-199303000-00018.

Boogaerts, J. G., Lafont, N. D., Declercq, A. G., Luo, H. C., Gravet, E. T., Bianchi, J. A., & Legros, F. J. (1994). Epidural administration of liposome associated bupivacaine for the management of postsurgical pain: A first study. *Journal of Clinical Anesthesia*, *6*(4), 315–320. https://doi.org/10.1016/0952-8180(94)90079-5.

Bourne, E., Wright, C., & Royse, C. (2010). A review of local anesthetic cardiotoxicity and treatment with lipid emulsion. *Local and Regional Anesthesia*, *3*(1), 11–19. https://doi.org/10.2147/lra.s8814.

Bozzuto, G., & Molinari, A. (2015). Liposomes as nanomedical devices. *International Journal of Nanomedicine*, *10*, 975. https://doi.org/10.2147/IJN.S68861.

Bramlett, K., Onel, E., Viscusi, E. R., & Jones, K. (2012). A randomized, double-blind, dose-ranging study comparing wound infiltration of DepoFoam bupivacaine, an extended-release liposomal bupivacaine, to bupivacaine HCl for postsurgical analgesia in total knee arthroplasty. *The Knee*, *19*(5), 530–536. https://doi.org/10.1016/j.knee.2011.12.004.

Brummett, C. M., & Williams, B. A. (2011). Additives to local anesthetics for peripheral nerve blockade. *International Anesthesiology Clinics*, *49*(4), 104–116. https://doi.org/10.1097/AIA.0b013e31820e4a49.

Çağdaş, M., Sezer, A. D., & Bucak, S. (2014). Liposomes as potential drug carrier systems for drug delivery. In *Application of nanotechnology in drug delivery* InTech. https://doi.org/10.5772/58459.

Campos, J. H., Seering, M., & Peacher, D. (2019). Is the role of liposomal bupivacaine the future of analgesia for thoracic surgery? An update and review. *Journal of Cardiothoracic and Vascular Anesthesia*. https://doi.org/10.1053/j.jvca.2019.11.014.

Chou, R., Gordon, D. B., de Leon-Casasola, O. A., Rosenberg, J. M., Bickler, S., Brennan, T., … Wu, C. L. (2016). Management of postoperative pain: A clinical practice guideline from the American Pain Society, the American Society of Regional Anesthesia and Pain Medicine, and the American Society of Anesthesiologists' Committee on Regional Anesthesia, Executive Commission. *Journal of Pain*, *17*(2), 131–157. https://doi.org/10.1016/j.jpain.2015.12.008.

Coppens, S. J. R., Zawodny, Z., Dewinter, G., Neyrinck, A., Balocco, A. L., & Rex, S. (2019). In search of the Holy Grail: Poisons and extended release local anesthetics. *Best Practice & Research. Clinical Anaesthesiology*, *33*(1), 3–21. https://doi.org/10.1016/j.bpa.2019.03.002.

Damjanovska, M., Cvetko, E., Hadzic, A., Seliskar, A., Plavec, T., Mis, K., … Stopar Pintaric, T. (2015). Neurotoxicity of perineural vs intraneural-extrafascicular injection of liposomal bupivacaine in the porcine model of sciatic nerve block. *Anaesthesia*, *70*(12), 1418–1426. https://doi.org/10.1111/anae.13189.

Dasta, J., Ramamoorthy, S., Patou, G., & Sinatra, R. (2012). Bupivacaine liposome injectable suspension compared with bupivacaine HCl for the reduction of opioid burden in the postsurgical setting. *Current Medical Research and Opinion*, *28*(10), 1609–1615. https://doi.org/10.1185/03007995.2012.721760.

FDA. (2015). *Supplemental NDA (sNDA) for EXPAREL FDA letter*. Retrieved from https://www.fda.gov/downloads/drugs/guidancecomplianceregula%0Atoryinformation/enforcementactivitiesbyfda/warninglettersandnoticeofviolationletterstopharmaceuticalcompanies/uc%0Am477250.pdf.

Gabriel, R. A., & Ilfeld, B. M. (2019). An updated review on liposome bupivacaine. *Current Anesthesiology Reports*, *9*(3), 321–325. https://doi.org/10.1007/s40140-019-00327-y.

Gadsden, J., & Long, W. J. (2016). Time to analgesia onset and pharmacokinetics after separate and combined administration of liposome bupivacaine and bupivacaine HCl: Considerations for clinicians. *Open Orthopaedics Journal*, *10*(1), 94–104. https://doi.org/10.2174/1874325001610010094.

Gan, T. J. (2017). Poorly controlled postoperative pain: Prevalence, consequences, and prevention. *Journal of Pain Research*. https://doi.org/10.2147/JPR.S144066. Dove Medical Press Ltd.

Golf, M., Daniels, S. E., & Onel, E. (2011). A phase 3, randomized, placebo-controlled trial of DepoFoam® bupivacaine (extended-release bupivacaine local analgesic) in bunionectomy. *Advances in Therapy*, *28*(9), 776–788. https://doi.org/10.1007/s12325-011-0052-y.

Gorfine, S. R., Onel, E., Patou, G., & Krivokapic, Z. V. (2011). Bupivacaine extended-release liposome injection for prolonged postsurgical analgesia in patients undergoing hemorrhoidectomy: A multicenter, randomized, double-blind, placebo-controlled trial. *Diseases of the Colon and Rectum*, *54*(12), 1552–1559. https://doi.org/10.1097/DCR.0b013e318232d4c1.

Grant, G. J., & Bansinath, M. (2001). Liposomal delivery systems for local anesthetics. *Regional Anesthesia and Pain Medicine*, *26*(1), 61–63. https://doi.org/10.1053/rapm.2001.19166.

Groban, L. (2003). Central nervous system and cardiac effects from long-acting amide local anesthetic toxicity in the intact animal model. *Regional Anesthesia and Pain Medicine*, *28*(1), 3–11. https://doi.org/10.1053/rapm.2003.50014.

Haas, E., Onel, E., Miller, H., Ragupathi, M., & White, P. F. (2012). A double-blind, randomized, active-controlled study for post-hemorrhoidectomy pain management with liposome bupivacaine, a novel local analgesic formulation. *The American Surgeon*, *78*(5), 574–581. https://doi.org/10.1177/000313481207800540.

Hadzic, A., Minkowitz, H. S., Melson, T. I., Berkowitz, R., Uskova, A., Ringold, F., … Ilfeld, B. M. (2016). Liposome bupivacaine femoral nerve block for postsurgical analgesia after total knee arthroplasty. *Anesthesiology*, *124*(6), 1372–1383. https://doi.org/10.1097/ALN.0000000000001117.

He, Y., Qin, L., Huang, Y., & Ma, C. (2020). Advances of nano-structured extended-release local anesthetics. *Nanoscale Research Letters*, *15*(1). https://doi.org/10.1186/s11671-019-3241-2.

Heimburger, D. C., McClave, S. A., Gramlich, L. M., & Merritt, R. (2010). The intersociety professional nutrition education consortium and American board of physician nutrition specialists: What have we learned? *Journal of Parenteral and Enteral Nutrition*, *34*(6 Suppl), 21–29. https://doi.org/10.1177/0148607110372393.

Hogan, Q. (2008). Pathophysiology of peripheral nerve injury during regional anesthesia. *Regional Anesthesia and Pain Medicine*, *33*(5), 435–441. https://doi.org/10.1016/j.rapm.2008.03.002.

Hu, D., Onel, E., Singla, N., Kramer, W. G., & Hadzic, A. (2013). Pharmacokinetic profile of liposome bupivacaine injection following a single administration at the surgical site. *Clinical Drug Investigation*, *33*(2), 109–115. https://doi.org/10.1007/s40261-012-0043-z.

Hussain, N., McCartney, C. J. L., Neal, J. M., Chippor, J., Banfield, L., & Abdallah, F. W. (2018). Local anaesthetic-induced myotoxicity in regional anaesthesia: A systematic review and empirical analysis. *British Journal of Anaesthesia*, *121*(4), 822–841. https://doi.org/10.1016/j.bja.2018.05.076.

Hutchins, J. L., Kesha, R., Blanco, F., Dunn, T., & Hochhalter, R. (2016). Ultrasound-guided subcostal transversus abdominis plane blocks with liposomal bupivacaine vs. non-liposomal bupivacaine for postoperative pain control after laparoscopic hand-assisted donor nephrectomy: A prospective randomised observer-blinded study. *Anaesthesia*, *71*(8), 930–937. https://doi.org/10.1111/anae.13502.

Ilfeld, B. M. (2017). Continuous peripheral nerve blocks: An update of the published evidence and comparison with novel, alternative analgesic modalities. *Anesthesia and Analgesia*, *124*(1), 308–335. https://doi.org/10.1213/ANE.0000000000001581.

Ilfeld, B. M., Gabriel, R. A., & Eisenach, J. C. (2018). Liposomal bupivacaine infiltration for knee arthroplasty: Significant analgesic benefits or just a bunch of fat? *Anesthesiology*, *129*(4), 623–626. https://doi.org/10.1097/ALN.0000000000002386.

Ilfeld, B. M., Malhotra, N., Furnish, T. J., Donohue, M. C., & Madison, S. J. (2013). Liposomal bupivacaine as a single-injection peripheral nerve block: A dose-response study. *Anesthesia and Analgesia*, *117*(5), 1248–1256. https://doi.org/10.1213/ANE.0b013e31829cc6ae.

Ilfeld, B. M., Viscusi, E. R., Hadzic, A., Minkowitz, H. S., Morren, M. D., Lookabaugh, J., & Joshi, G. P. (2015). Safety and side effect profile of liposome bupivacaine (Exparel) in peripheral nerve blocks. *Regional Anesthesia and Pain Medicine*, *40*(5), 572–582. https://doi.org/10.1097/AAP.0000000000000283.

Jorfeldt, L., Löfström, B., Pernow, B., Persson, B., Wahren, J., & Widman, B. (1968). The effect of local anaesthetics on the central circulation and respiration in man and dog. *Acta Anaesthesiologica Scandinavica*, *12*(4), 153–169. https://doi.org/10.1111/j.1399-6576.1968.tb00420.x.

Joshi, G., Patou, G., & Kharitonov, V. (2015). The safety of liposome bupivacaine following various routes of administration in animals. *Journal of Pain Research*, *8*, 781. https://doi.org/10.2147/JPR.S85424.

Kim, S., Kim, T., & Murdanhe, S. (2012). *Sustained-release liposomal anesthetic compositions*. United States Patent No: US8182835B2. Washington, DC, U.S. Retrieved from https://patents.google.com/patent/US8182835B2/en?q=sustained+release+LIPOSOMAL+ANESTHETIC&oq=sustained+release+LIPOSOMAL+ANESTHETIC.

Kim, S., Turker, M. S., Chi, E. Y., Sela, S., & Martin, G. M. (1983). Preparation of multivesicular liposomes. *Biochimica et Biophysica Acta (BBA) – Biomembranes*, *728*(3), 339–348. https://doi.org/10.1016/0005-2736(83)90504-7.

King, C. H., Beutler, S. S., Kaye, A. D., & Urman, R. D. (2017). Pharmacologic properties of novel local anesthetic agents in anesthesia practice. *Anesthesiology Clinics*, *35*(2), 315–325. https://doi.org/10.1016/j.anclin.2017.01.019.

Knudson, R. A., Dunlavy, P. W., Franko, J., Raman, S. R., & Kraemer, S. R. (2016). Effectiveness of liposomal bupivacaine in colorectal surgery: A pragmatic nonsponsored prospective randomized double blinded trial in a community hospital. *Diseases of the Colon and Rectum*, *59*(9), 862–869. https://doi.org/10.1097/DCR.0000000000000648.

Lachiewicz, P. F., Lee, G. C., Pollak, R. A., Leiman, D. G., Hu, J., & Sah, A. P. (2020). HTX-011 Reduced pain and opioid use after primary total knee arthroplasty: Results of a randomized phase 2b. *The Journal of Arthroplasty*, *35*(10), 2843–2851. https://doi.org/10.1016/j.arth.2020.05.044.

Malik, O., Kaye, A. D., Kaye, A., Belani, K., & Urman, R. D. (2017). Emerging roles of liposomal bupivacaine in anesthesia practice. *Journal of Anaesthesiology Clinical Pharmacology*. https://doi.org/10.4103/joacp.JOACP_375_15. Medknow Publications.

Malinovsky, J. M., Benhamou, D., Alafandy, M., Mussini, J. M., Coussaert, C., Couarraze, G., … Legros, F. J. (1997). Neurotoxicological assessment after intracisternal injection of liposomal bupivacaine in rabbits. *Anesthesia and Analgesia*, *85*(6), 1331–1336. https://doi.org/10.1097/00000539-199712000-00027.

Manna, S., Wu, Y., Wang, Y., Koo, B., Chen, L., Petrochenko, P., … Zheng, J. (2019). Probing the mechanism of bupivacaine drug release from multivesicular liposomes. *Journal of Controlled Release*, *294*(December 2018), 279–287. https://doi.org/10.1016/j.jconrel.2018.12.029.

Mantripragada, S. (2002). A lipid based depot (DepoFoam® technology) for sustained release drug delivery. *Progress in Lipid Research*, *41*(5), 392–406. https://doi.org/10.1016/S0163-7827(02)00004-8.

McAlvin, J. B., Padera, R. F., Shankarappa, S. A., Reznor, G., Kwon, A. H., Chiang, H. H., … Kohane, D. S. (2014). Multivesicular liposomal bupivacaine at the sciatic nerve. *Biomaterials*, *35*(15), 4557–4564. https://doi.org/10.1016/j.biomaterials.2014.02.015.

Mehran, R. J., Walsh, G. L., Zalpour, A., Cata, J. P., Correa, A. M., Antonoff, M. B., & Rice, D. C. (2017). Intercostal nerve blocks with liposomal bupivacaine: Demonstration of safety, and potential benefits. *Seminars in Thoracic and Cardiovascular Surgery*, *29*(4), 531–537. https://doi.org/10.1053/j.semtcvs.2017.06.004.

Mont, M. A., Beaver, W. B., Dysart, S. H., Barrington, J. W., & Del Gaizo, D. J. (2018). Local infiltration analgesia with liposomal bupivacaine improves pain scores and reduces opioid use after total knee arthroplasty: Results of a randomized controlled trial. *Journal of Arthroplasty*, *33*(1), 90–96. https://doi.org/10.1016/j.arth.2017.07.024.

Nadeau, M. H., Saraswat, A., Vasko, A., Elliott, J. O., & Vasko, S. D. (2016). Bupivacaine versus liposomal bupivacaine for postoperative pain control after augmentation mammaplasty: A prospective, randomized, double-blind trial. *Aesthetic Surgery Journal*, *36*(2), NP47–NP52. https://doi.org/10.1093/asj/sjv149.

Naseem, A., Harada, T., Wang, D., Arezina, R., Lorch, U., Onel, E., … Taubel, J. (2012). Bupivacaine extended release liposome injection does not prolong QTc interval in a thorough QT/QTc study in healthy volunteers. *Journal of Clinical Pharmacology*, *52*(9), 1441–1447. https://doi.org/10.1177/0091270011419853.

Neal, J. M., Bernards, C. M., Butterworth, J. F., Di Gregorio, G., Drasner, K., Hejtmanek, M. R., … Weinberg, G. L. (2010). ASRA practice advisory on local anesthetic systemic toxicity. *Regional Anesthesia and Pain Medicine*, *35*(2), 152–161. https://doi.org/10.1097/AAP.0b013e3181d22fcd.

Pacira Pharmaceuricals Inc. (2018). *Highlights of prescribing information, EXPAREL (bupivacaine liposome injectable suspension)*. Retrieved 7 May 2020, from https://www.exparel.com/hcp/prescriptioninformation.pdf.

Pacira Pharmaceuricals Inc. (n.d.). *Infiltration of exparel*. Retrieved 15 July 2020, from https://www.exparel.com/hcp/infiltration.

Padera, R., Bellas, E., Tse, J. Y., Hao, D., & Kohane, D. S. (2008). Local myotoxicity from sustained release of bupivacaine from microparticles. *Anesthesiology*, *108*(5), 921–928. https://doi.org/10.1097/ALN.0b013e31816c8a48.

Patel, M. A., Gadsden, J. C., Nedeljkovic, S. S., Bao, X., Zeballos, J. L., Yu, V., ... Bendtsen, T. F. (2020). Brachial plexus block with liposomal bupivacaine for shoulder surgery improves analgesia and reduces opioid consumption: Results from a multicenter, randomized, double-blind, controlled trial. *Pain Medicine, 21*(2), 387–400. https://doi.org/10.1093/pm/pnz103.

Raman, S., Lin, M., & Krishnan, N. (2018). Systematic review and meta-analysis of the efficacy of liposomal bupivacaine in colorectal resections. *Journal of Drug Assessment, 7*(1), 43–50. https://doi.org/10.1080/21556660.2018.1487445.

Rice, D. C., Cata, J. P., Mena, G. E., Rodriguez-Restrepo, A., Correa, A. M., & Mehran, R. J. (2015). Posterior intercostal nerve block with liposomal bupivacaine: An alternative to thoracic epidural analgesia. *Annals of Thoracic Surgery, 99*(6), 1953–1960. https://doi.org/10.1016/j.athoracsur.2015.02.074.

Rice, D., Heil, J. W., & Biernat, L. (2017). Pharmacokinetic profile and tolerability of liposomal bupivacaine following a repeated dose via local subcutaneous infiltration in healthy volunteers. *Clinical Drug Investigation, 37*(3), 249–257. https://doi.org/10.1007/s40261-017-0495-2.

Richard, B. M., Newton, P., Ott, L. R., Haan, D., Brubaker, A. N., Cole, P. I., ... Nelson, K. G. (2012). The safety of EXPAREL ® (bupivacaine liposome injectable suspension) administered by peripheral nerve block in rabbits and dogs. *Journal of Drug Delivery, 2012*, 1–10. https://doi.org/10.1155/2012/962101.

Richard, B. M., Ott, L. R., Haan, D., Brubaker, A. N., Cole, P. I., Nelson, K. G., ... Newton, P. E. (2011). The safety and tolerability evaluation of DepoFoam bupivacaine (bupivacaine extended-release liposome injection) administered by incision wound infiltration in rabbits and dogs. *Expert Opinion on Investigational Drugs, 20*(10), 1327–1341. https://doi.org/10.1517/13543784.2011.611499.

Richard, B. M., Rickert, D. E., Newton, P. E., Ott, L. R., Haan, D., Brubaker, A. N., ... Nelson, K. G. (2011). Safety evaluation of EXPAREL (DepoFoam bupivacaine) administered by repeated subcutaneous injection in rabbits and dogs: Species comparison. *Journal of Drug Delivery, 2011*, 1–14. https://doi.org/10.1155/2011/467429.

Sabesan, V. J., Shahriar, R., Petersen-Fitts, G. R., Whaley, J. D., Bou-Akl, T., Sweet, M., & Milia, M. (2017). A prospective randomized controlled trial to identify the optimal postoperative pain management in shoulder arthroplasty: Liposomal bupivacaine versus continuous interscalene catheter. *Journal of Shoulder and Elbow Surgery, 26*(10), 1810–1817. https://doi.org/10.1016/j.jse.2017.06.044.

Santamaria, C. M., Woodruff, A., Yang, R., & Kohane, D. S. (2017). Drug delivery systems for prolonged duration local anesthesia. *Materials Today (Kidlington), 20*(1), 22–31. https://doi.org/10.1016/j.mattod.2016.11.019.

Smoot, J. D., Bergese, S. D., Onel, E., Williams, H. T., & Hedden, W. (2012). The efficacy and safety of DepoFoam bupivacaine in patients undergoing bilateral, cosmetic, submuscular augmentation mammaplasty: A randomized, double-blind, active-control study. *Aesthetic Surgery Journal, 32*(1), 69–76. https://doi.org/10.1177/1090820X11430831.

Vandepitte, C., Kuroda, M., Witvrouw, R., Anne, L., Bellemans, J., Corten, K., ... Hadzic, A. (2017). Addition of liposome bupivacaine to bupivacaine HCl versus bupivacaine HCl alone for interscalene brachial plexus block in patients having major shoulder surgery. *Regional Anesthesia and Pain Medicine, 42*(3), 334–341. https://doi.org/10.1097/AAP.0000000000000560.

Viscusi, E. R., Candiotti, K. A., Onel, E., Morren, M., & Ludbrook, G. L. (2012). The pharmacokinetics and pharmacodynamics of liposome bupivacaine administered via a single epidural injection to healthy volunteers. *Regional Anesthesia and Pain Medicine, 37*(6), 616–622. https://doi.org/10.1097/AAP.0b013e318269d29e.

Viscusi, E. R., Sinatra, R., Onel, E., & Ramamoorthy, S. L. (2014). The safety of liposome bupivacaine, a novel local analgesic formulation. *Clinical Journal of Pain, 30*(2), 102–110. https://doi.org/10.1097/AJP.0b013e318288e1f6.

Yang, S., Abrahams, M. S., Hurn, P. D., Grafe, M. R., & Kirsch, J. R. (2011). Local anesthetic Schwann cell toxicity is time and concentration dependent. *Regional Anesthesia and Pain Medicine, 36*(5), 444–451. https://doi.org/10.1097/AAP.0b013e318228c835.

Ye, Q., Asherman, J., Stevenson, M., Brownson, E., & Katre, N. V. (2000). DepoFoam™ technology: A vehicle for controlled delivery of protein and peptide drugs. *Journal of Controlled Release, 64*(1–3), 155–166. https://doi.org/10.1016/S0168-3659(99)00146-7.

Chapter 28

Adverse events associated with analgesics: A focus on paracetamol use

Iwona Popiolek[a] and Grzegorz Porebski[b]

[a]Toxicology Clinic, University Hospital in Krakow, Krakow, Poland; [b]Department of Clinical and Environmental Allergology, Jagiellonian University Medical College, Krakow, Poland

Abbreviations

$[APAP]_P$	acetaminophen serum concentration
AC	activated charcoal
ALF	acute liver failure
ALP	alkaline phosphatase
ALT	alanine transaminase
APAP	N-acetyl-p-aminophenol, paracetamol, acetaminophen
AST	aspartate transaminase
ATP	adenosine triphosphate
DILI	drug induced liver injury
DRESS	drug reaction with eosinophilia and systemic symptoms
FDE	fixed drug eruption
GI	gastrointestinal
GSH	glutathione
GST	glutathione S-transferases
IV	intravenous
Ltx	liver transplantation
MPE	maculopapular eruption
NAC	N-acetylcysteine
NAPQI	N-acetyl-p-benzoquinone imine
NSAID	nonsteroidal antiinflammatory drugs
OTC	over the counter
ROS	reactive oxygen species
SJS/TEN	Stevens-Johnson syndrome/toxic epidermal necrolysis
SULT	sulfotransferase
UGT	uridine 5′-diphospho-glucuronosyltransferase
ULN	upper limit norm

Introduction

Paracetamol is worldwide used antipyretic and analgesic drug. It exhibits weak peripheral antiinflammatory and antiplatelet activity through the inhibition of prostaglandin and thromboxane synthesis (Thompson, Bundell, & Lucas, 2019). Despite of its excellent safety profile, excessive dose may cause toxicity targeted to liver and kidneys. Whereas hepatotoxicity due to paracetamol overdose is a well-known problem, hypersensitivity reactions induced by this drug are much less common in clinical practice. As the meaning of the title term "adverse event" is changing over the time, we follow here the definition of "adverse drug event", as it is described below in the section "Mini-dictionary of terms."

Epidemiology of APAP intoxication varies in different parts of the world. The countries of the highest number of publication describing acetaminophen use or overdose are the United States, the United Kingdom, Denmark, and Australia (Gunnell, Murray, & Hawton, 2000). Only in United States, APAP misuse cases are responsible for 26,000 hospitalizations, and 500 deaths per year (Agrawal & Khazaeni, 2020). Furthermore, the number of APAP intentional overdose incidence in

Treatments, Mechanisms, and Adverse Reactions of Anesthetics and Analgesics. https://doi.org/10.1016/B978-0-12-820237-1.00028-4
Copyright © 2022 Elsevier Inc. All rights reserved.

United States rise every year, especially in population 10–19 year old (Gummin et al., 2019). Easy access to this OTC drug combined with poor level of medical education creates the danger of accidental poisoning with this agent. Although the majority of patients experience mild adverse reactions, acetaminophen-related hepatotoxicity is generally estimated to account for approximately 45% of ALF cases in North America.

The strongest evidence on the prevalence of paracetamol hypersensitivity is difficult to obtain, as prescription data for OTC drugs are scarce. Nevertheless, the incidence of such reactions was estimated to be less than 1 per 100,000 in population of children under 15 years old, which is the common target group for this drug (Lee, 2017).

Mechanism of toxicity

Acetaminophen is easily and almost completely absorbed via the oral route and its maximum concentration is reached in 4 h after ingestion typically. The active substance from the tablet reaches central compartment after 45 min from the administration. Very high doses of paracetamol cause gastrointestinal irritating symptoms, but without elevated risk of GI bleeding (Bannwarth, 2004).

Scheme of APAP metabolism is presented on Fig. 1. Over 90% of absorbed acetaminophen goes to the reaction of the second phase with SULT and UGT resulting in conjugated products, eliminated by urine.

APAP hepatotoxicity occurs because of the formation of the noxious NAPQI metabolite through metabolism of the first phase. N-Acetyl-p-benzoquinone imine as a toxic compound is produced in hepatic and renal cytochromes, especially CYP2E1 but also CYP3A4, CYP2A6, CYP1A2 (Nelson, 2019). Then, in normal condition, NAPQI is quickly bounded by GSH forming nontoxic conjugates. In case of overdose, large amount of NAPQI is released into the cytoplasm, which cannot be detoxified by available glutathione. Subsequently, this reactive compound binds to –SH group from cysteine containing proteins, especially in mitochondria. This results in uncoupled mitochondria cytochrome reaction chain, breakdown of cellular respiration and activation the cascades of MAP kinases or other mechanisms leading to cellular necrosis. NAPQI forms also adducts proteins, which are released to the circulatory system and can be detected in serum. These products also may bind to nonsulfur compounds, for example mitochondrial proteins, promote forming of reactive oxygen species and finally decrease of ATP production resulting in cells death. These malfunctioning in mitochondrial metabolism led to acidosis and systemic hyperlactatemia. Histologically, the largest necrosis area is hepatic zone III, probably due to the highest cytochrome activity at this location. NAPQI is also produced in renal microsomes and the same mechanism is recognized as a cause for acute renal failure after APAP overdose independently from liver failure (Kennon-McGill & McGill, 2017).

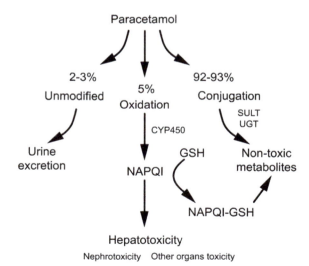

FIG. 1 Metabolic pathways of paracetamol. Diagram shows the pathophysiology of paracetamol toxicity. After absorption, a predominant part of the dose is conjugated in phase II pathway with sulfuric or glucuronic acid giving inactive metabolites excreted in the urine. A small percentage of drug is oxidized in microsomal systems of liver and kidney to NAPQI—a toxic metabolite, which can be neutralized by glutathione. A minority of the total dose is eliminated unmodified.

Clinical and laboratory manifestations of hepatotoxicity

The natural course of acute APAP toxicity can be divided into four stages. Firstly, the earliest major symptoms are from the gastrointestinal tract: nausea, vomiting, abdominal pain. Weakness, pallor, altered mental status, disturbance of consciousness, although rare, also may occur. In very massive overdose, metabolic acidosis with coma is possible, but other possible causes, for example mixed poisoning by popular combination with opioid, have to be excluded.

During the second phase, in next days after exposure, GI symptoms relieves, but liver damage develops. Hepatocellular indication parameters of acute liver failure such as serum aminotransferase activity are raising. The third phase starts usually after 2 days and represents the critical point—either hepatotoxicity will develop until acute liver failure with standard complications and a requirement for liver transplantation, or the symptoms get self-limiting and resolve completely. Death from acute liver failure occurs between 3 and 5 days after exposure. Acute APAP intoxication does not induce chronic liver failure.

Paracetamol induced liver injury is directly related with cellular necrosis. And for that reason, this type liver injury is associated with a large increase in aminotransferases activity, ALT and AST. During first 24 h all results are usually in normal range, except cases of very massive overdoses. Hepatotoxicity is defined as an elevation of AST or ALT over 1000 UI/l. This cut-off was established as a marker of significant hepatocytes' necrosis. Another widely available laboratory tests like prothrombin time, bilirubin level, arterial gasometry, creatinine level, or total blood count in the majority of cases have supportive role, but they are vital during qualification for urgent liver transplantation during severe intoxications. Another risk predictor, reserved only for those, who were treated by NAC is the multiplication result of ALT and serum paracetamol concentration (Wong & Graudins, 2017).

Risk factors of liver injury

The important factor for assessing hepatotoxicity is obviously ingested dose (Burns, Friedman, & Larson, 2020). This determines the amount of produced NAPQI, when the reactions of the second phase are saturated. The next key risk factor is the time from ingestion to start of treatment ("time to NAC"). NAC restores the substrate for NAPQI quick neutralization before necrosis begins (Wong & Graudins, 2017). The highest risk of hepatotoxic outcome occurs when the production of NAPQI is increased while detoxification mechanisms are impaired. Furthermore, greater risk of hepatotoxicity is associated with serum paracetamol concentration above the treatment thresholds on the paracetamol nomogram, the early increase in prothrombin time and chronic liver disease (Wong & Graudins, 2017). Both obesity and NAFLD are associated with increased CYP2E1 activity rising the risk. Ethanol is CYP1A2 substrate, so may compete with APAP lowering its toxicity, but an induced activity of cytochromes worsens prognosis during APAP overdose in alcoholic patients.

Children below 5 years are less susceptible for acetaminophen poisoning because of immaturity of the liver microsomal system, while in the opposite, older age is correlated with worse outcome because of depleted GSH stores. Finally, also genetically predisposition may alter the prognosis—for example patients with Gilbert syndrome have a high risk of liver injury after APAP overdose (Esteban & Pérez-Mateo, 1999). The summary of risk factors of liver injury after paracetamol poisoning is presented in Table 1.

TABLE 1 Risk factors paracetamol-induced hepatotoxicity divided into groups related to exposure, patient's medical history and laboratory results.

Risk of hepatotoxicity	Risk factors related to intoxication circumstances	Risk factors related to information from patient's history and examination	Risk factors related to laboratory findings at admission
Higher	– High dose of paracetamol ($\geq$40 g) – Co-ingestion of other hepatotoxic substances – Multiple overdoses	– Chronic alcohol intake – Malnutrition – Chronic liver disease – Obesity – Gilbert's syndrome	– Elevated ALT, AST during admission – Prolonged prothrombin time at admission – ALT $\times$ $[APAP]_P > 10{,}000$ mg/L $\times$ IU/L
Lower	– "Time to NAC" shorter than 8 h – Early GI tract decontamination	– Simultaneous alcohol consumption	– $[APAP]_P$ within normal range – Prothrombin time within normal range

The table summarizes factors which used for risk assessment (Burns et al., 2020; Chiew et al., 2017; Chomchai & Chomchai, 2018; Wong & Graudins, 2017; Wong, Sivilotti, & Graudins, 2017; Yoon, Babar, Choudhary, Kutner, & Pyrsopoulos, 2016).

Management of acute poisoning

Risk estimation of liver toxicity after a single overdose can be obtained using Rumack-Matthew nomogram (Rumack, 1981). This is semilogarithmic graph of modeled APAP concentration during the time after ingestion (Fig. 2). Maximal APAP concentration in this model starts at 4 h after ingestion, when absorption is likely to be completed, at 150 mg/L ("treatment line"), and at 200 mg/L for "probable hepatotoxicity line." If $[APAP]_P$ in specific time after exposure is above treatment line, this is an indication for NAC administration. This evaluation is quick and easy to apply and it is widely used until today. The main disadvantage of this nomogram is that its accuracy diminishes after time-point 16 h.

Gastrointestinal tract decontamination should be performed after single ingestion as soon as possible, but it is the most effective within the first 2 h. Activated charcoal in single dose absorbs acetaminophen, however repeated doses of AC are not recommended (Chiew, Gluud, Brok, & Buckley, 2018).

Antidote treatment is based on N-acetylcysteine was (Chiew et al., 2018), which is most often given intravenously. IV regime includes 21-h infusion with variable rate: first hour 150 mg/kg/h; then 4 h of rate 12.5 mg/kg/h and rest with 6.25 mg/kg/h. If hepatotoxicity occurs, the last infusion is prolonged until symptoms improvement or liver transplantation. Recently, it was proven that 12 h regime of NAC is noninferior to 21 h regime in hepatic injury prevention (Pettie et al., 2019). The most common side effects of NAC are nausea, and vomiting.

APAP is a small compound, which can be easily removed by means of extracorporeal elimination. Indications for this treatment include: (1) paracetamol serum concentration >1000 mg/L (6620 µmol/L) and an antidote is not administered; (2) paracetamol serum concentration >700 mg/L, the antidote is not administered and patient has one of following symptoms: altered mental status, metabolic acidosis with hyperlactatemia; (3) patient receives antidote but paracetamol serum concentration is over 900 mg/L and patient has one of following symptoms: altered mental status, metabolic acidosis with hyperlactatemia. For paracetamol poisoning, intermittent hemodialysis is recommended (Gosselin et al., 2014).

Patients with acute liver failure should be monitored and considered for liver transplantation. Criteria for LTx after paracetamol poisoning are: (1) Acidemia—pH below 7.3 or lactate over 3.0 mM after fluid resuscitation and (2) coagulopathy with INR over 6.5 or encephalopathy grade III or IV or creatinine level over 3.3 mg/dL. Patient with severe liver injury after paracetamol exposure should be managed in specialty medical centers (Yoon et al., 2016).

Chronic poisoning

Even a daily dose below 4 g/day, but repeated for a prolonged time may induce DILI (Zimmerman & Maddrey, 1995). The diagnosis of paracetamol-induced liver injury after chronic exposure is mostly based on exclusion of other causes, unless the history of excessive intake is clear. The latency period is several days after high therapeutic or supratherapeutic doses.

Some patients are more susceptible—especially those with a history of alcohol chronic consumption, malnutrition, and intensive weight loss. Chronic ethanol consumption induces CYP2E1, which increases the production of NAPQI

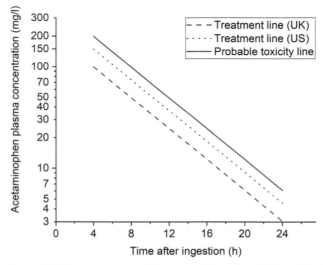

FIG. 2 Rumack-Matthew nomogram. Rumack-Matthew nomogram with treatment line and probable toxicity line. These lines are function graphs of $y = 200 \cdot 0{,}86^t$ *(dashed line)*, $y = 300 \cdot 0{,}86^t$ *(dotted line)*, and $y = 400 \cdot 0{,}86^t$ *(solid line)*. They are called: treatment line (United Kingdom), treatment line (United States), probable toxicity line, respectively.

(Nelson, 2019). Additionally, chronic alcoholics are at risk of malnutrition, with depleted hepatic glutathione stores and its impaired intracellular transportation. Finally, excessive alcohol consumption impairs the ability to correctly recognize danger and may delay seeking medical assistance.

Immediate hypersensitivity

Paracetamol-induced hypersensitivity reactions are heterogeneous in clinical presentation and underlying mechanisms. When immediate reactions to paracetamol occur in subjects cross-reacting to NSAIDs (more details—see the section "Application to other areas"), the suggested mechanism is related to its pharmacological effect, namely weak inhibition of cyclooxygenase-1, which results in alteration in arachidonic acid metabolism followed by overproduction of cysteinyl leukotrienes (Thompson et al., 2019). Whereas, an immediate hypersensitive reaction is reported in patient who is tolerant of NSAIDs, it implies alternative mechanism that stems from immune—IgE-dependent reaction. Such reaction is selective in nature and relies on production of drug-specific IgE at initial exposures in susceptible individuals. Subsequent treatments result in binding drug to IgE attached to its high-affinity receptors on mastocytes and basophils, degranulation of these cells and releasing vasoactive mediators, such as histamine and tryptase (Lee, 2017). Finally, when considering the mechanisms of immediate hypersensitivity, one should keep in mind, that different paracetamol commercial formulations commonly contain a number of excipients, like preservatives or colorings, which may be a cause of hypersensitivity by themselves (Lee, 2017).

The listed above mechanisms may be responsible for multiple clinical presentations of hypersensitivity, involving skin, mucosa, gastrointestinal tract, cardiovascular, and respiratory systems. They develop usually within 1 h after drug ingestion. The most common manifestations are urticaria and angioedema, bronchospasm with breathlessness. Hypotension, tachycardia, and anaphylactic shock are much less common. Abdominal pain, nausea, vomiting, and diarrhea are also possible. The severity ranges from the local symptom or sign to anaphylaxis. Obviously, previous tolerance of paracetamol does not protect from hypersensitivity reaction at subsequent exposure. Although immediate hypersensitivity to paracetamol is thought to be low, clinicians should be aware of it, particularly because of potentially severe outcomes including anaphylaxis (Thompson et al., 2019).

It is also worth to mention, that maternal paracetamol use during pregnancy has been accused of increasing the risk of childhood asthma. Meta-analyses dedicated to this issue showed increased odds of wheezing and asthma in infants, but the point of criticism was high heterogeneity across the analyzed surveys. Therefore, however existing data suggest an association between prenatal use of paracetamol and childhood asthma, there is a need for more well-designed, prospective, interventional studies to confirm it (Castro-Rodriguez, Forno, Rodriguez-Martinez, & Celedón, 2016).

Delayed hypersensitivity

Delayed reactions to paracetamol arise after T-cells response, which involves various effector cells responsible for hypersensitivity type IVa–IVc according to revised Gell-Coombs classification (Lee, 2017). Paracetamol is known to induce mild exanthemas (maculopapular eruption, MPE; fixed drug eruption, FDE) and severe cutaneous adverse reactions (Stevens-Johnson syndrome/toxic epidermal necrolysis, SJS/TEN; drug reaction with eosinophilia and systemic symptoms, DRESS) (Lee et al., 2019).

MPE is one of the most common delayed drug hypersensitivity reactions. Clinical picture corresponds to morbilliform exanthem, which spreads on the trunk and limbs about 7 days after the start of drug administration. Its immunologic mechanism is driven by specific CD4 cells, and to the lesser extend CD8 cells releasing perforin and granzyme B. Paracetamol often induces also FDE: solitary, well-demarcated, red to brown macules, which occur within a few hours after drug intake anywhere on the skin with a predilection to lips, groin, genitalia, hands, and feet. Skin signs are mediated by drug-specific CD8 T cells, which are located intraepidermally. DRESS is characterized by widespread maculopapular exanthema with accompanying peripheral blood eosinophilia, low-grade fever and organ dysfunction, usually liver injury. Symptoms typically develop within 2–6 weeks of treatment initiation. Pathologically, DRESS is related to eosinophilic inflammation with IL-5 and IL-4 predominance. Paracetamol may be also a culprit drug in SJS/TEN, but induces relatively mild symptoms. SJS/TEN starts with influenza-like symptoms followed by target-like skin lesions, blisters, mucosal ulcerations, conjunctivitis, and laboratory abnormalities (particularly anemia and lymphopenia). Loss of fluids and electrolytes due to extensive skin detachment may result in severe complications including hypovolemic shock and multiple organs failure. Therefore, SJS/TEN is potentially life-threatening condition with mortality rate reaching 45% in the most severe cases. Cell-mediated cytotoxicity plays a key role in its pathogenesis. Massive apoptosis of the keratinocytes is mediated by Fas/Fas ligand interaction and perforin, granulysin, and granzyme B released from cytotoxic CD8 cells and NK cells. The latency period between treatment initiation and onset of symptoms in SJS/TEN ranges from 4 to 28 days, but for paracetamol is usually short—a few days (Demoly et al., 2014; Lee, 2017; Porebski, Gschwend-Zawodniak, & Pichler, 2011).

Diagnosis and management of paracetamol-induced hypersensitivity reactions

A suspected drug has to be properly identified, as paracetamol has many trade names. Anamnesis should involve detailed information about time relationships between drug intake and symptoms (when they started, how long they persist), subsequent exposures to paracetamol (substances tolerated on the next occasion after the episode of hypersensitivity is unlikely to be a culprit), other drugs taken with paracetamol at the same time (other possible causative agents, single- or cross-reactions to NSAIDs). Also reaction induced by excipients, instead of an active substance, has to be taken into account. In case of SJS/TEN, it may happen that influenza-like prodromes are treated by antipyretic (e.g., paracetamol), but in fact a reaction is induced by a drug taken previously, what may result in causative misdiagnosis. Differential diagnosis at anamnesis includes, first of all skin symptoms of common diseases such as viral infections or chronic urticaria (Lee, 2017).

Skin prick test can be applied to detect IgE-mediated paracetamol allergy, but its sensitivity is low. A negative result may be also due to non-IgE-dependent mechanism of immediate hypersensitivity. Blood testing for specific IgE to paracetamol has been used in research, but it performed poorly (de Paramo et al., 2000; Thompson et al., 2019). Delayed reading of intradermal tests and patch tests with paracetamol may be helpful in diagnosing delayed reactions. These tests are specific, but not sensitive. The lymphocyte transformation test is an *in vitro* assay recommended in delayed reactions, but again it is used mostly for research purposes (Lee, 2017; Porebski et al., 2011).

Considering the above, drug provocation test with paracetamol turns out a reference diagnostic tool for formal causal diagnosis of hypersensitivity. Provocation should be carried out under supervision of the experienced clinicians in the setting that ensure the safety of patients re-exposed to paracetamol. Protocols can be individually adjusted depending on the severity of the initial hypersensitivity reaction. The initial dose usually ranges from 1/10 to 1/4 of the therapeutic dose and is followed by incremental doses with the maximal dose established at adult therapeutic dose. If hypersensitivity symptoms occur, the test is stopped and considered positive. Challenges are contraindicated during pregnancy, in patients with severe hypersensitivity reaction in anamnesis (SJS/TEN, DRESS, severe anaphylaxis, severe organ specific reactions, e.g., hepatitis), or with severe concomitant diseases (Kidon et al., 2018).

In the management of hypersensitivity reactions, suspected drugs should be stopped until the causal diagnostic work-up is performed. Later on, avoidance of substance confirmed as a culprit drug is necessary. This general rules apply also to hypersensitivity to paracetamol. Immediate reactions are usually treated with antihistamines, and depending on leading symptoms and their severity, with inhaled beta-mimetics, corticosteroids, oxygen, intravenous fluids, and intramuscular adrenaline. Therapy of delayed reactions is based on symptomatic treatment involving systemic corticosteroids and management in intensive care units in the most severe manifestations (e.g., SJS/TEN). As a patient hypersensitive to paracetamol often tolerates other antipyretics and painkillers, an alternative drug should be proposed, when its safety is confirmed by oral challenge. Management of cross-reactivity between paracetamol and NSAIDs is discussed more in details in the next section (Lee, 2017).

Applications to other areas

Acetaminophen poisoning is an example of intrinsic DILI. This type of dose related liver injury may also occur after exposure to high doses of other agents, for example niacin, aspirin, cocaine, IV amiodarone, anabolic steroids, cyclosporine, valproic acid, statins, IV methotrexate, and other cancer chemotherapy (Andrade et al., 2019; Hoofnagle & Björnsson, 2019). Exclusive ALT elevation after drug exposure might be also a phenotype of idiosyncratic DILI (Andrade et al., 2019). Non-APAP, idiosyncratic DILI is the second most common etiology of ALF and accounts for 10%–20% of ALF in developed countries (Stravitz & Lee, 2019).

In everyday practice, immediate hypersensitivity to paracetamol is often considered in conjunction with NSAIDs hypersensitivity, as paracetamol and NSAIDs have similar indications and mechanisms of action. Cyclooxygenases inhibition by NSAIDs accounts for their cross-reactive reactions, while paracetamol is only a weak inhibitor of cyclooxygenases. In clinic, it is reflected by the fact that up to 90% of NSAIDs hypersensitive patients tolerate paracetamol up to 500 mg. They usually tolerate also preferential and selective cyclooxygenase-2 inhibitors (such as meloxicam and celecoxib, respectively). Only a minority of hypersensitive patients (2%–10%) cross-react to paracetamol and NSAIDs. On the other hand, a patient with selective IgE-dependent allergy to paracetamol, may tolerate all NSAIDs. From a clinical point of view, hypersensitivity reactions related to IgE and to cyclooxygenase inhibition are often indistinguishable. Therefore, after positive oral challenge with paracetamol, a subsequent provocation with aspirin, playing a role of a NSAIDs representative, is necessary. If aspirin is tolerated, patients must avoid paracetamol, but will be able to tolerate NSAIDs. Patients presenting hypersensitivity reactions in both provocation tests, with paracetamol and aspirin, pose a management dilemma. They may be challenged under medical supervision with celecoxib, as a potential safe alternative (Gabrielli, Langlois, & Ben-Shoshan, 2018; Kowalski et al., 2011; Thompson et al., 2019).

Other agents of interest

Paracetamol is a specific compound which has the direct hepatotoxicity effect. Prevalence of liver injury due to other non-opioid analgesic drugs is not so high, but it also should be taken into account. In case of opioids, only combination these drugs with paracetamol increase the probability of liver injury (*LiverTox*, 2012).

Acetylsalicylic acid—Long-term therapy with high doses may result in elevations of ALT levels. Risk of hepatotoxicity rises when dose exceeds 100 mg/kg. This condition is often clinically asymptomatic. Acetylsalicylic acid is also associated with the occurrence of Rey's syndrome. This severe condition usually develops in children and young adults and clinical picture include metabolic acidosis, hepatic encephalopathy, and coma with high mortality. Nowadays, thanks to increasing awareness, this syndrome is incidental (*LiverTox*, 2012).

Diclofenac—15% of patients during chronic therapy have elevated serum aminotransferases levels, but clinically detectable liver function impairment is rare. Interestingly, diclofenac is able to cause clinical states very similar to chronic hepatitis due to its long latency period (Sriuttha, Sirichanchuen, & Permsuwan, 2018).

Celecoxib is able to induce both hepatocellular and cholestatic liver injury including prolonged jaundice and pruritus, but its prevalence is incidental. The mechanism of injury probably has an immune background (*LiverTox*, 2012).

Mini-dictionary of terms

Adverse drug event. An injury resulting from a medical occurrence related to a drug and including medication errors, hypersensitivity reactions and overdoses.

Anaphylaxis. An acute and potentially life-threating systemic hypersensitivity reaction caused by mediators released from mastocytes and basophils.

Antidote. A substance used for reversing the effect of the toxin. Antidotes can be specific, with a particular way of acting, for example naloxone for opioid overdose, or unspecific such as activated carbon which binds a large group of toxins preventing them from absorption to central compartment.

Extracorporeal elimination. It is a way of accelerating the elimination of a drug or toxin from the bloodstream using special tools that cleanse the blood outside the patient's body. Hemodialysis, hemoperfusion, and hemofiltration are the main methods of extracorporeal elimination. They differ in the efficiency of removing various substances from the blood. Each procedure requires a dialysis catheter or other vascular access that provides proper blood flow through the filter or absorption column.

Acute hepatic failure. It is rapid onset of symptoms of liver failure with coagulopathy, encephalopathy, and jaundice developed in short period in the absence of chronic liver disease. It is caused by paracetamol, hepatic ischemia, hepatitis (viral and autoimmune) and other hepatotoxic agents, drugs, industrial, or natural compounds. This condition has a 30% mortality (Stravitz & Lee, 2019).

Nomogram. It is a graph prepared for quick, simple and approximate determination (without any calculation) of the variable value. Such an example is the Rumack-Matthew nomogram, where, knowing the time from abuse and the concentration of paracetamol, the information about the risk of hepatotoxicity can be easily obtained.

Drug hypersensitivity reactions. Drug-induced reactions clinically resembling allergy, which are divided into **immune (allergic)** and **nonimmune (nonallergic)** reactions. Evidence for the immunological mechanism (either drug-specific T cells or antibodies) is demonstrated in allergic reactions. Whereas in nonallergic reactions, immunological mechanism is not confirmed or another background (e.g., nonspecific histamine release from mastocytes and basophiles, alternation in arachidonate metabolism related to cyclooxygenase inhibition) is suggested.

Immediate and nonimmediate drug hypersensitivity reactions. They are defined depending on the time of onset. Immediate reactions occur within 1–6 h after drug intake (e.g., urticaria, angioedema, bronchospasm). Nonimmediate reactions take place any time as from 1 h after exposition to drug, but usually after many days of treatment (e.g., maculopapular exanthema, Stevens-Johnson syndrome).

Key facts of paracetamol-induced toxicity and hypersensitivity

- Paracetamol in great majority is conjugated with sulfuric and glucuronic acids, but 5% of the absorbed dose is oxidized mostly by CYP2E1 producing toxic metabolite NAPQI.
- Acute poisoning presents four clinical stages: I—gastrointestinal symptoms; II—development of liver injury with a resolution of GI symptoms; III—clinically apparent liver failure; IV—outcome: death, liver transplantation or full recovery.
- *N*-Acetylcysteine administration, according to as an antidote is the most important intervention after paracetamol overdose. NAC provides –SH groups for glutathione regeneration. GSH reacts with NAPQI forming nontoxic conjugates.
- Evaluation of hepatotoxicity risk and antidote treatment indications after a single overdose can be obtain from Rumack-Matthew nomogram.
- Chronic poisoning may occur even with supratherapeutic doses, ingested for a prolonged time in specific, susceptible populations.
- The majority of patients hypersensitive to NSAIDs tolerate paracetamol up to 500 mg.
- Immediate reactions, including anaphylaxis can be manifestations of selective hypersensitivity to paracetamol.
- Paracetamol is known to induce SJS/TEN with a mild clinical course.

Summary points

- Paracetamol poisoning is common all over the world and in some regions is an important public health problem.
- Mechanism of hepatotoxicity is related to the presence of *N*-acetyl-*p*-benzoquinone imine, which impairs mitochondrial metabolism leading to hepatocytes necrosis.
- Nausea and vomiting are early symptoms of overdose, followed by an increase in ALT and AST levels.
- Management of acute poisoning relies on gastrointestinal tract decontamination with activated charcoal, antidote (*N*-acetylcysteine) IV administration according to Rumack-Matthew nomogram, and when necessary, extracorporeal removal of paracetamol and liver transplantation.
- Outcomes of paracetamol poisoning are very different—from complete recovery to death.
- Hypersensitivity to paracetamol involve immediate reactions such as urticaria, bronchospasm, anaphylaxis, and delayed reactions, including the most the most severe one—SJS/TEN.
- As skin tests and *in vitro* assays perform poorly, a supervised oral provocation is usually required for the diagnosis of hypersensitivity reactions to paracetamol.
- Management of paracetamol hypersensitivity is based on long-term drug avoidance and finding a safe alternative antipyretics.

References

Agrawal, S., & Khazaeni, B. (2020). *Acetaminophen toxicity. In StatPearls.* StatPearls Publishing(2020). *http://www.ncbi.nlm.nih.gov/books/NBK441917/.*

Andrade, R. J., Aithal, G. P., Björnsson, E. S., Kaplowitz, N., Kullak-Ublick, G. A., Larrey, D., et al. (2019). EASL clinical practice guidelines: Drug-induced liver injury. *Journal of Hepatology, 70*(6), 1222–1261. https://doi.org/10.1016/j.jhep.2019.02.014.

Bannwarth, B. (2004). Gastrointestinal safety of paracetamol: Is there any cause for concern? *Expert Opinion on Drug Safety, 3*(4), 269–272. https://doi.org/10.1517/14740338.3.4.269.

Burns, M. J., Friedman, S. L., & Larson, A. M. (2020). *Acetaminophen (paracetamol) poisoning in adults: Pathophysiology, presentation, and evaluation. In J. Grayzel (Ed.), UpToDate*(2020). https://www.uptodate.com/contents/acetaminophen-paracetamol-poisoning-in-adults-pathophysiology-presentation-and-evaluation.

Castro-Rodriguez, J. A., Forno, E., Rodriguez-Martinez, C. E., & Celedón, J. C. (2016). Risk and protective factors for childhood asthma: What is the evidence? *Journal of Allergy and Clinical Immunology: In Practice, 4*, 1111–1122.

Chiew, A. L., Gluud, C., Brok, J., & Buckley, N. A. (2018). Interventions for paracetamol (acetaminophen) overdose. *Cochrane Database of Systematic Reviews.* https://doi.org/10.1002/14651858.CD003328.pub3.

Chiew, A. L., Isbister, G. K., Kirby, K. A., Page, C. B., Chan, B. S. H., & Buckley, N. A. (2017). Massive paracetamol overdose: An observational study of the effect of activated charcoal and increased acetylcysteine dose (ATOM-2). *Clinical Toxicology, 55*(10), 1055–1065. https://doi.org/10.1080/15563650.2017.1334915.

Chomchai, S., & Chomchai, C. (2018). Being overweight or obese as a risk factor for acute liver injury secondary to acute acetaminophen overdose. *Pharmacoepidemiology and Drug Safety, 27*(1), 19–24. https://doi.org/10.1002/pds.4339.

de Paramo, B., Gancedo, Q., Cuevas, M., Camo, I., Martin, J., & Cosmes, E. (2000). Paracetamol (acetaminophen) hypersensitivity. *Annals of Allergy, Asthma & Immunology, 85*, 508–511.

Demoly, P., Adkinson, N. F., Brockow, K., Castells, M., Chiriac, A. M., Greenberger, P. A., et al. (2014). International consensus on drug allergy. *Allergy, 69*, 420–437.

Esteban, A., & Pérez-Mateo, M. (1999). Heterogeneity of paracetamol metabolism in Gilbert's syndrome. *European Journal of Drug Metabolism and Pharmacokinetics, 24*(1), 9–13. https://doi.org/10.1007/BF03190005.

Gabrielli, S., Langlois, A., & Ben-Shoshan, M. (2018). Prevalence of hypersensitivity reactions in children associated with acetaminophen: A systematic review and meta-analysis. *International Archives of Allergy and Immunology, 176*(2), 106–114.

Gosselin, S., Juurlink, D. N., Kielstein, J. T., Ghannoum, M., Lavergne, V., Nolin, T. D., et al. (2014). Extracorporeal treatment for acetaminophen poisoning: Recommendations from the EXTRIP workgroup. *Clinical Toxicology, 52*(8), 856–867. https://doi.org/10.3109/15563650.2014.946994.

Gummin, D. D., Mowry, J. B., Spyker, D. A., Brooks, D. E., Beuhler, M. C., Rivers, L. J., et al. (2019). 2018 Annual report of the American Association of Poison Control Centers' National Poison Data System (NPDS): 36th annual report. *Clinical Toxicology, 57*(12), 1220–1413. https://doi.org/10.1080/15563650.2019.1677022.

Gunnell, D., Murray, V., & Hawton, K. (2000). Use of paracetamol (acetaminophen) for suicide and nonfatal poisoning: Worldwide patterns of use and misuse. *Suicide & Life-Threatening Behavior, 30*(4), 313–326.

Hoofnagle, J. H., & Björnsson, E. S. (2019). Drug-induced liver injury—Types and phenotypes. *New England Journal of Medicine, 381*(3), 264–273. https://doi.org/10.1056/NEJMra1816149.

Kennon-McGill, S., & McGill, M. (2017). Extrahepatic toxicity of acetaminophen: Critical evaluation of the evidence and proposed mechanisms. *Journal of Clinical and Translational Research.* https://doi.org/10.18053/jctres.03.201703.005.

Kidon, M., Blanca-Lopez, N., Gomes, E., Terreehorst, I., Tanno, L., Ponvert, C., et al. (2018). EAACI/ENDA position paper: Diagnosis and management of hypersensitivity reactions to non-steroidal anti-inflammatory drugs (NSAIDs) in children and adolescents. *Pediatric Allergy and Immunology, 29*(5), 469–480.

Kowalski, M. L., Makowska, J. S., Blanca, M., Bavbek, S., Bochenek, G., Bousquet, J., et al. (2011). Hypersensitivity to nonsteroidal anti-inflammatory drugs (NSAIDs) – Classification, diagnosis and management: Review of the EAACI/ENDA and GA2LEN/HANNA. *Allergy, 66*, 818–829.

Lee, Q. U. (2017). Hypersensitivity to antipyretics: Pathogenesis, diagnosis, and management. *Hong Kong Medical Journal, 23*, 395–403.

Lee, S.-Y., Nam, Y. H., Koh, Y.-I., Kim, S. H., Kim, S., Kang, H.-R., et al. (2019). Phenotypes of severe cutaneous adverse reactions caused by nonsteroidal anti-inflammatory drugs. *Allergy, Asthma & Immunology Research, 11*(2), 212–221.

LiverTox: Clinical and research information on drug-induced liver injury. (2012). National Institute of Diabetes and Digestive and Kidney Diseases. (2012). *http://www.ncbi.nlm.nih.gov/books/NBK547852/.*

Nelson, L. (Ed.). (2019). *Goldfrank's toxicologic emergencies.* (11th ed.). McGraw-Hill Education.

Pettie, J. M., Caparrotta, T. M., Hunter, R. W., Morrison, E. E., Wood, D. M., Dargan, P. I., et al. (2019). Safety and efficacy of the SNAP 12-hour acetylcysteine regimen for the treatment of paracetamol overdose. *EClinicalMedicine, 11*, 11–17. https://doi.org/10.1016/j.eclinm.2019.04.005.

Porebski, G., Gschwend-Zawodniak, A., & Pichler, W. J. (2011). In vitro diagnosis of T cell-mediated drug allergy. *Clinical and Experimental Allergy, 41*(4), 461–470.

Rumack, B. H. (1981). Acetaminophen overdose. 662 cases with evaluation of oral acetylcysteine treatment. *Archives of Internal Medicine, 141*(3), 380–385. https://doi.org/10.1001/archinte.141.3.380.

Sriuttha, P., Sirichanchuen, B., & Permsuwan, U. (2018). Hepatotoxicity of nonsteroidal anti-inflammatory drugs: A systematic review of randomized controlled trials. *International Journal of Hepatology, 2018*, 1–13. https://doi.org/10.1155/2018/5253623.

Stravitz, R. T., & Lee, W. M. (2019). Acute liver failure. *The Lancet, 394*(10201), 869–881. https://doi.org/10.1016/S0140-6736(19)31894-X.

Thompson, G., Bundell, C., & Lucas, M. (2019). Paracetamol allergy in clinical practice. *Australian Journal of General Practice, 48*(4), 216–219.

Wong, A., & Graudins, A. (2017). Risk prediction of hepatotoxicity in paracetamol poisoning. *Clinical Toxicology, 55*(8), 879–892. https://doi.org/10.1080/15563650.2017.1317349.

Wong, A., Sivilotti, M. L. A., & Graudins, A. (2017). Accuracy of the paracetamol-aminotransferase multiplication product to predict hepatotoxicity in modified-release paracetamol overdose. *Clinical Toxicology, 55*(5), 346–351. https://doi.org/10.1080/15563650.2017.1290253.

Yoon, E., Babar, A., Choudhary, M., Kutner, M., & Pyrsopoulos, N. (2016). Acetaminophen-induced hepatotoxicity: A comprehensive update. *Journal of Clinical and Translational Hepatology, 4*(2), 131–142. https://doi.org/10.14218/JCTH.2015.00052.

Zimmerman, H. J., & Maddrey, W. C. (1995). Acetaminophen (paracetamol) hepatotoxicity with regular intake of alcohol: Analysis of instances of therapeutic misadventure. *Hepatology, 22*(3), 767–773.

Part IV

Novel and nonpharmacological aspects and treatments

Chapter 29

Acronychia pedunculata leaves and usage in pain

U.G. Chandrika[a] and W.M.K.M. Ratnayake[b]

[a]*Department of Biochemistry, Faculty of Medical Sciences, University of Sri Jayewardenepura, Nugeggoda, Sri Lanka,* [b]*Department of Pharmaceutical and Cosmetic Sciences, Faculty of Health Sciences, CINEC Campus, Malabe, Sri Lanka*

Abbreviations

AF-EEAL	alkaloid fraction of ethanol extract of *A. pedunculata* leaves
CMC	carboxymethyl cellulose
COX-2	cyclooxygenase-2
EEAL	70% ethanol extract of *A. pedunculata* leaves
NSAIDs	nonsteroidal anti-inflammatory drugs
WHO	World Health Organization

Introduction

Pain is one of the vital alert systems in the human body about the unpleasant sensation that occurs in the body. It helps to identify the various stimuli of intensities, which can be potentially harmful to the body tissue. Hence, pain is an essential signal which brings patients as well as physician attention in most of the disease conditions. Pain can be a somatogenic pain, which is caused by physical injury or psychological distress, which occurs due to some disturbance in mind. Further, based on the duration, pain can be classified as acute, subacute or chronic pain.

Although pain is a body protective mechanism, uncontrolled pain can reduce the quality of life by increasing the stress, decreasing appetite, disrupting sleep, causing anxiety and depression, etc. Hence, it is essential to manage the pain appropriately by treating with painkilling agents. There are four main pain treatment groups for muscle pain, neuropathic pain, inflammatory pain, and mechanical pain. The main drug groups are nonsteroidal anti-inflammatory drugs (NSAIDs), opioid analgesics, adjuvant medications like muscle relaxants and anxiolytics. However, many of these allopathic drugs are showing some adverse effects such as gastric irritations, which are leading to gastric ulcers, drug addiction, etc. Hence, there is a trend to have taken indigenous drugs, which are based on plant materials.

Traditional medicinal system in Sri Lanka

Like most of the other regions of the worldwide, Sri Lanka also has its medicinal systems within their cultural framework. Ancient people in Sri Lanka practiced their therapeutic method, and there is evidence from the period of King Ravana. According to the World Health Organization (WHO), about 80% of the world population still relies mainly on plant-based traditional medicine system (Patel, Patel, & Prajapati, 2013), and it is accessible within the society due to their abundant availability, low cost, and fewer side effects. Traditional medicine in Sri Lanka comprises four different systems; i.e. Ayurveda, Siddha, Unani and Deshiya chikista. Ayurveda and Siddha systems came to Sri Lanka from India along with several cultural waves, Unani system through the Arabs and Deshiya Chikitsa or indigenous medicine originated in Sri Lanka probably during prehistoric times and developed in its way. In Ayurveda, mostly herbal preparations are used as therapeutic agents for curing disease, and in the Siddha system, predominantly mineral preparations are used. The Unani system differs from the other two in its fundamental concepts. Deshiya Chikitsa uses mostly herbal preparations as in Ayurveda (Weragoda, 1980). Among the native flora of Sri Lanka, more than 1400 plants are used in traditional medicine (Wijesundera, 2004), and they are claimed as a treatment for certain diseases. The folk medicinal knowledge of the

Treatments, Mechanisms, and Adverse Reactions of Anesthetics and Analgesics. https://doi.org/10.1016/B978-0-12-820237-1.00029-6
Copyright © 2022 Elsevier Inc. All rights reserved.

322 **PART | IV** Novel and nonpharmacological aspects and treatments

remedies has been transferred from one generation to the next through traditional healers, knowledgeable elders or ordinary people, mostly without any written documents.

Typically, the plants that are being used for medicinal purposes are evergreen in nature and are grown in the backyards of houses by most people. Napagoda, Sundarapperuma, Fonseka, Amarasiri, and Gunaratna (2018) has carried out a study to assess the significance and contribution of medicinal plants in inflammatory conditions by collecting the data from 458 volunteers aged above 30 years, in Gampaha district. Out of the total participants, 51% claimed the use of medicinal plants for the treatment of inflammatory conditions such as fever, cough, asthma, swellings, and pain in the joints. The 66% of participants claimed that they use these herbal preparations at the initial stage of the disease before using for any other medications. Another 19% have mentioned that they are taking herbal medicine simultaneously with other drugs. Remaining 12%, use herbal therapeutics as a last resort, when other treatment methods have failed (Napagoda et al., 2018). These people diagnose the inflammatory conditions by their signs and symptoms rather than specific laboratory tests.

Acronychia pedunculata as a Sri Lankan medicinal plant

Acronychia pedunculata is a small evergreen plant that can range in size from a shrub to a large tree and is generally known as "Ankenda" in Sinhala, claw-flowered laurel in English and "Kattukanni" in Tamil (Jayaweera, 1982). It is widely distributed in Sri Lanka, especially in montane and rain forest understory, gaps, and fringes (Ashton et al., 1997), up to 5000 ft elevation (Jayaweera, 1982). It also found in Indonesia, Malaysia, Southern China, Hong Kong, Philippine islands (Jayaweera, 1982; Pathmasiri et al., 2005; Su et al., 2003).

Leaves are simple and oval or oblong shape with acute base. They are approximately 7.5–12.5 cm in length and have opposite or some alternate arrangement. And also, they are glabrous and shining dark green. The trunk has pale smooth bark and glabrous young branches. Flowers are regular, polygamous and pale yellowish-green in color with pedicels 1.8 cm long. They have loosely arranged in pyramidal, divaricate, corymbose cymes on long straight axillary peduncles. Four sepals are fused into four lobes. Lobes are short and broad. There are strap-shaped four petals with acute, inflexed at the tip and hairy within at the base. Anthers are versatile, the disc large and tomentose. The ovary is superior, tomentose, sunk on top of the disc. Style is concise. Stigma is four-lobed. Fruits are indehiscent, globular, 1.2–1.8 cm long, glabrous, rough with immersed glands, harder in the center but with no distinct stone and four-chambered.

Leaves, roots, barks and fruits of this plant have been used in folk medicine for the treatment of diarrhea, cough, asthma, ulcers, itchy skin, scales, pain, swellings, rheumatism and disorders with the involvement of the inflammatory process (Han, Pathmasiri, Bohlin, & Janson, 2004; Pathmasiri et al., 2005; Wu et al., 1989). Also, these are used as an antipyretic, antihemorrhagic and an aphrodisiac agent (Han et al., 2004).

In Sri Lankan traditional medicinal system, leaves, barks and roots of *A. pedunculata* are used for the preparation of various local applications as well as orally administered conventional drugs. "Ankenda thailaya," "Ratha thailaya," "Neel-dyaadee thailaya" are some of the examples for commonly used traditional preparations with *A. pedunculata*. Some of these preparations are externally applied, and some are taken internally as a purgative.

Chemistry and biological activities of *A. pedunculata*

The first phytochemical study of this plant was reported in 1979 by De Silva and his co-workers. They have isolated furoquinoline alkaloids, kokusaginine and evolitrine, from *A. pedunculata* leaves and timber, respectively. They found that kokusaginine showed 0.1% yield in the leaves, and evolitrine gave a yield of 0.05% from the wood (De Silva, De Silva, Mahendran, & Jennings, 1979). Although kokusaginine and evolitrine are the major alkaloids present in Sri Lankan *A. pedunculata*, the Indian variety of it collected from Madras (Chennai) does not contain these alkaloids (De Silva et al., 1979). Hence, these two phytochemicals may be considered as chemotaxonomic markers of *A. pedunculata* species of Sri Lanka. Ten years later, a new arylketone, 1-[2′,4′-dihydroxy-3′,5′-di-(3″-methyl-2″-butenyl)-6′ methoxy]phenylethanone was isolated together with acronylin, acrovestone, bergapten, β-amyrin and three furoquinoline alkaloids from the root bark of *A. pedunculata* by Kumar and co-workers (Kumar, Karunaratne, Sanath, & Meegalle, 1989). In the same year, Wu and co-workers have shown that methanol extract of the stem and root bark of *A. pedunculata* has significant cytotoxicity in the human KB tissue culture assay. They isolated a compound named as, acrovestone, as a cytotoxic principle in the early extract by bioassay directed fractionation (Wu et al., 1989).

Su et al. (2003) have isolated five new acetophenones from stem bark and root of this plant. They were namely acronyculatins A, B, C, D and E. In addition to these new compounds; the same research group isolated a known compound acrovestone. The antioxidant and antityrosinase activities of these five acetophenones and acrovestone were evaluated.

All new acetophenone compounds were found to be inactive at 500 μM in the DPPH assay as well as in antityrosinase activity assay. Acrovestone exhibited weak scavenging activity with IC_{50} value of 493 μM, compared to the reference compound, vitamin E (IC_{50}, 8.3 μM). Also, it showed weak inhibitory activity with an IC_{50} value of 333 μM, compared to the reference compound, kojic acid (IC_{50} value 125 μM) (Su et al., 2003). Sy and Brown has isolated a new aryl ketone, 1-[2', 4'-dihydroxy-3'-(3''-methylbut-2''-enyl)-5'-(1'''-ethoxy-3'''-methylbutyl)-6'-methoxy] phenylethanone from leaves of *A. pedunculata* (Sy & Brown, 1999). No biological activities of this new compound have been reported so far. In 2005, Pathmasiri and co-workers, isolated 1-[2, 4-dihydroxy-6-methoxy-3,5-bis(3- methylbut-2-en-1-yl)phenyl]ethanone and a new aryl ketone, named acrovestenol as in vitro cyclooxygenase-2 (COX-2) inhibitory principles from a dichloromethane extract of the bark of *A. pedunculata* by a bioassay-guided fractionation procedure. Pathmasiri et al. (2005), showed that acrovestenol inhibited COX-2 with IC_{50} value of 142.0 ± 2.15 μM, compared to the COX-2 inhibitory reference compound with an IC_{50} value of 11.3 ± 1.12 μM. And also their study showed that the compound 1-[2,4-dihydroxy-6-methoxy-3,5-bis (3-methylbut-2-en-1-yl) phenyl] ethanone, inhibited COX-2 catalyzed prostaglandin synthesis with 68% at a concentration of 500 μM (Pathmasiri et al., 2005). As COX-2 enzyme is involved in the production of prostaglandins that promote the inflammatory process, this study provided the scientific evidence for anti-inflammatory effect *A. pedunculata* bark used in traditional medicine for inflammatory conditions. Essential oil from aerial parts (branches, leaves and fruits) of *A. pedunculata* has been isolated, and 34 compounds were identified, accounting for 92.8% of the oil. The major constituents were α-pinene (57.4%) and (E)-β-caryophyllene (13.6%). The study of Lesueur and co-workers have shown that the essential oil of *A. pedunculata* possesses broad-spectrum antimicrobial activity against various bacteria, particularly *Salmonella enterica* and *Staphylococcus epidermidis* (Lesueur et al., 2008). Three new acetophenone diamers, acropyrone, acropyranol A, and acropyranol B were isolated from the trunk bark by Kouloura and co-workers. Also, the same research group isolated four known acetophenone diamers, acrovestone, acrovestenol, acrofolione A, acrofolione B and the acetophenone monomer acronyline and four furoquinoline alkaloids. These compounds were subjected to MTS assay against human DU145 prostate cancer and A2058 melanoma cells. Among the isolated compounds acrovestone and acrovestenol exhibited cytototoxicity (Kouloura et al., 2012).

Further, Kozaki and co-workers worked on chloroform soluble fractions of the methanolic extract of the dried leaves and twigs of this plant, and they could isolate three new acetophenone monomers (acronyculatin F, acronyculatin G, acronyculatin H) together with five known compounds, 1-[2',4'-dihydroxy-3',5'-di-(3''-methyl-2''-butenyl)-6' methoxy] phenylethanone, acronyculatin E and a mixture of β-sistosterol, stigmasterol and sesamin. There are reports that prenylated acetophenone dimers exhibited cytotoxic activities against human cancer cell lines (Kouloura et al., 2012). Further, Kozaki, Takenaka, Mizushina, Yamaura, and Tanahashi (2014) worked on the inhibitory activity of the isolated compounds on mammalian DNA polymerases using calf polymerase α and ray polymerase β as a representative DNA replicative polymerase and DNA repair-related polymerase, respectively. They found that acronyculatin F was the strongest polymerase α and β inhibitors among the compounds (Kozaki et al., 2014).

Anti-inflammatory activity of *A. pedunculate* leaves

As *A. pedunculata* leaves are used in both aqueous and oily traditional medicinal preparation, our research team worked on anti-inflammatory activity of aqueous and 70% ethanol extracts of it by using different *in-vivo* inflammatory models (Ratnayake, Suresh, Abeysekara, Salim, & Chandrika, 2019). It had been investigated effects of varying doses of both types of extracts.

It is found that 70% ethanol extract of *A. pedunculata* leaves (EEAL) has significant anti-inflammatory activity than the aqueous extract of it. Further studies on 70% ethanol extract of *A. pedunculata* leaves also carried out to find optimum anti-inflammatory dose on carrageenan-induced rat hind paw oedema model, which is a widely used model to evaluate acute anti-inflammatory activity.

Carrageenan-induced rat hind paw oedema model has several advantages, including the ability to detect orally acting anti-inflammatory agents in the acute phase (Owoyele, Adebukola, Funmilayo, & Soladoye, 2008). According to Di Rosa (1972), this test was instrumental in the discovery of the anti-inflammatory activity of indomethacin, which is one of the most initial NSAID. Carrageenan is the sulfated polysaccharide obtained from a sea-weed (Necas & Bartosikova, 2013). It is a widely used phlogistic agent that shows signs and symptoms of inflammation (Biradar, Kangralkar, Mandavkar, Thakur, & Chougule, 2010). The development of oedema in the paw of the rat after the subcutaneous injection of carrageenan is a multimediated phenomenon that liberates a diversity of mediators. It is believed to be a biphasic event (Patel et al., 2013; Vasudevan, Gunnam, & Parle, 2006). The initial phase observed during the first 2 h is attributed to the release of histamine and serotonin (Vasudevan et al., 2006). The second phase of oedema is due to the release of prostaglandins, proteases and lysozyme (Vasudevan et al., 2006). These chemical substances cause an increase in vascular permeability, thereby promoting the accumulation of fluid in tissues that accounts for oedema (Umukoro & Ashorobi, 2006). It has been reported that the second phase of oedema is sensitive to most of the clinically useful anti-inflammatory agents (Panda & Khambat, 2013). Thus, this model is vital to find new anti-inflammatory compounds.

According to our previous studies, the all the test doses of 70% ethanol extracts of *A. pedunculata* leaves in 0.5% carboxymethyl cellulose (CMC), i.e., 100, 200, 300 and 500 mg/kg b.w, significantly ($P < .05$) reduced paw oedema as compared to the negative control which received only 0.5% CMC (Ratnayake et al., 2019). Hence, all the test doses of EEAL have significant acute anti-inflammatory activity. Among the test doses, it has been found that 200 mg/kg b.w. dose is the minimum effective dose with maximum oedema inhibition (78% inhibition). It was 89% for indomethacin (5 mg/kg b.w.) which is the positive control. Interestingly, both of plant extract and indomethacin showed their maximum percentage reduction of oedema at the 5th h.

Identification of anti-inflammatory fractions and active compounds of *A. pedunculate* leaves

According to our previous studies, it has been shown that there is an enhancement anti-inflammatory activity of alkaloid fraction of *A. pedunculata* leaves (AF-EEAL). Further studies using activity-guided fraction studies, we were able to isolate the alkaloid responsible for anti-inflammatory activity as evolitrine, which is a furoquinoline alkaloids (Fig. 1).

Our previous studies were also showed that evolitrine has an activity enhancement on anti-inflammatory activity (Fig. 2). Hence, it was identified as an active compound in *A. pedunculate* leaves. However, the activity is less than that of indomethacin (5 mg/kg b.w), which inhibits oedema by 86% when compared with 89% by evolitrine (50 mg/kg b.w.)

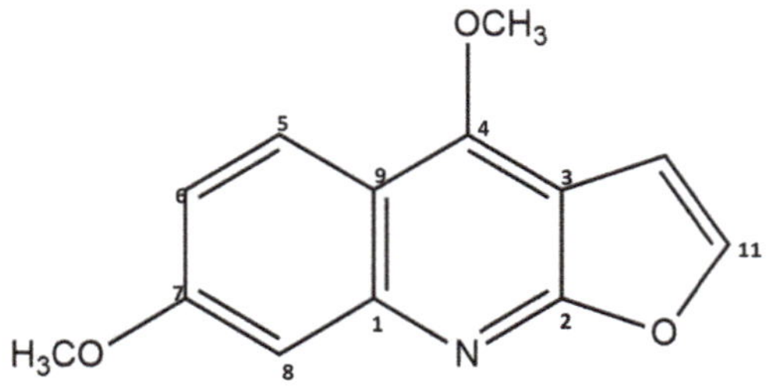

FIG. 1 The structure of the evolitrine.

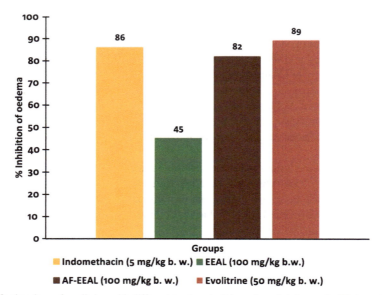

FIG. 2 Comparison of the effective dose of evolitrine with different treatment of *A. pedunculata* leaves in Wistar rats ($n = 6$/group) (*EEAL*, ethanol extract of *A. pedunculata* leaves; *AF-EEAL*, alkaloid fraction of ethanol extract of *A. pedunculata* leaves).

Analgesic activity of *A. pedunculate* leaves

As pain is one of the common signs in inflammation, infection, tissue damage, and most of the diseases, our research team have worked on the antinociceptive activity of *A. pedunculata* leaves. They used the acetic acid-induced writhing test, which is an *in-vivo* model to evaluate the peripheral analgesic activity. They also found that evolitrine has enhanced antinociceptive activity. The evolitrine (50 mg/kg b.w.) inhibits writhes by 63% compared with 25% by crude ethanol extract (200 mg/kg) at four times the dose of evolitrine (Fig. 3). Thus, there is a considerable enhancement of activity in evolitrine when compared with ethanol extract. Further, acetylsalicylic acid inhibits writhes by 55% by a dose at two times the dose of evolitrine. Thus, evolitrine has a higher antinociceptive activity than acetylsalicylic acid.

Pain is one of the cardinal sign of the inflammation and prostaglandin E_2 (PGE_2) is one of the chemical mediators, which plays a crucial role in inflammatory pain. It is produced as a result of inflammation both in peripheral inflamed tissues and in the spinal cord. Our previous studies have shown that 70% ethanol extract of *A. pedunculata* leaves have PGE_2 inhibition action on adjuvant-induced arthritis rat model (Ratnayake et al., 2019). The difference between the reduction of PGE_2 level by 200 mg/kg b. w. of EEAL and 20 mg/kg b. w. of celecoxib which is a well-known analgesic, were insignificant ($P > .05$). As the PGE_2 levels affect the progression of inflammation, anti-inflammatory and antinociceptive activity of 70% ethanol extract of *A. pedunculata* leaves, may be due to its ability to reduce the level of PGE_2.

Other areas of applications of *A. pedunculata*

In addition to the applications in traditional medicinal system *A. pedunculate* leaves are used locally as a food flavoring and for various other purposes. The young leaves are used as a condiment. The leaves are put in stimulating baths. This is probably due to the action of essential oil. The crushed leaves and fruits have a citrus smell. A fragrant oil from the stem and leaves is used in cosmetic perfumery. The bark is used for caulking boats and toughening nets.

Other agents of interest

Evolitrine was also identified as the major analgesic and anti-inflammatory compound in the *A. pedunculata* leaves, various other compounds have also isolated. i.e., acronylin, acrovestone, bergapten, β-amyrin, arylketone, 1-[2′,4′-dihydroxy-3′,5′-di-(3″-methyl-2″-butenyl)-6′-methoxy]phenylethanone, acrovestone, acronyculatins A, B, C, D, and E, acronyculatin F, acronyculatin G, acronyculatin H (Kouloura et al., 2012; Kumar et al., 1989; Su et al., 2003; Wu et al., 1989). There is scientific evidence of these compounds for various other medicinal activities such as broad-spectrum antimicrobial, anticancer, cytotoxicity and polymerase inhibitory activities.

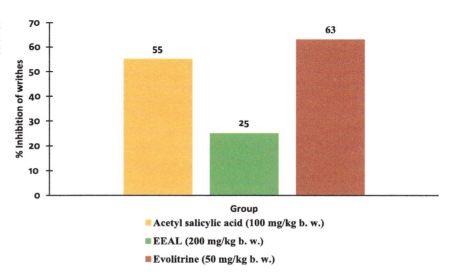

FIG. 3 Comparison of inhibition of writhes on antinociceptive activity of evolitrine with ethanol extract of *A. pedunculata* leaves (EEAL) and acetylsalicylic acid ($n = 6$/group).

Conclusion

The *A. pedunculata* leaves have the potential to control inflammatory and pain conditions through multiple mechanisms. The scientific observations provide some valuable support for the therapeutic application of leaves of *A. pedunculate* claimed in the traditional medicine system in Sri Lanka. The evolitrine is the major known anti-inflammatory and analgesic compound present in the leaves of *A. pedunculata*.

Mini-dictionary of terms

Anti-inflammatory drugs: A drug or substance that reduces inflammation (redness, swelling, and pain) in the body. Anti-inflammatory agents block certain substances in the body that cause inflammation.
Carrageenan: It is a family of linear sulfated polysaccharide that is extracted from red edible seaweeds.
Inflammation: A localized reaction that produces redness, warmth, swelling, and pain as a result of infection, irritation, or injury. Inflammation can be external or internal.
Nonsteroidal anti-inflammatory drugs (NSAIDs): They are the members of a drug class that reduces pain, decreases fever, prevents blood clots and in higher doses, decreases inflammation.
Prostaglandin: The prostaglandins are a group of lipids made at the sites of tissue damage or infection that are involved in dealing with injury and illness. They control processes such as inflammation, blood flow, the formation of blood clots and the induction of labor.
Traditional medicine: It is the total of the knowledge, skill, and practices based on the theories, beliefs, and experiences indigenous to different cultures, whether explicable or not, used in the maintenance of health as well as in the prevention, diagnosis, improvement, or treatment of physical and mental illness.

Key facts

There are potentials for the development of new analgesic and anti-inflammatory drugs from medicinal plants, and it is crucial to prove scientific evidence of their activities.

Acronychia pedunculata leaves have scientifically well-identified anti-inflammatory and analgesic activities, and evolitrine is the major anti-inflammatory and analgesic compound in its leaves.

Summary

Acronychia pedunculate is one of the commonly used medicinal plants in Sri Lankan traditional medicine as a treatment of asthma, itchy skin, cough, pain, swelling, and other disorders with the involvement of the inflammatory processes.

From scientific *in-vivo* studies, evolitrine was identified as the major anti-inflammatory and analgesic compound present in the leaves of *A. pedunculata*.

Hence, evolitrine has the potential to be incorporated into novel treatment agents against diseases associated with inflammatory and pain conditions.

A list of other novel plant-based agents used in analgesia or anesthesia

Linalool
Linalyl acetate
Beta-caryophyllene
Valerenic acid
Verbenol
Nerolidol
Myrtenol
Harman
Norharman
Hamol
Harmalol
Harmaline

Tetrahydroharman
Apigenin
Wogonin
Orientin
Isoorientin
Glabrol
Kaempferol
Vitexin
Chrysin

References

Ashton, M. S., Gunatilleke, S., De Zoysa, N., Dassanayake, M. D., Gunatilleke, N., & Wijesundera, S. (1997). *A field guide to the common trees and shrubs of Sri Lanka* (p. 333). Sri Lanka: WHT Publications (Pvt.) Limited.

Biradar, S., Kangralkar, V. A., Mandavkar, Y., Thakur, M., & Chougule, N. (2010). Antiinflammatory, anti-arthritic, analgesic and anti-convulsant activity of Cyperus essential oils. *International Journal of Pharmacy and Pharmaceutical Sciences, 2*(4), 112–115.

De Silva, L. B., De Silva, U. L. L., Mahendran, M., & Jennings, R. (1979). Kokusaginnine and evolitrine from *Acronychia pedunculata. Phytochemistry, 18,* 1255–1256.

Di Rosa, M. (1972). Biological properties of carrageenan. *Journal of Pharmacy and Pharmacology, 24,* 89–102.

Han, X., Pathmasiri, W., Bohlin, L., & Janson, J. C. (2004). Isolation of high purity 1-[2′, 4′- dihydroxy-3′, 5′-di- -(3″-methylbut-2″-enyl)-6′ methoxy] phenylethanone from *Acronychia pedunculata* (L.) Miq. By high speed counter current chromatography. *Journal of Chromatography A, 1022,* 213–216.

Jayaweera, D. M. A. (1982). *Medicinal plants (indigenous and exotice) used in Ceylon—Part I–V.* Sri Lanka: The National Science Foundation.

Kouloura, E., Halabalaki, M., Lallemand, M. C., Nam, S., Jove, R., Litaudon, M., et al. (2012). Cytotoxic prenylated acetophenone diamers from *Acronychia pedunculata. Journal of Natural Products, 75,* 1270–1276.

Kozaki, S., Takenaka, Y., Mizushina, Y., Yamaura, T., & Tanahashi, T. (2014). Three acetophenones from *Acronychia pedunculata. Journal of Natural Medicines, 68,* 421–426.

Kumar, V., Karunaratne, V., Sanath, M. R., & Meegalle, K. (1989). 1-[2′,4′-dihydroxy-3′,5′- di- -(3″-methyl-2″-butenyl)-6′ methoxy]phenylethanone from *Acronychia pedunculate* 290 root bark. *Phytochemistry, 28*(4), 1278–1279.

Lesueur, D., Serra, D., De, R., Bighelli, A., Hoi, T. M., Thai, T. H., et al. (2008). Composition and anti-microbial activity of the essential oil of *Acronychia pedunculate* (L.) Miq. From Vietnam. *Natural Product Research, 22*(5), 393–398.

Napagoda, M. T., Sundarapperuma, T., Fonseka, D., Amarasiri, S., & Gunaratna, P. (2018). An ethnobotanical study of the medicinal plants used as anti-inflammatory remedies in Gampaha District-Western Province, Sri Lanka. *Scientifica, 1,* 1–24.

Necas, J., & Bartosikova, L. (2013). Carrageenan: A review. *Veterinární Medicína, 58*(4), 187–205.

Owoyele, B. V., Adebukola, O. M., Funmilayo, A. A., & Soladoye, A. O. (2008). Anti-inflammatory activities of ethanolic extract of *Carica papaya* leaves. *Inflammopharmacology, 16,* 168–173.

Panda, V. S., & Khambat, P. D. (2013). In-vivo anti-inflammatory activity of *Garcinia indica* fruit rind (kokum) in rats. *The Journal of Phytopharmacology, 2*(5), 8–14.

Patel, H., Patel, B., & Prajapati, A. (2013). Evaluation of anti-inflammatory activity of *Cucurbita maxima* fruit. *Journal of Advance Pharmaceutical Research and Bioscience, 1*(1), 28–31.

Pathmasiri, W., El-Seedi, H. R., Han, X., Janson, J. C., Huss, U., & Bohlin, L. (2005). Aryl ketones from *Acronychia pedunculata* with cyclooxygenase-2 inhibitory effects. *Chemistry and Biodiversity, 2,* 463–469.

Ratnayake, W. M. K. M., Suresh, T. S., Abeysekara, A. M., Salim, N., & Chandrika, U. G. (2019). Acute anti-inflammatory and anti-nociceptive activities of crude extracts, alkaloid fraction and evolitrine from *Acronychia pedunculata* leaves. *Journal of Ethnopharmacology, 238,* 111827.

Su, C., Kuo, P., Wang, M., Liou, M., Damu, A. G., & Wu, T. (2003). Acetophenone derivatives from *Acronychia pedunculata. Journal of Natural Products, 66,* 990–993.

Sy, L., & Brown, G. D. (1999). 1- [2′, 4′-dihydroxy-3′-(3″-methylbut-2″-enyl)-5′-(1‴- ethoxy-3‴-methylbutyl)-6′-methoxy]phenylethanone from. *Phytochemistry, 52,* 681–683.

Umukoro, S., & Ashorobi, R. B. (2006). Evaluation of anti-inflammatory and membrane stabilizing property of aqueous leaf extract of *Momordica charantia* in rats. *African Journal of Biomedical Research, 9,* 119–124.

Vasudevan, M., Gunnam, K. K., & Parle, M. (2006). Anti-nociceptive and anti-inflammatory properties of *Daucus carota* seeds extract. *Journal of Health Science, 52*(5), 598–606.

Weragoda, P. B. (1980). The traditional system on medicine in Sri Lanka. *Journal of Ethnopharmacology, 2,* 71–73.

Wijesundera, D. S. A. (2004). Inventory, documentation and medicinal plant research in Sri Lanka. *Medicinal Plant Research in Asia, 1,* 184–195.

Wu, T., Wang, M., Jong, T., McPhail, A. T., McPhail, D. R., & Lee, K. (1989). X-ray crystal structure of acrovestone, a cytotoxic principle from *Acronychia pedunculata. Journal of Natural Products, 52*(6), 1284–1289.

Chapter 30

Adansonia digitata and its use in neuropathic pain: Prostaglandins and beyond

Aboyeji Lukuman Oyewole[a,b,c], Abdulmusawwir O. Alli-Oluwafuyi[c,d], Abdulrazaq Bidemi Nafiu[c,e], and Aminu Imam[f]

[a]*Neurophysiology, Inflammation and Tropical Diseases Unit, Department of Physiology, Faculty of Basic Medical Sciences, University of Ilorin, Ilorin, Nigeria,* [b]*Bioresearch Hub Laboratory, Ilorin, Nigeria,* [c]*Biotranslational Research Group, University of Ilorin, Ilorin, Nigeria,* [d]*Department of Pharmacology and Therapeutics, Faculty of Basic Clinical Sciences, University of Ilorin, Ilorin, Nigeria,* [e]*Departments of Human Physiology, Federal University Dutse, Dutse, Jigawa, Nigeria,* [f]*Neuroscience Unit, Department of Anatomy, Faculty of Basic Medical Sciences, University of Ilorin, Ilorin, Nigeria*

Abbreviations

5HT-2/5HT-3	Serotonin-2/Serotonin-3
AMPA	α-amino-3-hydroxy-5-methyl-4-isoxazolepropionic acid
BDNF	brain-derived neurotrophic factor
Cav.	T-type calcium channels
COX-2	cyclooxygenase
HIV-AIDS	human immunodeficiency virus-acquired immunodeficiency syndrome
mRNA	messenger ribonucleic acid
Nav.	sodium ion channel
NF-κB	nuclear factor kappa B
NMDA	*N*-methyl-D-aspartate
NO·	oxidative nitrogen radical
NR2A	NMDA receptor subunit 2A
NR2B	NMDA receptor subunit 2B
Nrf2	nuclear factor erythroid 2
PGE2	prostaglandin E2
ROS	reactive oxygen species
TRPV1	transient receptor potential family

Introduction

Brief description of *Adansonia digitata*

Adansonia digitata commonly called "baobab" or "African baobab" tree is a long living angiosperm classified as a member of Bombacoideae, a subfamily of Malvaceae (Fig. 1A). At a tender age (within 32 days) in the wild, the stem of baobab seedling could grow to the height of 30–45 cm and its diameter ranges from 1 to 1.3 cm (Assogbadjo et al., 2011). Depending on its genetic variation, baobab seedlings start with simple leaves and may produce these for years. However, baobab seedlings of West Africa origin appeared to produce compound leaves (Fig. 1B) earlier than other seedlings, especially when compared to those from South-Eastern African heritage (Cuni Sanchez et al., 2011; Wickens & Lowe, 2008). African baobab grows to be the largest tree among its eight Adansonia species. A well grown baobab tree could be more than 20 m in height and girth (Baum, 1995; Mullin, 2003; Wickens, 1982). Its root system is usually about 2 m below the

---

Treatments, Mechanisms, and Adverse Reactions of Anesthetics and Analgesics. https://doi.org/10.1016/B978-0-12-820237-1.00030-2
Copyright © 2022 Elsevier Inc. All rights reserved.

330 PART | IV Novel and nonpharmacological aspects and treatments

FIG. 1 *Adansonia digitata* and its various parts. (A) A seedless African baobab tree (B) leaves (C) root (D) bark (E) pericarp-enclosed fruits and its fruit pulps (F) fruit pulps and its seeds.

soil (Fig. 1C) while its bark (Fig. 1D) thickness varies with geographical distribution as baobab in dryland areas has thicker bark than those in subhumid areas. Though its shape varies from place to place, all baobab trees produce similar woody pericarp-enclosed fruits that are either spherical or ovoid in shape (Fig. 1E); (Kempe, Neinhuis, & Lautenschläger, 2018). The fruits are closely packed light beige fruit pulps that surround the seeds and appear chalky when dried (Fig. 1F) (Kempe et al., 2018).

Distribution

The geographical origin of the baobab tree remains subject of debate though phytogeographical studies suggested Madagascar as the origin for the genus Adansonia as six of its eight species are endemic to this region (Armstrong, 1983; Aubréville, 1975; Baum, 1995; Wickens & Lowe, 2008). From its geographical origin, large mammals (including humans) and sea contributed to the dispersal of baobab seeds across the world (Baum, 1995; Wickens, 1982; Wickens & Lowe, 2008). Today, the presence of Africa baobab tree is documented not only in Africa (Wickens, 1982) but also in South America (Boivin & Fuller, 2009), Asia (Armstrong, 1979; Maheshwari, 1971; Sidibe & Williams, 2002; Vandercone et al., 2004), and Australia (Baum, Small, & Wendel, 1998) among others. Wherever they are, baobab trees contribute significantly to the different aspects of people's lives in the hosting communities.

Significance of *Adansonia digitata*
Relevance among local folks

Baobab is a multipurposeful tree with diverse usefulness depending on its geographical location. Interestingly, every part of baobab has been reported to be useful by the locals. From the edible parts of baobab (leaf, fruit, and seed) which contribute

TABLE 1 The use of Baobab tree across different region.

S/No.	Part of Baobab	Purpose	Region	References
1	Seed	✓ Roasted—Eaten as snacks ✓ Grinded into flour—Use as thickening agent ✓ Fermented—Use as flavor enhancers ✓ Medicinal purpose	Burkina Faso and some other Africa countries	Gebauer, El-Siddig, and Ebert (2002) and Parkouda et al. (2015)
2	Pulp	✓ Fresh—Fruit juice ✓ Dried pulp—Fruit juice and mixing porridge ✓ Medicinal purpose	East Africa	Assogbadjo et al. (2011)
3	Leaf	✓ Fresh—eaten as vegetable ✓ Dried—use as seasoning/sauces ✓ Fermented—Local brews fermenting agent ✓ Medicinal purpose	West African, Tanzania, Malawi and Zimbabwe	Assogbadjo et al. (2011) and Gebauer et al. (2002, 2014)
4	Bark	✓ Rope making ✓ Fish trap production ✓ Medicinal purpose		Wickens (1982)
5	Stem	✓ Use as thatching for roofs ✓ As fodder ✓ Medicinal purpose		Wickens (1982)
6	Root	✓ Root tuber—eaten ✓ Medicinal purpose	Malawi	Ajibesin et al. (2007), Bamalli et al. (2014), and Kamanula et al. (2018)
7	Whole baobab tree (sacred tree)	✓ Religion (sacrifices are done underneath it to please gods)	Africa	Kamatou, Vermaak, and Viljoen (2011)

to nutrition of African folks (Assogbadjo et al., 2008) to the nonfood parts (bark, fibers, fodder, and timber) sold (Gebauer et al., 2016) as raw materials for other use. Table 1 summarizes the use of baobab across various Africa localities which are mainly directed toward nutrition, medicine, construction, and religious purposes.

Medicinal importance

Baobab tree is a commercially important plant that is popular for its medicinal use worldwide (Van Wyk, 2015). As shown in Table 1, every part of baobab has its medicinal value. The ethnopharmacological use of baobab has been extensively reviewed elsewhere (Kamatou et al., 2011). Baobab roots are used as an antimalarial agent in the Southeastern part of Nigeria (Ajibesin et al., 2007). Also, it has been used for its beneficial effects on memory and central nervous-related disease (Gelfland, 1985), antidiarrheal (Neuwinger, 2000), wound healing (Inngjerdingen et al., 2004), anti-HIV-AIDS, antimalarial (Chinsembu, 2015), antisickling (Amujoyegbe et al., 2016), and antidiabetic effects (Chinsembu, 2019). Its use as a beverage and food supplement is notable in the United States (Kinghorn et al., 2011) and parts of Southern Africa (Welcome & Van Wyk, 2019). It is also being developed for its cosmeceutical potential (Rahul et al., 2015). Of significance is the medicinal value of *A. digitata* on various forms of pain (Fabiyi et al., 1993; Kamatou et al., 2011; Owoyele & Bakare, 2018; Ramadan, Harraz, & El-Mougy, 1994). Unlike most plants, all parts of *A. digitata* plant are shown to have helpful effects on pain perception. Thus, the plant might be the cheapest and better candidate for the treatment of pain when fully developed. From its flower to the root, the effects of every part of *A. digitata* on pain sensation seems to be connected to its phytochemical constituents, which are known to be rich in pain-resolving flavonoids and minerals.

Composition of *Adansonia digitata* plant
Phytochemical constituents

The nutritional and medicinal properties of *A. digitata* have been well investigated and reported. Notably, virtually all parts of the plant possess one or more biological effects, which may be nutritional, medicinal or toxic. While biological effects may be common to more than one part of the plant, several other effects are unique to the particular plant part. The biological effects of plants are intrinsically and inextricably linked to the chemical constituents of the plant. A wide variety of chemical constituents have been detected and identified (Table 2). Polyphenols, flavonoids, mono- and poly-unsaturated fatty acids, and organic acids are some of the secondary metabolites identified in this plant. Considerable differences also occur between parts. For instance, in the study by Braca et al., 2018, qualitative differences in the chemical components of *Adansonia* fruit and leaf were reported (Braca et al., 2018). While feruloylquinic acid and organic acids such as citric acid was detected in the fruit, these compounds were not detected in the leaf. In the same vein, the leaf contained more polyphenols than the fruits. Specifically, compounds such as catechin, epicatechin, vitexin were detected in leaf but not in fruits. Whereas several other compounds including procyanidin dimer I, tiliroside I/II were detected in both fruit and seed. Similarly, in the study by Sena et al., 1998, though there were no qualitative differences in the amino-acid constituents between the *Adansonia* leaf and fruit, marked quantitative difference in the relative amounts of the amino-acids was observed (Sena et al., 1998). Thus, it is reasonable that these chemical constituents may underlie the similarities and differences in effects observed among the different plant parts.

Mineral components

Animals and humans alike require regular and adequate intake of about 22 mineral elements for optimal physiological function and health. While some of these are required in relatively larger amounts and thus referred to as macronutrients, others such as Zn^{2+} and Cu^{2+} are only required in minute or trace amounts because higher concentrations can be dangerous (Welch & Graham, 2004) and referred to as micronutrients. Mineral elements are obtained from animal and plant sources and form an integral part of human nutrition. The importance of these minerals become evident in deficiencies—plants, animals, and humans all show deficiencies to individual elements (Soetan, Olaiya, & Oyewole, 2010).

Over nine minerals including Ca^{2+}, K^+, Fe^{2+}, Mg^{2+}, Zn^{2+} have consistently been detected in different parts of the *A. digitata* plant (Table 3). While the qualitative composition of minerals across the different parts of the plant appear similar, considerable difference exist in their relative quantities. Foods of plant origin generally have high potassium and magnesium content while low in sodium (Martínez-Ballesta et al., 2010). Unsurprisingly, in the fruit and seed, potassium is the most abundant mineral element representing about 74% of total mineral content in the fruit while Ca^{2+} and Mg^{2+} are the next abundant with 12.5% and 9.9%, respectively. However, calcium is exceedingly the most abundant mineral in the leaf.

Impact of *Adansonia digitata* on pain
Effect on pain sensation

During the primitive age, the drive to resolve the unpleasantness associated with pain might be a part of compelling factors for humans to seek relief in medicinal plants within their environment. For a plant whose existence is dated about 1000–2000 years back (Magaia et al., 2013), information on when and how humans realized *A. digitata* plant for its analgesic medicinal value remains elusive. However, an increasing number of scientific studies on various extracts from *A. digitata* plant had confirmed its analgesic properties as popularly used among folks of different geographical areas where the plant domicile.

Interestingly, every part of the *A. digitata* plant appeared to be a potent pain killer. From its fruit pulp (Ramadan et al., 1994), seed and its oil (Kamatou et al., 2011), leaf (Fabiyi et al., 1993), bark (Owoyele & Bakare, 2018) to its root, *A. digitata* plant appears to be an excellent candidate for the development of a cheap and potent analgesic drug for various types of pain. The reported analgesic effects of *A. digitata* plant is limited neither by the diversity nor multifaceted nature of pain. Extracts from different parts of the plant were documented to favorably decrease pain perception associated with noxious thermal, chemical, and mechanical stimulus (Owoyele & Bakare, 2018). Also, specific types of pain such as neuropathic pain (Bakare, Oyewole, & Owoyele, 2019) and some symptomatic pain had all been shown to be mitigated by extracts from different parts of *A. digitata* plant. The effective dose however seems to vary from method of extraction to the part of the plant in consideration.

TABLE 2 Chemical and proximate composition of *Adansonia digitata*.

Plant parts	Chemical/ proximate components	Classes of chemical/ proximate contents	Extraction media	Isolated compound	Amount	References
Seeds			Ethylacetate fraction	mg/g		Adeoye et al. (2019)
				Apigenin	7.60	
				Quercetin	6.19	
				Luteolin	4.54	
				Gallic acid	2.35	
				Caffeic acid	2.29	
				Rutin	1.07	
			Chloroform fraction	Caffeic acid	0.17	
				Chlorogenic acid	0.09	
				Gallic acid	0.06	
Seeds				%		Mariod, Mirghani, and Hussein (2017)
				Saturated fatty acids	31.3	
				Monounsaturated	37.0	
				Polyunsaturated	31.7	
				Oleic acid	35.8	
				Linoleic	30.7	
				Palmitic	24.2	
Seeds				Arabinogalactans		Zahid et al. (2017)
Fruit				%		Ghoneim et al. (2016)
				Carb	76.0%	
				Phenols	6.5%	
				Protein	3.1%	
				Fat	0.25%	
Fruit		Organic acids		Citric acid		Braca et al. (2018)
				Tiliroside		
				Catechin		
		Phenolics		Procyanidin dimer I		
				Procyanidin dimer II		
				Feruloylquinic acid		
				Procyanidin trimer I		
				Procyanidin trimer II		
				Procyanidin tetramer		
				Quercetin 3-*O*-glucoside		
				Kaempferol 3-*O*-galactoside		
		Tiliroside isomer		Tiliroside I		
				Tiliroside II		
				Kaempferol		

Continued

334 PART | IV Novel and nonpharmacological aspects and treatments

TABLE 2 Chemical and proximate composition of *Adansonia digitata*—cont'd

Plant parts	Chemical/ proximate components	Classes of chemical/ proximate contents	Extraction media	Isolated compound	Amount	References
Leaf		Phenolics		Procyanidin dimer I		
				Procyanidin dimer II		
				Procyanidin trimer I		
				Procyanidin trimer II		
				Catechin		
				Epicatechin		
				Apigenin *O*-pentoside		
				Quercetin glycoside		
				Vitexin/isovitexin		
				Kaempferol glycoside		
				Rutin		
				Quercetin pentoside,,		
				Quercetin 3-hydroxy-3-methylglutaryl-*O*-hexoside		
				Kaempferol glycoside I		
				Kaempferol glycoside II,		
				Kaempferol 3-*O*-glucoside		
				Quercetin		
				Tiliroside I		
				Tiliroside II		
Fruit				%		
				Water	11.0 ± 5.0	Stadlmayr et al. (2013)
				Protein	2.5 ± 0.5	
				Fat	0.5 ± 0.3	
				Carbohydrate	74.9	
				Fiber	6.2 ± 1.7	
				Ash	4.9 ± 0.7	
				Vitamin C (mg)	273.0 ± 100	
Fruit pulp				Citric acid	25.7 ± 2.1	Magaia et al. (2013)
				Tartaric acid	ND	
				Malic acid (g/kg)	1.6 ± 0.2	
				Succinic acid	0.1 ± 0.0	
Fruit pulp				Hydroxycinnamic acid		Li et al. (2017)
				Glycosides (HAGs)		
				Iridoid glycosides (IGs)		
				Phenylethanoid glycosides		

TABLE 2 Chemical and proximate composition of *Adansonia digitata*—cont'd

Plant parts	Chemical/ proximate components	Classes of chemical/ proximate contents	Extraction media	Isolated compound	Amount	References
Fruit pulp				Epicatechin		Kamatou et al. (2011)
				Procyanidin B2		
				Procyanidin B5		
Fruit shell				Alkaloids		Seukep et al. (2013)
				Flavonoids		
				Phenols		
				Triterpenes		
				Sterols		
				Saponins		
Fruit shell	Polyphenols	Phenolic acids		Protocatechuic acid		Ismail et al. (2019)
				p-Hydroxy-benzoic acid		
				p-Hydroxycinnamic acid (*p*-Coumaric acid)		
				Chlorogenic acid		Ismail et al. (2019)
				Vanillic acid		
				Dihydrocaffeic acid		
	Flavonoids	Flavanols		Proanthocyanidin		
				D-(+) Catechin		
				(−)-Epicatechin		
				Procyanidin tetramer		
				Epicatechin derivative		
				Proanthocyanidin		
				(−)-Epiafzelechin		
		Flavonols		Dihydroquercetin-3-hexoside		
				Quercetin-3-rhamnoside		
				Quercetin 3-*O*-α-L-rhamnopyranosyl (1-N 2) [α-L-rhamnopyranosyl (1-N 6)]-β-D-glucopyranoside		
				Kaempferol-3-*O*-rhamnosylrutinoside		
				Rutin (isomer 1)		
				Rutin (isomer 2)		
				Dihydroquercetin		
				Isoquercetin (Quercetin 3-*O*-glucoside)		
				Kaempferol-3-glucoside-2″-*p*-coumaroyl (isomer 1)		
				Kaempferide I		
				Kaempferol		
				Kaempferol-3-glucoside-2″-*p*-coumaroyl (isomer 2)		
				Kaempferol-3-glucoside-2″-*p*-coumaroyl (isomer 3)		

Continued

TABLE 2 Chemical and proximate composition of *Adansonia digitata*—cont'd

Plant parts	Chemical/ proximate components	Classes of chemical/ proximate contents	Extraction media	Isolated compound	Amount	References
				Isorhamnetin		
				Quercetin		
		Flavone		Apigenin		
		Organic acids		Glucosyringic acid		
				Betulinic acid		
		Hydroxy fatty acids		Trihydroxy-octadecadienoic acid		
				Trihydroxy-octadecenoic acid		
				Dihydroxy-octadecenoic acid I		
				Hydroxyoctadeca-trienoic acid		
		Other organic compounds		Limonin		
				Saponin derivatives		
Leaf		Amino acids dry weight		mg/g		
				Aspartate	12.4	Sena et al. (1998)
				Glutamate	14.4	
				Serine	4.5	
				Glycine	5.4	
				Histidine	2.4	
				Arginine	8.3	
				Threonine	4.3	
				Alanine	6.8	
				Proline	6.4	
				Tyrosine	4.3	
				Valine	7.0	
				Methionine	1.5	
				Cysteine	2.4	
				Isoleucine	6.1	
				Leucine	9.8	
				Phenylalanine	4.6	
				Lysine	6.4	
				Tryptophan	3.4	
				Total protein	112.0	
Fruit				mg/g		
				Aspartate	15.9	Sena et al. (1998)
				Glutamate	38.8	
				Serine	8.8	
				Glycine	7.0	
				Histidine	3.8	
				Arginine	19.3	

TABLE 2 Chemical and proximate composition of *Adansonia digitata*—cont'd

Plant parts	Chemical/ proximate components	Classes of chemical/ proximate contents	Extraction media	Isolated compound	Amount	References
				Threonine	5.7	
				Alanine	7.2	
				Proline	7.5	
				Tyrosine	4.7	
				Valine	9.1	
				Methionine	1.6	
				Cysteine	3.6	
				Isoleucine	6.5	
				Leucine	10.9	
				Phenylalanine	8.4	
				Lysine	8.7	
				Tryptophan	2.8	
				Total protein	170.0	
Leaf			Petroleum ether	Phytosterols		Shri et al. (2004)
			Chloroform			
			Benzene			
			Methanol	Glycosides		
				Phytosterols		
				Saponins		
				Protein and amino acids		
				Phenolic compounds and tannins		
				Gums and mucilage		
				Flavanoids		
			Aqueous	Glycosides		
				Phytosterols		
				Saponins		
				Protein and amino acids		
				Phenolic compounds and tannins		
				Gums and mucilage		
				Flavanoids		
			Drug powder	Glycosides		
				Phytosterols		
				Saponins		
				Protein and amino acids		
				Phenolic compounds and tannins		
				Gums and mucilage		
				Flavanoids		

Continued

338   PART | IV Novel and nonpharmacological aspects and treatments

TABLE 2 Chemical and proximate composition of *Adansonia digitata*—cont'd

Plant parts	Chemical/proximate components	Classes of chemical/proximate contents	Extraction media	Isolated compound	Amount	References
Fruit pulp					%	Magaia et al. (2013)
		Dry matter			89.8 ± 0.1	
		Protein			2.4 ± 0.0	
		Fat			0.5 ± 0.1	
		Ash			5.5 ± 0.1	
Kernel		Dry matter			93.6 ± 0.0	
		Protein			36.7 ± 0.9	
		Fat			35.0 ± 0.2	
		Ash			7.7 ± 0.0	
AD		Dry matter			88.0	
		Glucose			3.0	
		Fructose			3.0	
		Sucrose			4.3	
Bark		Total phenolics (mgGAE/g)			2.43 ± 0.04	Mulaudzi et al. (2013)
		Flavonoids (mgCAE/g)			0.25 ± 0.06	
		Gallotannin (mgGAE/g)			2.85 ± 0.11	
		Condensed tannins (%LCE)			0.03 ± 0.03	
		Condensed tannins (%LCE)			0.03 ± 0.03	

Effect on neuropathic pain

Neuropathic pain is a chronic condition resulting from damage or disease within the part of the nervous system meant for processing of pain signals (Baron, 2000). Colloca et al. (2017) also described it as a chronic pain disorder caused by a lesion or disease of the somatosensory system (Colloca et al., 2017). Lesions or diseases affect the spinal cord, the brain as well as peripheral nerves particularly small unmyelinated C-fibers and myelinated A fibers. While cerebrovascular diseases such as stroke, Parkinson's, spinal cord trauma and demyelinating diseases such as multiple sclerosis may cause central neuropathic pain, diabetes, and cancer chemotherapy may induce peripheral neuropathic pain. Regardless of its location (central or peripheral neuropathy), neuropathic pain has always involved nerves transmitting within the nociceptive pathways. Other classical causes of neuropathic pain include postherpetic neuralgia, "channelopathies" or hereditary anomalies. Clinically, neuropathic pain may present with loss of response to tactile stimuli accompanied by hyperalgesia—enhanced sensitivity to pain, and/or allodynia—pain sensation elicited by nonnoxious stimuli such as light touch.

TABLE 3 Minerals composition of *Adansonia digitata*.

S/No.	Part of *A. digitata*	Calcium	Iron	Magnesium	Phosphorus	Potassium	Sodium	Zinc	Copper	Manganese	References
					mg/100 g of *A. digitata*						
1	Fruit	60.0–295.0	4.23–9.3	90.0–232.0	49.79–51.0	1240.0–1730.0	14.3–27.9	0.47–2.8	0.55–1.6	0.39–0.6	Eromosele, Eromosele, and Kuzhkuzha (1991), Lockett, Calvert, and Grivetti (2000), Osman (2004), and Stadlmayr et al. (2013)
2	Leaf	1800.0–17,500.0	30.0–93.1	246.0–1013.5	115.0–875.0	686.0	15.1	1.24–22.4	1.02–1.6	6.6–35.5	Barminas, Charles, and Emmanuel (1998), Catarino et al. (2019), Lockett et al. (2000), and Prentice et al. (1993)
3	Seed	264.0–410.0	4.35–6.4	270.0–278.0	678.0	910.0	28.3	4.29–5.2	1.19–2.6		Lockett et al. (2000) and Osman (2004)

340 **PART | IV** Novel and nonpharmacological aspects and treatments

Studies in animal models and human subjects have expanded our understanding of neuropathic pain. In response to nerve injury, there is an increase in the transmission of sensory signals and a decrease in inhibitory signaling emanating from inhibitory interneurons and descending nociceptive pathways. This imbalance results in a shift in pain threshold favoring hyperexcitability of pain sensing pathways. Several signaling molecules and receptors have been identified to play a role in the development of neuropathic pain. These changes occur in the injured primary afferents as well as adjoining noninjured neurons.

The expression and activity of several ion channels particularly sodium, calcium and potassium channels in primary afferents are altered in neuropathic pain. Several sodium channels are expressed in nerves and mediate the changes in conductance that underlie sensory transmission of mechanical and nociceptive signals. Nav1.3 appears to be the major sodium channel overexpressed and thus implicated in neuropathic pain. Conversely, Nav1.7, Nav1.8 and Nav1.9 are sodium channels downregulated in most neuropathic models and mostly associated with spontaneous firing in noninjured nerve fibers (Rogers et al., 2006). T-type Calcium channels Cav1.1, Cav1.2 and Cav1.3 have been proposed to play important roles in the sensitization of peripheral nerves to noxious stimuli in neuropathic pain (Bourinet, Francois, & Laffray, 2016). These channels lower the firing threshold of afferent pain fibers, promote synaptic release of neurotransmitters from afferent neurons and contribute to chronic pain possibly via posttranslational modifications (Bourinet et al., 2016). The transient receptor potential family of nonselective cation channels including TRPV1, TRPA1 are expressed at nerve terminals and along axons of nociceptors and transduce thermal, mechanical and chemical stimuli thus playing a central role in nociception (Basso & Altier, 2017). These channels are presumed to play an important role in neuropathic pain. For instance, TRPV1 is highly expressed in diabetic neuropathy and antagonism of these channels relieves painful neuropathy. The development of cold and heat hypersensitivity in neuropathic pain is also believed to be dependent upon TRPA1 (Basso & Altier, 2017). TRPM8 is involved in chemotherapy-induced neuropathic pain These channels have all emerged as potential therapeutic targets in management of neuropathic pain.

Aside from changes that occur in primary afferents, postsynaptic modifications occur in second-order neurons dramatically lowering the threshold for responding to mechano-thermal stimuli thereby constituting the central sensitization that occurs in neuropathic pain. Excitatory neuromodulators involved in this central sensitization include N-methyl-D-aspartate (NMDA) and α-amino-3-hydroxy-5-methyl-4-isoxazolepropionic acid (AMPA) receptors.

In addition, descending pathways such as α2-receptor mediated noradrenergic inhibition and 5HT-2/5HT-3 receptor-mediated serotonin enhancement which are involved in centrally-mediated pain inhibition are dysregulated.

Potential mechanism of actions for *A. digitata* extract, from prostaglandins and beyond

Prostaglandin E2 pathway

Following reports of the analgesic effects of *A. digitata* particularly in neuropathic pain, efforts to understand the potential mechanisms have been undertaken. Bakare and his colleagues in 2019 showed that *A. digitata* bark extract could be used to manage neuropathic pain, and they implicated the involvement of prostaglandin E2 in this notable role (Bakare et al., 2019). Implicating the involvement of prostaglandin E2 alone might suggest that *A. digitata* bark extract could only mitigate neuropathic pain of inflammatory etiological origin, and studies evaluating the effect of other parts of *A. digitata* plant (pulp, fruit, leaf and root) on neuropathic pain are scarce in the literature.

Moving from known to unknown, Bakare et al. (2019) opined that *A. digitata* lessen neuropathic pain via reduction of prostaglandin E2 level (Bakare et al., 2019). Their submission though showed a likely direction for the contributing mechanism but failed to provide the precise events leading to the decrease in prostaglandin E2. For instance, it is not clear whether the reduction in prostaglandin E2 level was due to decrease or inhibition of cyclooxygenase, an enzyme necessary for its production, or due to increase breakdown of prostaglandin E2 level, or owing to decrease in oxidative stress and tissue damage. Once produced however, prostaglandins (particularly prostaglandin E2) acts on peripheral sensory neurons and as well as centrally to sensitize nociceptive pathways. *A. digitata* is believed to alter the availability of prostaglandin E2 by decreasing its level thereby limiting its pathological impacts, diminishing its sensitization and stimulation effects on sensory neurons. Based on its phytochemical and minerals constituents, there are other potential mechanisms of actions that could be hypothesized for *A. digitata* for possible future study (Fig. 2).

Exploring *A. digitata* plant (Fig. 2,1A) for its phytochemical constituents showed abundant presence of polyphenols, flavonoids and amino acid compounds (Fig. 2,1B–C), as well as minerals and trace elements (Fig. 2,1E and F); (Braca et al., 2018; Ismail et al., 2019; Kamanula et al., 2018; Li et al., 2017; Magaia et al., 2013). Flavonoids are polyphenolic compounds found in virtually every part of most plants. Currently, more than $10,000$ unique flavonoids have been isolated and profiled for varied biological activities which including antinociceptive effect (Hossain et al., 2017). Remarkably, *A.*

digitata plant is rich in diverse classes of flavonoids which include flavones, flavanols, flavonols, flavanones, isoflavones and anthocyanins. The presence and abundance of these flavonoids is sometimes limited by the geographical location and the part of *A. digitata* in consideration. Previous physio-pharmacological investigations of these flavonoids and minerals showed diverse therapeutic benefits on neuropathic pain. Based on their neuropathic pain-linked evidences, of interest to the present review among the constituents of *A. digitata* are flavones (apigenin, vitexin), flavanols (catechin, epicatechin), flavonols (kaempferol, quercetin, rutin I & II,); (Fig. 2,2A–C), Mg^{2+} and Zn^{2+} (Fig. 2,2E and F).

Inflammatory and oxidative stress pathways

Following lesion or disease of somatosensory neurons, damage- and pathogen-related molecules are released at the primary site of injury or infection leading to initiation of inflammation process as a means of restoring body homeostasis. Basically, these damage- and pathogen-associated molecular patterns bind to pattern recognition receptors on tissue-resident immune cells (dendritic cells, macrophages, etc.) causing their activation, secretion of chemoattractant and expression of NF-κB. The expressed NF-κB further governs the expression of other genes for COX-2 and pro-inflammatory cytokines. This elevates the production of COX-2 and inflammation promoting cytokines. COX-2 enzyme catalyzes the production of prostagladin E2 which contributes to neural sensitization leading to increased nociception in neuropathic pain. *A. digitata* decreases the level of prostagladin E2 to relieve neuropathic pain (Bakare et al., 2019). However, the action of *A. digitata* extract on neuropathic may be more than mere decrease in prostaglandin E2. *A. digitata* plant is rich in apigenin, vitexin, catechin, epicatechin, kaempferol, quercetin and rutin flavonoids (Table 2). These flavonoids have been shown in the literature to impact pain-relieving effect on neuropathic pain. Briefly, all these aforementioned flavonoids exhibit antiinflammatory effects via inhibition/downregulation of NF-κB (Carvalho et al., 2019; Gao et al., 2016; Jiang et al., 2019; Lee et al., 2018; Lin & Lin, 1997; Patil et al., 2016; Tian et al., 2016), decreased recruitment of phagocytic cell (quercetin), impedes pro-inflammatory cytokine production (rutin), reduced NO•, COX-2 and prostaglandin E2 (PGE2) production (epicatechin and apigenin) and increased antiinflammatory cytokine while inhibiting pro-inflammatory cytokines production (vitexin); (Borghi et al., 2013, 2018; Carvalho et al., 2019; Guazelli et al., 2013; Patil et al., 2016; Zhong et al., 2012). It is possible that one or some of these flavonoids in *A. digitata* extract act in the same way to impact pain relieving-effects in neuropathic pain (Fig. 2,3A).

Also, during the above-described inflammation process, activated phagocytic cell usually release large amount of reactive oxygen and nitrogen species, primarily to kill the invading agent. The quantity of these radicals is exaggerated in inflammation process causing damage of healthy tissue (oxidative stress), expression of NF-κB and consequently initiate vicious cycle of inflammation. Summing up, the injured area is expanded, neuronal sensitization is increased and evoked pain is worsening. Strikingly, *A. digitata* plant has been reported for its antioxidant properties (Braca et al., 2018; Irondi et al., 2017) including in neuropathic pain study (Bakare et al., 2019). However, studies relating the precise antioxidant compound (in *A. digitata*) with specific events in neuropathic pain are scarce. Flavonoids such vitexin, rutin, and quercetin mediate improved oxidative defense through upregulation of Nrf2 expression (Carvalho et al., 2019; Lu et al., 2018; Tian et al., 2016), an important transcription factor that controls antioxidant defense genes, while apigenin was reported to scavenge hydroxyl radicals. There are high chances that vitexin, rutin, quercetin, and apigenin in *A. digitata* extract could also perform the same function (Fig. 2,3B). Study focusing on this will fill the current gap in knowledge.

Peripheral and central sensitization pathways

Both inflammation and oxidative stress do not only release mediators (prostagladin E2, cytokines, histamine, etc.) that cause neural sensitization but also induce varied molecular changes on the injured primary afferent neuron. The most important of these changes is up-and-down regulation of certain genes for ion channels associated with the receptors on the injured neuron. For instance, there is upregulation of T-type calcium channels (Cav1.1, Cav1.2 and Cav1.3), sodium channels (Nav 1.3) and the nonselective cation channels for transient receptor potential family (TRPV1, TRPA1 and TRPM8). On the other hand, there is downregulation of genes for certain sodium channels (Nav 1.7, Nav 1.8 and Nav 1.9) and Zn transporter (Transporter 1—leading to increased intracellular zinc ion). Elevated intracellular zinc ions precipitates increased conversion of pro-BDNF to mature-BDNF which in turn activates TrkB receptor that decreases expression of K^+-CL^- cotransporter and induce neuropathic pain in vivo (Kitayama et al., 2016). Together, there is decrease in firing threshold of affected receptors, increase in impulsive firing of surrounding noninjured neurons, increase in conductance and spontaneous/ectopic firing of the injured primary afferent neuron. To the best of our knowledge, study showing the impact of *A. digitata* on receptors' ions channels or their genes is not available in the literature. However, reports on similar flavonoids present in *A. digitata* extract were shown to affect certain ions channels. Precisely, vitexin

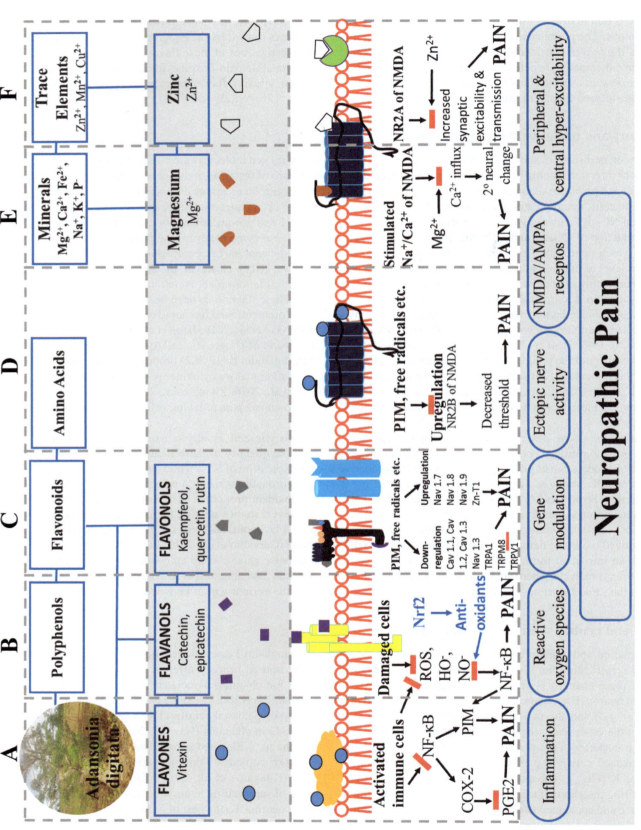

FIG. 2 Hypothesized mechanisms for *A. digitata*-mediated pain-relieving effects in Neuropathic Pain. (1A–F) Flavonoids and mineral composition of African Baobab plant. (2A–C) Specific subclasses of flavonoids detected in *A. digitata* plant that are relevant to neuropathic pain. (2E and F) Prominent mineral constituents of *A. digitata* plant that are relevant to neuropathic pain. (3A–F) Represent theorized endogenous mechanisms of African Baobab plant on neuropathic pain. The *RED bar* indicates possible suppression or inhibition of the flavonoids and minerals on endogenous pathway while the *BLUE arrows* represent flavonoids stimulation of pathway leading to mitigation of pain. Disease or damage of somatosensory neuron leads to activation of immune cells upregulate the expression of NF-κB and production of reactive oxygen and nitrogen species (ROS, NO˙). This leads to increased production of cyclooxygenase (COX-2) and pro-inflammatory mediators (PIM). These lead to modification of ion channels (Cav 1.1, Cav. 1.2, Cav. 1.3, Nav 1.3, etc.) on the pain receptor that are found on the injured afferent neuron. Also, the ROS and PIM cause modification NMDA receptor at the spinal cord. Taken together, there is decreased pain threshold, increased peripheral and central sensitization, and increased pain perception. Flavonoids and mineral components of *A. digitata* may block strategic events (*RED bar*) in these pathways as well as upregulate Nrf2 expression to increase antioxidant enzymes production to mop up ROS, NO˙ and HO˙.

and quercetin suppress the expression of TRPV1 (Chen et al., 2016; Gao et al., 2016). Since *A. digitata* extract has vitexin and quercetin flavonoids, it is possible the plant could inhibit the expression of mRNA for TRPV1. Study focusing on this is wanting in the literature (Fig. 2,3C).

Away from the injured primary afferent neuron in neuropathic pain, *N*-methyl-D-aspartate (NMDA) receptors on the second order neurons in the spinal cord are modified to shift from its normal firing threshold to a lower threshold. This shift from relatively high firing threshold to lower firing threshold constitute important aspect of central sensitization that occur in neuropathic pain. Again, study that investigate effect of *A. digitata* extract on central sensitization or NMDA receptor is rare in the literature. Though, vitexin was reported to downregulate expression of subunit 2B associated with glutamate ionotropic receptor NMDA. This subunit mediates increase in synaptic excitability and transmission of pain signals in neuropathic pain (Zhong et al., 2012). The possibility of *A. digitata* doing the same worth investigating (Fig. 2,3D).

Ions-mediated pathways

Beyond flavonoids constituents of *A. digitata*, mineral components of the plant extract may also contribute to its pain-relieving effects in neuropathic pain. Most prominent among the electrolytes in *A. digitata* (Seed, fruit, leaf and bark) plant are Mg, Ca, Fe, Zn and Mn. A few of these electrolytes were reported to have a relationship with neurodegeneration and neuropathic pain. For instance, magnesium, an abundant mineral in *A. digitata* was reported to exert analgesic effects in both humans (Koinig et al., 1998; Tramer et al., 1996) and animals (Begon et al., 2000; Takano et al., 2000). Magnesium modulates neuropathic pain by blocking spinal-mediated central sensitization that usually generates and maintains hypersensitivity phenomenon in chronic pain. Basically, exogenous magnesium treatment blocks the sodium/calcium ion channel of NMDA receptors and consequently prevents influx of extracellular calcium ions into the neural cell (Fawcett, Haxby, & Male, 1999). The blockage of extracellular calcium ions prevents secondary neuronal changes that lead to central sensitization associated with neuropathic pain. Based on these literature evidences, positioning the magnesium ion component of *A. digitata* plant as a contributor in its analgesic effects may not be a wide speculation (Fig. 2,3E).

Going further, in experimentally, induced-neuropathic pain, exogenous zinc ion was shown to bind with NR2A subunit of NMDA receptors, causing allosteric inhibition of the receptors that decreases synaptic excitability and transmission of pain signals which together result to pain-relieving effects (Nozaki et al., 2011). Also, zinc ion was reported to mitigate neuropathic pain via inhibition of acid-sensing ion channel (Lakhan & Avramut, 2012). Thus, there is a chance that the pain-relieving effects of *A. digitata* extract is not only due to its richness in flavonoids but also partly due to its abundance in zinc ions content (Fig. 2,3F).

Applications to other areas

A. digitata possesses a wide range of phytochemical constituents and in addition to its ameliorative effects in neuropathic pain, it has been evaluated for its therapeutic effects in several other disease areas. The discussion and table below highlight what we think are the most interesting areas for further exploration (Table 4).

Metabolic diseases

A. digitata appears to possess potential in metabolic diseases. Early studies had evaluated the modulatory effect of *A. digitata* extracts on key enzymes involved in metabolism that are therapeutic targets for diabetes mellitus. *A. digitata* extracts potently inhibited pancreatic amylase (Sudha, Zinjarde, Bhargava, & Kumar, 2011) and alpha-glucosidase enzyme (Irondi et al., 2017). In vivo, *A. digitata* administration reduced hyperglycemia in a streptozotocin-induced diabetes model. The hypoglycemic effect was further demonstrated in a high-sugar/high-fat diet model of metabolic syndrome. In this study, aqueous extracts of *A. digitata* lowered elevated glycemic and lipidemic levels (Suliman, Osman, Abdoon, Saad, & Khalid, 2020). In a clinical study where the effect of *A. digitata* on glycemic response was investigated, ingestion of *A. digitata* significantly lowered glycemic response that followed white bread ingestion, an effect associated with inhibition of digestible starch (Coe, Clegg, Armengol, & Ryan, 2013). Put together, these studies suggest the potential of *A. digitata* as an antidiabetic plant. More studies are however required. Standardized plant extracts that would enable objective evaluation of the plant extracts would need to be developed. Furthermore, extensive preclinical studies would be required to determine the optimal dose for hypoglycemic efficacy and the safety profile of the plant during ingestion. Eventually, human studies evaluating the safety and efficacy of *A. digitata* extracts in diabetic patients would have to be conducted. Drug discovery programs identifying and developing active hypoglycemic constituents are also recommended.

344 PART | IV Novel and nonpharmacological aspects and treatments

TABLE 4 Studies highlighting therapeutic potential of *Adansonia digitata*.

Therapeutic area	In silico	In vitro studies	In vivo studies	Human studies	Main findings
Diabetes/ metabolic syndrome		Cicolari et al. (2020), Ismail et al. (2019), and Sudha et al. (2011)	Suliman et al. (2020), Ebaid, Bashandy, Alhazza, Hassan, and Al-Tamimi (2019), and Braca et al. (2018)	Coe et al. (2013)	*A. digitata* extracts have been reported to possess hypoglycemic effects in model diabetic rats. It also lowers glucose excursion in humans during a meal. The hypoglycemic effects are related to its inhibitory enzyme on starch digestive enzymes such as glucosidase and amylase
Malaria	Adeoye et al. (2019)	Krishnappa, Elumalai, Dhanasekaran, and Gokulakrishnan (2012)	Adeoye and Bewaji (2018) and Musila et al. (2013)		*A. digitata* extracts have been demonstrated to suppress plasmodial development in murine model of malaria
Neuroprotection			Atuadu et al. (2021) and Shehu, Magaji, Yau, and Ahmed (2019)		*A. digitata* extracts protect against lead toxicity and improve key parameters in depression model
Hypertension			Ntchapda et al. (2020) and Ghoneim et al. (2016)		*A. digitata* extracts lowered blood pressure in animal models of hypertension
Antiviral		Hudson et al. (2000) and Anani et al. (2000)			*A. digitata* has potent virucidal properties

Malaria

The presence of sesquiterpene terpenoids in *A. digitata* suggests the plant may possess antimalarial efficacy. Aqueous extract of the stem bark was found effective in a chemosuppressive study of parasitemia (Adeoye & Bewaji, 2018; Musila, Dossaji, Nguta, Lukhoba, & Munyao, 2013). No clinical studies have yet been undertaken to evaluate the therapeutic potential in humans. Thus, more studies are still required before the antimalarial potential of *A. digitata* can be realized. Preclinical studies that evaluate the efficacy in chloroquine-sensitive and chloroquine-resistant strains of plasmodium would need to be conducted. Exploratory studies that also determine the active ingredients and the potential mechanism of action are still required. These studies may culminate in the design and conduct of safety and efficacy human studies evaluating the potential of *A. digitata* in human malaria patients.

Neuroprotection

Neurodegenerative and neurological disorders affect the quality of life of patients and current drugs do not fully address the pathophysiology of these conditions. Neuroprotective agents have been reported to provide additional benefits in diseases such as addiction, Alzheimer's and depression. In a murine model of depression, *A. digitata* extract was found to elevate the neuroprotective brain-derived neurotrophic factor and this was associated with improvement in behavioral measurements. Furthermore, in lead-induced neurotoxicity, *A. digitata* extract ameliorated the neurological impairments induced by the metal. Together, these studies indicate that there is indeed potential in exploring the therapeutic utility of *A. digitata* in neurodegenerative and neurologic disorders. Several questions need to be answered? Is the neuroprotection offered by *A. digitata* extensible to more brain-related disorders such as addiction, Parkinsonism and psychiatric illnesses? A human study has previously reported the satiety effect of *A. digitata* (Garvey, Clegg, & Coe, 2017) which suggest a rationale for its further evaluation in addiction and feeding disorders. Furthermore, what component(s) in *A. digitata* are responsible for neuroprotection and to what extent is this attributable to the agents reaching the brain. More human studies would then be needed to provide more evidence of safety in human consumption and efficacy for therapy in brain-related disorders.

Antiviral

Very early studies evaluated the antiviral potential of *A. digitata* extracts and these reports show that *A. digitata* possesses potent antiviral activity against viruses such as herpes simplex and polio virus (Anani et al., 2000; Hudson et al., 2000). With the recent coronavirus pandemic and the discovery of several potential antiviral components in plants (Gyebi, Ogunro, Adegunloye, Ogunyemi, & Afolabi, 2020), it may be worthwhile to evaluate the potential of *A. digitata* against the severe acute respiratory syndrome-2 virus (SARS2).

Other agents of interest

A wide variety of plants, plant constituents and existing therapeutic drugs have been evaluated and documented to possess efficacy in different models of neuropathic pain. Majority of the studies that have evaluated the therapeutic potential in humans are placebo-controlled, double-blinded randomized controlled clinical trials providing substantive evidence of efficacy. From such studies, evidence supports the therapeutic potential of oil extracts of *Citrullus colocynthis* (Bitter apple), different formulations of cannabis and cannabis-derived oils and nanocurcumin (see Table 5). However, more studies would be required to evaluate them against standard therapy. Meanwhile, a host of interesting results from clinical trials have thrown up more promising candidates for clinical evaluation. These include cannabidiol, moringin, beta-sitosterol among others (see Table 2).

Mini-dictionary of terms

- **Cosmeceuticals**—are cosmetics with biologically active compounds purported to have medical/pharmaceutical benefits.
- **Excitability**—is the ability of being quickly roused into state of action or excitement.
- **Flavonoids**—are polyphenolic natural compounds found in various plants that partly constitute humans' and animals' diets.
- **Homeostasis**—is the capacity to maintain relatively steady internal conditions irrespective changes outside the body.
- **Neuropathy**—dysfunction or injury of nerve(s) that result to pain, tingling, and numbness in the affected area.

TABLE 5 Showing agents evaluated for clinical efficacy in neuropathy.

S/N	Agent	Remarks	References
1	Topical *Citrullus colocynthis* (bitter apple) extract oil	Showed no efficacy in chemotherapy-induced neuropathy but good efficacy in painful diabetic neuropathy	Rostami, Mosavat, Heydarirad, Arbab Tafti, and Heydari (2019) and Heydari, Homayouni, Hashempur, and Shams (2016)
	Nanocurcumin	Improved diabetic sensorimotor polyneuropathy	Asadi et al. (2019)
	Low-dose vaporized cannabis	Showed efficacy in	Wilsey et al. (2013)
	Nutmeg extracts	Not efficacious in painful diabetic neuropathy after 2 months administration	Motilal and Maharaj (2013)
	St. John's Wort	No efficacy in painful polyneuropathy	Sindrup, Madsen, Bach, Gram, and Jensen (2001)

346 **PART | IV** Novel and nonpharmacological aspects and treatments

- **Noxious**—painful, very unpleasant, or harmful
- **Inflammation**—is one of the different biological responses of the body to injured cells, irritants or pathogens.
- **Phytochemicals**—are chemicals of plant origin
- **Phytogeographical**—relating to phytogeography—it is geographic distribution of various plant species that explain the origin, mode of dispersal, evolution and impact of precise plants on their geographic location.
- **Radical**—is an unpaired electron mostly formed when a single covalent bond breaks.
- **Receptors**—are mainly protein that receive and transduce signals capable of producing a biological activity or response
- **Transcription**—is the first step toward achieving gene expression in which an enzyme, RNA polymerases copies DNA based gene to make RNA molecules

Key facts

A. digitata attenuates pain sensations including neuropathic pain.
A. digitata decreases neural sensitizing molecules such as prostaglandin E2 and radicals to suppress neuropathic pain.
Inhibition or downregulation of NF-κB expression signaling during inflammation attenuates neuropathic pain.

Summary points

- Diverse flavonoids and mineral constituents of *A. digitata* plant seems to account for its efficacy in the treatment of neuropathic pain.
- Decreased prostaglandin E2 level, inhibition of acid-sensing ion channel and NR2A subunit of NMDA as well as selective suppression of mRNA for NF-κB, Nrf2, TRPV1, and subunit 2B of NMDA appeared to be the contributing mechanisms of *A. digitata* plant.
- When fully developed, *A. digitata* stands to offer a cheap, tolerable, effective, and less adversely prone drug for neuropathic pain managements.
- Future studies should be directed toward better our understanding on the precise downstream events that lead to pain-relieving effects during the treatments of neuropathic pain with *A. digitata* extract.

References

Adeoye, A., & Bewaji, C. O. (2018). Chemopreventive and remediation effect of *Adansonia digitata* L. Baobab (Bombacaceae) stem bark extracts in mouse model malaria. *Journal of Ethnopharmacology, 210,* 31–38.

Adeoye, A. O., et al. (2019). Molecular docking analysis of apigenin and quercetin from ethylacetate fraction of *Adansonia digitata* with malaria-associated calcium transport protein: An in silico approach. *Heliyon, 5*(9). https://doi.org/10.1016/j.heliyon.2019.e02248, e02248.

Ajibesin, K. K., et al. (2007). Ethnobotanical survey of Akwa Ibom State of Nigeria. *Journal of Ethnopharmacology, 115*(3), 387–408. https://doi.org/10.1016/j.jep.2007.10.021.

Amujoyegbe, O. O., et al. (2016). Ethnomedicinal survey of medicinal plants used in the management of sickle cell disorder in southern Nigeria. *Journal of Ethnopharmacology, 185,* 347–360. https://doi.org/10.1016/j.jep.2016.03.042.

Anani, K., Hudson, J. B., De Souza, C., Akpagana, K., Tower, G. H. N., Arnason, J. T., et al. (2000). Investigation of medicinal plants of Togo for antiviral and antimicrobial activities. *Pharmaceutical Biology, 38*(1), 40–45.

Armstrong, P. (1979). The history, natural history and distribution of Adansonia: A plant genus of the Indian Ocean. The Indian Ocean in focus. In *International Conference on Indian Ocean Studies, Perth, WA, 15–22 August 1979. Section I. Environment & Resources. University of Western Australia, Nedlands.* Available at: http://www.museum.wa.gov.au/maritime-archaeology-db/bibliography/history-natural-history-and-distribution-adansonia-plant-genus-indian-ocean-littoral.

Armstrong, P. (1983). The disjunct distribution of the genus Adansonia L. *National Geographical Journal of India, 29,* 142–163.

Asadi, S., Gholami, M. S., Siassi, F., Qorbani, M., Khamoshian, K., & Sotoudeh, G. (2019). Nano curcumin supplementation reduced the severity of diabetic sensorimotor polyneuropathy in patients with type 2 diabetes mellitus: A randomized double-blind placebo-controlled clinical trial. *Complementary Therapies in Medicine, 43,* 253–260.

Assogbadjo, A. E., et al. (2008). Folk classification, perception, and preferences of baobab products in West Africa: Consequences for species conservation and improvement. *Economic Botany, 62*(1), 74–84. https://doi.org/10.1007/s12231-007-9003-6.

Assogbadjo, A. E., et al. (2011). Natural variation in fruit characteristics, seed germination and seedling growth of *Adansonia digitata* L. in Benin. *New Forests, 41*(1), 113–125. https://doi.org/10.1007/s11056-010-9214-z.

Atuadu, V., Benneth, B. A., Oyem, J., Esom, E., Mba, C., Nebo, K., et al. (2021). *Adansonia digitata* L. leaf extract attenuates lead-induced cortical histoarchitectural changes and oxidative stress in the prefrontal cortex of adult male Wistar rats. *Drug Metabolism and Personalized Therapy, 36*(1), 63–71 (ahead-of-print).

Aubréville, A. (1975). Essais de géophylétique de Bombacacées. *Adansonia, 15*(2), 5–64.

Bakare, A. O., Oyewole, A. L., & Owoyele, B. V. (2019). Prostaglandin E2 and oxidative defense system contributed to anti-nociception action of aqueous *Adansonia digitata* bark extract in induced neuropathic pain in Wistar rats. *Oriental Pharmacy and Experimental Medicine, 19*(3), 287–298. https://doi.org/10.1007/s13596-019-00373-1.

Bamalli, Z., et al. (2014). Baobab tree (*Adansonia digitata* L) parts: Nutrition, applications in food and uses in ethno-medicine—A review. *Annals of Nutritional Disorders & Therapy, 1*(3), 1–9.

Barminas, J. T., Charles, M., & Emmanuel, D. (1998). Mineral composition of non-conventional leafy vegetables. *Plant Foods for Human Nutrition, 53*(1), 29–36. https://doi.org/10.1023/A:1008084007189.

Baron, R. (2000). Peripheral neuropathic pain: From mechanisms to symptoms. *Clinical Journal of Pain.* https://doi.org/10.1097/00002508-200006001-00004.

Basso, L., & Altier, C. (2017). Transient receptor potential channels in neuropathic pain. *Current Opinion in Pharmacology, 32*, 9–15. https://doi.org/10.1016/j.coph.2016.10.002.

Baum, D. A. (1995). A systematic revision of Adansonia (Bombacaceae). *Annals of the Missouri Botanical Garden, 82*(3), 440–471. Available at: www.jstor.org/stable/2399893.

Baum, D. A., Small, R. L., & Wendel, J. F. (1998). Biogeography and floral evolution of baobabs (Adansonia, Bombacaceae) as inferred from multiple data sets. *Systematic Biology, 47*(2), 181–207.

Begon, S., et al. (2000). Magnesium and MK-801 have a similar effect in two experimental models of neuropathic pain. *Brain Research, 887*(2), 436–439. https://doi.org/10.1016/S0006-8993(00)03028-6.

Boivin, N., & Fuller, D. Q. (2009). Shell middens, ships and seeds: Exploring coastal subsistence, maritime trade and the dispersal of domesticates in and around the ancient Arabian peninsula. *Journal of World Prehistory.* https://doi.org/10.1007/s10963-009-9018-2.

Borghi, S. M., et al. (2013). Vitexin inhibits inflammatory pain in mice by targeting TRPV1, oxidative stress, and cytokines. *Journal of Natural Products, 76*(6), 1141–1146. https://doi.org/10.1021/np400222v.

Borghi, S. M., et al. (2018). The flavonoid quercetin inhibits titanium dioxide (TiO_2)-induced chronic arthritis in mice. *Journal of Nutritional Biochemistry, 53*, 81–95. https://doi.org/10.1016/j.jnutbio.2017.10.010.

Bourinet, E., Francois, A., & Laffray, S. (2016). T-type calcium channels in neuropathic pain. *Pain, 157*(2), S15–S22. https://doi.org/10.1097/j.pain.0000000000000469.

Braca, A., et al. (2018). Phytochemical profile, antioxidant and antidiabetic activities of *Adansonia digitata* l. (baobab) from Mali, as a source of health-promoting compounds. *Molecules, 23*(12). https://doi.org/10.3390/molecules23123104.

Carvalho, T. T., et al. (2019). The granulopoietic cytokine granulocyte colony-stimulating factor (G-CSF) induces pain: Analgesia by rutin. *Inflammopharmacology, 27*(6), 1285–1296. https://doi.org/10.1007/s10787-019-00591-8.

Catarino, L., et al. (2019). Edible leafy vegetables from West Africa (Guinea-Bissau): Consumption, trade and food potential. *Food, 8*(10). https://doi.org/10.3390/foods8100493.

Chen, L., et al. (2016). Neuroprotective effects of vitexin against isoflurane-induced neurotoxicity by targeting the TRPV1 and NR2B signaling pathways. *Molecular Medicine Reports, 14*(6), 5607–5613. https://doi.org/10.3892/mmr.2016.5948.

Chinsembu, K. C. (2015). Plants as antimalarial agents in sub-Saharan Africa. *Acta Tropica, 152*, 32–48. https://doi.org/10.1016/j.actatropica.2015.08.009.

Chinsembu, K. C. (2019). Diabetes mellitus and nature's pharmacy of putative antidiabetic plants. *Journal of Herbal Medicine, 15*. https://doi.org/10.1016/j.hermed.2018.09.001.

Cicolari, S., Dacrema, M., Tsetegho Sokeng, A. J., Xiao, J., Atchan Nwakiban, A. P., Di Giovanni, C., et al. (2020). Hydromethanolic extracts from *Adansonia digitata* L. edible parts positively modulate pathophysiological mechanisms related to the metabolic syndrome. *Molecules, 25*(12), 2858.

Coe, S. A., Clegg, M., Armengol, M., & Ryan, L. (2013). The polyphenol-rich baobab fruit (*Adansonia digitata* L.) reduces starch digestion and glycemic response in humans. *Nutrition Research, 33*(11), 888–896.

Colloca, L., et al. (2017). Neuropathic pain. *Nature Reviews. Disease Primers, 3*, 1–20. https://doi.org/10.1038/nrdp.2017.2.

Cuni Sanchez, A., et al. (2011). Variation in baobab seedling morphology and its implications for selecting superior planting material. *Scientia Horticulturae, 130*(1), 109–117. https://doi.org/10.1016/j.scienta.2011.06.021.

Ebaid, H., Bashandy, S. A., Alhazza, I. M., Hassan, I., & Al-Tamimi, J. (2019). Efficacy of a methanolic extract of *Adansonia digitata* leaf in alleviating hyperglycemia, hyperlipidemia, and oxidative stress of diabetic rats. *BioMed Research International, 2019*. https://doi.org/10.1155/2019/2835152.

Eromosele, I. C., Eromosele, C. O., & Kuzhkuzha, D. M. (1991). Evaluation of mineral elements and ascorbic acid contents in fruits of some wild plants. *Plant Foods for Human Nutrition, 41*(2), 151–154. https://doi.org/10.1007/BF02194083.

Fabiyi, J. P., et al. (1993). Traditional therapy of dracunculiasis in the state of Bauchi – Nigeria. *Dakar Médical, 38*(2), 193–195.

Fawcett, W., Haxby, E., & Male, D. (1999). Magnesium: Physiology and pharmacology. *British Journal of Anaesthesia, 83*(2), 302–320. https://doi.org/10.1093/bja/83.2.302.

Gao, W., et al. (2016). Quercetin ameliorates paclitaxel-induced neuropathic pain by stabilizing mast cells, and subsequently blocking PKCε-dependent activation of TRPV1. *Acta Pharmacologica Sinica, 37*(9), 1166–1177. https://doi.org/10.1038/aps.2016.58.

Garvey, R., Clegg, M., & Coe, S. (2017). The acute effects of baobab fruit (*Adansonia digitata*) on satiety in healthy adults. *Nutrition and Health, 23*(2), 83–86.

Gebauer, J., El-Siddig, K., & Ebert, G. (2002). Baobab (*Adansonia digitata* L.): A review on a multipurpose tree with promising future in the Sudan. *Gartenbauwissenschaft, 67*(4), 155–160.

Gebauer, J., et al. (2014). Der Baobab (*Adansonia digitata* L.): Wildobst aus Afrika für Deutschland und Europa?! *Erwerbs-Obstbau, 56*(1), 9–24. https://doi.org/10.1007/s10341-013-0197-8.

Gebauer, J., et al. (2016). Africa's wooden elephant: The baobab tree (*Adansonia digitata* L.) in Sudan and Kenya: A review. *Genetic Resources and Crop Evolution, 63*(3), 377–399. https://doi.org/10.1007/s10722-015-0360-1.

Gelfland, M., et al. (1985). *The traditional medical practitioner in Zimbabwe: His principles of practice and pharmacopoeia.* Mambo Press.

Ghoneim, M. A. M., et al. (2016). Protective effect of *Adansonia digitata* against isoproterenol-induced myocardial injury in rats. *Animal Biotechnology, 27*(2), 84–95. https://doi.org/10.1080/10495398.2015.1102147.

Guazelli, C. F. S., et al. (2013). Quercetin-loaded microcapsules ameliorate experimental colitis in mice by anti-inflammatory and antioxidant mechanisms. *Journal of Natural Products, 76*(2), 200–208. https://doi.org/10.1021/np300670w.

Gyebi, G. A., Ogunro, O. B., Adegunloye, A. P., Ogunyemi, O. M., & Afolabi, S. O. (2020). Potential inhibitors of coronavirus 3-chymotrypsin-like protease (3CLpro): An in silico screening of alkaloids and terpenoids from African medicinal plants. *Journal of Biomolecular Structure and Dynamics,* 1–19 (just-accepted).

Heydari, M., Homayouni, K., Hashempur, M. H., & Shams, M. (2016). Topical C itrullus colocynthis (bitter apple) extract oil in painful diabetic neuropathy: A double-blind randomized placebo-controlled clinical trial: A double-blind randomized placebo-controlled clinical trial. *Journal of Diabetes, 8*(2), 246–252.

Hossain, M. S., et al. (2017). In vivo screening for analgesic and anti-inflammatory activities of syngonium podophyllum L.: A remarkable herbal medicine. *Annual Research & Review in Biology, 16*(3), 1–12. https://doi.org/10.9734/ARRB/2017/35692.

Hudson, J. B., Anani, K., Lee, M. K., De Souza, C., Arnason, J. T., & Gbeassor, M. (2000). Further Iinvestigations on the antiviral activities of medicinal plants of Togo. *Pharmaceutical Biology, 38*(1), 46–50.

Inngjerdingen, K., et al. (2004). An ethnopharmacological survey of plants used for wound healing in Dogonland, Mali, West Africa. *Journal of Ethnopharmacology, 92*(2-3), 233–244. https://doi.org/10.1016/j.jep.2004.02.021.

Irondi, E. A., et al. (2017). Blanching influences the phenolics composition, antioxidant activity, and inhibitory effect of *Adansonia digitata* leaves extract on α-amylase, α-glucosidase, and aldose reductase. *Food Science & Nutrition, 5*(2), 233–242. https://doi.org/10.1002/fsn3.386.

Ismail, B. B., et al. (2019). Characterizing the phenolic constituents of baobab (*Adansonia digitata*) fruit shell by LC-MS/QTOF and their in vitro biological activities. *Science of the Total Environment, 694*, 133387. https://doi.org/10.1016/j.scitotenv.2019.07.193.

Jiang, J., et al. (2019). Vitexin suppresses RANKL-induced osteoclastogenesis and prevents lipopolysaccharide (LPS)-induced osteolysis. *Journal of Cellular Physiology, 234*(10), 17549–17560. https://doi.org/10.1002/jcp.28378.

Kamanula, M., et al. (2018). Mineral and phytochemical composition of baobab (*Adansonia digitata* l.) root tubers from selected natural populations of Malawi. *Malawi Medical Journal, 30*(4), 250–255. https://doi.org/10.4314/mmj.v30i4.7.

Kamatou, G. P. P., Vermaak, I., & Viljoen, A. M. (2011). An updated review of *Adansonia digitata*: A commercially important African tree. *South African Journal of Botany, 77*(4), 908–919. https://doi.org/10.1016/j.sajb.2011.08.010.

Kempe, A., Neinhuis, C., & Lautenschläger, T. (2018). *Adansonia digitata* and Adansonia gregorii fruit shells serve as a protection against high temperatures experienced during wildfires. *Botanical Studies, 59*(1). https://doi.org/10.1186/s40529-018-0223-0.

Kinghorn, A. D., et al. (2011). The classical drug discovery approach to defining bioactive constituents of botanicals. *Fitoterapia, 82*(1), 71–79. https://doi.org/10.1016/j.fitote.2010.08.015.

Kitayama, T., et al. (2016). Down-regulation of zinc transporter-1 in astrocytes induces neuropathic pain via the brain-derived neurotrophic factor – K+-Cl− co-transporter-2 signaling pathway in the mouse spinal cord. *Neurochemistry International, 101*, 120–131. https://doi.org/10.1016/j.neuint.2016.11.001.

Koinig, H., et al. (1998). Magnesium sulfate reduces intra- and postoperative analgesic requirements. *Anesthesia and Analgesia, 87*(1), 206–210. https://doi.org/10.1097/00000539-199807000-00042.

Krishnappa, K., Elumalai, K., Dhanasekaran, S., & Gokulakrishnan, J. (2012). Larvicidal and repellent properties of *Adansonia digitata* against medically important human malarial vector mosquito *Anopheles stephensi* (Diptera: Culicidae). *Journal of Vector Borne Diseases, 49*, 86–90.

Lakhan, S. E., & Avramut, M. (2012). Matrix metalloproteinases in neuropathic pain and migraine: Friends, enemies, and therapeutic targets. *Pain Research and Treatment, 2012*. https://doi.org/10.1155/2012/952906.

Lee, H. N., et al. (2018). Anti-inflammatory effect of quercetin and galangin in LPS-stimulated RAW264.7 macrophages and DNCB-induced atopic dermatitis animal models. *International Journal of Molecular Medicine, 41*(2), 888–898. https://doi.org/10.3892/ijmm.2017.3296.

Li, X., et al. (2017). Application of a computer-assisted structure elucidation program for the structural determination of a new terpenoid aldehyde with an unusual skeleton. *Magnetic Resonance in Chemistry, 55*(3), 210–213. https://doi.org/10.1002/mrc.4466.Application.

Lin, Y. L., & Lin, J. K. (1997). (−)-Epigallocatechin-3-gallate blocks the induction of nitric oxide synthase by down-regulating lipopolysaccharide-induced activity of transcription factor nuclear factor-κB. *Molecular Pharmacology, 52*(3), 465–472. https://doi.org/10.1124/mol.52.3.465.

Lockett, C. T., Calvert, C. C., & Grivetti, L. E. (2000). Energy and micronutrient composition of dietary and medicinal wild plants consumed during drought. Study of rural Fulani, Northeastern Nigeria. *International Journal of Food Sciences and Nutrition, 51*(3), 195–208. https://doi.org/10.1080/09637480050029700.

Lu, Y., et al. (2018). Vitexin attenuates lipopolysaccharide-induced acute lung injury by controlling the Nrf2 pathway. *PLoS One, 13*(4), 1–12. https://doi.org/10.1371/journal.pone.0196405.

Magaia, T., et al. (2013). Dietary fiber, organic acids and minerals in selected wild edible fruits of Mozambique. *Springerplus, 2*(1), 1–8. https://doi.org/10.1186/2193-1801-2-88.

Maheshwari, J. K. (1971). The baobab tree: Disjunctive distribution and conservation. *Biological Conservation, 4*(1), 57–60. https://doi.org/10.1016/0006-3207(71)90058-9.

Mariod, A. A., Mirghani, M. E. S., & Hussein, I. (2017). *Adansonia digitata* Baobab seed oil. In A. A. Mariod, M. Elwathig, & H. I. Hussein (Eds.), *Unconventional oilseeds and new oil sources* (pp. 267–271). Academic Press.

Martínez-Ballesta, M. C., et al. (2010). Minerals in plant food: Effect of agricultural practices and role in human health. A review. *Agronomy for Sustainable Development, 30*(2), 295–309.

Motilal, S., & Maharaj, R. G. (2013). Nutmeg extracts for painful diabetic neuropathy: A randomized, double-blind, controlled study. *The Journal of Alternative and Complementary Medicine, 19*(4), 347–352.

Mulaudzi, R. B., et al. (2013). Anti-inflammatory and mutagenic evaluation of medicinal plants used by Venda people against venereal and related diseases. *Journal of Ethnopharmacology, 146*(1), 173–179. https://doi.org/10.1016/j.jep.2012.12.026.

Mullin, L. J. (2003). *Historic trees of Zimbabwe.* CBC Publishing. Available at: https://www.nhbs.com/historic-trees-of-zimbabwe-book.

Musila, M. F., Dossaji, S. F., Nguta, J. M., Lukhoba, C. W., & Munyao, J. M. (2013). In vivo antimalarial activity, toxicity and phytochemical screening of selected antimalarial plants. *Journal of Ethnopharmacology, 146*(2), 557–561.

Neuwinger, H. D. (2000). *African traditional medicine: A dictionary of plant use and applications with supplement : Search system for diseases. 2000.* Medpharm Scientific Publishers. Available at: https://books.google.com.ng/books/about/African_traditional_medicine.html?id=LopFAQAAIAAJ&redir_esc=y.

Nozaki, C., et al. (2011). Zinc alleviates pain through high-affinity binding to the NMDA receptor NR2A subunit. *Nature Neuroscience, 14*(8), 1017–1022. https://doi.org/10.1038/nn.2844.

Ntchapda, F., Bonabe, C., Atsamo, A. D., Kemata Azambou, D. R., Bekono Fouda, Y., Imar Djibrine, S., … Theophile, D. (2020). Effect of aqueous extract of Adansonia digitata stem bark on the development of hypertension in L-NAME-induced hypertensive rat model. *Evidence-based Complementary and Alternative Medicine.* https://doi.org/10.1155/2020/3678469, 3678469.

Osman, M. A. (2004). Chemical and nutrient analysis of baobab (*Adansonia digitata*) fruit and seed protein solubility. *Plant Foods for Human Nutrition, 59*(1), 29–33. https://doi.org/10.1007/s11130-004-0034-1.

Owoyele, B. V., & Bakare, A. O. (2018). Analgesic properties of aqueous bark extract of *Adansonia digitata* in Wistar rats. *Biomedicine and Pharmacotherapy, 97*, 209–212. https://doi.org/10.1016/j.biopha.2017.10.079.

Parkouda, C., et al. (2015). Biochemical changes associated with the fermentation of baobab seeds in Maari: An alkaline fermented seeds condiment from western Africa. *Journal of Ethnic Foods, 2*(2), 58–63. https://doi.org/10.1016/j.jef.2015.04.002.

Patil, R. H., et al. (2016). Anti-inflammatory effect of Apigenin on LPS-induced pro-inflammatory mediators and AP-1 factors in human lung epithelial cells. *Inflammation, 39*(1), 138–147. https://doi.org/10.1007/s10753-015-0232-z.

Prentice, A., et al. (1993). The calcium and phosphorus intakes of rural Gambian women during pregnancy and lactation. *British Journal of Nutrition, 69*(3), 885–896. https://doi.org/10.1079/bjn19930088.

Rahul, J., et al. (2015). *Adansonia digitata* L. (baobab): A review of traditional information and taxonomic description. *Asian Pacific Journal of Tropical Biomedicine, 5*(1), 79–84. https://doi.org/10.1016/S2221-1691(15)30174-X.

Ramadan, A., Harraz, F. M., & El-Mougy, S. A. (1994). Anti-inflammatory, analgesic and antipyretic effects of the fruit pulp of *Adansonia digitata. Fitoterapia, 65*(5), 418–422.

Rogers, M., et al. (2006). The role of sodium channels in neuropathic pain. *Seminars in Cell & Developmental Biology, 17*, 571–581. https://doi.org/10.1016/j.semcdb.2006.10.009.

Rostami, N., Mosavat, S. H., Heydarirad, G., Arbab Tafti, R., & Heydari, M. (2019). Efficacy of topical Citrullus colocynthis (bitter apple) extract oil in chemotherapy-induced peripheral neuropathy: A pilot double-blind randomized placebo-controlled clinical trial. *Phytotherapy Research, 33*(10), 2685–2691.

Sena, L. P., et al. (1998). Analysis of nutritional components of eight famine foods of the republic of Niger. *Plant Foods for Human Nutrition, 52*(1), 17–30. https://doi.org/10.1023/A:1008010009170.

Seukep, J. A., et al. (2013). Antibacterial activities of the methanol extracts of seven Cameroonian dietary plants against bacteria expressing MDR phenotypes. *Springerplus, 2*(1), 1–8. https://doi.org/10.1186/2193-1801-2-363.

Shehu, A., Magaji, M. G., Yau, J., & Ahmed, A. (2019). Methanol stem bark extract of *Adansonia digitata* ameliorates chronic unpredictable mild stress-induced depression-like behavior: Involvement of the HPA axis, BDNF, and stress biomarkers pathways. *Journal of Basic and Clinical Physiology and Pharmacology, 30*(3).

Shri, V. T., et al. (2004). Studies of pharmacognostical profiles of *Adansonia Digitata* linn. *Ancient Science of Life, XXVI*(2), 60–64.

Sidibe, M., & Williams, J. T. (2002). In N. Hughes, N. Haq, & R. Smith (Eds.), *Baobab (Adansonia digitata L.) propagation manual.* Southampton, UK: International Centre for Underutilised Crops. Cover.

Sindrup, S. H., Madsen, C., Bach, F. W., Gram, L. F., & Jensen, T. S. (2001). St. John's wort has no effect on pain in polyneuropathy. *Pain, 91*(3), 361–365.

Soetan, K. O., Olaiya, C. O., & Oyewole, O. E. (2010). The importance of mineral elements for humans, domestic animals and plants—A review. *African Journal of Food Science, 4*(5), 200–222.

Stadlmayr, B., et al. (2013). Nutrient composition of selected indigenous fruits from sub-Saharan Africa. *Journal of the Science of Food and Agriculture, 93*(11), 2627–2636. https://doi.org/10.1002/jsfa.6196.

Sudha, P., Zinjarde, S. S., Bhargava, S. Y., & Kumar, A. R. (2011). Potent α-amylase inhibitory activity of Indian Ayurvedic medicinal plants. *BMC Complementary and Alternative Medicine, 11*(1), 5.

Suliman, H. M., Osman, B., Abdoon, I. H., Saad, A. M., & Khalid, H. (2020). Ameliorative activity of *Adansonia digitata* fruit on high sugar/high fat diet-simulated metabolic syndrome model in male Wistar rats. *Biomedicine & Pharmacotherapy, 125*, 109968.

Takano, Y., et al. (2000). Antihyperalgesic effects of intrathecally administered magnesium sulfate in rats. *Pain, 84*(2–3), 175–179. https://doi.org/10.1016/S0304-3959(99)00207-9.

Tian, R., et al. (2016). Rutin ameliorates diabetic neuropathy by lowering plasma glucose and decreasing oxidative stress via Nrf2 signaling pathway in rats. *European Journal of Pharmacology, 771*, 84–92. https://doi.org/10.1016/j.ejphar.2015.12.021.

Tramer, M., et al. (1996). Role of magnesium sulfate in postoperative analgesia. *Anesthesiology, 84*, 340–347.

Van Wyk, B. E. (2015). A review of commercially important African medicinal plants. *Journal of Ethnopharmacology, 176*, 118–134. https://doi.org/10.1016/j.jep.2015.10.031.

Vandercone, R., et al. (2004). The status of the baobab (*Adansonia digitata* L.) in Mannar Island, Sri Lanka. *Current Science, 87*(12), 1709–1713. Available at: https://www.cabdirect.org/cabdirect/abstract/20053009933.

Welch, R. M., & Graham, R. D. (2004). Breeding for micronutrients in staple food crops from a human nutrition perspective. *Journal of Experimental Botany, 55*(396), 353–364.

Welcome, A. K., & Van Wyk, B. E. (2019). An inventory and analysis of the food plants of southern Africa. *South African Journal of Botany, 122*, 136–179. https://doi.org/10.1016/j.sajb.2018.11.003.

Wickens, G. E. (1982). *The baobab – Africa's upside-down tree. 47* (p. 2). Kew Bull. Available at: https://baobabstories.com/en/wickens-g-e-the-baobab-africas-upside-down-tree/.

Wickens, G. E., & Lowe, P. (2008). *The baobabs: Pachycauls of Africa, Madagascar and Australia* (1st ed.). https://doi.org/10.1007/978-1-4020-6431-9. Springer Netherlands.

Wilsey, B., Marcotte, T., Deutsch, R., Gouaux, B., Sakai, S., & Donaghe, H. (2013). Low-dose vaporized cannabis significantly improves neuropathic pain. *The Journal of Pain, 14*(2), 136–148.

Zahid, A., et al. (2017). Arabinogalactan proteins from baobab and Acacia seeds influence innate immunity of human keratinocytes in vitro. *Journal of Cellular Physiology, 232*(9), 2558–2568. https://doi.org/10.1002/jcp.25646.

Zhong, Y., et al. (2012). Anti-inflammatory activity of lipophilic epigallocatechin gallate (EGCG) derivatives in LPS-stimulated murine macrophages. *Food Chemistry, 134*(2), 742–748. https://doi.org/10.1016/j.foodchem.2012.02.172.

Chapter 31

Andrographis paniculata standardized extract (ParActin) and pain

Rafael A. Burgos, Pablo Alarcón, and Juan L. Hancke

Laboratory of Inflammation Pharmacology, Institute of Pharmacology and Morphophysiology, Universidad Austral de Chile, Valdivia, Chile

Abbreviations

ATF3	activating transcription factor 3
Cox2	cyclooxygenase 2
GR	glutathione reductase
GSH-Px	glutathione peroxidase
HDAC1	histone deacetylases1
HO-1	hem-oxygenase 1
HPLC	high-performance liquid chromatography
IL-1β	interleukin-1β
IL-6	interleukin-6
Keap1	Kelch-like ECH-associated protein1
LPS	lipopolysaccharide
MMP13	metalloproteinase 13
NET	neutrophils extracellular traps
NF-kB	nuclear factor-kappa B
NO	nitric oxide
Nrf2	nuclear factor-erythroid 2 p45-related factor 2
PGE2	prostaglandin E2
PMA	phorbol myristate acetate
RA-FLS	rheumatoid arthritis fibroblast-like Synoviocytes
ROS	reactive oxygen species
STAT3	signal transducer and activator of transcription 3
TH2	T helper cell type 2
TNF-α	tumor necrosis factor-α

Introduction

Andrographis paniculata: Classification and composition

Andrographis paniculata (Burm; Acanthaceae) is a medicinal plant widely used as a traditional herbal medicine in Bangladesh, China, Hong Kong, India, Pakistan, the Philippines, Malaysia, Indonesia, and Thailand (Akbar, 2011; Hossain, Urbi, Sule, & Hafizur Rahman, 2014). The whole plant (leaves and roots) is used as a co-adjuvant for different diseases in Asia and Europe (Akbar, 2011).

An analysis of the chemical composition of the leaves and stems of *A. paniculata* showed that it is abundant in diterpenoids and 2-oxygenated flavonoids including andrographolide, a bitter principle with derivates such as neoandrographolide, 14-deoxy-11,12-didehydroandrographolide, and 14-deoxyandrographolide, among others (Chao & Lin, 2010). Andrographolide can be isolated from the aerial part of the plant by extraction with alcohol or with alkaline solutions. The amount of andrographolide in leaves and stems of *A. paniculata* has been quantitatively determined by UV spectrophotometry after separation on silica gel plates and High-Performance Liquid Chromatography (HPLC). The stem was

Treatments, Mechanisms, and Adverse Reactions of Anesthetics and Analgesics. https://doi.org/10.1016/B978-0-12-820237-1.00003-X
Copyright © 2022 Elsevier Inc. All rights reserved.

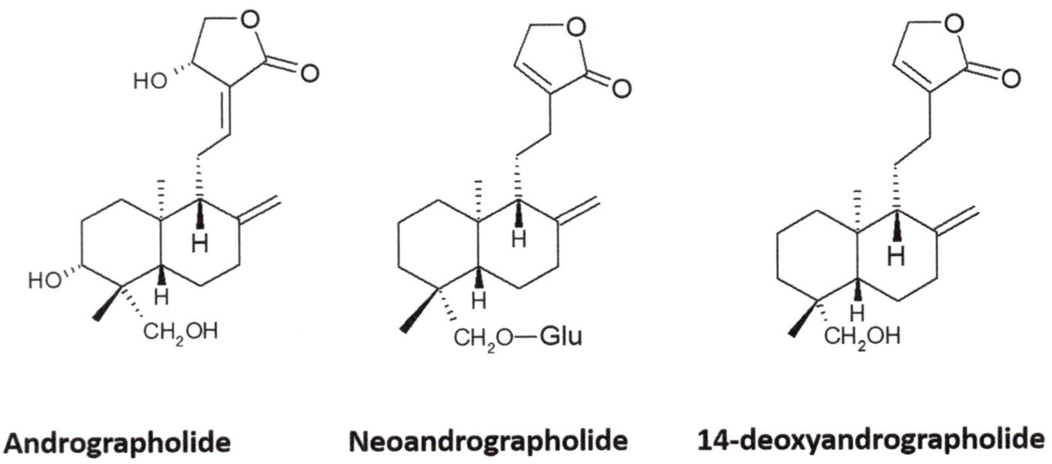

352 PART | IV Novel and nonpharmacological aspects and treatments

Andrographolide **Neoandrographolide** **14-deoxyandrographolide**

FIG. 1 Scheme of chemical structures of andrographolide and its derivates present in ParActin.

found to contain $0.2\% \pm 0.02\%$ andrographolide content, the seeds contained $0.13\% \pm 0.01\%$; the roots contained $0.44\% \pm 0.01\%$; and the leaves contained $2.39\% \pm 0.008\%$ (Sharma, Lal, & Handa, 1992). The leaves contain more than 2% andrographolide before the plant blossoms; thereafter, the content decreases to less than 0.5% (Tang & Eisenbrand, 1992). Andrographolide is soluble in acetone, methanol, chloroform, and ether, and sparingly soluble in water (Meng, 1981).

The most important andrographolide-like compounds, characterized by Balmain and Connolly (1973), are: 14-deoxy-11,12-didehydroandrographolide, with an average content in the leaf of 0.1%, 14-deoxyandrographolide (0.02%), and 14-deoxy-11-oxoandrographolide (0.12%) (Balmain & Connolly, 1973).

ParActin is a patented dried extract of *A. paniculata* standarized to 50% (w/w) of total andrographolides, of which 3%–6% (w/w) is 14-deoxyandrographolide, and approximately 0.2%–0.8% w/w is neoandrographolide (Fig. 1).

Preclinical antiinflammatory and analgesic effects of *A. paniculata* and andrographolide contained in ParActin

A. paniculata and andrographolide have been shown to have multiple properties in preclinical studies. For example, in a tumor induced by xenograft in a mouse model, the co-administration of *A. paniculata* (1600 mg/kg) with chemotherapeutics (cisplatin and 5-fluorouracil) showed an anti-metastatic effect, reducing the number of proliferative cancer cells in tumor sections and leading to an increase in apoptotic cancer cells in the tumor (Yue et al., 2019). Andrographolide can modulate the innate and adaptive immune responses by regulating macrophage phenotypic polarization induced by LPS or IL-4, inhibiting both M1 and M2 cytokine expression and decreasing the IL-12/IL-10 ratio (the ratio of M1/M2 polarization) (Wang et al., 2010). Andrographolide has been proposed as an anti-arthritis agent, due to its ability to promote chondrocyte proliferation and prevent apoptosis in pathologic conditions in an osteoarthritis mouse model. It also reduced interleukin-1β (IL-1β) via the miR-27-3p/MMP13 axis (Chen, Luo, & Chen, 2020) and nitric oxide (NO) production (Jin et al., 2017). In addition, andrographolide inhibited the production of NO, tumor necrosis factor-α (TNF-α), and cyclooxygenase 2 (COX2) expression via blockage of nuclear factor-kappa B (NF-κB) in lipopolysaccharide (LPS)-activated peritoneal macrophages in a dose-dependent manner (Gupta, Mishra, Singh, & Ganju, 2018). It is known that neutrophils play a key role in the onset of arthritis. Andrographolide has also been shown to reduce neutrophils infiltration in a complete Freund's adjuvant-induced arthritis rat model (Li et al., 2019). Furthermore, andrographolide accelerates neutrophil apoptosis in the presence of LPS and reduces neutrophil extracellular traps (NET)-formation induced by phorbol 12-myristate 13-acetate (PMA) (Li et al., 2019), therefore reducing inflammation. Andrographolide reduces the production and release of proinflammatory mediators such as TNF-α, IL-6, IL-1β (Li et al., 2017), NO, and PGE2 (Lee, Chang, Chung, & Lee, 2011), in LPS-stimulated RAW264.7 cells. Interestingly, the activation of the immune system and release of cytokines or other proinflammatory molecules have an effect on the sensory neurons, altering the spinal cord-brain axis through sensory neuron-expressed proinflammatory cytokine receptors (e.g., IL-1bR, TNF-αR, and IL-6R) (Vanderwall & Milligan, 2019). Another effect described for andrographolide is its capacity to relieve pain. For instance, in allodynia in mice with spared injured sciatic nerves, andrographolide was able to reduce mechanically induced pain (Wang et al., 2018). Andrographolide reduced the mechanical and thermal hyperalgesia of the sciatic nerve exposed to VIH-gp120 plus 2′,3′-dideoxycytidine injection through P2X7 downregulation and decreasing the production of IL-1β, thereby relieving the allodynia-associated pain (Yi et al., 2018). In addition, in a model of nitroglycerin-induced migraine, *A. paniculata* extract counteracted nitroglycerin-induced hyperalgesia, reducing mRNA levels of interleukin-6 (IL-6) and TNF-α in medulla and brain areas specific (mesencephalon) to migraine pain (Greco et al., 2016).

The other metabolites that compose ParActin have been described in terms of their biological activities. For instance, neoandrographolide has also been described to possess antiinflammatory effects. Neoandrographolide inhibited LPS-stimulated NO and prostaglandin E2 (PGE2) production (Liu, Wang, Ji, & Ge, 2007), PMA-induced reactive oxygen species production, and LPS-induced TNF-α production (Liu, Wang, & Ji, 2007) in murine macrophages. 14-deoxyandrographolide and andrographolide interact with several residues of COX2, suggesting an inhibitory effect on COX2 in inflammatory processes (Jiao et al., 2019). Moreover, 14-Deoxyandrographolide also reduced paw edema induced by egg white in rats (Liu, Yan, Zheng, Jia, & Han, 2020), demonstrating that andrographolide and its derivates possess antiinflammatory effects.

Another property described by andrographolide is its effect on the respiratory system. For instance, in a mouse model of ovalbumin-induced asthma, andrographolide improved allergic airway inflammation, reducing IL-4, IL-5, and IL-13 levels in TH2 cells. Eotaxin, and serum immunoglobulins were reduced in the bronchoalveolar lavage (Bao et al., 2009). Using the same model, other authors have described the reduction in the level of TNF-α, IL-1β, inflammasome activation, and ROS production (Peng et al., 2016). In addition, andrographolide reduces the expression of IL-6 in pulmonary dysfunction induced by cigarette smoke in a mouse model (Xia et al., 2019).

Previous studies have described the antiinflammatory activity of andrographolide in the intestine. In a mouse model of ulcerative colitis induced by dextran sulfate sodium, andrographolide suppressed the proinflammatory response, reducing TNF-α, IL-1β, and IL-6 levels (Kim et al., 2019), and inducible nitric oxide synthase (iNOS) and COX2 expression (Jing, Wang, & Xu, 2019).

Finally, in most cases, the origin of the pain is inflammatory, regardless of the etiology. With the elucidation of the role of inflammatory cytokines, there is now a clear understanding of the pathways by which many antiinflammatory drugs can alleviate inflammation and relieve pain (Maroon, Bost, & Maroon, 2010), making ParActin a promising candidate in the relief of pain derived from inflammation in pathologies such as asthma and ulcerative colitis, and other autoimmune diseases such as rheumatoid arthritis and multiple sclerosis.

Mechanism of action of andrographolide, the active principle of ParActin in inflammation and pain

In recent years there has been great scientific interest in the study and/or discovery of multitarget drugs (or drugs with polypharmacological profiling) due to their advantages as potential therapeutic solutions to diseases of complex etiology such as idiopathic pain (Turnaturi, Arico, Ronsisvalle, Pasquinucci, & Parenti, 2016). Andrographolide has been proposed as a multitarget molecule with neuroprotective and antiinflammatory effects (Kishore et al., 2017).

Proinflammatory stimuli of microbial origin or cytokines such as IL-1β and TNF-α, act through toll-like receptors, such as the IL-1 receptor family or TNF receptor family, respectively. Activation of all these receptor families triggers the activation of several signaling pathways that lead to the expression of several proteins associated with inflammation and pain (Chen et al., 2018; Yu, Sun, Wang, & Yan, 2016).

Effect of andrographolide on the NF-κB pathway

One of the most studied mechanisms of the antiinflammatory effect of andrographolide is the inhibition of the nuclear transcriptional factor-κB response (Fig. 2). The NF-κB family of transcription factors consists of five members, p50, p52, p65 (RelA), c-Rel, and RelB which can exist in the cytoplasm in the homo- or heterodimer form (Hayden & Ghosh, 2008). In unstimulated cells, NF-κB is found in the cytosol, inactivated by an inhibitor of the κB (IκB) molecule. Activation of the IκB kinase (IKK) complex targets IκB for degradation, thus releasing NF-κB and activating (or, in some cases, repressing) target gene transcription (Simmonds & Foxwell, 2008). NF-κB regulates the expression of several proinflammatory genes such as TNF-α, IL-1, IL-6, and COX2 (Chen et al., 2018), as well as chemokines, inducible nitric oxide synthase, and matrix metalloproteinases (Simmonds & Foxwell, 2008). In neutrophils, andrographolide acts as a negative regulator, since it interferes with NF-κB-DNA binding induced by PAF and N-formyl-methionyl-leucyl-phenylalanine (fMLP) and reduces the expression of COX2 (Hidalgo et al., 2005). In addition, andrographolide produces a covalent adduct with reduced cysteine 62 of p50, preventing the binding of NF-κB and transcriptional activity (Xia et al., 2004). Other authors have described how andrographolide increases the AMPK phosphorylation, indicating that andrographolide

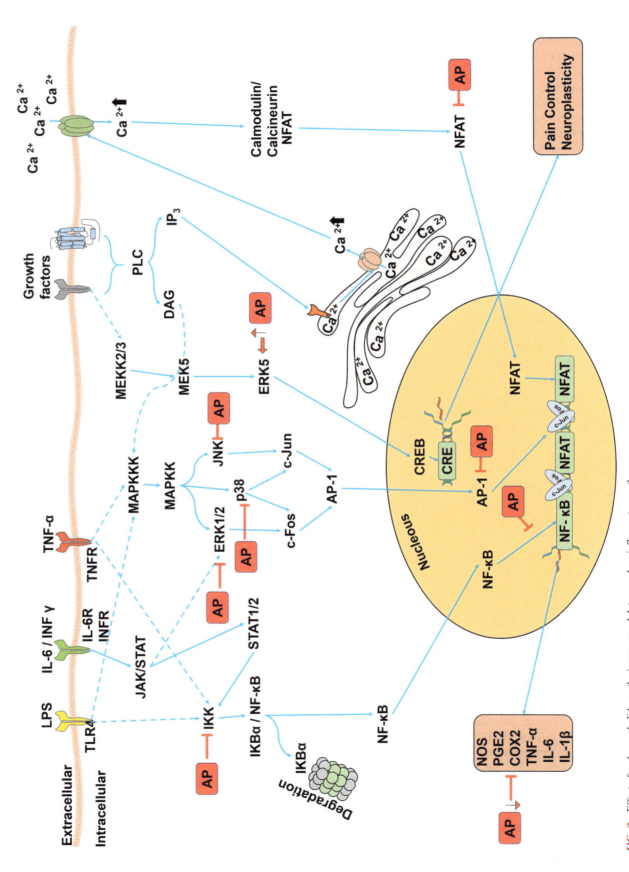

FIG. 2 Effect of andrographolide on the immunomodulatory and proinflammatory pathways.

is able to interfere with IκBα phosphorylation and its degradation (Kim et al., 2019). Moreover, there is evidence that andrographolide inhibits TNF-α-induced IKK activation and IκBα phosphorylation in endothelial cells (Chao et al., 2011) (Fig. 2). The NF-κB pathway is involved in pain, e.g., neuropathic pain (Xu et al., 2014). In fact, andrographolide decreases neuropathic pain in rodents (Wang et al., 2018), and ParActin reduces pain in conditions associated with upper respiratory tract infections (URTI) or joint inflammation in OA.

Effects of Andrographolide on MAPK and AP-1

Andrographolide reduces the phosphorylation of mitogen-activated protein kinases (MAPK) such as extracellular signal-regulated protein kinase 1/2 (ERK1/2), p38, and c-jun N-terminal kinase (JNK) in LPS-stimulated RAW264.7 cells (Li, He, et al., 2017) or TNF-α stimulated rheumatoid arthritis fibroblast-like synoviocytes (RA-FLS) (Li, Tan, Wang, & Li, 2017) (Fig. 2). It has also been shown that TNF-α-induced AP-1 activation occurs via the ERK 1/2 and p38 MAPK pathways (Yuan, Meng, Liao, & Lian, 2018). AP-1 has a central role in immune cells and is composed of four protein subfamilies combined in homo- or heterodimers: Jun (c-Jun, JunB, JunD), Fos (c-Fos, FosB, Fra1, Fra2), ATF-activating transcription factor (ATF2, LRF1/ATF3, BATF, JDP1, JDP2), and Maf (musculoaponeurotic fibrosarcoma; c-Maf, MafB, MafA. Mafg/f/k, and Nrl) (Atsaves, Leventaki, Rassidakis, & Claret, 2019).

Andrographolide inhibits the activation of activator protein-1 (AP-1), another transcriptional factor with pleiotropic effects, via the Src/MAPK pathway (Yuan et al., 2018). In neurons from dorsal root ganglia tissue, the activating transcription factor 3 (ATF3)/AP-1 pathway is increased in neuropathy and is positively correlated with the severity of hyperalgesia (Hsieh, Chiang, Lue, & Hsieh, 2012). Moreover, AP-1 and histone deacetylase 1 (HDAC1) expression is correlated with microglial activation in patients with neuropathic pain (Shimoyama et al., 2019), indicating that inhibition of this pathway can support the pain-controlling effect of andrographolide.

In addition, AP-1 is activated by the MAPK (Karin, 1995) and/or JAK/STAT pathways (Horvai et al., 1997). Andrographolide reduces the phosphorylation of STAT 1/2, thereby interfering with the JAK/STAT pathway in an influenza virus-induced inflammation murine model (Ding et al., 2017) (Fig. 2). In a murine model of pulmonary dysfunction induced by cigarette smoke, andrographolide reduced the signal transducer and activator of the transcription 3 (STAT3) pathway by interfering with the expression of SOCS1 and SOCS3 signaling (Lee et al., 2011). The JAK/STAT pathway is activated by cytokines or growth factors and is involved in heightened pain sensitivity under inflammatory conditions (Busch-Dienstfertig & Gonzalez-Rodriguez, 2013).

AP-1 also interacts with other nuclear factors such as NF-κB, CBP/p300, and nuclear factor of activated T-cells (NFAT) (Foletta, Segal, & Cohen, 1998). NFAT is a family of nuclear transcription factors with multiple effects on immune cells. Several proinflammatory stimuli induce calcium mobilization from intracellular stores (reticulum endoplasmic or mitochondrial) or extracellular sources, resulting in activation of many intracellular enzymes including the calcium- and calmodulin-dependent phosphatase calcineurin, a major upstream regulator of NFAT proteins. NFAT is involved in neuropathic pain in the dorsal horn (DH) and dorsal root ganglion (DRG) (Huang et al., 2018). Andrographolide interferes with NFAT activation via JNK/MAPK in Jurkat cells (Human acute T cells leukemia) stimulated by PMA/Ionomycin; this reduced IL-2 production (Carretta et al., 2009), thereby contributing to pain relief. Furthermore, it has been described that andrographolide reduces ERK5 phosphorylation in Jurkat cells treated with anti-CD3/Ionomycin (Carretta et al., 2009) (Fig. 2). The ERK 5 pathway is activated upstream by MAPK kinase 5 (MEK 5); downstream, ERK5 activates cyclic adenosine monophosphate (cAMP)-response element-binding protein (CREB). Interestingly, several studies have described that ERK 5 activation in the DRG (mainly in microglia cells) and the spinal cord takes part in mediating the transduction of pain signals and contributes to hyperalgesia and allodynia after peripheral inflammation or nerve injuries (neuropathic pain) (Mizushima et al., 2007; Obata et al., 2007; Xiao, Zhang, Cheng, & Zhang, 2008; Yu et al., 2016).

Antioxidant effect of andrographolide and role of the Nrf2/keap1 pathway

It is known that ROS such as superoxide anions induce inflammatory pain (Maioli et al., 2015). Andrographolide decreases radical oxygen species (ROS) production in neutrophils (Shen, Chen, & Chiou, 2002) (Fig. 3). In addition, andrographolide inhibits the oxidative biomarkers 8-isoprostane, 8-OHdG, and 3-nitrotyrosine, and promotes antioxidant enzyme glutathione peroxidase (GSH-Px) and glutathione reductase (GR) activities (Guan et al., 2013). Andrographolide increases the expression of antioxidant enzymes such as superoxide dismutase, catalase, glutathione reductase, glutathione peroxidase, glutathione-S-transferase (GST), and reduced glutathione (GSH), as well as glutathione disulfite (GSSG) concentrations (Das, Gautam, Dey, Maiti, & Roy, 2009). One of the transcription factors that regulate the status of cellular redox is nuclear factor-erythroid 2 p45-related factor 2 (Nrf2) (Hayes & Dinkova-Kostova, 2014). The inhibition of ROS is

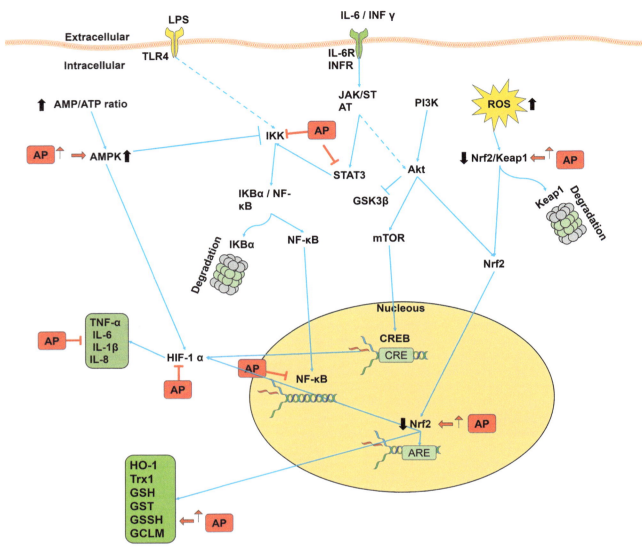

FIG. 3 Antioxidant and antiinflammatory effect of andrographolide on the different pathways.

associated with the inhibition of NADPH oxidase activation, up-regulation of hem-oxygenase-1 (HO-1) and glutamate-cysteine ligase modifier subunit (GCLM) expression, and elevation of GSH content through the PI3K/Akt/AP-1 and PI3K/Akt/Nrf2 pathways (Lu et al., 2014) (Fig. 3). Interestingly, it has been described that andrographolide increases the rate of nuclear translocation of Nrf2, inducing antioxidant response (Yu et al., 2010) and upregulating the expression of genes that control antioxidant, detoxicant, and antiinflammatory activity (Guan et al., 2013).

A functional Nrf2–Kelch-like ECH-associated protein1 (Keap1) axis is essential for protection against a plethora of diseases that involve oxidative stress and inflammation (Cuadrado et al., 2019). Furthermore, activation and nuclear accumulation of Nrf2 is associated with acute and chronic inflammatory pain relief (Staurengo-Ferrari et al., 2018), emulating the effects of NSAIDs on inflammatory pain (Cashman, 1996). Moreover, activation of the Keap1/Nrf2/Hem-oxygenase 1 (HO-1) axis (i.e., sulforaphane) enhances the analgesic effects of opioids (McDonnell, Leanez, & Pol, 2017; Redondo, Chamorro, Riego, Leanez, & Pol, 2017), which supports the fact that Nrf2 activators are potential candidates to relieve pain from inflammation.

Nrf2 induces transcription of cytoprotective enzymes, such as thioredoxin 1 (Trx1) (Tanito, Agbaga, & Anderson, 2007), and leads to an increase in hypoxia-inducible factor-1a (HIF-1α) levels (Malec et al., 2010), thereby inducing expression of the glycolytic genes and metabolic reprogramming (Bi et al., 2020). Andrographolide suppresses HIF-1α in human endothelial cell lines, activating Nrf2/HO-1 (Lin et al., 2017) (Fig. 3), and reduces the expression of HDAC1

(Shi, Zhang, Zheng, Lu, & Ji, 2017). Andrographolide attenuates hypoxia-induced invasiveness, reducing MMP-1, MMP-3, and MMP-9 expression of RA-FLSs, via inhibition of the interference of HIF-1α with DNA binding (Li, Qin, & Du, 2015). All these findings support the idea that andrographolide can control inflammatory pain.

Pharmacokinetics and metabolism

The pharmacokinetic parameters of A. paniculata dried powder in healthy volunteers administered in tablets (three times per day for 3 days), in doses equivalent to 32.6 mg, (total/day) of andrographolide, have been analyzed (Pholphana et al., 2016). In steady state, andrographolide shows a C_{max}, T_{max}, $t_{1/2}$, Cl/F, and area under the curve (AUC) of plasma concentration of 32.41 ng/mL, 0.78 and 2.65 h, 762.51 (L/h), and 55.23 h × ng/mL, respectively (Pholphana et al., 2016). Another study showed that an A. paniculata dried extract administered in a dose of 3 × 4 tablets/day, (about 1 mg andrographolide/kg/day), a concentration of 660 ng/mL at 24 h approximately, was registered in steady-state plasma concentration (Panossian et al., 2000). Andrographolide is quickly absorbed by the oral route but is intensively metabolized and less than 10% is eliminated via the urine after administration (Panossian et al., 2000). The absolute bioavailability (F) reported is 2.67%, indicating a rapid first-pass metabolism to the sulfate metabolite, and it is excreted into the intestine through the bile. In addition, the efflux by P-glycoprotein in the terminal intestine and biliary excretion could explain the low F (Ye et al., 2011). The hydroxyl group at C-19 is the major site for glucuronidation of andrographolide in human liver microsomal samples (Tian et al., 2015), which facilitates its excretion via the urine (Cui, Qiu, & Yao, 2005). In addition, the presence of several sulfate metabolites and a cysteine S-conjugate of andrographolide has also been detected in human urine (Cui, Qiu, Wang, & Yao, 2004). Potential interaction with the liver metabolism of other analgesic drugs such as morphine should be considered (Uchaipichat, 2018).

Clinical pharmacology and side effects

A. paniculata has been used successfully for the treatment of upper respiratory tract infections. It reduces the frequency and intensity of cough, expectoration, and nasal discharge (Hu et al., 2017). Furthermore, an analgesic effect has been observed including the reduction of sore throat, earache (Caceres, Hancke, Burgos, Sandberg, & Wikman, 1999; Thamlikitkul et al., 1991), and headache (Saxena et al., 2010). Oral administration of 1.8 g/day of A. paniculata for 8 weeks has been reported to be useful in the therapy of mild-to-moderate ulcerative colitis, leading to the observation of clinical remission and improvement in mucosal healing (Sandborn et al., 2013). In relapsing-remitting multiple sclerosis patients treated with interferon β-1a, ParActin at a dose of 170 mg/day, after 1 year, a significant improvement in fatigue was reported as compared with placebo. Four participants that received the placebo had new T2 lesions or new gadolinium-enhancing lesions at 12 months compared to one in the intervention group (Bertoglio et al., 2016). Moreover, in not active progressive multiple sclerosis, 140 mg/day of andrographolide for 2 years was shown to reduce brain atrophy and disability progression (Ciampi et al., 2020).

In rheumatoid arthritis, after 14 weeks of treatment with 90 mg/day ParActin extract, symptoms associated with joint inflammation were reduced (Burgos et al., 2009). In moderate knee osteoarthritis patients, both 300 mg/day and 600 mg/day of ParActin, reduced WOMAC pain score and stiffness score after 28 days of treatment (Hancke, Srivastav, Caceres, & Burgos, 2019). In addition, the therapy improved quality of life (SF36) and fatigue (FACIT-fatigue scale) (Hancke et al., 2019).

Authors of previous studies have stated that A. paniculata is safe, and does not cause mutations (Burgos et al., 1997; Chandrasekaran et al., 2009). In addition, A. paniculata is apparently safe for multiple oral dosing (Suriyo et al., 2017) and few side effects have been reported, such as mild headache, nasal discharge, epigastric pain, (Coon & Ernst, 2004; Poolsup, Suthisisang, Prathanturarug, Asawamekin, & Chancharoen, 2004), vertigo, dizziness, drowsiness, malaise, skin rash, urticaria (Hu et al., 2017), nausea, emesis, and diarrhea, but none have required further treatment (Wagner et al., 2015).

Contraindications and potential interactions

The co-administration of andrographolide with naproxen, nabumetone, or etoricoxib can alter the pharmacokinetic parameters, reducing the $t_{1/2}$ and C_{max} of these NSAIDs (Balap, Atre, Lohidasan, Sinnathambi, & Mahadik, 2016; Balap, Lohidasan, Sinnathambi, & Mahadik, 2017a, 2017b). A. paniculata extract and andrographolide accelerated the metabolism of tolbutamide, increasing CYP2C6/11 expression (Chen et al., 2013). However, there have been differences reported when A. paniculata or andrographolide has been used with theophylline. For instance, the co-treatment of andrographolide with theophylline was shown to increase the clearance of theophylline, while the co-administration of an ethanolic extract of *A. paniculata* with theophylline increases the level of theophylline due to its interaction with the substrate of CYP1A2 enzyme (Chien, Wu, Lee, Lin, & Tsai, 2010). Andrographolide increases the levels of C_{max}, $t_{1/2}$, and $AUC_{0-\infty}$ of warfarin (Zhang, Zhang, Wang, & Zhao, 2018). In addition, andrographolide and *A. paniculata* can affect the expression and activity of the cytochrome P450 (CYP) family, specifically CYP2C9 and CYP3A4 (Pekthong et al., 2009), therefore a potential interaction with cyclosporine A, barbiturate, non-steroidal antiinflammatory drugs, and warfarin should be considered.

Ongoing clinical trials in inflammation and pain

Currently, there are some clinical trials evaluating the efficacy of *A. paniculata* or andrographolide in different health conditions. For instance, an ongoing clinical trial is aiming to evaluate the effects of *A. paniculata* on palliative management of advanced oesophageal cancer (phase 3, NCT04196075). A randomized clinical trial is investigating the effects of *A. paniculata* and other compounds (magnesium, parthenium, andrographis, co-enzyme Q10, and riboflavin) in migraine prophylaxis. Another randomized controlled clinical trial is evaluating a novel dentifrice that contains andrographolide and its modulation of biochemical mediators of inflammation in generalized chronic periodontitis patients (CTRI/2019/09/021058).

The multitarget characteristics of andrographolide, as the main component of ParActin, allow its use in different acute and chronic inflammatory diseases associated with pain. ParActin is a therapeutic alternative or complementary treatment to traditional pain medicines. In addition, ParActin is well tolerated and does not lead to gastric mucosal alteration, unlike classical NSAIDs (Saranya, Geetha, & Selvamathy, 2011; Thakur, Soni, Rai, Chatterjee, & Kumar, 2014), making it safe for long term use.

Applications to other areas

ParActin has potential to be used in several diseases such as multiple sclerosis, ulcerative colitis, Duchenne muscular dystrophy, Alzheimer, Parkinson, asthma, and cancer.

Other agents of interest

There are several botanical extracts proposed with pain relieving activity, being the more common plants: *Acorus calamus, Ananas comosus, Artemisia dracunculus, Butea monosperma, Boswellia serrata, Citrullus colocynthis, Curcuma longa, Crocus sativus, Elaeagnus angustifolia, Harpagophytum procumbens, Ginkgo biloba, Mitragyna speciosa, Momordica charantia, Nigella sativa, Ocimum sanctum, Phyllanthus amarus, Pterodon pubescens, Punica granatum, Rubia cordifolia, Salvia officinalis, and Sesamum indicum.* Additionally, other natural active ingredients include avocado/soybean unsaponifiables, chili pepper, grape seed, green tea, peanuts, pine bark, rose hip, and soy, have been used for inflammation and pain control. Also, the potential benefits of cannabis-based medicine (herbal cannabis, plant-derived or synthetic, as tetrahydrocannabinol/cannabidiol) in chronic neuropathic pain, has shown some efficacy, nevertheless before cannabinoids can enter the broader market with patient-driven and industry-driven hype, high-quality clinical evidence is needed. For botanicals, absence of robust clinical data with isolated or standardized extract is still lacking, in order to recommend these ingredients, for pain symptom management.

Mini-dictionary of terms

NET-formation: a specific response of neutrophils, generated by different stimuli originating from microorganisms (viral, bacterial, fungal, and parasitic) or proinflammatory agents. The NETs are compounds formed by a network of DNA-histone fibers decorated with cytosolic proteins with bactericidal properties whose function is to prevent the spread of invading microorganisms. Exacerbation of this response can generate proinflammatory effects in the host.

Inflammasome: a multi-protein complex activated in response to infectious microbes and molecules derived from the host proteins of immune system cells such as macrophages. Activation of inflammasomes promotes proteolytic cleavage, maturation, and secretion of proinflammatory cytokines such as IL-1β and IL-18.

Herbal extract: compounds or substances extracted from a plant (leaves, stems, and/or roots) using different solvents (water, alcohols, or other organic solvents) alone or in a combination. A herbal extract can be standardized with respect to its active principles.

Andrographolide: is a principal bitter labdane diterpenoid obtained from *A. paniculata* and presents as a standardized dried extract in ParActin. It has multitarget properties such as immunomodulatory, antitumoral, antioxidant, hepatoprotective, and antiinflammatory, among others.

Andrographolide sulfate: is an andrographolide-derivative, found in human urine with sulfate substitution in the 3rd carbon position of diterpene.

14-Deoxyandrographolide: the second most abundant andrographolide derivate in ParActin with biological activities. This derivative does not possess a hydroxyl group in the 14th carbon.

Neoandrographolide: the third most abundant, non-bitter, andrographolide derivative present in ParActin. It has similar antiinflammatory activities to andrographolide. This derivative contains a glucoside group.

Key facts

- *paniculata* is a medicinal plant commonly mostly used in traditional Chinese and Ayurvedic medicine for the treatment of common cold, diarrhea, fever due to several infective causes, jaundice, as a health tonic for the liver and cardiovascular health, and as an antioxidant.
- ParActin is a patented dried extract of *A. paniculata* standardized to 50% (w/w) of total andrographolides, of which 3%–6% (w/w) is 14-deoxyandrographolide, and about 0.2%–0.8% (w/w) is neoandrographolide.
- ParActin reduces inflammation and pain in patients with rheumatoid arthritis and knee osteoarthritis.
- Andrographolide shows mild side effects but none of which require further treatment.
- ParActin shows potential interaction with warfarin, theophylline, and NSAIDs.

Summary points

- *A. paniculata* is a medicinal herb commonly used in Asia for pain control and inflammation in different diseases.
- Andrographolide is the principal compound in ParActin and its mechanism of action in the inflammatory process is to inhibit the activation of NF-kB and reduce iNOS and COX-2 expression.
- ParActin is a patented dried extract of *A. paniculata* standardized to 50% (w/w) of total andrographolides, of which 3%–6% (w/w) is 14-deoxyandrographolide, and about 0.2%–0.8% (w/w) is neoandrographolide.
- ParActin has been clinically used in diseases such as multiple sclerosis, rheumatoid arthritis, and knee osteoarthritis.
- ParActin has few adverse effects, allowing its long-term use in chronic diseases.

References

Akbar, S. (2011). Andrographis paniculata: A review of pharmacological activities and clinical effects. *Alternative Medicine Review, 16*(1), 66–77.

Atsaves, V., Leventaki, V., Rassidakis, G. Z., & Claret, F. X. (2019). AP-1 transcription factors as regulators of immune responses in cancer. *Cancers (Basel), 11*(7). https://doi.org/10.3390/cancers11071037.

Balap, A., Atre, B., Lohidasan, S., Sinnathambi, A., & Mahadik, K. (2016). Pharmacokinetic and pharmacodynamic herb-drug interaction of *Andrographis paniculata* (Nees) extract and andrographolide with etoricoxib after oral administration in rats. *Journal of Ethnopharmacology, 183,* 9–17. https://doi.org/10.1016/j.jep.2015.11.011.

Balap, A., Lohidasan, S., Sinnathambi, A., & Mahadik, K. (2017a). Herb-drug interaction of *Andrographis paniculata* (Nees) extract and andrographolide on pharmacokinetic and pharmacodynamic of naproxen in rats. *Journal of Ethnopharmacology, 195,* 214–221. https://doi.org/10.1016/j.jep.2016.11.022.

Balap, A., Lohidasan, S., Sinnathambi, A., & Mahadik, K. (2017b). Pharmacokinetic and Pharmacodynamic interaction of Andrographolide and standardized extract of *Andrographis paniculata* (Nees) with Nabumetone in Wistar rats. *Phytotherapy Research, 31*(1), 75–80. https://doi.org/10.1002/ptr.5731.

Balmain, A., & Connolly, J. D. (1973). Minor diterpenoid constituents of *Andrographis paniculata* Nees. *Journal of the Chemical Society, Perkin Transactions, 1,* 1247–1251. https://doi.org/10.1039/P19730001247.

Bao, Z., Guan, S., Cheng, C., Wu, S., Wong, S. H., Kemeny, D. M., et al. (2009). A novel antiinflammatory role for andrographolide in asthma via inhibition of the nuclear factor-kappaB pathway. *American Journal of Respiratory and Critical Care Medicine, 179*(8), 657–665. https://doi.org/10.1164/rccm.200809-1516OC.

Bertoglio, J. C., Baumgartner, M., Palma, R., Ciampi, E., Carcamo, C., Caceres, D. D., et al. (2016). *Andrographis paniculata* decreases fatigue in patients with relapsing-remitting multiple sclerosis: A 12-month double-blind placebo-controlled pilot study. *BMC Neurology, 16,* 77. https://doi.org/10.1186/s12883-016-0595-2.

Bi, Z., Zhang, Q., Fu, Y., Wadgaonkar, P., Zhang, W., Almutairy, B., et al. (2020). Nrf2 and HIF1alpha converge to arsenic-induced metabolic reprogramming and the formation of the cancer stem-like cells. *Theranostics, 10*(9), 4134–4149. https://doi.org/10.7150/thno.42903.

360 PART | IV Novel and nonpharmacological aspects and treatments

Burgos, R. A., Caballero, E. E., Sanchez, N. S., Schroeder, R. A., Wikman, G. K., & Hancke, J. L. (1997). Testicular toxicity assessment of *Andrographis paniculata* dried extract in rats. *Journal of Ethnopharmacology*, *58*(3), 219–224. https://doi.org/10.1016/s0378-8741(97)00099-8.

Burgos, R. A., Hancke, J. L., Bertoglio, J. C., Aguirre, V., Arriagada, S., Calvo, M., et al. (2009). Efficacy of an *Andrographis paniculata* composition for the relief of rheumatoid arthritis symptoms: A prospective randomized placebo-controlled trial. *Clinical Rheumatology*, *28*(8), 931–946. https://doi.org/10.1007/s10067-009-1180-5.

Busch-Dienstfertig, M., & Gonzalez-Rodriguez, S. (2013). IL-4, JAK-STAT signaling, and pain. *JAKSTAT*, *2*(4). https://doi.org/10.4161/jkst.27638, e27638.

Caceres, D. D., Hancke, J. L., Burgos, R. A., Sandberg, F., & Wikman, G. K. (1999). Use of visual analogue scale measurements (VAS) to assess the effectiveness of standardized *Andrographis paniculata* extract SHA-10 in reducing the symptoms of common cold. A randomized double blind-placebo study. *Phytomedicine*, *6*(4), 217–223. https://doi.org/10.1016/S0944-7113(99)80012-9.

Carretta, M. D., Alarcon, P., Jara, E., Solis, L., Hancke, J. L., Concha, I. I., et al. (2009). Andrographolide reduces IL-2 production in T-cells by interfering with NFAT and MAPK activation. *European Journal of Pharmacology*, *602*(2–3), 413–421. https://doi.org/10.1016/j.ejphar.2008.11.011.

Cashman, J. N. (1996). The mechanisms of action of NSAIDs in analgesia. *Drugs*, *52*(Suppl 5), 13–23. https://doi.org/10.2165/00003495-199600525-00004.

Chandrasekaran, C. V., Thiyagarajan, P., Sundarajan, K., Goudar, K. S., Deepak, M., Murali, B., et al. (2009). Evaluation of the genotoxic potential and acute oral toxicity of standardized extract of *Andrographis paniculata* (KalmCold). *Food and Chemical Toxicology*, *47*(8), 1892–1902. https://doi.org/10.1016/j.fct.2009.05.006.

Chao, C. Y., Lii, C. K., Tsai, I. T., Li, C. C., Liu, K. L., Tsai, C. W., et al. (2011). Andrographolide inhibits ICAM-1 expression and NF-kappaB activation in TNF-alpha-treated EA.hy926 cells. *Journal of Agricultural and Food Chemistry*, *59*(10), 5263–5271. https://doi.org/10.1021/jf104003y.

Chao, W. W., & Lin, B. F. (2010). Isolation and identification of bioactive compounds in *Andrographis paniculata* (Chuanxinlian). *Chinese Medicine*, *5*, 17. https://doi.org/10.1186/1749-8546-5-17.

Chen, L., Deng, H., Cui, H., Fang, J., Zuo, Z., Deng, J., et al. (2018). Inflammatory responses and inflammation-associated diseases in organs. *Oncotarget*, *9*(6), 7204–7218. https://doi.org/10.18632/oncotarget.23208.

Chen, H. W., Huang, C. S., Liu, P. F., Li, C. C., Chen, C. T., Liu, C. T., et al. (2013). *Andrographis paniculata* extract and Andrographolide modulate the hepatic drug metabolism system and plasma Tolbutamide concentrations in rats. *Evidence-based Complementary and Alternative Medicine*, *2013*, 982689. https://doi.org/10.1155/2013/982689.

Chen, S., Luo, Z., & Chen, X. (2020). Andrographolide mitigates cartilage damage via miR-27-3p-modulated matrix metalloproteinase13 repression. *The Journal of Gene Medicine*. https://doi.org/10.1002/jgm.3187, e3187.

Chien, C. F., Wu, Y. T., Lee, W. C., Lin, L. C., & Tsai, T. H. (2010). Herb-drug interaction of *Andrographis paniculata* extract and andrographolide on the pharmacokinetics of theophylline in rats. *Chemico-Biological Interactions*, *184*(3), 458–465. https://doi.org/10.1016/j.cbi.2010.01.017.

Ciampi, E., Uribe-San-Martin, R., Carcamo, C., Cruz, J. P., Reyes, A., Reyes, D., et al. (2020). Efficacy of andrographolide in not active progressive multiple sclerosis: A prospective exploratory double-blind, parallel-group, randomized, placebo-controlled trial. *BMC Neurology*, *20*(1), 173. https://doi.org/10.1186/s12883-020-01745-w.

Coon, J. T., & Ernst, E. (2004). *Andrographis paniculata* in the treatment of upper respiratory tract infections: A systematic review of safety and efficacy. *Planta Medica*, *70*(4), 293–298. https://doi.org/10.1055/s-2004-818938.

Cuadrado, A., Rojo, A. I., Wells, G., Hayes, J. D., Cousin, S. P., Rumsey, W. L., et al. (2019). Therapeutic targeting of the NRF2 and KEAP1 partnership in chronic diseases. *Nature Reviews. Drug Discovery*, *18*(4), 295–317. https://doi.org/10.1038/s41573-018-0008-x.

Cui, L., Qiu, F., Wang, N., & Yao, X. (2004). Four new andrographolide metabolites in human urine. *Chemical & Pharmaceutical Bulletin (Tokyo)*, *52*(6), 772–775. https://doi.org/10.1248/cpb.52.772.

Cui, L., Qiu, F., & Yao, X. (2005). Isolation and identification of seven glucuronide conjugates of andrographolide in human urine. *Drug Metabolism and Disposition*, *33*(4), 555–562. https://doi.org/10.1124/dmd.104.001958.

Das, S., Gautam, N., Dey, S. K., Maiti, T., & Roy, S. (2009). Oxidative stress in the brain of nicotine-induced toxicity: Protective role of *Andrographis paniculata* Nees and vitamin E. *Applied Physiology, Nutrition, and Metabolism*, *34*(2), 124–135. https://doi.org/10.1139/H08-147.

Ding, Y., Chen, L., Wu, W., Yang, J., Yang, Z., & Liu, S. (2017). Andrographolide inhibits influenza A virus-induced inflammation in a murine model through NF-kappaB and JAK-STAT signaling pathway. *Microbes and Infection*, *19*(12), 605–615. https://doi.org/10.1016/j.micinf.2017.08.009.

Foletta, V. C., Segal, D. H., & Cohen, D. R. (1998). Transcriptional regulation in the immune system: All roads lead to AP-1. *Journal of Leukocyte Biology*, *63*(2), 139–152. https://doi.org/10.1002/jlb.63.2.139.

Greco, R., Siani, F., Demartini, C., Zanaboni, A., Nappi, G., Davinelli, S., et al. (2016). *Andrographis paniculata* shows anti-nociceptive effects in an animal model of sensory hypersensitivity associated with migraine. *Functional Neurology*, *31*(1), 53–60. https://doi.org/10.11138/fneur/2016.31.1.053.

Guan, S. P., Tee, W., Ng, D. S., Chan, T. K., Peh, H. Y., Ho, W. E., et al. (2013). Andrographolide protects against cigarette smoke-induced oxidative lung injury via augmentation of Nrf2 activity. *British Journal of Pharmacology*, *168*(7), 1707–1718. https://doi.org/10.1111/bph.12054.

Gupta, S., Mishra, K. P., Singh, S. B., & Ganju, L. (2018). Inhibitory effects of andrographolide on activated macrophages and adjuvant-induced arthritis. *Inflammopharmacology*, *26*(2), 447–456. https://doi.org/10.1007/s10787-017-0375-7.

Hancke, J. L., Srivastav, S., Caceres, D. D., & Burgos, R. A. (2019). A double-blind, randomized, placebo-controlled study to assess the efficacy of *Andrographis paniculata* standardized extract (ParActin(R)) on pain reduction in subjects with knee osteoarthritis. *Phytotherapy Research*, *33*(5), 1469–1479. https://doi.org/10.1002/ptr.6339.

Hayden, M. S., & Ghosh, S. (2008). Shared principles in NF-kappaB signaling. *Cell*, *132*(3), 344–362. https://doi.org/10.1016/j.cell.2008.01.020.

Hayes, J. D., & Dinkova-Kostova, A. T. (2014). The Nrf2 regulatory network provides an interface between redox and intermediary metabolism. *Trends in Biochemical Sciences, 39*(4), 199–218. https://doi.org/10.1016/j.tibs.2014.02.002.

Hidalgo, M. A., Romero, A., Figueroa, J., Cortes, P., Concha, I. I., Hancke, J. L., et al. (2005). Andrographolide interferes with binding of nuclear factor-kappaB to DNA in HL-60-derived neutrophilic cells. *British Journal of Pharmacology, 144*(5), 680–686. https://doi.org/10.1038/sj.bjp.0706105.

Horvai, A. E., Xu, L., Korzus, E., Brard, G., Kalafus, D., Mullen, T. M., et al. (1997). Nuclear integration of JAK/STAT and Ras/AP-1 signaling by CBP and p300. *Proceedings of the National Academy of Sciences of the United States of America, 94*(4), 1074–1079. https://doi.org/10.1073/pnas.94.4.1074.

Hossain, M. S., Urbi, Z., Sule, A., & Hafizur Rahman, K. M. (2014). *Androgrphis paniculata* (Burm. F.) Wall. Ex Nees: A review of ethnobotany, phytochemistry, and pharmacology. *Scientific World Journal, 2014.* https://doi.org/10.1155/2014/274905, 274905.

Hsieh, Y. L., Chiang, H., Lue, J. H., & Hsieh, S. T. (2012). P2X3-mediated peripheral sensitization of neuropathic pain in resiniferatoxin-induced neuropathy. *Experimental Neurology, 235*(1), 316–325. https://doi.org/10.1016/j.expneurol.2012.02.013.

Hu, X. Y., Wu, R. H., Logue, M., Blondel, C., Lai, L. Y. W., Stuart, B., et al. (2017). *Androgrphis paniculata* (Chuan Xin Lian) for symptomatic relief of acute respiratory tract infections in adults and children: A systematic review and meta-analysis. *PLoS One, 12*(8). https://doi.org/10.1371/journal.pone.0181780, e0181780.

Huang, W., Huang, J., Jiang, Y., Huang, X., Xing, W., He, Y., et al. (2018). Oxaliplatin regulates chemotherapy induced peripheral neuropathic pain in the dorsal horn and dorsal root ganglion via the calcineurin/NFAT pathway. *Anti-Cancer Agents in Medicinal Chemistry, 18*(8), 1197–1207. https://doi.org/10.2174/1871520618666180525091158.

Jiao, J., Yang, Y., Wu, Z., Li, B., Zheng, Q., Wei, S., et al. (2019). Screening cyclooxygenase-2 inhibitors from *Androgrphis paniculata* to treat inflammation based on bio-affinity ultrafiltration coupled with UPLC-Q-TOF-MS. *Fitoterapia, 137,* 104259. https://doi.org/10.1016/j.fitote.2019.104259.

Jin, P., Wiraja, C., Zhao, J., Zhang, J., Zheng, L., & Xu, C. (2017). Nitric oxide nanosensors for predicting the development of osteoarthritis in rat model. *ACS Applied Materials & Interfaces, 9*(30), 25128–25137. https://doi.org/10.1021/acsami.7b06404.

Jing, M., Wang, Y., & Xu, L. (2019). Andrographolide derivative AL-1 ameliorates dextran sodium sulfate-induced murine colitis by inhibiting NF-kappaB and MAPK signaling pathways. *Oxidative Medicine and Cellular Longevity, 2019,* 6138723. https://doi.org/10.1155/2019/6138723.

Karin, M. (1995). The regulation of AP-1 activity by mitogen-activated protein kinases. *The Journal of Biological Chemistry, 270*(28), 16483–16486. https://doi.org/10.1074/jbc.270.28.16483.

Kim, N., Lertnimitphun, P., Jiang, Y., Tan, H., Zhou, H., Lu, Y., et al. (2019). Andrographolide inhibits inflammatory responses in LPS-stimulated macrophages and murine acute colitis through activating AMPK. *Biochemical Pharmacology, 170,* 113646. https://doi.org/10.1016/j.bcp.2019.113646.

Kishore, V., Yarla, N. S., Bishayee, A., Putta, S., Malla, R., Neelapu, N. R., et al. (2017). Multitargeting Andrographolide and its natural analogs as potential therapeutic agents. *Current Topics in Medicinal Chemistry, 17*(8), 845–857. https://doi.org/10.2174/1568026616666160927150452.

Lee, K. C., Chang, H. H., Chung, Y. H., & Lee, T. Y. (2011). Andrographolide acts as an antiinflammatory agent in LPS-stimulated RAW264.7 macrophages by inhibiting STAT3-mediated suppression of the NF-kappaB pathway. *Journal of Ethnopharmacology, 135*(3), 678–684. https://doi.org/10.1016/j.jep.2011.03.068.

Li, Y., He, S., Tang, J., Ding, N., Chu, X., Cheng, L., et al. (2017). Andrographolide inhibits inflammatory cytokines secretion in LPS-stimulated RAW264.7 cells through suppression of NF-kappaB/MAPK signaling pathway. *Evidence-based Complementary and Alternative Medicine, 2017.* https://doi.org/10.1155/2017/8248142, 8248142.

Li, G. F., Qin, Y. H., & Du, P. Q. (2015). Andrographolide inhibits the migration, invasion and matrix metalloproteinase expression of rheumatoid arthritis fibroblast-like synoviocytes via inhibition of HIF-1alpha signaling. *Life Sciences, 136,* 67–72. https://doi.org/10.1016/j.lfs.2015.06.019.

Li, Z. Z., Tan, J. P., Wang, L. L., & Li, Q. H. (2017). Andrographolide benefits rheumatoid arthritis via inhibiting MAPK pathways. *Inflammation, 40*(5), 1599–1605. https://doi.org/10.1007/s10753-017-0600-y.

Li, X., Yuan, K., Zhu, Q., Lu, Q., Jiang, H., Zhu, M., et al. (2019). Andrographolide ameliorates rheumatoid arthritis by regulating the apoptosis-NETosis balance of neutrophils. *International Journal of Molecular Sciences, 20*(20). https://doi.org/10.3390/ijms20205035.

Lin, H. C., Su, S. L., Lu, C. Y., Lin, A. H., Lin, W. C., Liu, C. S., et al. (2017). Andrographolide inhibits hypoxia-induced HIF-1alpha-driven endothelin 1 secretion by activating Nrf2/HO-1 and promoting the expression of prolyl hydroxylases 2/3 in human endothelial cells. *Environmental Toxicology, 32*(3), 918–930. https://doi.org/10.1002/tox.22293.

Liu, J., Wang, Z. T., & Ji, L. L. (2007). In vivo and in vitro antiinflammatory activities of neoandrographolide. *The American Journal of Chinese Medicine, 35*(2), 317–328. https://doi.org/10.1142/S0192415X07004849.

Liu, J., Wang, Z. T., Ji, L. L., & Ge, B. X. (2007). Inhibitory effects of neoandrographolide on nitric oxide and prostaglandin E2 production in LPS-stimulated murine macrophage. *Molecular and Cellular Biochemistry, 298*(1–2), 49–57. https://doi.org/10.1007/s11010-006-9349-6.

Liu, L., Yan, Y., Zheng, L., Jia, H., & Han, G. (2020). Synthesis and structure antiinflammatory activity relationships studies of andrographolide derivatives. *Natural Product Research, 34*(6), 782–789. https://doi.org/10.1080/14786419.2018.1501689.

Lu, C. Y., Yang, Y. C., Li, C. C., Liu, K. L., Lii, C. K., & Chen, H. W. (2014). Andrographolide inhibits TNFalpha-induced ICAM-1 expression via suppression of NADPH oxidase activation and induction of HO-1 and GCLM expression through the PI3K/Akt/Nrf2 and PI3K/Akt/AP-1 pathways in human endothelial cells. *Biochemical Pharmacology, 91*(1), 40–50. https://doi.org/10.1016/j.bcp.2014.06.024.

Maioli, N. A., Zarpelon, A. C., Mizokami, S. S., Calixto-Campos, C., Guazelli, C. F., Hohmann, M. S., et al. (2015). The superoxide anion donor, potassium superoxide, induces pain and inflammation in mice through production of reactive oxygen species and cyclooxygenase-2. *Brazilian Journal of Medical and Biological Research, 48*(4), 321–331. https://doi.org/10.1590/1414-431X20144187.

Malec, V., Gottschald, O. R., Li, S., Rose, F., Seeger, W., & Hanze, J. (2010). HIF-1 alpha signaling is augmented during intermittent hypoxia by induction of the Nrf2 pathway in NOX1-expressing adenocarcinoma A549 cells. *Free Radical Biology & Medicine, 48*(12), 1626–1635. https://doi.org/10.1016/j.freeradbiomed.2010.03.008.

Maroon, J. C., Bost, J. W., & Maroon, A. (2010). Natural antiinflammatory agents for pain relief. *Surgical Neurology International*, *1*, 80. https://doi.org/10.4103/2152-7806.73804.

McDonnell, C., Leanez, S., & Pol, O. (2017). The induction of the transcription factor Nrf2 enhances the antinociceptive effects of delta-opioid receptors in diabetic mice. *PLoS One*, *12*(7). https://doi.org/10.1371/journal.pone.0180998, e0180998.

Meng, Z. M. (1981). Studies on the structure of the adduct of andrographolide with sodium hydrogen sulfite (author's transl). *Yao Xue Xue Bao*, *16*(8), 571–575.

Mizushima, T., Obata, K., Katsura, H., Sakurai, J., Kobayashi, K., Yamanaka, H., et al. (2007). Intensity-dependent activation of extracellular signal-regulated protein kinase 5 in sensory neurons contributes to pain hypersensitivity. *The Journal of Pharmacology and Experimental Therapeutics*, *321*(1), 28–34. https://doi.org/10.1124/jpet.106.116749.

Obata, K., Katsura, H., Mizushima, T., Sakurai, J., Kobayashi, K., Yamanaka, H., et al. (2007). Roles of extracellular signal-regulated protein kinases 5 in spinal microglia and primary sensory neurons for neuropathic pain. *Journal of Neurochemistry*, *102*(5), 1569–1584. https://doi.org/10.1111/j.1471-4159.2007.04656.x.

Panossian, A., Hovhannisyan, A., Mamikonyan, G., Abrahamian, H., Hambardzumyan, E., Gabrielian, E., et al. (2000). Pharmacokinetic and oral bioavailability of andrographolide from *Andrographis paniculata* fixed combination Kan Jang in rats and human. *Phytomedicine*, *7*(5), 351–364. https://doi.org/10.1016/S0944-7113(00)80054-9.

Pekthong, D., Blanchard, N., Abadie, C., Bonet, A., Heyd, B., Mantion, G., et al. (2009). Effects of *Andrographis paniculata* extract and Andrographolide on hepatic cytochrome P450 mRNA expression and monooxygenase activities after in vivo administration to rats and in vitro in rat and human hepatocyte cultures. *Chemico-Biological Interactions*, *179*(2–3), 247–255. https://doi.org/10.1016/j.cbi.2008.10.054.

Peng, S., Gao, J., Liu, W., Jiang, C., Yang, X., Sun, Y., et al. (2016). Andrographolide ameliorates OVA-induced lung injury in mice by suppressing ROS-mediated NF-kappaB signaling and NLRP3 inflammasome activation. *Oncotarget*, *7*(49), 80262–80274. https://doi.org/10.18632/oncotarget.12918.

Pholphana, N., Panomvana, D., Rangkadilok, N., Suriyo, T., Puranajoti, P., Ungtrakul, T., et al. (2016). *Andrographis paniculata*: Dissolution investigation and pharmacokinetic studies of four major active diterpenoids after multiple oral dose administration in healthy Thai volunteers. *Journal of Ethnopharmacology*, *194*, 513–521. https://doi.org/10.1016/j.jep.2016.09.058.

Poolsup, N., Suthisisang, C., Prathanturarug, S., Asawamekin, A., & Chancharoen, U. (2004). *Andrographis paniculata* in the symptomatic treatment of uncomplicated upper respiratory tract infection: Systematic review of randomized controlled trials. *Journal of Clinical Pharmacy and Therapeutics*, *29*(1), 37–45. https://doi.org/10.1046/j.1365-2710.2003.00534.x.

Redondo, A., Chamorro, P. A. F., Riego, G., Leanez, S., & Pol, O. (2017). Treatment with sulforaphane produces antinociception and improves morphine effects during inflammatory pain in mice. *The Journal of Pharmacology and Experimental Therapeutics*, *363*(3), 293–302. https://doi.org/10.1124/jpet.117.244376.

Sandborn, W. J., Targan, S. R., Byers, V. S., Rutty, D. A., Mu, H., Zhang, X., et al. (2013). *Andrographis paniculata* extract (HMPL-004) for active ulcerative colitis. *The American Journal of Gastroenterology*, *108*(1), 90–98. https://doi.org/10.1038/ajg.2012.340.

Saranya, P., Geetha, A., & Selvamathy, S. M. (2011). A biochemical study on the gastroprotective effect of andrographolide in rats induced with gastric ulcer. *Indian Journal of Pharmaceutical Sciences*, *73*(5), 550–557. https://doi.org/10.4103/0250-474X.99012.

Saxena, R. C., Singh, R., Kumar, P., Yadav, S. C., Negi, M. P., Saxena, V. S., et al. (2010). A randomized double blind placebo controlled clinical evaluation of extract of *Andrographis paniculata* (KalmCold) in patients with uncomplicated upper respiratory tract infection. *Phytomedicine*, *17*(3–4), 178–185. https://doi.org/10.1016/j.phymed.2009.12.001.

Sharma, A., Lal, K., & Handa, S. S. (1992). Standardization of the Indian crude drug kalmegh by high pressure liquid chromatographic determination of andrographolide. *Phytochemical Analysis*, *3*(3), 129–131. https://doi.org/10.1002/pca.2800030308.

Shen, Y. C., Chen, C. F., & Chiou, W. F. (2002). Andrographolide prevents oxygen radical production by human neutrophils: Possible mechanism(s) involved in its antiinflammatory effect. *British Journal of Pharmacology*, *135*(2), 399–406. https://doi.org/10.1038/sj.bjp.0704493.

Shi, L., Zhang, G., Zheng, Z., Lu, B., & Ji, L. (2017). Andrographolide reduced VEGFA expression in hepatoma cancer cells by inactivating HIF-1alpha: The involvement of JNK and MTA1/HDCA. *Chemico-Biological Interactions*, *273*, 228–236. https://doi.org/10.1016/j.cbi.2017.06.024.

Shimoyama, S., Furukawa, T., Ogata, Y., Nikaido, Y., Koga, K., Sakamoto, Y., et al. (2019). Lipopolysaccharide induces mouse translocator protein (18 kDa) expression via the AP-1 complex in the microglial cell line, BV-2. *PLoS One*, *14*(9). https://doi.org/10.1371/journal.pone.0222861, e0222861.

Simmonds, R. E., & Foxwell, B. M. (2008). Signalling, inflammation and arthritis: NF-kappaB and its relevance to arthritis and inflammation. *Rheumatology (Oxford)*, *47*(5), 584–590. https://doi.org/10.1093/rheumatology/kem298.

Staurengo-Ferrari, L., Badaro-Garcia, S., Hohmann, M. S. N., Manchope, M. F., Zaninelli, T. H., Casagrande, R., et al. (2018). Contribution of Nrf2 modulation to the mechanism of action of analgesic and anti-inflammatory drugs in pre-clinical and clinical stages. *Frontiers in Pharmacology*, *9*, 1536. https://doi.org/10.3389/fphar.2018.01536.

Suriyo, T., Pholphana, N., Ungtrakul, T., Rangkadilok, N., Panomvana, D., Thiantanawat, A., et al. (2017). Clinical parameters following multiple oral dose administration of a standardized *Andrographis paniculata* capsule in healthy Thai subjects. *Planta Medica*, *83*(9), 778–789. https://doi.org/10.1055/s-0043-104382.

Tang, W., & Eisenbrand, G. (1992). *Andrographis paniculata* (Burm. F.) Nees. In *Chinese drugs of plant origin: Chemistry, pharmacology, and use in traditional and modern medicine* (pp. 97–103). Berlin, Heidelberg: Springer Berlin Heidelberg.

Tanito, M., Agbaga, M. P., & Anderson, R. E. (2007). Upregulation of thioredoxin system via Nrf2-antioxidant responsive element pathway in adaptive-retinal neuroprotection in vivo and in vitro. *Free Radical Biology & Medicine*, *42*(12), 1838–1850. https://doi.org/10.1016/j.freeradbiomed.2007.03.018.

Thakur, A. K., Soni, U. K., Rai, G., Chatterjee, S. S., & Kumar, V. (2014). Protective effects of *Androgrphis paniculata* extract and pure andrographolide against chronic stress-triggered pathologies in rats. *Cellular and Molecular Neurobiology, 34*(8), 1111–1121. https://doi.org/10.1007/s10571-014-0086-1.

Thamlikitkul, V., Dechatiwongse, T., Theerapong, S., Chantrakul, C., Boonroj, P., Punkrut, W., et al. (1991). Efficacy of *Androgrphis paniculata*, Nees for pharyngotonsillitis in adults. *Journal of the Medical Association of Thailand, 74*(10), 437–442.

Tian, X., Liang, S., Wang, C., Wu, B., Ge, G., Deng, S., et al. (2015). Regioselective glucuronidation of andrographolide and its major derivatives: Metabolite identification, isozyme contribution, and species differences. *The AAPS Journal, 17*(1), 156–166. https://doi.org/10.1208/s12248-014-9658-8.

Turnaturi, R., Arico, G., Ronsisvalle, G., Pasquinucci, L., & Parenti, C. (2016). Multitarget opioid/non-opioid ligands: A potential approach in pain management. *Current Medicinal Chemistry, 23*(40), 4506–4528. https://doi.org/10.2174/0929867323666161024151734.

Uchaipichat, V. (2018). In vitro inhibitory effects of major bioactive constituents of *Androgrphis paniculata, Curcuma longa* and *Silybum marianum* on human liver microsomal morphine glucuronidation: A prediction of potential herb-drug interactions arising from andrographolide, curcumin and silybin inhibition in humans. *Drug Metabolism and Pharmacokinetics, 33*(1), 67–76. https://doi.org/10.1016/j.dmpk.2017.10.005.

Vanderwall, A. G., & Milligan, E. D. (2019). Cytokines in pain: Harnessing endogenous anti-inflammatory signaling for improved pain management. *Frontiers in Immunology, 10*, 3009. https://doi.org/10.3389/fimmu.2019.03009.

Wagner, L., Cramer, H., Klose, P., Lauche, R., Gass, F., Dobos, G., et al. (2015). Herbal medicine for cough: A systematic review and meta-analysis. *Forschende Komplementärmedizin, 22*(6), 359–368. https://doi.org/10.1159/000442111.

Wang, H. C., Tsay, H. S., Shih, H. N., Chen, Y. A., Chang, K. M., Agrawal, D. C., et al. (2018). Andrographolide relieved pathological pain generated by spared nerve injury model in mice. *Pharmaceutical Biology, 56*(1), 124–131. https://doi.org/10.1080/13880209.2018.1426614.

Wang, W., Wang, J., Dong, S. F., Liu, C. H., Italiani, P., Sun, S. H., et al. (2010). Immunomodulatory activity of andrographolide on macrophage activation and specific antibody response. *Acta Pharmacologica Sinica, 31*(2), 191–201. https://doi.org/10.1038/aps.2009.205.

Xia, H., Xue, J., Xu, H., Lin, M., Shi, M., Sun, Q., et al. (2019). Andrographolide antagonizes the cigarette smoke-induced epithelial-mesenchymal transition and pulmonary dysfunction through antiinflammatory inhibiting HOTAIR. *Toxicology, 422*, 84–94. https://doi.org/10.1016/j.tox.2019.05.009.

Xia, Y. F., Ye, B. Q., Li, Y. D., Wang, J. G., He, X. J., Lin, X., et al. (2004). Andrographolide attenuates inflammation by inhibition of NF-kappa B activation through covalent modification of reduced cysteine 62 of p50. *Journal of Immunology, 173*(6), 4207–4217. https://doi.org/10.4049/jimmunol.173.6.4207.

Xiao, C., Zhang, L., Cheng, Q. P., & Zhang, L. C. (2008). The activation of extracellular signal-regulated protein kinase 5 in spinal cord and dorsal root ganglia contributes to inflammatory pain. *Brain Research, 1215*, 76–86. https://doi.org/10.1016/j.brainres.2008.03.065.

Xu, Y. Q., Jin, S. J., Liu, N., Li, Y. X., Zheng, J., Ma, L., et al. (2014). Aloperine attenuated neuropathic pain induced by chronic constriction injury via anti-oxidation activity and suppression of the nuclear factor kappa B pathway. *Biochemical and Biophysical Research Communications, 451*(4), 568–573. https://doi.org/10.1016/j.bbrc.2014.08.025.

Ye, L., Wang, T., Tang, L., Liu, W., Yang, Z., Zhou, J., et al. (2011). Poor oral bioavailability of a promising anticancer agent andrographolide is due to extensive metabolism and efflux by P-glycoprotein. *Journal of Pharmaceutical Sciences, 100*(11), 5007–5017. https://doi.org/10.1002/jps.22693.

Yi, Z., Ouyang, S., Zhou, C., Xie, L., Fang, Z., Yuan, H., et al. (2018). Andrographolide inhibits mechanical and thermal hyperalgesia in a rat model of HIV-induced neuropathic pain. *Frontiers in Pharmacology, 9*, 593. https://doi.org/10.3389/fphar.2018.00593.

Yu, A. L., Lu, C. Y., Wang, T. S., Tsai, C. W., Liu, K. L., Cheng, Y. P., et al. (2010). Induction of heme oxygenase 1 and inhibition of tumor necrosis factor alpha-induced intercellular adhesion molecule expression by andrographolide in EA.hy926 cells. *Journal of Agricultural and Food Chemistry, 58*(13), 7641–7648. https://doi.org/10.1021/jf101353c.

Yu, L. N., Sun, L. H., Wang, M., & Yan, M. (2016). Research progress of the role and mechanism of extracellular signal-regulated protein kinase 5 (ERK5) pathway in pathological pain. *Journal of Zhejiang University. Science. B, 17*(10), 733–741. https://doi.org/10.1631/jzus.B1600188.

Yuan, M., Meng, W., Liao, W., & Lian, S. (2018). Andrographolide antagonizes TNF-alpha-induced IL-8 via inhibition of NADPH oxidase/ROS/NF-kappaB and Src/MAPKs/AP-1 Axis in human colorectal cancer HCT116 cells. *Journal of Agricultural and Food Chemistry, 66*(20), 5139–5148. https://doi.org/10.1021/acs.jafc.8b00810.

Yue, G. G., Li, L., Lee, J. K., Kwok, H. F., Wong, E. C., Li, M., et al. (2019). Multiple modulatory activities of *Androgrphis paniculata* on immune responses and xenograft growth in esophageal cancer preclinical models. *Phytomedicine, 60*, 152886. https://doi.org/10.1016/j.phymed.2019.152886.

Zhang, X., Zhang, X., Wang, X., & Zhao, M. (2018). Influence of andrographolide on the pharmacokinetics of warfarin in rats. *Pharmaceutical Biology, 56*(1), 351–356. https://doi.org/10.1080/13880209.2018.1478431.

Chapter 32

Capsaicin: Features usage in diabetic neuropathic pain

Kongkiat Kulkantrakorn

Division of Neurology, Department of Internal Medicine, Faculty of Medicine, Thammasat University, Klong Luang, Pathumthani, Thailand

Abbreviations

DM   diabetic mellitus
PDN   painful diabetic neuropathy
TRPV1 transient receptor potentials vanilloid subfamily member 1

Introduction

Capsaicin

Capsaicin, the naturally extracted ingredient from hot chili peppers, acts as a ligand of the transient receptor potential cation channel of the vanilloid subfamily member 1 (TRPV1). TRPV1 plays a role in somatic and visceral peripheral nerve inflammation, modulating nociceptive inputs to the spinal cord and brain stem. After binding to its receptor, it works by releasing substance P from sensory nerve fibers and after repeated applications, depletes neurons of substance P (Rumsfield & West, 1991). Therefore, it initially increases sensitivity to noxious stimuli and this is followed by reduced sensitivity. After repetitive use, persistent desensitization will occur (Derry & Moore, 2012). As a result, the analgesic properties are mediated via defunctionalization of nociceptor fibers which is a temporary loss of neuronal membrane potential, the inability to transport neurotrophic factors leading to reversible retraction of epidermal and dermal nerve fiber terminals (Anand & Bley, 2011). This finding is confirmed by assessing the intraepidermal nerve fiber density from a skin biopsy. Capsaicin also exerts TRPV receptor-independent pathway and involves other neurotransmitters such as serotonin, neuropeptide, substance P, and somatostatin. Exploration of these mechanisms may help in further therapeutic indications (Patowary, Pathak, Zaman, Raju, & Chattopadhyay, 2017).

Topical capsaicin and other capsaicinoids have long been used as conventional medicine for pain relief in many syndromes such as osteoarthritis (Laslett & Jones, 2014), painful diabetic neuropathy (Tandan, Lewis, Krusinski, Badger, & Fries, 1992), postherpetic neuralgia (Moon et al., 2017), rheumatoid arthritis (Mason, Moore, Derry, Edwards, & McQuay, 2004), painful skin disorders (pruritus, psoriasis, and mastectomy), bladder disorders, and cluster headache (Rumsfield & West, 1991). It is available in several concentrations, ranging from 0.025% to 8%, and formulations such as gel, lotion, and patch. The 0.075% capsaicin is the most commonly used and widely available concentration. High concentration, at 8%, has been developed for localized neuropathic pain syndrome such as postherpetic neuralgia and HIV neuropathy. The data on diabetic neuropathy is favorable, but the potential side effect is concerning. With a very high dose (8%) patch (NGX-4010) for a much shorter exposure time for postherpetic neuralgia or HIV neuropathy, it yielded similar results and the nerve endings could recover themselves after 24 weeks in healthy volunteers (Kennedy et al., 2010). The low systemic exposure and very rapid elimination half-life of capsaicin after NGX-4010 patch administration ensure its favorable safety profile (Babbar et al., 2009; Kennedy et al., 2010; Üçeyler & Sommer, 2014).

Diabetic neuropathy

Diabetic neuropathy (DN) is one of the late common complications of diabetes mellitus, with a prevalence of about 30%. If left untreated, patients will have an increased risk of injury due to insensate feet. The prevalence of painful diabetic neuropathy (PDN) is about 20% in patients with type 2 diabetes and 5% in those with type 1 (Davies, Williams, & Taylor,

---

Treatments, Mechanisms, and Adverse Reactions of Anesthetics and Analgesics. https://doi.org/10.1016/B978-0-12-820237-1.00032-6
Copyright © 2022 Elsevier Inc. All rights reserved.

2006). The diagnosis of PDN is a clinical one, which relies on the patient's description of pain. The symptoms are distal, symmetrical, often associated with nocturnal exacerbations, and commonly described as prickling, deep aching, sharp, like an electric shock, and burning with hyperalgesia and frequently allodynia upon examination (Tesfaye et al., 2010)(Dyck et al., 2011). The symptoms are usually associated with the clinical signs of peripheral neuropathy, although occasionally in acute painful PDN, the symptoms may occur in the absence of signs. Having PDN has a significant negative effect on the quality of life, especially the physical aspect, and a significantly worse trajectory of quality of life outcomes over time as well as long-term increased total costs (daCosta DiBonaventura, Cappelleri, & Joshi, 2011). Based on the quality of life assessment by using the SF-36 questionnaire, PDN patients had poorer physical function than patients with other chronic neurological illnesses or than the general diabetic population, and comparable to patients with diabetic foot ulcers (Kulkantrakorn & Lorsuwansiri, 2013).

Common painful symptoms are burning, prickling, aching, sharp pain attacks, allodynia, cramping, and gnawing (Baron, Tölle, Gockel, Brosz, & Freynhagen, 2009; Petrikonis et al., 2010). These symptoms are always used in rating scales and standard pain questionnaires to assess the frequency and severity of painful symptoms and treatment response. On the neuropathic pain scale, sharp pain was the most common symptom and itching was the least common. Almost all patients had more than one type of pain, in the moderate pain range. In the short-form McGill Pain Questionnaire (SFMPQ), the sensory score, affective score, and the present pain score were in moderate severity (Kulkantrakorn & Lorsuwansiri, 2013). Moreover, each type of pain is believed to be caused by different pathophysiological mechanisms, therefore, each neuropathic pain medication might have a different effect on sensory symptoms (Maier et al., 2010).

There are several oral medications for treating painful diabetic neuropathy such as pregabalin, gabapentin, tricyclic antidepressants (TCAs), opioids, and anticonvulsants. They are moderately effective in pain control (Dosenovic et al., 2017), but are associated with clinically significant side effects such as nausea, vomiting, somnolence, dizziness, and ataxia (Duehmke et al., 2017). Topical treatment to alleviate severe pain is an interesting option to avoid these unpleasant systemic effects.

Clinical data of capsaicin in painful diabetic neuropathy

Clinical trials on topical capsaicin are summarized in Table 1.

Early clinical studies

Early data on topical capsaicin for PDN was reported in the 1980s and with successful results, especially in burning pain (Basha & Whitehouse, 1991). Subsequent short-term study (Tandan et al., 1992) showed remarkable benefits without significant side effects on sensory thresholds after 32 weeks of use in patients with preexisting neuropathic sensory impairment. Warm and vibration thresholds were unchanged but cold thresholds were reduced in both the capsaicin and the control group.

Tandan et al. conducted an 8-week randomized vehicle-controlled study with 0.075% topical capsaicin in 22 PDN subjects. It showed benefit in overall clinical improvement of pain status. In a follow-up open-label study, approximately 50% of subjects reported improved pain control or were cured, and 25% each were unchanged or worse. The burning sensation was common but it subsided over time (Tandan et al., 1992). A larger study in 277 subjects confirmed its efficacy in reducing pain with substantial improvement in walking, working, participation in recreational activities, sleeping, and other daily activities, which resulted in a better quality of life (Capsaicin Study Group, 1992). Metaanalysis in 1994 confirmed a favorable odds ratio (OR) of 2.74 (95% CI = 1.73, 4.32) in pain relief when comparing capsaicin with placebo in diabetic neuropathy (Zhang & Li Wan Po, 1994).

Comparative studies

An 8-week, double-blind, comparative study between topical capsaicin and oral amitriptyline was also performed on 235 patients with PDN. In terms of pain improvement, topical capsaicin and oral amitriptyline had similar efficacy. At the end of the study, 76% of patients in each group experienced less pain, with a mean reduction in the intensity of more than 40%. In both groups, the interference with daily activities from pain had improved significantly. Most patients in the amitriptyline group had at least one systemic side effect, including somnolence, and neuromuscular and cardiovascular adverse events, while there were none in the capsaicin group (Biesbroeck et al., 1995). Stinging and burning are common in the first week of capsaicin therapy but many patients can tolerate it well (Rains & Bryson, 1995). Even though its efficacy is modest at best, occasional patients appear to have a very good result, and these unusual cases may not be reflected by clinical trials

TABLE 1 Selected clinical trials on topical capsaicin for painful diabetic neuropathy (PDN).

Author (year)	Study drug/ comparator	Trial design	Duration (week)	No. of patients	Clinical significance	Outcome
The Capsaicin Study Group (1991)	0.075% cream vs placebo	Double blind, vehicle controlled	8	252	Yes	Improved in pain intensity, pain relief No systemic AE
Capsaicin Study Group (1992)	0.075% cream vs placebo	Double blind, vehicle controlled	8	277	Yes	Improvement in pain status, daily activities, QoL
Tandan et al. (1992)	0.075% cream vs placebo	Double blind, vehicle controlled	8	22	Yes	Improvement in pain status, intensity and severity
Kulkantrakorn et al. (2013)	0.025% capsaicin gel vs placebo	Double blind, randomized, cross over	8 weeks each arm, 4 week washout	33	No	No difference with placebo in pain relief, QoL Minor skin AE
Kulkantrakorn et al. (2019)	0.075% capsaicin lotion vs placebo	Double blind, randomized, cross over	8 weeks each arm, 4 weeks washout	42	No	No difference with placebo in pain relief, QoL Minor skin AE
Biesbroeck et al. (1995)	0.075% cream vs oral amitryptyline	Double blind, randomized controlled	8	235	Yes	Equal efficacy in pain reduction and QoL More systemic AE in amitryptyline group
Kiani et al. (2015)	0.075% capsaicin cream vs topical clonidine	Double blind, randomized controlled	12	139	Yes	Equal efficacy in pain relief, more skin complication in capsaicin group
Kiani et al. (2015)	0.075% capsaicin cream vs 2% amitriptyline cream	Double blind, randomized controlled	12	102	Yes	Equal efficacy in pain relief, more skin complication in capsaicin group
Simpson et al. (2017)	8% capsaicin patch vs placebo	Double blind, randomized, placebo controlled trial	12	369	Yes	Improvement in pain control, sleep More application site reaction
Vinik et al. (2016)	8% capsaicin patch add on vs standard of care alone	Open label, randomized, trial	52	468	Yes	Improvement in pain control, no worsening in sensory perception testing or neurological deficit More application site pain

AE, adverse event; *QoL*, quality of life.

(Watson, 1994). A 12-week, randomized, double-blind, comparative study with topical clonidine gel showed comparable efficacy in pain relief, but the capsaicin group had more skin complications (Kiani, Sajedi, Nasrollahi, & Esna-Ashari, 2015). Two percent amitriptyline cream has been shown to provide significant pain relief, similar to 0.075% capsaicin cream in a randomized double-blind study, but the capsaicin group had more skin adverse events (Kiani, Ahmad Nasrollahi, Esna-Ashari, Fallah, & Sajedi, 2015).

After the introduction of new antiepileptic drugs to neuropathic pain indication, especially gabapentin, oral medication became more commonly used than topical preparation (Duby, Campbell, Setter, White, & Rasmussen, 2004). Accumulating evidence has revealed limited efficacy for oral agents with significant systemic side effects (Alam, Sloan, & Tesfaye, 2020). New formulations and higher concentrations of topical capsaicin have been restudied with promising results (Anand & Bley, 2011).

Modern-day studies

Low dose topical capsaicin

Cochrane review in 2009 in six studies of low dose (0.075%) capsaicin cream compared with placebo cream revealed the number needed to treat (NNT) for at least 50% pain relief was 6.6 (Derry, Lloyd, Moore, & McQuay, 2009).

Lower concentration, at 0.025%, was studied to circumvent the local skin side effect. A 20-week, crossover, randomized, double-blinded, placebo-controlled trial failed to show significant pain relief, but it was well tolerated (Kulkantrakorn, Lorsuwansiri, & Meesawatsom, 2013). This can be explained by the patient's characteristics and their compliance. The PDN subjects who were enrolled were more intractable and had often been treated with more than one neuropathic pain medication. Therefore, the incremental benefit from capsaicin may not have been evident. The depth of skin penetration to the action site is also concentration dependent. Lower concentration preparation penetrates skin less and might not reach the skin nerve endings. Frequency of use, up to 3–4 times a day, also limits compliance.

A recent clinical trial using 0.075% capsaicin lotion compared with placebo in a crossover double-blinded study (Kulkantrakorn et al., 2019) also reported no significant improvement in pain control with either intention to treat or per-protocol analysis. Nevertheless, capsaicin lotion was well tolerated but local skin reactions were still common. The randomized, comparative study of capsaicin and topical turpentine oil for PDN showed similar efficacy when compared to baseline. There was no statistical difference between the two groups (Musharraf, Ahmad, & Yaqub, 2017). Considering the different outcomes in modern-day trials and previous studies in the 1990s, different capsaicin formulations, and patient characteristics may play more important roles than we had originally thought.

High dose topical capsaicin

An 8% capsaicin patch was developed to increase its efficacy and improve compliance with a single application. Pharmacokinetic data in patients with peripheral neuropathic pain showed low systemic exposure and a very rapid elimination half-life after local application for 60 or 90 min. Treatment on the feet caused lower systemic exposure than on the trunk (Babbar et al., 2009).

Two studies which compared a single application of a high dose (8%) capsaicin patch with a placebo patch showed the NNT for at least 30% pain relief was 12. Local skin reactions were common and tolerable but they may lead to withdrawal. The number needed to harm (NNH) for repeated low dose application was 2.5. They concluded that capsaicin, either as a repeated application of a low dose (0.075%) cream, or a single application of a high dose (8%) patch may provide a degree of pain relief to some patients with painful neuropathic conditions with a rare systemic side effect (Derry et al., 2009).

Therefore, an expert consensus has included this high concentration patch in the guideline for localized neuropathic pain. It can be used as a monotherapy or in combination with other, oral drugs to minimize drug interaction (Pickering, Martin, Tiberghien, Delorme, & Mick, 2017).

In 2012, Cochrane review of topical capsaicin (low concentration, <1%) could not conclude its efficacy in the treatment of neuropathic pain, due to insufficient data. Local skin irritation was mild and transient but may lead to withdrawal. Systemic side effects were rare (Derry & Moore, 2012). Recent Cochrane review of topical high concentration (8% patch) capsaicin for chronic neuropathic pain suggested a substantial level of pain relief when compared with much lower concentration cream (0.075%) in PDN, postherpetic neuralgia, and HIV neuropathy. However, the quality of evidence was moderate or very low. Its effect was similar to other chronic pain therapies (Derry, Rice, Cole, Tan, & Moore, 2017). In nondiabetic neuropathic pain patients with dynamic mechanical allodynia (DMA), 8% capsaicin patch also showed superior efficacy over pregabalin, in terms of reducing the intensity and area of DMA and the number of patients with complete resolution of DMA (Cruccu et al., 2018).

A dose-ranging study of capsaicin patch at 0.625% (50 $\mu g/cm^2$) 1.25% (100 $\mu g/cm^2$) compared to conventional 0.075% capsaicin cream or placebo patch in 60 patients with peripheral neuropathic pain revealed that 0.625% patch was as effective as 0.075% cream with significant improvement in pain. The patient global impression of change was favorable in the 0.625% patch only; other groups were not clearly different. However, a minority of patients in this study had PDN. High dropout rate and local skin reaction were still common in all groups (Moon et al., 2017).

EFNS guidelines on the pharmacological treatment of neuropathic pain: 2009 revision recommended duloxetine, gabapentin, pregabalin, a tricyclic antidepressant, and venlafaxine ER as first-line treatments for PDN. Opioids and tramadol were recommended as second- or third-line treatment. Unfortunately, topical capsaicin was not mentioned for PDN treatment. It was recommended as a second- or third-line treatment for postherpetic neuralgia (Attal et al., 2010). In contrast, the American Academy of Neurology guideline supported the use of 0.075% topical capsaicin four times a day in level B recommendation among other agents such as gabapentin, venlafaxine, duloxetine, amitriptyline, opioids, etc. Pregabalin is the only agent in their level A recommendation (Bril et al., 2011).

Applications to other areas

Osteoarthritis is common worldwide and often treated with oral paracetamol, oral or topical NSAIDs, or oral COXIBs. Oral therapy is sometimes associated with serious adverse events. Therefore, topical preparation is an interesting alternative and topical NSAIDs have become the first-line treatment in mild patients. For the past decade, capsaicin has been studied in many randomized controlled trials. They showed a good safety profile and efficacy in reducing osteoarthritis pain of the hand, knee, hip, or shoulder, and capsaicin was recommended at level B recommendation (Guedes, Castro, & Brito, 2018). Although there is no trial which directly compared capsaicin and topical NSAIDs, they were compared with placebo. The network metaanalysis indicates equal efficacy and effect size for pain relief between these 2 compounds (Persson, Stocks, Walsh, Doherty, & Zhang, 2018).

In Dermatology, pruritus is one of the difficult-to-treat symptoms. Topical capsaicin is often used as an adjuvant in patients who do not respond to other conventional agents. Treatment success in many itch-causing conditions was reported such as psoriasis, lichen simplex chronicus, prurigo nodularis, pruritus ani, pruritus of hemodialysis, aquagenic pruritus, alopecia areata, and other neuropathic itch conditions (Andersen, Arendt-Nielsen, & Elberling, 2017; Boyd, Shea, & Patterson, 2014). The formulations, dosages, and treatment duration are varied according to the indications (Boyd et al., 2014).

Other emerging indications were postmastectomy pain syndrome (Larsson, Ahm Sørensen, & Bille, 2017), painful radiculopathies (Baron et al., 2017), migraine (Cianchetti, 2010), fibromyalgia (Casanueva, Rodero, Quintial, Llorca, & González-Gay, 2013), and irritant-induced, unexplained chronic cough (Ternesten-Hasséus, Johansson, & Millqvist, 2015).

In conclusion, capsaicin is a unique compound acting on the TRPV1 receptor. Topical formulations in various concentrations have been tested in many peripheral neuropathic pain and other pain conditions. Topical capsaicin has been tested over the past several decades for the treatment of painful diabetic neuropathy which is the most common cause of neuropathic pain. In painful diabetic neuropathy, it is useful and has no systemic side effects. Therefore, it is often used as an adjunctive therapy or in patients who are intolerant to oral medications. Local skin reaction is common, especially in higher concentrations.

Clinical studies in other causes of neuropathic pain have been explored as well, such as postherpetic neuralgia and HIV neuropathy. The new formulation, route of administration, and higher concentration of capsaicin are promising in the future for patients with intractable pain.

Other agents of interest

Topical lidocaine

Lidocaine, widely used as a local anesthetic, has also been studied in many neuropathic pain conditions. Due to its sodium channel blocking property, it can block the pain transmission for ectopic discharge and stabilize membrane potentials resulting in neuropathic pain relief. It is included in many international guidelines as an option for postherpetic neuralgia (Pickering & Lucchini, 2020). Several short-term noncomparative studies in patients with PDN showed significant pain relief with a 5% lidocaine patch (Yang, Fang, Xiang, & Yang, 2019). A randomized, comparative study with pregabalin showed similar efficacy in terms of pain score and reduction of allodynia in patients with PHN or PDN, but the lidocaine group had substantially fewer adverse events and a lower discontinuation rate (Baron et al., 2009a). When used as add-on therapy with pregabalin in patients with PHN and painful PDN who failed to respond to monotherapy, it provides additional pain control and is safe and well-tolerated (Baron et al., 2009b). In the network metaanalysis, all interventions (topical capsaicin, lidocaine medicated plaster, amitriptyline, pregabalin, and gabapentin) were equally effective when compared

370 **PART | IV** Novel and nonpharmacological aspects and treatments

to placebo. Topical lidocaine may be associated with fewer and less clinically significant adverse events than are systemic medications (Wolff, Bala, Westwood, Kessels, & Kleijnen, 2010).

Vitamin and supplements

Alpha-lipoic acid (thioctic acid) is a potent antioxidant which can neutralize free radicals from oxidative stress in diabetic condition. It has been proven in many animal experiments and clinical studies, in both intravenous and oral forms. A metaanalysis from short-term studies supports the safety and efficacy of intravenous treatment at 300–600 mg/day for 2–4 weeks to improve the nerve conduction study and positive neuropathic symptoms (Han, Bai, Liu, & Hu, 2012). Later studies in the oral form provided similar results (Nguyen & Takemoto, 2018). It has been proposed to use as a monotherapy and combination therapy with other medications, vitamins, or supplements, based on the oxidative stress hypothesis in diabetes mellitus (Mostacci, Liguori, & Cicero, 2018).

Vitamin B group is essential in many biochemical pathways in the nervous system. In particular, thiamine (B1), pyridoxine (B6), and cobalamin (B12) are often used as supplements for patients with neuropathy. In metformin-treated diabetic patients, vitamin B12 level is sometimes low and may warrant the monitoring of its level or the use of B12 supplement to prevent or treat the neuropathy (Runeberg, Higbea, Weideman, & Alvarez, 2020). Nevertheless, metaanalysis for diabetic neuropathy does not support the use in diabetic neuropathy, especially vitamin B12 (Jayabalan & Low, 2016). Benfotiamine which is a highly absorbed vitamin B1 form, alone or in combination with lipoic acid works synergistically to reverse the pathomechanism of diabetic neuropathy (Varkonyi et al., 2017). The use of PDN has been suggested as an add-on with other supplements or neuropathic pain medications on the basis of pathogenetically based therapy (Javed, Petropoulos, Alam, & Malik, 2015). Most supplements are quite safe, except high dose pyridoxine which may cause sensory ataxic neuronopathy or neuropathy and worsen neuropathic conditions (Kulkantrakorn, 2014).

Many plant-based agents have been studied for anesthesia and analgesia for the past several decades. They were summarized in Table 2. The recent surge in medical cannabis usage leads to renewing interests in other plant-based agents as

TABLE 2 List of selected plant-based agents for analgesia and anesthesia.

Group	Potential mechanism	Usage
Aconitum alkaloids	Sodium channel blocker	Antineuritic Analgesic
Aspirin	Inhibition of arachidonic pathway	Antiinflammatory Analgesic
Cannabinoids	Cannabinoid receptor modulation	Analgesic
Capsaicin	Vanilloid receptor agonist	Analgesic
Cocaine and derivatives	Sodium channel blocker	Anesthetic
Flavonoids	Sodium channel blocker	Analgesic Antiinflammatory Anxiolytic
Ginsenosides	Opioid system modulation, calcium channel blockage, sympathomimetic	Analgesic CNS stimulant
Mitragynine	Partial opioid receptor agonist	Analgesic
Opioid and its derivatives	Opioid receptor agonist	Analgesic
Polygodial aesquiterpenes	Modulation of opioid, serotonin system	Antinociceptive
Salvinorin A	Kappa opioid receptor agonist	Analgesic
Stilbenoids	Sodium channel blocker	Analgesic Antiinflammatory
Terpenoids	Many voltage and ligand gated ion channel modulation	Analgesic Anesthetic Anxiolytic

well. Accumulating evidence supports medical cannabis usage in many pain conditions, but the side effects and long-term safety data are still required. Other promising groups are flavonoids, terpenoids, and new alkaloids (Table 2).

Mini-dictionary of terms

Neuropathic pain: Pain caused by a lesion or disease of the somatosensory nervous system.
Painful diabetic polyneuropathy: Symmetric, painful, length-dependent, polyneuropathy caused by diabetic mellitus. It is mainly involved small nerve fiber.
Capsaicinoids: A class of compounds found in members of the capsicum family. It activates the vanilloid receptor. The most common capsaicinoid is capsaicin.

Key facts

- Painful diabetic neuropathy (PDN) is common and disabling.
- Topical capsaicin is the important treatment option for pain relief for PDN and can be used as monotherapy or combination therapy with systemic medication.
- Higher concentration and different formulation of capsaicin is promising in neuropathic pain and other painful conditions.

Summary points

- Painful diabetic neuropathy (PDN) is the most common cause of peripheral neuropathic pain.
- Because current oral medications provide modest efficacy with systemic side effects, local therapy is an interesting therapeutic option for PDN.
- Capsaicin acts on the TRPV1 receptor in nerve endings and leads to modulation in pain pathways.
- Topical capsaicin in low concentration has been proved effective in clinical trials and is widely accepted in many guidelines.
- Local skin reaction is common, but systemic side effects are very rare.
- A very high concentration capsaicin formulation is promising in PDN and other localized neuropathic pain.

References

Alam, U., Sloan, G., & Tesfaye, S. (2020). Treating pain in diabetic neuropathy: Current and developmental drugs. *Drugs*, *80*(4), 363–384. https://doi.org/10.1007/s40265-020-01259-2.

Anand, P., & Bley, K. (2011). Topical capsaicin for pain management: Therapeutic potential and mechanisms of action of the new high-concentration capsaicin 8% patch. *British Journal of Anaesthesia*, *107*(4), 490–502. https://doi.org/10.1093/bja/aer260.

Andersen, H. H., Arendt-Nielsen, L., & Elberling, J. (2017). Topical capsaicin 8% for the treatment of neuropathic itch conditions. *Clinical and Experimental Dermatology*, *42*(5), 596–598. https://doi.org/10.1111/ced.13114.

Attal, N., Cruccu, G., Baron, R., Haanpää, M., Hansson, P., Jensen, T. S., & Nurmikko, T. (2010). EFNS guidelines on the pharmacological treatment of neuropathic pain: 2010 revision. *European Journal of Neurology*, *17*(9), 1113–e1188. https://doi.org/10.1111/j.1468-1331.2010.02999.x.

Babbar, S., Marier, J. F., Mouksassi, M. S., Beliveau, M., Vanhove, G. F., Chanda, S., & Bley, K. (2009). Pharmacokinetic analysis of capsaicin after topical administration of a high-concentration capsaicin patch to patients with peripheral neuropathic pain. *Therapeutic Drug Monitoring*, *31*(4), 502–510. https://doi.org/10.1097/FTD.0b013e3181a8b200.

Baron, R., Mayoral, V., Leijon, G., Binder, A., Steigerwald, I., & Serpell, M. (2009a). Efficacy and safety of 5% lidocaine (lignocaine) medicated plaster in comparison with pregabalin in patients with postherpetic neuralgia and diabetic polyneuropathy: Interim analysis from an open-label, two-stage adaptive, randomized, controlled trial. *Clinical Drug Investigation*, *29*(4), 231–241. https://doi.org/10.2165/00044011-200929040-00002.

Baron, R., Mayoral, V., Leijon, G., Binder, A., Steigerwald, I., & Serpell, M. (2009b). Efficacy and safety of combination therapy with 5% lidocaine medicated plaster and pregabalin in post-herpetic neuralgia and diabetic polyneuropathy. *Current Medical Research and Opinion*, *25*(7), 1677–1687. https://doi.org/10.1185/03007990903048078.

Baron, R., Treede, R. D., Birklein, F., Cegla, T., Freynhagen, R., Heskamp, M. L., ... Maihöfner, C. (2017). Treatment of painful radiculopathies with capsaicin 8% cutaneous patch. *Current Medical Research and Opinion*, *33*(8), 1401–1411. https://doi.org/10.1080/03007995.2017.1322569.

Baron, R., Tölle, T. R., Gockel, U., Brosz, M., & Freynhagen, R. (2009). A cross-sectional cohort survey in 2100 patients with painful diabetic neuropathy and postherpetic neuralgia: Differences in demographic data and sensory symptoms. *Pain*, *146*(1–2), 34–40. https://doi.org/10.1016/j.pain.2009.06.001.

Basha, K. M., & Whitehouse, F. W. (1991). Capsaicin: A therapeutic option for painful diabetic neuropathy. *Henry Ford Hospital Medical Journal*, *39*(2), 138–140.

372 PART | IV Novel and nonpharmacological aspects and treatments

Biesbroeck, R., Bril, V., Hollander, P., Kabadi, U., Schwartz, S., Singh, S. P., … Bernstein, J. E. (1995). A double-blind comparison of topical capsaicin and oral amitriptyline in painful diabetic neuropathy. *Advances in Therapy, 12*(2), 111–120.

Boyd, K., Shea, S. M., & Patterson, J. W. (2014). The role of capsaicin in dermatology. *Progress in Drug Research, 68*, 293–306. https://doi.org/10.1007/978-3-0348-0828-6_12.

Bril, V., England, J., Franklin, G., Backonja, M., Cohen, J., Del Toro, D., … Zochodne, D. (2011). Evidence-based guideline: Treatment of painful diabetic neuropathy: Report of the American Academy of Neurology, the American Association of Neuromuscular and Electrodiagnostic Medicine, and the American Academy of Physical Medicine and Rehabilitation. *Neurology, 76*(20), 1758–1765. https://doi.org/10.1212/WNL.0b013e3182166ebe.

Capsaicin Study Group. (1992). Effect of treatment with capsaicin on daily activities of patients with painful diabetic neuropathy. Capsaicin Study Group. *Diabetes Care, 15*(2), 159–165. https://doi.org/10.2337/diacare.15.2.159.

Casanueva, B., Rodero, B., Quintial, C., Llorca, J., & González-Gay, M. A. (2013). Short-term efficacy of topical capsaicin therapy in severely affected fibromyalgia patients. *Rheumatology International, 33*(10), 2665–2670. https://doi.org/10.1007/s00296-012-2490-5.

Cianchetti, C. (2010). Capsaicin jelly against migraine pain. *International Journal of Clinical Practice, 64*(4), 457–459. https://doi.org/10.1111/j.1742-1241.2009.02294.x.

Cruccu, G., Nurmikko, T. J., Ernault, E., Riaz, F. K., McBride, W. T., & Haanpää, M. (2018). Superiority of capsaicin 8% patch versus oral pregabalin on dynamic mechanical allodynia in patients with peripheral neuropathic pain. *European Journal of Pain, 22*(4), 700–706. https://doi.org/10.1002/ejp.1155.

daCosta DiBonaventura, M., Cappelleri, J. C., & Joshi, A. V. (2011). A longitudinal assessment of painful diabetic peripheral neuropathy on health status, productivity, and health care utilization and cost. *Pain Medicine, 12*(1), 118–126. https://doi.org/10.1111/j.1526-4637.2010.01012.x.

Davies, M. B. S., Williams, R., & Taylor, A. (2006). The prevalence, severity, and impact of painful diabetic peripheral neuropathy in type 2 diabetes. *Diabetes Care, 29*(7), 1518–1522.

Derry, S., Lloyd, R., Moore, R. A., & McQuay, H. J. (2009). Topical capsaicin for chronic neuropathic pain in adults. *Cochrane Database of Systematic Reviews*, (4), CD007393. https://doi.org/10.1002/14651858.CD007393.pub2.

Derry, S., & Moore, R. A. (2012). Topical capsaicin (low concentration) for chronic neuropathic pain in adults. *Cochrane Database of Systematic Reviews*, (9), CD010111. https://doi.org/10.1002/14651858.cd010111.

Derry, S., Rice, A. S., Cole, P., Tan, T., & Moore, R. A. (2017). Topical capsaicin (high concentration) for chronic neuropathic pain in adults. *Cochrane Database of Systematic Reviews, 1*(1), CD007393. https://doi.org/10.1002/14651858.CD007393.pub4.

Dosenovic, S., Jelicic Kadic, A., Miljanovic, M., Biocic, M., Boric, K., Cavar, M., … Puljak, L. (2017). Interventions for neuropathic pain: An overview of systematic reviews. *Anesthesia and Analgesia, 125*(2), 643–652. https://doi.org/10.1213/ane.0000000000001998.

Duby, J. J., Campbell, R. K., Setter, S. M., White, J. R., & Rasmussen, K. A. (2004). Diabetic neuropathy: An intensive review. *American Journal of Health-System Pharmacy, 61*(2), 160–173. quiz 175-166 https://doi.org/10.1093/ajhp/61.2.160.

Duehmke, R. M., Derry, S., Wiffen, P. J., Bell, R. F., Aldington, D., & Moore, R. A. (2017). Tramadol for neuropathic pain in adults. *Cochrane Database of Systematic Reviews, 6*, CD003726. https://doi.org/10.1002/14651858.CD003726.pub4.

Dyck, P. J., Albers, J. W., Andersen, H., Arezzo, J. C., Biessels, G.-J., Bril, V., … Toronto Expert Panel on Diabetic Neuropathy. (2011). Diabetic polyneuropathies: Update on research definition, diagnostic criteria and estimation of severity. *Diabetes/Metabolism Research and Reviews, 27*(7), 620–628. https://doi.org/10.1002/dmrr.1226.

Guedes, V., Castro, J. P., & Brito, I. (2018). Topical capsaicin for pain in osteoarthritis: A literature review. *Reumatología Clínica, 14*(1), 40–45. https://doi.org/10.1016/j.reuma.2016.07.008.

Han, T., Bai, J., Liu, W., & Hu, Y. (2012). A systematic review and meta-analysis of α-lipoic acid in the treatment of diabetic peripheral neuropathy. *European Journal of Endocrinology, 167*(4), 465–471. https://doi.org/10.1530/eje-12-0555.

Javed, S., Petropoulos, I. N., Alam, U., & Malik, R. A. (2015). Treatment of painful diabetic neuropathy. *Therapeutic Advances in Chronic Disease, 6*(1), 15–28. https://doi.org/10.1177/2040622314552071.

Jayabalan, B., & Low, L. L. (2016). Vitamin B supplementation for diabetic peripheral neuropathy. *Singapore Medical Journal, 57*(2), 55–59. https://doi.org/10.11622/smedj.2016027.

Kennedy, W. R., Vanhove, G. F., Lu, S.-P., Tobias, J., Bley, K. R, Walk, D., … Selim, M. M. (2010). A randomized, controlled, open-label study of the long-term effects of NGX-4010, a high-concentration capsaicin patch, on epidermal nerve fiber density and sensory function in healthy volunteers. *Journal of Pain, 11*(6), 579–587. https://doi.org/10.1016/j.jpain.2009.09.019.

Kiani, J., Ahmad Nasrollahi, S., Esna-Ashari, F., Fallah, P., & Sajedi, F. (2015). Amitriptyline 2% cream vs. capsaicin 0.75% cream in the treatment of painful diabetic neuropathy (double blind, randomized clinical trial of efficacy and safety). *Iranian Journal of Pharmaceutical Research, 14*(4), 1263–1268.

Kiani, J., Sajedi, F., Nasrollahi, S. A., & Esna-Ashari, F. (2015). A randomized clinical trial of efficacy and safety of the topical clonidine and capsaicin in the treatment of painful diabetic neuropathy. *Journal of Research in Medical Sciences, 20*(4), 359–363.

Kulkantrakorn, K. (2014). Pyridoxine-induced sensory ataxic neuronopathy and neuropathy: Revisited. *Neurological Sciences, 35*(11), 1827–1830. https://doi.org/10.1007/s10072-014-1902-6.

Kulkantrakorn, K., Chomjit, A., Sithinamsuwan, P., Tharavanij, T., Suwankanoknark, J., & Napunnaphat, P. (2019). 0.075% capsaicin lotion for the treatment of painful diabetic neuropathy: A randomized, double-blind, crossover, placebo-controlled trial. *Journal of Clinical Neuroscience, 62*, 174–179. https://doi.org/10.1016/j.jocn.2018.11.036.

Kulkantrakorn, K., & Lorsuwansiri, C. (2013). Sensory profile and its impact on quality of life in patients with painful diabetic polyneuropathy. *Journal of Neurosciences in Rural Practice, 4*(3), 267–270. https://doi.org/10.4103/0976-3147.118766.

Kulkantrakorn, K., Lorsuwansiri, C., & Meesawatsom, P. (2013). 0.025% capsaicin gel for the treatment of painful diabetic neuropathy: A randomized, double-blind, crossover, placebo-controlled trial. *Pain Practice, 13*(6), 497–503. https://doi.org/10.1111/papr.12013.

Larsson, I. M., Ahm Sørensen, J., & Bille, C. (2017). The post-mastectomy pain syndrome—A systematic review of the treatment modalities. *The Breast Journal, 23*(3), 338–343. https://doi.org/10.1111/tbj.12739.

Laslett, L. L., & Jones, G. (2014). Capsaicin for osteoarthritis pain. *Progress in Drug Research, 68*, 277–291. https://doi.org/10.1007/978-3-0348-0828-6_11.

Maier, C., Baron, R., Tölle, T. R., Binder, A., Birbaumer, N., Birklein, F., ... Treede, D.-R. (2010). Quantitative sensory testing in the German Research Network on Neuropathic Pain (DFNS): Somatosensory abnormalities in 1236 patients with different neuropathic pain syndromes. *Pain, 150*(3), 439–450. https://doi.org/10.1016/j.pain.2010.05.002.

Mason, L., Moore, R. A., Derry, S., Edwards, J. E., & McQuay, H. J. (2004). Systematic review of topical capsaicin for the treatment of chronic pain. *British Medical Journal, 328*(7446), 991. https://doi.org/10.1136/bmj.38042.506748.EE.

Moon, J. Y., Lee, P. B., Kim, Y. C., Lee, S. C., Nahm, F. S., & Choi, E. (2017). Efficacy and safety of 0.625% and 1.25% capsaicin patch in peripheral neuropathic pain: Multi-center, randomized, and semi-double blind controlled study. *Pain Physician, 20*(2), 27–35.

Mostacci, B., Liguori, R., & Cicero, A. F. (2018). Nutraceutical approach to peripheral neuropathies: Evidence from clinical trials. *Current Drug Metabolism, 19*(5), 460–468. https://doi.org/10.2174/1389200218666171031145419.

Musharraf, M. U., Ahmad, Z., & Yaqub, Z. (2017). Comparison of topical capsaicin and topical turpentine oil for treatment of painful diabetic neuropathy. *Journal of Ayub Medical College, Abbottabad, 29*(3), 384–387.

Nguyen, N., & Takemoto, J. K. (2018). A case for alpha-lipoic acid as an alternative treatment for diabetic polyneuropathy. *Journal of Pharmacy & Pharmaceutical Sciences, 21*(1), 177–191. https://doi.org/10.18433/jpps30100.

Patowary, P., Pathak, M. P., Zaman, K., Raju, P. S., & Chattopadhyay, P. (2017). Research progress of capsaicin responses to various pharmacological challenges. *Biomedicine & Pharmacotherapy, 96*, 1501–1512. https://doi.org/10.1016/j.biopha.2017.11.124.

Persson, M. S. M., Stocks, J., Walsh, D. A., Doherty, M., & Zhang, W. (2018). The relative efficacy of topical non-steroidal anti-inflammatory drugs and capsaicin in osteoarthritis: A network meta-analysis of randomised controlled trials. *Osteoarthritis and Cartilage, 26*(12), 1575–1582. https://doi.org/10.1016/j.joca.2018.08.008.

Petrikonis, K., Sčiupokas, A., Samušytė, G., Janušauskaitė, J., Sulcaitė, R., & Vaitkus, A. (2010). Importance of pain evaluation for more accurate diagnosis of painful diabetic polyneuropathy. *Medicina (Kaunas), 46*(11), 735–742.

Pickering, G., & Lucchini, C. (2020). Topical treatment of localized neuropathic pain in the elderly. *Drugs & Aging, 37*(2), 83–89. https://doi.org/10.1007/s40266-019-00739-9.

Pickering, G., Martin, E., Tiberghien, F., Delorme, C., & Mick, G. (2017). Localized neuropathic pain: An expert consensus on local treatments. *Drug Design, Development and Therapy, 11*, 2709–2718. https://doi.org/10.2147/dddt.S142630.

Rains, C., & Bryson, H. M. (1995). Topical capsaicin. A review of its pharmacological properties and therapeutic potential in post-herpetic neuralgia, diabetic neuropathy and osteoarthritis. *Drugs & Aging, 7*(4), 317–328. https://doi.org/10.2165/00002512-199507040-00007.

Rumsfield, J. A., & West, D. P. (1991). Topical capsaicin in dermatologic and peripheral pain disorders. *DICP, 25*(4), 381–387. https://doi.org/10.1177/106002809102500409.

Runeberg, H. A., Higbea, A. M., Weideman, R. A., & Alvarez, C. A. (2020). Evaluation of vitamin B(12) monitoring in veterans with type 2 diabetes on metformin therapy. *Journal of Pharmacy Practice*. https://doi.org/10.1177/0897190019899260. 897190019899260.

Simpson, D. M., Robinson-Papp, J., Van, J., Stoker, M., Jacobs, H., Snijder, R. J., et al. (2017). Capsaicin 8% patch in painful diabetic peripheral neuropathy: A randomized, double-blind, placebo-controlled study. *Journal of Pain, 18*(1), 42–53. https://doi.org/10.1016/j.jpain.2016.09.008.

Tandan, R., Lewis, G. A., Krusinski, P. B., Badger, G. B., & Fries, T. J. (1992). Topical capsaicin in painful diabetic neuropathy. Controlled study with long-term follow-up. *Diabetes Care, 15*(1), 8–14.

Ternesten-Hasséus, E., Johansson, E. L., & Millqvist, E. (2015). Cough reduction using capsaicin. *Respiratory Medicine, 109*(1), 27–37. https://doi.org/10.1016/j.rmed.2014.11.001.

Tesfaye, S., Boulton, A. J. M., Dyck, P. J., Freeman, R., Horowitz, M., Kempler, P., ... Toronto Diabetic Neuropathy Expert Group. (2010). Diabetic neuropathies: Update on definitions, diagnostic criteria, estimation of severity, and treatments. *Diabetes Care, 33*(10), 2285–2293. https://doi.org/10.2337/dc10-1303.

The Capsaicin Study Group. (1991). Treatment of painful diabetic neuropathy with topical capsaicin. A multicenter, double-blind, vehicle-controlled study. *Archives of Internal Medicine, 151*(11), 2225–2229. https://doi.org/10.1001/archinte.151.11.2225.

Üçeyler, N., & Sommer, C. (2014). High-dose capsaicin for the treatment of neuropathic pain: What we know and what we need to know. *Pain and Therapy, 3*(2), 73–84. https://doi.org/10.1007/s40122-014-0027-1.

Varkonyi, T., Korei, A., Putz, Z., Martos, T., Keresztes, K., Lengyel, C., ... Kempler, P. (2017). Advances in the management of diabetic neuropathy. *Minerva Medica, 108*(5), 419–437. https://doi.org/10.23736/s0026-4806.17.05257-0.

Vinik, A. I., Perrot, S., Vinik, E. J., Pazdera, L., Jacobs, H., Stoker, M., ... Katz, N. (2016). Capsaicin 8% patch repeat treatment plus standard of care (SOC) versus SOC alone in painful diabetic peripheral neuropathy: A randomised, 52-week, open-label, safety study. *BMC Neurology, 16*(1), 251. https://doi.org/10.1186/s12883-016-0752-7.

Watson, C. P. (1994). Topical capsaicin as an adjuvant analgesic. *Journal of Pain and Symptom Management, 9*(7), 425–433. https://doi.org/10.1016/0885-3924(94)90198-8.

Wolff, R. F., Bala, M. M., Westwood, M., Kessels, A. G., & Kleijnen, J. (2010). 5% lidocaine medicated plaster in painful diabetic peripheral neuropathy (DPN): A systematic review. *Swiss Medical Weekly, 140*(21–22), 297–306.

Yang, X. D., Fang, P. F., Xiang, D. X., & Yang, Y. Y. (2019). Topical treatments for diabetic neuropathic pain. *Experimental and Therapeutic Medicine, 17* (3), 1963–1976. https://doi.org/10.3892/etm.2019.7173.

Zhang, W. Y., & Li Wan Po, A. (1994). The effectiveness of topically applied capsaicin. A meta-analysis. *European Journal of Clinical Pharmacology, 46* (6), 517–522. https://doi.org/10.1007/bf00196108.

Chapter 33

Cola nitida and pain relief

L.D. Adedayo[a,b], O. Bamidele[a], and S.A. Onasanwo[b]

[a]*Neurophysiology Unit, Department of Physiology, College of Health Sciences, Bowen University, Iwo, Nigeria,* [b]*Neuroscience and Oral Physiology Unit, Department of Physiology, Faculty of Basic Medical Sciences, University of Ibadan, Ibadan, Nigeria*

Abbreviations

C	Cola
COX	cyclooygenase
MECN	methanol extract of *Cola nitida*
NSAIDs	non-steriodal anti-inflammatory drugs
TNF-α	tumor necrosis factor alpha

Introduction on *Cola nitida*

Cola nitida (family: *Sterculiaceae*), is a dicotyledonous plant with two subspecies, *alba* and *rubra*, the colorations are white and reddish respectively (Ibu et al., 1986). They originated from West Africa and have been introduced to all the forest areas of Central Africa (Chukwu, Odiete, & Briggs, 2006). They are grown in commercial quantities in Nigeria, Ghana, Côte d'Ivoire, and Sierra Leone and also in India, Brazil, and Jamaica in reasonable quantities (Lovejoy, 1980). *C. nitida* is commonly referred to as kola nut in the English language, *Obi* in the Yoruba Language, and *Gworo* in the Hausa language of the Western and Northern parts of Nigeria respectively.

Brief review on *Cola nitida*

Cola nitida like the coffee berry and tea leaf has been in use since ancient times. In West African traditions, it is chewed individually or in a social setting to restore vitality and ease hunger pangs. Kola nuts are also becoming popular in the Arab world, where they were first introduced during the trans-Saharan trade. Kola nuts are essential components of the West African traditional spiritual practice and religion which is more prominent in Nigeria (Aina, 2004). A pharmacist from Georgia, named John Pemberton in the 1800s took extracts of kola and Cocoa, mixed them with sugar and other ingredients, together with carbonated water to manufacture the first cola soft drink. The accountant named it "Coca-Cola" after tasting it. In 1904, Coca-Cola stops using kola in its original recipe.

Socio-cultural values and uses

In the Yoruba culture of western Nigeria, it is used as a gesture of friendship and is vital in traditional ceremonies as a symbol of peace and wishes for the goodwill of the people. Similarly, the sharing of kola nuts is a necessary prerequisite to business dealings that involve a strict etiquette in presenting, dividing, and eating the fruits. Proposals of marriage may be made by a young man's presentation of kola nuts to the prospective bride's father and her acceptance or refusal may be conveyed by a reciprocal gift of nuts, with the meaning depending upon the quality and color. Kola nuts presented by the bride's family signify fertility, productivity, prosperity, contentment, and desire for the union (Johnson & Johnson, 1976; Sundstrom, 1966).

It is used in divination and to know and learn the mind or intent of a god for healing the sick or against barrenness. In some areas, especially Eastern Nigeria, it is a component of an oath-taking process and well appreciated by the Igbo people in Nigeria. The possession and use of kola nuts may be a symbol of wealth and prestige (Hauenstein, 1974; Lovejoy, 1980).

Treatments, Mechanisms, and Adverse Reactions of Anesthetics and Analgesics. https://doi.org/10.1016/B978-0-12-820237-1.00033-8
Copyright © 2022 Elsevier Inc. All rights reserved.

376 PART | IV Novel and nonpharmacological aspects and treatments

Kola nuts are used as a religious object and sacred offering during prayers, ancestor veneration, and significant life events, such as naming ceremonies, weddings, and funerals (Fama, 2004). They are also used in a traditional divination system called Obi divination. For this use, only kola nuts divided into four lobes are suitable. They are cast upon a special wooden board and the resulting patterns are read by a trained diviner.

Distribution of *Cola nitida*

West Africa was the original habitant of *Cola nitida*, they are distributed along the coast of Sierra Leone to the Republic of Benin with Côte d'Ivoire and Ghana maintaining the highest occurrence and variability in their forest zones. It was reported that *Cola nitida* was cultivated eastwards through Nigeria toward Cameroon and Congo around 1900, and spread as far as Senegal, from the Western part of Nigeria (Opeke, 1992). *Cola nitida* is cultivated virtually by Western Africa (Voelcker, 1935). *Cola* species have been planted in tropical South and Central America, the West Indies, East of Mauritius, Sri Lanka, and Malaysia. Slaves' trade from West Africa was the channel for the spread of this species (Russell, 1955).

Composition of *Cola nitida*

It is composed of caffeine (2.0%–3.5%), theobromine (1.0%–2.5%), theophylline, phenolics, phlobaphens, epicatechin, D-catechin, tannic acid, sugar, cellulose, and water (Sonibare, Soladoye, Esan, & Sonibare, 2009). It has been reported that kola nuts are made up of 13.5% water, 9.5% crude protein, 1.4% fat, 45% sugar and starch, 7% cellulose, 3.8% tannin, and ash of 3% (Purseglove, 1968). The caffeine constituents of *C. nitida* seem to explain its use in energy tonics and are also attributed to the enhancement of male fertility. It has been reported that phytochemical screening of methanol extract of *C. nitida* seeds contains flavonoids, steroids, saponins, tannins, anthraquinones, terpenoids, and alkanoids (Adedayo et al., 2019). Chromatogram analysis of kola nuts revealed the presence of phenolic constituents in quantities that are higher than those typical for many fruits. Apples for instance contain 0.1–2.0/100 g fresh weight of polyphenolic compounds (Van Buren, 1970), compared with in excess of 4.0/100 g fresh weight in kola nuts. The study from Odebode (1996) showed that the two species differed markedly in the amount of total phenol and that differences also existed between different color variants as shown below. The total phenol content was greater in *C. nitida* than *C. acuminata*. In *Cola nitida*, the quantity of total phenol in red nuts was three times that of white and pink nuts; but in *Cola acuminata* the difference was not significant. This investigation supports the general view that *Cola nitida* is more astringent than *Cola acuminata*, because astringency is related to the phenolic content of fruits (Odebode, 1996) (Table 1).

Atawodi, Mende, Pfundstein, Preussmann, and Spiegelhalder (1995) analyzed both *C. acuminata* and *C. nitida* for their content of primary and secondary amines and assessed for their relative methylating potential. Seeds of both species contained high quantities of both primary and secondary amines. The methylating activity was significantly higher in kola nuts (170–490 mg/kg) than has ever been reported for a fresh plant product. The authors urge that the possible role of kola nut chewing in human cancer etiology should be explored in nations where kola nuts are voraciously consumed as stimulants. Ibu et al. (1986) reported that both species of Cola induced significant increases in gastric acid secretion. They advised sufferers from peptic ulcers to avoid eating kola nuts.

Uses of *Cola nitida*

Kola nuts contain some bioactive compounds which are powerful stimulants such as caffeine, alkaloids, and theobromine that counteract fatigue, suppress thirst and hunger, and enhances intellectual performance (Nickalls, 1986). More also, the

TABLE 1 Total phenolics in Cola species (mg/100 g fresh weight).

Color	Cola nitida	Cola acuminate
White	4.45	3.37
Pink	6.12	4.17
Red	9.09	–

The table shows the analysis of the total phenolics in two cola species which are *Cola nitida* and *Cola acuminate* take from Odebode (1996).
(Source: Odebode, A. C. (1996): Phenolic compounds in the kola nut (Cola nitida and Cola acuminata) (Sterculiaceae) in Africa. *Revista de Biología Tropical, 44*, 513–515.)

twigs are used as chewing sticks to clean the teeth and gums and its characteristics bitter taste made it effective for refreshing the mouth (Lewis & Elvin-Lewis, 1985). In Europe and the United States of America, they are chiefly used as flavoring agents for cola drinks which are refreshing or stimulating substitutes for tea and coffee (Irvine, 1956). Also, it has been reported that kola nuts are the source of alkaloids in pharmaceutical preparations (Opeke, 1992).

Traditional uses of *Cola nitida*

Burkill (1995) reported that traditionally, the leaves, twigs, flowers, fruits, and bark of *C. nitida* and *C. acuminata* are components of preparations used as a tonic remedy for dysentery, coughs, diarrhea, vomiting, and chest pain. Also, it has effective in folk medicine as an aphrodisiac and appetite suppressant. It is also used to treat morning sickness, migraine headache, and indigestion (Esimone, Adikwu, Nworu, Okoyeand, & Odimegwu, 2007), wounds and inflammation have been cure through the dermal application of *C. nitida* (Newall, Anderson, & Philipson, 1996). Pregnant women use *Cola nitida* to control vomiting, and also it is used as a principal stimulant to keep awake and withstand fatigue by drivers, students, and other 20 menial workers (Chukwu et al., 2006).

Pharmacological potentials of *Cola nitida*

Several studies have reported the pharmacological activities of *C. nitida* such as antiinflammatory, antibacterial, antidiabetic, antifungal, anticancer, antioxidant, hepato-protective effects (Kamatenesi-Mugisha & Oryem-Origa, 2005: Deeepa & Rajendran, 2007; Kisangau, Hosea, Joseph, & Lyaruu, 2007). Also, Farook and Atlee (2011) reported the hemolytic potential, larvicidal activity, and anthelmintic activity. Pain relief activity, central nervous system activity, sexual impotence, and erectile dysfunction, and hypolipidemic studies have been reported (Anthoney, Jackie, & Ngule, 2013; Anthoney, Ngule, & Obey, 2014; Arivoli & Tennyson, 2012; Hosahally et al., 2012; Joshi, Gopal, & Byregowda, 2011). In recent times, Adedayo et al. (2019) reported the antiinflammatory, antinociceptive potentials and its mechanism(s) of action via a cholinergic pathway.

Pain and nociception

An unpleasant emotional and sensory experience associated with actual or potential tissue damage, or description in terms of such damage is referred to as pain (Ibu et al., 1986). The entire nervous system, especially in the brain undergoes extensive processing of pain psychology and changes in its perception (D'Mello & Dickenson, 2008).

Nociception is essential for being aware of and reacting to potentially or actually damaging stimuli in the environment. Specialized sensory neuronal receptors detect noxious chemical, thermal and mechanical stimuli (Lee & Anderson, 2005). Electrical signals generated at these sites are essential amplified and transmitted further to the central nervous system's higher center in order to produce a systemic response aimed at self-preservation (Coderre, Vaccarino, & Melzack, 1990).

A body defense mechanism is being offered by pain and is a signal of a problem, particularly when in an acute state and may result in chronic where it outlasts any potential for healing and becomes modified centrally (Gould et al., 2002). Niv and Kreitler (2001) have reported that pain can poise an adverse effect on total well-being by suppressing physical and emotional function.

Causes of pain

Painful experiences are facilitated by several reasons. It may be felt due to the following: tissue necrosis, inflammation, infection, stretching of the tissue, chemical, or burn. Pain from skeletal muscles ensues from ischemia or hemorrhage. In the gastrointestinal tract, nociception may result from inflammation of the mucosa or from distention or spasms of the muscle. As a result of its cause, pain may be sudden and short-term, and the activity is marked primarily by a reflex withdrawal (Gould et al., 2002).

Pain can be classified as acute or chronic. Acute pain is known for its short duration and the cause often identifiable such as disease and trauma. But chronic pain persists after healing is expected to be complete, or ensue from a pre-existing chronic disease. Also, it may be modified by psychological factors, and the necessary attention to these is required in the management of pain (Rahman, D'Mello, & Dickenson, 2008).

Classification of pain

Pain can be classified into the following modalities:

a. First/fast pain (localized, acute, that appears in 0.1 s after the painful stimulus and ends with its action), accompanied by a flexion reflex and it is a consequence of the stimulation of A-delta myelinated fibers.
b. Secondary pain—slow, chronic, diffuse, that persists longer than the stimulus; it is accompanied by reflex muscular contraction and is a result of unmyelinated C fibers.

Woolf and co-workers (Woolf & King, 1989) reported that an essential conceptual breakthrough in understanding the physiology of pain is the recognition that the pain that occurs after most types of noxious stimulation is usually protective and distinct from pain ensuing from over damage to tissues or nerve and is the first type of pain termed physiologic pain.

Physiologic pain can also be called nociceptive pain because it is only stimulated when intense noxious stimuli threaten to injure tissue. High stimulus threshold well localized and transient are common features. Also, demonstrates a stimulus-response relationship similar to those of the other somatosensory (Woolf et al., 1995).

Nociceptive processing

Nociception is the physiologic component of pain, which involves the processes of transduction, transmission, and modulation of neural signals generated in response to an external noxious stimulus. It is an integral component of the physiologic process that results in the conscious perception of pain when carried to completion. In its simple modality, the pathway can be considered as a three-neuron chain, with the first-order neuron originating in the periphery and projecting to the spinal cord, the second-order neuron ascending the spinal cord, and the third-order neuron projecting to the cerebral cortex. At a complex level, the pathway involves a network of branches and communications with other sensory neurons and descending inhibitory neurons from the midbrain that regulate the afferent transmission of painful stimuli (Willis & Coggeshall, 1991).

Pathologic pain

The traditional stimulus-response model of physiologic pain is conceptually appealing and has laid the foundation for a more comprehensive understanding of nociceptive pathways. However, it must be recognized that physiologic pain cannot exist alone in the clinical setting. In most situations, the noxious stimulus is not transient and may be associated with serious tissue inflammation and nerve injury. In a situation like this, the classic "hard-wired system becomes less relevant, and dynamic changes in the processing of noxious input called pathologic pain (because it implies that the tissue damage has already occurred) or clinical pain, as ongoing discomfort and are characteristics of the patient's clinical symptomatology. Pathologic pain may manifest itself in several ways: spontaneous pain that may be dull, burning, or stabbing (causalgia); exaggerated pain in response to a noxious stimulus (hyperalgesia); and pain produced by a stimulus that is not normally noxious called allodynia (Woolf & Chong, 1993). Pathologic pain may arise from injury to a variety of tissue types invoking distinct neural mechanisms, and it is often further classified into inflammatory pain (involving somatic or visceral structures) or neuropathic pain (involving lesions of the nervous system). In addition, it is useful to characterize clinical pain from a temporal perspective and make the distinction between recently occurring (acute) and long-lasting (chronic) pain.

Peripheral mechanisms of sensory transmission

At the periphery is where sensory experience initiation begins, where the peripheral terminals of primary afferent fibers respond to a myriad of stimuli and translate this information into the dorsal horn of the spinal cord, where the central ends of these fibers terminate. $A\beta$-fibers, $A\delta$ fibers, and C-fibers are the three main types of sensory fibers in the peripheral nervous system. Each has different properties allowing them to respond to and transmit different types of sensory information. $A\beta$-fibers are large in diameter and highly myelinated, thus allowing them to conduct action potentials in fast response from their peripheral to central terminals. These fibers have low activation thresholds and normally respond to light touch and are responsible for conveying tactile sensory information. $A\delta$ fibers are smaller in diameter and thinly myelinated, and they also possess higher activation thresholds. They respond to both thermal and mechanical stimuli. C-fibers are the smallest type of primary afferents and are unmyelinated, thus, conduction is in the slowest modality. They have the greatest thresholds for activation and therefore detect selectively nociceptive or "painful" stimuli. Collectively, both $A\delta$ - and

C-fiber can be termed as nociceptors or "pain fibers," responding to noxious stimuli which may be in the form of mechanical, thermal, or chemical (Park, Chung, & Chung, 2000).

Cola nitida as a therapeutic agent for pain relief

Studies from Adedayo and co-workers (Adedayo et al., 2019) reported that extracts of *Cola nitida* exhibit analgesic activity at doses ranging from (50–200 mg/kg) using formalin-induced paw licking test as an assessment model which is better than well-known analgesic drug aspirin (Figs. 1 and 2; Table 2).

Furthermore, as shown above, the results obtained from the acetic acid-induced writhing model showed that *Cola nitida* possesses antinociceptive properties with a lower number of writhes obtained from the study. Lowering the number of writhes is caused by the inhibition of prostaglandin synthesis (Loganayaki, Siddhuraju, & Manian, 2012). To further investigate the involvement of pain pathways in the antinociceptive effects of *Cola nitida*, formalin-induced nociception was used. The formalin test was used for the assessment of the mechanistic study because of its advantages which include its ability to mimic human clinical pain conditions, sensitivity to mild analgesics, and non-steroidal antiinflammatory agents (Hunskaar & Hole, 1987; Tjolsen, Berge, Hunskaar, Rosland, & Hole, 1992). Formalin-induced test is a pain-related licking response of the injected paw in two distinct phases. The first phase is presented by neurogenic pain caused by direct chemical stimulation of nociceptors. The second phase is evidenced by inflammatory pain triggered by a fusion of stimuli-inflammation of the peripheral tissues and mechanisms of central sensitization (Tjolsen et al., 1992). Centrally acting analgesic drugs inhibit both phases of the formalin test, while peripherally acting analgesics restrict only the late phase responses. The first phase finding of the formalin test confirms the central antinociceptive effect of Methanol Extract *Cola nitida* (MECN). The late phase response as the antinociceptive effect observed in the acetic acid-induced writhing test is due to this inhibition of the inflammatory mediators (Oku, Ueda, Iinuma, & Ishiguro, 2005).

In the formalin-induced nociception assessments of the pathway involved in the antinociception effects of *Cola nitida* extracts, the opioid receptor blocker naloxone, cholinergic antagonist atropine, and beta-adrenergic receptor blocker propranolol were used. However, only atropine was able to reverse the antinociception offered by *Cola nitida* which is evidenced that the mechanism by which *Cola nitida* extract offers nociceptive potential seems to be mediated through cholinergic pathway but not through beta-adrenergic and opioid pathways.

It has been reported that flavonoids have been found to suppress the intracellular Ca^{2+} ion elevation in a dose-dependent manner, as well as possible release of pro-inflammatory mediators such as TNF-α (Kempuraj et al., 2005). Also, a study from Annegowda and co-workers showed that flavonoids may enhance the amount of endogenous serotonin or may act together with 5-HT2 and 5-HT3 receptors, which may be involved in the mechanism of central pain relief activity (Annegowda, Mordi, Ramanathan, & Mansor, 2010). There are also reports on the role of flavonoids in analgesic activity principally by targeting prostaglandins (Ramesh et al., 1998).

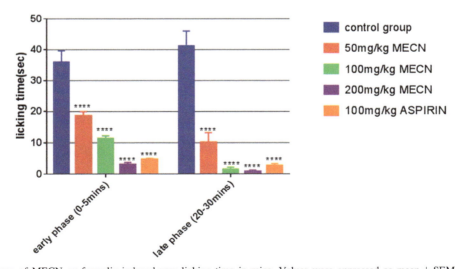

FIG. 1 Effect of doses of MECN on formalin-induced paw licking time in mice. Values were expressed as mean ± SEM, $N = 6$ in each group. ****$P < 0.0001$, compared to control. *(Source: Adedayo, L. D., Ojo, A. O., Awobajo, F. O., Adeboye, B. A., Adebisi, J. A., Bankole, T. J., Ayilara, G. O., Bamidele, O., Aitokhuehi, N. G., Onasanwo, S. A. (2019): Methanol extract of Cola nitida ameliorates inflammation and nociception in experimental animals, Neurobiology of Pain, 5, 100027.)*

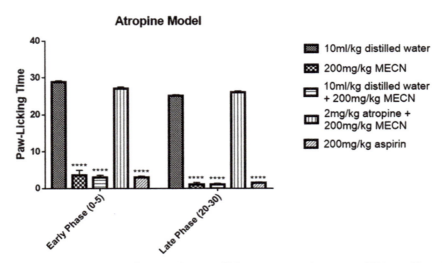

FIG. 2 Effect of Atropine on antinociceptive action of *cola nitida* extract. Values are expressed as mean ± SEM, $n = 6$ in each group. ****$P < 0.001$, compared with the control. *(Source: Adedayo, L. D., Ojo, A. O., Awobajo, F. O., Adeboye, B. A., Adebisi, J. A., Bankole, T. J., Ayilara, G. O., Bamidele, O., Aitokhuehi, N. G., Onasanwo, S. A. (2019): Methanol extract of Cola nitida ameliorates inflammation and nociception in experimental animals, Neurobiology of Pain, 5, 100027.)*

Furthermore, flavonoids have the capability to inhibit eicosanoid biosynthesis, such as prostaglandins which are involved in different immunological responses and are the products of the cyclooxygenase and lipoxygenase pathways (Jothimanivannan, Kumar, & Subramanian, 2010). Tannins are also believed to have a contribution to antinociceptive activity (Ramprasath, Shanthi, & Sachdanandam, 2006). Therefore, it can be agreed that cyclooxygenase (COX) inhibitory activity alongside antioxidant activity may reduce the production of arachidonic acid from phospholipids or may reduce the enzyme system responsible for the synthesis of prostaglandins and finally relieve pain-sensation.

Moreover, methanol extract of *Cola nitida* showed significant analgesic activity in the study reported by Adedayo et al. (2019) which may be due to its high flavonoid content and other phytochemical constituents present in the plant.

Several studies have shown that stimulation of neuronal nicotinic receptors has been known to produce analgesic effects both in human and experimental animals (Umana, Daniele, & McGehee, 2013). Also, the injection of anticholinesterase inhibitors in formalin tests in rats was found to produce an antinociceptive effect (Yoon, Choi, & Jeong, 2003). Antagonism of muscarinic receptors also has been shown to reverse the antinociceptive activities of anticholinesterase inhibitors (Tamaddonfard, Zanbouri, & Mojtahedin, 2009).

However, abolition of the methanol extract of *Cola nitida* induced-analgesia in the presence of this antagonist further validates central and peripheral mechanisms of analgesic action of the extract of *Cola nitida*.

The bioactive constituents of *Cola nitida* and some experimental studies reported on pain can validate the analgesic potential of *Cola nitida* and can be potential therapeutic targets for pain relief associated with the muscarinic receptors in the future.

TABLE 2 Effect of methanol extract of *Cola nitida* on the acetic acid-induced writhing model in mice.

Groups	Mean number of writhing	% Inhibition
Control	41 ± 3.96	
50 mg/kg MECN group	11.6 ± 0.75[d]	71.7%
100 mg/kg MECN group	6.4 ± 1.86[d]	84.39%
200 mg/kg MECN group	0 ± 0[d]	100%
Reference group (Aspirin)	0 ± 0[d]	100%

Values are expressed as mean ± SEM, $N = 6$ in each group, d$P < 0.0001$ compared with the control. The table shows the mean ± SEM of 5 sets of analysis of *Cola nitida* extract of various doses and aspirin taken from Adedayo et al. (2019).
(Source: Adedayo, L. D., Ojo, A. O., Awobajo, F. O., Adeboye, B. A., Adebisi, J. A., Bankole, T. J., Ayilara, G. O., Bamidele, O., Aitokhuehi, N. G., Onasanwo, S. A. (2019): Methanol extract of Cola nitida ameliorates inflammation and nociception in experimental animals, *Neurobiology of Pain, 5,* 100027.)

Application to other areas

Cola nitida is highly essential in socio-cultural values in Nigeria and West Africa. It is chewed individually or in a social setting such as marriages, burial rites, naming ceremonies to restore vitality and ease hunger pangs. They are also used as a component of spiritual practice and religion in West Africa especially Nigeria which attached more divinity to kola nuts.

Similarly, the sharing of kola nuts is a necessary prerequisite to business dealings that involve a strict etiquette in presenting, dividing, and eating the fruits which they believe will foster a good relationship amongst the party involved.

It is used in divination and to know and learn the mind or intent of a god for healing the sick or against barrenness in women and solution for the predicament. Also, in some areas, especially Eastern Nigeria, it is a component of an oath-taking process and well appreciated by the Igbo people in Nigeria. In Northern Nigeria, they chewed kola nuts as a daily food requirement for alertness, and this as created source of commercial ventures between the North and South of Nigeria, the west climate conditions favors its growth and therefore it is planted in South Western Nigeria massively. In general, the social-cultural values of *Cola nitida* cannot be overemphasized. It is consumed by many to withstand fatigue and keep them awake especially if the task is at hand, also is administered to pregnant women to control vomiting. *Cola nitida* is used to prepare a tonic as a remedy for dysentery, coughs, diarrhea, vomiting, and chest complaints and has been used in folk medicine as an aphrodisiac and an appetite suppressant.

Other agents of interest

Tetracarpidium conophorum which is popularly called walnut is a plant of potential, the study by Onasanwo, Babatunde, and Faborode (2016) has reported the analgesic potential of the plant; however, more work needs to be done on the mechanistic pathway by which the plant mediates pain. More also, the bioactive components and isolation of the various compounds that carry out all the aforementioned properties should be identified. Furthermore, the pharmacological evidence of *Cola nitida* needed urgent attention to unfolding various bioactive compounds responsible for the potential that has been reported in the literature.

Also, the use of antipsychotic drugs as an antiinflammatory agent has been reported by Adedayo et al. (2017) Though more work is needed to be done to identify various mechanism(s) through which this action is mediated. Little or no information on its activity on nociception is another area that needs to research. Though, work is ongoing on a generation of antipsychotic drugs on memory performance in rodents. But the research needs aids to facilitate the work and give it a more scientific approach for discovery in Nigeria.

Currently, we are trying to of identification of bioactive components of some nuts that are peculiar to West Africa and are on the extinction list. Resources to carry out this extensive project are the challenge. However, the best we can do to source for funding on these plants product is on top on-going. Isolations of compounds are what we are proposing with all the previous works that we have done in past but research funding has always been a challenge at this juncture. We want to improve on medicinal plants that have lesser untold effects when compared with NSAIDs and with much efficacy in the nearest future. The potentials of this medicinal plant have been reported in the literatures by many researchers.

Mini-dictionary

Cola nitida: A dicotyledonous plant with two subspecies, *alba* and *rubra*, the colorations are white and reddish respectively, the English name is Kola nut.

Obi-divination: Obi is a Yoruba name for *Cola nitida*, the Yoruba people of Western Nigeria believe in the traditional spiritualism of kola nut and its implication is social-cultural values.

Chromatogram: Results obtained from Gas chromatography-mass spectrophotometry (GC–MS) revealing the various compounds in the plant.

Somatosensory: Relating to or being sensing stimuli activity originating elsewhere than in the special senses (ears, eye) and conveying information about the state of the body properly and its immediate environment.

Sensory fibers: They are three types namely: Aß-fibers, Aδ fibers, and C-fibers. They have different properties allowing them to respond to and transmit different types of sensory information.

Aß-fibers: Large in diameter and highly myelinated, thus allowing them to conduct action potentials in fast response from their peripheral to central terminals.

Aδ fibers: Smaller in diameter and thinly myelinated, and they also possess higher activation thresholds.

C-fibers: The smallest type of primary afferents and are unmyelinated, thus, conduction is in the slowest modality.

MECN: Methanol extract of *Cola nitida*.

382 PART | IV Novel and nonpharmacological aspects and treatments

Paw licking time: The duration of time the animals use to lick their paws after exposure to formalin.
Atropine: Cholinergic antagonist.

Key facts

- *Cola nitida* lowers visceral nociception has seen in the reduced number of writhes in acetic acid-induced model
- Antinociceptive mechanism of *Cola nitida* potential which is evidenced through cholinergic pathway via atropine as the agent of the antagonism
- There is an abundance of *Cola nitida* in western Nigeria which produce in larger quantity for other parts of the country for usage
- The social-cultural values of kola nut cannot be overemphasized
- Numerous bioactive compounds have been identified in kola nut, especially caffeine with a larger percentage and needs more exploration on their biological importance.
- Classification of pain base on their modalities have been identified

Summary points

- *Cola nitida* is a plant belonging to the genus Cola and family *Sterculiaceae*, the plant is native to West Africa and can be found throughout the forest areas of West and Central Africa.
- It is cultivated in commercial quantities in Nigeria, Ghana, Côte d'Ivoire, and Sierra Leone and also to some extent in India, Brazil, and Jamaica.
- *Cola nitida* contains some bioactive constituents, this contributed to its usage in phytomedicine and potential therapeutic agents in the relief of various ailments in which pain is one of them.
- The study shows that the extract of Cola nitida possesses the analgesic potential and the mechanism by which the antinociception is conceived seems to involve the cholinergic pathway.
- The scientific findings from the studies carried-out justified the folk story narrated about the efficacies of the plant in West Africa.
- The untold adverse effects of some antinociceptive drugs can be ameliorated by unfolding various mechanism(s) by which Cola nitida mediates pain.

References

Adedayo, L. D., Ojo, A. O., Awobajo, F. O., Adeboye, B. A., Adebisi, J. A., Bankole, T. J., et al. (2019). Methanol extract of *Cola nitida* ameliorates inflammation and nociception in experimental animals. *Neurobiology of Pain, 5*, 100027.

Adedayo, L. D., Olawuyi, D. A., Ojo, A. O., Bamidele, O., Onasanwo, S. A., & Ayoka, A. O. (2017). The role of aripriprazole (an anti-psychotic drug) in the resolution of acute peripheral inflammation in male Wistar rats. *Journal of Pharmaceutical Research International, 17*(6), 1–8.

Aina, A. (2004). Man as a socio-rational being. In E. Ifie (Ed.), *Sir (Dr.) Lambert Eradiri: A legend in the oasis of the Niger* (pp. 224–227). Ibadan: Oputoru Books.

Annegowda, H. V., Mordi, M. N., Ramanathan, S., & Mansor, S. M. (2010). Analgesic and antioxidant properties of ethanolic extract of *Terminalia catappa* l. Leaves. *International Journal of Pharmacology, 6*, 910–915.

Anthoney, S. T., Jackie, O., & Ngule, C. M. (2013). In vitro control of selected pathogenic organisms by Vernonia adoensis roots. *International Journal of Pharmacy and Life Sciences, 4*(8), 2855–2859.

Anthoney, S. T., Ngule, C. M., & Obey, J. (2014). In vitro antibacterial activity of methanolic-aqua extract of Tragia brevipes leaves. *International Journal of Pharmacy and Life Sciences, 0976-7126. 5*(2), 3289–3294.

Arivoli, S., & Tennyson, S. (2012). Larvicidal efficacy of Strychnos nuxvomica Linn.(Loganiaceae) leaf extracts against the filarial vector Culex quinquefanciatus say (Diptera:Culicidae). *World Journal of Zoology, 7*(1), 06–11.

Atawodi, S. E., Mende, P., Pfundstein, B., Preussmann, R., & Spiegelhalder, B. (1995). Nitrosatable amines and nitrosamide formation in natural stimulants: *Cola acuminata, C. nitida* and *Garcinia cola*. *Food and Chemical Toxicology, 33*, 625–630.

Burkill, H. M. (1995). *The useful plants of west tropical Africa. Vol. 3* (2nd ed., pp. 522–527). London: Royal Kew Botanical Gardens, Kew.

Chukwu, L. O., Odiete, W. O., & Briggs, L. S. (2006). Basal metabolic regulatory responses and rhythmic activity of mammalian heart to aqueous kolanut extracts. *African Journal of Biotechnology, 5*, 484–486.

D'Mello, R., & Dickenson, A. H. (2008). Spinal cord mechanisms of pain. *British Journal of Anaesthesia, 101*(1), 8–16.

Coderre, T. J., Vaccarino, A. L., & Melzack, R. (1990). Central nervous system plasticity in the tonic pain response to subcutaneous formalin injection. *Brain Research, 535*, 155–158.

Deeepa, N., & Rajendran, N. N. (2007). Antibacterial and anti-fungal activities of various extracts of *Acanthospermum hispidum* DC. *Journal of Natural Remedies, 7*(2), 225–229.

Esimone, C. O., Adikwu, M. U., Nworu, C. S., Okoyeand, F. B. C., & Odimegwu, D. C. (2007). Adaptogenic potentials of *Camellia sinensis* leaves, *Garcinia kola* and Kolanitida seeds. *Scientific Research and Essay, 2*(7), 232–237.

Fama Aina Adewale-Somadhi, Chief. (2004). *Practioner's handbook for the Ifa professional* (p. 1). Ile Orunmila Communications.

Farook, S. M., & Atlee, W. C. (2011). Antidiabetic and hypolipidemic potential of Tragia involucrata Linn. In streptozotocin-nicotinamide induced type II diabetic rats. *International Journal of Pharmacy and Pharmaceutical Sciences, 3*(4), 103–109.

Gould, T. J., Rowe, W. B., Heman, K. L., Mesches, M. H., Young, D. A., Rose, G. M., et al. (2002). Effects of hippocampal lesions on patterned motor learning in the rat. *Brain Research Bulletin, 58*, 581–586.

Hauenstein, A. (1974). La noix de cola. Coutumes et rites de quelques ethaies de Cote d'Ivoire. *Anthropos, 69*, 457–493.

Hosahally, R. V., Sero, G., Sutar, P. S., Joshi, V. G., Sutar, K. P., & Karigar, A. A. (2012). Phytochemical and pharmacological evaluation of Tragia cannabinafor anti-inflammatory activity. *International Current Pharmaceutical Journal, 1*(8), 213–216.

Hunskaar, S., & Hole, K. (1987). The formalin test in mice: Dissociation between inflammatory and non-inflammatory pain. *Pain, 30*, 103–114.

Ibu, J. O., Iyama, A. C., Ijije, C. T., Ishmael, D., Ibeshi, M., & Nwokediuko, S. (1986). The effect of *Cola acuminata* and *Cola nitida* on gastric secretion. *Scandinavian Journal of Gastroenterology, 124*, 39–45.

Irvine, F. R. (1956). *Plants of the gold coast*. Oxford University Press.

Johnson, E. J., & Johnson, T. J. (1976). Economic plants in a rural Nigerian market. *Economic Botany, 30*, 375–381.

Joshi, C. G., Gopal, M., & Byregowda, S. M. (2011). Cytotoxic activity of Tragia involucrate Linn. Extracts. *American-Eurasian Journal of Toxicology Sciences, 3*(2), 67–69.

Jothimanivannan, C., Kumar, R. S., & Subramanian, N. (2010). Anti-inflammatory and analgesic activities ofethanol extract of aerial parts of justice gendarussa burm. *International Journal of Pharmaceutics, 6*, 278–283.

Kamatenesi-Mugisha, M., & Oryem-Origa, H. (2005). Traditional herbal remedies used in the management of sexual impotence and erectile dysfunction in western Uganda. *African Health Sciences, 5*(1), 40–49.

Kempuraj, D., Madhappan, B., Christodoulou, S., Boucher, W., Cao, J., Papadopoulou, N., … Theoharides, T. C. (2005). Flavonols inhibit proinflammatory mediator release, intracellular calcium ion levels and protein kinase C theta phosphorylation in human mast cells. *British Journal of Pharmacology, 145*, 934–944.

Kisangau, D. P., Hosea, K. M., Joseph, C. C., & Lyaruu, H. V. M. (2007). In vitro antimicrobial assay of plants used in traditional medicine in Bukoba rural district, Tanzania. *African Journal of Traditional, Complementary, and Alternative Medicines, 4*(4), 510–523.

Lee, J., & Anderson, R. (2005). Best evidence topic report. Effervescent agents for oesophageal food bolus impaction. *Emergency Medicine Journal, 22*(2), 123–124.

Lewis, W. H., & Elvin-Lewis, P. F. (1985). *Medical botany*. New York: Wiley and Sons.

Loganayaki, N., Siddhuraju, P., & Manian, S. (2012). Antioxidant, anti-inflammatory and anti-nociceptive effects of Ammannia baccifera L. (Lythracceae), a folklore medicinal plant. *Journal of Ethnopharmacology, 140*, 230–233.

Lovejoy, P. E. (1980). Kola in the history of West Africa. *Cahier d'Etudes Africaines, 77-78*, 97–134.

Newall, C. A., Anderson, L. A., & Philipson, J. D. (1996). *Herbal medicine: A guide for health care professionals* (pp. 199–200). London: The Phamarceutical Press.

Nickalls, R. W. D. (1986). W.F. Daniell (1817-1865): And the discovery that cola-nuts contain caffeine. *The Pharmaceutical Journal, 236*, 401–402.

Niv, D., & Kreitler, S. (2001). Pain and quality of life. *Pain Practice, 1*(2), 150–161.

Odebode, A. C. (1996). Phenolic compounds in the kola nut (*Cola nitida* and *Cola acuminata*) (Sterculiaceae) in Africa. *Revista de Biología Tropical, 44*, 513–515.

Oku, H., Ueda, Y., Iinuma, M., & Ishiguro, K. (2005). Inhibitory effects of xanthones from Guttiferae plants on PAF-induced hypotension in mice. *Planta Medica, 71*, 90–92.

Onasanwo, S. A., Babatunde, L. D., & Faborode, O. S. (2016). Anti-nociceptive and antiinflammatory potentials of fractions from the leaf extract of *Tetracarpidium conophorumin* rats and mice. *African Journal of Biomedical Research, 19*, 45–54.

Opeke, L. K. (1992). *Tropical tree crops*. Ibadan: Spectrum Books Ltd.

Park, S. K., Chung, K., & Chung, J. M. (2000). Effects of purinergic and adrenergic antagonists in a rat model of painful peripheral neuropathy. *Pain, 87*(2), 171–179.

Purseglove, J. W. (1968). *Tropical crops: Dicotyledons*. Longmans Green & Co Ltd. responses by prostaglandins and thromboxanes. J Clin Invest. 2001; 108: 15–23.

Rahman, W., D'Mello, R., & Dickenson, A. H. (2008). Peripheral nerve injury-induced changes in spinal α2-adrenoceptor-mediated modulation of mechanically evoked dorsal horn neuronal responses. *The Journal of Pain, 9*(4), 350–359.

Ramesh, M., Rao, Y. N., Rao, A. V., Prabhakar, M. C., Rao, C. S., Muralidhar, N., et al. (1998). Antinociceptive and anti-inflammatory activity of aflavonoid isolated from Caralluma attenuata. *Journal of Ethnopharmacology, 62*, 63–66.

Ramprasath, V. R., Shanthi, P., & Sachdanandam, P. (2006). Immunomodulatory and anti-inflammatory effects of *Semecarpus anacardium* Linn. Nut milk extract in experimental inflammatory conditions. *Biological & Pharmaceutical Bulletin, 29*(4), 693–700.

Russell, T. A. (1955). The kola of Nigeria and the Cameroons. *Tropical Agriculture (Trinidad), 32*, 210–240.

Sonibare, M., Soladoye, M., Esan, O., & Sonibare, O. (2009). Phytochemical and antimicrobial studies of four species of Cola Schott and Endl. (Sterculiaceae). *African Journal of Traditional, Complementary and Alternative Medicines, 6*, 518–525.

Sundstrom, L. (1966). The cola nut functions in west African social life. In *28. Studia Ethnographia Upsaliensia* (pp. 135–146). Stockholm: Almqvist and Wiksell.

Tamaddonfard, E., Zanbouri, A., & Mojtahedin, A. (2009). Role of central muscarinic cholinergic receptors in the formalin-induced pain in rats. *Indian Journal of Pharmacology, 41*(3), 147.

Tjolsen, A., Berge, O. G., Hunskaar, S., Rosland, J. H., & Hole, K. (1992). The formalin test: An evaluation of the method. *Pain, 51*, 5–17.

Umana, I. C., Daniele, C. A., & McGehee, D. S. (2013). Neuronal nicotinic receptors as analgesic targets: It's a winding road. *Biochemical Pharmacology, 86*(8), 1208–1214.

Van Buren, J. (1970). Fruit phenolics. In A. C. Hulme (Ed.), *Vol. 1. The biochemistry of fruits and their products* (pp. 269–304). London: Academic Press.

Voelcker, O. J. (1935). Cotyledon colour in kola. *Tropical Agriculture, 12*, 231–234.

Willis, W. D., & Coggeshall, R. E. (1991). *Sensory mechanisms of the spinal cord.* New York: Plenum Press.

Woolf, C. J., & Chong, M. S. (1993). Preemptive analgesia—Treating postoperative pain by preventing the establishment of central sensitization. *Anesthesia and Analgesia, 77*, 1–18.

Woolf, C. J., & King, A. E. (1989). Subthresholdcomponents of the cutaneous mechanoreceptive fields of dorsal horn neurons in the rat lumbar spinal cord. *Journal of Neurophysiology, 62*, 907–916.

Woolf, C. J., Shortland, P., Reynolds, M. L., Ridings, J., Doubell, T. P., & Coggeshall, R. E. (1995). Central regenerative sprouting: The reorganization of the central terminals of myelinated primary afferents in the rat dorsal horn following peripheral nerve section or crush. *The Journal of Comparative Neurology, 360*, 121–134.

Yoon, M. H., Choi, J. I., & Jeong, S. W. (2003). Antinociception of intrathecal cholinesterase inhibitors and cholinergic receptors in rats. *Acta Anaesthesiologica Scandinavica, 47*, 1079–1084.

Chapter 34

Analgesic effects of Ephedra herb and ephedrine alkaloids-free Ephedra herb extract (EFE)

Sumiko Hyuga[a], Shunsuke Nakamori[b], Yoshiaki Amakura[c], Masashi Hyuga[d], Nahoko Uchiyama[e], Yoshinori Kobayashi[b], Takashi Hakamatsuka[e], Yukihiro Goda[g], Hiroshi Odaguchi[h], and Toshihiko Hanawa[f]

[a]Department of Clinical Research, Oriental Medicine Research Center of Kitasato University, Tokyo, Japan, [b]Department of Pharmacognosy, School of Pharmacy, Kitasato University, Tokyo, Japan, [c]Department of Pharmacognosy, College of Pharmaceutical Sciences, Matsuyama University, Matsuyama, Ehime, Japan, [d]Division of Biological Chemistry and Biologicals, National Institute of Health Sciences, Kawasaki, Kanagawa, Japan, [e]Division of Pharmacognosy, Phytochemisry, and Narcotics, National Institute of Health Sciences, Kawasaki, Kanagawa, Japan, [f]Department of Kampo Medicine, Oriental Medicine Research Center of Kitasato University, Tokyo, Japan, [g]National Institute of Health Sciences, Kawasaki, Kanagawa, Japan, [h]Oriental Medicine Research Center of Kitasato University, Tokyo, Japan

Abbreviations

EFE	ephedrine alkaloids-free Ephedra Herb extract
EGFR	epidermal growth factor receptor
EGFR-TKI	EGFR tyrosine kinase inhibitors
EHE	Ephedra Herb extract
EHE-Ts	EHE purchased from Tsumua & Co
EMCT	Ephedra Herb Macromolecule Condensed-Tannin
Eph	ephedrine
HGF	Hepatocyte growth factor
NSAIDs	nonsteroidal antiinflammatory agents
NSCLC	nonsmall cell lung cancer
PGE2	prostaglandin E2
Pse	pseudoephedrine
PSN	peripheral sensory nerves
TRPA1	transient receptor potential ankyrin 1
TRPM8	transient receptor potential melastatin1
TRPV-1	transient receptor potential vanilloid 1

Introduction

Japanese traditional medicine, Kampo medicine, has developed into Japan's unique medical system based on the medical theory and treatment technology that was derived from ancient China. The word Kampo (漢方) is derived of two words, Kam (漢) referring to one of the Chinese dynasties, the Han Dynasty, and Po (方) meaning therapeutic strategy. Kampo medicines are combined crude drugs defined by their synergistic effects and are named and used as a separate unit for treatment (Hijikata, 2006; Hyuga, Odaguchi, & Hanawa, 2019).

In 1976, Kampo extract products were incorporated into the national health insurance system in Japan. These are composed of Kampo formulae that have been used in Japan between the late 1800s and mid-1900s and are included in most modern reference texts (The Editing Committee for Dictionary of Kampo Medicine, 2020). Currently, the national health insurance covers 148 kinds of Kampo extract products. More than 80% of medical doctors in Japan administer them, alone or in combination with conventional Western medications.

Kampo medicines are used to treat various types of pain such as lumbago, stiff shoulder, arthralgia, rheumatoid, neuralgia, headache, menstrual, psychogenic, and others (Hanawa, 2020; Hijikata, 2006). Historically, a Kampo medicine,

Treatments, Mechanisms, and Adverse Reactions of Anesthetics and Analgesics. https://doi.org/10.1016/B978-0-12-820237-1.00034-X
Copyright © 2022 Elsevier Inc. All rights reserved.

"Tsusensan," was used by Dr. Seishu Hanaoka, a Japanese medical practitioner who successfully resected breast cancer under general anesthesia using Tsusensan in 1804 (Hijikata, 2006).

Kampo medicines containing Ephedra Herb, such as kakkonto, makyoyokukanto, maoto, eppikajutsuto, and yokuininto (http://mpdb.nibiohn.go.jp/stork/), are used to treat pain like lumbago, stiff shoulder, arthralgia, and rheumatoid (Hanawa, 2020), and Ephedra Herb has reported an analgesic effect (Hyuga et al., 2016; Jong-Chol, 2005; Nakamori et al., 2017, 2019).

Ephedra herb and ephedrine alkaloids

Ephedra Herb is defined as a dried stem of *Ephedra sinica* Stapf, *Ephedra intermedia* Schrenk et C. A. Meyer, or *Ephedra equisetina* Bunge (Ephedraceae) that contains more than 0.7% total alkaloids (ephedrine and pseudoephedrine) (*"The Japanese Pharmacopoeia. 17th edition,"* 2016). The herb is widely known for its diaphoretic, antipyretic, antitussive, and antiinflammatory actions (Harada, 1980).

In 1885, the principal component in Ephedra Herb was found and was named as ephedrine by Professor Nagayoshi Nagai who is known as the founder of Japanese modern pharmaceutical science (Nagai, 1892). This marked the beginning of scientific research in Kampo medicines. Subsequently, other alkaloids, such as pseudoephedrine, methylephedrine, methylpseudoephedrine, norephedrine, and norpseudoephedrine, were isolated. Ephedrine and pseudoephedrine are present in high content in Ephedra Herb and play a critical role in its pharmacological actions (Harada, 1980). Ephedrine alkaloids, whose structures are similar to that of adrenaline, stimulate adrenaline receptors and result in an expansion of the bronchi and elimination of nasal mucosal hyperemia. Furthermore, they have been reported to inhibit prostaglandin E2 biosynthesis (Kasahara, Hikino, Tsurufuji, Watanabe, & Ohuchi, 1985), and the antiinflammatory effect of Ephedra Herb has been thought to be attributable to them (Hikino, Konno, Takata, & Tamada, 1980).

However, it is known that ephedrine alkaloids present in Ephedra Herb induce hypertension, palpitations, insomnia, dysuria, and other side effects. Consequently, Kampo medicines containing Ephedra Herb should be administered cautiously to patients with circulatory impairment, hypertension, or renal impairment; who are physically fragile; or elderly. The U.S. Food and Drug Administration has prohibited the sale of ephedrine alkaloid-containing products, such as Ephedra Herb, because of multiple deaths reported due to excessive intake or inappropriate use (Ling, 2004). Furthermore, the World Anti-Doping Agency has designated ephedrine, pseudoephedrine, and methylephedrine as doping prohibited drugs; therefore, athletes should take Kampo medicines containing Ephedra Herb prudently (Chih-Wei, Szu-Yun, Guan-Qian, & Mei-Chich, 2018; Ros, Pelders, & De Smet, 1999).

Novel active ingredients, herbacetin-glycosides, in Ephedra herb and the analgesic effect of herbacetin, an active metabolite of herbacetin-glycosides

We have found a novel pharmacological action of Ephedra Herb that impaired hepatocyte growth factor (HGF)-induced cancer cell motility and growth by suppressing the phosphorylation of HGF receptor (c-Met) (Hyuga, Shiraishi, Hyuga, Goda, & Hanawa, 2011). The c-Met inhibitory activity of Ephedra Herb was independent of ephedrine alkaloids. The novel constituents discovered in Ephedra Herb were herbacetin 7-O-neohesperidoside and herbacetin 7-O-glucoside (Amakura et al., 2013). Herbacetin, the aglycone of the herbacetin-glycosides, inhibits HGF-induced cell migration and phosphorylation of c-Met (Hyuga et al., 2013), but the herbacetin-glycosides had no effects. These glycosides are hydrolyzed to herbacetin by the intestinal flora after oral administration. Therefore, herbacetin was an active metabolite of herbacetin-glycosides in Ephedra Herb.

Herbacetin inhibits c-Met tyrosine kinase along with Tropomyosin-Receptor-Kinase A (Trk A), Aurora kinase, and Fms-like tyrosine Kinase 3, and is a multikinase inhibitor (Hanawa et al., 2018).

Trk A is a nerve growth factor (NGF) receptor. NGF is an inflammatory mediator and NGF-Trk A signaling is involved in inducing pain and itch (Indo, 2010). Herbacetin inhibited the tyrosine kinase activity of Trk A in a concentration-dependent manner, and IC50 was 189 nM (Fig. 1A). The NGF-induced phosphorylation of TrkA in rat pheochromocytoma cell line, PC12 cells, was inhibited in a concentration-dependent manner of herbacetin (Fig. 1B). The expression of TrkA was downregulated 30 min after treatment of NGF but was recovered by herbacetin (Fig. 1B). The downregulation of TrkA was caused by NGF-induced phosphorylation, and the recovery of TrkA expression was caused by the inhibition of phosphorylation by herbacetin. Furthermore, NGF-induced neurite outgrowth of PC12 cells was suppressed in a concentration-dependent manner of herbacetin (Fig. 1B and C). Next, we investigated the analgesic effect of herbacetin by formalin test. Formalin-induced nociceptive behavior shows an early phase (first phase) and a late phase (second phase). Central analgesics, such as morphine, inhibit the pain associated with the first and second phases. However, nonsteroidal antiinflammatory agents (NSAIDs) such as aspirin and diclofenac are ineffective against the first phase, but suppress the pain associated with the second phase (Tjolsen, Berge, Hunskaar, Rosland, & Hole, 1992). Formalin (2.5%) was inoculated into the plantar surfaces of mice's left hind paws 90 min after intraperitoneal injection of 50–150 mg/kg of herbacetin and licking behavior was recorded. In this study, formalin-induced biphasic behavior-related pain was defined as the first phase (0–5 min) and the second phase (15–45 min). Herbacetin suppressed only the second phase of formalin-induced pain in a dose-dependent manner (Fig. 1D), suggesting its role in relieving inflammatory pain.

To date, the analgesic action of Ephedra Herb has been traditionally believed to be due to the antiinflammatory effect of ephedrine alkaloids, but possibly herbacetin-glycosides in Ephedra Herb also contribute to the analgesic action.

Development of ephedrine alkaloids-free Ephedra Herb extract (EFE)

Herbacetin has c-Met inhibitory activity and analgesic effect, and Ephedra Herb has ephedrine alkaloids-independent pharmacological actions. However, the c-Met inhibitory activity of Ephedra Herb could not be explained by herbacetin-

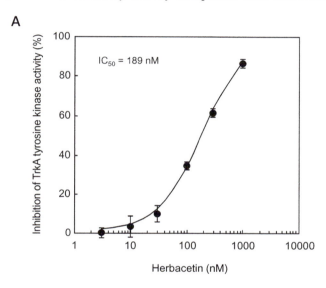

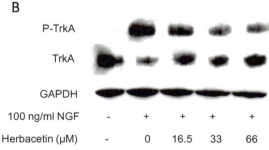

FIG. 1 Analgesic effect of herbacetin through inhibition of NGF-TrkA signaling. (A) Inhibitory effects of Trk A kinase activity of herbacetin by an in vitro Kinase Assay. Trk A tyrosine kinase activity was measured with an in vitro kinase assay. Briefly, a 25-μL mixture of 320 ng/mL recombinant TrkA kinase domain (Carna Biosciences Inc.), 1 μM CSKtide, and 75 μM ATP, 5 mM MgCl$_2$ and a buffer solution (20 mM HEPES, 0.01% Triton X-100, 2 mM DTT, pH 7.5) was incubated with herbacetin at room temperature for 60 min. The kinase reactions were terminated by the addition of a 60-μL termination buffer (QuickScout Screening Assist MSA; Carna Biosciences Inc.). The phosphorylated peptide was separated from the substrate peptide and quantified using a LabChip 3000 system (Caliper Life Sciences). The IC$_{50}$ was calculated using a four-parameter logistic model (Prism 5.0, GraphPad software). (B) The inhibitory effects of herbacetin on NGF-induced phosphorylation of TrkA in PC12 cells. PC12 cells (4 × 10^6 cells) were incubated for 30 min at 37°C in 4 mL RPMI-1640 containing: 50 ng/mL NGF (Recombinant human β-NGF, Globa bio-techne. Com., USA) and 0, 16.5, 33, 66 μM herbacetin. After the cells were washed three times with 4 mL cold phosphate-buffered saline without Ca and Mg [PBS(−)](Nacalai Tesque, Kyoto, Japan), they were treated with 300 μL Complete Lysis-M containing phosphatase inhibitor (Roche Diagnostics Co., Indianapolis, IN, USA) for 5 min in an ice bath. The lysates were collected and were analyzed by western blotting using an anti-p-TrkA monoclonal antibody (CST#4621), anti-Trk A poyclonal Ab (CST#2505), and ant-GAPDH polyclonal antibody (CST#25778).

(Continued)

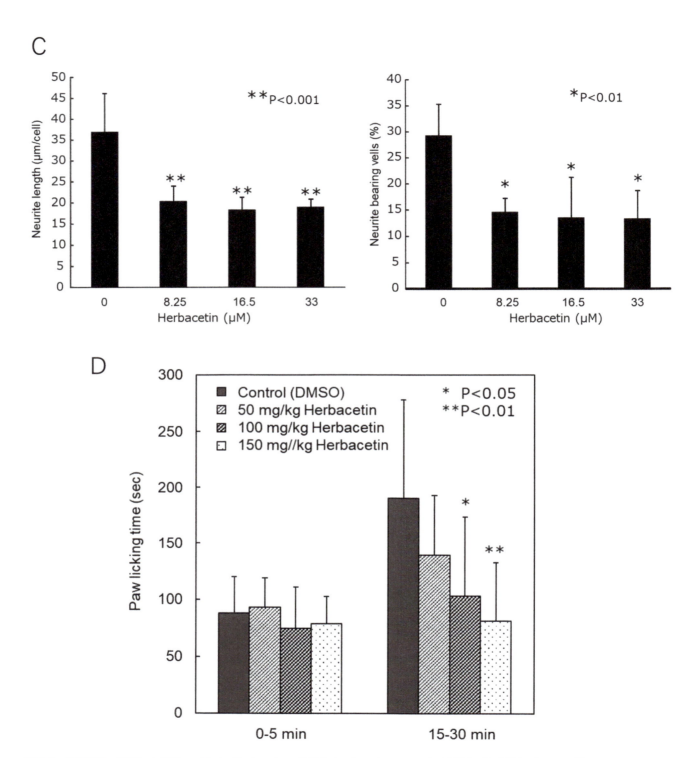

FIG. 1—CONT'D (C) The inhibitory effects of herbacetin on NGF-induced neurite outgrowth of PC12 cells. PC12 cells (1×10^5 cells) were incubated for 48 h at 37°C on 24-well type I collagen-coated cell culture plate (CORNING) in 400 μL 1% house serum-0.5% fetal calf serum-RPMI 1640 containing 10 ng/mL NGF and 0, 8.25, 16.5, 33 μM herbacetin. Moreover, the PC12 cells were photographed under phase-contrast microscopy. The neurite length and number of neurite-bearing cells on more than100 cells per one visual field were measured using Image analysis software, Image J, and each value was calculated from the average value of six fields of view. Data were analyzed by 1-way ANOVA. Significant differences between the control and treatment groups were determined by Dunnett's test using GraphPad Prism 7.03 (MDF Co. Ltd., Tokyo, Japan). *$P < 0.01$, **$P < 0.001$ vs control. (D): Analgesic effect of herbacetin on formalin-induced pain. ICR mice were injected intraperitoneally with 50, 100, 150 mg/kg of herbacetin. After 90 min, the mice were injected with 20 μL of 2.5% formalin into the plantar surface of the left hind paw. The amount of time that each animal spent licking the injection paw was recorded for 30 min in two phases, known as the first (0–5 min) and second (15–30 min) phases. Data were analyzed by 1-way ANOVA. Statistical significance was determined by Dunnett's test. *$P < 0.05$, **$P < 0.01$ vs control.

glycosides, because they have no effect on c-Met. Therefore, the nonalkaloid fraction of Ephedra Herb was presumed to have other ingredients with c-Met inhibitory activity.

It was previously thought that the removal of ephedrine alkaloids from Ephedra Herb would lead to a loss of its pharmacological actions, but our findings contradicted it. To reduce the adverse effects of Ephedra Herb, we utilized cation-exchange column chromatography to eliminate ephedrine alkaloids from Ephedra Herb extract (EHE), resulting in ephedrine alkaloids-free Ephedra Herb extract (EFE). The content of ephedrine alkaloids in EFE was less than 0.05 ppm, and the contents of 6-methoxykynurenic acid, 6-hydroxykynurenic acid, and *trans*-cinnamic acid in EFE were decreased as compared to that in EHE. 6-methoxykynurenic acid and 6-hydroxykynurenic acid might bind to the cation-exchange resin, as they contain a nitrogenous base in their structure. The *trans*-cinnamic acid might be bound to the resin possibly by the π-π interaction (Hyuga et al., 2016; Oshima et al., 2016).

EFE suppressed the HGF-induced motility of cancer cells by inhibiting the phosphorylation of c-Met at the EHE level. Moreover, EFE had analgesic and anti- influenza activities at the same level as EHE (Hyuga et al., 2016). However, EFE possibly loses the adrenergic receptor-mediated pharmacological actions such as antitussive action and amelioration of nasal obstruction.

Adverse effects of EHE and safety of EFE

In this chapter, we have described about the adverse effects of EHE such as excitation, insomnia, and arrhythmias, and safety of EFE clarified by the in vivo study and physician-initiated clinical trial.

The excitatory actions of EHE and EFE were examined in open-field and forced swim tests. The central stimulants such as caffeine induce a pronounced increase in locomotor activity in the open-field test and decrease immobility time in the forced swim test. The oral administration of EHE increased locomotor activity and shortened immobility time, whereas the administration of EFE had no effect. These results indicate that the excitatory action through central stimulation is present in EHE, but lacking in EFE (Takemoto et al., 2018).

The effects of EHE and EFE on sleep were investigated in the pentobarbital sleep test. EHE decreased sleep duration in a dose-dependent manner, whereas EFE showed no effect. The inhibitory action of EHE was attributed to the adrenergic action of ephedrine alkaloids, and the awakening effect of EHE was eliminated by the removal of ephedrine alkaloids (Takemoto et al., 2018).

The effect of EHE and EFE on cardiac electrophysiology was evaluated using electrocardiograms. The atrial fibrillation (i.e., irregular heart rhythm, absence of P waves, and appearance of f waves) was observed in the EHE-administered mice, but not EFE. These results suggested that this atrial fibrillation was induced by the stimulation of adrenaline β1 receptors by ephedrine alkaloids in EHE (Takemoto et al., 2018).

Our data indicated that EFE is free from adverse effects such as excitation, insomnia, and arrhythmias, and is a promising new botanical drug.

Next, the safety of EFE was assessed in comparison with EHE and water after repeated oral administration for 2 weeks. The dosages of EFE and EHE were converted to 50-fold of the human maximum dose of EHE. No significant differences were noted between the groups (Hyuga et al., 2016).

We conducted a clinical trial, a single-institution, double-blind, randomized, two-drug, two-stage, crossover comparative study, to verify the noninferiority of EFE's safety compared to that of EHE in humans. There were no significant differences in the incidences of adverse events, but the total number of adverse events was eight with EHE administration and three with EFE. Therefore, it was suggested that EFE is safer than EHE (Odaguchi et al., 2018).

Analgesic effects of EHE and EFE on formalin-induced pain

We examined the analgesic effect of EHE and EFE on formalin-induced pain. EHE was prepared by extracting Ephedra Herb, and EFE was prepared from EHE using cation-exchange column chromatography to eliminate ephedrine alkaloids (Oshima et al., 2016). Mice were orally administered water, 350 or 700 mg/kg EHE, or EFE for 3 days. On the third day, paw licking was induced in the mice by intraplantar injection of 2.5% formalin 6 h after administration of water, EHE, or EFE and the behavior was recorded for 30 min in the first phase (0–5 min) and the second phase (15–30 min). EHE and EFE showed no effects during the first phase but reduced paw-licking time in a dose-dependent manner during the second phase (Fig. 2A) (Hyuga et al., 2016). These results suggested that repeated oral administration of EHE or EFE had the analgesic effect against inflammatory pain.

The analgesic effect of single oral administrations of EHE was investigated at different time points from 0 to 24 h after administration. EHE purchased from Tsumura & Co (EHE-Ts) was used. Both first-phase and second-phase pain were transiently suppressed at 30 min after oral administration of 700 mg/kg EHE-Ts, but only second-phase pain was persistently suppressed 4–6 h after administration (Fig. 2B). These results showed that the single oral administration of EHE-Ts had an analgesic effect, which was biphasic, rapid transient and slow continuous (Nakamori et al., 2019).

Analgesic effects of EHE, EFE, ephedrine, and pseudoephedrine on formalin-induced pain

In this chapter, we have described what ingredients contribute to the biphasic analgesia of EHE.

We compared the analgesic effects of EFE and EHE at 30 min or 6 h after single oral administration. EHE was prepared by extracting Ephedra Herb, and EFE was prepared from EHE (Oshima et al., 2016). Thirty minutes after administration, EHE relieved both first-phase and second-phase pain in a dose-dependent manner, but EFE reduced only second-phase pain (Fig. 3A). Therefore, the analgesic effect of EHE at 30 min after oral administration against first-phase pain might be caused by ephedrine (Eph) and pseudoephedrine (Pse). Six hours after administration, both EHE and EFE reduced only second-phase pain in a dose-dependent manner (Fig. 3B) (Nakamori et al., 2019). Accordingly, it was clarified that EHE had biphasic analgesic effects, whereas EFE only had an analgesic effect against inflammatory pain 6 h after administration.

Then, we examined the contribution of Eph and Pse to the analgesic effects of EHE at 30 min and 6 h after administration. The dosages of Eph and Pse were based on their content in 700 mg/kg doses of EHE. Thirty minutes after administration, EHE and Eph significantly reduced both first- and second-phase pain at the same level, and EFE and Pse showed a decreasing tendency of only the second-phase pain (Fig. 3C) (Nakamori et al., 2019). These results suggested that the suppression of the first-phase pain at 30 min after administration of EHE was caused by Eph. The concentration of Eph in rat serum was reported to peak at 30 min after oral administration of EHE (Wei et al., 2014), and this report supports our results. We predict that Eph may express the analgesic action through a stimulating α2-adrenergic receptor in the central nervous system because the activation of α2-adrenergic receptor is responsible for sedation, analgesia, and sympatholytic effects (Giovannitti, et al., 2015). Six hours after administration, none of them had any effect on the first-phase pain, and Eph had almost no effect on the second-phase pain. EHE and EFE significantly suppressed second-phase pain, and Pse

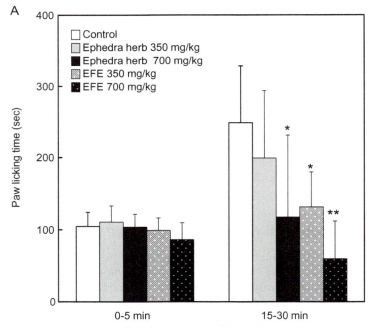

FIG. 2 Effects of EFE and Ephedra Herb extract on formalin-induced pain. (A) Effects of a repeat oral administration of EFE or EHE on formalin-induced pain. ICR mice were treated orally with water, 350 mg/kg EFE, 700 mg/kg EFE, or Ephedra Herb extract for 3 days. On the third day of treatment, formalin tests were performed 6 h after drug or placebo administration. The amount of time that each animal spent licking the injection paw was recorded for 30 min in two phases, known as the first (0–5 min) and second (15–30 min) phases. Statistical significance was determined by Dunnett's test. *$P < 0.05$ or **$P < 0.01$ vs control.

(Continued)

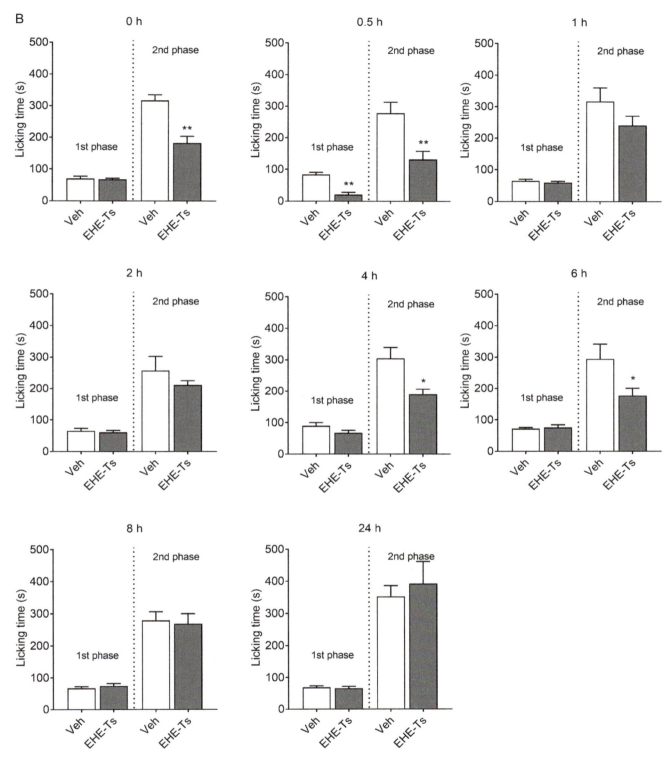

FIG. 2—CONT'D (B) The time-dependent changes of analgesic effect of EHE-Ts. ddY mice were orally administrated 700 mg/kg of EHE-Ts or water (Veh). After 0, 0.5, 1, 2, 4, 6, 8, and 24 h, the mice were injected with 10 μL of saline containing 2.5% formalin into the plantar surface of the left hind paw. Licking behaviors were observed for 45 min. First-phase paw-licking times were recorded 0–5 min after formalin injections, and second-phase paw-licking times were recorded 10–45 min after formalin injections. The data for each condition represents the means ± standard error. *(Panel A Reprinted from Hyuga, S., Hyuga, M., Oshima, N., Maruyama, T., Kamakura, H., Yamashita, T., et al. (2016). Ephedrine alkaloids-free Ephedra herb extract: A safer alternative to ephedra with comparable analgesic, anticancer, and anti-influenza activities.* Journal of Natural Medicines, 70*(3), 571–583. doi:10.1007/s11418-016-0979-z. Panel B Reproduced in part with permission from Nakamori, S., Takahashi, J., Hyuga, S., Yang, J., Takemoto, H., Maruyama, T., et al. (2019). Analgesic effects of Ephedra herb extract, ephedrine alkaloids-free Ephedra herb extract, ephedrine, and pseudoephedrine on formalin-induced pain.* Biological & Pharmaceutical Bulletin, 42*(9), 1538–1544. https://doi.org/10.1248/bpb.b19-00260 Copyright 2019 The Pharmaceutical Society of Japan).*

392 PART | IV Novel and nonpharmacological aspects and treatments

showed a decreasing tendency of second-phase pain (Fig. 3D). These results suggest that the suppression of second-phase pain by EHE is attributable to nonalkaloid constituents, although Pse may have partially contributed to the relief of second-phase pain.

Recently, we found high-molecular-mass condensed tannin, Ephedra Herb Macromolecule Condensed-Tannin (EMCT), in the nonalkaloid fraction of EHE and EFE. The weight-average molecular weight of EMCT was >45,000

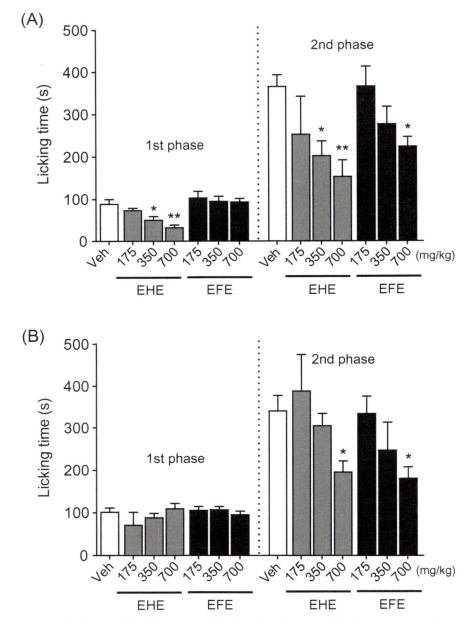

FIG. 3 Comparison of the analgesic effects of EHE, EFE, and ephedrine alkaloids at 30 min and 6 h after oral administration. (A and B) Comparison of the analgesic effects of EHE and EFE at 30 min or 6 h after oral administration. Mice were orally administrated 175–700 mg/kg of EHE and EFE. After 30 min (A) or 6 h (B), 10 μL of 2.5% formalin was injected into the plantar surfaces of their left hind paws. Licking behaviors were observed for 45 min. The data for each condition represent the means ± standard error. Statistical significance was determined with Dunnett's test; *$P < 0.05$ and **$P < 0.01$ vs the water group (Veh). (C and B) Comparison of the analgesic effects and EHE and EFE at 6 h after oral administration. Mice were orally administrated 700 mg/kg of EHE and EFE, 30 mg/kg of Eph, or 12 mg/kg of Pse. After 30 min

(Continued)

and the estimated content of EMCT in EHE and EFE was approximately 34% and 40.1%, respectively. Six hours after oral administration, EMCT significantly suppressed second-phase pain at the same level as EFE, but had no effect on first-phase pain (Yoshimura et al., 2020). Therefore, it is possible that the analgesic effect of EHE and EFE against inflammatory pain 6 h after administration was mainly caused by EMCT.

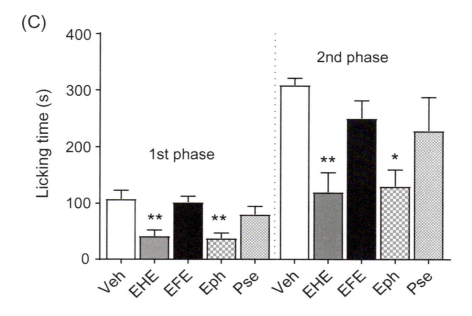

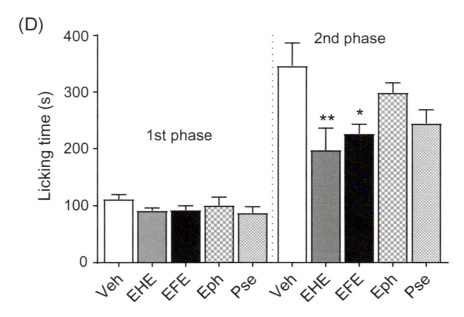

FIG. 3—CONT'D (C) or 6 h (B), 10 μL of 2.5% formalin was injected into the plantar surfaces of their left hind paws. Licking behaviors were observed for 45 min (0–5 min: first-phase pain; 10–45 min: second-phase pain). The data for each condition represent the means ± standard error. *(Panel A and B Reproduced in part with permission from Nakamori, S., Takahashi, J., Hyuga, S., Yang, J., Takemoto, H., Maruyama, T., et al. (2019). Analgesic effects of Ephedra herb extract, ephedrine alkaloids-free Ephedra herb extract, ephedrine, and pseudoephedrine on formalin-induced pain.* Biological & Pharmaceutical Bulletin, *42(9), 1538–1544. https://doi.org/10.1248/bpb.b19-00260 Copyright 2019 The Pharmaceutical Society of Japan.)*

However, herbacetin-glycosides, a nonalkaloid constituent, have a role in the analgesic effect of EHE and EFE against inflammatory pain, but are not the main active ingredients, as their content in EFE is 0.1% (Oshima et al., 2016).

Analgesic effect of EFE on pain in arthritis model mouse

Pain caused by age-related arthritis reduces the quality of life in the elderly and is one of the main causes of long-term care. Therefore, there is a social demand for the development of novel analgesics with a low physical burden on the elderly. EFE is considered a potential candidate, as it has almost no adverse effects such as hypertension, palpitations, insomnia, and dysuria. Therefore, we evaluated the analgesic effects of EFE against arthralgia in complete Freund's adjuvant-induced arthritis mouse model using the von Frey test (Chaplan, Bach, Pogrel, Chung, & Yaksh, 1994). EFE (175–700 mg/kg) significantly reduced the arthralgia in a dose-dependent manner 6 h after oral administration. The duration of analgesic effect with the oral administration of 700 mg/kg EFE once daily was 2 h, and from 6 to 8 h after administration, while the effect with the oral administration of 700 mg/kg EFE twice daily at 4 h intervals persisted 6 h and from 6 to 12 h after first administration (Fig. 4) (Nakamori et al., in preparation). The pain of arthritis becomes stronger at wake-up time, but the effects of almost all analgesics taken before bed do not continue until morning. If the arthritis patients take EFE 4 h before and just before bedtime, they may be able to control the wake-up pain (Fig. 4).

Reduction of capsaicin-induced pain via transient receptor potential vanilloid 1 (TRPV-1) by EHE

TRPV1 is a receptor of capsaicin, the pungent ingredient of the hot chili pepper (Frias & Merighi, 2016), and is a family of nonselective cation channels. It is activated by capsaicin, temperature (with threshold ~43°C), protons, and bacterial toxins, and is associated with thermoception and nociception. TRPV1, as a pain and heat sensor, is expressed at high levels in C fibers and is implicated in neurogenic inflammation and neuropathic pain. We investigated the involvement of TRPV1 in the analgesic effect of EHE. EHE significantly increased the intracellular Ca^{2+} concentration in stable mouse TRPV1-expressing transfected, mTRPV1/Flp-In293 cells, indicating that EHE directly activates mTRPV1. However, Eph had no effect on TRPV1. In addition, 30 min after intradermal injection of EHE into the plantar surface of the left hind paw of mice, the capsaicin-induced paw licking was suppressed (Fig. 5), suggesting that EHE elicits analgesic action by affecting TRPV1 function on peripheral sensory nerves (PSN) (Nakamori et al., 2017).

Intradermal injection of capsaicin contributes to the development of mechanical hyperalgesia through the activation of TRPV1 on sensory neurons. The activation of TRPV1 by capsaicin induces the releases of neuropeptides such as substance P, neurokinin A, and calcitonin-gene-related-peptide from the peripheral terminals of PSNs contributing to the onset of pain and inflammation. Thereafter, the neuropeptides' content in peripheral terminals is depleted, and the loss of membrane potential, the block of neurotrophic factor axonal trafficking, and a reversible retraction of cutaneous terminals are induced. As a result, the analgesic effect of capsaicin is developed (Frias & Merighi, 2016). Thus, the intradermal injection of EHE

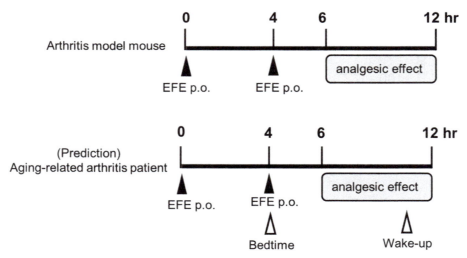

FIG. 4 Time until the onset of analgesic effect and duration of the analgesic effect after repeated administration of EFE.

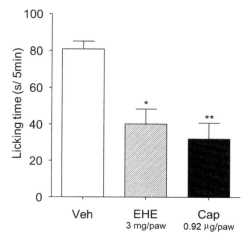

FIG. 5 Suppression of capsaicin-induced pain by i.d. administration of EHE. Mice were injected with 10 μL of vehicle (DMSO:Tween-80:physiological saline = 1:1:8) containing 3 mg/paw EHE or 0.92 μg/paw capsaicin (Cap) into the plantar surface of the left hind paw. After 30 min, 0.18 μg/paw capsaicin was injected into the same area. Licking behaviors were observed for 5 min. Data represent the mean ± standard error of 6 mice. Statistical significance was determined with Dunnett's test; *$P < 0.01$, and **$P < 0.001$ vs vehicle group (Veh). *(Modified from Nakamori, S., Takahashi, J., Hyuga, S., Tanaka-Kagawa, T., Jinno, H., Hyuga, M., et al. (2017). Ephedra herb extract activates/desensitizes transient receptor potential vanilloid 1 and reduces capsaicin-induced pain.* Journal of Natural Medicines, 71(1), 105–113. https://doi.org/10.1007/s11418-016-1034-9).

evokes similar analgesic action as that of capsaicin. We recently found that EFE increased the intracellular Ca^{2+} concentration in mTRPV1/Flp-In293 cells, suggesting that no-alkaloid-ingredients in EHE are involved in capsaicin-like analgesia.

Application to other areas

Recently, maoto, Kampo medicine containing Ephedra Herb (http://mpdb.nibiohn.go.jp/stork/), was reported to relieve the bone and joint pain associated with cancer treatment (Hayashi, 2014; Mantani & Oka, 2017).

This analgesic effect of maoto may be caused by Ephedra Herb. While, we previously reported that maoto suppresses cancer metastasis by inhibiting cancer motility (Hyuga, Hyuga, Yamagata, Yamagata, & Hanawa, 2004; Hyuga et al., 2007). And the prevention of cancer metastasis by maoto is due to inhibition of c-Met by Ephedra Herb (Hyuga et al., 2011). Therefore, these evidences suggested that maoto and Ephedra herb are useful in supporting cancer therapy.

We recently found the novel anticancer effect of Ephedra Herb on nonsmall cell lung cancer (NSCLC). The epidermal growth factor receptor (EGFR) tyrosine kinase inhibitors (EGFR-TKIs) are effective for NSCLC harboring EGFR-activating mutation, but the NSCLC acquires resistance to these treatments after a few years clinically. One of the causes of the resistance against EGFR-TKIs is a high level of c-Met amplification or c-Met protein overexpression/hyperactivation. Then, we examined the effect of EHE on EGFR-TKI resistant c-Met-overexpressed nonsmall cell lung cancer cell, H1993 cells. The combination of EHE and erlotinib, EGFR-TKI, was effective in inhibiting the growth of H1993 cells and its xenograft tumor, indicating that EHE contributed to the recovery from resistance to erlotinib of the cells. Moreover, EHE inhibited the phosphorylation of c-Met, and down-regulated the expression of c-Met through the promotion of endocytosis and degradation of c-Met and of EGFR and pEGFR. Similarly, EFE and EMCT promoted the downregulation of these receptors (Hyuga et al., 2020; Nishimura et al., 2016) (Fig. 6). Accordingly, the combination of EHE or EFE and EGFR-TKI may be useful for the treatment of EGFR-TKI resistant c-Met-overexpressed NSCLC patients.

Maoto is used to treat patients with the early stage of influenza virus infection in Japan (http://mpdb.nibiohn.go.jp/stork/). Ephedra Herb, EFE, and EMCT were reported to have an inhibitory action against infection with the influenza virus (Hyuga et al., 2016; Mantani, Andoh, Kawamata, Terasawa, & Ochiai, 1999; Yoshimura et al., 2020).

Therefore, Ephedra Herb and EFE may be excellent drugs to treat cancer and influenza virus infection and suppress the pain induced by these diseases. These crude drug extracts have various medicinal properties. These properties may be due to EMCT, a novel ingredient, in Ephedra Herb (Yoshimura et al., 2020). It regulates receptors on the cell surface, as EMCT being a tannin, has a high affinity for protein.

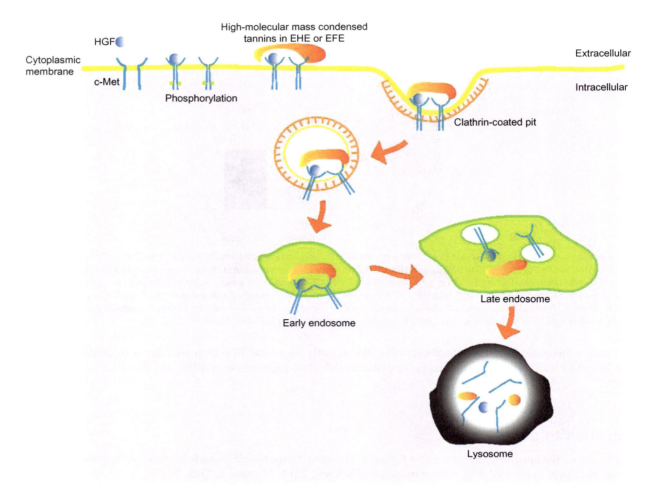

FIG. 6 The presumed mechanism of downregulation of c-Met by EHE and EFE. *(Reprinted from Hyuga, S., Hyuga, M., Amakura, Y., Yang, J., Mori, E., Hakamatsuka, T., et al. (2020). Effect of ephedra herb on erlotinib resistance in c-Met-overexpressing nonsmall-cell lung cancer cell line, H1993, through promotion of endocytosis and degradation of c-Met.* Evidence-based Complementary and Alternative Medicine, 2020, 7184114–7184129 pages.)

Conclusions and perspectives

The analgesic action of Ephedra Herb has been believed to be caused by ephedrine alkaloids, but we clarified that the analgesic effect remains even after the removal of ephedrine alkaloids from the Herb, and that it contains herbacetin-glycosides and EMCT associated with analgesic effect. The inhibition of NGF-TrkA signaling by herbacetin and the desensitization of TRPV-1 is possibly the molecular mechanisms. However, the detailed molecular mechanism of Ephedra Herb is still unknown. We think that EMCT plays an important role in the analgesia of Ephedra Herb, because oral administration of EMCT reduced both the formalin-induced second-phase pain and arthralgia in arthritis model mouse at the same level as EHE and EFE. EMCT may negatively modulate the nociception. The analgesic effect appears 6 h after EHE or EFE administration is as the intestinal absorption of EMCT may be slow. EMCT may be adsorbed from the intestine by transcytosis. EHE, EFE, and EMCT strongly induced endocytosis, which is associated with the first process of transcytosis, 4–8 h after cell treatment (Hyuga et al., 2020). Therefore, the time lag of 6 h until appearance of analgesic action after oral administration of EHE or EFE may arise in the process of transcytosis of EMCT.

In Japan, Ephedra Herb is classified into the list of raw materials exclusively used as pharmaceuticals and is approved as a pharmaceutical in Japan. Meanwhile, EFE is not approved as a pharmaceutical, because EFE is made by a modification to remove ephedrine-alkaloids from Ephedra Herb. Moreover, EFE is classified into the list of raw materials exclusively used as pharmaceuticals, since the raw material for EFE is Ephedra Herb. Therefore, EFE cannot be commercialized as food. We now try to obtain licensing approval for the therapeutic use of EFE in Japan. However, EMCT is unsuitable for the development of the drug, because the high-molecular-mass condensed tannins are unstable. Assumedly, EMCT is thought to be stable in EHE or EFE, because coexisting ingredients may protect it.

Other agents of interest

In this chapter, we have described about some Kampo medicines used for treating pain in Japan and their evidences.

1. Chemotherapy-induced peripheral neuropathy: Goshajinkigan (牛車腎気丸) has been reported to inhibit the progression of paclitaxel/carboplatin-induced peripheral neuropathy (Kaku et al., 2012), and the concomitant administration of docetaxel and goshajinkigan has been reported to prevent neuropathy in breast cancer patients treated with doceaxel (Abe et al., 2013). It was reported that the molecular mechanism of goshajinkigan was by suppressing the overexpression of transient receptor potential ankyrin 1 (TRPA1) and transient receptor potential melastatin 8 (TRPM8) (Kato et al., 2014; Mizuno et al., 2014).

   Furthermore, Motoo, Tomita, and Fujita (2020) reported that ninjin'yoeit (人参湯) reduces the incidence of oxaliplatin-induced cumulative peripheral neuropathy.

2. Chemotherapy-induced oral mucositis: Hangeshashinto (半夏心湯) has been demonstrated a safe and effective method for chemotherapy-induced oral mucositis (Kono et al., 2010; Matsuda et al., 2015; Taira, Fujiwara, Fukuhara, Koyama, & Takeuchi, 2020). Chemotherapy-induced oral mucositis is characterized by painful inflammation. Prostaglandin E2 (PGE2) is produced in the oral cavity during mucositis. Hangeshashinto and its active ingredients, shogaol, gingerol, wogonin, baicalein, baicalin, and berberine were reported to reduce PGE2 ptoduction (Kono et al., 2014). Furthermore, it was reported that gingerol and shogaol blocked Na^+ channel and inhibited the release of substance P (Hitomi et al., 2017).

3. Muscular pain: Shakuyakukanzoto (芍薬甘草湯) has been used by patients with muscle cramps and abdominal pains in Japan. Paeoniflorini in Paeoniae radix, which is a constituent crude drug of shakuyakukanzoto, was reported to a key antinociceptive compound. It increased noradrenaline release and activated α2-adrenergic receptor to modulate spinal nociceptive transmission in diabetic mice (Lee, Omiya, Yuzurihara, Kase, & Kobayashi, 2011).

4. Headache: Goshuyuto (茱萸湯) is effective for headache with chilling exposure. The efficacy of goshuyuto has been proven in randomized controlled trials (Odaguchi et al., 2006). Goreisan (五苓散) is effective for headache associated with bad weather (Ishida, 2013).

Mini-dictionary of terms

Adrenaline. A hormone secreted from the adrenal medulla and a neurotransmitter of the central nervous system.

Adrenaline was isolated for the first time from cow pancreas by Dr. Uenaka and Dr. Takamine in the USA 5 years after the discovery of ephedrine. Dr. Uenaka was one of the disciples of Professor Nagai.

c-Met. A receptor of HGF.

Expression of c-Met has been observed in a wide range of human malignancies, and HGF-mediated activation of c-Met in malignant cells stimulates migration, invasion, proliferation, and angiogenesis.

NSCLC. A type of lung cancer.

Lung cancer is the leading cause of cancer-related death in many countries around the world. NSCLC is accountable for approximately 85% of lung cancers.

Endocytosis. A process of transport of substances into the cells.

Endocytosis is a basic cellular function whereby cells internalize extracellular molecules via vesicular mechanisms.

Transcytosis. A type of transcellular transport of various macromolecules.

In epithelial cells, internalized molecules by endocytosis can be transported across the cells to the surface opposite. This process is called transcytosis.

Key facts

Key facts of formalin test

- Formalin test is widely used for evaluating analgesic effects.
- Formalin-induced nociceptive behavior shows an early phase (first phase) and a late phase (second phase).
- Pain in the first phase is thought to be caused by direct activation of type C fibers in nociceptive nerve endings, and substance P and bradykinin participate in the first phase.
- The second phase appears to be dependent on the combination of an inflammatory reaction in the peripheral tissue and functional changes in the dorsal horn of the spinal cord.

398 PART | IV Novel and nonpharmacological aspects and treatments

- Nonsteroidal antiinflammatory agents (NSAIDs) such as aspirin and diclofenac are ineffective against the first phase and suppress the pain associated with the second phase.
- Central analgesics, such as morphine, inhibit the pain associated with the first and second phases.

Key facts of Ephedra herb macromolecule condensed-tannin (EMCT)

- EMCT is a high molecular mass condensed tannin and is a primary active ingredient in EHE or EFE.
- The deduced structure of EMCT is constituted by mainly procyanidin B-type and partly procyanidin A-type, including pyrogallol- and catechol-type flavan 3-ols as extension and terminal units.
- The ratio of pyrogallol- and catechol-type is approximately 5:1, and the weight-average molecular weight based on the polystyrene standard was >45,000.
- EMCT is associated with c-Met inhibitory, analgesic, and antiinfluenza effects by EHE or EFE.
- The estimated content of EMCT in EHE and EFE was approximately 34% and 40.1%, respectively, and is higher than that of herbacetin-glycosides, whose content was 0.1%.

Summary points

- Kampo medicines containing Ephedra Herb are used to treat pain, and their analgesic action has been thought to be caused by the antiinflammatory effect of ephedrine alkaloids in Ephedra Herb.
- Ephedra Herb is known to have side effects, hypertension, palpitations, insomnia, and dysuria, to be induced by ephedrine alkaloids.
- Both the pharmacological actions and adverse effects of Ephedra Herb have been believed to be caused by ephedrine alkaloids.
- We have found that Ephedra Herb has an ephedrine alkaloids-independent pharmacological action, and developed ephedrine alkaloids-free Ephedra Herb extract (EFE) by eliminating ephedrine alkaloids from Ephedra Herb extract (EHE).
- EFE had the same analgesic effect as EHE.
- EFE may become a safe and useful analgesic drug.

References

Abe, H., Kawai, Y., Mori, T., Tomida, K., Kubota, Y., Umeda, T., et al. (2013). The Kampo medicine Goshajinkigan prevents neuropathy in breast cancer patients treated with docetaxel. *Asian Pacific Journal of Cancer Prevention*, *14*(11), 6351–6356. https://doi.org/10.7314/apjcp.2013.14.11.6351.

Amakura, Y., Yoshimura, M., Yamakami, S., Yoshida, T., Wakana, D., Hyuga, M., et al. (2013). Characterization of phenolic constituents from ephedra herb extract. *Molecules*, *18*(5), 5326–5334. https://doi.org/10.3390/molecules18055326.

Chaplan, S. R., Bach, F. W., Pogrel, J. W., Chung, J. M., & Yaksh, T. L. (1994). Quantitative assessment of tactile allodynia in the rat paw. *Journal of Neuroscience Methods*, *53*(1), 55–63. https://doi.org/10.1016/0165-0270(94)90144-9.

Chih-Wei, C., Szu-Yun, H., Guan-Qian, H., & Mei-Chich, H. (2018). Ephedra alkaloid contents of Chinese herbal formulae sold in Taiwan. *Drug Testing and Analysis*, *10*(2), 350–356. https://doi.org/10.1002/dta.2209.

Frias, B., & Merighi, A. (2016). Capsaicin, nociception and pain. *Molecules*, *21*(6). https://doi.org/10.3390/molecules21060797.

Giovannitti, J. A., Jr., Thoms, S. M., & Crawford, J. J. (2015). Alpha-2 adrenergic receptor agonists: A review of current clinical applications. *Anesthesia Progress*, *62*(1), 31–39. https://doi.org/10.2344/0003-3006-62.1.31.

Hanawa, T. (2020). Kampo. *The Japanese Journal of Clinical and Experimental Medicine*, *97*, 79–83.

Hanawa, T., Hyuga, S., Goda, Y., Hyuga, M., Amakura, Y., & Yoshimura, M. (2018). Multikinase inhibitors containing flavonoids or their glycosides as anticancer drugs, analgesics, and antipruritic drugs. In *Jpn. Kokai Tokkyo Koho, patent number 6264685*.

Harada, M. (1980). Pharmacological studies on Ephedra. *The Journal of Traditional Sino-Japanese Medicine*, *1*, 34–39.

Hayashi, A. (2014). Clinical experience with maoto: Managing the side effects of zoedronic acid hydrate—Bone pain and fever. *Pain and Kampo Medicine*, *24*, 149–151.

Hijikata, Y. (2006). Analgesic treatment with Kampo prescription. *Expert Review of Neurotherapeutics*, *6*(5), 795–802. https://doi.org/10.1586/14737175.6.5.795.

Hikino, H., Konno, C., Takata, H., & Tamada, M. (1980). Antiinflammatory principle of Ephedra herbs. *Chemical & Pharmaceutical Bulletin (Tokyo)*, *28*(10), 2900–2904. https://doi.org/10.1248/cpb.28.2900.

Hitomi, S., Ono, K., Terawaki, K., Matsumoto, C., Mizuno, K., Yamaguchi, K., et al. (2017). [6]-Gingerol and [6]-shogaol, active ingredients of the traditional Japanese medicine hangeshashinto, relief oral ulcerative mucositis-induced pain via action on Na(+) channels. *Pharmacological Research*, *117*, 288–302. https://doi.org/10.1016/j.phrs.2016.12.026.

Hyuga, S., Hyuga, M., Amakura, Y., Yang, J., Mori, E., Hakamatsuka, T., et al. (2020). Effect of ephedra herb on erlotinib resistance in c-Met-overexpressing nonsmall-cell lung cancer cell line, H1993, through promotion of endocytosis and degradation of c-Met. *Evidence-based Complementary and Alternative Medicine, 2020*, 7184114–7184129.

Hyuga, S., Hyuga, M., Nakanishi, H., Ito, H., Watanabe, K., Oikawa, T., et al. (2007). Maoto, Kampo medicine, suppresses the metastatic potential of highly metastatic osteosarcoma cells. *Journal of Traditional Medicine, 24*, 51–58.

Hyuga, S., Hyuga, M., Oshima, N., Maruyama, T., Kamakura, H., Yamashita, T., et al. (2016). Ephedrine alkaloids-free Ephedra herb extract: A safer alternative to ephedra with comparable analgesic, anticancer, and anti-influenza activities. *Journal of Natural Medicines, 70*(3), 571–583. https://doi.org/10.1007/s11418-016-0979-z.

Hyuga, S., Hyuga, M., Yamagata, S., Yamagata, T., & Hanawa, T. (2004). Mao-to, a Kampo medicine, inhibits motility of highly meastatic osteosarcoma cells. *Journal of Traditional Medicine, 21*, 174–181.

Hyuga, S., Hyuga, M., Yoshimura, M., Amakura, Y., Goda, Y., & Hanawa, T. (2013). Herbacetin, a constituent of ephedrae herba, suppresses the HGF-induced motility of human breast cancer MDA-MB-231 cells by inhibiting c-met and Akt phosphorylation. *Planta Medica, 79*(16), 1525–1530. https://doi.org/10.1055/s-0033-1350899.

Hyuga, S., Odaguchi, H., & Hanawa, T. (2019). The adverse effects of ephedrine alkaloids in Ephedra herb and development of a novel crude drug extract, EFE without the side effects. *Bioclinica, 8*, 118–122.

Hyuga, S., Shiraishi, M., Hyuga, M., Goda, Y., & Hanawa, T. (2011). Ephedrae herba, a major component of maoto, inhibits the HGF-induced motility of human breast cancer MDA-MB-231 cells through suppression of c-met tyrosine phosphorylation and c-met-expression. *Journal of Natural Medicines, 28*, 128–138.

Indo, Y. (2010). Nerve growth factor, pain, itch and inflammation: Lessons from congenital insensitivity to pain with anhidrosis. *Expert Review of Neurotherapeutics, 10*(11), 1707–1724. https://doi.org/10.1586/ern.10.154.

Ishida, K. (2013). [Kampo medicines as useful therapeutic agents in clinical practice of neurology: case reports & representative medicines]. *Rinsho Shinkeigaku, 53*(11), 938–941. https://doi.org/10.5692/clinicalneurol.53.938.

Jong-Chol, C. (2005). Position and types of Kampo medicines as analgesics. *Journal of Pain and Clinical Medicine, 5*, 42–48.

Kaku, H., Kumagai, S., Onoue, H., Takada, A., Shoji, T., Miura, F., et al. (2012). Objective evaluation of the alleviating effects of Goshajinkigan on peripheral neuropathy induced by paclitaxel/carboplatin therapy: A multicenter collaborative study. *Experimental and Therapeutic Medicine, 3*(1), 60–65. https://doi.org/10.3892/etm.2011.375.

Kasahara, Y., Hikino, H., Tsurufuji, S., Watanabe, M., & Ohuchi, K. (1985). Antiinflammatory actions of ephedrines in acute inflammations1. *Planta Medica, 51*(4), 325–331. https://doi.org/10.1055/s-2007-969503.

Kato, Y., Tateai, Y., Ohkubo, M., Saito, Y., Amagai, S. Y., Kimura, Y. S., et al. (2014). Gosha-jinki-gan reduced oxaliplatin-induced hypersensitivity to cold sensation and its effect would be related to suppression of the expression of TRPM8 and TRPA1 in rats. *Anticancer Drugs, 25*(1), 39–43. https://doi.org/10.1097/CAD.0000000000000022.

Kono, T., Kaneko, A., Matsumoto, C., Miyagi, C., Ohbuchi, K., Mizuhara, Y., et al. (2014). Multitargeted effects of hangeshashinto for treatment of chemotherapy-induced oral mucositis on inducible prostaglandin E2 production in human oral keratinocytes. *Integrative Cancer Therapies, 13*(5), 435–445. https://doi.org/10.1177/1534735413520035.

Kono, T., Satomi, M., Chisato, N., Ebisawa, Y., Suno, M., Asama, T., et al. (2010). Topical application of hangeshashinto (TJ-14) in the treatment of chemotherapy-induced oral mucositis. *World Journal of Oncology, 1*(6), 232–235. https://doi.org/10.4021/wjon263w.

Lee, K. K., Omiya, Y., Yuzurihara, M., Kase, Y., & Kobayashi, H. (2011). Antinociceptive effect of paeoniflorin via spinal alpha(2)-adrenoceptor activation in diabetic mice. *European Journal of Pain, 15*(10), 1035–1039. https://doi.org/10.1016/j.ejpain.2011.04.011.

Ling, A. M. (2004). FDA to ban sales of dietary supplements containing ephedra. *The Journal of Law, Medicine & Ethics, 32*(1), 184–186.

Mantani, N., Andoh, T., Kawamata, H., Terasawa, K., & Ochiai, H. (1999). Inhibitory effect of Ephedrae herba, an oriental traditional medicine, on the growth of influenza A/PR/8 virus in MDCK cells. *Antiviral Research, 44*(3), 193–200. https://doi.org/10.1016/s0166-3542(99)00067-4.

Mantani, N., & Oka, H. (2017). A case of paclitaxel-induced polyarthralgia effectively treated with Maoto. *Kampo & the Newest Therapy, 26*(3), 243–245.

Matsuda, C., Munemoto, Y., Mishima, H., Nagata, N., Oshiro, M., Kataoka, M., et al. (2015). Double-blind, placebo-controlled, randomized phase II study of TJ-14 (Hangeshashinto) for infusional fluorinated-pyrimidine-based colorectal cancer chemotherapy-induced oral mucositis. *Cancer Chemotherapy and Pharmacology, 76*(1), 97–103. https://doi.org/10.1007/s00280-015-2767-y.

Mizuno, K., Kono, T., Suzuki, Y., Miyagi, C., Omiya, Y., Miyano, K., et al. (2014). Goshajinkigan, a traditional Japanese medicine, prevents oxaliplatin-induced acute peripheral neuropathy by suppressing functional alteration of TRP channels in rat. *Journal of Pharmacological Sciences, 125*(1), 91–98. https://doi.org/10.1254/jphs.13244fp.

Motoo, Y., Tomita, Y., & Fujita, H. (2020). Prophylactic efficacy of ninjin'yoeito for oxaliplatin-induced cumulative peripheral neuropathy in patients with colorectal cancer receiving postoperative adjuvant chemotherapy: A randomized, open-label, phase 2 trial (HOPE-2). *International Journal of Clinical Oncology, 25*(6), 1123–1129. https://doi.org/10.1007/s10147-020-01648-3.

Nagai, N. (1892). Research of ingredients in Ephedra herb. *Journal of the Chemical Society of Japan, 13*. 29a-35.

Nakamori, S., Takahashi, J., Hyuga, S., Tanaka-Kagawa, T., Jinno, H., Hyuga, M., et al. (2017). Ephedra herb extract activates/desensitizes transient receptor potential vanilloid 1 and reduces capsaicin-induced pain. *Journal of Natural Medicines, 71*(1), 105–113. https://doi.org/10.1007/s11418-016-1034-9.

Nakamori, S., Takahashi, J., Hyuga, S., Yang, J., Takemoto, H., Maruyama, T., et al. (2019). Analgesic effects of Ephedra herb extract, ephedrine alkaloids-free Ephedra herb extract, ephedrine, and pseudoephedrine on formalin-induced pain. *Biological & Pharmaceutical Bulletin, 42*(9), 1538–1544. https://doi.org/10.1248/bpb.b19-00260.

Nishimura, Y., Hyuga, S., Takiguchi, S., Hyuga, M., Itoh, K., & Hanawa, T. (2016). Ephedrae herba stimulates hepatocyte growth factor-induced MET endocytosis and downregulation via early/late endocytic pathways in gefitinib-resistant human lung cancer cells. *International Journal of Oncology, 48*(5), 1895–1906. https://doi.org/10.3892/ijo.2016.3426.

Odaguchi, H., Sekine, M., Hyuga, S., Hanawa, T., Hoshi, K., Sasaki, Y., et al. (2018). A double-blind, randomized, crossover comparative study for evaluating the clinical safety of ephedrine alkaloids-free Ephedra herb extract (EFE). *Evidence-based Complementary and Alternative Medicine, 2018.* https://doi.org/10.1155/2018/4625358, 4625358.

Odaguchi, H., Wakasugi, A., Ito, H., Shoda, H., Gono, Y., Sakai, F., et al. (2006). The efficacy of goshuyuto, a typical Kampo (Japanese herbal medicine) formula, in preventing episodes of headache. *Current Medical Research and Opinion, 22*(8), 1587–1597. https://doi.org/10.1185/030079906X112769.

Oshima, N., Yamashita, T., Hyuga, S., Hyuga, M., Kamakura, H., Yoshimura, M., et al. (2016). Efficiently prepared ephedrine alkaloids-free Ephedra herb extract: A putative marker and antiproliferative effects. *Journal of Natural Medicines, 70*(3), 554–562. https://doi.org/10.1007/s11418-016-0977-1.

Ros, J. J., Pelders, M. G., & De Smet, P. A. (1999). A case of positive doping associated with a botanical food supplement. *Pharmacy World and Science, 21,* 44–46.

Taira, K., Fujiwara, K., Fukuhara, T., Koyama, S., & Takeuchi, H. (2020). The effect of hangeshashinto on oral mucositis caused by induction chemotherapy in patients with head and neck cancer. *Yonago Acta Medica, 63*(3), 183–187. https://doi.org/10.33160/yam.2020.08.007.

Takemoto, H., Takahashi, J., Hyuga, S., Odaguchi, H., Uchiyama, N., Maruyama, T., et al. (2018). Ephedrine alkaloids-free Ephedra herb extract, EFE, has no adverse effects such as excitation, insomnia, and arrhythmias. *Biological & Pharmaceutical Bulletin, 41*(2), 247–253. https://doi.org/10.1248/bpb.b17-00803.

The Editing Committee for Dictionary of Kampo Medicine. (2020). *The dictionary of Kampo medicine – Basic terms.* Medical Yukon Publishing Co. Ltd.

The Japanese pharmacopoeia. (2016) (17th ed.). Tokyo, Japan: Ministry of Health, Labour and Walfare.

Tjolsen, A., Berge, O. G., Hunskaar, S., Rosland, J. H., & Hole, K. (1992). The formalin test: An evaluation of the method. *Pain, 51*(1), 5–17. https://doi.org/10.1016/0304-3959(92)90003-t.

Wei, P., Huo, H. L., Ma, Q. H., Li, H. C., Xing, X. F., Tan, X. M., et al. (2014). Pharmacokinetic comparisons of five ephedrine alkaloids following oral administration of four different Mahuang-Guizhi herb-pair aqueous extracts ratios in rats. *Journal of Ethnopharmacology, 155*(1), 642–648. https://doi.org/10.1016/j.jep.2014.05.065.

Yoshimura, M., Amakura, Y., Hyuga, S., Hyuga, M., Nakamori, S., Maruyama, T., et al. (2020). Quality evaluation and characterization of fractions with biological activity from Ephedra herb extract and ephedrine alkaloids-free Ephedra herb extract. *Chemical & Pharmaceutical Bulletin (Tokyo), 68*(2), 140–149. https://doi.org/10.1248/cpb.c19-00761.

Chapter 35

Euphorbia bicolor (Euphorbiaceae) latex phytochemicals and applications to analgesia

Paramita Basu[a], Dayna L. Averitt[b], and Camelia Maier[b]

[a]*Department of Anesthesiology & Perioperative Medicine, Pittsburgh Center for Pain Research, and the Pittsburgh Project to End Opioid Misuse, University of Pittsburgh School of Medicine, Pittsburgh, PA, United States,* [b]*Department of Biology, Texas Woman's University, Denton, TX, United States*

Abbreviations

CAT	catalase
CCI	chronic constriction injury
CEF	chalcone-enriched fraction
CGRP	calcitonin gene-related peptide
CHS	chalcone synthase
CK	creatinine kinase
COX-2	cyclooxygenase 2
DRG	dorsal root ganglion
ER	estrogen receptor
ERα	estrogen receptor alpha
ERβ	estrogen receptor beta
G-CSF	granulocyte colony-stimulating factor
GPx	glutathione peroxidase
HO-1	heme oxygenase-1
i.p.	intraperitoneal
i.t.	intrathecal
IA	intra-articular
IL	interleukin
IL-1β	interleukin 1 beta
iNOS	inducible NO synthase
ipl	intraplantar
KO_2	superoxide anion donor
LPS	lipopolysaccharide
MCP-1	monocyte chemoattractant protein 1
MDA	malondialdehyde
MMP	matrix metalloproteinase
MPO	myeloperoxidase
MyoD	myoblast determination protein
NADPH	reduced nicotinamide adenine dinucleotide phosphate
NAG	N-acetyl-β-D-glucosaminidase
NF-κB	nuclear factor kappa B
NO − cGMP − PKG − K_{ATP} channel	nitric oxide–cyclic guanosine monophosphate–protein kinase G–ATP-sensitive potassium channel
Nrf2	nuclear factor erythroid-derived 2-like 2
OA	osteoarthritis
p.o.	per os
p38 MAPK	phosphorylated p38 mitogen-activated protein kinase
PGE2	prostaglandin E2

ROS	reactive oxygen species
RTX	resiniferatoxin
s.c.	subcutaneous
SGC	satellite glial cell
SNL	spinal nerve ligation
SOD	superoxide dismutase
TIMP	tissue inhibitors of metalloproteinase
TNFα	tumor necrosis factor alpha
TPA	12-O-tetradecanoylphorbol acetate
TRPA1	transient receptor potential cation channel subfamily A member 1
TRPM8	transient receptor potential cation channel subfamily M member 8
TRPV1	transient receptor potential V1 ion channel
UPLC-ESI-MS/MS	ultra-performance liquid chromatography-electrospray ionization tandem mass spectrometry

Introduction

The International Association for the Study of Pain defines pain as "an unpleasant sensory and emotional experience associated with, actual or potential tissue damage." Pain sensory transmission is typically initiated by the activation of the nerve endings of primary afferent fibers, which include unmyelinated C-fibers and myelinated Aβ and Aδ fibers, that send signals to the brain where the perception of pain is created. Pain perception occurs in three distinct stages: (1) activation of sensory neurons by noxious stimuli, (2) signal transmission to the spinal or medullary dorsal horn via the peripheral nervous system, and (3) further signal transmission to the thalamus and higher brain areas via the central nervous system. Pain can be acute or chronic based on duration; bone or joint, cutaneous, deep, or superficial, muscle or viscera based on location; and inflammatory, neuropathic, or nociceptive based on the cause or type (Fishman, 2012).

Chronic pain is a debilitating condition, which affects a significant amount of the population. According to a recent report, 10.6 million, or about 5% of the United States adult population, is affected by high-impact chronic pain (Pitcher, Von Korff, Bushnell, & Porter, 2019). The use of opioid analgesics has increased drastically in the management of chronic pain (Bruehl et al., 2013). However, many chronic pain patients on opioids suffer from physical dependence, tolerance, and addiction (Carter et al., 2014). Gastrointestinal side effects, including constipation, nausea, vomiting, as well as cardiovascular, endocrine, immune, musculoskeletal, respiratory, and central nervous system side effects, have led to the discontinuation of opioid therapy (Harris, 2008). In addition to opioids, several other pain therapeutics used to treat painful neuropathy are also accompanied by side effects, such as sedation. Moreover, opioids are not recommended for neuropathic pain by the American Academy of Neurology (Bril et al., 2011).

The negative side effects of opioid-based narcotics and other medications underscore the need for alternative, nonopioid bioactive compounds to treat pain. Several studies have reviewed the possible efficacy of dietary supplements, herbals, plant secondary metabolites, and nutraceuticals to treat pain with lesser side effects (Andrade & Valentão, 2018; Basu & Basu, 2020; de Souza Nascimento et al., 2013; Forouzanfar & Hosseinzadeh, 2018; Fürst & Zündorf, 2014; Ghasemian, Owlia, & Owlia, 2016; Herndon & Daniels, 2018; Lim & Kim, 2016; Maroon, Bost, & Maroon, 2010). Several genera of plants have a reported history of medicinal use in native populations. One genus that has several reports of plant species that have medicinal uses, including analgesia, is *Euphorbia* (*Euphorbiaceae*) (Gupta, Vishnoi, Wal, & Wal, 2013; Islam, Ara, Ahmad, & Uddin, 2019; Kumar & Saikia, 2016; Mali & Panchal, 2013; Mwine & Van Damme, 2011; Salehi et al., 2019; Seebaluck, Gurib-Fakim, & Mahomoodally, 2015; Zahidin et al., 2017). Here we provide a general overview of the medicinal properties of the *Euphorbia* species, then specifically discuss the analgesic properties of *E. bicolor* and its bioactive phytochemicals.

Medicinal properties of the *Euphorbia* species

The genus *Euphorbia* is the third largest genus of flowering plants with almost 2000 species (Govaerts, Frodin, & Radcliffe-Smith, 2000). Several *Euphorbia* species are known to possess medicinal properties, largely antiinflammatory and antioxidant properties (Kumar, Malhotra, & Kumar, 2010; Salehi et al., 2019; Shi, Su, & Kiyota, 2008; Vasas & Hohmann, 2014). One characteristic of the plants of the *Euphorbia* genus is the presence of a milky sap or latex (Govaerts et al., 2000; Horn et al., 2012). *Euphorbia* latex is known to contain biologically active natural products that can reduce pain. For instance, in phase I clinical trial, intrathecal administration of resiniferatoxin (RTX) isolated from *E. resinifera* induced analgesia in human cancer pain patients with no significant adverse effects (Mannes et al., 2010). RTX also strongly induced analgesic action in rats, clinical canine subjects, and one human subject with advanced cancer pain (Sapio et al., 2018). The latex of *E. antiquorum* is used for the treatment of toothache by the local people in Ben En National Park, Vietnam (Sam, Baas, & Kessler, 2008). Aqueous, dichloromethane-methanol, and petroleum ether extracts of *E. tirucalli* latex display analgesic properties in rats by reducing acetic acid-induced writhing and carrageenan-induced inflammatory edema (Prabha, Ramesh, Kuppast, & Mankani, 2008). Ethyl acetate fraction of the *E. royleana* latex extract induces analgesic activity in mice treated with acetic acid (Bani, Suri, Suri, & Sharma, 1997). The latex and leaf methanolic extracts of *E. helioscopia* display antinociceptive activity in mice models of chemical (acetic acid and formalin) and inflammatory (carrageenan) pain (Saleem, Ahmad, Ahmad, Hussain, & Bukhari, 2015). Besides latex extract, different organs' extracts of other *Euphorbia* species have also been reported to possess analgesic and/or antinociceptive properties (Das, Alam, Bhattacharjee, & Das, 2015; Gaur, Rana, Nema, Kori, & Sharma, 2009; Lanhers, Fleurentin, Dorfman, Mortier, & Pelt, 1991; Majid et al., 2015; Palit et al., 2018; Sdayria et al., 2018).

Medicinal properties of *E. bicolor* latex

E. bicolor Engelm. & A. Gray, commonly known as Snow-on-the-prairie, is a native plant to Southern USA, being found only in Arkansas, Louisiana, Oklahoma, and Texas. Recent preclinical studies have provided evidence for multiple medicinal properties of this previously unstudied plant (Basu et al., 2019; Basu, Hornung, Averitt, & Maier, 2019; Basu, Meza, Bergel, & Maier, 2019). These preclinical studies reported that the *E. bicolor* latex extract induced robust pain inhibition in rats via at least three distinct, nonopioid pain inhibition mechanisms: (1) targeting nociceptors (pain-sensing nerve endings), (2) modulating various inflammatory mediators, and (3) exerting antioxidant properties. These mechanisms are discussed in details below.

Effects on Nociceptors. *E. bicolor* latex extract was reported to induce long-lasting, nonopioid, peripheral analgesia in rat models of inflammatory pain (Basu, Tongkhuya, et al., 2019). While intraplantar injection of the extract induced transient pain behaviors, indicating an initial hyperalgesic activity of the extract, significant analgesia was observed for up to 3–4 weeks. Importantly, robust analgesia was observed in both male and female rats, while some pain medications, like morphine, are less effective in women (Averitt, Eidson, Doyle, & Murphy, 2019). Systemic injection of a broad-spectrum opioid receptor antagonist, naloxone, prior to administering *E. bicolor* latex extract did not alter extract-induced analgesia, indicating that extract-induced analgesia did not involve opioid receptors. Extract-induced analgesia was partly attenuated by blocking the transient receptor potential V1 ion channel (TRPV1), a pain-generating ion channel located on the nerve endings. The involvement of TRPV1 was further confirmed in primary neuronal culture in which pretreatment with TRPV1 blockers significantly decreased the extract-induced release of calcitonin gene-related peptide (CGRP) (Basu, Tongkhuya, et al., 2019). Together, these data suggest that *E. bicolor* latex extract contains phytochemicals that evoke peripheral analgesic activity which may be harnessed for pain management without encountering the central nervous system-mediated side effects.

Effects on Inflammatory Mediators. It was also reported that *E. bicolor* latex extract exerted antiinflammatory activity either by decreasing proinflammatory cytokines/chemokines or by increasing antiinflammatory cytokines/chemokines in a rat model of orofacial pain (Basu, Hornung, et al., 2019). *E. bicolor* latex extract significantly reduced proinflammatory cytokines interleukin (IL) 1 beta (IL-1β) and IL-17 at 1-h postinjection and IL-1 alpha at 6- and 24-h postinjection in male rats, which correlated with a reduction of pain behavior in males (Basu, Hornung, et al., 2019). Latex extract also reduced other cytokines, such as soluble intercellular adhesion molecule-1 and vascular endothelial growth factor, which are implicated in rheumatoid arthritis pathology, in male rats 6-h postinjection. In female rats, latex extract increased tissue inhibitors of metalloproteinase (TIMP) in the trigeminal ganglia at 72 h. This may be providing pain relief as intrathecal injection of TIMPs significantly reduce neuropathic pain (Kawasaki et al., 2008). Together, these data suggest that *E. bicolor* latex extract has the potential to decrease inflammatory orofacial pain and neuropathic pain.

Effects on Oxidative Stress Markers. Studies have reported that reactive oxygen species (ROS) can cause pain through the activation of TRPV1 (Ibi et al., 2008; Yoshida et al., 2006). Furthermore, TRPV1 can also be activated and potentiated by reduced nicotinamide adenine dinucleotide phosphate (NADPH) oxidase-generated ROS (Ding et al., 2016). *E. bicolor* latex extract evokes analgesia by reducing several oxidative stress biomarkers, including advanced oxidation protein products, ROS, and NADPH oxidase 4 expression (Basu, Hornung, et al., 2019). Latex extract exerted significantly higher ferric reducing activity compared to the standard butylated hydroxylated toluene, and also significantly scavenged several in vitro free radicals, such as 2,2-azino-bis-(3-ethylbenzothiazoline-6-sulphonic acid), 2,2-diphenyl-1-picrylhydrazyl, hydrogen peroxide, and nitric oxide, indicating pain-relieving properties of the extract by scavenging ROS. Together, these data indicate that *E. bicolor* latex extract modulates several oxidative stress biomarkers which can reduce pain.

Identification of *E. bicolor* latex phytochemicals

As discussed above, different species of *Euphorbia* are known to exert pain-relieving effects for various types of pain and via at least three distinct mechanisms (Bani et al., 1997; Basu, Hornung, et al., 2019; Basu, Tongkhuya, et al., 2019; Das et al., 2015; Gaur et al., 2009; Lanhers et al., 1991; Majid et al., 2015; Mannes et al., 2010; Palit et al., 2018; Prabha et al., 2008; Saleem et al., 2015; Sam et al., 2008; Sdayria et al., 2018). Here we identified the phytochemicals in the latex extract that are responsible for the pain-relieving properties of *E. bicolor*. Ultra-performance liquid chromatography-electrospray ionization tandem mass spectrometry (UPLC-ESI-MS/MS) analysis of the latex extract identified 11 candidate compounds that may contribute to *E. bicolor* latex extract-induced analgesia. The identified phytochemicals belong to the four major groups: coumestans (coumestrol), diterpenes (abietic acid and RTX), flavonoids (chalcone, kaempferol, naringenin, quercetin, and rutin), and isoflavones (biochanin A, daidzein, and genistein) (Fig. 1) (Basu, Tongkhuya, et al., 2019). Table 1 summarizes the structures of these latex phytochemicals. The following section gives details of the biosynthesis of four major groups of *E. bicolor* latex phytochemicals.

Coumestans, flavonoids, and isoflavones are synthesized through the phenylpropanoid pathway. Phenylpropanoids are derived from cinnamic acid, which is synthesized in a deamination reaction of phenylalanine by phenylalanine ammonia lyase (Ferrer, Austin, Stewart, & Noel, 2008). Isoflavones are synthesized from chalcone precursors, which are formed in a head-to-tail condensation of 4-coumaroyl CoA and three molecules of malonyl-CoA in a reaction catalyzed by chalcone synthase (CHS). CHS leads to the formation of tetrahydrochalcone, whereas chalcone reductase leads to the formation of trihydrochalcone, a second substrate for isoflavone synthesis. Chalcone isomerase catalyzes the formation of naringenin from tetrahydrochalcone. The first isoflavones are formed in a reaction catalyzed by isoflavone synthase (2-hydroxyisoflavanone synthase) in the presence of NADPH and molecular oxygen. In this reaction, the hydrogen radical at C3 is abstracted followed by migration of B-ring from C2 to C3 and subsequent hydroxylation of the C2 radical in naringenin and liquiritigenin to genistein and daidzein, respectively.

Genistein and daidzein serve as basic isoflavone skeletons and structural modifications and attachment of various oxygenated sides lead to the formation of other isoflavonoids. For instance, biochanin A is a 4′-methoxy derivative of genistein (Wang & Murphy, 1994), whereas coumestrol is derived from daidzein (Berlin, Dewick, Barz, & Grisebach, 1972; Dewick, Barz, & Grisebach, 1970). Furthermore, CHS catalyzes the condensation and subsequent intramolecular cyclization of

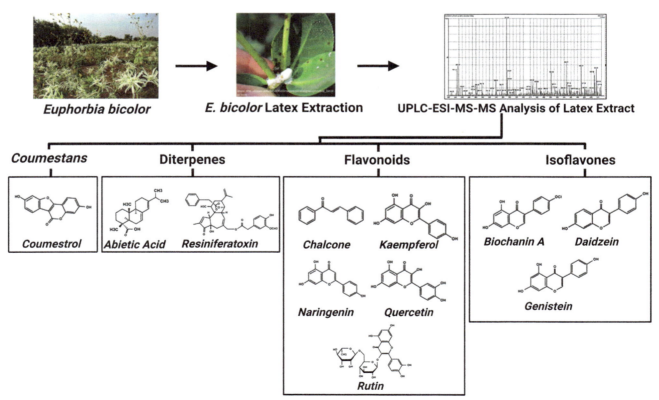

FIG. 1 *E. bicolor* latex phytochemicals identified by UPLC-ESI-MS/MS analysis.

TABLE 1 Structural description of *E. bicolor* latex phytochemicals.

Major group	Latex phytochemical	Structural description	References
Coumestans	Coumestrol	Coumestans with hydroxy substituents at positions 3 and 9	Dixon (1999)
Diterpene	Abietic acid	Abieta-7,13-diene substituted by a carboxy group at position 18	Veneziani, Ambrósio, Martins, Lemes, and Oliveira (2017)
	RTX	Three pharmacophoric regions: region A (4-hydroxy-3-methoxyphenyl), region B (connecting amide, ester or thiourea groups),region C (lipophilic side chains)	Szallasi and Blumberg (1989)
Isoflavones	Biochanin A	5,7-dihydroxy-4'-methoxy-isoflavone, methylated precursor of genistein	Kašparová, 2013
	Daidzein	7-hydroxyisoflavone substituted with an additional hydroxy group located at 4' position	Morissette, Litim, and Di Paolo (2018)
	Genistein	7-hydroxyisoflavone with additional hydroxy groups located at 5 and 4' positions	Morissette et al. (2018)
Flavonoids	Chalcone	Open-chain flavonoids with two aromatic rings (A and B) joined by a three-carbon α,β-unsaturated carbonyl system and without basic C ring from flavonoid backbone	Ninomiya and Koketsu (2013)
	Kaempferol	Tetrahydroxyflavone with four hydroxy groups located at 3, 5, 7 and 4' positions	Chen and Chen (2013)
	Naringenin	Flavanone with three hydroxy groups located at 4', 5, and 7 carbons positions	Morissette et al. (2018)
	Quercetin	Pentahydroxyflavone with four hydroxy groups located at 3, 5, 7, 3' and 4' positions	Bentz (2017)
	Rutin	Quercetin with the hydroxy group at position C-3 substituted with glucose and rhamnose sugar groups	Baliga et al. (2014)

Table is organized according to the alphabetical orders of major groups of latex phytochemicals.

three acetate units onto the *p*-coumaroyl-CoA. Following this reaction, the basic flavonoid backbone is modified by oxoglutarate-dependent $Fe2+/3+$ dioxygenases, hydroxylases, isomerases, and reductases, which leads to the formation of different flavonoid subclasses, including flavonols that are represented by kaempferol, quercetin, and rutin (a rutinose glycoside of quercetin) (Bowles, Isayenkova, Lim, & Poppenberger, 2005; Ferrer et al., 2008).

On the other hand, the addition of one unit of isopentenyl pyrophosphate to farnesyl pyrophosphate leads to the formation of geranylgeranyl-pyrophosphate, which is an intermediate in the biosynthesis of diterpenes (Davis & Croteau, 2000). Abietic acid is formed by the abietane skeleton via cyclization of the diterpenoid precursor geranylgeranyl diphosphate. The cyclization involves the formation of the intermediate (+)-copalyl diphosphate and is followed by the stepwise oxidation of the methyl group at the C18 position of the derived olefin abieta-7,13-diene (Funk & Croteau, 1994; LaFever, Vogel, & Croteau, 1994). RTX is a tigliane diterpenoid initially isolated from *E. resinifera*. Its biosynthesis involves the oxidation of casbene at the 5th position, indicating that casbene-5-oxidation is likely to be a conserved step in the biosynthesis of the majority of the diterpenoids produced by *Euphorbiaceae* (King, Brown, Gilday, Larson, & Graham, 2014).

Evidence of pain-relieving properties of the *E. bicolor* latex phytochemicals

The existing literature was searched for reports of analgesic, antiinflammatory, and antioxidant activities of the phytochemicals that have been thus far identified in *E. bicolor*. The following subsections report on the current published evidence of the effects of coumestans, diterpenes, flavonoids, and isoflavones specifically on pain, inflammation, and oxidative stress.

TABLE 2 Analgesic activities of *E. bicolor* latex coumestans.

Latex phytochemical	Animal or cell culture	Dose (animal—mg/kg; route of administration) or (in vitro—μM)	Effects	Reference
Coumestrol	Primary rat chondrocytes	1, 5, 10 μM 0.5, 1 μM	Counteracted 10 ng/mL IL-1β-induced proteoglycan loss, MMP-13, expressions of MMP-1, and MMP-3 Inhibited 10 ng/mL IL1β-induced increase in iNOS and COX-2 Downregulated PGE2 in presence of 10 ng/mL IL-1β Significantly decreased β-nerve growth factor in IL-1β treated chondrocytes Reduced levels of cytokines, chemokines, and catabolic growth factors in 10 ng/mL IL-1β treated chondrocytes	You et al. (2017)

Coumestans. Table 2 summarizes the pain-relieving activities of *E. bicolor* latex phytochemicals identified as coumestans. Coumestrol displays both antioxidant and antiinflammatory properties (Jin, Son, Min, Jung, & Choi, 2012). It also reduced lipopolysaccharide (LPS)-induced production of nitric oxide from a rat microglia cell line and exerted antiinflammatory activity (Jantaratnotai, Utaisincharoen, Sanvarinda, Thampithak, & Sanvarinda, 2013). Coumestrol-induced antiinflammatory activity involved inhibition of nitric oxide production, reduction of mRNA and protein expression of inducible NO synthase (iNOS), transcription factors governing iNOS, as well as reduction of LPS-induced chemokines and cytokines (Jantaratnotai et al., 2013).

Coumestrol can also counteract proinflammatory cytokines in a rat model of osteoarthritis (OA) (You et al., 2017). OA is a chronic degenerative joint disease that is prevalent in the elderly population, mainly caused by the progressive degeneration of the articular cartilage (Kean, Kean, & Buchanan, 2004). The upregulation of proinflammatory cytokines, such as IL-1β and tumor necrosis factor-alpha (TNFα), induce progressive degeneration of the articular cartilage by increasing the proteoglycan loss mediated by the expressions of matrix metalloproteinase (MMPs) and aggrecanases (Le Maitre, Pockert, Buttle, Freemont, & Hoyland, 2007). Coumestrol was reported to counteract IL-1β-induced proteoglycan loss, suppressed matrix-degrading enzymes, inhibited inflammatory mediators, such as iNOS and cyclooxygenase 2 (COX-2), downregulated prostaglandin E2 (PGE2), as well as counteracted levels of several cytokines, chemokines, and catabolic growth factors in primary rat chondrocytes (You et al., 2017). Furthermore, treatment of primary rat chondrocytes with IL-1β led to the production of β-nerve growth factor, which was decreased by coumestrol treatment (You et al., 2017). You et al. suggest that coumestrol might exert pain relieving effects by downregulating the proinflammatory cytokines and chemokines responsible for OA (You et al., 2017).

Diterpenes. Table 3 summarizes the pain-relieving activities of *E. bicolor* latex phytochemicals identified as diterpenes. Fernández et al. reported that oral or topical administration of abietic acid inhibited carrageenan-induced edema in male Wistar rats and 12-*O*-tetradecanoylphorbol acetate (TPA)-induced edema in male Swiss mice. Reduction in inflammation may be due to the phytochemical targeting macrophages as abietic acid also dose-dependently inhibited myeloperoxidase (MPO) activity, the production of PGE2, and the release of several cytokines (IL-1β, TNFα) in LPS-stimulated mouse macrophages (Fernández et al., 2001). However, abietic acid did not exert inhibitory activity on leukotriene C4 release, indicating that it did not affect the arachidonate pathways. Together, these findings show that topical or oral administration of abietic acid exerts antiinflammatory activity in acute inflammatory models, mainly via inhibition of macrophage release (Fernández et al., 2001).

RTX is a potent agonist for TRPV1 that triggers hyperalgesia followed by desensitization of the ion channel to produce long-lasting analgesia (Appendino & Szallasi, 1997; Mitchell et al., 2010; Ohbuchi et al., 2016; Wong & Gavva, 2009). Intrathecal administration of RTX in male Sprague-Dawley rats abolished cytokine TNF-induced mechanical allodynia and

TABLE 3 Analgesic activities of *E. bicolor* latex diterpenes.

Latex phytochemical	Animal or cell culture	Dose (animal—mg/ kg; route of administration) or (in vitro—μM)	Effects	References
Abietic acid	Male Wistar rats	50, 100 mg/kg; oral	Dose-dependently inhibited carrageenan-induced edema	Fernández et al. (2001)
	Male Swiss mice	0.25, 0.5, 1 mg/ear; topical	Dose-dependently inhibited TPA-induced edema	
	Peritoneal macrophages from male Swiss mice	1, 10, 100 μM	Inhibited MPO activity	
			Inhibited LPS-induced production of PGE2, IL-1β, TNFα	
			Inhibited calcium ionophore A23187-induced release of PGE2	
RTX	Dogs with clinically significant OA	10 μg; IA	Suppressed pain	Iadarola, Sapio, Raithel, Mannes, and Brown (2018)
			Improved gait, weight-bearing, daily activities for months or longer	
			Transcriptomic analyses showed several analgesic target genes in dog and human subjects	
	Male Sprague–Dawley rats	0, 0.1, 1 μg; perineural	Prevented SNL-induced hypersensitivity to mechanical, thermal, and cold stimuli	Koh et al. (2016)
			Decreased TRPM8 protein expression in DRG of rats with spinal nerve ligation	
			Did not decrease protein expressions of TRPV1, TRPA1	
		1.9 μg/kg; i.t.	RTX alone or with 2 or 20 ng TNF did not affect paw withdrawal thresholds	Leo et al. (2017)
			Alleviated TNF-induced mechanical allodynia and thermal hyperalgesia	
			RTX abolished TNF-mediated increase in TRPV1 protein expression and TRPV1-expressing neurons within DRG	
	Male Swiss mice	2 μg; i.t.	Pre- and posttreatment showed significant antiallodynic effects in the ipsilateral paw of neuropathic mice for 24 h	Parisi, Martins de Andrade, Torres Silva, and Silva (2017)
			The posttreatment group showed a significant reduction in TRPV1 expression in DRG	

Continued

TABLE 3 Analgesic activities of *E. bicolor* latex diterpenes—cont'd

Latex phytochemical	Animal or cell culture	Dose (animal—mg/ kg; route of administration) or (in vitro—μM)	Effects	References
	Male and female Sprague-Dawley rats	0.25 μg/100 μL; ipl	Significantly reduced full-thickness thermal injury-induced hyperalgesia in male and female rats at 2.5 h, 24 h, 1-week postinjection Significantly reduced full-thickness thermal injury-induced mechanical allodynia in male rats at 24 h and continued 1-week postinjection Reduced burn-injury induced spinal increase in Fos expression, CGRP, and substance P immunoreactivity	Salas, Clifford, Hayden, Iadarola, and Averitt (2017)

Table is organized according to the alphabetical orders of phytochemicals in the diterpenes group.

thermal hyperalgesia by decreasing TRPV1 protein expression and numbers of TRPV1-expressing neurons (Leo et al., 2017). On the other hand, intraplantar (ipl) administration of RTX induced analgesia in a rat model of burn injury by reversing burn injury-induced pain behaviors and reducing Fos expression (Salas et al., 2017). These findings suggest that local administration of RTX may provide long-lasting pain inhibition for burn patients (Salas et al., 2017). Administration of RTX via the intrathecal route in male Swiss mice (Parisi et al., 2017) or the perineural route in male Sprague-Dawley rats (Koh et al., 2016) was effective against neuropathic pain. Furthermore, in a canine intervention study, a single intra-articular (IA) injection of RTX induced a marked, long-lasting decrease in pain scores and also improved the use of the affected limb in the dog's daily activities, which lasted for 4 months or longer (Iadarola et al., 2018). Iadarola et al. also analyzed the canine dorsal root ganglia (DRG) transcriptome and provided some targets for analgesia, which are highly conserved in both protein sequence and expression level within the target tissues. The above study provides an important template for translational research in which a single injection of RTX might provide an effective intervention for human OA pain (Iadarola et al., 2018). Together, these data indicate that RTX can evoke robust analgesia in painful conditions induced by cancer, inflammation, burn, or nerve injury.

Flavonoids. Table 4 summarizes the analgesic activities of *E. bicolor* latex phytochemicals identified as flavonoids. Chalcone-enriched fraction (CEF) from stem-bark ethyl acetate extract of *Myracrodruon urundeuva*, containing three dimeric chalcones, urundeuvine A, B and C, inhibited acetic acid-induced abdominal contractions and nociception in female mice (Viana et al., 2003). Nociception was reversed by naloxone, indicating the involvement of the opioid system. Furthermore, CEF decreased thermal hyperalgesia and hind paw edema. Together, the findings suggest that CEF induces antiinflammatory as well as analgesic effects in female mice, but that the analgesic activity may be derived from targeting opioid receptors (Viana et al., 2003).

Kaempferol and naringenin have also been reported to have pain-relieving properties. Kaempferol showed antinociceptive activity in mice by reducing acetic acid-induced writhing in mice (Kim et al., 2015). Naringenin was effective against neuropathic pain, possibly through inhibiting neuroinflammation in male Sprague-Dawley rats (Hu & Zhao, 2014) as well as against diabetic neuropathy in male mice by reversing the hyperalgesia and allodynia and increasing the level of antioxidant enzyme superoxide dismutase (SOD) (Hasanein & Fazeli, 2014). Also, naringenin has been reported to prevent inflammatory pain and neurogenic inflammation in male mice by inhibiting inflammatory stimuli (carrageenan, capsaicin, CFA, or PGE2)-induced pain behavior, production of IL-33, TNFα, and IL-1β, and activation of nuclear factor kappa B (NF-κB) (Pinho-Ribeiro et al., 2016). Naringenin exerted its antiinflammatory activity by targeting the nitric oxide–cyclic guanosine monophosphate–protein kinase G–ATP-sensitive potassium channel ($NO-cGMP--PKG-K_{ATP}$ channel) signaling pathway (Pinho-Ribeiro et al., 2016). In male mice, the $NO-cGMP-PKG-K_{ATP}$ channel signaling pathway was targeted by naringenin to reduce superoxide anion donor (KO_2)-induced inflammatory pain (Manchope et al., 2016). Overall, these findings suggest that naringenin could be a good candidate for treating neuropathic and inflammatory pain conditions.

TABLE 4 Analgesic activities of *E. bicolor* latex flavonoids.

Latex Phytochemical	Animal	Dose (Animal— mg/kg; route of administration)	Effects	References
Chalcone-enriched fraction (CEF)	Female Swiss mice	5 and 10 mg/kg; intraperitoneal (i.p.) or per os (p.o.) 10 and 20 mg/kg; i.p. or p.o. 20 and 40 mg/kg; i.p. or oral	Inhibited acetic acid-induced abdominal constrictions; i.p. administration was more effective than p.o. and predominantly inhibited the second phase of the response i.p. but not p.o. increased thermal sensitivity in hot plate test Decreased carrageenan-induced paw edema	Viana, Bandeira, and Matos (2003)
Kaempferol	Mice, strain not specified	3 and 30 mg/kg; orally	Diminished acetic acid-induced writhing	Kim et al. (2015)
Naringenin	Male Wistar rats	20, 50, and 100 mg/kg; p.o.	Reversed formalin-induced hyperalgesia in streptozotocin-induced diabetic animals Reversed thermal hyperalgesia by increasing tail flicks Exerted antiallodynic effects Decreased SOD activity in diabetic animals	Hasanein and Fazeli (2014)
	Male Sprague-Dawley rats	50, 100, 200 mg/kg; i.t.	Attenuated thermal hyperalgesia and mechanical allodynia Inhibited SNL-induced activation of astrocytes and microglia Inhibited proinflammatory cytokines and chemokines, such as TNFα, IL-1β, MCP-1	Hu and Zhao (2014)
	Male Swiss mice	16.7, 50, or 150 mg/kg; p.o.	Inhibited KO$_2$-induced overt pain-like behavior, mechanical, thermal hyperalgesia Inhibited KO$_2$-induced MPO activity, lipid peroxidation, TNFα, IL-10, O$_2^-$ production, gp91phox (an NADPH oxidase component), cytokine (IL-33), COX-2 mRNA expression Increased Nrf2/HO-1 mRNA expression Activated the NO$-$cGMP$-$PKG$-$K$_{ATP}$ channel signaling pathway	Manchope et al. (2016)
	Male Swiss mice	16.7–150 mg/kg; p.o.	Inhibited acetic acid- and phenyl-p-benzoquinone-induced writhing Inhibited formalin-, capsaicin-, CFA-induced paw flinching and licking Inhibited carrageenan-, capsaicin-, PGE2- and CFA-induced mechanical hyperalgesia Inhibited carrageenan-induced oxidative stress, hyperalgesic cytokines (IL-33, TNFα, and IL-1β) production, and NF-κB activation Activated the analgesic NO-cyclic GMP-PKG-ATP sensitive K+ channel signaling pathway	Pinho-Ribeiro et al. (2016)

Continued

410 PART | IV Novel and nonpharmacological aspects and treatments

TABLE 4 Analgesic activities of *E. bicolor* latex flavonoids—cont'd

Latex Phytochemical	Animal	Dose (Animal—mg/kg; route of administration)	Effects	References
Quercetin		1–30 mg/kg; i.p.	Reduced muscle mechanical hyperalgesia	Borghi et al. (2016)
			Inhibited MPO, NAG, cytokine production, oxidative stress, COX-2, NF-κB activation in soleus muscle	
			Inhibited mRNA expressions of gp91phox, CK, MyoD in soleus muscle	
			Induced Nrf2 and HO-1 mRNA expression in soleus muscle	
			Inhibited spinal cord cytokine production, oxidative stress, glial cell activation	
		10, 30, 100 mg/g; i.p.	Reduced tumor-induced mechanical and thermal hyperalgesia	Calixto-Campos et al. (2015)
			Did not alter paw edema	
			Inhibited proinflammatory cytokines IL-1β and TNFα	
			Decreased neutrophil recruitment (MPO activity) and oxidative stress	
			Induced analgesic activity in an opioid-dependent pathway	
		10–60 mg/kg, i.p. or 100–500 mg/kg, p.o.	Inhibited acetic acid-induced pain and inhibited formalin-induced pain	Filho, Filho, Olinger, and de Souza (2008)
			Inhibited glutamate- and capsaicin-induced nociception	
			Exerted analgesic activity by interacting with L-arginine-nitric oxide, serotonin, and GABAergic systems	
	Male/female albino mice	100–300 mg/kg; i.p.	Increased tail-flick latency via D2-dopamine receptor and α2-adrenoreceptor	Naidu, Singh, and Kulkarni (2003)
	Male Sprague–Dawley rats	50 mg/kg; i.p.	Increased mechanical and thermal latencies	Yang et al. (2019)
			Inhibited enhanced P2X4 expression	
			Inhibited glial fibrillary acidic protein, a marker of SGCs	
			Decreased upregulation of p38 MAPK in DRG	
			Reduced P2X4 agonist adenosine triphosphate-activated currents in HEK293 cells transfected with P2X4 receptors	

TABLE 4 Analgesic activities of *E. bicolor* latex flavonoids—cont'd

Latex Phytochemical	Animal	Dose (Animal—mg/kg; route of administration)	Effects	References
Rutin	Male Swiss mice	10–100 mg/kg; i.p.	Reduced G-CSF-induced mechanical hyperalgesia	Carvalho et al. (2019)
			Rutin + morphine or rutin + indomethacin reduced G-CSF-induced pain	
			Inhibited NF-κB activation	
			Decreased mRNA expressions of Nrf2 and HO-1	
			Activated NO–cGMP–PKG–K_{ATP} channel signaling pathway to induce analgesia	
	Male Sprague-Dawley rats	5, 25, and 50 mg/kg; i.p.	Reduced mechanical hyperalgesia, thermal hyperalgesia, cold allodynia	Tian et al. (2016)
			Decreased NF-κB, IκBα, p-IκBα, IL-6 and TNFα, caspase-3 expression in DRG neurons	
			Decreased MDA and ROS levels	
			Increased H2S, Nrf2, and HO-1 in DRG neurons	
			Increased Na+, K+-ATPase activities in sciatic nerves	
			Partially increased antioxidant enzymes SOD, glutathione peroxidase (GPx), glutathione-S-transferase, and catalase (CAT) in sciatic nerves	

Table is organized according to the alphabetical orders of phytochemicals in the flavonoids group.

Quercetin induced analgesic activity in a mouse model of cancer pain by inhibiting mechanical and thermal hyperalgesia, inhibiting proinflammatory cytokines IL-1β and TNFα, neutrophil recruitment, and oxidative stress via an opioid-dependent pathway (Calixto-Campos et al., 2015). However, quercetin did not alter paw thickness or histological alterations, indicating that it induced analgesic activity without affecting tumor growth (Calixto-Campos et al., 2015). In diabetic neuropathic male rats, quercetin decreased the upregulation of P2X4 receptors in DRG satellite glial cells (SGCs) and consequently inhibited P2X4 receptor-mediated activation of phosphorylated p38 mitogen-activated protein kinase (p38 MAPK) (Yang et al., 2019). Furthermore, in male mice, quercetin exerted analgesic activity in different models of chemical and thermal nociception by interacting with L-arginine-nitric oxide, the serotonergic system, and the GABAergic system (Filho et al., 2008), and increased the latency of tail-flick in Albino mice by possibly involving dopaminergic and noradrenergic systems (Naidu et al., 2003). In addition, quercetin was also reported to reduce intense, acute, swimming-induced muscle pain in male Swiss mice by inhibiting both peripheral and spinal cord nociceptive mechanisms (Borghi et al., 2016). The study found that quercetin inhibited MPO and N-acetyl-β-D-glucosaminidase (NAG) in the soleus muscle, but not in the gastrocnemius muscle. Moreover, mRNA expression of creatinine kinase (CK) blood levels, myoblast determination protein (MyoD), and gp91phox were reduced, whereas mRNA expression of Nrf2 and HO-1 were induced. Cytokine production in both peripheral and spinal cord was inhibited, activation of glial fibrillary acidic protein, and ionized calcium-binding adapter molecule 1 mRNA expression in the spinal cord were also inhibited by quercetin (Borghi et al., 2016). Together, these data suggest that quercetin can reduce pain in a vast array of conditions, including cancer pain, diabetic neuropathy, muscle pain, chemical pain, and thermal pain, however analgesic activity may involve targeting the opioid system.

Similar to naringenin, rutin targeted the $NO - cGMP - PKG - K_{ATP}$ channel signaling pathway in male mice to induce analgesia in granulocyte colony-stimulating factor (G-CSF)-induced pain, inhibited NF-κB, and triggered the Nrf2/HO-1 pathway (Carvalho et al., 2019). To avoid infections and feverish conditions, G-CSF therapeutics aim to mobilize

412 PART | IV Novel and nonpharmacological aspects and treatments

hematopoietic stem cells from the bone marrow into circulation to increase the count of neutrophils in patients after chemotherapy (Bennett, Djulbegovic, Norris, & Armitage, 2013). Rutin did not affect the recruitment of leukocytes to the bloodstream, indicating that rutin reduced G-CSF-induced pain without inhibiting the G-CSF-induced mobilization of hematopoietic stem cells and neutrophils (Carvalho et al., 2019). Rutin was effective in reducing diabetic neuropathy in male Sprague-Dawley rats by reducing plasma glucose levels, diabetes-induced pain behaviors, attenuating several oxidative stress markers and proinflammatory cytokines, and increasing antioxidant enzymes (Tian et al., 2016). Based on these data, it can be concluded that rutin may serve as a good candidate for the treatment of diabetic neuropathy and G-CSF-induced pain. Overall, several flavonoids, including chalcone, kaempferol, naringenin, quercetin, and rutin have the potential to be developed into phytomedicines for pain management.

Isoflavones. Table 5 summarizes the analgesic activities of *E. bicolor* latex phytochemicals identified as isoflavones. Chundi et al. tested acute and chronic treatments with biochanin A in a rat model of diabetic neuropathy. Biochanin A was administered as either a single dose or administered for several days. The chronic treatment with biochanin A was more

TABLE 5 Analgesic activities of *E. bicolor* latex isoflavones.

Latex phytochemical	Animal	Dose (Animal— mg/kg; route of administration)	Effects	References
Biochanin A	Male albino Wistar rats	0.1, 1, 5 mg/kg; i.p.	Chronic treatment with a higher dose was more effective than acute treatment in reversing mechanical allodynia and hyperalgesia Exhibited better efficacy in reversing mechanical allodynia than mechanical hyperalgesia	Chundi et al. (2016)
Daidzein from *Pueraria lobata*	Male ddY mice	50, 100 mg/kg; i.p.	Inhibited acetic acid-induced writhing in a dose-dependent manner	Yasuda et al. (2005)
Genistein	Male C57BL/6 J mice	1, 3, 7.5, 15, or 30 mg/kg; subcutaneous (s.c.)	Reversed thermal hyperalgesia and mechanical allodynia at 15, 30 mg/kg Attenuated thermal and reversed allodynia at 7.5 mg/kg, while no effects recorded at 1, 3 mg/kg ER nonspecific antagonist ICI 182780 prevented only antiallodynic action ERα-specific antagonist MPP was ineffective in preventing anti-allodynic and anti-hyperalgesic actions ERβ-specific antagonist PHTPP prevented both anti-allodynic and anti-hyperalgesic actions Reversed mRNA over-expressions of IL-6 in the sciatic nerve and IL-1β in sciatic nerves, spinal cord, DRG Reversed IL-1β protein expression in sciatic nerve Abolished the NF-κB activation in sciatic nerve Reduced ROS and malondialdehyde (MDA) levels in CCI-operated paw tissues iNOS returned to sham-operated animal levels in spinal cord dorsal horn Repeated treatment increased GPx and CAT in neuropathic tissues	Valsecchi et al. (2008)

Table is organized according to the alphabetical orders of phytochemicals in the isoflavones group.

effective than acute in reversing mechanical allodynia and hyperalgesia. However, biochanin A treatment was more effective in reversing mechanical allodynia than mechanical hyperalgesia, indicating that biochanin A could be a good candidate to treat pain evoked by Aβ-fibers or low-threshold Aδ- and C-fibers, which also transmit nonpainful stimuli (Chundi et al., 2016).

Daidzein and genistein also have reported analgesic properties. Daidzein, one of the metabolites of *Pueraria lobata*, inhibited acetic acid-induced writhing in male mice (Yasuda et al., 2005). Genistein ameliorated painful neuropathy in male mice by two different mechanisms, exerting antiallodynic and antihyperalgesic actions in chronic sciatic nerve constriction-induced pain behavior, and reducing NF-κB, nitric oxide system, proinflammatory cytokines, and oxidative stress markers (Valsecchi et al., 2008). Since genistein is a phytoestrogen that binds to the estrogen receptor beta (ERβ) with higher affinity than to the estrogen receptor alpha (ERα) (An, Tzagarakis-Foster, Scharschmidt, Lomri, & Leitman, 2001; Kuiper et al., 1997), the study furthermore evaluated the antinociceptive activity of genistein by employing different ER-specific antagonists. The results showed that the ER antagonist ICI 182780 prevented only antiallodynic action. ERα-specific antagonist MPP did not prevent antiallodynic and antihyperalgesic actions of genistein, whereas ERβ-specific antagonist PHTPP prevented both antiallodynic and antihyperalgesic actions of genistein. This suggests that ERβ is mainly involved in the antinociceptive effect of genistein and genistein could also be effective in preventing and treating breast cancer-induced pain. In conclusion, the isoflavones biochanin A, daidzein, and genistein could serve as potential therapeutics for treating neuropathic pain.

Development of *E. bicolor* phytochemicals as phytomedicines for pain management

Based on the literature review, it is well evident that *E. bicolor* latex phytochemicals could serve as potential pain therapeutics in preventing, reducing, and treating a variety of acute and chronic pain conditions. Based on the preclinical studies discussed in this chapter, it can be speculated that biochanin A, genistein, naringenin, quercetin, and rutin could be effective for neuropathies, RTX and coumestrol for burn pain and osteoarthritis, and abietic acid, daidzein, dimeric chalcones and kaempferol for more general acute pain. CEF and quercetin appear to target the opioid system, so these phytochemicals would not be ideal for the treatment of chronic pain due to the potential negative side effects and abuse potential. Future studies should focus on the synergistic effects of these latex phytochemicals based on each of their discrete pain-relieving mechanisms to optimize their potential. Importantly, there are still many phytochemicals that have yet to be identified in the *E. bicolor* latex extract. The accumulation of work on *Euphorbia* phytochemicals in combination with future work discovering novel *Euphorbia* phytochemicals has the potential to lead to the development of neoteric phytomedicines. Pain medicine is currently in great need of new therapeutics with high potency and efficacy against severe or chronic pain, limited side effects, and curbed abuse potential. Historical efforts led to the development of several well-known pain medications currently in high demand; morphine from *Papaver somniferum* (poppy), aspirin from *Salix alba* (willow tree), cannabidiol from *Cannabis sativa* (cannabis), and capsaicin from *Capsicum annuum* (chili pepper). Continuing on this trajectory to find and develop novel plant phytochemicals that are highly effective against pain, but have nonopioid targets, may eventually address our current pain management needs.

Mini-dictionary of terms

Allodynia. Pain resulting from a stimulus that is usually not painful.

Chemokines. A class of cytokines that are chemotactic and attract white blood cells to the site of injury.

Cytokines. Small proteins secreted by cells, playing role in interactions and communications between cells. Proinflammatory cytokines are produced by activated macrophages and are involves in the upregulation of inflammatory reactions. Antiinflammatory cytokines control responses mediated by proinflammatory cytokines.

Hyperalgesia. An increased painful sensation, which occurs in response to noxious stimuli.

Intra-articular (IA). When administered directly into the joint.

Intraperitoneal (i.p.). When administered through the peritoneum, which is a thin transparent membrane that lines the wall of the peritoneal cavity and encloses abdominal organs, such as the stomach and intestines.

Intraplantar (ipl.). When administered within the sole of the foot.

Intrathecal (i.t.). When administered into the spinal canal or subarachnoid space reach the cerebrospinal fluid.

Perineural. When administered around a nerve or a bundle of nerve fibers.

Subcutaneous (s.c.). When administered into the subcutis, a skin layer present below the dermis and epidermis.

Key facts about coumestans

- Coumestrol was identified as the only coumestan
- Coumestrol exerts analgesic activity by downregulating several inflammatory mediators, including proinflammatory cytokines, and chemokines responsible for OA

Key facts about diterpenes

- Diterpenes identified were abietic acid and RTX
- Abietic acid exerts antiinflammatory activity in carrageenan-induced edema in rats and TPA-induced edema in mice
- Perineural RTX administration reduces thermal, mechanical, and especially cold hypersensitivity since it suppresses TRPM8 expression in DRG in SNL-induced animals
- RTX induces analgesic activity by reducing TNF-induced mechanical allodynia, thermal hyperalgesia, as well as TNF-induced increase in TRPV1 protein expression and TRPV1-expressing neurons
- RTX induces analgesia in a rat burn injury model by reducing pain behavior and Fos expression
- IA administration of RTX is effective against dog OA, which serves as a template for translational research in reducing OA pain

Key facts about flavonoids

- Flavonoids identified in *E. bicolor* latex extract were chalcone, kaempferol, naringenin, quercetin, and rutin
- Dimeric chalcones urundeuvine A, B, and C identified in CEF induces central and peripheral analgesic effects
- Kaempferol reduces acetic acid-induced writhing
- Naringenin reduces neuropathic pain either by increasing antioxidative enzymes or by inhibiting neuroinflammation
- Naringenin inhibits inflammatory pain by activating $NO - cGMP - PKG - K_{ATP}$ channel signaling pathway and inducing Nrf2/HO-1 pathway
- Quercetin reduces cancer pain by inhibiting proinflammatory cytokines and oxidative stress
- Quercetin reduces intense acute swimming-induced muscle pain by inhibiting several nociceptive markers at the peripheral and spinal cord
- Quercetin induces analgesia by interacting with either L-arginine-nitric oxide, serotonin, and GABAergic systems or dopaminergic and noradrenergic systems
- Quercetin inhibits neuropathic pain by inhibiting p38 MAPK
- Rutin induces analgesia in G-CSF-induced pain by activating $NO - cGMP - PKG - K_{ATP}$ channel signaling pathway and inhibiting NF-κB and triggering Nrf2/HO-1 pathway
- Rutin ameliorates diabetic neuropathy by modulating proinflammatory cytokines, oxidative stress markers, and electrophysiological measurements

Key facts about isoflavones

- Isoflavones identified in *E. bicolor* latex extract were biochanin A, daidzein, and genistein
- Biochanin A is more effective in reversing mechanical allodynia than mechanical hyperalgesia in a rodent neuropathic pain model
- Daidzein induces analgesia by reducing acetic acid-induced writhing
- Genistein decreases neuropathic pain by reducing NF-κB, nitric oxide system, proinflammatory cytokines, and oxidative stress markers

Summary points

- *E. bicolor* has not been studied before. The medicinal properties and phytochemical analysis of *E. bicolor* latex extract have been studied for the first time by our team
- *E. bicolor* latex extract induces nonopioid, long-lasting analgesia via TRPV1 in both male and female rats
- *E. bicolor* latex extract induces analgesia via reducing oxidative stress biomarkers and proinflammatory cytokines

- Eleven phytochemicals were identified so far in the *E. bicolor* latex extract, belonging to four major groups: coumestans (coumestrol), diterpenes (abietic acid and RTX), flavonoids (chalcone, kaempferol, naringenin, quercetin, and rutin), and isoflavones (biochanin A, daidzein, and genistein)
- Coumestrol exerts analgesic activity against OA by downregulating several inflammatory mediators, cytokines, and chemokines
- Oral or topical administration of abietic acid exerts antiinflammatory activity by inhibiting the action of macrophages in rodent acute inflammatory models
- RTX induces analgesia in rodent models of inflammation, nerve injury, and burn injury
- RTX also reduces OA pain in dogs, which provided a platform for testing the effects of RTX in humans
- Dimeric chalcones (urundeuvine A, B, and C) induces peripheral and central analgesic activities
- Kaempferol exerts antinociceptive activity by reducing acetic acid-induced writhing
- Naringenin induces analgesic activity against inflammatory or neuropathic pain either by activating $NO-cGMP--PKG-K_{ATP}$ channel signaling pathway or by inducing Nrf2/HO-1 pathway
- Quercetin induces analgesia against cancer pain, acute swimming-induced muscle pain, and neuropathic pain
- Quercetin-induced analgesia involves either L-arginine-nitric oxide, serotonin, and GABAergic systems or dopaminergic and noradrenergic systems
- Rutin induces analgesia against G-CSF-induced pain and neuropathic pain
- Biochanin A and genistein induce analgesia against neuropathic pain, whereas daidzein induces analgesia by reducing acetic acid-induced writhing

References

An, J., Tzagarakis-Foster, C., Scharschmidt, T. C., Lomri, N., & Leitman, D. C. (2001). Estrogen receptor beta-selective transcriptional activity and recruitment of coregulators by phytoestrogens. *The Journal of Biological Chemistry*, *276*(21), 17808–17814.

Andrade, P. B., & Valentão, P. (2018). Insights into natural products in inflammation. *International Journal of Molecular Sciences*, *19*(3), 644. https://doi.org/10.3390/ijms19030644.

Appendino, G., & Szallasi, A. (1997). Euphorbium: Modern research on its active principle, resiniferatoxin, revives an ancient medicine. *Life Sciences*, *60*(10), 681–696.

Averitt, D. L., Eidson, L. N., Doyle, H. H., & Murphy, A. Z. (2019). Neuronal and glial factors contributing to sex differences in opioid modulation of pain. *Neuropsychopharmacology*, *44*(1), 155–165.

Baliga, M. S., Saxena, A., Kaur, K., Kalekhan, F., Chacko, A., Venkatesh, P., et al. (2014). Polyphenols in the prevention of ulcerative colitis: Past, present and future. In R. R. Watson, V. R. Preedy, & S. Zibadi (Eds.), *Polyphenols in human health and disease* (pp. 655–663). San Diego: Academic Press (Chapter 50).

Bani, S., Suri, K. A., Suri, O. P., & Sharma, O. P. (1997). Analgesic and antipyretic properties of *Euphorbia royleana* latex. *Phytotherapy Research*, *11*(8), 597–599.

Basu, P., & Basu, A. (2020). *In vitro* and *in vivo* effects of flavonoids on peripheral neuropathic pain. *Molecules*, *25*(5). https://doi.org/10.3390/molecules25051171, 1171.

Basu, P., Hornung, R. S., Averitt, D. L., & Maier, C. (2019). *Euphorbia bicolor* (Euphorbiaceae) latex extract reduces inflammatory cytokines and oxidative stress in a rat model of orofacial pain. *Oxidative Medicine and Cellular Longevity*, *2019*, 8594375.

Basu, P., Meza, E., Bergel, M., & Maier, C. (2019). Estrogenic, antiestrogenic and antiproliferative activities of *Euphorbia bicolor* (*Euphorbiaceae*) latex extract and its phytochemicals. *Nutrients*, *12*(1). https://doi.org/10.3390/nu12010059.

Basu, P., Tongkhuya, S. A., Harris, T. L., Riley, A. R., Maier, C., Granger, J., et al. (2019). *Euphorbia bicolor* (*Euphorbiaceae*) latex phytochemicals induce long-lasting nonopioid peripheral analgesia in a rat model of inflammatory pain. *Frontiers in Pharmacology*, *10*(958). https://doi.org/10.3389/fphar.2019.00958.

Bennett, C. L., Djulbegovic, B., Norris, L. B., & Armitage, J. O. (2013). Colony-stimulating factors for febrile neutropenia during cancer therapy. *The New England Journal of Medicine*, *368*(12), 1131–1139.

Bentz, A. B. (2017). A review of quercetin: Chemistry, antioxident properties, and bioavailability. *Journal of Young Investigators*. Available online at https://www.jyi.org/2009-april/2017/10/15/a-review-of-quercetin-chemistry-antioxidant-properties-and-bioavailability.

Berlin, J., Dewick, P., Barz, W., & Grisebach, H. (1972). Biosynthesis of coumestrol in *Phaseolus aureus*. *Phytochemistry*, *11*(5), 1689–1693.

Borghi, S. M., Pinho-Ribeiro, F. A., Fattori, V., Bussmann, A. J., Vignoli, J. A., Camilios-Neto, D., et al. (2016). Quercetin inhibits peripheral and spinal cord nociceptive mechanisms to reduce intense acute swimming-induced muscle pain in mice. *PLoS One*, *11*(9), e0162267.

Bowles, D., Isayenkova, J., Lim, E.-K., & Poppenberger, B. (2005). Glycosyltransferases: Managers of small molecules. *Current Opinion in Plant Biology*, *8*(3), 254–263.

Bril, V., England, J., Franklin, G. M., Backonja, M., Cohen, J., Del Toro, D., et al. (2011). Evidence-based guideline: Treatment of painful diabetic neuropathy: Report of the American Academy of Neurology, the American Association of Neuromuscular and Electrodiagnostic Medicine, and the American Academy of Physical Medicine and Rehabilitation. *Neurology*, *76*(20), 1758–1765.

Bruehl, S., Apkarian, A. V., Ballantyne, J. C., Berger, A., Borsook, D., Chen, W. G., et al. (2013). Personalized medicine and opioid analgesic prescribing for chronic pain: Opportunities and challenges. *The Journal of Pain, 14*(2), 103–113.

Calixto-Campos, C., Corrêa, M. P., Carvalho, T. T., Zarpelon, A. C., Hohmann, M. S. N., Rossaneis, A. C., et al. (2015). Quercetin reduces Ehrlich tumor-induced cancer pain in mice. *Analytical Cellular Pathology, 2015*, 285708.

Carter, G. T., Duong, V., Ho, S., Ngo, K. C., Greer, C. L., & Weeks, D. L. (2014). Side effects of commonly prescribed analgesic medications. *Physical Medicine and Rehabilitation Clinics of North America, 25*(2), 457–470.

Carvalho, T. T., Mizokami, S. S., Ferraz, C. R., Manchope, M. F., Borghi, S. M., Fattori, V., et al. (2019). The granulopoietic cytokine granulocyte colony-stimulating factor (G-CSF) induces pain: Analgesia by rutin. *Inflammopharmacology, 27*(6), 1285–1296.

Chen, A. Y., & Chen, Y. C. (2013). A review of the dietary flavonoid, kaempferol on human health and cancer chemoprevention. *Food Chemistry, 138*(4), 2099–2107.

Chundi, V., Challa, S. R., Garikapati, D. R., Juvva, G., Jampani, A., Pinnamaneni, S. H., et al. (2016). Biochanin-A attenuates neuropathic pain in diabetic rats. *Journal of Ayurveda and Integrative Medicine, 7*(4), 231–237.

Das, B., Alam, S., Bhattacharjee, R., & Das, B. (2015). Analgesic and antiinflammatory activity of *Euphorbia antiquorum* Linn. *American Journal of Pharmacology and Toxicology, 10*, 46–55.

Davis, E. M., & Croteau, R. (2000). Cyclization enzymes in the biosynthesis of monoterpenes, sesquiterpenes, and diterpenes. In F. J. Leeper, & J. C. Vederas (Eds.), *Biosynthesis: Aromatic polyketides, isoprenoids, alkaloids* (pp. 53–95). Berlin, Heidelberg: Springer Berlin Heidelberg.

de Souza Nascimento, S., Desantana, J. M., Nampo, F. K., Ribeiro, E. A., da Silva, D. L., Araujo-Junior, J. X., et al. (2013). Efficacy and safety of medicinal plants or related natural products for fibromyalgia: A systematic review. *Evidence-based Complementary and Alternative Medicine, 2013*, 149468.

Dewick, P. M., Barz, W., & Grisebach, H. (1970). Biosynthesis of coumestrol in *Phaseolus aureus*. *Phytochemistry, 9*(4), 775–783.

Ding, R., Jiang, H., Sun, B., Wu, X., Li, W., Zhu, S., et al. (2016). Advanced oxidation protein products sensitized the transient receptor potential vanilloid 1 via NADPH oxidase 1 and 4 to cause mechanical hyperalgesia. *Redox Biology, 10*, 1–11.

Dixon, R. A. (1999). 1.28 – Isoflavonoids: Biochemistry, molecular biology, and biological functions. In S. D. Barton, K. Nakanishi, & O. Meth-Cohn (Eds.), *Comprehensive natural products chemistry* (pp. 773–823). Oxford: Pergamon.

Fernández, M. A., Tornos, M. P., García, M. D., de Las Heras, B., Villar, A. M., & Sáenz, M. T. (2001). Antiinflammatory activity of abietic acid, a diterpene isolated from *Pimenta racemosa* var. *grissea*. *Journal of Pharmacy and Pharmacology, 53*(6), 867–872.

Ferrer, J. L., Austin, M. B., Stewart, C., Jr., & Noel, J. P. (2008). Structure and function of enzymes involved in the biosynthesis of phenylpropanoids. *Plant Physiology and Biochemistry, 46*(3), 356–370.

Filho, A. W., Filho, V. C., Olinger, L., & de Souza, M. M. (2008). Quercetin: Further investigation of its antinociceptive properties and mechanisms of action. *Archives of Pharmacal Research, 31*(6), 713–721.

Fishman, S. M. (2012). *Bonica's management of pain.* Lippincott Williams & Wilkins.

Forouzanfar, F., & Hosseinzadeh, H. (2018). Medicinal herbs in the treatment of neuropathic pain: A review. *Iranian Journal of Basic Medical Sciences, 21*(4), 347–358.

Funk, C., & Croteau, R. (1994). Diterpenoid resin acid biosynthesis in conifers: Characterization of two cytochrome P450-dependent monooxygenases and an aldehyde dehydrogenase involved in abietic acid biosynthesis. *Archives of Biochemistry and Biophysics, 308*(1), 258–266.

Fürst, R., & Zündorf, I. (2014). Plant-derived antiinflammatory compounds: Hopes and disappointments regarding the translation of preclinical knowledge into clinical progress. *Mediators of Inflammation, 2014*, 146832.

Gaur, K., Rana, A., Nema, R., Kori, M., & Sharma, C. S. (2009). Antiinflammatory and analgesic activity of hydro-alcoholic leaves extract of *Euphorbia neriifolia* Linn. *Asian Journal of Pharmaceutical and Clinical Research, 2*, 26–29.

Ghasemian, M., Owlia, S., & Owlia, M. B. (2016). Review of antiinflammatory herbal medicines. *Advances in Pharmacological Sciences, 2016*, 9130979.

Govaerts, R., Frodin, D., & Radcliffe-Smith, A. (2000). *World checklist and bibliography of Euphorbiaceae and Pandaceae* (pp. 1–1622). Royal Bot. Garden, Kew. I-4.

Gupta, N., Vishnoi, G., Wal, A., & Wal, P. (2013). Medicinal value of *Euphorbia tirucalli*. *Systematic Reviews in Pharmacy, 4*, 40.

Harris, J. D. (2008). Management of expected and unexpected opioid-related side effects. *The Clinical Journal of Pain, 24*(Suppl 10), S8–s13.

Hasanein, P., & Fazeli, F. (2014). Role of naringenin in protection against diabetic hyperalgesia and tactile allodynia in male Wistar rats. *Journal of Physiology and Biochemistry, 70*(4), 997–1006.

Herndon, C. M., & Daniels, A. M. (2018). Nonopioid strategies for managing chronic pain. *Pharmacy Today, 24*(12), 47–61.

Horn, J. W., van Ee, B. W., Morawetz, J. J., Riina, R., Steinmann, V. W., Berry, P. E., et al. (2012). Phylogenetics and the evolution of major structural characters in the giant genus *Euphorbia* L. (*Euphorbiaceae*). *Molecular Phylogenetics and Evolution, 63*(2), 305–326.

Hu, C. Y., & Zhao, Y.-T. (2014). Analgesic effects of naringenin in rats with spinal nerve ligation-induced neuropathic pain. *Biomedical Reports, 2*(4), 569–573.

Iadarola, M. J., Sapio, M. R., Raithel, S. J., Mannes, A. J., & Brown, D. C. (2018). Long-term pain relief in canine osteoarthritis by a single intra-articular injection of resiniferatoxin, a potent TRPV1 agonist. *Pain, 159*(10), 2105–2114.

Ibi, M., Matsuno, K., Shiba, D., Katsuyama, M., Iwata, K., Kakehi, T., et al. (2008). Reactive oxygen species derived from NOX1/NADPH oxidase enhance inflammatory pain. *Journal of Neuroscience, 28*(38), 9486–9494.

Islam, M. S., Ara, H., Ahmad, K. I., & Uddin, M. M. (2019). A review on medicinal uses of different plants of *Euphorbiaceae* family. *Universal Journal of Pharmaceutical Research, 4*(1), 47–51.

Jantaratnotai, N., Utaisincharoen, P., Sanvarinda, P., Thampithak, A., & Sanvarinda, Y. (2013). Phytoestrogens mediated antiinflammatory effect through suppression of IRF-1 and pSTAT1 expressions in lipopolysaccharide-activated microglia. *International Immunopharmacology, 17*(2), 483–488.

Jin, S. E., Son, Y. K., Min, B. S., Jung, H. A., & Choi, J. S. (2012). Antiinflammatory and antioxidant activities of constituents isolated from *Pueraria lobata* roots. *Archives of Pharmacal Research, 35*(5), 823–837.

Kašparová, M. (2013). Phytoestrogens from red clover. *Praktický Lékar˘, 9*, 201–203.

Kawasaki, Y., Xu, Z. Z., Wang, X., Park, J. Y., Zhuang, Z. Y., Tan, P. H., et al. (2008). Distinct roles of matrix metalloproteases in the early- and late-phase development of neuropathic pain. *Nature Medicine, 14*(3), 331–336.

Kean, W. F., Kean, R., & Buchanan, W. W. (2004). Osteoarthritis: Symptoms, signs and source of pain. *Inflammopharmacology, 12*(1), 3–31.

Kim, S. H., Park, J. G., Sung, G. H., Yang, S., Yang, W. S., Kim, E., et al. (2015). Kaempferol, a dietary flavonoid, ameliorates acute inflammatory and nociceptive symptoms in gastritis, pancreatitis, and abdominal pain. *Molecular Nutrition & Food Research, 59*(7), 1400–1405.

King, A. J., Brown, G. D., Gilday, A. D., Larson, T. R., & Graham, I. A. (2014). Production of bioactive diterpenoids in the *Euphorbiaceae* depends on evolutionarily conserved gene clusters. *The Plant Cell, 26*(8), 3286–3298.

Koh, W. U., Choi, S.-S., Kim, J. H., Yoon, H. J., Ahn, H.-S., Lee, S. K., et al. (2016). The preventive effect of resiniferatoxin on the development of cold hypersensitivity induced by spinal nerve ligation: Involvement of TRPM8. *BMC Neuroscience, 17*(1), 38.

Kuiper, G. G., Carlsson, B., Grandien, K., Enmark, E., Haggblad, J., Nilsson, S., et al. (1997). Comparison of the ligand binding specificity and transcript tissue distribution of estrogen receptors alpha and beta. *Endocrinology, 138*(3), 863–870.

Kumar, S., Malhotra, R., & Kumar, D. (2010). *Euphorbia hirta*: Its chemistry, traditional and medicinal uses, and pharmacological activities. *Pharmacognosy Reviews, 4*, 58–61.

Kumar, A., & Saikia, D. (2016). *Euphorbia antiquorum* Linn: A comprehensive review of ethnobotany, phytochemistry and pharmacology. *Journal of Analytical & Pharmaceutical Research, 2.4*(2016), 00024.

LaFever, R. E., Vogel, B. S., & Croteau, R. (1994). Diterpenoid resin acid biosynthesis in conifers: Enzymatic cyclization of geranylgeranyl pyrophosphate to abietadiene, the precursor of abietic acid. *Archives of Biochemistry and Biophysics, 313*(1), 139–149.

Lanhers, M. C., Fleurentin, J., Dorfman, P., Mortier, F., & Pelt, J. M. (1991). Analgesic, antipyretic and antiinflammatory properties of *Euphorbia hirta*. *Planta Medica, 57*(3), 225–231.

Le Maitre, C. L., Pockert, A., Buttle, D. J., Freemont, A. J., & Hoyland, J. A. (2007). Matrix synthesis and degradation in human intervertebral disc degeneration. *Biochemical Society Transactions, 35*(4), 652–655.

Leo, M., Schulte, M., Schmitt, L. I., Schäfers, M., Kleinschnitz, C., & Hagenacker, T. (2017). Intrathecal resiniferatoxin modulates TRPV1 in DRG neurons and reduces TNF-induced pain-related behavior. *Mediators of Inflammation, 2017*, 2786427.

Lim, E. Y., & Kim, Y. T. (2016). Food-derived natural compounds for pain relief in neuropathic pain. *BioMed Research International, 2016*, 7917528.

Majid, M., Khan, M. R., Shah, N. A., Ul Haq, I., Farooq, M. A., Ullah, S., et al. (2015). Studies on phytochemical, antioxidant, antiinflammatory and analgesic activities of *Euphorbia dracunculoides*. *BMC Complementary and Alternative Medicine, 15*, 349.

Mali, P. Y., & Panchal, S. S. (2013). A review on phyto-pharmacological potentials of *Euphorbia thymifolia* L. *Ancient Science of Life, 32*(3), 165–172.

Manchope, M. F., Calixto-Campos, C., Coelho-Silva, L., Zarpelon, A. C., Pinho-Ribeiro, F. A., Georgetti, S. R., et al. (2016). Naringenin inhibits superoxide anion-induced inflammatory pain: Role of oxidative stress, cytokines, Nrf-2 and the NO-cGMP-PKG-KATP channel signaling pathway. *PLoS One, 11*(4), e0153015.

Mannes, A., Hughes, M., Quezado, Z., Berger, A., Fojo, T., Smith, R., et al. (2010). Resiniferatoxin, a potent TRPV1 agonist: Intrathecal administration to treat severe pain associated with advanced cancer—Case report. *The Journal of Pain, 11*(4), S43.

Maroon, J. C., Bost, J. W., & Maroon, A. (2010). Natural antiinflammatory agents for pain relief. *Surgical Neurology International, 1*, 80. https://doi.org/10.4103/2152-7806.73804.

Mitchell, K., Bates, B. D., Keller, J. M., Lopez, M., Scholl, L., Navarro, J., et al. (2010). Ablation of rat TRPV1-expressing Adelta/C-fibers with resiniferatoxin: Analysis of withdrawal behaviors, recovery of function and molecular correlates. *Molecular Pain, 6*, 94.

Morissette, M., Litim, N., & Di Paolo, T. (2018). Natural phytoestrogens: A class of promising neuroprotective agents for Parkinson disease. In G. Brahmachari (Ed.), *Discovery and development of neuroprotective agents from natural products* (pp. 9–61). Elsevier (Chapter 2).

Mwine, J., & Van Damme, P. (2011). Why do *Euphorbiaceae* tick as medicinal plants? A review of *Euphorbiaceae* family and its medicinal features. *Journal of Medicinal Plant Research, 5*, 652–662.

Naidu, P. S., Singh, A., & Kulkarni, S. K. (2003). D2-dopamine receptor and α 2, adrenoreceptor-mediated analgesic response of quercetin. *Indian Journal of Experimental Biology, 41*, 1400–1404.

Ninomiya, M., & Koketsu, M. (2013). Minor flavonoids (chalcones, flavanones, dihydrochalcones, and aurones). In *Natural products* (pp. 1867–1900). Springer.

Ohbuchi, K., Mori, Y., Ogawa, K., Warabi, E., Yamamoto, M., & Hirokawa, T. (2016). Detailed analysis of the binding mode of vanilloids to transient receptor potential vanilloid type I (TRPV1) by a mutational and computational study. *PLoS One, 11*(9), e0162543.

Palit, P., Mukherjee, D., Mahanta, P., Shadab, M., Ali, N., Roychoudhury, S., et al. (2018). Attenuation of nociceptive pain and inflammatory disorders by total steroid and terpenoid fraction of *Euphorbia tirucalli* Linn root in experimental *in vitro* and *in vivo* model. *Inflammopharmacology, 26*(1), 235–250.

Parisi, J. R., Martins de Andrade, A. L., Torres Silva, J. R., & Silva, M. L. (2017). Antiallodynic effect of intrathecal resiniferatoxin on neuropathic pain model of chronic constriction injury. *Acta Neurobiologiae Experimentalis (Wars), 77*(4), 317–322.

Pinho-Ribeiro, F. A., Zarpelon, A. C., Fattori, V., Manchope, M. F., Mizokami, S. S., Casagrande, R., et al. (2016). Naringenin reduces inflammatory pain in mice. *Neuropharmacology, 105*, 508–519.

Pitcher, M. H., Von Korff, M., Bushnell, M. C., & Porter, L. (2019). Prevalence and profile of high-impact chronic pain in the United States. *The Journal of Pain, 20*(2), 146–160.

Prabha, M. N. C., Ramesh, C. K., Kuppast, I. J., & Mankani, K. L. (2008). Studies on antiinflammatory and analgesic activities of *Euphorbia tirucalli L.* latex. *International Journal of Chemical Sciences*, 6(4), 1781–1787.

Salas, M. M., Clifford, J. L., Hayden, J. R., Iadarola, M. J., & Averitt, D. L. (2017). Local resiniferatoxin induces long-lasting analgesia in a rat model of full thickness thermal injury. *Pain Medicine*, 18(12), 2453–2465.

Saleem, U., Ahmad, B., Ahmad, M., Hussain, K., & Bukhari, N. I. (2015). Anti-nociceptive, antiinflammatory and anti-pyretic activities of latex and leaves methanol extract of *Euphorbia helioscopia*. *Asian Pacific Journal of Tropical Disease*, 5(4), 322–328.

Salehi, B., Iriti, M., Vitalini, S., Antolak, H., Pawlikowska, E., Kręgiel, D., et al. (2019). Euphorbia-derived natural products with potential for use in health maintenance. *Biomolecules*, 9(8), 337.

Sam, H. V., Baas, P., & Kessler, P. J. A. (2008). Traditional medicinal plants in Ben En National Park, Vietnam. *Blumea – Biodiversity, Evolution and Biogeography of Plants*, 53, 569–601.

Sapio, M. R., Neubert, J. K., LaPaglia, D. M., Maric, D., Keller, J. M., Raithel, S. J., et al. (2018). Pain control through selective chemo-axotomy of centrally projecting TRPV1+ sensory neurons. *The Journal of Clinical Investigation*, 128(4), 1657–1670.

Sdayria, J., Rjeibi, I., Feriani, A., Ncib, S., Bouguerra, W., Hfaiedh, N., et al. (2018). Chemical composition and antioxidant, analgesic, and antiinflammatory effects of methanolic extract of *Euphorbia retusa* in mice. *Pain Research & Management*, 2018, 4838413.

Seebaluck, R., Gurib-Fakim, A., & Mahomoodally, F. (2015). Medicinal plants from the genus *Acalypha* (*Euphorbiaceae*)—A review of their ethnopharmacology and phytochemistry. *Journal of Ethnopharmacology*, 159, 137–157.

Shi, Q.-W., Su, X.-H., & Kiyota, H. (2008). Chemical and pharmacological research of the plants in genus *Euphorbia*. *Chemical Reviews*, 108(10), 4295–4327.

Szallasi, A., & Blumberg, P. M. (1989). Resiniferatoxin, a phorbol-related diterpene, acts as an ultrapotent analog of capsaicin, the irritant constituent in red pepper. *Neuroscience*, 30(2), 515–520.

Tian, R., Yang, W., Xue, Q., Gao, L., Huo, J., Ren, D., et al. (2016). Rutin ameliorates diabetic neuropathy by lowering plasma glucose and decreasing oxidative stress *via* Nrf2 signaling pathway in rats. *European Journal of Pharmacology*, 771, 84–92.

Valsecchi, A. E., Franchi, S., Panerai, A. E., Sacerdote, P., Trovato, A. E., & Colleoni, M. (2008). Genistein, a natural phytoestrogen from soy, relieves neuropathic pain following chronic constriction sciatic nerve injury in mice: Anti-inflammatory and antioxidant activity. *Journal of Neurochemistry*, 107(1), 230–240.

Vasas, A., & Hohmann, J. (2014). *Euphorbia* diterpenes: Isolation, structure, biological activity, and synthesis (2008–2012). *Chemical Reviews*, 114(17), 8579–8612.

Veneziani, R. C. S., Ambrósio, S. R., Martins, C. H. G., Lemes, D. C., & Oliveira, L. C. (2017). Antibacterial potential of diterpenoids. In R. Atta-ur (Ed.), *Vol. 54. Studies in natural products chemistry* (pp. 109–139). Elsevier (Chapter 4).

Viana, G. S., Bandeira, M. A., & Matos, F. J. (2003). Analgesic and antiinflammatory effects of chalcones isolated from *Myracrodruon urundeuva allemao*. *Phytomedicine*, 10(2–3), 189–195.

Wang, H., & Murphy, P. A. (1994). Isoflavone content in commercial soybean foods. *Journal of Agricultural and Food Chemistry*, 42(8), 1666–1673.

Wong, G. Y., & Gavva, N. R. (2009). Therapeutic potential of vanilloid receptor TRPV1 agonists and antagonists as analgesics: Recent advances and setbacks. *Brain Research Reviews*, 60(1), 267–277.

Yang, R., Li, L., Yuan, H., Liu, H., Gong, Y., Zou, L., et al. (2019). Quercetin relieved diabetic neuropathic pain by inhibiting upregulated P2X4 receptor in dorsal root ganglia. *Journal of Cellular Physiology*, 234(3), 2756–2764.

Yasuda, T., Endo, M., Kon-no, T., Kato, T., Mitsuzuka, M., & Ohsawa, K. (2005). Antipyretic, analgesic and muscle relaxant activities of pueraria isoflavonoids and their metabolites from *Pueraria lobata* Ohwi-a traditional Chinese drug. *Biological & Pharmaceutical Bulletin*, 28(7), 1224–1228.

Yoshida, T., Inoue, R., Morii, T., Takahashi, N., Yamamoto, S., Hara, Y., et al. (2006). Nitric oxide activates TRP channels by cysteine S-nitrosylation. *Nature Chemical Biology*, 2(11), 596–607.

You, J. S., Cho, I. A., Kang, K. R., Oh, J. S., Yu, S. J., Lee, G. J., et al. (2017). Coumestrol counteracts interleukin-1β-induced catabolic effects by suppressing inflammation in primary rat chondrocytes. *Inflammation*, 40(1), 79–91.

Zahidin, N. S., Saidin, S., Zulkifli, R. M., Muhamad, I. I., Ya'akob, H., & Nur, H. (2017). A review of *Acalypha indica* L.(*Euphorbiaceae*) as traditional medicinal plant and its therapeutic potential. *Journal of Ethnopharmacology*, 207, 146–173.

Chapter 36

Analgesic properties and mechanisms of action of *Muntingia calabura* extracts: A review

Zainul Aminuddin Zakaria[a], Tavamani Balan[b], Mohd. Hijaz. Mohd. Sani[c], and Nurfuzillah Abdul Rani[b]

[a]*Department of Biomedical Sciences, Faculty of Medical and Health Sciences, University Putra Malaysia, Serdang, Selangor, Malaysia,* [b]*Department of Pharmaceutical Technology, Faculty of Pharmacy and Health Sciences, University Kuala Lumpur Royal College of Medicine Perak, Ipoh, Perak, Malaysia,* [c]*Department of Biomedical Sciences and Therapeutics, Faculty of Medicine and Health Sciences, University Malaysia Sabah, Kota Kinabalu, Sabah, Malaysia*

Abbreviations

AQMC	aqueous extract of *M. calabura*
AQPMC	aqueous partition of *M. calabura*
ASA	acetylsalicilic acid
CEMC	chloroform extract of *M. calabura*
cGMP	cyclic guanosine monophosphate
COX	cyclooxygenase
DPPH	2,2-diphenyl-1-picrylhydrazyl
EAPMC	ethyl acetate partition of *M. calabura*
HPLC	high-Performance Liquid Chromatography
LOX	lipoxygenase
L-NAME	NG-nitro-L-arginine methyl esters
L-NMMA	NG-monomethyl-L-arginine acetate
MB	methylene blue
MEMC	methanol extract of *M. calabura*
mRNA	messenger ribonucleic acid
NO	nitric oxide
NOS	nitric oxide synthase
NSAIDs	nonsteroidal antiinflammatory drugs
ORAC	oxygen radical absorbance capacity
PEPMC	petroleum ether partition of *M. calabura*
PMA	phorbol 12-myristate 13-acetate
PKA	protein kinase A
PKC	protein kinase C
T & CM	traditional and complementary medicine
TRPV1	transient receptor potential cation channel subfamily V member 1

Introduction

Pain is a discomforting sense caused by damaging stimuli and it is usually associated with the body's natural defense system, indicating the presence of tissue injury. Nociception is described as "the neural processes of encoding and processing noxious stimuli" that usually leads to pain (Zakaria et al., 2019). The pain sensation is aversive at the threshold. It serves as an important teaching signal that allows an organism to avoid injury, thus making it essential for survival (Ossipov, 2012).

Treatments, Mechanisms, and Adverse Reactions of Anesthetics and Analgesics. https://doi.org/10.1016/B978-0-12-820237-1.00036-3
Copyright © 2022 Elsevier Inc. All rights reserved.

In general, pain is modulated by various analgesic drugs. These drugs act via several mechanisms on the central and peripheral nervous system that serve to suppress pain signals (Kirkpatrick et al., 2015). Pain perception is mediated by numerous receptors and activation of many signaling cascades (Zakaria et al., 2018). The understanding behind the cascades activation and mediation of nociceptor sensitization is still at the preliminary level. Therefore, scientists around the world have always shown interest in exploring the complexity behind pain mediation and efforts are ongoing to develop new agents that target the pain pathway.

Furthermore, the available analgesic drugs are also associated with various adverse effects. For example, opioids cause constipation, nausea, vomiting, dizziness, sedation, dependency, and respiratory depression (Khansari, Sohrabi, & Zamani, 2013), while nonsteroidal antiinflammatory drugs (NSAIDs) are strongly associated with gastrointestinal bleeding and reduced platelet aggregation (Zakaria et al., 2019). Therefore, there is an urgent need for the development of antinociceptive agents with enhanced potency with promising pharmacological actions and safe with reduced or no side effects.

The realm of ethnopharmacology opens the path for the investigation of novel and safe therapeutic agents in pain management. The indigenous knowledge aids in the discovery and development of new agents from medicinal plants, eventually giving rise to the development of Traditional and complementary medicine (T & CM). Plants tend to be the promising source of a chemical substance that owns excellent therapeutic actions.

One of the medicinal plants that is in the limelight due to its potential to relieve pain is *Muntingia calabura* L (Fig. 1). *M. calabura* is known as "Jamaican cherry" and in Malaysia, the plant is more familiar with the name "kerukup siam" or "ceri kampung." It belongs to the family Muntingiacea and is widely cultivated in warm areas in India and Southeast Asia (Mahmood et al., 2014). *M. calabura* is well recognized for its medicinal benefit in Peruvian folklore. Traditionally, *M. calabura* is used to treat headaches, fever, and incipient cold. Scientifically, the plant has been reported to possess antitumor, antibacterial, antiinflammatory, antipyretic, antiulcer, antioxidant, and antiproliferative activities (Mahmood et al., 2014).

The antinociceptive activity of *M. calabura* has been well studied and its mechanism of action has been elucidated via researches that have been carried out by our team. It has been found that this plant possesses significant antinociceptive activity and there was the involvement of several pathways for the plant to exert its activity. This article is intended to review the antinociceptive activity of *M. calabura* from the previously published research articles to strengthen the stand of this plant as an alternative way to reduce pain.

Muntingia calabura L

M. calabura is the only species within the genus Muntingia. The plant is native to the Greater Antilles, tropical South America, Central America, southern Mexico, St. Vincent, and Trinidad. *M. calabura* is widely cultivated in warm areas of the Asian region (Chin, 1989). According to Peruvian folklore, the leaves of *M. calabura* (Fig. 2) are used to provide

FIG. 1 *Muntingia calabura* tree. The trees grow abundantly on the roadsides in Malaysia. This picture was captured in Serdang, Selangor.

FIG. 2 *Muntingia calabura* leaves and flower. The leaves are *dark green* color, alternate lanceolate, and minutely hairy on the surface. The flowers have *white* petals and prominent *yellow* stamens.

relief from gastric ulcers, reduce swelling of the prostate gland, and alleviate cold and headache while the flowers and strips of its bark are used as an antiseptic and as a wash to reduce swelling in the lower extremities (Morton, 1987; Zakaria, Mustapha, et al., 2007). The plant is also considered to be an antiseptic, antihysteric, antispasmodic, antidyspeptic, and diaphoretic agent (Kaneda et al., 1991; Nshimo, Pezzuto, Kinghorn, & Farnsworth, 1993; Perry, 1895). Besides that, the plant is used to treat mouth pimples, measles, and stomach ache in Mexico (Yasunaka et al., 2005) while the infusion of the flowers is used as a tranquillizer and tonic in Colombia (Kaneda et al., 1991; Perez-Arbealaez, 1975).

The establishment of the pharmacological value of *M. calabura* through meticulous scientific investigations has been done by researchers all over the world for the past 22 years. Various properties have been scientifically reported to be demonstrated by *M. calabura* leaves, including antiinflammatory (Jisha, Vysakh, Vijeesh, & Latha, 2019), antipyretic (Zakaria, Hazalin, et al., 2007; Zakaria et al., 2007), antiulcer (Balan et al., 2013), hepatoprotective (Zakaria, Mahmood, Mamat, Nasir, & Omar, 2018), antihypertensive (Shih, 2009), antioxidant (Sindhe, Bodke, & Chandrashekar, 2013; Zakaria et al., 2011), cytotoxic (Sufian, Ramasamy, Ahmat, Zakaria, & Yusof, 2013), antiproliferative (Zakaria et al., 2011), antihyperglycemic (Sindhe et al., 2013), and antibacterial activities (Sarojini & Mounika, 2018). In addition, various phytochemical constituents have been identified from various parts of *M. calabura* (Mahmood et al., 2014).

Antinociceptive activity of *Muntigia calabura* leaves extracts

The first attempt to study the antinociceptive activity of *M. calabura* was performed in 2006. Three separate reports evaluated the activity of aqueous extract of *M. calabura* (AQMC) leaves against acetic acid-induced abdominal constriction (Zakaria et al., 2006; Zakaria, Mustapha, et al., 2007; Zakaria, Somchit, Sulaiman, Mat Jais, & Fatimah, 2008) and hot plate test (Zakaria, Mustapha, et al., 2007). In these studies, the AQMC was prepared in the concentrations of 10%, 50% and 100% (equivalent to 27, 135 and 270 mg/kg) (Zakaria et al., 2006; Zakaria et al., 2008) and 1%, 5%, 10%, 50%, 100% (equivalent to 2.7, 13.5, 27, 135 and 270 mg/kg) (Zakaria, Mustapha, et al., 2007). Acetylsalicylic acid (ASA; 100 mg/kg) and 5 mg/kg morphine were used as the positive control for abdominal constriction and hot plate test, respectively. The injection of 0.6% acetic acid into the peritoneal cavity of mice causes stretching behavior or writhing. This test is a good tool to screen the pharmacological activity of compounds with analgesic properties, although some psychoactive agents are able to inhibit writhing. In all three studies, subcutaneous administration of the extract reduced the number of writhing in a concentration-dependent manner. The highest analgesic effect was produced by 100% AQMC at 83% (Zakaria, Mustapha, et al., 2007). Next, the antinociceptive effect of AQMC was determined using a hot plate test. This test is used to evaluate thermal-induced centrally mediated pain. Based on the findings, AQMC demonstrated a concentration-independent activity. The extract's analgesic effect was observed from 60 min until the end of the experiment (180 min) following administration of AQMC.

In addition, the effect of various temperatures (40, 60, 80, and 100°C) on 50% of AQMC's antinociceptive activity was also assessed using both the abdominal constriction and hot-plate tests. The findings demonstrated that 50% AQMC was stable against the effect of various temperatures as indicated by the presence of significant activity in both tests. It is well known that high temperatures can degrade the compounds present in the extract (Soquetta, Terra, & Bastos, 2018). Nevertheless, AQMC at various temperatures was still able to exert its activity in the hot plate test, indicating the presence of heat-stable compound(s) in the extract.

Zakaria and the team continued to build the antinociceptive profile of AQMC where they published another finding in the same year (Zakaria, Hazalin, et al., 2007). In this study, different concentrations of AQMC (10%, 50%, and 100%) were subjected to formalin test, a bi-phasic test used to evaluate centrally mediated neurogenic pain in the first phase and peripherally mediated inflammatory pain in the later phase. Based on the findings, AQMC at concentrations of 10%, 50%, and 100% (27, 135, and 270 mg/kg) produced a concentration-independent activity in both the early and late phases of nociception indicating the involvement of the central mechanism in the antinociception exerted by the extract.

In the following year, Zakaria et al. (2008) further studied the effect of pH and enzymes on the peripheral antinociceptive activity of AQMC. The same concentrations of AQMC (10%, 50%, and 100%) were subjected to an abdominal constriction test. The results showed that AQMC exerted significant antinociceptive activity in a concentration-dependent manner. In addition, 50% AQMC with pH 5.1 was pretreated against various pH (3, 5, 7, 9, 11, or 13) for 2 h before neutralizing back to pH 5.1 and administered to the mice, which was then subjected to the antinociceptive test. Furthermore, the 50% AQMC also was pretreated with 10% amylase, 10% lipase, or 1% protease for 2 h at 40°C before cooled to room temperature and then subjected to the antinociceptive assay. The acidic and alkaline conditions did not alter the antinociceptive activity of 50% AQMC. Furthermore, pretreatment with amylase, protease, lipase, or their combination did not result in any significant change in antinociception of AQMC. These results clearly proved that AQMC was able to resist the effect of extreme acidic and alkaline conditions as well as the chemical reactions of various enzymes. This could be due to the existence of AQMC's stable short-chain macromolecules or bioactive compounds.

While establishing the antinociceptive profile of AQMC, Zakaria and the team also studied the antinociceptive activity of the chloroform extract of *M. calabura* (CEMC) (Zakaria, Kumar, et al., 2007). CEMC, prepared in the concentrations of 10%, 50%, and 100%, was tested against acetic acid-induced abdominal constriction, formalin-induced paw licking, and the hot plate test. The extract exhibited a concentration-dependent activity with the highest concentration produced analgesic activity of more than 95%, while the 50% CEMC exerted equal effectiveness of the reference drug used, 100 mg/kg ASA. In the formalin test, CEMC showed significant activity in both phases whereby the activity was in a concentration-dependent manner in the early phase but not in the late phase. The reference drugs used in this test were morphine (5 mg/kg) for the early and late phase, and ASA (100 mg/kg) for the late phase. In the hot-plate test, CEMC showed inconsistent results throughout the entire experiment resulting in a concentration-independent activity On the whole, As CEMC was able to reduce the chemically and thermally induced nociception; thus, this indicates that CEMC was able to exert its effect both centrally and peripherally.

A few years later, Zakaria's team continued to explore the antinociceptive activity of *M. calabura* leaves, but this time using methanolic extract (MEMC) (Sani, Zakaria, Balan, Teh, & Salleh, 2012). In this study, 100, 250, and 500 mg/kg of MEMC were given to animals against three models of nociception, namely acetic acid-induced abdominal constriction, formalin-induced paw licking, and hot plate test. The extract exerted a significant dose-dependent antinociceptive activity in the abdominal constriction test. In the formalin test, only 250 and 500 mg/kg demonstrated significant dose-dependent analgesic activity in the early phase, whereas all doses produced significant antinociceptive activity in the late phase. Compared to the standard control, morphine (5 mg/kg) was significantly more effective than MEMC in both phases, while ASA (100 mg/kg) exhibited better activity compared to the extract in the late phase. The hot plate test, which evaluates central-mediated pain, demonstrated only MEMC at the highest dose (500 mg/kg) produced significant analgesic activity by prolonging the latency of response to heat stimuli by an average of 3 s. However, it was significantly lowered when compared to morphine (5 mg/kg), which increased latency by an average of 8 s.

In the following year, there was an attempt by Yusof et al. (2013) to fractionate MEMC and the fractions were subjected to antinociceptive activity. In this study, MEMC was partitioned to produce ethyl acetate (EAPMC), petroleum ether (PEPMC), and aqueous (AQPMC) partition. These partitions were tested against formalin-induced nociception in the doses of 100, 500, and 1000 mg/kg. The results obtained demonstrated that PEPMC has the most effective antinociception as compared to other partitions in the early and late phases.

On the other hand, in the year 2016, Zakaria, Mohd Sani, Abdul Kadir, Kek, and Salleh (2016) also had further partitioned the MEMC to PEPMC and studied its antinociceptive activity. PEPMC was subjected to acetic acid-induced abdominal constriction, formalin, and hot plate tests in the doses of 100, 250, and 500 mg/kg. The acetic-acid induced abdominal constriction test revealed significant and dose-dependent effects. Next, the result obtained from the formalin test showed a similar dose-dependent effect to acetic acid-induced abdominal constriction in both phases (centrally and peripherally). However, the activity (in both phases) was lower as compared to the reference drug, which was 5 mg/kg morphine. Meanwhile, in the hot plate test, morphine exerted the highest latency time as compared to other doses of PEPMC extract on the starting minutes.

Overall, it could be summarized that all the extracts of *M. calabura* exerted antinociceptive activity in the abdominal constriction, hot-plate, and formalin tests (Table 1). In the abdominal constriction test, pain is produced through an injection of an irritant into the peritoneal cavity which produces episodes of typical abdominal stretching movements (Le Bars, Gozariu, & Cadden, 2001). This behavior is said to be related to the hindrance in the peritoneal fluid's levels of prostaglandin and cytokines. Besides, this model also represents the stimulation of peripheral mechanisms as it is believed that phlogogen administration leads to a rise in cyclooxygenase (COX) and lipooxygenase (LOX) levels (Ikeda, Ueno, Naraba, & Oh-ishi, 2001). Thus, it could be deduced that the extracts of *M. calabura* could have contributed to the inhibition of prostaglandin because the nociceptive mechanism involves the processing or releasing of arachidonic acid metabolites through COX and biosynthesis of prostaglandin (Guimarães et al., 2010). The hot plate test involves elevation of the supraspinal nociceptive processing (Deuis, Dvorakova, & Vetter, 2017). It has been applied to evaluate the activity of various extracts of *M. calabura*. The results showed that the administration of these extracts possessed antinociception either centrally or peripherally or both. Besides, the effect of various temperatures, pH, and enzymes indicates that the extract may contain stable bioactive compounds that were able to resist the effect of these extreme temperatures, acidic, and alkaline conditions, and chemical reactions of various enzymes.

TABLE 1 Summary of antinociceptive profile studies that have been done on *M. calabura* extracts using various animal models.

Plant Extracts	Concentration/doses used	Animal Model	Mechanism of action	References
AQMC	10%, 50% and 100% 5%, 50% and 100% 1%, 5%, 10%, 50%, 100%	Abdominal constriction test	Peripheral	Zakaria et al. (2006) Zakaria et al. (2008) Zakaria, Mustapha, et al. (2007)
	1%, 5%, 10%, 50%, 100% 10%, 50%, and 100%	Hot plate test	Central	Zakaria, Mustapha, et al. (2007) Zakaria, Hassan, et al. (2007)
	10%, 50%, and 100%	Formalin test	Central and peripheral	Zakaria, Hazalin, et al. (2007)
CEMC	10%, 50%, and 100%	Abdominal constriction and formalin test	Central and peripheral	Zakaria, Kumar, et al. (2007)
	10%, 50%, and 100%	Hot plate test	Central	Zakaria, Kumar, et al., 2007, Zakaria, Hassan, et al. (2007)
MEMC	100, 250, and 500 mg/kg	Abdominal constriction, hot plate, and formalin test	Central and peripheral	Sani et al. (2012)
Fractions of MEMC PEPMC, EAPMC, AQPMC	100, 500, and 1000 mg/kg	Formalin test	Central and peripheral	Yusof et al., 2013
PEPMC	100, 250, and 500 mg/kg	Abdominal constriction test, hot plate test, and formalin test	Central and peripheral	Zakaria et al. (2016)

All the extracts of *M. calabura* were able to exert its antinociceptive activity both centrally and peripherally. *AQMC*: aqueous extract of *M. calabura*; *AQPMC*: aqueous partition of *M. calabura*; *CEMC*: chloroform extract of *M. calabura*; *MEMC*: methanol extract of *M. calabura*; *PEPMC*: petroleum ether partition of *M. calabura*; *EAPMC*: ethyl acetate partition of *M. calabura*.

424 PART | IV Novel and nonpharmacological aspects and treatments

TABLE 2 Summary of phytoconstituents present in *M. calabura* leaves extracts.

Extract	Pyhtoconstituents	Types of analysis	References
AQMC	Flavanoids (catechin, rutin, fisetin)	HPLC	Zakaria, Hazalin, et al. (2007)
MEMC	Flavanoids (rutin, quercitrin, fisetin)	HPLC	Zakaria et al. (2014)
PEPMC	5-hydroxy-3,7,8-trimethoxyflavone 3,7-dimethoxy-5-hydroxyflavone 2′,4′-dihydroxy-3′-methoxychalcone Calaburone	Isolation of compounds LC-MS NMR	Yusof et al. (2013)
PEPMC	Flavanoids (pinostrobin, flavanone)	HPLC	Zakaria et al. (2016)

The *M. calabura* extracts showed the presence of flavanoid-based compounds, which could have contributed to its antinociceptive effect. *AQMC*: aqueous extract of *M. calabura*; *HPLC*: high-performance liquid chromatography; *LC-MS*: liquid chromatography-mass spectrometry; *MEMC*: methanol extract of *M. calabura*; *NMR*: nuclear magnetic resonance; *PEPMC*: petroleum ether partition of *M. calabura*.

The effect of neurogenic pain at the early stage is from an extreme reaction to formalin's direct activity on nociceptors within the intraplantar zone (Giorno, Moreira, Rezende, & Fernandes, 2018). For the late phase, the result observed as an antiinflammatory mediated pain is due to the tonic response from the release of inflammatory mediators and activation of neurons in the dorsal horns of the spinal cords (Kendroud & Hanna, 2019). Therefore, the formalin test represents the irritating effect at the C-type sensory in the early phase and tonic inflammatory pain response in the later phase (Chen et al., 2014). Extracts of *M. calabura* were able to show a significant effect in the formalin test, indicating its involvement in both the early and late phases of nociception.

On the other hand, Zakaria, Hazalin, et al. (2007), Zakaria et al. (2014, 2016) and Yusof et al. (2013) have reported the presence of various phytoconstituents in the extracts of *M. calabura* as summarized in Table 2. Flavonoids are multitarget molecules and are well known for their analgesic, antiinflammatory, and antioxidant effects while having safe clinical and preclinical profiles (Ferraz et al., 2020). Thus, it is worth saying that the antinociceptive activity exerted by *M. calabura* extracts could be due to the presence of the flavanoid-based compounds and their synergistic actions.

Mechanisms of action underlying the antinociceptive activity of *Muntingia calabura* leaves extracts

The involvement of the opioid receptor system has been studied on AQMC, MEMC, and PEPMC (Sani et al., 2012; Zakaria et al., 2006; Zakaria et al., 2016). The antinociceptive of 50% AQMC in the hot plate test was significantly reversed by the pretreatment of a nonselective opioid receptor antagonist, naloxone (2 and 10 mg/kg), administered intraperitoneally (i.p.). The analgesic activity of 500 mg/kg MEMC and PEPMC was also significantly reversed in all pain assay models (abdominal constriction, formalin, and hot plate tests) when the animals were pretreated with 5 mg/kg naloxone. Further evaluation on the involvement of MEMC and PEPMC in opioid receptor modulation, β-funaltrexamine, naltrindole, and nor-binaltorphimine (μ-, δ- and κ-opioid receptors antagonist, respectively) was used to specifically determine the opioid subtype involved in the antinociceptive action of the extracts. Through this study, these results demonstrated the possible involvement of opioid receptors in the extracts' antinociceptive mode of action (Zakaria et al., 2014; Zakaria et al., 2016).

Various pain-related nonopioid receptors were also investigated to unravel the possible mechanism of action of *M. calabura* extracts. In the study by Zakaria, Hassan, et al. (2007), antinociceptive activity of CEMC was suggested to be involved through the modulation of GABAergic, α1-adrenergic, α2-adrenergic, and β-adrenergic receptors following the pretreatment of 10 mg/kg bicuculline, 10 mg/kg pindolol, 10 mg/kg phenoxybenzamine and 10 mg/kg yohimbine, respectively. In this study, the analgesic effect of the extract at the concentration of 10%, 50%, and 100% significantly reduced as the latency of distress recorded in the hot plate test reduces. Zakaria et al. (2008) demonstrated the possible involvement of AQMC (50%) in modulating adrenergic and muscarinic receptors in its antinociceptive activity based on the attenuation of activity following subcutaneous administration of pindolol (10 mg/kg) and atropine (5 mg/kg), respectively. The effect of various receptor antagonists on MEMC-induced antinociception assay was conducted by Zakaria et al. (2014) using the abdominal constriction test. Atropine (10 mg/kg, i.p.), caffeine (3 mg/kg, i.p.), pindolol (1 mg/kg. i.p.), haloperidol (20 mg/kg, i.p.), yohimbine (0.15 mg/kg, i.p) or prazosin (0.15 mg/kg, i.p), were pretreated to the animal before the administration of MEMC (500 mg/kg). Except for atropine and haloperidol, the antinociceptive effect of MEMC was significantly reversed indicating the possible involvement of adenosinergic, serotonergic, and noradrenergic receptor systems.

The involvement of potassium channels was evaluated by Zakaria et al. (2014, 2016). Intraperitoneal administration of 0.04 mg/kg apamin (small-conductance Ca^{2+}-activated K^+ channels), 10 mg/kg glibenclamide (an ATP sensitive K^+ channel inhibitor), 4 mg/kg tetraethylammonium chloride (a nonselective voltage dependant K^+ channel inhibitor), and 0.02 mg/kg charybdotoxin (an inhibitor of large-conductance Ca^{2+}-activated K^+ channels) significantly reversed MEMC (500 mg/kg). On the other hand, the antinociceptive activity of PEPMC (500 mg/kg) involved the modulation of all but large-conductance Ca^{2+}-activated K^+ channel.

The role of transient receptor potential cation channel subfamily V member 1 (TRPV1), glutamatergic (Sani et al., 2012), B_2 receptor, and protein kinase C (PKC) pathway (Zakaria et al., 2014) in the modulation of MEMC and PEPMC antinociceptive activity was also successfully demonstrated. Effect of pain induced by capsaicin (1.6 μg/paw, 50 μL, intraplantar) significantly reduced by all doses of MEMC (100, 250, and 500 mg/kg) while only 250 and 500 mg/kg PEPMC reduces paw-licking time, indicating the involvement of TRPV1 receptor system. Both MEMC and PEPMC in all three doses significantly reduced paw licking time in the glutamate-induced nociception test suggesting the involvement of glutamatergic receptors. In the B_2 and PKC pathways, only 250 and 500 mg/kg of MEMC and PEPMC significantly attenuated the pain induced by bradykinin and phorbol 12-myristate 13-acetate (PMA), indicating the possible involvement of B_2 receptor and PKC pathway.

The involvement of the L-arginine/nitric oxide/cyclic guanosine monophosphate (L-arginine/NO/cGMP) pathway in the antinociception of the extracts was demonstrated in the study by Zakaria et al. (2006, 2016) and Sani et al. (2012). The pretreatment by L-arginine, the precursor of nitric oxide (NO), reversed the effect of AQMC antinociceptive activity, while pretreatment by NG-nitro-L-arginine methyl esters (L-NAME), a competitive and reversible inhibitor of NO synthase (NOS), showed an improvement to reduce the number of abdominal constriction at 27 and 135 mg/kg dose of AQMC. NG-monomethyl-L-arginine acetate (L-NMMA) (an inhibitor of NOS) produced a reduction in the number of abdominal constrictions and enhancement of AQMC antinociception properties, regardless of the doses used. In addition, the results also showed that during the co-treatment of L-arginine either with methylene blue (MB) or L-NAME improved antinociceptive activity at all concentrations used but both co-treatments did not help in improving the antinociceptive activity of AQMC, respectively. Therefore, it has proved the involvement of the L-arginine/NO/cGMP pathway in the antinociception of AQMC (Zakaria et al., 2006). MEMC analgesic activity was attenuated by the pretreatment of L-arginine (20 mg/kg, i. p.), while L-NAME (20 mg/kg, i.p.) maintained the antinociceptive activity of MEMC. L-Arginine was found to reverse the L-NAME-induced antinociceptive activity. Synergistically, both L-NAME and L-arginine failed to affect the same activity (Sani et al., 2012). However, based on the study by Zakaria et al. (2016), the PEPMC antinociceptive activity did not involve the modulation of the L-arginine/NO/cGMP pathway.

Conclusion

Pain modulation is a complex process that involves numerous mediators and receptors. Based on the findings, it can be concluded that *M. calabura* is a potential antinociceptive agent that exerts its effect via interaction with several physiological pathways in the central and peripheral nervous system. Fig. 3 represents a schematic diagram showing the overall possible mechanisms of action of *M. calabura* extracts. *M. calabura* seems to modulate antinociception via L-arginine/NO/cGMP pathway and also acts as an agonist for opioids, potassium channels, serotonergic, adrenergic, adenosinergic, as well as cholinergic receptors. On the other hand, the extracts seem to inhibit TRPV1, PKC, prostaglandin, bradykinin, and glutamatergic receptors. All these activities could be associated with the synergistic effect contributed by the flavanoid-based phytoconstituents present in the extracts. Therefore, these findings strongly show that *M. calabura* has great potential as an antinociceptive agent and the knowledge gained from these studies can be utilized in the discovery of new drug agents.

Applications to other areas

In this chapter, we have reviewed the antinociceptive potential of *M. calabura* and its possible mechanisms of action. The role of opioid and nonopioid receptors and involvement of the cGMP pathway has also been highlighted. In this section, we describe other pharmacological activities of *M. calabura* including some of the activities that could be associated with the antinociception exerted by the plant. *M. calabura* possesses significant antioxidant and radical scavenging activities. The plant is rich in antioxidants such as phenols and flavonoids while its ability to attenuate harmful radicals has been observed via the 2,2-diphenyl-1-picrylhydrazyl (DPPH), oxygen radical absorbance capacity (ORAC), reducing power, ferric ion chelating assay, superoxide anion, and NO scavenging activity assays. A well correlation between antioxidant activity and total phenolic and flavonoid contents has been observed. On the other hand, the plant was also able to reduce inflammation and pyrexia, which was observed via the carrageenan-induced paw edema and yeast-induced pyrexia assays,

426 PART | IV Novel and nonpharmacological aspects and treatments

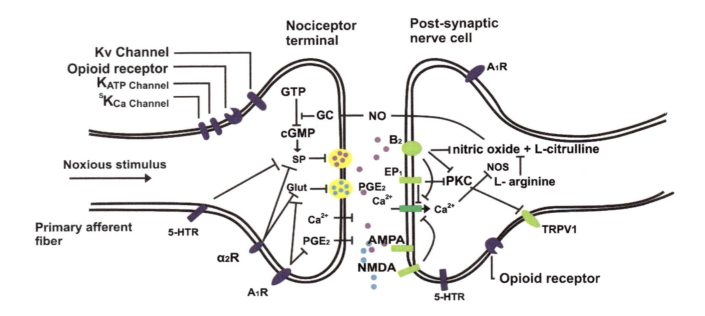

FIG. 3 The schematic diagram shows the possible mechanisms of the action exerted by *M. calabura* leaves. *M. calabura* shows antinociception both at central and peripheral levels and involves the modulation of various neurotransmitters and receptors. *5-HTR:* serotonergic receptor; *A₁R:* adenosine receptor; *AMPA:* α-amino-3-hydroxy-5-methyl-4-isoxazolepropionic acid; *B₂:* bradykinin receptor; *Ca²⁺:* calcium ion; *cGMP:* cyclic guanosine monophosphate; *EP₁:* prostaglandin receptor; *GC:* guanylate cyclase; *Glut:* glutamate; *GTP:* guanosine triphosphate; *K_{ATP}:* ATP-sensitive potassium channel; *ˢK_{Ca}:* small conductance calcium-activated potassium channel; *K_v:* voltage-gated potassium channel; *NMDA:* N-methyl-ᴅ-aspartate; *NO:* nitric oxide; *NOS:* nitric oxide synthase; *PGE₂:* prostaglandin E₂; *PKC:* protein kinase C enzyme; *TRPV1:* transient receptor potential cation channel subfamily V member 1; *α₂R:* α₂-adrenergic receptor.

respectively. Due to the strong antioxidant potential, *M. calabura* was also reported to exert significant antiulcer and hepatoprotective effect, which was partly modulated by the antiinflammatory ability of the plant. Furthermore, decreased activity of marker enzymes such as alanine transaminase, aspartate transaminase, creatinine phosphokinase, and lactate dehydrogenase in the heart tissues and serum of the rats proves the cardioprotective activity of *M. calabura* while the reduction in the mean systemic arterial pressure of the animal corresponds to the antihypertensive effect of the plant, which was mediated by the activation of nitric-oxide-dependent protein kinase G signaling pathways. In addition, *M. calabura* was found to exert a significant reduction in fasting blood glucose level, which was also associated with its antioxidant potential, hence offering protection against free radical-mediated damages. Besides, *M. calabura* also exerts antiproliferative and antibacterial effects, showing that the plant is able to impair and inhibit the growth of cancer cells as well as several microbial species that cause diseases in human. In a nutshell, extensive studies have been carried out on *M. calabura* and they were found to possess a broad range of pharmacological activities. Some of these activities are involved and associated with the pain relief pathways. Thus, this supports the suitability of *M. calabura* as a candidate for the treatment of neuropathic pain.

Other agents of interest

In this chapter, we have seen the potential use of the herbal plant, *M. calabura,* in the treatment of neuropathic pain. Besides *M. calabura,* there are also several other medicinal plants that have been found to be a promising agent that could be used in the management of pain, namely:

Syzygium jambos
Phyllanthus amarus
Papaver somniferum
Alpinia zerumbet
Pancratium maritimum
Zataria multiflora
Sempervivum tectorum
Salvia officinalis
Equisetum arvense
Physalis angulata

One of the plants that has caught the attention of researchers is *Physalis angulata. Physalis angulata* is a widespread indigenous herb from the family Solanaceae, mainly found in Africa, the Americas, and Asia. Traditionally used to treat sore throats, abdominal pain, as an analgesic and antirheumatic, scientists began studying the plant in 1987, with its analgesic properties evaluated for the first time in 2003. The chemical compounds, physalins, with potential antinociceptive and antiinflammatory activities have been isolated from *P. angulata.* Physalin B, D, F, and G isolated from the stem of the plant have been demonstrated to possessed consistent analgesic activities which are suggested to be via supraspinal pain inhibition, while only physalin F possesses antiinflammatory properties via inhibition of local edema and production of TNF-α. The compounds were also reported to have no sedative effect and did not affect the locomotor activity of the experimental animals. The study also reported the activity of physalin F and G to be longer-lasting when compared to morphine suggested its potential to be further developed into new analgesic drugs. Even though most natural product researches are targeted to isolate and identify the pure compound responsible for the activity, it is important to note that some plants exert promising activity in semipure or crude form. In line with this, the concentrated ethanolic extract from *P. angulata* was demonstrated to exert a more potent antinociceptive effect when compared to the purified compounds. The antinociceptive and antiinflammatory activities of this extract were via the modulation of tumor necrosis factor-α, interleukin-1B, COX-2, and iNOS mRNA based on the inflammatory analysis from the paw edema. The superiority of the extract over the purified compounds was believed to be due to synergistic actions of the bioactive constituents' mixture from the plant extracts. The result of this study suggested that direct purification of any new plant of interest might not be the best method to evaluate its pharmacological activity, while bioassay-guided isolation and purification could be a better pathway in discovering new plant-based agents. The pharmacological potency of both physalin and ant the concentrated extract showed a strong potential to be developed into phytomedicine; however, the current studies on its antinociceptive and antiinflammatory activities lack proper toxicity evaluation.

Mini-dictionary of terms

Antinociceptive: the activity of inhibiting or obstructing the detection of an injurious or painful stimulus by sensory neurons.

Writhing: large twisting movement in the abdomen due to unbearable pain.

Opioid drugs: class of pain-relieving drugs that act on an opioid receptor that belongs to the family of G protein-coupled receptors (GPCRs). These drugs give a strong pain-relieving effect. An example is a morphine.

Nonopioid drugs: pain-relieving drugs that act on other than opioid receptors. This includes selective and nonselective COX inhibitors and acetaminophen.

L-Arginine/nitric oxide/cyclic guanosine monophosphate: This is the pathway where nitric oxide is synthesized from L-arginine and mediates the cellular and intercellular communication by activating soluble guanylyl cyclase to produce cyclic GMP. This results in multiple effects, including the regulation of neuronal ion channels involved in the pain pathway.

Key facts of traditional and complementary medicine

- Traditional and complementary medicine (T&CM) is a crucial health resource which has many applications in the management and prevention of a variety of diseases.
- According to Global Report On Traditional And Complementary Medicine 2019 by World Health Organization (WHO), there are more countries that are recognizing the role of T&CM in their national health systems.

428 PART | IV Novel and nonpharmacological aspects and treatments

- One of the objectives of the WHO Traditional Medicine Strategy 2014–2023 is "To promote universal health coverage by integrating T&CM services into health care service delivery and self-health care".
- Thus, it is important to incorporate the use of herbal medicine in the attempt to seek a safer and effective alternative therapy in pain management.
- *M. calabura* has been proven to significantly reduce antinociception via several pathways, which makes it one of the promising candidates in the application of T&CM to improve global health.

Summary points

- This chapter focuses to present on the antinociceptive potency of *M. calabura* and its possible mechanisms of action involved have also been discussed.
- Various extracts of *M. calabura* showed significant antinociceptive activity against abdominal constriction, hot-plate, and formalin tests, showing that the extracts exert its activity either centrally or peripherally or via both.
- Moreover, it was also postulated that *M. calabura* may contain bioactive compounds which are stable and able to withstand extreme temperatures, acidic and alkaline conditions, and chemical reactions.
- Besides, *M. calabura* has been shown to modulate pain via L-arginine/NO/cGMP pathway, opioids, potassium channels, serotonergic, adrenergic, adenosinergic, as well as cholinergic receptors.
- Based on the findings demonstrated by the researchers, *M. calabura* is a T&CM which has a promising therapeutic effect in pain management and has the potential to be developed into health-related product as well as applied in the new drug discovery.

References

Balan, T., Sani, M. H., Suppaiah, V., Mohtarrudin, N., Suhaili, Z., Ahmad, Z., et al. (2013). Antiulcer activity of *Muntingia calabura* leaves involves the modulation of endogenous nitric oxide and non-protein sulfhydryl compounds. *Pharmaceutical Biology*, 52, 410–418.

Chen, Y., Kanju, P., Fang, Q., Lee, S. H., Parekh, P. K., Lee, W., et al. (2014). TRPV4 is necessary for trigeminal irritant pain and functions as a cellular formalin receptor. *Pain*, 155(12), 2662–2672.

Chin, W. Y. (1989). *A guide to the wayside trees of Singapore*. Singapore: BP Singapore Science Centre.

Deuis, J. R., Dvorakova, L. S., & Vetter, I. (2017). Methods used to evaluate pain behaviors in rodents. *Frontiers in Molecular Neuroscience*, 10, 284.

Ferraz, C. R., Carvalho, T. T., Manchope, M. F., Artero, N. A., Rasquel-Oliveira, F. S., Fattori, V., et al. (2020). Therapeutic potential of flavonoids in pain and inflammation: Mechanisms of action, pre-clinical and clinical data, and pharmaceutical development. *Molecules*, 25(3), 762.

Giorno, T., Moreira, I., Rezende, C. M., & Fernandes, P. D. (2018). New βN-octadecanoyl-5-hydroxytryptamide: Antinociceptive effect and possible mechanism of action in mice. *Scientific Reports*, 8(1), 10027.

Guimarães, A. G., Oliveira, G. F., Melo, M. S., Cavalcanti, S. C., Antoniolli, A. R., Bonjardim, L. R., et al. (2010). Bioassay-guided evaluation of antioxidant and antinociceptive activities of carvacrol. *Basic & Clinical Pharmacology & Toxicology*, 107(6), 949–957.

Ikeda, Y., Ueno, A., Naraba, H., & Oh-ishi, S. (2001). Involvement of vanilloid receptor VR1 and prostanoids in the acid-induced writhing responses of mice. *Life Sciences*, 69(24), 2911–2919.

Jisha, N., Vysakh, A., Vijeesh, V., & Latha, M. S. (2019). Anti-inflammatory efficacy of methanolic extract of *Muntingia calabura* L. leaves in carrageenan induced paw edema model. *Pathophysiology*, 26(3–4), 323–330.

Kaneda, N., Pezzuto, J. M., Soejarto, D. D., Kinghorn, A. D., Farnworth, N. R., Santisuk, T., et al. (1991). Plant anticancer agents, XLVIII. New cytotoxic flavonoids from *Muntingia calabura* roots. *Journal of Natural Products*, 54, 196–206.

Kendroud, S., & Hanna, A. (2019). Physiology, nociceptive pathways. In *StatPearls [internet]* StatPearls Publishing.

Khansari, M., Sohrabi, M., & Zamani, F. (2013). The usage of opioids and their adverse effects in gastrointestinal practice: A review. *Middle East Journal of Digestive Diseases*, 5(1), 5–16.

Kirkpatrick, D. R., McEntire, D. M., Hambsch, Z. J., Kerfeld, M. J., Smith, T. A., Reisbig, M. D., et al. (2015). Therapeutic basis of clinical pain modulation. *Clinical and Translational Science*, 8(6), 848–856.

Le Bars, D., Gozariu, M., & Cadden, S. W. (2001). Animal models of nociception. *Pharmacological Reviews*, 53(4), 597–652.

Mahmood, N. D., Nasir, N. L. M., Rofiee, M. S., Tohid, S. M., Ching, S. M., Teh, L. K., et al. (2014). *Muntingia calabura*: A review of its traditional uses, chemical properties, and pharmacological observations. *Pharmaceutical Biology*, 52(12), 1598–1623.

Morton, J. F. (1987). *Jamaica cherry. Fruits of warm climates* (pp. 65–69). Miami, FL: Julia F. Morton.

Nshimo, C. M., Pezzuto, J. M., Kinghorn, A. D., & Farnsworth, N. R. (1993). Cytotoxic constituents of *Muntingia calabura* leaves and stems collected in Thailand. *International Journal of Pharmacology*, 31, 77–81.

Ossipov, M. H. (2012). The perception and endogenous modulation of pain. *Scientifica*, 2012, 561761.

Perez-Arbealaez, E. (1975). *Plants Medicinales y Venenosas de Colombia* (p. 192). Medellin, Colombia: Hernando Salazar.

Perry, L. M. (1895). *Medicinal plants of east and Southeast Asia*. Cambridge MA: The Massachusetts Institute of Technology.

Sani, M. H., Zakaria, Z. A., Balan, T., Teh, L. K., & Salleh, M. Z. (2012). Antinociceptive activity of methanol extract of *Muntingia calabura* leaves and the mechanisms of action involved. *Evidence-based Complementary and Alternative Medicine, 2012*, 1–10.

Sarojini, S., & Mounika, B. (2018). *Muntingia Calabura* (Jamaica cherry): An overview. *PharmaTutor, 6*(11), 1–9.

Shih, C. D. (2009). Activation of nitric oxide/cGMP/PKG signaling cascade mediates antihypertensive effects of *Muntingia calabura* in anesthetized spontaneously hypertensive rats. *The American Journal of Chinese Medicine, 37*, 1045–1058.

Sindhe, M. A., Bodke, Y. D., & Chandrashekar, A. (2013). Antioxidant and *in vivo* anti-hyperglycemic activity of *Muntingia calabura* leaves extracts. *Der Pharmacia Letter, 5*, 427–435.

Soquetta, M. B., Terra, L. M., & Bastos, C. P. (2018). Green technologies for the extraction of bioactive compounds in fruits and vegetables. *CyTA Journal of Food, 16*(1), 400–412.

Sufian, A. S., Ramasamy, K., Ahmat, N., Zakaria, Z. A., & Yusof, M. I. M. (2013). Isolation and identification of antibacterial and cytotoxic compounds from the leaves of *Muntingia calabura* L. *Journal of Ethnopharmacology, 146*(1), 198–204.

Yasunaka, K., Abe, F., Nagayama, A., Okabe, H., Lozada, L., López, E., et al. (2005). Antibacterial activity of crude extracts from Mexican medicinal plants and purified coumarines and xanthones. *Journal of Ethnopharmacology, 97*, 293–299.

Yusof, M., Izwan, M., Salleh, M., Lay Kek, T., Ahmat, N., Azmin, N., et al. (2013). Activity-guided isolation of bioactive constituents with antinociceptive activity from *Muntingia calabura* L. leaves using the formalin test. *Evidence-based Complementary and Alternative Medicine*, 2013.

Zakaria, Z. A., Abdul Rahim, M. H., Roosli, R., Mohd Sani, M. H., Omar, M. H., Mohd Tohid, S. F., et al. (2018). Antinociceptive activity of methanolic extract of *Clinacanthus nutans* leaves: Possible mechanisms of action involved. *Pain Research & Management, 2018*, 9536406.

Zakaria, Z. A., Hassan, M. H., Nurul Aqmar, M. N. H., Abd Ghani, M., Mohd Zaid, S. N. H., Sulaiman, M. R., et al. (2007). Effects of various nonopioid receptor antagonists on the antinociceptive activity of *Muntingia calabura* extracts in mice. *Methods and Findings in Experimental and Clinical Pharmacology, 29*(8), 515–520.

Zakaria, Z. A., Hazalin, N. M. N., Zaid, S. M., Ghani, M. A., Hassan, M. H., Gopalan, H. K., et al. (2007). Antinociceptive, antiinflammatory and anti-pyretic effects of *Muntingia calabura* aqueous extract in animal models. *Journal of Natural Medicines, 61*(4), 443–448.

Zakaria, Z. A., Kumar, G. H., Zaid, S. N., Ghani, M. A., Hassan, M. H., Hazalin, N. A., et al. (2007). Analgesic and antipyretic actions of *Muntingia calabura* leaves chloroform extract in animal models. *Oriental Pharmacy and Experimental Medicine, 7*(1), 34–40.

Zakaria, Z. A., Mahmood, N. D., Mamat, S. S., Nasir, N., & Omar, M. H. (2018). Endogenous antioxidant and LOX-mediated systems contribute to the hepatoprotective activity of aqueous partition of methanol extract of *Muntingia calabura* L. Leaves against paracetamol intoxication. *Frontiers in Pharmacology, 8*, 982.

Zakaria, Z. A., Mohamed, A. M., Jamil, N. S., Rofiee, M. S., Hussain, M. K., Sulaiman, M. R., et al. (2011). *In vitro* antiproliferative and antioxidant activities of the extracts of *Muntingia calabura* leaves. *The American Journal of Chinese Medicine, 39*, 1–18.

Zakaria, Z. A., Mohd Sani, M. H., Abdul Kadir, A., Kek, T. L., & Salleh, M. Z. (2016). Antinociceptive effect of semi-purified petroleum ether partition of *Muntingia calabura* leaves. *Revista Brasileira de Farmacognosia, 26*(4), 408–419.

Zakaria, Z. A., Mustapha, S., Sulaiman, M. R., Jais, A. M. M., Somchit, M. N., & Abdullah, F. C. (2007). The antinociceptive action of aqueous extract from *Muntingia calabura* leaves: The role of opioid receptors. *Medical Principles and Practice, 16*(2), 130–136.

Zakaria, Z. A., Rahim, M. H. A., Mohd Sani, M. H. M., Omar, M. H., Ching, S. M., Kadir, A. A., et al. (2019). Antinociceptive activity of petroleum ether fraction obtained from methanolic extract of *Clinacanthus nutans* leaves involves the activation of opioid receptors and NO-mediated/cGMP-independent pathway. *BMC Complementary and Alternative Medicine, 19*(1), 79.

Zakaria, Z. A., Sani, M. H. M., Cheema, M. S., Kader, A. A., Kek, T. L., & Salleh, M. Z. (2014). Antinociceptive activity of methanolic extract of *Muntingia calabura* leaves: Further elucidation of the possible mechanisms. *BMC Complementary and Alternative Medicine, 14*(1), 63.

Zakaria, Z. A., Somchit, M. N., Sulaiman, M. R., Mat Jais, A. M., & Fatimah, C. A. (2008). Effects of various receptor antagonists, pH and enzymes on *Muntingia calabura* antinociception in mice. *Research Journal of Pharmacology, 2*(3), 31–37.

Zakaria, Z. A., Sulaiman, M. R., Jais, A. M. M., Somchit, M. N., Jayaraman, K. V., Balakhrisnan, G., et al. (2006). The antinociceptive activity of *Muntingia calabura* aqueous extract and the involvement of L-arginine/nitric oxide/cyclic guanosine monophosphate pathway in its observed activity in mice. *Fundamental & Clinical Pharmacology, 20*(4), 365–372.

Chapter 37

Resolving neuroinflammation and pain with maresin 1, a specialized pro-resolving lipid mediator

Victor Fattori[a], Camila R. Ferraz[a], Fernanda S. Rasquel-Oliveira[a], Tiago H. Zaninelli[a], Sergio M. Borghi[a,b], Rubia Casagrande[c], and Waldiceu A. Verri, Jr[a]

[a]Departament of Pathology, Center of Biological Sciences, Londrina State University, Londrina, Paraná, Brazil, [b]Center for Research in Health Science, University of Northern Paraná, Londrina, Paraná, Brazil, [c]Departament of Pharmaceutical Sciences, Center of Health Science, Londrina State University, Londrina, Paraná, Brazil

Abbreviations

AA	arachidonic acid
AT-SPM	aspirin-triggered specialized pro-resolving lipid mediator
CFA	complete Freund's Adjuvant
CGRP	calcitonin gene-related peptide
Chem23	chemerin receptor 23
COX	cyclooxygenase
DAMP	damage-associated molecular pattern
DHA	docosahexaenoic acid
DRG	dorsal root ganglion
EPA	eicosapentaenoic acid
ERK	extracellular signal-regulated kinase
i.p.	intraperitoneal
i.pl.	intraplantar
i.t.	intrathecal
i.v.	intravenous
IASP	International Association for the Study of Pain
LGR6	receptor 6 coupled to G protein-containing Leucine-rich repeats
LOX	lipoxygenase
LX	lipoxin
MAPK	mitogen-activated protein kinase
MaR	maresin
NALP1	NAcht, Leucine-rich repeat, and PYD domain-containing protein 1
NK	natural killer
NLRP3	nod-like receptor protein 3
PAF	platelet-activating factor
PD	protectin
PGE_2	prostaglandin E_2
PSD95	postsynaptic density protein 95
Rv	resolvin
SNL	spinal nerve ligation
SPM	specialized pro-resolving lipid mediator
TRPA1	transient receptor potential cation channel subfamily A member 1
TRPM8	transient receptor potential cation channel subfamily M member 8
TRPV1	transient receptor potential cation channel subfamily V member 1
TRPV2	transient receptor potential cation channel subfamily V member 2

Treatments, Mechanisms, and Adverse Reactions of Anesthetics and Analgesics. https://doi.org/10.1016/B978-0-12-820237-1.00037-5
Copyright © 2022 Elsevier Inc. All rights reserved.

TRPV3	transient receptor potential cation channel subfamily V member 3
TRPV4	transient receptor potential cation channel subfamily V member 4
UVB	ultraviolet B

Introduction

Inflammation is an essential and protective mechanism preserved during evolution. After tissue injury, for example, tissue-resident cells such as macrophages and dendritic cells recognize the damage-associated molecular patterns (DAMPs) and produce pro-inflammatory mediators (Pinho-Ribeiro, Verri, & Chiu, 2017; Verri et al., 2006). These molecules contribute to the inflammatory response by inducing the recruitment of leukocytes, such as neutrophils, to the inflammatory foci (Fattori, Amaral, & Verri, 2016). These pro-inflammatory mediators produced by neutrophil and resident cells sensitize or activate the peripheral nociceptors (Basbaum, Bautista, Scherrer, & Julius, 2009; Verri et al., 2006). After activation, nociceptors release neuropeptides such as substance P and calcitonin gene-related peptide (CGRP) to orchestrate neurogenic inflammation and the activity of immune cells in a context-dependent manner (Baral, Udit, & Chiu, 2019).

Two phenomena are recognized as the main drivers of persistent pathological pain: peripheral and central sensitization. Peripheral sensitization, which occurs by the lowering of the peripheral neuronal threshold (Fattori et al., 2017; Verri et al., 2006). Among other mediators, peripheral sensitization is mediated by bradykinin, prostaglandin E_2 (PGE_2), as well as pro-inflammatory cytokines released by peripheral immune cells, such as macrophages, neutrophils, and mast cells (Ferreira, 1972; Verri et al., 2006; Woolf, Allchorne, Safieh-Garabedian, & Poole, 1997). Central sensitization is mediated by cytokines, chemokines, and growth factors released by microglia, astrocytes, and oligodendrocytes in the spinal cord. Glial cell dysregulation (gliopathy) also contributes, in part, to the pathogenesis of chronic pain (Basbaum et al., 2009; Woolf & Salter, 2000; Zarpelon et al., 2016).

The unresolved acute inflammatory response can lead to chronic inflammation, causing an increase in tissue damage, tissue remodeling, and poor tissue healing (Ortega-Gomez, Perretti, & Soehnlein, 2013). These conditions are known to induce the transition to chronic and maladaptive inflammatory pain. Once thought to be a passive process, it is now recognized that resolution of acute inflammation is a time-dependent and active process and requires biosynthesis of specialized pro-resolving mediators (SPMs) (Fig. 1). Omega (ω)-3 polyunsaturated fatty acids docosahexaenoic acid (DHA) or eicosapentaenoic acid (EPA), or the ω-6 fatty acid arachidonic acid (AA) are biosynthesized to SPMs, such as resolvins (Rvs), protectins (PDs), maresins (MaRs), and lipoxins (LXs). SPMs are known for their antiinflammatory and pro-resolution effects in various models and divided according to their ω-fatty acid precursor (Bannenberg & Serhan, 2010; Fattori, Zaninelli, Rasquel-Oliveira, Casagrande, & Verri, 2020; Serhan, 2017). For instance, MaRs and PDs are DHA-derived, Rvs are either DHA (RvD-series) or EPA-derived (RvE-series), and the LXs are AA-derived (Fig. 2) (Chiang & Serhan, 2017; Serhan, 2017).

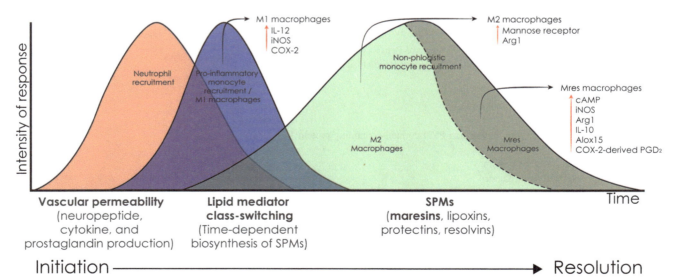

FIG. 1 Resolution process of acute inflammatory response. Time-dependent biosynthesis of SPMs that promotes active resolution of inflammation. SPMs might also stimulate the recruitment of resolution-phase macrophages (Mres). While sharing markers with both M1 (iNOS and COX-2, for example) and M2 macrophages (Arg1), Mres display a distinct signature controlled by cAMP with weaker bactericidal property than M1, which contributes to full resolution of inflammation (Bystrom et al., 2008; Stables et al., 2011).

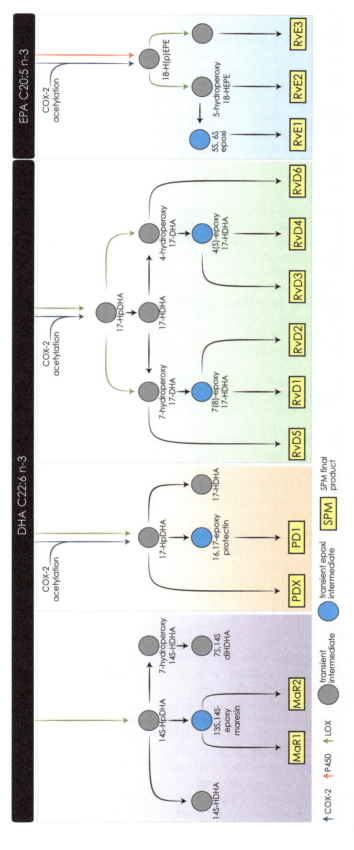

FIG. 2 Biosynthetic route for SPM production. Schematic figure with intermediates and precursors for the endogenous production of SPMs from omega-3 fatty acids EPA and DHA. *Red arrows* indicate the routes that are induced through the acetylation of COX-2, giving rise to the AT-SPMs.

Aspirin-triggered (AT) SPMs are another class of immunoresolvent molecules, which are produced from aspirin acetylation of COX-2. This acetylation changes its enzymatic activity from COX-like to lipoxygenase (LOX)-like. This mechanism not only contributes to the blockage of PGE_2 production but also allows the biosynthesis of SPM epimers, the AT-SPMs (Fig. 2) (Chiang & Serhan, 2017; Serhan, 2017; Serhan, Chiang, Dalli, & Levy, 2015). SPMs play an important role in the blockage of neutrophil infiltration and change of macrophage phenotype to contribute to the resolution of inflammation. For example, SPMs downregulate pro-inflammatory cytokines, such as IL-1β, IL-6, and TNF-α, to induce inflammation resolution and tissue regeneration (Bannenberg & Serhan, 2010; Fattori et al., 2020; Serhan, 2017). SPMs not only promote the resolution of inflammation but also show promising analgesic activity without immunosuppressive effects (Fattori et al., 2020). In this chapter, therefore, we focus on the DHA-derived SPM MaR1 and review the biosynthesis and its receptor and mechanism actions of MaR1 in controlling preclinical pain, and we highlight the therapeutic potential of MaR1 in pain management.

MaR1

MaRs (MaR1 and MaR2) are part of a family of molecules derived from the ω-3 fatty acid DHA. Fig. 2 highlight the routes for MaR1 production and other SPMs. MaR1 was initially described as produced by macrophages, thus the name macrophage mediators in resolving inflammation - Maresin-1 (Dalli et al., 2013; Serhan et al., 2009). The endogenous biosynthesis of MaR1 begins with the 14-lipoxygenation of DHA by the 12-LOX of macrophages. This reaction gives rise to the intermediate molecule 14S-HpDHA acid (14S-hydro (peroxy)-docosa-4Z, 7Z, 10Z, 12E, 16Z, 19Z-hexanoic) (Dalli et al., 2013). An enzymatic hydrolysis via nucleophilic attack at carbon 7 of 14S-HpDHA finally produces MaR1 (Dalli et al., 2013; Serhan et al., 2012). Another study demonstrates that the interaction between neutrophils and platelets is an alternative route for the endogenous biosynthesis of MaR1 (Abdulnour et al., 2014). Co-incubation of neutrophils with platelet-activating factor (PAF)-stimulated platelets or after incubation with 13S, 14S-epoxy-maresin, or 14S-HpDHA derived from platelets, leads to the production of MaR1, suggesting a transcellular and alternative route for the biosynthesis of this SPM (Abdulnour et al., 2014).

Part of the anti-inflammatory effect of MaR1 is related to inhibiting the recruitment of neutrophils to the inflammatory focus and changing the profile of macrophages from M1 to resolving macrophages (Fattori et al., 2020; Serhan, Dalli, Colas, Winkler, & Chiang, 2015). This change in immune cell phenotype is associated with the increase in apoptotic neutrophil efferocytosis, for example. In a spinal cord injury model, although MaR1 does not reduce the initial recruitment of neutrophils into the spinal cord, it increases the clearance of these cells and decreases the presence of these cells by 50% 7 days after the injury. Furthermore, MaR1 reduces Ly6C^{hi} macrophages (pro-inflammatory) and increases Ly6C^{low} macrophages (anti-inflammatory) (Francos-Quijorna et al., 2017).

To produce its immunoresolvent effects (ability to resolve inflammation without an immunosuppressive effect), MaR1 activates the LGR6 receptor (receptor 6 coupled to G protein-containing Leucine-rich repeats) in neutrophils and macrophages (Chiang, Libreros, Norris, De La Rosa, & Serhan, 2019). After activation of these cells, MaR1 promotes an increase in the efferocytosis and phagocytosis of *Escherichia coli* and zymosan particles (Chiang et al., 2019). It is worth mentioning that LGR6 is expressed by glial cells (microglia, astrocytes, and oligodendrocytes), peripheral immune cells (macrophages, neutrophils, NK cells, etc.), and neurons, which explains its beneficial effect in different models. However, as LGR6 was only recently been shown to be activated by MaR1, data validating MaR1/LGR6 signaling is still required to concepts developed before the demonstration of this ligand/receptor duet.

Analgesic effects of MaR1

Increasing attention has been paid to the analgesic effects of MaR1 (Fattori et al., 2020). Table 1 summarizes the mechanistic findings related to the analgesic effects of MaR1 in different experimental models of pain. The first demonstration of the MaR1 analgesic effect was provided by Prof. Charles Serhan and his group. Treatment with MaR1 reduces capsaicin-induced overt pain behavior and the mechanical threshold in vincristine-induced neuropathic pain (Serhan et al., 2012). These effects were attributed to the inhibition of the transient receptor potential cation channel subfamily V member 1 (TRPV1) activation in dorsal root ganglia (DRG) neurons by MaR1 (Serhan et al., 2012). Extending these results, our group demonstrated that MaR1 provides a long-lasting analgesic effect (5 days upon one single i.t. treatment). Using classical models of inflammatory pain, we demonstrated that MaR1 inhibits carrageenan- and complete Freund adjuvant (CFA)-induced mechanical and thermal pain in mice (Fattori et al., 2019). In the periphery, we demonstrate that MaR1 reduces TRPV1$^+$ DRG neuron activation as observed by lower calcium influx when compared to CFA-stimulated mice. Importantly, our group was the first to demonstrate that MaR1 blocks neuro-immune communication by inhibiting the release of calcitonin gene-related peptide (CGRP) by DRG neuron (Fattori et al., 2019), which was further confirmed in a recent

TABLE 1 MaR1 effects in experimental models of pain.

Model	Specie	Route	Dosage	Analgesic effects and related mechanisms	Reference
Capsaicin-induced inflammatory pain	Mouse	(a) i.pl. (b) in vitro	(a) 10 ng (b) 100 nM	(a) Inhibition of overt pain-like behavior (b) Inhibition of TRPV1 activation in DRG neurons	Serhan et al. (2012)
Vincristine-induced neuropathic pain		i.p.	40 ng	Inhibition of mechanical allodynia	
Carrageenan	Mouse	i.t.	(a) 10 ng (b) 3 ng/mL	(a) Reduction of mechanical and thermal hyperalgesia, spinal pro-inflammatory cytokine production, and NF-κB activation, and reduction of recruitment of neutrophils and macrophages to the paw skin	Fattori et al. (2019)
CFA				(a) Reduction of mechanical and thermal hyperalgesia, decreases astrocytes and microglia and NF-κB activation, and spinal pro-inflammatory cytokine release. In the periphery, reduction of the number of leukocytes close to CGRP[+] fibers and calcium influx in DRG neurons (b) Blockage of CGRP release by DRG neurons	
SNL-induced neuropathic pain	Rat	i.t.	100 ng	Inhibition of mechanical and thermal hyperalgesia, glia activation, NFκB, and pro-inflammatory cytokines; and increase in levels of PSD95 and synapsin II	Gao et al. (2018)
Noncompressive lumbar disc herniation-induced radicular pain	Rat	i.t.	100 ng	Inhibition of mechanical and thermal hyperalgesia, NLRP3 inflammasome and IL-1β and IL-18 production	Wang et al. (2020)
K/BxN serum transfer-induced inflammatory arthritis	Mouse	(a) i.p. (b) in vitro	(a) 100 ng (b) 3 ng/mL (c) 100 nM	(a) Inhibition of mechanical hyperalgesia and macrophage migration to DRG (b) Inhibition of TRPV1 activation in DRG neurons (c) Inhibition of CGRP release by DRG neurons	Allen et al. (2020)
Tibial bone fracture-induced pos-operative pain	Mouse	i.v. or i.t.	500 ng	Reduction of mechanical and cold allodynia	Zhang et al. (2018)

CFA: complete Freund's adjuvant; *CGRP*: calcitonin gene-related peptide; *DRG*: dorsal root ganglion; *i.pl.*: intraplantar; *i.p.*: intraperitoneal; *i.t.*: intrathecal; *i.v.*: intravenous; *PSD95*: postsynaptic density protein 95; *SNL*: spinal nerve ligation.

work (Allen et al., 2020). As consequence, we observed fewer CD11b$^+$ cells close to CGRP$^+$ fibers (Fattori et al., 2019). In the spinal cord, we observed reduced activation of microglia and astrocytes, and reduced activation of NF-κB and downstream cytokines TNF-α and IL-1β (Fattori et al., 2019). Using a rat model spinal nerve ligation (SNL)-induced neuropathic pain, treatment with MaR1 also attenuates mechanical and thermal hyperalgesia with reduced levels of pro-inflammatory cytokines and glia cells activation (Gao et al., 2018). In a model of noncompressive disc herniation in the lumbar segment of rats, treatment with MaR1 inhibits mechanical and thermal hyperalgesia as well as NLRP3 inflammasome-related proteins and downstream cytokines IL-1β and IL-18 (Wang et al., 2020). The Authors demonstrate that inhibition of caspase-1 activity leads to an increase in 14-HDHA (MaR1 precursor) and reduction of PGE$_2$ levels in the spinal cord, indicating MaR1 orchestrate resolution of neuroinflammation (Wang et al., 2020).

In a model of inflammatory arthritis induced by immunization with a serum of transgenic K/BxN mice, MaR1 negatively correlates with the persistence of pain observed after the disappearance of clinical signs and with an accumulation of M1 macrophages in DRG (Allen et al., 2020). Interestingly, in corroboration to our study, Allen and colleagues show that MaR1 displays a long-lasting analgesic effect as well. After repeated treatments, MaR1 presents 14 days of analgesia (Allen et al., 2020). In a model of postoperative pain after tibial bone fracture in mice, MaR1 delays the development of pain (Zhang et al., 2018). Additionally, treatment with MaR1 but not DHA reduces mechanical hyperalgesia and cold hyperalgesia induced by the model (Zhang et al., 2018). These results suggest a distinct potential for DHA and DHA-derived MaR1 analgesic effects, in favor of MaR1 regarding postoperative bone fracture pain management (Zhang et al., 2018). Altogether, it is likely that MaR1 controls neuroinflammation by blocking peripheral and central sensitization.

Clinical analgesic evidence of MaR1 and its precursors

The promising preclinical evidence of the analgesic effect of MaR1 raises the question of the possibility of its applicability on clinical stages. In this section, we focus on the studies that measured either MaR1 or its precursor levels in human subjects. Table 2 summarizes studies using ω-3 supplementation and Table 3 summarizes studies that measured MaR1 or its precursor levels in human samples. A comprehensive description of the analgesic role of SPMs and ω-3 supplementation in clinical studies can be found elsewhere (Fattori et al., 2020). One of the main challenges of translation studies relies on the fact that SPMs are unstable molecules that undergo local metabolic inactivation near the inflammation site, being active only on target cells in their immediate milieu (Serhan & Petasis, 2011). One approach that has been developed to solve this issue is the synthesis of SPMs analogs, with higher stability and comparable bioactivity, currently on different stages of

TABLE 2 The effect of ω-3 supplementation on MaR1 and its precursor levels in human subjects.

Condition	n	Treatment	Duration	Outcomes	Reference
Rheumatoid arthritis	38	*Schizochytrium* sp. oil (2.11 g DHA, daily)	10 weeks (cross-over)	Decline in the sum of tender and swollen joints, and ultrasound score (US-7) and increase in 17-HpDHA and 14-HpDHA levels	Dawczynski et al. (2018)
	15	8% EPA + 12% DHA w/w	Variable	Plasma MaR1 was increased when compared to healthy volunteers taking 2.4 g ω-3 EPA+ DHA supplementation for 4 weeks	Barden et al. (2016)
Psoriatic arthritis	6	8% EPA + 12% DHA w/w	Variable	Plasma MaR1 was increased when compared to healthy volunteers taking 2.4 g ω-3 EPA + DHA supplementation for 4 weeks	Barden et al. (2016)
Gout	4				
Pauci arthritis	4				
B27 spondyloarthritis	2				
Monoarthritis	2				
Other rheumatic-associated condition	5				

DHA: docosahexaenoic acid; *EPA*: eicosapentaenoic acid; ω-3: omega-3.

TABLE 3 Presence of MaR1 and its precursors in human samples.

Condition	n	Sample	Outcomes	Reference
Rheumatoid arthritis	5	Synovial fluid	Presence of MaR1 precursor 14-HpDHA	Dawczynski et al. (2018)
	9	Serum	MaR1 levels are increased in patients with inactive arthritis when compared to active arthritis	Jin et al. (2018)
Rheumatic diseases	36	Synovial fluid and plasma	Presence of MaR1 and its precursor 14-HpDHA	Barden et al. (2016)

clinical trials (ClinicalTrials.gov Identifier: NCT02342691; NCT00799552). Future studies will also determine whether local injection of SPMs might be suitable therapeutic approaches to overcome metabolic inactivation.

While MaR1 (among other SPMs) were identified in human synovial fluids of patients with arthritis (Giera et al., 2012), no clinical studies evaluating its effect on pain relief are available. Nonetheless, the presence of these mediators was negatively associated with pain score, indicating SPMs possibly limit pain (Barden et al., 2016). In fact, patients with active rheumatoid arthritis show reduced MaR1 serum levels when compared to healthy volunteers and patients with inactive disease (Jin et al., 2018). Interestingly, patients with an inactive phase of the disease present higher levels even when compared to healthy volunteers (Jin et al., 2018). These data indicate that MaR1 is likely to control disease status in humans.

As already mentioned, EPA and DHA are substrates for enzymes that give rise to several SPMs including MaR1. Therefore, it is rational to consider that ω-3 supplementation and improvement of clinical symptoms and inflammatory parameters can occur through resolution promoted by MaR1 activity. Patients with the different rheumatic diseases under ω-3 supplementation show increased levels of SPMs (including MaR1) and their precursors when compared to healthy volunteers that were also taking ω-3 supplementation (Barden et al., 2016). In addition to ω-3 capsules supplementation, seafood (fish, fish oils, and algae oils) comprise a very good and well-known source of ω-3 (Whelan & Rust, 2006). One study conducted by Dawczynski et al. (2018) aimed to investigate the clinical benefit of daily consumption of microalgae oil-enriched foods in 38 patients with rheumatoid arthritis (Dawczynski et al., 2018). This placebo-controlled, randomized double-blinded crossover study consisted of 10-week intervention period, 10-week washout period and another 10-week intervention period. Before and after each period, blood samples were collected, and disease activity assessed. Subjects consumed 8 g of *Schizochytrium* sp. oil (a daily DHA dose of 2.11 g), and the placebo group received 8 g commercially available sunflower oil. DHA-enriched products result in a significant lower number of tender joints, along with a strong increase in EPA and DHA levels and decrease in ω-6, as well as significant differences of ultrasound score-7. Moreover, duration of morning stiffness from the worsening changes observed in placebo. Furthermore, it was noticed an enhanced production of the Rvs, PDs, and MaR precursors 17-HpDHA and 14-HpDHA, what may explain the clinical improvements (Dawczynski et al., 2018).

Application to other areas

In this chapter, we review how MaR1 controls neuroinflammation and pain. As we addressed in the clinical section, imbalance on pro-resolution lipid mediator profile is one of the key aspects of chronic diseases. Psoriasis is a chronic disease with an increase in ω-6 fatty acid-oxidized derivatives, which shift the response to a pro-inflammatory profile (Sorokin et al., 2018; Sorokin et al., 2018). Blockade of neuropeptide release by neurons is shown to ameliorate inflammation in a psoriasis model (Riol-Blanco et al., 2014). In fact, we have demonstrated that treatment with MaR1 reduces UVB irradiation-induced skin damages, indicating it can be useful to treat other skin diseases as well (Cezar et al., 2019).

We also address how MaR1 silence neurons and the release of the neuropeptide CGRP as part of its analgesic effect. A growing body of evidence demonstrates how nociceptors control immune response, inflammation, and infection (Pinho-Ribeiro et al., 2017). MaR1 and other SPMs while limiting neutrophil recruitment and stimulate the recruitment of nonphlogistic monocytes. At the inflammatory foci, macrophages then phagocyte apoptotic neutrophils as well as micro-organisms to resolve inflammation and infection (Chiang & Serhan, 2017; Serhan, 2017; Serhan, Chiang, et al., 2015). In fact, during infections SPMs RvD1, RvD5, and PD1 show a temporal and differential regulation during infections and lower the antibiotic requirements for bacterial clearance, meaning that they work as immunoresolvent rather than immunosuppressant molecules (Chiang et al., 2012). For instance, RvD1, RvD5, and PD1 increase host response against

E. coli and lowers antibiotic requirement (Chiang et al., 2012). Similarly, treatment with MaR1 prevents sepsis-induced lethality by decreasing both local and systemic bacterial burden (Hao et al., 2019). Recent evidence demonstrates vagus nerve produces SPMs promoting, therefore, tissue homeostasis and infection resolution (Serhan, De La Rosa, & Jouvene, 2018). Among other SPMs, the human vagus nerve produces MaR1 and in accordance, vagotomy delays the resolution of self-limiting E. coli infection in mice (Dalli, Colas, Arnardottir, & Serhan, 2017). These data show the immunoresolvent properties of MaR1, which might contribute to control infection by promoting micro-organism clearance. Therefore, it might reduce the antibiotic requirements to treat infections, which can contribute to reducing antibiotic resistance and post-infectious sequelae.

Other agents of interest

Despite effective for a fraction of patients with rheumatic diseases and neuropathic pain, current therapies cause unwanted side effects, immunosuppression, or induce analgesic tolerance, and addiction (Fattori et al., 2020). Table 4 summarizes the main drugs used to treat inflammatory pain and their long-term use side effects. On the other hand, isolated SPMs show efficacy at very low doses and have been used to treat pain in experimental models without side effects (Chiang & Serhan, 2017; Fattori et al., 2020). For instance, MaR1 shows a long-lasting analgesic effect upon a single treatment (up to 5 days) (Fattori et al., 2019) and when administrated before the development of tactile allodynia, a single treatment with RvD1 produces an enduring analgesic effect (up to 30 days) (Huang, Wang, Serhan, & Strichartz, 2011). Moreover, by acting on the receptor Chem23, RvE1 blocks TNF-α-induced ERK phosphorylation in cultured DRG neurons, which contributes to its potent analgesic effect with a dose $1000\times$ lower than morphine in the formalin model of pain (Xu et al., 2010). AT-RvD1 is an aspirin-triggered SPM that also suppresses neuronal activation in a TRPV3-dependent manner without any effects on TRPV1, TRPV2, TRPV4, TRPM8, or TRPA1 (Bang, Yoo, Yang, Cho, & Hwang, 2012).

LXs are biosynthesized from arachidonic acid and were the first SPMs described (Serhan, Hamberg, & Samuelsson, 1984). Treatment with LXA_4 (intrathecal) reduces carrageenan-induced thermal hyperalgesia, while LXA_4 and AT-LXA_4 (intravenous) reduce edema and thermal hyperalgesia induced by carrageenan in rats. AT-LXA_4, moreover, prevents ATP-induced MAPK signaling on astrocytes in vitro, indicating a direct effect on this glial cell (Svensson, Zattoni, & Serhan, 2007). Similarly, LXA_4 reduces carrageenan-induced hyperalgesia (Abdelmoaty et al., 2013). In addition to the

TABLE 4 Main drugs used to treat inflammatory pain.

Class	Best known examples	Main analgesic mechanism of action	Main side effects (long-term use)
Weak analgesic	Metamizole Paracetamol	Multiple targets	Hepatotoxicity Nephrotoxicity
NSAID	Acetyl-salicylic acid Diclofenac Ibuprofen Indomethacin Ketoprofen Ketorolac Naproxen	Nonselective blockage of COX	GI bleeding Nephrotoxicity
	Celecoxib Etoricoxib Meloxicam Rofecoxib Valdecoxib	Selective blockage of COX-2	Nephrotoxicity Cardiotoxicity Ischemic or hemorrhagic stroke
Corticosteroid	Betamethasone Prednisone Prednisolone	Binds to glucocorticoid receptor and blocks NF-κB activation	GI bleeding Susceptibility to infections Myopathy Osteoporosis
Biological agent	IL-1ra	Antagonism of IL-1 receptor	Development of adaptive immunity Susceptibility to infections
	Etanercept	Soluble decoy receptor for TNF-α	

COX: cycloxygenase; GI: gastrointestinal; NSAID: nonsteroidal antiinflammatory drug; ra: receptor antagonist.

analgesic effect, AT-LXA$_4$ blocks morphine tolerance by blocking NALP1-derived IL-1β levels in vivo and in vitro (Tian et al., 2015). These effects are important considering the current struggle with opioid side effects, tolerance, and addiction. Altogether, these findings highlight the potential of different SPMs as promising therapeutic approaches to control pain and inflammation.

Mini-dictionary of terms

Allodynia: "Pain due to a stimulus that does not normally provoke pain" (official definition by the International Association for the Study of Pain – IASP).

Central sensitization: changes in the neuronal pain circuit and glial cell phenotype resulting in a reduction in the activation threshold and/or an increase in the magnitude of responsiveness of neurons at the spinal cord level.

Epimer: One of a pair of molecules (more specifically stereoisomers) that differ in the configuration of a single stereocenter.

Hyperalgesia: "Increased pain from a stimulus that normally provokes pain" (official definition by IASP).

Immunoresolvent: molecules with the ability to resolve inflammation without causing immunosuppression.

Peripheral sensitization: changes in the neuronal pain circuit and immune cell phenotype resulting in a reduction in the activation threshold and/or an increase in the magnitude of responsiveness at the peripheral ends of nociceptor neurons.

Key facts

Key facts of SPMs

- Imbalance between ω-3 and ω-6 fatty acids toward ω-6 production favors chronic disease development.
- SPMs actively resolve inflammation without immunosuppressive effects.
- SPMs resolve infections and lower antibiotic requirement.
- ω-3 supplementation reduces pain score in disease, such as rheumatoid arthritis.
- Vagus nerve produce SPMs, including MaR1, and limit infection.

Key facts of MaR1

- MaR1 is an agonist of LGR6.
- MaR1 selectively blocks TRPV1 activation in DRG neurons in vitro (it shows no effect over TRPA1).
- MaR1 shows analgesic effect in several experimental models and displays a long-lasting analgesic effect in inflammatory pain and arthritis.
- MaR1 blocks CGRP release by DRG neurons.
- Patients with active rheumatoid arthritis have lower MaR1 levels when compared to patients with inactive disease.

Summary points

- SPMs are a class of molecules biosynthesized from ω-3 (DHA and EPA) or ω-6 (AA) fatty acids that are actively produced during the resolution phase of inflammation.
- SPMs do not present immunosuppressive effects, which is a common side effect of standard of care analgesic drugs such as opioids, biological agents, and corticosteroids.
- MaR1 is a DHA-derived SPM that produces enduring analgesic and antiinflammatory effects.
- MaR1 silences nociceptors to reduce pain in part through the blockade of neuropeptide release and TRPV1 activation.
- MaR1 reduces the activation of spinal cord glial cells such as astrocytes and microglia to reduce pain.
- ω-3 supplementation increases levels of SPMs precursors, prevent the development of chronic diseases and reduces pain score in disease, such as rheumatoid arthritis.

Funding

We are thankful for the financial support of the Department of Science and Technology from the Science, Technology and Strategic Inputs Secretariat of the Ministry of Health (Decit/SCTIE/MS, Brazil) intermediated by the National Council for Scientific and Technological Development (CNPq, Brazil) with support of the Araucária Foundation and State Health Secretariat, Paraná (SESA-PR, Brazil; PPSUS Grant agreement041/2017, protocol 48.095); Programa de Apoio a Grupos de Excelência (PRONEX) grant supported by SETI/Araucária Foundation and MCTI/CNPq; and Paraná State Government (agreement 014/2017, protocol 46.843). V.F., T.H.Z., and F.S.R.-O. acknowledge PhD degree scholarship

440 **PART | IV** Novel and nonpharmacological aspects and treatments

from Coordination for the Improvement of Higher Education Personnel (CAPES, Brazil, finance code 001). C.R.F. acknowledges a CNPq PDJ postdoc fellowship. S.M.B. acknowledges a FUNADESP research fellowship. W.A.V.J. and R.C. acknowledge the CNPq Productivity fellowship.

References

Abdelmoaty, S., Wigerblad, G., Bas, D. B., Codeluppi, S., Fernandez-Zafra, T., El-Awady, E. S., et al. (2013). Spinal actions of lipoxin A4 and 17(R)-resolvin D1 attenuate inflammation-induced mechanical hypersensitivity and spinal TNF release. *PLoS One, 8*, e75543.

Abdulnour, R. E., Dalli, J., Colby, J. K., Krishnamoorthy, N., Timmons, J. Y., Tan, S. H., et al. (2014). Maresin 1 biosynthesis during platelet-neutrophil interactions is organ-protective. *Proceedings of the National Academy of Sciences of the United States of America, 111*, 16526–16531.

Allen, B. L., Montague-Cardoso, K., Simeoli, R., Colas, R. A., Oggero, S., Vilar, B., et al. (2020). Imbalance of pro-resolving lipid mediators in persistent allodynia dissociated from signs of clinical arthritis. *Pain, 161*, 2155–2166.

Bang, S., Yoo, S., Yang, T. J., Cho, H., & Hwang, S. W. (2012). 17(R)-resolvin D1 specifically inhibits transient receptor potential ion channel vanilloid 3 leading to peripheral antinociception. *British Journal of Pharmacology, 165*, 683–692.

Bannenberg, G., & Serhan, C. N. (2010). Specialized pro-resolving lipid mediators in the inflammatory response: An update. *Biochimica et Biophysica Acta, 1801*, 1260–1273.

Baral, P., Udit, S., & Chiu, I. M. (2019). Pain and immunity: Implications for host defence. *Nature Reviews. Immunology, 19*, 433–447.

Barden, A. E., Moghaddami, M., Mas, E., Phillips, M., Cleland, L. G., & Mori, T. A. (2016). Specialised pro-resolving mediators of inflammation in inflammatory arthritis. *Prostaglandins, Leukotrienes, and Essential Fatty Acids, 107*, 24–29.

Basbaum, A. I., Bautista, D. M., Scherrer, G., & Julius, D. (2009). Cellular and molecular mechanisms of pain. *Cell, 139*, 267–284.

Bystrom, J., Evans, I., Newson, J., Stables, M., Toor, I., Van Rooijen, N., et al. (2008). Resolution-phase macrophages possess a unique inflammatory phenotype that is controlled by camp. *Blood, 112*, 4117–4127.

Cezar, T. L. C., Martinez, R. M., Rocha, C. D., Melo, C. P. B., Vale, D. L., Borghi, S. M., et al. (2019). Treatment with maresin 1, a docosahexaenoic acid-derived pro-resolution lipid, protects skin from inflammation and oxidative stress caused by UVB irradiation. *Scientific Reports, 9*, 3062.

Chiang, N., Fredman, G., Backhed, F., Oh, S. F., Vickery, T., Schmidt, B. A., et al. (2012). Infection regulates pro-resolving mediators that lower antibiotic requirements. *Nature, 484*, 524–528.

Chiang, N., Libreros, S., Norris, P. C., De La Rosa, X., & Serhan, C. N. (2019). Maresin 1 activates LGR6 receptor promoting phagocyte immunoresolvent functions. *The Journal of Clinical Investigation, 129*, 5294–5311.

Chiang, N., & Serhan, C. N. (2017). Structural elucidation and physiologic functions of specialized pro-resolving mediators and their receptors. *Molecular Aspects of Medicine, 58*, 114–129.

Dalli, J., Colas, R. A., Arnardottir, H., & Serhan, C. N. (2017). Vagal regulation of group 3 innate lymphoid cells and the Immunoresolvent PCTR1 controls infection resolution. *Immunity, 46*, 92–105.

Dalli, J., Zhu, M., Vlasenko, N. A., Deng, B., Haeggstrom, J. Z., Petasis, N. A., et al. (2013). The novel 13S,14S-epoxy-maresin is converted by human macrophages to maresin 1 (MaR1), inhibits leukotriene A4 hydrolase (LTA4H), and shifts macrophage phenotype. *The FASEB Journal, 27*, 2573–2583.

Dawczynski, C., Dittrich, M., Neumann, T., Goetze, K., Welzel, A., Oelzner, P., et al. (2018). Docosahexaenoic acid in the treatment of rheumatoid arthritis: A double-blind, placebo-controlled, randomized cross-over study with microalgae vs. sunflower oil. *Clinical Nutrition, 37*, 494–504.

Fattori, V., Amaral, F. A., & Verri, W. A., Jr. (2016). Neutrophils and arthritis: Role in disease and pharmacological perspectives. *Pharmacological Research, 112*, 84–98.

Fattori, V., Hohmann, M. S. N., Rossaneis, A. C., Manchope, M. F., Alves-Filho, J. C., Cunha, T. M., et al. (2017). Targeting IL-33/ST2 signaling: Regulation of immune function and analgesia. *Expert Opinion on Therapeutic Targets, 21*, 1141–1152.

Fattori, V., Pinho-Ribeiro, F. A., Staurengo-Ferrari, L., Borghi, S. M., Rossaneis, A. C., Casagrande, R., et al. (2019). The specialised pro-resolving lipid mediator maresin 1 reduces inflammatory pain with a long-lasting analgesic effect. *British Journal of Pharmacology, 176*, 1728–1744.

Fattori, V., Zaninelli, T. H., Rasquel-Oliveira, F. S., Casagrande, R., & Verri, W. A. (2020). Specialized pro-resolving lipid mediators: A new class of non-immunosuppressive and non-opioid analgesic drugs. *Pharmacological Research, 151*, 104549.

Ferreira, S. H. (1972). Prostaglandins, aspirin-like drugs and analgesia. *Nature: New Biology, 240*, 200–203.

Francos-Quijorna, I., Santos-Nogueira, E., Gronert, K., Sullivan, A. B., Kopp, M. A., Brommer, B., et al. (2017). Maresin 1 promotes inflammatory resolution, neuroprotection, and functional neurological recovery after spinal cord injury. *The Journal of Neuroscience, 37*, 11731–11743.

Gao, J., Tang, C., Tai, L. W., Ouyang, Y., Li, N., Hu, Z., et al. (2018). Pro-resolving mediator maresin 1 ameliorates pain hypersensitivity in a rat spinal nerve ligation model of neuropathic pain. *Journal of Pain Research, 11*, 1511–1519.

Giera, M., Ioan-Facsinay, A., Toes, R., Gao, F., Dalli, J., Deelder, A. M., et al. (2012). Lipid and lipid mediator profiling of human synovial fluid in rheumatoid arthritis patients by means of LC-MS/MS. *Biochimica et Biophysica Acta, 1821*, 1415–1424.

Hao, Y., Zheng, H., Wang, R. H., Li, H., Yang, L. L., Bhandari, S., et al. (2019). Maresin1 alleviates metabolic dysfunction in septic mice: A (1)H NMR-based metabolomics analysis. *Mediators of Inflammation, 2019*, 2309175.

Huang, L., Wang, C. F., Serhan, C. N., & Strichartz, G. (2011). Enduring prevention and transient reduction of postoperative pain by intrathecal resolvin D1. *Pain, 152*, 557–565.

Jin, S., Chen, H., Li, Y., Zhong, H., Sun, W., Wang, J., et al. (2018). Maresin 1 improves the Treg/Th17 imbalance in rheumatoid arthritis through miR-21. *Annals of the Rheumatic Diseases, 77*, 1644–1652.

Ortega-Gomez, A., Perretti, M., & Soehnlein, O. (2013). Resolution of inflammation: An integrated view. *EMBO Molecular Medicine, 5*, 661–674.

Pinho-Ribeiro, F. A., Verri, W. A., Jr., & Chiu, I. M. (2017). Nociceptor sensory neuron-immune interactions in pain and inflammation. *Trends in Immunology, 38*, 5–19.

Riol-Blanco, L., Ordovas-Montanes, J., Perro, M., Naval, E., Thiriot, A., Alvarez, D., et al. (2014). Nociceptive sensory neurons drive interleukin-23-mediated psoriasiform skin inflammation. *Nature, 510*, 157–161.

Serhan, C. N. (2017). Treating inflammation and infection in the 21st century: New hints from decoding resolution mediators and mechanisms. *The FASEB Journal, 31*, 1273–1288.

Serhan, C. N., Chiang, N., Dalli, J., & Levy, B. D. (2015). Lipid mediators in the resolution of inflammation. *Cold Spring Harbor Perspectives in Biology, 7*, a016311.

Serhan, C. N., Dalli, J., Colas, R. A., Winkler, J. W., & Chiang, N. (2015). Protectins and maresins: New pro-resolving families of mediators in acute inflammation and resolution bioactive metabolome. *Biochimica et Biophysica Acta, 1851*, 397–413.

Serhan, C. N., Dalli, J., Karamnov, S., Choi, A., Park, C. K., Xu, Z. Z., et al. (2012). Macrophage proresolving mediator maresin 1 stimulates tissue regeneration and controls pain. *The FASEB Journal, 26*, 1755–1765.

Serhan, C. N., De La Rosa, X., & Jouvene, C. C. (2018). Cutting edge: Human Vagus produces specialized proresolving mediators of inflammation with electrical stimulation reducing proinflammatory eicosanoids. *Journal of Immunology, 201*, 3161–3165.

Serhan, C. N., Hamberg, M., & Samuelsson, B. (1984). Lipoxins: Novel series of biologically active compounds formed from arachidonic acid in human leukocytes. *Proceedings of the National Academy of Sciences of the United States of America, 81*, 5335–5339.

Serhan, C. N., & Petasis, N. A. (2011). Resolvins and protectins in inflammation resolution. *Chemical Reviews, 111*, 5922–5943.

Serhan, C. N., Yang, R., Martinod, K., Kasuga, K., Pillai, P. S., Porter, T. F., et al. (2009). Maresins: Novel macrophage mediators with potent antiinflammatory and proresolving actions. *The Journal of Experimental Medicine, 206*, 15–23.

Sorokin, A. V., Domenichiello, A. F., Dey, A. K., Yuan, Z. X., Goyal, A., Rose, S. M., et al. (2018). Bioactive lipid mediator profiles in human psoriasis skin and blood. *The Journal of Investigative Dermatology, 138*, 1518–1528.

Sorokin, A. V., Norris, P. C., English, J. T., Dey, A. K., Chaturvedi, A., Baumer, Y., et al. (2018). Identification of proresolving and inflammatory lipid mediators in human psoriasis. *Journal of Clinical Lipidology, 12*, 1047–1060.

Stables, M. J., Shah, S., Camon, E. B., Lovering, R. C., Newson, J., Bystrom, J., et al. (2011). Transcriptomic analyses of murine resolution-phase macrophages. *Blood, 118*, e192–e208.

Svensson, C. I., Zattoni, M., & Serhan, C. N. (2007). Lipoxins and aspirin-triggered lipoxin inhibit inflammatory pain processing. *The Journal of Experimental Medicine, 204*, 245–252.

Tian, Y., Liu, M., Mao-Ying, Q. L., Liu, H., Wang, Z. F., Zhang, M. T., et al. (2015). Early single aspirin-triggered Lipoxin blocked morphine antinociception tolerance through inhibiting NALP1 inflammasome: Involvement of PI3k/Akt signaling pathway. *Brain, Behavior, and Immunity, 50*, 63–77.

Verri, W. A., Jr., Cunha, T. M., Parada, C. A., Poole, S., Cunha, F. Q., & Ferreira, S. H. (2006). Hypernociceptive role of cytokines and chemokines: Targets for analgesic drug development? *Pharmacology & Therapeutics, 112*, 116–138.

Wang, Y. H., Li, Y., Wang, J. N., Zhao, Q. X., Wen, S., Wang, S. C., et al. (2020). A novel mechanism of specialized Proresolving lipid mediators mitigating radicular pain: The negative interaction with NLRP3 Inflammasome. *Neurochemical Research*.

Whelan, J., & Rust, C. (2006). Innovative dietary sources of n-3 fatty acids. *Annual Review of Nutrition, 26*, 75–103.

Woolf, C. J., Allchorne, A., Safieh-Garabedian, B., & Poole, S. (1997). Cytokines, nerve growth factor and inflammatory hyperalgesia: The contribution of tumour necrosis factor alpha. *British Journal of Pharmacology, 121*, 417–424.

Woolf, C. J., & Salter, M. W. (2000). Neuronal plasticity: Increasing the gain in pain. *Science, 288*, 1765–1769.

Xu, Z. Z., Zhang, L., Liu, T., Park, J. Y., Berta, T., Yang, R., et al. (2010). Resolvins RvE1 and RvD1 attenuate inflammatory pain via central and peripheral actions. *Nature Medicine, 16*, 592–597 (1p following 597).

Zarpelon, A. C., Rodrigues, F. C., Lopes, A. H., Souza, G. R., Carvalho, T. T., Pinto, L. G., et al. (2016). Spinal cord oligodendrocyte-derived alarmin IL-33 mediates neuropathic pain. *The FASEB Journal, 30*, 54–65.

Zhang, L., Terrando, N., Xu, Z. Z., Bang, S., Jordt, S. E., Maixner, W., et al. (2018). Distinct analgesic actions of DHA and DHA-derived specialized pro-resolving mediators on post-operative pain after bone fracture in mice. *Frontiers in Pharmacology, 9*, 412.

Chapter 38

Therapeutic role of naringenin to alleviate inflammatory pain

Marília F. Manchope[a], Camila R. Ferraz[a], Sergio M. Borghi[a,c], Fernanda Soares Rasquel-Oliveira[a], Anelise Franciosi[a], Julia Bagatim-Souza[a], Amanda M. Dionisio[a], Rubia Casagrande[b], and Waldiceu A. Verri, Jr[a]

[a]*Department of Pathology, Center of Biological Sciences, Londrina State University, Londrina, Paraná, Brazil,* [b]*Department of Pharmaceutical Sciences, Center of Health Sciences, Londrina State University, Londrina, Paraná, Brazil,* [c]*Center for Research in Health Sciences, University of Northern Paraná, Londrina, Paraná, Brazil*

Abbreviations

ARE	antioxidant responsive elements
ATP	adenosine triphosphate
CAT	catalase
CCI	chronic constriction injury
cGMP	cyclic guanosine monophosphate
DAMPs	damage-associated molecular patterns
DENV	dengue virus
GPx	glutathione peroxidase
GSH	reduced glutathione
HCV	hepatitis C virus
HO-1	heme oxygenase-1
IL	interleukin
LPS	lipopolysaccharide
miRNA	micro-RNA
MSU	monosodium urate
NFκB	nuclear factor κB
NO	nitric oxide
Nrf2	nuclear factor-erythroid 2-related factor 2
OH	hydroxyl group
PAMPs	pathogen-associated molecular patterns
PKG	protein kinase G
ROS	reactive oxygen species
SOD	superoxide dismutase
TiO$_2$	titanium dioxide
TNFα	tumor necrosis factor-alpha
TRP channels	transient receptor potential channels

Introduction

The nervous system identifies and interprets a broad spectrum of biological, chemical, mechanical, and thermal stimuli. Nociceptors sense what is happening in the periphery through receptors that modulate pain transmission (Basbaum, Bautista, Scherrer, & Julius, 2009). During tissue injury or infection, toll-like receptors are activated by pathogen- and/ or damage-associated molecular patterns (PAMPs and/or DAMPs) inducing NFκB activation and upregulation of inflammatory gene expression and lately inflammatory mediators release. There is also the production of reactive oxygen species (ROS) leading to a disbalance between oxidant and antioxidant systems causing oxidative stress (Betteridge, 2000). Nociceptors are sensitized by inflammatory mediators such as ROS, prostaglandins, pro-inflammatory cytokines such as tumor

Treatments, Mechanisms, and Adverse Reactions of Anesthetics and Analgesics. https://doi.org/10.1016/B978-0-12-820237-1.00038-7
Copyright © 2022 Elsevier Inc. All rights reserved.

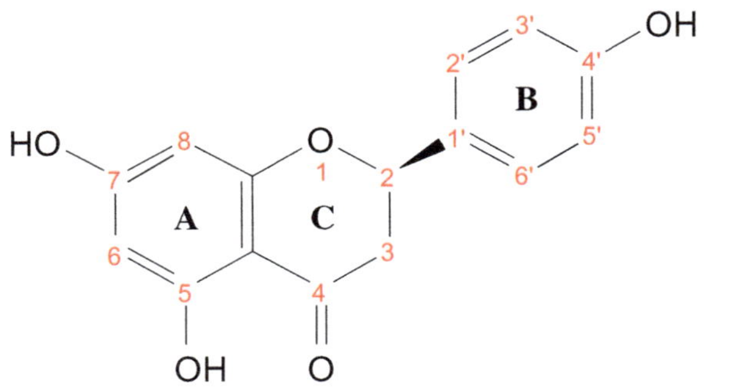

444 PART | IV Novel and nonpharmacological aspects and treatments

FIG. 1 Naringenin chemical structure. Naringenin chemical structure has three hydroxyls (OH) groups at positions 4' of B ring and 5 and 7 of A ring. The OH group located at the B ring allows naringenin to stabilize hydroxyl, peroxyl, and peroxynitrite radicals, generating stability to naringenin through resonance. This direct antioxidant activity of naringenin explains in part the antioxidant chemical properties of naringenin.

TABLE 1 Common food sources of naringenin.

Food Description	mg of naringenin/100 mL or g	References
Dried Mexican oregano	372	Lin, Mukhopadhyay, Robbins, and Harnly (2007)
Commercial grapefruit juice	4.20	Gattuso, Barreca, Gargiulli, Leuzzi, and Caristi (2007)
Citrus paradisi (grapefruit) juice	2.70	Gattuso et al. (2007)
Almonds	0.09	Teets, Minardi, Sundararaman, Hughey, and Were (2009)

Amount of naringenin in mg per 100 mL or g of food.

necrosis factor-alpha (TNFα), interleukin (IL)-1β, and pro-inflammatory chemokines (Basbaum et al., 2009). Thus, modulating the sensory neuron sensibilization is an emerging trend in pain treatment.

Analgesic drugs available for inflammatory pain relief are based on mitigating inflammatory mediators' production in the periphery by nonsteroidal antiinflammatory drugs, and also opioids that act on the central nervous system in cases with difficulty in pain management (Kurella, Bennett, & Chertow, 2003). Opioids and other drugs such as dipyrone have a peripheral analgesic mechanism, which induces sensory neuron hyperpolarization decreasing nociceptive transmission (Cunha et al., 2010; Sachs, Cunha, & Ferreira, 2004). However, those analgesic drugs have side effects such as gastric lesions, hepatotoxicity, nephrotoxicity, and also opioids can lead to addiction and even opioid hyperalgesia (Kurella et al., 2003).

This brief introduction on pain and analgesic mechanisms in inflammation indicates that compounds targeting inflammatory pain mechanisms and/or sharing active analgesic mechanisms with clinically available drugs would be of potential clinical relevance. That is the case of naringenin, a trihydroxyflavanone with substitutions at positions 4' of B ring and 5' and 7' of A ring. (Fig. 1) found in different food sources (Table 1) (Gattuso et al., 2007; Lin et al., 2007; Teets et al., 2009). Naringenin analgesic and antiinflammatory activities were described in varied preclinical painful inflammatory models, which will be discussed herein, and also, the naringenin clinical data and safety, other applications, and other agents of interest. Figs. 2 and 3 summarize analgesic mechanisms of naringenin.

Naringenin actions on transient receptor potential (TRP) channels and inflammatory pain relief

TRP channels act as sensors of temperature and/or natural and endogenous substances that activate sensory neurons through calcium influx reviewed elsewhere (Julius, 2013). Activation of TRP channels in nociceptors induces spontaneous pain like-behavior in the case of TRPV1 activation by capsaicin, and TRPA1 activation by formalin in its neurogenic phase of behavioral testing, and ROS (Julius, 2013). Moreover, TRP channels are more easily activated during inflammatory painful conditions due to inflammatory mediators' activation and sensitization of sensory neurons (Julius, 2013). Thus, downmodulating TRP channels activity is analgesic.

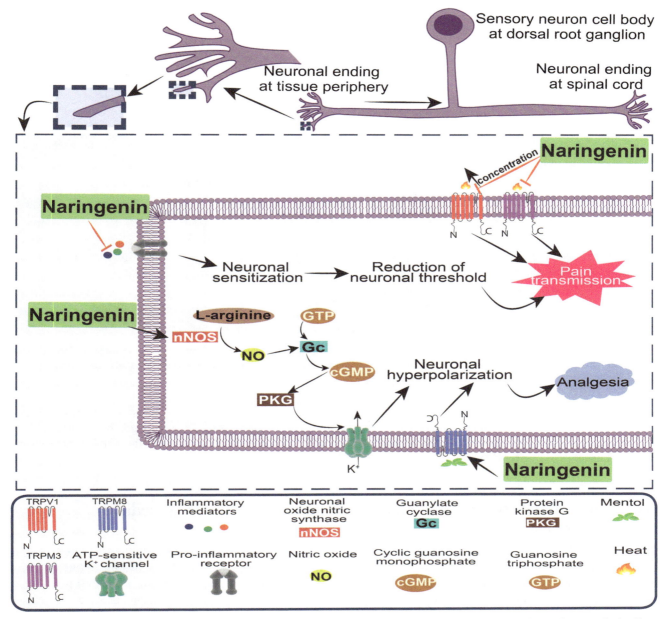

FIG. 2 Naringenin induces analgesia through sensory neuron modulation. Sensory neurons have channels responsible for sensing several stimuli e.g. chemical, thermal, and mechanical. Naringenin blocks TRPM3 and in higher concentration TRPV1. Naringenin inhibits the release of inflammatory mediators which are responsible for neuronal sensitization. Naringenin induces analgesia by activating the NO/cGMP/PKG/ATP-sensitive K^+ pathway. Naringenin activates TRPM8 which is responsible for menthol analgesic activity.

Naringenin inhibits the neurogenic phase of formalin, capsaicin- and superoxide anion-induced flinches, which are models involving TRPA1 and TRPV1 activation (Manchope et al., 2016; Pinho-Ribeiro et al., 2016). Electrophysiologic studies show that naringenin does not block TRPA1 activation and blocks TRPM3 and partially blocks TRPV1 in higher concentrations on HEK cells (Straub et al., 2013). TRPM3 is activated by noxious heat (Vriens et al., 2011). Naringenin inhibits thermal hyperalgesia induced by lipopolysaccharide (LPS) (Pinho-Ribeiro et al., 2016) and superoxide anion (Manchope et al., 2016). Thus, it is reasonable to expect that this thermal analgesia is related to the TRPM3 block by naringenin (Straub et al., 2013). On the opposite side, TRPM8 is activated by menthol and induces analgesia in Complete Freund's adjuvant-induced inflammatory pain (Liu et al., 2013). Naringenin activates TRPM8 in HEK cells (Straub et al., 2013) suggesting that not only inhibiting TRP channels that induce pain, but also activating TRP channels that induce

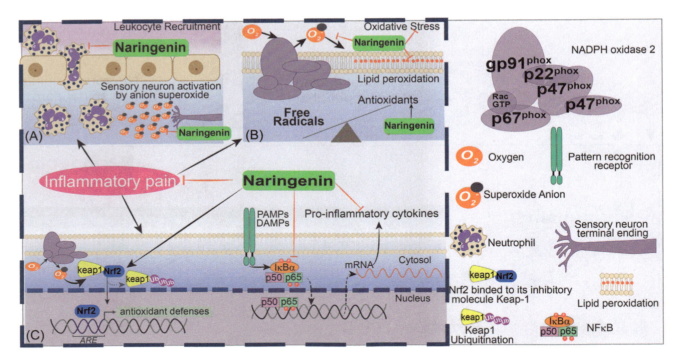

FIG. 3 Naringenin peripherical analgesic mechanism in inflammatory pain. Naringenin inhibits inflammatory pain through modulating oxidative and immunoinflammatory events. Naringenin inhibits leukocyte recruitment and nociceptor sensory neuron activation by free radical (A); oxidative stress (B), modulates transcription factors NFκB and Nrf2 and downstream pathways (C).

analgesia are mechanistically important for naringenin analgesia. However, more data are needed to determine whether TRPM8 activation by naringenin has an analgesic role *in vivo* in inflammatory pain. Thus, naringenin analgesic activity involves modulation of TRP channels during inflammatory painful conditions. Fig. 2 summarizes the activity of naringenin in TRP channels.

Naringenin inhibits the production of endogenous cytokines that mediate inflammatory pain

Tissue-resident cells expressing pattern recognition receptor (PRR) are the first to recognize PAMPs and DAMPs including tissue macrophages, mast cells, and NK cells, but also neurons innervating tissue barriers. The activation of PRRs leads to intracellular signaling resulting in the activation of master inflammatory transcription factors such as NFκB that orchestrates the production of inflammatory mediators, encompassing ROS, lipid mediators, and inflammatory cytokines and chemokines. It is interesting that these molecules contribute to inflammatory pain. For instance, chemokine CXCL1 (IL-8 homolog in mice) chemoattracts neutrophils, which in turn produce PGE_2 that sensitizes nociceptor neurons (Cunha et al., 2008; Verri et al., 2006).

Naringenin has consistently been shown to inhibit neutrophil recruitment, suggesting the importance of this action to its analgesic effect in varied models of inflammation. This activity is in line with the inhibition of cytokine production by naringenin, which are molecules that can chemoattract neutrophils and also sensitize nociceptive sensory neurons. Naringenin inhibits the production of IL-33 in carrageenan and zymosan inflammatory pain (Bussmann et al., 2019; Pinho-Ribeiro, Zarpelon, Fattori, et al., 2016), which is an essential cytokine to trigger and potentiate the production of other hyperalgesic cytokines (Zarpelon et al., 2013). In fact, naringenin inhibits TNFα and IL-1β in inflammatory painful condition induced by carrageenan (Pinho-Ribeiro, Zarpelon, Fattori, et al., 2016), superoxide anion (Manchope et al., 2016), LPS (Pinho-Ribeiro, Zarpelon, Mizokami, et al., 2016), zymosan (Bussmann et al., 2019), monosodium urate (MSU) (Ruiz-Miyazawa et al., 2018), and diabetic neuropathic pain (Singh, Bansal, Kuhad, Kumar, & Chopra, 2020). Specific biologics and genetic studies demonstrated that inhibiting cytokines, their production, and receptors reduce pain (Verri et al., 2006). Therefore, the consistent inhibitory effect of naringenin over cytokine production is an important analgesic mechanism (Fig. 3).

Naringenin targets oxidative stress to reduce inflammatory pain

Oxidative stress occurs due to an uncontrolled and excessive generation of ROS causing an imbalance between oxidants (free radicals) and endogenous antioxidant systems (Betteridge, 2000). The imbalance between free radicals and antioxidants is a hallmark in inflammatory painful conditions (Manchope et al., 2016; Pinho-Ribeiro, Zarpelon, Fattori, et al., 2016; Ruiz-Miyazawa et al., 2018). ROS induce spontaneously and evoked nociceptive behavior. Spontaneous pain induced by ROS occurs through TRPA1 (Julius, 2013) and evoked pain caused by ROS has the contribution of the TNFα/TNFR1 signaling (Yamacita-Borin et al., 2015) and NFκB activation (Pinho-Ribeiro, Fattori, et al., 2016) to induce mechanical hyperalgesia, demonstrating that oxidative stress induces pain by directly activating nociceptor sensory neurons and by inflammation. Thus, considering that oxidative stress during inflammation accounts for pain, its inhibition contributes to analgesia.

Nevertheless, an important action of naringenin is the antioxidant activity related to its antioxidant chemical groups which are responsible for donating an electron and directly stabilizing free radicals such as superoxide anion, hydroxyl radical, and also, chelate heavy metals and by resonance, naringenin stays stable after electron donation (Cavia-Saiz et al., 2010; Hernandez-Aquino & Muriel, 2018). Thus, reducing oxidative stress is a contributing analgesic mechanism of naringenin.

During inflammation, leukocytes are a major source of superoxide anion produced by the multimeric complex of NADPH oxidase 2 which has two catalytic domains, gp91^{phox} and p22^{phox} (Mittal, Siddiqui, Tran, Reddy, & Malik, 2014), and naringenin inhibits gp91^{phox} mRNA expression in inflammatory pain induced by superoxide anion itself (Manchope et al., 2016), zymosan (Bussmann et al., 2019), MSU (Ruiz-Miyazawa et al., 2018), and Titanium dioxide (TiO$_2$) (Manchope et al., 2018). Also, naringenin inhibits xanthine oxidase activity in a cell-free condition (Cavia-Saiz et al., 2010), suggesting a direct effect and not related to gene expression. Oxidative damage occurs by the interaction of free radicals with biomolecules such as membrane lipids (Betteridge, 2000) and lipoperoxidation is amenable by naringenin in painful conditions (Manchope et al., 2016, 2018; Pinho-Ribeiro, Zarpelon, Mizokami, et al., 2016; Singh et al., 2020). Thus, naringenin indirectly downmodulates oxidative stress-reducing oxidative damage, and the expression and activity of enzymes involved in ROS production, which account for its analgesic activity in inflammatory pain.

Nrf2 binding to antioxidant responsive elements (ARE) drives the expression of antioxidant and detoxification enzymes and contributes to analgesia and antiinflammatory actions (Staurengo-Ferrari et al., 2018). Heme-oxygenase-1(HO-1) is one of several enzymes controlled by Nrf2. HO-1 metabolizes the heme group in biliverdin and carbon monoxide (antioxidant molecules). HO-1 induction inhibits spontaneous pain behavior in the inflammatory phase of formalin but not the neurogenic (Rosa et al., 2008), and naringenin induces Nrf2/HO-1 antioxidant signaling in inflammatory pain such as induced by superoxide anion (Bussmann et al., 2019; Manchope et al., 2016; Ruiz-Miyazawa et al., 2018). Naringenin upregulates tissue antioxidants in inflammatory pain including reduced glutathione (GSH) that is regulated by Nrf2 (Bussmann et al., 2019; Manchope et al., 2016; Pinho-Ribeiro, Zarpelon, Mizokami, et al., 2016; Ruiz-Miyazawa et al., 2018). Thus, naringenin modulates oxidative stress through scavenging activity, balancing the levels of oxidant and antioxidant molecules, and inducing Nrf2 and downstream pathways.

Analgesic pathways actively induced by naringenin

Nociceptor hyperpolarization diminishes painful sensation and is a peripheral mechanism of action of drugs such as morphine, dipyrone, and NO donors which inhibit ongoing prostaglandin E$_2$-induced neuronal sensitization and mechanical hyperalgesia (Cunha et al., 2010; Sachs et al., 2004). In terms of peripheral analgesic mechanisms, morphine and dipyrone induce neuronal nitric oxide synthase to produce low levels of NO. NO increases the levels of cGMP produced by guanylate cyclase. The second messenger cGMP activates neuronal PKG, which increases potassium influx in ATP-sensitive-K$^+$ channels and leads to neuronal hyperpolarization diminishing painful sensation (Cunha et al., 2010; Sachs et al., 2004). Naringenin induces analgesia through activation of NO/cGMP/PKG/ATP-sensitive K$^+$ pathway reducing neuronal pain transmission in inflammatory pain models (Manchope et al., 2016; Pinho-Ribeiro, Zarpelon, Fattori, et al., 2016).

Antiinflammatory cytokines such as IL-10 and TGF-β are produced during inflammatory conditions to counterbalance and avoid the exaggerated inflammatory processes by limiting the production of inflammatory cytokines (Verri et al., 2006). Also, those cytokines induce analgesia. For instance, IL-10 inhibits carrageenan-induced inflammatory pain (Verri et al., 2006) and TGF-β1 alleviates neuropathic pain induced by nerve injury (Echeverry et al., 2009). Stimulating endogenous antiinflammatory cytokines is an interesting analgesic approach since, in principle, it does not induce overdose with immunosuppressive consequences. Naringenin increases IL-10 production zymosan arthritis (Bussmann et al., 2019). On the other hand, in inflammatory pain induced by superoxide anion naringenin decreases the production of IL-10

448 PART | IV Novel and nonpharmacological aspects and treatments

(Manchope et al., 2016) demonstrating that naringenin's upregulation of IL-10 depends on the inflammatory painful stimulus. Naringenin increases the production of TGF-β in the knee joint of mice with gout arthritis. Thus, naringenin actively induces the activation of the analgesic pathway NO/cGMP/PKG/ATP-sensitive K^+ that reduces neuronal depolarization as well as may increase analgesic cytokines IL-10 and TGF-β depending on disease context.

Naringenin modulates transcription factors and miRNA

Nrf2 and NFκB are the main transcription factors modulated by naringenin. Naringenin induces antioxidant response indirectly through activation of Nrf2 regulating antioxidant responses. Nrf2 binds to ARE resulting in transcriptional gene induction of, for instance, HO-1, nicotinamide adenine dinucleotide phosphate quinone oxidoreductase, effector glutamyl cysteinyl ligase enzyme, glutathione peroxidase (GPx), and glutathione reductase (Bussmann et al., 2019; Fan, Pan, Zhu, & Zhang, 2017; Hernandez-Aquino & Muriel, 2018; Manchope et al., 2016; Staurengo-Ferrari et al., 2018).

ROS activate the NFκB pathway and reciprocally its pro-inflammatory signaling induces the expression of pro-oxidant components (Li & Karin, 1999; Pahl, 1999). Through the inhibition of NFκB activation, naringenin contributes efficiently to counteract a positive feedback loop involving pro-oxidative metabolism (Bussmann et al., 2019; Ferraz et al., 2020; Hernandez-Aquino & Muriel, 2018; Manchope et al., 2018; Pinho-Ribeiro, Zarpelon, Fattori, et al., 2016). NFκB inhibition accounts for blocking the production of pro-inflammatory/hyperalgesic cytokines (Ferraz et al., 2020; Hernandez-Aquino & Muriel, 2018). Moreover, Nrf2 activation indirectly inhibits NFκB activity (Pan, Wang, Wang, Zhu, & Mao, 2012) since both transcription factors compete for the same transcriptional co-activator p300; thus, Nrf2 activation inhibits nuclear p65-p300 interaction (Kim, Lee, Shin, & Lee, 2013). Therefore, naringenin inhibits NFκB activation directly by interfering with NFκB activation and by activating Nrf2.

Naringenin also modulates protein synthesis by targeting miRNAs (Hernandez-Aquino & Muriel, 2018). *miR-17-3p* targets mRNAs coding for superoxide dismutase (SOD) and GPx, inhibiting their protein synthesis, and naringenin downregulates *miR-17-3p*, thus, allowing the codification of SOD and GPx that belong to the endogenous antioxidant system (Curti et al., 2017; Hernandez-Aquino & Muriel, 2018). *miR-223* regulates inflammatory response in neutrophils and it is upregulated in the early phase of the secondary damage at the spinal cord after an injury (Izumi et al., 2011). Naringenin efficiently reduces *miR-223* upregulation and neutrophil activity in the spinal cord injury in a rat model (Shi et al., 2016). Thus, demonstrating a broad spectrum of modulation of gene expression by naringenin involving important genes controlled by NFκB, Nrf2, and miRNAs. Table 2 summarizes preclinical data of naringenin analgesic activity in inflammatory pain.

Clinical applicability and safety

To our knowledge, there is no report evaluating the effect of sole naringenin on inflammatory pain in humans so far. According to the clinicaltrials.gov database, there are 11 clinical trials registered involving treatment with naringenin or naringin (its glycoside form). Except for one, all studies test naringenin/naringin associated with other compounds or as a complex food supplement (e.g., natural juice), limiting clinical conclusions specifically about naringenin/naringin effects themselves. One specific trial (identifier: NCT03582553), still in early phase I, aims to evaluate the safety, tolerability, and bioavailability of an extract of *Citrus sinensis* containing naringenin and naringin (150–900 mg) in humans.

Previous clinical studies have shown antiinflammatory and/or antioxidant effects after naringenin intake (Table 3). Snyder et al. (2011) demonstrated that orange juice consumption (a great source of flavonoids including naringenin) results in decreased serum levels of lipoprotein oxidation and increased antioxidant capacity values, independently of sugar and vitamin C content (Snyder et al., 2011). Comparably, regular intake of orange juice by patients with chronic hepatitis C induced augmented serum antioxidant capacity (evaluated by lipoperoxidation assay) and decreased inflammation parameters (C-reactive protein) and blood cholesterol (Goncalves et al., 2017). In another randomized placebo-controlled study, overweight individuals taking Sinetrol-XPur (900 mg/day), a citrus polyphenolic extract, had improvement of oxidative status without adverse effects after 12 weeks of intervention (Dallas et al., 2014). Thirty hypercholesterolemic subjects ingesting naringin (400 mg/capsule/day) had significant improvements in the erythrocyte antioxidant enzymes SOD and catalase (CAT) activities (Jung et al., 2003). Furthermore, 103 prediabetic individuals presented beneficial outcomes after supplementation with Eriomin, composed of eriocitrin (70%), hesperidin (5%), and naringin (4%), at doses ranging from 200 to 800 mg/day. Results show lower levels of IL-6, TNFα, and lipid peroxidation, and enhanced antioxidant capacity, reversing the prediabetic condition in 24% of the subjects (Ribeiro et al., 2019).

Improvement of clinical symptoms was reported as well. A randomized, double-blind, placebo-controlled, crossover study enrolling 36 postmenopausal women evaluated the efficacy of hop extract capsules, containing 100 mg of 8-prenylnaringenin, for relief of menopausal discomforts. Tools used for the evaluation of patients were the Kuppermann

TABLE 2 Preclinical data of naringenin analgesic activity in inflammatory pain.

Model	Dose	Molecular targets	Tissues/cells	References
Carrageenan paw inflammation	50 mg/kg	↓ Oxidative stress, NFκB activation, TNF-α, IL-1β, and IL-33 production, and triggering NO/cGMP/PKG/ATP- sensitive K$^+$ activation	Paw skin	Pinho-Ribeiro, Zarpelon, Fattori, et al. (2016)
Capsaicin-induced spontaneous pain behavior	50 mg/kg	N/A	Paw skin	Pinho-Ribeiro, Zarpelon, Fattori, et al. (2016)
CFA-induced spontaneous and evoked pain behavior	50 mg/kg	N/A	Paw skin	Pinho-Ribeiro, Zarpelon, Fattori, et al. (2016)
Formalin-induced spontaneous pain behavior (neurogenic and inflammatory phases)	50 mg/kg	N/A	Paw skin	Pinho-Ribeiro, Zarpelon, Fattori, et al. (2016)
LPS paw inflammation	50 mg/kg	↓ Oxidative stress, NFκB activation, TNF-α, IL-1β, IL-6, and IL-12 production	Paw skin	Pinho-Ribeiro, Zarpelon, Mizokami, et al. (2016)
MSU-induced gout arthritis	150 mg/kg	↓ Oxidative stress, NFκB activation, TNF-α, IL-1β, IL-6, IL-17, and IL-10 production, preproET-1, Nlrp3, ASC, caspase-1, and pro-IL-1β mRNA expression ↑ TGF- β production, Nrf2, and HO-1 mRNA expression	Knee joint	Ruiz-Miyazawa et al. (2018)
Superoxide anion-triggered paw inflammation	50 mg/kg	↓ Oxidative stress, TNF-α production, COX-2, preproET-1, IL-33 mRNA expression, and triggering NO/cGMP/PKG/ATP- sensitive K$^+$ activation, Nrf2 and HO-1 mRNA expression	Paw skin	Manchope et al. (2016)
TiO$_2$ prosthesis component-induced chronic arthritis	50 mg/kg	↓ Oxidative stress, NFκB activation, IL-33, TNF-α, pro-IL-1β, IL-6, and gp91^{phox} mRNA expression	Knee joint	Manchope et al. (2018)
Zymosan-induced arthritic pain	50 mg/kg	↓ Oxidative stress, NFκB activation, TNF-α, IL-1β, and IL-33 production, preproET-1, Nlrp3, ASC, caspase-1, and pro-IL-1β mRNA expression ↑ IL-10 production, Nrf2, and HO-1 mRNA expression	Knee joint	Bussmann et al. (2019)

N/A: not available data.

index, Menopause Rating Scale, and Visual analog scale. A significant reduction in menopausal complaints was observed after 16 weeks (Erkkola et al., 2010).

There is still no evidence of inflammatory pain relief provided exclusively by naringenin intake by humans. However, this was not disproved. Robust preclinical data support the analgesic effect of naringenin, which merits further clinical investigation.

Applications to other areas

Naringenin therapeutic activities also encompass nephroprotective, neuroprotective, and antimicrobial effects (Zaidun, Thent, & Latiff, 2018). Renal diseases such as glomerulonephritis, tubulointerstitial nephritis, and chronic renal failure involve oxidative stress with an increase in ROS production by leukocytes, vascular cells, glomerular cells, and interstitial

450 PART | IV Novel and nonpharmacological aspects and treatments

TABLE 3 Clinical studies involving naringenin or naringin consumption.

Condition	n	Treatment	Duration	Outcomes	References
Healthy subjects	16	Placebo (ascorbic acid and sugar equivalent to orange juice); placebo plus hesperidin; placebo plus hesperidin, luteolin, and naringenin; and orange juice (positive control)	4 weeks	Treatment with flavonoids found in orange juice increase antioxidant capacity and lower lipoprotein oxidation values independent of sugar and ascorbic acid content	Snyder et al. (2011)
Individuals with chronic hepatitis C	63	Orange juice (500 mL/day)	8 weeks	Levels of total cholesterol, LDL-cholesterol, CRP, and parameters of oxidative stress decreased	Goncalves et al. (2017)
Healthy overweight volunteers	95	2 capsules of Sinetrol-XPur/ day (each containing 90% of total polyphenols (expressed as catechin), at least 20% of total flavanones (expressed as naringin) and between 1% and 3% of natural caffeine)	12 weeks	Waist and hip circumference, abdominal fat, inflammatory markers, and oxidative stress was reduced, as well as increased superoxide dismutase and glutathione activity	Dallas et al. (2014)
Hypercholesterolemic subjects	60	Naringin (400 mg/capsule/ day)	8 weeks	Erythrocyte superoxide dismutase and catalase activities significantly increased	Jung et al. (2003)
Prediabetes patients	103	200–800 mg/day of Eriomin (70% eriocitrin, 5% hesperidin, 4% naringin, and 1% didymin)	12 weeks	Benefited glycemic control, reduced systemic inflammation and oxidative stress, and reversed the prediabetic condition in 24% of the evaluated patients	Ribeiro, Ramos, Manthey, and Cesar (2019)
Menopausal women	36	Hop extract (100 mg 8-prenylnaringenin per day)	16 weeks	Reduced discomforts and complaints associated with the menopause as measured by Kupperman Index, Menopause Rating Scale, and a multifactorial Visual Analogue Scale	Erkkola et al. (2010)

Clinical studies available at clinicaltrials.gov database. Note: No clinical trial with sole naringenin and/or naringin was found.

renal cells (Galle, 2001). Naringenin improves renal function in cisplatin-induced nephrotoxicity by preventing renal lipid peroxidation, restoring endogenous (GSH), and enzymatic antioxidant system (CAT, SOD, and GPx). Thus, naringenin nephroprotective activity is due related to its antioxidant activity (Badary, Abdel-Maksoud, Ahmed, & Owieda, 2005). By the mechanisms observed for naringenin in the cisplatin nephrotoxicity model, it is possible that Nrf2 induction is involved (Bussmann et al., 2019; Hernandez-Aquino & Muriel, 2018; Manchope et al., 2016, 2018).

Neuronal tissue has a long-life span and a high oxygen consumption rate; thus it needs an appropriate antioxidant system to avoid oxidative damage. In chronic neurodegenerative diseases such as Alzheimer's disease, senile plaques and neurofibrillary tangles trigger oxidative stress in the central nervous system (Zhao & Zhao, 2013). Naringenin improves Aβ-induced impairment of learning and memory in mice decreasing lipid peroxidation levels in the hippocampus, demonstrating the relevance of antioxidant activity of naringenin for neuroprotection (Ghofrani et al., 2015).

Antibacterial and antiviral effects of naringenin seem also relevant. Naringenin has antibacterial activity against *Escherichia coli* and *Bacillus subtilis* (Ulanowska, Majchrzyk, Moskot, Jakóbkiewicz-Banecka, & Węgrzyn, 2007). In arboviruses, naringenin was reported to have direct virucidal activity against type-2 Dengue virus (DENV)-2 (Zandi et al., 2011), inhibited the infection by all four DENV strains in a hepatocyte lineage cell (Frabasile et al., 2017), and prevented Zika virus infection in human cells possibly by acting on viral replication or assembly of viral particles (Cataneo et al., 2019). Naringenin can interfere with hepatitis C virus (HCV) infection by reducing the bioavailability of lipids necessary for HCV's successful cellular replication (Hernandez-Aquino & Muriel, 2018; Nahmias et al., 2008). Altogether these data rank naringenin as a pleiotropic compound that has antibacterial and antiviral effects in diseases that are also accompanied by pain and inflammation in which all these activities may, eventually, be useful together.

Other agents of interest

Phenolic compounds are a large group of plant molecules with analgesic, antiinflammatory, and antioxidant activities due to their structure with electron-donating ability (Balasundram, Sundram, & Samman, 2006; Ferraz et al., 2020). This session discusses the analgesic, antiinflammatory, and antioxidant activity of diosmin, which is another citrus flavonoid already used for vascular disorders treatment. In addition, Table 4 exemplifies the analgesic activity of other plant-based agents such as flavonoids (quercetin), dihydroxybenzoic acid derivatives (vanillic acid), and diterpenes (kaurenoic acid)

TABLE 4 Other plant-based agents with analgesic activity.

Molecule	Model	Dose	Molecular targets	Tissues/ Cells	References
Quercetin	Zymosan-induced arthritis pain	100 mg/kg	↓ Oxidative stress, gp91^{phox}, prepoET-1, and COX-2 mRNA expression, NF-κB activation, TNF-α, and IL-1β production ↑GSH levels, HO-1 and Nrf2 mRNA expression	Knee joint	Guazelli et al. (2018)
	Carrageenan paw inflammation	100 mg/kg	↓ Oxidative stress and IL-1β production ↑GSH levels	Paw skin	Valério et al. (2009)
	Formalin-induced spontaneous pain behavior	100 mg/kg	Analgesia through 5-HT$_{1A}$ receptor activation	Paw skin	Martinez et al. (2009)
Kaurenoic Acid	Carrageenan-induced pain	10 mg/kg	↓ TNF-α and IL-1β production, and triggering NO/cGMP/PKG/ATP- sensitive K$^+$ activation.	Paw skin	Mizokami et al. (2012)
	CFA-induced inflammatory pain	1 mg/kg	N/A	Paw skin	Dalenogare et al. (2019)
	Thermal latency on tail-flick test	1 mg/kg	Opioids receptor activation	Tail	Dalenogare et al. (2019)
	Formalin-induced pain	15 mg/kg	N/A	Paw skin	Montiel-Ruiz et al. (2020)
Vanillic Acid	Carrageenan paw inflammatory pain	30 mg/kg	↓ Oxidative stress, lipid oxidation, TNF-α, IL-1β, and IL-33 production, NF-κB activation ↑ GSH levels	Paw skin	Calixto-Campos et al. (2015)
	Acetic acid-induced abdominal pain	100 mg/kg	Analgesia through 5-HT$_3$ receptor activation ↓ Lipid oxidation	Plasma	Yrbas, Morucci, Alonso, and Gorzalczany (2015)
	Formalin-induced spontaneous pain behavior	100 mg/kg	N/A	Paw skin	Morucci, Lopez, Mino, Ferraro, and Gorzalczany (2012)

N/A: not available data.

452 PART | IV Novel and nonpharmacological aspects and treatments

(Calixto-Campos et al., 2015; Dalenogare et al., 2019; Guazelli et al., 2018; Martinez et al., 2009; Mizokami et al., 2012; Montiel-Ruiz et al., 2020; Morucci et al., 2012; Valério et al., 2009; Yrbas et al., 2015).

Preclinically diosmin has analgesic, antiinflammatory, and antioxidant effects in chronic constriction injury (CCI)-induced neuropathic pain, LPS-induced inflammatory pain, and peritonitis in mice (Bertozzi et al., 2017; Fattori et al., 2020). Diosmin inhibits CCI-induced neuropathic pain by inducing the analgesic pathway NO/cGMP/PKG/ATP- sensitive K^+ pathway. In neuropathic pain induced by CCI, IL-33 is produced by oligodendrocytes at the spinal cord. IL-33 further induces the production of TNFα and IL-1β in the spinal cord and activates astrocytes and microglia (Zarpelon et al., 2016). Diosmin modulates the expression of inflammatory cytokines reducing the expression of IL-33, TNFα, and IL-1β, and activation of oligodendrocytes, astrocytes, and microglia at the spinal cord in CCI-induced neuropathic pain (Bertozzi et al., 2017).

In terms of inflammatory pain, diosmin inhibits inflammatory pain and peritonitis induced by LPS through diminishing leukocyte recruitment to paw skin and peritoneum. Recruited leukocytes are a major source of ROS, and diosmin inhibits ROS production and superoxide anion production in peritoneal recruited cells and paw tissue. Moreover, diosmin increases antioxidant response by increasing GSH levels in paw skin. Thus, diosmin inhibits oxidative stress in LPS-induced inflammatory pain and peritonitis. Furthermore, diosmin inhibits the production of TNFα, IL-1β, and IL-6 in the paw skin and peritoneum, and NFκB activation in paw skin and recruited leukocytes to the peritoneum (Fattori et al., 2020). Thus, flavonoids derived from citrus fruits such as naringenin and diosmin have analgesic, antiinflammatory activities by inducing neuronal hyperpolarization, reducing leukocyte recruitment, oxidative stress, neuroinflammation, modulating cytokines, and NFκB.

Mini-dictionary of terms

Hyperalgesia: "Increased pain sensation from a stimulus that normally elicits pain. Hyperalgesia reflects increased pain on suprathreshold stimulation" as defined by The International Association for the Study of Pain (IASP).

Nociception: "The neural process of encoding noxious stimuli" as defined by IASP.

Nociceptor: "A high-threshold sensory neuron of the peripheral somatosensory nervous system that is capable of transducing and encoding noxious stimuli" as defined by IASP.

Oxidative stress: disbalance between oxidants levels (also called free radicals) and antioxidant defenses leading to cell damage.

TRPA1: Transient receptor potential of the ankyrin subfamily member 1 is a cation channel involved in transduction of cold temperatures in mammals, mechanic and irritant chemical stimulus, such as allyl isothiocyanate (AITC).

TRPM3: Transient receptor potential of the melastatin subfamily member 3 is a cation channel related to the transduction of thermal stimuli, such as noxious heat, also is activated by pregnenolone sulfate.

TRPM8: Transient receptor potential of the melastatin subfamily member 8 is a cation channel which transduces cold stimulus and also plays a role in menthol inducing analgesia.

TRPV1: Transient receptor potential of the vanilloid subtype member 1 is calcium-permeable channel which transduce a plethora of stimulus such as noxious heat, capsaicin, osmotic and mechanical changes.

Key facts of inflammatory pain

- Inflammatory pain is involved in several kinds of painful conditions such as arthritic gout, rheumatoid arthritis, infection, and others.
- Recruited leukocytes are a major source of ROS and oxidative stress upon a stimulus inducing inflammatory pain.
- ROS induce spontaneous pain behavior through neuronal activation.
- Inflammatory pain is induced by pro-inflammatory cytokines and prostanoids released mainly by tissue-resident cells and recruited leukocytes.
- Sequentially, there is a hierarchical inflammatory mediators release involving chemokines, cytokines, and prostanoids, which lately induces primary afferent nociceptor sensitization.
- NFκB is a major transcriptional factor involved in the expression of inflammatory mediators and enzymes involved in inflammatory pain.

Summary points

- Naringenin modulates TRP channels.
- Naringenin has analgesic activity in inflammatory pain models.

- Naringenin induces neuronal hyperpolarization triggering the analgesic pathway NO/cGMP/PKG/ATP- sensitive K^+.
- Naringenin induces antioxidant responses directly scavenging ROS or indirectly through inhibiting the production of free radicals and inducing Nrf2 and downstream antioxidant gene expression.
- Naringenin inhibits pro-inflammatory cytokine release in inflammatory painful conditions.
- Naringenin inhibits NFκB activation in inflammatory painful conditions.

Funding

This work was supported by Brazilian grants from PPSUS funded by Decit/SCTIE/MS intermediated by CNPq with support of Fundação Araucária and SESA-PR (agreement 041/2017); CAPES (finance code 001); CNPq; and Programa de Apoio a Grupos de Excelência (PRONEX) grant supported by SETI/Fundação Araucária and MCTI/CNPq, and Governo do Estado do Paraná (agreement 014/2017).

References

Badary, O. A., Abdel-Maksoud, S., Ahmed, W. A., & Owieda, G. H. (2005). Naringenin attenuates cisplatin nephrotoxicity in rats. *Life Sciences, 76*(18), 2125–2135. https://doi.org/10.1016/j.lfs.2004.11.005.

Balasundram, N., Sundram, K., & Samman, S. (2006). Phenolic compounds in plants and Agri-industrial by-products: Antioxidant activity, occurrence, and potential uses. *Food Chemistry, 99*(1), 191–203. https://doi.org/10.1016/j.foodchem.2005.07.042.

Basbaum, A. I., Bautista, D. M., Scherrer, G., & Julius, D. (2009). Cellular and molecular mechanisms of pain. *Cell, 139*(2), 267–284. https://doi.org/10.1016/j.cell.2009.09.028.

Bertozzi, M. M., Rossaneis, A. C., Fattori, V., Longhi-Balbinot, D. T., Freitas, A., Cunha, F. Q., et al. (2017). Diosmin reduces chronic constriction injury-induced neuropathic pain in mice. *Chemico-Biological Interactions, 273*, 180–189. https://doi.org/10.1016/j.cbi.2017.06.014.

Betteridge, D. J. (2000). What is oxidative stress? *Metabolism, 49*(2 Suppl 1), 3–8. https://doi.org/10.1016/s0026-0495(00)80077-3.

Bussmann, A. J. C., Borghi, S. M., Zaninelli, T. H., Dos Santos, T. S., Guazelli, C. F. S., Fattori, V., et al. (2019). The citrus flavanone naringenin attenuates zymosan-induced mouse joint inflammation: Induction of Nrf2 expression in recruited CD45(+) hematopoietic cells. *Inflammopharmacology, 27*(6), 1229–1242. https://doi.org/10.1007/s10787-018-00561-6.

Calixto-Campos, C., Carvalho, T. T., Hohmann, M. S. N., Pinho-Ribeiro, F. A., Fattori, V., Manchope, M. F., et al. (2015). Vanillic acid inhibits inflammatory pain by inhibiting neutrophil recruitment, oxidative stress, cytokine production, and NFκB activation in mice. *Journal of Natural Products, 78*(8), 1799–1808. https://doi.org/10.1021/acs.jnatprod.5b00246.

Cataneo, A. H. D., Kuczera, D., Koishi, A. C., Zanluca, C., Silveira, G. F., Arruda, T. B., et al. (2019). The citrus flavonoid naringenin impairs the in vitro infection of human cells by Zika virus. *Scientific Reports, 9*(1), 16348. https://doi.org/10.1038/s41598-019-52626-3.

Cavia-Saiz, M., Busto, M. D., Pilar-Izquierdo, M. C., Ortega, N., Perez-Mateos, M., & Muñiz, P. (2010). Antioxidant properties, radical scavenging activity and biomolecule protection capacity of flavonoid naringenin and its glycoside naringin: A comparative study. *Journal of the Science of Food and Agriculture, 90*(7), 1238–1244. https://doi.org/10.1002/jsfa.3959.

Cunha, T. M., Roman-Campos, D., Lotufo, C. M., Duarte, H. L., Souza, G. R., Verri, W. A., et al. (2010). Morphine peripheral analgesia depends on activation of the PI3Kgamma/AKT/nNOS/NO/KATP signaling pathway. *Proceedings of the National Academy of Sciences of the United States of America, 107*(9), 4442–4447. https://doi.org/10.1073/pnas.0914733107.

Cunha, T. M., Verri, W. A., Schivo, I. R., Napimoga, M. H., Parada, C. A., Poole, S., et al. (2008). Crucial role of neutrophils in the development of mechanical inflammatory hypernociception. *Journal of Leukocyte Biology, 83*(4), 824–832. https://doi.org/10.1189/jlb.0907654.

Curti, V., Di Lorenzo, A., Rossi, D., Martino, E., Capelli, E., Collina, S., et al. (2017). Enantioselective modulatory effects of Naringenin enantiomers on the expression levels of miR-17-3p involved in endogenous antioxidant defenses. *Nutrients, 9*(3). https://doi.org/10.3390/nu9030215.

Dalenogare, D. P., Ferro, P. R., De Pra, S. D. T., Rigo, F. K., de David Antoniazzi, C. T., de Almeida, A. S., et al. (2019). Antinociceptive activity of Copaifera officinalis Jacq. L oil and kaurenoic acid in mice. *Inflammopharmacology, 27*(4), 829–844. https://doi.org/10.1007/s10787-019-00588-3.

Dallas, C., Gerbi, A., Elbez, Y., Caillard, P., Zamaria, N., & Cloarec, M. (2014). Clinical study to assess the efficacy and safety of a citrus polyphenolic extract of red orange, grapefruit, and orange (Sinetrol-XPur) on weight management and metabolic parameters in healthy overweight individuals. *Phytotherapy Research, 28*(2), 212–218. https://doi.org/10.1002/ptr.4981.

Echeverry, S., Shi, X. Q., Haw, A., Liu, H., Zhang, Z. W., & Zhang, J. (2009). Transforming growth factor-beta1 impairs neuropathic pain through pleiotropic effects. *Molecular Pain, 5*, 16. https://doi.org/10.1186/1744-8069-5-16.

Erkkola, R., Vervarcke, S., Vansteelandt, S., Rompotti, P., De Keukeleire, D., & Heyerick, A. (2010). A randomized, double-blind, placebo-controlled, cross-over pilot study on the use of a standardized hop extract to alleviate menopausal discomforts. *Phytomedicine, 17*(6), 389–396. https://doi.org/10.1016/j.phymed.2010.01.007.

Fan, R., Pan, T., Zhu, A. L., & Zhang, M. H. (2017). Anti-inflammatory and anti-arthritic properties of naringenin via attenuation of NF-kappaB and activation of the heme oxygenase HO-1/related factor 2 pathway. *Pharmacological Reports, 69*(5), 1021–1029. https://doi.org/10.1016/j.pharep.2017.03.020.

Fattori, V., Rasquel-Oliveira, F. S., Artero, N. A., Ferraz, C. R., Borghi, S. M., Casagrande, R., et al. (2020). Diosmin treats lipopolysaccharide-induced inflammatory pain and peritonitis by blocking NF-kappaB activation in mice. *Journal of Natural Products, 83*(4), 1018–1026. https://doi.org/10.1021/acs.jnatprod.9b00887.

Ferraz, C. R., Carvalho, T. T., Manchope, M. F., Artero, N. A., Rasquel-Oliveira, F. S., Fattori, V., et al. (2020). Therapeutic potential of flavonoids in pain and inflammation: Mechanisms of action, pre-clinical and clinical data, and pharmaceutical development. *Molecules, 25*(3). https://doi.org/10.3390/molecules25030762.

Frabasile, S., Koishi, A. C., Kuczera, D., Silveira, G. F., Verri, W. A., Jr., Duarte Dos Santos, C. N., et al. (2017). The citrus flavanone naringenin impairs dengue virus replication in human cells. *Scientific Reports, 7*, 41864. https://doi.org/10.1038/srep41864.

Galle, J. (2001). Oxidative stress in chronic renal failure. *Nephrology, Dialysis, Transplantation, 16*(11), 2135–2137. https://doi.org/10.1093/ndt/16.11.2135.

Gattuso, G., Barreca, D., Gargiulli, C., Leuzzi, U., & Caristi, C. (2007). Flavonoid composition of Citrus juices. *Molecules, 12*(8), 1641–1673. https://doi.org/10.3390/12081641.

Ghofrani, S., Joghataei, M. T., Mohseni, S., Baluchnejadmojarad, T., Bagheri, M., Khamse, S., et al. (2015). Naringenin improves learning and memory in an Alzheimer's disease rat model: Insights into the underlying mechanisms. *European Journal of Pharmacology, 764*, 195–201. https://doi.org/10.1016/j.ejphar.2015.07.001.

Goncalves, D., Lima, C., Ferreira, P., Costa, P., Costa, A., Figueiredo, W., et al. (2017). Orange juice as dietary source of antioxidants for patients with hepatitis C under antiviral therapy. *Food & Nutrition Research, 61*(1), 1296675. https://doi.org/10.1080/16546628.2017.1296675.

Guazelli, C. F. S., Staurengo-Ferrari, L., Zarpelon, A. C., Pinho-Ribeiro, F. A., Ruiz-Miyazawa, K. W., Vicentini, F., et al. (2018). Quercetin attenuates zymosan-induced arthritis in mice. *Biomedicine & Pharmacotherapy, 102*, 175–184. https://doi.org/10.1016/j.biopha.2018.03.057.

Hernandez-Aquino, E., & Muriel, P. (2018). Beneficial effects of naringenin in liver diseases: Molecular mechanisms. *World Journal of Gastroenterology, 24*(16), 1679–1707. https://doi.org/10.3748/wjg.v24.i16.1679.

Izumi, B., Nakasa, T., Tanaka, N., Nakanishi, K., Kamei, N., Yamamoto, R., et al. (2011). MicroRNA-223 expression in neutrophils in the early phase of secondary damage after spinal cord injury. *Neuroscience Letters, 492*(2), 114–118. https://doi.org/10.1016/j.neulet.2011.01.068.

Julius, D. (2013). TRP channels and pain. *Annual Review of Cell and Developmental Biology, 29*, 355–384. https://doi.org/10.1146/annurev-cellbio-101011-155833.

Jung, U. J., Kim, H. J., Lee, J. S., Lee, M. K., Kim, H. O., Park, E. J., et al. (2003). Naringin supplementation lowers plasma lipids and enhances erythrocyte antioxidant enzyme activities in hypercholesterolemic subjects. *Clinical Nutrition, 22*(6), 561–568. https://doi.org/10.1016/s0261-5614(03)00059-1.

Kim, S.-W., Lee, H.-K., Shin, J.-H., & Lee, J.-K. (2013). Up-down regulation of HO-1 and iNOS gene expressions by ethyl pyruvate via recruiting p300 to Nrf2 and depriving it from p65. *Free Radical Biology & Medicine, 65*, 468–476. https://doi.org/10.1016/j.freeradbiomed.2013.07.028.

Kurella, M., Bennett, W. M., & Chertow, G. M. (2003). Analgesia in patients with ESRD: A review of available evidence. *American Journal of Kidney Diseases, 42*(2), 217–228. https://doi.org/10.1016/S0272-6386(03)00645-0.

Li, N., & Karin, M. (1999). Is NF-kappaB the sensor of oxidative stress? *The FASEB Journal, 13*(10), 1137–1143. http://www.ncbi.nlm.nih.gov/pubmed/10385605.

Lin, L. Z., Mukhopadhyay, S., Robbins, R. J., & Harnly, J. M. (2007). Identification and quantification of flavonoids of Mexican oregano (*Lippia graveolens*) by LC-DAD-ESI/MS analysis. *Journal of Food Composition and Analysis, 20*(5), 361–369. https://doi.org/10.1016/j.jfca.2006.09.005.

Liu, B., Fan, L., Balakrishna, S., Sui, A., Morris, J. B., & Jordt, S. E. (2013). TRPM8 is the principal mediator of menthol-induced analgesia of acute and inflammatory pain. *Pain, 154*(10), 2169–2177. https://doi.org/10.1016/j.pain.2013.06.043.

Manchope, M. F., Artero, N. A., Fattori, V., Mizokami, S. S., Pitol, D. L., Issa, J. P. M., et al. (2018). Naringenin mitigates titanium dioxide (TiO2)-induced chronic arthritis in mice: Role of oxidative stress, cytokines, and NFkappaB. *Inflammation Research, 67*(11–12), 997–1012. https://doi.org/10.1007/s00011-018-1195-y.

Manchope, M. F., Calixto-Campos, C., Coelho-Silva, L., Zarpelon, A. C., Pinho-Ribeiro, F. A., Georgetti, S. R., et al. (2016). Naringenin inhibits superoxide anion-induced inflammatory pain: Role of oxidative stress, cytokines, Nrf-2 and the NO-cGMP-PKG-KATPChannel signaling pathway. *PLoS One, 11*(4). https://doi.org/10.1371/journal.pone.0153015, e0153015.

Martinez, A. L., Gonzalez-Trujano, M. E., Aguirre-Hernandez, E., Moreno, J., Soto-Hernandez, M., & Lopez-Munoz, F. J. (2009). Antinociceptive activity of Tilia americana var. mexicana inflorescences and quercetin in the formalin test and in an arthritic pain model in rats. *Neuropharmacology, 56*(2), 564–571. https://doi.org/10.1016/j.neuropharm.2008.10.010.

Mittal, M., Siddiqui, M. R., Tran, K., Reddy, S. P., & Malik, A. B. (2014). Reactive oxygen species in inflammation and tissue injury. *Antioxidants & Redox Signaling, 20*(7), 1126–1167. https://doi.org/10.1089/ars.2012.5149.

Mizokami, S. S., Arakawa, N. S., Ambrosio, S. R., Zarpelon, A. C., Casagrande, R., Cunha, T. M., et al. (2012). Kaurenoic acid from Sphagneticola trilobata inhibits inflammatory pain: Effect on cytokine production and activation of the NO-cyclic GMP-protein kinase G-ATP-sensitive potassium channel signaling pathway. *Journal of Natural Products, 75*(5), 896–904. https://doi.org/10.1021/np200989t.

Montiel-Ruiz, R. M., Cordova-de la Cruz, M., Gonzalez-Cortazar, M., Zamilpa, A., Gomez-Rivera, A., Lopez-Rodriguez, R., et al. (2020). Antinociceptive effect of Hinokinin and Kaurenoic acid isolated from Aristolochia odoratissima L. *Molecules, 25*(6). https://doi.org/10.3390/molecules25061454.

Morucci, F., Lopez, P., Mino, J., Ferraro, G., & Gorzalczany, S. (2012). Antinociceptive activity of aqueous extract and isolated compounds of Lithrea molleoides. *Journal of Ethnopharmacology, 142*(2), 401–406. https://doi.org/10.1016/j.jep.2012.05.009.

Nahmias, Y., Goldwasser, J., Casali, M., van Poll, D., Wakita, T., Chung, R. T., et al. (2008). Apolipoprotein B-dependent hepatitis C virus secretion is inhibited by the grapefruit flavonoid naringenin. *Hepatology, 47*(5), 1437–1445. https://doi.org/10.1002/hep.22197.

Pahl, H. L. (1999). Activators and target genes of Rel/NF-kappaB transcription factors. *Oncogene, 18*(49), 6853–6866. https://doi.org/10.1038/sj.onc.1203239.

Pan, H., Wang, H., Wang, X., Zhu, L., & Mao, L. (2012). The absence of Nrf2 enhances NF-κB-dependent inflammation following scratch injury in mouse primary cultured astrocytes. *Mediators of Inflammation, 2012*. https://doi.org/10.1155/2012/217580, 217580.

Pinho-Ribeiro, F. A., Zarpelon, A. C., Fattori, V., Manchope, M. F., Mizokami, S. S., Casagrande, R., et al. (2016). Naringenin reduces inflammatory pain in mice. *Neuropharmacology, 105*, 508–519. https://doi.org/10.1016/j.neuropharm.2016.02.019.

Pinho-Ribeiro, F. A., Zarpelon, A. C., Mizokami, S. S., Borghi, S. M., Bordignon, J., Silva, R. L., et al. (2016). The citrus flavonone naringenin reduces lipopolysaccharide-induced inflammatory pain and leukocyte recruitment by inhibiting NF-κB activation. *Journal of Nutritional Biochemistry, 33*, 8–14. https://doi.org/10.1016/j.jnutbio.2016.03.013.

Ribeiro, C. B., Ramos, F. M., Manthey, J. A., & Cesar, T. B. (2019). Effectiveness of Eriomin(R) in managing hyperglycemia and reversal of prediabetes condition: A double-blind, randomized, controlled study. *Phytotherapy Research, 33*(7), 1921–1933. https://doi.org/10.1002/ptr.6386.

Rosa, A. O., Egea, J., Lorrio, S., Rojo, A. I., Cuadrado, A., & Lopez, M. G. (2008). Nrf2-mediated haeme oxygenase-1 up-regulation induced by cobalt protoporphyrin has antinociceptive effects against inflammatory pain in the formalin test in mice. *Pain, 137*(2), 332–339. https://doi.org/10.1016/j.pain.2007.09.015.

Ruiz-Miyazawa, K. W., Borghi, S. M., Pinho-Ribeiro, F. A., Staurengo-Ferrari, L., Fattori, V., Fernandes, G. S. A., et al. (2018). The citrus flavanone naringenin reduces gout-induced joint pain and inflammation in mice by inhibiting the activation of NFκB and macrophage release of IL-1β. *Journal of Functional Foods, 48*, 106–116. https://doi.org/10.1016/j.jff.2018.06.025.

Sachs, D., Cunha, F. Q., & Ferreira, S. H. (2004). Peripheral analgesic blockade of hypernociception: Activation of arginine/NO/cGMP/protein kinase G/ATP-sensitive K^+ channel pathway. *Proceedings of the National Academy of Sciences of the United States of America, 101*(10), 3680–3685. https://doi.org/10.1073/pnas.0308382101.

Shi, L. B., Tang, P. F., Zhang, W., Zhao, Y. P., Zhang, L. C., & Zhang, H. (2016). Naringenin inhibits spinal cord injury-induced activation of neutrophils through miR-223. *Gene, 592*(1), 128–133. https://doi.org/10.1016/j.gene.2016.07.037.

Singh, P., Bansal, S., Kuhad, A., Kumar, A., & Chopra, K. (2020). Naringenin ameliorates diabetic neuropathic pain by modulation of oxidative-nitrosative stress, cytokines and MMP-9 levels. *Food & Function, 11*(5), 4548–4560. https://doi.org/10.1039/c9fo00881k.

Snyder, S. M., Reber, J. D., Freeman, B. L., Orgad, K., Eggett, D. L., & Parker, T. L. (2011). Controlling for sugar and ascorbic acid, a mixture of flavonoids matching navel oranges significantly increases human postprandial serum antioxidant capacity. *Nutrition Research, 31*(7), 519–526. https://doi.org/10.1016/j.nutres.2011.06.006.

Staurengo-Ferrari, L., Badaro-Garcia, S., Hohmann, M. S. N., Manchope, M. F., Zaninelli, T. H., Casagrande, R., et al. (2018). Contribution of Nrf2 modulation to the mechanism of action of analgesic and anti-inflammatory drugs in pre-clinical and clinical stages. *Frontiers in Pharmacology, 9*, 1536. https://doi.org/10.3389/fphar.2018.01536.

Straub, I., Mohr, F., Stab, J., Konrad, M., Philipp, S. E., Oberwinkler, J., et al. (2013). Citrus fruit and fabacea secondary metabolites potently and selectively block TRPM3. *British Journal of Pharmacology, 168*(8), 1835–1850. https://doi.org/10.1111/bph.12076.

Teets, A. S., Minardi, C. S., Sundararaman, M., Hughey, C. A., & Were, L. M. (2009). Extraction, identification, and quantification of flavonoids and phenolic acids in electron beam-irradiated almond skin powder. *Journal of Food Science, 74*(3), C298–C305. https://doi.org/10.1111/j.1750-3841.2009.01112.x.

Ulanowska, K., Majchrzyk, A., Moskot, M., Jakóbkiewicz-Banecka, J., & Węgrzyn, G. (2007). Assessment of antibacterial effects of flavonoids by estimation of generation times in liquid bacterial cultures. *Biologia, 62*(2), 132–135.

Valério, D.a., Georgetti, S. R., Magro, D.a., Casagrande, R., Cunha, T. M., Vicentini, F. T. M. C., et al. (2009). Quercetin reduces inflammatory pain: Inhibition of oxidative stress and cytokine production. *Journal of Natural Products, 72*(11), 1975–1979. https://doi.org/10.1021/np900259y.

Verri, W. A., Cunha, T. M., Parada, C. A., Poole, S., Cunha, F. Q., & Ferreira, S. H. (2006). Hypernociceptive role of cytokines and chemokines: Targets for analgesic drug development? *Pharmacology & Therapeutics, 112*(1), 116–138. https://doi.org/10.1016/j.pharmthera.2006.04.001.

Vriens, J., Owsianik, G., Hofmann, T., Philipp, S. E., Stab, J., Chen, X., et al. (2011). TRPM3 is a nociceptor channel involved in the detection of noxious heat. *Neuron, 70*(3), 482–494. https://doi.org/10.1016/j.neuron.2011.02.051.

Yamacita-Borin, F. Y., Zarpelon, A. C., Pinho-Ribeiro, F. A., Fattori, V., Alves-Filho, J. C., Cunha, F. Q., et al. (2015). Superoxide anion-induced pain and inflammation depends on TNFα/TNFR1 signaling in mice. *Neuroscience Letters, 605*, 53–58. https://doi.org/10.1016/j.neulet.2015.08.015.

Yrbas, M. L., Morucci, F., Alonso, R., & Gorzalczany, S. (2015). Pharmacological mechanism underlying the antinociceptive activity of vanillic acid. *Pharmacology, Biochemistry, and Behavior, 132*, 88–95. https://doi.org/10.1016/j.pbb.2015.02.016.

Zaidun, N. H., Thent, Z. C., & Latiff, A. A. (2018). Combating oxidative stress disorders with citrus flavonoid: Naringenin. *Life Sciences, 208*, 111–122. https://doi.org/10.1016/j.lfs.2018.07.017.

Zandi, K., Teoh, B., Sam, S., Wong, P., Mustafa, M., & AbuBakar, S. (2011). In vitro antiviral activity of Fisetin, Rutin and Naringenin against dengue virus type-2. *Journal of Medicinal Plants Research, 5*(23), 5534–5539 (research paper).

Zarpelon, A. C., Cunha, T. M., Alves-Filho, J. C., Pinto, L. G., Ferreira, S. H., McInnes, I. B., et al. (2013). IL-33/ST2 signalling contributes to carrageenin-induced innate inflammation and inflammatory pain: Role of cytokines, endothelin-1 and prostaglandin E2. *British Journal of Pharmacology, 169*(1), 90–101. https://doi.org/10.1111/bph.12110.

Zarpelon, A. C., Rodrigues, F. C., Lopes, A. H., Souza, G. R., Carvalho, T. T., Pinto, L. G., et al. (2016). Spinal cord oligodendrocyte-derived alarmin IL-33 mediates neuropathic pain. *The FASEB Journal, 30*(1), 54–65. https://doi.org/10.1096/fj.14-267146.

Zhao, Y., & Zhao, B. (2013). Oxidative stress and the pathogenesis of Alzheimer's disease. *Oxidative Medicine and Cellular Longevity, 2013*, 316523. https://doi.org/10.1155/2013/316523.

Chapter 39

Analgesic properties of plants from the genus *Solanum* L. (Solanaceae)

F.J.R. Paumgartten[a], G.R. de Souza[a], A.J.R. da Silva[b], and A.C.A.X. De-Oliveira[a]

[a]*Department of Biological Sciences, National School of Public Health, Oswaldo Cruz Foundation, Rio de Janeiro, RJ, Brazil,* [b]*Institute of Research on Natural Products, Federal University of Rio de Janeiro, Rio de Janeiro, RJ, Brazil*

Abbreviations

AP	acetaminophen
CGA	chlorogenic acid
COX	cyclooxygenase
CNS	central nervous system
LPS	lipopolysaccharides
NAC	*N*-acetyl-cysteine
NSAIDs	nonsteroidal antiinflammatory drugs
PG	prostaglandins
SA	salicylic acid

Introduction

The origins of pain pharmacology can be traced back to a millenary use of medicinal plants such as the opium poppy (*Papaver somniferum*) and the willow tree (*Salix alba* L.) to ease aches and pains. Morphine, a benzylisoquinoline alkaloid abundant in the dried juice of poppy seedpod (opium), is the prototype of the opioid central-acting analgesics. If morphine and related narcotic opium alkaloids (codeine, thebaine) are the roots of a pharmacological class of opioid analgesics, the willow tree bark is the primary source of another class of widely used pain-relief agents, the NSAIDs.

The bark and leaves of willow trees are rich in the phenolic glycoside salicin that, in the intestines, is converted into salicyl alcohol that is further oxidized to a phenolic acid, the active analgesic principle. Acetylsalicylic acid or aspirin (a prodrug) is an ester of salicylic acid that is less irritant to the stomach than the parent acid compound. In 1971, John Vane deciphered the mode by which aspirin exerts its analgesic, antipyretic and antiinflammatory actions, and demonstrated that SA inhibited the cyclooxygenase-catalyzed synthesis of prostaglandins that are the key mediators of inflammation, pain, and fever (Mahdi, 2010). The elucidation of the mode of action of aspirin opened the door for the development of a class of nonsteroidal antiinflammatory (NSAID) and analgesic drugs to which was added, 20 years later, a novel subclass of selective COX-2 inhibitors (Botting, 2010). Currently, both nonselective and selective COX-2 inhibitors play a key role in the therapeutic armamentarium available to relieve pain and reduce inflammation.

Opium poppy and willow tree illustrate how important medicinal plants are as potential sources of new analgesic drugs. Over one-hundred plant species of a number of botanical families and genera were reported to have analgesic and/or antiinflammatory and/or anesthetic activities (Almeida, Navarro, & Barbosa-Filho, 2001; Anilkumar, 2010; Sengupta, Sheorey, & Hinge, 2012; Uritu et al., 2018), and different types of phytochemicals may account for these therapeutic properties (Tsuchiya, 2017). Table 1 lists a set of phytochemicals and plants that are potential sources of novel plant-based analgesic and/or anesthetic agents useful for the clinical management of pain.

In the following lines, we address the experimental evidence suggesting that plants belonging to the *Solanum* genus might also be a source of novel and safer pain relief medicines. Moreover, the information available on the therapeutical potential of crude extracts of *Solanum* plants provides a scientific basis to corroborate their use to alleviate moderate aches of different types. As highlighted by the WHO's Declaration of Alma-Ata in 1978, reiterated several times thereafter, in

---

Treatments, Mechanisms, and Adverse Reactions of Anesthetics and Analgesics. https://doi.org/10.1016/B978-0-12-820237-1.00039-9
Copyright © 2022 Elsevier Inc. All rights reserved.

458 PART | IV Novel and nonpharmacological aspects and treatments

TABLE 1 Phytochemicals with pain-relieving properties, a source of novel plant-based analgesics and anesthetics.

Phytochemical	Plant (family)	Pharmacological/clinical evidence
Analgesics		
Cannabinoids, cannabis extracts, Δ^9tetra-hidrocannabinol, cannabidiol (CBD), synthetic cannabinoids	*Cannabis sativa* (*Cannabaceae*)	Meta-analyses indicate that there is moderate evidence that cannabinoids relieve chronic noncancer pain. Studies suggest, however, that cannabinoids are not effective for acute pain [a,b]
Opioids, morphine, codeine, thebaine, papaverine, and other opiate alkaloids.	*Papaver somniferum* (*Papaveraceae*)	Analgesic opioids are ligands to μ opioid (MOP) receptors in the brain and peripheral nervous system where they act in an agonist manner [c]
Salicilin/salicylic acid (SA)	*Salix alba* (*Salicaceae*) (SA is a phytohormone found in plants)	Analgesic, Antiinflammatory, and antipyretic effects arising from the inhibition of cyclooxygenases (COX 1 and 2)
Flavonoids/polyphenols Quercetin, resveratrol, narigenin, kaempferol, apigenin, fisetin, luteolin, hydroxycinnamates	Products of plant secondary metabolism. Ubiquitious presence in plants.	Analgesic Antiinflammatory and antioxidant activities in rodent assays and Antiinflammatory-like actions in in vitro tests[d]
Carnosol (CS), carnosic acid (CA), phenolic diterpes	*Rosmarinus officinalis, Salvia* spp. (*Lamiaceae*)	Antioxidant and Antiinflammatory actions. In silico docking suggests that CS and CA act on microsomal prostaglandin E synthase-1 (mPGES-1) and arachidonate 5-lipoxygenase (5-LO) thereby inhibiting the formation of proinflammatory eicosanoids[e]
Steroidal saponins; verniosides, vernioside B2	*Vernonia amygdalina, Vernonia condensata, Vernonia polyanthes* (*Asteraceae*)	Antiinflammatory and analgesic effects in rodent assays[f]
Anesthetics		
Cocaine (alkaloid)	*Erythroxylon coca* (and other *Erythroxylaceae* species)	Cocaine and related amino esters and amino amides are local anesthetics. They inhibit sensory and motor neuron conduction by acting on voltage-gated Na^+ ion channels in the axonal plasma membrane[g]
Spilanthol (aliphatic amide)	*Acmella oleracea* (syn *Spilanthes acmella*) (*Asteraceae*)	Condiment widely used in Amazon culinary. It causes tongue tingling and was shown to have local anesthetic properties[h]
Terpenoids	Essential oils of plants	Eugenol and thymol derivatives (e.g., enal, propinal) have general anesthetic properties. Propofol is a synthetic short-acting intravenous general anesthetic structurally related to phenolic terpenoids[i]
Eugenol	*Syzygium aromaticum* (*Myrtaceae*) *Myristica fragrans* (*Myristicaceae*) *Cinnamomum verum* (*Lauraceae*) *Ocimum basilicum* (*Lamiaceae*)	
Thymol	*Thymus vulgaris* (*Lamiaceae*)	
Pipercallosidine, piperine, valeramide, and other amides	*Ottonia anisum* (*Piperaceae*) (methanolic extract from roots)	Local anesthetic activity after intradermal injection in guinea pigs[j]
Aconitum and delphinium alkaloids, lappaconitine, aconitine, bulleyaconitine A	*Aconitum bulleyanum* (*Ranunculaceae*)	Strongly reduced Na^+ current (voltage clamp experiments) thereby blocking motor and sensory nervous conduction in rat sciatic nerve[k]

TABLE 1 Phytochemicals with pain-relieving properties, a source of novel plant-based analgesics and anesthetics—cont'd

Phytochemical	Plant (family)	Pharmacological/clinical evidence
Valerian extracts and valerenic acid	*Valeriana officinalis* (*Caprifoleaceae*)	Anesthetic and sedative effects possibly mediated by modulation of GABA$_A$ receptor function[l]
Capsaicin	*Capsicum* spp. (*Capsiceae*)	Neuropeptide releasing agent (causes depletion of substance P and tachykinins) selective for primary sensory peripheral neurons. Studies suggest that it may be useful for neuropathic pain[m]

[a]Johal H, Devji T, Chang Y, Simone J, Vannabouathong C, Bhandari M. Cannabinoids in Chronic Non-Cancer Pain: A Systematic Review and Meta-Analysis. Clin Med Insights Arthritis Musculoskelet Disord. 2020;13:1179544120906461.
[b]Stevens AJ, Higgins MD. A systematic review of the analgesic efficacy of cannabinoid medications in the management of acute pain. Acta Anaesthesiol Scand. 2017;61(3):268–280.
[c]Pathan H, Williams J. Basic opioid pharmacology: an update. Br J Pain. 2012;6(1):11–16.
[d]Maione F, Cantone V, Pace S, et al. Anti-inflammatory and analgesic activity of carnosol and carnosic acid in vivo and in vitro and in silico analysis of their target interactions. Br J Pharmacol. 2017;174(11):1497–1508.
[e]Wang J, Song H, Wu X, et al. Steroidal Saponins from Vernonia amygdalina Del. and Their Biological Activity. Molecules. 2018;23(3):579.
[f]Calatayud J, González A; History of the Development and Evolution of Local Anesthesia Since the Coca Leaf. Anesthesiology 2003;98(6):1503–1508.
[g]Rondanelli M, Fossari F, Vecchio V, et al. Acmella oleracea for pain management. Fitoterapia. 2020;140:104419.
[h]López KS, Marques AM, Moreira DL, et al. Local Anesthetic Activity from Extracts, Fractions and Pure Compounds from the Roots of Ottonia anisum Spreng. (Piperaceae). An Acad Bras Cienc. 2016;88(4):2229–2237.
[i]Wang CF, Gerner P, Wang SY, Wang GK. Bulleyaconitine A isolated from aconitum plant displays long-acting local anesthetic properties in vitro and in vivo. Anesthesiology. 2007;107(1):82–90.
[j]Yuan CS, Mehendale S, Xiao Y, Aung HH, Xie JT, Ang-Lee MK. The gamma-aminobutyric acidergic effects of valerian and valerenic acid on rat brainstem neuronal activity. Anesth Analg. 2004;98(2):353–358.
[k]Ferraz CR, Carvalho TT, Manchope MF, et al. Therapeutic Potential of Flavonoids in Pain and Inflammation: Mechanisms of Action, Pre-Clinical and Clinical Data, and Pharmaceutical Development. Molecules. 2020;25(3):762.
[l]Reiner GN, Delgado-Marín L, Olguín N, et al. GABAergic pharmacological activity of propofol related compounds as possible enhancers of general anesthetics and interaction with membranes. Cell Biochem Biophys. 2013;67(2):515–525.
[m]Anand P, Bley K. Topical capsaicin for pain management: therapeutic potential and mechanisms of action of the new high-concentration capsaicin 8% patch. Br J Anaesth. 2011;107(4):490–502.

many developing countries a large part of their underprivileged population still relies on the use of medicinal plants and traditional practices for primary healthcare including management of pain and inflammatory conditions (WHO, 1998).

Solanum genus L. (Solanaceae)

The genus *Solanum* comprises over 1250 species (including mainly herbs and shrubs, and rarely trees) that are widely distributed throughout tropical and subtropical regions of the world. It is the largest genus within the *Solanaceae* family and one of the broadest and most diverse genera among the flowering plants or Angiosperms (Kaunda & Zhang, 2019). Plants of this genus are known to be potentially poisonous, depending on their content of solanine, a toxic steroidal glycoalkaloid, and some, such as potato (*S. tuberosum*), tomato (*S. lycopersicum*), and eggplant (*S. melongena*), are important food crops.

Antinociceptive effects of *Solanum* spp. extracts

A number of experimental studies showed that extracts of a variety of *Solanum* spp. (Fig. 1) present a broad range of pharmacological properties including antiinflammatory and analgesic activities.

As shown in Table 2, extracts (predominantly aqueous, hydro-alcoholic, and ethanolic extracts) from 12 *Solanum* species have been reported to exert antinociceptive effects in rodent assays. The tested extracts were obtained mainly from the plant aerial parts (mostly leaves) and, in a minority of cases, from roots, tubers (potatoes), or seeds (Table 2). *S. nigrum* (black nightshade) and *S. torvum* (turkey berry) with five and three studies, respectively, were the *Solanum* species most investigated for analgesic activity.

Antinociceptive effects of *Solanum* plant extracts were consistently demonstrated by a set of tests which are highly sensitive to centrally acting (opioid-like) analgesic agents (hot plate test, tail-flick methods) as well as by assays used to screen nonnarcotic analgesics (e.g., NSAIDs), agents that generally possess antiinflammatory and/or antipyretic activities in addition to pain-relieving properties (abdominal writhing test, formalin sub-plantar injection test/2nd phase responses).

Fig. 2 depicts the analgesic effects of aqueous extracts (ethyl acetate fraction) of leaves of *S. paniculatum* (SPOE) and *S. torvum* (STOE) in the mouse abdominal writhing assay. The antinociceptive response to an oral dose of

460 PART | IV Novel and nonpharmacological aspects and treatments

FIG. 1 Plants of *Solanum* genus with analgesic properties. Leaves and flowers of *S. paniculatum* (on the left) and *S. torvum* (on the right). The two plants are widely used in traditional medicine in Brazil and elsewhere. The aqueous extracts from leaves of these two plants were described to possess analgesic, antiinflammatory, antioxidant, and liver-protective effects in rodent assays.

TABLE 2 Studies reporting analgesic and antiinflammatory properties of extracts of plants from the genus *Solanum* L (Solanaceae).

			Antinociceptive assays			
Species	Plant part used	Extract/solvent/ compounds	Rodent (administration route)	Test	Related pharmacological activity	Reference
S. nigrum	Leaves	Aqueous extract	Rats (ip injection)	Pressure-induced pain, Thermally-induced pain	Antiinflammatory: egg albumin-induced paw edema	Wannang, Anuka, Kwanashie, and Auta (2006)
S. nigrum	Leaves (air dried)	Chloroform extract Aqueous extract	♂ Mice (Balb/c) (sc injection)	Abdominal writhing test (0.6% acetic acid) Hot plate test	NR[a]	Zakaria et al. (2006) Zakaria et al. (2009)
			Rats (Sprague Dawley) (sc injection)	Formalin test (1st and 2nd phase responses)	Antiinflammatory: carrageenan-induced paw edema Antipyretic: brewer's yeast pyrexia test	
S. nigrum	Fruit (dried)	Ethanolic extract	♂♀ Mice (Swiss) (oral, gavage)	Abdominal writhing (0.7% acetic acid)	NR	Karmakar et al. (2010)
S. nigrum	Leaves Seeds	Aqueous extract Methanolic extract	♀ Mice (Swiss) (oral, gavage)	Hot plate test Tail flick test	NR	Ibrahim et al. (2018)

Analgesic properties of *Solanum* genus plants Chapter | 39 461

TABLE 2 Studies reporting analgesic and antiinflammatory properties of extracts of plants from the genus *Solanum* L (Solanaceae)—cont'd

Species	Plant part used	Extract/solvent/ compounds	Antinociceptive assays		Related pharmacological activity	Reference
			Rodent (administration route)	Test		
S. torvum	Leaves (air dried)	Aqueous extract (90 °C)	♂♀ Mice (Swiss)	Abdominal writhing test (1.0% (v/v) acetic acid).	–	Ndebia et al. (2006)
			Rats (Wistar)	Hind paw pressure-induced pain	Antiinflammatory: carrageenan-induced paw edema	
S. torvum	Aerial parts	Ethanolic extract	♂♀ Mice (Swiss) (oral, gavage)	Abdominal writhing (0.6% acetic acid)	NR	Acharyya and Khatun (2018)
			♂♀ Rats (Wistar) (oral, gavage)	Tail immersion method (55 °C) Formalin-induced paw licking test		
S. torvum	Leaves (shade dried)	Ethanolic extract	♂♀ Mice (Swiss) (oral, gavage)	Abdominal writhing test (1.0% acetic acid) Hot plate test Formalin (sub-plantar injection) test Tail flick (radiant heat)	NR	Kar (2019)
			♂♀ Rats (Wistar) (oral, gavage)	Paw pressure pain test	Antiinflammatory: Carrageenan-induced paw edema	
S. tuberosum	Tubers (peeled off potatoes)	Ethanolic extract	♂ Mice (ICR) (oral, gavage)	Formalin-induced paw licking test Abdominal writhing (3% acetic acid) test Hot plate test	Antiinflammatory: Carrageenan-induced paw edema Formalin-induced paw edema Arachidonic acid–induced ear edema	Choi and Koo (2005)
S .tuberosum	Tubers (potato peel)	Potato peel extract (solvent not reported)	♂ Rats (Wistar) (oral, gavage)	Hot plate test	Antiinflammatory: carrageenan-induced paw edema	Wahyudi, Ramadhan, Wijaya, Ardhani, and Utami (2020)

Continued

462 PART | IV Novel and nonpharmacological aspects and treatments

TABLE 2 Studies reporting analgesic and antiinflammatory properties of extracts of plants from the genus *Solanum* L (Solanaceae)—cont'd

Species	Plant part used	Extract/solvent/ compounds	Antinociceptive assays		Related pharmacological activity	Reference
			Rodent (administration route)	Test		
S. melongena	Leaves	Leaf juice after pressing, evaporated	♂♀ Mice (Swiss) (oral, gavage)	Abdominal writhing (0.7% (v/v) acetic acid)	NR	Mutalik, Paridhavi, Rao, and Udupa (2003)
S. melongena	Leaves (air dried)	Ethanolic extract	♂♀ Mice (Swiss) (oral, gavage)	Abdominal writhing (acetic acid)	NR	Das, Gohain, and Das (2017)
			♂♀ Rats (albino) (oral, gavage)	Tail-flick test (radiant heat) Hot plate test		
S. acanthodes	Fruit	Ethanolic extract	Mice (oral, gavage)	Formalin intraplantar injection test (1st and 2nd phase responses) Glutamate intraplantar injection test	NR	Bento, Azevedo, Luiz, Moura, and Santos (2004)
S. trilobatum	Roots	Toluene:water CFA (Concentrated Alkaloid Fraction)	♂♀ Mice (Swiss) (oral, gavage)	Hot plate test; Abdominal writhing (0.7% (v/v) acetic acid)	NR	Annamalai, Khosa, and Hemalatha (2009)
			♂♀ Rats (Wistar) (oral, gavage)	Formalin (sub-plantar) induced pain (1st and 2nd phase responses)		
S. cernuum		Dichloromethane crude extract, Compounds: (1) cycloeucalenone; (2) 24-oxo-31-norcycloartanone	Mice (oral, gavage)	Abdominal writhing (0.8% (v/v) acetic acid)	NR	Lopes et al. (2014)
			Rats (oral, gavage)		Antiinflammatory: carrageenan-induced paw edema Inhibition of COX-2 expression (compound 2)	

TABLE 2 Studies reporting analgesic and antiinflammatory properties of extracts of plants from the genus *Solanum* L (Solanaceae)—cont'd

Species	Plant part used	Extract/solvent/ compounds	Rodent (administration route)	Test	Related pharmacological activity	Reference
			Antinociceptive assays			
S. incanum	Roots	Dichloromethane extract	Rats (Wistar) (oral, gavage)	Formalin-induced pain (paw)	Antiinflammatory: Carrageenan-induced paw edema	Mwonjoria et al. (2014)
S. incanum	Roots	Petroleum ether/ methanol (flavonoids rich fraction)	♂♀ Mice (Swiss) (sc injection)	Formalin-induced pain (paw)	NR	Enoc et al. (2018)
S. violaceum	Leaves	Methanolic extract	♂♀ Mice (Swiss) (oral, gavage)	Abdominal writhing (0.6% (v/v) acetic acid); Hot plate test	NR	Mahaldar et al. (2016)
S. sisymbriifolium	Leaves	Methanolic extract	♂ Mice (Swiss) (oral, gavage)	Hot plate test Tail immersion test Formalin (sub-plantar injection) test (1_{st} and 2nd phase responses) Abdominal writhing (0.7% (v/v) acetic acid)	NR	Nasrin et al. (2017)
S. viarum	Leaves (shade dried)	Ethanolic extract	♂♀Rats (Wistar) (oral, gavage)	Hot plate test Tail-flick test (thermal stimulus)	Antipyretic: brewer's yeast pyrexia test	Kausar and Singh (2018)
S. paniculatum	Leaves	Aqueous extract, ethyl-acetate fraction	♂ Mice (Swiss) (oral, gavage)	Abdominal writhing (0.6% (v/v) acetic acid)	Protective effects against paracetamol-induced liver injury	de Souza et al. (2019)

[a]*NR: not reported.*

S. paniculatum aqueous extract as high as 300 mg/kg body wt was similar to that produced by 180 mg/kg of acetaminophen (Fig. 2A). The aqueous extracts from leaves of *S. torvum* (STOE) were somewhat more potent than those of *S. paniculatum* (SPOE), so that doses as high as 150 mg/kg resulted in an antinociceptive response comparable to or even slightly more pronounced than that induced by 90 mg/kg of acetaminophen (Fig. 2B). Moreover, when mice were concomitantly treated with acetaminophen (AP 90 mg/kg) and *S. torvum* extract (STOE 150 mg/kg), the magnitude of the analgesic response (combined treatment) indicated that the effects of these drugs are likely to be additive (Fig. 2B).

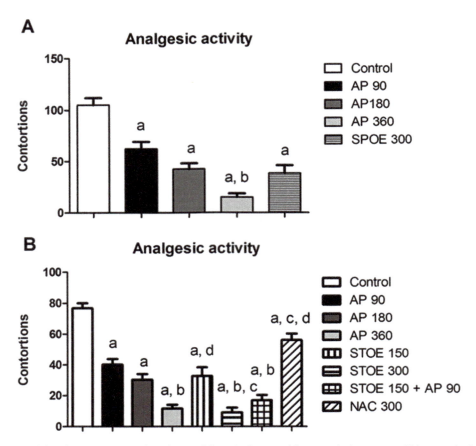

FIG. 2 Antinociceptive activity of aqueous extracts from leaves of *S. paniculatum* and *S. torvum* in the mouse writhing test. Antinociceptive effects of aqueous extracts (ethyl acetate fraction) from leaves of *S. paniculatum* (SPOE 300 mg/kg po) and *S. torvum* (STOE 150, 300 mg/kg po), paracetamol (AP 90, 180, 360 mg/kg ip), paracetamol (90 mg/kg ip) plus STOE (150 mg/kg po), and *N*-acetyl-cysteine (NAC 300 mg/kg po). Contortions were counted (0–40 min) after mice were injected with acetic acid 0.6% (v/v), ip. Bar height is means ± SD. Means ($N = 5$ to 7 mice per group) were compared by ANOVA and Bonferroni's test, $P < 0.05$, and differences are indicated by letters as follows; (A): a $\neq$ control, b $\neq$ other treated groups. (B): bars not sharing the same letters differ between them. *(Panel A: reproduced from de Souza, G. R., De-Oliveira, A. C. A. X., Soares, V., Chagas, L. F., Barbi, N. S., Paumgartten, F. J. R., et al. (2019). Chemical profile, liver protective effects and analgesic properties of a* Solanum paniculatum *leaf extract.* Biomedicine & Pharmacotherapy, *110, 129–138. Panel B: unpublished data by Paumgartten, de Souza, da Silva & De-Oliveira.)*

Antiinflammatory activity of *Solanum* spp. extracts

In some cases, studies on the analgesic properties of *Solanum* spp. (*S. nigrum, S. torvum, S. cernum,* and *S. incarnum*) showed that extracts also possessed antiinflammatory activity (carrageenan-, formalin- or arachidonic acid-induced paw edema assays) (Table 2). A number of additional investigations that did not evaluate directly whether they possessed analgesic properties have revealed that extracts from leaves, fruits, or roots of *Solanum* spp. present antiinflammatory activity. The antiinflammatory *Solanum* spp. extracts include those from roots of *S. erianthum* (Chen et al., 2013), from *S. alatum* (Lin, Lin, Yang, & Shieh, 1995), from the whole-plant (Aryaa & Viswanathswamy, 2017), and berries of *S. nigrum* (Ravi, Saleem, Patel, Raamamurthy, & Gauthaman, 2009), berries of *S. khasianum* (Jarrald et al., 2008), tubers of *S. tuberosum* (Ahmad, Ahmad, Keservani, & Sharma, 2016), aerial parts and a β-sitosterol isolated from *S. lingustrinum* (Delporte et al., 1998), leaves of *S. scabrum* (Ogunnaike et al., 2013), leaves of *S. surattense* (Reddy, Rao, Vangoori, & Sundharam, 2014), fruits from *S. xanthocarpum* (Datta, Gosh, Das, & Nath, 2018), leaves of *S. anomalum* (Okokon, Davies, Amazu, & Umoh, 2017), leaves (da-Costa et al., 2015) of *S. lycocarpum* and fruits of *S. lycopersicum* (Li, Deng, Liu, Loewen, & Tasao, 2014). Topical application of extracts of leaves of *S. corymbiflorum* and *S. paranense* was shown to decrease inflammatory response in a croton oil-induced ear edema assay in mice (Piana et al., 2016; Piana, Camponogara, Boligon, & Oliveira, 2017). Moreover, in vitro tests indicated that steroidal saponins (glycosides) isolated from *S. nigrum* berries inhibited the macrophage (RAW 246.7 cells) production of nitric oxide, IL-6, and IL-1β elicited by stimulation with LPS (Wang, Xiang, Yi, & He, 2017).

The antinociceptive effects of *Solanum* plant extracts were generally more pronounced in rodent assays involving inflammatory pain/hyperalgesic stimuli (abdominal writhing and formalin tests) than in tests involving thermal stimuli (hot plate, tail immersion). This fact allied to the overall evidence from repeated reports of antiinflammatory activities of *Solanum* spp. extracts is consistent with an NSAID-like (mediated by COX inhibition) analgesic property. Antinociceptive effects of *S. sisymbriifolium* extracts (400 mg/kg) on the hot plate and tail immersion tests were reported to be reversed by naloxone (2 mg/kg ip), and the authors suggested that the contribution of a modulation of an opioid system by compounds contained in the extracts could not be entirely ruled out (Nasrin, Khandaker, Akter, & Imam, 2017). The authors, however, did not investigate whether naloxone (a nonselective competitive opioid receptor antagonist) would also reduce the effect of *S. sisymbriifolium* extracts on acetic acid-induced abdominal writhings and formalin-induced paw licking behaviors. If extract constituents with opioid-like actions do contribute to the overall antinociceptive activity, a partial reduction of behavioral responses of mice to acetic acid and formalin stimuli is expected to occur following treatment with opioid antagonists. No evidence of opioid-like action of *Solanum* crude extracts, however, has been provided.

Type of analgesia produced by *Solanum* spp. extracts

Overall, findings described in Table 2 and those from additional studies on rodent models of aseptic inflammation indicated that extracts from several *Solanum* plants used in traditional medicine for different purposes present analgesic and antiinflammatory properties.

Evidence of a possible adjunctive antipyretic activity is scanty. As shown in Table 2, only three studies suggested that *S. nigrum* (aqueous/chloroform extracts) and *S. viarum* (ethanolic extract) leaf extracts have antipyretic activity (against brewer's yeast-caused pyrexia in rats), and relatively modest reductions of fever were noted after oral administration of leaf extracts of the two plants (Kausar & Singh, 2018; Zakaria et al., 2006, 2009).

As aforementioned, there are two main classes of analgesics, opioid drugs, and NSAIDs, such as salicylates and other inhibitors of COX, including the most recent selective COX-2 blockers. Paracetamol, or acetaminophen, and metamizole, also known as dipyrone, are two of the most commonly consumed analgesics that do not exactly fit into one of these two main pharmacological classes of painkillers.

Although sharing analgesic and antipyretic activities with NSAIDs, unlike most of these drugs, acetaminophen, and metamizole present only weak antiinflammatory activities. The exact mode of action of acetaminophen remains yet to be elucidated. Notwithstanding not reducing tissue inflammation, acetaminophen is believed to inhibit COX-mediated PG synthesis particularly at the Central Nervous System (for reviews, see Ghanem, Pérez, Manautou, & Mottino, 2016; Graham, Davies, Day, Mohamudally, & Scott, 2013). The mode of action of metamizole (which has also a spasmolytic action besides the analgesic and antipyretic effects) is complex and not completely understood either. It most probably involves not only the inhibition of a central COX-3, but also an adjuvant participation of PG-independent pathways (Ferreira, 1980; Jasiecka, Maślanka, Jaroszewski, 2014).

The facts that several extracts from the aerial parts and roots of *Solanum* spp. listed in Table 2 present both analgesic and antiinflammatory properties, and that similar extracts from other *Solanum* plants were also demonstrated to reduce tissue inflammation, point toward an aspirin- or NSAID-like mode of action for the analgesia elicited by *Solanum* spp. Additional investigations, however, are needed to determine whether *Solanum* extracts and/or their constituents do in fact inhibit COX-mediated PG synthesis. Studies are also lacking to corroborate their use in traditional medicine as remedies against fever.

Chemical constituents of active extracts

Steroidal saponins and alkaloids, steroidal glycosides, anthocyanins, phenylpropanoid amides, terpenes, flavonoids, lignans, sterols, phenolic acid, hydroxycinnamates, and coumarins are amidst the chemical constituents of *Solanum* plants. Plant extracts are, as a rule, immensely complex multiconstituent mixtures. Very few pharmacological studies, however, have adequately analyzed the qualitative and quantitative chemical composition of tested botanical materials. Moreover, vegetable extracts used in phytotherapy are often poorly standardized, and it is well known that their content of active (s) compound (s) may vary depending on the plant genetic background, climate conditions, and soil in which it was grown, the phytopathology status, and, of course, the extraction methods, among other factors.

Phenolic acids and polyphenols (including flavonoids) are products of plant secondary metabolism that are generally abundant in *Solanum* spp. (aqueous and alcoholic extracts) extracts where they are found combined with mono- and polysaccharides.

466 **PART | IV** Novel and nonpharmacological aspects and treatments

TABLE 3 Quantitative determination of major phenolic compounds in the aqueous extract (ethyl acetate fraction) from *S. paniculatum* leaves showing analgesic properties.

Compounds	mg/g (mean ± SD)
Hydroxycinnamates (total)	76.29 ± 0.01
Flavonoids (total)	18.01 ± 0.04
Phenolics (total)	94.30 ± 0.05
Monocaffeoylquinic acid (total)	7.89 ± 0.010
Monocoumaroylquinic acid (total)	8.33 ± 0.003
Dicaffeoylquinic acid (total)	2.62 ± 0.004
Chlorogenic acid	2.36 ± 0.010
Robinobioside of quercetin	6.94 ± 0.028
Rutinoside of quercetin (rutin)	1.99 ± 0.004
Ramnoside of quercetin	1.27 ± 0.002

(Adapted from de Souza, G. R., De-Oliveira, A. C. A. X., Soares, V., Chagas, L. F., Barbi, N. S., Paumgartten, F. J. R., et al. (2019). Chemical profile, liver protective effects and analgesic properties of a Solanum paniculatum leaf extract. *Biomedicine & Pharmacotherapy, 110*, 129–138.)

An analgesic aqueous extract (ethyl acetate partition) from *S. paniculatum* leaves was reported to contain (tentatively identified by mass spectroscopy) a set of 35 flavonoids, esters of hydroxycinnamic acid, and isomers of chlorogenic acid, such as monocaffeoylquinic acids, monocoumaroylquinic acids, quercetin robinobioside, rutin, hexoside of hydroxyluteolin, dicaffeoyl quinic acids, quercetin, and hexoside of methylquercetin (de Souza et al., 2019). The quantification (analysis by HPLC-DAD) of major phenolic compounds found in this analgesic extract of *S. paniculatum* leaves is depicted in Table 3 (Fig. 3).

When administered orally, the glycosylated flavonoids (hydrophilic compounds) are poorly absorbed (by passive diffusion) in the small intestine. Nonetheless, glycosylated flavonoids are extensively hydrolyzed by bacteria found in the colon microflora so that the corresponding aglycones (more lipophilic compounds) are absorbed (Hollman, 2004). The absorption of quercetin glycosides (i.e., those derived from glucose) also occurs in the small intestines through a Na^+-dependent glucose co-transporter (Hollman, 2004). Therefore, it is plausible to assume that the quercetin from glycosylated quercetin compounds found in the *Solanum* extracts is to a great extent bioavailable.

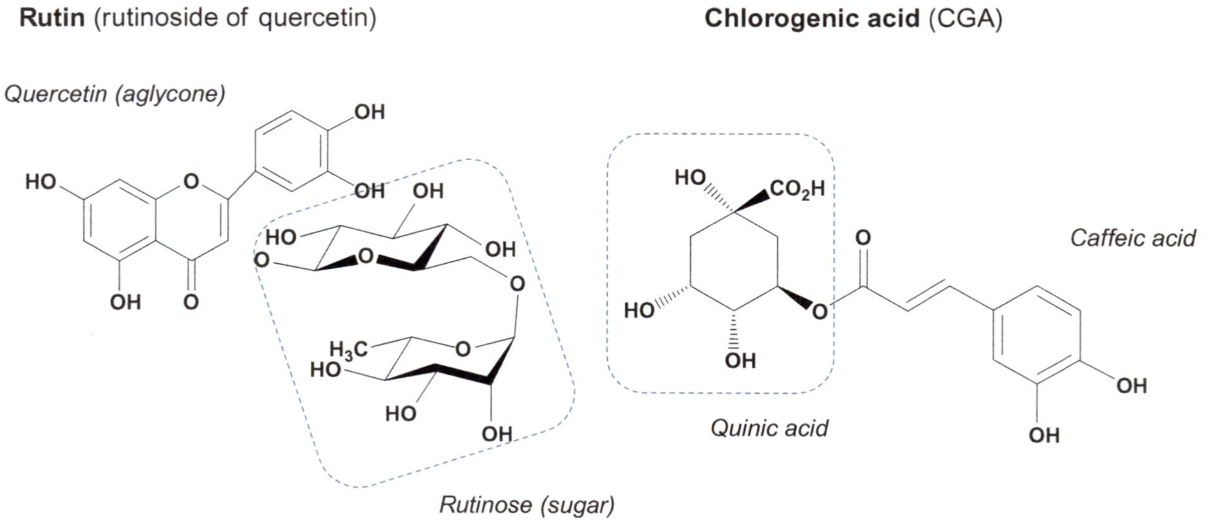

FIG. 3 Phenolic compounds found in *Solanum* spp. leaf aqueous extracts. Quercetin, rutin, and chlorogenic acid are constituents of extracts from *Solanum* spp. with analgesic and antiinflammatory properties. Rutin is a flavonol glycoside the aglycone of which is quercetin, while chlorogenic acid (CGA) is an ester of caffeic and quinic acid.

Flavonoids such as quercetin and rutin (a flavonol glycoside) may be at least partially responsible for the analgesic and antiinflammatory properties (and other pharmacological activities as well) of the *Solanum* plant extracts.

Experimental evidence has been provided that quercetin, given either by ip or by oral routes, reduces inflammatory pain in mouse models of nociception and inflammation (Valério et al., 2009) and, administered by ip injection, this flavonoid was shown to attenuate Ehrlich tumor-induced cancer pain in mice (Calixto-Campos et al., 2015). Antinociceptive activity of rutin was also demonstrated in mouse assays (Ganeshpurkar & Saluja, 2017). Chlorogenic acid, the ester of caffeic and (−)-quinic acid, is another major constituent of *Solanum* plant extracts that have been described to have analgesic and antiinflammatory activities when administered (by gavage) to rats (dos Santos, Almeida, Lopes, & Souza, 2006). No evidence that CGA or quercetin also inhibited LPS-elicited fever in rats was found (dos Santos, Almeida, Lopes, & de Souza, 2006; Kanashiro, Machado, Malvar, Aguiar, & Souza, 2008). It is of note that hydroxycinnamic acid derivatives (hydroxycinnamates) present in *Solanum* spp. and in plants from other genera and families were also reported to possess antiinflammatory properties (Taofiq, González-Paramás, Barreiro, & Ferreira, 2017).

In conclusion, several major phenolic constituents of *Solanum* spp. have antinociceptive and antiinflammatory activities and may contribute to the overall analgesic and antiinflammatory properties of plant crude extracts, as evidenced by rodent tests.

Other pharmaco-toxicological activities of interest

Phenolic compounds and polyphenols/flavonoids that are abundant in *Solanum* spp. are known to be strong natural antioxidants (Kiokias, Proestos, & Oreopoulou, 2020; Lin et al., 2016; Panche, Diwan, & Chandra, 2016) and, thus, it is not surprising that extracts from plants of this genus were reported to present antioxidant properties as well. Antioxidant activity has been found in several extracts tested for analgesic and/or antiinflammatory properties, such as *S. paniculatum* (de Souza et al., 2019), *S. lycocarpum* (Da Costa et al., 2015), *S. nigrum* (Karmakar, Tarafder, Sadhu, Biswas, & Shill, 2010), and many others.

In addition, extracts from several *Solanum* spp. were described to prevent or ameliorate hepatocellular damage induced by a variety of liver toxicants. Extracts from leaves of *S. surattense* (Parvez et al., 2019), *S. trilobatum* (Shahjahan, Sabitha, Jainu, & Shyamala Devi, 2004), and fruits of *S. xanthocarpum* (Gupta et al., 2011) were demonstrated to protect against acute liver injury induced by CCl_4 in rats. *S. xanthocarpum* fruit extract was described to prevent liver damage caused by a combination of antituberculosis drugs (isoniazid, rifampicin, and pyrazinamide) in rats (Hussain et al., 2012), while extracts from *S. alatum* (Lin et al., 2000), from leaves of *S. fastigiatum* (Sabir & Rocha, 2008) and from leaves of *S. paniculatum* (de Souza et al., 2019) were reported to protect against paracetamol-caused hepatotoxicity in mice. Moreover, it was described that extracts from leaves of *S. muricatum* protect the mouse liver against ethanol-induced injury (Hsu, Lin, Hsu, Chen, & Chen, 2018) and that extracts from *S. nigrum* alleviate hepatic fibrosis in rats fed a high fat and alcohol diet (Tai et al., 2016), and hepatic inflammation in mice fed a high-fat diet (Chang et al., 2017).

Application to other areas

Overall, data from rodent studies suggest that crude extracts from aerial parts and/or roots of several plants of the *Solanum* genus exhibited not only analgesic and antiinflammatory properties but also antioxidant and liver-protective effects. These findings seem to support their use in traditional medicine to treat slight to moderate pain and inflammatory conditions. Furthermore, the liver protective activity of some extracts (e.g., *S. paniculatum* leaf aqueous extract) was demonstrated to confer effective protection against paracetamol-elicited hepatic injury comparable to that of *N*-acetyl-cysteine, a classical antidote to paracetamol-caused hepatotoxic effects (de Souza et al., 2019). Aqueous extracts (ethyl acetate partitioned fractions) of *S. paniculatum, S. torvum,* and of other plants of the *Solanum* genus, therefore, seem to be plant-derived materials that could be used in the development of new phytotherapeutic drugs combining analgesic and liver protective activities. Since analgesic effects of these plant extracts and those of paracetamol were shown to be additive, a phytotherapeutic drug formulation combining them is likely to give rise to effective and safer medications for moderate pain relief.

Mini-dictionary of terms

Analgesia: absence or reduced capacity of feeling pain while in a conscious state.

Antinociception: blockage of behavioral responses to potentially painful stimuli. Animal tests measure antinociceptive effects.

Formalin (licking) test: Measures the rodent response (licking/biting) to a formalin sc injection into the hind paw; the early response reflects an activation of peripheral nociceptors, while the late response corresponds to inflammatory pain.

Flavonoids: Phenolic compounds (polyphenols) produced by plant secondary metabolism which were reported to exhibit antioxidant and other biological activities.

Antioxidants: substances that inhibit oxidation processes and can prevent or delay cell damage produced by free radicals or molecules with unpaired electrons.

Cyclooxygenases (COX-1, COX-2, COX-3): enzymes that mediate the synthesis of prostanoids, including thromboxane and prostaglandins from arachidonic acid. Of the three isoforms, COX-2 is an inducible form the expression of which is rapidly enhanced in several cell types in response to proinflammatory molecules.

Nonsteroidal antiinflammatory drugs (NSAID): Class of drugs that inhibit COX and reduce pain, fever, and inflammation. Nonselective COX inhibitors (aspirin) also prevent blood clotting. COX-2 selective inhibitors present a lower risk of gastrointestinal bleeding side effects but enhance risks of thrombosis.

Glycosides: compounds (e.g., flavonol glycosides) in which sugar molecules are linked to a noncarbohydrate moiety (aglycone, e.g., quercetin) via a glycosidic bound.

Key facts of *Solanum* analgesia

- *Solanum* sp. is the largest genus within the Solanaceae family and one of the broadest of angiosperms.
- Extracts of aerial parts and roots of various *Solanum* spp. are widely used in traditional medicine for several therapeutic purposes.
- Plants used in traditional medicine are a potential source of analgesic and antiinflammatory compounds.
- Underprivileged people from the developing world still rely on the use of plants for primary healthcare.
- Evidence on the analgesic/antiinflammatory activity of these plants is relevant to corroborate this popular use.

Other agents of interest

In this chapter, we described the evidence suggesting that extracts of *Solanum* spp. and their constituents have analgesic properties. It is of note that a variety of plants belonging to other genera of Solanaceae family, and from other plant families (e.g., Liliaceae, Caesalpinaceae, Phyllanthaceae, Malvaceae, Lamiaeceae, Labiatae, Asteraceae, and many others) have also been described to exhibit analgesic and/or antiinflammatory properties. Moreover, several glycosides with analgesic properties found in *Solanum* plants were also isolated from plants from other genera and families.

Summary points

- This chapter addresses the evidence on the analgesic/antiinflammatory properties of *Solanum* plants.
- Antinociceptive and antiinflammatory activities of *Solanum* spp. extracts were consistently found by rodent assays.
- Flavonol glycosides (e.g., rutin) and hydroxycinnamates are among the most abundant constituents of *Solanum* spp. analgesic extracts.
- Quercetin, rutin, CGA, and other hydoxycinnamates found in *Solanum* plants have been shown to possess analgesic and/or antiinflammatory activities.
- Antioxidant and liver protective activities have also been found in *Solanum* spp. extracts with analgesic properties.
- Analgesic effects of paracetamol and *S. torvum* leaf extract were demonstrated to be additive.
- Overall, available evidence seems to support the use of several *Solanum* spp. extracts for the treatment of moderate pain in traditional medicine practices.
- Ethyl acetate partitioned fraction of aqueous extracts of *S. paniculatum* or *S. torvum* (and possibly of other plants of *Solanum* genus), alone or associated with lower doses of paracetamol, can be developed as new, effective, and safer therapeutic options to treat moderate pain.

References

Acharyya, S., & Khatun, B. (2018). Antimicrobial and analgesic activity of *Solanum torvum*. *Haya: The Saudi Journal of Life Sciences, 3*(6), 459–464.

Ahmad, M. F., Ahmad, S. M., Keservani, R. K., & Sharma, A. K. (2016). Anti-inflammatory activity of tuber extracts of *Solanum tuberosum* in male albino rats. *National Academy Science Letters, 39*, 421–425.

Almeida, R. N., Navarro, D. S., & Barbosa-Filho, J. M. (2001). Plants with central analgesic activity. *Phytomedicine, 8*(4), 310–322.

Anilkumar, M. (2010). Ethnomedicinal plants as antiinflammatory and analgesic agents. In D. Chattopadhyay (Ed.), *Ethnomedicine: A source of complementary therapeutics* (pp. 267–293). Kerala: Research Signpost (Chapter 10).

Annamalai, P., Khosa, R. L., & Hemalatha, S. (2009). Evaluation of analgesic potential of *Solanum trilobatum* roots. *Iranian Journal of Pharmaceutical Research, 8*(4), 269–273.

Aryaa, A., & Viswanathswamy, A. H. (2017). Effect of *Solanum nigrum* Linn on acute and sub-acute models of inflammation. *Journal of Young Pharmacists, 9*(4), 566–570.

Bento, A. F., Azevedo, M. S., Luiz, A. P., Moura, J. A., & Santos, A. R. S. (2004). Atividade antinociceptiva do extrato etanólico do fruto de Solanum acanthodes Hook.f. em camundongos. *Revista Brasileira de Farmacognosia, 4*(Supp 1), 9–10.

Botting, R. M. (2010). Vane's discovery of the mechanism of action of aspirin changed our understanding of its clinical pharmacology. *Pharmacological Reports, 62,* 518–525.

Calixto-Campos, C., Corrêa, M. P., Carvalho, T. T., Zarpelon, A. C., Hohmann, M. S. N., Rossaneis, A. C., et al. (2015). Quercetin reduces Ehrlich tumor-induced cancer pain in mice. *Analytical Cellular Pathology, 2015.* 1155/2015/285708.

Chang, J.-J., Chung, D.-J., Lee, Y.-J., Wen, B.-H., Jao, H.-Y., & Wang, C.-J. (2017). *Solanum nigrum* polyphenol extracts inhibit hepatic inflammation, oxidative stress, and lipogenesis in high-fat-diet-treated mice. *Journal of Agricultural and Food Chemistry, 65*(42), 9255–9265.

Chen, Y.-C., Lee, H.-Z., Chen, H.-C., Wen, C.-L., Kuo, Y.-H., & Wang, G.-J. (2013). Anti-inflammatory components from the root of *Solanum erianthum. International Journal of Molecular Sciences, 14,* 12581–12592.

Choi, E., & Koo, S. (2005). Anti-nociceptive and anti-inflammatory effects of the ethanolic extract of potato (*Solanum tuberosum*). *Food and Agricultural Immunology, 16*(1), 29–39.

Da-Costa, G. A. F., Morais, M. G., Saldanha, A. A., Silva, I. C. A., Aleixo, A. A., Ferreira, J. M. S., et al. (2015). Antioxidant, antibacterial, cytotoxic, and Antiinflammatory potential of the leaves of *Solanum lycocarpum* A. St. Hil. (Solanaceae). *Evidence-based Complementary and Alternative Medicine, 2015.* https://doi.org/10.1155/2015/315987.

Das, M., Gohain, K., & Das, S. (2017). Evaluation of central and peripheral activities of Solanum melongena ethanolic leaf extract in experimental animals. *International Journal of Pharmaceutical Sciences and Research, 8*(3), 1168–1172.

Datta, M., Ghosh, R., Das, L., & Nath, S. (2018). Evaluation of anti-inflammatory activity of methanolic extract of fruits of *Solanum xanthocarpum* Schrad & Wendl in albino rats. *International Research Journal of Pharmacy and Medical Sciences, 1*(6), 1–5.

de Souza, G. R., De-Oliveira, A. C. A. X., Soares, V., Chagas, L. F., Barbi, N. S., Paumgartten, F. J. R., et al. (2019). Chemical profile, liver protective effects and analgesic properties of a *Solanum paniculatum* leaf extract. *Biomedicine & Pharmacotherapy, 110,* 129–138.

Delporte, C., Backhouse, N., Negrete, R., Salinas, P., Rivas, P., Cassels, B. K., et al. (1998). Antipyretic, hypothermic and antiinflammatory activities and metabolites from *Solanum ligustrinum* Lood. *Phytotherapy Research, 12,* 118–122.

dos Santos, M. D., Almeida, M. C., Lopes, N. P., & de Souza, G. E. (2006). Evaluation of the anti-inflammatory, analgesic and antipyretic activities of the natural polyphenol chlorogenic acid. *Biological & Pharmaceutical Bulletin, 29*(11), 2236–2240.

Enoc, W. N., Daisy, M. G. N., Wilbroda, O. A., Alphonse, W. N., Joseph, N. J. N., & Maina, M. J. K. (2018). Antinociceptive and Antiinflammatory effects of flavonoids rich fraction of *Solanum incanum* (Lin) root extracts in mice. *Journal of Phytopharmacology, 7*(4), 399–403.

Ferreira, S. H. (1980). Peripheral analgesia: Mechanism of the analgesic action of aspirin-like drugs and opiate-antagonists. *British Journal of Clinical Pharmacology, 10*(Suppl 2), 237S–245S.

Ganeshpurkar, A., & Saluja, A. K. (2017). The pharmacological potential of rutin. *Saudi Pharmaceutical Journal, 25*(2), 149–164.

Ghanem, C. I., Pérez, M. J., Manautou, J. E., & Mottino, A. D. (2016). Acetaminophen from liver to brain: New insights into drug pharmacological action and toxicity. *Pharmacological Research, 109,* 119–131.

Graham, G. G., Davies, M. J., Day, R. O., Mohamudally, A., & Scott, K. F. (2013). The modern pharmacology of paracetamol: Therapeutic actions, mechanism of action, metabolism, toxicity and recent pharmacological findings. *Inflammopharmacology, 21*(3), 201–232.

Gupta, R. K., Hussain, T., Panigrahi, G., Das, A., Singh, G. N., Sweety, K., et al. (2011). Hepatoprotective effect of *Solanum xanthocarpum* fruit extract against CCl_4 induced acute liver toxicity in experimental animals. *Asian Pacific Journal of Tropical Medicine, 4*(12), 964–968.

Hollman, P. C. H. (2004). Absorption, bioavailability, and metabolism of flavonoids. *Pharmaceutical Biology, 42*(Suppl), 74–83.

Hsu, J.-Y., Lin, H.-H., Hsu, C.-C., Chen, B.-C., & Chen, J.-H. (2018). Aqueous extract of Pepino (*Solanum muriactum* Ait) leaves ameliorate lipid accumulation and oxidative stress in alcoholic fatty liver disease. *Nutrients, 10*(7), 931.

Hussain, T., Gupta, R. K., Sweety, K., Khan, M. S., Hussain, M. S., Arif, M., et al. (2012). Evaluation of antihepatotoxic potential of Solanum xanthocarpum fruit extract against antitubercular drugs induced hepatopathy in experimental rodents. *Asian Pacific Journal of Tropical Biomedicine, 2*(6), 454–460.

Ibrahim, A., Alnaeli, G., Misbah, A., Abuzanoona, A., Emhemed, S., Almahdi, M., et al. (2018). The evaluation of analgesic activity of *Solanum nigrum. Libyan Journal of Medical Research, 12*(2), 1–10.

Jarald, E. E., Edwin, S., Saini, V., Deb, L., Gupta, V. B., Wate, S. P., et al. (2008). Anti-inflammatory and anthelmintic activities of *Solanum khasianum* Clarke. *Natural Product Research, 22*(3), 269–274.

Jasiecka, A., Maślanka, T., & Jaroszewski, J. J. (2014). Pharmacological characteristics of metamizole. *Polish Journal of Veterinary Sciences, 17*(1), 207–214.

Kanashiro, A., Machado, R. R., Malvar, D. C., Aguiar, F. A., & Souza, G. E. (2008). Quercetin does not alter lipopolysaccharide-induced fever in rats. *The Journal of Pharmacy and Pharmacology, 60*(3), 357–362.

Kar, P. K. (2019). Analgesic and anti-inflammatory properties of ethanolic leaf extract of plant *Solanum torvum. World Journal of Pharmaceutical Research, 8*(11), 756–765.

Karmakar, U. K., Tarafder, U. K., Sadhu, S. K., Biswas, N. M., & Shill, M. C. (2010). Biological investigations of dried fruit of *Solanum nigrum* Linn. *Stamford Journal of Pharmaceutical Sciences, 3*(1), 38–45.

Kaunda, J. S., & Zhang, Y.-J. (2019). The genus *Solanum*: An ethnopharmacological, phytochemical and biological properties review. *Natural Products and Bioprospecting, 9*, 77–137.

Kausar, M., & Singh, B. K. (2018). Pharmacological evaluation of Solanum viarum dunal leaves extract for analgesic and antipyretic activities. *Journal of Drug Delivery and Therapeutics, 8*(4), 356–361.

Kiokias, S., Proestos, C., & Oreopoulou, V. (2020). Phenolic acids of plant origin—A review on their antioxidant activity in vitro (O/W emulsion systems) along with their in vivo health biochemical properties. *Food, 9*(4), 534.

Li, H., Deng, Z., Liu, R., Loewen, S., & Tsao, R. (2014). Bioaccessibility, in vitro antioxidant activities and in vivo anti-inflammatory activities of a purple tomato (*Solanum lycopersicum* L.). *Food Chemistry, 159*, 353–360.

Lin, S.-C., Chung, T.-C., Ueng, T.-H., Lin, Y.-H., Hsu, S.-H., Chiang, C.-L., et al. (2000). The hepatoprotective effects of *Solanum alatum* Moench. On acetaminophen-induced hepatotoxicity in mice. *The American Journal of Chinese Medicine, 28*(1), 105–114.

Lin, C.-C., Lin, W.-C., Yang, S.-R., & Shieh, D.-E. (1995). Anti-inflammatory and hepatoprotective effects of Solanum alatum. *The American Journal of Chinese Medicine, 23*(1), 65–69.

Lin, D., Xiao, M., Zhao, J., Li, Z., Xin, B., Li, X., et al. (2016). An overview of plant phenolic compounds and their importance in human nutrition and management of Type 2 diabetes. *Molecules, 21*(10), 1374.

Lopes, L. C., de Carvalho, J. E., Kakimore, M., Vendramini-Costa, D. B., Medeiros, M. A., Spindola, H. M., et al. (2014). Pharmacological characterization of *Solanum cernuum* Vell.: 31-norcycloartanones with analgesic and anti-inflammatory properties. *Inflammopharmacology, 22*(3), 179–185.

Mahaldar, K., Saifuzzaman, M., Irin, T., Barman, A. K., Islam, M. K., Rahman, M. M., et al. (2016). Analgesic, anthelmintic and toxicity studies of Solanum violaceum Linn. Leaves. *Oriental Pharmacy and Experimental Medicine, 2016*. https://doi.org/10.1007/s13596-016-0227-9.

Mahdi, J. G. (2010). Medicinal potential of willow: A chemical perspective of aspirin discovery. *Journal of Saudi Chemical Society, 14*, 317–322.

Mutalik, S., Paridhavi, K., Rao, C. M., & Udupa, N. (2003). Antipyretic and analgesic effect of leaves of *Solanum malongena* Linn. In rodents. *Indian Journal of Pharmacology, 35*, 312–315.

Mwonjoria, J. K., Ngeranwa, J. J., Githinji, C. G., Kahiga, T., Kariuki, H. N., & Waweru, F. N. (2014). Suppression of nociception by *Solanum incanum* (Lin.) Diclomethane root extract is associated anti-inflammatory activity. *Journal of Phytopharmacology, 3*(3), 156–162.

Nasrin, T., Khandaker, M., Akter, S., & Imam, M. Z. (2017). Antinociceptive activity of methanol extract of leaves of *Solanum sisymbriifolium* in heat and chemical-induced pain. *Journal of Applied Pharmaceutical Science, 7*(11), 142–146.

Ndebia, E. J., Kamgang, R., & Nkeh-Chungag Anye, B. N. (2006). Analgesic and anti-inflammatory properties of aqueous extract from leaves of *Solanum torvum* (Solanaceae). *African Journal of Traditional, Complementary, and Alternative Medicines, 4*(2), 240–244.

Ogunnaike, B. F., Okutachi, I. R., Anucha, E. S., Gbodi, O. O., Shokunbi, O. S., & Onajobi, F. D. (2013). Comparative anti-inflammatory activities of *Jatropha curcas, Ocimum gratissimum* and *Solanum scabrum* leaves. *Journal of Natural Product and Plant Resources, 3*(1), 59–66.

Okokon, J. E., Davies, K. O., Amazu, L. U., & Umoh, E. E. (2017). Anti-inflammatory activity of leaf extract of *Solanum anomalum*. *Journal of Herbal Drugs, 7*(4), 243–249.

Panche, A. N., Diwan, A. D., & Chandra, S. R. (2016). Flavonoids: An overview. *Journal of Nutritional Science, 5*, e47.

Parvez, M. K., Al-Dosari, M. S., Arbab, A. H., Alam, P., Alsaid, M. S., & Khan, A. A. (2019). Hepatoprotective effect of *Solanum surattense* leaf extract against chemical-induced oxidative and apoptotic injury in rats. *BMC Complementary and Alternative Medicine, 19*(1), 154.

Piana, M., Camponogara, C., Boligon, A. A., Machado, M. M., Brum, T. F., Oliveira, S. M., et al. (2016). Topical anti-inflammatory activity of *Solanum corymbiflorum* leaves. *Journal of Ethnopharmacology, 179*, 16–21.

Piana, M., Camponogara, C., Boligon, A. A., & Oliveira, S. M. (2017). *Solanum paranense* extracts and solanine present anti-inflammatory activity in an acute skin inflammation model in mice. *Evidence-based Complementary and Alternative Medicine, 2017*, 4295680.

Ravi, V., Saleem, T. S. M., Patel, S. S., Raamamurthy, J., & Gauthaman, K. (2009). Anti-inflammatory effect of methanolic extract of *Solanum nigrum* Linn berries. *International Journal of Applied Research in Natural Products, 2*(2), 33–36.

Reddy, R. V. R., Rao, K. U. M., Vangoori, Y., & Sundharam, J. M. (2014). Evaluation of diuretic and anti-inflammatory property of ethanolic extract of *Solanum surattense* in experimental animal models. *International Journal of Pharmacy and Pharmaceutical Sciences, 6*(1), 387–393.

Sabir, S. M., & Rocha, J. B. (2008). Antioxidant and hepatoprotective activity of aqueous extract of Solanum fastigiatum (false "Jurubeba") against paracetamol-induced liver damage in mice. *Journal of Ethnopharmacology, 120*(2), 226–232.

Sengupta, R., Sheorey, S. D., & Hinge, M. A. (2012). Analgesic and anti-inflammatory plants: An updated review. *International Journal of Pharmaceutical Sciences Review and Research, 12*(2), 114–119.

Shahjhan, M., Sabitha, K. E., Jainu, M., & Shyamala Devi, C. S. (2004). Effect of *Solanum trilobatum* against carbon tetrachloride induced hepatic damage in albino rats. *The Indian Journal of Medical Research, 120*(3), 194–198.

Tai, C.-J., Choong, C.-Y., Shi, Y.-C., Lin, Y.-C., Wang, C.-W., Lee, B.-H., et al. (2016). *Solanum nigrum* protects against hepatic fibrosis via suppression of hyperglycemia in high-fat/ethanol diet-induced rats. *Molecules, 21*(3), 269.

Taofiq, O., González-Paramás, A. M., Barreiro, M. F., & Ferreira, I. C. (2017). Hydroxycinnamic acids and their derivatives: Cosmeceutical significance, challenges and future perspectives, a review. *Molecules, 22*(2), 281.

Tsuchiya, H. (2017). Anesthetic agents of plant origin: A review of phytochemicals with anesthetic activity. *Molecules, 22*(8), 1369.

Uritu, C. M., Mihai, C. T., Stanciu, G. D., Dodi, G., Alexa-Stratulat, T., Luca, A., et al. (2018). Medicinal plants of the family Lamiaceae in pain therapy: A review. *Pain Research & Management, 2018*, 7801543.

Valério, D. A., Georgetti, S. R., Magro, D. A., Casagrande, R., Cunha, T. M., Vicentini, F. T. M., et al. (2009). Quercetin reduces inflammatory pain: Inhibition of oxidative stress and cytokine production. *Journal of Natural Products, 72*(11), 1975–1979.

Wahyudi, I. A., Ramadhan, F. R., Wijaya, R. I. K., Ardhani, R., & Utami, T. W. (2020). Analgesic, antiinflammatory and anti-biofilm-forming activity of potato (*Solanum tuberosum* L.) peel extract. *Indonesian Journal of Cancer Chemoprevention, 11*(1), 30–35.

Wang, Y., Xiang, L., Yi, X., & He, X. (2017). Potential antiinflammatory steroidal saponins from the berries of *Solanum nigrum* L. (European black nightshade). *Journal of Agricultural and Food Chemistry, 65*(21), 4262–4272.

Wannang, N. N., Anuka, J. A., Kwanashie, H. O., & Auta, A. (2006). Analgesic and Antiinflammatory activities of the aqueous leaf extract of *Solanum nigrum* Linn (Solanaceae) in rats. *Nigerian Journal of Pharmaceutical Research, 5*(1). https://doi.org/10.4314/njpr.v5i1.53546.

World Health Organization. (1998). *Regulatory situation of herbal medicines. A worldwide review* (pp. 1–45). Geneva: WHO. WHO/TRM/98.1.

Zakaria, Z. A., Gopalan, H. K., Zainal, H., Pojan, N. H. M., Morsid, N. A., Aris, A., et al. (2006). Antinociceptive, Antiinflammatory and antipyretic effects of *Solanum nigrum* chloroform extract in animal models. *Yakugaku Zasshi, 126*(11), 1171–1178.

Zakaria, Z. A., Sulaiman, M. R., Morsid, N. A., Aris, A., Zainal, H., Pojan, N. H. M., et al. (2009). Antinociceptive, antiinflammatory and antipyretic effects of *Solanum nigrum* aqueous extract in animal models. *Methods and Findings in Experimental and Clinical Pharmacology, 31*(2), 81–88.

Chapter 40

Dietary constituents act as local anesthetic agents: Neurophysiological mechanism of nociceptive pain

Mamoru Takeda and Yoshihito Shimazu

Laboratory of Food and Physiological Sciences, Department of Life and Food Sciences, School of Life and Environmental Sciences, Azabu University, Sagamihara, Kanagawa, Japan

Abbreviations

Ach M$_2$R	acethylcholine muscarinic 2 receptor
ASIC2	acid-sensing channel 2
C1-C2	upper cervical spinal cord
CAM	complementary alternative medicine
Cav	voltage-gated Ca^{2+} channels
CGA	chlorogenic acid
DA	decanoic acid
DHA	docosahexaenoic acid
DRG	dorsal root ganglion
EPSP	excitatory postsynaptic potential
IA	fast-inactivating transient Kv currents
ID	slow-inactivating transient Kv currents
IK	slow-inactivating sustained Kv currents
JOR	jaw-opening reflex
Kv	voltage-gated K^+ channels
Nav	voltage-gated Na^+ channels
NS	nociceptive-specific
PBN	parabrachial nucleus
REV	resveratrol
SpVc	spinal trigeminal nucleus caudalis
TG	trigeminal ganglion
TMJ	temporomandibular joint
TRP	transient receptor potential
TRPA1	transient receptor potential ankyrin
TTX	tetrodotoxin
WDR	wide-dynamic range

Introduction

Medicinal plants (e.g., cocaine) have been used since ancient times for relieving pain caused by injuries and disease (Chidiac et al., 2011). Complementary alternative medicine (CAM) therapies, such as herbal medicines and acupuncture, are often used in pain management, especially after a failure of conventional Western medicine (Rao et al., 1999; Konvicka, Meyer, & Roberson, 2008; Rosenberg et al., 2008) or when adverse side effects are a concern (Shir et al., 2001; Ernst, 2003). Previous studies reported that various dietary constituents can potentially affect protective biological mechanisms, such as those in the cardiovascular, neural, and anticancer systems (Frémont, 2000; Pervaiz, 2003). In these recent studies, various dietary constituents and supplements could potentially affect the excitability of the peripheral and central

Treatments, Mechanisms, and Adverse Reactions of Anesthetics and Analgesics. https://doi.org/10.1016/B978-0-12-820237-1.00040-5
Copyright © 2022 Elsevier Inc. All rights reserved.

TABLE 1 Possible molecular targets for modulation of excitability of primary afferents by dietary constituents under in vitro conditions.

Dietary constituents	Molecular targets	Effects	References
• *Polyphenol*			
• Resveratrol	• TRPA1	Inhibition	Yu et al. (2013)
	• Nav	Inhibition	Kim et al. (2005)
	• Kv	Facillitation	Grannados-Soto et al. (2002)
• Chrologenic acid	• ASIC	Inhibition	Qu et al. (2014)
	• Kv	Facillitation	Liu et al. (2016)
• Genistein	• Nav	Inhibition	Liu et al. (2004)
	• Kv	Inhibition	Liu et al. (2004)
• Cathechin	• Nav	Inhibition	Kim et al. (2009)
• Quercetin	• Nav	Inhibition	Wallace et al. (2006)
• *Carotenoid*			
• Lutein	• TRPA1	Inhibition	Horváth et al. (2005)
			Hórvath et al. (2012)
• *Fatty acids*			
• Dcosahexanoic acid	• TRPA1	Inhibition	Motter and Ahern (2012)
	• Nav	Inhibition	Xiao et al. (1995)
• Eicosapentanoic acid	• Nav	Inhibition	Xiao et al. (1995)
• Decanoic acid	• Ach M_2R	Facillitation	Gwynne et al. (2004)

Abbreviations: *ASIC 2*, acid-sensing channel 2; *Ach M_2R*, acetylcholine muscarinic 2 receptor; *Kv*, voltage-gated K^+ channels; *Nav*, voltage-gated Na^+ channels; *TRPA1*, transient receptor potential ankyrin 1.

nervous systems. In Table 1, we summarize the possible molecular targets for the modulation of the neuronal excitability of primary sensory afferent neurons by dietary constituents under in vitro experimental conditions. For example, many researchers showed that under in vitro conditions, various chemical substances, including polyphenol compounds, carotenoids, and fatty acids, modify the excitability in nociceptive neurons in both the peripheral and central nervous systems via various voltage-gated Na^+ and K^+ (Nav and Kv) channels, transient receptor potential (TRP) family, and G-protein-coupled channels.

The majority of currently used anesthetic agents is derived from natural products such as plants (Tsuchiya, 2017). Local anesthetics reversibly block Nav channels that are responsible for initiation in sensory nerve terminals and conduction of action potentials from the peripheral nervous system to the central nervous system (Fozzard, Lee, & Lpkind, 2005). The trigeminal nervous system has unique structures and functions for the processing of orofacial nociception as well as non-noxious sensations. We recently investigated whether local administration of some dietary constituents attenuated the excitability of trigeminal nociceptive neurons using in vivo electrophysiological experiments, based on the data obtained from previous in vitro experiments (Table 1). The present review focuses on the mechanisms by which dietary constituents might modulate nociceptive neuronal excitability associated with trigeminal pain, based on our experimental in vivo data. In addition, we discuss the physiological contribution of dietary constituents as CAM agents on nociceptive pain.

Ascending pain pathway in trigeminal system

This review focuses on the potential contribution of dietary constituents to the relief of nociceptive pain. Based on recent data demonstrating the effect of dietary constituents on the excitability of nociceptive neurons in a trigeminal pain rat model, we will introduce the trigeminal pain pathway and the general physiological properties of the two types of nociceptive neurons.

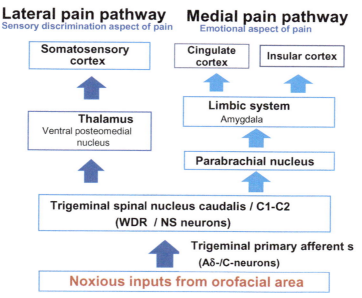

FIG. 1 The trigeminal nociceptive sensory pathway is divided into a lateral and a medial system. Although the lateral pain system conveys the information for the sensory-discriminative aspect of pain, the medial pain system conveys information on the emotional aspect of pain. Orofacial noxious inputs are conveyed to the spinal trigeminal nucleus caudalis (SpVc) and upper cervical dorsal horn (C1–2). SpVc and C1–2 neurons project axons to the somatosensory cortex via the ventral posteromedial nucleus in the thalamus (lateral pathway) and also to both the anterior cingulate cortex and the insular cortex via the limbic system, such as the amygdala (medial pathway).

As shown in Fig. 1, the trigeminal nociceptive sensory pathway is divided into two systems, the lateral and the medial systems. The lateral pain system conveys information underlying sensory discrimination of pain, including its location, intensity, and quality, while the medial pain system conveys the emotional aspect of pain (Iwata, Takeda, Oh, & Shinoda, 2017; Takeda, Matsumoto, et al., 2011). Orofacial tissues, such as the oral mucosal membrane, tongue, tooth pulp, and temporomandibular joint (TMJ), are innervated by small-diameter Aδ-fibers and unmyelinated C-fibers. Noxious orofacial signals are transmitted via trigeminal ganglion (TG) neurons to second-order neurons in the spinal trigeminal nucleus caudalis (SpVc) and upper cervical spinal cord (C1-C2). These SpVc/C1-C2 are important signaling stations for trigeminal nociceptive inputs generated by tissue inflammation and damage. SpVc/C1–2 nociceptive neurons include two types, nociceptive-specific (NS) and wide-dynamic range (WDR) neurons. NS neurons respond only to noxious stimulation of receptive fields, and thus possibly send location-related information to higher centers. In contrast, WDR neurons respond to noxious and nonnoxious stimulation; graded nociceptive stimulation to the most sensitive area of the receptive field induces increased firing frequency in proportion to stimulus intensity. Therefore, WDR neurons act physiologically to transmit noxious stimulus-intensity information from peripheral tissue to a higher center of the central nervous system. Central projection neurons in both the SpVc and C1-C2 send axons to the ventromedial posterior and medial thalamic nuclei and the parabrachial nucleus (PBN), whereby the ventromedial posterior transmits nociceptive information to primary and secondary somatosensory cortical neurons. Recent anatomical studies have demonstrated that projections from the SpVc and C1-C2 to the PBN are much denser compared with those to the ventromedial posterior (Al-Khater and Todd, 2009). The PBN projects to the limbic system, specifically the amygdala, the anterior cingulate corex, and the insular cortex (Vogt, 2005), which contributes to the medial pain pathway system and concerns motivational and affective aspects of pain.

Sensory transduction and noxious transmission

Fig. 2 shows the generally accepted mechanism of nociceptive sensory signaling, which depends on the following four processes: (i) transduction of external stimuli from peripheral terminals; (ii) action potential generation; (iii) action potential propagation along axons; and (iv) transmission to central terminals that form the presynaptic elements of the first synapses in sensory pathways in the central nervous system (Takeda, Matsumoto, et al., 2011; Harriott and Gold, 2009). If noxious mechanical stimulation is applied to a peripheral tissue, such as skin, mechanosensitive ion channels (transient receptor potential family) open in the free nerve endings of noxious primary afferent neurons, and the resulting depolarization creates a generator potential. When this generator potential reaches a certain threshold, Nav and Kv channels open, resulting in the firing of action potentials. The action potentials are then conducted through the primary afferent

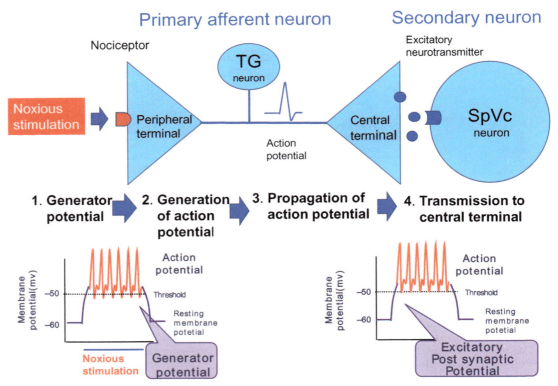

FIG. 2 Four general processes involved in nociceptive sensory signaling. When noxious mechanical stimulation is applied to the tissue, the generator potential in the peripheral terminal of trigeminal ganglion (TG) neurons is activated. This depolarization (generator potential) further opens voltage-dependent sodium and potassium channels, generating action potentials. Action potentials are discharged through primary afferent TG neurons to the central terminal in nociceptive neurons in the spinal trigeminal nucleus caudalis (SpVc). When action potentials are conducted to the central terminal of the SpVc, presynaptic voltage-dependent calcium channels open, leading to the release of neurotransmitters into the synaptic cleft, which then binds to postsynaptic receptors, activating excitatory postsynaptic potentials. If the amplitude of excitatory postsynaptic potential (EPSP) exceeds the action potential threshold, a barrage of action potentials is conducted to higher centers in the pain pathway, and pain is perceived.

fibers, Aδ- and C-fibers, to the central terminals in NS and WDR nociceptive neurons, located in the SpVc and C1-C2. Presynaptic Cav channels are activated by the propagation of action potentials in the central terminals and subsequent release of excitatory neurotransmitters (e.g., glutamate) into the synaptic cleft and binding to postsynaptic ionotropic glutamate receptors, which activates excitatory postsynaptic potentials (EPSP). If the EPSP amplitude exceeds the action potential threshold, a barrage of action potentials is conducted to higher centers in both lateral and medial pain pathways and pain is perceived.

Possible molecular targets for local anesthetic agents

When anesthetic agents are applied locally on peripheral tissues, these agents inhibit the action potentials in the peripheral terminals of primary afferents involved in nociceptive stimulation.

Generator potential in the nociceptive terminals

Concerning the development of generator potential in the free nerve ending of nociceptive primary afferents, it can generally be assumed that there are several possible molecular targets for mechanotransduction channels in mammals, including nociception channels, for example, TRP ankyrin 1 (TRPA1) and acid-sensing channels 2 (ASIC2). TRPA1 is a member of the TRP ion channel family which is implicated in thermosensation and mechanosensation (McKemy et al., 2002; Walker, Willingham, & Zuker, 2000; Tracey et al., 2003). Recent reports indicate that ASIC2 might be a mammalian cutaneous mechanoreceptor (Borzan, Zhao, Mayer, & Raja, 2010; Kang et al., 2012; Price et al., 2001). ASIC2 also belong to the family of proton-gated cation channels and act as critical pH sensors, which is important for nociception (Alvarez de la Rosa, Zhang, Shao, White, & Canessa, 2000; Deval et al., 2008).

Action potential in the nociceptive terminals

When the amplitude of the generator potential reaches the membrane potential threshold of Nav and Kv channels, an action potential is generated in the nerve terminals and conducted by nociceptive afferents to the central terminal in the nociceptive neurons in the SpVc. Generally, Nav channels are composed of a core α-subunit associated with one or more regulatory β-subunits (β1-β4). (Catterall, 2012). Catterall (2012) described that the α-subunit is not only a selectively permeable pore for Na$^+$ ions, but also contains the binding or receptor site that can cause occlusion of the pore, resulting in the blockade of Nav channels (Catterall, 2012). At least nine distinct α-subunits (Nav1.1-Nav1.9) have been cloned from mammalian Nav channels (Catterall, 2012). Only Nav 1.7, Nav 1.8, and Nav 1.9 are expressed in nociceptive neurons. Based on their sensitivity to tetrodotoxin (TTX), Nav channel subtypes are divided into TTX-sensitive (Nav1.1-Nav1.4, Nav1.6, and Nav 1.7) and TTX-resistant (Nav1.5, Nav1.8, Nav1.9) neurons (Kim and Chung, 1999;). Since Nav1.8 and Nav1.9 isoforms were found in small-diameter (e.g., Aδ- and C-fibers) dorsal root ganglia (DRG) and TG neurons, it can be assumed that Nav1.8 and Nav1.9 play an important role in pain transmission; therefore, these channels are implicated as targets for anesthetic and analgesic drugs.

In contrast, Kv channels are known to be one of the most important physiological regulators of membrane potential in the excitable tissue, including nociceptors. Nociceptive primary afferent neurons, such as DRG and TG neurons, express three distinct classes of K$^+$ currents in varying quantities: slow-inactivating sustained (K-current; I_K), fast-inactivating transient (A-current; I_A), and slow-inactivating transient (D-current; I_D) currents, based on their inactivation of kinetics and drug sensitivities to tetraethylammonium, 4-aminopyridine, and a-dendrotoxin, respectively (Matsumoto, Yoshida, Takahashi, Saiki, & Takeda, 2010; Takeda, Tanimoto, Ikeda, Kadoi, & Matsumoto, 2004). Actually, previous studies indicated that peripheral nerve injury and inflammation markedly reduce the density of Kv channels. These changes implicate hyperexcitability of nociceptive primary afferent neurons that results in the development of neuropathic/inflammatory pain (Everill & Kocsis, 1999; Kim, Choi, Rim, & Cho, 2002; Park, Choi, & Cho, 2003). Since the opening of K$^+$ channels leads to hyperpolarization of the cell membrane and a resultant decrease in cell excitability, several types of Kv channels have been proposed as potential target candidates for attenuated signals in pain pathways (Lawson, 2006; Takeda, Matsumoto, et al., 2011; Takeda, Tsuboi, et al., 2011).

Peripheral mechanism for potential candidates of dietary constituents as local anesthetic agents

Modulation of generator potential by dietary constituents

Resveratrol: Resveratrol (REV; *trans*-3,4,5-trihydroxystilbene) is a natural polyphenolic compound found in a large number of plants that form part of the human diet, including peanuts, mulberries, grapes, and red wine. REV has several beneficial biological actions, including antioxidative, antiinflammatory, neuroprotective, anticancer, and cardioprotective effects (Takeda, Takehana, & Shimazu, 2018). Recently, Meng et al. (2015) demonstrated that mechanical stimuli induce mechanosensitive currents in vitro via mechanosensitive channels, such as TRPA1, to trigger mechanotransduction in the trigeminal ganglion neurons innervating the inner walls of the anterior eye chamber. In addition, TRPA1 channels have been shown to modulate mechanotransduction in primary sensory neurons (Kwan, Glazer, Corey, Rice, & Stucky, 2009). REV is a potent inhibitor of TRPA1 channels in vitro and in vivo (Yu et al., 2013), suggesting that REV attenuates the generator potential and inhibits action potential firing via the mechanical transduction process. Actually, under in vivo conditions, we have demonstrated that local injection of REV into the peripheral receptive field suppresses neuronal firing rates of SpVc neurons responding to nociceptive mechanical stimulation in a dose-dependent manner, partially via inhibition of TRP channels in the nociceptive nerve terminals of trigeminal ganglion neurons (Shimazu et al., 2016). Therefore, it is likely that REV inhibits both the generator potential and action potential firing via the mechanical transduction process.

Lutein: Lutein (LT) is a naturally occurring carotenoid present in various food products, such as fruits and leafy green vegetables (Alves-Rodrigues & Shao, 2004). Lutein inhibits stimulation-induced responses to TRPA1, but not TRP vanilloid 1 (TRPV1), on cell bodies and peripheral terminals of sensory neurons in vitro and in vivo (Hórvath et al., 2012). Previously, Kwan et al. (2009) demonstrated that TRPA1 channels modulate mechanotransduction via the generator potential in primary sensory neurons (Kwan et al., 2009). Recently, we tested whether systemic administration of LT inhibits hyperexcitability of nociceptive SpVc WDR neurons responding to noxious stimulation in complete Freund's adjuvant-induced inflamed rats (Syoji et al., 2018). We found that systemic administration of LT attenuates inflammatory hyperalgesia associated with hyperexcitability of nociceptive SpVc WDR neurons via the inhibition of prostaglandin E$_2$ production by suppression of peripheral cyclooxygenase-2 signaling. It can be assumed that LT could attenuate the

478 PART | IV Novel and nonpharmacological aspects and treatments

amplitude of the generator potential and subsequent firing rates of trigeminal ganglion neurons. However, further study is needed to clarify the hypothesis that local injection of LT into the peripheral receptive field suppresses the excitability of SpVc neurons as described in a previous study (Shimazu et al., 2016).

Chlorogenic acid: Chlorogenic acid (5-caffeoylquinic acid, CGA) is a natural polyphenolic compound found in many plant species consumed by humans, including fruits and vegetables. ASIC2 might be a mammalian cutaneous mechano-receptor (Price et al., 2001; Borzan et al., 2010; Kang et al., 2012). Since CGA was recently shown to inhibit neuronal firing at the DRG via blocking ASIC2 and modulating the neuronal excitability of DRG neurons (Qu, Liu, QiU, Li, & Hu, 2014), CGA presumably inhibits the nociceptive nerve terminal of TG neurons via modulation of ASIC2. ASIC2 also belongs to the family of proton-gated cation channels and acts as critical pH sensors, predominantly expressed in the nociceptor where they might be important for nociception (Alvarez de la Rosa et al., 2002; Deval et al., 2008). We recently demonstrated that local injection of CGA into the peripheral receptive field dose-dependently suppressed the excitability of SpVc neurons responding to noxious mechanical stimulation, possibly via the modulation of ASIC2 in the nociceptive nerve terminal of trigeminal ganglion neurons (Kakita et al., 2018). Therefore, local administration of CGA possibly inhibits the excitability of SpVc WDR neurons by inhibiting the development of generator potential after noxious mechanical stimulation in the peripheral terminal of TG neurons.

Modulation of action potential by dietary constituents

Resveratrol: REV modulates Na^+ and K^+ currents in primary afferent sensory neurons, which are associated with the generation of action potentials (Kim et al., 2005; Grannados-Soto et al., 2002). Specifically, Kim et al. (2005) showed that REV predominantly inhibits Na^+ currents in acutely dissociated DRG neurons, indicating that REV inhibits the generation of action potentials. In addition, REV inhibits both TTX-S and TTX-R Na^+ currents in acutely dissociated DRG neurons, and TTX-R Na^+ channels appear to be selectively expressed in nociceptive (small- and medium-sized) DRG neurons, corresponding to $A\delta$-/C-primary afferent trigeminal ganglion neurons (Cummins et al., 1999; Takeda et al., 2006). Previously, we investigated whether local subcutaneous administration of REV attenuates mechanical stimulation-induced excitability of SpVc WDR neuron activity in rats in vivo. As a result, local subcutaneous administration of REV into the orofacial skin dose-dependently and significantly reduced the mean number of SpVc WDR neurons firing in response to both nonnoxious and noxious mechanical stimuli, reaching maximal inhibition of discharge frequency in response to both stimuli within 5 min. These inhibitory effects were no longer evident after approximately 20 min. Taken together, the results suggest that REV inhibits the excitability of peripheral terminals of trigeminal afferent nerves by modulating generation (initiating action potentials) mechanisms, as well as transduction (generator potential).

Docosahexaenoic acid: Docosahexaenoic acid (DHA) is one of the major polyphenolic saturated fatty acids in marine fish oil and the human central nervous system. Previous studies suggest that DHA modulates the neuronal excitability of the peripheral and central nervous systems via TRP channels and voltage-dependent ion channels (Motter & Ahern, 2012; Xiao, Kang, Morgan, & Leaf, 1995). Since DHA is a potent inhibitor of TRPA1 in vitro and in vivo (Motter & Ahern, 2012), we speculate that DHA may attenuate the amplitude of generator potentials responding to noxious mechanical stimulation in the peripheral terminal of primary afferents. In addition, DHA modulates the Nav currents in DRG neurons that are associated with the generation of action potentials (Xiao et al., 1995). Taken together, these findings strongly suggest that local DHA administration may suppress the generation of action potentials associated with the peripheral nervous system. Since the jaw-opening reflex (JOR), induced by electrical stimulation of orofacial tissues (e.g., tooth pulp and tongue), is a valid index for reflex response to noxious stimuli (trigeminal nociceptive reflex), the JOR threshold has been used as an indicator of the intensity of the stimulus applied to orofacial tissues (Kokuba, Takehna, Oshima, Shimazu, & Takeda, 2017; Mitome, Takehana, Oshima, Shimazu, & Takeda, 2018). Using an in vivo nociceptive jaw-opening reflex rat model, we recently reported the following observations: (i) Noxious stimulation-evoked JOR amplitude was dose-dependently inhibited by local administration of DHA; (ii) The maximum inhibition of the amplitude of these JOR responses was seen within approximately 10 min, and these inhibitory effects were reversed after approximately 40 min (Mitome et al., 2018). Taken together, these findings suggest that DHA attenuates the nociceptive JOR by blocking Nav channels and strongly supports the idea that DHA is a potential therapeutic agent.

Chlorogenic acid: Recently, Zhang et al. (2014) showed that CGA modulates the neuronal excitability of TG neurons via Kv channels. In addition, a rat formalin test showed that REV induced peripheral antinociception in vivo via the opening of several Kv channels (Grannados-Soto, Argulles, & Ortiz, 2002). Since the opening of Kv channels leads to hyperpolarization of the resting membrane potential and, in turn, decreased neuronal excitability, several types of Kv channels have been proposed as potential target candidates for pain therapy (Takeda, Tsuboi, et al., 2011). Liu et al. (2016) also recently reported a possible analgesic effect of CGA via the modulation of Kv channel activation and inactivation under inflammatory conditions, providing a novel molecular and ionic mechanism underlying antiinflammatory pain with CGA administration. Recently, under in vivo conditions, Kakita et al. (2017) reported that (i) the mean firing rate of SpVc WDR neurons in response to both nonnoxious and noxious mechanical stimuli was dose-dependently reduced by local injection of CGA into the receptive field; (ii) CGA application-induced inhibition of the discharge frequency in response to both non-noxious and noxious mechanical stimuli was reversible (within 30 min); and (iii) local injection of the vehicle had no significant effect on SpVc WDR neuron activity evoked by nonnoxious or noxious mechanical or pinch stimulation. These findings suggest that local CGA injection into the peripheral receptive field suppresses the excitability of SpVc neurons, possibly via activation of Kv channels in the nociceptive nerve terminal of trigeminal ganglion neurons. Therefore, it is likely that local injection of CGA could contribute to the treatment of trigeminal nociceptive pain.

Decanoic acid: The dietary constituent and fatty acid, decanoic acid (DA), performs several biological actions. In a study using guinea pig duodenum and jejunum, Gwyme et al. (2004) demonstrated that DA-induced motor activity is reversibly abolished by a muscarinic antagonist, suggesting that DA may act as an acetylcholine muscarinic 2 receptor ($AchM_2R$) agonist. In fact, the $AchM_2R$ family is involved in a large number of important physiological functions in both the central and peripheral nervous systems, including nociception (Wess et al., 2003). Previously, Bernardini, Sauer, and Haberberger (2001) reported that in electrophysiological experiments using a rat skin-nerve preparation, $AchM_2R$ M_2 exerted an inhibitory or desensitizing effect on the peripheral terminal of C-nociceptors. This desensitization can be explained by the fact that, apart from lowering the intracellular cAMP concentration, $AchM_2R$ also affects low-threshold Kv channels, which builds up a hyperpolarization force (Pan and Williams, 1994). There is evidence that $AchM_2R$ mRNA and immunoreactivity are found in small- and medium-diameter TG neurons (Dussor et al., 2004; Noguchi et al., 2017). In agreement with these findings, we recently showed that acute DA application induces short-term mechanical hypoalgesia. This effect was mainly due to suppression of the excitability of TG and SpVc WDR and NS neurons via M_2 $AchM_2R$-induced membrane hyperpolarization (Nakajima et al., 2018; Noguchi et al., 2017). These observations support the idea of DA as a potential therapeutic CAM for the treatment of trigeminal nociception.

Functional significance for dietary constituents as local anesthetic agents

Recently, there has been increased interest in the use of CAM for the treatment of acute and chronic pain symptoms. Patients with pain frequently turn to CAM for attenuation of pain when other Western medical treatments are ineffective (Rosenberg et al., 2008). Therefore, it is reasonable to speculate that the development and discovery of new dietary constituents acting as analgesic drugs without side effects would lead to a contribution to CAM. The Nav channel blocker, lidocaine, is the most representative and widely used form of local anesthetic. Previously, we compared the magnitude of inhibition of SpVc WDR nociceptive transmission between REV and lidocaine under in vivo conditions (Shimazu et al., 2016). The mean magnitude of inhibition of SpVc neuronal discharge frequency responding to noxious stimulation was almost equal between 10 mM REV and 37 mM lidocaine (=1% lidocaine), suggesting that the potency of the inhibitory effect of REV on nociceptive transmission was fourfold higher than that of lidocaine. Similarly, we also examined the local application of DHA for the nociceptive jaw-opening reflex (Mitome et al., 2018). We found that the mean magnitude of inhibition was almost equal between 25 mM DHA and 37 mM lidocaine (=1% lidocaine), suggesting that the potency of the inhibitory effect of nociceptive jaw-opening reflex was 1.5-fold higher for DHA than 1% lidocaine. Therefore, it can be assumed that among both polyphenolic compounds, REV and polyunsaturated fatty acids, DHA might be a candidate for replacing local analgesic drugs by inhibiting both the generation and/or conduction of action potentials in the trigeminal primary afferents to the central pain pathway. The results of these studies suggested that the potency of the inhibitory effect of these local analgesic drugs was as follows: REV > DHA > lidocaine.

Alternatively, in several types of trigeminal nerve injury/inflammatory models (Takeda, Tsuboi, et al., 2011), blocking Kv channels contributed to the improvement of peripheral nerve injury-induced mechanical allodynia (Kitagawa et al., 2006; Nakagawa et al., 2010; Takeda, Matsumoto, et al., 2011). These common changes contribute to the incremental spike discharge and prolonging the duration of action potentials of TG neurons in the neuropathic pain rat model (Takeda, Tsuboi, et al., 2011). These results raised the possibility that Kv channel openers could be potential therapeutic agents for preventing trigeminal neuropathic/inflammatory pain. Previously, we observed significantly greater inhibition of SpVc WDR neuronal discharge frequency by CGA for noxious than nonnoxious stimuli. This finding supports the idea of CGA as a local anesthetic agent for treating trigeminal nociceptive pain. We also compared the magnitude of inhibition of SpVc WDR nociceptive neurons between CGA and 1% lidocaine, a Nav channel blocker, and a well-known local anesthetic agent. We found that the potency of the inhibitory effect of nociceptive transmission was fourfold higher for CGA than 1% lidocaine (Kakita et al., 2017), suggesting that the potency of the inhibitory effect is the following: CGA = REV > lidocaine.

Finally, we made an original ointment for the application of DA to the natural cutaneous tissue. DA is a saturated medium-chain fatty acid with a 10-carbon backbone and a lipophilic agent. The original ointment containing DA had the following composition: 30% decanoic acid, 15% caffeine, 4% D-mannitol, 31% calcium carbonate, and 20% solvent medium (alcohol, olive oil, and macrogol ointment) (Noguchi et al., 2017). When applied to rat natural cutaneous tissue, this DA ointment induced mechanical hypoalgesia in the absence of inflammation/neuropathy (Noguchi et al., 2017). As a result, its inhibitory effect was mainly due to the suppression of excitability of the SpVc WDR and NS neurons via the inhibition of peripheral AchM$_2$R, possibly via TG neurons (Nakajima et al., 2018). Therefore, it can be speculated that DA-containing ointment may effectively reduce clinical pain, such as injection-related pain during blood sampling and other clinical treatment. However, further study is needed to clarify this hypothesis.

Future direction

In this review, we describe the current state of knowledge for dietary constituents as local anesthetic agents. Based on previous in vivo and in vitro studies by our group, they inhibit not only the generation of action potentials, but also the development of generator potential in the peripheral nociceptive afferent terminals. In the future, we are planning to examine whether a mixture of dietary constituents that block both generator and action potential could synergistically attenuate nociception. Table 1 introduces possible molecular targets for the modulation of excitability of primary afferents by dietary constituents under in vitro conditions. Many dietary constituents are of plant origin. Polyphenol compounds, such as genistein, catechin, and quercetin, have a strong inhibitory effect on the Nav channel activity (Liu et al., 2004; Kim et al., 2009; Wallace, Baczko, Jones, Fercho, & Light, 2006). Thus, we are planning to explore their potential local anesthetic effect on the trigeminal nociceptive neuronal activity under in vivo conditions and to compare this inhibitory effect with that of lidocaine.

Finally, the pain has very complicated sensory-discriminative, motivational, and affective aspects (Fig. 1). The sensory-discriminative aspect of pain is analogous to a nonnoxious sensation and thought to be involved in the discrimination of pain features, such as its location, intensity, and quality (lateral system in the pain pathway) (Iwata et al., 2017). In contrast, the motivational and affective aspects of pain are believed to be related to emotional and autonomic responses due to long-lasting, intense noxious stimuli (medial system in the pain pathway) (Iwata et al., 2017). Although most research on REV and nociceptive and pathological pain has focused on the lateral system in the pain pathways, such as the sensory-discriminative aspect of pain (Kokuba et al., 2017; Sekiguchi et al., 2016; Takehana et al., 2016, 2017), future studies should focus on the effect of REV on the medial system in the pain pathway (an emotional aspect of pain). Concerning the mechanisms underlying the effects of dietary constituents on pain, many researchers have mainly focused on the lateral pain system, covering the sensory-discriminative aspect of pain. In the future, we aim to investigate whether pretreatment with dietary constituents attenuates the nocifensive response in the medial pain pathway system related to the motivational and affective aspects of pain.

Concluding remarks

Our knowledge on the mechanisms underlying the modulatory effects of dietary constituents on the excitability of nociceptive neurons associated with pain relief can be summarized as follows: (i) REV, CGA, LT, and DHA inhibit the nociceptive neuronal excitability by modulating the generator potential, such as in the TRP channel and ASIC2 channels; (ii) REV and DHA inhibit the action potential in nociceptive neurons by suppressing Nav channels; (iii) CGA, REV, and DA inhibit the action potential in nociceptive neurons by facilitating the opening of Kv channels (Fig. 3). In conclusion, our in vivo experimental studies provide evidence that, when locally applied on peripheral tissues, dietary constituents inhibit not only the generation of the action potential, but also the development of generator potential in the peripheral terminal of nociceptive primary afferents after nociceptive stimulation. Our results suggest that dietary constituents, such as polyphenolic compounds, carotenoid, and fatty acids, could contribute to the development of analgesic drugs for treating clinical nociceptive pain, including orofacial pain, with fewer and less severe side effects.

Mini-dictionary of terms

- **Complementary and alternative medicine**

    Complementary alternative medicine (CAM) therapies, such as herbal medicines and acupuncture, are often used in pain management, especially after the failure of conventional Western medicine or when adverse side effects are a concern.

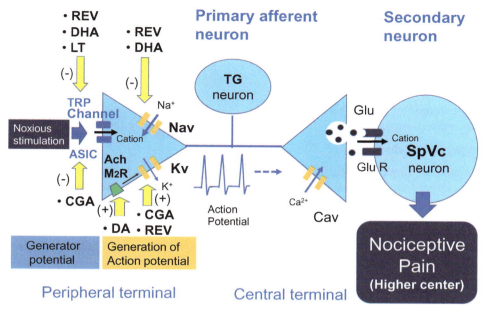

FIG. 3 Possible molecular target for dietary constituents as local anesthetic agents. When locally applied on peripheral tissues, dietary constituents (RES, CGA, and DHA) inhibit the development of both generator potential and action potential in the peripheral terminal of primary afferents after nociceptive stimulation. However, DA inhibits only the development of action potential in the peripheral terminal of primary afferents after nociceptive stimulation. LT inhibits only the development of generator potential in the peripheral terminal of primary afferents after nociceptive stimulation. *ASIC*, acid-sensing ion channel; *Ach M₂R*, acetylcholine muscarinic receptor 2; *Cav*, voltage-gated calcium channel; *CGA*, chlorogenic acid; *DA*, decanoic acid; *Glu*, glutamate; *Glu R*, glutamate receptor; *Kv*, voltage-gated potassium channel; *LT*, lutein; *Nav*, voltage-gated sodium channel; *REV*, resveratrol; *SpVc*, spinal trigeminal nucleus caudalis; *TG*, trigeminal ganglion; *TRP*, transient receptor potential. (+), excitation; (−), inhibition.

- **Polyphenolic compounds**

  Polyphenolic compounds have several beneficial biological actions, including antioxidative, antiinflammatory, neuroprotective, anticancer, and cardioprotective effects. Polyphenols comprise a large variety of molecules, including some with one unique phenol ring or typically more than one. According to both the number of phenol rings and the structural elements binding these rings to one another, polyphenols can be divided into the following groups: flavonoids (e.g., catechin, genistein), phenolic acids (e.g., chlorogenic acids), stilbenes (e.g., resveratrol), and lignans.

- **Carotenoid**

  Carotenoids are important dietary nutrients with antioxidant potential. They are lipophilic substances derived from the same C40 isoprenoid skeleton. These fat-soluble yellow pigments are abundantly present in fruits and leafy green vegetables. One of the most prevalent carotenoids is lutein.

- **Fatty acids**

  Fatty acids are classified as saturated fatty acids with no double bonds (e.g., C10, decanoic acid) or as unsaturated fatty acids with double or triple bonds. The latter are divided into monounsaturated fatty acids with only one double bond and polyunsaturated fatty acids with two or more double bonds. Polyunsaturated fatty acids are further divided into omega-3 fatty acids, represented by α-linoleic acid, eicosatetraenoic acid (EPA), and docosahexaenoic acid (DHA), and omega-6 fatty acids, represented by linoleic acid, γ-linoleic acid, and arachidonic acid. DHA has several beneficial biological actions, including antioxidative, antiinflammatory, neuroprotective, anticancer, and cardioprotective effects.

- **Generator potential**

  Generator potential (or receptor potential) refers to stimulus-induced changes in the membrane potential of sensory receptors, especially primary sensory receptors, such as caused by free nerve endings of peripheral neurons. When the magnitude of the generator potential reaches the activation threshold for Na⁺ channels, an action potential is generated in the sensory nerve.

- **Action potential**

  A brief fluctuation in membrane potential is caused by the rapid opening and closing of voltage-gated sodium and potassium ion channels. Action potentials rise like a wave, or spike, cause all-or-none rapid depolarization, and create electrical signals along axons to transfer information from one place to another in the nervous system.

- **Voltage-gated sodium channels**

  A membrane protein forming a pore that is permeable to Na^+ and gated by depolarization of the membrane. Sodium currents play an important role in generating action potentials and controlling their firing rates.

- **Voltage-gated potassium channels**

  A membrane protein forming a pore that is permeable to K^+ and gated by depolarization of the membrane. Voltage-gated potassium currents are important physiological regulators of membrane potentials, action potential shape, and firing adaptation in excitable tissues, including nociceptive sensory neurons.

- **Nociceptive pain**

  Sensation induced by activation of some receptor selective for potentially harmful stimuli. Normally, pain is felt when signals originating in thinly myelinated (Aδ-) and unmyelinated (C-) nociceptive afferents reach a conscious brain. Typical examples of nociceptive pain include a pinprick or a stubbed toe. The sensation felt (pain) matches the stimulus (noxious).

- **Spinal trigeminal nucleus caudalis**

  Orofacial tissues are innervated by small-diameter Aδ-fibers and unmyelinated C-fibers of the trigeminal nerve. Noxious orofacial signals are transmitted via the trigeminal afferent nerve (trigeminal ganglion) neurons to second-order neurons in the spinal trigeminal nucleus and upper cervical spinal cord and then transmitted to higher centers in the pain pathway. The spinal trigeminal nucleus serves as an important relay station for the transmission of orofacial sensory information. It is functionally subdivided from rostral to caudal: oralis, interpolaris, and caudalis.

Key facts

1. Recently, we have investigated whether local administration of some dietary constituents attenuates the excitability of trigeminal nociceptive neurons using in vivo electrophysiological experiments, based on data obtained from previous in vitro experiments.
2. REV, CGA, LT, and DHA inhibit the nociceptive neuronal excitability via the modulation of generator potential, such as in TRPA1 and ASIC2 channels.
3. REV and DHA inhibit the action potential in nociceptive neurons via the suppression of Nav channels.
4. CGA, REV, and DA inhibit the action potential in the nociceptive neuron by facilitating the opening of Kv channels.
5. The results of our in vivo electrophysiological study suggest that the potency of the inhibitory effect of local analgesic drugs on the excitability of nociceptive primary afferents is as follows: REV = CGA > DHA > lidocaine.

Summary points

- Recent reports have described the use of CAM, such as herbal medicines and acupuncture, for the treatment of clinical pain.
- We have investigated whether local administration of some dietary constituents attenuates the excitability of trigeminal nociceptive neurons using in vivo electrophysiological experiments, based on data obtained from previous in vitro experiments.
- When locally applied on peripheral tissues, anesthetic agents inhibit not only the generation of the action potential, but also the development of generator potential in the peripheral terminal of primary afferents after noxious mechanical stimulation.
- Polyphenolic compounds (REV and CGA) and fatty acids (DA and DHA) modulate the neuronal excitability of the peripheral nervous system via Nav, Kv, TRPA1, ASIC channels, and Ach M_2R
- Dietary constituents might contribute to the development of analgesic drugs for treating clinical pain, including orofacial pain, with fewer and less severe side effects.

References

Al-Khater, K. M., & Todd, A. J. (2009). Collateral projections of neurons in laminae I, III, and IV of rat spinal cord to thalamus, periaqueductal gray matter, and lateral parabrachial area. *Journal of Comparative Neurology, 515*, 629–646.

Alvarez de la Rosa, D., Zhang, P., Shao, D., White, F., & Canessa, C. M. (2000). Functional implications of the localization and activity of acid-sensitive channels in rats peripheral nervous system. *Proceedings of the National Academy of Sciences United States of America, 99*, 2326–2331.

Alves-Rodrigues, A., & Shao, A. (2004). The science behind lutein. *Toxicology Letters, 150*, 57–83.

Bernardini, N., Sauer, S. K., Haberberger, R., Fischer, M. J., & Reech, P. W. (2001). Excitatory nicotinic and desensitizing muscarinic (M2) effects on C-nociceptors in isolated rat skin. *Journal of Neuroscience, 21*, 3295–3302.

Borzan, J., Zhao, C., Mayer, R. A., & Raja, S. N. (2010). A role for acid-sensing ion channel 3, but not acid-sensing ion channels 2, in sensing dynamic mechanical stimuli. *Anesthesiology, 113*, 647–654.

Catterall, W. A. (2012). Voltage-gated sodium channels at 60: Structure, function and pathophysiology. *Journal of Physiology, 590*, 2577–2589.

Chidiac, E., Kaddoum, R. N., & Fuleihan, S. F. (2011). Special article: Mandragora: Anesthetic of the ancients. Anesthetic of the antients. *Anethesia & Analgesia, 115*, 1437–1441.

Cummins, T. R., Dib-Hajj, S. D., Black, J. A., Akopian, A. N., Wood, J. N., & Waxman, S. G. (1999). A novel persistent tetrodotoxin-resistant sodium current in SNS-null and wild-type small primary sensory neurons. *The Journal of Neuroscience, 19*, RC43.

Deval, E., Noel, J., Lay, N., Alloui, A., Diochot, S., Friend, V., et al. (2008). ASIC3, sensor of acidic and primary inflammatory pain. *EMBO Journal, 27*, 3047–3055.

Dussor, G. O., Helesic, G., Hargreaves, K. M., & Flores, C. M. (2004). Cholinergic modulation of nociceptive responses in vivo and neuropeptide release in vitro at the level of primary sensory neuron. *Pain, 107*, 22–32.

Ernst, E. (2003). Complementary medicine. *Current Opinion in Rheumatology, 15*, 151–155.

Everill, B., & Kocsis, J. D. (1999). Reduction in potassium currents in identified cutaneous afferent dorsal root ganglion neurons after axotomy. *Journal of Neurophysiology, 82*, 700–708.

Fozzard, H. A., Lee, P. J., & Lpkind, G. M. (2005). Mechanism of local anesthetic action on voltage-gated sodium channels. *Current Pharmaceutical Design, 11*, 2671–2686.

Frémont, L. (2000). Biological effects of resveratrol. *Life Sciences, 66*, 663–673.

Grannados-Soto, V., Argulles, C. F., & Ortiz, M. I. (2002). The peripheral antinociceptive effect of resveratrol is associated with activation of potassium channels. *Neuropharmacology, 43*, 917–923.

Gwynne, R. M., Thomas, E. A., Goh, S. M., & Bornstein, J. C. (2004). Segmentation induced by intraluminal fatty acid in isolated Guinea-pig duodenum and jejenum. *Journal of Physiology, 556*, 557–569.

Harriott, A. M., & Gold, M. S. (2009). Contribution of primary afferent channels to neuropathic pain. *Current Pain and Headache Reports, 13*, 197–207.

Horváth, G., Kemény, Á., Barthó, L., Molnár, P., Deli, J., Szente, L., et al. (2005). Effects of some natural carotenoids on TRPA1- and TRPV1-induced neurogenic inflammatory processes in vivo in the mouse skin. *Journal of Molecular Neuroscience, 56*. 113–21.

Hórvath, G., Szoke, E., Kemeny, A., Bagoly, T., Deli, J., Szente, L., et al. (2012). Lutein inhibits the function of the transient receptor potential A1 ion channel in different in vitro and in vivo models. *Journal of Molecular Neuroscience, 4*, 1–9.

Iwata, K., Takeda, M., Oh, S., & Shinoda, M. (2017). Neurophysiology of orofacial pain. In C. S. Farah, et al. (Eds.), *Contemporary oral medichine* Springer International Publishing.

Kakita, K., Tsubouchi, H., Adachi, M., Takehana, S., Shimazu, Y., & Takeda, M. (2017). Local subcutaneous injection of chrologenic acid inhibits the nociceptive spinal trigeminal nucleus caudalis neurons in rats. *Neuroscience Research, 134*, 49–55.

Kakita, K., Tsubouchi, H., Adachi, M., Takehana, S., Shimazu, Y., & Takeda, M. (2018). Local subcutaneous injection of chrologenic acid inhibits the nociceptive spinal trigeminal nucleus caudalis neurons in rats. *Neuroscience Research, 134*, 49–55.

Kang, S., Jang, J. H., Price, M. P., Gautam, M., Benson, C. J., Geong, H. G., et al. (2012). Simultaneous disruption of mouse ASIC1a, AsSIC2 and ASIC3 genes enhances cutaneous mechanosensitivity. *PLoS One, 7*, e35225.

Kim, D. S., Choi, J. O., Rim, H. D., & Cho, H. J. (2002). Down regulation of voltage gated potassium channels α gene expression in dorsal root ganglion following chronic constriction injury of rat sciatic nerve. *Molecular Brain Research, 105*, 146–152.

Kim, H. C., & Chung, M. K. (1999). Voltage-dependent sodium and calcium currents in acutely isolated adult rat trigeminal root ganglion neurons. *Journal of Neurophysiology, 81*, 1123–1134.

Kim, H. I., Kim, T. H., & Song, J. H. (2005). Resveratrol inhibits Na+ currents in rat dorsal root ganglion neurons. *Brain Research, 1045*, 134–141.

Kim, T. H., Lim, J. M., Kim, S. S., Kim, J., Park, M., & Song, J. H. (2009). Effect of (−) epigallocatechin-3-gallate on Na (+)currents in rat dorsal root ganglion neurons. *European Journal of Pharmacology, 604*, 20–26.

Kokuba, S., Takehna, S, Oshima, K., Shimazu, Y., & Takeda, M. (2017). Systemic administration of the dietary constituents resveratrol inhibits the nociceptive jaw-opening reflex in rats via the endogenous opioid system. *Neuroscience Research, 119*, 1–6.

Konvicka, J. J., Meyer, T. A., McDavid, A. J., & Roberson, C. R. (2008). Complementary/alternative medicine use among chronic pain clinic patients. *Journal of Perianesthesia Nursing, 23*, 17–23.

Kwan, K. Y., Glazer, J. M., Corey, D. P., Rice, F. L., & Stucky, C. L. (2009). TRPA1 modulates mechanotransduction in cutaneous sensory neurons. *The Journal of Neuroscience, 29*, 4808–4819.

Lawson, K. (2006). Potassium channel as targets for the management of pain. *Central Nervous System Agents in Medicinal Chemistry, 6*, 119–128.

Liu, L., Yang, T., & Simon, S. A. (2004). The protein tyrosine kinase inhibitor, genistein, decreases excitability of nociceptive neurons. *Pain, 112*, 131–141.

Liu, F., Lu, X. W., Zhang, Y. J., Kou, L., Song, N., Wu, M. K., et al. (2016). Effect of chlorogenic acid on voltage-gated pottasium channels of trigeminal ganglion neuron in an inflammatory environments. *Brain Research Bulletin, 127*, 119–125.

Matsumoto, S., Yoshida, S., Takahashi, M., Saiki, C., & Takeda, M. (2010). The roles of ID, IA and IK in the electrophysiological functions of small-diameter rat trigeminal ganglion neurons. *Current Molecular Pharmacology, 3*, 30–36.

McKemy, D. D., Nuhausser, W. M., & Julius, D. (2002). Identification of a cold receptor reveals a general role for TRP in thermosensation. *Nature, 416*, 52–58.

Meng, Q., Fang, P., Hu, Z., Ling, Y., & Liu, H. (2015). Mechanotransduction of trigeminal ganglion neurons innervating inner walls of rat anterior eye chamber. *American Journal of Physiology. Cell Physiology, 309*, C1–C10.

Mitome, K., Takehana, S., Oshima, K., Shimazu, Y., & Takeda, M. (2018). Local anesthetic effect of docosahexanoic acid on the nociceptive jaw-opening reflex in rats. *Neuroscience Research, 137*, 30–35.

Motter, A. L., & Ahern, G. P. (2012). TRPA1 is a polyunsaturated fatty acid sensor in mammals. *PLoS One, 7*, e38439.

Nakagawa, K., Takeda, M., Tsuboi, Y., Kondo, M., Kitagawa, J., Matsumoto, S., et al. (2010). Alteration of primary afferent activity following inferior alveolar nerve transection in rats. *Molecular Pain, 6*, 9.

Nakajima, R., Uehara, A., Takehana, S., Akama, Y., Shimazu, Y., & Takeda, M. (2018). Decanoic acid attenuates the excitability of nociceptive trigeminal primary and secondary neuron associated with hypoalgesia. *Journal of Pain Research, 11*, 2867–2876.

Noguchi, Y., Matsuzawa, N., Akama, Y., Sekiguchi, K., Takehana, S., Shimazu, Y., et al. (2017). Dietary constituents, decanoic acid suppresses the excitability of nociceptive trigeminal neuronal activity associated with hypoalgesia via muscarinic M_2 receptor signaling. *Molecular Pain, 13*, 1744806917710779.

Pan, Z. Z., & Williams, J. T. (1994). Muscarine hyperpolarizes a subpopulation of neurons by activating an M2 muscarinic receptor in rat nucleus raphe magnus in vitro. *Journal of Neuroscience, 14*, 1332–1338.

Park, S. Y., Choi, J. Y., Lee, Y. S., Cho, H., & Kim, D. S. (2003). Down regulation of voltage gated potassium channel α gene expression by axotomy and neurotrophins in rat dorsal root ganglia. *Molecular Cell, 16*, 256–259.

Pervaiz, S. (2003). Resveratrol: From grapevines to mammalian biology. *FASEB Journal, 17*, 1975–1985.

Price, M. P., McIIwrath, S. I., Xie, J., Cheng, C., Qiao, J., Tar, T. E., et al. (2001). The DRASIC cation channel contributes to the detection of cutaneous touch and acid stimuli in mice. *Neuron, 32*, 1071–1083.

Qu, Z.-W., Liu, T.-T., QiU, C.-Y., Li, J.-D., & Hu, W.-P. (2014). Inhibition of acid-sensing ion channels by chlorogenic acid in rat dorsal root ganglion neurons. *Neuroscience Letters, 567*, 35–39.

Rao, J. K., Mihaliak, K., Kroenke, K., Bradley, J., Tierney, W. M., & Weinberger, M. (1999). Use of complementary therapies for arthritis among patients of rheumatologists. *Annals of Internal Medicine, 131*, 409–416.

Rosenberg, E. I., Genao, I., Chen, I., Mechaber, A. J., Wood, J. A., Faselis, C. J., et al. (2008). Complementary and alternative medicine use by primary care patients with chronicpain. *Pain Medicine, 9*, 1065–1072.

Sekiguchi, K., Takehana, S., Shibuya, E., Matsuzawa, N., Hidaka, H., Kanai, Y., et al. (2016). Resveratrol attenuates inflammation-induced hyperexcitability of trigeminal spinal nucleus caudalis neurons associated with hyperalgesia in rats. *Molecular Pain, 12*. 1744806916643082.

Shimazu, Y., Shibuya, E., Takehana, S., Sekiguchi, K., Oshima, K., Kamata, H., et al. (2016). Local administration of resveratrol inhibits excitability of nociceptive wide-dynamic range neurons in rat spinal trigeminal nucleus caudalis. *Brain Research Bulletin, 124*, 262–268.

Shir, Y., Raja, S. N., Weissman, C. S., Campbell, J. N., & Seltzer, Z. (2001). Consumption of soy diet before nerve injury preempts the development of neuropathic pain in rats. *Anesthesiology, 95*, 1238–1244.

Syoji, Y., Kobayashi, R., Miyamura, N., Hirohara, T., Kubota, Y., Uotsu, N., et al. (2018). Suppression of hyperexcitability of trigeminal nociceptive neurons associated with inflammatory hyperalgesia following systemic administration lutein via inhibition of cyclooxygenase-2 cascade signaling. *Journal of Inflammation, 15*, 24.

Takeda, M., Tanimoto, T., Ikeda, M., Kadoi, J., & Matsumoto, S. (2004). Activation of GABAB receptor inhibits the excitability of rat small diameter trigeminal root ganglion neurons. *Neuroscience, 123*, 491–505.

Takeda, M., Tanimoto, T., Ikeda, M., Nasu, M., Kadoi, J., Yoshida, S., et al. (2006). Enhanced excitability of rat trigeminal root ganglion neurons via decrease in A-type potassium currents following temporomandibular joint inflammation. *Neuroscience, 138*, 621–630.

Takeda, M., Tsuboi, Y., Kitagawa, J., Nakagawa, K., Iwata, K., & Matsumoto, S. (2011). Potassium channels as a potential therapeutic target for trigeminal neuropathic and inflammatory pain. *Molecular Pain, 7*, 5.

Takeda, M., Matsumoto, S., Sessle, B. J., Shinoda, M., & Iwata, K. (2011). Peripheral and central mechanisms of trigeminal neuropathic and inflammatory pain. *Journal of Oral Biosciences, 53*, 318–329.

Takeda, M., Takehana, S., & Shimazu, Y. (2018). The polyphenolic compound resveratrol attenuates pain: Neurophysiological mechanisms. In R. Watson, V. Preedy, & S. Zibadi (Eds.), *Polyphenol: Mechanisms of action in human health and disease* (2nd ed., pp. 237–247). Academic Press.

Takehana, S., Sekiguchi, K., Inoue, M., Kubota, Y., Ito, Y., Yui, K., et al. (2016). Systemic administration of resveratrol suppress the nociceptive neuronal activity of spinal trigeminal nucleus caudalis in rats. *Brain Research Bulletin, 120*, 117–122.

Takehana, S., Kubota, Y., Uotsu, N., Yui, K., Iwata, K., Shimazu, Y., et al. (2017). The dietary constituent resveratrol suppresses nociceptive transmission via NMDA receptor. *Molecular Pain, 13*. 1744806917697010.

Tracey, W. D., Jr., Wilson, R. I., Laurent, G., & Benzer, S. (2003). Painless, Drosophila gene essential for nociception. *Cell, 113*, 261–273.

Tsuchiya, H. (2017). Anesthetic agents of plant origin: A review of phytochemicals with anesthetic activity. *Molecules, 22*, 1369.

Vogt, B. A. (2005). Pain and emotion interactions in subregions of the cingulate gyrus. *Nature Review Neuroscience, 6*, 533–544.

Walker, R. G., Willingham, A. T., & Zuker, C. S. (2000). A Drosophila mechanosensory transduction channel. *Science, 287*, 2229–22234.

Wallace, C. H., Baczko, I., Jones, L., Fercho, M., & Light, P. E. (2006). Inhibition of cardiac voltage-gated channels by grape polyphenol. *British Journal of Pharmacology*, *149*, 657–665.

Wess, J., Duttaroy, A., Gomeza, J., Zhang, W., Yamada, M., Felder, C. C., et al. (2003). Muscarinic receptor subtypes mediating central and peripheral antinociception studied with muscarinic receptor knockout mice: A review. *Life Sciences*, *72*, 2047–2054.

Xiao, Y.-F., Kang, J. X., Morgan, J. P., & Leaf, A. (1995). Blocking effects of polyunsaturated fatty acids on Na^+ channels of neonatal rat ventricular myocytes. *Proceedings of the National Academy of Sciences of the United States of America*, *92*, 11000–11004.

Yu, L., Wang, S., Kogure, Y., Yamamoto, S., Noguchi, K., & Dai, Y. (2013). Modulation of TRP channels by resveratrol and other stilbenoids. *Molecular Pain*, *9*, 3.

Zhang, Y.-J., Lu, X.-W., Song, N., Kou, L., Wu, M.-K., Liu, F., et al. (2014). Chlorogenic acid alters the voltage-gated potassium currents of trigeminal ganglion neurons. *International Journal of Oral Science*, *6*, 233–240.

Chapter 41

Pain response following prenatal stress and its modulation by antioxidants

Che Badariah Abd Aziz[a], Asma Hayati Ahmad[a], and Hidani Hasim[b]

[a]Department of Physiology, School of Medical Sciences, Universiti Sains Malaysia, Kubang Kerian, Kelantan, Malaysia, [b]Department of Chemical Pathology, School of Medical Sciences, Universiti Sains Malaysia, Kubang Kerian, Kelantan, Malaysia

Abbreviations

5-HT	serotonin
ACTH	adrenocorticotropic hormone
BDNF	brain-derived neurotrophic factor
CAT	catalase
CRH	corticotropin-releasing hormone
GABA	gamma-aminobutyric acid
GSH	glutathione
GPx	glutathione peroxidase
HIV	human immunodeficiency virus
HPA axis	hypothalamic-pituitary-adrenal axis
MDA	malondialdehyde
NMDA	N-methyl-D-aspartate
ROS	reactive oxygen species
SOD	superoxide dismutase

Introduction

Prenatal stress refers to the exposure of an expectant mother to stressors which include medical illness, social or emotional problems, or other types of stressful events (Markham & Koenig, 2011). Prenatal stress is linked to poor birth outcomes such as abortion, reduced birth weight, and premature delivery (Mulder et al., 2002). In addition, prenatal stress may result in abnormalities of behavior, cognition, emotion, and psychosocial characteristics of the offspring. The abnormalities may be attributed to altered brain development during prenatal stress.

Studies show an increased incidence of depression and schizophrenia in the children of prenatally stressed mothers (Dorrington et al., 2014; Weinstock, 2008). An increasing number of human and animal studies have reported that pain responses of the offspring were modulated after exposure to prenatal stress (Abd Aziz, Ahmad, Mohamed, & Wan Yusof, 2013; Butkevich, Barr, Mikhailenko, & Otellin, 2006; Davis, Glynn, Waffarn, & Sandman, 2011).

To determine the associated mechanisms responsible for the alteration in brain function following prenatal stress, studies have utilized various methods to induce stress in animals, including repeated restraint, hypoxia, electric shocks, deprivation of food, and exposure to noise, light, or extreme temperature. Exposure to prenatal stress will activate the hypothalamic-pituitary-adrenal (HPA) axis and stimulate the release of adrenocorticotropic hormone (ACTH) and glucocorticoid. A report has shown that restraint stress applied three times daily from day 11 of pregnancy until delivery was associated with a significantly increased level of fecal corticosterone on day 20 of pregnancy compared to day 10 (before restraint stress), (95% CI, 14.034 to 325.775), t (17) = 2.300, $p < 0.034$ (Fig. 1; Ahmad Suhaimi, 2019). The differences were not significant in the control and treated groups. 11β-hydroxysteroid dehydrogenase (11β-HSD), available in the placenta, has a role in converting glucocorticoid; cortisol (in human) and corticosterone (in rats), into inactive metabolites such as cortisone and 11-dehydrocorticosterone (Yang, 1997). The enzyme protects the fetus from the transfer of a large amount of corticosterone through the placenta. However, when glucocorticoid is produced in excess, the inactivation process is not adequate and more cortisol/corticosterone enter the fetal circulation.

Treatments, Mechanisms, and Adverse Reactions of Anesthetics and Analgesics. https://doi.org/10.1016/B978-0-12-820237-1.00041-7
Copyright © 2022 Elsevier Inc. All rights reserved.

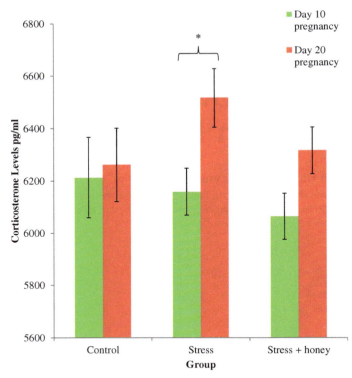

FIG. 1 Corticosterone level in the feces of maternal rats ($n = 6$ rats in each group). Data are presented as mean ± SEM. *$P < 0.05$ Statistical comparison of corticosterone level on day 10 and day 20 in the stress group. Restraint stress was applied three times daily from day 11 until delivery. *(From Ahmad Suhaimi, S. Q. (2019). Spinal cord mechanism in the modulation of nociceptive response in the male rats offspring exposed to prenatal stress. MSc Thesis. Universiti Sains Malaysia, Kubang Kerian, with permission.)*

During prenatal stress, glucocorticoid may directly stimulate the placenta to release corticotropin-releasing hormone (CRH) which results in activation of the fetal HPA axis (Challis, Matthews, Gibb, & Lye, 2000). In addition, there is stimulation of the sympathetic system and adrenal medulla that leads to increased catecholamine levels. One of the effects of catecholamines is vasoconstriction which reduces placental blood flow and decreases nutrient transfer to the fetus. The stress results in increased release of fetal glucocorticoid, which in turn would increase activation of excitatory amino acid receptors and increase the level of calcium in the cells that ultimately contribute to the production of reactive oxygen species (Zhu et al., 2004).

Oxidative stress

Reactive oxygen species (ROS) produce adverse modifications of the cell which results in oxidative damage of the structures in various organs (Birben, Sahiner, Sackesen, Erzurum, & Kalayci, 2012). Studies utilizing various types of stressors to induce prenatal stress have demonstrated the presence of oxidative stress in the brain, spinal cord, and liver of offspring (Table 1). The oxidative stress may result in neuronal loss in the affected brain and spinal cord (Abd Aziz et al., 2019; Charil, Laplante, Vaillancourt, & King, 2010).

Malondialdehyde (MDA), a product of lipid peroxidation, was found to be increased while antioxidants were lower in the offspring exposed to prenatal stress (Table 1). Although the reduced antioxidant level is one of the parameters to suggest the presence of oxidative stress, a higher level of antioxidants following prenatal stress has been reported, suggesting that a compensatory mechanism has occurred to combat the oxidative stress (Paidi et al., 2014; Vega et al., 2016).

Oxidative stress may contribute to neuronal damage by altering signaling molecules and DNA damage. A study by Chen, Zhang, and Wang (2019) showed that prenatal stress induced DNA changes in the fetal organs, with lasting effect in the adult offspring. Another study reported that prenatal stress resulted in reduced mRNA and protein expressions of factors that are involved in the neurogenesis during development such as brain-derived neurotrophic factor (BDNF), β-catenin, and Notch (Fatima, Srivastav, Ahmad, & Mondal, 2019). The altered neurogenesis may contribute to alteration in mood, behavior, and pain responses in the offspring.

TABLE 1 Experimental studies using various types of stress and the parameters of oxidative stress investigated.

Authors	Type of prenatal stress	Structures investigated in the offspring/types of animals	Parameters of oxidative stress
Ajarem, Al-Basher, Allam, and Mahmoud (2017)	Nicotine exposure	Cerebellum and medulla oblongata in mice	GSH lower SOD lower MDA higher
Paidi, Schjoldager, Lykkesfeldt, and Tveden-Nyborg (2014)	Vitamin C deficiency	Whole brain in guinea pig	SOD higher[a] MDA higher
Soleimani, Goudarzi, Abrari, and Lashkarbolouki (2016)	Lead and ethanol exposure	Hippocampus in rats	GPX lower SOD lower CAT lower MDA higher
Abd Aziz et al. (2019)	Repeated physical restraint	Spinal cord in rats	GSH lower SOD lower CAT lower MDA higher
Al-Amin, Sultana, Sultana, Rahman, and Reza (2016)	Lipopolysaccharide exposure	Whole brain and liver in rats	GSH lower SOD lower CAT lower MDA higher
Vega et al. (2016)	Low protein diet	Fetal liver in rats	MDA higher SOD higher[a] GPx higher[a] ROS higher

[a]Higher antioxidant levels may suggest a compensatory mechanism has occurred to combat the oxidative stress in fetus. The lower level of antioxidants may suggest "defense exhaust", when the cell cannot produce more antioxidants because of severe or prolonged stress. GSH, glutathione; GPx, glutathione peroxidase; SOD, superoxide dismutase.
Almost all the parameters were compared with the control group which did not receive any stress. Higher malondialdehyde (MDA) level was shown in all studies.

Following exposure to prenatal stress, there is altered neurogenesis in the central nervous system which may include the neuronal network involved in pain processing. The altered neuronal network may modulate the pain responses in the offspring. Clinical reports have demonstrated that prenatal stress may lead to increased pain response in newborns (Davis et al., 2011), more somatic complaints in toddlers (Wolff et al., 2010), and chronic pain in adult offspring (Grégoire, Jang, Szyf, & Stone, 2020) (Fig. 2).

The increase in pain responses is accompanied by alteration of both structure and function of the nervous system. The altered function was shown by an imaging study done by Scheinost et al. (2016) that demonstrated reduced functional connectivity between structures in the brain such as the amygdala and thalamus. In addition, reduced volume of the structures such as the prefrontal cortex and medial temporal lobe in the central nervous system of neonates following prenatal stress has also been reported (Buss, Davis, Muftuler, Head, & Sandman, 2010; Scheinost et al., 2016). In the brain of adult offspring, other changes seen include reduced dendritic branching with a reduced number of synapses (Barros, Duhalde-Vega, Caltana, Brusco, & Antonelli, 2006). The increased pain responses in the offspring may be contributed by alteration of GABAergic and serotonergic neurons (Butkevich, Mikhailenko, Vershinina, Otellin, & Aloisi, 2011; Lussier & Stevens, 2016), increased expression of fos neurons (Butkevich et al., 2006), and reduction in the number of normal neurons (Abd Aziz et al., 2019).

Human and animal studies have shown that oxidative stress may lead to the alteration of pain responses. A linear correlation has been found between oxidative stress biomarkers and pain responses in full term and premature babies following painful procedures such as heel prick (Perrone et al., 2017) while a significant decrease in opioid antinociception was directly correlated with oxidative damage in the cortex, striatum, and midbrain of the aging brain (Raut & Ratka, 2009). If oxidative stress contributes to the changes in pain response of the offspring following exposure to prenatal stress, administration of antioxidant would prevent or inhibit the mechanism involved in the modulation of responses.

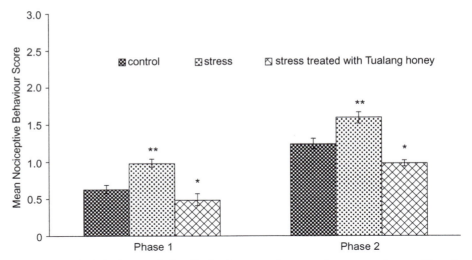

FIG. 2 Comparison of mean nociceptive behavior score in the offspring from control, stress, and stress-treated groups. The columns represent the mean nociceptive behavior score in each phase: phase 1 (0–10 min) and phase 2 (15–60 min). **$P < 0.001$ a significant difference between stress and other groups, *$P < 0.05$ shows a significant difference between stress group and stress treated with Tualang honey. *(From Hasim, H. (2018). The effects of Tualang honey on nociceptive responses in the thalamus of prenatally stressed male rats offspring. Thesis. Universiti Sains Malaysia, Kubang Kerian, with permission.)*

Antioxidant

Oxidation is a chemical reaction that generates free radicals which are harmful to various cells. Antioxidants have the capability to inhibit the reaction and thus protect the cells from damage. There are two types of antioxidants: enzymatic and non-enzymatic antioxidants. The former works by degrading and removing free radicals, while non-enzymatic antioxidants interrupt free radical chain reactions. Superoxide dismutase (SOD), glutathione peroxidase (GSHPx), and catalase (CAT) are examples of enzymatic antioxidants, while vitamin C, vitamin A, vitamin E, and phenolic compounds are included under the other category.

Phenolic compounds are further subdivided into subgroups based on their chemical structures namely phenolic acids, tannins, coumarins, flavonoids, and others. Considerable attention has been focused on flavonoids because of their potential benefits to human health (Maleki, Crespo, & Cabanillas, 2019). Apart from antioxidant effects, a flavonoid also exhibits antinociceptive property. Pandurangan, Krishnappan, Subramanian, and Subramanyan (2014) revealed that the flavonoid quercetin reduced pain responses in animal models of pain including formalin test, acetic acid-induced writhing test, and tail-flick test. The antinociceptive effects are mediated by several mechanisms including action on opioid receptors and involvement of serotonergic and GABAergic systems. Accordingly, reduction of the antinociceptive effects was demonstrated by administration of naloxone (opioid receptor antagonist), ondansetron (serotonin receptor antagonist), or bicuculine (GABA receptor antagonist) (Pandurangan et al., 2014). In addition, flavonoid compounds have the ability to suppress the release of kinins and other inflammatory mediators (Maleki et al., 2019) which are involved in pain response following formalin-induced inflammation (Table 2).

There are a few subclasses of flavonoids including flavones, flavanols, flavanones, isoflavones, flavonols, and anthocyanidins (Kumar & Pandey, 2013). Flavonoids can be found in onions, apples, grapes, soy products, vegetables, honey, and others. Generally, the flavonoid intake in the diet may be influenced by age, physical activity, type of diet, and supplements taken by the individual (Vanhees et al., 2013).

Honey

Honey is an example of a supplement that is rich in flavonoids. Honey is a natural product that has a viscous liquid consistency. Its color ranges from golden to amber. It is produced by honeybees and other insects (Crane, 1991). Different types of honey are available worldwide and they are consumed for various health problems. In Malaysia, the types of honey include Gelam, Nenas, Kelulut, and Tualang. The antinociceptive and antioxidant properties of honey have been demonstrated by a number of studies (Abd Aziz, Ismail, Hussin, & Mohamed, 2014; Abd Aziz, Ismail, Iberahim, Mohamed, & Kamaruljan, 2014; Kishore, Halim, Syazana, & Sirajudeen, 2011). Among the many types of Malaysian honey, Tualang

TABLE 2 List of benefits of flavonoids.

	Benefits of flavonoids
Antiinflammatory	• Inhibition of protein kinase • Inhibition of NF-Kb • Inhibition of phosphodiesterases • Inhibits release of kinin and other inflammatory mediators
Antimicrobial	• Improves wound healing • Combats infection • Maintain CD4 cell count
Antinociceptive	• Possible actions on opioid, serotonergic, and GABAergic receptors • Through antioxidant and antiinflammatory properties
Antioxidant	• Reduces oxidant level • Improves antioxidant level/activity • Improves oxidative stress
Cardioprotective effects	• Reduces systemic blood pressure • Decreases size of atherosclerosis and • Inhibits activation of platelet and formation of thrombosis
Neuroprotective effects	• Decreases neuronal death • Improves number of normal neurons • Through antioxidant and antiinflammatory properties

The list summarizes the benefits of flavonoids that have been discussed in the review.

honey has been shown to possess greater antioxidant activity as well as higher free radical scavenging activities (Khalil, Alam, Moniruzzaman, Sulaiman, & Gan, 2011; Kishore et al., 2011). Mohamed, Sirajudeen, Swamy, Yaacob, and Sulaiman (2010) reported that the total phenolic content in Tualang honey is 251.7 ± 7.9 mg gallic acid/kg honey. The result is expressed as mg gallic acid/kg honey because gallic acid was used in the calibration.

Antinociceptive effects of Tualang honey in the rat offspring following prenatal stress have been shown by Abd Aziz et al. (2013). The study investigated whether the administration of Tualang honey to the pregnant dams had a significant effect on the pain behavior of the offspring. It was demonstrated that the offspring from Tualang honey-treated group had a lower nociceptive behavior score in the formalin test compared to the untreated group. The results suggest that facilitatory modulation of the pain pathway that occurs in the offspring following exposure to prenatal stress is reduced by Tualang honey.

Tualang honey administration to the pregnant rats may allow certain substances in the honey to be delivered to the fetus via the placenta. Schroder-van der Elst, van der Heide, Rokos, Morreale de Escobar, and Kohrle (1998) reported that flavonoid was discovered in fetal tissues including the central nervous system after infusion of flavonoids in pregnant dams. The flavonoids which are the antioxidants probably contribute to reduced oxidative stress and nociceptive behavior score in the adult rat offspring following exposure to prenatal stress (Abd Aziz et al., 2019). This finding was well substantiated by Vanhees et al. (2013) who showed that flavonoid administration is also associated with increased expression of antioxidant genes that confer protection to the DNA from damage induced by oxidative stress in the adult offspring. Clinical studies have also supported a close association of oxidative stress with complicated pregnancies such as preeclampsia, diabetes, and hypoxia. Oxidative stress may lead to neuronal damage, which gives rise to decreased number of normal neurons in the brain of the rat offspring. Tualang honey given during prenatal stress provides antioxidants to the fetus that will protect the neural structures and possibly reverse the detrimental effects in the central nervous system (Fig. 3).

Protective effects of other antioxidants in the prenatally stressed offspring

Resveratrol

Resveratrol is a natural product that is present in grapes, peanuts, and red wines. Its antioxidant property is contributed by its phenolic content. A few experimental studies have reported regarding the beneficial effects of resveratrol. Madhyastha, Sahu, and Rao (2014) reported the presence of oxidative stress in the brain of the rat offspring following early or late gestational stress. The oxidative stress was decreased by the administration of resveratrol. Improved neurogenesis and

492 PART | IV Novel and nonpharmacological aspects and treatments

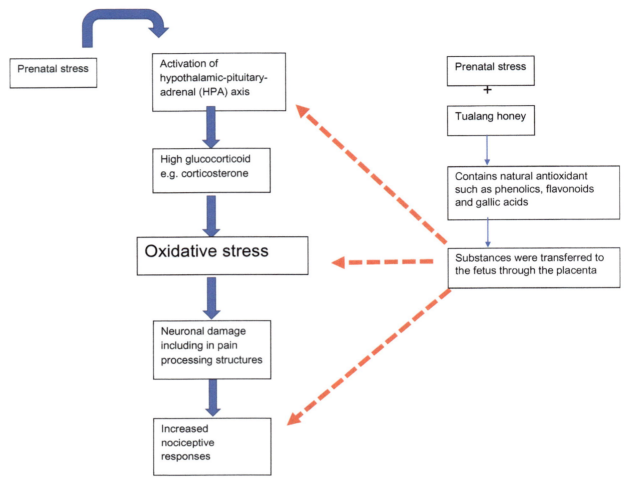

FIG. 3 A simplified diagram showing possible effects of Tualang honey administration during prenatal stress in pregnant dams that prevent alteration of nociceptive responses in adult offspring. ⬇ Stimulate;, Stimulate; ◀▬▬▬ Inhibit, Inhibit. *(From Hasim, H. (2018). The effects of Tualang honey on nociceptive responses in the thalamus of prenatally stressed male rats offspring. Thesis. Universiti Sains Malaysia, Kubang Kerian, with permission.)*

expression of BDNF in the brain of offspring were also seen following its administration (Madhyastha, Sahu, & Rao, 2013). The use of this antioxidant was also associated with improvement in learning and memory abilities of the rat offspring following prenatal stress (Sahu, Madhyastha, & Rao, 2013). These data suggest the potential protective effects of resveratrol against prenatal stress-induced impaired neurogenesis which might also be beneficial to modulate pain response in the offspring.

Spirulina

Spirulina is a filamentous microalga that is rich in vitamins and phenolic acids. In prenatal stress secondary to lead poisoning, supplementation of pregnant and lactating rats with spirulina (*A. platensis*), protected the offspring from oxidative stress, renal damages, and alteration of hematological parameters (Gargouri, Soussi, Akrouti, Magné, & El Feki, 2018).

Vitamins

Another study also showed that vitamin E administration in pregnant dams exposed to ethanol-induced oxidative stress did not only reduce DNA damage and brain atrophy in the offspring but its administration was also associated with inhibition of histopathological and hormonal alterations in the testis (Shirpoor et al., 2009).

Apart from vitamin E, vitamin C has been reported to improve the health status of offspring. A clinical study conducted in infants born to smokers revealed impaired respiratory function as shown by alteration of respiratory flows, respiratory compliance, and tidal volumes in the infants, resulting in increased childhood asthma and hospital admission for respiratory infections (McEvoy et al., 2014). Supplementation of vitamin C to pregnant smokers was associated with improvement in the health condition of the offspring. In comparison with the vitamins, honey has various antioxidants including flavonoids, phenolic acids, and also vitamins (vitamin C and E) which may work synergistically to give excellent antioxidant effects (Vallianou, Gounari, Àlex, Panagos, & Kazazis, 2014) compared to the individual vitamin which may work alone. All the studies discussed have demonstrated the role of antioxidants in preventing the pathological changes in fetal development following prenatal stress.

Applications to other areas

Apart from the aforementioned antinociceptive effects and protective effects against structural and functional alteration of various systems due to prenatal stress, several studies have also documented other therapeutic benefits of antioxidants.

Various studies have investigated the use of antioxidants in the treatment or prevention of various diseases including cardiovascular disease (CVD), diabetes, and acting as antimicrobial. Oxidative stress and inflammation are reported to contribute to the development of the diseases. Antioxidants such as flavonoids have anti-inflammatory properties and have the potential to inhibit the development as well as retard the disease process. The anti-inflammatory properties of flavonoids are contributed by multiple factors including their ability to inhibit protein kinase (Hou & Kumamoto, 2010) and transcription factors, such as NF-κB (Peng, Huang, Cheng, & Liou, 2018). In addition, they may inhibit phosphodiesterases that regulate various cell functions during inflammation (Guo et al., 2018).

Role on cardiovascular disease

Various cardioprotective effects of flavonoids have been shown in animal and human studies. The beneficial effects seen associated with flavonoid consumption include reducing systemic blood pressure, decreasing the size of atherosclerosis, and inhibiting activation of platelet and formation of thrombosis (Hayek et al., 1997; Pearson et al., 2002; Ried, Sullivan, Fakler, Frank, & Stocks, 2010). Management of risk factors of CVD might assist in the treatment or prevention of the development of CVD.

Role in diabetes mellitus

Another risk factor for CVD is diabetes mellitus, a disease that is characterized by hyperglycemia with other metabolic abnormalities. The disease is associated with the oxidation of glucose with generation of free radicals. The abundant free radicals may act on the tissues in various structures including the nervous system that may lead to polyneuropathy (Obrosova et al., 2007). Supplementation with antioxidants may be beneficial in preventing the generation of oxidants, removing free radicals if they are formed, and improving the antioxidant activity (Bajaj & Khan, 2012).

Role as antimicrobial agent

Flavonoids have important anti-inflammatory properties and might be beneficial in combating infection and improving immune status. Honey that has several antioxidants including flavonoids has antibacterial action including controlling the growth of *P. aeruginosa*, as demonstrated by Khoo, Halim, Singh, and Mohamad (2010). The healing effects are seen both in animals, such as in the treatment of alkali injury of rabbit's eye (Bashkaran et al., 2011), and in humans, such as in post-tonsillectomy patients (Mat Lazim, Abdullah, & Salim, 2013). In asymptomatic HIV-infected inmates, administration of 40 and 60 g daily of Tualang honey for 6 months resulted in improvement in viral load while maintaining CD4 count (Wan Yusuf et al., 2018). Postulated mechanisms of Tualang honey in giving rise to the improvement are detailed in Table 3.

Other agents of interest

Apart from antioxidants, there are drugs that can be given depending on the type of stress. If the patient is suffering from medical illness such as depression or anxiety, appropriate treatment can be given accordingly. However, their effects on oxidative stress and pain responses in the offspring may need to be evaluated further. Furthermore, medical practitioners must also evaluate the possible teratogenic effects on the developing fetus. The drugs include:

494 PART | IV Novel and nonpharmacological aspects and treatments

TABLE 3 Postulated mechanisms of phenolic compounds in Tualang honey that prevent an increase in viral load and a decrease in white cell count in asymptomatic HIV inmates.

Asymptomatic HIV patients	Components in Tualang honey	Postulated mechanisms	Tualang honey + asymptomatic HIV patients
Increasing trend of viral load	Caffeic acid Quercetin Kaempferol	Inhibits HIV integrase	Suppression of HIV proliferation
	Chrysin Apigenin	Inhibits HIV replication	
Decreasing trend of (i) CD4 (ii) CD8 (iii) T cells (iv) Neutrophils (v) Lymphocytes	Flavonoids	Mitogenic effect Inhibit pyroptosis Antiinflammatory, (inhibit Caspase-1) and antioxidant effects	Maintaining the white cell counts

From Wan, Y. (2019). Effects of Tualang honey on clinical parameters and quality of life in asymptomatic Human Immunodeficiency Virus (HIV) infected inmates. PhD Thesis. Universiti Sains Malaysia, Kubang Kerian, with permission.

Buspirone
Fluoxetine
Aminooxyacetic acid (AOAA)

Buspirone (5-HT1A receptor agonist) and fluoxetine (antidepressant) are among the drugs that have been investigated in experimental studies. Buspirone is an anxiolytic agent that is used as a second-line drug for the treatment of generalized anxiety disorder. Serotonergic system may be modulated during prenatal stress hence contributing to alteration of pain response and mood disturbance in the offspring. The use of buspirone was found to be helpful to reduce pain behavior due to formalin-induced inflammation and depressive behavior in male rat offspring (Butkevich et al., 2011).

In the clinical setting, sometimes buspirone is added to fluoxetine (antidepressant) to increase efficacy in treating depression. Butkevich, Mikhailenko, Vershinina, and Barr (2019) have shown that administration of buspirone, fluoxetine, or a combination of both drugs to the pregnant dams with prenatal stress was associated with inhibition of pain responses to noxious thermal stimulus and formalin-induced inflammation in the rat offspring. However, the combination is rarely used in pregnant mothers as it may give rise to serious side effects such as hallucination, seizure, incoordination, and gastrointestinal symptoms.

Apart from enhanced peripherally induced pain, prenatal stress may also be associated with chronic abdominal pain in the offspring (Wang et al., 2018). Wang et al. (2018) showed that there was an increased pain response due to colorectal distension in the adult offspring following prenatal stress suggesting the presence of visceral hypersensitivity. The heightened response was inhibited by the administration of aminooxyacetic acid (AOAA), an inhibitor for cystathionine-β-synthase (CBS), to the offspring. CBS is one of the endogenous enzymes that have an important role in the development of visceral hyperalgesia (Xu et al., 2009). However, administration to pregnant dams has not been tested.

Mini-dictionary of terms

Reactive oxygen species—unstable molecule that has oxygen and reacts with other molecules.
Oxidative stress—an abnormal state due to the presence of free radicals or imbalance between the level of antioxidants and oxidants.
Brain-derived neurotrophic factor—a protein that is a member of the neurotrophin family of growth factors.
Antinociceptive—the action or process that inhibits painful sensation.
Antioxidant—a substance that prevents oxidation of other molecules.
Acetic acid-induced nociception—a nociceptive experiment assessing the pain behavior due to acetic acid injection.
Hot plate test—a nociceptive experiment assessing the pain behavior due to contact with a hot surface in rodents.
Formalin test—a nociceptive experiment assessing the pain behavior of rodents after injection of formalin in the hind paw.

Tail flick test—a nociceptive experiment assessing tail-flick latency in response to a noxious thermal stimulus which is usually applied on the tail.

Key facts of prenatal stress

- Prenatal stress refers to the exposure of an expectant mother to stressors which include medical illness, social or emotional problems, or due to other stressful events
- Prenatal stress is linked to poor birth outcomes such as abortion, reduced birth weight, and premature delivery, as well as behavioral, cognitive, and emotional changes in the offspring
- Pain responses in the human and rat offspring are increased following exposure to prenatal stress
- The increased pain response is due to changes in the structure and function of the nervous system
- The structural damage in the central nervous system of prenatally-stressed offspring is partly due to oxidative stress

Key facts of antioxidants

- Enzymatic antioxidant degrades and removes free radicals, while non-enzymatic antioxidant interrupts free radical chain reactions.
- Flavonoids are antioxidants that have anti-inflammatory and antinociceptive properties.
- The flavonoids which are present in honey include catechin, luteolin, kaempferol, apigenin, and naringenin.
- The phenolic acids, vitamins, and flavonoid contents in honey contribute to its antioxidant property.
- The antioxidant property contributes to pain modulatory effects in the offspring following exposure to prenatally stress.

Summary points

- Prenatal stress is associated with the release of stress hormones including catecholamines and glucocorticoids.
- Glucocorticoids alter the metabolism in the fetus and may lead to oxidative stress.
- Oxidative stress in the central nervous system of prenatally stressed offspring may result in cell damage, gene alteration, and increased nociceptive responses.
- Antioxidant administration to pregnant dams during prenatal stress may lead to reduced pain response in the offspring in the formalin test.
- Antioxidants present such as flavonoids can be transferred through the placenta and have the ability to improve oxidative stress in the fetus as well as prevent neuronal damage, thus protecting from altered pain responses.

References

Abd Aziz, C. B., Ahmad, R., Mohamed, M., & Wan Yusof, W. N. (2013). The effects of Tualang honey intake during prenatal stress on pain responses in the rat offsprings. *European Journal of Integrative Medicine, 5*(4), 326–331.

Abd Aziz, C. B., Ahmad Suhaimi, S. Q., Hasim, H., Ahmad, A. H., Long, I., & Zakaria, R. (2019). Effects of Tualang honey in modulating nociceptive responses at the spinal cord in offspring of prenatally stressed rats. *Journal of Integrative Medicine, 17*(1), 66–70.

Abd Aziz, C. B., Ismail, C. A. N., Hussin, C. M. C., & Mohamed, M. (2014). The antinociceptive effects of Tualang honey in male Sprague-Dawley rats: a preliminary study. *Journal of Traditional and Complementary Medicine, 4*(4), 298–302.

Abd Aziz, C. B., Ismail, C. A. N., Iberahim, M. I., Mohamed, M., & Kamaruljan, S. (2014). Effects of different doses of Tualang honey on pain behavior in rats with formalin-induced inflammation. *Journal of Physiological and Biomedical Sciences, 27*(2), 39–43.

Ahmad Suhaimi, S. Q. (2019). *Spinal cord mechanism in the modulation of nociceptive response in the male rats offspring exposed to prenatal stress.* MSc Thesis: Universiti Sains Malaysia, Kubang Kerian.

Ajarem, J. S., Al-Basher, G., Allam, A. A., & Mahmoud, A. M. (2017). Camellia sinensis prevents perinatal nicotine-induced neurobehavioral alterations, tissue injury, and oxidative stress in male and female mice newborns. *Oxidative Medicine and Cellular Longevity, 5985219.* https://doi.org/10.1155/2017/5985219.

Al-Amin, M. M., Sultana, R., Sultana, S., Rahman, M. M., & Reza, H. M. (2016). Astaxanthin ameliorates prenatal LPS-exposed behavioral deficits and oxidative stress in adult offspring. *BMC Neuroscience, 17*(1), 11.

Bajaj, S., & Khan, A. (2012). Antioxidants and diabetes. *Indian Journal of Endocrinology and Metabolism, 16*(Suppl 2), S267–S271. https://doi.org/10.4103/2230-8210.104057.

Barros, V. G., Duhalde-Vega, M., Caltana, L., Brusco, A., & Antonelli, M. C. (2006). Astrocyte-neuron vulnerability to prenatal stress in the adult rat brain. *Journal of Neuroscience Research, 83*(5), 787–800.

Bashkaran, K., Zunaina, E., Bakiah, S., Sulaiman, S. A., Sirajudeen, K., & Naik, V. (2011). Anti-inflammatory and antioxidant effects of Tualang honey in alkali injury on the eyes of rabbits: Experimental animal study. *BMC Complementary and Alternative Medicine, 11*(1), 90. https://doi.org/10.1186/1472-6882-11-90.

Birben, E., Sahiner, U. M., Sackesen, C., Erzurum, S., & Kalayci, O. (2012). Oxidative stress and antioxidant defense. *World Allergy Organization Journal, 5*(1), 9–19.

Buss, C., Davis, E. P., Muftuler, L. T., Head, K., & Sandman, C. A. (2010). High pregnancy anxiety during mid-gestation is associated with decreased gray matter density in 6-9-year-old children. *Psychoneuroendocrinology, 35*(1), 141–153.

Butkevich, I. P., Barr, G. A., Mikhailenko, V. A., & Otellin, V. A. (2006). Increased formalin-induced pain and expression of fos neurons in the lumbar spinal cord of prenatally stressed infant rats. *Neuroscience Letters, 403*(3), 222–226.

Butkevich, I. P., Mikhailenko, V. A., Vershinina, E. A., & Barr, G. A. (2019). Differences between the prenatal effects of fluoxetine or Buspirone alone or in combination on pain and affective behaviors in prenatally stressed male and female rats. *Frontiers in Behavioral Neuroscience, 11*(13), 125. https://doi.org/10.3389/fnbeh.2019.00125.

Butkevich, I. P., Mikhailenko, V. A., Vershinina, E. A., Otellin, V. A., & Aloisi, A. M. (2011). Buspirone before prenatal stress protects against adverse effects of stress on emotional and inflammatory pain-related behaviors in infant rats: Age and sex differences. *Brain Research, 1419*, 76–84.

Challis, J. R., Matthews, S. G., Gibb, W., & Lye, S. J. (2000). Endocrine and paracrine regulation of birth at term and preterm. *Endocrine Reviews, 21*(5), 514–550.

Charil, A., Laplante, D. P., Vaillancourt, C., & King, S. (2010). Prenatal stress and brain development. *Brain Research Reviews, 65*(1), 56–79. https://doi.org/10.1016/j.brainresrev.2010.06.002.

Chen, X., Zhang, L., & Wang, C. (2019). Prenatal hypoxia-induced epigenomic and transcriptomic reprogramming in rat fetal and adult offspring hearts. *Scientific Data, 6*(1), 238. https://doi.org/10.1038/s41597-019-0253-9.

Crane, E. (1991). Honey from honeybees and other insects. *Ethology Ecology and Evolution, 3*(sup1), 100–105.

Davis, E. P., Glynn, L. M., Waffarn, F., & Sandman, C. A. (2011). Prenatal maternal stress programs infant stress regulation. *Journal of Child Psychology and Psychiatry, 52*(2), 119–129.

Dorrington, S., Zammit, S., Asher, L., Evans, J., Heron, J., & Lewis, G. (2014). Perinatal maternal life events and psychotic experiences in children at twelve years in a birth cohort study. *Schizophrenia Research, 152*(1), 158–163. https://doi.org/10.1016/j.schres.2013.11.006.

Fatima, M., Srivastav, S., Ahmad, M. H., & Mondal, A. C. (2019). Effects of chronic unpredictable mild stress induced prenatal stress on neurodevelopment of neonates: Role of GSK-3β. *Scientific Reports, 9*(1), 1305. https://doi.org/10.1038/s41598-018-38085-2.

Gargouri, M., Soussi, A., Akrouti, A., Magné, C., & El Feki, A. (2018). Ameliorative effects of Spirulina platensis against lead-induced nephrotoxicity in newborn rats: Modulation of oxidative stress and histopathological changes. *Experimental Clinical Sciences Journal, 17*, 215–232. https://doi.org/10.17179/excli2017-1016.

Grégoire, S., Jang, S. H., Szyf, M., & Stone, L. S. (2020). Prenatal maternal stress is associated with increased sensitivity to neuropathic pain and sex-specific changes in supraspinal mRNA expression of epigenetic- and stress-related genes in adulthood. *Behavioural Brain Research, 380*. https://doi.org/10.1016/j.bbr.2019.112396.

Guo, Y. Q., Tang, G. H., Lou, L. L., Li, W., Zhang, B., Liu, B., et al. (2018). Prenylated flavonoids as potent phosphodiesterase-4 inhibitors from Morus alba: Isolation, modification, and structure-activity relationship study. *European Journal of Medicinal Chemistry, 144*, 758–766.

Hayek, T., Fuhrman, B., Vaya, J., Rosenblat, M., Belinky, P., Coleman, R., et al. (1997). Reduced progression of atherosclerosis in apolipoprotein E-deficient mice following consumption of red wine, or its polyphenols quercetin or catechin, is associated with reduced susceptibility of LDL to oxidation and aggregation. *Arteriosclerosis, Thrombosis, and Vascular Biology, 11*, 2744–2752.

Hou, D. X., & Kumamoto, T. (2010). Flavonoids as protein kinase inhibitors for cancer chemoprevention: Direct binding and molecular modeling. *Antioxidants & Redox Signaling, 13*, 691–719.

Khalil, M., Alam, N., Moniruzzaman, M., Sulaiman, S., & Gan, S. (2011). Phenolic acid composition and antioxidant properties of Malaysian honeys. *Journal of Food Science, 76*(6), C921–C928.

Khoo, Y. T., Halim, A. S., Singh, K. K. B., & Mohamad, N. A. (2010). Wound contraction effects and antibacterial properties of Tualang honey on full-thickness burn wounds in rats in comparison to hydrofibre. *BMC Complementary and Alternative Medicine, 10*(1), 48. https://doi.org/10.1186/1472-6882-10-48.

Kishore, R. K., Halim, A. S., Syazana, M. S., & Sirajudeen, K. N. (2011). Tualang honey has higher phenolic content and greater radical scavenging activity compared with other honey sources. *Nutrition Research, 31*(4), 322–325.

Kumar, S., & Pandey, A. K. (2013). Chemistry and biological activities of flavonoids: An overview. *The Scientific World Journal, 162750*.

Lussier, S. J., & Stevens, H. E. (2016). Delays in GABAergic interneuron development and behavioral inhibition after prenatal stress. *Developmental Neurobiology, 76*(10), 1078–1091.

Madhyastha, S., Sahu, S. S., & Rao, G. (2013). Resveratrol improves postnatal hippocampal neurogenesis and brainderived neurotrophic factor in prenatally sressed rats. *International Journal of Developmental Neuroscience, 31*, 580–585.

Madhyastha, S., Sahu, S. S., & Rao, G. (2014). Resveratrol for prenatal-stress- induced oxidative damage in growing brain and its consequences on survival of neurons. *Journal of Basic and Clinical Physiology and Pharmacology, 25*(1), 63–72.

Maleki, S. J., Crespo, J. F., & Cabanillas, B. (2019). Anti-inflammatory effects of flavonoids. *Food Chemistry, 299*, 125124. https://doi.org/10.1016/j.foodchem.2019.125124.

Markham, J. A., & Koenig, J. I. (2011). Prenatal stress: Role in psychotic and depressive diseases. *Psychopharmacology, 214*(1), 89–106.

Mat Lazim, N., Abdullah, B., & Salim, R. (2013). The effect of Tualang honey in enhancing post tonsillectomy healing process. An open labelled prospective clinical trial. *International Journal of Pediatric Otorhinolaryngology, 77*(4), 457–461.

McEvoy, C. T., Schilling, D., Clay, N., Jackson, K., Go, M. D., Spitale, P., et al. (2014). Vitamin C supplementation for pregnant smoking women and pulmonary function in their newborn infants: A randomized clinical trial. *Journal of the American Medical Association, 311*(20), 2074–2082.

Mohamed, M., Sirajudeen, K., Swamy, M., Yaacob, M., & Sulaiman, S. (2010). Studies on the antioxidant properties of Tualang honey of Malaysia. *African Journal of Traditional, Complementary, and Alternative Medicines, 7*(1), 59–63.

Mulder, E. J., Robles de Medina, P. G., Huizink, A. C., Van den Bergh, B. R., Buitelaar, J. K., & Visser, G. H. (2002). Prenatal maternal stress: Effects on pregnancy and the (unborn) child. *Early Human Development, 70*(1–2), 3–14.

Obrosova, I. G., Drel, V. R., Oltman, C. L., Mashtalir, N., Tibrewala, J., Groves, J. T., et al. (2007). Role of nitrosative stress in early neuropathy and vascular dysfunction in streptozotocin-diabetic rats. *American Journal of Physiology, 293*, E1645–E1655.

Paidi, M. D., Schjoldager, J. G., Lykkesfeldt, J., & Tveden-Nyborg, P. (2014). Prenatal vitamin C deficiency results in differential levels of oxidative stress during late gestation in foetal Guinea pig brains. *Redox Biology, 2*, 361–367.

Pandurangan, K., Krishnappan, V., Subramanian, V., & Subramanyan, R. (2014). Antinociceptive effect of certain dimethoxy flavones in mice. *European Journal of Pharmacology, 727*, 148–157.

Pearson, D. A., Paglieroni, T. G., Rein, D., Wun, T., Schramm, D. D., Wang, J. F., et al. (2002). The effects of flavanol-rich cocoa and aspirin on ex vivo platelet function. *Thrombosis Research, 106*, 191–197.

Peng, H. L., Huang, W. C., Cheng, S. C., & Liou, C. J. (2018). Fisetin inhibits the generation of inflammatory mediators in interleukin-1β-induced human lung epithelial cells by suppressing the Nf-κb and Erk1/2 pathways. *International Immunopharmacology, 60*, 202–210.

Perrone, S., Bellieni, C. V., Negro, S., Longini, M., Santacroce, A., Tataranno, M. L., et al. (2017). Oxidative stress as a physiological pain response in full-term newborns. *Oxidative Medicine and Cellular Longevity, 3759287*. https://doi.org/10.1155/2017/3759287.

Raut, A., & Ratka, A. (2009). Oxidative damage and sensitivity to nociceptive stimulus and opioids in aging rats. *Neurobiology of Aging, 30*(6), 910–919.

Ried, K., Sullivan, T., Fakler, P., Frank, O. R., & Stocks, N. P. (2010). Does chocolate reduce blood pressure? A meta-analysis. *BMC Medicine, 8*, 39. https://doi.org/10.1186/1741-7015-8-39.

Sahu, S. S., Madhyastha, S., & Rao, G. M. (2013). Neuroprotective effect of resveratrol against prenatal stress induced cognitive impairment and possible involvement of Na+, K+-ATPase activity. *Pharmacology Biochemistry and Behavior, 103*(3), 520–525.

Scheinost, D., Kwon, S. H., Lacadie, C., Sze, G., Sinha, R., Constable, R. T., et al. (2016). Prenatal stress alters amygdala functional connectivity in preterm neonates. *Neuroimage Clinical, 12*, 381–388.

Schroder-van der Elst, J. P., van der Heide, D., Rokos, H., Morreale de Escobar, G., & Kohrle, J. (1998). Synthetic flavonoids cross the placenta in the rat and are found in fetal brain. *The American Journal of Physiology, 274*(2 Pt 1), E253–E256.

Shirpoor, A., Minassian, S., Salami, S., Khadem-Ansari, M. H., Ghaderi-Pakdel, F., & Yeghiazaryan, M. (2009). Vitamin E protects developing rat hippocampus and cerebellum against ethanol-induced oxidative stress and apoptosis. *Food Chemistry, 113*(1), 115–120.

Soleimani, E., Goudarzi, I., Abrari, K., & Lashkarbolouki, T. (2016). The combined effects of developmental lead and ethanol exposure on hippocampus dependent spatial learning and memory in rats: Role of oxidative stress. *Food and Chemical Toxicology, 96*, 263–272.

Vallianou, N. G., Gounari, P., Àlex, S. R., Panagos, J., & Kazazis, C. (2014). Honey and its anti-inflammatory, anti-bacterial and anti-oxidant properties. *Biology, Medicine*. https://doi.org/10.4172/2327-5146.1000132.

Vanhees, K., van Schooten, F. J., van Waalwijk van Doorn-Khosrovani, S. B., van Helden, S., Munnia, A., Peluso, M., et al. (2013). Intrauterine exposure to flavonoids modifies antioxidant status at adulthood and decreases oxidative stress-induced DNA damage. *Free Radical Biology & Medicine, 57*, 154–161.

Vega, C. C., Reyes, L. A., Rodríguez-González, G. L., Bautista, C. J., Vázquez-Martínez, M., Larrea, F., et al. (2016). Resveratrol partially prevents oxidative stress and metabolic dysfunction in pregnant rats fed a low protein diet and their offspring. *Journal of Physiology, 594*(5), 1483–1499.

Wan Yusuf, W. N., Wan Mohammad, W., Gan, S. H., Mustafa, M., Abd Aziz, C. B., & Sulaiman, S. A. (2018). Tualang honey ameliorates viral load, CD4 counts and improves quality of life in asymptomatic human immunodeficiency virus infected patients. *Journal of Traditional and Complementary Medicine, 9*(4), 249–256. https://doi.org/10.1016/j.jtcme.2018.05.003.

Wang, H. J., Xu, X., Xie, R. H., Rui, Y. Y., Zhang, P. A., Zhu, X. J., et al. (2018). Prenatal maternal stress induces visceral hypersensitivity of adult rat offspring through activation of cystathionine-β-synthase signaling in primary sensory neurons. *Molecular Pain, 14*. https://doi.org/10.1177/1744806918777406.

Weinstock, M. (2008). The long-term behavioural consequences of prenatal stress. *Neuroscience and Biobehavioral Reviews, 32*(6), 1073–1086.

Wolff, N., Darlington, A.S., Hunfeld, J, Verhulst, F., Jaddoe, V., Hofman, A., Passchier, J., & Tiemeier, H. (2010). Determinants of somatic complaints in 18-month-old children: The generation R study. Journal of Pediatric Psychology, 35(3), 306–316.

Xu, G. Y., Winston, J. H., Shenoy, M., Zhou, S., Chen, J. D., & Pasricha, P. J. (2009). The endogenous hydrogen sulfide producing enzyme cystathionine-beta synthase contributes to visceral hypersensitivity in a rat model of irritable bowel syndrome. *Molecular Pain, 5*, 44. https://doi.org/10.1186/1744-8069-5-44.

Yang, K. (1997). Placental 11 beta-hydroxysteroid dehydrogenase: Barrier to maternal glucocorticoids. *Reviews of Reproduction, 2*(3), 129–132.

Zhu, Z., Li, X., Chen, W., Zhao, Y., Li, H., Qing, C., et al. (2004). Prenatal stress causes gender-dependent neuronal loss and oxidative stress in rat hippocampus. *Journal of Neuroscience Research, 78*(6), 837–844. https://doi.org/10.1002/jnr.20338.

Chapter 42

Physical activity and exercise in the prevention of musculoskeletal pain in children and adolescents

Pablo Molina-García[a], Patrocinio Ariza-Vega[b], and Fernando Estévez-López[c]

[a]*Biohealth Research Institute, Physical Medicine and Rehabilitation Service, Virgen de las Nieves University Hospital, Granada, Spain,* [b]*Department of Physiotherapy, University of Granada, Granada, Spain,* [c]*Department of Pediatrics, Wilhelmina Children's Hospital, University Medical Center Utrecht, Utrecht University, Utrecht, the Netherlands*

Abbreviations

BMI   body mass index
MSKP   musculoskeletal pain

Background

Musculoskeletal pain (MSKP) in childhood and adolescence has increased significantly in the last decade and now is similar to the figures in adults. The overall prevalence of MSKP in children and adolescents ranges from 4% to 40% and, MSKP is a major health concern in this population (King et al., 2011). The most common sites of musculoskeletal pain are the low back, knee, and neck. Concretely, low back pain seems the most affected area reaching a prevalence of up to 54% (Cruz, 2015). In children and adolescents, MSKP negatively impacts on quality of life and school ability. MSKP supposes a huge economic impact on the health systems around the world (Groenewald, Essner, Wright, Fesinmeyer, & Palermo, 2014). The experience of MSKP during childhood or adolescence is strongly associated with the presence of pain in adulthood. For instance, a previous episode of low back pain was a consistent risk factor for a new episode of low back pain in young adults (Øiestad et al., 2020). Similarly, having low back pain and pain in other regions during young adulthood has demonstrated to be risk factors for low back pain and sciatica later in life (Parreira, Maher, Steffens, Hancock, & Ferreira, 2018). In a large cohort, 4 out of 10 adolescents with knee pain in late adolescence still experienced pain 5 years later, severe enough to negatively impact their quality of life, sports participation, and even affecting their choice of job or education (Rathleff et al., 2019). All this emphasizes the importance of preventing MSKP from the early stages of life to avoid its progression to adulthood, as well as the development of more severe musculoskeletal disorders.

Physical activity in the prevention of musculoskeletal pain

In the transition from childhood to adulthood, it is often speculated that a physically active life may protect against MSKP yet the evidence in this regards is unclear. Indeed, inconsistent results have emerged from a recent systematic review on the longitudinal association of physical activity during childhood and adolescence with new episodes of low back pain (Øiestad et al., 2020). An umbrella review studying the risk factors for the development of low back pain and sciatica in more than 50,000 adults determined unclear results for physical activity participation during the lifespan (Parreira et al., 2018). Physical activity was demonstrated neither to be a risk factor for non-specific neck pain in children and adolescents nor young adults (Huguet et al., 2016; Jahre, Grotle, Smedbråten, Dunn, & Øiestad, 2020). Evidence is such as contradictory that some studies found negative effects whereas other studies found positive effects of physical activity in the prognosis of certain MSKPs (Fig. 1).

Treatments, Mechanisms, and Adverse Reactions of Anesthetics and Analgesics. https://doi.org/10.1016/B978-0-12-820237-1.00042-9
Copyright © 2022 Elsevier Inc. All rights reserved.

500 PART | IV Novel and nonpharmacological aspects and treatments

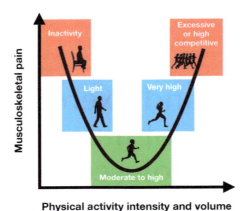

FIG. 1 Theoretical association of musculoskeletal pain and physical activity in u-shaped curve.

The nature of this discrepancy seems to be in the volume and intensity in which physical activity is performed. For instance, participation at an elite level and very high volumes of physical activity in youth are associated with low back pain and higher risk of knee and hip osteoarthritis and its symptoms (pain among them) in later adulthood (Beynon, Hebert, Lebouef-Yde, & Walker, 2019; Calvo-Muñoz, Kovacs, Roqué, Fernández, & Calvo, 2018; Tran et al., 2016). On the other hand, lifetime participation in organized sport or other types of leisure-time physical activity reduce the risk of developing chronic low back pain by 11%–16% in both men and women, with highly active people (i.e., $\geq$3–4 times or 2–4 h per week) reaching the greater profits (Shiri & Falah-Hassani, 2017). In a large cohort study in children, the relationship between physical activity and spinal pain demonstrated a strong dependency on the intensity (Franz et al., 2017). Those who engaged in the highest intensity of physical activity were more likely to experience some types of spinal pain, while moderate intensity showed a protective effect. Light intensity did not demonstrate any association. A possible explanation of this finding is that high demanding physical activity practiced in high volumes increases the risk of suffering traumatic injuries, which lead to the onset of acute and chronic MSKP (Franz et al., 2017).

All this evidence suggests that the relationship between physical activity and MSKP might follow a U-shaped curve, with more pain in inactive people as well high-performance practitioners or excessively active people, whereas a reduction of pain in moderate-to-high intensity and volume practitioners. However, further research is necessary to corroborate this hypothesis, including high-quality measures of pain and physical activity (e.g., accelerometry).

Physical activity recommendations

We recommend to follow the internationally accepted physical activity guidelines for pre-schoolers, children, and adolescents, which overall encourage to do moderate-to-vigorous physical activity daily (Table 1).

TABLE 1 Physical activity recommendations for children and adolescents.

Preschool-aged children (ages 3 through 5 years old)

- To be physically active throughout the day to enhance growth and development.
- Adult caregivers should encourage active play that includes a variety of activity types.

Children and Adolescents (ages 6 through 17 years old)
- 60 min or more of moderate-to-vigorous physical activity daily.
- Most of the 60 min or more per day should be either moderate- or vigorous-intensity aerobic physical activity and should include vigorous-intensity physical activity on at least 3 days a week.
- As part of their 60 min or more of daily physical activity, children and adolescents should include muscle- and bone-strengthening physical activity on at least 3 days a week.

Physical fitness components in the prevention of musculoskeletal pain

Cardiorespiratory fitness and aerobic/anaerobic training

Cardiorespiratory fitness has been studied as a prognostic factor in the development of MSKP. Three high-quality and large population studies reported conflictive results in the association between cardiorespiratory fitness and spine pain (Lardon, Leboeuf-Yde, & Le Scanff, 2015). One study found that greater cardiorespiratory fitness was associated with an increased likelihood of low back pain in boys and was not significantly associated in girls (Perry, Straker, O'sullivan, Smith, & Hands, 2009). Other study discovered a protective effect of cardiorespiratory fitness in the experience of neck pain in elementary school children (Cardon et al., 2004). The third study demonstrated that the associations between CRF and low back pain disappeared after taking into account back muscle endurance (Bo Andersen, Wedderkopp, & Leboeuf-Yde, 2006). A systematic review analyzed the effectiveness of aerobic exercise in a pediatric population with chronic pain and found inconclusive results in the only three available randomized controlled trials (Kichline & Cushing, 2019).

Exercise–induced hypoalgesia effects of aerobic exercise seem to vary depending on the exercise intensity. In one study, four pain induction procedures were used to test the acute effects of different intensities of aerobic exercise in a young population (Naugle, Naugle, Fillingim, Samuels, & Riley, 2014). The authors found that both moderate and vigorous aerobic exercise reduced pain ratings in heat stimuli, with vigorous exercise producing a large effect, while neither exercise intensities influenced pressure pain ratings. This intensity dependence might be behind the contradictory results in pain reduction found in aerobic interventions, indicating that the hypoalgesia would be only achieved with moderate-to-vigorous intensity. Moreover, exercise might reduce pain in some specific dimensions such as heat stimuli, which is often omitted in the questionnaires most commonly used in previous studies.

Further research has also demonstrated that the positive effects of cardiorespiratory fitness are explained by the fact that fitter young also have greater endurance in the core musculature providing functional support for different types of physical activity (Bo Andersen et al., 2006). Therefore, the aerobic activity enhancing this protective core strengthening should be identified to maximize the effects of cardiorespiratory fitness training in the prevention of low back pain.

Aerobic/anaerobic training recommendations

Overall, we found controversial effects of having optimal cardiorespiratory fitness or develop aerobic/anaerobic training in the prevention of MSKP in children and adolescents. It is widely known that cardiorespiratory fitness is a powerful marker of health in childhood. All physical activity guidelines recommend to engage in moderate to vigorous aerobic exercise. The general recommendations proposed in the lasted guides of fitness training in youth are depicted in Table 2 (Faigenbaum, Lloyd, Oliver, & American College of Sports Medicine, 2019).

Muscular strength and resistance training

Muscle strength and endurance have been also identified as relevant fitness components in the prevention of MSKP. Muscle strength is the force generated by the contraction of certain muscles, whereas muscle endurance refers to the capacity of these muscles to sustain a contraction force for a long period of time. A systematic review and meta-analysis revealed that only quadriceps and hip abductors strength were risk factors for future anterior knee pain, whereas strength in hamstrings, hip extensors, internal and external rotators was not (Neal et al., 2019). Similarly, lower quadriceps strength was associated with an increased risk of symptomatic, pain among them, and functional deterioration in knee osteoarthritis (Culvenor, Ruhdorfer, Juhl, Eckstein, & Elin Øiestad, 2017).

TABLE 2 General principles for maximizing aerobic and anaerobic gains in children and adolescents.

- At least 3 sessions per week.
- Session duration of 5–60 min, depending on the training modality. High-Intensity Interval Training (HIIT) as short as 5 min have demonstrated success in improving adolescents' cardiorespiratory fitness (Logan, Harris, Duncan, & Schofield, 2014), whereas team sports are often played for at least 60 min at high intensity are also demonstrate benefits (Wrigley, Drust, Stratton, Scott, & Gregson, 2012). Sessions that last 40 to 60 min are generally the most successful.
- ≥85% maximal heart rate (HRmax) effort work intensity.
- Global movement patterns such as running, object manipulation or jumps.
- Mixed-intermittent, ranging from <10 s to 4 min and from structured interval work to relatively unstructured games.

PART | IV Novel and nonpharmacological aspects and treatments

At the spine region, poorer muscle endurance in lower limbs, measured as the capacity to maintain isometric position such as squat or a hip extensor test, were identified as risk factors for low back pain in adolescents (Potthoff, De Bruin, Rosser, Humphreys, & Wirth, 2018). The core musculature understood as muscles that support the lumbo-pelvic-hip arrangement which helps to stabilize the spine, has demonstrated an important role in the prevention of low back pain and neck pain. Two systematics reviews agreed by finding that muscle endurance seems to have a major role compared to strength (Lardon et al., 2015; Potthoff et al., 2018), which is logical assuming that core musculature has a stabilization role and requires slow-twitch fibers for sustained contraction rather than fast fibers for powerful contractions. Results from cross-sectional studies covered in these two reviews consistently demonstrate that children and adolescents with low back pain have weaker back and abdominal muscle endurance compared with those free of pain.

It has been reported that activation, size, strength, and endurance of the deep cervical muscles may change in patients with chronic neck pain. In fact, previous studies have demonstrated a large association between deep cervical musculature weakness and neck pain (Falla, Jull, & Hodges, 2004; Javanshir et al., 2011). Some experts propose that a healthy cervical musculature free of pain is characterized by a modulation in the activation of superficial cervical muscles (e.g., sternocleidomastoid and anterior scalene) and deep cervical muscles (e.g., longus capitis or longus colli), avoiding the overactivation of superficial and inactivation of deep muscles (Blomgren, Strandell, Jull, Vikman, & Röijezon, 2018). Stretch overactivated muscles and strengthen inhibited muscles is a proposed treatment of this muscle imbalance. In workplace environments, self-perceived muscular tension in the cervical region demonstrated to be a risk factor for neck pain in a recent meta-analysis, while other factors that have been traditionally considered relevant such as longer duration of computer device use or break time were not associated (Jun, Zoe, Johnston, & O'Leary, 2017). This muscular tension occurs in superficial cervical muscles of workers, and cohort studies in the same population demonstrate that deep cervical muscles endurance has a preventive effect in the development of neck pain (Shahidi, Curran-Everett, & Maluf, 2015; Wigaeus Tornqvist, Hagberg, Hagman, Hansson Risberg, & Toomingas, 2009). Several systematic reviews and meta-analyses have confirmed this observation in exercise-based trials, demonstrating that cervical, shoulder, and scapulothoracic strengthening, and endurance exercises have positive effects in realizing chronic neck pain (Amiri Arimi, Mohseni Bandpei, Javanshir, Rezasoltani, & Biglarian, 2017; Blomgren et al., 2018; Gross et al., 2015). A special emphasis is placed on deep cervical muscle endurance training at low intensity and focused on improving neuromuscular coordination and cervical posture as the most effective exercise therapy.

Resistance training recommendations

Muscle strength in lower limbs and muscle endurance in the core and deep cervical muscles seems important risk factors for MSKP. Then, it is justified that enhancing a proper strength development in the early stages of life could be beneficial to prevent MSKP and injuries. Based on the guide from Stricker, Faigenbaum, and McCambridge (2020), Table 3 shows our recommendations on resistance training for children and adolescents.

Flexibility and mobility training

The terms mobility and flexibility are often used interchangeably, but they do not mean exactly the same. Flexibility has a more anatomical point of view and refers to the range of motion of one or more joints in a given action. On the other hand,

---

TABLE 3 General guideline for youth resistance training.

- Begin with 1–2 sets of 8–12 repetitions using a low resistance training intensity (i.e., ≤60% 1 repetition maximum -RM-) as proper technique is developed.
- As strength skills competence improves, it is reasonable to increase weight in 5%–10% increments and reduce the number of repetitions.
- The program can be progressed to 2–4 sets of 6–12 repetitions with a low to moderate training intensity (≤80% 1 RM).
- Multijoint exercises, such as weightlifting, always focus on completing all repetitions with proper technique to achieve proper motor control development.
- Include global muscle groups, with special emphasis on core muscles.
- Perform exercise through the full range of motion with proper technique.
- Sessions should last, at least, 20–30 min and be performed 2–3 times per week on non-consecutive days to have improvements.
- Keep the resistance training stimulus effective and enjoyable by periodically varying the exercises, sets, and repetitions.
- Use dynamic warm-up exercises integrated into the training session followed by cool-down periods with appropriate stretching techniques.
- Youth resistance training programs should be technique driven and consistent with the needs, abilities, and maturity level of the participants.

mobility is understood as a functional concept and refers to the ability to perform global movements in a free and fluid manner. We could say that mobility represents a bridge between flexibility and the ability to perform motor skills. Theoretically, optimal joint flexibility (e.g., ankle dorsiflexion) allows adequate mobility in functional tasks (e.g., squat pattern). which decreases excessive stress and helps to prevent the onset of MSKP (Clark & Lucett, 2014). However, the literature is not fully conclusive in this regard.

Global flexibility tests, such as the sit and reach or standing toe touch, have overall demonstrated to be no associated with both low back pain and neck pain in young (Potthoff et al., 2018; Wirth, Potthoff, Rosser, Humphreys, & De Bruin, 2018). Some authors point that both high and low levels of spinal flexion and hamstring flexibility could increase spinal strain during functional activities, and then might not be recommended (Perry et al., 2009). Another clear example demonstrating that excess mobility is not beneficial became evident in the joint hypermobility syndrome (Bravo & Wolff, 2006). However, three independent meta-analyses agreed by suggesting that having joint hypermobility is not a risk factor for the onset of MSKP in youth (Huguet et al., 2016; McCluskey, ÓKane, Hann, Weekes, & Rooney, 2012; Pourbordbari, Riis, Jensen, Olesen, & Rathleff, 2019).

In the lumbar spine, sagittal flexibility demonstrated markedly controversial results in the development of spine pain in young (Potthoff et al., 2018). One prospective study demonstrated that an elevated lumbar sagittal flexibility combined with low lumbar extension strength was a risk factor for low back pain (Sjölie & Ljunggren, 2001). Panjabi's stability model states that the lumbar spine is a structure that requires stability in a neutral and physiological zone to support functional tasks with a minimum mechanical stress on the vertebral structures (Hoffman & Gabel, 2013). Then, it is logical to assume that excessive lumbar mobility could not be beneficial in the prevention of MSKP. Panjabi's model has been extended by another author as Graig Cook through the joint by joint approach (Fig. 2) (Cook, 2010). All these theories have in common that a harmonious relationship between subjacent stable and mobile joints has to be present to maximize human movement and reduce musculoskeletal injuries (Hoffman & Gabel, 2013). Based on this concept, and knowing that the lumbar spine is a complex with a stabilization function, subjacent thoracic spine and hip joints should demonstrate further profits in the prevention of low back pain. Although there is still no a sufficient level of evidence to establish conclusions, there are some studies demonstrating the positive effects of the thoracic spine and hip mobility in the prevention of low back pain (Potthoff et al., 2018; Roach et al., 2015; Yang, Kim, Park, & Kim, 2015).

Something similar is observed in the knee, which is also considered a joint that mainly requires stability. A systematic review demonstrated that hip and ankle range of motion are risk factors for medial tibial stress syndrome, one of the most common causes of exercise-induced knee pain (Winkelmann, Anderson, Games, & Eberman, 2016). Furthermore, a meta-analysis found that decreased ankle ROM was moderately associated with increased knee abduction angle, which

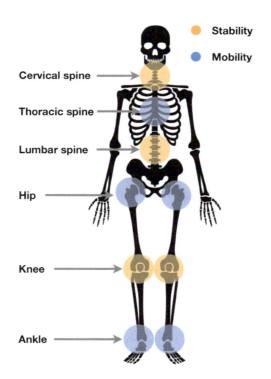

FIG. 2 Joint-by-joint theory.

TABLE 4 Recommendations to incorporate flexibility training in the warm up and cool down.

Dynamic warm-up recommendations:
- It should take less than 15 min
- Follow the **RAMP** approach (The Warm-Up: Maximize Performance and Improve Long-Term Athletic Development (English Edition) eBook: Jeffreys, Ian: Amazon.es: Tienda Kindle, n.d.): (1) **R**aise the body temperature, (2) **A**ctivate key muscles, (3) **M**obilize key joint and (4) **P**otentiate the movement skills that will be used in the session/game.
- According to the joint-by-joint approach, emphasize the mobility drills in those joints that have a mobility role in functional movements (e.g., ankle, hip, and thoracic spine).
- Perform dynamic stretches of large-muscle-group through functional movements such as shoulder circles, squat, or lunge.
- Six to eight movement patterns are enough for beginners.
- Prepare dynamic, diverse, and engaging warm-ups that create an expectation for the subsequent session.
- Self-myofascial release technique such as foam roller can be introduced since do not affect the performance and increase the flexibility during the practice (Cheatham, Kolber, Cain, & Lee, 2015).
- In the presence of joint hypermobility, the recommendations should be the focus on located muscular strength training to gain stability.

Cool down recommendations:
- Static stretching and PNF may negatively influence the performance (Kallerud & Gleeson, 2013), so that it is better to introduce them in the cool down. Furthermore, these techniques are more effective with high muscle temperature.
- PNF seems to have greater flexibility improvements compare to static stretching (Wicke, Gainey, & Figueroa, 2014).
- Self-myofascial release has been demonstrated to reduce post-training muscle pain (Cheatham et al., 2015).

is considered to be a risk factor for knee pain and injuries such as anterior cruciate ligament (Cronström, Creaby, Nae, & Ageberg, 2016).

Flexibility training recommendations

The role of global flexibility during youth in the prevention of MSKP is questioned. Evidence seems to point to increase flexibility on some specific joints as the most reasonable strategy to prevent the onset of MSKP, via the maintenance of an adequate biomechanics between body structures. Various training modalities can be used to increase flexibility levels both in the warm up and cool down. Static and dynamic stretching, proprioceptive neuromuscular facilitation (PNF), or self-myofascial release are some examples. We recommend active warm-ups that include dynamic stretching or self-myofascial release that raises body temperature, activates key muscle groups, mobilizes joints, potentiates performance, and develops desired movement skills. In the cool down, we recommend using self-myofascial release, static stretching, or PNF since can improve the flexibility as well as help in the recovering process (Table 4).

Risk factors for MSKP and the preventive role of physical activity and exercise

Biomechanical factors: An integrative approach

Nowadays, the sport sciences field is inconceivable to study muscular strength or flexibility in an isolated manner. Thus, the previous sections should be interpreted together with this section. The human movement works as a synchronized orchestrate that encompasses the muscular and myofascial system together with articular and skeletal components, all controlled by the central nervous system in perfect harmony. An optimal neuromuscular control, structural alignment, and force-couple, and muscular length-tension relationship would allow to perform functional activities in an efficient and safe manner (Sahrmann, Caldwell, & Bloom, 2002). All this complex perspective of the human movement system can be summarized as the importance of having a proper posture and movement biomechanics in our daily life. A body structure presenting a non-optimal posture or biomechanics leads to muscle overload/tension, microtraumas, and inflammation that initiates what is known as the cumulative injury and pain cycle (Clark & Lucett, 2014).

Both sitting and standing posture seem to be risk factors in the development of MSKP, but the evidence is not fully conclusive yet. Suboptimal cervical and thoracic spine postures during sitting have demonstrated to be related to the presence of neck pain, low back pain, and overall MSKP in adolescents, yet these associations depend on the posture measure used and vary between sexes (Potthoff et al., 2018; Prins, Crous, & Louw, 2008; Wirth et al., 2018). This fact is especially important considering the worrying numbers of sedentarism worldwide that affect all age ranges including childhood and adolescence. Non-neutral posture alignment in stance has been also concordantly stated as a risk factor for different MSKP, but results are again inconsistent. Studies in adolescents have demonstrated that alterations in the lumbo-pelvis position are a risk factor for low back pain but are not for overall MSKP in adolescence (Potthoff et al., 2018; Wirth et al., 2018). Confirming this, a meta-analysis in adults only including imaging measures of lumbo-pelvic posture (e.g., MRI or radiographs) demonstrated that the loss of a physiological lumbar lordosis is strongly related to the presence of low back pain, which was corroborated in a longitudinal study (Chun, Lim, Kim, Hwang, & Chung, 2017). A cohort study in children revealed that having a more aligned posture of the head in a standing position was related to a lower prevalence of neck pain and medical assistance in later life (Dolphens et al., 2012).

Biomechanical impairments during functional tasks such as walking or running are also suggested to lead to the onset of MSKP. For instance, common lower extremity impairments such as excessive foot pronation, increased knee valgus, and increased lumbo-pelvic movement can lead to several complications such as plantar fasciitis, shin splints, anterior knee pain, and low back pain (Clark & Lucett, 2014). Moreover, upper extremity impairments (usually characterized by rounded shoulders, forward head, and suboptimal scapulothoracic and glenohumeral arthrokinematics) may generate rotator cuff impingement, shoulder instability, biceps tendinitis, or headaches (Clark & Lucett, 2014). A recent consensus statement has evidenced that important biomechanical factors are involved in the pathogenesis of patellofemoral pain (Powers, Witvrouw, Davis, & Crossley, 2017). Hip adduction and internal rotation, differences in tibiofemoral kinematics in all three planes of motion, impaired quadriceps function, and excessive rotation of the tibia accompanies subtalar joint pronation are some examples. Individuals with chronic low back pain seem to exhibit higher global trunk muscle activity compared to asymptomatic peers (Ghamkhar & Kahlaee, 2015). The changed muscle activity pattern seems to be a compensatory strategy to enhance the reduced spinal stability or to compensate for the inactivation of more intrinsic core musculature in individuals with chronic low back pain.

The engagement in physical activities and reduction of sedentary behaviors from early childhood is crucial to develop a healthy posture and a natural movement biomechanics. For instance, physical activity has been demonstrated to be associated with a better spine posture in both sitting and standing (Sullivan et al., 2011; Wyszyńska et al., 2016), and foot mechanics during walking in children and adolescents (Riddiford-Harland et al., 2015). The previous meta-analysis has also demonstrated that well-designed exercise program significantly improves the movement proficiency in youth and can reduce the injury risk (Morgan et al., 2013; Steffen et al., 2013). All this evidence suggests an additional pathway in which the role of exercise in the prevention of MSKP comes from postural and biomechanical improvement during youth. In fact, the last guides for youth fitness support it and highlight that the acquisition of motor skills is a key developmental milestone to minimize injury and prevent MSKP (Faigenbaum et al., 2019).

Pediatric obesity

Although it has been assumed that the presence of overweight/obesity is associated with the onset of MSKP, the literature is inconsistent. Systematic reviews and meta-analyses of cross-sectional studies evidence that children and adolescents with overweight/obesity have 26% higher prevalence of overall MSKP than children with normal-weight and in some areas as the low back, this higher incidence is up to 42% (Paulis, Silva, Koes, & Van Middelkoop, 2014). However, longitudinal studies testing an increased body mass index (BMI) during childhood and adolescence found no significant relationship with the onset of low back nor neck pain (Beynon et al., 2019; Jahre et al., 2020). This relationship seems more evident in the knee joint, with several cohort studies demonstrating that a higher BMI status during childhood is associated with knee pain and the development of osteoarthritis in adulthood (Antony, Jones, Jin, & Ding, 2016).

The multi-factorial nature of both obesity and MSKP likely explains some of these conflicting results. However, there are mainly two strong pathomechanisms (Fig. 3) that justify the relationship between having overweight/obesity during childhood and adolescence and develop MSKP:

- Systemic inflammation: Adipose tissue accumulation increases proinflammatory cytokines and decreases anti- inflammatory adipokines, which enhances a systemic inflammation environment (Roffey et al., 2014). An inflamed tissue over time in this population can lead to pain or amplify the intensity by decreasing the neural excitation thresholds (Roffey et al., 2014).
- Biomechanical factors: recent systematic reviews have demonstrated that the presence of overweight/obesity in childhood and adolescence is associated with biomechanical alterations in basic situations such as static body posture (submitted article) and walking (Molina-Garcia et al., 2019). Some of these alterations are the presence of flatfoot, genu valgum both in static and dynamic situations, lumbar hyperlordosis, longer and wider stance phase, greater joint load in the lower limbs during stance phase and greater calf activation in the push-off phase. Furthermore, these biomechanical alterations have been also observed in adulthood, demonstrating a continuation in the life course (Runhaar, Koes, Clockaerts, & Bierma-Zeinstra, 2011). All these biomechanical characteristics could play a major role in the onset

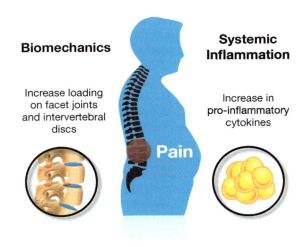

FIG. 3 Patho-mechanisms leading to musculoskeletal pain in pediatric obesity.

of MSKP and more severe disorders in the long terms such as osteoarthrirtis (Antony et al., 2016; Molina-Garcia et al., 2019). Importantly, all these alterations occur in a static situation or walking, where peak loads of 1–1.2 times the bodyweight are supported, but the consequences could be even more harmful in demanding activities such as running or jumping, where peak loads of 2–5 times the bodyweight are supported (Cavanagh & Lafortune, 1980; Jacob, 2001).

Exercise and physical activity seem to be especially relevant in the prevention of MSKP in this population. A recent meta-analysis has demonstrated that physical activity interventions reverse the inflammatory state normally observed in children and adolescents with overweight/obesity (Han et al., 2019). From a biomechanical point of view, some physical fitness components such as cardiorespiratory fitness, muscular strength, or speed agility were associated with a more aligned posture of upper and lower limbs in children with overweight/obesity (Molina-Garcia et al., 2020). Moreover, exercise interventions in this population have demonstrated positive effects both in body posture and the biomechanics of walking (Horsak et al., 2019; Molina-Garcia et al., 2020, 2020). With the available evidence, physical activity and exercise seem a promising therapy in the prevention of MSKP in pediatric obesity via inflammatory and biomechanical improvements, although it is a topic that still requires extensive research.

Psychosocial factors

There is increasing evidence demonstrating that psychosocial factors play an important role in the pathogenesis of MSKP. In a systematic review investigating the risk factors for the first episode of neck pain, a depressed mood, and high role conflict were detected as the strongest factors, while perceived empowering leadership and high perceived social climate demonstrated a protective role (Kim, Wiest, Clark, Cook, & Horn, 2018). Similarly, other systematic reviews demonstrated that depression and distress were the most common psychosocial factors influencing the experience of upper quadrant musculoskeletal pain in children and adolescents (Prins et al., 2008). Psychological stress and depression have also been raised as important risk factors in the development of low back pain and sciatica both in adolescence and adulthood (Minghelli, 2017; Parreira et al., 2018). The mechanism underlying the association between psychological factors and MSKP is not completely understood, but some researchers point to genetic components or a weakened ability to deal with MSKP in people with psychological symptoms (Parreira et al., 2018; Pinheiro et al., 2016).

Increasing physical activity and reducing sedentary behavior have been proposed to have positive effects on structure and function of the brain through different mechanisms. For instance, it seems that participation in physical activity might enhance mental health via the release of endorphins (Dishman & O'Connor, 2009), increases in brain-derived neurotrophic factor (Cotman, Berchtold, & Christie, 2007), and growth of new capillaries (Kleim, Cooper, & VandenBerg, 2002), which in turn might enhance the structural and functional composition of the brain. In a systematic review and meta-analysis including 114 studies, we recently found that physical activity has a small but significant positive effect on adolescents' mental health, while sedentary behavior was negatively associated with mental health in children and adolescents (Rodriguez-Ayllon et al., 2019). Concretely, participation in team sports demonstrated the most consistent positive correlation with mental health in children and adolescents, emphasizing the importance of the social role of physical activity. Another important finding of this review is that positive results in overall mental health were found in young meeting the physical activity recommendations of at least 60 min/day. Mental health improvement induced by physical activity may be an additional pathway in the prevention of MSKP from early life, but further research is necessary to corroborate this hypothesis.

Although physical activity recommendations may seem easy to reach, to change habits and routines is challenging. For instance, the habits and routines of children and adolescents are often linked with those of their parents. Thus, involve parents in the consolidation of physical activity routine should be a key strategy to follow. In addition, a behavioral change needs to be kept along the time and the motivation to change has to be explored on the basis of individual differences (e.g., values, preferences, and needs). In children and adolescents, a first step may be to include physical activity in their activities of daily living such as walking to the school/high school or practicing sports chosen by themselves (according to their interests). In those cases with special needs, the supervision and monitoring of a sport trainer would be recommended.

Atypical brain development

The literature often speculates that brain neurobiological features (central structures) that deviate from typical development are important risk factors for developing MSKP. It should be noted that a potential sensitized brain does not exclude a possible role of the peripheral nervous system in MSKP (as discussed in previous sections). According to the hypothesis of central sensitization, MSKP may be experienced at lowered thresholds when brain development deviates from the typical features (e.g., structure and function).

Although atypical brain development is often suggested as a risk factor for MSKP, most of the previous literature was cross-sectional and conducted in adults. Thus, whether the atypical features of the brain are risk factors or consequences of MSKP is unclear. In children and adolescents, there is a paucity of research on the association between brain features and MSKP. However, studies in other types of pain are available in the past literature. For instance, in comparison to controls, adolescents with abdominal pain have lower gray matter volumes in the anterior cingulate cortex and prefrontal cortex (Bhatt et al., 2019).

Collectively, the atypical brain is often speculated as a risk factor for developing MSKP in children and adolescents. However, this hypothesis remains to be corroborated. It should be also noted that reverse causality or bidirectionality of the association of brain development with MSKP is plausible. Thus, research testing whether atypical brain development is a risk factor or consequence of MSKP is of interest. In this context, Table 5 presents recommendations for designing studies on this topic.

Applications to other areas

In this chapter, we have reviewed the current evidence about the beneficial role of physical activity and exercise in the prevention of MSKP in children and adolescents. Because MSKP is one of the many bodily symptoms that cannot be adequately explained by organic pathology, this chapter may provide valuable knowledge not only to understand chronic MSKP in children and adolescents but also other persistent physical symptoms (disabling fatigue, headaches, and abdominal pain).

TABLE 5 Research recommendations for disentangling whether atypical brain development is a risk factor or consequence of musculoskeletal pain.

- Measure MSKP as a phenotype that is continuous (e.g., number of pain sites) instead of binary (patient vs. control).
- Design population-based research to enhance the generalisability of the findings.
- Include the most common neuroimaging features (e.g., structural, diffusion, and functional magnetic resonance imaging -MRI-).
- Consider carefully which confounders may be potentially involved in the association under study, which may help to solve problems with poor replicability of neuroimaging findings.

Other agents of interest

This chapter summarizes the role that physical activity and exercise have in the prevention of musculoskeletal pain in children and adolescents. Because physical activity or exercise programs are not widely implemented currently, more scalable solutions are urgently needed. Thus, it seems of interest to develop digital health solutions (those delivered through the Internet and smartphone applications) for promoting physical activity in children and adolescents at increased risk of developing chronic widespread musculoskeletal pain.

Mini-dictionary of terms

Musculoskeletal pain: Refers to pain in the muscles, bones, ligaments, and tendons and is a known consequence of cumulative trauma injury, repetitive strain, or overuse.

Physical activity: It is any bodily movement that increases energy expenditure above resting energy expenditure.

Physical exercise: It is a subset of physical activity that is planned, structured, repetitive, and purposeful that might have a different effect on mental health depending on the constituent elements (e.g., competitive vs. non-competitive).

Sedentary behavior: It is defined as any waking behavior characterized by an energy expenditure ≤ 1.5 metabolic equivalents (METs), while in a sitting, reclining or lying posture.

Cardiorespiratory fitness: A measure of the functional capabilities of the heart, lungs and muscles, relative to the demands of specific exercise routines such as running or cycling. VO_2 max or peak is the most common measure.

Resistance training: A type of strength-building exercise program that requires the body muscle to exert a force against some form of resistance, such as weight, stretch bands, water, or immovable objects.

Flexibility: Refers to the range of motion of one or more joints in a given action.

Mobility: Refers to the ability to perform global movements in a free and fluid manner.

Biomechanics: The application of mechanical principles in the study of human movement.

Psychosocial factors: Refers to influences that affect a person psychologically or socially.

Key facts

- MSKP is highly prevalent in childhood and adolescence and continues into adulthood with more severe consequences.
- Physical activity and exercise have been identified as key actions in the prevention and treatment of MSKP in children and adolescents but a review of the literature is needed to corroborate it.

Summary points

- Moderate-to-high volumes and intensities of physical activity seem to have the greatest benefits to reduce MSKP, whereas inactivity or an excessive practice could be harmful.
- The role of cardiorespiratory fitness in preventing MSKP is contradictory. Moderate and vigorous aerobic exercise have demonstrated the largest hypoalgesia effects. These positive effects might be leaded by a greater endurance in the core musculature.
- Muscle strength and endurance are associated with the presence of MSKP. Endurance seems to have a major role compared to strength in the prevention of MSKP, with special emphasis in stabilizing muscles.
- Excessive flexibility could be more harmful than beneficial in the development of MSKP. Optimal flexibility in key joints to support functional mobility may have positive effects in the prevention of some pain types such as low back and knee pain.
- Body posture and movement biomechanics are proposed as risk factors in the onset of MSKP, although evidence is not fully conclusive. Physical activity and exercise lead to the natural development of posture and movement biomechanics, contributing to minimize injury and prevent MSKP.
- Pediatric obesity may lead to the development of MSKP via biomechanical alterations and systemic inflammation. Exercise interventions have demonstrated improvements in these two pato-mechanisms and could be promising therapies to treat MSKP in this population.
- Psychosocial factors and atypical brain development are important risk factor to developing MSKP. Recent evidence support that physical activity has a positive effect in the mental health status and brain structure of children and adolescents, which in turn could help in the prevention of MSKP.

References

Amiri Arimi, S., Mohseni Bandpei, M. A., Javanshir, K., Rezasoltani, A., & Biglarian, A. (2017). The effect of different exercise programs on size and function of deep cervical flexor muscles in patients with chronic nonspecific neck pain. *American Journal of Physical Medicine and Rehabilitation*. https://doi.org/10.1097/PHM.0000000000000721.

Antony, B., Jones, G., Jin, X., & Ding, C. (2016). Do early life factors affect the development of knee osteoarthritis in later life: A narrative review. *Arthritis Research & Therapy*, *18*(1), 202. https://doi.org/10.1186/s13075-016-1104-0.

Beynon, A. M., Hebert, J. J., Lebouef-Yde, C., & Walker, B. F. (2019). Potential risk factors and triggers for back pain in children and young adults. A scoping review, part I: Incident and episodic back pain. In *Chiropractic and manual therapies*. https://doi.org/10.1186/s12998-019-0280-9.

Bhatt, R. R., Gupta, A., Labus, J. S., Zeltzer, L. K., Tsao, J. C., Shulman, R. J., et al. (2019). Altered brain structure and functional connectivity and its relation to pain perception in girls with irritable bowel syndrome. *Psychosomatic Medicine*. https://doi.org/10.1097/PSY.0000000000000655.

Blomgren, J., Strandell, E., Jull, G., Vikman, I., & Röijezon, U. (2018). Effects of deep cervical flexor training on impaired physiological functions associated with chronic neck pain: A systematic review 11 medical and health sciences 1103 clinical sciences. *BMC Musculoskeletal Disorders*. https://doi.org/10.1186/s12891-018-2324-z.

Bo Andersen, L., Wedderkopp, N., & Leboeuf-Yde, C. (2006). Association between back pain and physical fitness in adolescents. *Spine*. https://doi.org/10.1097/01.brs.0000224186.68017.e0.

Bravo, J. F., & Wolff, C. (2006). Clinical study of hereditary disorders of connective tissues in a chilean population: Joint hypermobility syndrome and vascular Ehlers-Danlos syndrome. *Arthritis and Rheumatism*. https://doi.org/10.1002/art.21557.

Calvo-Muñoz, I., Kovacs, F. M., Roqué, M., Fernández, I. G., & Calvo, J. S. (2018). Risk factors for low Back pain in childhood and adolescence: A systematic review. *Clinical Journal of Pain*. https://doi.org/10.1097/AJP.0000000000000558.

Cardon, G., De Bourdeaudhuij, I., De Clercq, D., Philippaerts, R., Verstraete, S., & Geldhof, E. (2004). Physical fitness, physical activity, and self-reported back and neck pain in elementary schoolchildren. *Pediatric Exercise Science*. https://doi.org/10.1123/pes.16.2.147.

Cavanagh, P. R., & Lafortune, M. A. (1980). Ground reaction forces in distance running. *Journal of Biomechanics*. https://doi.org/10.1016/0021-9290(80)90033-0.

Cheatham, S. W., Kolber, M. J., Cain, M., & Lee, M. (2015). The effects of self-myofascial release using a foam roll or roller massager on joint range of motion, muscle recovery, and performance: A systematic review. *International Journal of Sports Physical Therapy*, *10*(6), 827–838. Retrieved from http://www.ncbi.nlm.nih.gov/pubmed/26618062%0Ahttp://www.pubmedcentral.nih.gov/articlerender.fcgi?artid=PMC4637917.

Chun, S. W., Lim, C. Y., Kim, K., Hwang, J., & Chung, S. G. (2017). The relationships between low back pain and lumbar lordosis: A systematic review and meta-analysis. *The Spine Journal*. https://doi.org/10.1016/j.spinee.2017.04.034.

Clark, M. A., & Lucett, S. C. (2014). In Lippincot Williams & Wilkins (Ed.), *NASM essentials of corrective exercise training* (2nd ed.). Baltimore: Lippincot Williams & Wilkins.

Cook, G.,, E. (2010). *Movement functional movement systems: Screening, assessment and corrective strategies*. Aptos, CA.

Cotman, C. W., Berchtold, N. C., & Christie, L. A. (2007). Exercise builds brain health: Key roles of growth factor cascades and inflammation. *Trends in Neurosciences*. https://doi.org/10.1016/j.tins.2007.06.011.

Cronström, A., Creaby, M. W., Nae, J., & Ageberg, E. (2016). Modifiable factors associated with knee abduction during weight-bearing activities: A systematic review and meta-analysis. *Sports Medicine*. https://doi.org/10.1007/s40279-016-0519-8.

Cruz, E. R. L. (2015). Epidemiology of non-specific Back pain in children and adolescents: A systematic review of observational studies. *Journal of Novel Physiotherapies*. https://doi.org/10.4172/2165-7025.1000266.

Culvenor, A. G., Ruhdorfer, A., Juhl, C., Eckstein, F., & Elin Øiestad, B. (2017). Knee extensor strength and risk of structural, symptomatic, and functional decline in knee osteoarthritis: A systematic review and meta-analysis. *Arthritis Care and Research*. https://doi.org/10.1002/acr.23005.

Dishman, R. K., & O'Connor, P. J. (2009). Lessons in exercise neurobiology: The case of endorphins. *Mental Health and Physical Activity*. https://doi.org/10.1016/j.mhpa.2009.01.002.

Dolphens, M., Cagnie, B., Coorevits, P., Vanderstraeten, G., Cardon, G., D'Hooge, R., et al. (2012). Sagittal standing posture and its association with spinal pain: A school-based epidemiological study of 1196 flemish adolescents before age at peak height velocity. *Spine*. https://doi.org/10.1097/BRS.0b013e3182408053.

Faigenbaum, A. D., Lloyd, R. S., Oliver, J. L., & American College of Sports Medicine. (2019). Essentials of youth fitness. In *Essentials of youth fitness* Human Kinetics.

Falla, D. L., Jull, G. A., & Hodges, P. W. (2004). Patients with neck pain demonstrate reduced electromyographic activity of the deep cervical flexor muscles during performance of the craniocervical flexion test. *Spine*. https://doi.org/10.1097/01.brs.0000141170.89317.0e.

Franz, C., Møller, N. C., Korsholm, L., Jespersen, E., Hebert, J. J., & Wedderkopp, N. (2017). Physical activity is prospectively associated with spinal pain in children (CHAMPS study-DK). *Scientific Reports*. https://doi.org/10.1038/s41598-017-11762-4.

Ghamkhar, L., & Kahlaee, A. H. (2015). Trunk muscles activation pattern during walking in subjects with and without chronic low back pain: A systematic review. *PM and R*. https://doi.org/10.1016/j.pmrj.2015.01.013.

Groenewald, C. B., Essner, B. S., Wright, D., Fesinmeyer, M. D., & Palermo, T. M. (2014). The economic costs of chronic pain among a cohort of treatment-seeking adolescents in the United States. *The Journal of Pain*. https://doi.org/10.1016/j.jpain.2014.06.002.

Gross, A., Kay, T. M., Paquin, J. P., Blanchette, S., Lalonde, P., Christie, T., et al. (2015). Exercises for mechanical neck disorders. *Cochrane Database of Systematic Reviews*. https://doi.org/10.1002/14651858.CD004250.pub5.

Han, Y., Liu, Y., Zhao, Z., Zhen, S., Chen, J., Ding, N., et al. (2019). Does physical activity-based intervention improve systemic proinflammatory cytokine levels in overweight or obese children and adolescents? Insights from a meta-analysis of randomized control trials. *Obesity Facts*, *12*(6), 653–668. https://doi.org/10.1159/000501970.

Hoffman, J., & Gabel, P. (2013). Expanding Panjabi's stability model to express movement: A theoretical model. *Medical Hypotheses*. https://doi.org/10.1016/j.mehy.2013.02.006.

Horsak, B., Schwab, C., Baca, A., Greber-Platzer, S., Kreissl, A., Nehrer, S., et al. (2019). Effects of a lower extremity exercise program on gait biomechanics and clinical outcomes in children and adolescents with obesity: A randomized controlled trial. *Gait and Posture*, *70*(February), 122–129. https://doi.org/10.1016/j.gaitpost.2019.02.032.

Huguet, A., Tougas, M. E., Hayden, J., McGrath, P. J., Stinson, J. N., & Chambers, C. T. (2016). Systematic review with meta-analysis of childhood and adolescent risk and prognostic factors for musculoskeletal pain. *Pain*. https://doi.org/10.1097/j.pain.0000000000000685.

Jacob, H. A. C. (2001). Forces acting in the forefoot during normal gait—An estimate. *Clinical biomechanics*. https://doi.org/10.1016/S0268-0033(01)00070-5.

Jahre, H., Grotle, M., Smedbråten, K., Dunn, K. M., & Øiestad, B. E. (2020). Risk factors for non-specific neck pain in young adults. A systematic review. *BMC Musculoskeletal Disorders*, *21*(1), 366. https://doi.org/10.1186/s12891-020-03379-y.

Javanshir, K., Rezasoltani, A., Mohseni-Bandpei, M. A., Amiri, M., Ortega-Santiago, R., & Fernández-De-Las-Peñas, C. (2011). Ultrasound assessment of bilateral longus colli muscles in subjects with chronic bilateral neck pain. *American Journal of Physical Medicine & Rehabilitation*. https://doi.org/10.1097/PHM.0b013e31820173e5.

Jun, D., Zoe, M., Johnston, V., & O'Leary, S. (2017). Physical risk factors for developing non-specific neck pain in office workers: A systematic review and meta-analysis. *International Archives of Occupational and Environmental Health*. https://doi.org/10.1007/s00420-017-1205-3.

Kallerud, H., & Gleeson, N. (2013). Effects of stretching on performances involving stretch-shortening cycles. *Sports Medicine*. https://doi.org/10.1007/s40279-013-0053-x.

Kichline, T., & Cushing, C. C. (2019). A systematic review and quantitative analysis on the impact of aerobic exercise on pain intensity in children with chronic pain. *Children's Health Care*. https://doi.org/10.1080/02739615.2018.1531756. Routledge.

Kim, R., Wiest, C., Clark, K., Cook, C., & Horn, M. (2018). Identifying risk factors for first-episode neck pain: A systematic review. *Musculoskeletal Science & Practice*. https://doi.org/10.1016/j.msksp.2017.11.007.

King, S., Chambers, C. T., Huguet, A., MacNevin, R. C., McGrath, P. J., Parker, L., et al. (2011). The epidemiology of chronic pain in children and adolescents revisited: A systematic review. *Pain*. https://doi.org/10.1016/j.pain.2011.07.016.

Kleim, J. A., Cooper, N. R., & VandenBerg, P. M. (2002). Exercise induces angiogenesis but does not alter movement representations within rat motor cortex. *Brain Research*. https://doi.org/10.1016/S0006-8993(02)02239-4.

Lardon, A., Leboeuf-Yde, C., & Le Scanff, C. (2015). Is back pain during childhood or adolescence associated with muscle strength, muscle endurance or aerobic capacity: Three systematic literature reviews with one meta-analysis. *Chiropractic and Manual Therapies*. https://doi.org/10.1186/s12998-015-0065-8.

Logan, G. R. M., Harris, N., Duncan, S., & Schofield, G. (2014). A review of adolescent high-intensity interval training. *Sports Medicine*. https://doi.org/10.1007/s40279-014-0187-5.

McCluskey, G., ÓKane, E., Hann, D., Weekes, J., & Rooney, M. (2012). Hypermobility and musculoskeletal pain in children: A systematic review. *Scandinavian Journal of Rheumatology*. https://doi.org/10.3109/03009742.2012.676064.

Minghelli, B. (2017). *Low back pain in childhood and adolescent phase: Consequences, prevalence and risk factors—A revision*. https://doi.org/10.4172/2165-7939.1000351.

Molina-Garcia, P., Migueles, J. H., Cadenas-Sanchez, C., Esteban-Cornejo, I., Mora-Gonzalez, J., Rodriguez-Ayllon, M., et al. (2019). A systematic review on biomechanical characteristics of walking in children and adolescents with overweight/obesity: Possible implications for the development of musculoskeletal disorders. *Obesity Reviews*, *20*(7), 1033–1044. https://doi.org/10.1111/obr.12848.

Molina-Garcia, P., Miranda-Aparicio, D., Molina-Molina, A., Plaza-Florido, A., Migueles, J. H., Mora-Gonzalez, J., et al. (2020). Effects of exercise on plantar pressure during walking in children with overweight/obesity. *Medicine & Science in Sports & Exercise*, *52*(3), 654–662. https://doi.org/10.1249/MSS.0000000000002157.

Molina-Garcia, P., Mora-Gonzalez, J., Migueles, J. H., Rodriguez-Ayllon, M., Esteban-Cornejo, I., Cadenas-Sanchez, C., et al. (2020). Effects of exercise on body posture, functional movement, and physical fitness in children with overweight/obesity. *Journal of Strength and Conditioning Research*. https://doi.org/10.1519/JSC.0000000000003655.

Molina-Garcia, P., Plaza-Florido, A., Mora-Gonzalez, J., Torres-Lopez, L. V., Vanrenterghem, J., & Ortega, F. B. (2020). Role of physical fitness and functional movement in the body posture of children with overweight/obesity. *Gait & Posture*. https://doi.org/10.1016/j.gaitpost.2020.04.001.

Morgan, P. J., Barnett, L. M., Cliff, D. P., Okely, A. D., Scott, H. A., Cohen, K. E., et al. (2013). Fundamental movement skill interventions in youth: A systematic review and meta-analysis. *Pediatrics*, *132*(5), e1361–e1383. https://doi.org/10.1542/peds.2013-1167.

Naugle, K. M., Naugle, K. E., Fillingim, R. B., Samuels, B., & Riley, J. L. (2014). Intensity thresholds for aerobic exercise-induced hypoalgesia. *Medicine and Science in Sports and Exercise*. https://doi.org/10.1249/MSS.0000000000000143.

Neal, B. S., Lack, S. D., Lankhorst, N. E., Raye, A., Morrissey, D., & Van Middelkoop, M. (2019). Risk factors for patellofemoral pain: A systematic review and meta-analysis. *British Journal of Sports Medicine*. https://doi.org/10.1136/bjsports-2017-098890.

Øiestad, B. E., Hilde, G., Tveter, A. T., Peat, G. G., Thomas, M. J., Dunn, K. M., et al. (2020). Risk factors for episodes of back pain in emerging adults. A systematic review. *European Journal of Pain (United Kingdom)*. https://doi.org/10.1002/ejp.1474.

Parreira, P., Maher, C. G., Steffens, D., Hancock, M. J., & Ferreira, M. L. (2018). Risk factors for low back pain and sciatica: An umbrella review. *The Spine Journal*. https://doi.org/10.1016/j.spinee.2018.05.018.

Paulis, W. D., Silva, S., Koes, B. W., & Van Middelkoop, M. (2014). Overweight and obesity are associated with musculoskeletal complaints as early as childhood: A systematic review. *Obesity Reviews*. https://doi.org/10.1111/obr.12067.

Perry, M., Straker, L., O'sullivan, P., Smith, A., & Hands, B. (2009). Fitness, motor competence, and body composition are weakly associated with adolescent back pain. *The Journal of Orthopaedic and Sports Physical Therapy.* https://doi.org/10.2519/jospt.2009.3011.

Pinheiro, M. B., Ferreira, M. L., Refshauge, K., Maher, C. G., Ordoñana, J. R., Andrade, T. B., et al. (2016). Symptoms of depression as a prognostic factor for low back pain: A systematic review. *The Spine Journal.* https://doi.org/10.1016/j.spinee.2015.10.037.

Potthoff, T., De Bruin, E. D., Rosser, S., Humphreys, B. K., & Wirth, B. (2018). A systematic review on quantifiable physical risk factors for non-specific adolescent low back pain. *Journal of Pediatric Rehabilitation Medicine.* https://doi.org/10.3233/PRM-170526.

Pourbordbari, N., Riis, A., Jensen, M. B., Olesen, J. L., & Rathleff, M. S. (2019). Poor prognosis of child and adolescent musculoskeletal pain: A systematic literature review. *BMJ Open.* https://doi.org/10.1136/bmjopen-2018-024921.

Powers, C. M., Witvrouw, E., Davis, I. S., & Crossley, K. M. (2017). Evidence-based framework for a pathomechanical model of patellofemoral pain: 2017 patellofemoral pain consensus statement from the 4th international patellofemoral pain research retreat, Manchester, UK: Part 3. *British Journal of Sports Medicine.* https://doi.org/10.1136/bjsports-2017-098717.

Prins, Y., Crous, L., & Louw, Q. (2008). A systematic review of posture and psychosocial factors as contributors to upper quadrant musculoskeletal pain in children and adolescents. *Physiotherapy Theory and Practice.* https://doi.org/10.1080/09593980701704089.

Rathleff, M. S., Holden, S., Straszek, C. L., Olesen, J. L., Jensen, M. B., & Roos, E. M. (2019). Five-year prognosis and impact of adolescent knee pain: A prospective population-based cohort study of 504 adolescents in Denmark. *BMJ Open.* https://doi.org/10.1136/bmjopen-2018-024113.

Riddiford-Harland, D. L., Steele, J. R., Cliff, D. P., Okely, A. D., Morgan, P. J., Jones, R. A., et al. (2015). Lower activity levels are related to higher plantar pressures in overweight children. *Medicine and Science in Sports and Exercise, 47*(2), 357–362. https://doi.org/10.1249/MSS.0000000000000403.

Roach, S. M., San Juan, J. G., Suprak, D. N., Lyda, M., Bies, A. J., & Boydston, C. R. (2015). Passive hip range of motion is reduced in active subjects with chronic low back pain compared to controls. *International Journal of Sports Physical Therapy.*

Rodriguez-Ayllon, M., Cadenas-Sánchez, C., Estévez-López, F., Muñoz, N. E., Mora-Gonzalez, J., Migueles, J. H., et al. (2019). Role of physical activity and sedentary behavior in the mental health of preschoolers, children and adolescents: A systematic review and meta-analysis. *Sports Medicine.* https://doi.org/10.1007/s40279-019-01099-5.

Roffey, D. M., Simon Dagenais, D. C., Ted Findlay, D. O., Marion, T. E., McIntosh, G., Shamji, M. F., et al. (2014). Obesity, weight loss, and low back pain: An overview for primary care providers—Part 1. *Journal of Current Clinical Care, 4*(4). Retrieved from file:///F:/Služba izpit/neimenovano.pdf.

Runhaar, J., Koes, B. W., Clockaerts, S., & Bierma-Zeinstra, S. M. (2011). A systematic review on changed biomechanics of lower extremities in obese individuals: A possible role in development of osteoarthritis. *Obesity Reviews, 12*(12), 1071–1082. https://doi.org/10.1111/j.1467-789X.2011.00916.x.

Sahrmann, S., Caldwell, N., & Bloom, C. (2002). *Diagnosis and treatment of movement impairment syndromes.* Mosby.

Shahidi, B., Curran-Everett, D., & Maluf, K. S. (2015). Psychosocial, physical, and neurophysiological risk factors for chronic neck pain: A prospective inception cohort study. *The Journal of Pain.* https://doi.org/10.1016/j.jpain.2015.09.002.

Shiri, R., & Falah-Hassani, K. (2017). Does leisure time physical activity protect against low back pain? Systematic review and meta-analysis of 36 prospective cohort studies. *British Journal of Sports Medicine.* https://doi.org/10.1136/bjsports-2016-097352.

Sjölie, A. N., & Ljunggren, A. E. (2001). The significance of high lumbar mobility and low lumbar strength for current and future low back pain in adolescents. *Spine.* https://doi.org/10.1097/00007632-200112010-00019.

Steffen, K., Emery, C. A., Romiti, M., Kang, J., Bizzini, M., Dvorak, J., et al. (2013). High adherence to a neuromuscular injury prevention programme (FIFA 11+) improves functional balance and reduces injury risk in Canadian youth female football players: A cluster randomised trial. *British Journal of Sports Medicine.* https://doi.org/10.1136/bjsports-2012-091886.

Stricker, P. R., Faigenbaum, A. D., & McCambridge, T. M. (2020). Resistance training for children and adolescents. *Pediatrics, 145*(6). https://doi.org/10.1542/peds.2020-1011.

Sullivan, P. B. O., Smith, A. J., Beales, D. J., Straker, L. M., O'Sullivan, P. B., Smith, A. J., et al. (2011). Association of biopsychosocial factors with degree of slump in sitting posture and self-report of back pain in adolescents: A cross-sectional study. *Physical Therapy, 91*(4), 470–483. https://doi.org/10.2522/ptj.20100160.

The Warm-Up: Maximize Performance and Improve Long-Term Athletic Development (English Edition) eBook: Jeffreys, Ian: Amazon.es: Tienda Kindle. (n.d.). Retrieved July 13, 2020, from https://www.amazon.es/Warm-Up-Maximize-Performance-Long-Term-Development-ebook/dp/B07MHJZMYJ.

Tran, G., Smith, T. O., Grice, A., Kingsbury, S. R., McCrory, P., & Conaghan, P. G. (2016). Does sports participation (including level of performance and previous injury) increase risk of osteoarthritis? A systematic review and meta-analysis. *British Journal of Sports Medicine.* https://doi.org/10.1136/bjsports-2016-096142.

Wicke, J., Gainey, K., & Figueroa, M. (2014). A comparison of self-administered proprioceptive neuromuscular facilitation to static stretching on range of motion and flexibility. *Journal of Strength and Conditioning Research.* https://doi.org/10.1519/JSC.0b013e3182956432.

Wigaeus Tornqvist, E., Hagberg, M., Hagman, M., Hansson Risberg, E., & Toomingas, A. (2009). The influence of working conditions and individual factors on the incidence of neck and upper limb symptoms among professional computer users. *International Archives of Occupational and Environmental Health.* https://doi.org/10.1007/s00420-009-0396-7.

Winkelmann, Z. K., Anderson, D., Games, K. E., & Eberman, L. E. (2016). Risk factors for medial tibial stress syndrome in active individuals: An evidence-based review. *Journal of Athletic Training.* https://doi.org/10.4085/1062-6050-51.12.13.

Wirth, B., Potthoff, T., Rosser, S., Humphreys, B. K., & De Bruin, E. D. (2018). Physical risk factors for adolescent neck and mid back pain: A systematic review. *Chiropractic and Manual Therapies.* https://doi.org/10.1186/s12998-018-0206-y.

Wrigley, R., Drust, B., Stratton, G., Scott, M., & Gregson, W. (2012). Quantification of the typical weekly in-season training load in elite junior soccer players. *Journal of Sports Sciences*. https://doi.org/10.1080/02640414.2012.709265.

Wyszyńska, J., Podgórska-Bednarz, J., Drzał-Grabiec, J., Rachwał, M., Baran, J., Czenczek-Lewandowska, E., et al. (2016). Analysis of relationship between the body mass composition and physical activity with body posture in children. *BioMed Research International*, *2016*(1851670), 1–10. https://doi.org/10.1155/2016/1851670.

Yang, S. R., Kim, K., Park, S. J., & Kim, K. (2015). The effect of thoracic spine mobilization and stabilization exercise on the muscular strength and flexibility of the trunk of chronic low back pain patients. *Journal of Physical Therapy Science*. https://doi.org/10.1589/jpts.27.3851.

Chapter 43

Linking aerobic exercise and childhood pain alleviation: A narrative

Tiffany Kichline[a], Adrian Ortega[a], and Christopher C. Cushing[b]

[a]Clinical Child Psychology Program, University of Kansas, Lawrence, KS, United States; [b]Clinical Child Psychology Program, Schiefelbusch Institute for Life Span Studies, University of Kansas, Lawrence, KS, United States

Abbreviations

EIH	exercise-induced hypoalgesia
mHealth	mobile health
NDMA	N-methyl-D-aspartate
OT	occupational therapy
PT	physical therapy
RVM	rostral ventromedial medulla

Introduction

Chronic pain is defined as pain lasting more than 3 months and is estimated to influence approximately 11%–38% of children (King et al., 2011; Perquin et al., 2000). Chronic pediatric pain has been shown to impact a variety of domains in youths' lives, including their functioning and physical activity levels (Long, Palermo, & Manees, 2008; Palermo, 2000). One important area of concern is an aerobic activity. It is estimated that up to 75% of youth with chronic pain withdraw from sports participation and experience reductions in activities such as running, walking, gym class, and playing (Hunfeld et al., 2002; Konijnenberg et al., 2005). Given widely disseminated physical activity targets for health promotion and disease prevention (Lee et al., 2012; USDHHS, 2018), low rates of aerobic activity in youth with chronic pain increase the risk for morbidity.

Aerobic activity levels in youth with chronic pain

The World Health Organization (2010) and the US Department of Health and Human Services (2018) recommend that youth engage in 60 min of aerobic activity (e.g., tennis, swimming, and brisk walking). However, the literature suggests that youth with chronic pain do not meet these recommendations (Kichline, Cushing, Ortega, Friesen, & Schurman, 2019; Long et al., 2008). Further, one study found that adolescents with chronic pain engaged in even lower levels of physical activity than their healthy counterparts (Long et al., 2008). This suggests that pediatric chronic pain patients are at higher risk for negative health outcomes related to inactivities, such as stroke, type 2 diabetes, cancer, and overweight or obesity over the long term (Lee et al., 2012). In addition, there is some literature that suggests that general negative health outcomes related to inactivity (e.g., obesity) can relate to worse pain outcomes. For example, pediatric chronic pain patients with overweight or obesity report worse pain severity (Bonilla, Wang, & Saps, 2011; Tsiros et al., 2014; Wilson, Samuelson, & Palermo, 2010). However, more research is needed to better understand the causal relationship between overweight and obesity and increased pain severity. Further, the causal mechanisms underlying chronic pain and reduced activity levels remain unclear. Taken together, the literature suggests that increasing aerobic exercise in youth with chronic pain could importantly benefit their health and potentially reduce chronic pain symptoms.

Treatments, Mechanisms, and Adverse Reactions of Anesthetics and Analgesics. https://doi.org/10.1016/B978-0-12-820237-1.00043-0
Copyright © 2022 Elsevier Inc. All rights reserved.

Effect of aerobic exercise on pediatric chronic pain intensity

Youth with chronic pain often experience diminished physical and role functioning due to their condition (Palermo, 2000). In other words, youth with chronic pain can develop secondary physical limitations and functional disabilities, such as the inability to accomplish tasks of daily living, in addition to receding from individual and group physical activities (Palermo, 2000, 2012). Physical functioning is a multifaceted construct including perceived disability, physiological functioning (e.g., fitness, functional capacity), and physical activity (Wilson & Palermo, 2012). However, increases in pain-related activity limitations in youth have been moderately correlated with decreased levels of physical activity as assessed objectively via accelerometry (Long et al., 2008; Wilson & Palermo, 2012). While there is variation in how youth will experience pain-related functional disability, measuring an individual's physical activity can still serve as an indicator for understanding the impact of chronic pain in youth (Wilson & Palermo, 2012). For example, the assessment of physical activity is important for several reasons, including (a) to evaluate the extent to which pain may interfere with an individual's functioning, (b) to determine the level of need a patient may require to accomplish everyday tasks (and consequently the level of burden for caregivers), (c) to assess for improvements in treatment outcomes during and following clinical interventions, and on a more exploratory note, and (d) to study the individual- and group-level differences in exercise among youth with various pain types.

The treatment of pediatric chronic pain often includes multimodal and multidisciplinary approaches (Odell & Logan, 2013). Given the impact of pain on functional disability, the major goals of many intensive multidisciplinary pediatric pain rehabilitation programs are to improve physical functioning and reintegration into daily, typical life activities for youth with chronic pain (Odell & Logan, 2013). A major component of these pain treatment programs is aerobic exercise embedded within physical therapy (PT) or occupational therapy (OT). Exercise is a commonly used treatment for pain populations such as fibromyalgia, back pain, chronic fatigue syndrome, and arthritis (Bement & Sluka, 2016). It is important to note that providing exercise therapy is consistent with the CDC's recommendation of providing non-opioid and non-pharmacological treatments for managing chronic pain (Dowell, Haegerich, & Chou, 2016).

Exercise-induced hypoalgesia (EIH) is the process by which acute exercise reduces pain sensitivity (Naugle, Fillingim, & Riley, 2012). Mechanisms underlying the process by which exercise produces hypoalgesia are still unclear. However, research has identified key areas and processes involved in EIH. Aerobic exercise is theorized to reduce pain sensitivity through a variety of mechanisms influencing opioid and non-opioid systems such as serotonergic systems, changes in N-methyl-D-aspartate (NDMA) receptors, and endocannabinoid systems (Koltyn, Brellenthin, Cook, Sehgal, & Hillard, 2014; Lima, Abner, & Sluka, 2017; Rice et al., 2019). Several studies have found that aerobic exercise stimulates the release of beta-endorphins which can attenuate the perception of pain (Naugle et al., 2012; Thorén, Floras, Hoffmann, & Seals, 1990). Structurally, the rostral ventromedial medulla (RVM) is a brain region area for pain modulation (Lima et al., 2017). Nuclei within the RVM are involved in pain inhibitory systems as well as in the modulation of motor movements (Porreca, Ossipov, & Gebhart, 2002), demonstrating the importance of this brain structure for understanding EIH. Consistent physical activity reduces the facilitation of nociception in NMDA receptors embedded in the RVM (Sluka, Danielson, Rasmussen, & Dasilva, 2012). Of note, the current literature is relatively premature and most of the mechanistic research linking exercise and reductions in pain has been conducted in animal models, healthy adults, or adults with chronic pain (Lima et al., 2017; Rice et al., 2019) which might not translate to youth, as many neurobehavioral processes are still developing during this time. Further, one study suggested that individuals with chronic pain may report different levels of pain sensitivity following exercise (Rice et al., 2019). It is possible that there may be other unknown mechanisms that could contribute to variability in EIH across pain populations. Nonetheless, EIH research in animal models and adults has provided useful avenues for future research in this area.

Current literature on effect of aerobic exercise on chronic pediatric pain intensity

There is some adult literature that suggests that aerobic exercise reduces chronic pain intensity (Busch, Schachter, Peloso, & Bombardier, 2002; Jansen, Viechtbauer, Lenssen, Hendriks, & de Bie Rob, 2011). For example, meta-analyses in adults with chronic pain found a small effect of aerobic exercise on pain intensity (Busch et al., 2002; Busch, Barber, Overend, Peloso, & Schachter, 2007). However, there are few studies that have examined the effect of aerobic exercise in children with chronic pain conditions. Examining the effects of aerobic exercise in children is important because childhood is a time of rapid changes in cognitive and physical development, which could result in a different relationship between exercise and pain intensity. For instance, levels of physical activity decrease from childhood to adolescence (Nader, Bradley, Houts, McRitchie, &

O'Brien, 2008), which might impact physical functioning and even potentially serve as a risk factor for chronic pain development (Sluka, O'Donnell, Danielson, & Rasmussen, 2013; Wilson & Palermo, 2012).

The current literature on the effect of aerobic exercise on chronic pediatric pain conditions is relatively mixed. Some studies found that increased aerobic exercise related to reduced pain intensity outcomes (Fanucchi, Stewart, Jordaan, & Becker, 2009; Kashikar-Zuck et al., 2010; Long et al., 2008). For example, one randomized control trial found that pediatric patients with low back pain that engaged in an 8-week aerobic exercise program reported reduced levels of pain intensity compared to a no-intervention group. Reductions in pain intensity were observed at 1 month and 3-month follow-up for the aerobic exercise group (Fanucchi et al., 2009). Kashikar-Zuck et al. (2010) also found that juvenile fibromyalgia patients that engaged in higher levels of physical activity reported lower levels of pain intensity. On the other hand, some studies found no association between physical activity levels and chronic pain intensity (Bohr, Nielsen, Müller, Karup Pedersen, & Andersen, 2015). For example, Bohr et al. (2015) examined objective levels of physical activity in youth with juvenile idiopathic arthritis and found no association between increased levels of physical activity and pain intensity. Of note, Bohr et al. (2015) had participants rate their level of pain intensity as an aggregated based on the last week. In other studies, participants rated their level of pain in the moment or day-to-day (Fanucchi et al., 2009; Kashikar-Zuck et al., 2010). Taken together, the studies suggest variability in the relationship between aerobic activity and pediatric chronic pain intensity.

A recent meta-analysis aimed to clarify the effect of aerobic activity on pain intensity in youth with chronic pain conditions by aggregating the literature (Kichline & Cushing, 2019). To reduce inaccuracies influenced by the recall and social desirability bias, subjective reports (i.e., self-report) of aerobic activity were not included in the review (Dhurandhar et al., 2015; Sallis & Saelens, 2000). Therefore, the meta-analysis consisted of studies in (1) pre–post-test comparisons and randomized control trials and (2) longitudinal actigraphy studies that objectively measured physical activity levels. Results indicated a small effect of aerobic exercise interventions ($g = -0.39$, 95% confidence interval [CI] $= -0.65$ to -0.14) on reducing chronic pain intensity (Kichline & Cushing, 2019). However, these findings were determined to be inconclusive due to the high risk of biases and low methodological quality of the studies. Specifically, the meta-analysis found a large risk of bias regarding potential failure to blind outcome assessment, personnel, and participants. Thus, the results revealed an important need for increased randomized control trials to better understand the causal relationships. Future use of randomized control trials would improve the quality of literature and allow for researchers to better understand the causal relationship between aerobic exercise and chronic pain (Kichline & Cushing, 2019). Based on these current limitations, the evidence on aerobic activity reducing pain intensity is preliminary and relatively inconclusive. The following sections focus on specific preliminary findings related to aerobic activity and pain intensity and identify areas in need of further research (Table 1).

Effect of aerobic activity across pediatric chronic pain conditions

The current literature on the effect of aerobic activity and pain intensity primarily focuses on chronic pediatric pain patients with fibromyalgia or juvenile arthritis (Kichline & Cushing, 2019). This focus may be because treatment recommendations for joint-based or musculoskeletal-based pain complaints often specifically include aerobic exercise and physical therapy (Kashikar-Zuck & Ting, 2014). Treatment for other pediatric chronic pain conditions, such as headache, and abdominal pain, often include recommendations to increase functioning and day-to-day activity, but focus less on aerobic activity levels (Chiou & Nurko, 2010; O'Brien, Kabbouche, & Hershey, 2012; Odell & Logan, 2013). However, there are a few studies that suggest that aerobic exercise may reduce pain intensity in different pediatric chronic pain conditions (Long et al., 2008; Rabbitts, Holley, Karlson, & Palermo, 2014). For example, Long et al. (2008) found that increased aerobic activity was related to reduced chronic pain intensity in youth with general chronic pain. Specifically, study participants included pediatric patients with primary pain locations in the head and neck, shoulder, abdomen, back, and extremities for at least 3 months, 3 days per week with moderate intensity (Long et al., 2008). Additionally, Rabbitts et al. (2014) found that increased levels of activity predicted lower levels of pain intensity at the end of the day for participants that reported having general chronic pain over the course of 3 months, but were not diagnosed with arthritis or other medical diagnoses (Rabbitts et al., 2014). However, one study in youth with chronic abdominal pain found mixed associations between aerobic activity and chronic pain intensity (Kichline et al., 2019). Taken together, though some studies suggest aerobic activity may reduce pain intensity across a variety of pediatric pain conditions, results are far from being conclusive. Currently, there are limited studies to identify and trends or differences in the effect of aerobic exercise across pediatric chronic pain conditions. Examining individual differences in the effect of aerobic exercise on chronic pain could provide more information (Fig. 1).

516 PART | IV Novel and nonpharmacological aspects and treatments

TABLE 1 Examples of aerobic exercise interventions in youth with chronic pain.

Author	Exercise intervention	Frequency
Dodds, Bjornson, Sweeney, and Narayanan (2016)	180 min of physical activity intervention with aerobic exercises including cycling, swimming, and recreational activities	Weekdays for 5 weeks
Elnaggar and Elshafey (2016)	Physical activity program with aerobic exercises including fast walking exercises and water kicking exercises	Frequency not report. Duration was 12 weeks
Epps et al. (2005) Land	60 min training program with aerobic exercises including static biking, step machine, walking, skipping, hopping, and bunny jumps	16 sessions over 2 weeks
Epps et al. (2005) Aquatic	60 min training program with aerobic activities underwater including jogging on the spot, scissor kicks, star jumps, bobbing, and jumping	16 sessions over 2 weeks
Fanucchi et al. (2009)	40–45 min exercise program with aerobic exercises including walking, lunges, squats	8 sessions over 8 weeks
Klepper (1999)	60-min commercial exercise tape with aerobic exercises including uncomplicated dance steps on the floor, stepping up and down 4-in. platforms, activities while sitting on large vinyl balls	24 sessions over 8 weeks
Mendonça et al. (2013)	Pilates and conventional exercise program consisting of stretching, flexibility, and aerobic movements	50 min 2 times a week/ 24 weeks
Olsen et al. (2013)	6 h of physical and occupational therapy with an emphasis on aerobic training	~4 weeks
Stephens et al. (2008)	30-min aerobic, cardio-dance, and boxing movement program	3 times per week/ 12 weeks
Tarakci, Yeldan, Baydogan, Olgar, and Kasapcopur (2012)	Comprehensive land-based home and physical therapist-lead exercise program including strengthening, stretching, postural, and functional exercises	4 times for 20–45 min/12 weeks
van Dijk-Lokkart et al. (2016)	High intensity cardiorespiratory and muscle strength training	2 times for 45 min/12 weeks

Examples of aerobic exercise interventions in previous studies examining aerobic exercise and chronic pain intensity.

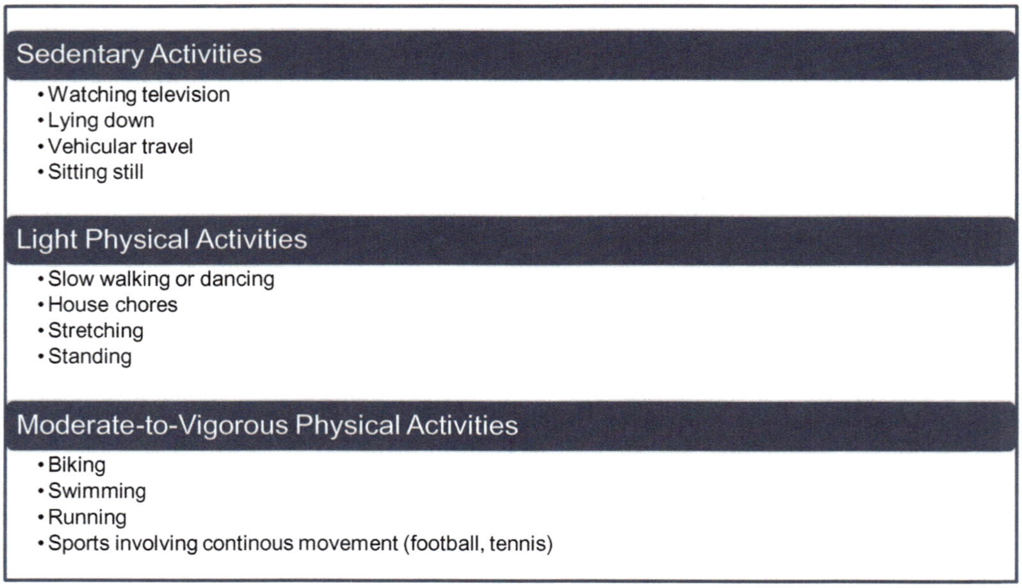

Sedentary Activities
- Watching television
- Lying down
- Vehicular travel
- Sitting still

Light Physical Activities
- Slow walking or dancing
- House chores
- Stretching
- Standing

Moderate-to-Vigorous Physical Activities
- Biking
- Swimming
- Running
- Sports involving continous movement (football, tennis)

FIG. 1 Activity types classified by intensity. Examples of varying activity types by intensity.

Individual differences and the effect of aerobic exercise

Currently, most aerobic exercise studies in chronic pain examine the relationship between aerobic activity and pain intensity on a trait-level or macrotemporal processes, which are processes that are stable or slow to change (e.g., higher trait-level of physical activity) (Dunton, 2018). However, physical activity is often characterized on a state-level or microtemporal process, which are processes that change on a short timescale (e.g., across minutes, hours, days) (Dunton, 2018). Investigating the impact of physical activity on a microtemporal level in addition to the macrotemporal level could provide valuable information on the relationship between physical activity and pain intensity. Specifically, identifying microtemporal processes could improve our understanding of the day-to-day fluctuations between physical activity and pain intensity above and beyond its macrotemporal processes. For example, examining microtemporal processes would allow us to answer questions such as, "Does experience increased levels of physical activity *than typical for oneself* relate to reduced pain intensity?" as well as whether momentary changes in a physical activity produce immediate or lagged reductions in pain intensity in free-living environments. In addition, identifying the microtemporal effects of physical activity on pain intensity could also lead to more precise predictive models for individuals (e.g., do increased levels of physical activity *than typical for oneself* relate to reduced pain intensity *across individuals*?). Further, one might be better able to elucidate the microtemporal association of pain intensity on physical activity.

Results from a recent observational study on youth with chronic abdominal pain suggest that it may be important to consider microtemporal level in addition to macrotemporal relationships when investigating the association between exercise and pain intensity (Kichline et al., 2019). Findings indicated that the microtemporal relationship between physical activity and pain intensity may differ based on the individual. Put in other words, when some participants had a particularly active day, they reported more pain intensity. On the other hand, when some participants had a particularly active day, they reported decreased pain intensity. Thus, these results suggest that the effect of exercise on a microtemporal level could be different based on the individual. One possibility is that individuals who experienced reductions in chronic pain intensity may have been consistently engaging in physical activity prior to the study and thus are experiencing a long-term benefit of reduced pain intensity associated with aerobic activity. Individuals who experienced increases in chronic pain intensity after engaging in more aerobic activity than usual may experience initial discomfort, leading to a heterogeneous response to aerobic activity (Rice et al., 2019). However, this study was limited by its observational design (Kichline et al., 2019). Further microtemporal studies are necessary to better understand state-level fluctuations between physical activity and pain intensity.

Aerobic exercise and strength training in pediatric chronic pain

Within multimodal treatment programs, aerobic exercise is just a type of exercise to alleviate chronic pain in children. Other forms of exercise include strength training in the forms of isometric exercise and dynamic resistance exercise (Naugle et al., 2012). Isometric exercises involve static muscle contractions without any joint movement and dynamic resistance exercises incorporate muscle contractions with joint movements (Naugle et al., 2012). Some research has investigated the effect of aerobic activity in combination with neuromuscular strength training exercises on chronic pediatric pain intensity (Kashikar-Zuck et al., 2018). Neuromuscular training consists of exercises derived from an injury prevention standpoint that focuses on posture, strength, balance, and functional movement (Tran et al., 2016). Kashikar-Zuck et al. (2018) aimed to examine the effect of FIT Teens, a new intervention that integrated neuromuscular strength training and aerobic exercise with cognitive behavioral therapy, on chronic pediatric pain intensity. Findings indicated that participants that engaged in FIT Teens (i.e., neuromuscular training, aerobic activity, and cognitive behavioral therapy) reported lower levels of pain intensity compared to participants that only received cognitive behavioral therapy (Kashikar-Zuck et al., 2018). Thus, results suggest that engaging in both aerobic activity and strength training contributes to reduced pain intensity, above and beyond cognitive behavioral therapy. However, it is unclear if the reduction in pain intensity can be attributed to aerobic activity, strength training, or the combination of aerobic activity and strength training. Therefore, future studies should aim to clarify the effects of strength training and aerobic activity, separately and combined, to better understand the important treatment components. In addition, the differential effects of isometric and dynamic resistance exercises on reductions in chronic pain have yet to be clarified.

Conclusion

The current literature suggests that aerobic activity may be an effective treatment modality for reducing pediatric chronic pain intensity. However, the overall effect of aerobic exercise on pain intensity remains inconclusive due to the low

518 PART | IV Novel and nonpharmacological aspects and treatments

methodological quality and high risks of bias in these studies. Thus, the results highlight the need for additional randomized control trials that investigate the relationship between aerobic exercise and chronic pain intensity. In addition, increased sample sizes could allow us to better understand how aerobic activity impacts chronic pain intensity across pediatric pain conditions. Further, future research could identify microtemporal relationships between aerobic activity and pain intensity, and ultimately improve treatment by providing precise models tailored on an individual level.

Application to other areas

Utilizing mobile health (mHealth), which are defined as the use of technology to function as an active treatment ingredient, such as health behavior interventions (Cushing & Steele, 2010; van Heerden, Tomlinson, & Swartz, 2012), to deliver aerobic exercise interventions could address some limitations associated with current face-to-face interventions for youth with chronic pain. Specifically, current face-to-face interventions are often expensive, time-intensive, and burdensome for pediatric pain patients. The majority of face-to-face physical activity interventions require patients to come in for multiple sessions over the course of months (Kichline & Cushing, 2019). Further, face-to-face interventions may also decrease accessibility to large segments of populations (Cushing, 2017). Prior research indicates that pediatric chronic pain patients have significant limitations to pediatric care due to geographic distance from treatment centers (Palermo, 2013). Specifically, one study found that chronic pediatric pain patients waited for 6 months on average before seeing a pain provider (Palermo et al., 2019). Thus, increasing access to health interventions, such as physical activity interventions, is likely important for pediatric pain populations and improved health outcomes.

mHealth interventions could address the limitations of face-to-face interventions could have important research and clinical implications. First, mHealth interventions could reduce cost and burden for participants and potentially increase researchers' feasibility to conduct randomized control trials, therefore, ultimately improving our knowledge about the effect of aerobic exercise and pediatric chronic pain intensity. Second, if researchers find that aerobic activity significantly reduces chronic pain intensity, then mHealth interventions could improve accessibility to large segments of the pediatric pain population. For example, 88% of teens own or have access to cell phones (Lenhart, 2015); thus, this suggests that an mHealth aerobic activity intervention would likely increase accessibility. In addition, the use of mHealth interventions could reduce the burden and cost for the participant, increasing the feasibility of delivering and continuing treatment. Therefore, if aerobic activity interventions effectively reduce chronic pain intensity, utilizing mHealth interventions could ultimately improve its treatment efficacy.

Previous studies have utilized mHealth interventions to change activity levels in healthy populations (Bond et al., 2014; Militello, Melnyk, Hekler, Small, & Jacobson, 2016). For example, one research group utilized an mHealth text messaging intervention that included automated goal setting, passive objective monitoring of sedentary time, and text messages with prompts and feedback (Bond et al., 2014). mHealth physical activity interventions could potentially be applied in youth with chronic pain. Future studies may consider further investigating and developing mHealth aerobic exercise interventions for chronic pediatric pain to improve our understanding of aerobic activity on chronic pain intensity.

Other agents of interest

Psychological treatment of pediatric chronic pain typically employs cognitive-behavioral approaches, which use cognitive strategies, such as cognitive restructuring, and behavioral strategies, such as behavioral activation and scheduling activities to reduce anxiety and depression (Odell & Logan, 2013). Previous literature indicates that improved psychosocial outcomes, such as lower levels of depression and anxiety, are related to better pain outcomes (Linton & Shaw, 2011; Norris, Carroll, & Cochrane, 1992; Severeijns, Vlaeyen, van den Hout, & Weber, 2001). For example, cognitive-behavioral therapy often includes aerobic activity as a form of behavioral activation (Dimidjian, Barrera, Martell, Muñoz, & Lewinsohn, 2011). Pediatric pain patients that engage in aerobic exercise may ultimately be engaging in behavioral activation, which has been shown to improving anxiety and depression (Dimidjian et al., 2011). This reduction in anxiety and depressive symptoms may ultimately reduce pain intensity (Asmundson & Katz, 2009). In addition, engaging in scheduled aerobic activity could reduce pediatric pain intensity because it serves as a distraction for youth with chronic pediatric pain. Previous literature indicates that distraction is related to reduced pain intensity in youth (Walker et al., 2006). Further, aerobic exercise has been shown to increase hippocampal volume, which has been shown to be related to reduced anxiety and depressive symptoms (Cha et al., 2016; Den Ouden et al., 2018). It is possible that this increase in hippocampal volume and its related effects on depression and anxiety could relate to reduced pain intensity. Thus, cognitive and behavioral mechanisms may also underlie EIH modulation of chronic pain, although more research is needed on this topic in youth with chronic pain.

Mini-dictionary of terms

- *Aerobic activity*. Aerobic activity is defined as physical activity that includes low, moderate, to high intensity. When engaged in aerobic activity, one's heart rate increases and should exhibit some difficulty engaging in conversation while performing the aerobic activity. Examples of aerobic activity include brisk walking, tennis, swimming, and running.
- *Chronic pain*. Chronic pain is defined as pain that is persistent or recurrent that lasts longer than 3 months.
- *Exercise-induced hypoalgesia*. The process by which acute exercise reduces pain sensitivity.
- *Macrotemporal processes*. Macrotemporal processes provide information about stable or slow-to-change relationships.
- *Microtemporal processes*. Microtemporal processes provide information about state-level factors that fluctuate or change on a smaller timescale (e.g., minutes, hours, days).

Key facts

- The World Health Organization and Center for Disease Control and Prevention recommend that youth engage in 60 min of physical activity daily.
- Chronic pediatric pain conditions include a variety of pain areas including headache, abdominal pain, amplified musculoskeletal pain syndrome (AMPS), juvenile arthritis, and complex regional pain syndrome (CRPS).
- Youth with chronic pain do not engage in recommended activity levels.
- The pain response to activity is likely variable depending on the individual.
- Aerobic exercise is theorized to reduce pain sensitivity through a variety of mechanisms influencing opioid and non-opioid systems such as serotonergic systems, changes in N-methyl-D-aspartate (NDMA) receptors, and endocannabinoid systems.

Summary points

- This chapter focuses on aerobic activity levels and the effect of aerobic exercise on chronic pain intensity.
- Youth with chronic pain do not engage in the recommended levels of aerobic activity. Thus, they are more likely to have negative health-related outcomes, such as obesity, diabetes, and heart disease.
- Aerobic exercise may reduce chronic pain through a variety of processes including reduced facilitation of pain through opioid and non-opioid systems including such as serotonergic systems, changes in N-methyl-D-aspartate (NDMA) receptors, endocannabinoid systems, and beta-endorphins.
- Preliminary literature suggests that aerobic activity reduces pediatric chronic pain intensity.
- However, the overall effect of aerobic exercise on pain intensity remains inconclusive due to the low methodological quality and high risks of bias in these studies.
- There is a clear need for increased randomized control trials examining the effect of aerobic exercise on chronic pain intensity.

References

Asmundson, G. J., & Katz, J. (2009). Understanding the co-occurrence of anxiety disorders and pain: State-of-the-art. *Depression and Anxiety, 26,* 888–901. https://doi.org/10.1002/da.20600.

Bement, M., & Sluka, K. (2016). Exercise-induced analgesia: An evidence-based review. In K. Sluka (Ed.), *Mechanisms and management of pain for the physical therapist* (2nd ed., pp. 177–201). Seattle, WA: Wolters Kuwer, IASP Press.

Bohr, A. H., Nielsen, S., Müller, K., Karup Pedersen, F., & Andersen, L. B. (2015). Reduced physical activity in children and adolescents with juvenile idiopathic arthritis despite satisfactory control of inflammation. *Pediatric Rheumatology, 13*(1), 57–66. https://doi.org/10.1186/s12969-015-0053-5.

Bond, D. S., Thomas, J. G., Raynor, H. A., Moon, J., Sieling, J., Trautvetter, J., et al. (2014). B-MOBILE—A smartphone-based intervention to reduce sedentary time in overweight/obese individuals: A within-subjects experimental trial. *PLoS One, 9*(6). https://doi.org/10.1371/journal.pone.0100821, e100821.

Bonilla, S., Wang, D., & Saps, M. (2011). Obesity predicts persistence of pain in children with functional gastrointestinal disorders. *International Journal of Obesity, 35*(4), 517–521. https://doi.org/10.1038/ijo.2010.245.

Busch, A. J., Barber, K. A. R., Overend, T. J., Peloso, P. M. J., & Schachter, C. L. (2007). Exercise for treating fibromyalgia syndrome. *Cochrane Database of Systematic Reviews, 4.* https://doi.org/10.1002/14651858.CD003786.pub2.

Busch, A., Schachter, C. L., Peloso, P. M., & Bombardier, C. (2002). Exercise for treating fibromyalgia syndrome (Cochrane review). *Cochrane Database of Systematic Reviews, 3,* 1–36.

Cha, J., Greenberg, T., Song, I., Blair Simpson, H., Posner, J., & Mujica-Parodi, L. R. (2016). Abnormal hippocampal structure and function in clinical anxiety and comorbid depression. *Hippocampus, 26*(5), 545–553. https://doi.org/10.1002/hipo.22566.

Chiou, E., & Nurko, S. (2010). Management of functional abdominal pain and irritable bowel syndrome in children and adolescents. *Expert Review of Gastroenterology & Hepatology, 4*(3), 293–304. https://doi.org/10.1586/egh.10.28.

Cushing, C. C. (2017). eHealth applications in pediatric psychology. In M. C. Roberts, & R. G. Steele (Eds.), *Handbook of pediatric psychology* (5th ed., pp. 201–211). New York: The Guilford Press.

Cushing, C. C., & Steele, R. G. (2010). A meta-analytic review of eHealth interventions for pediatric health promoting and maintaining behaviors. *Journal of Pediatric Psychology, 35*(9), 937–949. https://doi.org/10.1093/jpepsy/jsq023.

Den Ouden, L., Kandola, A., Suo, C., Hendrikse, J., Costa, R. J. S., Watt, M. J., et al. (2018). The influence of aerobic exercise on hippocampal integrity and function: Preliminary findings of a multi-modal imaging analysis. *Brain Plasticity, 4*(2), 211–216. https://doi.org/10.3233/bpl-170053.

Dhurandhar, N. V., Schoeller, D., Brown, A. W., Heymsfield, S. B., Thomas, D., Sørensen, T. I. A., et al. (2015). Energy balance measurement: When something is not better than nothing. *International Journal of Obesity, 39*(7), 1109–1113. https://doi.org/10.1038/ijo.2014.199.

Dimidjian, S., Barrera, M., Martell, C., Muñoz, R. F., & Lewinsohn, P. M. (2011). The origins and current status of behavioral activation treatments for depression. *Annual Review of Clinical Psychology, 7*(1), 1–38. https://doi.org/10.1146/annurev-clinpsy-032210-104535.

Dodds, C. B., Bjornson, K. F., Sweeney, J. K., & Narayanan, U. G. (2016). The effect of supported physical activity on parental-reported sleep qualities and pain severity in children with medical complexity. *Journal of Pediatric Rehabilitation Medicine, 9*(3), 195–206.

Dowell, D., Haegerich, T. M., & Chou, R. (2016). CDC guideline for prescribing opioids for chronic pain-United States, 2016. *Journal of the American Medical Association, 315*(15), 1624–1645. https://doi.org/10.1001/jama.2016.1464.

Dunton, G. F. (2018). Sustaining health-protective behaviors such as physical activity and healthy eating. *Journal of the American Medical Association, 320*(7), 639–640. https://doi.org/10.1001/jama.2018.6621.

Elnaggar, R. K., & Elshafey, M. A. (2016). Effects of combined resistive underwater exercises and interferential current therapy in patients with juvenile idiopathic arthritis: A randomized controlled trial. *American Journal of Physical Medicine & Rehabilitation, 95*(2), 96–102.

Epps, H., Ginnelly, L., Utley, M., Southwood, T., Gallivan, S., Sculpher, M., et al. (2005). Is hydrotherapy cost-effective? A randomised controlled trial of combined hydrotherapy programmes compared with physiotherapy land techniques in children with juvenile idiopathic arthritis. *Health Technology Assessment (Winchester, England), 9*(39), iii–iv.

Fanucchi, G. L., Stewart, A., Jordaan, R., & Becker, P. (2009). Exercise reduces the intensity and prevalence of low back pain in 12-13 year old children: A randomised trial. *The Australian Journal of Physiotherapy, 55*(2), 97–104. https://doi.org/10.1016/S0004-9514(09)70039-X.

Hunfeld, J. A. M., Perquin, C. W., Hazebroek-Kampschreur, A. A. J. M., Passchier, J., Van Suijlekom-Smit, L. W. A., & Van der Wouden, J. C. (2002). Physically unexplained chronic pain and its impact on children and their families: The mother's perception. *Psychology and Psychotherapy: Theory, Research and Practice, 75*(3), 251–260. https://doi.org/10.1348/147608302320365172.

Jansen, M. J., Viechtbauer, W., Lenssen, A. F., Hendriks, E. J. M., & de Bie Rob, A. A. (2011). Strength training alone, exercise therapy alone, and exercise therapy with passive manual mobilisation each reduce pain and disability in people with knee osteoarthritis: A systematic review. *Journal of Physiotherapy, 57*(1), 11–20. https://doi.org/10.1016/S1836-9553(11)70002-9.

Kashikar-Zuck, S., Black, W. R., Pfeiffer, M., Peugh, J., Williams, S. E., Ting, T. V., et al. (2018). Pilot randomized trial of integrated cognitive-behavioral therapy and neuromuscular training for juvenile fibromyalgia: The FIT teens program. *The Journal of Pain, 19*(9), 1049–1062. https://doi.org/10.1016/j.jpain.2018.04.003.

Kashikar-Zuck, S., Flowers, S. R., Verkamp, E., Ting, T. V., Lynch-Jordan, A. M., Graham, T. B., et al. (2010). Actigraphy-based physical activity monitoring in adolescents with juvenile primary fibromyalgia syndrome. *The Journal of Pain, 11*(9), 885–893. https://doi.org/10.1016/j.jpain.2009.12.009.

Kashikar-Zuck, S., & Ting, T. V. (2014). Juvenile fibromyalgia: Current status of research and future developments. *Nature Reviews Rheumatology, 10*(2), 89–96. https://doi.org/10.1038/nrrheum.2013.177.

Kichline, T., & Cushing, C. C. (2019, April). A systematic review and quantitative analysis on the impact of aerobic exercise on pain intensity in children with chronic pain. *Children's Health Care, 48*, 244–261. https://doi.org/10.1080/02739615.2018.1531756.

Kichline, T., Cushing, C. C., Ortega, A., Friesen, C., & Schurman, J. V. (2019). Associations between physical activity and chronic pain severity in youth with chronic abdominal pain. *Clinical Journal of Pain, 35*(7), 618–624. https://doi.org/10.1097/AJP.0000000000000716.

King, S., Chambers, C. T., Huguet, A., Mac Nevin, R. C., McGrath, P. J., Parker, L., et al. (2011). The epidemiology of chronic pain in children and adolescents revisited: A systematic review. *Pain, 152*(12), 2729–2738. https://doi.org/10.1016/j.pain.2011.07.016.

Klepper, S. E. (1999). Effects of an eight-week physical conditioning program on disease signs and symptoms in children with chronic arthritis. *Arthritis Care & Research, 12*(1), 52–60.

Koltyn, K. F., Brellenthin, A. G., Cook, D. B., Sehgal, N., & Hillard, C. (2014). Mechanisms of exercise-induced hypoalgesia. *The Journal of Pain, 15*(12), 1294–1304. https://doi.org/10.1016/j.jpain.2014.09.006.

Konijnenberg, A. Y., Uiterwaal, P. M., Kimpen, J. L. L., Van Der Hoeven, J., Buitelaar, J. K., & De Graeff, E. R. (2005). Children with unexplained chronic pain: Substantial impairment in everyday life. *Archives of Disease in Childhood, 90*(7), 680–686. https://doi.org/10.1136/adc.2004.056820.

Lee, I. M., Shiroma, E. J., Lobelo, F., Puska, P., Blair, S. N., Katzmarzyk, P. T., et al. (2012). Effect of physical inactivity on major non-communicable diseases worldwide: An analysis of burden of disease and life expectancy. *Lancet, 380*(9838), 219–229. https://doi.org/10.1016/S0140-6736(12)61031-9.

Lenhart, A. (2015). *Teens, Social Media & Technology Overview*, 2015. http://www.pewinternet.org/2015/04/09/teens-social-media-technology-2015/.

Lima, L. V., Abner, T. S. S., & Sluka, K. A. (2017). Does exercise increase or decrease pain? Central mechanisms underlying these two phenomena. *Journal of Physiology, 595*(13), 4141–4150. https://doi.org/10.1113/JP273355.

Linton, S. J., & Shaw, W. S. (2011). Impact of psychological factors in the experience of pain. *Physical Therapy, 91*(5), 700–711.

Long, A. C., Palermo, T. M., & Manees, A. M. (2008). Brief report: Using actigraphy to compare physical activity levels in adolescents with chronic pain and healthy adolescents. *Journal of Pediatric Psychology, 33*(6), 660–665. https://doi.org/10.1093/jpepsy/jsm136.

Mendonça, T. M., Terreri, M. T., Silva, C. H., Neto, M. B., Pinto, R. M., Natour, J., et al. (2013). Effects of Pilates exercises on health-related quality of life in individuals with juvenile idiopathic arthritis. *Archives of Physical Medicine and Rehabilitation, 94*(11), 2093–2102.

Militello, L. K., Melnyk, B. M., Hekler, E., Small, L., & Jacobson, D. (2016). Correlates of healthy lifestyle beliefs and behaviors in parents of overweight or obese preschool children before and after a cognitive behavioral therapy intervention with text messaging. *Journal of Pediatric Health Care, 30*(3), 252–260. https://doi.org/10.1016/j.pedhc.2015.08.002.

Nader, P. R., Bradley, R. H., Houts, R. M., McRitchie, S. L., & O'Brien, M. (2008). Moderate-to-vigorous physical activity from ages 9 to 15 years. *Journal of the American Medical Association, 300*(3). https://doi.org/10.1001/jama.300.3.295.

Naugle, K. M., Fillingim, R. B., & Riley, J. L. (2012). A meta-analytic review of the hypoalgesic effects of exercise. *The Journal of Pain, 13*, 1139–1150. https://doi.org/10.1016/j.jpain.2012.09.006.

Norris, R., Carroll, D., & Cochrane, R. (1992). The effects of physical activity and exercise training on psychological stress and well-being in an adolescent population. *Journal of Psychosomatic Research, 36*(1), 55–65.

O'Brien, H. L., Kabbouche, M. A., & Hershey, A. D. (2012). Treating pediatric migraine: An expert opinion. *Expert Opinion on Pharmacotherapy, 13*(7), 959–966. https://doi.org/10.1517/14656566.2012.677434.

Odell, S., & Logan, D. E. (2013). Pediatric pain management: The multidisciplinary approach. *Journal of Pain Research, 6*, 785–790. https://doi.org/10.2147/JPR.S37434.

Olsen, M. N., Sherry, D. D., Boyne, K., McCue, R., Gallagher, P. R., & Brooks, L. J. (2013). Relationship between sleep and pain in adolescents with juvenile primary fibromyalgia syndrome. *Sleep, 36*(4), 509–516.

Palermo, T. M. (2000). Impact of recurrent and chronic pain on child and family daily functioning: A critical review of the literature. *Journal of Developmental and Behavioral Pediatrics, 21*(1), 58–69. https://doi.org/10.1097/00004703-200002000-00011.

Palermo, T. M. (2012). *Cognitive-behavioral therapy for chronic pain in children and adolescents.* New York: Oxford University Press.

Palermo, T. M. (2013). Remote management of pediatric pain. In G. F. Gebhart, & R. F. Schmidt (Eds.), *Encyclopedia of pain* (pp. 3389–3393). https://doi.org/10.1007/978-3-642-28753-4_4967.

Palermo, T. M., Slack, M., Zhou, C., Aaron, R., Fisher, E., & Rodriguez, S. (2019). Waiting for a pediatric chronic pain clinic evaluation: A prospective study characterizing waiting times and symptom trajectories. *The Journal of Pain, 20*(3), 339–347. https://doi.org/10.1016/j.jpain.2018.09.009.

Perquin, C. W., Hazebroek-Kampschreur, A. A. J. M., Hunfeld, J. A. M., Bohnen, A. M., Van Suijlekom-Smit, L. W. A., Passchier, J., et al. (2000). Pain in children and adolescents: A common experience. *Pain, 87*(1), 51–58. https://doi.org/10.1016/S0304-3959(00)00269-4.

Porreca, F., Ossipov, M. H., & Gebhart, G. F. (2002). Chronic pain and medullary descending facilitation. *Trends in Neurosciences, 25*(6), 319–325. https://doi.org/10.1016/s0166-2236(02)02157-4.

Rabbitts, J. A., Holley, A. L., Karlson, C. W., & Palermo, T. M. (2014). Bidirectional associations between pain and physical activity in adolescents. *Clinical Journal of Pain, 30*(3), 251–258. https://doi.org/10.1097/AJP.0b013e31829550c6.

Rice, D., Nijs, J., Kosek, E., Wideman, T., Hasenbring, M. I., Koltyn, K., et al. (2019). Exercise-induced hypoalgesia in pain-free and chronic pain populations: State of the art and future directions. *The Journal of Pain, 20*(11), 1249–1266. https://doi.org/10.1016/j.jpain.2019.03.005.

Sallis, J. F., & Saelens, B. E. (2000). Assessment of physical activity by self-report: Status, limitations, and future directions. *Research Quarterly for Exercise and Sport, 71*, 1–14. https://doi.org/10.1080/02701367.2000.11082780.

Severeijns, R., Vlaeyen, J. W., van den Hout, & Weber, W. E. (2001). Pain catastrophizing predicts pain intensity, disability, and psychological distress independent of the level of physical impairment. *The Clinical Journal of Pain, 17*(2), 165–172.

Sluka, K. A., Danielson, J., Rasmussen, L., & Dasilva, L. F. (2012). Exercise-induced pain requires nmda receptor activation in the medullary raphe nuclei. *Medicine and Science in Sports and Exercise, 44*(3), 420–427. https://doi.org/10.1249/MSS.0b013e31822f490e.

Sluka, K. A., O'Donnell, J. M., Danielson, J., & Rasmussen, L. A. (2013). Regular physical activity prevents development of chronic pain and activation of central neurons. *Journal of Applied Physiology, 114*(6), 725–733. https://doi.org/10.1152/japplphysiol.01317.2012.

Stephens, S., Feldman, B. M., Bradley, N., Schneiderman, J., Wright, V., Singh-Grewal, D., et al. (2008). Feasibility and effectiveness of an aerobic exercise program in children with fibromyalgia: Results of a randomized controlled pilot trial. *Arthritis Care & Research: Official Journal of the American College of Rheumatology, 59*(10), 1399–1406.

Tarakci, E., Yeldan, I., Baydogan, S. N., Olgar, S., & Kasapcopur, O. (2012). Efficacy of a land-based home exercise programme for patients with juvenile idiopathic arthritis: A randomized, controlled, single-blind study. *Journal of Rehabilitation Medicine, 44*(11), 962–967.

Thorén, P., Floras, J. S., Hoffmann, P., & Seals, D. R. (1990). Endorphins and exercise: Physiological mechanisms and clinical implications. *Medicine and Science in Sports and Exercise, 22*(4), 417–428. https://doi.org/10.1249/00005768-199008000-00001.

Tran, S. T., Thomas, S., DiCesare, C., Pfeiffer, M., Sil, S., Ting, T. V., et al. (2016). A pilot study of biomechanical assessment before and after an integrative training program for adolescents with juvenile fibromyalgia. *Pediatric Rheumatology, 14*(1), 53. https://doi.org/10.1186/s12969-016-0103-7.

Tsiros, M. D., Buckley, J. D., Howe, P. R. C., Walkley, J., Hills, A. P., & Coates, A. M. (2014). Musculoskeletal pain in obese compared with healthy-weight children. *The Clinical Journal of Pain, 30*(7), 583–588. https://doi.org/10.1097/AJP.0000000000000017.

US Department of Health and Human Services. (2018). *Physical activity guidelines for Americans* (2nd ed.). Washington, DC.

van Dijk-Lokkart, Braam, K. I., van Dulmen-den Broeder, E., Kaspers, G. J., Takken, T., & Huisman, J. (2016). Effects of a combined physical and psychosocial intervention program for childhood cancer patients on quality of life and psychosocial functioning: Results of the QLIM randomized clinical trial. *Psycho-oncology, 25*(7), 815–822.

van Heerden, A., Tomlinson, M., & Swartz, L. (2012). Point of care in your pocket: A research agenda for the field of m-health. *Bulletin of the World Health Organization, 90*(5), 393–394. https://doi.org/10.2471/BLT.11.099788.

Walker, L. S., Williams, S. E., Smith, C. A., Garber, J., Van Slyke, D. A., & Lipani, T. A. (2006). Parent attention versus distraction: Impact on symptom complaints by children with and without chronic functional abdominal pain. *Pain, 1*, 43–52.

Wilson, A. C., & Palermo, T. M. (2012). Physical activity and function in adolescents with chronic pain: A controlled study using actigraphy. *The Journal of Pain, 13*(2), 121–130. https://doi.org/10.1016/j.jpain.2011.08.008.

Wilson, A. C., Samuelson, B., & Palermo, T. M. (2010). Obesity in children and adolescents with chronic pain: Associations with pain and activity limitations. *Clinical Journal of Pain, 26*(8), 705–711. https://doi.org/10.1097/AJP.0b013e3181e601fa.

World Health Organization. (2010). *Global recommendations on physical activity for health.*

Chapter 44

Physical activity and exercise in the management of chronic widespread musculoskeletal pain: A focus on fibromyalgia

Thomas Davergne[a,*], Fernando Estévez-López[b,*], Ana Carbonell-Baeza[c,†], and Inmaculada C. Álvarez-Gallardo[c,†]

[a]Sorbonne Université, INSERM, Pierre Louis Institute of Epidemiology and Public Health (IPLESP), Paris, France, [b]Department of Pediatrics, Wilhelmina Children's Hospital, University Medical Center Utrecht, Utrecht University, Utrecht, the Netherlands, [c]Department of Physical Education, Faculty of Education Sciences, University of Cádiz, Cádiz, Spain

Abbreviations

EULAR	EUropean ALliance of Associations for Rheumatology
NICE	The National Institute for Health Care
FITT	frequency, intensity, time and type of exercise
VO₂R	oxygen uptake reserve
HRR	heart rate reserve
RM	repetition maximum
RPE	rated perceived exertion

Musculoskeletal pain

Musculoskeletal pain includes muscle or joint disorders in different parts of the body, such as the neck, shoulders, hips, and knees. New evidence shows that all musculoskeletal pain disorders share common biopsychosocial risk profiles for pain and disability. Similarly, clinical practice guidelines are converging toward common recommendations, regardless of the body region (Caneiro et al., 2019; Lin et al., 2020). Following these recommendations, health professionals are encouraged to:

- provide management addressing physical activity and/or exercise,
- screen for biopsychosocial factors and health comorbidities,
- embrace patient-centered communication,
- educate beyond words using active learning approaches,
- coach toward self-management,
- unless specifically indicated (e.g., red flag condition), offer evidence-informed non-surgical care prior to surgery, and
- facilitate continuation or resumption of work.

Physical activity in musculoskeletal pain: A historical perspective

In the last decades, the dominant view about physical activity in musculoskeletal pain has changed dramatically. The common advice for pain management has changed from suggesting to pace down activity and rest to increase physical activity and reduce sedentary behaviors. Currently, physical activity and exercise programs are increasingly promoted

---

* Thomas Davergne and Fernando Estévez-López equally contributed as the first author to this chapter.

† Ana Carbonell-Baeza and Inmaculada C. Álvarez-Gallardo equally contributed as the last author to this chapter.

Treatments, Mechanisms, and Adverse Reactions of Anesthetics and Analgesics. https://doi.org/10.1016/B978-0-12-820237-1.00044-2
Copyright © 2022 Elsevier Inc. All rights reserved.

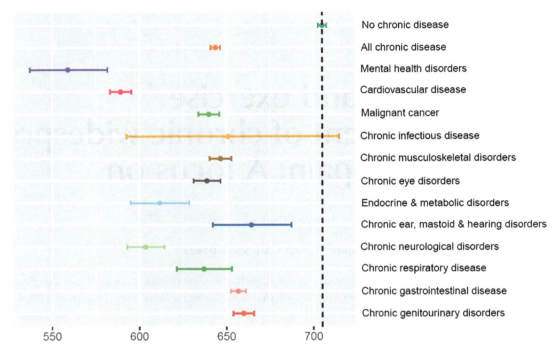

FIG. 1 Geometric mean moderate activity in minutes per week on 96,706 participants with and without chronic diseases (Barker et al., 2019). (© The Author(s) 2019. Published by Oxford University Press on behalf of the International Epidemiological Association. This is an Open Access article distributed under the terms of the Creative Commons Attribution License (http://creativecommons.org/licenses/by/4.0/), which permits unrestricted reuse, distribution, and reproduction in any medium, provided the original work is properly cited.)

and offered in various health care systems and national organizations (such as the NICE) and international organizations (such as the EUropean ALliance of Associations for Rheumatology (EULAR)) (Lin et al., 2020).

Levels of physical activity in musculoskeletal pain

Despite the benefits of physical activity, people with chronic musculoskeletal disorders often avoid physical activity to prevent pain, fatigue, or flares. In particular, peers from the general population engage in physical activity in around 60 min/week more than people with chronic musculoskeletal disorders do (Barker et al., 2019). Fig. 1 shows that, as other chronic conditions, patients with chronic musculoskeletal disorders are significantly less active than healthy subjects (Barker et al., 2019).

Interventions to enhance physical activity in musculoskeletal pain: Current evidence and novel approaches

Two systematic reviews evaluated the effectiveness of the interventions focused on increasing physical activity in adults with a chronic musculoskeletal disorder (Marley et al., 2017; Oliveira et al., 2016). In general, the available evidence highlights that, in people with chronic musculoskeletal pain, physical activity interventions lead to little or no difference in objectively measured physical activity. A small effect was observed when physical activity was measured by questionnaires but mostly based on low-quality studies. Because both objectively and subjectively measured physical activities have their own strengths and limitations, it is advised to include both types of measures (Estevez-Lopez et al., 2016; Munguía-Izquierdo et al., 2019; Pulido-Martos et al., 2020).

New technologies might play a promising role in promoting physical activity with apps and activity trackers. A systematic review of interventions aiming to increase regular physical activity with a wearable activity tracker was performed in patients with rheumatic and musculoskeletal diseases (Davergne, Pallot, Dechartres, Fautrel, & Gossec, 2019). Meta-analysis based on seven studies ($n = 463$) showed at the end of the intervention a mean difference over the comparator group of 1520 steps/per day compared with groups without trackers. Activity trackers may be an effective option to increase physical activity in this at-risk population but with uncertainties for long-term effect.

Physical exercise in musculoskeletal pain

Physical exercise is considered the cornerstone of treatment for musculoskeletal pain and is recommended by clinical guidelines for a variety of conditions (Geenen et al., 2018a; Lin et al., 2020). Exercise-based treatments have shown a small to moderate effect on pain in musculoskeletal disorders. However, evidence is based on large variability in exercise prescriptions and on the low quality of studies (Geneen et al., 2017). In inflammatory arthritis and osteoarthritis, a multidisciplinary task force including professionals and patient representatives conducted a systematic literature review of systematic reviews (Geenen et al., 2018a). The results of this task force are summarized in Table 1. Overall, physical exercise may be an acceptable intervention for people with chronic pain, with minimal negative effects (increase in pain or muscle soreness, which would have subsided after a few weeks of intervention) (Geneen et al., 2017).

General advice to design physical exercise programs for chronic musculoskeletal pain

Musculoskeletal disorders are considered chronic when pain lasts longer than the normal tissue healing time, usually 3 months. The chronicity is generally linked with secondary pathologies, such as, for instance, anxiety and a persistent state of high pain reactivity called central sensitization (Booth et al., 2017). It is suggested that exercise should focus on the primary disorder as well as secondary pathologies for a bigger chance of success. A consensus on exercise for chronic musculoskeletal disorders is converging toward individualized and supervised exercise. These exercises should be based on the patient's presentation, goals, and preferences. Exercise should be perceived as safe and non-threatening to avoid fostering unnecessary associations between physical activity and pain. To reconceptualize pain-related fear through exercise, evidence from a narrative view suggests adopt the following goals treatment (Smith et al., 2019):

- figure out what the patient understands about pain,
- challenge unhelpful beliefs about pain and exercises,
- enhance self-efficacy in completing exercises,
- provide safety cues,
- provide advice on suitable levels of pain, i.e., searching for tolerable pain, not absence of pain, and
- provide advice on exercise modification.

The weight of evidence in favor of exercise has been provided by studies of aerobic and resistance exercise, yet there is considerable uncertainty about how best to apply the results to exercise prescription. Tables 2 and 3 display the recommendations from (Booth et al., 2017). Table 4 summarizes the previous literature in the field.

Physical activity in fibromyalgia

Fibromyalgia is a disease characterized by widespread musculoskeletal pain accompanied by other symptoms (e.g., fatigue and signs of depression). People with fibromyalgia spent more time in sedentary behaviors (Segura-Jiménez et al., 2015) and are less physically active than their peers from the general population (Joustra, Zijlema, Rosmalen, & Janssens, 2018; Segura-Jiménez et al., 2015). Only about 20% of the people with fibromyalgia fulfill the recommendation for 150 min/week of moderate-to-vigorous physical activity in bouts of at least 10 min/bout and 16% achieved the recommendation for 10,000 steps/day (Segura-Jiménez et al., 2015). Furthermore, women with fibromyalgia who are highly sedentary (Segura-Jimenez et al., 2015) and low physically active (Joustra et al., 2018; Segura-Jimenez et al., 2015) show greater pain and disease severity than those who are low sedentary and high physically active (Segura-Jimenez et al., 2015). Thus, despite having fibromyalgia, it is important to engage in physical activity. A feasible strategy to achieve health benefits in this condition might be to first, replace inactivity with light physical activity, and eventually increase physical activity to moderate intensity levels (Gavilán-Carrera et al., 2019).

To date, scarce research has examined the effect of interventions focused on increasing lifestyle physical activity on pain or disease severity in people with fibromyalgia. Fontaine, Conn, and Clauw (2010) observed that a 12-week program produced a 54% increase in the average number of steps taken per day (from 3800 at baseline to 5800 at the final, approximately). Compared to an educational group, these participants significantly reduced their disease severity by 18% and pain by 35% (Fontaine et al., 2010). However, opposite findings have been also found (Kaleth, Slaven, & Ang, 2014). Thus, future studies should focus on analyzing the effects of lifestyle physical activity interventions.

526 PART | IV Novel and nonpharmacological aspects and treatments

TABLE 1 Overview of systematic reviews of randomized trials regarding physical activity and exercise: Treatment modality and disease, the direction of effect and the quality of the evidence according to GRADE in a patient with rheumatoid arthritis, axial spondyloarthritis, and osteoarthritis.

Treatment modality disease	Reviews (n)	Direction of effect	GRADE quality of evidence
General exercise			
RA	5	o/+	⊕⊕
SpA	6	+	⊕⊕
OA-general	6	+	⊕⊕⊕
OA-hand/wrist	4	o/+	⊕⊕
OA-hip/knee	11	+	⊕⊕⊕
OA-hip	11	o/+	⊕⊕
OA-knee	18	+	⊕⊕⊕
OA-foot/ankle	2	+	⊕⊕
Aerobic exercise			
RA	3	o/+	⊕⊕
OA-general	3	+	⊕⊕⊕
OA-hip/knee	2	o/+	⊕⊕
OA-hip	1	o	⊕
OA-knee	9	+	⊕⊕⊕
Strength and resistance			
RA	2	o/+	⊕⊕
OA-general	3	+	⊕⊕⊕
OA-hand/wrist	2	o/+	⊕⊕
OA-hip/knee	4	+	⊕⊕⊕
OA-hip	3	+	⊕⊕
OA-knee	14	+	⊕⊕⊕
Tai chi, yoga, qigong, whole-body vibration			
RA	3	?/+	⊕
OA-general	6	o/+	⊕ to ⊕⊕
OA-hand/wrist	3	+	⊕
OA-hip/knee	1	o/+	⊕⊕
OA-knee	12	o/+	⊕ to ⊕⊕

Diseases without a review were excluded from the table. The direction of effect: +, positive; o, no; −, negative;?, unclear (effect equivocal) or a combination thereof meaning that different reviews reached different conclusions about the effect of treatment. GRADE: ++++, high; +++, moderate; ++ low; + to ++, very low to low; + very low. GRADE: Grades of Recommendation, Assessment, Development and Evaluation; OA; osteoarthritis; RA, rheumatoid arthritis; SpA, spondyloarthritis.

Aerobic exercise in fibromyalgia

Aerobic exercise is the most widely studied type of exercise in fibromyalgia. Bidonde et al. (2017) reviewed the effects of aerobic exercises versus control conditions. The authors observed that aerobic exercise, in comparison to the control group, resulted in low to moderate benefits in pain intensity and symptoms severity, corresponding to a decrease of pain intensity of 11% and symptoms severity of 8%, respectively. In this systematic review, the duration of aerobic exercises ranged from

TABLE 2 General guidelines for aerobic and resistance exercise prescriptions using the FIIT principle (frequency, intensity, time, and type) for people with chronic musculoskeletal pain.

Aerobic exercise	Intensity HR equivalents: low intensity: 40 < 55% HRmax; moderate intensity: 55 < 70% HRmax; high intensity: 70 < 90% HRmax
Frequency	$\geq$ 2 times/week; $\geq$ 6 weeks
Intensity	Low intensity (RPE 8–10) to moderate intensity (RPE 11–13). Higher intensity (RPE 14–16) for goals involving more demanding work, sport, or recreation where tolerated
Time	20–60 min and < 20 min with exercise intolerance. Consider shorter intervals interspersed with other exercise modalities (e.g., 3 × 7 min walking separated by resistance exercise)
Type	Modalities involving continuous and rhythmic exercises that engage major muscle groups but do not exacerbate symptoms (walking, jogging, swimming, dancing, etc.)
Progression	Commence RPE 8–10 grading to RPE 11–13 as tolerance increases; RPE $\geq$14 for high-intensity training. Increase duration before intensity (e.g., for treadmill walking increase duration and walking speed before incline)
Resistance exercise	Intensity 1RM equivalents: low intensity: 40 < 60% 1RM; moderate intensity: 60–70% 1RM; high intensity: $\geq$70% 1RM
Frequency	2–3 times/week; $\geq$ 6 weeks
Intensity	Low-intensity (RPE 8–10) to moderate-intensity exercise (RPE 11–13). For more demanding work, sport or recreation consider high-intensity training (RPE 14–16)
Time	For low- to moderate-intensity exercise 1–2 sets of 15–20 reps reduced/adapted for exercise intolerance. For high-intensity exercise 1–2 sets of 8–12 reps
Type	Modalities that engage muscles of affected body part(s) and/or major muscle groups (weight-bearing activity, free weights, floor exercise; machines, resistance bands, motor control exercise, etc.) that do not exacerbate symptoms
Progression	Commence RPE 8–10 grading to 11–13 as tolerance and function increases; RPE $\geq$14 for higher intensity training. Increase reps before load; commence floor exercise with short holds and higher reps and increase hold duration before exercise difficulty. For functional exercise commence at a level specific to the patient's presentation and increase reps before load

ex, exercise; HRmax, age-predicted hearth rate maximum; 1RM, 1 repetition maximum; RPE, rating of perceived exertion assessed using Borg 6–20 RPE scale; reps, repetitions.

TABLE 3 Key points concerning exercise prescription for chronic musculoskeletal pain.

- Understanding contemporary pain biology and 'explaining pain' are key competencies required for biopsychosocial treatment
- Frequently reassure patients that it is safe to move/pace up despite their symptoms
- Exercise prescription should be time, as opposed to pain, contingent using a tolerable/not tolerable dichotomy
- Have ready-made responses to flare-ups can reduce the severity
- Exercise should be individualized, enjoyable, related to patient goals, and with a level of supervision specific to the patient
- Many patients with chronic musculoskeletal pain will respond to lower exercise dosage than recommended for healthy individuals (i.e. graded low to moderate intensity)
- Closely observe and monitor exercise practice, seek and provide feedback, and correct poor technique
- Encourage patients to self-monitor exercise (diaries, activity trackers, etc.)
- Place emphasis on developing/restoring movement confidence and quality

TABLE 4 Characteristics of the studies testing the effectiveness of physical exercise in the management of pain- or severity-related outcomes in musculoskeletal pain.

Reference	Design, analyses	Condition	Intervention	Control group	Outcome(s): findings
Geneen et al. (2017)	Overview of Cochrane SRs of RCTs	Chronic pain in adults, including rheumatoid arthritis, osteoarthritis, fibromyalgia, low back pain, intermittent claudication, dysmenorrhoea, mechanical neck disorder, spinal cord injury, post-polio syndrome, and patellofemoral pain.	Aerobic, strength, flexibility, range of motion, core, balance training programs, yoga, Pilates, and tai chi	No exercise or minimal intervention as defined by the authors (usual care, waiting list control, placebo/sham treatment, other treatment, or a combination of treatments (as long as the effect of exercise could be measured distinctly))	Based on 16 SRs, the qualitative synthesis was that, at end of intervention and follow-up, exercise had a small to moderate beneficial effect on pain severity in most SRs. Only three SRs found no statistically significant changes in pain from any intervention. Results were inconsistent across interventions and follow-up. The quality of the evidence was low, largely due to small sample sizes and potentially underpowered studies. Adherence to exercises was the same in both groups Most of the adverse events were: increased soreness or muscle pain, which reportedly subsided after a few weeks of the intervention. Only one review reported death separately from other adverse events: the intervention was protective against death (based on the available evidence), though did not reach statistical significance

Geenen et al. (2018b)	A EULAR multidisciplinary task force including professionals and patient representatives conducted a systematic literature review of SRs	Inflammatory arthritis and osteoarthritis	General exercise, aerobic exercise, strength, and resistance training programs, Tai Chi, yoga, qigong, and whole-body vibration	No exercise or minimal intervention, such as advice during a consultation, use of brochures, and e-health psychoeducation	Based on 88 SRs or RCTs, the qualitative synthesis was that, at end of the intervention, general exercise, aerobic exercise, and strength and resistance training had a beneficial effect on pain in most of the included works. However, results were inconsistent with some SRs concluding no effects. The quality of studies was low to moderate Qualitative synthesis showed for tai chi, yoga, qigong, and whole-body vibration had an unclear effect on pain The quality of the evidence was low to very low
Smith et al. (2017)	SR of RCTs	Chronic musculoskeletal diseases	Exercises that was advised to be purposively painful, or where the pain was allowed or tolerated	Non-painful exercises(use therapeutic exercises that were pain-free)	In the short term ($\leq$3 months from randomization), based on 3 trials including 215 participants, MA showed a small beneficial short-term effect in favor of painful exercises (SMD -0.27 (95% CI -0.54 to -0.05) with moderate-quality evidence The quality of the evidence was low due to trial design and low participant numbers For pain in the medium ($>$3 and $<$ 12 months from randomization, 1 trial with 40 participants) and long term ($\geq$12 months from randomization, 3 trials with 215 participants), there was no significant difference The quality of the evidence was low to moderate due to trial design and low participant numbers

Continued

Reference	Design, analyses	Condition	Intervention	Control group	Outcome(s): findings
Regnaux et al. (2015)	Cochrane SR of RCTs	Knee or hip osteoarthritis	High-intensity physical activity or exercise program	Low-intensity physical activity or exercise program	Immediately at the end of the exercise program, based on 4 trials including 313 participants, MA showed a small beneficial short-term effect in favor of high-intensity exercises (MD -0.84 (95% CI -1.63 to −0.04). The quality of the evidence was low. At the mid-term (from 6 to 16 weeks), based on 2 trials including 199 participants, showed non-significant effects on pain (MD -0.82, 95% CI -1.90 to 0.26). The quality of the evidence was low. In the long-term (40 weeks), based on 1 trial including 139 participants, MA showed a small beneficial effect in favor of high-intensity exercise (MD -1.33, 95% CI -2.56 to −0.10). The quality of evidence was low
O'Keeffe, Hayes, McCreesh, Purtill, and O'Sullivan (2017)	SR of RCTs	Low back pain, neck pain, knee pain, and shoulder pain	Group-based physiotherapy exercise programs (mainly consisted of exercise, education, and relaxation)	Individual physiotherapy exercise programs (mainly consisted of exercise, education, and passive modalities)	In the short term (≤3 months from randomization), based on 12 trials including 480 participants, MA showed no statistically differences for pain intensity (MD = −0.39, 95% CI −3.07 to 2.28) or disability (MD = −0.02, 95% CI −0.21 to 0.16) between both groups. At the mid-term (>3

and < 12 months), based on seven studies with 361 participants, MA showed no statistically differences for pain intensity between both groups (MD = −1.41, 95% CI −4.48 to 1.66). A small, yet statistically significant, difference was found for disability between individual and group physiotherapy including exercise in 8 trials including 414 participants (MD = −0.20, 95% CI −0.41 to 0.00) At the long-term (≥12 months), based on 7 studies including 307 participants, MA showed that trials including exercise were more beneficial in group-based interventions in comparison to individual interventions (MD = −3.51, 95% CI −6.14 to −0.88).). A small, yet statistically significant, difference was found for disability between individual and group physiotherapy including exercise in 7 trials including 343 participants (MD = −0.14, 95% CI −0.30 to 0.01) A sensitivity analysis based on the 'high-quality' studies only (i.e., with a PEDro score > 6/10) showed no significant differences for pain

Continued

TABLE 4 Characteristics of the studies testing the effectiveness of physical exercise in the management of pain- or severity-related outcomes in musculoskeletal pain—cont'd

Reference	Design, analyses	Condition	Intervention	Control group	Outcome(s): findings
Hurley et al. (2018)	A mixed methods Cochrane SR	Participants aged 45 years or older, with a clinical or self-reported diagnosis of chronic osteoarthritis of hip or knee	NA	NA	Result of the patients' perspectives from qualitative synthesis, based on 12 trials including 6–29 participants indicates ways to improve the delivery of exercise interventions included: provide better information and advice about the safety and value of exercise; provide exercise tailored to individual's preferences, abilities and needs; challenge inappropriate health beliefs and provide better support. The quality of evidence was generally good
Marley et al. (2017)	SR of RCTs and quasi-RCTs	Individuals with persistent musculoskeletal pain: osteoarthritis (13 trials) and persistent low back pain (7 trials)	All interventions that had a clear aim of increasing physical activity		In the short term c $\leq$ 3 months from randomization) based on 9 trials including 1096 participants, MA showed no statistically differences for subjective measure of physical activity (SMD 0.24, 95% CI -0.07, 0.55). The quality of the evidence was downgraded from high to very low. At mid-term (>3 months, <6 months post randomization), based on 9 trials including 1309 participants, MA showed a small significant beneficial effect (SMD 0.25, 95% CI 0.01, 0.48). The quality of the evidence was downgraded from high to

| | | | | | moderate. In the long term (>6 months post randomization), based on 11 trials including 1872 participants, MA showed small significant beneficial effect (SMD 0.21, 95% CI 0.08, 0.33). The quality of the evidence was downgraded from high to moderate For measures of objective physical activity: interventions were ineffective based on very low to low quality evidence |
| Oliveira et al. (2016) | SR of RCTs or quasi-RCTs | Chronic (persistent or episodic pain lasting more than 3 months) musculoskeletal pain (around the axial skeleton (neck or low back) or large peripheral joints (hip, knee, or shoulder)) | any intervention of physical activity promotion, provided that it aimed to increase activity or fitness levels, delivered individually or to a group of patients, under clinician supervision or home-based, including physical activity coaching or counseling, cognitive-behavioral therapies, and feedback using electronic devices | No/minimal intervention: waiting list, placebo, and education (including information on pain management and self-care strategies) | In the short term (≤3 months follow up), based on 6 trials including 420 participants, MA showed a non- significant effect (SMD 0.34, 95% CI -0.09, 0.77) between physical activity interventions and no/ minimal interventions on objectively measured physical activity For intermediate follow up (>3, <6 months follow up), based on 5 trials including 347 participants, MA showed no statistically differences (SMD 0.03, 95% CI -0.20, 0.26) on objectively measured physical activity. For long-term follow up (>6 months follow up), based on 2 trials including 105 participants, MA showed a non- significant effect (SMD 0.22, 95% CI -0.17, 0.60) on objectively measured physical activity. The overall quality of evidence was considered to be low for all time points assessed |

Continued

Reference	Design, analyses	Condition	Intervention	Control group	Outcome(s): findings
Davergne et al. (2019)	SR of RCTs	patients with rheumatic and musculoskeletal diseases: lower-extremity osteoarthritis, low-back pain, or chronic inflammatory rheumatic diseases (i.e., spondyloarthritis, rheumatoid arthritis, psoriatic arthritis, or juvenile arthritis)	Any intervention using wearable activity tracker to increase physical activity (i.e., any device designed to be worn on the user's body, using accelerometers, with or without altimeters or other sensors to track the wearer's movements and/or biometric data)	Same intervention without tracker or waiting list	Immediately at the end if intervention, based on 7 trials including 463 patients, MA showed a large effect on mean daily steps (mean $\pm$ SD study duration 13.9 $\pm$ 6.9 weeks; SMD 0.83, 95% CI 0.29, 1.38), corresponding to a mean difference over the comparator group of 1520 steps per day [95% CI 580, 2460]). In studies with a prolonged follow-up after the end of the intervention period, no significant results were found for mean daily steps, based on 2 trials. Quality of evidence was not assessed by the authors
Eisele, Schagg, Krämer, Bengel, and Göhner (2019)	SR of RCTs, cluster-RCTs or quasi-RCTs	Patients with chronic (for at least three months) musculoskeletal conditions	Intervention aimed at improving physical activity levels adherence, contained at least one behavior change technics (as defined by Michie) targeting any kind of activity and were delivered by health professionals within any setting	Usual care, minimal intervention (not specified by the authors), placebo or no intervention	For medium-term follow-ups (3–6 months), based on 15 trials including 2138 participants, MA showed was a small, significant overall effect (SMD = 0.20, 95% CI 0.08–0.33, $p < 0.01$) in favor of the intervention in physical activity level collected by questionnaires. For long-term follow-ups (7–12 months), based on 7 trials including 1484 participants, MA showed no effect in physical activity (pooled

SMD = 0.13, 95% CI -0.02 to 0.28, $p = 0.09$)
There was a small and significant overall effect (SMD = 0.29, 95% CI 0.19–0.40, $p < 0.001$) for interventions using >8 BCTs and no effect (SMD = 0.08, 95% CI -0.11-0.27, $p = 0.41$) for interventions using <8 behavior change technics Methodological quality of the included original studies varied between moderate to high (18 studies with PEDro-Scores between 6 and 8) and poor (4 studies with scores of 5)

SMD, standardized mean difference; MD, mean difference; 95% CI, 95% confidence interval; RCT, randomized controlled trials; SR, systematic review; MA, meta-analysis; EULAR, the European League Against Rheumatism.

536 PART | IV Novel and nonpharmacological aspects and treatments

6 to 24 weeks, with exercise frequency of three times per week in most studies and an average intervention time of 35 min (minimum-maximum: 20–60). Walking was the most popular intervention examined in this review (Bidonde et al., 2017). The quality of evidence was low, mainly due to the risk of bias. Moreover, evidence on long-term effects (24–208 weeks post-intervention) shows persistent benefits on pain, but not on symptoms severity. In all, aerobic exercise seems to be well tolerated by the patients, with few adverse events (Andrade, Dominski, & Sieczkowska, 2020; Bidonde et al., 2017).

Water-based and land-based aerobic exercise showed an equivalent effect on pain (Hauser et al., 2010). Swimming and walking programs had similar beneficial effects on pain and symptoms severity. In addition, both swimming and walking appear to be a well-tolerated modality of aerobic exercise by patients with fibromyalgia (Fernandes, Jennings, Nery Cabral, Pirozzi Buosi, & Natour, 2016).

Resistance training in fibromyalgia

Several reviews showed that resistance training has positive effects on pain, tenderness with a reduction of the number of tender points in people with fibromyalgia (Andrade, de Azevedo Klumb Steffens, Sieczkowska, Peyré Tartaruga, & Torres Vilarino, 2018; Busch et al., 2013). Several studies have analyzed if a resistance training can be more effective on pain or symptoms severity than relaxation (Larsson et al., 2015; Silva et al., 2019) or flexibility training (Gavi et al., 2014), with contradictory results. In this sense, a progressive resistance training twice a week for 15 weeks improved symptoms severity (9%), current pain intensity (23%), pain disability (11%), and pain acceptance (9%) compared with relaxation therapy (Larsson et al., 2015). However, Silva et al. (2019) did not observe differences between the effects of a resistance training twice a week for 12 weeks and a relaxation intervention (Sophrology) on pain and symptoms severity. Compared with flexibility exercise, a progressive resistance training twice a week for 16 weeks showed greater improvements in pain, but there were no differences between the effects of both treatments on severity symptoms (Gavi et al., 2014).

Multicomponent exercise in fibromyalgia

Multicomponent exercise comprises all programs which combine two or more types of exercise (aerobic, resistance, balance, flexibility). Regarding randomized clinical trials published over the last decade that compared multicomponent exercises with usual care control (García-Martínez, De Paz, & Márquez, 2012; Paolucci et al., 2015; Sañudo et al., 2010), it can be observed that exercise decreased pain an average of 12% (6%–20%) and symptoms severity of 14% (10%–21%). These programs were supervised group interventions consisting of 60-min sessions, performed 2–5 times per week for 5–24 weeks which included aerobic, resistance, flexibility, and relaxation. Exercise intensity varied from light to moderate. Aerobic exercises ranged from 60% to 70% maximum heart rate (up to 85% maximum heart rate in one of them (García-Martínez et al., 2012)), 1–3 sets of 8–10 repetitions was used for resistance training and stretching holding the position for 30–60 s was the most common. A review (Bidonde et al., 2019) about multicomponent exercise found an improvement of 5% on pain and 7% on symptoms severity; however, authors reflected the high variability of exercise types used and exercise dosages (frequency, intensity, time, and duration).

Multicomponent exercise interventions seem to be effective for individuals with fibromyalgia. Yet, the evidence showed small to moderate effects with considerable variation in the way interventions were designed and delivered (Bidonde et al., 2019). This situation makes it difficult to have a conclusion about what combination of exercise components is more effective in people with fibromyalgia.

Mind-body exercise in fibromyalgia

Mind-body therapies have increased in popularity over the last decade as a treatment for fibromyalgia. For instance, a meta-analysis indicated that Tai Chi exerts a significantly greater effect on reducing the severity and pain score at 12–16 weeks than standard care (Cheng et al., 2019), being the Yang style Tai Chi the most practiced intervention. Furthermore, Tai chi intervention Yang style program twice weekly for 12 weeks had a positive effect on pain (Jones et al., 2012) and severity (Jones et al., 2012; Wang et al., 2010) compared with the educational group. In Table 5, the effects of mind-body interventions compared with usual care can be observed (Bravo, Skjaerven, Espart, Guitard Sein-Echaluce, & Catalan-Matamoros, 2019; Wong et al., 2018).

TABLE 5 Characteristics of randomized clinical trials studies testing the effectiveness of physical exercise interventions (aerobic, resistance, multicomponent and mind-body) in the management of pain- or severity-related outcomes in fibromyalgia.

Reference	Design, analyses	Sample	Intervention				Outcome(s)[a]: findings	Adherence (A, mean, range) and side effects (SE)
			Type	Frequency, length	Duration	Intensity		
Aerobic training								
Andrade, Zamunér, Forti, Tamburús, and Silva (2019)	RCT, Intention-to-treat analyses	$n_{ex} = 27$ $n_c = 27$	Aquatic physical training in a heated pool (30 ± 2 °C)	2 day/week, 16 weeks	45′: 5′ stretching, 5′ walking exercises and lateral displacement, 30′ aerobic exercise protocol, 5′ relaxation	From 80% VAT HR to 110% VAT HR	Pressure pain threshold: increased (10% in Ex) Pain (VAS): decreased (7% in Ex) Severity: decreased (18% in Ex)	A: not reported SE: not reported
Kayo, Peccin, Sanches, and Trevisani (2012)	RCT Intention-to-treat analyses[b]	$n_{ex} = 30$ $n_c = 30$	Supervised walking either outdoors or indoors in a gymnasium, depending on the weather	3 day/week, 16 weeks	60′: 5–10′ stretching of a warm-up period, 25–50′ walking and 5′ cool down period	From 40 to 50% and progressed to 60–70% HRR	Pain (VAS)*: no effects Pain (SF-36): decreased (57% in Ex) Severity: no effects	A: not reported SE: no adverse effect
Sañudo et al. (2010)	RCT Intention-to-treat analyses	$n_{ex} = 22$ $n_c = 21$	Supervised aerobic exercise: walking/running	2 day/week, 16 weeks	45–60′: 10′ warm-up activities; 15–20′ steady-state aerobic exercise at 60% to 65% of HRmax; 15′ interval training at 75%–80% HRmax that included aerobic dance and jogging; 5–10 'cool-down activities	From 60% to 80% age-adjusted HRmax	Severity*: decreased (14% in Ex)	A: 89% SE: 1 unable to exercise after an injury, but it is unclear whether the injury was related to participation in the exercise

Continued

Reference	Design, analyses	Sample	Intervention				Outcome(s): findings	Adherence (A, mean, range) and side effects (SE)
			Type	Frequency, length	Duration	Intensity		
Resistance training								
Kayo et al. (2012)	RCT Intention-to-treat analyses[b]	$n_{ex} = 30$ $n_c = 30$	Muscle-strengthening exercises	3 day/week, 16 weeks	11 free active exercises in the standing, sitting, and lying positions -First 2 weeks: 3 sets of 10 repetitions, exercises were without load −3-16 weeks: repetitions were increased to 15 After week 5, load was included to the exercises	Exercise load and intensity were increased every 2 weeks, according to the patient's tolerance and by following the Borg Scale	Pain (VAS)*: no effects Pain (SF-36): decreased (41% in Ex) Severity: no effects	A: not reported SE: no adverse effect
Assumpção et al. (2018)	RCT Efficacy analysis	$n_{ex} = 16$ $n_c = 14$	Supervised resistance training program	2 day/week, 12 weeks	40': A series of 8 repetitions of 9 resistance exercises with dumbbells (upper limbs) and shin pads (lower limbs)	In the first 2 sessions, no load was used. Subsequently, 0.5 kg was added each week if the patient identified the effort as slightly intense on the Borg scale (score = 13).	Pressure pain threshold: no effects Tender point count: no effects Pain (VAS): no effects Severity: no effects	A: not reported SE: no adverse effect
Multicomponent training								
Sañudo et al. (2010)	RCT Efficacy analysis	$n_{ex} = 18$ $n_c = 20$	Supervised multicomponent exercise: aerobic (walking, running or dancing), strength, and flexibility	2 day/week, 24 weeks	45–60': 10' warm-up, 10–15' aerobic, 15–20' strength training, 10' flexibility cold-down	AE: 65–70% HRmax (220-age) ST: 1 set of 8–10 reps 8 muscle groups (circuit) at a level that the patients could easily manage F: 8–9 exercises stations, each 1 set of 3 reps, holding the position for 30 s	Pain (SF-36): decreased (6.7% in Ex) Severity*: decreased (10.25% in Ex)	A: 85% SE: not reported

García-Martínez et al. (2012)	RCT Efficacy analysis	$n_{ex} = 14$ $n_c = 14$	Supervised multicomponent exercise: aerobic, strengthening, and stretching	3 day/week, 12 weeks	60': 10' warm-up, 20' aerobic, 20' stretching and strength, 10' cool-down	AE: 60–70% -75-85% Hrmax ST: not reported	Pain (SF-36): decreased (20.8% in Ex) Severity: decreased (21.9% in Ex)	A: not reported SE: not reported
Paolucci et al. (2015)	RCT Efficacy analysis	$n_{ex} = 19$ $n_c = 18$	Supervised multicomponent: low impact aerobic training, agility training and balance exercises, postural exercises, strength training, and stretching, breathing exercises and relaxation	2 day/week, 5 weeks	60': 10' aerobic; 20' agility, balance and postural exercise; 30' resistance exercises, stretching, and breathing exercises	AE: 60% HRmax (220-age) ST: 3 sets of 10 rep Stretching 30–60 s 3 rep	Severity: decreased (9,9% in Ex)	A: not reported SE: no adverse events
Mind-body therapies								
Wong et al. (2018)	RCT Efficacy analysis	$n_{ex} = 18$ $n_c = 19$	10 forms from the classic Yang style of Taichi	3 day/week, 12 weeks	55': 10' warm up, 40' practice Taichi and 5' cool-down	40%–50% HRR	Pain (VAS): decreased (29% in Ex)	A: 97.3% ± 1.4% SE: no adverse effects or injuries
Bravo et al. (2019)	RCT Efficacy analysis	$n_{ex} = 20$ $n_c = 21$	Basic Body Awareness Therapy	2 day/week, 5 weeks	First 2 individual sessions of 60' 8 group sessions of 90': Basic Body Awareness Therapy movements, dropsy massage, and 15' for sharing reflections about the experiences during the session within the group	Not reported	Pain (VAS)*: decreased (13% in Ex)	A: not reported SE: no adverse events

Note: A, adherence; AE, aerobic; C, control group; Ex, exercise group; FIQ, fibromyalgia impact questionnaire; F: flexibility; HR, heart rate; HR_{max}, maximum heart rate; HRR, heart rate reserve; RCT, randomized controlled trial; s, second; SE, side effects; ST, strength; VAS, visual analog scale; VAT, ventilatory anaerobic threshold.

[a]*Most of the studies included other outcomes, however, only pain-related or global disease severity/impact outcomes are showed. An asterisk (*) indicates that the outcome was the main outcomes of the study.*
[b]*The study also reported efficacy analyses, however, only results from intention-to-treat analyses are displayed.*

Exergames in fibromyalgia

Recently, a new modality of exercise has been proven in people with fibromyalgia, the exergames. In particular, two studies have observed that two sessions per week (1 h per session) during 8 weeks or 24 weeks have positive effects on pain compared with control group (Collado-Mateo, Dominguez-Munoz, Adsuar, Garcia-Gordillo, & Gusi, 2017; Villafaina, Collado-Mateo, Domínguez-Muñoz, Fuentes-García, & Gusi, 2019). The exercise intervention was based on an exergame, the VirtualEx-FM, focused on mobility, postural control, upper and lower limbs coordination, aerobic fitness, and strength (Collado-Mateo et al., 2017; Villafaina et al., 2019).

General exercise recommendations

Evidence-based frequency, intensity, time and type of exercise (FITT) in the management of fibromyalgia are described below (Andrade et al., 2018, 2020; Pescatello, Arena, Riebe, & Thompson, 2013):

- *Frequency:* 2–3 times/week, with a minimum of 24–48 h between sessions. Begin with 1–2 days/week and gradually progress to 2–3 if the patient presents very low physical function and severe symptomatology.
- *Intensity:* Begin with light and gradually progress to moderate intensity.
  - Aerobic: From 30% to 39% and progressed to 40%–59% VO_2R or HRR.
  - Resistance: 40%–80% 1-RM. If the individual cannot complete at least 3 repetitions easily and without pain, it is advised the intensity must be reduced to a level where no pain is experienced.
  - Flexibility: the stretch should be held to the point of tightness or slight discomfort.
- *Time:* for 30–60 min in each session
  - Aerobic: begin with 10 min/day and progress to a total of 30–60 min/day as soon as tolerated.
  - Resistance: 10–20 repetitions per muscle group, increasing to 2–3 sets.
  - Flexibility: Initially hold the stretch for 10–30 s. If it is well tolerated, progress to holding each stretch for up to 60 s.
- *Type:*

  - Aerobic: Low-impact (e.g., water exercise, walking, dance, and other aerobic movement to music, swimming, cycling) to minimize the pain that may be caused by exercise.
  - Resistance: Elastic bands, dumbbells, cuff/ankle weights, weight machines, or bodyweight exercises. For resistance in the water, use devices to manipulate turbulence (velocity, surface area). Circuits, where muscle groups are alternated, can be a good option.
  - Flexibility: Static stretches (passive and/or active), for all major muscle tendon groups.

People with fibromyalgia are part of a heterogeneous group (Estévez-López et al., 2017) and the same person could present a great variability with flares in symptoms over time. Exercise prescriptions must be individualized, based on the patient's baseline testing, and adapted daily according to their symptoms. Physical function, the severity of pain and fatigue and other fibromyalgia symptoms, tolerance to exercise, and exercise-induced pain should be taken into account. It is essential to prescribe recovery times between workouts, and to plan the progression of exercises carefully (Cerrillo-Urbina, García-Hermoso, Sánchez-López, & Martínez-Vizcaíno, 2015). The lack of compliance to physical exercise in this population at the beginning of a program often depends on the appearance of pain after exercise (Busch et al., 2011).

There is no evidence that high-intensity exercise is more effective than moderate exercise for this population, and therefore it seems more appropriate to recommend physical exercise at low–moderate intensity (Bidonde, Busch, Bath, & Milosavljevic, 2014). The intensity of exercise must be individually adjusted each session depending on the perception of pain. The ratings of perceived exertion (RPE: CR-10 scale) could be an easy way for monitoring intensity (Soriano-Maldonado et al., 2014), as well as a Visual Analog Scale for monitoring pain in sessions.

Most of the studies of physical exercise interventions in people with fibromyalgia include group activities (Andrade et al., 2020). Individual strategies do not seem to produce psychological improvements, require more effort and perseverance, and result in a higher dropout rate (Cerrillo-Urbina et al., 2015). Thus, it seems reasonable to suggest that, in order to increase compliance, group strategies should be adopted in the implementation of physical exercise programs. Moreover, group interventions should be compatibles with individualization of exercise and exercise adaptions to patients' preferences, thus helping exercise adherence.

People with fibromyalgia should be encouraged to be more physically active participating in physical exercise programs and with daily active behaviors. Long-term exercise program seems to be essential to help facilitate effective lifestyle changes for this population group.

Effectiveness of physical exercise in fibromyalgia: A summary of the evidence

Table 5 summarizes the previous randomized controlled trial in the field. Furthermore, a recent systematic review with a meta-analysis of randomized controlled trials highlighted that exercise was moderately effective for lowering fatigue and had small effects on enhancing sleep quality in fibromyalgia (Estévez-López et al., 2020).

Applications to other areas

In this chapter, we have reviewed the current evidence about the beneficial role of physical activity and physical exercise in chronic musculoskeletal pain, in general, and fibromyalgia, in particular. Because the characteristics of these populations are similar to other diseases such as rheumatic diseases (e.g., Sjogren's syndrome) and mental health disease (e.g., depression), this chapter may provide valuable knowledge not only to chronic musculoskeletal pain populations but also to other populations with similar characteristics.

Other agents of interest

This chapter shows that physical activity and exercise in the management of chronic widespread musculoskeletal pain. However, physical activity or exercise programs are not widely implemented in clinical settings, which highlights the need of more scalable solutions. Moreover, the use of behavior change technics (e.g., goal setting, self-monitoring, and problem solving) could be an interesting option to increase physical activity levels in people with chronic musculoskeletal diseases (Eisele et al., 2019). In this context, digital health interventions (those delivered through the Internet and smartphone applications) promoting behavioral change have an unique potential to bridge this critical gap in service delivery (Palermo, de la Vega, Murray, Law, & Zhou, 2020). Thus, it is of interest to develop digital health solutions promoting people with chronic widespread musculoskeletal pain to engage in physical activity or exercise. Due to the paucity of research in this promising field, future research is warranted.

Mini-dictionary of terms

Catastrophizing (as a pain-related cognition): the tendency to focus on and magnify pain experiences and to feel helpless during pain episodes.

Exercise: It is a subset of physical activity that is planned, structured, repetitive, and purposeful that might have a different effect on mental health depending on the constituent elements (e.g., competitive vs. non-competitive).

Fibromyalgia severity: the impact that fibromyalgia has on people daily life by means of either the fibromyalgia impact questionnaire (total score) or the modified 2011 preliminary criteria questionnaire (polysymptomatic distress scale).

Objective measurements: Information based on clinical observations (e.g., tender points) or performance-based tests (e.g., neuropsychological or physical tests).

Physical activity: It is any bodily movement that increases energy expenditure above resting energy expenditure.

Sedentary behavior: It is defined as any waking behavior characterized by an energy expenditure ≤ 1.5 metabolic equivalents (METs), while in a sitting, reclining, or lying posture.

Subjective assessments: information self-reported by people by means of questionnaires; also referred to as patient-reported outcomes.

Key facts

- The common advice for pain management has changed from suggesting to pace down activity and rest to increase physical activity and reduce sedentary behaviors.
- The mechanisms underlying the beneficial effects of physical activity on musculoskeletal pain remain to be understood.
- Despite the benefits of physical activity, people with chronic musculoskeletal disorders often avoid physical activity to prevent pain, fatigue, and flares.
- People with musculoskeletal pain may also experience unhelpful cognitions in relation to physical activity such as, for instance, fear of movement (kinesiophobia) and pain catastrophizing.

542 **PART | IV** Novel and nonpharmacological aspects and treatments

Summary points

- In people with chronic musculoskeletal pain, physical activity is associated with better health-related outcomes.
- In people with chronic musculoskeletal pain, physical activity interventions lead to little or no difference in objectively measured physical activity.
- Physical exercise is considered the cornerstone of treatment for musculoskeletal pain.
- Aerobic exercise is the most widely studied the type of exercise in fibromyalgia, showing benefits on pain. Resistance training is also associated with more favorable scores on pain-related outcomes.
- There is no evidence of a superiority of water-based over land-based aerobic exercise effect on pain.
- Mind-body therapies have increased in popularity over the last decade as a treatment for fibromyalgia.

References

Andrade, A., de Azevedo Klumb Steffens, R., Sieczkowska, S. M., Peyré Tartaruga, L. A., & Torres Vilarino, G. (2018). A systematic review of the effects of strength training in patients with fibromyalgia: Clinical outcomes and design considerations. *Advances in Rheumatology, 58*(1), 36. https://doi.org/10.1186/s42358-018-0033-9.

Andrade, A., Dominski, F. H., & Sieczkowska, S. M. (2020). What we already know about the effects of exercise in patients with fibromyalgia: An umbrella review. *Seminars in Arthritis and Rheumatism, 000.* https://doi.org/10.1016/j.semarthrit.2020.02.003.

Andrade, C. P., Zamunér, A. R., Forti, M., Tamburús, N. Y., & Silva, E. (2019). Effects of aquatic training and detraining on women with fibromyalgia: Controlled randomized clinical trial. *European Journal of Physical and Rehabilitation Medicine, 55*(1), 79–88.

Assumpção, A., Matsutani, L. A., Yuan, S. L., Santo, A. S., Sauer, J., Mango, P., et al. (2018). Muscle stretching exercises and resistance training in fibromyalgia: Which is better? A three-arm randomized controlled trial. *European Journal of Physical and Rehabilitation Medicine, 54*(5), 663–670.

Barker, J., Smith Byrne, K., Doherty, A., Foster, C., Rahimi, K., Ramakrishnan, R., et al. (2019). Physical activity of UK adults with chronic disease: Cross-sectional analysis of accelerometer-measured physical activity in 96 706 UK biobank participants. *International Journal of Epidemiology, 48*(4), 1167–1174. https://doi.org/10.1093/ije/dyy294.

Bidonde, J., Busch, A. J., Bath, B., & Milosavljevic, S. (2014). Exercise for adults with fibromyalgia: An umbrella systematic review with synthesis of best evidence. *Current Rheumatology Reviews, 10*(1), 45–79. https://doi.org/10.2174/1573403X10666140914155304.

Bidonde, J., Busch, A. J., Schachter, C. L., Overend, T. J., Kim, S. Y., Góes, S. M., et al. (2017). Aerobic exercise training for adults with fibromyalgia. *Cochrane Database of Systematic Reviews, 2017*(6). https://doi.org/10.1002/14651858.CD012700.

Bidonde, J., Busch, A. J., Schachter, C. L., Webber, S. C., Musselman, K. E., Overend, T. J., et al. (2019). Mixed exercise training for adults with fibromyalgia (review). *Cochrane Database of Systematic Reviews, 2019*(9). https://doi.org/10.1002/14651858.CD013419.

Booth, J., Moseley, G. L., Schiltenwolf, M., Cashin, A., Davies, M., & Hübscher, M. (2017). Exercise for chronic musculoskeletal pain: A biopsychosocial approach. *Musculoskeletal Care, 15*(4), 413–421. https://doi.org/10.1002/msc.1191.

Bravo, C., Skjaerven, L. H., Espart, A., Guitard Sein-Echaluce, L., & Catalan-Matamoros, D. (2019). Basic body awareness therapy in patients suffering from fibromyalgia: A randomized clinical trial. *Physiotherapy Theory and Practice, 35*(10), 919–929. https://doi.org/10.1080/09593985.2018.1467520.

Busch, A. J., Webber, S. C., Brachaniec, M., Bidonde, J., Bello-Haas, V. D., Danyliw, A. D., et al. (2011). Exercise therapy for fibromyalgia. *Current Pain and Headache Reports, 15*(5), 358–367. https://doi.org/10.1007/s11916-011-0214-2.

Busch, A. J., Webber, S. C., Richards, R. S., Bidonde, J., Schachter, C. L., Schafer, L. A., et al. (2013). Resistance exercise training for fibromyalgia. *The Cochrane Database of Systematic Reviews, 12*(12). https://doi.org/10.1002/14651858.CD010884, CD010884.

Caneiro, J. P., Roos, E. M., Barton, C. J., O'Sullivan, K., Kent, P., Lin, I., et al. (2019). It is time to move beyond body region silos' to manage musculoskeletal pain: Five actions to change clinical practice. *British Journal of Sports Medicine.* https://doi.org/10.1136/bjsports-2018-100488. BMJ Publishing Group.

Cerrillo-Urbina, A. J., García-Hermoso, A., Sánchez-López, M., & Martínez-Vizcaíno, V. (2015). Effect of exercise programs on symptoms of fibromyalgia in peri-menopausal age women: A systematic review and meta-analysis of randomized controlled trials. *Myopain, 23*(1–2), 56–70. https://doi.org/10.3109/10582452.2015.1083640.

Cheng, C.-A., Chiu, Y.-W., Wu, D., Kuan, Y.-C., Chen, S.-N., & Tam, K.-W. (2019). Effectiveness of tai chi on fibromyalgia patients: A meta-analysis of randomized controlled trials. *Complementary Therapies in Medicine, 46*, 1–8. https://doi.org/10.1016/j.ctim.2019.07.007.

Collado-Mateo, D., Dominguez-Munoz, F. J., Adsuar, J. C., Garcia-Gordillo, M. A., & Gusi, N. (2017). Effects of exergames on quality of life, pain, and disease effect in women with fibromyalgia: A randomized controlled trial. *Archives of Physical Medicine and Rehabilitation, 98*(9), 1725–1731. https://doi.org/10.1016/j.apmr.2017.02.011.

Davergne, T., Pallot, A., Dechartres, A., Fautrel, B., & Gossec, L. (2019). Use of wearable activity trackers to improve physical activity behavior in patients with rheumatic and musculoskeletal diseases: A systematic review and meta-analysis. *Arthritis Care & Research, 71*(6), 758–767. https://doi.org/10.1002/acr.23752.

Eisele, A., Schagg, D., Krämer, L. V., Bengel, J., & Göhner, W. (2019). Behaviour change techniques applied in interventions to enhance physical activity adherence in patients with chronic musculoskeletal conditions: A systematic review and meta-analysis. *Patient Education and Counseling.* https://doi.org/10.1016/j.pec.2018.09.018.

Estevez-Lopez, F., Alvarez-Gallardo, I. C., Segura-Jimenez, V., Soriano-Maldonado, A., Borges-Cosic, M., Pulido-Martos, M., et al. (2016). The discordance between subjectively and objectively measured physical function in women with fibromyalgia: Association with catastrophizing and self-efficacy cognitions. The al-Ándalus project. *Disability and Rehabilitation*, 1–9. https://doi.org/10.1080/09638288.2016.1258737.

Estévez-López, F., Maestre-Cascales, C., Russell, D., Álvarez-Gallardo, I. C., Rodriguez-Ayllon, M., Hughes, C. M., et al. (2020). Effectiveness of exercise on fatigue and sleep quality in fibromyalgia: A systematic review and meta-analysis of randomised trials. *Archives of Physical Medicine and Rehabilitation*. https://doi.org/10.1016/j.apmr.2020.06.019.

Estévez-López, F., Segura-Jiménez, V., Álvarez-Gallardo, I. C., Borges-Cosic, M., Pulido-Martos, M., Carbonell-Baeza, A., ... Delgado-Fernández, M. (2017). Adaptation profiles comprising objective and subjective measures in fibromyalgia: The al-Ándalus project. *Rheumatology*, 56(11), 2015–2024. https://doi.org/10.1093/rheumatology/kex302.

Fernandes, G., Jennings, F., Nery Cabral, M. V., Pirozzi Buosi, A. L., & Natour, J. (2016). Swimming improves pain and functional capacity of patients with fibromyalgia: A randomized controlled trial. *Archives of Physical Medicine and Rehabilitation*, 97(8), 1269–1275. https://doi.org/10.1016/j.apmr.2016.01.026.

Fontaine, K. R., Conn, L., & Clauw, D. J. (2010). Effects of lifestyle physical activity on perceived symptoms and physical function in adults with fibromyalgia: Results of a randomized trial. *Arthritis Research & Therapy*, 12(2), R55. https://doi.org/10.1186/ar2967.

García-Martínez, A. M., De Paz, J. A., & Márquez, S. (2012). Effects of an exercise programme on self-esteem, self-concept and quality of life in women with fibromyalgia: A randomized controlled trial. *Rheumatology International*, 32(7), 1869–1876. https://doi.org/10.1007/s00296-011-1892-0.

Gavi, M. B. R. O., Vassalo, D. V., Amaral, F. T., Macedo, D. C. F., Gava, P. L., Dantas, E. M., et al. (2014). Strengthening exercises improve symptoms and quality of life but do not change autonomic modulation in fibromyalgia: A randomized clinical trial. *PLoS One*, 9(3). https://doi.org/10.1371/journal.pone.0090767, e90767.

Gavilán-Carrera, B., Segura-Jiménez, V., Mekary, R. A., Borges-Cosic, M., Acosta-Manzano, P., Estévez-López, F., et al. (2019). Substituting sedentary time with physical activity in fibromyalgia and the association with quality of life and impact of the disease: The al-Ándalus project. *Arthritis Care & Research*, 71(2), 281–289. https://doi.org/10.1002/acr.23717.

Geenen, R., Overman, C. L., Christensen, R., Åsenlöf, P., Capela, S., Huisinga, K. L., et al. (2018a). EULAR recommendations for the health professional's approach to pain management in inflammatory arthritis and osteoarthritis. *Annals of the Rheumatic Diseases*, 77(6), 797–807. https://doi.org/10.1136/annrheumdis-2017-212662.

Geenen, R., Overman, C. L., Christensen, R., Åsenlöf, P., Capela, S., Huisinga, K. L., et al. (2018b). EULAR recommendations for the health professional's approach to pain management in inflammatory arthritis and osteoarthritis. *Annals of the Rheumatic Diseases*, 77(6), 797–807. https://doi.org/10.1136/annrheumdis-2017-212662.

Geneen, L. J., Moore, R. A., Clarke, C., Martin, D., Colvin, L. A., & Smith, B. H. (2017). Physical activity and exercise for chronic pain in adults: An overview of cochrane reviews. *Cochrane Database of Systematic Reviews*. https://doi.org/10.1002/14651858.CD011279.pub3. John Wiley and Sons Ltd.

Hauser, W., Klose, P., Langhorst, J., Moradi, B., Steinbach, M., Schiltenwolf, M., et al. (2010). Efficacy of different types of aerobic exercise in fibromyalgia syndrome: A systematic review and meta-analysis of randomised controlled trials. *Arthritis Research & Therapy*, 12(3), R79. https://doi.org/10.1186/ar3002.

Hurley, M., Dickson, K., Hallett, R., Grant, R., Hauari, H., Walsh, N., ... Oliver, S. (2018). Exercise interventions and patient beliefs for people with hip, knee or hip and knee osteoarthritis: a mixed methods review. *Cochrane Database of Systematic Reviews*, 4(4), CD010842. https://doi.org/10.1002/14651858.CD010842.pub2.

Jones, K. D., Sherman, C. A., Mist, S. D., Carson, J. W., Bennett, R. M., & Li, F. (2012). A randomized controlled trial of 8-form Tai chi improves symptoms and functional mobility in fibromyalgia patients. *Clinical Rheumatology*, 31(8), 1205–1214. https://doi.org/10.1007/s10067-012-1996-2.

Joustra, M. L., Zijlema, W. L., Rosmalen, J. G. M., & Janssens, K. A. M. (2018). Physical activity and sleep in chronic fatigue syndrome and fibromyalgia syndrome: Associations with symptom severity in the general population cohort lifelines. *Pain Research & Management*, 2018. https://doi.org/10.1155/2018/5801510.

Kaleth, A. S., Slaven, J. E., & Ang, D. C. (2014). Does increasing steps per day predict improvement in physical function and pain interference in adults with fibromyalgia? *Arthritis Care & Research*, 66(12), 1887–1894. https://doi.org/10.1002/acr.22398.

Kayo, A. H., Peccin, M. S., Sanches, C. M., & Trevisani, V. F. M. (2012). Effectiveness of physical activity in reducing pain in patients with fibromyalgia: A blinded randomized clinical trial. *Rheumatology International*, 32(8), 2285–2292. https://doi.org/10.1007/s00296-011-1958-z.

Larsson, A., Palstam, A., Löfgren, M., Ernberg, M., Bjersing, J., Bileviciute-Ljungar, I., et al. (2015). Resistance exercise improves muscle strength, health status and pain intensity in fibromyalgia—A randomized controlled trial. *Arthritis Research & Therapy*, 17(1), 161. https://doi.org/10.1186/s13075-015-0679-1.

Lin, I., Wiles, L., Waller, R., Goucke, R., Nagree, Y., Gibberd, M., et al. (2020). What does best practice care for musculoskeletal pain look like? Eleven consistent recommendations from high-quality clinical practice guidelines: Systematic review. *British Journal of Sports Medicine*. https://doi.org/10.1136/bjsports-2018-099878.

Marley, J., Tully, M. A., Porter-Armstrong, A., Bunting, B., O'Hanlon, J., Atkins, L., et al. (2017). The effectiveness of interventions aimed at increasing physical activity in adults with persistent musculoskeletal pain: A systematic review and meta-analysis. *BMC Musculoskeletal Disorders*. https://doi.org/10.1186/s12891-017-1836-2. BioMed Central Ltd.

Munguía-Izquierdo, D., Pulido-Martos, M., Acosta, F. M., Acosta-Manzano, P., Gavilán-Carrera, B., Rodriguez-Ayllon, M., et al. (2019). Objective and subjective measures of physical functioning in women with fibromyalgia: What type of measure is associated most clearly with subjective well-being? *Disability and Rehabilitation*, 1–8. https://doi.org/10.1080/09638288.2019.1671503.

O'Keeffe, M., Hayes, A., McCreesh, K., Purtill, H., & O'Sullivan, K. (2017). Are group-based and individual physiotherapy exercise programmes equally effective for musculoskeletal conditions? A systematic review and meta-analysis. *British Journal of Sports Medicine, 51*(2), 126–132. https://doi.org/10.1136/bjsports-2015-095410.

Oliveira, C. B., Franco, M. R., Maher, C. G., Christine Lin, C. W., Morelhão, P. K., Araújo, A. C., et al. (2016). Physical activity interventions for increasing objectively measured physical activity levels in patients with chronic musculoskeletal pain: A systematic review. *Arthritis Care and Research, 68*(12), 1832–1842. https://doi.org/10.1002/acr.22919.

Palermo, T. M., de la Vega, R., Murray, C., Law, E., & Zhou, C. (2020). A digital health psychological intervention (WebMAP Mobile) for children and adolescents with chronic pain: Results of a hybrid effectiveness-implementation stepped-wedge cluster randomized trial. *Pain, 161*(12), 2763–2774. https://doi.org/10.1097/j.pain.0000000000001994.

Paolucci, T., Vetrano, M., Zangrando, F., Vulpiani, M. C., Grasso, M. R., Trifoglio, D., et al. (2015). MMPI-2 profiles and illness perception in fibromyalgia syndrome: The role of therapeutic exercise as adapted physical activity. *Journal of Back and Musculoskeletal Rehabilitation, 28*(1), 101–109. https://doi.org/10.3233/BMR-140497.

Pescatello, L. S., Arena, R., Riebe, D., & Thompson, P. D. (2013). In L. S. Pescatello (Ed.), *ACSM's guidelines for exercise testing and prescription* (9th ed.). Wolters Kluwer Health.

Pulido-Martos, M., Luque-Reca, O., Segura-Jiménez, V., Álvarez-Gallardo, I. C., Soriano-Maldonado, A., Acosta-Manzano, P., et al. (2020). Physical and psychological paths toward less severe fibromyalgia: A structural equation model. *Annals of Physical and Rehabilitation Medicine, 63*(1), 46–52. https://doi.org/10.1016/j.rehab.2019.06.017.

Regnaux, J. P., Lefevre-Colau, M. M., Trinquart, L., Nguyen, C., Boutron, I., Brosseau, L., et al. (2015). High-intensity versus low-intensity physical activity or exercise in people with hip or knee osteoarthritis. *Cochrane Database of Systematic Reviews*. https://doi.org/10.1002/14651858. CD010203.pub2. John Wiley and Sons Ltd.

Sañudo, B., Galiano, D., Carrasco, L., Blagojevic, M., de Hoyo, M., & Saxton, J. (2010). Aerobic exercise versus combined exercise therapy in women with fibromyalgia syndrome: A randomized controlled trial. *Archives of Physical Medicine and Rehabilitation, 91*(12), 1838–1843. https://doi.org/10.1016/j.apmr.2010.09.006.

Segura-Jiménez, V., Álvarez-Gallardo, I. C., Estévez-López, F., Soriano-Maldonado, A., Delgado-Fernández, M., Ortega, F. B., et al. (2015). Differences in sedentary time and physical activity between female patients with fibromyalgia and healthy controls: The al-Ándalus project. *Arthritis and Rheumatology, 67*(11). https://doi.org/10.1002/art.39252.

Segura-Jimenez, V., Borges-Cosic, M., Soriano-Maldonado, A., Estevez-Lopez, F., Alvarez-Gallardo, I. C., Herrador-Colmenero, M., et al. (2015). Association of sedentary time and physical activity with pain, fatigue, and impact of fibromyalgia: The al-Andalus study. *Scandinavian Journal of Medicine & Science in Sports*. https://doi.org/10.1111/sms.12630.

Silva, H. J.d. A., Assunção Júnior, J. C., de Oliveira, F. S., Oliveira, J. M.d. P., Figueiredo Dantas, G. A., Lins, C. A.d. A., et al. (2019). Sophrology versus resistance training for treatment of women with fibromyalgia: A randomized controlled trial. *Journal of Bodywork and Movement Therapies, 23*(2), 382–389. https://doi.org/10.1016/j.jbmt.2018.02.005.

Smith, B. E., Hendrick, P., Bateman, M., Holden, S., Littlewood, C., Smith, T. O., et al. (2019). Musculoskeletal pain and exercise—Challenging existing paradigms and introducing new. *British Journal of Sports Medicine*. https://doi.org/10.1136/bjsports-2017-098983. BMJ Publishing Group.

Smith, B. E., Hendrick, P., Smith, T. O., Bateman, M., Moffatt, F., Rathleff, M. S., et al. (2017). Should exercises be painful in the management of chronic musculoskeletal pain? A systematic review and meta-analysis. *British Journal of Sports Medicine*. https://doi.org/10.1136/bjsports-2016-097383. BMJ Publishing Group.

Soriano-Maldonado, A., Ruiz, J. R., Álvarez-Gallardo, I. C., Segura-Jiménez, V., Santalla, A., & Munguía-Izquierdo, D. (2014). Validity and reliability of rating perceived exertion in women with fibromyalgia: Exertion-pain discrimination. *Journal of Sports Sciences*, 1–8. https://doi.org/10.1080/02640414.2014.994661.

Villafaina, S., Collado-Mateo, D., Domínguez-Muñoz, F. J., Fuentes-García, J. P., & Gusi, N. (2019). Benefits of 24-week exergame intervention on health-related quality of life and pain in women with fibromyalgia: A single-blind, randomized controlled trial. *Games for Health Journal, 8*(6), 380–386. https://doi.org/10.1089/g4h.2019.0023.

Wang, C., Schmid, C. H., Rones, R., Kalish, R., Yinh, J., Goldenberg, D. L., et al. (2010). A randomized trial of tai chi for fibromyalgia. *New England Journal of Medicine, 363*(8), 743–754. https://doi.org/10.1056/NEJMoa0912611.

Wong, A., Figueroa, A., Sanchez-Gonzalez, M. A., Son, W. M., Chernykh, O., & Park, S. Y. (2018). Effectiveness of Tai Chi on cardiac autonomic function and symptomatology in women with fibromyalgia: A randomized controlled trial. *Journal of Aging and Physical Activity, 26*(2), 214–221. https://doi.org/10.1123/japa.2017-0038.

Chapter 45

Spinal cord stimulation and limb pain

Timothy Sowder, Usman Latif, Edward Braun, and Dawood Sayed

Department of Anesthesiology, University of Kansas Medical Center, Kansas City, KS, United States

Abbreviations

10KHz	high frequency 10,000 Hz
CRPS	chronic regional pain syndrome
DRG	dorsal root ganglion
FBSS	failed back surgery syndrome
FDA	Food and Drug Administration
LF	low frequency
PDN	painful diabetic neuropathy
RCT	randomized controlled trial
SCS	spinal cord stimulation
VAS	visual analog scale
WDR	wide-dynamic-range

History

Stimulation of the spinal cord was first described as a potential treatment for chronic pain by Shealy et al. in a landmark paper more than 50 years ago (Shealy, Mortimer, & Reswick, 1967). Originally described as dorsal column stimulation, this form of neuromodulation is now described as spinal cord stimulation (SCS) to reflect the varying forms and strategies of treatments that have emerged since then (Verrills, Sinclair, & Barnard, 2016). The United States Food and Drug Administration (FDA) approved SCS for the treatment of neuropathic limb pain in 1989. SCS offers the ability to treat resistant pain syndromes as a low-risk intervention that carries minimal side effects, is relatively safe, and can potentially be reversed (Deer et al., 2014b; Song, Popescu, & Bell, 2014).

Historically, the majority of SCS was grounded in paresthesia-based stimulation. The goal was to replace pain sensation with paresthesias that required mapping of stimulation to the painful areas during SCS trials and implantations (Deer et al., 2014). Altered pain processing due to masking of pain sensations by paresthesia sensations was thought to result in improved pain control; however, a significant portion of patients may not tolerate the paresthesias, or find them uncomfortable or undesirable. In addition, conventional SCS therapy may potentially result in undesirable sensations or changes in efficacy with positional changes (Verrills et al., 2016).

Conventional tonic SCS

Conventional SCS devices are generally capable of stimulation in the range of 2–1200 Hz but are typically utilized in the 40–60 Hz range. With SCS used for other conditions, patients undergo a trial where leads are percutaneously placed in the epidural space and connected to an external generator. Patients who have substantial pain benefit with a one-week trial may proceed to permanent implantation with internalized leads and an internal pulse generator. Leads are typically placed through a paramedian incision and anchored before being tunneled to a second, often gluteal, pocket incision for connection to the internal pulse generator. Lead placement is typically performed utilizing epidural needles, except when paddle leads are employed, and it generally requires laminotomy or a partial open approach.

The core tenet of conventional tonic SCS is that patients experience non-painful paresthesias instead of chronic painful sensations in the dermatomal distribution of their typical pain. Electrical stimulation at the contacts on the leads results in

Treatments, Mechanisms, and Adverse Reactions of Anesthetics and Analgesics. https://doi.org/10.1016/B978-0-12-820237-1.00045-4
Copyright © 2022 Elsevier Inc. All rights reserved.

action potentials which reduce pathological excitation and increase inhibition (Bicket, Dunn, & Ahmed, 2016). Action potentials propagate antidromically into the dorsal horn. Aß fibers synapse with wide-dynamic-range (WDR) neurons which synapse and release inhibitory transmitters such as GABA and adenosine (Ahmed, Yearwood, De Ridder, & Vanneste, 2018). The dorsal horn is affected by orthodromic potentials, which induce inhibition through the serotonergic and noradrenergic pathways. The degree of pain control seems to be most strongly related to the level of activation of the descending pain inhibitory pathway.

Dorsal root ganglion stimulation

Advances in recent years have resulted in additional placement strategies and modalities for stimulation. Dorsal root ganglion (DRG) stimulation is a very targeted form of SCS where the leads are placed at the DRG to specifically target particular painful dermatomes. The DRG is known to play roles in both nociceptive and neuropathic pain (Wall & Devor, 1983). Success has been demonstrated with focal areas of pain in the limb such as the foot. (Liem et al., 2015). The DRG consists of primary afferent sensory neuron cell bodies (Liem, van Dongen, Huygen, Staats, & Kramer, 2016). There is research to suggest that patients with neuropathic pain have increased spontaneous nerve firing in the DRG. It is believed that DRG stimulation may reduce nociceptive input to the spinal cord by restoring some of the filtering functions of the DRG (Liem, 2015). DRG stimulation has been utilized to treat limb pain attributable to chronic regional pain syndrome (CRPS), and studies have shown greater pain relief compared to traditional SCS (Liem et al., 2016).

High frequency 10 kHz SCS

High frequency 10 kHz (10 kHz) SCS is based on research demonstrating that stimulation at 10,000 Hz results in paresthesia-free pain relief. 10KhZ SCS was approved for use in Europe and Australia in 2011 and the United States in 2015. The FDA has approved 10 kHz SCS for the treatment of chronic refractory pain of the trunk and/or limbs. Kapural et al. published in their 2015 SENZA randomized controlled trial (RCT) study that 10Khz SCS treatment is superior to conventional SCS (Kapural et al., 2015).

Preclinical studies have demonstrated that high-frequency SCS can result in selective blockade of larger motor fibers (Tanner, 1962). In addition, Cuellar, Alataris, Walker, Yeomans, and Antognini (2013) demonstrated that high-frequency SCS at the dura mater overlying WDR neurons at frequencies ranging from 2 to 100 kHz resulted in blockade of conduction related to noxious peripheral stimulation. 10 kHz SCS can inactivate sodium channels along with several nodes of Ranvier. By blocking large-diameter fibers from generating action potentials and activating medium and small diameter fibers that decrease WDR cell signaling, neuropathic pain signaling can be diminished (Ahmed et al., 2018).

Burst SCS

Burst SCS provides stimulation in a series of five quick bursts using conventional frequency parameters (De Ridder, Plazier, Kamerling, Menovsky, & Vanneste, 2013). With a burst frequency of 40 Hz, a train of five 1000 μs pulses at a frequency of 500 Hz and a reduced amplitude are employed to result in little to no paresthesia. This is demonstrated in a study by De Ridder et al., which showed paresthesia present in 92% of those receiving conventional tonic SCS and 17% of those receiving Burst SCS (De Ridder, Vanneste, Plazier, van der Loo, & Menovsky, 2010). Burst SCS reduces the firing of WDR and high-threshold neurons (Ahmed et al., 2018). Unlike conventional tonic SCS, Burst SCS does not appear to inhibit neuronal firing through effects on GABA release and activation of GABA receptors.

Research has shown that Burst SCS may improve sleep cycles and depressive symptoms via the anti-inflammatory IL-10. The subthreshold firing of Aβ fibers suggests that pain may be modulated via the gate control mechanism despite the absence of paresthesia. Burst SCS has been shown to provide better pain relief than conventional SCS at various follow-up intervals as well as improved performance on measures such as attention to pain and pain changes (De Ridder et al., 2013). Encephalogram analysis in the Burst SCS group revealed more activation of the dorsal anterior cingulate and dorsolateral prefrontal cortex compared to the conventional tonic SCS group (De Ridder et al., 2013). Burst SCS affects the medial, lateral, and descending pathway whereas conventional tonic SCS primarily influences the descending pathway.

Conventional tonic SCS has historically targeted chronic neuropathic limb pain with limited ability to target axial back pain (Roulaud et al., 2015). Newer treatments such as DRG SCS, 10 kHz SCS, and Burst SCS have demonstrated improved ability to target limb pain (de Vos, Bom, Vanneste, Lenders, & de Ridder, 2014; Kapural et al., 2015; Liem et al., 2015). Liem et al. demonstrated that DRG SCS reduced leg pain scores by 62% (74.6–28.7) and foot pain by 80% (81.4–22.0) (Liem et al., 2015).

Ischemic pain

SCS therapy was first demonstrated to show benefit for ischemic pain in the 1960s. Despite evidence supporting the role of SCS in ischemic pain as well as refractory angina pectoris, it is typically utilized late in the treatment algorithm. It is possible that earlier intervention may result in improved outcomes (Deer & Raso, 2006).

Chronic SCS is believed to modify the alpha-1 adrenergic pathway, resulting in a sympatholytic and vasodilatative effect (Kinfe, Pintea, & Vatter, 2016). This results in improved microcirculation in ischemic tissue. Nitric oxide and calcitonin gene-related peptide further amplify the vasodilatative, endothelial-protective, and angiogenic effects. Other chemicals of interest in this process include acetylcholine, adenosine, and substance P. Animal models suggest that improved microcirculation is primarily due to reduced sympathetic outflow combined with a release of vasodilatative transmitters (Bernardini, Neuhuber, Reeh, & Sauer, 2004). Ischemic patients undergoing treatment with SCS have been observed to have reduced ulceration and oxygen demand with resultant pain reduction and improved perfusion status (Deer et al., 2014a).

Patient selection

Indications

SCS has the ability to provide significant pain relief for many chronic pain conditions including a variety of limb pain etiologies. The goal of stimulator therapy is the reduction of pain and/or the improvement of functionality by more than 50% (Atkinson et al., 2011). To meet the criteria for a spinal cord stimulator, patients should have:

- An understanding of and interest in pursuing stimulator therapy
- Suboptimal response to conservative therapy, such as failed physical and behavioral therapy and treatment with opioid analgesics, non-steroidal anti-inflammatory drugs, anticonvulsants, and antidepressants
- No significant psychological comorbidities identified
- Definitive surgical correction (further) not identified
- No contraindication to surgical approach

Failure of conservative treatment

Before initiating a trial of SCS therapy, conservative treatment modalities should be thoroughly employed. For instance, over-the-counter analgesics, physical therapeutic modalities and home exercise, cognitive-behavioral therapies, and injection therapies should have failed to provide adequate pain relief. In addition, the risks of further surgical intervention should be thought to outweigh potential benefits.

Psychological screening

The experience of pain includes the presence of psychological and social factors. The purpose of psychological screening is to identify factors that may have a negative effect on pain coping ability, such as depression, anxiety, and maladaptive behaviors. A psychological screening assessment is usually performed before offering SCS therapy, in part because it is mandated by most insurance plans.

The examination, performed by a psychologist, usually consists of a social and family history, in addition to substance abuse history and past psychological diagnoses and treatments. The psychological evaluation also covers expectations of benefit from the stimulator (Jamison & Craig, 2011).

Identification of psychosocial issues is not an absolute contraindication to the pursuit of stimulator therapy, but rather an opportunity to formulate a treatment plan to address challenges and optimize the chances of a favorable response to treatment. A study of the predictive value of psychological screening showed that patients without psychiatric disorders improved 1.5 times more on the Oswestry Disability Index and 2.5 times more on the Pain Catastrophizing Scale (Fama et al., 2016). In addition, patients experienced significantly improved outcomes when depression was identified and treated (Fama et al., 2016). A psychological evaluation performed by a mental health specialist not only satisfies an insurance requirement but is also a tool to treat patients holistically.

Medical comorbidities

Evaluation of comorbid medical conditions is an important step before initiating SCS. General contraindications include an unresolved psychiatric disorder, persistent infection, uncontrolled diabetes or hypertension, and uncontrolled bleeding disorders (De La Cruz et al., 2015). Where possible, optimization of these conditions provides an opportunity to move forward with a stimulator trial. A failed SCS trial may be addressed by considering alternate stimulator therapy options. Anticipated need for MRI exams based on medical comorbidities is an important consideration when deciding which stimulator therapy is the best match for a patient.

Screening trial

Conducting an SCS trial before permanent implantation of the device is common practice (Bruehl, 2015). The purpose is to predict whether the device will provide substantial pain relief in the postoperative period (Song et al., 2014). In a letter to healthcare providers, the FDA currently recommends trial stimulation to confirm satisfactory pain relief before implanting a spinal cord stimulator (US Food and Drug Administration, 2020). The recommendation is based on a review of published literature as well as reporting of adverse events. An SCS trial usually lasts 5–7 days, and a successful trial is usually defined as a pain reduction of more than 50% and/or functional improvement based on instruments such as the Visual Analog Scale (VAS), McGill Pain Questionnaire, and Oswestry Disability Index (Deer, Mekhail, et al., 2014a). It is also recommended that the patient be provided with the manufacturer's patient labeling and educational materials specific to the device planned for implantation (US Food and Drug Administration, 2020) (Fig. 1).

Outcomes

Failed back surgical syndrome

Lower limb pain associated with failed back surgical syndrome (FBSS) is one of the most common indications for SCS. The predominately neuropathic nature of limb pain in this condition makes it easier to achieve pain relief compared to back pain, which may have a stronger nociceptive component that is less responsive to SCS. Several initial case reports and series showed promise in treating this difficult-to-treat condition. Stratified meta-analysis of early data on patients with predominately leg pain showed an average pain reduction of 54% with conventional SCS (Taylor, Desai, Rigoard, & Taylor, 2014). In the first RCT comparing SCS to repeat lumbar spine surgery, 60 patients with persistent or recurrent radicular pain after

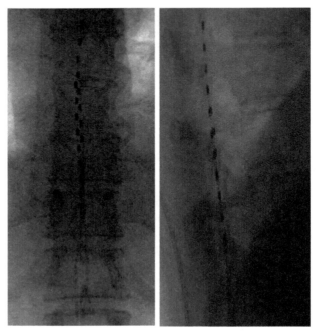

FIG. 1 Anteroposterior (A) and lateral (B) fluoroscopic images of two 8-contact spinal cord stimulator leads placed in the thoracic spine to treat CRPS of the lower extremity. *(Images courtesy of T. Sowder, M.D.)*

one or more lumbosacral spine surgeries were randomized to receive either SCS or reoperation (North, Kidd, Farrokhi, & Piantadosi, 2005). Nine of 19 (47%) patients in the SCS group compared to 3 of 26 (12%) patients in the reoperation group had 50% or greater pain relief and patient satisfaction, resulting in a significant ($p < 0.01$) difference between the two groups. Kumar et al. conducted a trial comparing conservative medical management (CMM) to SCS in patients with FBSS (Kumar et al., 2007). At 6 months, 24 of 52 (48%) SCS patients compared to 4 of 48 (9%) CMM patients achieved >50% reduction of leg pain ($p < 0.001$). Compared with the CMM group, the SCS group experienced significantly improved leg and back pain relief, quality of life, and functional capacity, as well as greater treatment satisfaction ($p \leq 0.05$).

More recently, novel SCS waveforms have proven effective in this patient population. An RCT by Kapural et al. randomized 198 patients to receive either high frequency (HF) 10 kHz SCS or conventional low frequency (LF) SCS to treat chronic back and leg pain (Kapural et al., 2015). At 3-months, 83% of the 10 kHz group had >50% relief of leg pain compared with 55% of the LF-SCS group ($p < 0.001$). Responder rates were sustained at 12 and 24 months in both groups but remained higher in the 10 kHz group (79% and 73% vs 51% and 49% for LF-SCS, $p < 0.001$). The 10 kHz group also reported a larger decrease in leg pain of 70% compared to 49% in the LF-SCS group. Similar outcomes were noted in prospective, single-arm trials by Van Buyten and Al-Kaisy (Al-Kaisy et al., 2014; Van Buyten, Al-Kaisy, Smet, Palmisani, & Smith, 2013). The use of HF SCS to treat upper limb pain associated with FBSS was assessed by Amirdalfen et al. In a single-arm study of 45 patients (Amirdelfan et al., 2020). The primary endpoint of ≥50% pain relief at 3 months was achieved in 86.7% of patients. At the 12-month assessment, 95.0% of patients reported ≥50% pain relief from baseline with a significant reduction in mean VAS scores from a baseline of 7.1 ± 0.3 cm to 1.0 ± 0.2 cm ($p < 0.001$) (Amirdelfan et al., 2020). Burst stimulation has also shown efficacy in treating FBSS (De Ridder et al., 2010; Schu et al., 2014). In the SUNBURST RCT by Deer et al., burst stimulation achieved superior pain relief compared to tonic stimulation for limb pain ($p < 0.045$) and was preferred by 68.2% of patients (Deer et al., 2018).

Complex regional pain syndrome

Limb pain from CRPS has also been treated with SCS. Patient selection is important and the Budapest criteria can help establish a clinical diagnosis of CRPS (Table 1). In an RCT by Kemler et al., 54 patients with CRPS were randomized 2:1 to SCS and physical therapy or physical therapy alone (Kemler, de Vet, Barendse, van den Wildenberg, & van Kleef, 2008). Twenty-four of 36 participants in the SCS group had a successful trial and underwent permanent implant. The SCS group had a mean VAS reduction of 2.4 cm at six months compared to an increase of 0.2 cm in the group that completed physical therapy alone ($p < 0.001$). There were also more patients with a score of 6 ("much improved") for the global perceived effect, which was much higher in the SCS group than in the control group (39% vs. 6%, $p = 0.01$). Despite diminishing relief at 2 and 5 years follow up, there were persistent improvements in global perceived effect ($p = 0.02$) and

---

TABLE 1 Budapest criteria for CRPS (Harden, Bruehl, Stanton-Hicks, & Wilson, 2007).

1. Continuing pain, which is disproportionate to any inciting event

2. Must report at least one symptom in three of the following four categories:

   Sensory: Hyperesthesia and/or allodynia

   Vasomotor: Temperature asymmetry and/or skin color changes and/or skin color asymmetry

   Sudomotor/edema: Edema and/or sweating changes and/or sweating asymmetry

   Motor/trophic: Decreased range of motion and/or motor dysfunction (weakness, tremor, dystonia) and/or trophic changes (hair, nail, skin)

3. Must display at least one sign at time of evaluation in *two or more* of the following categories:

   Sensory: Evidence of hyperalgesia (to pinprick) and/or allodynia (to light touch and/or temperature sensation and/or deep somatic pressure and/or joint movement)

   Vasomotor: Evidence of temperature asymmetry (>1 °C) and/or skin color changes and/or asymmetry

   Sudomotor/edema: Evidence of edema and/or sweating changes and/or sweating asymmetry

   Motor/trophic: Evidence of decreased range of motion and/or motor dysfunction (weakness, tremor, dystonia) and/or trophic changes (hair, nail, skin)

4. There is no other diagnosis that better explains the signs and symptoms

550 PART | IV Novel and nonpharmacological aspects and treatments

pain relief ($p = 0.06$) in patients with SCS compared to physical therapy alone, and 95% of patients with an SCS implant indicated they would repeat the treatment for the same result (Kemler et al., 2000, 2008; Kemler, De Vet, Barendse, Van Den Wildenberg, & Van Kleef, 2004). Small studies investigating the use of novel SCS waveforms to treat CRPS have demonstrated efficacy with various frequencies (500–10,000 Hz), though larger RCTs are needed (Al-Kaisy et al., 2014; Gill, Asgerally, & Simopoulos, 2019; Kriek, Groeneweg, Stronks, de Ridder, & Huygen, 2017).

The DRG has emerged as a target for neuromodulation with initial studies reporting efficacy in treating CRPS. In a small case series by Van Buyten et al., 6 of 8 patients reported >50% relief at 12 months and some patients noted improvement in edema, trophic skin changes, and function (Van Buyten, Smet, Liem, Russo, & Huygen, 2015). The subsequent ACCURATE RCT randomized 152 patients with CRPS of the lower extremity to either conventional SCS or DRG stimulation. The number of patients with greater than 50% pain relief was statistically higher in the DRG group at 3 months (81.2% vs 55.7, $p < 0.001$) and 12 months (74.2% vs 53.0%, $p = 0.005$) compared to the SCS group (Deer et al., 2017).

Diabetic neuropathy

Treatment of painful diabetic neuropathy (PDN) refractory to conventional medical therapy with SCS has yielded positive results. An RCT by De Vos et al. randomized patients to best conventional medical practice with or without SCS. The SCS group reported a decrease of 42 mm in mean VAS score with 65% reporting at least 50% relief compared to no VAS improvement and only 5% achieving at least 50% relief in the control group ($p < 0.001$). Assessed by SF-MPQ and EuroQoL 5D questionnaires, the SCS group also had reduced pain and improved health and quality of life after 6 months of treatment (de Vos et al., 2014). In a similar RCT, Slangen noted treatment success in 59% of SCS patients compared to 7% of control patients at 6 months ($p < 0.01$), with success defined as ≥50% pain relief or "(very) much improved" for pain and sleep on the Patient Global Impression of Change scale (Slangen et al., 2014). Two-year follow-up data indicated that analgesic effects of SCS were sustained over that period (van Beek et al., 2015). Studies of novel SCS waveforms in the treatment of PDN are ongoing (Mekhail et al., 2020).

Ischemic pain

As noted previously, SCS has positive effects on ischemia and pain in ischemic limb conditions. A Cochrane review of 6 trials ($N = 444$) compared SCS to conservative treatment alone for non-reconstructable chronic critical leg ischemia. The review noted significantly higher limb salvage after 12 months in the SCS group ($RR = 0.71$) and more prominent pain relief in the SCS group where the patients required significantly less analgesics (Ubbink & Vermeulen, 2013). Small studies have shown benefit in other ischemic limb conditions including Raynaud's phenomenon, thrombangiitis obliterans (Buerger's Disease), and systemic sclerosis (Augustinsson, Linderoth, & Mannheirmer, 1992; Donas, Schulte, Ktenidis, & Horsch, 2005; Francaviglia, Silvestro, Maiello, Bragazzi, & Bernucci, 1994; Provenzano et al., 2011; Robaina, Dominguez, Díaz, Rodriguez, & de Vera, 1989; Swigris, Olin, & Mekhail, 1999; Vaquer Quiles et al., 2009).

Applications to other areas

This chapter has reviewed the use of spinal cord stimulation for both neuropathic and ischemic limb pain. This is not intended to be a comprehensive list of indications for SCS as there are other limb pain conditions that may be responsive to SCS. The variable pathology of pain syndromes highlights the importance of a thoughtful evaluation of each individual patient when considering SCS treatment. There is data to support the use of SCS for phantom limb pain (Aiyer, Barkin, Bhatia, & Gungor, 2017), peripheral polyneuropathy (Galan et al., 2019), post-herpetic neuralgia (Lin, Lin, Lao, & Chen, 2019), and spinal cord or nerve root injury. As indications continue to expand, more research is needed to evaluate the efficacy of SCS in these and other less prevalent limb pain syndromes.

Other agents of interest

In this chapter, we have described detailed mechanisms of how SCS works in treating a variety of limb pain syndromes. Other considerations for those with refractory limb pain include variations of existing SCS technology including combined frequencies or waveforms. As the field continues to evolve, new waveforms like differential target multiplexed SCS (Fishman et al., 2020) or currently untested frequencies may prove beneficial. In addition, peripheral nerve stimulation can be used to treat some forms of limb pain and may be especially helpful in those with contraindications to SCS.

Mini-dictionary of terms

Spinal cord stimulation: use of an electrical lead placed in the epidural space to dose the spinal cord with electricity to treat refractory pain syndromes.

Conventional tonic SCS: initial iteration of SCS which typically utilizes the 40–60 Hz range to produce non-painful paresthesias instead of chronic painful sensations in the area of pain.

10 kHz: high-frequency stimulation at 10,000 Hz, resulting in paresthesia-free pain relief.

Burst SCS: stimulation in a series of five quick bursts using 40 Hz, a train of five 1000 μs pulses at a frequency of 500 Hz, and a reduced amplitude are employed to result in little to no paresthesia.

DRG stimulation: the targeted form of SCS where the leads are placed at the DRG to specifically target particular painful dermatomes.

Key facts of spinal cord stimulation and limb pain

- Stimulation of the spinal cord was first described as a potential treatment for chronic pain by Shealy et al. in a landmark paper more than 50 years ago
- Historically, the majority of SCS was grounded in paresthesia-based stimulation where the goal was to replace pain sensation with paresthesias that required mapping of stimulation to the painful areas during SCS trials and implantations
- Novel waveforms like high frequency 10 kHz and burst SCS, have changed this paradigm and offer patients pain relief without paresthesias
- Dorsal root ganglion (DRG) stimulation is a very targeted form of SCS where the leads are placed at the DRG to specifically target particular painful dermatomes
- In ischemic pain, SCS is believed to modify the alpha-1 adrenergic pathway, resulting in a sympatholytic and vasodilatative effect which results in improved microcirculation in ischemic tissue

Summary points

- Spinal cord stimulation (SCS) can treat resistant limb pain syndromes as a relatively safe, low-risk intervention with minimal side effects
- Before initiating a trial of SCS therapy, conservative treatment modalities should be thoroughly employed
- A psychological screening assessment is usually performed before offering SCS therapy to identify factors that may have a negative effect on pain coping ability, such as depression, anxiety, and maladaptive behaviors
- General contraindications to SCS implantation include an unresolved psychiatric disorder, persistent infection, uncontrolled systemic disease, and coagulopathy. Where possible, optimization of these conditions provides an opportunity to move forward with a stimulator trial
- Conducting an SCS trial before permanent implantation of the device is common practice and recommended by the FDA to predict whether the device will provide substantial pain relief after permanent implantation
- SCS has shown efficacy in limb pain associated with failed back surgery syndrome, complex regional pain syndrome, diabetic neuropathy and other neuropathic conditions, and ischemic pain conditions
- Indications for SCS continue to expand and further research is needed to evaluate efficacy in additional limb pain syndromes

References

Ahmed, S., Yearwood, T., De Ridder, D., & Vanneste, S. (2018). Burst and high frequency stimulation: Underlying mechanism of action. *Expert Review of Medical Devices, 15*(1), 61–70. https://doi.org/10.1080/17434440.2018.1418662.

Aiyer, R., Barkin, R. L., Bhatia, A., & Gungor, S. (2017). A systematic review on the treatment of phantom limb pain with spinal cord stimulation. *Pain Management, 7*(1), 59–69. https://doi.org/10.2217/pmt-2016-0041.

Al-Kaisy, A., Van Buyten, J. P., Smet, I., Palmisani, S., Pang, D., & Smith, T. (2014). Sustained effectiveness of 10 kHz high-frequency spinal cord stimulation for patients with chronic, low back pain: 24-month results of a prospective multicenter study. *Pain Medicine, 15*(3), 347–354. https://doi.org/10.1111/pme.12294.

Amirdelfan, K., Vallejo, R., Benyamin, R., Yu, C., Yang, T., Bundschu, R., et al. (2020). High-frequency spinal cord stimulation at 10 khz for the treatment of combined neck and arm pain: Results from a prospective multicenter study. *Neurosurgery, 87*(2), 176–185. https://doi.org/10.1093/neuros/nyz495.

Atkinson, L., Sundaraj, S. R., Brooker, C., O'Callaghan, J., Teddy, P., Salmon, J., et al. (2011). Recommendations for patient selection in spinal cord stimulation. *Journal of Clinical Neuroscience, 18*(10), 1295–1302. https://doi.org/10.1016/j.jocn.2011.02.025.

Augustinsson, L. E., Linderoth, B., & Mannheirmer, C. (1992). Spinal cord stimulation in various ischaemic conditions. In L. Illis (Ed.), *Spinal cord dysfunction. III: Functional stimulation* (pp. 272–295). Oxford, UK: Oxford Medical Publications.

Bernardini, N., Neuhuber, W., Reeh, P. W., & Sauer, S. K. (2004). Morphological evidence for functional capsaicin receptor expression and calcitonin gene-related peptide exocytosis in isolated peripheral nerve axons of the mouse. *Neuroscience, 126*(3), 585–590. https://doi.org/10.1016/j.neuroscience.2004.03.017.

Bicket, M. C., Dunn, R. Y., & Ahmed, S. U. (2016). High-frequency spinal cord stimulation for chronic pain: Pre-clinical overview and systematic review of controlled trials. *Pain Medicine, 17*(12), 2326–2336. https://doi.org/10.1093/pm/pnw156.

Bruehl, S. (2015). Complex regional pain syndrome. *BMJ, 351*, h2730. https://doi.org/10.1136/bmj.h2730.

Cuellar, J. M., Alataris, K., Walker, A., Yeomans, D. C., & Antognini, J. F. (2013). Effect of high-frequency alternating current on spinal afferent nociceptive transmission. *Neuromodulation, 16*(4), 318–327. discussion 327 https://doi.org/10.1111/ner.12015 (Epub 2012 Dec 17).

De La Cruz, P., Fama, C., Roth, S., Haller, J., Wilock, M., Lange, S., et al. (2015). Predictors of spinal cord stimulation success. *Neuromodulation, 18*(7), 599–602. discussion 602 https://doi.org/10.1111/ner.12325.

De Ridder, D., Plazier, M., Kamerling, N., Menovsky, T., & Vanneste, S. (2013). Burst spinal cord stimulation for limb and back pain. *World Neurosurgery, 80*(5), 642–649. e641 https://doi.org/10.1016/j.wneu.2013.01.040.

De Ridder, D., Vanneste, S., Plazier, M., van der Loo, E., & Menovsky, T. (2010). Burst spinal cord stimulation: Toward paresthesia-free pain suppression. *Neurosurgery, 66*(5), 986–990. https://doi.org/10.1227/01.Neu.0000368153.44883.B3.

de Vos, C. C., Bom, M. J., Vanneste, S., Lenders, M. W., & de Ridder, D. (2014). Burst spinal cord stimulation evaluated in patients with failed back surgery syndrome and painful diabetic neuropathy. *Neuromodulation, 17*(2), 152–159. https://doi.org/10.1111/ner.12116.

de Vos, C. C., Meier, K., Zaalberg, P. B., Nijhuis, H. J., Duyvendak, W., Vesper, J., et al. (2014). Spinal cord stimulation in patients with painful diabetic neuropathy: A multicentre randomized clinical trial. *Pain, 155*(11), 2426–2431. https://doi.org/10.1016/j.pain.2014.08.031.

Deer, T. R., Krames, E., Mekhail, N., Pope, J., Leong, M., Stanton-Hicks, M., et al. (2014). The appropriate use of neurostimulation: New and evolving neurostimulation therapies and applicable treatment for chronic pain and selected disease states. Neuromodulation appropriateness consensus committee. *Neuromodulation, 17*(6), 599–615. discussion 615 https://doi.org/10.1111/ner.12204.

Deer, T. R., Levy, R. M., Kramer, J., Poree, L., Amirdelfan, K., Grigsby, E., et al. (2017). Dorsal root ganglion stimulation yielded higher treatment success rate for complex regional pain syndrome and causalgia at 3 and 12 months: A randomized comparative trial. *Pain, 158*(4), 669–681. https://doi.org/10.1097/j.pain.0000000000000814.

Deer, T. R., Mekhail, N., Provenzano, D., Pope, J., Krames, E., Leong, M., et al. (2014a). The appropriate use of neurostimulation of the spinal cord and peripheral nervous system for the treatment of chronic pain and ischemic diseases: The neuromodulation appropriateness consensus committee. *Neuromodulation, 17*(6), 515–550. discussion 550 https://doi.org/10.1111/ner.12208.

Deer, T. R., Mekhail, N., Provenzano, D., Pope, J., Krames, E., Thomson, S., et al. (2014b). The appropriate use of neurostimulation: Avoidance and treatment of complications of neurostimulation therapies for the treatment of chronic pain. Neuromodulation appropriateness consensus committee. *Neuromodulation, 17*(6), 571–597. discussion 597–578 https://doi.org/10.1111/ner.12206.

Deer, T. R., & Raso, L. J. (2006). Spinal cord stimulation for refractory angina pectoris and peripheral vascular disease. *Pain Physician, 9*(4), 347–352.

Deer, T., Slavin, K. V., Amirdelfan, K., North, R. B., Burton, A. W., Yearwood, T. L., et al. (2018). Success using neuromodulation with BURST (SUNBURST) study: Results from a prospective, randomized controlled trial using a novel burst waveform. *Neuromodulation, 21*(1), 56–66. https://doi.org/10.1111/ner.12698.

Donas, K. P., Schulte, S., Ktenidis, K., & Horsch, S. (2005). The role of epidural spinal cord stimulation in the treatment of Buerger's disease. *Journal of Vascular Surgery, 41*(5), 830–836. https://doi.org/10.1016/j.jvs.2005.01.044.

Fama, C. A., Chen, N., Prusik, J., Kumar, V., Wilock, M., Roth, S., et al. (2016). The use of preoperative psychological evaluations to predict spinal cord stimulation success: Our experience and a review of the literature. *Neuromodulation, 19*(4), 429–436. https://doi.org/10.1111/ner.12434.

Fishman, M. A., Calodney, A., Kim, P., Slezak, J., Benyamin, R., Rehman, A., et al. (2020). Prospective, multicenter feasibility study to evaluate differential target multiplexed spinal cord stimulation programming in subjects with chronic intractable Back pain with or without leg pain. *Pain Practice, 20*(7), 761–768. https://doi.org/10.1111/papr.12908 (Epub 2020 Jun 2).

Francaviglia, N., Silvestro, C., Maiello, M., Bragazzi, R., & Bernucci, C. (1994). Spinal cord stimulation for the treatment of progressive systemic sclerosis and Raynaud's syndrome. *British Journal of Neurosurgery, 8*(5), 567–571. https://doi.org/10.3109/02688699409002949.

Galan, V., Chang, P., Scowcroft, J., Li, S., Staats, P., & Subbaroyan, J. (2019). A prospective clinical trial to assess high frequency spinal cord stimulation (HF-SCS) at 10 kHz in the treatment of chronic intractable pain from peripheral polyneuropathy (PPN). In *Paper presented at the 22nd annual meeting of the North American Neuromodulation Society (NANS), Las Vegas, NV*.

Gill, J. S., Asgerally, A., & Simopoulos, T. T. (2019). High-frequency spinal cord stimulation at 10 khz for the treatment of complex regional pain syndrome: A case series of patients with or without previous spinal cord stimulator implantation. *Pain Practice, 19*(3), 289–294. https://doi.org/10.1111/papr.12739.

Harden, R. N., Bruehl, S., Stanton-Hicks, M., & Wilson, P. R. (2007). Proposed new diagnostic criteria for complex regional pain syndrome. *Pain Medicine, 8*(4), 326–331.

Jamison, R. N., & Craig, K. D. (2011). Psychological assessment of persons with chronic pain. In M. E. Lynch, K. D. Craig, & P. W. H. Peng (Eds.), *Clinical pain management: A practical guide* (pp. 81–91). Oxford: Wiley-Blackwell.

Kapural, L., Yu, C., Doust, M. W., Gliner, B. E., Vallejo, R., Sitzman, B. T., et al. (2015). Novel 10-kHz high-frequency therapy (HF10 therapy) is superior to traditional low-frequency spinal cord stimulation for the treatment of chronic back and leg pain: The SENZA-RCT randomized controlled trial. *Anesthesiology, 123*(4), 851–860. https://doi.org/10.1097/aln.0000000000000774.

Kemler, M. A., Barendse, G. A., van Kleef, M., de Vet, H. C., Rijks, C. P., Furnée, C. A., et al. (2000). Spinal cord stimulation in patients with chronic reflex sympathetic dystrophy. *The New England Journal of Medicine, 343*(9), 618–624. https://doi.org/10.1056/nejm200008313430904.

Kemler, M. A., De Vet, H. C., Barendse, G. A., Van Den Wildenberg, F. A., & Van Kleef, M. (2004). The effect of spinal cord stimulation in patients with chronic reflex sympathetic dystrophy: Two years' follow-up of the randomized controlled trial. *Annals of Neurology, 55*(1), 13–18. https://doi.org/10.1002/ana.10996.

Kemler, M. A., de Vet, H. C., Barendse, G. A., van den Wildenberg, F. A., & van Kleef, M. (2008). Effect of spinal cord stimulation for chronic complex regional pain syndrome type I: Five-year final follow-up of patients in a randomized controlled trial. *Journal of Neurosurgery, 108*(2), 292–298. https://doi.org/10.3171/jns/2008/108/2/0292.

Kinfe, T. M., Pintea, B., & Vatter, H. (2016). Is spinal cord stimulation useful and safe for the treatment of chronic pain of ischemic origin? A review. *The Clinical Journal of Pain, 32*(1), 7–13. https://doi.org/10.1097/ajp.0000000000000229.

Kriek, N., Groeneweg, J. G., Stronks, D. L., de Ridder, D., & Huygen, F. J. (2017). Preferred frequencies and waveforms for spinal cord stimulation in patients with complex regional pain syndrome: A multicentre, double-blind, randomized and placebo-controlled crossover trial. *European Journal of Pain, 21*(3), 507–519. https://doi.org/10.1002/ejp.944.

Kumar, K., Taylor, R. S., Jacques, L., Eldabe, S., Meglio, M., Molet, J., et al. (2007). Spinal cord stimulation versus conventional medical management for neuropathic pain: A multicentre randomised controlled trial in patients with failed back surgery syndrome. *Pain, 132*(1–2), 179–188. https://doi.org/10.1016/j.pain.2007.07.028.

Liem, L. (2015). Stimulation of the dorsal root ganglion. *Progress in Neurological Surgery, 29*, 213–224. https://doi.org/10.1159/000434673.

Liem, L., Russo, M., Huygen, F. J., Van Buyten, J. P., Smet, I., Verrills, P., et al. (2015). One-year outcomes of spinal cord stimulation of the dorsal root ganglion in the treatment of chronic neuropathic pain. *Neuromodulation, 18*(1), 41–48. discussion 48-49 https://doi.org/10.1111/ner.12228.

Liem, L., van Dongen, E., Huygen, F. J., Staats, P., & Kramer, J. (2016). The dorsal root ganglion as a therapeutic target for chronic pain. *Regional Anesthesia and Pain Medicine, 41*(4), 511–519. https://doi.org/10.1097/aap.0000000000000408.

Lin, C. S., Lin, Y. C., Lao, H. C., & Chen, C. C. (2019). Interventional treatments for postherpetic neuralgia: A systematic review. *Pain Physician, 22*(3), 209–228.

Mekhail, N. A., Argoff, C. E., Taylor, R. S., Nasr, C., Caraway, D. L., Gliner, B. E., et al. (2020). High-frequency spinal cord stimulation at 10 kHz for the treatment of painful diabetic neuropathy: Design of a multicenter, randomized controlled trial (SENZA-PDN). *Trials, 21*(1), 87. https://doi.org/10.1186/s13063-019-4007-y.

North, R. B., Kidd, D. H., Farrokhi, F., & Piantadosi, S. A. (2005). Spinal cord stimulation versus repeated lumbosacral spine surgery for chronic pain: A randomized, controlled trial. *Neurosurgery, 56*(1), 98–106. discussion 106-107 https://doi.org/10.1227/01.neu.0000144839.65524.e0.

Provenzano, D. A., Nicholson, L., Jarzabek, G., Lutton, E., Catalane, D. B., & Mackin, E. (2011). Spinal cord stimulation utilization to treat the microcirculatory vascular insufficiency and ulcers associated with scleroderma: A case report and review of the literature. *Pain Medicine, 12*(9), 1331–1335. https://doi.org/10.1111/j.1526-4637.2011.01214.x.

Robaina, F. J., Dominguez, M., Díaz, M., Rodriguez, J. L., & de Vera, J. A. (1989). Spinal cord stimulation for relief of chronic pain in vasospastic disorders of the upper limbs. *Neurosurgery, 24*(1), 63–67. https://doi.org/10.1227/00006123-198901000-00010.

Roulaud, M., Durand-Zaleski, I., Ingrand, P., Serrie, A., Diallo, B., Peruzzi, P., et al. (2015). Multicolumn spinal cord stimulation for significant low back pain in failed back surgery syndrome: Design of a national, multicentre, randomized, controlled health economics trial (ESTIMET study). *Neurochirurgie, 61*(Suppl 1), S109–S116. https://doi.org/10.1016/j.neuchi.2014.10.105.

Schu, S., Slotty, P. J., Bara, G., von Knop, M., Edgar, D., & Vesper, J. (2014). A prospective, randomised, double-blind, placebo-controlled study to examine the effectiveness of burst spinal cord stimulation patterns for the treatment of failed back surgery syndrome. *Neuromodulation, 17*(5), 443–450. https://doi.org/10.1111/ner.12197.

Shealy, C. N., Mortimer, J. T., & Reswick, J. B. (1967). Electrical inhibition of pain by stimulation of the dorsal columns: Preliminary clinical report. *Anesthesia and Analgesia, 46*(4), 489–491.

Slangen, R., Schaper, N. C., Faber, C. G., Joosten, E. A., Dirksen, C. D., van Dongen, R. T., et al. (2014). Spinal cord stimulation and pain relief in painful diabetic peripheral neuropathy: A prospective two-center randomized controlled trial. *Diabetes Care, 37*(11), 3016–3024. https://doi.org/10.2337/dc14-0684.

Song, J. J., Popescu, A., & Bell, R. L. (2014). Present and potential use of spinal cord stimulation to control chronic pain. *Pain Physician, 17*(3), 235–246.

Swigris, J. J., Olin, J. W., & Mekhail, N. A. (1999). Implantable spinal cord stimulator to treat the ischemic manifestations of thromboangiitis obliterans (Buerger's disease). *Journal of Vascular Surgery, 29*(5), 928–935. https://doi.org/10.1016/s0741-5214(99)70221-1.

Tanner, J. A. (1962). Reversible blocking of nerve conduction by alternating-current excitation. *Nature, 195*, 712–713. https://doi.org/10.1038/195712b0.

Taylor, R. S., Desai, M. J., Rigoard, P., & Taylor, R. J. (2014). Predictors of pain relief following spinal cord stimulation in chronic back and leg pain and failed back surgery syndrome: A systematic review and meta-regression analysis. *Pain Practice, 14*(6), 489–505. https://doi.org/10.1111/papr.12095.

Ubbink, D. T., & Vermeulen, H. (2013). Spinal cord stimulation for non-reconstructable chronic critical leg ischaemia. *Cochrane Database of Systematic Reviews, 2013*(2), Cd004001. https://doi.org/10.1002/14651858.CD004001.pub3.

United States Food and Drug Administration. (2020). *Conduct a trial stimulation period before implanting a spinal cord stimulator (SCS) - Letter to health care providers*. Retrieved from https://www.fda.gov/medical-devices/letters-health-care-providers/conduct-trial-stimulation-period-implanting-spinal-cord-stimulator-scs-letter-health-care-providers.

van Beek, M., Slangen, R., Schaper, N. C., Faber, C. G., Joosten, E. A., Dirksen, C. D., et al. (2015). Sustained treatment effect of spinal cord stimulation in painful diabetic peripheral neuropathy: 24-month follow-up of a prospective two-center randomized controlled trial. *Diabetes Care, 38*(9), e132–e134. https://doi.org/10.2337/dc15-0740.

Van Buyten, J. P., Al-Kaisy, A., Smet, I., Palmisani, S., & Smith, T. (2013). High-frequency spinal cord stimulation for the treatment of chronic back pain patients: Results of a prospective multicenter European clinical study. *Neuromodulation, 16*(1), 59–65. discussion 65-56 https://doi.org/10.1111/ner.12006.

Van Buyten, J. P., Smet, I., Liem, L., Russo, M., & Huygen, F. (2015). Stimulation of dorsal root ganglia for the management of complex regional pain syndrome: A prospective case series. *Pain Practice, 15*(3), 208–216. https://doi.org/10.1111/papr.12170.

Vaquer Quiles, L., Blasco González, L., Asensio Samper, J., Villanueva Pérez, V. L., López Alarcón, M. D., & De Andrés Ibáñez, J. (2009). Epidural neurostimulation of posterior funiculi for the treatment of Buerger's disease. *Neuromodulation, 12*(2), 156–160. https://doi.org/10.1111/j.1525-1403.2009.00203.x.

Verrills, P., Sinclair, C., & Barnard, A. (2016). A review of spinal cord stimulation systems for chronic pain. *Journal of Pain Research, 9*, 481–492. https://doi.org/10.2147/jpr.S108884.

Wall, P. D., & Devor, M. (1983). Sensory afferent impulses originate from dorsal root ganglia as well as from the periphery in normal and nerve injured rats. *Pain, 17*(4), 321–339. https://doi.org/10.1016/0304-3959(83)90164-1.

Chapter 46

Effectiveness of neural mobilization on pain and disability in individuals with musculoskeletal disorders

Carlos Romero-Morales[a], César Calvo-Lobo[b], David Rodríguez-Sanz[b], Daniel López-López[c], Marta San Antolín[d], Victoria Mazoteras-Pardo[e], Eva María Martínez-Jiménez[b], Marta Losa-Iglesias[f], and Ricardo Becerro-de-Bengoa-Vallejo[b]

[a]*Physiotherapy Department, Faculty of Sport Sciences, Universidad Europea de Madrid, Madrid, Spain,* [b]*Faculty of Nursing, Physiotherapy and Podiatry, Universidad Complutense de Madrid, Madrid, Spain,* [c]*Research, Health and Podiatry Group, Department of Health Sciences, Faculty of Nursing and Podiatry, Universidade da Coruña, Ferrol, Spain,* [d]*Department of Psychology, Universidad Europea de Madrid, Madrid, Spain,* [e]*Research Group "ENDOCU", Department of Nursing, Physiotherapy and Occupational Therapy, Faculty of Physiotherapy and Nursing of Toledo, University of Castilla-La Mancha, Ciudad Real, Spain,* [f]*Department of Nursing, Faculty of Health Sciences, Universidad Rey Juan Carlos, Alcorcón, Spain*

Abbreviations

CP	cervicobrachial pain
CTS	carpal tunnel syndrome
LBP	low back pain
MTrP	myofascial trigger points
NG	neurodynamic gliding
NM	neural mobilization
NS	neurodynamic stretching
PNS	peripheral nervous system
PPT	pain pressure threshold
ROM	range of motion
SLR	straight leg raise
TTS	tarsal tunnel syndrome

Introduction

The 2016 Global Burden of Disease reported that musculoskeletal conditions comprised the second highest contributor to disability worldwide (Global, regional, and national incidence, prevalence, and years lived with disability for 328 diseases and injuries for 195 countries, 1990–2016: a systematic analysis for the Global Burden of Disease Study 2016, 2017). Patients are characterized by pain, weakness, stiffness, mental disorders, a decrease in functionality which often resulted in chronic health disturbances that impact negatively on social relationships (Fig. 1; Blyth et al., 2019). The prevalence of musculoskeletal conditions was comparable with cardiovascular and respiratory alterations, resulting in an estimated 213 billion dollars (Blyth et al., 2019). The maintaining of musculoskeletal health is essential for social and functional autonomy and also to the increase of productive life years (Schofield et al., 2015).

Prior research support that the peripheral nervous system (PNS) is involved in every syndrome or injury related to the musculoskeletal system, such as low back-related leg pain or neck-related arm pain (Leaver et al., 2013; Schafer, Hall, & Briffa, 2009). The PNS also plays an important role in neuropathies caused by entrapment locations, such as sciatica (Neto et al., 2017), plantar heel pain, carpal tunnel syndrome, epicondylalgia (Basson et al., 2017), or even tension-type headache (Ferragut-Garcias et al., 2017).

Treatments, Mechanisms, and Adverse Reactions of Anesthetics and Analgesics. https://doi.org/10.1016/B978-0-12-820237-1.00046-6
Copyright © 2022 Elsevier Inc. All rights reserved.

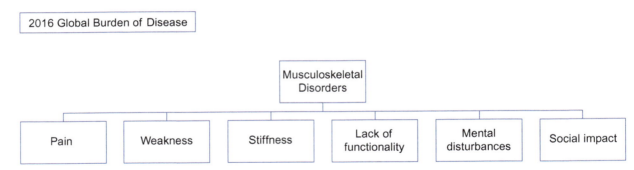

FIG. 1 Musculoskeletal disorders features.

In 1995 Shacklock argued that the nervous system approaches by manual therapy could be an interesting method for the treatment of pain (Shacklock, 1995). This approach was based on the mechanical stimulation of the pain physiology pathways by the treatment of neural tissues and structures surrounding the nervous system (e.g., muscle, fascia, connective tissues) (Shacklock, 1995). Thus, neural mobilization (NM) techniques were employed for the diagnosis and treatment of the mechanical and physiological properties of the PNS (Shacklock, 1995). Several studies reported the benefits of the NM in both healthy (Waldhelm et al., 2019) and individuals with musculoskeletal disorders (De-la-Llave-Rincon et al., 2012; Villafane, Silva, & Fernandez-Carnero, 2011). NM techniques are based on manual mobilizations and exercises focused on restoring the homeostasis inside and around the nervous system (Coppieters & Butler, 2008). Shacklock described that NM approaches were: neurodynamic gliding (NG) was defined as the mobilization of the nerve alternating sliding mobilizations of at least two joints tensioning the peripheral nervous system while the other movement at the same time decreasing the nervous system tension". Neurodynamic stretching (NS) approach was described as "the simultaneous movements of two joints that one movement load the nervous system while the other movement of the other joint increases the load tension of the nervous system" (Fig. 2) (Beltran-Alacreu, Jimenez-Sanz, Carnero, & La Touche, 2015; Butler, 1991; Shacklock, 1995).

Currently, research developed in animal and human models reported the benefits of the NB on the decreasing of pain and hyperalgesia (Song et al., 2006), intraneural edema of the median nerve in subjects with carpal tunnel syndrome (Schmid et al., 2012), for the improvement of the cervical range of motion (ROM) with non-differences respect to ibuprofen in individuals with cervicobrachial pain (Calvo-Lobo et al., 2018), an increase of the glial cells and neural growth factor responses (Santos et al., 2012), a decreasing of the electromyographic activity in the biceps brachii on stroke patients with spasticity (Castilho et al., 2012), facilitating nerve regeneration involving nerve growth factor and myelin protein zero in individuals after sciatic nerve injury (da Silva et al., 2015). Currently, evidence support that NM could be effective in structural nerve disturbances, for example, a situation in which the cross-sectional area was involved for nerve compression, such as entrapment syndromes (Kantarci et al., 2014). In addition, individuals with musculoskeletal disorders experimented sensitivity alterations and could be related to neuropathic or nociceptive pain, defined as "pain caused by a lesion or disease of the somatosensory nervous system"(Treede et al., 2008) Therefore, NM interventions are employed as a treatment for different musculoskeletal disorders (Neto et al., 2017).

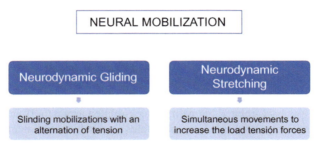

FIG. 2 Neural mobilization techniques.

Headache and neck pain

Considering the NB techniques developed in the upper quadrant, Ferragut-Garcias et al. (2017) carried out a study in individuals diagnosed with a tension-type headache to evaluate a program based on soft tissue management and NM. Tension-type headache was described as the most prevalent primary headache disorder worldwide and was related to the presence of musculoskeletal disorders, such as the presence of latent or active myofascial trigger points (MTrP) in masseter, temporalis, and sternocleidomastoid muscles (Do et al., 2018). The authors reported that the treatments based on soft tissue and NM were beneficial in pain pressure threshold (PPT) variables in temporal muscles and supraorbital region and for a decrease of maximal pain intensity and the frequency of the episodes. These findings may conclude that the combination of two manual therapy interventions – including NM – could be beneficial to pain-relieving and headache symptoms. NM has also been employed in combination with cervical tractions in individuals with cervical nerve root pathology in a case report developed by Savva and Giakas (2013). The study reflects an improvement of the pain intensity and the neck functionality tested by the neck disability index with a 4-week follow up which means that the patient was able to develop the job tasks without pain and limitations. Neck pain was defined as a common musculoskeletal condition presented in 45% of the general population (Fejer, Kyvik, & Hartvigsen, 2006). Current evidence reports that individuals with neck pain reported several muscle disorders, such as in the upper trapezius, scalene, sternocleidomastoid, and suboccipital muscles. Fernandez-Carnero et al. (2019) performed a randomized clinical trial to determine the immediate effect of an NM technique in individuals with chronic neck pain compared with a sham intervention. The results exhibit the effectiveness of the neural tensioner technique on conditioned pain modulation, but no significant differences were observed between groups for the pain intensity, neck disability index, and cervical ROM. These findings support the benefits of the NM for improving endogenous pain modulation in subjects with chronic neck pain. Regarding the neurophysiological effects of the NM techniques, a study conducted by Beltran-Alacreu et al. (2015) was developed to assess the short-term mechanical hypoalgesic effects in asymptomatic individuals comparing the two NM techniques, NG and NS. Pain pressure threshold (PPT) of the masseter and temporalis muscles were evaluated to demonstrate the hypoalgesic effects by activation of the descending inhibitory pain for the NG group.

Cervicobrachial pain

Cervicobrachial pain (CP) was described as a common condition that affects 83 in 100,000 people in the course of their lives. Currently, the research described the CP by the presence of pain in the neck which is referred to as the arm. This condition was estimated at approximately 868 million US dollars per year, increasing day after day. Magnetic resonance imaging was defined as a gold standard to diagnose CP reporting a positive correlation between pathological and clinical findings (Calvo-Lobo et al., 2018). The classical pain symptoms were caused by the presence of musculoskeletal disturbances and neuropathic irradiation due to nerve anatomical injuries (Salt, Wright, Kelly, & Dean, 2011). Non-steroidal anti-inflammatory drugs (NSAIDs), such as oral ibuprofen were described as the first option for the treatment of patients with CP. In addition, physiotherapy techniques were also recommended for these patients. Several authors reported the effectiveness of NM techniques to relieve the CP symptoms through the activation of the descending nervous system pain pathway (Beneciuk, Bishop, & George, 2009; Coppieters, Stappaerts, Wouters, & Janssens, 2003). Calvo-Lobo et al. (2018) developed a randomized clinical trial to study the effectiveness between the median nerve NM and cervical lateral glide intervention (group 1) and oral ibuprofen (group 2) in individuals with CP. Predictably, oral ibuprofen was effective for the reduction of pain intensity and disability during 6-weeks with respect to the NM therapy but, for the ROM variable, no statistical differences were reported taking into consideration the possible oral ibuprofen adverse effects (Fig. 3). Rodriguez-Sanz et al. (2018) also considered effective the NM techniques for individuals with CP in the study developed to demonstrate the effectiveness of median nerve NM in this population. The authors reported the benefits of an NM approach to the reduction of pain compared with a waiting-list group in patients with CP. These findings evidence that NM techniques had a positive impact on the reduction the peripheral and central neurogenic pain in patients with CP. Median nerve NM technique has been also compared with the oral ibuprofen treatment in individuals with CP reporting both interventions positive results for the reduction of pain intensity (Sanz et al., 2018). Based on these findings, NM could be effective as a non-pharmacological option for the treatment of CP patients, without the NSAIDs' adverse effects. In this line, Kim, Chung, and Jung (2017) reported that NM added to a cervical traction technique was effective in relieving pain, recovery from a neck disability, a ROM improvement, and deep flexor muscles endurances in individuals with disturbances in the cervical nerve roots, such as cervical radiculopathy. In addition, Ayub, Osama, and Ahmad (2019) performed a double-blinded randomized clinical trial and compared the effectiveness between passive NM and active NM techniques

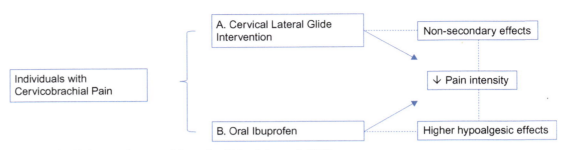

FIG. 3 Cervical lateral glide intervention vs. oral ibuprofen (Calvo-Lobo et al., 2018).

in a 12-treatment sessions program. Results showed that both active and passive NM approaches were effective in the management of cervical radiculopathy with no differences between groups for the pain rating scale, neck disability and cervical ROM variables.

Median, cubital, and radial nerves entrapment syndromes

Regarding the upper limb, lateral elbow pain is one of the most common musculoskeletal conditions reported by people who work with a computer. Pain and a decreasing ROM and functionality were the most common symptoms described by this population. An NM of the radial nerve employing 8 series of 8 oscillations during 3 times was effective to an improvement of pain symptoms in individuals with lateral elbow pain. The authors argued that a neurodynamic approach could be effective to restore the dynamic balance between the neural tissues and surrounding structures (Arumugam, Selvam, & MacDermid, 2014).

Carpal tunnel syndrome (CTS) was considered the most common peripheral entrapment nerve condition related to a compression of the median nerve at the carpal tunnel. Currently, research reported a prevalence of 3.72% in the US population and it affects women more than men. Several non-pharmacological options were available to the management of the CTS, such as passive mobilizations, thermotherapy, electrotherapy, and NM. De-la-Llave-Rincon et al. (2012) showed the effectiveness of soft tissue mobilizations and a neurodynamic technique in women with chronic CTS diagnosed by electromyography and clinical reproduction of clinical symptoms. The soft tissue approach included: myofascial release, soft tissue mobilization, stretching, and cross-fiber friction manipulations in the soft tissue structures related with the median nerve—scalene, pectoralis minor and pronator muscles, transverse carpal ligament, and the palmar aponeurosis— Immediately the soft tissue application authors developed a nerve slider intervention to the median nerve. The combination of these two manual therapy techniques reported a decrease in pain intensity but did not improve widespread pain sensitivity in women with chronic CTS. Supported by the current literature, it can be seen that the combination of diverse manual therapy techniques including the NM approach produced better improvements to relieving the neuropathic mechanical and sensitive clinical symptoms, mainly the pain intensity. In this line, median nerve mobilization was also compared with carpal bone mobilization and a non-intervention group in individuals with CTS and exhibits that no differences between neurodynamic interventions were observed. However, the authors reported a clear positive trend between the treatment and no intervention groups encouraging the manual therapy approach in populations who suffer CTS (Tal-Akabi & Rushton, 2000). In addition, Oskouei, Talebi, Shakouri, and Ghabili (2014) showed the effectiveness of the median nerve NM combined with a conventional routine of physiotherapy (treatment group) compared with a conventional routine physiotherapy group (control) in individuals with CTS in a randomized clinical trial. Both groups showed an improvement for the pain intensity and symptoms severity scale and significant differences were observed for an improvement of the distal motor latency and functional scale for the NM plus the conventional physiotherapy intervention group. Regarding the nerve sliding techniques, Schmid et al. (2012) reported a reduction in intraneural edema and an improvement in pain and functionality in patients with CTS. Based on these findings, evidence-based manual approaches support that non-surgery intervention could be the first option to relieving the neural entrapment symptoms and for the improvement of the functionality capacities in the individuals with CTS. Moreover, it's considered necessary that these neurodynamic techniques were developed under an adequate clinical reasoning umbrella. Different NM approaches produce different amounts of excursion and tension in the nervous system. Coppieters, Hough, and Dilley (2009) reported that sliding neural techniques were related to the largest excursion while the neural tensioning approach induced with the smaller nerve excursion induced by the individual's movements of the neck or elbow. Thus, it's plausible to conclude that different NM approaches produced different physiological and therapeutic effects.

Cubital tunnel syndrome was considered the second most common entrapment neuropathies after CTS. Caused by compression, traction, and pressure in the ulnar nerve at the cubital tunnel. Commonly experimented in individuals who work in large periods of elbow flexion—in elbow flexed position authors estimated that narrows the canal 55% -. Therefore, this syndrome was related to an increase of intraneural pressure in this position and, if a wrist extension and a shoulder abduction were added, the intraneural pressure at the cubital tunnel increase by 6 times (Oskay et al., 2010). Clinical symptoms were described by the patients by paresthesia alongside the arm (fourth and fifth finger), elbow pain, and weakness of grip and pinch maneuver. Oskay et al. (2010) in a case series report described that a non-pharmacological treatment with an 8-week rehabilitation program that included NMs—sliding and tensioning interventions – could be beneficial for patients with moderate and mild cubital tunnel syndrome. Regarding the radial nerve mobilization, Villafañe, Silva, Bishop, and Fernandez-Carnero (2012) reported the effectiveness of the sliding mobilization of the proximal-distal nerve for the decreasing of pain intensity and increasing tip pinch strength in subjects with thumb carpometacarpal osteoarthritis. In addition, neurodynamic techniques for the upper limb could be effective in other conditions, for example, Godoi et al. (2010) supported that NM was effective for the reduction of the myoelectric activity in biceps brachii muscles in post-stroke individuals or Kim, Cha, and Ji (2016) exhibit that a median nerve NM have a positive effect on recovery from delayed onset soreness.

Low back pain

Low back pain (LBP) was defined as one of the most health disability musculoskeletal conditions worldwide. Almost 75% of the population experience one episode of LBP in their lives. Currently, research demonstrated that LBP causes a very high cost to the healthcare systems that influences the personal, work, and social environments of this population. For example, a total cost was up to 77 billion dollars in the United States with an average cost of 11.151 dollars per individual. Exercise and manual therapy have been described as effective evidence-based interventions to relieving pain and disability in this population (Hahne et al., 2017). Based on the current effectiveness of different manual therapy approaches combined with each other, Butler's neural tissue mobilization with mulligan's bent leg raise produced an improvement in pain and straight leg raise (SLR) range immediately after the application but this effect was not observed with 24 h of follow up in individuals with low backache (Tambekar, Sabnis, Phadke, & Bedekar, 2016). The NM effects have been also compared with an electrotherapy program in subjects with LBP. Both groups exhibit benefits on pain and functional disability and may be employed as self-treatment added to a supplementary program in patients with LBP (Kurt, Aras, & Buker, 2020). Related to LBP many patients developed radiating pain and radicular symptoms. Lumbar radiculopathy was defined as a result of an irritated lumbar nerve trunk in consequence of a herniated lumbar disc. Current research support that physical therapy was the first option evidence-based for the management of lumbar radiculopathy (Plaza-Manzano et al., 2019). A recent study combined NM with a motor control exercises program in subjects with lumbar radiculopathy to assess pain, disability, and pressure sensitivity variables. The therapist carried out the neural slider technique into the sciatic nerve of the affected side over 3 sets of 10 repetitions. Authors exhibit that the NM with a motor control exercises program did not improve pain, function, and pressure sensitivity variables but subjects showed changes in neural mechano-sensitivity and SLR (Plaza-Manzano et al., 2019).

Lower limb and neurodynamic techniques

Regarding the relationship between neurodynamic and performance, in Aksoy et al. (2020) performed a randomized double-blinded study to investigate the effects of femoral nerve mobilization and sciatic nerve mobilization in vertical and horizontal jumps. NM isn't only effective for the treatment of pain and functionality, it has been demonstrated that both femoral and sciatic NM were effective to increase an immediate vertical jump in adults. In addition, neurodynamic techniques (neural sliders and tensioners) were effective for increasing hamstring flexibility (Sharma, Balthillaya, Rao, & Mani, 2016). In addition, Coppieters et al. (2015) support that different neurodynamic approaches for the lower limb developed different neural excursions for the sciatic nerve assessed by high-resolution ultrasound imaging. The authors explained while the results of their study and current research that the mobilization of adjacent joints could be an impact on nerve biomechanics due to the continuity of the nervous system.

Patellofemoral pain syndrome was described as a common knee disturbance that presents pain on the anterior knee aspect while the subjects performed activities (e.g. going up or downstairs) Current literature support that patellofemoral pain symptoms could be related to peripheral nerve damage concluding that nerve disorders might be another factor to the development of the patellofemoral symptoms added to the most habitual mechanisms, such as overuse or biomechanical features (Huang, Shih, Chen, & Ma, 2015). Femoral nerve mobilization has been also employed as a predictor tool to consider if an individual with patellofemoral pain should receive an NM treatment or not. In addition, the authors showed that these individuals responded positively to a femoral nerve mobilization within 2-weeks (Huang et al., 2015). Regarding the peroneal nerve, Villafañe, Pillastrini, and Borboni (2013) performed a case report on a 24 years old patient with peroneal nerve paralysis. Foot disturbances (leg and foot pain, sensory alterations, a decrease in ankle dorsiflexion function) were related to musculoskeletal and neural syndromes or processes. Peroneal nerve paralysis was considered the most common neural condition in the lower limb and was related to the foot drop. The authors developed a combination of manual therapy interventions (spinal and fibular manipulation, soft tissue treatment of the hamstring and psoas muscles) with an NM sliding technique of the sciatic nerve in 3-sessions over 1-week period. The patient showed a pain decreasing, and an increase in the left lower limb function, ROM, and strength. Based on these findings, the effectiveness of NM combined with manual therapy interventions has demonstrated a robust clinical evidence-based also in the lower limb improving the motor control, biomechanics, axoplasmic flow, neural connective layers elasticity, and for the reduction on pain symptoms and sensitivity of the neural system (Villafañe et al., 2013). Furthermore, future research could be interesting for the NM of the peroneal nerve in anterior cruciate surgeries or ankle sprains conditions. Another peripheral neuropathy could be managing with NM; for example, individuals with diabetic peripheral neuropathy presented an altered neural sensitivity, pain, and impaired proprioception, gait and balance disturbances, and a decrease in ankle dorsiflexion ROM. In addition, this condition affected more than 50% of type II diabetic patients and patients with hyperglycemia, resulting in an increase of fluids between nerve layers (Boyd, Nee, & Smoot, 2017; Goyal, Esht, & Mittal, 2019). Current evidence in rats showed that NM was effective in alleviating mechanical allodynia processed due to diabetic neuropathy (Zhu, Tsai, Chen, & Hung, 2018).

Regarding the effectiveness of neural manipulation approaches, Martins et al. (2019) developed a comparison between neural gliding and tensioning techniques and support that both interventions showed similar positive effects for hamstring flexibility in the mobilized limb, however, tensioning was more effective for the non-mobilized limb. Moreover, the authors report that the neural gliding approach was superior for pressure pain and heat thresholds.

Considering the foot disorders, tarsal tunnel syndrome (TTS) was defined as an entrapment neuropathy of the posterior tibial nerve that comprises: the medial calcaneal, medial plantar, and lateral plantar branches placed in the flexor retinaculum behind the medial malleolus of the ankle. Kavlak and Uygur reported that patients with TTS who were instructed for the development of a self-nerve mobilization intervention added to a conservative program have positive effects on 2-point discrimination test and Tinel sign over with a 6-weeks follow up (Kavlak & Uygur, 2011). In this line, the study performed by Saban, Deutscher, and Ziv (2014) exhibits that the use of NM in an exercise program was positive in plantar heel pain.

Clinical applications

Strong evidence supported that NM is effective for the management of nerve-related musculoskeletal disorders. Current research exhibit that NM intervention added to a physical therapy program have immediate mechanical hypoalgesic effects in individuals with cervicobrachial pain, carpal and cubital tunnel syndromes, low back pain, and lower limb musculoskeletal disturbances related to neural tissue conditions. This chapter may be helpful for researchers and clinicians regarding the treatment of the pain features from an evidence-based neurodynamic approach in patients with musculoskeletal disorders and evidence the potential of this non-pharmacological intervention, without adverse secondary effects of drugs (Fig. 4).

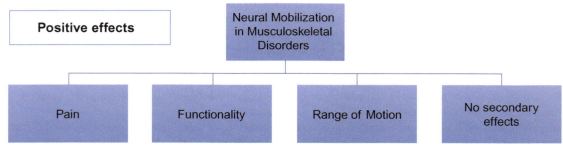

FIG. 4 Positive effects of neural mobilization in musculoskeletal disorders.

Application to other areas

Neurodynamic application may be also beneficial in other areas, for example, patients with fibromyalgia or rheumatoid arthritis. Fibromyalgia was described as a chronic condition involving widespread pain, fatigue, sleep disorders, and cognitive symptoms. Individuals with fibromyalgia described pain as the most frequent feature reported being catastrophizing an important variable to a better understanding of the pain experience. Torres et al. (2015) argued that NM has been effective for decreasing pain, neural biomechanics, and fatigue in patients with fibromyalgia. Neural exercises for the median, femoral and saphenous nerves during 4–8 weeks could be also beneficial on pain in patients with rheumatoid arthritis (Lau et al., 2019). Moreover, neurodynamic interventions have been demonstrated positive effects on performance in healthy individuals, for example, neural exercises targeting the sciatic nerve improve specific joints ROM, vertical and horizontal jump, or hamstrings flexibility (Aksoy et al., 2020; Sharma et al., 2016).

Other agents of interest

The use of the NM techniques should be considered not only for the management or intervention but also as a diagnostic tool. Neural provocation test was effective to complement the physical examination in patients with musculoskeletal disorders who potentially were related to neural disturbances (based on the previous research, most bone, muscle, and soft tissue conditions were associated to varying degrees with the nerve system). These mobilizations were focused to manifest the mechanical stress to irritate the neural tissues. Zamorano argued that neural test could generate in the trunk nerves: length increase, blood flow restriction, compression mechanical forces, and nerve mobilization (Zamorano, 2013). Response for the neural tests comprised: pain, paresthesia, dysesthesia, muscle spam, or a ROM restriction. Therefore, the inclusion of a neural provocation test is recommended in physical examinations or diagnosis protocols.

Mini-dictionary of terms

- **Neurodynamic:** The continuum of the different parts of the nervous system and their relationship with the musculoskeletal system.
- **Neural mobilization:** manual intervention with the objective of move and stretching the neural tissues to their surrounding structures.
- **Musculoskeletal system:** The integration of bone, muscle, cartilage, tendons, ligaments, and soft tissues which form the locomotor system.
- **Nervous system:** considered as an "organ" that involves the peripheral and the central nervous system.
- **Hypoalgesia**: defined as a sensitivity decreasing due to a painful stimulus.
- **Paresthesia**: abnormal sensation perceived on skin. Described as "needles or burning feeling" all over the skin temporally.

Key facts of musculoskeletal disorders

- Musculoskeletal conditions were considered the second highest contributor to disability worldwide.
- Musculoskeletal disorders resulting in an estimated 213 billion dollars in 2011.
- Features include pain, mental disorders, a decrease of functionality often resulted in chronic health disturbances.
- The peripheral and central nervous systems are involved in every syndrome or injury related to the musculoskeletal system.
- Manual therapy and exercise programs have been demonstrated benefits in musculoskeletal conditions.

Summary points

- This chapter focuses on the effectiveness of neural mobilization (NM) techniques in musculoskeletal conditions.
- NM interventions have been demonstrated immediate mechanical hypoalgesic effects related to musculoskeletal disorders.
- NM approaches have been useful added to a manual therapy or exercise program.
- NM has been demonstrated benefits for neck and cervicobrachial pain, median, cubital, and radial tunnel syndromes.
- Neurodynamic interventions have been demonstrated benefits on pain in individuals diagnosed with low back pain.
- NM may be had positive effects on pain in peripheral neuropathies, such as diabetic peripheral neuropathy.

References

Aksoy, C. C., et al. (2020). The immediate effect of neurodynamic techniques on jumping performance: A randomised double-blind study. *Journal of Back and Musculoskeletal Rehabilitation, 33*(1), 15–20.

Arumugam, V., Selvam, S., & MacDermid, J. C. (2014). Radial nerve mobilization reduces lateral elbow pain and provides short-term relief in computer users. *The Open Orthopaedics Journal, 8*, 368–371. https://pubmed.ncbi.nlm.nih.gov/25352930.

Ayub, A., Osama, M., & Ahmad, S. (2019). Effects of active versus passive upper extremity neural mobilization combined with mechanical traction and joint mobilization in females with cervical radiculopathy: A randomized controlled trial. *Journal of Back and Musculoskeletal Rehabilitation, 32*(5), 725–730.

Basson, A., et al. (2017). The effectiveness of neural mobilization for neuromusculoskeletal conditions: A systematic review and meta-analysis. *The Journal of Orthopaedic and Sports Physical Therapy, 47*(9), 593–615.

Beltran-Alacreu, H., Jimenez-Sanz, L., Carnero, J. F., & La Touche, R. (2015). Comparison of hypoalgesic effects of neural stretching vs neural gliding: A randomized controlled trial. *Journal of Manipulative and Physiological Therapeutics, 38*(9), 644–652.

Beneciuk, J. M., Bishop, M. D., & George, S. Z. (2009). Effects of upper extremity neural mobilization on thermal pain sensitivity: A sham-controlled study in asymptomatic participants. *The Journal of Orthopaedic and Sports Physical Therapy, 39*(6), 428–438.

Blyth, F. M., et al. (2019). The global burden of musculoskeletal pain-where to from here? *American Journal of Public Health, 109*(1), 35–40. https://pubmed.ncbi.nlm.nih.gov/30495997.

Boyd, B. S., Nee, R. J., & Smoot, B. (2017). Safety of lower extremity Neurodynamic exercises in adults with diabetes mellitus: A feasibility study. *The Journal of Manual & Manipulative Therapy, 25*(1), 30–38.

Butler, D. J. M. (1991). *Mobilizaion of the nervous system*. London: Churchill.

Calvo-Lobo, C., et al. (2018). Is pharmacologic treatment better than neural mobilization for Cervicobrachial pain? A randomized clinical trial. *International Journal of Medical Sciences, 15*(5), 456–465. https://pubmed.ncbi.nlm.nih.gov/29559834.

Castilho, J., et al. (2012). Analysis of electromyographic activity in spastic biceps brachii muscle following neural mobilization. *Journal of Bodywork and Movement Therapies, 16*(3), 364–368. http://www.sciencedirect.com/science/article/pii/S1360859211002014.

Coppieters, M. W., & Butler, D. S. (2008). Do 'sliders' slide and 'tensioners' tension? An analysis of Neurodynamic techniques and considerations regarding their application. *Manual Therapy, 13*(3), 213–221.

Coppieters, M. W., Hough, A. D., & Dilley, A. (2009). Different nerve-gliding exercises induce different magnitudes of median nerve longitudinal excursion: An in vivo study using dynamic ultrasound imaging. *The Journal of Orthopaedic and Sports Physical Therapy, 39*(3), 164–171.

Coppieters, M. W., Stappaerts, K. H., Wouters, L. L., & Janssens, K. (2003). The immediate effects of a cervical lateral glide treatment technique in patients with neurogenic cervicobrachial pain. *The Journal of Orthopaedic and Sports Physical Therapy, 33*(7), 369–378.

Coppieters, M. W., et al. (2015). Excursion of the sciatic nerve during nerve mobilization exercises: An in vivo cross-sectional study using dynamic ultrasound imaging. *The Journal of Orthopaedic and Sports Physical Therapy, 45*(10), 731–737.

da Silva, J. T., et al. (2015). Neural mobilization promotes nerve regeneration by nerve growth factor and myelin protein zero increased after sciatic nerve injury. *Growth Factors, 33*(1), 8–13. https://doi.org/10.3109/08977194.2014.953630.

De-la-Llave-Rincon, A. I., et al. (2012). Response of pain intensity to soft tissue mobilization and neurodynamic technique: A series of 18 patients with chronic carpal tunnel syndrome. *Journal of Manipulative and Physiological Therapeutics, 35*(6), 420–427.

Do, T. P., et al. (2018). Myofascial trigger points in migraine and tension-type headache. *The Journal of Headache and Pain, 19*(1), 84. https://pubmed.ncbi.nlm.nih.gov/30203398.

Fejer, R., Kyvik, K. O., & Hartvigsen, J. (2006). The prevalence of neck pain in the world population: A systematic critical review of the literature. *European Spine Journal : Official Publication of the European Spine Society, the European Spinal Deformity Society, and the European Section of the Cervical Spine Research Society, 15*(6), 834–848. https://pubmed.ncbi.nlm.nih.gov/15999284.

Fernandez-Carnero, J., et al. (2019). Neural tension technique improves immediate conditioned pain modulation in patients with chronic neck pain: A randomized clinical trial. *Pain Medicine (Malden, Mass.), 20*(6), 1227–1235.

Ferragut-Garcias, A., et al. (2017). Effectiveness of a treatment involving soft tissue techniques and/or neural mobilization techniques in the management of tension-type headache: A randomized controlled trial. *Archives of Physical Medicine and Rehabilitation, 98*(2), 211–219. e2.

Global, Regional, and National Incidence, Prevalence, and Years Lived with Disability for 328 Diseases and Injuries for 195 Countries, 1990–2016: A Systematic Analysis for the Global Burden of Disease Study 2016. (2017). *Lancet (London, England), 390*(10100), 1211–1259.

Godoi, J., et al. (2010). Electromyographic analysis of biceps brachii muscle following neural mobilization in patients with stroke. *Electromyography and Clinical Neurophysiology, 50*(1), 55–60.

Goyal, M., Esht, V., & Mittal, A. (2019). A study protocol on nerve mobilization induced diffusion tensor imaging values in posterior tibial nerve in healthy controls and in patients with diabetic neuropathy-multigroup pretest posttest design. *Contemporary Clinical Trials Communications, 16*, 100451. https://pubmed.ncbi.nlm.nih.gov/31650071.

Hahne, A. J., et al. (2017). Individualized physical therapy is cost-effective compared with guideline-based advice for people with low Back disorders. *Spine, 42*(3), E169–E176.

Huang, B.-Y., Shih, Y.-F., Chen, W.-Y., & Ma, H.-L. (2015). Predictors for identifying patients with patellofemoral pain syndrome responding to femoral nerve mobilization. *Archives of Physical Medicine and Rehabilitation, 96*(5), 920–927.

Kantarci, F., et al. (2014). Median nerve stiffness measurement by shear wave Elastography: A potential sonographic method in the diagnosis of carpal tunnel syndrome. *European Radiology, 24*(2), 434–440.

Kavlak, Y., & Uygur, F. (2011). Effects of nerve mobilization exercise as an adjunct to the conservative treatment for patients with tarsal tunnel syndrome. *Journal of Manipulative and Physiological Therapeutics, 34*(7), 441–448.

Kim, M.-K., Cha, H.-G., & Ji, S. G. (2016). The initial effects of an upper extremity neural mobilization technique on muscle fatigue and pressure pain threshold of healthy adults: A randomized control trial. *Journal of Physical Therapy Science, 28*(3), 743–746.

Kim, D.-G., Chung, S. H., & Jung, H. B. (2017). The effects of neural mobilization on cervical radiculopathy patients' pain, disability, ROM, and deep flexor endurance. *Journal of Back and Musculoskeletal Rehabilitation, 30*(5), 951–959.

Kurt, V., Aras, O., & Buker, N. (2020). Comparison of conservative treatment with and without neural mobilization for patients with low Back pain: A prospective, randomized clinical trial. *Journal of Back and Musculoskeletal Rehabilitation.*

Lau, Y. N., et al. (2019). A brief report on the clinical trial on neural mobilization exercise for joint pain in patients with rheumatoid arthritis. *Zeitschrift für Rheumatologie, 78*(5), 474–478.

Leaver, A. M., et al. (2013). Characteristics of a new episode of neck pain. *Manual Therapy, 18*(3), 254–257.

Martins, C., et al. (2019). Neural gliding and neural tensioning differently impact flexibility, heat and pressure pain thresholds in asymptomatic subjects: A randomized, parallel and double-blind study. *Physical Therapy in Sport : Official Journal of the Association of Chartered Physiotherapists in Sports Medicine, 36*, 101–109.

Neto, T., et al. (2017). Effects of lower body quadrant neural mobilization in healthy and low Back pain populations: A systematic review and meta-analysis. *Musculoskeletal Science & Practice, 27*, 14–22.

Oskay, D., et al. (2010). Neurodynamic mobilization in the conservative treatment of cubital tunnel syndrome: Long-term follow-up of 7 cases. *Journal of Manipulative and Physiological Therapeutics, 33*(2), 156–163.

Oskouei, A. E., Talebi, G. A., Shakouri, S. K., & Ghabili, K. (2014). Effects of neuromobilization maneuver on clinical and electrophysiological measures of patients with carpal tunnel syndrome. *Journal of Physical Therapy Science, 26*(7), 1017–1022. https://pubmed.ncbi.nlm.nih.gov/25140086.

Plaza-Manzano, G., et al. (2019). Effects of adding a neurodynamic mobilization to motor control training in patients with lumbar radiculopathy due to disc herniation: A randomized clinical trial. *American Journal of Physical Medicine & Rehabilitation.*

Rodriguez-Sanz, D., et al. (2018). Effects of median nerve neural mobilization in treating cervicobrachial pain: A randomized waiting list-controlled clinical trial. *Pain Practice : The Official Journal of World Institute of Pain, 18*(4), 431–442.

Saban, B., Deutscher, D., & Ziv, T. (2014). Deep massage to posterior calf muscles in combination with neural mobilization exercises as a treatment for heel pain: A pilot randomized clinical trial. *Manual Therapy, 19*(2), 102–108.

Salt, E., Wright, C., Kelly, S., & Dean, A. (2011). A systematic literature review on the effectiveness of non-invasive therapy for cervicobrachial pain. *Manual Therapy, 16*(1), 53–65.

Santos, F. M., et al. (2012). Neural mobilization reverses behavioral and cellular changes that characterize neuropathic pain in rats. *Molecular Pain, 8*, 57. https://pubmed.ncbi.nlm.nih.gov/22839415.

Sanz, D. R., et al. (2018). Effectiveness of median nerve neural mobilization versus Oral ibuprofen treatment in subjects who suffer from cervicobrachial pain: A randomized clinical trial. *Archives of Medical Science, 14*(4), 871–879.

Savva, C., & Giakas, G. (2013). The effect of cervical traction combined with neural mobilization on pain and disability in cervical radiculopathy. A case report. *Manual Therapy, 18*(5), 443–446.

Schafer, A., Hall, T., & Briffa, K. (2009). Classification of low Back-related leg pain–a proposed Patho-mechanism-based approach. *Manual Therapy, 14*(2), 222–230.

Schmid, A. B., et al. (2012). Effect of splinting and exercise on intraneural edema of the median nerve in carpal tunnel syndrome—An MRI study to reveal therapeutic mechanisms. *Journal of Orthopaedic Research : Official Publication of the Orthopaedic Research Society, 30*(8), 1343–1350.

Schofield, D. J., et al. (2015). Lost productive life years caused by chronic conditions in Australians aged 45–64 years, 2010–2030. *The Medical Journal of Australia, 203*(6), 260. e1–6.

Shacklock, M. (1995). Neurodynamics. *Physiotherapy, 81*(1), 9–16. http://www.sciencedirect.com/science/article/pii/S0031940605670241.

Sharma, S., Balthillaya, G., Rao, R., & Mani, R. (2016). Short term effectiveness of neural sliders and neural tensioners as an adjunct to static stretching of hamstrings on knee extension angle in healthy individuals: A randomized controlled trial. *Physical Therapy in Sport : Official Journal of the Association of Chartered Physiotherapists in Sports Medicine, 17*, 30–37.

Song, X.-J., et al. (2006). Spinal manipulation reduces pain and hyperalgesia after lumbar intervertebral foramen inflammation in the rat. *Journal of Manipulative and Physiological Therapeutics, 29*(1), 5–13.

Tal-Akabi, A., & Rushton, A. (2000). An investigation to compare the effectiveness of carpal bone mobilisation and Neurodynamic mobilisation as methods of treatment for carpal tunnel syndrome. *Manual Therapy, 5*(4), 214–222.

Tambekar, N., Sabnis, S., Phadke, A., & Bedekar, N. (2016). Effect of Butler's neural tissue mobilization and Mulligan's bent leg raise on pain and straight leg raise in patients of low Back ache. *Journal of Bodywork and Movement Therapies, 20*(2), 280–285.

Torres, J. R., et al. (2015). Results of an active neurodynamic mobilization program in patients with fibromyalgia syndrome: A randomized controlled trial. *Archives of Physical Medicine and Rehabilitation, 96*(10), 1771–1778.

Treede, R.-D., et al. (2008). Neuropathic pain: Redefinition and a grading system for clinical and research purposes. *Neurology, 70*(18), 1630–1635.

Villafañe, J. H., Pillastrini, P., & Borboni, A. (2013). Manual therapy and neurodynamic mobilization in a patient with peroneal nerve paralysis: A case report. *Journal of Chiropractic Medicine, 12*(3), 176–181. https://pubmed.ncbi.nlm.nih.gov/24396318.

Villafañe, J. H., Silva, G. B., Bishop, M. D., & Fernandez-Carnero, J. (2012). Radial nerve mobilization decreases pain sensitivity and improves motor performance in patients with thumb carpometacarpal osteoarthritis: A randomized controlled trial. *Archives of Physical Medicine and Rehabilitation, 93*(3), 396–403. https://doi.org/10.1016/j.apmr.2011.08.045.

Villafane, J. H., Silva, G. B., & Fernandez-Carnero, J. (2011). Short-term effects of neurodynamic mobilization in 15 patients with secondary thumb carpometacarpal osteoarthritis. *Journal of Manipulative and Physiological Therapeutics, 34*(7), 449–456.

Waldhelm, A., et al. (2019). Acute effects of neural gliding on athletic performance. *International Journal of Sports Physical Therapy, 14*(4), 603–612. https://pubmed.ncbi.nlm.nih.gov/31440411.

Zamorano, E. (2013). *Movilización Neuromeníngea*. Madrid: Médica Panamericana, D.L.

Zhu, G.-C., Tsai, K.-L., Chen, Y.-W., & Hung, C.-H. (2018). Neural mobilization attenuates mechanical allodynia and decreases proinflammatory cytokine concentrations in rats with painful diabetic neuropathy. *Physical Therapy, 98*(4), 214–222.

Chapter 47

Virtual reality and applications to treating neck pain

M. Razeghi, I. Rezaei, and S. Bervis

Physical Therapy Department, School of Rehabilitation Sciences, Shiraz University of Medical Sciences, Shiraz, Iran

Abbreviations

ACC   anterior cingulate cortex
HMD   head mounted display
NP   neck pain
PAG   periaqueductal gray matter
ROM   range of motion
VR   virtual reality
SS   simulator sickness

Introduction

Virtual reality (VR) applications have been growing quickly in a variety of healthcare areas including rehabilitation assessment and training. This provides a virtual environment to perform rehabilitation-related activities for patients in a similar manner to what they experience in the real life. VR provides an interactive environment between the virtual world and the user. Different technologies such as HMD, cyber-gloves, and head mouse extreme has been applied to bring the subject into the virtual world. Furthermore, VR may immerse the users by engaging their senses, like sight and hearing (Sato et al., 2012).

One of the fundamental components of rehabilitation is therapeutic exercise programs. Gaining the best outcome from an exercise protocol needs feedback, motivation, and repetition.

Staying engaged during exercises needs motivation. Repeating a set of exercises may cause patients to withdraw from the rehabilitation protocol due to inadequate motivation. VR can provide and enhance this motivation by rewarding the users with their best score. Feedback is also an essential component of motor learning. Motivation may facilitate feedback as the knowledge of results and performance, making learning and performance being achieved (Holden, 2005). It may also enable the users to adapt a load of exercise to his/her ability without becoming bored. Virtual rehabilitation has been suggested as an alternative to face-to-face conventional physiotherapy reducing the cost and frequency of clinic visits (Holden, 2005).

"Distraction", or the act of drawing attention away, is another effective mechanism being proposed to relieving pain when using VR. This directs the limited sources of attention away from painful medical conditions and consequently modifies pain perception (Gold, Belmont, & Thomas, 2007). Immersion and multisensory experiences in VR make it more effective in distracting the mind from unpleasant feelings as pain, stress, etc.

Neck pain and VR-based rehabilitation

Neck pain (NP) is a common musculoskeletal disorder with an annual prevalence of 30%–50% (Sarig-Bahat, Weiss, & Laufer, 2010). Nearly 70% of people experience NP in their lifetime (Hoy, Protani, De, & Buchbinder, 2010). Aging, traumas, and prolonged poor posture during working with computer or sleeping are some of the reasons for NP and stiffening (Hakala, Rimpelä, Saarni, & Salminen, 2006; Hoy et al., 2010; Sakakibara, Miyao, Kondo, & Yamada, 1995).

Treatments, Mechanisms, and Adverse Reactions of Anesthetics and Analgesics. https://doi.org/10.1016/B978-0-12-820237-1.00047-8
Copyright © 2022 Elsevier Inc. All rights reserved.

Patients with NP demonstrate different impairments such as muscle weakness, numbness, limitation of motion, and alterations in speed, accuracy, and smoothness of neck movements, leading to "functional limitation" (Sjölander, Michaelson, Jaric, & Djupsjöbacka, 2008; Williams, Sarig-Bahat, Williams, Tyrrell, & Treleaven, 2017). As a consequence, it may lead to disability and fear of movement (Sarig Bahat, Takasaki, Chen, Bet-Or, & Treleaven, 2015).

Sensorimotor impairments are common in NP patients, causing pain chronic and recurrent. Neck region mechanoreceptors are important contributors in the sense of position and movement and postural control (Kavounoudias, Gilhodes, Rodillo, & Rodillo, 1999). Alteration in these mechanoreceptors' function is associated with dizziness, headache, and balance and postural control impairments (Kristjansson & Treleaven, 2009).

Accurate evaluation of the neck region warrants effective treatment by monitoring the patients' signs and symptoms at the beginning and through monitoring their progression and efficacy of treatment during the intervention.

Exercise therapy is the main component of the rehabilitation program in NP. Common neck exercise protocols include muscle stretching, muscle strengthening, stabilization, and ROM exercises (Gross et al., 2016; Häkkinen, Salo, Tarvainen, Wiren, & Ylinen, 2007). Proprioception training as gaze stability, eye-head coordination, repositioning training are important parts of exercise therapy protocols in NP patients as well (Rezaei, Razeghi, Ebrahimi, Kayedi, & Rezaeian Zadeh, 2019).

VR has been widely employed to effectively prescribe exercise protocols. It allows the users to precisely conduct a repeated number of neck movements with greater motivation, compliance and provides real-time feedback to correct the movement errors, by the focus on kinematic and movement sense training (Rezaei et al., 2019; Sarig Bahat et al., 2018). Using the distraction technique while executing the exercise in the form of playing a serious game makes it an effective tool in NP rehabilitation. Additionally, in the field of neck evaluation, VR devices are a valid reliable tool to assess cervical kinematics (Sarig-Bahat, Weiss, & Laufer, 2009).

VR systems design in NP rehabilitation

To develop an effective and attractive tool in the assessment and treatment of NP patients, VR systems must possess several characteristics. It includes safety, motivation, the possibility of performing a controlled number of repetitions, providing different kinds of feedback, and implementing a multi-sensory environment to immerse the user. While simplicity, attractiveness, and pleasure aspect of gamification should be taken into account, emphasis should be placed on exercise protocol guidelines. The neck movements should follow the principle of static and dynamic control over the range, velocity, accuracy, hardness, and energy expenditure.

Several VR applications in NP rehabilitation have been introduced so far. Some are designed in Android and iOS mobile platforms simply to alarm the user when bad cervical posture is found or to provide instructions to further correct postural misalignment (Wang et al., 2015). Others apply advanced game techniques to provide a more real attractive environment in NP assessment and treatment.

Kramer et al. (2009) introduce a VR system to detect proprioceptive deficits in NP by employing cervico-kinesthetics tests. Users are requested to track precisely a globe with a diamond by an HMD which is controlled by the head and neck

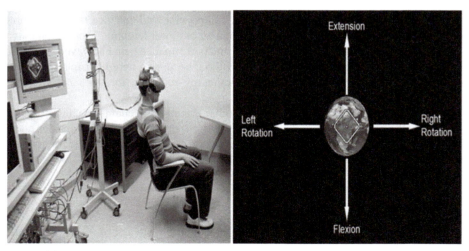

FIG. 1 Experimental setup and the scene that users see in the system developed by Kramer et al. The user immerses into a virtual environment by a head mouse device and tracks the globe precisely with the diamond by head and neck movements. *(With permission from Kramer, M., Honold, M., Hohl, K., Bockholt, U., Rettig, A., Elbel, M., & Dehner, C. (2009). Reliability of a new virtual reality test to measure cervicocephalic kinaesthesia. Journal of Electromyography and Kinesiology, 19(5), e353–e361, Figs. 1, 2, p. e355.)*

FIG. 2 The virtual scene of exergame developed by Sarig-Bahat et al. The user flies the red airplane to reach yellow targets appearing in four directions in three modules of range, velocity, and accuracy of motion. *(With permission from Sarig Bahat, H., Takasaki, H., Chen, X., Bet-Or, Y., & Treleaven, J. (2015). Cervical kinematic training with and without interactive VR training for chronic neck pain—A randomized clinical trial. Manual Therapy, 20(1), 68–78, Fig. 1, p. 69.)*

movements. The preciseness of joint position sense and movement is evaluated as static and dynamic joint re-positioning error (Fig. 1) (Kramer et al., 2009).

Sarig-Bahat et al. (2009, 2010) developed another VR system using HMD. Different scenarios of flying a red airplane in different directions to reach or follow targets appearing on the scene have been employed to assess and train head and neck movements in terms of ROM, velocity, and accuracy in the advanced version (Fig. 2) (Sarig Bahat et al., 2015).

"PlayMancer" has been presented as an exergame for chronic pain rehabilitation. The user plays the game by movements of different parts of the body while wearing a motion suit. A set of infrared cameras capture the motion of the body and muscle activation level is also measure to provide feedback on muscular tension (Jansen-Kosterink et al., 2013).

Harvie et al. (2015) proposed a VR system that provides information on the real movement, then transforming into virtual understate or overstate virtual movements through an HMD to give visual proprioceptive feedback to the user to increase neck mobility (Fig. 3) (Harvie et al., 2015).

"Motor Offset Visual Illusion" is another VR system invented by Harvie et al. (2020) to modulate the kinematic information for altering the body movement perception, including the neck of the user. The hypothesis behind the design of the system is the assumption that non-painful stimuli, as proprioception information, can affect pain perception, manipulating non-aching stimuli which may reduce pain intensity (Harvie et al., 2020).

Kim, Jeon, and Moon (2017) developed a VR exergame for neck training by providing an attractive visual and auditory virtual environment. The game requires the user to follow the generated signals that appeared in different directions based on background music (Kim et al., 2017).

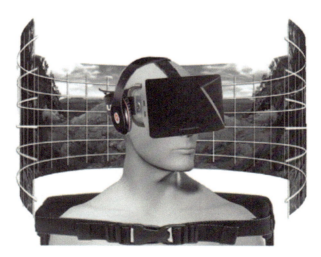

FIG. 3 The "PlayMancer". It is a 360 virtual-template with six different scenes provided by HMD and headphones. The VR system tricks the user by understating the neck range of motion. This can encourage the user to increase neck mobility. *(With permission from Harvie, D. S., Broecker, M., Smith, R. T., Meulders, A., Madden, V. J., & Moseley, G. L. (2015). Bogus visual feedback alters onset of movement-evoked pain in people with neck pain. Psychological Science, 26(4), 385–392, Fig. 1, p. 3.)*

For both assessing and training neck movements, Mihajlovic, Popovic, Brkic, and Cosic (2018) introduced another VR system. This system uses two separate parts, namely a system set-up and calibration and exergame, making individualization of neck exercise program possible based on each user ROM. The user should follow a static, jerky jumping, and moving target, a butterfly and trap it, while information on the neck kinematic characteristics, reaction time, peak and average velocity, jerk intensity, and smoothness of movements can be gathered (Fig. 4) (Mihajlovic et al., 2018).

Rezaei et al. (2019) designed a video game named "Cervigame." A "Head Mouse Extreme" attached above the laptop screen detects reflexive markers fixed on the users' face to transport head movements into the VR environment. This game requires the user to move a rabbit attempting to reach carrots in different directions without hitting the obstacles. The movements of the rabbits are controlled by head movements. The head and neck should move the rabbit, while the subject's eyes are stable on the monitor. This could train gaze stability, eye–head coordination, and position and movement sense. The game includes 50 different stages of movement in two modes of unidirectional and two-directional, progressing from easy to hard. The difficulty of stages increased by the level of obstacles (Fig. 5) (Rezaei et al., 2019).

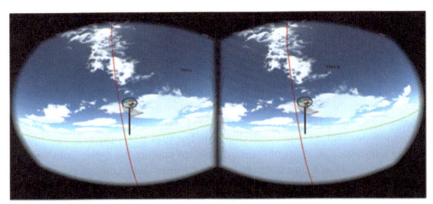

FIG. 4 Two game scene (classic and realistic virtual environment) including butterfly and a net. The user chases the butterfly by neck movement with the net in different directions. *(With permission from Mihajlovic, Z., Popovic, S., Brkic, K., & Cosic, K. (2018). A system for head-neck rehabilitation exercises based on serious gaming and virtual reality.* Multimedia Tools and Applications, 77(15), 19113–19137, Fig. 6&7, p. 19121–19122.)

FIG. 5 The virtual scene of the Cervigame in different stages. The user moves the rabbit by head movements to reach carrots without hitting obstacles in (A) unidirectional game for right and left neck rotation, (B) unidirectional game for neck flexion and extension, (C) two-directional game. *(With permission from Rezaei, I., Razeghi, M., Ebrahimi, S., Kayedi, S., & Rezaeian Zadeh, A. (2019). A novel virtual reality technique (Cervigame®) compared to conventional proprioceptive training to treat neck pain: A randomized controlled trial.* Journal of Biomedical Physics & Engineering, 9(3), 355–366, Fig. 1, p. 358.)

Clinical efficacy of VR in neck disorders

VR technology has been employed to alter some clinical manifestations in NP patients. This mainly includes pain, disability, ROM, and postural control.

Pain

VR has been employed as a pain relief method as soon as 1998 (Hoffman, Doctor, Patterson, Carrougher, & Furness III, 2000). The usage of the technique as an analgesic intervention has been reported afterward (Wiederhold, Gao, Sulea, & Wiederhold, 2014).

There are reports that even a single session of VR training reduces NP (Sarig-Bahat et al., 2010). Significant pain relief is documented after 4–6 sessions of VR training in short (Jansen-Kosterink et al., 2013; Rezaei et al., 2019; Sarig Bahat et al., 2015) and intermediate-term (Rezaei et al., 2019).

Some research showed that VR application can increase pain-free ROM through manipulating visual proprioceptive feedback (Harvie et al., 2015). However, the hypothesis was not supported in NP patients (Harvie et al., 2020).

The superiority of VR compared to the control group receiving treatments other than VR has been reported (Rezaei et al., 2019; Sarig Bahat et al., 2018). However, it is not obvious that VR application poses an added value of superiority when comparing a combination group of kinematic plus VR training versus kinematic training (Sarig Bahat et al., 2015).

Several underlying mechanisms have been proposed for the beneficial effects of VR on pain alleviation. "Distraction" is the main mechanism affecting pain perception. Pain is a perceptual response to protect the body from injuries which mainly results from contextual, psychological, and sensory inputs (Harvie et al., 2015). Factors such as attention, cognition, emotion, expectation, motivation, and memory are influential in pain perception (Tabor, Thacker, Moseley, & Körding, 2017).

Being engaged in a game draws the user s' attention away from pain because playing needs the users' attention, concentration, and multi-sensory integration.

There might be a higher threshold of pain perception by occupying pathways of proprioception sense which is rather concentrated on a peripheral sense other than pain when focusing on the game (Harvie et al., 2015).

Another possible mechanism is the "gate control theory of Melzak". The activity of the central nervous system regions related to attention, emotion, and memory affect the level of pain perception. Painful signals transmission is blocked from traveling through the sensory pathways, when other signals are transmitted depending on the level of subjects' attention, emotion, and past experiences related to pain (Melzack & Wall, 1965).

Furthermore, the capacity of attention is limited (McCaul & Malott, 1984). When the user draws attention to the game, more capacity is occupied, remaining less space for pain. The "multiple resources theory of Wickens" describes that different sensory system resources are in action separately. It may explain how a higher pain relief would be achieved when employing more sensory sources of attention like visual, auditory, and tactile sensation by VR games (Li, Montaño, Chen, & Gold, 2011).

Neuroimaging studies have proposed parts of the brain cortex as regions responsible for pain modulation called the pain matrix. This includes ACC, insula, thalamus, and the primary (S1) and secondary (S2) somatosensory cortices. Furthermore, reduction of the brain activity in parts of the cortex including the insula, thalamus, S2, S1, and ACC (midcingulate part) has been claimed for pain relief when the user engages in a VR game (Gold et al., 2007; Hoffman et al., 2007). Another study has reported a decreased activity in the pain matrix and increased activity of orbitofrontal and ACC (perigenual part) (Bantick et al., 2002). The increased activity of the two latter regions may suggest an intercortical or descending pain control modulation of pain while using VR.

Cingulo-frontal cortex activity associated with attention processing is proposed to facilitate PAG and consequently pain inhibition (Valet et al., 2004) PAG, as a part of the midbrain induces spinal-level pain inhibition via descending fibers. It receives signals from cortices processing attention and emotion. This could explain how attentional-based treatments as VR may modulate pain perception by increasing perigenual ACC and orbitofrontal and consequently PAG activity and decreasing pain matrix activity. In addition, emotions involved in VR in a pleasant environment can modulate pain perception by activating the amygdala, facilitating PAG, and reducing pain (Gold et al., 2007).

VR games require the users to involve in neck exercises that facilitate muscle education. Improvement of muscle function and coordination leads to better neck region motor control and an enhanced eyes, neck and vestibular inter-neural connection and consequently a pain (Sarig Bahat et al., 2015).

A higher skin temperature and lower heart rate are positive signs of autonomic system activation, showing the relaxation effect of VR and justifying pain reduction (Wiederhold et al., 2014).

Kinematic impairments

Different aspects of neck movements are reported to impair after NP including range, velocity, and accuracy.

Neck ROM limitation is one of the common impairments found in NP patients (Sjölander et al., 2008). Controversy exists in this regard. Although pieces of evidence showed an increase in the neck ROM with understating visual-proprioceptive feedback of the neck rotation degree in the VR environment (Harvie et al., 2015), VR training with augmented feedback in a study did not support this finding (Harvie et al., 2020).

The neck ROM was reported to increase even after one session of VR including repetitive kinematic training (Sarig-Bahat et al., 2009, 2010) and after 4–6 session of such VR kinematic training (Jansen-Kosterink et al., 2013; Sarig Bahat et al., 2015). However, no improvement was found in the neck in NP subjects by Sarig Bahat et al. (2018). The controversy might be related to the normal baseline ROM values of the participants, the type of exergame, and the study design (Sarig Bahat et al., 2018).

The velocity of movements has been mentioned as a better indicator of kinematic impairments in NP patients even with mild to moderate disability and normal neck ROM (Descarreaux, Passmore, & Cantin, 2010). VR is reported to the peak and means movement velocity improvements in short and intermediate-term after 4 weeks of VR training (Sarig Bahat et al., 2015, 2018) and when compared to the control group (Sarig Bahat et al., 2018).

Deficiency in the neck movement accuracy is another impairment found in NP patients (Sjölander et al., 2008). VR games induced significant improvement in the movement accuracy after 4–6 session (Sarig Bahat et al., 2015, 2018) when compared to the control group (Sarig Bahat et al., 2018).

The possible VR-induced kinematic improvement has been attributed to a better motor control, eyes, head and vestibular coordination, and motor learning. "Distraction" effect, relaxation, enhanced motivation, and decreased fear of movement are proposed to further explain these changes.

Disability

Pieces of evidence fairly support disability reduction by employing VR in NP subjects (Jansen-Kosterink et al., 2013; Sarig Bahat et al., 2015, 2018) and when compared to the control group (Rezaei et al., 2019; Sarig Bahat et al., 2018). Better deep and superficial neck muscle control and coordination, improved capability to move the neck more rapidly (in a greater range), less neck segments stress in daily activities, advanced fine motor control and neural relation between neck, visual, and vestibular system are possible justifications for disability reduction (Sarig Bahat et al., 2015). Further studies are needed to elaborate on the topic.

Postural control and balance impairment

An effective and timely postural control needs appropriate integration of visual, vestibular, and proprioceptive information processing (Vuillerme & Pinsault, 2009). It is more complex in the neck region, as indicated by a higher number of muscle spindles (Treleaven, 2008). Muscle spindles of this region have a complex reflex connection with vestibular and visual systems and play a more crucial role in postural control. Evidence is in favor of postural control impairments in NP clients (Silva & Cruz, 2013). Therefore, neck muscle deficiency may potentially impair postural control.

VR has been shown to significantly enhancing balance in step-up (Sarig Bahat et al., 2015) and Y-balance test (Rezaei et al., 2019) in the short and intermediate-term. Improving neck muscle coordination might be speculation for balance improvements. Furthermore, VR intervention seems to facilitate reflex responses of neck muscles. This may play an essential role in the interaction of reflexes of the head, neck, eyes, and vestibular system.

Pain relief and disability reduction further contribute to improved postural control and balance due to the tuning effects of VR on reducing muscle spindle sensitivity reportedly increased by NP (Thunberg et al., 2001; Treleaven, 2008).

Factors affecting the efficacy of VR-based treatment in NP

Factors determining the quality of VR-based treatment effectiveness include the type of technology used in the VR design, the amount of sense of presence while putting the user in a virtual environment, and the appropriate time and frequency of VR interventions.

There are pieces of evidence that immersive VR technology with HMD and high technology increases the benefits of VR on pain reduction compared to non-immersive or low technology HMD (Li et al., 2017). Employing higher gamification technology to provide a more realistic environment would enhance the user's motivation to achieve treatment goals.

Higher technology as an HMD and realistic game design have been reported to create a greater sense of presence in a virtual environment leading to a greater attenuation of pain (Mihajlovic et al., 2018). However, these results have not yet been evaluated in NP. There is a need for future studies in this population to justify such effects.

To achieve an optimal treatment goal, a proper time and frequency of VR interventions are important. Although further future studies are needed to fully confirm, 4–8 sessions, each 20–30 min long, over 4–5 weeks is recommended (Sarig Bahat et al., 2015, 2018).

VR application as an assessment tool in NP

VR technology has been applied as an evaluation tool to monitor the outcome of therapeutic interventions. This covers a wide range of measuring neck ROM, speed and smoothness and accuracy of motion, and reaction time by static and dynamic kinematic tests (Sarig-Bahat et al., 2009, 2010).

Neck motion measurements encounter some difficulties in the determination of neck landmarks, 3-D motion, and participation of multi-joints. Technologies such as motion tracker and HMD used in the VR systems are reported to overcome these difficulties by 3-D instantaneous determination of neck position by employing gyroscopes and magnetometers. There are pieces of evidence that VR systems have been valid and reliable in neck kinematic assessments (Chang et al., 2019; Sarig-Bahat et al., 2009, 2010). Furthermore, reporting the movement analysis by VR systems decreases errors in reading the values by the assessors. It also provides a self-evaluating tool to giving users accurate feedbacks to monitor their progression (Chang et al., 2019).

The idea behind VR-based assessments is to evaluate the spontaneous responses of users to move their head and neck in response to an applied stimulus within a VR environment as occurring in real-life tasks. In contrast to most of the conventional methods requiring head and neck voluntary movements in response to assessor verbal commands, VR seems to be a more functional approach. VR would motivate and engage users to perform the tests with major attempt and more compliance. This makes VR-based assessment more precise and valid when evaluating the maximum users' capabilities (Sarig-Bahat et al., 2009, 2010).

Movement velocity in terms of peak, mean, and time to peak values may be evaluated while the user attempts rapid head and neck movements to catch a target. Evaluating the time the user reaches a target from the onset of the object's appearance could precisely determine the reaction time.

The movement smoothness and accuracy can be determined by measuring the deviation of the user's head and neck movements from a predetermined pathway in VR while tracking a mobile target (Mihajlovic et al., 2018; Sarig Bahat et al., 2015).

Disadvantages of VR application in NP

Despite the beneficial effects of using VR in NP patients, some concerns have been raised for unwanted side effects mainly as "SS."

Neck movement during the application of VR, especially with immersion in the virtual environment, involves the inter-activation of vestibular, visual, and cervical motor and sensory control systems (Sato et al., 2012). This may cause SS, characterized by pallor, cold sweating, nausea, and vomiting. SS has been reported to affect approximately 80%–95% of VR users especially during immersion with HMD. Since dizziness is one of the common complaints in people with NP, using VR in this population may exacerbate SS side effects (Tyrrell, Sarig-Bahat, Williams, Williams, & Treleaven, 2018).

Rate of dizziness, the presence of visual disturbance, exposure time, age, and female gender are among factors increasing the possibility of SS after VR application (Tyrrell et al., 2018).

It is rather to screen people with NP before prescribing VR-based treatments.

Reducing the exposer time to approximately 15 min (Tyrrell et al., 2018), using high HMD technology with greater sampling rate frequency and less latency between real motion and tracking device display are recommended as measures reducing the possibility of SS (Treleaven et al., 2015).

Two points are worthy-mentioning. Firstly, patients may experience the same signs and symptoms of SS even during real neck exercise, and, secondly, VR-related SS has been mainly rated as mild. On the other hand, there are pieces of evidence that VR exercise could decrease dizziness in NP subjects (Tyrrell et al., 2018). Further studies are needed to expand the knowledge of predisposing factors for SS in NP during VR Application.

A relatively high cost and complex technology applied in VR systems can restrict the use of VR (Sarig Bahat et al., 2018). Developing more user-friendly and low-cost VR applications and systems could ensure the usage of VR in NP.

Application of VR may have psychological side effects such as desensitization in long-term usage.

Conclusions

VR possesses the capability of assessment and treatment in NP with the main benefits of motivation and distraction. There is a growing tendency of employing VR through attractive games, making it a desirable rehabilitation tool in exercise testing and prescription in NP patients.

Applications to other areas

VR has the potential to be employed as a telerehabilitation tool. High cost, long distances to reach, and long waiting period for medical services and treatment sessions make telerehabilitation a warranted alternative. VR-based rehabilitation seems to provide a good telerehabilitation tool. This technology authorized therapists to remotely follow their patients' progression from home. Furthermore, the possibility of rehabilitation use at home may provide self-management even without therapists' supervision. This can emphasize motivation to continue exercises at home and self-monitoring by the patients. It turns rehabilitation into an affordable low-cost treatment as well.

Other agents of interest

VR has drawn a big deal of attention in the sport industry, being employed by athletes, coaches, and other sport-related professionals to evaluate sports performance. It has been demonstrated that interactive immersive VR applications could improve performance, physiological, and psychological outcomes. Factors related to the athlete, VR environment factors, task factors, and the non-VR environment factors are considered influential to ultimate outcomes. Variables such as the presence of others in the virtual environment, competitiveness, task autonomy, immersion, attentional focus, and feedback are of utmost importance in this regard.

Mini-dictionary of terms

NP: considered as pain in an area confined to the superior nuchal line and root of spine of the scapula with or without referred pain to the head or upper extremities.
VR: a 2D or 3D computer-based technology that provides a simulated interactive environment.
Head-mounted display: a technique commonly used in VR system by employing gyroscope, accelerometer, and magnetometer that transport the user to the virtual world and track the head motions.
Gamification: the process of turning an activity or task into a game or something resembling a game.
Exergame: the process of turning an activity or task into a game or something resembling a game.
Telerehabilitation: is the delivery of rehabilitation services over telecommunication networks and the Internet.
Immersion: in game design is that a game creates a spatial presence when the player starts to feel that he/she is there in the world created by game.

Key facts of VR

- There are reports that even a single session of VR training reduces pain and limitation of motion in NP patients.
- "Distraction" is an effective mechanism being proposed to relieving pain when using VR.
- VR has been shown to significantly enhancing balance in the short and intermediate term in NP patients.
- "Simulating sickness" is the most important unwanted side effects when using VR in NP patients.
- VR technology has been applied as an evaluation tool to monitor the outcome of therapeutic interventions.

Summary points

- VR application has drawn attention in a variety of healthcare areas including rehabilitation assessment and training.
- Applying exercise protocols need providing feedback, motivation, and repetition which are accessible through VR.
- VR systems are considered effective when they provide safety, motivation, the possibility of performing a controlled number of repetition, providing different kinds of feedback, and implementing a multi-sensory environment to immerse the user.
- VR technology has been employed to alter some clinical manifestations in NP patients, mainly includes pain, disability, ROM, and postural control.

- Factors determining the quality of VR-based treatment effectiveness include the type of technology used in the VR design, the amount of sense of presence while putting the user in a virtual environment, and the appropriate time and frequency of VR interventions.

References

Bantick, S. J., Wise, R. G., Ploghaus, A., Clare, S., Smith, S. M., & Tracey, I. (2002). Imaging how attention modulates pain in humans using functional MRI. *Brain*, *125*(2), 310–319.

Chang, K.-V., Wu, W.-T., Chen, M.-C., Chiu, Y.-C., Han, D.-S., & Chen, C.-C. (2019). Smartphone application with virtual reality goggles for the reliable and valid measurement of active craniocervical range of motion. *Diagnostics*, *9*(3), 71.

Descarreaux, M., Passmore, S. R., & Cantin, V. (2010). Head movement kinematics during rapid aiming task performance in healthy and neck-pain participants: The importance of optimal task difficulty. *Manual Therapy*, *15*(5), 445–450.

Gold, J. I., Belmont, K. A., & Thomas, D. A. (2007). The neurobiology of virtual reality pain attenuation. *Cyberpsychology & Behavior*, *10*(4), 536–544.

Gross, A., Paquin, J.-P., Dupont, G., Blanchette, S., Lalonde, P., Cristie, T., et al. (2016). Exercises for mechanical neck disorders: A Cochrane review update. *Manual Therapy*, *24*, 25–45.

Hakala, P. T., Rimpelä, A. H., Saarni, L. A., & Salminen, J. J. (2006). Frequent computer-related activities increase the risk of neck–shoulder and low back pain in adolescents. *The European Journal of Public Health*, *16*(5), 536–541.

Häkkinen, A., Salo, P., Tarvainen, U., Wiren, K., & Ylinen, J. (2007). Effect of manual therapy and stretching on neck muscle strength and mobility in chronic neck pain. *Journal of Rehabilitation Medicine*, *39*(7), 575–579.

Harvie, D. S., Broecker, M., Smith, R. T., Meulders, A., Madden, V. J., & Moseley, G. L. (2015). Bogus visual feedback alters onset of movement-evoked pain in people with neck pain. *Psychological Science*, *26*(4), 385–392.

Harvie, D. S., Smith, R. T., Moseley, G. L., Meulders, A., Michiels, B., & Sterling, M. (2020). Illusion-enhanced virtual reality exercise for neck pain: A replicated single case series. *The Clinical Journal of Pain*, *36*(2), 101–109.

Hoffman, H. G., Doctor, J. N., Patterson, D. R., Carrougher, G. J., & Furness, T. A., III. (2000). Virtual reality as an adjunctive pain control during burn wound care in adolescent patients. *Pain*, *85*(1–2), 305–309.

Hoffman, H. G., Richards, T. L., Van Oostrom, T., Coda, B. A., Jensen, M. P., Blough, D. K., et al. (2007). The analgesic effects of opioids and immersive virtual reality distraction: Evidence from subjective and functional brain imaging assessments. *Anesthesia & Analgesia*, *105*(6), 1776–1783.

Holden, M. K. (2005). Virtual environments for motor rehabilitation. *Cyberpsychology & Behavior*, *8*(3), 187–211.

Hoy, D., Protani, M., De, R., & Buchbinder, R. (2010). The epidemiology of neck pain. *Best Practice & Research. Clinical Rheumatology*, *24*(6), 783–792.

Jansen-Kosterink, S. M., Huis In 't Veld, R. M., Schönauer, C., Kaufmann, H., Hermens, H. J., & Vollenbroek-Hutten, M. M. (2013). A serious exergame for patients suffering from chronic musculoskeletal back and neck pain: A pilot study. *Games for Health Journal*, *2*(5), 299–307.

Kavounoudias, A., Gilhodes, J., Rodillo, R., & Rodillo, J. (1999). De la regulación del balance a la orientación del cuerpo: dos metas para el tratamiento de la información propioceptivo del músculo. *Cerebro Res De Exp*, *124*(1), 80–88.

Kim, W., Jeon, I., & Moon, J. (2017). *Nexercise VR: A VR-based exergame for neck exercise*. Available at SSRN 3448344.

Kramer, M., Honold, M., Hohl, K., Bockholt, U., Rettig, A., Elbel, M., et al. (2009). Reliability of a new virtual reality test to measure cervicocephalic kinaesthesia. *Journal of Electromyography and Kinesiology*, *19*(5), e353–e361.

Kristjansson, E., & Treleaven, J. (2009). Sensorimotor function and dizziness in neck pain: Implications for assessment and management. *Journal of Orthopaedic & Sports Physical Therapy*, *39*(5), 364–377.

Li, A., Montaño, Z., Chen, V. J., & Gold, J. I. (2011). Virtual reality and pain management: Current trends and future directions. *Pain Management*, *1*(2), 147–157.

Li, L., Yu, F., Shi, D., Shi, J., Tian, Z., Yang, J., et al. (2017). Application of virtual reality technology in clinical medicine. *American Journal of Translational Research*, *9*(9), 3867.

McCaul, K. D., & Malott, J. M. (1984). Distraction and coping with pain. *Psychological Bulletin*, *95*(3), 516.

Melzack, R., & Wall, P. D. (1965). Pain mechanisms: A new theory. *Science*, *150*(3699), 971–979.

Mihajlovic, Z., Popovic, S., Brkic, K., & Cosic, K. (2018). A system for head-neck rehabilitation exercises based on serious gaming and virtual reality. *Multimedia Tools and Applications*, *77*(15), 19113–19137.

Rezaei, I., Razeghi, M., Ebrahimi, S., Kayedi, S., & Rezaeian Zadeh, A. (2019). A novel virtual reality technique (Cervigame®) compared to conventional proprioceptive training to treat neck pain: A randomized controlled trial. *Journal of Biomedical Physics & Engineering*, *9*(3), 355–366.

Sakakibara, H., Miyao, M., Kondo, T.-A., & Yamada, S. Y. (1995). Overhead work and shoulder-neck pain in orchard farmers harvesting pears and apples. *Ergonomics*, *38*(4), 700–706.

Sarig Bahat, H., Croft, K., Carter, C., Hoddinott, A., Sprecher, E., & Treleaven, J. (2018). Remote kinematic training for patients with chronic neck pain: A randomised controlled trial. *European Spine Journal*, *27*(6), 1309–1323.

Sarig Bahat, H., Takasaki, H., Chen, X., Bet-Or, Y., & Treleaven, J. (2015). Cervical kinematic training with and without interactive VR training for chronic neck pain—a randomized clinical trial. *Manual Therapy*, *20*(1), 68–78.

Sarig-Bahat, H., Weiss, P. L., & Laufer, Y. (2009). Cervical motion assessment using virtual reality. *Spine*, *34*(10), 1018–1024.

Sarig-Bahat, H., Weiss, P. L. T., & Laufer, Y. (2010). Neck pain assessment in a virtual environment. *Spine*, *35*(4), E105–E112.

Sato, K., Fukumori, S., Miyake, K., Obata, D., Gofuku, A., & Morita, K. (2012). Pain in perspective. In S. Ghosh (Ed.), *A novel application of virtual reality for pain control: Virtual reality-mirror visual feedback therapy* (pp. 237–254). IntechOpen.

574 PART | IV Novel and nonpharmacological aspects and treatments

Silva, A. G., & Cruz, A. L. (2013). Standing balance in patients with whiplash-associated neck pain and idiopathic neck pain when compared with asymptomatic participants: A systematic review. *Physiotherapy Theory and Practice, 29*(1), 1–18.

Sjölander, P., Michaelson, P., Jaric, S., & Djupsjöbacka, M. (2008). Sensorimotor disturbances in chronic neck pain—Range of motion, peak velocity, smoothness of movement, and repositioning acuity. *Manual Therapy, 13*(2), 122–131.

Tabor, A., Thacker, M. A., Moseley, G. L., & Körding, K. P. (2017). Pain: A statistical account. *PLoS Computational Biology, 13*(1), e1005142.

Thunberg, J., Hellström, F., Sjölander, P., Bergenheim, M., Wenngren, B.-I., & Johansson, H. (2001). Influences on the fusimotor-muscle spindle system from chemosensitive nerve endings in cervical facet joints in the cat: Possible implications for whiplash induced disorders. *Pain, 91*(1), 15–22.

Treleaven, J. (2008). Sensorimotor disturbances in neck disorders affecting postural stability, head and eye movement control. *Manual Therapy, 13*(1), 2–11.

Treleaven, J., Battershill, J., Cole, D., Fadelli, C., Freestone, S., Lang, K., et al. (2015). Simulator sickness incidence and susceptibility during neck motion-controlled virtual reality tasks. *Virtual Reality, 19*(3–4), 267–275.

Tyrrell, R., Sarig-Bahat, H., Williams, K., Williams, G., & Treleaven, J. (2018). Simulator sickness in patients with neck pain and vestibular pathology during virtual reality tasks. *Virtual Reality, 22*(3), 211–219.

Valet, M., Sprenger, T., Boecker, H., Willoch, F., Rummeny, E., Conrad, B., et al. (2004). Distraction modulates connectivity of the cingulo-frontal cortex and the midbrain during pain—An fMRI analysis. *Pain, 109*(3), 399–408.

Vuillerme, N., & Pinsault, N. (2009). Experimental neck muscle pain impairs standing balance in humans. *Experimental Brain Research, 192*(4), 723–729.

Wang, Q., Chen, W., Timmermans, A. A., Karachristos, C., Martens, J.-B., & Markopoulos, P. (2015). Smart rehabilitation garment for posture monitoring. In *Paper presented at the 2015 37th annual international conference of the IEEE engineering in medicine and biology society (EMBC)* (pp. 5736–5739).

Wiederhold, B. K., Gao, K., Sulea, C., & Wiederhold, M. D. (2014). Virtual reality as a distraction technique in chronic pain patients. *Cyberpsychology, Behavior and Social Networking, 17*(6), 346–352.

Williams, G., Sarig-Bahat, H., Williams, K., Tyrrell, R., & Treleaven, J. (2017). Cervical kinematics in patients with vestibular pathology vs. patients with neck pain: A pilot study. *Journal of Vestibular Research, 27*(2–3), 137–145.

Chapter 48

Virtual reality induced analgesia and dental pain

Elitsa Veneva[a], Ani Belcheva[a], and Ralitsa Raycheva[b]

[a]Department of Pediatric Dentistry, Faculty of Dental Medicine, Medical University – Plovdiv, Plovdiv, Bulgaria, [b]Department of Social medicine and Public Health, Faculty of Public Health, Medical University – Plovdiv, Plovdiv, Bulgaria

Abbreviations

CCLAD   computer-controlled local anesthetic delivery system
LA   local anesthesia
SADE   sensory adapted dental environment
VR   virtual reality

Introduction

Pain management is one of the most important aspects in the care for dental patients. Progress in the field of local anesthesia is critical to the development of dental science, allowing the profession a therapeutic approach that would otherwise be impossible.

The strategies to relieve dental pain during treatment are mainly divided by the use of a chemical agent into two groups: non-pharmacological and pharmacological strategies, summarized in Table 1. As pain has both physiological and psychological components, often a combination of techniques is required to achieve painless dental treatment.

Local anesthesia infiltration is a common method to relieve procedural dental pain during treatment. Nevertheless, the injection of a local anesthetic agent is not painless itself, thereby inducing anxiety and often fear in patients. Potential methods and means to reduce injection discomfort during local anesthesia are introduced in Table 2.

In addition to the effective injection method, the introduction of highly engaging forms of distraction is being studied in order to divert attention from the stress factor—the needle. The implementation of audio-visual stimulation via virtual reality during medical procedures led to the introduction of a new term in the literature—"virtual anesthesia" (Gold, Kant, Kim, & Rizzo, 2005).

Distraction as an approach to relieve injection discomfort

A successful and atraumatic approach to local anesthesia requires control over the anxiety and pain associated with the procedure. Various techniques are implemented to guide the patient's behavior. Verbal communication and the tell-show-do techniques are widely used methods of diverting a patient's attention from the pain stimuli. Thus, the most common practices for distraction and anxiety reduction during local anesthesia remain verbal.

The rationale behind pain reduction caused by distraction, although not fully understood, is based on the principle that different activities of the central nervous system can play a significant role in the sensory perception of pain stimuli (Li, Montaño, Chen, & Gold, 2011). Distracting factors can be inactive or passive form.

Active distraction agents promote patient involvement including several sensory components. An example of that are interactive toys and virtual reality games controlled by the patient. Although active strategies are thought to be more effective than passive ones, other studies show that passive distraction may be as effective or even better because active distractions are too demanding for children (Attar & Baghdadi, 2014; Dahlquist et al., 2007).

Passive ones are delivered to the patient who is experiencing them without active actions required from their side. Two forms of passive distraction, widely used and studied in dentistry, are audio (different categories of music) and audio-visual distraction (TV, tablet, and two-dimensional video glasses) (Aitken, Wilson, Coury, & Moursi, 2002; Attar & Baghdadi,

TABLE 1 Strategies to reduce pain during dental treatment.

Non-pharmacological methods	Pharmacological methods
Distraction Verbal distraction Audio-visual stimulation Sensory-adapted dental environment Hypnosis	**Diffuse mechanism** Topical anesthesia (lidocaine patches) Intranasal anesthesia
Electronic dental analgesia	**Vascular-diffuse mechanism** Computer- controlled local anesthetic delivery systems Jet injections
Laser analgesia	**Infiltration mechanism** Local anesthetic injection

2014; El-Sharkawi, El-Housseiny, & Aly, 2012; Marwah, Raju, & Prabhakar, 2007; Ram et al., 2010). Music remains the preferred choice of general practitioners and pediatric dentists, compared to other forms of anxiety management available due to the availability of the method.

Passive distraction can also be achieved by using a virtual reality device in passive mode. Virtual reality devices or three-dimensional glasses offer a range of animation videos available on the internet. During invasive medical procedures like local anesthesia, it is required that patients watch videos that do not promote any movement from the viewer.

Virtual reality devices in dental pain management

A virtual reality device is a head-mounted device that uses advanced technologies to create virtual environments of sights, sounds, and movement, allowing patients to be immersed in an interactively simulated world to reduce pain (Gold et al., 2005). VR devices have a wide field of view and three-dimensional display that projects images right in front of the user's eyes. The stereoscopic head-mounted display provides separate images for each eye. Virtual reality headsets can either have a dedicated internal display or can use the display in a smartphone. In the following case, the VR content is projected from the screen of the inserted smartphone through lenses that act as a stereoscope. Fig. 1 illustrates an example of a smartphone-

TABLE 2 Potential methods and means to reduce injection discomfort during local anesthesia.

Step	Reducing pain related to local anesthetic injection
1.	**Warming** of local anesthetic agent
2.	**Buffering** of local anesthetic agent
3.	**Distraction** Verbal distraction Audio-visual stimulation (VR, TV, tablet, music) Sensory-adapted dental environment Hypnosis
4.	**Injection site pre-conditioning** Vibrotactile devices Cooling of the injection site (ice popsicles)
5.	**Combination of anesthetics** Using a topical LA prior to injection
6.	**Injection technique** Slow injection (1 mL/min) Pulse injection (3 s injecting, 1 s rest) Two-step injection (injecting a drop, retracting the needle, then injecting again after few seconds).

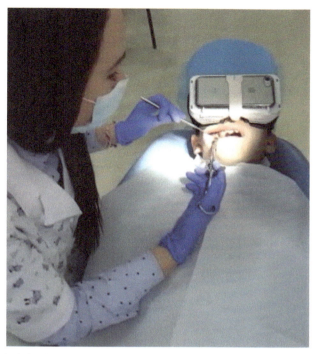

FIG. 1 Virtual reality-induced analgesia during local anesthetic infiltration in a pediatric dental patient. A smartphone (iPhone 7, Apple) is attached to a VR device (*Noon VR, FX Gear*). The VR device is projecting a relaxing forest walk while the patients receive local anesthetic infiltration.

mediated virtual reality device in use during local anesthesia. Fig. 2 shows a screenshot of the actual image projected to the patient's eyes.

VR devices not only show potentially attractive audio-visual stimuli but also exclude all other visual stimuli in the environment that may affect the patient. The lack of eye contact with the syringe during local anesthesia assisted by a VR device is beneficial, especially in children.

Despite some limitations, researchers in this field have successfully demonstrated its feasibility and satisfaction in reducing pain associated with medical interventions (Agarwal, 2017; Li et al., 2011; Malloy & Milling, 2010). Clinical trials for the application of the technique in the dental environment (Attar & Baghdadi, 2014; Bansal, Jain, Tyagi, & Jain, 2018; El-Sharkawi et al., 2012; Nuvvula, Alahari, Kamatham, & Challa, 2014), albeit a few, report a subjectively and objectively assessed reduction in pain and anxiety during dental procedures in both adults and pediatric patients (Asl Aminabadi, Erfanparast, Sohrabi, Ghertasi Oskouei, & Naghili, 2012; Bagattoni, D'Alessandro, Sadotti, Alkhamis, & Piana, 2018; Furman et al., 2009; Hoffman et al., 2001; Mitrakul, Asvanund, Arunakul, & Paka-Akekaphat, 2015; Nuvvula et al., 2014; Ram et al., 2010; Wiederhold, Gao, & Wiederhold, 2014).

FIG. 2 A screenshot of the image, projected on the virtual reality device through a smartphone. A stereoscopic video of a relaxing walk through a forest is projected to the patient's eyes before and during local anesthetic infiltration.

Safety considerations

The use of audio-visual distraction via virtual reality devices is limited by some conditions:

- Patients with vision requiring correction with optical glasses.
- Children with a history of photosensitive epileptic or non-epileptic seizures.
- Patients with recent trauma to the eye or face, which makes it difficult to comfortably use a virtual reality device.

During a longer procedure, some patients might experience headache, nausea, or just feel uncomfortable with the image displayed. Removal of the device is necessary also when the patient is curious and has a desire to stay aware during the treatment performed.

A restriction of the method of VR-assisted local anesthesia is the inability of the operator to see the patient's eyes during injection or treatment. Thus, early signs of indisposition in the patient could be missed. Constant verbal contact should be held with the patient to avoid complications.

Another limitation of the method is the unavailability of eyeglasses for children with small face/head size and a reported questionable contraindication below the age of seven years (Nuvvula et al., 2014; Robson, 2010). Also when used during longer dental procedures, the dental clinician might experience inconvenience when treating the patient because of the usually bulky shape of the device mounted just above the oral cavity.

An important factor to be considered is the disinfection of the device, which should be performed thoroughly after each patient.

Effectiveness of VR in reduction of injection discomfort

The use of virtual reality devices for distraction during dental treatment in children has gained popularity in the last 20 years with the development of technology. The applicability and efficacy of the method to reduce fear and pain during conservative treatment have been confirmed in several clinical studies (Kaur et al., 2015; Mitrakul et al., 2015; Ram et al., 2010).

From the available scientific literature, contradictory results are reported only by the scientific team Aitken et al. (2002), which identified the technology as ineffective in reducing anxiety, pain, or uncooperative behavior during conservative dental treatment in children (Aitken et al., 2002).

There are few studies in the scientific literature on the use of VR devices to control pain and anxiety during local anesthesia in pediatric dental patients (Asvanund, Mitrakul, Juhong, & Arunakul, 2015; Bansal et al., 2018; El-Sharkawi et al., 2012; Khanapurkar, Nagpal, Lamba, Choudhari, & Hotwani, 2018; Nuvvula et al., 2014), but all concluded that the use of audio-visual stimulation had a positive effect on the acceptance of local anesthesia manipulation.

The study by Bansal et al. in 2018 examines the efficiency of a VR device using an attached mobile phone. Their results confirm the efficacy of the method in children, although they do not specify what type of injection technique they use (Bansal et al., 2018).

Asnavund et al. in 2015 found the method of VR-distraction is successful during the application of local anesthesia in both jaws, as well as in children with previous experience (Asvanund et al., 2015). According to their results, the use of virtual reality has helped to place the LA, leading to lower values of subjective pain, pain-related behavior, and heart rate compared to the LA by the traditional method without VR.

Amal Al Khotani et al. in a 2016 study confirmed the efficacy of the method for reducing subjective pain, anxiety, and physical distress during local anesthesia in children, compared with a control group of patients receiving conventional LA manipulation (Al-Khotani, Bello, & Christidis, 2016).

A common drawback of the available studies could be considered the use of a topical anesthetic before injection of local anesthesia. This is a prerequisite for bias in the results, as it is unclear whether the pain reduction is entirely due to VR or dose-dependent on the topical anesthetic. The lack of control in individual cases and operators for pre-drying of the mucosa, the amount of topical anesthetic used, the anesthetic agent, and time for its action make the scientific conclusions inapplicable.

The current review of the literature defines the use of a virtual reality device as an effective and applicable non-pharmacological method for reducing injection pain and anxiety in children (Asvanund et al., 2015; Attar & Baghdadi, 2014; Bansal et al., 2018; El-Sharkawi et al., 2012; Fakhruddin, Hisham, & Gorduysus, 2015; Ghaderi, Banakar, & Rostami, 2013; Khanapurkar et al., 2018; Nuvvula et al., 2014).

Given the progress and increasing availability of technology for clinical use, although with some limitations, virtual reality has the potential to be a useful addition to modern pediatric dental practice.

Applications to other areas

The proven beneficial effect of virtual reality distraction could be applied to reduce anxiety associated with invasive medical procedures of short duration such as the ones that include needle injection, as the sight of a needle is a strong factor determining the psychological component of pain.

Other agents of interest

In this chapter, we have described the detailed mechanisms of action behind the new term "virtual anesthesia." Other engaging forms of distraction include:

Sensory-Adapted Dental Environment (SADE) a specially designed space that offers various forms of sensory stimulation, such as lights, sounds, aromas and music, tactile surfaces, moving images, and more, aiming to promote a positive experience. Results the few studies available show that behavioral and psychophysiological parameters for relaxation are significantly improved in SADE compared to the conventional dental environment (Cermak et al., 2015; Shapiro, Melmed, Sgan-Cohen, Eli, & Parush, 2007; Shapiro, Melmed, Sgan-Cohen, & Parush, 2009).

Hypnosis can serve as an effective addition to local anesthesia and in some cases as an alternative to pain control in dentistry. Adequate application of the method could increase the cooperation of patients and reduce the pain and anxiety associated with dental manipulations, especially in childhood (Al-Harasi, Ashley, Moles, Parekh, & Walters, 2017; Burghardt, Koranyi, Magnucki, Strauss, & Rosendahl, 2018; Oberoi et al., 2018).

Mini-dictionary of terms

Virtual reality device: An advanced technology that creates virtual environments of sights, sounds, and movement, allowing patients to be immersed in an interactively simulated world. The VR device headset has a wide field of view and a three-dimensional display that projects images right in front of the user's eyes.

Key facts of the use of VR devices in dental pain management

- Virtual reality devices are non-pharmacological means aiming to distract the patient from pain stimuli.
- The virtual environment of sights, sounds, and movement immerses the patient in an interactively simulated world and excludes all other visual stimuli in the dental environment that may act as a stress factor and affect the patient.
- VR devices are head-mounted and can use either a dedicated internal display or the display of an inserted smartphone.
- The use of virtual reality devices is limited by some health conditions and should be performed with care regarding the wellbeing of the patient, disinfection protocols, and possible side effects like nausea and headache.
- Constant verbal contact should be kept with the patient to confirm their wellbeing during the VR-assisted treatment.
- Dental professionals might experience difficulty in working on a patient wearing a VR headset.
- For maximum comfort for the patient during VR-assisted LA infiltration, a topical anesthetic agent should be used on the injection site before needle penetration.
- Topical anesthetic before the use of VR for LA should be avoided in the protocol of clinical trials aiming to evaluate the efficacy of pain reduction due to VR.
- During invasive dental procedures such as local anesthesia, it is required that patients watch videos that do not promote any movement from the viewer to avoid trauma.

Summary points

- The rationale behind pain reduction caused by distraction, although not fully understood, is based on the principle that different activities of the central nervous system can play a significant role in the sensory perception of pain stimuli.
- Virtual reality devices are implemented as a non-pharmacological method for reducing pain and anxiety in patients during painful dental procedures.
- VR-assisted local anesthesia offers a passive distraction to divert the patient's attention from the painful event.
- Despite some limitations of the method, researchers have successfully demonstrated its feasibility and satisfaction in reducing pain associated with dental procedures.
- The use of VR during local anesthesia in dental patients can increase patient cooperativeness and promote the dentist-patient relationship.
- Pictures of VR-assisted local anesthesia can be found in Figs. 1 and 2.

References

Agarwal, N. (2017). Effectiveness of two topical anaesthetic agents used along with audio visual aids in paediatric dental patients. *Journal of Clinical and Diagnostic Research.* https://doi.org/10.7860/JCDR/2017/23180.9217.

Aitken, J. C., Wilson, S., Coury, D., & Moursi, A. M. (2002). The effect of music distraction on pain, anxiety and behavior in pediatric dental patients. *Pediatric Dentistry, 24*(2), 114–118.

Al-Harasi, S., Ashley, P. F., Moles, D. R., Parekh, S., & Walters, V. (2017). Hypnosis for children undergoing dental treatment. *Cochrane Database of Systematic Reviews, 6.* https://doi.org/10.1002/14651858.CD007154.pub3, CD007154.

Al-Khotani, A., Bello, L. A., & Christidis, N. (2016). Effects of audiovisual distraction on children's behaviour during dental treatment: A randomized controlled clinical trial. *Acta Odontologica Scandinavica, 74*(6). https://doi.org/10.1080/00016357.2016.1206211.

Asl Aminabadi, N., Erfanparast, L., Sohrabi, A., Ghertasi Oskouei, S., & Naghili, A. (2012). The impact of virtual reality distraction on pain and anxiety during dental treatment in 4-6 year-old children: A randomized controlled clinical trial. *Journal of Dental Research, Dental Clinics, Dental Prospects, 6*(4), 117–124. https://doi.org/10.5681/joddd.2012.025.

Asvanund, Y., Mitrakul, K., Juhong, R., & Arunakul, M. (2015). Effect of audiovisual eyeglasses during local anesthesia injections in 5- to 8-year-old children. *Quintessence International, 46*(6), 513–521. https://doi.org/10.3290/j.qi.a33932.

Attar, R. H., & Baghdadi, Z. D. (2014). Comparative efficacy of active and passive distraction during restorative treatment in children using an iPad versus audiovisual eyeglasses: A randomised controlled trial. *European Archives of Paediatric Dentistry, 16*(1), 1–8. https://doi.org/10.1007/s40368-014-0136-x.

Bagattoni, S., D'Alessandro, G., Sadotti, A., Alkhamis, N., & Piana, G. (2018). Effects of audiovisual distraction in children with special healthcare needs during dental restorations: A randomized crossover clinical trial. *International Journal of Paediatric Dentistry, 28*(1), 111–120. https://doi.org/10.1111/ipd.12304.

Bansal, A., Jain, S., Tyagi, P., & Jain, A. (2018). Effect of virtual reality headset using smart phone device on pain and anxiety levels during local anesthetic injection in children with 6-10 years of age. *Paripex-Indian Journal of Research, 7*(6). https://doi.org/10.36106/paripex.

Burghardt, S., Koranyi, S., Magnucki, G., Strauss, B., & Rosendahl, J. (2018). Non-pharmacological interventions for reducing mental distress in patients undergoing dental procedures: Systematic review and meta-analysis. *Journal of Dentistry, 69,* 22–31. https://doi.org/10.1016/J.JDENT.2017.11.005.

Cermak, S. A., Stein Duker, L. I., Williams, M. E., Dawson, M. E., Lane, C. J., & Polido, J. C. (2015). Sensory adapted dental environments to enhance oral care for children with autism spectrum disorders: A randomized controlled pilot study. *Journal of Autism and Developmental Disorders, 45*(9), 2876–2888. https://doi.org/10.1007/s10803-015-2450-5.

Dahlquist, L. M., Mckenna, K. D., Jones, K. K., Dillinger, L., Weiss, K. E., & Ackerman, C. S. (2007). Active and passive distraction using a head-mounted display helmet: Effects on cold pressor pain in children. *Health Psychology, 26*(6), 794–801. https://doi.org/10.1037/0278-6133.26.6.794.

El-Sharkawi, H. F. A., El-Housseiny, A. A., & Aly, A. M. (2012). Effectiveness of new distraction technique on pain associated with injection of local anesthesia for children. *Pediatric Dentistry, 34*(2), e35–e38.

Fakhruddin, K. S., Hisham, E. B., & Gorduysus, M. O. (2015). Effectiveness of audiovisual distraction eyewear and computerized delivery of anesthesia during pulp therapy of primary molars in phobic child patients. *European Journal of Dentistry, 9*(4), 470–475. https://doi.org/10.4103/1305-7456.172637.

Furman, E., Jasinevicius, T. R., Bissada, N. F., Victoroff, K. Z., Skillicorn, R., & Buchner, M. (2009). Virtual reality distraction for pain control during periodontal scaling and root planing procedures. *The Journal of the American Dental Association, 140*(12), 1508–1516. https://doi.org/10.14219/jada.archive.2009.0102.

Ghaderi, F., Banakar, S., & Rostami, S. (2013). Effect of pre-cooling injection site on pain perception in pediatric dentistry: "A randomized clinical trial". *Dental Research Journal, 10*(6), 790–794.

Gold, J. I., Kant, A. J., Kim, S. H., & Rizzo, A. (2005). Virtual anesthesia: The use of virtual reality for pain distraction during acute medical interventions. *Seminars in Anesthesia, Perioperative Medicine and Pain, 24*(4), 203–210. https://doi.org/10.1053/j.sane.2005.10.005.

Hoffman, H. G., Garcia-Palacios, A., Patterson, D. R., Jensen, M., Furness, T., & Ammons, W. F. (2001). The effectiveness of virtual reality for dental pain control: A case study. *Cyberpsychology & Behavior, 4*(4), 527–535. https://doi.org/10.1089/109493101750527088.

Kaur, R., Jindal, R., Dua, R., Mahajan, S., Sethi, K., & Garg, S. (2015). Comparative evaluation of the effectiveness of audio and audiovisual distraction aids in the management of anxious pediatric dental patients. *Journal of the Indian Society of Pedodontics and Preventive Dentistry, 33*(3). https://doi.org/10.4103/0970-4388.160357.

Khanapurkar, P. M., Nagpal, D. I., Lamba, G., Choudhari, P., & Hotwani, K. (2018). Effect of virtual reality distraction on pain and anxiety during local anesthesia injection in children—A randomized controlled cross-over clinical study. *Journal of Advanced Medical and Dental Sciences Research, 6* (11). https://doi.org/10.21276/jamdsr.

Li, A., Montaño, Z., Chen, V. J., & Gold, J. I. (2011). Virtual reality and pain management: Current trends and future directions. *Pain, 1*(2), 147–157. https://doi.org/10.2217/pmt.10.15.Virtual.

Malloy, K. M., & Milling, L. S. (2010). The effectiveness of virtual reality distraction for pain reduction: A systematic review. *Clinical Psychology Review, 30*(8), 1011–1018. https://doi.org/10.1016/j.cpr.2010.07.001.

Marwah, N., Raju, O., & Prabhakar, A. (2007). A comparison between audio and audiovisual distraction techniques in managing anxious pediatric dental patients. *Journal of the Indian Society of Pedodontics and Preventive Dentistry, 25*(4), 177. https://doi.org/10.4103/0970-4388.37014.

Mitrakul, K., Asvanund, Y., Arunakul, M., & Paka-Akekaphat, S. (2015). Effect of audiovisual eyeglasses during dental treatment in 5-8 year-old children. *European Journal of Paediatric Dentistry, 16*(3), 239–245.

Nuvvula, S., Alahari, S., Kamatham, R., & Challa, R. R. (2014). Effect of audiovisual distraction with 3D video glasses on dental anxiety of children experiencing administration of local analgesia: A randomised clinical trial. *European Archives of Paediatric Dentistry, 16*(1), 43–50. https://doi.org/10.1007/s40368-014-0145-9.

Oberoi, J., Panda, A., Bhatia, R., Garg, I., Soni, S., & Lecturer, E. (2018). Effect of hypnosis during administration of local anesthesia in pediatric patients—A pilot study. *International Journal of Oral Care and Research, 6*(2), 39–43.

Ram, D., Shapira, J., Holan, G., Magora, F., Cohen, S., & Davidovich, E. (2010). Audiovisual video eyeglass distraction during dental treatment in children. *Quintessence International, 41*(8), 673–679.

Robson, W. (2010). Retrieved from http://www.audioholics.com/editorials/warning-3d-video-hazardous-to-your-health.

Shapiro, M., Melmed, R. N., Sgan-Cohen, H. D., Eli, I., & Parush, S. (2007). Behavioural and physiological effect of dental environment sensory adaptation on children's dental anxiety. *European Journal of Oral Sciences, 115*(6), 479–483. https://doi.org/10.1111/j.1600-0722.2007.00490.x.

Shapiro, M., Melmed, R. N., Sgan-Cohen, H. D., & Parush, S. (2009). Effect of sensory adaptation on anxiety of children with developmental disabilities: A new approach. *Pediatric Dentistry, 31*(3), 222–228.

Wiederhold, M. D., Gao, K., & Wiederhold, B. K. (2014). Clinical use of virtual reality distraction system to reduce anxiety and pain in dental procedures. *Cyberpsychology, Behavior and Social Networking, 17*(6), 359–365. https://doi.org/10.1089/cyber.2014.0203.

Chapter 49

Vibrotactile devices, DentalVibe, and local anesthesia

Elitsa Veneva[a], Ani Belcheva[a], and Ralitsa Raycheva[b]

[a]*Department of Pediatric Dentistry, Faculty of Dental Medicine, Medical University – Plovdiv, Plovdiv, Bulgaria;* [b]*Department of Social Medicine and Public Health, Faculty of Public Health, Medical University – Plovdiv, Plovdiv, Bulgaria*

Abbreviations

DV DentalVibe
LA local anesthesia

Introduction

Providing painless dental treatment is crucial to reducing fear and anxiety, facilitating treatment, developing patient confidence, and accepting future treatment. Performing local anesthesia is a vital part of this, although it remains one of the most difficult aspects of dentistry, especially in pediatric dental practice. The fear associated with experiencing needle penetration and the sensation of swelling soft tissues is common factors causing patients and dental clinicians to experience anxiety regarding the use of infiltration local anesthesia.

Recent progress in the field of local anesthesia has led to the development of newer agents, delivery devices, and modification in injection techniques. A growing number of studies are focusing on the non-pharmacological methods to reduce discomfort associated with local anesthesia, such as warming or buffering of local anesthetic agents, cooling the injection site, and applying vibration or pressure from high-tech devices to relieve pain before or after injection of local anesthesia. In addition to the effective injection method, the introduction of highly engaging forms of distraction is studied to divert attention from the stress factor through means of audio-visual stimulation and virtual reality, which led to the introduction of a new term in the literature—"virtual anesthesia". Other methods aim to replace infiltrative analgesia with one supplied by physical means, such as laser analgesia.

Non-pharmacological approach to injection discomfort

Distraction is a widely used method of diverting a child's attention from pain stimuli. The main practices for distraction and anxiety reduction are verbal. The mechanism of action, although fully understood, is based on the principle that different activities of the central nervous system can play a significant role in the sensory perception of pain stimuli (Li, Montaño, Chen, & Gold, 2011). Distracting factors can be inactive or passive form, with the active ones promoting patient involvement, including several sensory components.

Vibrotactile devices offer passive distraction and their action is expressed in the principle of masking the pain of the needle penetration by applying pressure, vibration, micro oscillations, or a combination of them. Applied physical stimuli alter or interfere with pain signals by closing the neural gate of the cerebral cortex, thus reducing the sensation of pain due to distraction.

Melzack and Wall's gate control theory of pain

In their work Melzack and Wall describe a transmission station (a "gate") in the spinal cord that influences the flow of nerve impulses to the brain. The role of presynaptic inhibition in accentuating or filtering afferent inputs can be described as switching between inputs when there are competing ones (Melzack, 1996). For example, the presynaptic inhibition of cutaneous nociceptive fibers can be achieved by switching from the input of pain to an input of touching. Vibrotactile devices use pressure, vibration, micro oscillations, or a combination of them to block the nociceptive input of needle penetration.

Treatments, Mechanisms, and Adverse Reactions of Anesthetics and Analgesics. https://doi.org/10.1016/B978-0-12-820237-1.00049-1
Copyright © 2022 Elsevier Inc. All rights reserved.

Fear of dental injections

Dental injection phobia is a subtype of a group of phobias about blood, injury, and injections. Milgrom believes that the fear of injections in general, including pain and fear of injury, are the main aspects of the fear of injectable local anesthesia in the dental practice (Milgrom, Coldwell, Getz, Weinstein, & Ramsay, 1997). The vast majority of children report a strong fear of injections, although with age the fear of needles decreases, probably due to cognitive maturation or the development of coping behavior in dealing with pain (Majstorovic & Veerkamp, 2004). In addition, the authors found a strong association between blood-injury-injection phobia and dental anxiety (Vika, Skaret, Raadal, Öst, & Kvale, 2008). In addition, dental anxiety and injection pain appear to be strongly correlated, with many anxious patients reporting increased and prolonged pain perception (van Wijk & Makkes, 2008). Weisman shows that inadequate analgesia for invasive medical procedures in young children may reduce the effect of future anesthetic procedures (Weisman, Bernstein, & Schechter, 1998). Similarly, it has been shown that previous negative experiences with inadequate local infiltrative anesthesia can lead to behavioral problems in subsequent treatment sessions (Versloot, Veerkamp, & Hoogstraten, 2008).

Reduction of discomfort during local anesthesia

The discomfort of the local anesthetic injection is related to both the penetration of the needle and the tension created by the dilation of the tissues as the anesthetic solution penetrates. It is hypothesized that vibration from the devices not only relieves the pain of the needle prick but could also reduce infiltration pain by physically facilitating the distribution of anesthetic fluid.

Vibrotactile devices for dental use

VibraJect is a battery-powered device that attaches to a standard anesthetic syringe, causing the syringe and its adjacent needle to vibrate during infiltration. There are currently only two clinical trials of the device involving children and the results are contradictory. While the results of Roeber et al. in 2011. Roeber, Wallace, Rothe, Salama, and Allen (2011) indicate that the device does not contribute to the reduction of discomfort under local anesthesia, Chaundhry et al. in 2015 proved its effectiveness but in a study with three times smaller sample size (Chaudhry, Shishodia, Singh, & Tuli, 2015). A consideration about the use of the VibraJect can be the fact that the whole syringe vibrates during infiltration which could cause stress in both patient and operator.

The DentalVibe Comfort Injection System is a wireless handheld device that gently stimulates sensory receptors at the injection site. The advantage is that it vibrates on the tissues before the needle penetrates, during the injection, and a few seconds after, which implies masking the unpleasant sensations. The DentalVibe method has been evaluated as easily applicable, successful, and well-accepted also by pediatric dental patients (Jayanthi et al., 2015; Nanitsos, Vartuli, Forte, Dennison, & Peck, 2009; Tandon et al., 2018). The downside is that it is not directly attached to the syringe, so both hands of the operator are involved. To overcome the inconvenience, the device has a lightweight design with angulation such as a dental mirror and light, for easier retraction and visibility.

Accupal is a wireless device, similar to DentalVibe, that applies not only vibrations but also pressure to the injection site. At present, the device is not yet available for sale, which limits the possibilities for conducting a clinical trial.

Effectiveness in reduction of injection discomfort

Predominant results confirm the effectiveness of vibrotactile stimulation in reducing injection pain and confirm patient's preference for it over traditional infiltration anesthesia (Chaudhry et al., 2015; Ching, Finkelman, & Loo, 2014; Jayanthi et al., 2015; Nanitsos et al., 2009; Shilpapriya et al., 2015; Tung, Carillo, Udin, Wilson, & Tanbonliong, 2018) with the exception of reports from several author teams which demonstrate the opposite (Erdogan, Sinsawat, Pawa, Rintanalert, & Vuddhakanok, 2018; Roeber et al., 2011; Şermet Elbay et al., 2016).

Despite the potential that the method suggests, the results of clinical trials are insufficient to establish the vibrotactile method as effective in reducing the pain during the infiltrative technique of anesthesia in children (Chaudhry et al., 2015; Davoudi et al., 2016; Elbay et al., 2015). Additional studies are needed to assess the clinical adequacy of the vibrotactile devices to establish the possibility of performing local anesthesia with reduced discomfort.

Evaluation of the efficacy of vibrotactile devices

The patient's physical, psychological and cognitive development can affect the experience of pain and anxiety during local anesthesia. The difficulty in determining pain in children is exacerbated by the fact that it is difficult to distinguish between behavior that results solely from pain and behavior that results from fear and a mixture of other factors (Beyer, McGrath, & Berde, 1990; Versloot, Veerkamp, & Hoogstraten, 2004).

As the subjective assessment of pain in patients, especially pediatric ones, may affect the reliability of the results, the efficacy of vibrotactile devices should be examined in a complex way. In future trials, no topical anesthetic should be used before injection assisted with vibrotactile devices as the results could be thus distorted. Subjective and objective measurements should be engaged such as evaluation of dental fear before injection and heart rate of the patient throughout the visit. Both subjective anxiety and pain felt during LA placement should be examined, as well as pain-related behavior during the procedure. Patients with no previous experience with local anesthesia should be enrolled in studies due to the possibility of a preconceived notion of the procedure associated with memories of a previous experience. Well-developed protocols of clinical trials should be accessible in public databases before patient inclusion to help establish a better understanding of the nature of results (Veneva, Cholakova, Raycheva, & Belcheva, 2019).

Applications to other areas

The discomfort of the local anesthetic injection is related to both the penetration of the needle and the tension created by the dilation of the tissues as the anesthetic solution penetrates. Vibration from vibrotactile devices has been found to relieve needle puncture pain, but could also reduce infiltration pain by physically facilitating the distribution of anesthetic fluid. Although it has been suggested that this may stimulate the induction of an anesthetic effect, a study by Shaefer et al. found that the use of DV did not affect the time of onset of anesthesia ($P > .05$) (Shaefer, Lee, & Anderson, 2017).

Other agents of interest

In this chapter, we have described detailed mechanisms of vibrotactile devices that work to achieve a less painful local anesthetic injection. Other devices intended to minimize infiltration pain are:

Jet injections are based on the principle of using a mechanical energy source to generate pressure, allowing a thin stream of anesthetic fluid of sufficient strength to penetrate the soft tissues (Arapostathis, Dabarakis, Coolidge, Tsirlis, & Kotsanos, 2010).

Computer-controlled local anesthetic delivery devices allow constant velocity and pressure of the anesthetic solution and are found to reduce the injection pain and the time of onset of the anesthetic effect (Alamoudi et al., 2016; Kämmerer et al., 2015; Thoppe-Dhamodharan et al., 2015).

Mini-dictionary of terms

Vibrotactile device: A devices offer passive distraction and their action is expressed in the principle of masking the pain of the needle penetration by applying pressure, vibration, micro oscillations, or a combination of them. Applied physical stimuli alter or interfere with pain signals by closing the neural gate of the cerebral cortex, thus reducing the sensation of pain due to distraction.

Key facts of vibrotactile devices for dental use

- Vibrotactile devices are a non-pharmacological mean aiming to reduce the discomfort on infiltrative local anesthesia.
- They are developed on the principle of masking the pain of needle penetration by applying pressure, vibration, micro oscillations, or a combination of them.
- The motion created by the action of the devices is speculated to also promote the distribution of the anesthetic agent, thus reducing the sensation of swelling.
- The Vibraject device should be attached to a syringe causing it to vibrate during the injection. However, the device is not commercially marketed though may be useful in comparing mechanisms of action when investigating new devices.
- The DentalVibe and Accupal devices are used separately from the syringe, enabling conditioning of the tissues before injection.
- For maximum comfort of the patient during infiltration, a topical anesthetic agent should be used on the injection site before needle penetration.

586 PART | IV Novel and nonpharmacological aspects and treatments

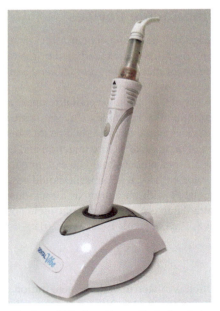

FIG. 1 The DentalVibe Comfort Injection System. The DentalVibe is a vibrotactile device, designed with the shape of a cheek retractor and equipped with a light for easier visualization of the operative field.

Summary points

- Fear associated with experiencing needle penetration and sensation of swelling soft tissues are common factors causing anxiety during infiltration of local anesthesia.
- Vibrotactile devices offer a passive distraction to interfere with pain signals from the injection by closing the neural gate of the cerebral cortex, thus reducing the sensation of pain.
- Their aim is to mask the pain of needle penetration by applying pressure, vibration, micro oscillations, or a combination of them.
- A reduced sensation of soft tissue swelling could be expected due to the vibration created by the action of devices like DentalVibe.
- Pictures of the vibrotactile device DentalVibe can be found in Figs. 1 and 2.

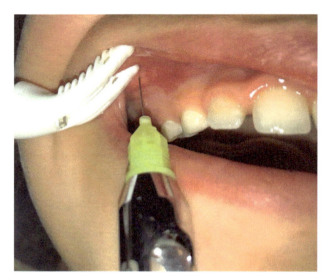

FIG. 2 The DentalVibe Comfort Injection System in use during local anesthesia. The DentalVibe is placed on the mucosa before infiltration in working mode. The needle should be inserted near one of the two prongs. After infiltration and retraction of the needle, the device should still vibrate for another 5 s.

References

Alamoudi, N. M., Baghlaf, K. K., Elashiry, E. A., Farsi, N. M., El Derwi, D. A., & Bayoumi, A. M. (2016). The effectiveness of computerized anesthesia in primarjy mandibular molar pulpotomy: A randomized controlled trial. *Quintessence International, 47*(3), 217–224. https://doi.org/10.3290/j.qi.a34977.

Arapostathis, K. N., Dabarakis, N. N., Coolidge, T., Tsirlis, A., & Kotsanos, N. (2010). Comparison of acceptance, preference, and efficacy between jet injection INJEX and local infiltration anesthesia in 6 to 11 year old dental patients. *Anesthesia Progress, 57*(1), 3–12. https://doi.org/10.2344/0003-3006-57.1.3.

Beyer, J. E., McGrath, P. J., & Berde, C. B. (1990). Discordance between self-report and behavioral pain measures in children aged 3-7 years after surgery. *Journal of Pain and Symptom Management, 5*(6), 350–356.

Chaudhry, K., Shishodia, M., Singh, C., & Tuli, A. (2015). Comparative evaluation of pain perception by vibrating needle (Vibraject TM) and conventional syringe anesthesia during various dental procedures in pediatric patients: A short study. *International Dental & Medical Journal of Advanced Research, 1*(1), 1–5. https://doi.org/10.15713/ins.idmjar.5.

Ching, D., Finkelman, M., & Loo, C. Y. (2014). Effect of the DentalVibe injection system on pain during local anesthesia injections in adolescent patients. *Pediatric Dentistry, 36*(1), 51–55.

Davoudi, A., Rismanchian, M., Akhavan, A., Nosouhian, S., Bajoghli, F., Haghighat, A., ... Jahadi, S. (2016). A brief review on the efficacy of different possible and nonpharmacological techniques in eliminating discomfort of local anesthesia injection during dental procedures. *Anesthesia, Essays and Researches, 10*(1), 13–16. https://doi.org/10.4103/0259-1162.167846.

Elbay, M., Şermet Elbay, Ü., Yıldırım, S., Uğurluel, C., Kaya, C., Baydemir, C., ... Baydemir, C. (2015). Comparison of injection pain caused by the DentalVibe injection system versus a traditional syringe for inferior alveolar nerve block anaesthesia in paediatric patients. *European Journal of Paediatric Dentistry, 16*(2), 123–128.

Erdogan, O., Sinsawat, A., Pawa, S., Rintanalert, D., & Vuddhakanok, S. (2018). Utility of vibratory stimulation for reducing intraoral injection pain. *Anesthesia Progress, 65*(2), 95–99. https://doi.org/10.2344/anpr-65-02-01.

Jayanthi, M., Shilpapriya, M., Reddy, V., Sakthivel, R., Selvaraju, G., & Vijayakumar, P. (2015). Effectiveness of new vibration delivery system on pain associated with injection of local anesthesia in children. *Journal of the Indian Society of Pedodontics and Preventive Dentistry, 33*(3), 173. https://doi.org/10.4103/0970-4388.160343.

Kämmerer, P. W., Schiegnitz, E., von Haussen, T., Shabazfar, N., Kämmerer, P., Willershausen, B., ... Daubländer, M. (2015). Clinical efficacy of a computerised device (STA™) and a pressure syringe (VarioJect INTRA™) for intraligamentary anaesthesia. *European Journal of Dental Education, 19*(1), 16–22. https://doi.org/10.1111/eje.12096.

Li, A., Montaño, Z., Chen, V. J., & Gold, J. I. (2011). Virtual reality and pain management: Current trends and future directions. *Pain, 1*(2), 147–157. https://doi.org/10.2217/pmt.10.15.Virtual.

Majstorovic, M., & Veerkamp, J. S. J. (2004). Relationship between needle phobia and dental anxiety. *Journal of Dentistry for Children, 71*(3), 201–205.

Melzack, R. (1996). Gate control theory. *Pain Forum, 5*(2), 128–138. https://doi.org/10.1016/S1082-3174(96)80050-X.

Milgrom, P., Coldwell, S. E., Getz, T., Weinstein, P., & Ramsay, D. S. (1997). Four dimensions of fear of dental injections. *Journal of the American Dental Association (1939), 128*(6), 756–766.

Nanitsos, E., Vartuli, R., Forte, A., Dennison, P., & Peck, C. (2009). The effect of vibration on pain during local anaesthesia injections. *Australian Dental Journal, 54*(2), 94–100. https://doi.org/10.1111/j.1834-7819.2009.01100.x.

Roeber, B., Wallace, D. P., Rothe, V., Salama, F., & Allen, K. D. (2011). Evaluation of the effects of the VibraJect attachment on pain in children receiving local anesthesia. *Pediatric Dentistry, 33*(1), 46–50.

Şermet Elbay, Ü., Elbay, M., Yıldırım, S., Kaya, E., Kaya, C., Uğurluel, C., & Baydemir, C. (2016). Evaluation of the injection pain with the use of DentalVibe injection system during supraperiosteal anaesthesia in children: A randomised clinical trial. *International Journal of Paediatric Dentistry, 26*(5), 336–345. https://doi.org/10.1111/ipd.12204.

Shaefer, J. R., Lee, S. J., & Anderson, N. K. (2017). A vibration device to control injection discomfort. *Compendium of Continuing Education in Dentistry, 38*(6), e5–e8.

Shilpapriya, M., Jayanthi, M., Reddy, V. N., Sakthivel, R., Selvaraju, G., & Vijayakumar, P. (2015). Effectiveness of new vibration delivery system on pain associated with injection of local anesthesia in children. *Journal of the Indian Society of Pedodontics and Preventive Dentistry, 33*(3), 173–176. https://doi.org/10.4103/0970-4388.160343.

Tandon, S., Kalia, G., Sharma, M., Mathur, R., Rathore, K., & Gandhi, M. (2018). Comparative evaluation of mucosal vibrator with topical anesthetic gel to reduce pain during administration of local anesthesia in pediatric patients: An in vivo study. *International Journal of Clinical Pediatric Dentistry, 11*(4), 261–265. https://doi.org/10.5005/jp-journals-10005-1523.

Thoppe-Dhamodharan, Y. K., Asokan, S., John, B. J., Pollachi-Ramakrishnan, G., Ramachandran, P., & Vilvanathan, P. (2015). Cartridge syringe vs computer controlled local anesthetic delivery system: Pain related behaviour over two sequential visits – A randomized controlled trial. *Journal of Clinical and Experimental Dentistry, 7*(4), e513–e518. https://doi.org/10.4317/jced.52542.

Tung, J., Carillo, C., Udin, R., Wilson, M., & Tanbonliong, T. (2018). Clinical performance of the DentalVibe® injection system on pain perception during local anesthesia in children. *Journal of Dentistry for Children, 85*(2), 51–57.

van Wijk, A. J., & Makkes, P. C. (2008). Highly anxious dental patients report more pain during dental injections. *British Dental Journal, 205*(3), E7. https://doi.org/10.1038/sj.bdj.2008.583.

Veneva, E., Cholakova, R., Raycheva, R., & Belcheva, A. (2019). Efficacy of vibrotactile device DentalVibe in reducing injection pain and anxiety during local anaesthesia in paediatric dental patients: A study protocol for a randomised controlled clinical trial. *BMJ Open, 9.* https://doi.org/10.1136/bmjopen-2019-029460.

Versloot, J., Veerkamp, J., & Hoogstraten, J. (2008). Dental anxiety and psychological functioning in children: Its relationship with behaviour during treatment. *European Archives of Paediatric Dentistry, 9*(Suppl 1), 36–40.

Versloot, J., Veerkamp, J. S. J., & Hoogstraten, J. (2004). Assessment of pain by the child, dentist, and independent observers. *Pediatric Dentistry, 26*(5), 445–449.

Vika, M., Skaret, E., Raadal, M., Öst, L.-G., & Kvale, G. (2008). Fear of blood, injury, and injections, and its relationship to dental anxiety and probability of avoiding dental treatment among 18-year-olds in Norway. *International Journal of Paediatric Dentistry, 18*(3), 163–169. https://doi.org/10.1111/j.1365-263X.2007.00904.x.

Weisman, S. J., Bernstein, B., & Schechter, N. L. (1998). Consequences of inadequate analgesia during painful procedures in children. *Archives of Pediatrics & Adolescent Medicine, 152*(2), 147–149.

Chapter 50

Cooled radiofrequency ablation as a treatment for knee osteoarthritis

Antonia F. Chen[a] and Eric J. Moorhead[b]

[a]Department of Orthopaedic Surgery, Brigham and Women's Hospital, Harvard Medical School, Boston, MA, United States, [b]Global Clinical Affairs, Avanos Medical, Alpharetta, GA, United States

Abbreviations

CI	confidence interval
CRFA	cooled radiofrequency ablation
GPE	global perceived effect
HA	hyaluronic acid
NRS	numeric rating scale
NSAID	nonsteroidal antiinflammatory drug
OA	osteoarthritis
OKS	Oxford Knee Score
PRP	platelet-rich plasma
RF	radiofrequency
RFA	radiofrequency ablation
SD	standard deviation
TKA	total knee arthroplasty
VAS	visual analog scale
WOMAC	Western Ontario and McMaster Universities Osteoarthritis Index

Introduction

Affecting an estimated 14 million people in the United States, osteoarthritis (OA) of the knee is the leading cause of chronic knee pain (Deshpande et al., 2016). Injury or overuse of the knee joint may lead to OA, which results from progressive degradation of the cartilage. Knee OA is a common cause of pain and disability (Gupta, Huettner, & Dukewich, 2017; Jamison & Cohen, 2018), and its prevalence is increasing in conjunction with the aging population and rising obesity rates (Charlesworth, Fitzpatrick, Perera, & Orchard, 2019). Many patients with knee OA will be refractory to both nonpharmacological and pharmacological conservative approaches, and many will undergo surgery. However, up to 25% of patients may not be eligible for surgery due to comorbidities or they may be unwilling to undergo surgery (Gossec et al., 2011). Denervation approaches, such as cooled radiofrequency ablation (CRFA), are emerging options that have demonstrated efficacy in the treatment of knee pain caused by OA (Bellini & Barbieri, 2015; Chen et al., 2020a, 2020b; Davis et al., 2018, 2019; Hunter, Davis, Loudermilk, Kapural, & DePalma, 2020).

Nonpharmacological and pharmacological conservative treatments

The goals of OA treatment are to reduce symptoms (e.g., relieve pain and inflammation) and to improve or maintain mobility, function, and health-related quality of life (Bannuru et al., 2019; Gupta et al., 2017). Treatment of knee OA pain varies depending on clinical assessments and patient comorbidities. Most treatment approaches address pain symptoms, but are not curative and do not modify the natural history or progression of OA (Charlesworth et al., 2019). Nonpharmacological management of knee OA includes lifestyle modifications (exercise, diet, and weight loss), assistive walking devices, and physical therapy. Pharmacological interventions include analgesics, nonsteroidal antiinflammatory drugs (NSAIDs), and intra-articular injections. Although used extensively as an analgesic, acetaminophen is not a recommended treatment

Treatments, Mechanisms, and Adverse Reactions of Anesthetics and Analgesics. https://doi.org/10.1016/B978-0-12-820237-1.00050-8
Copyright © 2022 Elsevier Inc. All rights reserved.

589

and has been found to have little to no efficacy in individuals with OA (Bannuru et al., 2019; da Costa et al., 2017). In addition, the use of acetaminophen is restricted in patients with hepatic impairment or active liver disease because of the potential for hepatotoxicity (Charlesworth et al., 2019). NSAIDs, such as diclofenac, have demonstrated efficacy in the treatment of knee OA (Bannuru et al., 2015; Jung et al., 2018). However, NSAIDs should be used with caution due to the risk of adverse events (e.g., gastrointestinal [GI] bleeding, renal impairment, and hypersensitivity) (Fine, 2013; Pelletier, Martel-Pelletier, Rannou, & Cooper, 2016). NSAIDs are contraindicated in patients with comorbidities, such as peptic ulcer disease (Bozimowski, 2015), or in patients who are on chronic anticoagulation. Opioids also cause a number of adverse effects and are no longer recommended for the treatment of knee OA due to growing concerns related to opioid use disorder and addiction (Fuggle et al., 2019; Swiontkowski & Heckman, 2019; Trang et al., 2015). A retrospective cohort study showed a significant increase in opioid prescriptions between 2003 and 2009 for Medicare beneficiaries with knee OA (Wright, Katz, Abrams, Solomon, & Losina, 2014). In many patients, pain due to severe OA of the knee is not reliably responsive to conservative therapies (Conaghan et al., 2015; Crawford, Miller, & Block, 2013). In light of the escalating opioid crisis, safe and efficacious nonopioid alternatives are needed for patients with knee pain due to OA.

Minimally invasive and surgical approaches

Intra-articular injections encompass a wide range of medications, such as antiinflammatory corticosteroids, platelet-rich plasma (PRP) solutions, and viscosupplementation (hyaluronic acid (HA); Altman, Manjoo, Fierlinger, Niazi, & Nicholls, 2015; Chevalier et al., 2010; McAlindon & LaValley, 2017; Meheux, McCulloch, Lintner, Varner, & Harris, 2016). Steroid injections provide short-lasting pain relief (Zeng et al., 2019). Results from a randomized clinical trial demonstrated greater cartilage loss and no significant difference in knee pain with 40 mg of intra-articular triamcinolone compared with placebo at 2 years (McAlindon et al., 2017). In addition, long-term use of steroid injections is associated with numerous complications, including crystal-induced synovitis, fat necrosis, cartilage damage, and infection (Gupta et al., 2017; Hepper et al., 2009; Jüni et al., 2015; Zeng et al., 2019). Like steroid injections, HA injections only provide short-term pain relief, demonstrating a maximum clinical durability of 6 months (Chevalier et al., 2010). HA injections are not recommended for the treatment of knee OA (Jevsevar, 2013). PRP injections have shown mixed results for treating knee OA pain. One study showed no superiority of PRP injections when compared to both HA and saline injections (Filardo et al., 2015). Although some review articles state that PRP injections can result in a significant clinical improvement up to 12 months post injection (Meheux et al., 2016), more recent review articles have called the lack of standardization among PRP products into question and the high risk of bias in previously conducted clinical trials (Gato-Calvo, Magalhaes, Ruiz-Romero, Blanco, & Burguera, 2019).

Surgery is reserved for eligible patients with knee OA who are refractory to conservative treatments. Surgical approaches include arthroscopic procedures [arthroscopic lavage and cartilage debridement—which is not recommended in the setting of knee OA (Jevesar et al., 2013)], osteotomies, and open unicompartmental or total knee arthroplasties (TKA; Kamaruzaman, Kinghorn, & Oppong, 2017). Even after undergoing TKA, which is the terminal treatment for knee OA, up to 44% of patients have reported experiencing persistent postsurgical pain (Wylde, Hewlett, Learmonth, & Dieppe, 2011). Therefore, effective nonsurgical options are needed to treat refractory pain in patients with knee OA before proceeding to surgical management.

Denervation treatment in knee osteoarthritis

The knee joint is innervated by several branches of the genicular nerves, arising from the femoral, common peroneal, saphenous, tibial, and obturator nerves (Oladeji & Cook, 2019). Ablative procedures target sensory nerves of the peripheral nervous system that relay pain signals to the central nervous system. It is believed that ablation results in attenuation of pain as the nerve structure heals (Farrell, Gutierrez, & Desai, 2017). These procedures have been investigated for the treatment of pain associated with knee OA, and include cryoablation, radiofrequency ablation (RFA), and CRFA. Various clinical evidence supporting the use of these methods in the treatment of knee OA are available, and Table 1 shows selected outcome measures used in studies to assess the clinical benefits.

Cryoablation

Cryoablation (also known as cryoneurolysis) is performed using a novel device to produce lesions in the peripheral nervous tissue by applying low temperatures to the targeted site, causing temporary destruction of the nerve structure and blockade of pain transmission (Miller, Block, & Malfait, 2018). A retrospective chart review evaluated the benefit of perioperative

TABLE 1 Outcome measures used in knee osteoarthritis.

Measure	Description
NRS	11-point scale that captures the amount of index knee pain at different timepoints
OKS	12 items (pain, 5 items; function, 7 items), each item worth 0–4 points
	Range: 0–48 points
	Categories: excellent (>41), good (34–41), fair (27–33), and poor (<27)
Pain VAS	Continuous scale comprised a line, 100 mm in length, anchored by two descriptors, one for each symptom extreme
	Score: ranges from 0 ("no pain") to 100 ("worst imaginable pain")
WOMAC	24 items [pain (5 items), stiffness (2 items), and physical function (17 items)]
	Range: 0–98 points

NRS, numeric rating scale; *OKS*, Oxford Knee Score; *VAS*, visual analog scale. References: Dowsey and Choong (2013); van der Wees et al. (2017); Davis et al. (2018).

cryoneurolysis prior to TKA in patients with knee OA (Dasa et al., 2016). Postoperative outcomes were compared between a treatment group of 50 patients who received cryoneurolysis 5 days before TKA and a control group of 50 TKA patients who did not receive cryoneurolysis. Compared with the control group, the treatment group had a lower proportion of patients with a length of stay of 2 days or more (6% vs 67%, $P < .0001$). The treatment group also required 45% less opioids during the first 12 weeks of surgery ($P < .0001$), had greater reduction in the symptom subscale score of the Knee Injury and Osteoarthritis Outcome Score (KOOS) from baseline at 6 weeks ($P = .0037$) and 12 weeks ($P = .0011$) of follow-up, and had reduced pain intensity scores as measured by the Patient-Reported Outcomes Measurement Information System (PROMIS; $P < .0001$).

A randomized, double-blind, sham-controlled, multicenter trial evaluated cryoneurolysis for reduction of pain symptoms associated with knee OA (Radnovich et al., 2017). A total of 180 patients with mild-to-moderate knee OA were randomized 2:1 to cryoneurolysis targeting the infrapatellar branch of the saphenous nerve ($n = 121$) or sham treatment ($n = 59$). Patients treated with cryoneurolysis had a greater decrease in the Western Ontario and McMaster Universities Osteoarthritis Index (WOMAC) pain score, WOMAC total score, and pain Visual Analog Scale (VAS) score at 30 days. The improvement in WOMAC total score was greater at 90 days with cryoneurolysis than in sham treatment, but no differences were observed at 60 days. In addition, no differences in pain VAS scores were observed between cryoneurolysis and sham treatment at 60 and 90 days.

Radiofrequency ablation

RFA is a minimally invasive technique that uses a needle electrode inserted through the skin, applying high-frequency electrical current to destroy the target tissue and disrupt the nociceptive supply to pain-provoking structures (Fig. 1; Farrell et al., 2017). Conventional RFA relies on ionic heat generated by the continuous flow of radiofrequency (RF) current through the needle electrode to cause local tissue destruction, resulting in nerve ablation through a process of coagulative necrosis, collagen destruction, and axonal degeneration (Orhurhu, Urits, Grandhi, & Abd-Elsayed, 2019). Several conditions, related to electrode structure and behavior as well as sources of heat conduction and transfer, can cause problems with maintaining current flow and thermal conductivity; these potential failure points can have a detrimental effect on the outcomes observed with conventional RFA and require careful monitoring (Ball, 2014). A major limitation with conventional

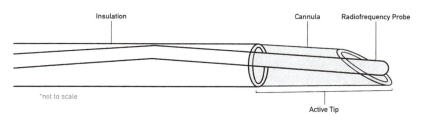

FIG. 1 Standard radiofrequency probe. *(Reproduced with permission from Avanos Medical, Inc.)*

RFA occurs when extended periods of ionic heating cause desiccation and charring of tissues surrounded by the probe (Kapural & Deering, 2020). As a result, the charred tissue prevents thermal transfer to the adjacent tissue and limits the size of lesions created by the electrode. Lesions formed are elliptical, following the shape of the exposed electrode, which can make it more difficult to reach the desired nerve targets.

Clinical trials of radiofrequency ablation

A prospective, observational, single-arm, longitudinal study investigated the efficacy of genicular nerve RFA in 25 patients with advanced OA of the knee (Santana Pineda et al., 2017). The primary endpoint was the change in knee pain VAS from baseline. At least 50% improvement in pain was reported in 88% (22/25) of patients at 1 month, 64% (16/25) at 6 months, and 32% (8/25) at 12 months. At 6 months, patient satisfaction scores were reported as "very good" in 64% (16/25) of patients, "good" in 20% (5/25), "average" in 4% (1/25), and "poor" in 8% (2/25). Study results also showed reduced analgesic use following RFA treatment. No serious adverse events were reported.

A prospective, single-arm study investigated genicular nerve RFA treatment in 49 patients with chronic knee pain due to OA (Kirdemir, Catav, & Alkaya Solmaz, 2017). Significantly reduced pain was observed with RFA treatment as measured by decreases in VAS scores from baseline at 1, 4, and 12 weeks ($P < .01$ at all timepoints). Similarly, RFA treatment resulted in decreases in WOMAC scores (including the total score and subscores for pain, stiffness, and functioning) from baseline at 1, 4, and 12 weeks, indicating significantly reduced pain and improved function over time ($P < .01$ at all timepoints). No procedure-related complications were reported.

A randomized, sham-controlled trial investigated RFA of the genicular nerve in 38 patients with chronic knee OA pain (Choi et al., 2011). Results showed that RFA significantly reduced knee pain as measured by VAS scores at 4 weeks ($P < .001$) and at 12 weeks ($P < .001$) compared with sham. In all, 10 of the 17 (59%) patients in the RFA group achieved at least 50% pain reduction at 12 weeks, compared with no patients in the control group ($P < .001$). The Oxford Knee Score (OKS) and patient satisfaction as measured by the Global Perceived Effect (GPE) were also significantly improved in the RFA group ($P < .001$ for both outcomes). No adverse events were reported during the 3-month follow-up period.

Another randomized trial compared the efficacy of genicular nerve RFA ($n = 37$) with an intra-articular injection composed of bupivacaine, morphine, and betamethasone ($n = 36$) in patients with chronic knee OA (Sari et al., 2018). The primary endpoint was pain intensity with functional status as a secondary endpoint compared at baseline, 1-month, and 3-month follow-ups. The RF group reported a significant reduction in pain VAS score ($P < .001$) than did the intra-articular injection group at 1-month and 3-month follow-ups. The RF group also had significantly lower WOMAC scores than did the intra-articular injection group at 1-month follow-up ($P < .001$). No adverse events were reported in either treatment group.

Other studies of radiofrequency ablation

A retrospective/prospective case-series chart review analyzed the efficacy of genicular nerve RFA in 26 patients with chronic knee OA pain (Iannaccone, Dixon, & Kaufman, 2017). The study population consisted of patients who had received conservative treatment (physical therapy, steroid or HA injection, oral analgesics) prior to RFA. At 3 months, a 67% improvement in pain from baseline was reported, with an average 0–10 pain score of 2.9. Of the patients with pain relief at 3 months, 95% had pain relief at the 6-month follow-up and reported a 64% improvement in pain from baseline, with an average 0–10 pain score of 3.3. No serious adverse events were reported.

A retrospective chart review evaluated the efficacy of genicular nerve RFA in 48 patients with chronic refractory knee pain due to OA (Konya, Akin Takmaz, Basar, Baltaci, & Babaoglu, 2020). Significant reductions in pain VAS scores ($P < .001$) and WOMAC scores (including total score and subscores for pain, stiffness, and functioning; $P < .01$) from baseline were observed at 1, 3, and 6 months postprocedure. Quality-of-life scores were also significantly improved at all timepoints ($P < .05$). No serious complications were reported.

Jamison and Cohen (2018) conducted a systematic literature review investigating patient selection, anatomic targets, treatment parameters, and effectiveness of RF techniques to treat chronic knee pain. The review consisted of 9 clinical trials evaluating knee RFA for OA and persistent postsurgical pain and included a total of 592 patients. Studies included one randomized, placebo-controlled trial, one randomized, controlled trial evaluating RFA as an add-on therapy, four comparative-effectiveness studies, two randomized trials comparing different techniques and treatment paradigms, and one nonrandomized, controlled trial. The authors found significant benefits for both pain reduction and functional improvement lasting between 3 and 12 months, with questionable utility for prognostic blocks. Considerable variation was found in the described neuroanatomy, neural targets, radiofrequency techniques, and selection criteria. Overall findings supported RFA of the knee as a viable and effective treatment option, providing significant benefits to well-selected patients lasting at least 3 months.

Summary of radiofrequency ablation

Previous studies have shown evidence of pain relief with conventional RFA in patients with knee OA. Conventional RFA has demonstrated improved outcomes compared with an intra-articular injection of anesthetic and steroid combined with morphine. Clinical benefits were observed at 1 and 3 months post procedure. Inherent limitations associated with RFA may negatively affect the ability to attain effective denervation, given the smaller surface area of treatment. Longer-term studies with active comparators are needed to evaluate the durability of treatment outcomes with RFA.

Cooled radiofrequency ablation

Approved by the US Food and Drug Administration (FDA) in 2017 for the treatment of knee OA, a specific version of CRFA (COOLIEF) uses a RF probe cooled by circulated water through the probe tip to moderate temperatures at the tissue-tip interface (Fig. 2; Ball, 2014; Kapural & Deering, 2020). Although this technology is similar to conventional (thermal) RF, this newer CRFA technology can overcome limitations inherent to standard RF probes by minimizing charring and delivering more energy to generate larger lesions (Fig. 3). In addition, preclinical studies have demonstrated that CRFA probes deliver more cellular-level damage to the nerve structure (Zachariah et al., 2020). Although the probe operates at a set temperature of 60°C, temperatures in the tissues reach 80°C. Genicular nerve targets for ablation (Fig. 4), informed by anatomical studies (Tran, Agur, & Peng, 2020), show a high degree of nerve variability. As such, producing larger lesions via CRFA can accomplish a more successful denervation.

Clinical trials of cooled radiofrequency ablation

This CRFA received FDA clearance based on the results of a prospective, randomized, crossover study investigating the efficacy and safety of genicular nerve CRFA compared with intra-articular steroid injections (Davis et al., 2018). The study included 151 patients with knee OA unresponsive to conservative modalities. Pain relief with CRFA was superior to that obtained with intra-articular steroids at all time periods (1, 3, and 6 months following treatment). At the 6-month follow-up, 74% of the CRFA group had at least 50% pain relief [as evaluated by the Numeric Rating Scale (NRS) score] versus 16% of the intra-articular steroid group. Complete reduction of pain was reported in 22% of the CRFA group versus 4% of the intra-articular steroid group. Improvements in knee function (as measured by change in the OKS) and GPE were also superior in the CRFA group. In addition, patients in the CRFA group had a greater mean change in nonopioid (NSAID) analgesic use. No treatment-related serious adverse events were reported in either group.

A follow-up of the Davis 2018 cohort reported the 12-month outcomes in 52 participants originally treated with CRFA and showed similar improvements in knee pain and function at follow-up (Davis et al., 2019). At least 50% pain reduction was reported in 65% of patients at 12 months, with a mean reduction of 4.3 ± 2.7 points in the NRS score ($P < .0001$). Overall, 75% of patients reported improvement in the perceived effects of treatment.

Hunter et al. (2020) reported follow-up results at 18 and 24 months in the original cohort from the pivotal trial. A total of 33 patients were included in the extended outcome study ($n = 19$ from the CRFA group and $n = 14$ from the crossover arm). At 18 months, the mean NRS pain score was 3.1 ± 2.7, the mean OKS was 47.2 ± 8.1, and 80% reported an improvement in the perceived effect of treatment (GPE). At 24 months, the mean NRS pain score was 3.6 ± 2, the mean

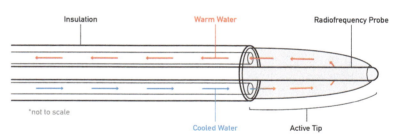

FIG. 2 Cooled radiofrequency probe. *(Reproduced with permission from Avanos Medical, Inc.)*

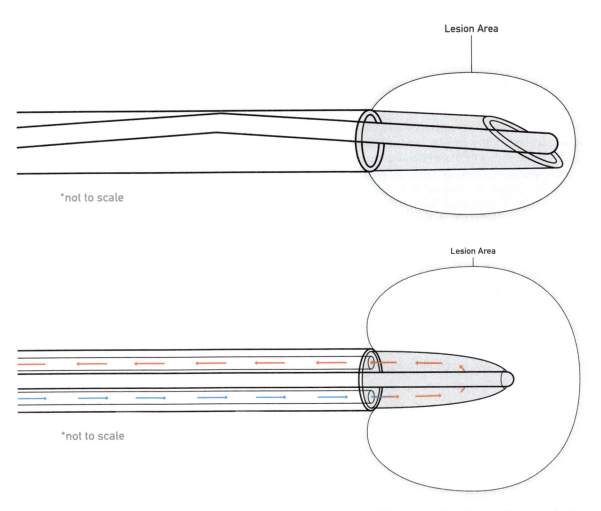

FIG. 3 Lesion size and shape created by standard radiofrequency probes (top) and cooled radiofrequency probes (bottom). *(Reproduced with permission from Avanos Medical, Inc.)*

OKS was 46.8 ± 10.3, and 67% reported an improvement in GPE. No adverse events were reported at 18 and 24 months post CRFA.

A multicenter, randomized study ($n = 158$) compared the efficacy and safety of CRFA ($n = 87$) with a single intraarticular HA injection ($n = 88$) in patients with OA-related knee pain (Chen et al., 2020a). The proportion of patients achieving at least 50% reduction in the NRS pain score was significantly higher in the CRFA group than in the HA group (71.1% vs 37.8%; $P < .0001$). The mean NRS ± standard deviation (SD) at 6 months was significantly lower in the CRFA group than in the HA group (2.7 ± 2.3 vs 4.5 ± 2.7; $P < .0001$). The mean WOMAC score improvement at 6 months from baseline was also significantly greater in the CRFA group than in the HA group (48.2% vs 22.6%; $P < .0001$). More patients in the CRFA group reported improvement in GPE at 6 months than in the HA group (72.4% vs 40.2%; $P < .0001$). The incidence of adverse events was similar between treatment groups, and no serious adverse events were reported in either group.

In the 12-month follow-up to this study, subjects receiving CRFA continued to report improved clinical outcomes (Chen et al., 2020b). At 12 months, 65.2% of subjects in the CRFA cohort reported ≥50% pain relief from baseline. Those treated with CRFA had a mean NRS pain score of 2.8 ± 2.4, compared to a baseline of 6.9 ± 0.8, representing a 4.1 decrease in the NRS pain score ($P < .0001$). Subjects within this cohort also saw a 46.2% improvement in the total WOMAC score. These data suggest that CRFA has a clinical durability of 12 months for a majority of subjects undergoing the procedure.

Overall, there is a robust amount of clinical evidence supporting the use of CRFA in the management of chronic knee pain caused by OA. Subjects report diminished pain scores with a clinical durability often lasting 12 months, with some patients reporting pain relief lasting 24 months (Fig. 5).

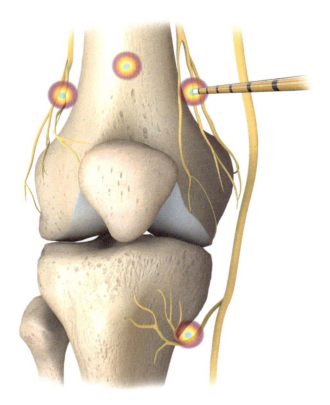

FIG. 4 Genicular nerve location and needle placement for cooled radiofrequency ablation. *(Reproduced with permission from Avanos Medical.)*

Other studies of cooled radiofrequency ablation

A case-series report at a single center investigated the efficacy of CRFA of the genicular nerve in nine patients with knee OA (Bellini & Barbieri, 2015). No adverse events were reported. The primary endpoint was the pain VAS score with the WOMAC score as a secondary endpoint. Improvement in pain VAS scores were observed from a mean of 8.0 ± 1.9 at baseline to mean scores of 2.0 ± 0.5, 2.3 ± 0.7, 2.1 ± 0.5, and 2.2 ± 0.2 at 1, 3, 6, and 12 months, respectively. Improvement in WOMAC scores were also observed from a mean of 88 ± 1.9 at baseline to mean scores of 20 ± 2.0, 22 ± 0.5, 21 ± 1.7, and 20 ± 1.0 at 1, 3, 6, and 12 months, respectively. No complications were reported.

A retrospective, cross-sectional survey evaluated CRFA of the genicular nerve in 33 patients (52 treated knees) with chronic knee pain due to OA (McCormick et al., 2017). At 6 months, 35% [95% confidence interval (CI): 22%–48%] of patients achieved a successful outcome measured by a composite of at least a 50% reduction in the NRS score, a reduction of 3.4 or more points in the Medication Quantification Scale III score, and the Patient Global Impress of Change (PGIC) score consistent with "very much improved/improved." Complete pain relief was attained in 19% (95% CI: 10%–33%) of the patients at a minimum follow-up of 6 months. No serious adverse events were reported.

A retrospective electronic chart review evaluated the long-term effectiveness of CRFA in patients with chronic knee pain (Kapural, Lee, Neal, & Burchell, 2019). The study included 275 consecutive patients who had undergone a geniculate nerve block at a single-site pain practice between July 1, 2014 and July 1, 2017. Of these 275 patients, 205 (75%) received CRFA and 89% (183/205) returned for follow-up. Reductions in pain VAS scores were observed from a baseline of 8.5 to 2.2 after the geniculate local anesthetic block, and to 4.2 after CRFA. After treatment with CRFA, more than 50% pain relief occurred in 65% of the patients [mean duration: 12.5 months (range 0–35 months)], a decrease in the pain VAS score of at least 2 was observed in 77%, and no pain at all was reported in 14% of the patients. Similar pain relief was achieved in patients who underwent a repeated procedure ($P = .402$). No difference in outcomes after geniculate CRFA was found between patients who had TKA ($n = 21$) and maintained chronic knee pain and those who had no prior surgery ($P = .542$).

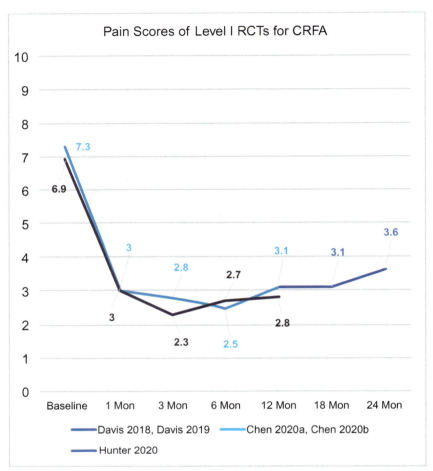

FIG. 5 Pain scores from level 1 randomized clinical trials of CRFA.

Summary of cooled radiofrequency ablation

The use of CRFA in the treatment of knee OA has been cleared by the FDA and is supported by a substantial body of evidence that demonstrates extended clinical durability compared with other treatments. Patients with refractory knee OA pain currently have limited nonsurgical options. With emerging data supporting its use in OA-related knee pain, CRFA offers a safe and efficacious option that fills a gap in the treatment algorithm to achieve significant pain relief and improved function without surgery or opioids.

Applications to other areas

In this chapter, we have reviewed the clinical evidence supporting ablative procedures in the treatment of knee osteoarthritis (OA). We have described in detail the limitations with existing nonpharmacological, pharmacological, and surgical approaches to treating knee OA. In the following section, we described how denervation using ablative procedures, including cooled radiofrequency ablation (CRFA), can target sensory nerves of the knee joint that relay pain signals to the central nervous system. We then summarized the emerging data supporting the use of CRFA in OA-related knee pain. This body of evidence may have implications in the use of CRFA to treat pain affecting other regions of the body and caused by other conditions.

Other ablative procedures of interest

In this chapter, we have described detailed mechanisms of how cooled radiofrequency ablation works. Other ablative procedures include

Conventional radiofrequency ablation.
Cryoablation.
Chemical neurolysis.

Conventional radiofrequency ablation and cryoablation are also discussed in this chapter. The literature on chemical neurolysis is very limited in the treatment of knee OA.

Mini-dictionary of terms

Ablation: Minimally invasive technique that involves the destruction of tissue, targeting sensory nerves of the peripheral nervous system that relay pain signals to the central nervous system.
Cooled radiofrequency ablation (CRFA): Procedure that uses a radiofrequency probe cooled by circulated water through the probe tip to moderate temperatures at the tissue-tip interface. CRFA is used to cause local tissue destruction, resulting in nerve ablation.
Cryoablation: Procedure that uses a device to produce lesions in the peripheral nervous tissue by applying low temperatures to the targeted site, temporarily destroying the nerve structure and blocking pain transmission.
Knee osteoarthritis: Degenerative joint disease caused by injury or overuse of the knee joint. Knee osteoarthritis is the leading cause of chronic knee pain.
Radiofrequency ablation (RFA): Procedure that uses a needle electrode inserted through the skin, applying high-frequency electrical current to destroy the target tissue and disrupt nociceptive supply to pain-provoking structures.

Key facts of cooled radiofrequency ablation

- Cooled radiofrequency ablation (CRFA) treatment (COOLIEF) was cleared by the US Food and Drug Administration in 2017 for treatment of knee osteoarthritis.
- The use of CRFA can minimize charring and deliver more energy to generate larger lesions and more cellular-level damage to the nerve structure.
- In a multicenter, randomized study, 71.1% of patients treated with CRFA achieved at least 50% reduction in the Numerical Rating Scale pain score versus 37.8% of those treated with a single intra-articular hyaluronic acid injection.
- In a randomized, crossover study, pain relief with CRFA was superior to that obtained with intra-articular steroids at 1, 3, and 6 months following treatment.
- At 6 months, complete reduction of pain was reported in 22% of the CRFA group versus 4% of the intra-articular steroid group.

Summary points

- Denervation approaches, such as cooled radiofrequency ablation (CRFA), are an emerging option that have demonstrated efficacy in the treatment of knee pain caused by osteoarthritis (OA).
- Radiofrequency ablation (RFA) has demonstrated improved outcomes compared with an intra-articular injection of anesthetic and corticosteroid combined with morphine.
- RFA is associated with inherent limitations, which may negatively affect the ability to attain effective denervation.
- CRFA was approved by the US Food and Drug Administration in 2017 and may overcome limitations inherent to standard radiofrequency probes to generate larger lesions and more cellular-level damage to the nerve structure.
- The use of CRFA in the treatment of knee OA is supported by a substantial body of evidence, which demonstrates extended clinical durability when compared with other treatments.

References

Altman, R. D., Manjoo, A., Fierlinger, A., Niazi, F., & Nicholls, M. (2015). The mechanism of action for hyaluronic acid treatment in the osteoarthritic knee: A systematic review. *BMC Musculoskeletal Disorders, 16*, 321.

Ball, R. D. (2014). The science of conventional and water-cooled monopolar lumbar radiofrequency rhizotomy: An electrical engineering point of view. *Pain Physician, 17*(2), E175–E211.

Bannuru, R. R., Osani, M. C., Vaysbrot, E. E., Arden, N. K., Bennell, K., Bierma-Zeinstra, S. M. A., ... McAlindon, T. E. (2019). OARSI guidelines for the non-surgical management of knee, hip, and polyarticular osteoarthritis. *Osteoarthritis and Cartilage, 27*(11), 1578–1589.

Bannuru, R. R., Schmid, C. H., Kent, D. M., Vaysbrot, E. E., Wong, J. B., & McAlindon, T. E. (2015). Comparative effectiveness of pharmacologic interventions for knee osteoarthritis: A systematic review and network meta-analysis. *Annals of Internal Medicine, 162*(1), 46–54.

Bellini, M., & Barbieri, M. (2015). Cooled radiofrequency system relieves chronic knee osteoarthritis pain: The first case-series. *Anaesthesiology Intensive Therapy, 47*(1), 30–33.

Bozimowski, G. (2015). A review of nonsteroidal anti-inflammatory drugs. *AANA Journal, 83*(6), 425–433. Retrieved from: https://www.ncbi.nlm.nih.gov/pubmed/26742337.

Charlesworth, J., Fitzpatrick, J., Perera, N. K. P., & Orchard, J. (2019). Osteoarthritis—A systematic review of long-term safety implications for osteoarthritis of the knee. *BMC Musculoskeletal Disorders, 20*(1), 151.

Chen, A. F., Khalouf, F., Zora, K., DePalma, M., Kohan, L., Guirguis, M., ... Lyman, J. (2020a). Cooled radiofrequency ablation compared with a single injection of hyaluronic acid for chronic knee pain: A multicenter, randomized clinical trial demonstrating greater efficacy and equivalent safety for cooled radiofrequency ablation. *Journal of Bone & Joint Surgery, 102*, 1501–1510.

Chen, A. F., Khalouf, F., Zora, K., DePalma, M., Kohan, L., Guirguis, M., ... Lyman, J. (2020b). Cooled radiofrequency ablation provides extended clinical utility in the management of knee osteoarthritis: 12-month results from a prospective, multi-center, randomized, cross-over trial compating cooled radiofrequency ablation to a single hyaluronic acid injection. *BMC Musculoskeletal Disorders, 21*(1), 363.

Chevalier, X., Jerosch, J., Goupille, P., van Dijk, N., Luyten, F. P., Scott, D. L., ... Pavelka, K. (2010). Single, intra-articular treatment with 6 ml hylan G-F 20 in patients with symptomatic primary osteoarthritis of the knee: A randomised, multicentre, double-blind, placebo controlled trial. *Annals of the Rheumatic Diseases, 69*(1), 113–119.

Choi, W. J., Hwang, S. J., Song, J. G., Leem, J. G., Kang, Y. U., Park, P. H., & Shin, J. W. (2011). Radiofrequency treatment relieves chronic knee osteoarthritis pain: A double-blind randomized controlled trial. *Pain, 152*(3), 481–487.

Conaghan, P. G., Peloso, P. M., Everett, S. V., Rajagopalan, S., Black, C. M., Mavros, P., ... Taylor, S. D. (2015). Inadequate pain relief and large functional loss among patients with knee osteoarthritis: Evidence from a prospective multinational longitudinal study of osteoarthritis real-world therapies. *Rheumatology (Oxford), 54*(2), 270–277.

Crawford, D. C., Miller, L. E., & Block, J. E. (2013). Conservative management of symptomatic knee osteoarthritis: A flawed strategy? *Orthopedic Reviews (Pavia), 5*(1), e2.

da Costa, B. R., Reichenbach, S., Keller, N., Nartey, L., Wandel, S., Juni, P., & Trelle, S. (2017). Effectiveness of non-steroidal anti-inflammatory drugs for the treatment of pain in knee and hip osteoarthritis: A network meta-analysis. *Lancet, 390*(10090), e21–e33.

Dasa, V., Lensing, G., Parsons, M., Harris, J., Volaufova, J., & Bliss, R. (2016). Percutaneous freezing of sensory nerves prior to total knee arthroplasty. *The Knee, 23*(3), 523–528.

Davis, T., Loudermilk, E., DePalma, M., Hunter, C., Lindley, D., Patel, N., ... Kapural, L. (2018). Prospective, multicenter, randomized, crossover clinical trial comparing the safety and effectiveness of cooled radiofrequency ablation with corticosteroid injection in the management of knee pain from osteoarthritis. *Regional Anesthesia and Pain Medicine, 43*(1), 84–91.

Davis, T., Loudermilk, E., DePalma, M., Hunter, C., Lindley, D. A., Patel, N., ... Kapural, L. (2019). Twelve-month analgesia and rescue, by cooled radiofrequency ablation treatment of osteoarthritic knee pain: Results from a prospective, multicenter, randomized, cross-over trial. *Regional Anesthesia and Pain Medicine.* https://doi.org/10.1136/rapm-2018-100051.

Deshpande, B. R., Katz, J. N., Solomon, D. H., Yelin, E. H., Hunter, D. J., Messier, S. P., ... Losina, E. (2016). Number of persons with symptomatic knee osteoarthritis in the US: Impact of race and ethnicity, age, sex, and obesity. *Arthritis Care & Research (Hoboken), 68*(12), 1743–1750.

Dowsey, M. M., & Choong, P. F. (2013). The utility of outcome measures in total knee replacement surgery. *International Journal of Rheumatology, 2013*, 506518.

Farrell, M. E., Gutierrez, G., & Desai, M. J. (2017). Demonstration of lesions produced by cooled radiofrequency neurotomy for chronic osteoarthritic knee pain: A case presentation. *PM & R : The Journal of Injury, Function, and Rehabilitation, 9*(3), 314–317.

Filardo, G., Di Matteo, B., Di Martino, A., Merli, M. L., Cenacchi, A., Fornasari, P., ... Kon, E. (2015). Platelet-rich plasma intra-articular knee injections show no superiority versus viscosupplementation: A randomized controlled trial. *The American Journal of Sports Medicine, 43*(7), 1575–1582.

Fine, M. (2013). Quantifying the impact of NSAID-associated adverse events. *The American Journal of Managed Care, 19*(14 Suppl), s267–s272.

Fuggle, N., Curtis, E., Shaw, S., Spooner, L., Bruyere, O., Ntani, G., ... Cooper, C. (2019). Safety of opioids in osteoarthritis: Outcomes of a systematic review and meta-analysis. *Drugs & Aging, 36*(Suppl. 1), 129–143.

Gato-Calvo, L., Magalhaes, J., Ruiz-Romero, C., Blanco, F. J., & Burguera, E. F. (2019). Platelet-rich plasma in osteoarthritis treatment: Review of current evidence. *Therapeutic Advances in Chronic Disease, 10.* 2040622319825567.

Gossec, L., Paternotte, S., Maillefert, J. F., Combescure, C., Conaghan, P. G., Davis, A. M., ... OARSI-OMERACT Task Force Total Articular Replacement as Outcome Measure in OA. (2011). The role of pain and functional impairment in the decision to recommend total joint replacement in hip and knee osteoarthritis: An international cross-sectional study of 1909 patients. Report of the OARSI-OMERACT Task Force on total joint replacement. *Osteoarthritis and Cartilage, 19*(2), 147–154.

Gupta, A., Huettner, D. P., & Dukewich, M. (2017). Comparative effectiveness review of cooled versus pulsed radiofrequency ablation for the treatment of knee osteoarthritis: A systematic review. *Pain Physician, 20*(3), 155–171.

Hepper, C. T., Halvorson, J. J., Duncan, S. T., Gregory, A. J., Dunn, W. R., & Spindler, K. P. (2009). The efficacy and duration of intra-articular corticosteroid injection for knee osteoarthritis: A systematic review of level I studies. *The Journal of the American Academy of Orthopaedic Surgeons, 17* (10), 638–646.

Hunter, C., Davis, T., Loudermilk, E., Kapural, L., & DePalma, M. (2020). Cooled radiofrequency ablation treatment of the genicular nerves in the treatment of osteoarthritic knee pain: 18- and 24-month results. *Pain Practice, 20*(3), 238–246.

Iannaccone, F., Dixon, S., & Kaufman, A. (2017). A review of long-term pain relief after genicular nerve radiofrequency ablation in chronic knee osteoarthritis. *Pain Physician, 20*(3), E437–E444.

Jamison, D. E., & Cohen, S. P. (2018). Radiofrequency techniques to treat chronic knee pain: A comprehensive review of anatomy, effectiveness, treatment parameters, and patient selection. *Journal of Pain Research, 11*, 1879–1888.

Jevesar, D., Brown, G. A., Jones, D. L., Matzin, E. G., Manner, P., Mooar, P., ... Stovitz, S. (2013). Treatment of osteoarthritis of the knee. *American Academy of Orthopedic Surgeons* (2nd ed.). Retrieved from https://www.orthoguidelines.org/topic?id=1005.

Jevsevar, D. S. (2013). Treatment of osteoarthritis of the knee: Evidence-based guideline, 2nd edition. *The Journal of the American Academy of Orthopaedic Surgeons, 21*(9), 571–576.

Jung, S. Y., Jang, E. J., Nam, S. W., Kwon, H. H., Im, S. G., Kim, D., ... Sung, Y. K. (2018). Comparative effectiveness of oral pharmacologic interventions for knee osteoarthritis: A network meta-analysis. *Modern Rheumatology, 28*(6), 1021–1028.

Jüni, P., Hari, R., Rutjes, A. W., Fischer, R., Silletta, M. G., Reichenbach, S., & da Costa, B. R. (2015). Intra-articular corticosteroid for knee osteoarthritis. *Cochrane Database of Systematic Reviews, 10*. CD005328.

Kamaruzaman, H., Kinghorn, P., & Oppong, R. (2017). Cost-effectiveness of surgical interventions for the management of osteoarthritis: A systematic review of the literature. *BMC Musculoskeletal Disorders, 18*(1), 183.

Kapural, L., & Deering, J. P. (2020). A technological overview of cooled radiofrequency ablation and its effectiveness in the management of chronic knee pain. *Pain Management*. https://doi.org/10.2217/pmt-2019-0066.

Kapural, L., Lee, N., Neal, K., & Burchell, M. (2019). Long-term retrospective assessment of clinical efficacy of radiofrequency ablation of the knee using a cooled radiofrequency system. *Pain Physician, 22*(5), 489–494.

Kirdemir, P., Catav, S., & Alkaya Solmaz, F. (2017). The genicular nerve: Radiofrequency lesion application for chronic knee pain. *Turkish Journal of Medical Sciences, 47*(1), 268–272.

Konya, Z. Y., Akin Takmaz, S., Basar, H., Baltaci, B., & Babaoglu, G. (2020). Results of genicular nerve ablation by radiofrequency in osteoarthritis-related chronic refractory knee pain. *Turkish Journal of Medical Sciences, 50*(1), 86–95.

McAlindon, T. E., & LaValley, M. P. (2017). Long-term intra-articular steroid injections and knee cartilage-reply. *JAMA, 318*(12), 1185–1186.

McAlindon, T. E., LaValley, M. P., Harvey, W. F., Price, L. L., Driban, J. B., Zhang, M., & Ward, R. J. (2017). Effect of intra-articular triamcinolone vs saline on knee cartilage volume and pain in patients with knee osteoarthritis: A randomized clinical trial. *JAMA, 317*(19), 1967–1975.

McCormick, Z. L., Korn, M., Reddy, R., Marcolina, A., Dayanim, D., Mattie, R., ... Walega, D. R. (2017). Cooled radiofrequency ablation of the genicular nerves for chronic pain due to knee osteoarthritis: Six-month outcomes. *Pain Medicine, 18*(9), 1631–1641.

Meheux, C. J., McCulloch, P. C., Lintner, D. M., Varner, K. E., & Harris, J. D. (2016). Efficacy of intra-articular platelet-rich plasma injections in knee osteoarthritis: A systematic review. *Arthroscopy, 32*(3), 495–505.

Miller, R. E., Block, J. A., & Malfait, A. M. (2018). What is new in pain modification in osteoarthritis? *Rheumatology (Oxford), 57*(Suppl_4), iv99–iv107.

Oladeji, L. O., & Cook, J. L. (2019). Cooled radio frequency ablation for the treatment of osteoarthritis-related knee pain: Evidence, indications, and outcomes. *The Journal of Knee Surgery, 32*(1), 65–71.

Orhurhu, V., Urits, I., Grandhi, R., & Abd-Elsayed, A. (2019). Systematic review of radiofrequency ablation for management of knee pain. *Current Pain and Headache Reports, 23*(8), 55.

Pelletier, J. P., Martel-Pelletier, J., Rannou, F., & Cooper, C. (2016). Efficacy and safety of oral NSAIDs and analgesics in the management of osteoarthritis: Evidence from real-life setting trials and surveys. *Seminars in Arthritis and Rheumatism, 45*(4 Suppl), S22–S27.

Radnovich, R., Scott, D., Patel, A. T., Olson, R., Dasa, V., Segal, N., ... Metyas, S. (2017). Cryoneurolysis to treat the pain and symptoms of knee osteoarthritis: A multicenter, randomized, double-blind, sham-controlled trial. *Osteoarthritis and Cartilage, 25*(8), 1247–1256.

Santana Pineda, M. M., Vanlinthout, L. E., Moreno Martin, A., van Zundert, J., Rodriguez Huertas, F., & Novalbos Ruiz, J. P. (2017). Analgesic effect and functional improvement caused by radiofrequency treatment of genicular nerves in patients with advanced osteoarthritis of the knee until 1 year following treatment. *Regional Anesthesia and Pain Medicine, 42*(1), 62–68.

Sari, S., Aydin, O. N., Turan, Y., Ozlulerden, P., Efe, U., & Kurt Omurlu, I. (2018). Which one is more effective for the clinical treatment of chronic pain in knee osteoarthritis: Radiofrequency neurotomy of the genicular nerves or intra-articular injection? *International Journal of Rheumatic Diseases, 21* (10), 1772–1778.

Swiontkowski, M., & Heckman, J. D. (2019). Research in musculoskeletal pain management: Time to focus. *The Journal of Bone and Joint Surgery. American Volume, 101*(7), 571.

Tran, J., Agur, A., & Peng, P. (2020). Revisiting the anatomical evidence supporting the classical landmark of genicular nerve ablation. *Regional Anesthesia and Pain Medicine, 45*(5), 393–394.

Trang, T., Al-Hasani, R., Salvemini, D., Salter, M. W., Gutstein, H., & Cahill, C. M. (2015). Pain and poppies: The good, the bad, and the ugly of opioid analgesics. *The Journal of Neuroscience, 35*(41), 13879–13888.

van der Wees, P. J., Wammes, J. J., Akkermans, R. P., Koetsenruijter, J., Westert, G. P., van Kampen, A., ... Schreurs, B. W. (2017). Patient-reported health outcomes after total hip and knee surgery in a Dutch University Hospital Setting: Results of twenty years clinical registry. *BMC Musculoskeletal Disorders, 18*(1), 97.

Wright, E. A., Katz, J. N., Abrams, S., Solomon, D. H., & Losina, E. (2014). Trends in prescription of opioids from 2003-2009 in persons with knee osteoarthritis. *Arthritis Care & Research (Hoboken)*, *66*(10), 1489–1495.

Wylde, V., Hewlett, S., Learmonth, I. D., & Dieppe, P. (2011). Persistent pain after joint replacement: Prevalence, sensory qualities, and postoperative determinants. *Pain*, *152*(3), 566–572.

Zachariah, C., Mayeux, J., Alas, G., Adesina, S., Mistretta, O. C., Ward, P. J., ... Washington, A. (2020). Physiological and functional responses of water-cooled versus traditional radiofrequency ablation of peripheral nerves in rats. *Regional Anesthesia & Pain Management*, *45*(10), 792.

Zeng, C., Lane, N. E., Hunter, D. J., Wei, J., Choi, H. K., McAlindon, T. E., ... Zhang, Y. (2019). Intra-articular corticosteroids and the risk of knee osteoarthritis progression: Results from the osteoarthritis initiative. *Osteoarthritis and Cartilage*, *27*(6), 855–862.

Chapter 51

Nonpharmacologic analgesic therapies: A focus on photobiomodulation, acustimulation, and cryoanalgesia (ice) therapy

Roya Yumul[a,b], Ofelia L. Elvir Lazo[a], and Paul F. White[a,c]

[a]*Department of Anesthesiology, Cedars-Sinai Medical Center, Los Angeles, CA, United States;* [b]*Department of Anesthesiology, UCLA, Charles R. Drew University of Medicine and Science, Los Angeles, CA, United States;* [c]*The White Mountain Institute, The Sea Ranch, CA, United States*

Abbreviations

APS	American Pain Society
ASA	American Society of Anesthesiologists
ATP	adenosine 5′-triphosphate
CDC	Center for Disease Control and Prevention
CNS	central nervous system
EA	electroacupuncture
FDA	Federal Drug Administration
HILT	high-intensity laser therapy
LASER	light amplification by stimulated emission of radiation
LED	light-emitting diode
LLLT	low-level laser therapy
mRNA	messenger ribonucleic acid
OA	osteoarthritis
PBMT	photobiomodulation therapy
PENS	percutaneous electrical nerve stimulation
PNS	peripheral nerve stimulation
PONV	postoperative nausea and vomiting
RSD	reflex sympathetic dystrophy
TAES	transcutaneous acupoint stimulation
TCM	traditional Chinese medicine
TEA	transcutaneous electrical acupoints
TENS	transcutaneous electrical nerve stimulation
TMJ	temporomandibular joint
TNFα	tumor necrosis factor alpha

Introduction

The search for safe and effective pain treatments is ongoing. In the last decades, the use of opioid analgesic for postoperative pain management and for treating chronic pain has contributed to opioid dependence in many patients throughout the world. This is contributing to the global opioid epidemic. Opioid dependence continues growing despite the economic, social, and cultural issues. Lack of evidence on long-term effectiveness and safety concerns speaks to the need to consider alternative treatments of opioids for chronic pain. The 2016 Center for Disease Control and Prevention (CDC) guidelines for prescribing opioids for chronic pain recommend that nonopioid therapy is preferred for the treatment of chronic pain. A multimodal pharmacological (nonopioid) and nonpharmacological approach has been suggested for the management of

602 PART | IV Novel and nonpharmacological aspects and treatments

acute and chronic pain (White, 2017). In 2016, the American Pain Society (APS), in collaboration with the American Society of Anesthesiologists (ASA), has published a new set of guidelines for managing postoperative pain supporting the use of nonpharmacological modalities such as Photobiomodulation Therapy (PBMT) (White, 2017). The nonpharmacologic analgesic therapies which will be described in this chapter are PBMT, Acustimulation, and Cryoanalgesia.

Photobiomodulation therapy

Cold (nonsurgical) laser therapy is a type of PBMT that has been used clinically for over 50 years for the management of acute and chronic pain therapy. It has demonstrated to be effective when used as an adjuvant to both pharmacologic analgesics and physicotherapy, in enhancing analgesia and accelerating healing (de Freitas & Hamblin, 2016).

History

In 1903 Dr. Nils Finsen used concentrated light radiation to treat lupus vulgaris and was awarded a Nobel Prize for his contribution in 1903. Albert Einstein published his laser theory development, where he used the term "LASER" (Light Amplification by Stimulated Emission of Radiation) in 1916. Endre Mester, who applied low-level laser light to treat people with skin ulcers in 1967, is recognized as the grandfather of laser therapy (Mester et al., 1968). In 1987 companies were selling lasers to treat pain, accelerate wound healing, and reduce injuries without enough scientific evidence at that time (Avci et al., 2013). The Federal Drug Administration (FDA) approved the first cold laser for treating pain in 2002, and since then, Low-Level Laser Therapy (LLLT) was widely used in the United States. The latest development in cold laser therapy is the use of High-Intensity Laser Therapy (HILT) with the first scientific publication in 2009 (Santamato et al., 2009).

Mechanisms of PBMT

PBMT is a nonpharmacological intervention that triggers positive physiological effects. More specifically, it increases microcirculation, stimulation of the mitochondrial respiratory chain and function, increases Adenosine $5'$-triphosphate (ATP) synthesis, and factors that may influence the metabolism of various pathologies (Tomazoni et al., 2017). PBMT also regulates inflammatory mediators such as tumor necrosis factor-alpha (TNF-α), interleukins, and prostaglandin (Aver Vanin et al., 2016), contributing to pain relief and tissue repair (Chow, Johnson, Lopes-Martins, & Bjordal, 2009). It stimulates fibroblast proliferation and the synthesis of type I and III procollagen messenger ribonucleuc acid (mRNA) which facilitates wound revascularization and accelerates bone healing (de Freitas & Hamblin, 2016). Laser therapy promotes cells to absorb the released photons of light and converts the energy into ATP (Table 1). It then augments the production of antioxidants and heat shock proteins (de Freitas & Hamblin, 2016). The light-absorbing components of the cell are named chromophores (or photo-acceptor) and are contained within the mitochondria and cellular membrane. This allows the absorption of light in the near-infrared due to the heme and copper centers. These absorption peaks are generally in the red (600–700 nm) and near-infrared (760–940 nm) spectral regions. PBMT could increase the speed of recovery from conduction block or inhibit the A-delta- and C-fiber transmission by having a direct stimulatory effect on nerve structures (Chow, Armati, Laakso, Bjordal, & Baxter, 2011). Laser therapy can disturb neuronal physiology, modifying axonal flow and cytoskeleton organization (Chow & Armati, 2016). HILT has photochemistry effects that stimulate oxidation of mitochondria and ATP creation by delivering high energy output inside tissues. HILT can cause quick absorption of edema and removal of exudates through increased metabolism and blood circulation (Choi et al., 2017). These laser-induced modifications are entirely reversible with no residual nerve damage or side effects.

PBMT treatment parameters

PBMT is an easily administered, painless, noninvasive, and effective alternative therapy for a wide variety of medical and surgical conditions (e.g., arthritis, plantar fasciitis, and back pain) (Tables 2; White, Elvir Lazo, Galeas, & Cao, 2017; White, Elvir Lazo, & Yumul, 2019). The therapeutic effects of PBMT depend on multiple parameters (Table 2; Fulop et al., 2010). The findings in the peer-reviewed literature showed contradiction due to the lack of a complete description of parameters used. Laser and light-emitting diode (LED) categories utilized for treating pain belong to classes III and IV. LLLT devices (class III) use a red beam or near-infrared with a wavelength between 600 and 1000 nm (nm), and power from 5 to 500 mW (Tables 1–3). HILT devices, (class IV) with a power >1 W and longer wavelengths >800 nm (e.g., 1275–1280 nm) have a more powerful laser beam (20–75 W) (Alayat, Aly, Elsayed, & Fadil, 2017; Dundar, Turkmen,

TABLE 1 Terminology and definitions used in laser science and laser safety issues.

Term	Description	Comments
Photobiomodulation	The new term for LLLT, cold laser, laser therapy, and HILT	A form of light therapy that utilizes nonionizing light sources, including lasers, light-emitting diodes, and/or broadband light
Electromagnetic radiation	The radiant energy is released by certain electromagnetic processes	Includes radio waves, microwaves, infrared radiation, visible light, ultraviolet radiation, X-rays, and gamma rays
Electromagnetic spectrum	The range of wavelengths and frequencies of electromagnetic radiation	
Chromophore	The part of the molecule that captures light energy	
Cytochrome c	A component of the electron transport chain in mitochondria	
Laser wavelength (nm)	Wavelength refers to the physical distance between crests of successive waves in the laser beam	Medical laser wavelengths: Visible: 400–700 nm Near-infrared: 650 nm to 1350
Near-infrared (NIR) window	Therapeutic or optical window: The range of wavelengths from 650 to 1350 nm where light has its maximum depth of penetration into biological tissue	
Laser pulse energy (J)	Pulsed mode: the energy of laser pulse is a more reliable parameter for reporting the power of these lasers	Energy is measured in Joules
Continuous-wave	A laser runs at constant power output for a defined period of time (60–80 s)	
Laser power (W)	The rate at which energy is generated by the laser Radiant energy emitted per unit of time Laser power of 1 W: 1 J of energy in 1 s	
Pulse duration (ms) or pulse width	The temporal length of laser pulse; that is, the time during which the laser emits energy	
Peak power	Power level during an individual laser pulse	Peak power = pulse energy/pulse duration
Laser beam area orspot size (cm²)	Diameter of the laser beam on the target. By changing the distance of the laser head from the skin, the laser beam spot size will vary Area of the beam on a surface. Beam area will depend on the treatment head used and the distance the treatment head is held from the surface	

Continued

604 PART | IV Novel and nonpharmacological aspects and treatments

TABLE 1 Terminology and definitions used in laser science and laser safety issues—cont'd

Term	Description	Comments
Treatment area(cm^2)	Area of the skin surface that is treated	
Fluence (J/cm^2) Dose of energy or energy density	Quantity of energy (J) delivered to the treated area (cm^2)	Fluence = Energy/Area
Frequency (hertz [Hz]) (repetition rate)	Laser pulses are emitted periodically at a pulse rate, such as 10–40 pulses/sec. The Class III superpulsed lasers, e.g., Multi Radiance Medical Technology (Solon, Ohio), produce "high powered light (905 nm) in billionth of a second" pulses to create a high photon density. This type of LLLT device is alleged to generate a total power of up to 50,000 mW (comparable to a HILT device) Pulses per second = Hz	
Irradiance or power density (W/cm^2)	The output power of the laser is divided by the beam area. The power intensity at the surface of the skin	
Dose = Irradiance × exposure time (J/cm^2)	Total energy delivered is divided by the total treatment area.	
Absorption coefficient = absorbed energy/ laser energy	The amount of absorbed energy versus the total used energy	One of the most important optical features of the target tissue is its ability to absorb laser light

TABLE 1 Terminology and definitions used in laser science and laser safety issues—cont'd

Term	Description	Comments
Thermal relaxation time (TRT)	The time it takes for an object to cool down from 100°C to 50°C Rule: A smaller object cools faster than a larger object of the same material and shape, which means that the smaller target has a shorter TRT This fact is important when the tissue needs to be heated to the desired temperature at a certain fluence setting. If the pulse width is too long, the tissue will start cooling itself via thermal conduction prior to the completion of a pulse, causing a negative clinical effect	The second parameter that should be taken into consideration when estimating the TRT is the shape of the target tissue. A sphere (e.g., skin cells) having 360 degrees of cooling surface area cools faster than a cylinder (e.g., hair follicles). This allows a hair follicle to retain its heat while the skin cells can cool much more efficiently. For this reason, parameters can be selected to destroy the follicle without causing damage to the skin. For targeting smaller structures, shorter pulse duration and higher fluence are recommended
Laser safety issues	Eye protection	All persons in the operating room must wear safety eyewear. Light from the laser can cause severe corneal and retinal damage to the unprotected eye. Eyewear must have side shields and be worn over prescription glasses
	Reflection	Laser light is easily reflected, and care must be taken to ensure the beam is not directed toward shiny surfaces
	Electrical hazard	The interior of the laser machine contains high voltages and exposed invisible laser radiation. Only technicians trained in electrical and laser safety are authorized to perform internal maintenance

Modified from White, P., Elvir Lazo, O., & Yumul, R. (2019). Cold laser therapy for acute and chronic pain management: A comparison of low-level and high-intensity laser therapy devices. *Anesthesiology News*, 65–77.

Toktas, Ulasli, & Solak, 2015). With higher wavelengths, a deeper tissue penetration into the muscle and soft tissue is permitted because of reduced absorption of the laser beam by both melanin and hemoglobin (Fig. 1 and Table 2). As a result, it allows the treatment of degenerative joint conditions, musculoskeletal disorders, and greater tissue regions (White et al., 2019; Fig. 1).

The optimal laser dose needed to generate the required photo-stimulating effect on specific body regions is still unknown. If the stimuli are weak (underdosing), it generates little or no effect on cellular function. Moderate-to-strong laser stimuli positively enhance cellular function, while excessively strong stimuli (overdosing) can suppress cellular function and further damage the tissue due to the adverse effects of overheating of the tissue (Huang et al., 2015). The ability of a laser beam to penetrate tissue is determined by wavelength (Huang et al., 2015; Fig. 1 and Tables 1 and 2). Superficial tissue layers reduce the laser light effect by 50%–90% due to absorption into the subcutaneous tissues and the remaining energy penetrating into the deeper tissue layers.

Diodes with wavelengths between 400 and 700 nm transfer light energy to the epidermal and subdermal tissue layers (<1 cm), suitable for treating superficial wounds and skin disorders (Joensen et al., 2012). Diodes with wavelengths between 820 and 904 nm transmit light energy from 2 to 4 cm beyond the skin interface. Wavelengths (>1000 nm) are suitable for treating deep soft tissue disorders (e.g., muscle, joints, ligaments, deep fascia) (Enwemeka et al., 2004).

How is PBMT administered?

Laser can be applied in both short bursts (120–150 μs) or continuous at 60-s intervals, with a short duty cycle (0.1%) to avoid superficial tissue damage by harmful thermal accumulation. The pulsed release of light allows higher doses to reach deeper tissues, principally at short-pulsed, and low repetition rates (Gross et al., 2013). The administration of PBMT is a simple "point-and-shoot" technique at the affected tissue area(s) and/or nerves. The laser technician must be proficient in laser safety practices, knowledgeable in anatomy and physical therapy, and/or kinesiology. Once recognizing the painful region, the laser beam is pointed straight at the overlying skin area. Low-powered lasers used for LLLT can be positioned from direct contact to 3–5 in. from the skin surface. Superficial treatments usually last 5–20 min. The laser probe of HILT

606 PART | IV Novel and nonpharmacological aspects and treatments

TABLE 2 Characteristics of cold laser devices used for the administration of low-level laser therapy (LLLT) and high-intensity laser therapy (HILT) and clinical applications.

	LLLT	HILT
Laser class	I, Im, II, III	IV
Wavelength	600–980 nm	660–1280 nm
Power	<1 W	1–75 W
Penetration abilities	Low (<2 cm)	Deep (5–15 cm)
Temperature changes	<1.0°C	Low thermal accumulation
Clinical applications for acute and chronic pain	Acute and chronic pain related to the herpes virus	Bactericidal effects
	Bactericidal effects	Chronic lumbar radiculopathy
	Carpal tunnel syndrome	Fibromyalgia-related pain
	Dental pain	Hemophilic arthropathy
	Fibromyalgia-related pain	Low back pain
	Headache	Myofascial pain syndrome
	Low back pain	Neck pain
	Mucositis-associated pain	Opioid dependence
	Musculoskeletal back pain	Osteoarthritis
	Neck pain	Pain related to the herpes virus
	Neuropathic pain	Postburn pruritus
	Opioid dependence	Postoperative pain
	Osteoarthritis	Shoulder pain
	Pain due to acute muscle injury	Note: Several of the applications listed for LLLT have also been successfully treated with HILT. Although some of these applications have not been formally studied using a HILT device, there is no reason to expect that these conditions would not respond as well or even better compared with LLLT
	Plantar fasciitis	
	Postoperative pain	
	Shoulder pain	
	Trigeminal neuralgia	
	Wound repair	

Modified from White, P. F., Elvir Lazo, O. L., Galeas, L., & Cao, X. (2017). Use of electroanalgesia and laser therapies as alternatives to opioids for acute and chronic pain management. *F1000Research, 6*, 2161. Copyright permission has been obtained.

devices must be placed to a bigger distance from the skin to avoid treated area overheating, usually 8–12 in. from the skin surface. The painful body area(s) receives a series of 60 s. treatments located approximately 3–6 in. separately to cover completely the symptomatic area (distance varies on the laser beam width).

With the HILT devices, the width of the laser beam varies from 2.5–5 cm; reducing the treatment time to 60–90 s at each point will avoid "overstimulating" the tissue, which can result in tissue fatigue and liberation of muscle enzymes. Every treatment session can last from 10 to 45 min, depending on the affected area and/or the number of symptomatic painful areas. To optimize the management of a tendon or joint, the patients may be asked to perform simple range of motion exercises immediately before, during, and after the laser treatment session. The analgesia obtained after a single laser treatment session lasts for 2–4 days and can be extended for up to 7–10 days or longer with recurring sessions. Selected acute injuries frequently resolve following 1–2 treatment sessions, chronic pain conditions improve after 8–12 treatment sessions throughout a 3–4 week period, followed by a maintenance treatment every 3–5 weeks.

TABLE 3 Commercially available cold laser systems.

Brand name	Wattage	Wavelength (nm)
Apollo	0.5–20 W	810
Aspen Summit	1–60 W	810–980
Avant	0.66–1.4 W	808
BioLase Epic	10 W	940
BioLight Aura PTL	5 mW	635
BTL 6000 High Intensity Laser	10 W	1064
CureWave HILT	1–44 W	1280
Cutting Edge	1.1–3.3 W	810
Erchonia XLR8 laser	7.5–20 mW laser diodes	635
Ga-Al-As laser	50 mW	809
Infrared laser IR810/400 probe	400 mW	808
Irradia Mid-Laser	120 mW	660
	20 W	904
K-Laser	6–20 W	660–980
Light Force/LiteCure	10–25 W	810/980
Lumix Series 2–3	45–250 W	810–910
Multiradiance LaserStim	7.5 mW	660
	25 W	905
Nd:YAG laser	0.25 W	1064
Nexus	10 W	810–980
Phoenix Thera-Lase	37–75 W	1275
Pilot Diode Laser	9 W	810–980
R650/50 probe	50 mW	658
TerraQuant	15–50 W	660–905
THOR Laser	200 mW	660
	2 W	810

Modified from White, P., Elvir Lazo, O., & Yumul, R. (2019). Cold laser therapy for acute and chronic pain management: A comparison of low-level and high-intensity laser therapy devices. *Anesthesiology News*, 65–77. Copyright permission has been obtained.

In patients with chronic pain, it has been seen that HILT and Acustimulation presented similar pain-relieving effects when using needles at either acupoints and dermatomes, but HILT showed longer-lasting favorable effects (White, Elvir Lazo, Galeas, et al., 2017). When used for postoperative pain treatment and faster wound healing, the lasers beams are targeted at the surgical incision sites, painful trigger points and acupoint sites correlating with the painful areas. Also, HILT is used to treat the paraspinous region of the spinal column, peripheral nerves innervating areas of numbness and pain (e.g., neuropathic pain), and/or broad soft tissue coverage to decrease identified pain and inflammation, thus stimulating tissue and peripheral nerve healing (de Andrade, Bossini, & Parizotto, 2016).

PBMT side effects

In contrast to pharmacotherapy, laser therapy has minor contraindications and has gained great patient acceptance due to the low incidence of side effects. The only well-known side effect of cold laser therapy is transient skin discoloration (redness,

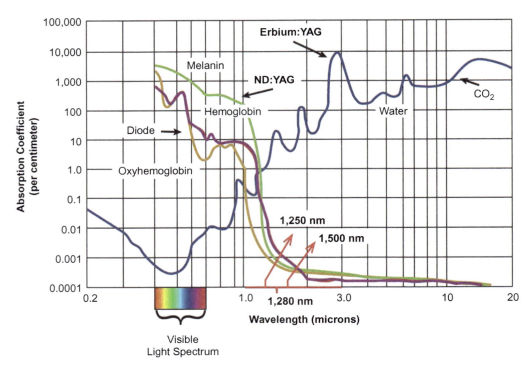

FIG. 1 This graph illustrates the effect of different wavelengths of the laser beam on the absorption of infrared light by water, hemoglobin, oxyhemoglobin, and melanin. High-intensity laser therapy (HILT) devices like Phoenix Thera-lase and CureWave, which function at longer wavelengths (>1250 nm), have markedly reduced absorption by melanin, hemoglobin, and oxyhemoglobin compared with HILT devices like LightForce (LiteCure; 980 nm). *(Modified from White, P. F., Elvir Lazo, O. L., Galeas, L., & Cao, X. (2017). Use of electroanalgesia and laser therapies as alternatives to opioids for acute and chronic pain management. F1000Research, 6, 2161. Copyright permission has been obtained.)*

hypo, or hyper pigmentation), and a "warming sensation" if the hand-held laser head comes near the skin's surface (<8 in.). Special protective goggles should be worn by both patient and operator during the treatment sessions to avoid direct exposure of the laser beam with the eye. These goggles should block the wavelength of the laser system in use and can be purchased from laser safety manufacturers.

PBMT clinical applications

Acute and chronic pain related to herpes simplex and herpes zoster infections

Decreased pain, viral titer, recurrence of herpes labialis, and postherpetic neuralgia, aided healing of skin lesions, improving quality of life.

Acute dental and oral-associated pain: Decreases pain, soft tissue swelling, and antiinflammatory meds after oral surgery, improves orthodontic treatment and prevents relapse.

Acute joint and soft tissue injuries: Delays the onset, and improves the postexercise fatigue in athletes. It is successfully effective for acute lateral epicondylitis, tendinopathies, and acute joint injuries.

Acute postoperative pain: LLLT reduced postoperative pain. HILT showed to be an effective therapy for weaning off opioid-dependent patients after surgery off from their opioid medications (White, Elvir-Lazo, & Hernandez, 2017).

Burn-related pain: HILT reduced postburn pruritus and pain, improving functionality and quality of life.

Chronic neck pain: PBMT reduced acute and chronic neck pain, improving range of motion, and functionality.

Chronic low back pain: PBMT reduced musculoskeletal back pain, improving functionality and disability.

Carpal tunnel syndrome: Reduced pain, symptom severity, and improved functionality.

Diabetic peripheral neuropathic pain: Decreased neuropathic pain, accelerated tissue repair, improving quality of life.

Fibromyalgia-related pain: Reduced pain, accelerating tissue healing, and improving quality of life.

Headache pain: Reduced pain of chronic headaches.

Hemophilia-induced arthritis: HILT reduced pain, gait functioning, and functional capacity.

Mucositis, leukoplakia and pain-related to radiation therapy: Decreased pain.

Neuropathic pain, allodynia, and radiculopathy: Decreased pain sensitivity, improved neurological function, and quality of life.

Osteoarthritis (OA): Decreased pain, enhanced physical disability/activity, and extended the time to needing surgery for knee replacement.

Pain due to trigeminal neuralgia: Reduced chronic neuropathic pain.

Plantar fasciitis: PBMT reduced pain, improved functionality, and quality of life. However, LLLT was more beneficial in regards to the short and long-term.

Temporomandibular Joint (TMJ) related pain: LLLT decreased pain, improved range of motion, and tension headaches.

Treatment of opioid dependency: Multimodal pain therapy as such as nonopioid and nonpharmacologic analgesic therapies like PBMT should be considered for patients with chronic pain (White, 2017).

Applications to other areas

Androgenetic alopecia, skin conditions ophthalmic, neurological, neurodegenerative, and psychiatric disorders (Hamblin, 2018; Hipskind et al., 2019).

Conclusion

PBMT gained high patient acceptance and satisfaction in acute and chronic pain therapy due to its efficacy and absence of drug-related side effects. The usage of PBMT as part of multimodal pain management continues to be underutilized by the medical multidisciplinary community due to the lack of education in medical training of benefits of this nontraditional therapy, high cost of equipment, and low reimbursement by 3rd party payers despite the fact it is very safe and effective compared to long term use of common pharmaceutical therapies.

Acustimulation

Traditional Chinese medicine (TCM) in western countries has been used widely and effectively to treat several pain disorders such as low back pain and migraines (Fung, 2009). The Chinese medical acupuncture theory suggests that Yin and Yang are two opposing and complementary forces that coexist in nature and interact to regulate the flow of "vital energy," identified as Qi. Qi flows throughout a network of channels called meridians (12; 6 Yin and 6 Yang), which bring Qi from the internal organs to the skin surface. Acupoints are the body's surface sites in which the Qi of organs and meridians assembled to act as target and response points of treatment (Deng & Shen, 2013).

Acupoints can be stimulated to modulate the body physiology to correct the imbalance and restore the body to normal health, such as reducing postoperative nausea and vomiting (PONV), improving pain, lowering blood pressure, avoiding arrhythmic recurrences, and improving the overall quality of life (Deng & Shen, 2013; Li et al., 2015; Wang, Kain, & White, 2008). Acupoints can be stimulated using moxibustion, mechanical stimulation of the needle, electrical current stimulation of electroacupuncture (EA), and the radiated stimulation of laser acupuncture (Li et al., 2015). Acustimulation within the central nervous system (CNS) generates the release of neurotransmitters and endogenous opioid-like substances and activates c-fos when inducing direct inhibitory effects on peripheral nerves, which reduces acute pain input into the CNS. In chronic pain patients, laser-induced changes in the spinal cord produce longer-term suppression of pain in the CNS (Li et al., 2015).

Acupuncture

Acupuncture intervention consists of applying pressure, heat, needling, and electrical stimulation to specific acupuncture points. Acupuncture and/or acupressure has shown to be effective in reducing a variety of pain such as cancer pain (also decreasing the analgesics use), early stages of labor, migraine headaches, pancreatic pain, musculoskeletal pain, endometriosis, pain in children with sickle cell disease, knee osteoarthritis pain, dysmenorrhea and the short-term management of low back pain (He et al., 2020; Kwon, Pittler, & Ernst, 2006; Wang et al., 2008). Chronic nociceptive low back pain showed a long-term pain-relieving effect using needle acupuncture, decreased analgesic intake, and improvement in quality of sleep (Carlsson & Sjölund, 2001). Psychosocial factors for pain results such as patients' beliefs about acupuncture and the therapeutic relationship are significant predictors of treatment and for facilitating good clinical outcomes (Wang et al., 2008).

Laser acupuncture has shown long term effectiveness for treating musculoskeletal conditions, for the relief of chronic tension headaches, improved pain after pediatric percutaneous kidney biopsies, and can be an effective alternative to conventional needling in patients who are needle phobic or applied to complicated points (Fernandes, Duarte Moura, Da Silva, De Almeida, & Barbosa, 2017; Oates et al., 2017).

Acupuncture moxibustion has been reported to mitigate all types of pain such as primary dysmenorrhea, Reflex Sympathetic Dystrophy (RSD) in the early stages, knee osteoarthritis, TMJ disturbance syndrome, soft tissue injury, heel pain, swelling pain, and pain related to herpes zoster (Deng & Shen, 2013).

Applications to other areas

Pregnancy-induced hyperemesis, chemotherapy, PONV.
    Major postoperative symptoms (e.g., reduced the time of first flatus, defecation, ambulation, and hospital stay).
    Decreased blood pressure, improved dyspnea and fatigue, and reduced insomnia.

Conclusion

Acustimulation in all its modalities should be widely used as an adjuvant to multimodal analgesic approach for the management of a variety of acute and chronic pain syndromes.

The treatments are practical and easy to perform. Nevertheless, higher quality clinical evidence is required to establish efficacy.

Electroanalgesia

This term is used for a variety of different types of neuromodulation therapies, including Transcutaneous Electrical Nerve Stimulation (TENS), Transcutaneous Acupoint Stimulation (TAES), ultrasound-guided acupotomy, Electro-acupuncture (EA), Peripheral Nerve Stimulation (PNS), and Percutaneous Electrical Nerve Stimulation (PENS). PNS may help to select candidates for implantation of a spinal cord stimulator in patients who experience significant improvement in pain and disability (Chakravarthy, Nava, Christo, & Williams, 2016).

TENS applied to acupoints has successfully reduced pain and improved patient satisfaction during small procedures (Lisón et al., 2017). TENS reduced pain and improved physical activity in a variety of chronic pain syndromes for example temporal-mandibular disorders, myofascial pain syndrome, neck/low back pain, osteoarthritis/gonarthritis, abdominal/pelvic pain, and pain (and opioid consumption) in mechanically ventilated patients (White, Elvir Lazo, Galeas, et al., 2017). Needleless Transcutaneous Electrical Acupoints (TEA) improved major postoperative symptoms and pain by enhancing vagal and suppressing sympathetic activities (Zhang et al., 2018). However, there are some studies that have reported that TAES/TENS offered low-quality evidence or no significant advantage supporting its effectiveness (Page et al., 2016).

EA has effectively treated pain in chronic rotator cuff disease, fibromyalgia, and chronic low back pain (Page et al., 2016; Salazar, Stein, Marchese, Plentz, & Pagnussat, 2017; White, Elvir Lazo, Galeas, et al., 2017). Also reduced inflammatory mediators' levels, intraoperative anesthetic/analgesic requirements, and residual postoperative cognitive dysfunction (Zhang et al., 2017). PENS reduced the pain level of chronic low back pain, sciatica, myofascial chronic neck pain, postsurgical pain, neck pain, knee osteoarthritis, headaches, peripheral neuropathic pain, fibromyalgia, and enhanced exercise therapy (Qi et al., 2016; White, Elvir Lazo, Galeas, et al., 2017). Acupotomy EA was effective for pain control and functional recovery in patients with knee osteoarthritis (Ding et al., 2016).

The duration of pain relief after each electroanalgesia treatment session is fairly short-lived (<48 h), after a series of treatments there is some "cumulative" benefit, consequently, maintenance therapy is typically required. Regardless of the peer-reviewed literature support, clinical value, minimally invasive application, and characteristics (Table 4) of these techniques have failed to gain widespread clinical acceptance due to time-consuming therapy sessions and extremely low insurance reimbursement compared to other medical procedures.

Cryoanalgesia (ice) therapy

Surface cooling utilizing ice has been used for centuries as a pain relief technique. Cryoanalgesia therapy is a noninvasive, simple, cheap, and convenient option, with a very low frequency of adverse side effects. It is considered a safe method that has a strong short-term analgesic effect in several painful conditions, principally in the musculoskeletal system. Cryotherapy is applied over the skin of a painful and/or inflamed area generating a reduction in the temperature, slowing

TABLE 4 Comparison of typical characteristics and applications of transcutaneous and percutaneous electroanalgesia techniques for treating acute and chronic pain.

	Transcutaneous electroanalgesia techniques	Percutaneous electroanalgesia techniques
Examples	Transcutaneous electrical nerve stimulation (TENS) Acupoint-like transcutaneous electrical nerve stimulation	Electroacupuncture (EA) Percutaneous electrical nerve stimulation (PENS) Peripheral nerve stimulation
Application	Noninvasive Cutaneous pads (disposable)	Minimally invasive Acupuncture needles inserted through the skin
Duration of pain relief	Limited short-term benefits (<24 h)	Short-term and some longer-term benefits (24–72 h)
Cost	Low (self-administered)	High (personnel required)
Time required to perform techniques	Minimal	More labor-intensive and time-consuming to perform
Applications for acute and chronic pain management	Acute postoperative pain, low back pain, neck pain, osteoarthritis/gonarthritis, abdominal/pelvic pain, myofascial pain syndrome, chronic rotator cuff diseases, and temporomandibular disorders	Acute postsurgical pain, low back pain, sciatica, neck pain, knee osteoarthritis, osteoarthritis/gonarthritis, headaches, peripheral neuropathic pain, and fibromyalgia

Modified from White, P. F., Elvir Lazo, O. L., Galeas, L., & Cao, X. (2017). Use of electroanalgesia and laser therapies as alternatives to opioids for acute and chronic pain management. *F1000Research, 6*, 2161. Copyright permission has been obtained.

the pain signal transmission by reducing the velocity of nerve conduction in C and A-delta fibers consequently elevating the pain threshold and producing an antinociceptive effect (Attia & Hassan, 2017). The physiological effects of cryotherapy also include vasoconstriction and reductions in blood flow, inflammation, edema, muscle spasm, and metabolic demand, therefore, reducing the oxygen demand of cells and limiting the tissue free radical production. When the tissue temperature changes, the pain receptors called thermoreceptors (temperature-sensitive nerve endings) are activated and can block nociception within the spinal cord (Keskin, Özdemir, Uzun, & Güler, 2017; Malanga, Yan, & Stark, 2015).

Cryoanalgesia administration

Cryoanalgesia therapy has several administering choices including first-generation like crushed ice in a plastic bag, cold or gel packs; second-generation with circulating ice water with or without compression; and third-generation advanced computer-assisted devices with specific continuous controlled-temperature (11°C) for prolonged use of time. There is no standardization or precise technique of crucial factors such as the time period, duration, application mode, and cold agent for cryotherapy administration. There have been some general descriptions such as:

(a) during the perioperative period applied an ice pack for a continuous 120 min after surgery over the incision and once the ice had melted the pack was replaced as requested by staff and patients. Or after knee surgery applied 3 ice packs (2 anterior and one posterior) for 15 min on arrival to PACU and again on arrival to the ward, then 2 h and 4 h after surgery.
(b) For ankle pain applied 2 cm^{-3} cm of ice in a plastic bag.
(c) For pain related to mucositis held ice chips in the mouth.
(d) In root canal procedure cold saline irrigation at 2.5°C as final irrigant.
(e) For mandibular third molars removal applied intermittent cryotherapy for 30 min during the first postoperative days.
(f) For oral mucosa pain after dental injection applied ice for 1 min.
(g) For postpartum perineal pain applied for 10 min a single ice-pack after spontaneous vaginal birth (Hindocha, Manhem, Bäckryd, & Bågesund, 2019).

Clinical applications

Cryotherapy has reduced pain tolerance at the ankle, pain at the puncture site of Arterio-Venous Fistula in adults and children, pain rheumatoid arthritis, pain for vaccination in children, pain associated with mucositis, pain and scrotal edema, orthopedic trauma, sports injuries, acute low back pain, acute musculoskeletal injury, postpartum perineal pain after spontaneous vaginal

612 PART | IV Novel and nonpharmacological aspects and treatments

birth, pain after root canal procedure, teeth restoration/removal and dental injection, facial swelling/trismus related pain, post mandibular third molars removal, postoperative pain (hernioplasty, knee arthroplasty, gynecological, craniotomy, eyelid edema, and facial ecchymosis) (Guillot et al., 2014; Idayu Mat Nawi, Lei Chui, Wan Ishak, & Hsien Chan, 2018; Peres et al., 2017). It has also shown to prevent or decrease edema after injury and to reduce pain medication due to the corticosteroid and nonsteroidal antiinflammatory drug dose-sparing effects in renal patients. However, some authors suggested creating a standard randomized controlled trial to conclude the effectiveness of cryotherapy (Dantas et al., 2019).

Conclusion

Cryoanalgesia therapy is noninvasive and has been used as a complementary therapeutic option to reduce pain. The low frequency of its adverse effects makes it simple, convenient, and cheap to use. Ice application is a safe technique with a strong short-term analgesic effect in a wide range of painful conditions.

Other agents of interest

Pain is a complex phenomenon, in part due its multidimensional and subjective experience, which consists of sensory, affective, behavioral, and cognitive elements. In this chapter, we have described photobiomodulation, acustimulation, and cryoanalgesia (ice) therapy, but we would like to list some of the other nonpharmacologic analgesic therapies/tools used for pain management.

Mindfulness meditation (MM) is a nonelaborative, nonjudgmental awareness of the present-moment experience, which improves mechanisms supporting cognitive control, positive mood, emotion regulation, and acceptance. Buddhist monks for thousands of years have stated that MM can significantly alter the subjective experience of pain and reduce it. In the past three decades, scientists studied the processes underlying MM induced pain relief and health improvements, which engage multiple unique brain mechanisms that attenuate the subjective experience of pain in experimental and clinical settings. This mental practice teaches how to slow down racing thoughts, let go of negativity, and calm both mind and body by fully focusing on "the now." Such an exercise allows for acknowledgement and acceptance of thoughts, feelings, and sensations without judgment. MM improves many cognitive and health outcomes such as anxiety, depression, stress, pain, sleep, immunity, cognition, self-regulation, and affective and social functioning (Zeidan & Vago, 2016). MM can be practiced anywhere; it is inexpensive and low risk; it reduces opioid requirement and overall healthcare costs, and it improves the quality of life in patients with chronic pain conditions.

Cognitive behavioral interventions include a complex clinical interaction and makes use of a wide range of techniques. This intervention focuses on modifying cognition and behavior with the objective of decreasing pain sensations. This treatment includes counseling, relaxation training, deep breathing, guided imagery, positive reframing, and specific mindfulness interventions. Adding this technique to a multimodal treatment reduces pain scores, pain medication used, PACU length of stay, and healthcare costs. It also improves the function and quality of life in patients with chronic pain conditions (Frances, Shawyer, Cayoun, Enticott, & Meadows, 2020).

Music therapy (MT) is a beneficial intervention that has been demonstrated to reduce pain perception when used as part of a multidisciplinary pain therapy. It has been shown to be effective in reducing anxiety and pain, supporting relaxation, decreasing sedation medication used during procedures, and improving patient comfort and satisfaction (Bechtold, Perez, Puli, & Marshall, 2006; Sonke et al., 2015). The effects of music on the brain, such as its neurochemical changes and powerful emotional qualities have already been studied by many researchers and is an area of further investigation. MT may promote mental health and healing by activating self-repair mechanisms found in the brain and body. During MT, patients select music they find especially satisfying. MM is readily available, low risk, inexpensive, and does not require intense training by staff. It also reduces costs of medical care with few, if any, side effects (Bernatzky, Presch, Anderson, & Panksepp, 2019).

Other techniques used as an adjuvant therapy to control pain, such as biofeedback techniques, virtual reality distraction analgesia, massage therapy, aroma therapy, and distraction therapy, should be incorporated as part of multimodal pain management, to reduce pharmacotherapy induced adverse effects and improve patient care and satisfaction.

Mini-dictionary of terms

Laser: An acronym created from Light Amplification by Stimulated Emission of Radiation, defined by Albert Einstein.
Laser wavelength: It refers to the physical distance between crests of successive waves in the laser beam.
Acupoints: Specific body points are used in acustimulation (e.g., acupuncture) where pressure or needle is applied as a treatment to correct the imbalance and restore the body's normal health.

TENS: Transcutaneous Electrical Nerve Stimulation.
PNS: Peripheral Nerve Stimulation.

Key facts of nonpharmacologic analgesic therapies

- Cold laser therapy has been available since the 1960s, but it was not approved by the Food and Drug Administration (FDA) until 2002.
- The administration of cold laser therapy is simple, painless, noninvasive, and effective in a wide range of acute and chronic pain conditions.
- PBMT continues gaining high patient acceptance and satisfaction due to its efficacy, efficacious nonpharmaceuticals, and absence of side effects.
- Acustimulation can be used in children and adults.
- Cryoanalgesia Therapy or ice application is noninvasive, simple, cheap, convenient, and a safe method with a strong short-term analgesic effect in several painful conditions.

Summary points

- This chapter describes three nonpharmacologic analgesic therapies: photobiomodulation therapy (PBMT) or cold laser, acustimulation, and cryoanalgesia therapy.
- 80% of patients after elective surgery experience mild-to-severe pain, and in certain settings may conduce to chronic pain in 40% of these cases.
- Disproportionate reliance on opioid analgesics for treating postoperative pain plays an important role in opioid dependence.
- PBMT light therapy with an advantage over other antiinflammatory and pain medications uses lasers to provide analgesic, intracellular photochemical changes, and bio-stimulating effects, to modulate the antiinflammatory response, promoting tissue regeneration and wound healing.
- Acupoints are the body's surface spots, in which the Qi of organs and meridians are assembled, and act as target points and response points of treatment.
- Novel nonpharmacologic therapies like PBMT, acustimulation, and cryoanalgesia therapy could prove to be safe and cost-effective adjuncts to pharmacotherapy and physical therapy for managing acute and chronic pain.
- The interest among patients in nonpharmacologic therapies behoves pain specialists and anesthesiologists to be familiar with the benefits of these complementary treatments.

References

Alayat, M. S. M., Aly, T. H. A., Elsayed, A. E. M., & Fadil, A. S. M. (2017). Efficacy of pulsed Nd:YAG laser in the treatment of patients with knee osteoarthritis: A randomized controlled trial. *Lasers in Medical Science, 32*(3), 503–511.

Attia, A. A. M., & Hassan, A. M. (2017). Effect of cryotherapy on pain management at the puncture site of arteriovenous fistula among children undergoing hemodialysis. *International Journal of Nursing Sciences, 4*(1), 46–51.

Avci, P., Gupta, A., Sadasivam, M., Vecchio, D., Pam, Z., Pam, N., & Hamblin, M. R. (2013). Low-level laser (light) therapy (LLLT) in skin: Stimulating, healing, restoring. *Seminars in Cutaneous Medicine and Surgery, 32*(1), 41–52.

Aver Vanin, A., De Marchi, T., Tomazoni, S. S., Tairova, O., Leão Casalechi, H., de Tarso Camillo de Carvalho, P., … Leal-Junior, E. C. (2016). Pre-exercise infrared low-level laser therapy (810nm) in skeletal muscle performance and postexercise recovery in humans, what is the optimal dose? A randomized, double-blind, placebo-controlled clinical trial. *Photomedicine and Laser Surgery, 34*(10), 473–482.

Bechtold, M. L., Perez, R. A., Puli, S. R., & Marshall, J. B. (2006). Effect of music on patients undergoing outpatient colonoscopy. *World Journal of Gastroenterology, 12*(45), 7309–7312. https://doi.org/10.3748/wjg.v12.i45.7309.

Bernatzky, G., Presch, M., Anderson, M., & Panksepp, J. (2019). Emotional foundations of music as a non-pharmacological pain management tool in modern medicine. *Neuroscience and Biobehavioral Reviews, 35*(9), 1989–1999. https://doi.org/10.1016/j.neubiorev.2011.06.005.

Carlsson, C. P., & Sjölund, B. H. (2001). Acupuncture for chronic low back pain: A randomized placebo-controlled study with long-term follow-up. *The Clinical Journal of Pain, 17*(4), 296–305.

Chakravarthy, K., Nava, A., Christo, P. J., & Williams, K. (2016). Review of recent advances in peripheral nerve stimulation (PNS). *Current Pain and Headache Reports, 20*(11), 60.

Choi, H.-W., Lee, J., Lee, S., Choi, J., Lee, K., Kim, B.-K., & Kim, G.-J. (2017). Effects of high intensity laser therapy on pain and function of patients with chronic back pain. *Journal of Physical Therapy Science, 29*(6), 1079–1081.

Chow, R. T., & Armati, P. J. (2016). Photobiomodulation: Implications for anesthesia and pain relief. *Photomedicine and Laser Surgery, 34*(12), 599–609.

Chow, R. T., Johnson, M. I., Lopes-Martins, R. A. B., & Bjordal, J. M. (2009). Efficacy of low-level laser therapy in the management of neck pain: A systematic review and meta-analysis of randomised placebo or active-treatment controlled trials. *Lancet, 374*, 1897–1908.

Chow, R., Armati, P., Laakso, E. L., Bjordal, J. M., & Baxter, G. D. (2011). Inhibitory effects of laser irradiation on peripheral mammalian nerves and relevance to analgesic effects: A systematic review. *Photomedicine and Laser Surgery, 29*(6), 365–381.

Dantas, L. O., Moreira, R. D. F. C., Norde, F. M., Mendes Silva Serrao, P. R., Alburquerque-Sendín, F., & Salvini, T. F. (2019). The effects of cryotherapy on pain and function in individuals with knee osteoarthritis: A systematic review of randomized controlled trials. *Clinical Rehabilitation, 33*(8), 1310–1319.

de Andrade, A. L. M., Bossini, P. S., & Parizotto, N. A. (2016). Use of low level laser therapy to control neuropathic pain: A systematic review. *Journal of Photochemistry and Photobiology. B, Biology, 164*, 36–42.

de Freitas, L. F., & Hamblin, M. R. (2016). Proposed mechanisms of photobiomodulation or low-level light therapy. *IEEE Journal of Selected Topics in Quantum Electronics, 22*(3), 7000417.

Deng, H., & Shen, X. (2013). The mechanism of moxibustion: Ancient theory and modern research. *Evidence-based Complementary and Alternative Medicine: Ecam, 2013*, 379291.

Ding, Y., Wang, Y., Shi, X., Luo, Y., Gao, Y., & Pan, J. (2016). Effect of ultrasound-guided acupotomy vs electro-acupuncture on knee osteoarthritis: A randomized controlled study. *Journal of Traditional Chinese Medicine, 36*(4), 450–455.

Dundar, U., Turkmen, U., Toktas, H., Ulasli, A. M., & Solak, O. (2015). Effectiveness of high-intensity laser therapy and splinting in lateral epicondylitis; a prospective, randomized, controlled study. *Lasers in Medical Science, 30*(3), 1097–1107.

Enwemeka, C. S., Parker, J. C., Dowdy, D. S., Harkness, E. E., Sanford, L. E., & Woodruff, L. D. (2004). The efficacy of low-power lasers in tissue repair and pain control: A meta-analysis study. *Photomedicine and Laser Surgery, 22*(4), 323–329.

Fernandes, A. C., Duarte Moura, D. M., Da Silva, L. G. D., De Almeida, E. O., & Barbosa, G. A. S. (2017). Acupuncture in temporomandibular disorder myofascial pain treatment: A systematic review. *Journal of Oral & Facial Pain and Headache, 31*(3), 225–232.

Frances, S., Shawyer, F., Cayoun, B., Enticott, J., & Meadows, G. (2020). Study protocol for a randomized control trial to investigate the effectiveness of an 8-week mindfulness-integrated cognitive behavior therapy (MiCBT) transdiagnostic group intervention for primary care patients. *BMC Psychiatry, 20* (1), 7. 1–13 https://doi.org/10.1186/s12888-019-2411-1.

Fulop, A. M., Dhimmer, S., Deluca, J. R., Johanson, D. D., Lenz, R. V., Patel, K. B., … Enwemeka, C. S. (2010). A meta-analysis of the efficacy of laser phototherapy on pain relief. *The Clinical Journal of Pain, 26*(8), 729–736.

Fung, P. C. (2009). Probing the mystery of Chinese medicine meridian channels with special emphasis on the connective tissue interstitial fluid system, mechanotransduction, cells durotaxis and mast cell degranulation. *Chinese Medicine, 4*(10), 1–6.

Gross, A. R., Dziengo, S., Boers, O., Goldsmith, C. H., Graham, N., Lilge, L., … White, R. (2013). Low level laser therapy (LLLT) for neck pain: A systematic review and meta-regression. *The Open Orthopaedics Journal, 7*, 396–419.

Guillot, X., Tordi, N., Mourot, L., Demougeot, C., Dugué, B., Prati, C., & Wendling, D. (2014). Cryotherapy in inflammatory rheumatic diseases: A systematic review. *Expert Review of Clinical Immunology, 10*(2), 281–294.

Hamblin, M. R. (2018). Photobiomodulation for traumatic brain injury and stroke. *Journal of Neuroscience Research, 96*(4), 731–743.

He, Y., Guo, X., May, B. H., Zhang, A. L., Liu, Y., Lu, C., … Zhang, H. (2020). Clinical evidence for association of acupuncture and acupressure with improved cancer pain: A systematic review and meta-analysis. *JAMA Oncology, 6*(2), 271–278.

Hindocha, N., Manhem, F., Bäckryd, E., & Bågesund, M. (2019). Ice versus lidocaine 5% gel for topical anaesthesia of oral mucosa—A randomized cross-over study. *BMC Anesthesiology, 19*(1), 227.

Hipskind, S. G., Grover, F. L., Fort, T. R., Helffenstein, D., Burke, T. J., Quint, S. A., … Hurtado, T. (2019). Pulsed transcranial red/near-infrared light therapy using light-emitting diodes improves cerebral blood flow and cognitive function in veterans with chronic traumatic brain injury: A case series. *Photobiomodulation, Photomedicine, and Laser Surgery, 37*(2), 77–84.

Huang, Z., Ma, J., Chen, J., Shen, B., Pei, F., & Kraus, V. B. (2015). The effectiveness of low-level laser therapy for nonspecific chronic low back pain: A systematic review and meta-analysis. *Arthritis Research & Therapy, 17*, 360.

Idayu Mat Nawi, R., Lei Chui, P., Wan Ishak, W. Z., & Hsien Chan, C. M. (2018). Oral cryotherapy: Prevention of oral mucositis and pain among patients with colorectal cancer undergoing chemotherapy. *Clinical Journal of Oncology Nursing, 22*(5), 555–560.

Joensen, J., Ovsthus, K., Reed, R. K., Hummelsund, S., Iversen, V. V., Lopes-Martins, R.Á. B., & Bjordal, J. M. (2012). Skin penetration time-profiles for continuous 810 nm and superpulsed 904 nm lasers in a rat model. *Photomedicine and Laser Surgery, 30*(12), 688–694.

Keskin, C., Özdemir, Ö., Uzun, İ., & Güler, B. (2017). Effect of intracanal cryotherapy on pain after single-visit root canal treatment. *Australian Endodontic Journal, 43*(2), 83–88.

Kwon, Y. D., Pittler, M. H., & Ernst, E. (2006). Acupuncture for peripheral joint osteoarthritis: A systematic review and meta-analysis. *Rheumatology (Oxford, England), 45*(11), 1331–1337.

Li, F., He, T., Xu, Q., Lin, L.-T., Li, H., Liu, Y., … Liu, C.-Z. (2015). What is the acupoint? A preliminary review of acupoints. *Pain Medicine, 16*(10), 1905–1915.

Lisón, J. F., Amer-Cuenca, J. J., Piquer-Martí, S., Benavent-Caballer, V., Biviá-Roig, G., & Marín-Buck, A. (2017). Transcutaneous nerve stimulation for pain relief during office hysteroscopy: A randomized controlled trial. *Obstetrics and Gynecology, 129*(2), 363–370.

Malanga, G. A., Yan, N., & Stark, J. (2015). Mechanisms and efficacy of heat and cold therapies for musculoskeletal injury. *Postgraduate Medicine, 127* (1), 57–65.

Mester, E., Ludány, G., Sellyei, M., Szende, B., Gyenes, G., & Tota, G. J. (1968). Studies on the inhibiting and activating effects of laser beams. *Langenbecks Archiv Fur Chirurgie, 322*, 1022–1027.

Oates, A., Benedict, K. A., Sun, K., Brakeman, P. R., Lim, J., & Kim, C. (2017). Laser acupuncture reduces pain in pediatric kidney biopsies: A randomized controlled trial. *Pain, 158*(1), 103–109.

Page, M. J., Green, S., Mrocki, M. A., Surace, S. J., Deitch, J., McBain, B., ... Buchbinder, R. (2016). Electrotherapy modalities for rotator cuff disease. *Cochrane Database of Systematic Reviews, 6*, CD012225 (1469-493X (Electronic)).

Peres, D., Sagawa, Y., Jr., Dugué, B., Domenech, S. C., Tordi, N., & Prati, C. (2017). The practice of physical activity and cryotherapy in rheumatoid arthritis: Systematic review. *European Journal of Physical and Rehabilitation Medicine, 53*(5), 775–787.

Qi, L., Tang, Y., You, Y., Qin, F., Zhai, L., Peng, H., & Nie, R. (2016). Comparing the effectiveness of electroacupuncture with different grades of knee osteoarthritis: A prospective study. *Cellular Physiology and Biochemistry: International Journal of Experimental Cellular Physiology, Biochemistry, and Pharmacology, 39*(6), 2331–2340.

Salazar, A. P. D. S., Stein, C., Marchese, R. R., Plentz, R. D. M., & Pagnussat, A. D. S. (2017). Electric stimulation for pain relief in patients with fibromyalgia: A systematic review and meta-analysis of randomized controlled trials. *Pain Physician, 20*(2), 15–25.

Santamato, A., Solfrizzi, V., Panza, F., Tondi, G., Frisardi, V., Leggin, B. G., ... Fiore, P. (2009). Short-term effects of high-intensity laser therapy versus ultrasound therapy in the treatment of people with subacromial impingement syndrome: A randomized clinical trial. *Physical Therapy, 89*(7), 643–652.

Sonke, J., Pesata, V., Arce, L., Carytsas, F. P., Zemina, K., & Jokisch, C. (2015). The effects of arts-in-medicine programming on the medical-surgical work environment. *Arts & Health, 7*(1), 27–41. https://doi.org/10.1080/17533015.2014.966313.

Tomazoni, S. S., Frigo, L., Dos Reis Ferreira, T. C., Casalechi, H. L., Teixeira, S., de Almeida, P., ... Leal-Junior, E. C. P. (2017). Effects of photobiomodulation therapy and topical non-steroidal anti-inflammatory drug on skeletal muscle injury induced by contusion in rats-part 1: Morphological and functional aspects. *Lasers in Medical Science, 32*(9), 2111–2120.

Wang, S.-M., Kain, Z. N., & White, P. (2008). Acupuncture analgesia: I. The scientific basis. *Anesthesia and Analgesia, 106*(2), 602–610.

White, P. F. (2017). What are the advantages of non-opioid analgesic techniques in the management of acute and chronic pain? *Expert Opinion on Pharmacotherapy, 18*(4), 329–333.

White, P. F., Elvir-Lazo, O. L., & Hernandez, H. (2017). A novel treatment for chronic opioid use after surgery. *Journal of Clinical Anesthesia, 40*, 51–53.

White, P. F., Elvir Lazo, O. L., Galeas, L., & Cao, X. (2017). Use of electroanalgesia and laser therapies as alternatives to opioids for acute and chronic pain management. *F1000Research, 6*, 2161.

White, P., Elvir Lazo, O., & Yumul, R. (2019). Cold laser therapy for acute and chronic pain management: A comparison of low-level and high-intensity laser therapy devices. *Anesthesiology News, 65*–77.

Zeidan, F., & Vago, D. R. (2016). Mindfulness meditation-based pain relief: A mechanistic account. *Annals of the New York Academy of Sciences, 1373*(1), 114–127. https://doi.org/10.1111/nyas.13153.

Zhang, Q., Li, Y.-N., Guo, Y.-Y., Yin, C.-P., Gao, F., Xin, X., ... Wang, Q.-J. (2017). Effects of preconditioning of electro-acupuncture on postoperative cognitive dysfunction in elderly: A prospective, randomized, controlled trial. *Medicine, 96*(26), e7375.

Zhang, B., Xu, F., Hu, P., Zhang, M., Tong, K., Ma, G., ... Chen, J. D. Z. (2018). Needleless transcutaneous electrical acustimulation: A pilot study evaluating improvement in post-operative recovery. *The American Journal of Gastroenterology, 113*(7), 1026–1035.

Chapter 52

New coping strategies and self-education for chronic pain management: E-health

Victoria Mazoteras-Pardo[a], Marta San Antolín[b], Daniel López-López[c], Ricardo Becerro-de-Bengoa-Vallejo[d], Marta Losa-Iglesias[e], Carlos Romero-Morales[f], David Rodríguez-Sanz[d], Eva María Martínez-Jiménez[d], and César Calvo-Lobo[d]

[a]Research Group "ENDOCU", Department of Nursing, Physiotherapy and Occupational Therapy, Faculty of Physiotherapy and Nursing of Toledo, University of Castilla-La Mancha, Ciudad Real, Spain, [b]Department of Psychology, Universidad Europea de Madrid, Madrid, Spain, [c]Research, Health and Podiatry Group, Department of Health Sciences, Faculty of Nursing and Podiatry, Universidade da Coruña, Ferrol, Spain, [d]Faculty of Nursing, Physiotherapy and Podiatry, Universidad Complutense de Madrid, Madrid, Spain, [e]Department of Nursing, Faculty of Health Sciences, Universidad Rey Juan Carlos, Alcorcón, Spain, [f]Physiotherapy Department, Faculty of Sport Sciences, Universidad Europea de Madrid, Madrid, Spain

Abbreviations

Apps applications
IASP International Association for the Study of Pain

Introduction

Society has always been interested in issues such as Health, Disease, and Pain, but today, in addition, a new culture of prevention is being established in our area (Matarín, 2015).

The democratization of information, technological innovations produced by the progression of scientific knowledge, and of course, the great acceptance of new technologies has created greater awareness for the population regarding Health Promotion and Disease Prevention (Perroy, 2016; Silva, Rodrigues, de la Torre Díez, López-Coronado, & Saleem, 2015; Teo, Ng, & White, 2017).

On the one hand, these healthier living conditions and the progress in Public Health and health care have led to a demographic change, which leads to an increase in life expectancy, in the prevalence of chronic diseases, and obviously, in pain and morbidity. The pain, when it lasts more than 3 months, can be considered a chronic disease (International Association for the Study of Pain (IASP), 2012).

Currently, persistent pain is very prevalent. This supposes a physical, emotional, and socioeconomic burden for the person in pain or at risk of any type of pain and for their families or caregivers. Therefore, it has a great impact at the health level (Choinière et al., 2010; Pompili et al., 2012; Schopflocher, Taenzer, & Jovey, 2011).

On the other hand, the aforementioned demographic change raises the need to generate new organizational models and the provision of services, in which people are more responsible and active in caring for their health (Teo et al., 2017). Examples of coping strategy devices or protocols can be seen in Table 1. Effective pain management is essential, and clearly, people with pain or risk of any type of pain should be involved in decisions about proposed pain management interventions (Heath & Philip, 2020).

In this sense, the relationship between health and technology is growing, since it is part of a new culture of prevention-control, allowing the exchange of experiences and knowledge, quickly and easily. The new technologies for Health support information systems that promote participation and intervention for the diagnosis, treatment, prevention, and evaluation of diseases, research and continuous training of professionals (Bradbury, Watts, Arden-Close, Yardley, & Lewith, 2014; Kampmeijer, Pavlova, Tambor, Golinowska, & Groot, 2016; Marceglia & Conti, 2017).

If we combine the high prevalence of chronic pain and the importance of effective management with the benefits of digital health, the benefits can be enormous, both for the person, their family environment, and the health system. The current health system aims to tackle the chronic pain epidemic (Heath & Philip, 2020; Portelli & Eldred, 2016).

Treatments, Mechanisms, and Adverse Reactions of Anesthetics and Analgesics. https://doi.org/10.1016/B978-0-12-820237-1.00052-1
Copyright © 2022 Elsevier Inc. All rights reserved.

618 PART | IV Novel and nonpharmacological aspects and treatments

TABLE 1 Other coping strategy devices or protocols.

Other coping strategy devices or protocols

Ahmed, W., & Mohammed, B. (2019). Nursing students' stress and coping strategies during clinical training in KSA. *Journal of Taibah University Medical Sciences, 14*(2), 116–122. https://doi.org/10.1016/j.jtumed.2019.02.002

Banerjee, Y., Akhras, A., Khamis, A. H., Alsheikh-Ali, A., & Davis, D. (2019). Investigating the relationship between resilience, stress-coping strategies, and learning approaches to predict academic performance in undergraduate medical students: Protocol for a proof-of-concept study. *JMIR Research Protocols, 8*(9), e14677. https://doi.org/10.2196/14677

Berra, E., Muñoz, S. I., Vega, C.Z., Rodríguez, A. S., & Gómez, G. (2014). Emotions, stress and coping in adolescents from the Lazarus and Folkman model. *Intercontinental Journal of Psychology and Education, 16*(1), 37–57

Failo, A., Beals-Erickson, S. E., & Venuti, P. (2018). Coping strategies and emotional well-being in children with disease-related pain. *Journal of child health care: For professionals working with children in the hospital and community, 22*(1), 84–96. https://doi.org/10.1177/1367493

Karaca, A., Ünsal, G., Asik, E., Keser, I., Ankarali, H., & Merih, Y. D. (2018). Development and assessment of a coping scale for infertile women in Turkey. *African journal of reproductive health, 22*(3), 13–23. https://doi.org/10.29063/ajrh2018/v22i3.2

Mazoteras Pardo, V., Losa Iglesias, M. E., López Chicharro, J., & Becerro de Bengoa Vallejo, R. (2017). The QardioArm app in the assessment of blood pressure and heart rate: Reliability and validity study. *JMIR mHealth and uHealth, 5*(12), e198. https://doi.org/10.2196/mhealth.8458

Saetes, S., Hynes, L., McGuire, B. E., & Caes, L. (2017). Family resilience and adaptive coping in children with juvenile idiopathic arthritis: Protocol for a systematic review. *Systematic reviews, 6*(1), 221. https://doi.org/10.1186/s13643-017-0619-z

Slepecky, M., Kotianova, A., Prasko, J., Majercak, I., Gyorgyova, E., Kotian, M., Zatkova, M., Tonhajzerova, I., Chupacova, M., & Popelkova, M. (2017). Coping, schemas, and cardiovascular risks—Study protocol. *Neuropsychiatric Disease and Treatment, 13*, 2599–2605

Wesner, A. C., Behenck, A., Finkler, D., Beria, P., Guimarães, L., Manfro, G. G., Blaya, C., & Heldt, E. (2019). Resilience and coping strategies in cognitive behavioral group therapy for patients with panic disorder. *Archives of Psychiatric Nursing, 33*(4), 428–433

The epidemic of chronic pain

According to the IASP (2012), pain is defined as an unpleasant experience in the face of real or potential tissue damage. This experience is sensory and harbors the physiology and affective nature of pain.

Chronic pain is considered a disease, with a frequency of more than 5 days per week for more than 3 months continuously or intermittently. Most of the time it is difficult to treat (Heath & Philip, 2020).

It is divided into two main categories:

– Nociceptive pain considers a warning signal caused by actual or potential damage to nonneuronal tissue that triggers the activation of nociceptors in a normally functioning nervous system.
– Neuropathic pain is a clinical description of pain believed to be damaged by damage from a lesion or disease of the somatosensory nervous system that is confirmed by diagnostic tests.

In turn, pain can be classified according to the area of injury, duration, and diagnosis, considering that the different types of pain are compatible. The classification of pain divides it into seven types: Primary chronic pain; Chronic cancer pain; Chronic postsurgical or posttraumatic pain; Chronic neuropathic pain; Orofacial pain and headache; Chronic visceral pain, and Chronic musculoskeletal pain (Fig. 1).

This concept is very relevant since for many people it causes discomfort throughout their lives (Lynch, 2011). In the world, chronic and/or poorly managed pain poses a great problem for the subject, the community, and the health system (Choinière et al., 2010; Pompili et al., 2012; Schopflocher et al., 2011).

The quality of life of people with chronic pain is affected day by day, and generally, it also affects mood and even develops mental problems (Lerman, Rudich, Brill, Shalev, & Shahar, 2015).

Regarding the prevalence of pain, it has been proven that it is proportional to the age of the person. Sixty-five percent of those over 65 years have chronic pain and this number increases in people of the same age who live in residences for the elderly (Breivik, Collett, Ventafridda, Cohen, & Gallacher, 2006; Hadjistavropoulos, Hunter, & Dever Fitzgerald, 2009; Lynch, 2011).

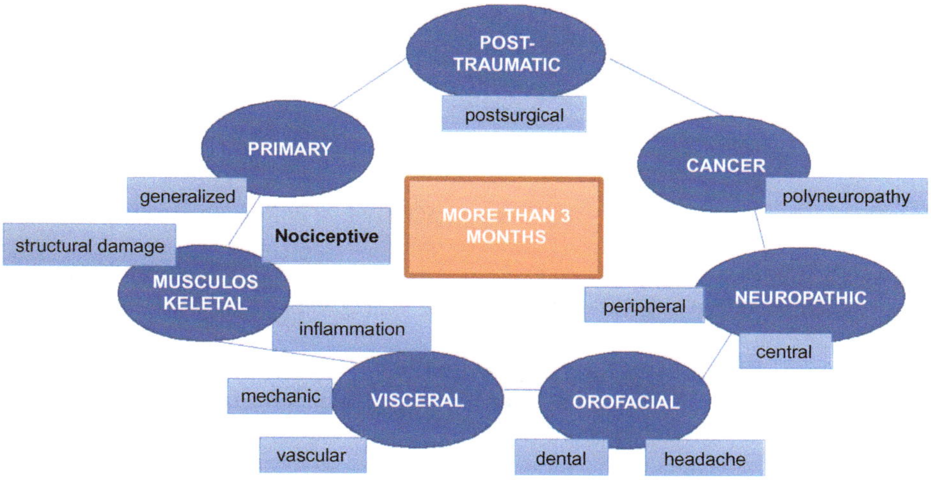

Chronic pain and E-health **Chapter | 52** 619

FIG. 1 Main types of chronic pain.

Effective chronic pain management

Proper pain management should be one of the fundamental cares and rights of all patients, regardless of their cause or type of pain (Heath & Philip, 2020; Jarzyna et al., 2011; Jungquist et al., 2020; Fig. 2).

To effectively manage pain, it is assumed that pain is very subjective and multidimensional and has sensory, cognitive, and affective components (IASP, 2012).

This must be done in a comprehensive, multidimensional, and interprofessional manner, considering all the spiritual, cultural, biological, and social conditions and circumstances of the individual (Association of Ontario, 2013). There are many factors to consider other than pain, such as whether there is a disability, cognitive status, age, ethnicity, level of education, beliefs, communication skills, or previous pain experiences (Curry Narayan, 2010; Lewis Ramos & Eti, 2019).

Apart from interprofessional, the different professionals must work in continuous contact not only with the sick but also with their caregivers, relatives, and different members of their area. They must adapt to the different rhythms of life, always trying, as far as possible, to promote the self-care of the subject to control their pain (Clare, Andiappan, MacNeil, Bunton, & Jarrett, 2013; The British Pain Society, 2019; Zhou & Thompson, 2008).

Also, pain managers should be trained in pain assessment. This includes knowledge of pain and communication techniques, having the ability to recognize pain, even if the subject is not able to express it, and even having an environment with adequate resources (Czarnecki et al., 2011; Herr, Coyne, McCaffery, Manworren, & Merkel, 2011). As mentioned, pain is subjective and some people are unable to describe their discomfort or do not use the appropriate words to express their pain (IASP, 2012).

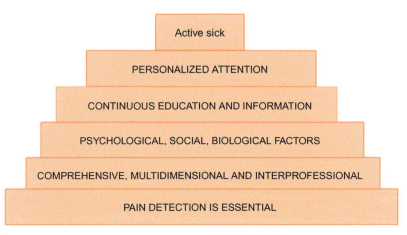

FIG. 2 Effective chronic pain management.

620 PART | IV Novel and nonpharmacological aspects and treatments

However, it has been found that most of the time, pain is managed inappropriately regardless of age. Furthermore, not all centers have appropriate context and resources based on scientific evidence to manage pain. Pain care is also not always available (American Society of Anesthesiologists, 2010; Loeser & Schatman, 2017).

Pain detection

To achieve effective pain management, pain detection is essential (Heath & Philip, 2020; Schofield, O'Mahony, Collett, & Potter, 2008).

For the healthcare professional to detect pain or the risk of pain, they do not have to assume that the patient or caregiver is going to explain it voluntarily, but they have to ask about the pain directly (American Medical Directors Association [AMDA], 2012; The British Pain Society, 2019).

When the person can communicate, expressing it for himself is the most effective way to assess pain, since as mentioned, pain is a multidimensional and subjective phenomenon (Association of Ontario, 2013; Mills, Torrance, & Smith, 2016; Vargovich et al., 2019). It is important to understand that the inability to describe pain does not mean that the person is not experiencing it. Assessment of pain in people who are unable to express it is essential for adequate care (Herr et al., 2011; IASP, 2012).

Listed below are questions that may be appropriate to detect pain in any population that can express it (AMDA, 2012, p. 8):

a. "Do you feel any pain or pain now?"
b. "Does any part of your body hurt?"
c. "Do you have any discomfort?"
d. "Have you taken any pain medication?"
e. "Do you have any pain or discomfort that continues overnight?"
f. "Have you had problems with any of your normal day-to-day activities?"
g. "How intense is the pain?"

It should be borne in mind that in the health consultation, pain should be assessed: In the visit to any health professional, whatever the reason; After any change in the person's state of health, especially after a chronic disease; and lastly, before, during, and after any medical intervention (Mills et al., 2016).

It has been seen that early detection by the interprofessional team is very relevant because the diagnosis of neuropathic pain may require further investigation to facilitate its early management. Thus, neuropathic pain can be a challenge in diagnosis for the interprofessional team. There are neuropathic pain screening questionnaires that incorporate the person's signs and symptoms (Wong, Hui, Chung, & Wong, 2014).

As for the way to assess pain, it must be comprehensive and systematic collecting, among others, the background of previous pain, the sensory characteristics of pain (area, intensity, when it increases or decreases), how it affects pain on a daily basis (work, leisure, rest) and the psychological impact of pain (Herr et al., 2011; Lewis Ramos & Eti, 2019; Schofield et al., 2008; The British Pain Society, 2019; Fig. 2).

Although many healthcare workers participate in pain assessment and control, both directly and indirectly, nurses are considered to have a leading role in pain detection. This is because the infirmary has a lot of direct daily contact with the sick.

Therefore, this position is crucial to pain detection, monitoring, and comprehensive evaluation of the patient's pain experience. If the professional knows the patient, they will carry out a more complete and individualized assessment of pain (Heath & Philip, 2020; Mills et al., 2016; Schneiderhan, Clauw, & Schwenk, 2017).

Tools to assess chronic pain

It is important to prevent, anticipate, and manage pain whenever possible. To assess pain, it is relevant to use tools that are easily used and interpreted by the subject and their caregivers. Today, various translated and validated tools are available in different languages (Gélinas et al., 2019; Herr, Bjoro, & Decker, 2006; Lichtner et al., 2014).

The tools can be one-dimensional or multidimensional, but all comprehensive. The comprehensive assessment determines both the quality and the severity of the pain. If they are one-dimensional, they focus only on one aspect of pain such as intensity, and if the tool is multidimensional, they value more aspects, resulting in a more complete assessment.

The characteristics that a tool to assess pain must meet are (Association of Ontario, 2013; Booker & Herr, 2016; Fig. 3):

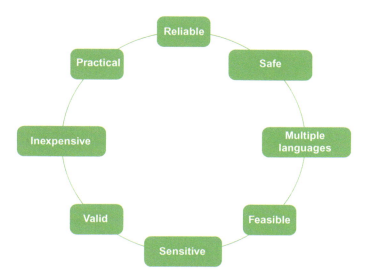

FIG. 3 Pain tools requirements.

- Reliable, providing consistent and reliable scores, regardless of time, institution, or by whom the measure is administered.
- Valid, that measures what it should measure, achieve the objective for which the tool was invented.
- Sensitive, which detects fluctuations in pain levels according to evolution and interventions carried out.
- Feasible, which can be handled and understood easily and quickly, without much prior knowledge and training. Also, make it a tool that likes patients, healthcare professionals, and researchers use it.
- Practical, which evaluates, whenever possible, the different types of pain. Furthermore, the tool must be culturally adapted to the population to which it is applied. Often a person's beliefs about pain affect whether they will seek help to alleviate it and what strategies they will accept to manage it (Curry Narayan, 2010; Lewis Ramos & Eti, 2019). Difficulties arise when a person makes decisions based on wrong beliefs due to a lack of understanding and incomplete knowledge about pain.
- Available in multiple languages or easily translatable.
- Easy to obtain, reproduce and distribute.
- What can be disinfected to ensure safe and hygienic general use.

The selected tool will be the one that adapts to all the person's conditions and the sociological, cultural, biological, and psychological factors. Pain management tools should aim to reduce pain intensity and aim to improve function, sleep, and overall quality of life (Curry Narayan, 2010; Gélinas, 2016; Herr et al., 2006; Lichtner et al., 2014; Schofield et al., 2008; Zhou & Thompson, 2008). Once the tool has been chosen, it must be adequately explained to the person who is going to apply it (AMDA, 2012; The British Pain Society, 2019).

Current nonpharmacological interventions in chronic pain

Nonpharmacological interventions are often used in conjunction with pharmacological interventions to control pain (Fig. 4).

Professionals should explore the conditions of the person suffering from pain, such as age, health status, type of pain, as well as their abilities, cultural concerns, and beliefs about the use of complementary or alternative forms of care (Association of Ontario, 2013; Curry Narayan, 2010; Schofield et al., 2008).

It has been shown that nonpharmacological interventions should be considered in conjunction with pharmacological interventions to reduce pain, improve sleep, mood, and general well-being. Evidence confirms the efficacy of nonpharmacological care when used alone or in combination with pharmacological interventions. The main alternatives were physical and psychological, among which are cognitive-behavioral therapy, massages, exercises, relaxation, and fundamental, the education of people over 65 (Cheatle, 2016; Eccleston, Morley, & Williams, 2013).

But, currently, pain care has been added through new technologies (Keogh, Rosser, & Eccleston, 2010).

622 PART | IV Novel and nonpharmacological aspects and treatments

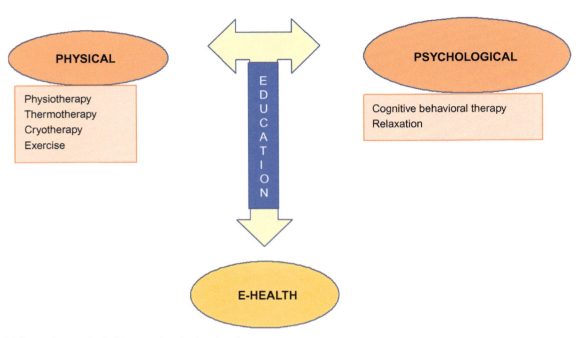

FIG. 4 Main nonpharmacological interventions in chronic pain.

The "E-health" refers to the application of communication and information technologies in all those aspects that concern health and its management, providing health services, regardless of where the patients are located, the professionals, equipment, or medical history (Anderson, Burford, & Emmerton, 2016; Bradbury et al., 2014; Kampmeijer et al., 2016; Fig. 5).

Within the previous one, there is the "M-health," which is the branch that uses wireless devices such as smartphones, laptops, watches, tablets, patient monitoring devices, or digital personal assistants in the healthcare environment at a clinical and training level, assistance, and research.

Therefore, examples of M-health are all the interventions that use these technologies to monitor health indicators broadly, those that educate the population or health professionals, those that provide an appointment to a patient for a consultation, or those that give you instructions for your lifestyle in real-time. Thus, M-health encompasses both the education,

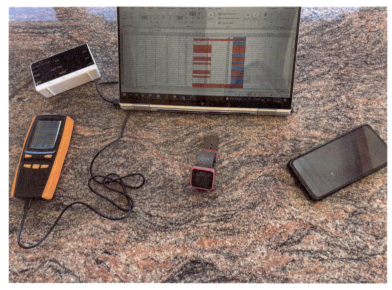

FIG. 5 Examples of E-health.

diagnosis, treatment, control, and monitoring of subjects and their administrative efforts favoring Health Promotion and Disease Prevention. In addition, when health professionals communicate, these with patients or among the latter, they also make use of M-health, by means of text messages on the smartphone, email, or video calls (Bradbury et al., 2014; de Jongh, Gurol-Urganci, Vodopivec-Jamsek, Car, & Atun, 2012; Perroy, 2016).

The splendor that such a concept has experienced brings with it the development of applications for smartphones (apps) related to health, which have the great advantage that once created, they can be used an infinite number of times by the entire population without the need to go to a health consultation while they are new working tools for the professional (Anderson et al., 2016; Omboni, Caserini, & Coronetti, 2016; Paschou, Sakkopoulos, & Tsakalidis, 2013; Silva et al., 2015; Teo et al., 2017).

E-health in chronic pain

If we approach these technologies and information, they are a new model of Health Education in chronic pain. On the web, there is a wide variety of content related to chronic pain, created by pain professionals. These resources are increasing, and with it, access to information about pain and painful conditions through the Internet has also increased (Gordon & Crouch, 2019; Polomano, Droog, Purinton, & Cohen, 2007; Walsh & Volsko, 2008).

The main advantage of these resources is to allow relevant and updated information to be accessed from anywhere quickly and efficiently. Disadvantages may include that the information available is not safe, valid, and accurate (Corcoran, Haigh, Seabrook, & Schug, 2009; Martorella et al., 2017; Powell, Inglis, Ronnie, & Large, 2011).

As for electronic devices, these have also been applied as tools to measure and manage pain, in a personalized and valid way and other characteristics already explained in the corresponding section. Apps, electronic diaries, and personal digital assistants are examples that allow the storage of large amount of data on pain, its emotions, and related behaviors, offering a new professional-healthcare patient relationship in the control of symptoms. This data can be instantly saved and allows professional feedback remotely in real-time without being susceptible to bias, transcription, and memory errors (Keogh, 2013; Ranney, Duarte, Baird, Patry, & Green, 2016). Thus, through electronic devices, patients can receive specific treatment for their pain, without the need to go to the health consultation, reducing the economic costs and associated risks of attending traditional hospital settings (Borrelli & Ritterband, 2015; Du et al., 2011; Portelli & Eldred, 2016; Pronovost, Peng, & Kern, 2009).

Also, the apps can be included in the treatment of chronic pain, promoting self-care of the subject through personal goals, objectives, and activities that are evaluated over time (Du, Liu, Cai, Hu, & Dong, 2020; Rosser & Eccleston, 2011; Whitehead & Seaton, 2016).

There are studies that ensure that these are easy to use and provide the patient a high level of compliance and better coping with chronic pain. In addition, these new tools also allow connecting different professionals, achieving multidisciplinary, comprehensive, and interprofessional chronic pain management (Marceglia & Conti, 2017).

Another type of device is remote sensing sensors, to manage chronic pain anywhere through movements, behaviors, and activities. They are wireless, discrete, and nonintrusive (Keogh, 2013).

Likewise, there has been an expansion in virtual social networks (Matarín, 2015). These have been shown to offer very useful content regarding the beliefs and expectations of treating the patient with chronic pain since they share experiences and feelings. Thus, there are regulated groups made up of subjects with pain where they provide common support and offer advice by connecting different platforms. By increasing the use of social networks, it implies a greater demand for health care (Musich, Wang, Slindee, Kraemer, & Yeh, 2019; Pronovost et al., 2009).

Digital Health provides new opportunities and advantages in chronic pain management (Fig. 6), but, nevertheless, it is not one of the most studied chronic diseases, such as hypertension, obesity, diabetes, or cardiovascular disease (Portelli & Eldred, 2016).

Clinical applications

E-health should be considered as an effective tool for the management of chronic pain, both by subjects with pain and by healthcare professionals.

It can be used together with other pharmacological and nonpharmacological treatments, achieving better health results.

Also, these technologies have become a new model of Health education and promote self-care of the population (Du et al., 2020; Gordon & Crouch, 2019; Keogh, 2013; Musich et al., 2019; Pronovost et al., 2009; Ranney et al., 2016).

FIG. 6 Advantages of E-health in chronic pain management.

Application to other areas

In addition to chronic pain, other prevalent chronic diseases such as hypertension, diabetes mellitus, cardiovascular diseases, or obesity could be benefited by the inclusion of E-health into a multidimensional and comprehensive intervention (Du et al., 2020; Portelli & Eldred, 2016; Sarwar et al., 2018; Silva et al., 2015).

Other agents of interest

Another alternative for patients with chronic pain is emotional control through E-health. These patients could have a self-reported psychological questionnaire on their smart device and thus, professionals could relate their results to other clinical symptoms and signs. An association has been seen between psychological factors and central sensitization in these patients (San-Antolín et al., 2020).

Mini-dictionary of terms

E-health: This term may be defined as the communication and information technologies used for health.
Chronic pain: This term may be defined as pain from any part of the body that lasts more than 3 months continuously or intermittently, with a frequency of more than 5 days per week.
Chronic pain management: This term can be defined as the detection, measurement, monitoring, and control of symptoms and signs of chronic pain.
M-health: This is the branch of E-health that uses wireless devices in the setting of clinical and health care training, assistance, and research.

Key facts of chronic pain

- Chronic pain can be considered an epidemic, a very prevalent chronic disease.
- Chronic pain affects biological, social, economic, and psychological levels.
- Detection and effective management of chronic pain are crucial.

- Pain management must be comprehensive, multidimensional, and interprofessional.
- Nonpharmacological interventions should be included in the management of patients with chronic pain.
- In chronic pain management, the use of E-health is essential.

Summary points

- This chapter focuses on effective chronic pain management through E-health.
- Currently, in the world, poorly managed pain presents great problems for the subject, the community, and the health system.
- The management of chronic pain must be personalized, comprehensive, multidimensional, and interprofessional. E-health fulfills these characteristics.
- Nonpharmacological interventions are also very effective in managing chronic pain, in combination or not with pharmacological.
- People with chronic pain should have an active position in monitoring and controlling it.
- The tools that control chronic pain must be fundamentally reliable, valid, sensitive, feasible, safe, and practical, among others. E-health can meet these conditions.
- The E-health has additional advantages in the management of chronic pain. It improves knowledge, increases compliance, saves healthcare costs, and promotes self-care of the patient, whether applied alone or in combination with pharmacological treatment.
- More research on E-health and pain is needed to improve the knowledge of detection, management, and monitoring of chronic pain, as well as the validity and safety of these technologies.

References

American Medical Directors Association (AMDA) (2012). *Pain management in the long-term care setting.* Columbia, MD: AMDA.

American Society of Anesthesiologists Task Force on Chronic Pain Management, & American Society of Regional Anesthesia and Pain Medicine (2010). Practice guidelines for chronic pain management: An updated report by the American Society of Anesthesiologists Task Force on Chronic Pain Management and the American Society of Regional Anesthesia and Pain Medicine. *Anesthesiology, 112*(4), 810–833. https://doi.org/10.1097/ALN.0b013e3181c43103.

Anderson, K., Burford, O., & Emmerton, L. (2016). Mobile health apps to facilitate self-care: A qualitative study of user experiences. *PLoS One, 11*(5), e0156164. https://doi.org/10.1371/journal.pone.0156164.

Association of Ontario (2013). *Assessment and management of pain* (3rd ed.). Toronto, ON: Registered Nurses' Association of Ontario.

Booker, S. Q., & Herr, K. A. (2016). Assessment and measurement of pain in adults in later life. *Clinics in Geriatric Medicine, 32*(4), 677–692. https://doi.org/10.1016/j.cger.2016.06.012.

Borrelli, B., & Ritterband, L. M. (2015). Special issue on eHealth and mHealth: Challenges and future directions for assessment, treatment, and dissemination. *Health Psychology : Official Journal of the Division of Health Psychology, American Psychological Association, 34S,* 1205–1208.

Bradbury, K., Watts, S., Arden-Close, E., Yardley, L., & Lewith, G. (2014). Developing digital interventions: A methodological guide. *Evidence-based Complementary and Alternative Medicine: Ecam, 2014,* 561320. https://doi.org/10.1155/2014/561320.

Breivik, H., Collett, B., Ventafridda, V., Cohen, R., & Gallacher, D. (2006). Survey of chronic pain in Europe: Prevalence, impact on daily life, and treatment. *European Journal of Pain, 10*(4), 287–333. https://doi.org/10.1016/j.ejpain.2005.06.009.

Cheatle, M. D. (2016). Biopsychosocial approach to assessing and managing patients with chronic pain. *The Medical Clinics of North America, 100*(1), 43–53. https://doi.org/10.1016/j.mcna.2015.08.007.

Choinière, M., Dion, D., Peng, P., Banner, R., Barton, P. M., Boulanger, A., … Ware, M. (2010). The Canadian STOP-PAIN project—Part 1: Who are the patients on the waitlists of multidisciplinary pain treatment facilities? *Canadian Journal of Anesthesia, 57,* 539–548.

Clare, A., Andiappan, M., MacNeil, S., Bunton, T., & Jarrett, S. (2013). Can a pain management programme approach reduce healthcare use? Stopping the revolving door. *British Journal of Pain, 7*(3), 124–129. https://doi.org/10.1177/2049463713484907.

Corcoran, T. B., Haigh, F., Seabrook, A., & Schug, S. A. (2009). The quality of internet-sourced information for patients with chronic pain is poor. *The Clinical Journal of Pain, 25*(7), 617–623. https://doi.org/10.1097/AJP.0b013e3181a5b5d5.

Curry Narayan, M. (2010). Culture's effects on pain assessment and management. *American Journal of Nursing, 110*(4), 38–47.

Czarnecki, M. L., Turner, H. N., Collins, P. M., Doellman, D., Wrona, S., & Reynolds, J. (2011). Procedural pain management: A position statement with clinical practice recommendations. *Pain Management Nursing, 12*(2), 95–111. https://doi.org/10.1016/j.pmn.2011.02.003.

de Jongh, T., Gurol-Urganci, I., Vodopivec-Jamsek, V., Car, J., & Atun, R. (2012). Mobile phone messaging for facilitating self-management of long-term illnesses. *The Cochrane Database of Systematic Reviews, 12*(12), CD007459.

Du, S., Liu, W., Cai, S., Hu, Y., & Dong, J. (2020). The efficacy of e-health in the self-management of chronic low back pain: A meta analysis. *International Journal of Nursing Studies, 106,* 103507. https://doi.org/10.1016/j.ijnurstu.2019.10.350.

Du, S., Yuan, C., Xiao, X., Chu, J., Qiu, Y., & Qian, H. (2011). Self-management programs for chronic musculoskeletal pain conditions: A systematic review and meta-analysis. *Patient Education and Counseling, 85*(3), e299–e310. https://doi.org/10.1016/j.pec.2011.02.021.

Eccleston, C., Morley, S. J., & Williams, A. C. (2013). Psychological approaches to chronic pain management: Evidence and challenges. *British Journal of Anaesthesia, 111*(1), 59–63. https://doi.org/10.1093/bja/aet207.

Gélinas, C. (2016). Pain assessment in the critically ill adult: Recent evidence and new trends. *Intensive & Critical Care Nursing, 34*, 1–11. https://doi.org/10.1016/j.iccn.2016.03.001.

Gélinas, C., Joffe, A. M., Szumita, P. M., Payen, J. F., Bérubé, M., Shahiri, T. S., ... Puntillo, K. A. (2019). A psychometric analysis update of behavioral pain assessment tools for noncommunicative, critically ill adults. *AACN Advanced Critical Care, 30*(4), 365–387. https://doi.org/10.4037/aacnacc2019952.

Gordon, N. P., & Crouch, E. (2019). Digital information technology use and patient preferences for internet-based health education modalities: Cross-sectional survey study of middle-aged and older adults with chronic health conditions. *JMIR Aging, 2*(1), e12243. https://doi.org/10.2196/12243.

Hadjistavropoulos, T., Hunter, P., & Dever Fitzgerald, T. (2009). Pain assessment and management in older adults: Conceptual issues and clinical challenges. *Canadian Psychology/Psychologie Canadienne, 50*(4), 241–254. https://doi.org/10.1037/a0015341.

Heath, L., & Philip, A. (2020). Chronic pain care: Time for excellence. *Family Medicine and Community Health, 8*(2), e000285. https://doi.org/10.1136/fmch-2019-000285.

Herr, K., Bjoro, K., & Decker, S. (2006). Tools for assessment of pain in nonverbal older adults with dementia: A state-of-the-science review. *Journal of Pain and Symptom Management, 31*(2), 170–192. https://doi.org/10.1016/j.jpainsymman.2005.07.001.

Herr, K., Coyne, P. J., McCaffery, M., Manworren, R., & Merkel, S. (2011). Pain assessment in the patient unable to self-report: Position statement with clinical practice recommendations. *Pain Management Nursing, 12*(4), 230–250. https://doi.org/10.1016/j.pmn.2011.10.002.

International Association for the Study of Pain (IASP) (2012). *Desirable characteristics of national pain strategies: Recommendations by the International Association for the Study of Pain.* .

Jarzyna, D., Jungquist, C. R., Pasero, C., Willens, J. S., Nisbet, A., Oakes, L., ... Polomano, R. C. (2011). American Society of Pain Management Nursing guidelines on monitoring for opioid-induced sedation and respiratory depression. *Pain Management Nursing, 12*(3), 118–145.

Jungquist, C. R., Quinlan-Colwell, A., Vallerand, A., Carlisle, H. L., Cooney, M., Dempsey, S. J., ... Polomano, R. C. (2020). American Society for Pain Management Nursing guidelines on monitoring for opioid-induced advancing sedation and respiratory depression: Revisions. *Pain Management Nursing, 21*(1), 7–25. https://doi.org/10.1016/j.pmn.2019.06.007.

Kampmeijer, R., Pavlova, M., Tambor, M., Golinowska, S., & Groot, W. (2016). The use of e-health and m-health tools in health promotion and primary prevention among older adults: A systematic literature review. *BMC Health Services Research, 16*(Suppl. 5), 290.

Keogh, E. (2013). Developments in the use of e-health for chronic pain management. *Pain Management, 3*(1), 27–33. https://doi.org/10.2217/pmt.12.70.

Keogh, E., Rosser, B. A., & Eccleston, C. (2010). E-health and chronic pain management: Current status and developments. *Pain, 151*(1), 18–21. https://doi.org/10.1016/j.pain.2010.07.014.

Lerman, S. F., Rudich, Z., Brill, S., Shalev, H., & Shahar, G. (2015). Longitudinal associations between depression, anxiety, pain, and pain-related disability in chronic pain patients. *Psychosomatic Medicine, 77*(3), 333–341. https://doi.org/10.1097/PSY.0000000000000158.

Lewis Ramos, V., & Eti, S. (2019). Assessment and management of chronic pain in the seriously ill. *Primary Care, 46*(3), 319–333. https://doi.org/10.1016/j.pop.2019.05.001.

Lichtner, V., Dowding, D., Esterhuizen, P., Closs, S. J., Long, A. F., Corbett, A., & Briggs, M. (2014). Pain assessment for people with dementia: A systematic review of systematic reviews of pain assessment tools. *BMC Geriatrics, 14*, 138. https://doi.org/10.1186/1471-2318-14-138.

Loeser, J. D., & Schatman, M. E. (2017). Chronic pain management in medical education: A disastrous omission. *Postgraduate Medicine, 129*(3), 332–335. https://doi.org/10.1080/00325481.2017.1297668.

Lynch, M. (2011). The need for a Canadian pain strategy. *Pain Research & Management, 16*(2), 77–80.

Marceglia, S., & Conti, C. (2017). A technology ecosystem for chronic pain: Promises, challenges, and future research. *mHealth, 3*, 6. https://doi.org/10.21037/mhealth.2017.02.03.

Martorella, G., Boitor, M., Berube, M., Fredericks, S., Le May, S., & Gélinas, C. (2017). Tailored web-based interventions for pain: Systematic review and meta-analysis. *Journal of Medical Internet Research, 19*(11), e385. https://doi.org/10.2196/jmir.8826.

Matarín, T. M. (2015). Redes sociales en prevención y promoción de la salud. Una revisión de la actualidad. *Revista Española de Comunicación en Salud, 6*(1), 62–69.

Mills, S., Torrance, N., & Smith, B. H. (2016). Identification and management of chronic pain in primary care: A review. *Current Psychiatry Reports, 18*(2), 22. https://doi.org/10.1007/s11920-015-0659-9.

Musich, S., Wang, S. S., Slindee, L., Kraemer, S., & Yeh, C. S. (2019). Association of resilience and social networks with pain outcomes among older adults. *Population Health Management, 22*(6), 511–521. https://doi.org/10.1089/pop.2018.0199.

Omboni, S., Caserini, M., & Coronetti, C. (2016). Telemedicine and M-health in hypertension management: Technologies, applications and clinical evidence. *High Blood Pressure & Cardiovascular Prevention, 23*(3), 187–196. https://doi.org/10.1007/s40292-016-0143-6.

Paschou, M., Sakkopoulos, E., & Tsakalidis, A. (2013). easyHealthApps: E-health apps dynamic generation for smartphones & tablets. *Journal of Medical Systems, 37*(3), 9951. https://doi.org/10.1007/s10916-013-9951-6.

Perroy, A. C. (2016). M-health in an age of e-health. Promises, challenges and liabilities. *Annales Pharmaceutiques Françaises, 74*(6), 421–430. https://doi.org/10.1016/j.pharma.2016.03.002.

Polomano, R. C., Droog, N., Purinton, M. C., & Cohen, A. S. (2007). Social support web-based resources for patients with chronic pain. *Journal of Pain & Palliative Care Pharmacotherapy, 21*(3), 49–55.

Pompili, M., Forte, A., Palermo, M., Stefani, H., Lamis, D. A., Serafini, G., … Girardi, P. (2012). Suicide risk in multiple sclerosis: A systematic review of current literature. *Journal of Psychosomatic Research, 73*(6), 411–417. https://doi.org/10.1016/j.jpsychores.2012.09.011.

Portelli, P., & Eldred, C. (2016). A quality review of smartphone applications for the management of pain. *British Journal of Pain, 10*(3), 135–140. https://doi.org/10.1177/2049463716638700.

Powell, J., Inglis, N., Ronnie, J., & Large, S. (2011). The characteristics and motivations of online health information seekers: Cross-sectional survey and qualitative interview study. *Journal of Medical Internet Research, 13*(1), e20. https://doi.org/10.2196/jmir.1600.

Pronovost, A., Peng, P., & Kern, R. (2009). Telemedicine in the management of chronic pain: A cost analysis study. *Canadian Journal of Anaesthesia, 56*(8), 590–596. https://doi.org/10.1007/s12630-009-9123-9.

Ranney, M. L., Duarte, C., Baird, J., Patry, E. J., & Green, T. C. (2016). Correlation of digital health use and chronic pain coping strategies. *mHealth, 2*, 35. https://doi.org/10.21037/mhealth.2016.08.05.

Rosser, B. A., & Eccleston, C. (2011). Smartphone applications for pain management. *Journal of Telemedicine and Telecare, 17*(6), 308–312. https://doi.org/10.1258/jtt.2011.101102.

San-Antolín, M., Rodríguez-Sanz, D., López-López, D., Romero-Morales, C., Carbajales-Lopez, J., Becerro-de-Bengoa-Vallejo, R., … Calvo-Lobo, C. (2020). Depression levels and symptoms in athletes with chronic gastrocnemius myofascial pain: A case-control study. *Physical Therapy in Sport, 43*, 166–172. https://doi.org/10.1016/j.ptsp.2020.03.002.

Sarwar, C., Vaduganathan, M., Anker, S. D., Coiro, S., Papadimitriou, L., Saltz, J., … Butler, J. (2018). Mobile health applications in cardiovascular research. *International Journal of Cardiology, 269*, 265–271. https://doi.org/10.1016/j.ijcard.2018.06.039.

Schneiderhan, J., Clauw, D., & Schwenk, T. L. (2017). Primary care of patients with chronic pain. *JAMA, 317*(23), 2367–2368. https://doi.org/10.1001/jama.2017.5787.

Schofield, P., O'Mahony, S., Collett, B., & Potter, J. (2008). Guidance for the assessment of pain in older adults: A literature review. *British Journal of Nursing, 17*(14), 914–918. https://doi.org/10.12968/bjon.2008.17.14.30659.

Schopflocher, D., Taenzer, P., & Jovey, R. (2011). The prevalence of chronic pain in Canada. *Pain Research & Management, 16*(6), 445–450. https://doi.org/10.1155/2011/876306.

Silva, B. M., Rodrigues, J. J., de la Torre Díez, I., López-Coronado, M., & Saleem, K. (2015). Mobile-health: A review of current state in 2015. *Journal of Biomedical Informatics, 56*, 265–272. https://doi.org/10.1016/j.jbi.2015.06.003.

Teo, C. H., Ng, C. J., & White, A. (2017). What do men want from a health screening mobile app? A qualitative study. *PLoS One, 12*(1), e0169435. https://doi.org/10.1371/journal.pone.0169435.

The British Pain Society (2019). *Guidelines for pain management programmes for adults. (2019). www.britishpainsociety.org.*

Vargovich, A. M., Schumann, M. E., Xiang, J., Ginsberg, A. D., Palmer, B. A., & Sperry, J. A. (2019). Difficult conversations: Training medical students to assess, educate, and treat the patient with chronic pain. *Academic Psychiatry, 43*(5), 494–498. https://doi.org/10.1007/s40596-019-01072-4.

Walsh, T. M., & Volsko, T. A. (2008). Readability assessment of internet-based consumer health information. *Respiratory Care, 53*(10), 1310–1315.

Whitehead, L., & Seaton, P. (2016). The effectiveness of self-management mobile phone and tablet apps in long-term condition management: A systematic review. *Journal of Medical Internet Research, 18*(5), e97. https://doi.org/10.2196/jmir.4883.

Wong, C. S., Hui, G. K., Chung, E. K., & Wong, S. H. (2014). Diagnosis and management of neuropathic pain. *Pain Management, 4*(3), 221–231. https://doi.org/10.2217/pmt.14.7.

Zhou, Y., & Thompson, S. (2008). Quality assurance for interventional pain management procedures in private practice. *Pain Physician, 11*(1), 43–55.

Chapter 53

Postoperative pain management: Truncal blocks in obstetric and gynecologic surgery

Pelin Corman Dincer

Department of Oral & Maxillofacial Surgery, Faculty of Dentistry, Marmara University, Istanbul, Turkey

Abbreviations

ERAS	enhanced recovery after surgery
etCO$_2$	end-tidal carbon dioxide
IH-II	iliohypogstric-ilioinguinal
LAST	local anesthetic systemic toxicity
QLB	quadratus lumborum block
SpO$_2$	peripheral oxygen saturation
TAP	transversus abdominis plane
TFP	transversalis fascia plane

Introduction

A growing number of people need surgery every day; thus, postoperative pain treatment is critical. In the preoperative period, pain treatment options should be discussed with the patient, and they should be started promptly to achieve an effective result. It is known that adequate pain relief is associated with early mobilization, fewer thromboembolic complications, and early return of bowel functions.

Untreated or undertreatment of acute pain may lead to chronic pain; therefore, multimodal analgesia protocols are being used in clinics (Fletcher et al., 2015). Analgesic treatments with minimal side effects are desired. To avoid the unwanted effects of systemically administered opioid drugs, drugs with different action mechanisms are added to treatment protocols. Additive and/or synergistic effects of the administered drugs result in fewer complications, faster discharge times, and reduced hospital costs (Pogatzki-Zahn, Segelcke, & Schug, 2017).

Enhanced recovery after surgery (ERAS) protocols aim to improve outcomes after surgery. Because various types and locations of incisions can result in different pain conditions, analgesic efficacy is not equivalent in different surgical settings (Pogatzki-Zahn et al., 2017). According to the surgery, perioperative care protocols are developed and implemented by the ERAS group (Ljungqvist, 2014). One of the aspects of ERAS protocols is to use opioid-sparing multimodal perioperative analgesia methods, which include regional analgesia techniques.

Regional analgesia and anesthesia can be achieved by central neuraxial blocks or by peripheral nerve blocks. Landmark-based techniques are generally used, but portable ultrasound devices have gained popularity over nerve stimulators and landmark techniques. Direct visualization of the area and vasculature enables high success rates and fewer complications.

Therapeutic interventions with less risk of adverse effects but with equivalent or more efficacy than blocks already being used are being sought. Ultrasound images of anatomical muscle planes and the visualization of spreading local anesthetic drugs ensure successful trunk blocks (Chakraborty, Khemka, & Datta, 2016). According to the surgery site, several truncal blocks are described, and their efficacy in postoperative pain management is being compared to other blocks.

Unlike peripheral nerve blocks, truncal blocks do not require direct visualization of the nerves and neural plexus; the local anesthetic drug is injected into the targeted plane and spreads along the plane, making them simple to perform and easy to learn when compared to other regional techniques.

This chapter provides information about the truncal blocks that are used in obstetric and gynecologic operations.

Treatments, Mechanisms, and Adverse Reactions of Anesthetics and Analgesics. https://doi.org/10.1016/B978-0-12-820237-1.00053-3
Copyright © 2022 Elsevier Inc. All rights reserved.

Preparation for the truncal block

After discussing the benefits, complications, and alternative postoperative pain management techniques with the patient before the operation, written consent must be obtained. As in any other regional procedure, before initiating the block, the patient's heart rate, blood pressure, SpO_2, respiratory rate, and, if applicable, $etCO_2$ levels must be monitored. A running intravenous cannula must be placed, resuscitation equipment and drugs, i.e., oxygen supply, airways, bag valve mask devices, laryngoscopes, endotracheal intubation tubes, suction, defibrillator, epinephrine, phenylephrine, atropine, etc. must be within reach. Although local anesthetic toxicity (LAST) is rare with truncal blocks, emergency medicine, including 20% lipid solutions and LAST protocols, should also be available.

Before donning sterile gloves and, if intuitions protocols require it, a sterile gown, the potential landmarks must be checked and verified with the ultrasound device. A marking pen can be used to label the injection points. A high-frequency linear transducer (6–18 MHz) is efficient, but if adipose tissue is too thick or a more posterior approach is planned, then a curved transducer (2–5 MHz) might be needed. Depending on the adipose tissue thickness, the depth settings will be adjusted between 3 and 5 cm (Chakraborty et al., 2016). Skin preparation is done with an antiseptic solution as in any peripheral block. Sterile equipment (50–100 mm, 20–22-gauge peripheral block needle and, if necessary, peripheral nerve catheter, syringes containing local anesthetic drug(s), another syringe containing 1–2 mL saline or local anesthetic drug for hydrodissection and checking the right position of the block needle, sterile gauze or sponges, sterile bag for the probe and ultrasound gel) must be placed in a sterile place.

An anesthesiologist with sufficient knowledge and skills of the block must be accompanied by a health professional who will assist the anesthesiologist throughout the procedure, particularly if a complication arises. Nonanesthesiologist physicians like emergency physicians, critical care specialists, and surgeons also perform regional anesthesia in settings outside the operating rooms (Pawa & El-Boghdadly, 2018).

Patient selection

The spread of local anesthetics in the facial plane enables truncal blocks to be used in various abdominal surgeries. Upper, lower, posterior, or unilateral abdominal surgeries, including laparoscopic surgery and cesarean section, may benefit from these blocks.

When informing the patient about postoperative pain treatment options, truncal blocks must also be discussed, especially when neuraxial blocks are contraindicated. In patients who are obese or who have sleep apnea disease, systematically applied opioids are used with caution, which may lead to undertreatment. In patients, when epidural catheter placement fails or when opioid-free anesthesia techniques are used, truncal blocks may provide efficient analgesia in a multimodal analgesia treatment.

Contraindications

As in any nerve block technique, patients' refusal to the block, infection at the injection site, and allergy to the local anesthetic to be used are absolute contraindications.

Anatomical variations, loss of tissue integrity due to previous or current surgeries, anticoagulant drug therapies, neurologic diseases, and lack of skill and knowledge of the anesthesiologist can be encountered among the relative contraindications.

Abdominal wall

The abdominal wall has two vertical muscles: the rectus abdominis and the pyramidalis, and three flat muscles: the external abdominal oblique, the internal abdominal oblique, and the transversus abdominis. These three flat muscles form a broad aponeurosis, which forms the rectus sheath, and in the midline, it fuses into the linea alba. In the area above the arcuate line, both the anterior and posterior rectus sheath, below the arcuate line, just the anterior rectus sheath is formed by the aponeurosis (Fig. 1).

The sensory innervation of the skin, muscles, and parietal peritoneum of the anterior abdominal wall is done by the anterior rami of T7-L1 nerves (McDonnell et al., 2007). In the plane between the internal oblique and transversus abdominis muscles, the intercostal, subcostal, iliohypogastric and ilioinguinal nerves exist, and these nerves and branches form upper and lower transversus abdominis plane plexuses and rectus sheath plexus (Koh & Lee, 2018; Mishra & Mishra, 2016).

Truncal blocks in OB/GYN **Chapter | 53** 631

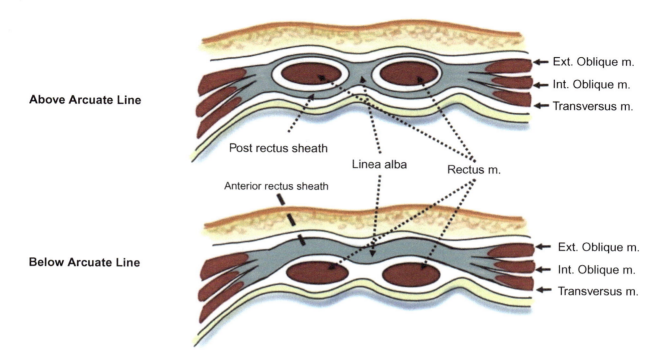

FIG. 1 Anterior and posterior rectus sheath.

Truncal blocks of the abdominal wall

Effective pain treatment after operations is essential. Advances in technology and pharmacology enable physicians to use new techniques that are more effective but with fewer complications. Table 1, in short, summarizes the pain treatment methods used for gynecologic and obstetric operations.

Transversus abdominis plane block

In this technique, a local anesthetic drug is injected into the "transversus abdominis plane" between the internal oblique and transversus abdominis muscles. This block can be used bilaterally in laparoscopic and open gynecologic operations as well as in cesarean sections. It can be performed preoperatively, perioperatively, or postoperatively. Patients who have general or spinal anesthesia may benefit from this block, but visceral pain will not be affected, so it must be used as an adjunct to

TABLE 1 Perioperative analgesic treatment methods used for gynecologic and obstetric operations.

Neuraxial block	Spinal block
	Epidural block
Truncal blocks	Transversus abdominis plane block
	Quadratus lumborum block
	Rectus sheath block
	Ilioinguinal-iliohypogastric nerve block
	Transversalis fascia plane block
	Erector spinae block
	Lumbar paravertebral block
Wound infiltration	

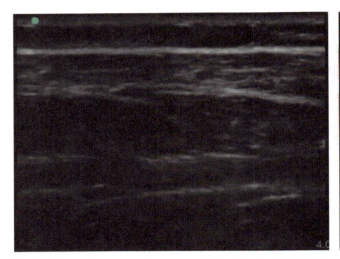
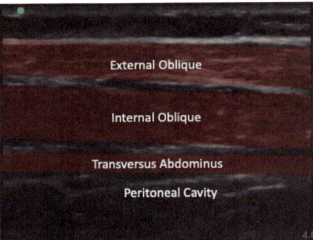

FIG. 2 Ultrasound anatomy of transversus abdominis plane (TAP) block.

multimodal analgesia. Without catheter placement, the transversus abdominis plane (TAP) block lasts for 24–48 h (Abdallah, Laffey, Halpern, & Brull, 2013).

Rafi first described this block in 2001 as a landmark-based technique (Rafi, 2001). In the blind (landmark) technique, Petit's lumbar triangle, which lies between the costal margin, iliac crest, external oblique, and latissimus dorsi muscles, is determined. When the needle passes the external oblique and internal oblique muscles, two pops will be felt, and the local anesthetic drug will be injected (Fig. 2). The thinning of the internal oblique aponeurosis in obese and pregnant patients may lead to missing the second pop feeling.

Surgeon-assisted TAP block is relatively safe as the surgeon's hand guides the needle or the catheter in the right position (Zhong et al., 2013).

Laparoscopic-assisted TAP block is faster and as efficacious as an ultrasound-guided block. In laparoscopic surgeries, when there is no ultrasound machine available, this method can be used (Ravichandran et al., 2017).

Robot-assisted TAP block is done in robot-assisted radical prostatectomy and has been shown to achieve better pain scores and fewer narcotic analgesic requirements than local anesthetic infiltration (Shahait et al., 2019).

Ultrasound-guided TAP block was introduced by Hebbart and et al. in 2007 (Hebbard, Fujiwara, Shibata, & Royse, 2007). To get the best image, the probe must be moved and tilted. Simulation-based training is beneficial in mastering ultrasound-guided blocks (Park et al., 2020). The learning curve for the TAP block was studied, and a 90% success rate was achieved after 16 blocks, and within 20 blocks, all residents were able to perform the block (Vial et al., 2015).

Several approaches are described with ultrasound guidance. The patient must be positioned in the lateral decubitus or supine position according to the approach type. The internal oblique muscle is the thickest of the three muscles visualized.

a. *Posterior approach*: the probe is placed on the lateral abdominal wall, in the mid-axillary line between the iliac crest and costal margin, and moved posteriorly. With an in-plane technique, the needle is advanced until the needle tip is seen in the plane between the internal oblique abdominal muscle and transversus abdominis muscle. After negative aspiration, the local anesthetic drug is injected, and the elliptical separation of the fascial layers is visualized using ultrasound. This approach blocks thoracolumbar nerves within the plane and the drug used reaches paravertebral space by spreading around the quadratus lumborum muscle (Tran, Bravo, Leurcharusmee, & Neal, 2019).
b. *Subcostal approach (Upper TAP)*: the probe is placed parallel to the subcostal margin near the xyphoid process. Subcutaneous tissue, rectus abdominis, and transversus abdominis muscles are visualized, and the needle is advanced with an in-plane technique. The local anesthetic is injected between the posterior rectus sheath and the transversus abdominis muscle. If analgesia is required for a more extended period, a catheter can be placed with this approach (Hebbard, Barrington, & Vasey, 2010). The anterior abdominal wall between the xyphoid process and the anterosuperior iliac spine is affected by this approach.
c. *Lateral approach*: The needle is inserted into the anterior axillary line and advanced to reach the TAP. It covers mainly the T10 to L1 dermatomes (Tran et al., 2019). For the cesarean section, this approach provides less pain relief than the posterior approach (Faiz et al., 2018).

d. *Anterior approach*: The insertion point lies medial to the anterior iliac spine. The deep circumflex iliac artery is near the TAP.

e. *Multiple injection (dual) approach*: subcostal and lateral approaches are combined for better results.

Lateral approaches are insufficient for effective pain treatment in surgeries above the umbilicus level (Abrahams, Derby, & Horn, 2016). Tran et al. suggest that if a long-acting opioid drug is used in cesarean delivery, then the TAP block will be futile to perform (Tran et al., 2019). Open hysterectomy patients may benefit from lateral and posterior TAP block, while for laparoscopic hysterectomies, no benefit has been shown (Zhou et al., 2018).

Injectates

Ropivacaine, bupivacaine, and levobupivacaine can be used. In pregnant women, the drug calculation must be based on lean body weight (Tran et al., 2019).

It is recommended to use large volumes as the widespread of the drug, i.e., 15–30 mL per side, and low concentration of local anesthetic drugs, i.e., 0.125%–0.25% (Tran et al., 2019).

Aly et al. compared epidural analgesia with TAP block achieved by using 0.5 mL/kg 0.25% bupivacaine for lower abdominal surgeries and concluded that epidural analgesia is superior to TAP (Aly, Talaat, Menshawi, & Mohammed, 2020).

Complications

The TAP block is relatively safe when done under direct visualization of the region. Nevertheless, complications related to the TAP block have also been reported. Finally, intravascular injection, allergic reactions, nerve injury, intramuscular injection, and infection can be seen in any peripheral block. In pregnancy, the drugs' pharmacokinetics may be altered, making the patient prone to LAST (Griffiths et al., 2013).

A transient abdominal motor block was observed in an elderly patient. The authors suggested that caution must be taken in patients with weak abdominal walls (Bortolato, Ori, & Freo, 2015). Femoral nerve blockage, hematoma, intraperitoneal injection, and visceral organ trauma, like liver laceration and bowel perforation, are less likely when a blunt needle is used (Abrahams et al., 2016; Lancaster & Chadwick, 2010).

Quadratus lumborum block

Blanco first described this block as a variant of the TAP block and then named it as quadratus lumborum block (QLB; Blanco, Ansari, Riad, & Shetty, 2016). It is an interfascial plane block located in the posterior part of the abdominal wall. Several researchers have modified their approaches in order to achieve the maximum spread of the local anesthetic drug with better outcomes.

It may provide visceral analgesia and can be used in cesarean section and total abdominal hysterectomies (Blanco et al., 2016; Yousef, 2018). A recent study revealed that in laparoscopic hysterectomy, total opioid consumption is not reduced with QLB (Hansen et al., 2021). In cesarean surgeries, QLB provides better analgesia than TAP block and, when combined with multimodal analgesia, provides adequate analgesia compared to spinal anesthesia (Blanco et al., 2016; Blanco, Ansari, & Girgis, 2015).

This block can be done preoperatively, intraoperatively, postoperatively, even in PACU and ICU. The patient can be under general anesthesia or regional anesthesia.

The quadratus lumborum muscle is the deepest abdominal muscle and lies between the crista iliaca, 12th rib, and the L1-L5 lumbar vertebra's transverse processes. Layers of fascia transversalis and fascia thoracolumbaris cover the muscles, communicate with another fascia, and aid in the spread of the injectate. Due to thoracic paravertebral and lumbar plexus spread, visceral analgesia can be achieved (Akerman, Pejčić, & Veličković, 2018; Ueshima, Otake, & Lin, 2017). If a longer duration of analgesia is required, a catheter can be placed (Visoiu & Yakovleva, 2013).

In this block, an ultrasound device, color Doppler, pressure monitors, and peripheral nerve stimulator usage are beneficial as the vicinity of lumbar arteries, large vessels, nerves, and lumbar plexus may lead to complications (Ueshima et al., 2017). The duration of the block is approximately 24–48 h (Murouchi, Iwasaki, & Yamakage, 2016).

The patient's position varies according to the physician's preference, patient mobility, and planned approach. Supine with or without a lateral tilt, lateral, sitting, and prone positions can be used (Elsharkawy, El-Boghdadly, & Barrington, 2019). Below the most commonly used positions are mentioned.

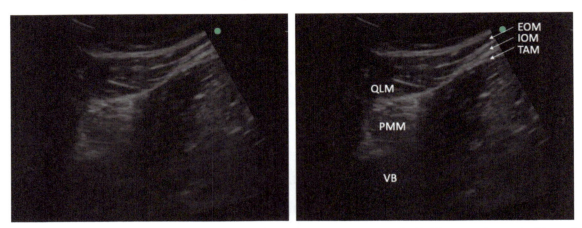

FIG. 3 Ultrasound anatomy of quadratus lumborum block (QLB). *QLM*, quadratus lumborum muscle; *PMM*, psoas major muscle; *VB*, vertebral body; *EOM*, external oblique muscle; *IOM*, internal oblique muscle; *TAM*, transversus abdominis muscle.

a. *Anterior QLB (transmuscular block) (QL3 block)*: The patient lies in the lateral position, and the probe is placed above the crista iliaca. The needle is advanced anteromedially, and the drug is injected into the plane between the quadratus lumborum and psoas muscle (Fig. 3). A less peritoneal breach is expected with this method (Ueshima et al., 2017).

Subcostal (paramedian sagittal oblique) QLB: It is also an anterior block. *The probe is* placed 3 cm lateral to the L2 vertebra. The needle is advanced laterally until it reaches the targeted area.

"Shamrock sign" can be viewed in this approach, corpus vertebra as the stem, erector spinae muscle as the posterior leaf, psoas major muscle as the anterior leaf, and quadratus lumborum muscle as the lateral leaf (Børglum et al., 2013).

b. *Lateral QLB (QL1 block)*: The patient lies in the supine position, the probe is placed in Petit's triangle and moved posteriorly to visualize quadratus lumborum muscle. The drug is injected into the anterolateral border of the quadratus lumborum, where it contacts the transversalis fascia.

c. *Posterior QLB (QL2 block)*: The patient lies in the supine position, and the injectate is given between quadratus lumborum and erector spinae muscles.

d. *Intramuscular QLB (QL4 block)*: The patient lies in the supine position, and the injectate is given into the quadratus lumborum muscle.

Injectates

Ropivacaine, bupivacaine and levobupivacaine can be used per side in low concentration (0.2%–0.5%, 0.1%–0.5%, 0.25%–0.5% respectively) and high volumes, i.e., 20–25 mL (Elsharkawy et al., 2019; Fernandes et al., 2021). An optimal dose and concentration are yet to be found. Bilateral application of 15–30 mL 0.125%–0.375% local anesthetic is proposed (Akerman et al., 2018). In laparoscopic ovarian surgery, 20 mL of 0.375% ropivacaine per side was effective and long-lasting (Murouchi et al., 2016).

Dexamethasone and epinephrine can be added to local anesthetic to ease nausea and vomiting, and it also prolongs the duration of the block (Akerman et al., 2018; Elsharkawy et al., 2019).

Complications

In a patient with bleeding diathesis, this block is not recommended (Elsharkawy et al., 2019). As this is a deep block visceral injury, peritoneal or pleural breach, hematoma, infection, inadvertent neural injury, vascular injection, LAST can be encountered. Motor block and lower limb weakness may accompany a successful block. Hypotension and tachycardia are also reported (Sa et al., 2018). The kidneys are nearby, and visualizing them during the injection is essential.

Rectus sheath block

Since its first description in 1899 by Schleich, this block has been used mostly for analgesia and anesthesia in specific cases (Quek & Phua, 2014). The aponeurosis of the external oblique, internal oblique, and transversus abdominis muscles form the anterior and posterior rectus sheaths (Fig. 1). It is located under the superficial fascia (Fig. 4). Small epigastric arteries

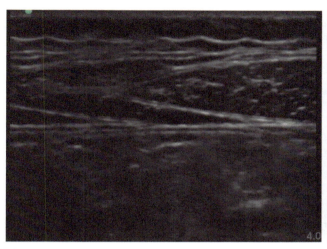 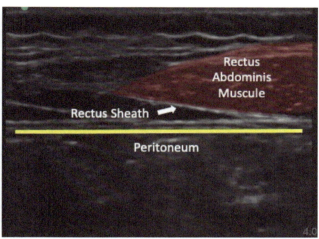

FIG. 4 Ultrasound anatomy of rectus sheath block.

and 9–11th intercostal nerves course are located in the plane between the rectus abdominis muscle and the posterior rectus sheath.

The bilateral rectus sheath block provides analgesia from the xyphoid process to the symphysis pubis. This block is used as a part of multimodal analgesia treatments as it provides somatic pain relief. This anterior wall block is useful for postoperative analgesia in umbilical surgeries; major gynecological and laparoscopic surgeries may also benefit from this block (Azemati & Khosravi, 2005; Crosbie, Massiah, Achiampong, Dolling, & Slade, 2012). A combination of TAP and rectus sheath blocks can be used according to the place of the incision, but safe anesthetic drug dosage must be calculated (Abdelsalam & Mohamdin, 2016).

This block provides denser analgesia, but for a shorter period than the TAP block; therefore, it is advised to place a catheter for more prolonged pain relief (Quek & Phua, 2014). Lower limb weakness, hemodynamic changes are not encountered; thus, early mobilization and fewer complications are seen.

It can be done while the patient is still under anesthesia, sedation, or awake.

Traditional blind technique: the needle is advanced through the muscle and fascia of the anterior abdominal wall.

Ultrasound-assisted technique: The linear probe is placed lateral to the umbilicus in a transverse position when the patient is lying in the supine position. Both in-plane and out-of-plane techniques can be used. In obese patients, out-of-plane can be much easier to apply. Injectate spread is viewed between the rectus muscle and the posterior rectus sheath.

Surgical technique: the block is done under direct visualization of the muscle.

Injectate

Bilateral 10–20 mL of local anesthetic drug (ropivacaine 0.25%–0.5%, bupivacaine 0.25%–0.5%, or levobupivacaine 0.25%–0.5%) administration provides sufficient postoperative analgesia (Fernandes et al., 2021; Jin, Li, Tan, Ma, & Lu, 2015).

Complication

Peritoneal breaching, infection, visceral organ puncture, and vascular puncture leading to rectus sheath hematoma or intravenous administration of local anesthetics can be encountered.

Iliohypogastric and ilioinguinal nerve block

The ilioinguinal and iliohypogastric (IH-II) nerves are branches of the first lumbar nerve. They emerge from the lateral border of the psoas major muscle. They coarse between the transverse abdominis and the internal oblique muscles (Fig. 5). Their sensory innervation of the lower abdominal wall's skin makes the nerve block useful for pain treatment after cesarean sections (Bell et al., 2002). This block is also used as an adjunct to other analgesia techniques.

Preoperative and postoperative injection of 30 mL of 0.5% bupivacaine +5 μg/mL epinephrine revealed that after female reproductive tract operations, pain scores and narcotic analgesic usage were not affected (Wehbe et al., 2008),

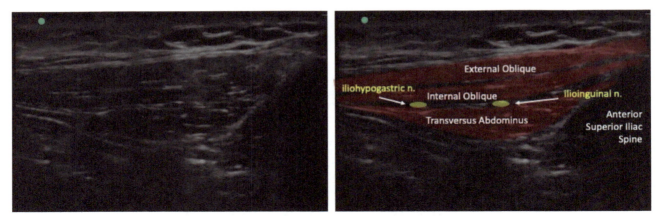

FIG. 5 Ultrasound anatomy of iliohypogastric and ilioinguinal (IH-II) nerve block.

but in nonlaparoscopic gynecologic surgery, postoperative bilateral block achieved by injecting 15 mL of a mixture of 5 mg/mL ropivacaine and 0.5 μg/kg clonidine decreased morphine consumption (Oriola et al., 2007).

The ultrasound probe is placed on the imaginary line that joins the umbilicus and anterior superior iliac spine while the patient is lying in the supine position. With an in-plane approach, after visualizing the nerves, a local anesthetic is injected, and the spread between the transversus abdominis and internal oblique muscle planes is seen. Color Doppler is used before injection to verify the nerves and the vessels.

Injectate

Bilateral 10–15 mL of local anesthetic (Ropivacaine 0.25%–0.5%, Bupivacaine 0.25%–0.5%, or Levobupivacaine 0.25%–0.5%) is injected (Chakraborty et al., 2016; Fernandes et al., 2021).

Complication

Intraperitoneal injection, intravascular injection, quadriceps paralysis, femoral nerve block, and bowel perforation can be seen (Fernandes et al., 2021; Oriola et al., 2007).

Transversalis fascia plane block

The ilioinguinal and iliohypogastric nerves are blocked where they course between the fascia of the transversus abdominis muscle (thoracolumbar fascia) and the transversal fascia.

TAP may not cover the L1 dermatome, and variations may exist in the course of iliohypogastric and ilioinguinal nerves (Choudhary, Mishra, & Jadhav, 2018). However, variations are less in the site of transversalis fascia plane (TFP) block, and analgesic treatment of surgeries involving L1 dermatome benefit from this block (Choudhary et al., 2018). Post cesarean pain treatment may also benefit from this block (Tulgar & Serifsoy, 2019).

The probe (linear or curvilinear) is placed on the midaxillary line when the patient is in the lateral or supine position. The probe is then moved posteriorly and visualizes the aponeurosis formed by transversus abdominis and internal oblique muscles, and between the transversus abdominis and the extraperitoneal fascia, the transversalis fascia is seen. The needle is advanced with an in-plane technique. Hydro dissection can be used for correct needle tip position, and local anesthetic spread is seen over the anterior surface of the quadratus lumborum (Hebbard, 2009).

Injectate

A low concentration of 20 mL local anesthetic with or without epinephrine is used.

Complication

Motor weakness, block failure, LAST, peritoneal penetration, liver trauma can be seen with this block (Hebbard, 2009; Lee, Goetz, & Gharapetian, 2015).

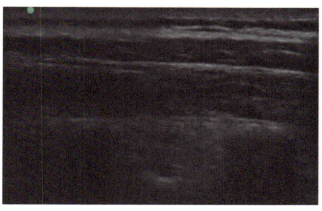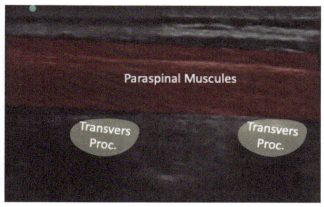

FIG. 6 Ultrasound anatomy of paravertebral block (PVB).

Lumbar paravertebral block

It is also called a *Paravertebral lumbar sympathetic block, paravertebral nerve block, paraspinal epidural blocks* (Fig. 6). Spinal nerves are blocked when they come out of the intervertebral foramina. A single level block will affect five unilateral somatic dermatomal levels and eight sympathetic dermatomes (Mitchell et al., 2019). For the cesarean section, T10-L2 levels should be blocked.

The patient can be sitting, prone, or in a lateral decubitus position. Nerve stimulation and color Doppler can also be used to enhance safety.

Conventional Landmark-Guided Technique: Insertion point is 2–2.5 cm lateral to the spinous process. The needle is advanced perpendicular to the skin until contacting the transverse process. With a "pop" the costotransverse ligament is penetrated, and loss of air resistance happens.

Ultrasound-Guided Technique: Linear probe is used to locate the paravertebral space. Although the *parasagittal in-plane approach* leads to less risk for epidural spread than the *transverse in-plane approach*, needle visualization is more inadequate. The needle is advanced to the costotransverse ligament, and the drug is injected with direct visualization (Mitchell et al., 2019). The out-of-plane technique can also be used (Batra, Krishnan, & Agarwal, 2011).

Injectate

Bilateral 2–7 mL of local anesthetic with or without epinephrine is injected, duration of action is approximately 9–12 h (Mitchell et al., 2019). For labor analgesia, 15 mL of 0.375% bupivacaine with 1:400,000 epinephrine and a pudendal block is successfully used (Suelto & Shaw, 1999).

Complication

The epidural or intrathecal spread of the drug, unintentional epidural or intrathecal injections, pleural puncture and pneumothorax, spinal cord injury, vascular puncture, hypotension, pulmonary hemorrhage, contralateral harlequin, and ipsilateral Horner's syndrome can be seen.

Erector spina plane block

Forero first described it in 2016 (Forero, Adhikary, Lopez, Tsui, & Chin, 2016). Spinalis, longissimus, and iliocostalis muscles form the erector spinae muscle.

Sitting, prone, or lateral decubitus positions can be used depending on the patient's condition (awake or under anesthesia). The blind technique or fluoroscopy, or an ultrasound can be used (Kot et al., 2019). The probe (linear or curvilinear) is placed 3 cm lateral to the spinous process, and when the transverse process is identified, it is rotated to stand in a parasagittal plane. A local anesthetic is injected into the plane between the erector spinae muscles and the transverse processes of thoracic or lumbar vertebrae (Fig. 7).

It is effective in pain treatment and relatively easy to apply with low complication risk (Kot et al., 2019). It is mostly used as a component of multimodal analgesia treatment.

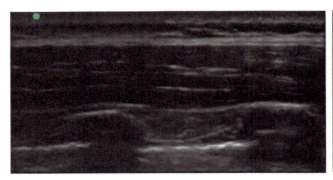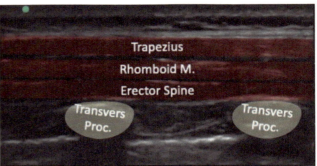

FIG. 7 Ultrasound anatomy of erector spinae (ESP) block.

It is a relatively new defined technique, and more research is still needed. Erector spina plane (ESP), compared to TAP block after abdominal hysterectomy, 20 mL of bupivacaine 0.375% plus 5 μg/mL adrenaline (1:200,000) in each side at the level of T9 provided better and longer analgesia with less morphine consumption (Kamel, Amin, & Ibrahem, 2020).

Injectate
Low concentration-high volume of local anesthetics are used.

Conclusion

Regional anesthesia techniques and approaches are still evolving. New drugs or old drugs with new indications are also being investigated to obtain better analgesia control with fewer systemic side effects.

Truncal blocks are more frequently used as the imaging technology enables them to be performed more precisely with real-time direct visualization of the drugs spread and anatomical structures.

Every person has a different pain sensation, therefore "one block does not fit" all the patients with similar surgeries. Pain treatment options must be planned before the surgery according to the surgery planned and the history of the patient.

The efficacy of the truncal blocks in different settings demonstrated by the researchers makes them to be used more frequently but optimal dosing protocols are still needed.

Application to other areas

Truncal blocks used in obstetric and gynecologic surgeries are not specific to these operations. All surgeries involving the abdominal wall may benefit the analgesic properties of these blocks. One must take into consideration of the surgical site and level of injection when applying the block as described in this chapter, as the obstetric and gynecologic surgeries mostly involve the lower abdominal wall. Truncal blocks can be used as an alternative to epidural analgesia especially when epidural approach is contraindicated. With the guidance of ultrasound, these blocks are done safely and control pain efficiently. Although they are mostly used for acute postoperative pain treatment, they can be useful in chronic pain treatment as well.

Other agents of interest

Epidural analgesia is considered the gold standard method after abdominal surgeries but the search for providing better analgesia with fewer complications is still ongoing. New approaches to already in use methods and novel drug combinations are being evaluated for this purpose. Robotic assistance, injection of the drugs from inside the abdomen, combining different blocks and timing of block performance (pre or postop).

Mini-dictionary of terms

Dermatome: It is an area of skin in which a single spinal nerve root is responsible for its sensation.
Local anesthetic drug: Drugs that create a loss of pain sensation without affecting consciousness.

Neuraxial block: Injection of local anesthetic drugs around the nerves of the central nervous system in the spinal column to achieve analgesia and anesthesia.

Peripheral nerve block: Injection of local anesthetic drug near a specific nerve or bundle of nerves to achieve analgesia and anesthesia.

Truncal block: Injection of local anesthetic drugs into muscle planes of the torso to achieve analgesia.

Key facts

Key facts of truncal blocks in obstetric and gynecologic surgery

- Truncal blocks provide efficient postoperative analgesia.
- Complications related to truncal blocks resemble other nerve blocks, caution must be taken when applying truncal blocks.
- Truncal blocks are widely used as an adjunct to multimodal analgesia. They reduce narcotic analgesic consumption.
- Wide accessibility to an ultrasound machine in the operating rooms leads to performing these blocks under direct visualization of the anatomical structure.
- Single shot, multiinjection, repeated injections, or catheter placement is possible. The technique must be planned before the operation.

Summary points

- Regional anesthesia techniques are still evolving due to technological improvements.
- Patients with comorbidities, elderly patients, emergency cases, and critically ill patients sometimes cannot get neuraxial pain management treatments.
- Truncal blocks when performed by direct visualization, i.e., real-time ultrasonography, have a lower complication and higher success rates as compared to traditional landmark techniques.
- Some of the truncal blocks are easy to learn and perform but a good knowledge of anatomy, pharmacology, and competence in ultrasound usage is essential.
- Truncal blocks can be used together with other neuraxial and truncal blocks, but maximum local anesthetic doses must be calculated.

References

Abdallah, F. W., Laffey, J. G., Halpern, S. H., & Brull, R. (2013). Duration of analgesic effectiveness after the posterior and lateral transversus abdominis plane block techniques for transverse lower abdominal incisions: A meta-analysis. *British Journal of Anaesthesia, 111*(5), 721–735. https://doi.org/10.1093/bja/aet214.

Abdelsalam, K., & Mohamdin, O. W. (2016). Ultrasound-guided rectus sheath and transversus abdominis plane blocks for perioperative analgesia in upper abdominal surgery: A randomized controlled study. *Saudi Journal of Anaesthesia, 10*(1), 25–28. https://doi.org/10.4103/1658-354X.169470.

Abrahams, M., Derby, R., & Horn, J. L. (2016). Update on ultrasound for truncal blocks: A review of the evidence. *Regional Anesthesia and Pain Medicine, 41*(2), 275–288. https://doi.org/10.1097/AAP.0000000000000372.

Akerman, M., Pejčić, N., & Veličković, I. (2018). A review of the quadratus lumborum block and ERAS [review]. *Frontiers in Medicine, 5*(44). https://doi.org/10.3389/fmed.2018.00044.

Aly, N. M., Talaat, S. M., Menshawi, M. A., & Mohammed, E. R. (2020). Bilateral transversus abdominis plane block versus epidural analgesia for postoperative pain relief in lower abdominal surgery. *QJM, 113*(Suppl. 1). https://doi.org/10.1093/qjmed/hcaa039.003.

Azemati, S., & Khosravi, M. B. (2005). An assessment of the value of rectus sheath block for postlaparoscopic pain in gynecologic surgery. *Journal of Minimally Invasive Gynecology, 12*(1), 12–15. https://doi.org/10.1016/j.jmig.2004.12.013.

Batra, R. K., Krishnan, K., & Agarwal, A. (2011). Paravertebral block. *Journal of Anaesthesiology Clinical Pharmacology, 27*(1), 5–11. https://www.ncbi.nlm.nih.gov/pubmed/21804697.

Bell, E. A., Jones, B. P., Olufolabi, A. J., Dexter, F., Phillips-Bute, B., Greengrass, R. A., et al. (2002). Iliohypogastric-ilioinguinal peripheral nerve block for post-cesarean delivery analgesia decreases morphine use but not opioid-related side effects. *Canadian Journal of Anaesthesia, 49*(7), 694–700. https://doi.org/10.1007/BF03017448.

Blanco, R., Ansari, T., & Girgis, E. (2015). Quadratus lumborum block for postoperative pain after caesarean section: A randomised controlled trial. *European Journal of Anaesthesiology, 32*(11), 812–818. https://doi.org/10.1097/EJA.0000000000000299.

Blanco, R., Ansari, T., Riad, W., & Shetty, N. (2016). Quadratus lumborum block versus transversus abdominis plane block for postoperative pain after cesarean delivery: A randomized controlled trial. *Regional Anesthesia and Pain Medicine, 41*(6), 757–762. https://doi.org/10.1097/AAP.0000000000000495.

Børglum, J., Moriggl, B., Jensen, K., Lønnqvist, P. A., Christensen, A. F., Sauter, A., et al. (2013). Ultrasound-guided transmuscular quadratus lumborum blockade. *British Journal of Anaesthesia, 111*(eLetters Suppl). https://doi.org/10.1093/bja/el_9919.

Bortolato, A., Ori, C., & Freo, U. (2015). Transient abdominal motor block after a transversus abdominis plane block in an elderly patient. *Canadian Journal of Anaesthesia, 62*(7), 837–838. https://doi.org/10.1007/s12630-015-0341-z.

Chakraborty, A., Khemka, R., & Datta, T. (2016). Ultrasound-guided truncal blocks: A new frontier in regional anaesthesia. *Indian Journal of Anaesthesia, 60*(10), 703–711. https://doi.org/10.4103/0019-5049.191665.

Choudhary, J., Mishra, A. K., & Jadhav, R. (2018). Transversalis fascia plane block for the treatment of chronic postherniorrhaphy inguinal pain: A case report. *A&A Practice, 11*(3), 57–59. https://doi.org/10.1213/XAA.0000000000000730.

Crosbie, E. J., Massiah, N. S., Achiampong, J. Y., Dolling, S., & Slade, R. J. (2012). The surgical rectus sheath block for post-operative analgesia: A modern approach to an established technique. *European Journal of Obstetrics, Gynecology, and Reproductive Biology, 160*(2), 196–200. https://doi.org/10.1016/j.ejogrb.2011.10.015.

Elsharkawy, H., El-Boghdadly, K., & Barrington, M. (2019). Quadratus lumborum block: Anatomical concepts, mechanisms, and techniques. *Anesthesiology, 130*(2), 322–335. https://doi.org/10.1097/ALN.0000000000002524.

Faiz, S. H. R., Alebouyeh, M. R., Derakhshan, P., Imani, F., Rahimzadeh, P., & Ghaderi Ashtiani, M. (2018). Comparison of ultrasound-guided posterior transversus abdominis plane block and lateral transversus abdominis plane block for postoperative pain management in patients undergoing cesarean section: A randomized double-blind clinical trial study. *Journal of Pain Research, 11*, 5–9. https://doi.org/10.2147/JPR.S146970.

Fernandes, H. D. S., Azevedo, A. S., Ferreira, T. C., Santos, S. A., Rocha-Filho, J. A., & Vieira, J. E. (2021). Ultrasound-guided peripheral abdominal wall blocks. *Clinics (São Paulo, Brazil), 76*, e2170. https://doi.org/10.6061/clinics/2021/e2170.

Fletcher, D., Stamer, U. M., Pogatzki-Zahn, E., Zaslansky, R., Tanase, N. V., Perruchoud, C., et al. (2015). Chronic postsurgical pain in Europe: An observational study. *European Journal of Anaesthesiology, 32*(10), 725–734. https://doi.org/10.1097/EJA.0000000000000319.

Forero, M., Adhikary, S. D., Lopez, H., Tsui, C., & Chin, K. J. (2016). The erector spinae plane block: A novel analgesic technique in thoracic neuropathic pain. *Regional Anesthesia and Pain Medicine, 41*(5), 621–627. https://doi.org/10.1097/AAP.0000000000000451.

Griffiths, J. D., Le, N. V., Grant, S., Bjorksten, A., Hebbard, P., & Royse, C. (2013). Symptomatic local anaesthetic toxicity and plasma ropivacaine concentrations after transversus abdominis plane block for caesarean section. *British Journal of Anaesthesia, 110*(6), 996–1000. https://doi.org/10.1093/bja/aet015.

Hansen, C., Dam, M., Nielsen, M. V., Tanggaard, K. B., Poulsen, T. D., Bendtsen, T. F., et al. (2021). Transmuscular quadratus lumborum block for total laparoscopic hysterectomy: A double-blind, randomized, placebo-controlled trial. *Regional Anesthesia and Pain Medicine, 46*(1), 25. https://doi.org/10.1136/rapm-2020-101931.

Hebbard, P. D. (2009). Transversalis fascia plane block, a novel ultrasound-guided abdominal wall nerve block. *Canadian Journal of Anesthesia, 56*(8), 618–620. https://doi.org/10.1007/s12630-009-9110-1.

Hebbard, P. D., Barrington, M. J., & Vasey, C. (2010). Ultrasound-guided continuous oblique subcostal transversus abdominis plane blockade: Description of anatomy and clinical technique. *Regional Anesthesia and Pain Medicine, 35*(5), 436–441. https://doi.org/10.1097/aap.0b013e3181e66702.

Hebbard, P., Fujiwara, Y., Shibata, Y., & Royse, C. (2007). Ultrasound-guided transversus abdominis plane (TAP) block. *Anaesthesia and Intensive Care, 35*(4), 616–617. https://www.ncbi.nlm.nih.gov/pubmed/18020088.

Jin, F., Li, X. Q., Tan, W. F., Ma, H., & Lu, H. W. (2015). Preoperative versus postoperative ultrasound-guided rectus sheath block for improving pain, sleep quality and cytokine levels of patients with open midline incisions undergoing transabdominal gynaecological operation: Study protocol for a randomised controlled trial. *Trials, 16*, 568. https://doi.org/10.1186/s13063-015-1096-0.

Kamel, A. A. F., Amin, O. A. I., & Ibrahem, M. A. M. (2020). Bilateral ultrasound-guided erector spinae plane block versus transversus abdominis plane block on postoperative analgesia after total abdominal hysterectomy. *Pain Physician, 23*(4), 375–382. https://www.ncbi.nlm.nih.gov/pubmed/32709172.

Koh, W. U., & Lee, J. H. (2018). Ultrasound-guided truncal blocks for perioperative analgesia. *Anesthesia and Pain Medicine, 13*(2), 128–142. https://doi.org/10.17085/apm.2018.13.2.128.

Kot, P., Rodriguez, P., Granell, M., Cano, B., Rovira, L., Morales, J., et al. (2019). The erector spinae plane block: A narrative review. *Korean Journal of Anesthesiology, 72*(3), 209–220. https://doi.org/10.4097/kja.d.19.00012.

Lancaster, P., & Chadwick, M. (2010). Liver trauma secondary to ultrasound-guided transversus abdominis plane block. *British Journal of Anaesthesia, 104*(4), 509–510. https://doi.org/10.1093/bja/aeq046.

Lee, S., Goetz, T., & Gharapetian, A. (2015). Unanticipated motor weakness with ultrasound-guided transversalis fascia plane block. *A&A Practice, 5*(7). https://journals.lww.com/aacr/Fulltext/2015/10010/Unanticipated_Motor_Weakness_with.6.aspx.

Ljungqvist, O. (2014). ERAS-enhanced recovery after surgery: Moving evidence-based perioperative care to practice. *JPEN Journal of Parenteral and Enteral Nutrition, 38*(5), 559–566. https://doi.org/10.1177/0148607114523451.

McDonnell, J. G., O'Donnell, B. D., Farrell, T., Gough, N., Tuite, D., Power, C., et al. (2007). Transversus abdominis plane block: A cadaveric and radiological evaluation. *Regional Anesthesia and Pain Medicine, 32*(5), 399–404. https://doi.org/10.1016/j.rapm.2007.03.011.

Mishra, M., & Mishra, S. P. (2016). Transversus abdominis plane block: The new horizon for postoperative analgesia following abdominal surgery. *Egyptian Journal of Anaesthesia, 32*(2), 243–247. https://doi.org/10.1016/j.egja.2015.12.003.

Mitchell, K. D., Smith, C. T., Mechling, C., Wessel, C. B., Orebaugh, S., & Lim, G. (2019). A review of peripheral nerve blocks for cesarean delivery analgesia. *Regional Anesthesia and Pain Medicine*. https://doi.org/10.1136/rapm-2019-100752.

Murouchi, T., Iwasaki, S., & Yamakage, M. (2016). Quadratus lumborum block: Analgesic effects and chronological ropivacaine concentrations after laparoscopic surgery. *Regional Anesthesia and Pain Medicine, 41*(2), 146–150. https://doi.org/10.1097/AAP.0000000000000349.

Oriola, F., Toque, Y., Mary, A., Gagneur, O., Beloucif, S., & Dupont, H. (2007). Bilateral ilioinguinal nerve block decreases morphine consumption in female patients undergoing nonlaparoscopic gynecologic surgery. *Anesthesia and Analgesia, 104*(3), 731–734. https://doi.org/10.1213/01.ane.0000255706.11417.9b.

Park, S. J., Kim, H. J., Yang, H. M., Yoon, K. B., Lee, K. Y., Ha, T., et al. (2020). Impact of simulation-based anesthesiology training using an anesthetized porcine model for ultrasound-guided transversus abdominis plane block. *The Journal of International Medical Research, 48*(3). https://doi.org/10.1177/0300060519896909. 300060519896909.

Pawa, A., & El-Boghdadly, K. (2018). Regional anesthesia by nonanesthesiologists. *Current Opinion in Anaesthesiology, 31*(5), 586–592. https://doi.org/10.1097/ACO.0000000000000643.

Pogatzki-Zahn, E. M., Segelcke, D., & Schug, S. A. (2017). Postoperative pain-from mechanisms to treatment. *Pain Reports, 2*(2), e588. https://doi.org/10.1097/PR9.0000000000000588.

Quek, K. H., & Phua, D. S. (2014). Bilateral rectus sheath blocks as the single anaesthetic technique for an open infraumbilical hernia repair. *Singapore Medical Journal, 55*(3), e39–e41. https://doi.org/10.11622/smedj.2014042.

Rafi, A. N. (2001). Abdominal field block: A new approach via the lumbar triangle. *Anaesthesia, 56*(10), 1024–1026. https://doi.org/10.1046/j.1365-2044.2001.02279-40.x.

Ravichandran, N. T., Sistla, S. C., Kundra, P., Ali, S. M., Dhanapal, B., & Galidevara, I. (2017). Laparoscopic-assisted tranversus abdominis plane (TAP) block versus ultrasonography-guided transversus abdominis plane block in postlaparoscopic cholecystectomy pain relief: Randomized controlled trial. *Surgical Laparoscopy, Endoscopy & Percutaneous Techniques, 27*(4), 228–232. https://doi.org/10.1097/SLE.0000000000000405.

Sa, M., Cardoso, J. M., Reis, H., Esteves, M., Sampaio, J., Gouveia, I., et al. (2018). Quadratus lumborum block: Are we aware of its side effects? A report of 2 cases. *Revista Brasileira de Anestesiologia, 68*(4), 396–399. https://doi.org/10.1016/j.bjan.2017.04.023 (Bloqueio do quadrado lombar: estamos cientes de seus efeitos colaterais? Relato de dois casos.).

Shahait, M., Yezdani, M., Katz, B., Lee, A., Yu, S. J., & Lee, D. I. (2019). Robot-assisted transversus abdominis plane block: Description of the technique and comparative analysis. *Journal of Endourology, 33*(3), 207–210. https://doi.org/10.1089/end.2018.0828.

Suelto, M. D., & Shaw, D. B. (1999). Labor analgesia with paravertebral lumbar sympathetic block. *Regional Anesthesia and Pain Medicine, 24*(2), 179–181. https://doi.org/10.1016/s1098-7339(99)90082-2.

Tran, D. Q., Bravo, D., Leurcharusmee, P., & Neal, J. M. (2019). Transversus abdominis plane block: A narrative review. *Anesthesiology, 131*(5), 1166–1190. https://doi.org/10.1097/ALN.0000000000002842.

Tulgar, S., & Serifsoy, T. E. (2019). Transversalis fascia plane block provides effective postoperative analgesia for cesarean section: New indication for known block. *Obstetric Anesthesia Digest, 39*(3). https://journals.lww.com/obstetricanesthesia/Fulltext/2019/09000/Transversalis_Fascia_Plane_Block_Provides.60.aspx.

Ueshima, H., Otake, H., & Lin, J. A. (2017). Ultrasound-guided quadratus lumborum block: An updated review of anatomy and techniques. *BioMed Research International, 2017*, 2752876. https://doi.org/10.1155/2017/2752876.

Vial, F., Mory, S., Guerci, P., Grandjean, B., Petry, L., Perrein, A., et al. (2015). Evaluating the learning curve for the transversus abdominal plane block: A prospective observational study. *Canadian Journal of Anaesthesia, 62*(6), 627–633. https://doi.org/10.1007/s12630-015-0338-7 (Evaluation de la courbe d'apprentissage du bloc du plan transverse abdominal: Etude prospective observationnelle.).

Visoiu, M., & Yakovleva, N. (2013). Continuous postoperative analgesia via quadratus lumborum block—An alternative to transversus abdominis plane block. *Paediatric Anaesthesia, 23*(10), 959–961. https://doi.org/10.1111/pan.12240.

Wehbe, S. A., Ghulmiyyah, L. M., Dominique El, K. H., Hosford, S. L., Ehleben, C. M., Saltzman, S. L., et al. (2008). Prospective randomized trial of iliohypogastric-ilioinguinal nerve block on post-operative morphine use after inpatient surgery of the female reproductive tract. *Journal of Negative Results in Biomedicine, 7*, 11. https://doi.org/10.1186/1477-5751-7-11.

Yousef, N. K. (2018). Quadratus lumborum block versus transversus abdominis plane block in patients undergoing total abdominal hysterectomy: A randomized prospective controlled trial. *Anesthesia, Essays and Researches, 12*(3), 742–747. https://doi.org/10.4103/aer.AER_108_18.

Zhong, T., Wong, K. W., Cheng, H., Ojha, M., Srinivas, C., McCluskey, S. A., et al. (2013). Transversus abdominis plane (TAP) catheters inserted under direct vision in the donor site following free DIEP and MS-TRAM breast reconstruction: A prospective cohort study of 45 patients. *Journal of Plastic, Reconstructive & Aesthetic Surgery, 66*(3), 329–336. https://doi.org/10.1016/j.bjps.2012.09.034.

Zhou, H., Ma, X., Pan, J., Shuai, H., Liu, S., Luo, X., et al. (2018). Effects of transversus abdominis plane blocks after hysterectomy: A meta-analysis of randomized controlled trials. *Journal of Pain Research, 11*, 2477–2489. https://doi.org/10.2147/JPR.S172828.

Index

Note: Page numbers followed by *f* indicate figures and *t* indicate tables.

A

Abdominal wall, 630
   truncal blocks of, 631–633, 631*t*
Aβ-fibers, 378–379, 381
Abietic acid, 405–406, 405*t*, 407–408*t*, 413–415
Ablation, 597
Accupal, 584
Acetaminophen (APAP), 46, 285–286, 309
   acute poisoning, 312
   applications, 314–315
   chronic poisoning, 312–313
   delayed hypersensitivity, 313–314
   diagnosis, 314
   drug provocation test, 314
   hepatotoxicity, 310–311
   immediate hypersensitivity, 313
   key facts of, 316
   management, 314
   mechanism, 310
   poisoning, 314
   risk factors of, 311
Acetophenones, 322–323
Acetylcholine muscarinic 2 receptor
      (AchM2R), 479
Acetylcholine theory, 157
Acetylsalicylic acid, 315, 457
Acid-sensing channels 2 (ASIC2), 476, 478
Acronychia pedunculata
   active compounds of, 324
   analgesic activity, 325
   antiinflammatory activity, 323–324
   antiinflammatory fractions, identification of,
      324
   applications, 325
   biological activities, 322–323
   chemistry, 322–323
   key facts, 326
   as Sri Lankan medicinal plant, 322
Acrovestone, 322–323
Activator protein-1 (AP-1), 355
Active cooling, 8
Active distraction, 575
Acupoints, 609–610, 612
Acupuncture, 609–610
Acustimulation, 609
   acupuncture, 609–610
   applications, 609–610
   electroanalgesia, 610, 611*t*
   overview, 610

Acute dental pain, 608
Acute hepatic failure, 315
Acute joint injuries, 608
Acute pain, 608
Acute poisoning, 312
Acute postoperative pain, 608
Adansonia digitata (Baobab), 343
   antiviral, 345
   central sensitization pathways, 341–343
   description of, 329–330
   distribution, 330
   inflammatory pathway, 341
   ions-mediated pathways, 343
   key facts, 346
   malaria, 344
   medicinal importance, 331
   metabolic diseases, 343
   mineral components, 332
   on neuropathic pain, 332–337
   neuroprotection, 344–345
   oxidative stress pathway, 341
   on pain sensation, 332–337
   peripheral pathway, 341–343
   phytochemical constituents, 332, 333–338*t*
   prostaglandin E2 pathway, 340–341
   therapeutic potential of, 344*t*
   use of, 330–331, 331*t*
Aδ fibers, 378–379, 381
Adrenaline, 23, 386, 397
Adrenocorticotropic hormone (ACTH), 286
Adverse drug event, 309, 315
Aerobic activity, 519
Aerobic/anaerobic training, 501
Aerobic exercise, pediatric chronic pain
      intensity
   conditions, 515–516
   effect of, 514–517
   individual differences, 517
   levels in, 513
   strength training, 517
Allodynia, 406–408, 412–413, 438–439, 609
Alpha-adrenergic receptors, 82
Alpha-2 agonist drugs, 85
Alpha-lipoic acid, 370
American Pain Society (APS), 601–602
American Society of Anesthesiologists Physical
      Status (ASA-PS) I and II patients, 23,
      134–136, 601–602
Amides, 153

Aminooxyacetic acid (AOAA), 494
Analgesia-breastfeeding issues, 213
   confounders, choice of, 215, 216–217*t*
   enrolment criteria, 213
   labor neuraxial blockade techniques, different
      accuracy in, 213–215, 216*t*
   limitations of, 213
   methodological flaws, 215–217
   new research opportunity, 218
   sampling strategies, 213, 214–215*t*
   study designs, 213
Analgesic Nociception Index (ANI), 34
Anaphylaxis, 313, 315
Andrographis paniculata, 359
   activator protein-1 (AP-1), 355
   analgesic effects, 352–353
   antioxidant effect of, 355–357
   classification, 351–352
   clinical pharmacology, 357
   composition, 351–352
   contraindications, 357–358
   in inflammation and pain, 358
   key facts, 359
   metabolism, 357
   on mitogen-activated protein kinases
      (MAPK), 355
   on nuclear factor-kappa B (NF-κB) pathway,
      353–355
   in ParActin, 352–353
   pharmacokinetics, 357
   preclinical antiinflammatory, 352–353
   side effects, 357
Andrographolide sulfate, 359
Ankenda. *See Acronychia pedunculata*
Anterior approach, 633
Anterior rectus sheath, 631*f*
Antiarrhythmic agents, 8–9
Antidote, 312, 315
Antiinflammatory cytokines, 447–448
Antiinflammatory drugs, 326
Antimicrobial agent, 493
Antinociception, 459–463, 460–463*t*, 467
Antinociceptive activity, 421–425
Antioxidants, 467–468
   of andrographolide, 355–357
Antioxidants, in prenatal stress, 490, 493
   antimicrobial agent, 493
   cardiovascular disease, 493
   diabetes mellitus, 493

643

644 Index

Antioxidants, in prenatal stress *(Continued)*
  honey, 490–491
  key facts of, 495
  resveratrol, 491–492
  spirulina, 492
  vitamins, 492–493
Antiviral potential, 345
Aqueous extract of *M. calabura* (AQMC),
    421–422, 423–424*t*, 424
Area under the concentration-time curve (AUC),
    23
Aspirin, 457
Asthma, 104
Atropine, 379, 380*f*, 382
Attention-Deficit/Hyperactivity Disorder
    (ADHD), 201
Audio-visual distraction, 578
Automatic control of anesthesia, 39–40
  analgesia, 36
  anesthetic agents, 39
  applications, 38
  clinical signs and physiological signals, in
      surgery, 34
  evaluation and supervisory phases, 38
  final feedback control phase, 38
  hypnosis, 33–36
  index analysis phase, 38
  modeling and control of, 35
  neuromuscular relaxation, 33–34, 36–37
  nociception, 33–34
  nonlinear analysis phase, 37–38
  propofol, 39
  signal decomposing, 37
  system biology phase, 38
Autonomic nervous system (ANS), 155
Axonal degeneration, 591–592

B

Baby-friendly hospital initiative, 222
Balance impairment, 570
Baobab. *See Adansonia digitata* (Baobab)
Baroreceptor reflexes, 134
Benfotiamine, 370
Benzodiazepines, 133–136
Benzodiazepines, preoperative opioid and,
    247–248, 258
  analgesia, increased complications risk,
      253–254
  application, 256
  differential opioid tolerance, 258
  gabapentinoids, 256
  liposomal bupivacaine, 258
  long-term postoperative outcomes, for
      patients, 255–256
  opioid-induced hyperalgesia, 258
  patient cohorts, characteristics of, 248–251
  perioperative outcomes, of patients, 252–253
  persistent opioid use, 258
  prescription, 248–251, 250–251*t*
  prevalence, 248
  short-term outcomes, 255
  tolerance, 258
Beta arrestins 2 (βarr2), 229, 231–232

Beta blockers, concomitant use of, 106
Binding immunoglobulin protein (BiP), 92
Biochanin A, 404–405, 405*t*, 412–415, 412*t*
Bispectral index (BIS), 23, 34, 39
Body mass index (BMI), 283
Brain development, 507
Brain, long-term effects of anesthesia on, 206
  aging brain, 203–204
  anesthetic neurotoxicity, 205
  cognitive and behavioral development
      in *Homo sapiens,* after childhood
      surgery, 200–202
  historical overview, 198–200
  neurodegeneration, 198, 199*t*
  on neuronal networks, 202
  neurotoxicity, of anesthetic agents, 197–198
  neurotransmitters signaling, during cerebral
      development, 196–197
  in nonhuman primates, 202–203
  postoperative cognitive dysfunction, 205
  surgical stress, effect of, 200
  synaptogenesis, 197
Breastfeeding, 223
  analgesia-breastfeeding issue, 213
  applications, 221–222
  breastfeeding initiation success (BIS),
      nonbiased indicators of, 220–221
  dyadic neurological competence, 218–220
  first 6 months of an infant's life, 213
  mother-baby's breastfeeding abilities,
      219–220*t*
Breastfeeding initiation success (BIS)
  measuring and grading, 221
  nonbiased indicators of, 220–221
Bupivacaine, 46, 633
Buprenorphine, 55, 58–60
  adverse effects, 56–57
  applications of, 58
  clinical use of, 57
  perioperative management, 57–58
  pharmacology, 55–56
  preparations, in United States, 55, 56*t*
Burn, 608
Burst spinal cord stimulation, 546, 551
Buspirone, 494

C

Caffeine, as analgesic adjuvant, 70
  active comparators, 66
  applications, 69–70
  clinical efficacy data, in acute pain trials,
      65–66
  and headache diseases, 67
  in humans, analgesic properties, 63–64
  and migraine, 67
  non-narcotic analgesics, 63
  NSAIDs, 66
  pain and functional impairment, 68
  pharmacokinetic properties, absence and
      presence of analgesics, 64–65
  preclinical models of pain, mechanism of
      action and effects, 63
  safety of, 68–69

5-caffeoylquinic acid (CGA), 478–479
Calcium antagonists, 106
Calcium displacement theory, 157
Capnography, 136
Capsaicin, 365
  applications, 369
  comparative studies, 366–368
  early clinical studies, 366
  high dose topical capsaicin, 368–369
  low dose topical capsaicin, 368
  painful diabetic neuropathy (PDN), 365–366,
      367*t*
Capsaicin induced inflammatory pain, 434–436,
    435*t*
Capsaicin induced pain, 444–446, 449*t*
Capsaicin-induced pain reduction, 394–395
Capsaicinoids, 365, 371
Cardiac effects, 104
Cardiorespiratory fitness, 501, 508
Cardiotoxicity, 30
Cardiovascular disease (CVD), 493
Carotenoid, 474*t*, 477–478, 481
Carpal tunnel syndrome (CTS), 558–559, 608
Carrageenan, 323–324, 326, 435*t*, 438–439
Carrageenan paw inflammation, 446–448, 449*t*
Cartilage damage, 590
CEF. *See* Chalcone-enriched fraction (CEF)
Celecoxib, 315
CEMC. *See* Chloroform extract of *M. calabura*
    (CEMC)
Centers for Disease Control and Prevention
    (CDC), 235–236, 247–248
Central nervous system (CNS), 81–82, 104, 166,
    195–196, 227–228, 609–610
Central sensitization, 432, 436, 439
  pathways, 341–343
Cervicobrachial pain (CP), 557–558
Cervigame., 568, 568*f*
CFA. *See* Complete Freund's Adjuvant (CFA)
C-fibers, 378–379, 381
Chalcone-enriched fraction (CEF), 404, 405*t*,
    408, 409–411*t*, 413–414
Chalcone synthase (CHS), 404–405
Charlson comorbidity index, 248–251
Chemokines, 403, 406, 413
Chinese medical acupuncture theory, 609
Chloroform extract of *M. calabura* (CEMC),
    422, 423*t*, 424
Chlorogenic acid, 474*t*, 478–479
Chloroprocaine, 77
  adverse effects, 77
  applications, chloroprocaine 2%, 77
  clinical application, 75–77
  comparative pharmacology, of local
      anesthetics, 74, 75*t*
  features and applications, 78
  history and controversies, 73
  mechanism of action, 74
  pharmacokinetics, of 2-chloroprocaine, 74
  pharmacology, 73–74
  structure and properties, 74
Chromatogram, 376
Chronic cancer pain, 618
Chronic Fatigue Syndrome, 267–268

Chronic inflammation, 432–434
Chronic leg ulcers, 176
Chronic low back pain, 608
Chronic neck pain, 608
Chronic neuropathic pain, 618
Chronic obstructive pulmonary
    disease (COPD), 104
Chronic pain, 402, 432, 608, 619–620, 619*f*, 624
  agents of, 624
  application, 624
  coping strategy devices/protocols, 617, 618*t*
  definition, 513, 519
  e-health in, 622*f*, 623
  epidemic of, 618
  nonpharmacological interventions in,
      621–623, 622*f*
  pain detection, 620
  pediatric (*see* Pediatric chronic pain intensity)
  tools, 620–621, 621*f*
  types of, 619*f*
  youth aerobic activity level with, 513
Chronic poisoning, 312–313
Chronic postsurgical pain, 618
Citrus sinensis, 448
Clonidine, 86
  adjuvants, regional anesthesia, 83
  antihypertensive medication, 81
  anti-shivering effect, 85
  applications, 85–86
  cardiovascular system, protection of, 83–84
  clinical applications, in perioperative period,
      82–85
  controlled hypotension, 84–85
  noradrenaline, 82
  pharmacodynamics, 81–82
  pharmacokinetics, 82
  postoperative agitation, 85
  pre-anesthetic medication, 82–83
  sedative effect, 81
c-Met, 386–389, 395, 396*f*, 397
Coagulative necrosis, 591–592
Cochrane meta-analysis, 222
Cognitive-behavioral therapy (CBT), for
      chronic pain and OUD, 238, 239*t*, 242
  acute pain, 238
  alternative therapies, 242
  catastrophizing, 240
  clinical complexity of patients with, 236
  decreased assertiveness, 241
  decreased distress tolerance, 238–240
  functional analysis worksheet, 241, 241*f*
  medication for OUD (MOUD), 235–237
  modules, 238
  necessary pharmacological platform,
      (MOUD), 237
  optimal medical management, 237
  patient inactivity, 238
  prevalence of, 235–236
  provider consideration, 236
Cola acuminata, 376–377
Cola nitida
  application, 381
  brief review on, 375
  composition, 376

distribution, 376
  in divination, 381
  key facts, 382
  nociception, 377
  overview, 375
  pain, 377
  pharmacological potentials, 377
  phenolics in, 376, 376*t*
  socio-cultural values, 375
  as therapeutic agent
      for pain relief, 379–380
  traditional uses, 377
  uses, 375–377
Cold (non-surgical) laser therapy, 602
Collagen destruction, 591–592
Combined spinal-epidural anesthesia (CSEA),
      281–282, 287
  headache, techniques of, 283–284
  for special patient groups, 283
Compartment nerve block, LA-dextran
      mixture
  analgesic effects, 19
  injection site, 19–21
  suitability, 18
  unintended spread of injected LA, 21–22
Complementary and alternative medicine
      (CAM), 473–474, 479–480, 482
Complete Freund's Adjuvant (CFA), 434–436,
      435*t*
Complex regional pain syndrome (CRPS),
      549–550
Computer-controlled local anesthetic delivery
      devices, 585
Constipation, opioids, 232–233
  application, 232
  enteric nervous system, GI function, 228–230
  on GI motility, 231
  opioid receptors, endogenous ligands and,
      229
  PAMORAs, 232
  tolerance development, 231–232
Conventional Landmark-Guided Technique,
      637
Conventional tonic spinal cord stimulation,
      545–546, 551
Cooled radio frequency ablation (CRFA), 593,
      594–595*f*, 595–597
  clinical trials of, 593–594, 596*f*
  overview, 596
Cooled radiofrequency probe, 593*f*
Cosmeceuticals, 331, 345
Cosyntropin, 286–287
Coumestans, 406, 406*t*, 414
Coumestrol, 404–406, 405–406*t*, 413–415
Cryoablation, 590–591, 597
Cryoanalgesia (ice) therapy, 610
  administration, 610–612
  clinical applications, 611
  overview, 611–612
Cryoneurolysis, 590–591
Crystal induced synovitis, 590
CTS. *See* Carpal tunnel syndrome (CTS)
Cubital tunnel syndrome, 559
Cyclooxygenase (COX), 46

Cyclooxygenases, 457, 465, 468
Cystathionine beta synthase (CBS), 262
Cystathionine gamma lyase (CGL), 262
Cytochrome P450 (CYP P450), 107, 136
Cytokines, 403, 406, 413

D

Daidzein, 404–405, 405*t*, 413–415
Dantrolene, 7–8
Decanoic acid, 479
Dental pain management
  injection discomfort
      distraction, 575–576
      during local anesthesia, 575, 576*t*
      virtual reality, effectiveness of, 578
      treatment, 575, 576*t*
      virtual reality devices, 576–577
      key facts of, 579
      safety considerations, 578
Dental treatment
  injections, fear of, 584
  vibrotactile devices, 584–585
DentalVibe comfort injection system, 584
14-Deoxyandrographolide, 351–353, 359
Depth of anesthesia (DOA), 35–36
Dermatome, 636, 638
Desflurane, 101
Dexmedetomidine, 86
Dextran, in regional anesthesia, 23–24
  adjuvant adrenaline, adverse effects of, 16–18
  adrenaline and dextran, as LA adjuvant, 16
  adrenaline toxicity, reduction, 18
  applications, 18–23
  epidural anesthesia, 15
  first appearance, as LA adjuvant, 18
  interfascial compartment nerve block, 15, 23
  low-molecular weight, 24
  mechanism, as LA adjuvant, 18
  minimum alveolar concentration (MAC), 24
  numerical rating scale (NRS), 24
  ropivacaine, 24
  ultrasound guidance, for nerve block, 24
  ultrasound technology, 15
DHA. *See* Docosahexaenoic acid (DHA)
Diabetic neuropathy (DN), 365–366, 550
Diabetic peripheral neuropathic pain, 608
Diclofenac, 315
Dietary constituents
  chlorogenic acid, 478–479
  decanoic acid, 479
  docosahexaenoic acid, 478
  functional significance, 479–480
  lutein, 477–478
  resveratrol, 477–478
Digital Health, 623
Dihydropyridine receptor (DHPR), 3–4
Dimeric chalcones, 408, 413–415
Disability, virtual reality (VR), 570
Disseminated intravascular coagulation (DIC),
      5, 12
Distraction, 565–566, 569–570, 572, 583
  injection discomfort, 575–576
Diterpenes, 406, 407–408*t*, 414

646 Index

Docosahexaenoic acid (DHA), 432–434, 433*f*, 436–437, 478–479
Dorsal root ganglion (DRG) stimulation, 546, 550–551
Drug-drug interactions, 106
Drug hypersensitivity reactions, 313–315
Drug induced liver injury (DILI), 312, 314
Drug reaction with eosinophilia and systemic symptoms (DRESS), 313–314
Duchenne Muscular Dystrophy (DMD), 267–268
Dynamic resistance exercises, 517

E

ED$_{95}$, 106–107
EFE. *See* Ephedrine alkaloids free Ephedra herb extract (EFE)
EHE. *See* Ephedra herb extract (EHE)
E-health, 622, 622*f*, 624
  in chronic pain management, 623, 624*f*
  clinical applications, 623
Eicosapentaenoic acid (EPA), 432–434, 433*f*, 436*t*, 437
Electroacupuncture (EA), 610
Electroanalgesia, 610, 611*t*
Electroconvulsive therapy (ECT), 143–144
Electroencephalography (EEG), 34
Electronic dental analgesia, 576*t*
EMCT. *See* Ephedra herb macromolecule condensed-tannin (EMCT)
Endocytosis, 395–397
Endogenous cytokines, 446
Enhanced recovery after surgery (ERAS), 629
Enteric nervous system (ENS), GI function, 228–229
Environmental effects, 105
EPA. *See* Eicosapentaenoic acid (EPA)
Ephedra herb extract (EHE), 396
  adverse effects of, 389
  analgesic effects, on formalin-induced pain, 389–394
  application, 395
  capsaicin-induced pain, reduction of, 394–395
  and ephedrine alkaloids, 386
  herbacetin-glycosides in, 386–387
Ephedra herb macromolecule condensed-tannin (EMCT), 392–393, 395–396, 398
Ephedrine alkaloids free Ephedra herb extract (EFE), 386
  analgesic effects
    on arthritis model mouse pain, 394
    on formalin-induced pain, 389–394
  development of, 387–389
  safety of, 389
Epidermal growth factor receptor-tyrosine kinase inhibitors (EGFR-TKIs), 395
Epidural anesthesia, 287
Epidural blood patch (EBP), 285–287
Epidural catheter, 273, 277–278, 284–285
  application, 277
  blocked epidural catheter, 275
  breakage of, 274–275

catheter migration to other locations, 274
insertion problems, 273–274
neuraxial anesthesia techniques, 277
removal difficulties, 275
spinal epidural abscess, 277
spinal epidural hematoma (SEH), 276
subarachnoid migration, 274
Epimer, 432–434, 439
Epinephrine, 106
Erector spina plane (ESP) block, 637–638, 638*f*
Esters, 153
Euphorbia bicolor
  development, 413
  effects
    on inflammatory mediators, 403
    on nociceptors, 403
    on oxidative stress markers, 403
  latex phytochemicals
    identification, 404–405
    pain-relieving properties, 405–413
    as phytomedicines, 413
    structural description, 404, 405*t*
  medicinal properties, 403
  pain-relieving properties, 405
    coumestans, 406, 406*t*, 414
    diterpenes, 406, 407–408*t*, 414
    flavonoids, 408, 409–411*t*, 414
    isoflavones, 412–414, 412*t*
    resiniferatoxin (RTX), 406–408, 407–408*t*
Euphorbia sp.
  E. antiquorum, 402–403
  E. bicolor (*see Euphorbia bicolor*)
  E. helioscopia, 402–403
  E. resinifera, 402–403
  E. royleana, 402–403
  medicinal properties, 402–403
European group of MH (EMHG), 261–262
Eutectic mixture of local anesthetics (EMLA), 161, 165
Evolitrine, 325, 325*f*
Exercise-induced hypoalgesia (EIH), 514, 519
Exercise therapy, 566
Exergame, 567–568, 567*f*, 570, 572
EXPAREL, 292
Extracorporeal elimination, 312, 315

F

Failed back surgical syndrome (FBSS), 548–549
Fat necrosis, 590
Fatty acids, 474*t*, 478, 481
Federal Drug Administration (FDA), 195–196, 205, 602
Femoral nerve mobilization, 559–560
Fentanyls, 98–99
  affinity of, 94–95
  applications, 97–98
  dual μOR/σ$_1$R ligands, 96–97
  opioid receptors, 92–93, 95–96
  role of, 90
  secondary pharmacology of, 90–92
Fibromyalgia, 608
  aerobic exercise in, 526–536
  exergames, 540

mind-body exercise in, 536–539
multicomponent exercise in, 536
physical exercise effectiveness in, 541
resistance training, 536
Fixed drug eruption (FDE), 313–314
Flavonoids, 340–341, 342*f*, 343, 345, 379–380, 408, 409–411*t*, 414, 466–468
  benefits of, 490, 491*t*
  cardioprotective effects of, 493
  honey, 490–491
  naringenin (*see* Naringenin)
Flavonol glycosides, 466*f*, 467–468
Flexibility, 502–504, 504*t*, 508
Fluoxetine, 494
Formalin, 444–447, 449*t*
Formalin-induced pain, 389–390
Formalin test, 379, 397–398, 460–463*t*, 465, 468, 494
Formula-feeding, 213, 222
Frequency, intensity, time and type of exercise (FITT), 540

G

Gabapentin, 286
Gabapentinoids, 47–48
Gamification, 566, 570, 572
Gamma-aminobutyric acid (GABA), 107, 132
Gastrointestinal (GI) tract, 227–229
Gate control theory, Melzack, 583
Gate control theory, Wall, 583
General anesthesia (GA), 195, 197–198
  epigenetic modulation, 198
Generalized convulsive status epilepticus (GCSE), 135
Generator potential, 476
Genistein, 404–405, 405*t*, 412*t*, 413–415
Glibenclamide, 269
Global perceived effect (GPE), 592
Glucocorticoid, 487–488, 495
Glucoronidation, 145
Glycoside, 464, 468
Glycosylated flavonoids, 466
G proteincoupled receptors (GPCRs), 229
Greater occipital nerve block (GONB), 286–287
Greenhouse effect, 173
Greenhouse technique, 173

H

Haloethers, 176
Halothane, 101
Hard-to-heal ulcers, 176
Hard-to-heal wounds, sevoflurane as prohealing agent, 175, 176*f*
Headache, 557, 608
Head-mounted display (HMD), 566–567, 567*f*, 570–572
Health workers, systemic adverse effects for, 173
Heme-oxygenase-1 (HO-1), 447–448
Hemophilia-induced arthritis, 609
Henderson-Hasselbach equation, 153
Hepatic effects, 104
Hepatotoxicity, 310

Herbacetin-glycosides
  active metabolite of, 386–387
  in *Ephedra* herb, 386–387
Herbal extract, 358
High frequency spinal cord stimulation (HF-SCS), 549
High-intensity laser therapy (HILT), 602
HMD. *See* Head-mounted display (HMD)
Homeostasis, 341, 345
Homocysteine (Hcy), 267–268
Honey, 490–491
Hot plate test, 494
H_2S-generating enzymes, 269
Hydrocortisone, 286
Hydrogen sulfide, 262, 264*f*
  and malignant hyperthermia (MH) syndrome, 263–264
  signaling, 267–268
Hydromorphone, 144
Hyperacusis, 282, 287
Hyperalgesia, 406–408, 411–413, 434–436, 435*t*, 438–439, 445–447, 452
Hypercapnia/hypercarbia, 12
Hyperhomocysteinemia (HHcy), 267–268
Hyperkalaemia, 12
Hypersensitivity, 316
  delayed, 313–314
  immediate, 313
  paracetamol induced hypersensitivity, 314
Hypnosis, 576*t*, 579
Hypoalgesia, 561

I

Ilioinguinal and iliohypogastric (IH-II) nerves block, 635–636, 636*f*
Immersion, 565, 571–572
Immunoresolvent, 432–434, 437–439
Infant breastfeeding assessment tool (IBFAT), 220
Inflammasome, 358
Inflammation, 326
  pain, 325
  symptoms, 323–324
Inflammation resolution, 432–434, 432*f*
Inflammatory mediators, 403
Inflammatory pain, 403, 408, 414, 432–436, 435*t*, 438–439
  drugs, 438, 438*t*
  key facts of, 452
  naringenin (*see* Naringenin)
Inflammatory pathway, 341
Inhaled anesthetics, 38
  blood-gas coefficients of, 102*t*
Injection discomfort
  distraction, 575–576
  during local anesthesia, 575, 576*t*
  non-pharmacological approach, 583
  VR, effectiveness of, 578
International Board Certified Lactation Consultants (IBCLCs), 220
International Headache Society, 282
International multicenter study on the long-term POCD (ISPOCD1), 203–204

Intra-articular (IA) injection, 406–408, 413, 589–590
Intramuscular QLB (QL4 block), 634
Intraperitoneal local anesthetic agents, postoperative pain, 31
  applications, 30
  bupivacaine, 27
  evaporation-based humidifiers, 31
  gas-humidifying devices, 31
  gynecologic surgery, 28
  laparoscopic appendectomy, 28, 30*t*
  laparoscopic cholecystectomy, 27–28, 29*t*
  laparoscopic surgery, 27
  lidocaine, 27
  micro-vibration humidifiers, 31
  parietal peritoneum, 27
  phrenic nerve pain prevention, instillation at hemidiaphragm, 28–30
  ropivacaine, 27
  safety of, 30
  S-isomer forms, 27
  visceral peritoneum, 27
Intraplantar injection, 403
Intrathecal injection, 403
Intravascular migration, 273
Intravenous paracetamol, 147–148
  in adult, pharmacokinetics, 142, 143*t*
  applications, 146–147
  breakthrough pain, 144
  chemical structure and pharmacokinetics, 139–146
  in children, pharmacokinetics, 145–146
  COX-2 inhibitors, 141
  COX pathway, inhibition of, 141
  endogenous cannabinoid system, modulation of, 141
  nitric oxide, inhibition of, 141–142
  non-superiority of IV paracetamol, 144
  in obstetrics, 144–145
  paracetamol pharmacokinetics, 140
  preemptive analgesic, for major oncology operations, 139
  side effect of, 146
  site of action, 141
  uses, in adult, 142–144
Intrinsic primary afferent neurons (IPANs), 228–229
In vitro contracture test (IVCT), 6, 7*f*, 261–262, 269
  procedure, 262, 263*f*
Ions-mediated pathways, 343
Ischemic pain, 547, 550–551
Isoflavones, 412–414, 412*t*
Isoflurane, 101
  adverse effects, 105
  agents of, 106–107
  application, 106
  calcium antagonists, 106
  cardiac effects, 104
  central nervous system effects, 104
  chemical structure, 102, 102*f*
  clinical effects, 103–105, 103*t*
  concomitant use of beta blockers, 106
  drug-drug interactions, 106

  environmental effects, 105
  epinephrine, 106
  hepatic effects, 104
  malignant hyperthermia, 105
  mechanism of action, 102–103
  metabolism, 105–106
  neuromuscular blocking agents, 106
  nitrous oxide, 106
  obstetric effects, 105
  opioids, 106
  pharmacokinetics, 102
  pharmacology, 102
  properties, 102
  renal effects, 105
  respiratory effects, 104
  skeletal and smooth muscle effects, 104
  teratogenic effects, 105
Isoflurane, adverse effects, 105
Isometric exercises, 517
9-item Patient Health Questionnaire (PHQ-9), 45
30-item State Trait Operation Anxiety (STOA), 45

J

Jamaican cherry. *See Muntingia calabura*
Jet injections, 585
Joint-by-joint theory, 503, 503*f*

K

Kaempferol, 404–405, 405*t*, 408, 409–411*t*, 413–415
Kampo medicine, 385–386, 395
K_{ATP} channels, 264
Kaurenoic acid, 451–452, 451*t*
K/BxN serum transfer induced inflammatory arthritis, 435*t*, 436
Kerukup siam. *See Muntingia calabura*
Kinematics, 566–567, 569
Knee osteoarthritis, 589, 596–597
  denervation treatment in, 590, 591*t*
  nonpharmacological and pharmacological conservative treatments, 589–590
Kola nut. *See Cola nitida*
Kv7 channel, paradoxical depolarizing activity of, 264–265, 267*f*

L

Labor analgesia, 220
  in guidelines and meta-analyses, 222
Laparoscopic-assisted TAP block, 632
L-Arginine, 425, 427
Laser acupuncture, 610
Laser analgesia, 576*t*
Laser therapy, 602
Laser wavelength, 602–605, 607–608, 612
Latching-audible-type-comfort-help (LATCH), 220
Lateral approach, 632
Lateral elbow pain, 558
Lateral QLB (QL1 block), 634
Leukocytes, 447, 452
Leukoplakia, 609

648 Index

Levobupivacaine, 23, 633
Lidocaine, 16–18, 23, 369–370
Lidocaine patch, 117–118
  GPE scale, 116
  LOHS, 117
  minimally invasive approach, 116
  placebo patch, 117
  postoperative analgesia, 116–117
  post-thoracotomy pain, 109
  PRISMA guideline, 110*f*
  regional techniques, 109
  search strategy, 110
  spirometry tests, 116
  systemic opioid administration, 109
  WOMAC index, 116–117
Light Amplification by Stimulated Emission of
  Radiation (LASER), 602, 612
Lipophilicity, 74
Lipopolysaccharide (LPS), 445–446, 449*t*, 452
Liposomal bupivacaine (LB), pain relief and
  adverse events, 291–292, 303–304
  administration, 301
  bupivacaine hydrochloride, 291
  cardiovascular system toxicity, 298
  central nervous system toxicity, 297
  compatibility, 299
  and DepoFoam, 292
  HTX-011, 302
  inflammation and wound healing, 298
  local adverse events, 298–299
  maternal/fetal toxicity, 299
  multimodal analgesia, 291
  muscle injury, 299
  neosaxitoxin, 302
  nerve injury, 299
  neuraxial use, 301
  peripheral nerve blocks and perineural use,
    300–301
  pharmacodynamics and mechanism
    of action, 294
  pharmacokinetics, 294–296, 296*t*
  placebo, 300
  release characteristics and stability, 294
  safety and adverse events, 297–299
  sucrose acetate isobutyrate extended-release
    bupivacaine (SABER-bupivacaine), 301
  surgical site infiltration, 300
Lipoxin (LX), 432–434, 438–439
Liver disease, 186
Local adverse effects, sevoflurane, 172
Local anesthetics (LAs), 15, 23, 45–46,
  476–477, 575–579, 576*t*, 630–632, 638
  discomfort, reduction of, 584–585
  systemic toxicity, 24
Local skin reaction, 368–369, 371
Long-term opioid therapy (LTOT), 235–237
Low back pain (LBP), 559
Lower limb and neurodynamic techniques,
  559–560
Low frequency spinal cord stimulation
  (LF-SCS), 549
Low-level laser therapy (LLLT), 602
Lumbar paravertebral block, 637, 637*f*
Lumbar spine, 503

Lumbo pelvic-hip arrangement, 502
Lumbo-pelvic movement, 504–505
Lutein, 477–478

M

Macrotemporal processes, 517, 519
Maculopapular eruption (MPE), 313–314
Malaria, *A. digitata*, 344
Malignant hyperthermia (MH), 12, 105,
  261–264
  applications, 11
  cellular hypermetabolic reaction, 3
  clinical manifestations, in MH crisis, 5–6
  diagnosis and differential diagnosis, 6–7
  halothane, 11
  perioperative care, anaesthesia, 10–11
  persulfidation of Kv7 channels in, 266–267,
    268*f*
  post-anaesthetic recovery room, 12
  pre-operative care, anaesthesia, 9
  soda lime, 12
  susceptible patients, 4–5, 11–12
  treatment, of MH crisis, 7–9
  triggering agents, 3, 4*f*, 12
  vaporizers, 12
Malondialdehyde (MDA), 488
Manual therapy, 556–562
Maoto, 386, 395
Maresin-1 (MaR1), 439
  analgesic effects, 434–436
  antiinflammatory effect, 434
  application, 437–438
  clinical analgesic evidence, 436–437
  key facts of, 439
  precursors, 436–437
Mayo Anesthesia Safety in Kids (MASK) Study,
  201
MECN. *See* Methanol extract of *Cola nitida*
  (MECN)
Melzack's Gate control theory, 583
Memantine, 127–128
  applications, 127
  chronic pain, 121–122, 124–127
  clinical evidence, in chronic pain,
    124–127
  clinical use considerations, 127
  fundamental pharmacology, 122–123
  N-methyl-D-aspartate (NMDA) receptors,
    121–122
  preclinical evidence, effect on pain, 124
MEMC. *See* Methanolic extract of *M. calabura*
  (MEMC)
Menopausal syndrome, 86
Mental health, 506–507
Mepivacaine, 222
Meridians, 609
Metabolic diseases, *A. digitata*, 343
Metabolism, 105–106
Methadone, 58–59
Methanol extract of *Cola nitida* (MECN),
  379, 379*f*, 381
Methanolic extract of *M. calabura* (MEMC),
  422, 423–424*t*, 424

Meyer–Overton hypothesis, 157–158
MH Association of the United States (MHAUS),
  261–262
M-health, 622–624
MicroRNA (miRNA), 269, 448
Microtemporal processes, 517, 519
Midazolam, 136
  advantages of, 131
  cardiovascular system, effects on, 134
  central nervous system, 133
  clinical uses and dosage, 134–135
  feature of, 131
  flumazenil, 136
  GABA$_A$ receptor, 132, 136
  mechanism of action, 132
  palliative care, 135–136
  pharmacokinetics, 132–133
  predictable pharmacokinetics, 131
  procedural premedication, 134–135, 134*t*
  procedural sedations, 131
  respiratory system, effects on, 134
Minimally invasive and surgical approaches,
  590
Minimum alveolar concentration (MAC),
  103–104, 106–107
Mitogen-activated protein kinases (MAPK), 355
Mobile health (mHealth), 518
Mobility, 502–504, 508
Monosodium urate (MSU), 446–447, 449*t*
μ-opioid receptor (μOR), 90, 181, 227–228
Morphine, 232, 457
Motor Offset Visual Illusion, 567
MSKP. *See* Musculoskeletal pain (MSKP)
Mucositis, 609
Multi-layer perceptron artificial neural networks
  (MLP-ANN), 36
Multiple injection (dual) approach, 633
Multiple resources theory of Wickens, 569
Multivesicular liposomes (MVLs), 291–292
Muntingia calabura, 420–421
  antinociceptive activity, 420–425
  applications, 425–426
  mechanisms of action, 424–425
Muscular strength, 501–502
Muscular tension, 502
Musculoskeletal pain (MSKP), 508, 541–542
  applications, 541
  atypical brain development, 507
  biomechanical factors, 504–505, 508
  cardiorespiratory fitness and aerobic/
    anaerobic training, 501, 508
  chronic musculoskeletal pain, 525, 528–535*t*
  clinical practice guidelines, 523
  fibromyalgia, physical activity in, 525
  flexibility and mobility training, 502–504,
    504*t*, 508
  interventions, physical activity, 524–525
  key facts, 508
  muscle/joint disorders, 523
  muscular strength and resistance training,
    501–502, 508
  pediatric obesity, 505–506
  physical activity, 499–500, 507, 523–524
  physical exercise, 525

psychosocial factors, 506–508
risk factors, 504–507
Myenteric plexus, 228–229

N

N-Acetylcysteine (NAC), 311–312, 316
N-acetyl-ᴘ-benzoquinone-imine (NAPQI)
  metabolism, 140, 310–313, 316
Naltrexone, 59
Nanostructured lipid carriers (NLCs), 155
N-arachidinoyl-phenolamine (AM404), 141
Naringenin, 404, 405*t*, 408, 409–411*t*, 411–415
  analgesic pathways, 447–448
  antibacterial effect, 450–451
  antiviral effect, 450–451
  chemical structure, 444*f*
  clinical applicability and safety, 448–449
  common food sources, 444, 444*t*
  endogenous cytokines, 446
  on inflammatory pain relief, 444–446
  miRNA, 448
  oxidative stress, 443–444, 446*f*, 447,
    449–450, 449*t*, 451*t*, 452
  transcription factors, 448
  on transient receptor potential (TRP)
    channels, 444–446, 452
Nausea, 282
Neck pain (NP), VR, 557, 566
  applications, 572
  as assessment tool, 571
  clinical efficacy, 569
  design, 566–568
  disability, 570
  disadvantages, 571
  factors affecting the efficacy of, 570–571
  key facts, 572
  kinematic impairments, 570
  pain, 569
  postural control and balance impairment, 570
  range of motion (ROM), 570
  rehabilitation, 566–568
Needle-through-needle technique (NTN), 283
Neoandrographolide, 351–353, 359
Nerve growth factor (NGF) receptor, 386–387
Neural manipulation approaches, 560
Neural mobilization (NM), 561
  application, 561
  carpal tunnel syndrome (CTS), 558–559
  cervicobrachial pain (CP), 557–558
  clinical applications, 560
  cubital tunnel syndrome, 559
  femoral nerve mobilization, 559–560
  headache, 557
  low back pain (LBP), 559
  lower limb and neurodynamic techniques,
    559–560
  manipulation approaches, 560
  musculoskeletal disorders, 561
  neck pain, 557
  patellofemoral pain syndrome, 559–560
  tarsal tunnel syndrome (TTS), 560
Neuraxial anesthesia, 287
  unintentional dural puncture (UDP) in,
    284–285

Neuraxial block, 630, 639
Neuraxial labor analgesia, 211
  childbirth pain, 211–212
  effects of, 218, 218*t*
  human rotational delivery, 211–212
Neurodynamic gliding (NG), 556
Neurodynamic stretching (NS), 556
Neurogenic pain, 424
Neuromuscular blocking agents, 106
Neuromuscular training, 517
Neuropathic pain, 176, 402–403, 406–408,
    413–415, 609, 618
  Adansonia digitata in (*see Adansonia digitata*
    (Baobab))
Neuropathy, 340, 345, 345*t*
Neuroprotection, *A. digitata*, 344–345
Neutrophils extracellular traps (NET), 352–353,
    358
Nfr2/keap1 pathway, role of, 355–357
Nitric oxide, 425, 427
Nitrous oxide, 106
N-methyl-D-aspartate (NMDA) receptors, 343,
    514, 519
Nociception, 377–378, 419, 452
Nociceptive pain, 618
Nociceptive specific (NS) neurons, 475
Nociceptive terminals
  action potential, 477
  generator potential, 476
Nociceptors, 403, 432, 437–439, 443–444, 452
Nomogram, 311–312, 315
Noncompressive lumbar disc herniation,
    434–436, 435*t*
Non-opioid analgesia, otolaryngology, 48, 48*t*
  acetaminophen (APAP), 46
  gabapentinoids, 47–48
  local anesthetics, 45–46
  nonsteroidal anti-inflammatory drugs
    (NSAID), 46–47
Nonopioid drugs, 427
Non-opioid intravenous anesthetics, 38
Nonpharmacological pain management
    interventions (NPMIs), 237
Nonsmall cell lung cancer (NSCLC), 395, 397
Nonsteroidal anti-inflammatory drugs
    (NSAIDs), 141, 143, 165, 285–286, 321,
    326, 457, 465, 468, 557–558, 589–590
Noxious transmission, 475–476
Nuclear factor-erythroid 2-related factor 2
    (Nrf2), 355–357, 446*f*, 447–450
Nuclear factor-kappa B (NF-κB) pathway,
    353–355
Nuclear factor of activated T cells (NFAT), 355
Numeric Rating Scale (NRS) score, 593–594

O

Obi. See Cola nitida
Obstetric effects, 105
Omega-3 supplementation, 436–437, 436*t*, 439
Opioid dependency, treatment of, 609
Opioid-induced bowel dysfunction (OBD),
    227–228
Opioid-induced constipation (OIC), 232–233,
    253

application, 232
enteric nervous system, GI function, 228–230
on GI motility, 231
opioid receptors, endogenous ligands and,
    229
PAMORAs, 232
tolerance development, 231–232
Opioid receptors, fentanyls, 99
  interactions with, 96
Opioid-sparing effect, 172
Opioid use disorder (OUD), CBT and chronic
    pain, 238, 239*t*, 242
  acute pain, 238
  alternative therapies, 242
  catastrophizing, 240
  clinical complexity of patients with, 236
  decreased assertiveness, 241
  decreased distress tolerance, 238–240
  functional analysis worksheet, 241, 241*f*
  medication for OUD (MOUD), 235–237
  modules, 238
  necessary pharmacological platform,
    (MOUD), 237
  optimal medical management, 237
  patient inactivity, 238
  prevalence of, 235–236
  provider consideration, 236
Opium poppy (*Papaver somniferum*), 457
Oral-associated pain, 608
Osteoarthritis (OA), 369, 406, 589, 609
Otolaryngology, 43, 50
  applications, 48
  corticosteroids, 49
  nasal surgery, 44
  non-opioid analgesia, 45–48
  oncologic surgery, 45
  oral cavity and oropharynx, 44
  otologic surgery, 44
  pain medication use in, 49
  patient pain experience in, 43–45, 49
  risk factors, 45
  visceral/soft tissue, 45
Oxford Knee Score (OKS), 592
Oxidative stress, 443–444, 446*f*, 447, 449–450,
    449*t*, 451*t*, 452, 488–489, 494
  markers, 403
  pathway, *A. digitata*, 341

P

Pain
  analgesic drugs, 420
  causes of, 377
  classification, 378
    nociceptive processing, 378
    pathologic pain, 378
  definition, 402
  detection, 620
  gate control theory, 583
  management, 413, 575
  modulation, 425
  overview, 321
  sensation, *A. digitata* on, 332–337
  sensory transmission, 402
  virtual reality (VR), 569

650 Index

Painful diabetic neuropathy (PDN), 365–366,
 367*t*, 368, 370–371
"Pain-OUD Dysfunction Cycle,", 238
Pain pressure threshold (PPT), 557
Paracetamol. *See* Intravenous paracetamol
ParActin, 352, 359
 active principle of, 353
 andrographolide in, 352–353
 in inflammation and pain, 353
Paraspinal epidural blocks, 637
Paravertebral lumbar sympathetic block, 637
Paravertebral nerve block, 637, 639
Paresthesia, 559, 561
Paresthesia-based stimulation, 545, 551
Passive distraction, 575–576
Patellofemoral pain syndrome, 559–560
Pathologic pain, 378
Patient Global Impress of Change (PGIC), 595
PDN. *See* Painful diabetic neuropathy
 (PDN)
Pediatric chronic pain intensity
 aerobic exercise, 517
 psychological treatment of, 518
 treatment, 514
Pediatric obesity, 505–506
Pentazocine, 99
PEPMC. *See* Petroleum ether partition of
 M. calabura (PEPMC)
Percutaneous Electrical Nerve Stimulation
 (PENS), 610
Periaqueductal gray matter (PAG), 569
Perioperative analgesia, inadequate, 252–253
Peripherally acting mu opioid receptor
 antagonists (PAMORAs), 227–228
Peripheral nerve block, 629, 639
Peripheral nerve stimulation (PNS), 610, 613
Peripheral nervous system (PNS), 555–556
Peripheral sensitization, 432, 436, 439
Persulfidation (S-sulphydration), 265–266, 269
Pethidine, 222
Petroleum ether partition of M. calabura
 (PEPMC), 422–424, 423–424*t*
Phantom limb pain (PLP), 124–127
Pharmacodynamics (PD) model, 35
Phenolic acids, 465, 467
Phenolic glycoside salicin, 457
Phosphodiesterase (PDEs), 263, 267–268
Photobiomodulation therapy (PBMT), 602–609
 administration of, 605–607
 applications, 608–609
 clinical applications, 608–609
 history, 602
 mechanisms of, 602, 603–605*t*
 overview, 609
 side effects, 607–608
 treatment parameters, 602–605, 606–607*t*,
  608*f*
Photophobia, 282, 287
Physalis angulata, 427
Physical activity, 499–500, 507–508
Physical activity, musculoskeletal pain,
 523–524
 fibromyalgia, 525
 interventions, 524–525

levels of, 524
Physiologic pain, 378
Phytochemicals, 332, 333–338*t*
Phytoconstituents, 424–425, 424*t*
Pin-prick test, 167
Plantar fasciitis, 609
Plant-based agents, 451–452, 451*t*
 for analgesia and anesthesia, 370–371, 370*t*
Plant-based analgesics, 458–459*t*
Plant-based anesthetics, 458–459*t*
PlayMancer, 567, 567*f*
PNS. *See* Peripheral nervous system (PNS)
Point-and-shoot technique, 605–606
Polyphenols, 465, 467
Post-anaesthetic recovery room (PACU), 11
Postdural headache, 68–69
Postdural puncture headache (PDPH), 281
 alternative therapies, 286–287
 application, 287
 characteristics of, 282
 epidural blood patch (EBP), 286
 history of, 281
 management, 285–287
 mechanism of, 282
 medical management, 285–286
 prophylaxis of, 284–285
 supportive treatment, 285
Posterior QLB (QL2 block), 634
Posterior rectus sheath, 631*f*
Post-operative cognitive dysfunction (POCD),
 203–205
Postoperative nausea and vomiting (PONV),
 134–135
Posttranslational mechanism, 269
Posttraumatic pain, 618
Postural control, 570
Potassium channels, 425, 426*f*
 H$_2$S molecular targets in MHS, 264, 265–266*f*
Pregabalin, 286
Premature ejaculation (PE), 185
Preoperative opioid, benzodiazepines and,
 247–248, 258
 analgesia, increased complications risk,
  253–254
 application, 256
 differential opioid tolerance, 258
 gabapentinoids, 256
 liposomal bupivacaine, 258
 long-term postoperative outcomes, for
  patients, 255–256
 opioid-induced hyperalgesia, 258
 patient cohorts, characteristics of, 248–251
 perioperative outcomes, of patients, 252–253
 persistent opioid use, 258
 prescription, 248–251, 250–251*t*
 prevalence, 248
 short-term outcomes, 255
 tolerance, 258
Preterm infant breastfeeding behavior scale
 (PIBBS), 220
Prilocaine, 162
 advantages, 151
 anesthesia duration, 160
 applications, 161

bupivacaine, 155
chemical structure of, 151, 152*f*
classification, of anesthetic drugs, 153
dental treatment procedures, 153–154
distribution, 159
elimination, 160
exceptional strategy, 154
intravenous regional anesthesia, 153–154
lidocaine, secondary amide analogue of,
 153–154
lipid solubility and protein binding, 158
local anesthetic drug (LAs), 151–152, 157
metabolism and excretion, 158–161
methemoglobinemia, 151, 160–161
perianal surgery, 156
pharmacology and mechanism action, of local
 anesthetic agent, 157–158
physicochemical and biological properties of,
 151, 152*t*
premature ejaculation (PE), 155
preparation method of, 157
spinal anesthesia, for ambulatory arthroscopic
 surgery, 156
structure, 152–153
TNS, 156–157
toluene derivative, 151
topical anesthesia, 155
toxicity, 160
tramadol, as nasal packing removal, 155
transient neurologic symptoms (TNS), 156
Primary chronic pain, 618
Primary human SKM (PHSKM) cells, 264–265
Propofol, 222
 automatic control of anesthesia
  (*see* Automatic control of anesthesia)
Propofol infusion syndrome (PRIS), 38
Proprioceptive neuromuscular facilitation
 (PNF), 504
Propylene glycol, 137
Prostaglandin E2 (PGE2), 322–326, 340–341
Proteins posttranslational modifications
 (PTMs), 265–266
Pruritus, 369
Pseudoephedrine, analgesic effects
 on formalin-induced pain, 389–394
Psoriasis, 437
Psychological screening, 547
Psychological stress, 506
Pueraria lobata, 412*t*, 413

Q

Quadratus lumborum block (QLB), 633–634,
 634*f*
Quality of life, 172
Quercetin, 404–405, 405*t*, 409–411*t*, 411,
 413–415, 451–452, 451*t*
Quinolines, 153

R

Radiation therapy, 609
Radiculopathy, 609
Radiofrequency ablation (RFA), 591–593,
 591*f*, 597

clinical trials of, 592
overview, 593
Randomized controlled trials (RCTs), 284, 537–539*t*, 541
Reactive oxygen species (ROS), 443–444, 447–450, 452, 488, 494
Rebound pain, 77
Rectus sheath block, 634–635, 635*f*
Refractory pain, 177
Regional analgesia, 629
Rehabilitation, virtual reality (VR), 566–568
Remote sensing sensors, 623
Renal dysfunction, 186
Renal effects, 105
Resiniferatoxin (RTX), 402–408, 407–408*t*, 415
Resistance training, 501–502, 508
Respiratory depression, 187
Respiratory effects, 104
Response Entropy (RE), 34
Resveratrol, 474*t*, 477–478, 491–492
Retigabine, 269
Rheumatoid arthritis, 357
Robot-assisted TAP block, 632
Ropivacaine, 633
Rostral ventromedial medulla (RVM), 514
RTX. *See* Resiniferatoxin (RTX)
Rumack-Matthew nomogram, 312, 312*f*, 316
Rutin, 404–405, 405*t*, 409–411*t*, 411–415
Ryanodinopathies, 4
RyR1 genetic mutations, 268–269

S

SCS. *See* Spinal cord stimulation (SCS)
Sedentary behavior, 505–508
Seizure disorders, 186
Self-medication, 69
Sensorimotor impairments, 566
Sensory-adapted dental environment (SADE), 576*t*, 579
Sensory fibers, 378–379, 381
Sensory neurons, 228–229
Sensory transduction, 475–476
Sensory transmission, 378–379
Sevoflurane
    agents of, 175–176
    analgesic effects of, haloethers, 166
    analgesic profile, 167, 172*f*
    antimicrobial agent for infected wounds, 174–175
    application, 174–175
    dosages and methods of administration, 173, 174*f*
    economic implications, 173
    greenhouse effect, 173
    health workers, systemic adverse effects for, 173
    local adverse effects, 172
    opioid-sparing effect, 172
    painful chronic wounds, 167–172, 168–171*t*
    painful medical conditions from wounds, 175
    patients, systemic adverse effects for, 172–173

peripheral analgesic effects of, haloethers, 166–167
    prohealing agent for hard-to-heal wounds, 175, 176*f*
    quality of life, 172
    safety issues of, 172–173
    systemic effects of, haloethers, 165–166, 166*f*
    topical analgesic, 166
    wounds and pains, 167–172
Shamrock sign, 634
Silver staining, 198
Simulator sickness (SS), 571–572
Single space technique, 283
Skeletal muscle (SKM), 104, 267–268
Sleep disturbance, 69
Smooth muscle effects, 104
SNL-induced inflammatory pain, 434–436, 435*t*
Soft tissue injuries, 608
Solanum spp., 459
    analgesia, 460–463*t*, 465, 468
    antiinflammatory activity of, 460–463*t*, 464–465
    antinociceptive effects of, 459–463
    application, 467
    chemical constituents, 465–467
    pharmaco-toxicological activities, 467
    S. acanthodes, 460–463*t*
    S. cernuum, 460–463*t*
    S. incanum, 460–463*t*
    S. melongena, 459, 460–463*t*
    S. nigrum, 459, 460–463*t*, 464–465, 467
    S. paniculatum, 459–463, 460*f*, 460–463*t*, 464*f*, 466–468, 466*t*
    S. sisymbriifolium, 460–463*t*, 465
    S. torvum, 459–463, 460*f*, 464*f*, 467–468
    S. trilobatum, 460–463*t*, 467
    S. tuberosum, 460–463*t*, 464
    S. viarum, 460–463*t*, 465
    S. violaceum, 460–463*t*
Solid lipid nanoparticles (SLNs), 155
Somatic pain, 176
Specialized pro-resolving mediators (SPMs), 432–434, 432–433*f*, 436–439
Spinal anesthesia, 281, 287
Spinal cord stimulation (SCS), 551
    applications, 550
    burst, 546, 551
    conventional tonic, 545–546, 551
    dorsal root ganglion (DRG) stimulation, 546, 551
    high frequency 10 kHz, 546, 551
    history, 545
    ischemic pain, 547, 551
    key facts of, 551
    outcomes
        complex regional pain syndrome (CRPS), 549–550
        diabetic neuropathy, 550
        failed back surgical syndrome (FBSS), 548–549
        ischemic pain, 550
    patient selection
        conservative treatment, failure of, 547
        indications, 547

medical comorbidities, 548
    psychological screening, 547
    screening trial, 548
Spinal trigeminal nucleus caudalis (SpVc), 475, 475–476*f*, 478, 482
Spirulina, 492
SPMs. *See* Specialized pro-resolving mediators (SPMs)
S-sulphydration, 265–266, 269
State Entropy (SE), 34
Steroid injections, 590
Stevens-Johnson syndrome (SJS), 313–314, 316
Strength training, 517
Subcostal approach (Upper TAP), 632
Subcostal (paramedian sagittal oblique) QLB, 634
Submucosal plexus, 228–229
Suicidal tendency, 187
Sumatriptan, 286
Superoxide anion, 445–448, 449*t*, 452
Surgeon-assisted TAP block, 632
Surgical Stress Index (SSI), 34
Systemic inflammation, 505

T

Tail flick test, 495
Tannins, 380
Tarsal tunnel syndrome (TTS), 560
Telerehabilitation, 572
Temporomandibular Joint (TMJ), 609
Teratogenic effects, 105
Tetracarpidium conophorum, 381
Theophylline, 286
Thermoreceptors, 611
3-Mercaptosulfutransferase (3-MST), 262
Tibial bone fracture, 435*t*, 436
Tinnitus, 282, 287
Titanium dioxide (TiO_2), 447, 449*t*
Topical sevoflurane
    painful chronic wounds, 167–172, 168–171*t*
    for painful conditions, 175
    safety issues of, 172–173
Total knee arthroplasties (TKA), 590
Tourette's syndrome, 86
Toxicity
    mechanism of, 310
    prilocaine, 160
    tramadol, 185–186
Traditional and complementary medicine (T&CM), 420, 427–428
Traditional blind technique, 635
Traditional Chinese medicine (TCM), 609
Tramadol, as analgesics
    acute myocardial infarction pain, 184
    acute renal pain, 185
    cancer pain, 183–184
    central analgesic mechanism, 181
    chemical synthesis, 181
    contraindications, 186–187
    dependence, withdrawal, abuse and tolerance, 186
    emergency department, 184
    5-HT2C receptors, 182

Tramadol, as analgesics *(Continued)*
　isomers of, 181, 182f
　as local anesthetic, 185
　neuropathic pain, 183
　osteoarthritis (OA) pain, 183
　peripheral local anesthetic mechanism, 181
　pharmacokinetics, 182
　postoperative pain, 184–185
　postoperative shivering, 185
　therapeutic uses, 183–185
　toxicity, adverse effects and management of, 185–186
　use/off-label use, 185
Transcription factor nuclear factor κB (NFκB), 443–444, 446f, 447–448, 449t, 452–453
Transcutaneous Acupoint Stimulation (TAES), 610
Transcutaneous Electrical Nerve Stimulation (TENS), 610, 613
Transcytosis, 396–397
Transient neurological symptoms (TNS), 77
Transient receptor potential (TRP), 167, 444–446, 452
Transient receptor potential of melastatin subfamily member 3 (TRPM3), 445–446, 452
Transient receptor potential of melastatin subfamily member 8 (TRPM8), 445–446, 445f, 452
Transient receptor potential vanilloid 1 (TRPV -1), 365, 369, 394–395, 403, 444–446, 445f, 452
Transversalis fascia plane (TFP) block, 636
Transversus abdominis plane (TAP) block, 631–633, 632f
Trifluoroacetic acid, 104
Trigeminal neuralgia, 609
Trigeminal pain, 474–475
Tropomyosin-Receptor-Kinase A (Trk A), 386–387

TRP ankyrin 1 (TRPA1), 476
TRPV1. *See* Transient receptor potential vanilloid 1 (TRPV -1)
Truncal block, in obstetric and surgery, 630–639
　erector spina plane (ESP) block, 637–638, 638f
　ilioinguinal and iliohypogastric (IH-II) nerves block, 635–636, 636f
　lumbar paravertebral block, 637, 637f
　patient selection, 630
　preparation for, 630–638
　quadratus lumborum block (QLB), 633–634, 634f
　rectus sheath block, 634–635, 635f
　transversalis fascia plane (TFP) block, 636
　truncal blocks of abdominal wall, 631–633, 631t
Tualang honey, 491, 492f, 493, 494t
Tuohy needle, 284, 287
Tuohy/spinal needles, 274, 277–278
2016 Center for Disease Control and Prevention (CDC) guidelines, 601–602
Type 1 ryanodine receptor (RyR1), 3–4

U

Ultrasound-guided acupotomy, 610
Ultrasound-guided TAP block, 632
Unintentional dural puncture (UDP), 282
　in neuroaxial anesthesia, 284–285
Unitary Hypothesis, 157–158
Upper limb pain, 558
Upper respiratory tract infections, 357
Urticaria, 313
U-shaped curve, 500, 500f

V

Vanillic acid, 451–452, 451t
VibraJect, 584

Vibrotactile devices, dental treatment, 584–585
　applications, 585
　effectiveness, 584
　efficacy evaluation, 585
Vincristine induced inflammatory pain, 434–436, 435t
Virtual anesthesia, 575, 579
Virtual reality devices, dental pain management, 576–577, 579
　key facts of, 579
　safety considerations, 578
Virtual reality-induced analgesia, 577f
Virtual reality (VR), neck pain, 566
　applications, 572
　as assessment tool, 571
　clinical efficacy, 569
　design, 566–568
　disability, 570
　disadvantages, 571
　factors affecting the efficacy of, 570–571
　key facts, 572
　pain, 569
　postural control and balance impairment, 570
　range of motion (ROM), 570
　rehabilitation, 566–568
Vitamin B, 370
Vitamins, 492–493
Voltage-gated sodium channels (VGSC), 74

W

Wall's Gate control theory, 583
Wide-dynamic range (WDR) neurons, 475
Willow tree *(Salix alba)*, 457
World Health Organization (WHO), 222
Writhing, 421–422, 427

Z

Zymosan, 446–447, 449t